M. Fukugita · A. Suzuki (Eds.)

Physics and Astrophysics of Neutrinos

Springer Japan KK

Professor Masataka Fukugita
Yukawa Institute for Theoretical Physics, Kyoto University, Sakyo-ku, Kyoto, 606-01 Japan;
Institute for Advanced Study, Princeton, NJ 08540, USA

Professor Atsuto Suzuki
Faculty of Science, Tohoku University, Aoba-ku, Sendai, 980 Japan

ISBN 978-4-431-67031-5 ISBN 978-4-431-67029-2 (eBook)
DOI 10.1007/978-4-431-67029-2

Originally published by Springer-Verlag Tokyo in 1994.

Preface

Ever since it was first introduced by Pauli, the neutrino has played a crucial role in the development of particle physics, from the early formulation of the four Fermi theory till the completion of the unified theory of electromagnetic and weak interactions by Glashow, Weinberg, and Salam. Its role is not over yet: we have good reason to anticipate that the neutrino will still be at the vanguard of particle physics beyond today's standard theory. Whether or not neutrinos are strictly massless is among the most important issues from this point of view. This is also the most important theme of recent experiments concerning neutrino physics: whilst a variety of topics have been explored using a variety of methods, the essential point of these experiments lies in the quest for a very small mass of the neutrino, which, if it happens to be discovered, should have vast implications for our understanding of particle physics, especially in that it probably implies new physics beyond the standard theory.

Current interest in the neutrino, in particular in the massive neutrino, has been magnified by the fact that it plays an important role in the astrophysical and cosmological environment. Increasing effort has recently been focused on the observation of neutrinos of astrophysical origin, which not only brings us new insights for astrophysics, but also provides a new method of exploring particle physics which cannot be carried out in laboratories. The detection of the neutrino burst from SN1987A was an epoch-making event in that it demonstrated an intimate interplay of astrophysics with particle physics. Recently, additional significant developments have been made in solar neutrino physics as well: thanks to the new experiments and theories developed over the last several years, the solution to the long-standing problem posed by Davis now seems to be emerging on the horizon, which lends us hope that the question as to whether the original problem should be ascribed to a nonstandard property of the neutrino will be resolved within the coming decade. Whether the massive neutrino is (a part of) dark matter in the universe is also a question of utmost importance in both fields of physics.

Against this background we have experienced a flurry of activity in neutrino physics and astrophysics research worldwide over the last decade. This was also true in Japan. From 1988 to 1990, research in Japan was, in fact, supported by the Ministry of Education with a significant amonut of funding under The Grant-in-Aid for Scientific Research on Priority Areas. This book is a product of this neutrino physics and astrophysics research done in Japan.

In order to document the outcome of this research and make it available to general readers in textbook format, we asked for contributions from our Japanese colleagues who are carrying out unique work in this field, the most typical and well-known of which is represented by the Kamiokande experiment. The topics covered also include the electron neutrino mass experiment, the double beta decay experiment, and the neutrino oscillation experiment, especially that using an emulsion technique, and R&D studies for new solar neutrino experiments, as well as subjects concerning atmospheric neutrinos, supernova neutrinos, and very high-energy neutrinos from astronomical objects.

The aim of the present book is to present a comprehensive account of this work, which cannot be found in journal papers, conference proceedings, or review articles, and to describe these subjects starting from elementary levels so that they are readily accessible to nonexperts, including graduate students. This book is also intended to serve as a handbook for researchers in the relevant fields. A general introduction was added to introduce the state-of-art theory of neutrino physics which is the basic background behind the experimental work described in this book.

It was not our intention to cover the entire range of subjects in this field; rather, we have concentrated on those to which Japanese physicists made significant contributions. Therefore, some important topics in this field are not included. We expect that readers will supplement the missing parts with other modern textbooks, which have rapidly become available to us in the last few years.

Finally, we would like to express our sincere thanks to Hirotaka Sugawara, Director of KEK, who served as the Principal Investigator of the Grant-in-Aid project; Jiro Arafune, Director of the Institute for Cosmic Ray Research; and Masatoshi Koshiba, the founding father of the Kamiokande project. We are greatly indebted to their enthusiastic engagement and continuous encouragement throughout the project. We also thank Masaki Mori for his invaluable assistance in the time-consuming editorial process. The greater part of the work described in this book was supported by a Grant-in-Aid for Scientific Research on Priority Areas of the Ministry of Education, Science and Culture (No. 029) for the fiscal years of 1988-1990; this publication was also supported in part by a Grant-in-Aid for Publication of Scientific Research Result (No. 3001) for the fiscal year 1993. We wish to thank the Ministry for its generous support, which made the publication of this book possible.

Kyoto-Sendai, 1 January 1994

Masataka Fukugita
Atsuto Suzuki

Guide to this book

The first article by M. Fukugita and T. Yanagida is a textbook-style review of the present status of the theory of neutrino physics. This article is intended to provide the background for the research presented in this volume. The key notions are explained in detail and derivations are given for fundamental equations, whenever possible. The knowledge of no more than elementary field theory is required for full comprehension of this article.

The next two articles describe the Kamiokande experiment. All details of the experiment, which were previously reported only as internal documents, are presented, starting from an elementary level, by K. Nakamura, T. Kajita, M. Nakahata, and A. Suzuki. A description of the detector concepts, geography and geology of the site, detector performance, and an explanation of the Monte-Carlo code is followed by a review of the physics output it has achieved. A historical account of the Kamiokande experiment is provided by A. Suzuki, one of the original researchers involved in the experiment.

The following three articles are reports by the representatives of groups which carried out neutrino mass experiments. T. Ohshima and H. Kawakami review the status and problems of current neutrino mass experiments using tritium beta decay, with particular emphasis given to their own experiment carried out at the Institute of Nuclear Study at the University of Tokyo. This experiment, together with those of Zürich and Los Alamos, marked the second generation of experiments after the Moscow result of a finite electron neutrino mass. H. Ejiri, who has been conducting a double beta decay experiment at the Kamioka mine using the "ELEGANTS" detector he developed, contributes a short summary of the present status of double beta decay experiments. K. Niwa presents a detailed account of unique neutrino oscillation experiments using a hybrid emulsion-counter system which are being carried out at Fermilab and CERN with his colleagues. The new experiments are expected to achieve a sensitivity that will result in a strong constraint on the hypothesis of the tau neutrino being cosmic dark matter.

Results from the atmospheric neutrino experiments conducted at Kamiokande and IMB have ignited controversy over whether or not the deficit of muon-neutrino flux can be ascribed to neutrino physics. T. Kajita, who has been engaged in the analysis of atmospheric neutrinos as a member of the Kamiokande group, examines the experimental procedure of the existing atmospheric neutrino experiments in detail and presents a review on this issue. This article is supplemented by a report by M. Honda explaining the key points of calculations of the atmospheric neutrino flux in the hope of clarifying the degree of reliability of the existing flux predictions and possible problems posed by these calculations.

Y. Totsuka, who is currently conducting the Kamiokande experiment, presents an overview of the detection methods for low-energy astrophysical neutrinos, in particular for solar neutrinos, with both real-time detection and radiochemical methods. The fundamentals of these detection methods are described in detail. This article is followed by two reports on low-energy neutrino detection using indium (by Y. Suzuki and K. Inoue) and cryogenic techniques (by T. Ebisu and T. Watanabe), which are still in the R&D stage.

The theory of supernova neutrinos is discussed by H. Suzuki, who has been carrying out a calculation of the neutrino flux from supernovae for several years. Following an elementary account of the dynamics of supernovae, details of his neutrino flux calculation and an extensive comparison with other work are presented.

The last two articles are dedicated to reviews of very high-energy gamma rays and neutrinos from astronomical objects. T. Kifune and M. Mori provide an extensive review of the status of very high-energy gamma ray observations and also discuss the future prospects of very high-energy neutrino detection. The article by F. Takahara is a textbook-style description of particle acceleration around active compact astronomical objects, in connection with very high-energy gamma ray/neutrino production.

Among a number of available textbooks related to the present subject we particularly recommend *Neutrino Astrophysics* by J. N. Bahcall (Cambridge University Press, 1988), especially for the astrophysics of solar neutrinos, and *Neutrino Physics*, edited by K. Winter (Cambridge University Press, 1991), which appeared during the preparation of this book, for physics of neutrino-hadron interactions, since these subjects are not covered in this book.

Contents

Observation of Atmospheric Neutrinos 559

T. Kajita

Calculation of Low-Energy Atmospheric Neutrino Fluxes 606

M. Honda

Experiments for Solar Neutrinos and Supernova Neutrinos 625

Y. Totsuka

Indium Solar Neutrino Experiments 694

Y. Suzuki and K. Inoue

Superheated Superconducting Granules for a Particle Detector 720

T. Ebisu and T. Watanabe

Physics of Neutrinos

Masataka Fukugita

Yukawa Institute for Theoretical Physics, Kyoto University

Tsutomu Yanagida

Department of Physics, Tohoku University

I. Introduction

1 Why neutrinos?

Ever since it was first proposed the neutrino has played a crucial role in the advancement of our understanding of particle physics. The neutrino was first conjectured by W. Pauli in a letter written in 1930 [1] as "verzweifeltes Ausweg" in order to save the energy conservation law in the nuclear beta-decay process, in which the emitted electron exhibits a continuous energy spectrum [2], as well as to explain the wrong statistics of ^{14}N and ^{6}Li — the neutron was not yet known at that time, and this latter problem was solved by its discovery. According to his conjecture this hypothetical particle is electrically neutral and has a mass much smaller than the proton. The particle was also supposed to have a spin of 1/2 and to respect the exclusion (Pauli) principle. Pauli argued that this neutral particle may have a finite, but small, magnetic moment no greater than $10^{-13}e$ cm (0.02 Bohr magneton) with a penetration power of at least 10 times greater than that of the gamma ray. He asked experimentalists participating in the conference in Tübingen whether such a particle could be found experimentally.

Three years after Pauli's conjecture a theory regarding the interaction of the neutrino with matter was formulated by E. Fermi [3] in a form in which the interaction is written as a product of four spinor fields, now called the four Fermi interaction. From the empirical electron energy spectrum he assumed that the mass of the neutrino is zero (this was also discussed by Perrin [4]). In Fermi's form of the interaction Hamiltonian five types of interactions are possible (Fermi chose the vector type in analogy to QED). Much of the work since then, for over a quarter of a century, was aimed at determining the type of interactions (e.g.,[5]–[7]). It was finally settled to take the form $V - A$ [8], following the discovery of parity violation [9]; namely, beta decay is described by

$$(1/\sqrt{2})G_F\bar{p}\gamma_\mu(1-\lambda\gamma_5)n\bar{e}\gamma^\mu(1-\gamma_5)\nu \tag{1}$$

In the present-day view this $(1-\gamma_5)$ structure is taken as being the first evidence which indicates that massless particles play a basic role in the weak interaction, and that the constant $\lambda \neq 1$ indicates that nucleons have structures which renormalise the strength of the axial coupling.

The existence of the neutrino was confirmed by Reines and Cowan (1956) [10] using an antineutrino beam from a nuclear pile, detecting $p+\bar{\nu} \to n+e^+$ followed by $e^+e^- \to \gamma\gamma$ and a γ ray emitted from neutron capture on Cd in a large mass scintillator. That the neutrino from the beta decay is left-handed $(V-A)$ was confirmed by an (e,ν) angular correlation measurement with ^{35}A [11] and a devised experiment which measured the γ circular polarisation emitted from the excited state of ^{152}Sm in the electron capture of ^{152}Eu [12]. A review was given of the development to this epoch by Pauli, himself [13].

The idea that the four Fermi interaction is a low-energy effective theory and that there exist more fundamental interactions behind the Fermi theory, is old [14, 15]. The evidence may be demonstrated by considering the fact that the cross section of $\bar{\nu}+p \to e^+ + n$ scattering, for example, violates the bound derived from unitarity if the incident energy becomes larger than 1 TeV, indicating a new physics at around or below this energy. This point is most clearly recognised when the "renormalisability" of the theory is taken to be the fundamental requisite for the self-consistency of the field theory. An important step towards a correct theory was made by Glashow [16], who gave the correct structure for the mixing of the photon and the Z^0-boson. Weinberg and Salam then presented, based on the gauge principle with the SU(2)×U(1) group, a definitive theory [17, 18], in which the mass of the intermediate bosons was predicted with the aid of the Higgs mechanism from the internal consistency of the theory. The full internal consistency of the Weinberg–Salam theory as a field theory was shown by the proof of renormalisability [19]; the correctness of the theory was evidenced by the discovery of neutral-current induced neutrino interactions [20]. The theory was then established by further explorations of neutrino-induced reactions [21] and finally by the discovery of the W and Z [22]. Today, the theory is being tested to quite high precision; we know that the nature is extremely close to what the Weinberg-Salam theory predicts, which, on the other hand, is almost frustrating for those people looking for something beyond the standard theory.

The aspect of neutrino physics that is of interest today arises from the belief that the physics of neutrino might yet lead to a new physics, if any, beyond the standard theory. The first point is that the neutrino is the only particle which is free from interactions stronger than the weak interaction, and that we may yet hope that a weaker interaction would be most apparent in the physics of neutrinos. The second point lies in the fact that the mass of the neutrino is either zero or anomalously small compared with the masses of other particles. We do not know of any symmetry that forbids the neutrino from having a finite mass. Should it be strictly massless, the most important task is to find the reason for this masslessness. On the other hand, there are reasonably convincing explanations for a small neutrino mass: the smallness is associated with a small lepton number violation; that is, the neutrino mass is controlled by a Majorana mass term rather than the Dirac mass term, which is responsible for the mass of other particles.

If we try to describe the coupling which gives rise to the Majorana mass term in terms of the fields that appear in the standard theory, we necessarily obtain an interaction of the dimension greater than four,

$$f\ \nu\nu\phi\phi, \tag{2}$$

where ϕ is the Weinberg–Salam's Higgs doublet [23]. When ϕ develops a vacuum expectation value, this interaction yields a lepton-number violating Majorana mass term. In fact, most of the models for the neutrino mass take essentially this form. In this expression the coupling constant f has the dimension of $[\text{mass}]^{-1}$; this interaction is regarded as an effective Lagrangian, just as in the case with (1), where G_F has the dimension of $[\text{mass}]^{-2}$. If we consider $\nu\phi \to \nu\phi$ scattering, the cross section does not decrease as $(\text{energy})^{-2}$, but remains constant. If we require this to satisfy the S-wave unitarity bound for energy smaller than the value corresponding to the Planck mass ($M_{pl} = 1.22 \times 10^{19}$GeV), we must have $f < 2\pi/M_{pl}$; that is, the neutrino mass must be $< 10^{-5}$eV. If it would happen to be found that the neutrino has a mass larger than this limit, there should be a new interaction below the Planck mass scale. This is the prime reason why we are so interested in the neutrino mass search.

2 Physics of massive neutrinos — general survey

The experimental effort carried out over the last two decades has established that the interaction of the neutrino with matter is just as described by the Weinberg-Salam theory. While an important task still remains to find any possible minuscule deviation of the ν interaction from that described in the Weinberg-Salam theory, current interest is focused more on the mass of the neutrino and its related phenomena.

If the neutrino indeed has a finite mass, many questions would open up concerning the properties of the neutrino: whether the neutrino is of the Dirac type or of the Majorana type (Majorana 1937 [24]) is perhaps the first question to ask. Generation mixing would also be a very important issue, which would provide a hint to our understanding of the mass problem in particle physics. We must ask whether the massive neutrino is stable or not. CP violation in the lepton sector might be substantially different from what we have seen in the quark sector. As we have discussed above, all of these questions may be closely related to the physics at a very high energy, much higher than which the present experiments can explore.

The dominant part of the current experimental effort with neutrinos is, in fact, concerned with the search for a finite neutrino mass. The most straightforward of them is the search for the electron neutrino mass with nuclear beta decay. The limit on the electron neutrino mass has been significantly updated almost every decade over the last half century.[1]

[1] Pauli conjectured the neutrino mass as not being larger than the electron mass. Earlier

The search for double beta decays also has a long history. Following a study by Goeppert-Meyer (1935) [31] for $(N,Z) \rightarrow (N,Z+2) + 2\bar{\nu} + 2e^-$, Furry (1939) [32] pointed out that neutrinoless double beta decays, $(N,Z) \rightarrow (N,Z+2) + 2e^-$, may take place if the neutrino is of the Majorana type. Interest in this process was revived with the advent of grand unification, a theory in which neutrinos are likely to appear as the Majorana type.

Pontecorvo (1957) [33] suggested that a neutrino may oscillate into its antipartner in the vacuum if the lepton number is not conserved, just like $K^0 - \bar{K}^0$ oscillation in the presence of CP violation. Oscillation among the different kinds of neutrinos was then proposed by Maki et al. (1962) [34]. Such a phenomenon may occur if some of the neutrinos are massive and if there is a mixing between the different neutrinos. There has been a persistent effort to search for neutrino oscillation using varieties of neutrino beams, i.e., those from nuclear piles, accelerators, and those of cosmic-ray origin (atmospheric neutrinos). Experiments with different beams explore different regions in mass-mixing parameter space.

Although we have occasionally seen some positive indications which suggest a finite mass from these experiments, most of them were soon excluded based on successive efforts. There have, however, been a few, if not convincing, which have not yet been completely excluded. It will be very important to either confirm or refute such indications, since, if confirmed, would immediately lead us to think about something beyond the standard theory.

Neutrinos are also important in the astrophysical environment. They are copiously produced in a high-temperature and/or high-density environment, and these neutrinos may often be used to probe the properties of neutrinos which can by no means be accessible by laboratory experiments. The representative examples are the neutrino mass limit from the cosmological mass density of the universe [35, 36], and the solar neutrino problem [37]–[39]. In particular, the latter indicates that either there is a serious departure from the standard understanding of the astrophysics of the Sun, or that the neutrino has an unexpected property in today's standard theory of particle physics. The recently started new solar-neutrino experiments [40, 41, 42], when combined with the old one, indicate that the latter is likely to be the case. We consider that the solar-neutrino problem is an issue which should be taken seriously, among others, in neutrino physics. The problem may point to the first evidence suggesting a new energy scale beyond the standard theory.

The plan of this review is as follows. In Chap. II the standard theory of the (electro)weak interaction and neutrinos is sketched; after giving a brief review of Weinberg–Salam's theory, various limits on the basic properties of neutrinos are

---

searches for the neutrino mass were made by Lyman (1939) [25] and Haxby et al. (1940) [26], resulting in $0\pm0.2m_e$ and $0\pm0.1m_e$, respectively, using $^{13}\mathrm{N} \rightarrow ^{13}\mathrm{C} + e^+ + \nu_e$. Konopinski (1947) extracted the limit on the neutrino mass ($m_\nu < 10$keV) from $^3\mathrm{H} \rightarrow {}^3\mathrm{He} + e^- + \bar{\nu}_e$ for the first time [27]. Since then, tritium has been used almost exclusively to study the electron neutrino mass. The limit obtained by Langer and Moffat (1952) [28] $m_\nu < 250$eV had been the best bound for long until Bergkvist [29] set a new limit ($m_{\nu_e} < 60$eV) in 1971 by his careful study. A surprise was announced by the Moscow ITEP group (1980) [30] who reported 17eV$< m_{\nu e} <$ 40eV. Several groups then initiated work to either confirm or to refute the ITEP result; it took a decade to rule out conclusively a finite mass in this range (see Sect. 6).

summarised. Several useful low-energy cross sections of the neutrino-induced reactions predicted by the standard theory are also given. Chap. III is given to a discussion of the elementary theory of the neutrino mass. After first discussing the equivalence of the Weyl and the Majorana neutrinos, various forms of the mass terms are considered. How a small neutrino mass is translated into physics at a large mass scale is also explained here. The quantisation of the Majorana field is presented in Chap. IV using the Weyl field basis. Particle states are defined and Feynman propagators are derived. As applications, a few characteristic processes involving Majorana neutrinos are calculated. One of them is the neutrinoless double beta decay; a brief review is given regarding the present status. Chap. V considers the problems in the presence of generations. The problem of the wave function phases is discussed in particular detail. Neutrino oscillation, neutrino conversion in matter, and the CP properties of neutrinos are also described in this chapter. Varieties of models for the neutrino mass are presented in Chap. VI. In this chapter possible implications of a small neutrino mass in the view of higher unification of particle theories are considered. The neutrino magnetic dipole moment and its physical implications are discussed in Chap. VII. In Chap. VIII topics concerning the baryon number generation in the Universe in connection to the neutrino problems are discussed.

Interplays between neutrino physics and cosmology are one of the important topics of neutrino physics, which have been paid much attention, particularly over the last decade. In this article, however, we mention this subject only to a minimal extent. We refer the reader to other review articles [43, 44] concerning this subject.

In this article we attempt to derive the fundamental equations and to explain the key notions in as much detail as the space permits. On the other hand, we do not try to cover extensive theoretical attempts towards particle physics modeling which have been presented in the literature; we must refer readers to other more specific review articles for these topics.

II. Standard theory of the weak interaction and properties of the neutrinos

3 Classical theory of the weak interaction

In the four Fermi interaction the neutrino interacts with the electron via the $V-A$ current of the form

$$\bar{e}\gamma_\mu(1-\gamma_5)\nu \ . \tag{1}$$

In modern field theory the left-handed and right-handed fields are regarded as being fundamental fields (this handedness is called chirality). They are given by

$$\psi_{\substack{L\\R}} = \frac{1\mp\gamma_5}{2}\psi \ , \tag{2}$$

where ψ is the conventional Dirac field, and ψ_L and ψ_R are two component fields which are called Weyl fermions [45]. ψ_L and ψ_R are eigenstates of the chiral transformation (Stech-Jensen transformation [6]) $\psi \to \gamma_5\psi$ with eigenvalues of -1 and $+1$, respectively. The weak current of form (1) is written as

$$\bar{\psi}_L\gamma_\mu\psi_L \ . \tag{3}$$

This means that the weak interaction conserves chirality, that is, it is invariant under a chiral transformation. The mass term $m\bar{\psi}\psi$, on the other hand, flips the chirality; it is written as

$$m(\bar{\psi}_R\psi_L + \bar{\psi}_L\psi_R) \ . \tag{4}$$

In the absence of mass the weak interaction theory is chiral invariant.

For leptons the fundamental fields are e_L, ν_L and e_R, and for quarks they are q_L and q_R with $q = u$ or d for the first-generation particles; e_R and q_R do not appear in the weak-interaction process. Whether ν_R exists in nature or not is unknown, since it decouples from the theory. The standard theory assumes that it does not exist.

The classical weak interaction Hamiltonian is given by

$$H_W = \frac{G_F}{\sqrt{2}}J_\mu^\dagger J_\mu \ , \tag{5}$$

where the current J_μ is given by

$$J_\mu = 2(\bar{\nu}_L\gamma_\mu e_L^- + \bar{u}_L\gamma_\mu d_L) \tag{6}$$

and G_F is the Fermi coupling constant. The right-handed field does not appear in this weak Hamiltonian.

The interaction of the form given in (5) suggests that the weak interaction has an underlying SU(2) symmetry, and that e_L and ν_L (u_L and d_L) form a doublet.

The non-appearance of e_R leads us to suppose that it is a singlet of SU(2), i.e.,

$$\psi_L = \begin{pmatrix} \nu_L \\ e_L \end{pmatrix} \ \mathbf{2}\ ; \quad \psi_R = e_R \ \mathbf{1}\ . \tag{7}$$

The first term of (6), for instance, is then written as

$$\bar{\psi}_L \gamma_\mu \tau_+ \psi_L \ \ (\tau_+ = \frac{\tau_1 + i\tau_2}{2}) \tag{8}$$

with $\vec{\tau}$ the Pauli matrices describing the SU(2) (weak isospin) of the weak interaction.

4 Weinberg-Salam theory of the electroweak interaction

4.1 SU(2) $\times$ U(1) Symmetry

We now sketch the Weinberg-Salam theory of the electroweak interaction [17, 18], the fundamental theory describing both weak and electromagnetic interactions. The theory is taken as a prototype of unified theories which we often refer to in this article. We review the theory only regarding its practical merit, and defer any theoretical detail to other textbooks (e.g., ref.[46]).

In the Weinberg-Salam theory the symmetry of the weak interaction is taken to be a gauge symmetry. Therefore, the SU(2) quantum number is regarded as the charge of a particle. For instance, the neutrino ν_L has a weak charge, $T = 1/2$ and $T_3 = 1/2$, while carrying no electromagnetic charge.

The most general unitary transformation for ψ_L and ψ_R in (7) is obviously U(2) $\times$ U(1). This, however, includes the lepton-number phase transformation, which we would like to exclude from gauge symmetry. Therefore, the most general desired symmetry is SU(2)$\times$ U(1); in fact, it turns out that this is the symmetry which correctly describes both electromagnetic (em) and weak interactions at the same time [16]. Let T_i ($i = 1 - 3$) be the three operators of SU(2). The operator Y, defined by

$$Y = Q - T_3, \quad Q = \begin{pmatrix} 0 & 0 \\ 0 & -1 \end{pmatrix}, \tag{9}$$

commutes with T_i (Q represents the em charge operator), and gives the desired U(1) operator. A consistent description is made if the U$(1)_Y$ charge is assigned to be $-1/2$ for ψ_L and -1 for ψ_R.

4.2 The Gauge fields

The Weinberg-Salam theory is a theory of gauge fields [47] which couple universally to weak charges of SU(2) ($\vec{T}$) and U(1) (Y). We have three gauge fields, $W^\pm$ and W_3, associated with the generator of SU(2) and one gauge field B for U$(1)_Y$.

The gauge principle dictates that the gauge fields interact with matter field ψ_L, ψ_R through the covariant derivative,

$$D_\mu = \partial_\mu + ig\,\vec{W}_\mu \vec{T} + ig' B_\mu Y \tag{10}$$

Regarding the charge assignment for ψ_L and ψ_R discussed above,

$$\mathcal{L} = i\bar{\psi}_L \not{D} \psi_L + i\bar{\psi}_R \not{D}' \psi_R \ , \tag{11}$$

where

$$D_\mu = \partial_\mu + ig\,\vec{W}_\mu \frac{\vec{\tau}}{2} - i\frac{g'}{2} B_\mu \tag{12}$$

$$D'_\mu = \partial_\mu - ig' B_\mu \tag{13}$$

from (10).

This uniquely determines the coupling of the gauge bosons to the matter fields. At this level all fields, including fermion fields, are massless, since there are no interactions that break the chiral symmetry.

4.3 Spontaneous symmetry breaking and the Higgs mechanism

In reality, SU(2) $\times$ U(1) symmetry must be broken, since only $U(1)_{em}$ charge seems to be conserved in nature. This is deeply related to the fact that only the photon among others is a massless gauge field.

There are generally two ways to break symmetry, explicit and spontaneous. We must take the latter for gauge theory, since otherwise symmetry breaking spoils the renormalisability of the entire theory. In order to break symmetry spontaneously we introduce a scalar field ϕ. In principle it may be in any representation other than the singlet of SU(2). It seems, however, only to be the simplest choice, ϕ being a doublet, which agrees with the experiment. The contribution from scalars with other representations, if any, must be small. We introduce

$$\phi = \begin{pmatrix} \phi^+ \\ \phi^0 \end{pmatrix} \quad \mathbf{2}\ , \quad Y = +1/2 \tag{14}$$

and give a vacuum expectation value to ϕ^0 in order to break SU(2) $\times$ $U(1)_Y$ down to $U(1)_{em}$. The coupling to the gauge field is given by

$$(D_\mu \phi)^\dagger (D^\mu \phi) \ , \tag{15}$$

where D_μ is given by (11) with $\vec{T} = \vec{t}\ /2$ and $Y = +1/2$:

$$D_\mu = \partial_\mu + i\,\vec{W}_\mu \frac{\vec{t}}{2} + i\frac{g'}{2} B_\mu \ . \tag{16}$$

We may assume a self-interaction of the scalar field (Higgs potential) with a negative mass squared,

$$V(\phi) = -\mu^2 \phi^\dagger \phi + \lambda (\phi^\dagger \phi)^2 \ . \tag{17}$$

The potential then takes a minimum at $\phi^\dagger \phi = \mu^2 / 2\lambda$, which leads to

$$\langle\phi^0\rangle = \frac{v}{\sqrt{2}}, \quad v = \sqrt{\mu^2/\lambda}\ . \tag{18}$$

When continuous symmetry is spontaneously broken, zero-mass Nambu-Goldstone bosons appear [48]. When the symmetry is a gauge symmetry, however, these zero mass modes appear in the same form as the gauge degrees of freedom and are absorbed into the gauge bosons to give their longitudinal mode, and hence masses [49].

4.4 Gauge-boson mixing

The Higgs ϕ^0 that develops a vacuum expectation is an SU(2) non-singlet and has a non-zero $U(1)_Y$ charge, but zero $U(1)_Q$ charge. It thus breaks $SU(2)\times U(1)_Y$, with the $U(1)_Q$ charge,

$$Q = T_3 + Y\ , \tag{19}$$

remaining unbroken. Therefore, the gauge field

$$A_\mu \propto \frac{1}{g}W_\mu^3 + \frac{1}{g'}B_\mu \tag{20}$$

which couples to

$$Q = \frac{1}{g}(gT_3) + \frac{1}{g'}(g'Y) \tag{21}$$

remains massless. The field orthogonal to (20),

$$Z_\mu \propto \frac{1}{g'}W_\mu^3 - \frac{1}{g}B_\mu \tag{22}$$

becomes massive (as seen explicitly in eq.(26) below). $(W^3, B) \to (A, Z)$ is an orthogonal transformation [16],

$$\begin{pmatrix} A_\mu \\ Z_\mu \end{pmatrix} = \begin{pmatrix} \cos\theta_W & \sin\theta_W \\ -\sin\theta_W & \cos\theta_W \end{pmatrix} \begin{pmatrix} B_\mu \\ W_\mu^3 \end{pmatrix}\ , \tag{23}$$

and the angle θ_W is called the Weinberg angle or the weak mixing angle,

$$\mathrm{tg}\ \theta_W = \frac{g'}{g}\ , \tag{24}$$

from (20) and (22). This angle is not determined within the theory.

Writing

$$\phi^0 = \frac{v + \phi'}{\sqrt{2}} \tag{25}$$

we can see that this vacuum expectation value gives mass to the gauge field, i.e.,

$$
\begin{aligned}
(D_\mu\phi)^\dagger(D_\mu\phi) &= |(\partial_\mu + ig\frac{\vec{\tau}}{2}\vec{W}'_\mu + i\frac{g'}{2}B_\mu)\begin{pmatrix} 0 \\ \frac{v+\phi'}{\sqrt{2}} \end{pmatrix}|^2 \\
&= \frac{1}{2}(\partial_\mu\phi')^2 + \left(\frac{g^2}{4}W_\mu^+W^{-\mu} + \frac{g^2+g'^2}{8}Z_\mu Z^\mu\right)v^2 \\
&\quad + \cdots
\end{aligned}
\tag{26}
$$

after rewriting W^3 and B with A and Z. This shows that the W_μ and Z_μ fields acquire masses of

$$
M_W^2 = \frac{g^2v^2}{4} \quad \text{and} \quad M_Z^2 = \frac{g^2+g'^2}{4}v^2 \ , \tag{27}
$$

respectively, whereas the A_μ field (photon) remains massless. We note that the relation

$$
\frac{M_W^2}{M_Z^2} = \frac{1}{1+(g'/g)^2} = \cos^2\theta_W \tag{28}
$$

is a consequence of ϕ being a doublet.

4.5 Fermion mass

The doublet ϕ field also couples to the leptons. The coupling pattern is unique and takes the form

$$
\mathcal{L}_Y = -f_e\overline{(\nu_e, e^-)_L}\begin{pmatrix} \phi^+ \\ \phi^0 \end{pmatrix} e_R^- + h.c. \tag{29}
$$

This is called the Yukawa coupling. The vacuum expectation value of ϕ also breaks chiral symmetry (the ϕ field carries chirality -2), giving to the electron a mass of

$$
m_e = f_e\langle\phi^0\rangle = \frac{f_e v}{\sqrt{2}} \ . \tag{30}
$$

4.6 Introduction of quarks

The introduction of quarks [50] is straightforward. We take

$$
\begin{aligned}
\psi_L = \begin{pmatrix} u \\ d \end{pmatrix}_L \quad &\mathbf{2}, \quad Y = 1/6 \\
u_R \quad &\mathbf{1}, \quad Y = 2/3 \\
d_R \quad &\mathbf{1}, \quad Y = -1/3,
\end{aligned}
\tag{31}
$$

so that (19) is satisfied. For the most economical case the Yukawa couplings read

$$\begin{aligned}\mathcal{L}_Y &= -f_e\overline{(\nu_e, e^-)_L}\begin{pmatrix}\phi^+\\ \phi^0\end{pmatrix}e_R^- \\ &\quad - f_d\overline{(u,d)_L}\begin{pmatrix}\phi^+\\ \phi^0\end{pmatrix}d_R - f_u\overline{(u,d)_L}\begin{pmatrix}\phi^{0*}\\ -\phi^-\end{pmatrix}u_R,\end{aligned} \tag{32}$$

where ϕ and its conjugate $\tilde{\phi} = i\tau_2\phi^*$ give masses to the d- and u-quarks, respectively.

4.7 Low-energy weak interactions

The couplings of the A_μ, Z_μ, W_μ gauge fields to matter fields are given in (10)–(13). To study a low-energy effective theory we write the operator

$$\mathcal{O}_\mu = g\,\vec{T}\vec{W}_\mu + g'YB_\mu \tag{33}$$

in (11) explicitly in terms of A_μ, Z_μ and $W_\mu^\pm = \frac{1}{\sqrt{2}}(W_\mu^1 \mp W_\mu^2)$:

$$\begin{aligned}\mathcal{O}_\mu &= g[\sin\theta_W Q A_\mu + \cos\theta_W (T_3 - \mathrm{tg}^2\theta_W Y)Z_\mu \\ &\quad + \frac{1}{\sqrt{2}}(T_1 - iT_2)W_\mu^- + \frac{1}{\sqrt{2}}(T_1 + iT_2)W_\mu^+].\end{aligned} \tag{34}$$

It is obvious that $g\sin\theta_W$ can be identified with the electric charge e,

$$e = g\sin\theta_W\ . \tag{35}$$

The Lagrangian describing the coupling to matter (11) is written

$$\mathcal{L} = -\bar{\psi}_L\mathcal{O}\psi_L - \bar{\psi}_R\mathcal{O}\psi_R\ . \tag{36}$$

The part where $W_\mu^\pm$ appear is

$$\mathcal{L}(W^\pm) = -\frac{g}{2\sqrt{2}}(J_\mu W^{+\mu} + J_\mu^\dagger W^{-\mu}) \tag{37}$$

with J_μ the current of (6). With this Lagrangian one can calculate the effective Hamiltonian describing scattering of a lepton and a quark, e.g., $e_L^- + u_L \to \nu_L + d_L$ with the $W^\pm$ boson exchanged, which reads

$$-iH_W^{\rm eff} = -ig^2\frac{1}{2}(\overline{\nu_L}\gamma_\mu e_L)\frac{1}{M_W^2 - k^2}(\overline{d_L}\gamma^\mu u_L)\ , \tag{38}$$

where k is the momentum transfer between the lepton and the quark. When $k^2 \ll M_W^2$, $H_W^{\rm eff}$ reduces to the classical four Fermi interaction with the identification

$$\frac{G_F}{\sqrt{2}} = \frac{g^2}{8M_W^2}. \tag{39}$$

The other part of (36) describes the electromagnetic and the weak neutral current interactions, which contain A_μ and Z_μ:

Table 1: Neutral current couplings.

	g_{iL}		g_{iR}
e_L	$-\frac{1}{2}+\sin^2\theta_W$	e_R	$\sin^2\theta_W$
ν_L	$+\frac{1}{2}$		
u_L	$+\frac{1}{2}-\frac{2}{3}\sin^2\theta_W$	u_R	$-\frac{2}{3}\sin^2\theta_W$
d_L	$-\frac{1}{2}+\frac{1}{3}\sin^2\theta_W$	d_R	$\frac{1}{3}\sin^2\theta_W$

$$\begin{aligned}\mathcal{L}(A,Z) &= -g\bar{\psi}_L[\sin\theta_W Q A_\mu + \cos\theta_W(T_3 - \mathrm{tg}^2\theta_W Y)Z_\mu]\psi_L \\ &\quad - g'\bar{\psi}_R Q[\cos\theta_W A_\mu - \sin\theta_W Z_\mu]\psi_R \qquad (40)\\ &= -eA^\mu J_\mu^{\mathrm{em}} - \frac{g}{\cos\theta_W}\frac{1}{2}J_\mu^{NC}Z^\mu \qquad (41)\end{aligned}$$

with

$$J_\mu^{\mathrm{em}} = Q\bar{\psi}\gamma_\mu\psi \qquad (42)$$

and

$$\begin{aligned}J_\mu^{NC} &= 2\sum_{L,R}\bar{\psi}_{L,R}(T_3 - \sin^2\theta_W Q)\gamma_\mu\psi_{L,R} \qquad (43)\\ &= 2\sum_i(g_{iL}\bar{\psi}_{Li}\gamma_\mu\psi_{Li} + g_{iR}\bar{\psi}_{Ri}\gamma_\mu\psi_{Ri}) \ , \qquad (44)\end{aligned}$$

where $g_{iL,R}$ is given explicitly in Table 1 for convenience.

A procedure similar to (38) leads to an effective Hamiltonian which is valid at low energies; the weak interaction Hamiltonian is summarised as

$$H_W^{\mathrm{eff}} = \frac{G_F}{\sqrt{2}}[J_\mu^\dagger J_\mu + \rho J_\mu^{NC}J_\mu^{NC}] \qquad (45)$$

with the current given in (6) and (44) and

$$\rho \equiv \frac{M_W^2}{M_Z^2\cos^2\theta_W} \ , \qquad (46)$$

which is unity due to (28). If there are several Higgs bosons ϕ_i with weak isospin $\vec{T}(\phi_i)$, ρ is given by

$$\rho = \frac{\sum_i[T(\phi_i)(T(\phi_i)-1) - T_3(\phi_i)^2]|\langle\phi_i\rangle|^2}{2\sum_i T(\phi_i)^2|\langle\phi_i\rangle|^2} \ . \qquad (47)$$

4.8 Introduction of the generation

We have so far discussed only the first generation of particles, $(\nu_e, e^-)_L$, e_R^-, $(u,d)_L$, u_R and d_R. There are, however, two additional generations (possibly more) in nature. It is easy to incorporate these three generations into the Weinberg-Salam theory. Let us denote the generation with superscripts $\alpha, \beta, \cdots$. The gauge coupling can be written in a form that is diagonal with respect to the generation:

$$i\bar{\psi}_L^\alpha \not{D}\psi_L^\alpha \ , \quad i\bar{\psi}_R^\alpha \not{D}\psi_R^\alpha \ . \tag{48}$$

The Yukawa coupling giving the matter field a mass is then

$$f^{\alpha\beta}\bar{\psi}_L^\alpha\psi_R^\beta\phi \ , \tag{49}$$

and the mass term is generally given by the matrix $M^{\alpha\beta}\bar{\psi}_L^\alpha\psi_R^\beta$ with non-vanishing off-diagonal elements ($M^{\alpha\beta} = f^{\alpha\beta}\langle\phi\rangle$). We note that there are three matrices corresponding to f_e, f_d and f_u, as in (32).

The matrix M can be diagonalised by a biunitary transformation,

$$V^{-1}MW = \widetilde{M}, \tag{50}$$

where $\widetilde{M}$ stands for a diagonal matrix. This can be easily shown by applying the polar decomposition theorem, which states that one can write M as

$$M = HU \tag{51}$$

with H a hermitian and U a unity matrix. Since H can be diagonalised by the unitary matrix V as

$$V^{-1}HV = \widetilde{M} \ , \tag{52}$$

we obtain (50) upon writing $U^{-1}V = W^{-1}$.

The eigenstate of gauge coupling, i.e., the eigenstate of the weak interaction (let us denote it by a prime) is given by the mass eigenstate as

$$\psi_L' = V\psi_L \ , \quad \psi_R' = W\psi_R \ . \tag{53}$$

The weak current is then written with the mass eigenstate as

$$\begin{aligned} J_\mu &\equiv \bar{\psi}_L'(Q=2/3)\gamma_\mu\bar{\psi}_L'(Q=-1/3) \\ &= \bar{\psi}_L(Q=2/3)U_{KM}\gamma_\mu\psi_L(Q=-1/3), \end{aligned} \tag{54}$$

where

$$U_{KM} = V^\dagger(Q=2/3) \ \ V(Q=-1/3) \ . \tag{55}$$

This quark mixing matrix U_{KM}, which was first introduced by Gell-Mann and Levy [50] and then by Cabbibo [52][2], is now called the Cabibbo [52] (when 2×2) or the Kobayashi-Maskawa [53] matrix (when 3×3).

In this manipulation four transformation matrices [$V(Q = 2/3)$, $V(Q = -1/3)$, $W(Q = 2/3)$, $W(Q = -1/3)$] appear. We emphasize, however, that only U_{KM} is physical in the standard Weinberg-Salam theory. In order to demonstrate this, we note that the original Lagrangian (kinetic and gauge parts) is invariant under three rotations with respect to the generation index:

$$\psi_L^\alpha \rightarrow S_\psi^{\alpha\beta}\psi_L^\beta , \qquad [\psi_L \equiv \begin{pmatrix} u \\ d \end{pmatrix}_L] \tag{56}$$

$$d_R^\alpha \rightarrow S_d^{\alpha\beta} d_R^\beta , \tag{57}$$

$$u_R^\alpha \rightarrow S_u^{\alpha\beta} u_R^\beta . \tag{58}$$

We can therefore diagonalise the Yukawa term $f_d^{\alpha\beta}\bar{\psi}_L^\alpha \phi d_R^\beta$ using (56) and (57); this means that we can take $V(Q = -1/3) = 1$ without any loss of generality. The freedom of rotation, (57) and (58), can further be used to choose W in eq.(53) to be unity. This makes only $U_{KM} = V^\dagger(Q = 2/3)$ physical. If there exist some new interactions that break invariance under (56–58), however, the relevant transformation matrices acquire physical meaning.

An $n \times n$ unitary matrix U_{KM} generally has $n(n+1)/2$ phases. In (54), however, n phase degrees of freedom can be absorbed into the phase of $\bar{\psi}_L(Q = 2/3)$, and additional $n-1$ phases into $\psi_L(Q = 1/3)$. We are therefore left with $(n-1)(n-2)/2$ physical phases in the matrix. For $n = 3$ we have only one physical phase, which is supposed to give rise to CP violation [53].

The situation with the lepton sector is different. If all the neutrino mass vanishes (or precisely degenerates), the unitary transformation for the neutrino $V(\nu)$ has no physical meaning. One can therefore always choose $V(\nu)$ so that it cancels against $V(l)$, i.e., $U_l = V^\dagger(\nu)V(l) = 1$. The lepton mixing becomes physically meaningful when neutrinos have finite masses. Extra phases may appear if the neutrinos are of the Majorana type (see Sect.4).

4.9 Experimental test

4.9.1 Electroweak parameters

The theory contains three fundamental parameters: g, g' and v. We traditionally represent these parameters in terms of $\alpha = e^2/4\pi$, G_F and $\sin^2\theta_W$. From the a.c. Josephson effect and the quantum Hall effect we have

$$\alpha^{-1} = 137.0359895(61) . \tag{59}$$

[2]Gell-Mann and Levy (1958) introduced this mixing angle to keep universality of the weak interaction. Cabbibo (1963) then carried out an analysis, showing that this mixing indeed describes well the weak interaction phenomenology. The mixing matrix in an orthogonal matrix form (the GIM form) was first introduced by Maki, Nakagawa and Sakata (1962) [34] within the framework of Sakata's composite model.

The Fermi coupling constant is defined by the muon decay constant, using

$$\Gamma(\mu \to e\nu\bar{\nu}) = \frac{G_\mu m_\mu^5}{192\pi^3} f(m_e^2/m_\mu^2)\left[1 + \frac{\alpha}{2\pi}(\frac{25}{4} - \pi^2)\right] , \tag{60}$$

where

$$f(x) = 1 - 8x + 8x^3 - x^4 - 12x^2 \ln x \tag{61}$$

is the phase-space volume factor. Note that the radiative correction is explicitly taken into account for the QED part [54], while the corrections due to W and Z^0 exchanges are absorbed into G_μ. From $\tau_\mu = \Gamma_\mu^{-1} = 2.197035(40)\mu$s we have

$$G_F \equiv G_\mu = 1.16637(2) \times 10^{-5}\text{GeV}^{-2}. \tag{62}$$

Here, the data are taken from ref.[55]. (In this review we adopt data from ref.[55], unless otherwise explicitly quoted.) The Weinberg angle $\sin^2\theta_W$ is determined from a variety of neutral-current effects as well as the physics with the W and Z bosons.

The test of the Weinberg–Salam model is to examine whether $\sin^2\theta_W$ is determined consistently from different effects. The most accurate determination of $\sin^2\theta_W$ comes from the gauge-boson masses. From (39) and (35),

$$M_W = \frac{(\pi\alpha/\sqrt{2}G_F)^{1/2}}{\sin\theta_W} , \tag{63}$$

and from (28)

$$M_Z = M_W/\cos\theta_W , \tag{64}$$

or

$$\sin^2\theta_W = 1 - \frac{M_W^2}{M_Z^2} . \tag{65}$$

The average of four experiments at LEP [56]–[59] yields $M_Z = 91.175 \pm 0.021$ GeV [60]. The error is only 0.03%. This M_Z gives $\sin^2\theta_W = 0.2122 \pm 0.0001$ using (63) and (65). This most precise W boson mass is given by UA2 [61] and CDF [62]. The UA2 group presented $M_W/M_Z = 0.8813 \pm 0.0041$. When combined with the LEP M_Z value, it yields $M_W = 80.35 \pm 0.37$ GeV. The W mass from the CDF group is $M_W = 79.91 \pm 0.39$ GeV. Using (61) one obtains $\sin^2\theta_W = 0.223 \pm 0.007$ from the UA2 data and 0.232 ± 0.008 from the CDF data. This may be compared with the most precise values from low-energy neutral-current processes, i.e., the values from the ratio [63] of $\nu_\mu + N \to \nu_\mu + X$ to $\nu_\mu + N \to \mu + X$ in deep inelastic neutrino scattering: $\sin^2\theta_W = 0.236 \pm 0.005 \pm 0.005$ (data from the CDHS) or $0.245 \pm 0.005 \pm 0.005$ (data from the CHARM-II experiment [64]). Here, the second error represents the theoretical uncertainties.

The scatter in the $\sin^2\theta_W$ values derived above are as large as 0.03. Most of the scatter, however, can be removed (to an accuracy of 0.005) by radiative corrections (we note, however, that there remains some systematic uncertainty which amounts to 0.01 among experiments, as shown above).

A full discussion of the radiative correction is too much involved to be reviewed here. We only give a very brief sketch of the central result. We first note that since the definitions of parameters $e, g, g', M_W, M_Z, v, \theta_W$ are renormalisation scheme dependent, we must fix the scheme in order to give an unambiguous meaning to the coupling constant. For α it is customary to take the value defined by Thomson scattering, and for G_F that by muon decay. The following two schemes are often used to define $\sin^2\theta_W$:

(i) $\overline{\mathrm{MS}}$ scheme [65]. The renormalised couplings g_r, g'_r are defined by subtracting $2/(4-n) + \log 4\pi - \gamma_E$ from g and g', and the Weinberg angle is defined by

$$\sin^2\hat{\theta}_W(\mu) = \frac{g_r'^2}{g_r^2 + g_r'^2}. \tag{66}$$

This definition of $\sin^2\theta_W$ depends on the renormalisation point μ, which is usually chosen to be M_Z.

(ii) On-shell scheme [66, 67, 68]. The Weinberg angle is defined by

$$\sin^2\theta_W = 1 - \frac{M_W^2}{M_Z^2}, \tag{67}$$

where the gauge-boson masses are their on-shell (physical) values.

When scheme (i) is adopted, (63) is modified to

$$M_W = \frac{(\pi\alpha/\sqrt{2}G_F)^{1/2}}{\sin\hat{\theta}_W(1-\Delta\hat{r}_W)^{1/2}}, \tag{68}$$

and (64) to

$$M_Z = \frac{M_W}{(1+\Delta\hat{\rho})^{1/2}\cos\theta_W}. \tag{69}$$

In (68) Δr_W depends weakly on the top quark mass m_t, and is dominantly ($\gtrsim$90%) given by the correction due to the running of α,

$$\alpha(M_Z)^{-1} = \alpha^{-1} - \frac{1}{3\pi}[3 \sum_{i=u,d,s,c,b} Q_i^2 \log(M_Z/m_i) + \sum_{i=e,\mu,\tau} \log(M_Z/m_i)] \simeq 128.4\ . \tag{70}$$

That is,

$$\frac{1}{(1-\Delta\hat{r}_W)} \simeq \alpha(M_Z)/\alpha \tag{71}$$

gives a correction of 0.067 which is compared with the values tabulated in Table 2; $\sin^2\hat{\theta}_W$ is only weakly dependent on m_t (-1% change as m_t increases from 100 GeV to 200GeV). On the other hand, $\Delta\hat{\rho}$ depends largely on m_t, as

$$\Delta\hat{\rho} \simeq \frac{3G_F m_t^2}{8\sqrt{2}\pi^2} = 0.0070(m_t/150\mathrm{GeV})^2 \tag{72}$$

Table 2: Renormalisation corrections $\Delta\hat{r}_W$ and Δr for $m_\phi = 250$ GeV.

m_t	$\Delta\hat{r}_W$	Δr
100 GeV	0.0695 (13)	0.0605 (13)
200 GeV	0.0719 (13)	0.0248 (13)

Table 3: Weinberg angle $\sin^2\theta_W$ derived from various experiments. $m_t = 150$ GeV and $m_\phi = 250$ GeV are assumed for the one loop results.

	tree	with one loop corrections $\overline{\text{MS}}$	on-shell
M_Z	0.2122 (1)	0.2326 (3)	0.2261 (3)
M_W/M_Z	0.2276 (53)	0.2339 (22)	0.2276
$\nu N \to \nu X$	0.240 (7)	0.237 (6)	0.232 (6)
$\nu_\mu e \to \nu_\mu e$	0.237(10)	0.237(10)	

for $m_t \gg M_W$. The mass of Z decreases as m_t increases.

With scheme (ii),

$$M_W = \frac{(\pi\alpha/\sqrt{2}G_F)^{1/2}}{\sin\theta_W(1-\Delta r)^{1/2}}, \tag{73}$$

while the relation $M_Z = M_W/\cos\theta_W$ is kept fixed. This means that M_Z/M_W, and hence $\sin^2\theta_W$, is strongly dependent on m_t, as is clear from (69) and (72) ($\sim -6\%$ change for $m_t = 100 \to 200$GeV); Δr also depends strongly on m_t (see Table 2). For $m_t \simeq 100$GeV and $M_\phi \simeq 250$GeV, for instance, the relation between the two angles is $\sin^2\hat{\theta}_W(M_Z) = 1.01\sin^2\theta_W$.

The phenomenology of neutral-current induced processes, as governed by (44), must be modified by taking into account process dependent radiative corrections. For (45) the overall strength is multiplied by $(1+\Delta\rho_i)$, and $\sin^2\theta_W$ is replaced by $(1+\Delta\kappa_i)\sin^2\theta_W$, where i is the process being considered. The Weinberg angle obtained from the Z, W phenomenology and from various neutral-current effects are presented in Table 3.

A most notable feature which gives evidence of the correctness of the electroweak theory, including the radiative correction, is that one can pin down the range for the top-quark mass consistently from $\sin^2\theta_W$ determined from various processes. The value quoted by Langacker [69] is

$$m_t = 149\begin{matrix}+21\\-27\end{matrix} \pm 16 \text{ GeV}. \tag{74}$$

The best value of the Weinberg angle, as quoted by Langacker, is

$$\sin^2 \hat{\theta}_W(M_Z) = 0.2326 \pm 0.0008 \tag{75}$$

$$\sin^2 \theta_W = 0.2261 \pm 0.0030 \ , \tag{76}$$

where most of the errors in (76) come from the uncertainty of the top-quark mass (2σ range is taken for m_t in these fits). Another analysis by Ellis and Fogli gives $m_t = 127^{+24}_{-30}$GeV and $\sin^2 \theta_W = 0.2273 \pm 0.0032$ [70]. The Weinberg angle from an analysis of the joint LEP groups is $\sin^2 \hat{\theta}_W(M_Z) = 0.2337 \pm 0.0014^{+0.0001}_{-0.0004}$ and $\sin^2 \theta_W = 0.2299^{+0.0047}_{-0.0048}$ [60]. All of the analyses almost agree at the one standard-deviation level. We adopt (75) or (76) in our article.

A test for the consistency of the theory is also made in the following way: if the average value of $\sin^2 \theta_W$ is inserted into (68) and (69), one obtains $\Delta r = 0.051^{+0.007}_{-0.010}$, which is compared with the predicted value $\Delta r = 0.0605(13)$ for $m_t = 100$GeV or $0.0248(13)$ for $m_t = 200$GeV. In the $\overline{\text{MS}}$ scheme, $\Delta \hat{r}_W = 0.060^{+0.009}_{-0.004}$ is obtained, which is compared with the prediction 0.0695(13)–0.0719(13) for $m_t = 100 - 200$GeV.

We note, however, that the radiative corrections discussed here are dominated (80–90%) by those of QED (as seen in (70) and (71) above) and of self-energy insertions from fermion loops; the vertex correction from W and Z is small. For a full consistency test for the one-loop radiative corrections of the Weinberg-Salam theory, the experimental error (e.g. that of the W mass) must be reduced by a factor 5 or so.

For reviews of experimental tests we refer the reader to [71] for earlier summaries, [72] for results before the LEP experiments and [56, 69, 70] for results which include the LEP experiments.

Other elementary parameters in the Weinberg-Salam theory

From the above results, we can calculate other elementary parameters that appear in the Weinberg-Salam theory. The two gauge couplings of the theory are: $\alpha_2 \equiv g^2/4\pi \simeq 1/30$ and $\alpha_1 \equiv g'^2/4\pi \simeq 1/100$. For the vacuum expectation value of the Higgs field, we have $v \simeq 247$GeV. The Yukawa couplings are given by $f_i \simeq 5.7 \times 10^{-6}(m_i/1\text{MeV})$, where m_i is the fermion mass. The ϕ^4 coupling constant is related to the Higgs mass as $\lambda \simeq (1/19)(m_\phi^2/m_W^2)$.

Higgs sector

The Higgs sector is the least understood part of the Weinberg-Salam theory. The mass of the Higgs particle (or equivalently the ϕ^4 coupling in the minimal version of the model) is unconstrained within the model. The Higgs sector may well be more complicated than has been described here, without changing the predictions of the gauge sector. The only constraint that we can derive on the Higgs sector is that on ρ of eq.(47); it is very close to unity within 1%, which means that Higgs particles that give a vacuum expectation value to the gauge bosons are completely dominated by an $SU(2)_L$ doublet.

A direct search for Higgs particles is made using $Z^0 \to \phi +$ "Z^0(virtual)", "Z^0" $\to \ell^+\ell^-$ and $\nu\bar{\nu}$. The Aleph group [73] (and subsequently the other three groups at LEP) excluded a Higgs particle in the mass range $0 \le m_\phi \le 41$GeV (95%C.L.) in their early experiment at LEP for the minimal version of the model. This excluded the simple possibility [74] that spontaneous symmetry breaking is caused by radiative corrections, starting from a flat effective potential (Coleman-Weinberg mechanism); the model predicts m_ϕ=9.7GeV. The experimental limit has now been raised to 58GeV (95%) [75]. The constraint, however, is substantially relaxed in non-minimal models, where the production cross section of ϕ^0 is model-dependent.

The search for Higgs particles and the exploration of the Higgs sector are among the most pressing issues for the next-generation high-energy accelerators. The upper bounds presently available on the mass of the Higgs particle are only suggestive. A bound $m_\phi \lesssim 1$TeV is inferred by requiring that the Higgs coupling remains weak, so that perturbation theory is applicable [76]. An almost identical limit ($\lambda \le 2\pi/3$ or $m_\phi \lesssim 1$TeV) is also inferred from tree unitarity [77]. Lattice calculations, exploiting the expected triviality of the scalar theory, set an upper bound on $\lambda = (m_\phi^2/m_W^2)(g^2/8)$, giving $m_\phi \lesssim 0.6-0.8$TeV [78]. None of these bounds seem to be compelling, however.

4.9.2 Cabibbo universality and quark mixing

Although the theory requires that the strengths of the leptonic weak interaction and those of hadronic weak interaction are equal, allowing for mixing of the form (54) for the quark sector [52]. Tests are usually made by comparing G_μ from μ decay with the strength of semileptonic weak interactions of hadrons. If the masses of neutrinos are all negligible compared with the muon mass, mixings in the leptonic sector, if any, cannot be detected in this comparison. The significance of this universality test also lies in the fact that it constrains possible exotic interactions which do not respect universality.

Expressing U_{KM} of (55) as

$$U_{KM} = \begin{bmatrix} U_{ud} & U_{us} & U_{ub} & \cdots \\ U_{cd} & U_{cs} & U_{cb} & \cdots \\ U_{td} & U_{ts} & U_{tb} & \cdots \\ \cdots & & & \end{bmatrix}, \tag{77}$$

each matrix element is determined from the relevant experiments, usually by assuming universality for the overall weak coupling. The test for universality is therefore translated into that for unitarity, i.e.,

$$|U_{ud}|^2 + |U_{us}|^2 + |U_{ub}|^2 = 1 \ , \tag{78}$$

etc.

A precise test of unitarity can be made only for the first row of (77), i.e., (78). For this test the determination of only elements U_{ud} and U_{us} is important, because

$|U_{ub}|^2$ is negligibly small ($< 5 \times 10^{-5}$) compared with the present accuracy of U_{ud} and U_{us} in the unitarity sum of (78).

The most precise determination of $|U_{ud}|$ is given by a comparison of the strength of the $0^+ \to 0^+$ nuclear Fermi transitions (pure superallowed beta decay) with that of muon decay. Since these transitions are caused solely by the vector current, and the vector current is conserved (up to isospin violation), the transition amplitude does not receive corrections from strong interactions, which are difficult to estimate. Since we are now interested in an accuracy of 1% or less, radiative corrections are essential for our purpose. It has been known for a long time that the radiative correction for nuclear beta decay diverges, in contrast to that for μ decay, and one needs a cutoff [79]. In the modern theory this cutoff is given by the gauge-boson mass; that is, radiative corrections contain the factor $(\alpha/2\pi)\ln(m_W/m_e)$, which is quite large.

The electromagnetic radiative corrections conventionally split into "outer" ($\delta_{\rm out}$) and "inner" ($\delta_{\rm in}$) parts [80]. The former is the conventional QED corrections for nuclear beta decays given by (B.12). The latter corrections comprise a part which is independent of the nuclei. We include corrections from W and Z^0 exchanges into $\delta_{\rm in}$. Using G_μ defined in eq.(60), we have

$$
\begin{aligned}
|U_{ud}|^2 &= \frac{\pi^3 \ln 2/m_e^5}{(ft)' G_\mu^2 (1+\delta_{\rm out}+\delta_{\rm in}(\beta)-\delta_{\rm in}(\mu))} \\
&= \frac{2984.4 \pm 0.1\ {\rm s}}{(ft)'(1+\delta_{\rm out}+\delta_{\rm in}(\beta)-\delta_{\rm in}(\mu))}. \qquad (79)
\end{aligned}
$$

Here $\delta_{\rm in}(\mu)$ is the correction from $W^\pm$ and Z^0 exchanges to μ-decay, which was implicitly absorbed into G_μ in (60), and

$$
\begin{aligned}
\delta_{\rm in}(\beta) - \delta_{\rm in}(\mu) &\simeq \frac{\alpha}{2\pi}[4\ln\frac{M_Z}{m_p} + \ln\frac{m_p}{m_A} + {\rm const}] \qquad (80) \\
&\simeq 0.0234 \pm 0.0018 \qquad (81)
\end{aligned}
$$

is the difference between the inner corrections for β decay and μ decay, where m_A is an effective mass which refers to the axial current [81]. The conventional ft value (f is the statistical weight factor with the spectral shape correction factor also taken into account and t is the half-life for the transition) is corrected for the effect arising from an isospin violation $(ft)' = (ft)(1-\Delta_c)$, where $\Delta_c \approx 0.3 - 0.6\%$ depending on the nuclei (configuration mixing and the difference in the wave functions between the initial and the final states). The most recent estimate by Hardy et al. [82] yielded

$$
(ft)'(1+\delta_{\rm out}) = 3073.3 \pm 3.53 \qquad (82)
$$

from eight $0^+ \to 0^+$ transitions between $T=1$ isomultiplets (^{14}O, ^{26m}Al, ^{34}Cl, ^{38m}K, ^{42}Sc, ^{46}V, ^{50}Mn, ^{54}Co), where $\delta_{\rm out} = 1.4 - 1.5\%$ includes 0.2% $Z\alpha^2$ and $Z^2\alpha^3$ corrections [83], which remove a sizable atomic number dependence of resulting ft value. From (82) one obtains

$$|U_{\rm ud}| = 0.9740 \pm 0.0010 \tag{83}$$

(Sirlin's earlier estimate was 0.9750 ± 0.0010 [83]). In this estimation the dominant sources of errors are those arising from the model-dependent part, i.e., those in Δ_c (0.1%) and radiative corrections for the axial part (0.15%). It thus seems not to be easy to reduce the error beyond the present level.

Another precise determination of $|U_{ud}|$ results from the neutron lifetime τ_n and g_A/g_V. The best experimental results are

$$g_A = 1.262 \pm 0.005 \text{ [84]} \tag{84}$$

and

$$\tau_n = 889.1 \pm 2.6\,\text{s [85]} \ . \tag{85}$$

Using the statistical weight factor of ref.[86],

$$|U_{ud}|^2 = \frac{4904.0 \pm 5.0\text{s}}{\tau_n[1 + 3(g_A/g_V)^2]} \ , \tag{86}$$

one obtains

$$|U_{ud}| = 0.9770 \pm 0.0035 \ , \tag{87}$$

which is consistent with (83) but having a somewhat larger error.

$|U_{us}|$ is determined from either K_{e3} decay or hyperon beta decays. K_{e3} is caused by the vector current, for which the effects of flavour SU(3) symmetry breaking enter in the second order [87]; the theoretical uncertainty is controlled better for this process. Taking into account the effect of isospin violation for $K^+ \to \pi^0 e\nu$ and $K^0 \to \pi^- e\nu$ with the aid of chiral perturbation theory (which predicts $f^{K^0\to\pi^-}(0) = 0.961 \pm 0.008$, $f_+^{K^+\to\pi^0}/f_+^{K^0\to\pi^-} = 1.022$), Leutwyler and Roos [88] obtained $|U_{us}| = 0.2196 \pm 0.0023$. Hyperon decays involve the axial current. Using a phenomenological model for symmetry breaking [89], $|U_{us}| = 0.222 \pm 0.003$ is quoted [55] from the WA2 data [90]. The PDG authors [55] adopt

$$|U_{us}| = 0.220 \pm 0.002 \ . \tag{88}$$

U_{ub} is a very small quantity. The PDG quotes

$$|U_{ub}| = 0.005 \pm 0.002 \ , \tag{89}$$

determined from $\Gamma(b \to u\ell\nu)/\Gamma(b \to c\ell\nu)$ (which gives $|U_{ub}/U_{cb}| = 0.09 \pm 0.04$) and $|U_{cb}|$ (see below).

The matrix elements that refer to the charm quark (the second row) are estimated more model-dependently and are rather poorly determined. The estimate of $|U_{cd}|(= 0.215 \pm 0.0016)$ uses di-muon productions in deep inelastic ν_μ and $\bar\nu_\mu$ scattering through charm productions off d quarks. $|U_{cs}| = 0.98 \pm 0.12$ is estimated by similar di-muon productions off strange quarks in the sea quark component of nucleons. $|U_{cb}| = 0.046 \pm 0.005$ is obtained from the lifetime of bottom mesons and

$\Gamma(b \to c\ell\nu)$ using model form factors. No direct information is available concerning U_{ti} (the third row).

The resulting matrix elements are

$$U_{KM} = \begin{pmatrix} 0.975 \pm 0.007 & 0.220 \pm 0.002 & 0.005 \pm 0.002 \\ 0.215 \pm 0.016 & 0.98 \pm 0.12 & 0.046 \pm 0.005 \\ \cdots & \cdots & \cdots \end{pmatrix} \tag{90}$$

The unitarity relation (78) is thus tested to an accuracy of

$$|U_{ud}|^2 + |U_{us}|^2 + |U_{ub}|^2 = 0.9970 \pm 0.0020 \ . \tag{91}$$

If $|U_{ud}|$ is replaced by (87), the right-hand side becomes 1.003 ± 0.005. The test for the unitarity constraint does not say much about the possible existence of the fourth generation, unless mixing with the first generation happens to be unexpectedly large. However, eq.(91) restricts quite stringently the couplings which are often expected in some non-standard models. An example of such a use can be seen in Sect.5.

The KM matrix for practical use is determined by imposing the unitarity constraints. A modern determination [55] reads

$$U_{KM} = \begin{pmatrix} 0.9747 - 0.9759 & 0.218 - 0.224 & 0.002 - 0.007 \\ 0.218 - 0.224 & 0.9735 - 0.9751 & 0.032 - 0.054 \\ 0.003 - 0.018 & 0.030 - 0.054 & 0.9985 - 0.9995 \end{pmatrix} . \tag{92}$$

A few parametrisations for the 3×3 KM matrix are used in the literature. We adopt

$$\begin{aligned} U_{KM} = & \begin{pmatrix} 1 & 0 & 0 \\ 0 & \cos\theta_2 & \sin\theta_2 \\ 0 & -\sin\theta_2 & \cos\theta_2 \end{pmatrix} \begin{pmatrix} \cos\theta_1 & \sin\theta_1 & 0 \\ -\sin\theta_1 & \cos\theta_1 & 0 \\ 0 & 0 & 1 \end{pmatrix} \\ & \times \begin{pmatrix} 1 & & \\ & 1 & \\ & & e^{i\delta} \end{pmatrix} \begin{pmatrix} 1 & 0 & 0 \\ 0 & \cos\theta_3 & \sin\theta_3 \\ 0 & -\sin\theta_3 & \cos\theta_3 \end{pmatrix} \\ = & \begin{pmatrix} c_1 & s_1c_3 & s_1s_3 \\ -s_1c_2 & c_1c_2c_3 - s_2s_3e^{i\delta} & c_1c_2s_3 + s_2c_3e^{i\delta} \\ s_1s_2 & -c_1s_2c_3 - c_2s_3e^{i\delta} & -c_1s_2s_3 + c_2c_3e^{i\delta} \end{pmatrix} , \end{aligned} \tag{93}$$

where $c_i = \cos\theta_i$ and $s_i = \sin\theta_i$.

The leptonic counterpart of unitarity for the KM matrix is $e-\mu-\tau$ universality. Writing the effective coupling constants for the $\ell\bar{\nu}_e W$ vertex, which refer to e, μ and τ as being $g^{(e)}, g^{(\mu)}, g^{(\tau)}$ i.e., $(G_\mu \propto g^{(e)}g^{(\mu)})$, the ratio $g^{(e)}/g^{(\mu)} = 1.0003 \pm 0.0016$, derived from $\pi \to e\nu(+\gamma)/\pi \to \mu\nu(+\gamma)$ [91], gives the best verification for $\mu - e$ universality (see Table 4 for other data). The most accurate test for $g^{(\tau)}/g^{(\mu,e)}$ is given by a comparison of the decay rate of $\tau \to \mu(e)\nu\bar{\nu}$ with μ decay. The data as of the beginning of 1992 [55] indicated that $g^{(\tau)}/g^{(\mu,e)} = 0.975 \pm 0.010$, the departure from universality being at a 2.5 standard-deviation level. The new mass

Table 4: $e - \mu - \tau$ universality tests.

	$g^{(e)}/g^{(\mu)}$		$g^{(\tau)}/g^{(e)}$
$\frac{\pi \to e\nu}{\pi \to \mu\nu}$	1.0003 ± 0.0016[91]		
$\frac{\tau \to e\nu_e\nu_\tau}{\tau \to \mu\nu_\mu\nu_\tau}$	0.996 ± 0.011[55]		
$\frac{W \to e\nu}{W \to \mu\nu}$	1.01 ± 0.04[92]	$\frac{W \to \pi\nu}{W \to e\nu}$	0.997 ± 0.070[93]
$\frac{\tau \to \pi\nu_\tau}{\pi \to e\nu_e}$			0.985 ± 0.026[94]
$\frac{\tau \to (e,\mu)\nu\bar{\nu}}{\mu \to e\nu\bar{\nu}}$			$g^{(\tau)}/g^{(e,\mu)} = 0.975 \pm 0.010$[55] $(0.998 \pm 0.012)^*$
$\frac{Z \to ee}{Z \to \mu\mu}$	1.001 ± 0.007[55]	$\frac{Z \to \tau\tau}{Z \to ee}$	0.997 ± 0.007[55]

Note– * The Aleph data are used for τ_τ [97], see text.

measurements with ARGUS [95] and with the Beijing Electron Spectrometer [96], however, have given $m_\tau = 1776.3 \pm 2.4 \pm 1.4$ and 1776.78 ± 0.5MeV, respectively, 6MeV smaller than the DELCO value (1783^{+3}_{-4}MeV) which had been adopted for quite a while. These new results for the τ mass half way remove the departure from universality. The ALEPH group at LEP then reported a new value for τ_τ, which is 3.5% (1.7σ) longer than the world average [97]. If this lifetime is used, we obtain $g^{(\tau)}/g^{(\mu,e)} = 0.998 \pm 0.012$, which is consistent with universality. The systematics problem in the τ lifetime measurements has yet to be settled, however.

5 Electromagnetic properties of the neutrino

5.1 Electromagnetic form factors

In order to discuss the electromagnetic properties of neutrinos, we define the form factors by

$$\begin{aligned}\langle \nu(p',\lambda')|J^{\rm em}_\mu|\nu(p,\lambda)\rangle &= \bar{\nu}(p',\lambda')[\gamma_\mu F_1(q^2) - \gamma_\mu\gamma_5 G_1(q^2) \\ &\quad + \sigma_{\mu\nu}q^\nu\{F_2(q^2) - \gamma_5 G_2(q^2)\}]\nu(p,\lambda) \qquad (94)\end{aligned}$$

with $q = p' - p$. F_1 and G_1 are charge form factors, and F_2 and G_2 are magnetic and electric dipole form factors, respectively. For the left-handed neutrino, (94) reduces to

$$\langle \nu_L' | J_\mu^{\rm em} | \nu_L \rangle = \bar{\nu}_L \gamma_\mu \{ F_1(q^2) + G_1(q^2) \} \nu_L \ , \tag{95}$$

where $F_1(0) + G_1(0)$ is the charge of the left-handed neutrino. For the right-handed neutrino the charge is $F_1(0) - G_1(0)$. Two-component (Weyl) neutrinos cannot have a magnetic or an electric dipole moment.

Since the Majorana neutrino is equivalent to the Weyl neutrino (we show this in Sect.5 and 4), the Majorana neutrino cannot have a dipole moment.[3]

If we rewrite (95) with the Majorana field χ (see (19) of Sect.5),

$$\langle \chi | J_\mu^{\rm em} | \chi \rangle = -\frac{1}{2}(F_1 + G_1)\bar{\chi}\gamma_\mu\gamma_5\chi \ . \tag{96}$$

The divergence is

$$\langle \chi | \partial^\mu J_\mu^{\rm em} | \chi \rangle = -(F_1 + G_1) m_\chi . \tag{97}$$

The electromagnetic current is not conserved unless $m_\chi = 0$ or $F_1 + G_1 = 0$. This shows that only neutral fermions ($F_1 + G_1 = 0$) can have a finite Majorana mass; quarks and charged leptons cannot have a Majorana mass.

In the following we discuss the electromagnetic properties of the neutrino. We refer the reader to ref. [98] for a classical survey of this subject.

5.2 Electric charge

Whether strict charge neutrality can be derived from the standard theory is an intriguing question. Let us define the charge in units of the charge of the gauge particle, or equivalently of the Higgs particle. At a naive level the prediction of the theory is only $Q_\nu - Q_e = 1$. $Q_\nu (= F_1(0) + G_2(0)) = 0$ is set by choosing $Y = -1/2$ for $\psi_L = (\nu, e^-)_L$ by hand. [If the neutrino has a Majorana mass, $Q_\nu = 0$ by the argument above.]

When one looks into the full internal consistency of the theory, however, a much stronger condition is obtained on the neutrino charge. The requirement is that the theory is free from any triangular anomaly [99]; the triangle diagram should be canceled among all fermions contained in the theory. If the existing neutrinos are only left-handed, the strict charge neutrality is derived straightforwardly from the $\gamma - \gamma - \gamma$ diagram (Fig.1) to be

$$Q_{\nu \rm em}^3 = 0 \ , \tag{98}$$

[3]This can be easily seen in the following way (see (19) of Sect.5). Let us write

$$\mathcal{L} = \bar{\chi}\sigma_{\nu\mu}(f + g\gamma_5)\chi \ .$$

Rewriting this in terms of $\chi^c = C\bar{\chi}^T$ we have

$$\begin{aligned} \mathcal{L} &= -\bar{\chi}^c\sigma_{\mu\nu}(f + g\gamma_5)\chi^c \\ &= -\bar{\chi}\sigma_{\mu\nu}(f + g\gamma_5)\chi \ , \end{aligned}$$

where the Majorana property $\chi^c = \chi$ is used in the last step. This proves that the dipole form factors f and g vanish.

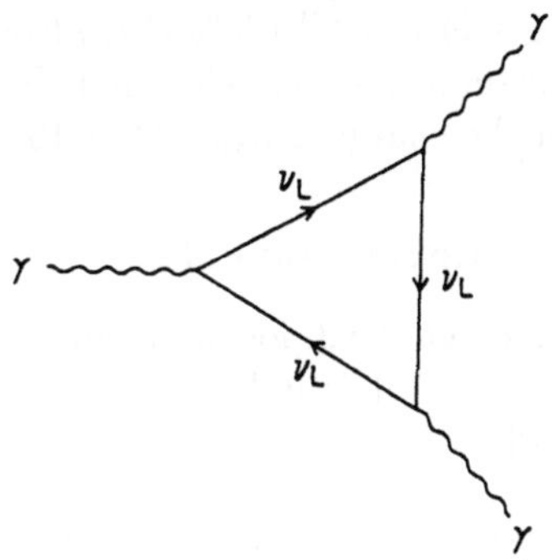

Figure 1: Three photon triangle diagram giving rise to anomaly. The contributions from other fermions are canceled between left-handed and right-handed fermion loops.

because the contributions from other fermions cancel between their left- and right-handed components [100].[4] This argument does not depend on whether the neutrino has a Majorana mass or not.

If there exists, however, a right-handed partner of the neutrino, the $\gamma-\gamma-\gamma$ diagram vanishes due to a cancellation between ν_L and ν_R, and the charge of the left-handed neutrino becomes free. It is easily shown that the neutrino (and also the neutron) may have a charge, while $Q_n + Q_\nu = Q_p + Q_{e^-} = 0$ should strictly hold based on the more general anomaly free conditions[5] (see Appendix A).

The experimental limit on Q_ν is

$$|Q_\nu| \leq (0.5 \pm 2.9) \times 10^{-21} . \qquad (68\% \text{ C.L.}) \tag{99}$$

This is obtained by assuming charge conservation in $n \to p + e^- + \bar{\nu}_e$ with the

[4]If one allows a parity violation in the electromagnetic interaction, the neutrino can have a non-vanishing electric charge, $Q_{\nu\text{em}} = \epsilon$. In this case, however, $\langle\phi\rangle \neq 0$ always breaks the electric charge conservation and gives rise to a finite photon mass [101]. The present upper limit on the photon mass $m_\gamma < 10^{-25}$eV [102] implies that $Q_{\nu\text{em}} = 10^{-26}$ [101].

[5]Constraints are derived from the anomaly free conditions for the WWB coupling and the BBB coupling. When these constraints are combined with the condition of the charge neutrality condition for the terms in (32), e.g., $Q_{u_L} - Q_{e_L} = 1$, $Q_{u_L} - Q_{d_L} = 1$ etc., we are finally led to

$$\begin{aligned} Q_e &= -1 + \epsilon , \\ Q_{\nu_L} &= Q_{\nu_R} = \epsilon , \\ Q_u &= 2/3 - \epsilon/3 , \\ Q_d &= -1/3 - \epsilon/3 . \end{aligned}$$

This means that

$$Q(\text{proton}) = 1 - \epsilon \qquad Q(\text{neutron}) = -\epsilon ,$$

i.e., strict neutrality holds for the hydrogen atom.

combined use of the measurement $|Q_p + Q_{e^-}| = 0.9 \pm 2.7 \times 10^{-21}$ (from an acoustic technique test by applying an alternating electric field to SF_6 gas in a cavity) [103] and the value of $Q_n = (-0.4 \pm 1.1) \times 10^{-21}$ (68% C.L.) from a null result from the deflection of a cold neutron beam in a strong electric field [104]. We add that the astrophysical limit derived on the charge is $|Q_\nu| < 2 \times 10^{-15}$ from SN1987A [105].

5.3 Magnetic and electric dipole moments

The Dirac neutrino may have a magnetic dipole moment. In the standard Weinberg-Salam theory with ν_R, for instance, it is induced by radiative corrections and is calculated to be [106, 107, 108]

$$\mu_{\nu_i} = \frac{3eG_F}{8\sqrt{2}\pi^2} m_{\nu_i} = 3 \times 10^{-19} \mu_B (\frac{m_\nu}{1\text{eV}}) \ , \tag{100}$$

where μ_B is the Bohr magneton,

$$\mu_B = e/2m_e = 5.788 \times 10^{-9}\text{eV} \cdot \text{Gauss}^{-1} = 1.93 \times 10^{-11} e \text{ cm}. \tag{101}$$

This means that a Dirac neutrino has a finite magnetic moment if $m_\nu \neq 0$. In a more general context the magnetic moment is not necessarily proportional to the neutrino mass, and even a massless neutrino may have a finite magnetic moment. This subject is discussed fully in Chap. VII.

A particle with a finite magnetic moment scatters off electrons with a cross section [109] of

$$\frac{d\sigma}{dE'_e} = \pi \left(\frac{\alpha^2}{m_e}\right) \mu_\nu^2 \left(\frac{1}{E'_e - m_e} - \frac{1}{E_\nu}\right) \tag{102}$$

with a recoil electron energy of E'_e or

$$\sigma \simeq \pi (\frac{\alpha}{m_e})^2 \mu_\nu^2 \log(q^2_{\max}/q^2_{\min}) \ , \tag{103}$$

where $q^2_{\max}$ and $q^2_{\min}$ are cutoffs ($q^2_{\min} \gg m_e^2$) of momentum squared at high- and low-momentum transfers.

In the original proposal for the neutrino, Pauli set the limit that μ_{ν_e} be smaller than $0.02\mu_B$ based on the absence of interactions with matter [1]. The detection of the neutrino interaction [10] immediately improved the limit to $\mu_{\nu_e} < 1 \times 10^{-9}\mu_B$ [110]. More recently, Kyuldjiev [111] analysed the $\bar{\nu}_e e$ scattering data by Reines et al. [112] and obtained $\mu_{\nu_e} < 1.5 \times 10^{-10}\mu_B$. This estimate has now been corrected to a slightly more modest limit of

$$\mu_{\nu_e} < 4 \times 10^{-10} \mu_B \tag{104}$$

using a revised estimate of the reactor antineutrino flux [113, 114]. A modern stopped muon experiment resulted in $\mu_{\nu_e} < 10.8 \times 10^{-10}\mu_B$ [115]. We note that a better limit is obtained more easily from a neutrino beam with a lower energy, since the weak interaction cross section increases as E_ν, whereas (103) stays constant.

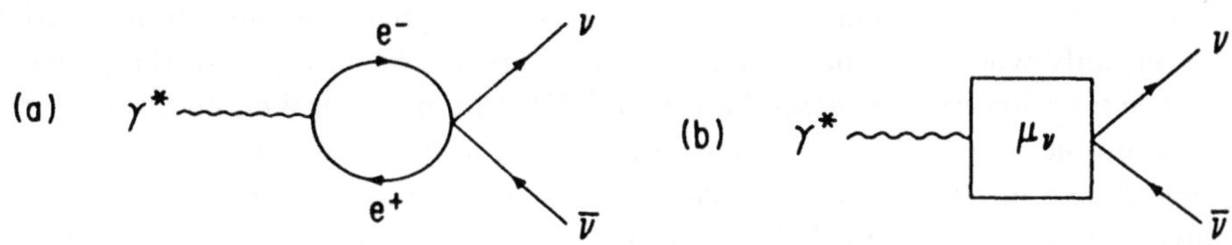

Figure 2: (a) Plasmon decay into a neutrino pair. (b) Extra stellar cooling process induced by neutrino magnetic moment.

This explains why the most recent limit with the stopped muon experiment improves only little the limit obtained by Cowan and Reines [110] 35 years ago , who used low-energy neutrino beams from a nuclear reactor. For ν_μ recent experiments [117, 116] give

$$\mu_{\nu_\mu} < 7.4 \times 10^{-10} \mu_B \quad . \tag{105}$$

The present laboratory limit on μ_{ν_τ} is

$$\mu_{\nu_\tau} < 4 \times 10^{-6} \mu_B \tag{106}$$

from $e^+e^- \to \nu\bar{\nu}\gamma$ [118].

A stronger limit is derived from the constraint against excess stellar cooling. In the stellar interior, a photon which has acquired a mass of the plasma frequency decays into a neutrino pair through virtual e^+e^- pair production, called plasmon decay (Fig. 2) [119]; this is often the chief mechanism of stellar cooling for post-main-sequence stars. The presence of a finite magnetic moment would enhance this effect and deprive the stellar core of nuclear energy, resulting in excess cooling, which would promote the evolution of the star, especially during later stages [98], [120]–[122]. A quantitative consideration yields a bound $\mu_{\nu_i} < 1 \times 10^{-11}\mu_B$ [123] on the largest of the magnetic moment of neutrinos from the number count of He-burning stars after helium ignition in open clusters. Blinnikov [124] derived $\mu_{\nu_i} < 0.5 \times 10^{-11}\mu_B$ from the luminosity function of cooling white dwarfs. An even stronger bound of $\mu_{\nu_i} < 0.3 \times 10^{-11}\mu_B$ was inferred by Raffelt [125] from an estimate of the critical mass for a helium flash. At this level, we are not very sure whether the limit does not depend on the details of stellar-evolution calculations and observational uncertainties. We may take

$$\mu(\nu_i) < (0.5 - 1) \times 10^{-11} \mu_B \tag{107}$$

as a conservative limit from astrophysics. A number of stronger limits are inferred from the cooling argument for supernova SN1987A [126]. We consider that we cannot take these limits too literally, since they depend on the details of the supernova theory, which have not been established as well as that for stationally burning stars. Only a weak bound is obtained on μ_ν from neutron-star cooling [127].

The Dirac neutrino may also have an electric dipole moment. It may arise, however, only when both parity and CP are broken. When $m_\nu = 0$, the phase of the wave function can be rotated away and CP cannot be broken and, hence, the electric dipole moment always vanishes. For $m_\nu \neq 0$, the neutrino may have an electric dipole moment. Since νe scattering experiments are made using neutrino beams with an energy much larger than the possible neutrino mass, neutrinos essentially behave as being left-handed, and neutrino scattering off electrons via electric dipole moment and that via magnetic dipole moment yield an identical result. Therefore, the limits on the magnetic moment are interpreted as being those on the electric dipole moment. With the conventional unit the experimental limit (104) is written

$$ed \leq 8 \times 10^{-21} e \text{ cm}. \tag{108}$$

5.4 Charge radius

The charge radius is defined by

$$\langle r^2 \rangle = 6 \frac{\partial}{\partial q^2} f(q^2)|_{q^2=0} \quad , \tag{109}$$

where $f(q^2) = F_1(q^2) + G_1(q^2)$ and $f(0) = 0$. The neutrino may generally have a finite $\langle r^2 \rangle$. Since the charge density of the neutrino is not positive definite, $\langle r^2 \rangle$ can be negative.

The calculation of the charge radius in the standard model has a confusing history [128]; if one calculates only vertex corrections, they are not gauge invariant and are generally divergent, and therefore physically meaningless. Finally, Degrassi et al. [129] identified $\langle r^2 \rangle$ as measured in experiments with a target-independent part of the radiative correction for neutrino scattering off leptons or hadrons. This physical (gauge-independent) $\langle r^2 \rangle$ contains not only radiative corrections to the $\nu\bar{\nu}\gamma$ vertex, but also terms arising from box diagrams. The calculated $\langle r^2 \rangle$ is sensitive to the top quark mass, and the result given in Table I of [122] may be represented as

$$\langle r^2 \rangle_{\nu_e} = [+(6.5 \pm 3.6) + 40(m_{t120}^2 - 1)] \times (10^{-17}\text{cm})^2 \quad , \tag{110}$$
$$\langle r^2 \rangle_{\nu_\mu} = [+(37.7 \pm 3.6) + 40(m_{t120}^2 - 1)] \times (10^{-17}\text{cm})^2 \quad , \tag{111}$$
$$\langle r^2 \rangle_{\nu_\tau} = [+(55.3 \pm 3.6) + 40(m_{t120}^2 - 1)] \times (10^{-17}\text{cm})^2 \tag{112}$$

to the accuracy of 10% for $90\text{GeV} \lesssim m_t \lesssim 180\text{GeV}$, where $m_{t120} = m_t/120\text{GeV}$, and the on-shell scheme of Sect.4.9.1 is used for renormalisation.

The effect of the $\langle r^2 \rangle$ term on νe elastic scattering is obviously absorbed into a shift of the vector coupling of the Z^0 exchange $g_V = g_L + g_R$ as $g_V + 2\delta$, where

$$2\delta = \frac{\sqrt{2}\pi\alpha}{3G_F} \langle r^2 \rangle = 2.38 \times 10^{30}\text{cm}^{-2} \langle r^2 \rangle \quad , \tag{113}$$

or $\sin^2\theta_W \rightarrow \sin^2\theta_W + \delta$. Therefore, one can estimate $\langle r^2 \rangle$ from νe scattering when $\sin^2\theta_W$ is known from other places.

If $\sin^2\theta_W$ is constrained to lie within 0.229 and 0.223, the $\nu_\mu e \to \nu_\mu e$ experiment by Ahrens et al. [116][6] gives

$$(-5.8 \sim -5.3) \times 10^{-32} < \langle r^2 \rangle_{\nu_\mu} < (0.08 \sim 1.3) \times 10^{-32}\text{cm}^2 \ . \tag{114}$$

If we use the data of the CHARM II collaboration [117], we have

$$(-0.67 \sim -0.77) \times 10^{-32} < \langle r^2 \rangle_{\nu_\mu} < (2.0 \sim 2.5) \times 10^{-32}\text{cm}^2 \ . \tag{115}$$

For ν_e, a LAMPF experiment [115] would yield

$$(-5.6 \sim -5.0) \times 10^{-32} < \langle r^2 \rangle_{\nu_e} < (9.6 - 10.2) \times 10^{-32}\text{cm}^2 \ . \tag{116}$$

In (114)–(115) two limits correspond to $\sin^2\theta_W = 0.229$ and 0.223, showing the dependence of the limit on the value of $\sin^2\theta_W$. Comparing (114) with (111), we may derive a limit on the top-quark mass, $m_t < 210$GeV.

We remark that the present experiment already indicates a value of $\langle r^2 \rangle$ of the order expected in the one-loop order of the standard theory, in contrast to the case with the magnetic dipole moment where an ample gap exists between the empirical limit and the one-loop prediction of the standard model.

6 Limit on the mass of neutrinos

While the neutrino is traditionally supposed to be massless, no exact symmetry is known that forbids neutrinos to have a finite mass. (It should be remembered that gauge symmetry forbids the photon having a finite mass.) Massive neutrinos can readily be accommodated into the Weinberg-Salam theory with a minimal modification, as described in the next chapter.

On the other hand, there is no compelling evidence at present for the neutrino to have a finite mass. The present laboratory limits on the neutrino masses are summarised as follows:

$$m_{\nu_e} < 7.3\text{eV} \ \ (95\%\text{C.L.}) \ , \tag{117}$$

$$m_{\nu_\mu} < 270 - 500\text{keV} \ \ (90\%\text{C.L.}) \ , \tag{118}$$

$$m_{\nu_\tau} < 31\text{MeV} \ \ (95\%\text{C.L.}) \ . \tag{119}$$

The modern ν_e mass measurements use exclusively the decay of tritium ${}^3\text{H} \to {}^3\text{He} + \bar{\nu}_e + e^-$, which has a reasonably small Q value ($Q = 18.6$ keV; Q-value is defined by the energy difference between the two neutral atoms), and look for a rounding of the electron spectrum at the end point (Kurie plot). For the allowed

[6] Ahrens et al. [116] remarked that their earlier publication contained errors in their formula. Their corrected formula, however, still contains errors of a factor $\sqrt{2}$. Here these errors are corrected to derive (114).

Table 5: Atomic and molecular corrections for tritium beta decays. P_0 is the probability staying in the ground state, and $\overline{\Delta E}$ is the average energy of excitations.

	T (atomic) [130]	HT (molecular) [132]	CH_3T [135]	CH_3T [136]	CH_3–CHT–CH_3 [136]	valine [135]
P_0	0.7023	0.5769	0.6122	0.607 ~0.623	0.598	0.6122
$\overline{\Delta E}$(eV)	27.21	29.27	30.30	—	—	30.12

or superallowed β transitions the transition matrix element is constant (i.e., non-relativistic; see (B.3) of Appendix B) and the transition rate Γ is determined solely by the phase-space volume factor,

$$d\Gamma \propto d^3p_e d^3p_\nu \delta(E_e + E_\nu - Q) \propto E_e p_e E_\nu p_\nu dE_e \quad . \tag{120}$$

For the massless neutrino, the shape factor of the electron spectrum is then given by

$$S(E_e) \equiv \frac{d\Gamma(E_e)/dE_e}{E_e p_e F(E_e, Z)} \propto (E_0 - E_e)^2 \quad , \tag{121}$$

where $E_0(= Q)$ is the maximum energy of the electron and $F(E_e, Z)$ is the Fermi function for the correction of the Coulomb wave-function distortion (B9). With a finite mass the shape factor is modified so as to be

$$S(E_e) \propto (E_0 + m_\nu - E_e)[(E_0 + m_\nu - E_e)^2 - m_\nu^2]^{1/2} \quad , \tag{122}$$

where $E_0 = Q - m_\nu$. In practice, this spectrum is rounded by the finite resolution of the spectrometer and of the energy loss at the tritium source. Radiative corrections (see B.11) change the end-point spectrum very little.

If one attempts to set a limit that is smaller than a few tens of eV, one meets a complication which arises from the fact that the β-decay process excites the helium atom and/or surrounding atoms [29]. For a free tritium atom, for instance, the shape factor is replaced by

$$S(E_e) = \sum_{n=1}^{\infty} P_{nS} S_{nS}(m_\nu; E_{nS}) + \int_0^\infty dE_\alpha P_\alpha(E_\alpha) S_\alpha(m_\nu : E_\alpha) \quad , \tag{123}$$

where P_{nS} (P_α) is the probability that the nS state (the continuum state) is excited and $S_{nS(\alpha)}$ is given by (122) with $E_0 \to E_0 - E_{nS(\alpha)}$. For a free hydrogen atom,

Table 6: Summary of ν_e mass experiments using tritium β decay.

Group [ref.]	Source	Corrections	fit of m_ν^2 (eV^2)	m_ν limit (eV) (at 95% C.L.)
ITEP (1980-1987) [27, 95]	valin	[135]	$+970 \pm 50 \pm 160$	$17 < m_\nu < 40$
Zürich (1986) [138]	T_2^+ implanted into C	CH_3T assumed [135]	-11 ± 63	< 18
INS (1987) [139]	arachidic acid $CH_3(CH_2)_{18}$ COOH	[135]	$+287 \pm 341$	< 32
Los Alamos (1987) [140]	molecular T_2	[133]	-57^{+453}_{-118}	< 27
Zürich (1990) [141]	tritiated octadecyltrichlorosilane	$CH_3-CHT-CH_3$ [135]	$-24 \pm 48 \pm 61$	< 11
INS (1991) [142]	arachidic acid	[135]	$-65 \pm 85 \pm 65$	< 13
Los Alamos (1991) [143]	molecular T_2	[132, 133, 134]	$-147 \pm 68 \pm 41$	< 9.3
Mainz (1992) [144]	frozen T_2		$-38.8 \pm 34.1 \pm 15.1$	< 7.2

$$P_{nS} = |\langle Z = 2, nS | Z = 1, 1S\rangle|^2 \tag{124}$$

in the impulse approximation, and is calculated for all states (e.g., $P_{1S} = 70.2\%, P_{2S} = 25\%, P_{3S} = 12.7\%, ..., P_{\rm cont} = 2.6\%$; mean excitation energy $\overline{\Delta E}$=27.21eV) [130, 131]. Accurate calculations have also been carried out for a T_2 molecule [132, 133, 134]. For convenience of making a tritium source, however, it is often embedded in organic molecules. The correction for surrounding molecules, especially for the shake-off process (ionisation), is substantially more complicated. Calculations for molecules were made in ref.[135, 136] using the wave function obtained by the molecular orbital method, and were claimed to be sufficiently accurate to extract the mass of the electron neutrino at a necessary level. The most important parameter that controls $S(E_e)$ is the probability remaining at the ground state and the mean excitation energy, which are given in Table 5.

Recent ν_e mass experiments are summarised in Table 6.[7] We conclude that the best current upper limit on ν_e is 7.3eV, obtained by the Mainz group, and that the claim of the ITEP group that m_{ν_e} is between 18 and 40eV was ruled out.

The use of internal bremsstrahlung electron capture (IBEC) $(Z, A) + e^- \rightarrow (Z-1, A) + \nu_e + \gamma$ [145] has been proposed to study the mass of the electron neutrino [146]. The photon (X-ray) emitted in the internal conversion has a spectrum of the form of (122) with E_e being replaced by the photon energy k_γ (the prefactor $E_e p_e$ must be replaced with k_γ, however). De Rújula has shown that the IBEC rate, which is otherwise too small, receives a resonance-like enhancement if k_γ is smaller than the energy of the characteristic X-rays. Although a few experiments have been

[7] Let us note that higher precision experiments all resulted in a negative m_ν^2 for their best fits. This suggests that a part of the rounding-off effects may not be quite controlled. It is also to be noted that the shape factor given by (122) does not fit the data quite well for a wide range ($\Delta E \simeq$1keV) of the spectrum; the origin of this distortion is not understood.

carried out [147]–[149] for ^{163}Ho, which has the smallest Q-value ($\simeq$ 2keV) among others, the rate turns out to be too small to yield any useful information concerning the neutrino mass; the limit obtained so far with this method is $m_{\nu_e} < 225$eV [149]. This method was not as successful as it was first hoped.

The electron capture rate of ^{163}Ho, itself, may also be used to obtain information concerning m_{ν_e} [150]. The obtained limit is $m_{\nu_e} < 500$eV [147, 148].

The limit on the ν_μ mass comes from the two-body kinematics of $\pi^+ \to \mu^+ + \nu_\mu$ decay. Abela et al. [151] measured the muon momentum from $\pi^+ \to \mu^+ + \nu_\mu$ decay at rest to an accuracy of $\Delta p_\mu / p_\mu = 28$ppm, and derived the limit $m_{\nu_\mu} <$ 0.25MeV at the 90% C.L. Since then, the pion mass measurement (using pionic $4f-3d$ transition in ^{24}Mg) has been improved [152] ($\Delta m_\pi / m_\pi = 3.8$ppm), and the updated limit is $m_{\nu_\mu} < 0.27$MeV at the 90% C.L. The accuracy for the muon mass is $\Delta m_\mu / m_\mu = 0.32$ppm, much better than other quantities. The limit appears to be set by the accuracy of the muon momentum measurement. However, we note that a continued effort to reduce the error of the muon momentum ($\Delta p_\mu / p_\mu = 18$ppm) has revealed that the pion mass from p_μ (with $m_{\nu_\mu} \geq 0$) is larger than that from a pionic X-ray measurement by 3.5σ [153]. This implies that the currently quoted limit may be marred by some systematic problems.

There is another type of experiment measuring p_π and p_μ for pion decay in flight [154]. The advantage is that the uncertainty in m_π enters in the result as a minor factor. The limit obtained with this method is $m_{\nu_\mu} < 500$keV at the 90% C.L. Taking account of the current uncertainty in the stopped pion measurement, this value perhaps represents a safe limit on the muon neutrino mass.

The best limit on the ν_τ mass is derived from the upper end of the invariant mass distribution of the 5π system in $\tau \to 5\pi + \nu_\tau$. With the maximum likelihood analysis the ARGUS group [155] obtained $m_{\nu_\tau} < 31$MeV at the 95% C.L. with a new τ mass from their measurement [95]. A recent result from CLEO is $m_{\nu_\tau} <$ 32.6MeV at the 95% C.L. [156].

We should also mention the bound on the neutrino mass from cosmology [35, 36]. In the early universe neutrinos were in equilibrium with other radiation. As the Universe expanded, neutrinos decoupled from equilibrium at around $T \sim$ 4MeV while being relativistic, unless m_ν is larger than a few MeV. The number density is thus prescribed by the Fermi distribution, and the density today should be $n_{\nu+\bar{\nu}} = (3\zeta(3)/4\pi^2)T_\nu^3 \simeq 110\mathrm{cm}^{-3}$ per neutrino species with $T_\nu = (4/11)^{1/3} T_\gamma \simeq$ 1.9K. Therefore, massive neutrinos contribute to the mass density of the Universe by an amount of $m_\nu n_{\nu+\bar{\nu}}$. By requiring that this does not exceed the critical density of the Universe,

$$\rho_{\mathrm{crit}} = 3H_0^2/8\pi G = 10.54 h^2 \mathrm{keV} \cdot \mathrm{cm}^{-3} \ , \tag{125}$$

where H_0 is the Hubble constant and $h = H_0/100\mathrm{km\ s^{-1}Mpc^{-1}}$ is usually taken to be between 0.5 and 1, one obtains

$$\sum_i m_{\nu_i} \lesssim 96 h^2 \mathrm{eV} \tag{126}$$

if the neutrinos are stable on the cosmological time scale. This limit is stronger by several orders of magnitude than the laboratory limits for ν_μ and ν_τ, thus leading us to believe that all three neutrinos have very small masses. Within the standard theory, the lifetime of neutrinos ($m_\nu \lesssim 1$MeV) should be longer than the age of the Universe. The possibility of unstable neutrinos will be discussed in Sects.28 and 45.

There are attempts to set a limit on m_{ν_e} based on the duration of the neutrino burst at the time of supernova SN1987A [157]. The limit derived varies from $\lesssim$6eV to $\lesssim$30eV. The better bound, however, is more model-dependent, and the conclusive bound is at best no better than the present laboratory limit.

7 Number of generations

If all neutrinos have light masses, the counting of neutrino species in high-energy experiments gives information concerning the number of generations.

An important constraint is derived from the total width of Z^0. The Z^0 decays into e^+e^-, $\mu^+\mu^-$, $\tau^+\tau^-$, $q\bar{q}$ ($\bar{q} = u, d, s, c, b$) and $\nu\bar{\nu}$'s, with the partial width to $\nu\bar{\nu}$ in particular,

$$\Gamma_{\nu\bar{\nu}} = \frac{1}{12\pi}\frac{G_F}{\sqrt{2}}M_Z^3 N_\nu \simeq 0.181 N_\nu \mathrm{GeV} \tag{127}$$

for the neutral-current coupling given in (44). The partial width of Z^0 for invisible channels measured at the LEP experiments yields

$$N_\nu = 2.99 \pm 0.04 \quad . \tag{128}$$

This result not only gives the number of neutrino species, but also means that any particles that carry SU(2)$_L$ and/or U(1)$_Y$ charge with a mass smaller than $\approx m_Z/2$ are excluded.

The constraint can also be inferred from a neutrino counting experiment with $e^+e^- \to \nu\bar{\nu} + \gamma$ [158]. The OPAL experiment at LEP yields $N_\nu = 3.0 \pm 0.4 \pm 0.2$ [159]. This experiment gives a constraint independent of that from Z^0 decay for some special cases; this process may also be sensitive to right-handed neutrino species, if they exist, and couple to e^+e^- through, e.g., a non-standard Higgs boson exchange.

Before these good laboratory limits became available, the constraint $N_\nu \lesssim 4$ was inferred based on the argument on primordial nucleosynthesis for light elements. This argument was first explicitly pointed out by Shvartsman [160] and a more quantitative estimate was initiated by Steigman et al. [161]. The constraint is derived from the fact that extra neutrino species increase the rate of the expansion of the early universe, and thus increase the temperature of decoupling T_β of the β equilibrium ($n + \nu_e \rightleftarrows p + e^-$ etc). This increases the neutron fraction, and thus the ^{4}He abundance by an amount of $\delta Y_p = [\partial Y_p/\partial(n/p)][\partial(n/p)/\partial T_\beta](\partial T_\beta/\partial g_*)(7/4)\delta N_\nu \simeq 0.012\delta N_\nu$, where Y_p is the mass fraction of helium and g_* is a relativistic degree of freedom that contributes to the expansion of the universe. The production of ^{4}He monotonically decreases as the baryon density decreases, but a lower limit is inferred on the baryon density from the abundance of ^{7}Li and an upper bound on the ^{3}He $+ d$ abundance.

Therefore, the observation of primordial ^{4}He ($Y_p \lesssim 0.245$) sets an upper limit on the number of neutrino species as well as the baryon density. Using modern observational data and accurate laboratory data for the neutron lifetime, a consistent determination of the baryon density from ^{4}He, ^{3}He, d and ^{7}Li can be made only with $N_\nu \leq 3.6$ [162].

8 Examples of neutrino scattering cross sections

8.1 ν-lepton scattering

The part of eq.(45) relevant to $\nu_e e^-$ scattering is

$$\begin{aligned} H_W^{\text{eff}} &= \frac{G_F}{\sqrt{2}}\{[\bar{e}\gamma_\mu(1-\gamma_5)\nu_e][\bar{\nu_e}\gamma^\mu(1-\gamma_5)e] \\ &\quad + 2\rho[\bar{\nu}\gamma_\mu\frac{1}{2}(1-\gamma_5)\nu][\bar{e}\gamma^\mu(g_V - g_A\gamma_5)e]\} \ , \end{aligned} \tag{129}$$

where $g_V = g_L + g_R = -1/2 + 2\sin^2\theta_W$ and $g_A = g_L - g_R = -1/2$. The first term is absent for $\nu_\mu e^-$ scattering. After the Fierz rearrangement of the first term, we have

$$H_W^{\text{eff}} = \frac{G_F}{\sqrt{2}}[\bar{\nu}\gamma_\mu(1-\gamma_5)\nu][\bar{e}\gamma^\mu(c_V - c_A\gamma_5)e] \tag{130}$$

$$\begin{array}{rcc} & \nu_e e^- & \nu_\mu e^- \\ c_V = & \rho g_V + 1 & \rho g_V \\ c_A = & \rho g_A + 1 & \rho g_A \end{array} \tag{131}$$

With the standard cross section calculation we obtain the recoil energy (E_e') distribution in the lab frame,

$$\frac{d\sigma}{dE_e} = \frac{2G_F^2 m_e}{\pi}\left[c_L^2 + c_R^2\left(1 - \frac{E_e'}{E_\nu}\right)^2 + c_L c_R \frac{m_e}{E_\nu}\frac{E_e'}{E_\nu}\right] \ , \tag{132}$$

where

$$\begin{aligned} c_L &= \frac{c_V + c_A}{2} &&= g_L + 1 \qquad (\text{for} \quad \nu_e) \\ & &&= g_L \qquad\quad\ \ (\text{for} \quad \nu_\mu) \\ c_R &= \frac{c_V - c_A}{2} &&= g_R \ . \end{aligned} \tag{133}$$

The inelasticity parameter $y = E_e/E_\nu$ is often used to display eq.(132). In the center of mass (CM) system

$$1 - y = \frac{1}{2}(1 + \cos\theta) \ , \tag{134}$$

Table 7: Coefficients which appear in neutrino-lepton scattering.

	c_L	c_R	c	α
$\nu_e e^-$	$1/2+\sin^2\theta_W$	$\sin^2\theta_W$	0.990	0.096
$\bar{\nu}_e e^-$	$\sin^2\theta_W$	$1/2+\sin^2\theta_W$	0.383	10.4
$\nu_\mu e^-$	$-1/2+\sin^2\theta_W$	$\sin^2\theta_W$	0.157	0.67
$\bar{\nu}_\mu e^-$	$\sin^2\theta_W$	$-1/2+\sin^2\theta_W$	0.129	1.49

Note– c_L and c_R are defined by eq.(133). The cross section is given by $\sigma = c \times 10^{-43}\text{cm}^2\ (E/10\ \text{MeV})$. The differential energy dependence is $d\sigma/dE_e \propto 1+\alpha(1-E_e/E_\nu)^2$.

with θ the CMS scattering angle between ν and e. When the third term of (132) is ignored ($E_\nu \gg m_e$), the total cross section is

$$\sigma = \frac{2G_F^2 m_e E_\nu}{\pi}\left[c_L^2 + \frac{1}{3}c_R^2\right] \quad . \tag{135}$$

For $\bar{\nu}_{e(\mu)}$ e scattering c_L and c_R are interchanged. For convenience c_L, c_R are tabulated in Table 7. With $\sin^2\theta_W = 0.225$, the νe scattering cross sections are

$$\sigma(\nu_e e^- \to \nu_e e^-) = 0.920 \times 10^{-43}(E_\nu/10\text{MeV})\text{cm}^2 \quad , \tag{136}$$
$$\sigma(\bar{\nu}_e e^- \to \bar{\nu}_e e^-) = 0.383 \times 10^{-43}(E_\nu/10\text{MeV})\text{cm}^2 \quad , \tag{137}$$
$$\sigma(\nu_\mu e^- \to \nu_\mu e^-) = 0.157 \times 10^{-43}(E_\nu/10\text{MeV})\text{cm}^2 \quad , \tag{138}$$
$$\sigma(\bar{\nu}_\mu e^- \to \bar{\nu}_\mu e^-) = 0.129 \times 10^{-43}(E_\nu/10\text{MeV})\text{cm}^2 \quad . \tag{139}$$

We note that the recoil electron has a soft energy spectrum, as can be seen in eq.(132); the observable cross section is significantly reduced if low-energy recoil electrons are discarded in the detector. An example is given in Table 8 for neutrinos from an ^{8}B decay source ($E_\nu \leq 14$MeV) with the spectrum given in ref.[39]. The detectable fraction is $\approx 30\%$, even if the recoil electron energy threshold is set at 5 MeV.

If $O(\alpha, G_F)$ radiative corrections are taken into account, (129) for $\rho = 1$ is modified to [163, 164, 165],

$$H_W^{\text{eff}} = \frac{G_F}{\sqrt{2}}\rho^{(\nu,\ell)}[\bar{\nu}\gamma_\mu(1-\gamma_5)\nu][\bar{e}\gamma^\mu(c'_V(q^2) - c'_A(q^2)\gamma_5)e] \quad , \tag{140}$$

where $\rho^{(\nu,\ell)} = 1 + O(g^2/\pi)$ and c'_V and c'_A are obtained by replacing

$$\sin^2\theta_W \longrightarrow \kappa^{(\nu,\ell)}(q^2)\sin^2\theta_W \quad , \tag{141}$$

with $\kappa^{(\nu,\ell)}(q^2) = 1 + O(g^2/\pi)$ (the superscript ℓ shows the explicit target dependence of the radiative correction). An explicit expression for $\rho^{(\nu,\ell)}$ is given by

Table 8: Neutrino electron scattering for ^{8}B solar neutrinos. Detection fractions are shown for the given threshold energy for recoil electrons.

E_{th}(MeV)	$\nu_e e$	$\nu_\mu e$
5	30.2%	26.5%
7.5	10.4%	8.9%
10	4.2%	3.6%
$\sigma_{\rm tot}({\rm cm})^2$	6.6×10^{-44}	1.0×10^{-44}

(3b)–(3d) of ref.[165], and $\kappa^{(\nu,\ell)}$ by (4a)–(4c) of ref.[165] and (A15), (B10), (C3), (A5) and (A6) of ref.[166]. The differential and total cross sections are explicitly worked out in [163]–[165]. The correction brings a 0.4–0.5% decrease in $\sin^2\theta_W$ in the $\overline{\rm MS}$ scheme for given cross section data compared with its tree level value. This correction is substantially smaller than that for ν hadron scattering, where it amounts to a few %. The reason for the smallness of this radiative correction is the same as that for the μ decay compared with that for the β decay. Major experiments for νe scattering are summarised in Table 9.

8.2 ν-nucleon scattering

The effective interaction is given by (I.1) with the Cabibbo angle, i.e.,

$$H_{\rm eff} = \frac{G_F}{\sqrt{2}}|U_{ud}|[\bar{e}\gamma_\mu(1-\gamma_5)\nu_e]\ [\bar{p}\gamma^\mu(g_V - g_A\gamma_5)n] + h.c., \tag{142}$$

where $g_V = 1$ and g_A is an effective axial vector coupling constant. For sufficiently low energy, a non-relativistic treatment for the nucleon in the second bracket gives $g_V \cdot \mathbf{1}(\mu = 0)$ and $-g_A\,\vec{\sigma}\ (\mu = 1-3)$. With an elementary calculation we have

$$\begin{aligned}\sigma(\bar{\nu}_e p \to e^+ n) &= \sigma(\nu_e n \to e^- p) \\ &= \frac{G_F^2 E_\nu^2}{\pi}|\cos^2\theta_c|^2\left[1 + 3(\frac{g_A}{g_V})^2\right] \qquad (143) \\ &= 9.23 \times 10^{-42}\left(\frac{E_\nu}{10{\rm MeV}}\right)^2 {\rm cm}^2\ , \qquad (144)\end{aligned}$$

where g_A/g_V=1.262 [84] and $\cos\theta_c = |U_{ud}| = 0.975$ are used. The differential cross section is almost flat; $d\sigma/d\theta \propto 1 - 0.103\cos\theta$ (see eq.(155) below).

For energy $E_\nu \gtrsim 50$ MeV the structure of nucleons must be taken into account. By writing the matrix element of the vector and the axial currents as

$$\langle N(p')|V_\mu|N(p)\rangle = \bar{u}(p')(\gamma_\mu F_V + \frac{1}{2m_N} i\sigma_{\mu\nu}q^\nu F_W)u(p), \tag{145}$$

Table 9: Major neutrino-lepton scattering experiments.

experiment [ref.]	ν beam	process	number of events	$< E_\nu >$	$\sin^2\theta_W$
Reines et al. [112]	Savannah River reactor	$\bar{\nu}_e e^- \to \bar{\nu}_e e^-$	~ 400	~ 3 MeV	0.29 ± 0.05
Irvine-LANL -Maryland [115]	LAMPF beam stop	$\nu_e e^- \to \nu_e e^-$	234	~ 35 MeV ~ 35 MeV	
CHARM [117]	CERN SPS WBB	$\nu_\mu e^- \to \nu_\mu e^-$ $\bar{\nu}_\mu e^- \to \bar{\nu}_\mu e^-$	83 112		} $0.211 \pm 0.035 \pm 0.011$
US-JAPAN [116]	BNL AGS	$\nu_\mu e^- \to \nu_\mu e^-$ $\bar{\nu}_\mu e^- \to \bar{\nu}_\mu e^-$	160 97	1.3 GeV 1.2GeV	} $0.195 \pm 0.018 \pm 0.013$
CHARM II [167]	CERN SPS WBB	$\nu_e e^- \to \nu_e e^-$ $\bar{\nu}_\mu e^- \to \bar{\nu}_\mu e^-$	2105 2215	24 GeV 19 GeV	} $0.237 \pm 0.007 \pm 0.007$
Kamiokande II [40]	solar ν	$\nu_e e^- \to \nu_e e^-$		~ 9.5 MeV	

$$\langle N(p')|A_\mu|N(p)\rangle = \bar{u}(p')(\gamma_\mu\gamma_5 F_A + \frac{1}{2m_N}\gamma_5 q_\mu F_p)u(p) \ , \tag{146}$$

where $q = p' - p$, the cross section is calculated [168] to give

$$\begin{aligned} &\frac{d\sigma}{dq^2}\begin{bmatrix} \nu_e + n & \to & p + e^- \\ \bar{\nu}_e + p & \to & n + e^+ \end{bmatrix} \\ &= \frac{G_F^2}{4\pi}\cos^2\theta_c\Big[(F_V + F_W \pm F_A)^2 + (F_V + F_W \mp F_A)^2 \\ &\quad \times \left(1 - \frac{q^2}{2E_\nu m_N}\right)^2 \\ &\quad + \left\{F_A^2 - (F_V + F_W)^2\right\}\frac{q^2}{2E_\nu^2} \\ &\quad + \left\{F_W^2\frac{(q^2 + 4m_N^2)}{4m_N^2} - 2(F_V + F_W)F_W\right\} \\ &\quad \times \left\{2 - \frac{q^2(m_N + 2E_\nu)}{2E_\nu^2 m_N}\right\}\Big] \ . \end{aligned} \tag{147}$$

Note that the second term of (146) gives a term that is proportional to the lepton mass, and is ignored in (147). From the conservation of the vector current we have

$$F_{V,W}(q^2) = F_{V,W}^{(p)}(q^2) - F_{V,W}^{(n)}(q^2) \ . \tag{148}$$

The parametrisation, taken conventionally, is

$$F_V(q^2) = \left[1 + \frac{q^2}{(0.84\text{GeV})^2}\right]^{-2} \ ,$$

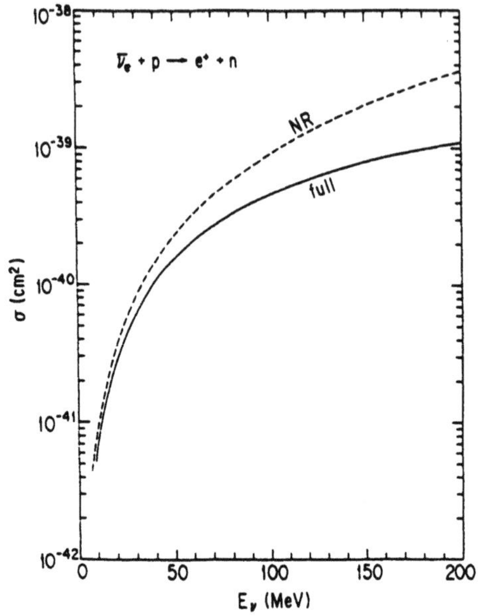

Figure 3: Cross section of $\bar{\nu}_e + p \to e^+ + n$ calculated from the full expression of eq. (147). The non-relativistic expression (NR) is also plotted for comparison.

$$
\begin{aligned}
F_W(0) &= (\mu_p - \mu_n)/\mu_N, \quad F_W(q^2) \propto F_V(q^2) \ , \\
F_A(0) &= (g_A/g_V)\left[1 + \frac{q^2}{(1.032\mathrm{GeV})^2}\right]^{-2} \ , \qquad (149)
\end{aligned}
$$

where $(\mu_p - \mu_n)/\mu_N = 4.706$ is the difference (i.e. isovector part) of the proton and neutron magnetic moments in units of the nuclear magneton, $\mu_N = e/2m_N$. The total and differential cross sections are shown for $\bar{\nu}_e + p \to n + e^+$ in Figs. 3 and 4. The forward peak develops as E_ν increases: e.g., $\sigma(0°)/\sigma(180°) = 1.7$ at E_ν=50MeV and 3.7 at E_ν=100MeV for $\bar{\nu}_e + p \to e^+ + n$.

For the neutral-current interaction the effective Hamiltonian is

$$
H_{\rm eff} = \frac{G_F}{\sqrt{2}}[\bar{\nu}\gamma^\mu(1-\gamma_5)\nu][\bar{N}\{(g_V\gamma_\mu + g_A\gamma_\mu\gamma_5)\frac{\tau_3}{2} - 2\sin\theta_W Q g_V\gamma_\mu\}N] \ , \qquad (150)
$$

where $N = (p, n)$. The cross section for $\overset{(-)}{\nu} p \to \overset{(-)}{\nu} p$ and $\overset{(-)}{\nu} n \to \overset{(-)}{\nu} n$ can then be obtained by replacing the form factors in (147) as

$$
\begin{aligned}
F_{V,W} &\longrightarrow \pm\frac{1}{2}F_{V,W} - \sin^2\theta_W F_{V,W}^{(p \text{ or } n)} \ , \\
F_A &\longrightarrow \pm\frac{1}{2}F_A \ . \qquad (151)
\end{aligned}
$$

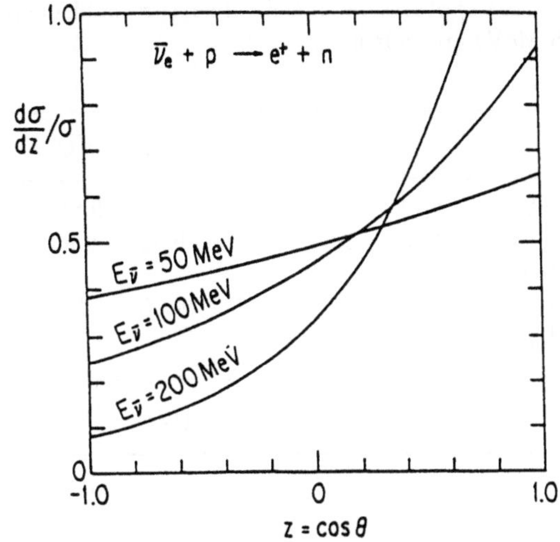

Figure 4: Normalised differential cross section of $\bar{\nu}_e + p \rightarrow e^+ + n$. θ is the CMS angle.

When the neutrino energy is larger than 1GeV, inelastic processes become increasingly important. In such situations, inclusive reactions, such as $\nu + N \rightarrow (\mu, \nu)$ + anything, are described well by scattering off quarks, $\nu + q \rightarrow \mu(\text{or } \nu) + q$, for which the cross section is given by (135) with c_L and c_R being replaced by appropriate factors given in Table 1. The νN cross section is then calculated as a convolution integral with the quark wave function in the nucleons [169]. High-energy neutrino scattering off nucleons has played a very important role in exploring the hadron structure. Although this is another very important use of neutrinos, and constitutes a major field in particle physics, we omit all such topics from the present review.

8.3 Neutrino-nucleus scattering

Neutrino-nucleus scattering (inverse beta decay) is of practical importance, since nuclei are often used as targets for the detection of low-energy neutrinos. The advantage of using a nuclear target lies in the fact that it is sensitive to a specific neutrino reaction, thus efficiently reducing the background noise. For very low-energy neutrinos, $E_\nu \lesssim$ 5MeV, this is practically the only method available today for neutrino detection.

For the low-energy limit of $\nu_e + (A, Z) \rightarrow e^- + (A, Z+1)$, only superallowed and allowed transitions give a significant contribution, and the cross section is given by

$$\sigma = \frac{G_F^2}{\pi} \cos^2\theta_c \sum_i |M_i|^2 \, E_e \, (E_e^2 - m_e^2)^{1/2} F(Z, E_e) \quad , \qquad (152)$$

where

Table 10: Neutrino (ν_e) captures in nuclei. The threshold energy for E_ν and cross section (for $E_\nu \sim 35$ MeV) are shown.

	n	^{2}H	^{12}C	^{16}O	^{37}Cl	^{56}Fe
E_{th}(MeV)	-0.78	1.44	17.34	15.43	0.814	4.57
σ ($\times 10^{-41}$cm^2)	12.0	3.5	1.46	0.52	9.90	17 – 69
σ/neutron ($\times 10^{-42}$cm^2)	120	35	2.4	0.65	4.9	5.6 – 23
ref.	—	[194]	[170]	[192]	[174]	[171]

$$|M_i|^2 = \langle 1\rangle_i^2 + g_A^2\langle\sigma\rangle_i^2 \tag{153}$$

and the summation i is taken over the levels under consideration (the isospin operator τ^+ is suppressed here) [see (B.16)] . For $\nu_e + n \to p + e^-$, $\langle 1\rangle^2 = 1$ and $\langle\vec{\sigma}\rangle^2 = 3$ and we recover (143).

We show examples of the reaction cross section in Table 10. They are much smaller than that expected for a free neutron times the number of neutrons in the nuclei, and are often even smaller than that for a single neutron.[8] The differential cross section has an angular dependence of

$$\frac{d\sigma}{d\theta} \propto 1 + \alpha\cos\theta \tag{154}$$

in the CMS, and α is

$$\alpha = \frac{\langle 1\rangle^2 - \frac{1}{3}g_A^2\langle\vec{\sigma}\rangle^2}{\langle 1\rangle^2 + g_A^2\langle\vec{\sigma}\rangle^2}\ , \tag{155}$$

as can be seen from (B.2) for the case of β decay.

For neutrino-nucleus scattering the ground-state to ground-state transition is reliably estimated from the ft value of the relevant electron capture process (or β decay), just by taking into account the spin weight factor $(2J_i+1)/(2J_f+1)$. For the transition to excited states, the Fermi transition $\langle 1\rangle^2$ is reliably calculated to an accuracy up to isospin violation [see (B.8)]. On the other hand, a reliable estimation is generally difficult for the Gamow-Teller (GT) matrix element $\langle\vec{\sigma}\rangle^2$, and the estimate generally suffers from a large uncertainty. There are, however, a few exceptional cases. The best known example is neutrino capture on ^{37}Cl; the Gamow-Teller matrix elements for $^{37}\mathrm{Cl} \to {}^{37}\mathrm{A}$ can be estimated from the β decay of the mirror process $^{37}\mathrm{Ca} \to {}^{37}\mathrm{K} + \beta^+$ (see Fig.5) [172]–[174].

[8] ^{18}O is an exceptional case; the neutrino cross section on ^{18}O is close to $2\times\sigma(\nu n)$.

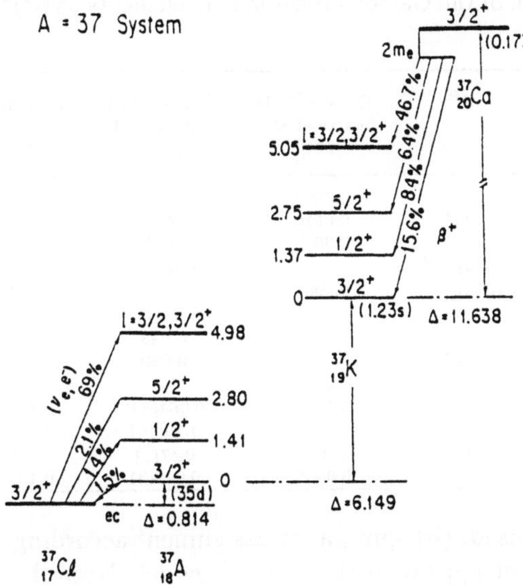

Figure 5: $A = 37$ ${}^{37}_{17}$Cl$-{}^{37}_{18}$A$-{}^{37}_{19}$K$-{}^{37}_{20}$Ca system. Energy levels and the Q-value Δ are in units of MeV. The branching fractions of ^{37}Ca $\rightarrow^{37}$K$+\beta^{+}$ are the values measured by Sextro et al. [174] and those for ^{37}Cl are approximate values for ^{37}Cl(ν_e, e^-) ^{37}A for $E_\nu \simeq 35$ MeV. Levels denoted by thick lines are the $I = 3/2$ isobaric analogue states.

For general cases, one may think of a calculation with the shell model to estimate the GT transition to the excited state. The shell model generally predicts a GT transition that is too large if the canonical value of g_A is used. For p-shell nuclei a systematic study of β-decays [175] shows that agreement is achieved with a "good wave function" [176] to an accuracy of 0.6–1.5 (in matrix element square) if g_A is reduced to a value of around ≈ 1 (quenching effect). The origin of this quenching effect has not been understood, though a number of possibilities have been discussed [177, 178]. For the $s-d$ shell, even with a "good wave function" [178], one fails to predict the matrix element within a factor of 2 (e.g., see Table 11 below). The shell-model calculation may be good only to the extent to predict global qualitative characteristics, and cannot be used for a quantitative prediction already for the $s-d$ shell nuclei, unless extra constraints are introduced from the outside.

Another general method often used is to utilise an empirical approximate proportionality between the GT matrix element and the (p, n) reaction cross section near to the forward direction [179]. Much effort has been invested in this work from its practical importance. Taddeuchi et al. [180] claim that the proportionality is good within an error of 5% for the same pair of nuclei, and that the proportionality holds within 20–50% for the more general case. For the ^{37}A system, however, where empirical knowledge exists concerning the matrix elements, the level-to-level

Table 11: Comparison of the Gamow-Teller matrix elements $g_A^2\langle\vec{\sigma}\rangle^2$ for ^{37}Cl$(\nu_e, e^-)^{37}$A.

^{37}Cl level (keV)	^{37}K level (keV)	J^P	^{37}Ca → $^{37}K\beta+$ Sextro et al. [174]	^{37}Ca → ^{37}Kβ Garcia et al. [174]	sd shell$^{(a)}$ [178]	(pn) Rapaport et al. [179]
0	0	3/2+	0.0493	0.0483	0.072	0.034
1410	1371	1/2+	0.0510	0.074(10)	0.034	$\leq$ 0.009
2796	2750	5/2+	0.0833	0.067(4)	0.273	
3170	3239	(5/2)+$^{(b)}$	—	0.004 (1)	0.058	0.23
3605	3622	(3/2)+$^{(b)}$	0.0907	0.075	0.157	
?	3840	?	0.118	0.094(5)	x$^{(c)}$	
?	4413	?	0.0464	0.043	x$^{(c)}$	
?	4495	1/2+	0.0894	0.060	0.356	0
?	4665	?	0.0561	—	x$^{(c)}$	
?	5016	?	0.1186	0.064(64)	x$^{(c)}$	
4980	5051	3/2+	3.083	3.10(10)	3.119	3.08
?	5120	1/2+	0.414	0.47(2)	x$^{(c)}$	?
5130	5320	3/2+	0.0868	0.048(3)	0.131	?

Note– (a) $g_A^{\rm eff} = 1$ is used; (b) spin-parity assignment according to the shell model; (c) levels which do not appear in the *sd* shell model. Note that the state at 5120 keV (^{37}K) is a particularly strong level.

agreement between the GT matrix elements determined from the mirror beta decay and those for the (p, n) reaction seems to be much worse than the accuracy quoted above. There is no theoretical basis for expecting this proportionality.

The accuracy of various estimates is demonstrated in Table 11 for the GT matrix element squared, $B(GT) = |\langle f|\ \vec{\sigma}\ |i\rangle|^2$, for the ^{37}Cl → ^{37}A transition. The reference values are taken from two experiments for the ^{37}Ca beta decay in ref.[174]. The situation would be much worse for heavier nuclei.

As emphasised above, ^{37}Cl, among the targets used or proposed for the solar neutrinos detection, is a fortunate example [181] in that about 65% of the expected capture cross section is given by the Fermi transition to $A(3/2^+, E = 4.98\text{MeV})$, for which a theoretical estimate is reliable, and reasonably reliable estimates are also available for the GT transitions for the other levels below 5–6MeV. ^{71}Ga is another fortunate example [182]. ^{71}Ga ($3/2^-$, gnd) → ^{71}Ge ($1/2^-$, gnd) has a sufficiently low threshold (Q=0.236MeV) to detect the copious *pp* solar neutrino component, yet the first excited state ($5/2^-$, 0.175 MeV) lies at an energy that is already high enough so that *pp* neutrinos contribute very little. Therefore, 87% of the total captures take place through the ground-to-ground transition (see Table 12 [183]). For the rest, an uncertainty may exist that is as large as by a factor of 2; however, the overall uncertainty in the total capture cross section remains small. For other proposed targets a reliable estimate of the capture cross section is far more difficult (perhaps except for light nuclei ^{7}Li and ^{11}B). For current estimates of the neutrino capture cross sections (for solar neutrinos), see refs.[184].

Nuclear targets can also be used to detect selectively the neutral-current induced neutrino reaction by measuring the γ-rays from the nuclear excitation, $\nu A \to \nu A^*$ followed by $A^* \to \gamma + A$ [185]–[189]. For low-energy nuclear exci-

Table 12: Solar neutrino capture rate on ^{71}Ga in units of SNU. This table is reconstructed from Krofcheck et al. [183] and Bahcall [39].

	$pp\nu$	other ν's	subtotal
ground state	71	44	115
excited states	0	17	17
(sub) total	71	61	132±20

tation, only the isovector axial-vector current part of (142) contributes, and the effective interaction is given by

$$H_{\text{eff}} \simeq \frac{G_F}{2\sqrt{2}} g_A \vec{\sigma} \tau_3 [\bar{\nu} \vec{\gamma} (1 - \gamma_5) \nu] \quad . \tag{156}$$

The cross section is then

$$\sigma = \frac{G_F^2}{\pi} \sum_i |M_i|^2 (E_\nu - E_i)^2 \quad , \tag{157}$$

where

$$|M_i|^2 = [\frac{1}{2} \langle \vec{\sigma}\, \tau \rangle]^2 \quad . \tag{158}$$

The matrix element is often estimated from the $ep \rightarrow e'p$ reaction or from the magnetic dipole transition by assuming $\langle \vec{\sigma} \tau_z \rangle \propto \langle \vec{\sigma} \rangle$.

Special considerations have been given to nuclear excitations of ^{6}Li : $1^+ \rightarrow 0^+$ (3.162 MeV) [186, 187] and ^{12}C : $0^+ \rightarrow 1^+$ (15.1 MeV) [186]–[188], [190, 191]. Although the 1^+ level of ^{12}C is too high for solar neutrinos, it is a good candidate to detect neutral-current induced neutrino reactions from a stellar collapse [190, 192]. ^{11}B has been proposed to be a good candidate target for solar-neutrino detection [189] (see Fig.6). This experiment is of particular interest, since the neutral-current induced reactions are not affected by neutrino oscillation or neutrino conversion in matter, the most plausible explanations for the solar neutrino problem. ^{13}C has also been considered by some authors for the same purpose [193].

A deuterium target offers a promising possibility to detect the neutral-current interaction through $\nu + d \rightarrow \nu + p + n$, with n identified by the characteristic γ-ray arising from its subsequent absorption into nuclei, e.g., $n + d \rightarrow {}^3\text{He} + \gamma$ (6.25MeV) or that from neutron capture on some nuclei doped into the detector. Calculations have been made for both $\nu + d \rightarrow \nu + p + n$ and $\nu + d \rightarrow e^- + p + p$ in ref. [194]–[197]. For these reactions one can calculate the reaction cross section to a reasonable accuracy (~10%). Figure 7 shows examples of calculations [197].

Estimations of the cross section for $E_\nu \gtrsim$ 50MeV is much more difficult, since forbidden transitions start to contribute significantly. One has to also retain the terms involving $O(E_\nu/M)$ and its higher orders in calculating the matrix elements. Yet, such a calculation is of practical importance for the detection of neutrinos of astrophysical origin. An elaborate shell-model calculation has been made for $\nu_e {}^{16}$O

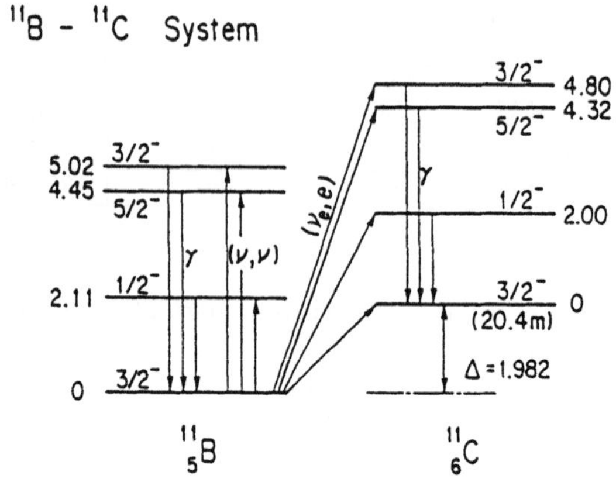

Figure 6: $^{11}_{5}$B$-^{11}_{6}$C mirror system.

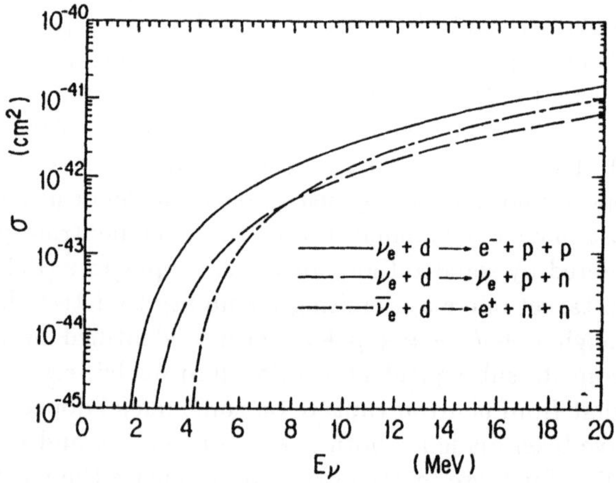

Figure 7: Neutrino reaction cross sections on deuterium. Figures are taken from calculations of Ref. [197].

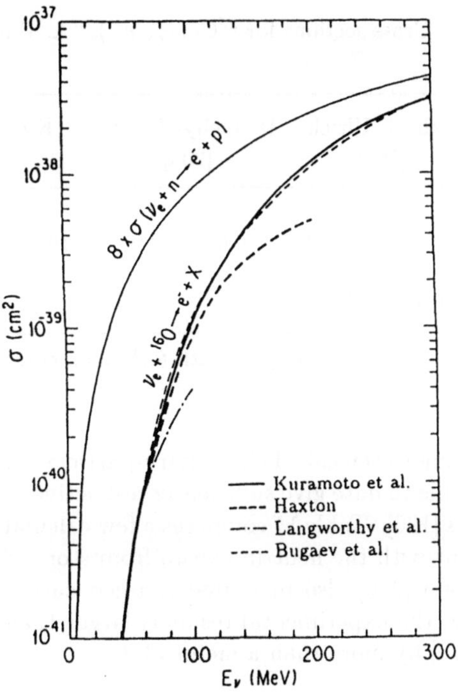

Figure 8: Neutrino reaction cross sections on ^{16}O (inclusive reactions). The calculations of Langworthy [199] and Haxton [198] use the shell model, and those of Bugaev et al. [200] and Kuramoto et al. [202] adopt a Fermi gas model. $8 \times \sigma(\nu_e + n \to e^- + p)$ is also plotted for comparison.

reactions in [198], taking account of all levels of ^{16}F states up to $2\hbar w(1\hbar w)$ excitations for positive (negative) energy states (see also [199] for an earlier calculation). The result seems to be valid up to 70 MeV. For higher energies (inclusive reaction) a Fermi gas model approximation would perhaps give a reasonable description [200]–[203]. The cross section for $\nu_e + {}^{16}\text{O} \to e^- + X$ is shown in Fig.8.

A calculation has also been made for ^{37}Cl $(\nu_e, e^-)^{37}$A up to E_ν = 300MeV [202]. The authors show that the contribution from forbidden transitions exceeds that from allowed transitions for $E \gtrsim 100$ MeV, and that the uncertainties of the calculation rapidly increases from $E_\nu \simeq$ 80MeV. For lower energies, the uncertainties are not too large for the reasons mentioned above. The cross section for ν_e from stopped μ^+ is estimated to a nominal accuracy of 25% (see Table 13 [184, 202, 204]).

Recently, neutrino-nucleus reaction cross sections were measured for the first time at LAMPF [205] and at ISIS (KARMEN experiment) [206, 207] of the Rutherford Laboratory for the ^{12}C target. The observed reactions are $\nu + {}^{12}\text{C} \to {}^{12}\text{N}(1^+,\ \text{gnd}) + e^-$ [205, 207] and $\nu + {}^{12}\text{C} \to \nu + {}^{12}\text{C}(1^+,\ 15.1\text{MeV})$ [206]. Here ^{12}N(1^+, gnd) and ^{12}C(1^+, 15.1 MeV), together with ^{12}B(1^+, gnd), form a $T = 1$ triad. For the reactions to this triad, one may apply the form-factor approach as is

Table 13: Cross sections for ^{37}Cl (ν_e, e^-) ^{37}A. Units are in cm^2.

Ref.	Bahcall-Ulrich [184]	Donnelly-Haxton [204]	Kuramoto et al. [202]
	7.2×10^{-41} (a)	9.9×10^{-41}	$(9.4 \pm 1.4) \times 10^{-41}$
	7.6×10^{-41} (b)		

Note– (a) without/(b) with forbidden corrections.

used for $\nu\, p$ scattering (often called elementary-particle approach), since the experimental data available to date give sufficiently restrictive information for determining the form factors [190]. Table 14 compares a few calculations [170, 190, 191, 208] with the experiment with the neutrino beam from stopped muons. More recently, the KARMEN group [206] also measured charged current transitions to excited states of ^{12}N. Here, the experimental result is larger than a calculation based on the shell model [170] by more than a factor of 3.

8.4 Coherent nuclear scattering

The neutral current induces coherent scattering of neutrinos off nuclei. For instance, scattering off a $T = 0$, $Z = N = A/2$ nucleus is described by the effective Hamiltonian with $g_V^{I=0} = g_V^P + g_V^A = -\sin^2\theta_W$. With the matrix element given by

$$\langle A(p')|V_\mu^{I=0}|A(p)\rangle = \frac{1}{(2\pi)^3}\frac{1}{\sqrt{2E'2E}}(p+p')_\mu\, g_V^{I=0} f^{I=0}(q^2) \quad , \tag{159}$$

where

$$f(q^2) = Ae^{-bq^2}, \qquad b = \frac{1}{6}\langle r^2\rangle \quad , \tag{160}$$

the cross section is

$$\frac{d\sigma}{dq^2} = \frac{G_F^2}{2\pi}(g_V^{I=0})^2 A^2 e^{-2bq^2}\left(1 - q^2\frac{2m_A E_\nu + m_A^2}{4m_A^2 E_\nu}\right) \quad , \tag{161}$$

where $q = p' - p$ and $m_A = Am_N$. For a small q,

$$\frac{d\sigma}{d\cos\theta} = \frac{G_F^2}{2\pi}\left(g_V^{I=0}\right)^2 A^2 E_\nu^2(1+\cos\theta) \quad , \tag{162}$$

and we have

$$\sigma \simeq 1.68 \times 10^{-42}\sin^4\theta_W A^2\left(\frac{E_\nu}{10\mathrm{MeV}}\right)^2 \mathrm{cm}^2 \quad . \tag{163}$$

Table 14: Cross section of $\nu+$ ^{12}C reactions (in units of 10^{-42}cm^2). ν_e+ $^{12}\mathrm{C} \to e^-+$ $^{12}\mathrm{N}_{\mathrm{gnd}}$.

Donnelly [170]	Fukugita et al. [190]	Minz et al. [208]	expt.-LAMPF [205]	expt.-ISIS [206]
12$^{(a)}$	9.2	9.0	10.5±1.0±1.0	8.0±0.75± 0.75

Note– (a) 9.4 is quoted in [205].

ν_e+ $^{12}\mathrm{C} \to \nu_e+$ $^{12}\mathrm{C}^*$ (15.1) plus $\bar{\nu}_\mu+$ $^{12}\mathrm{C} \to \bar{\nu}_\mu+$ $^{12}\mathrm{C}^*$(15.1)

Fukugita et al. [190]	Pourkaviani et al. [191]	expt.-ISIS [206]
10.4$^{(b)}$	10.7$^{(b)}$	9.5 ± 1.8 ± 1.35

Note– (b) Fukugita et al. give $\sigma(\nu_e\mathrm{C})/\sigma(\bar{\nu}_\mu\mathrm{C})$=1.84, whereas Pourkaviani et al. give 0.84 for the same ratio.

For $E_\nu \simeq$ 10MeV and $A = 50$, $\sigma = 2.1 \times 10^{-40}$cm^2 yields the mean free path length $\lambda = 780\mathrm{m}/(\rho/10^{13}\mathrm{g\ cm}^{-3})$. This value is compared with $\lambda = 2 \times 10^4\mathrm{m}/(\rho/10^{13}\mathrm{g\ cm}^{-3})$ for νe scattering. This provides a mechanism for a neutrino trap in the core of protoneutron stars after supernova core collapse, and plays an important role in supernova explosions [209]. The large cross section of (163) is also one of the basic principles for cryogenic detection for low-energy neutrinos [210] (and for Dirac neutrino like dark matter particles in the Galactic halo [211]).

8.5 Refractive effects

Neutrinos propagating through matter receive a refractive effect due to their interaction with matter. If the interaction gives extra energy V, the energy (E)–momentum (k') relation is

$$E = \sqrt{k'^2 + m^2} + V \quad . \tag{164}$$

The refractive index n is defined by

$$\Psi = \exp(i\ n\ \vec{k}\vec{x} - iEt) \quad , \tag{165}$$

where k and E satisfy the dispersion relation in vacuum, $k^2 + m^2 = E^2$. Since k' is identified with nk,

$$(nk)^2 + m^2 = E^2 - 2EV \ . \tag{166}$$

Namely,

$$n = 1 - \frac{EV}{k^2} \ . \tag{167}$$

For an interaction of the form (130),

$$V = \sqrt{2} G_F c_V n_e \tag{168}$$

$$= \pm \sqrt{2} G_F (\pm 1/2 + 2 \sin^2 \theta_W) n_e \tag{169}$$

with n_e the electron density when matter is at rest (only the $\mu = 0$ component contributes, i.e. $\langle \bar{e} \gamma_\mu e \rangle = \langle e^\dagger e \rangle = n_e \delta_{\mu 0}$; also note that $\bar{\nu}_e \gamma^0 (1 - \gamma_5) \nu_e = 2 \nu_{e_L}^\dagger \nu_{e_L}$). The $\pm$ sign in the parenthesis refers to $\nu_e(+)$ and $\nu_{\mu,\tau}(-)$ and that in front to $\nu(+)$ and $\bar{\nu}(-)$. Although the deviation of n from unity is very small, $\Delta n \sim 7.6 \times 10^{-19} (\rho/100\text{g cm}^{-3})(E/10\text{MeV})^{-1}$, this causes an important effect on the neutrino propagation if the neutrino has a small mass, as discussed in Sect.26.

Similarly, for interactions with protons and neutrons we have from Table 1,

$$V = \pm \sqrt{2} G_F (1/2 - 2 \sin^2 \theta_W) n_p \mp \sqrt{2} G_F (1/2) n_n \ , \tag{170}$$

where the $\pm$ signs refer to ν(upper) and $\bar{\nu}$(lower) for all three neutrinos.

The phase of the neutrino induced by this coherent effect rotates by 2π when it propagates typically $2\pi/V \sim 1.6 \times 10^9 \text{cm} (\rho/1gH)^{-1}$ (gH stands for gramme hydrogen equivalent). This is an effect first order in G_F and is greatly shorter than the interaction length for the low energy neutrino $l \sim 1/\sigma(\nu_e e \to \nu_e e) n_e \sim 1.8 \times 10^{19} \text{cm} \ (E/10 \text{ MeV})^{-1} (\rho/1\text{gH})^{-1}$.

The refractive effect in the general high-temperature background, such as in the early universe, and higher-order effects are discussed in refs.[212, 213, 214].

III. Massive neutrinos

9 Massive neutrinos –overview

The mass term connects a left-handed (LH) field to its right-handed (RH) partner, i.e., it is the term which flips the chirality of a particle. There are two possible mass terms, the Dirac mass term and the Majorana mass term,

$$m(\bar{\psi}_R\psi_L + h.c.) \tag{1}$$

and

$$M(\overline{\psi^c_L}\psi_L + h.c.), \tag{2}$$

where $\psi^c = C\gamma^0\psi^*$ is the charge-conjugated field of ψ and $\psi^c_L \equiv (\psi_L)^c = \frac{(1+\gamma_5)}{2}\psi^c$ has a right-handed chirality. The former is the term $m\bar{\psi}\psi$ giving a mass to quarks and charged leptons. The latter obviously violates lepton-number conservation by two units, and makes a particle and its antiparticle indistinguishable. (The classical experiment by Davis [215], who showed that $\bar{\nu}+\ {}^{37}\mathrm{Cl} \to {}^{37}\mathrm{A} + e^-$ does not take place, is interpreted to be by a virtue of the right-handed helicity of $\bar{\nu}$; this test does not preclude the possibility of the Majorana neutrino.) As discussed in Sect.5, a Majorana mass term is allowed only for the neutrino. When neutrinos have a Majorana mass term, they are usually called the "Majorana neutrino". Since this terminology is not quite accurate, a more precise definition is given in this chapter.

The Majorana field is originally a name given to some two-component (massless) fermion fields [24], constructed from the Dirac field by imposing the reality condition (Majorana condition). Another familiar representation of the two-component filed is the Weyl field [45], defined by the chirality projection of the Dirac field, $\chi_\pm = \frac{1\pm\gamma_5}{2}\psi$ (equivalently the Dirac field is given by two Weyl fields). It is shown that the Majorana field is unitary equivalent to the Weyl field. If there exists only one Weyl field χ, the Majorana mass would be the only mass allowed for the field; this necessarily breaks the lepton number. If there are two Weyl fields $\chi^{(1)}$ and $\chi^{(2)}$, there may exist a mass term which conserves the lepton number defined by $L = L^{(1)} - L^{(2)}$ with $L^{(i)}$ referring to $\chi^{(i)}$. In particular, if $\chi^{(2)}_L = \chi^{(1)c}_L$, the mass is called the Dirac mass.

This implies that there are two ways to extend the standard electroweak theory so as to incorporate massive neutrinos. If there exists ν_R in addition to ν_L one can construct a Dirac mass term and treat the neutrino mass in parallel to the mass term for other charged particles. If only the ν_L exists, the Majorana mass is the only possible form to give the neutrino a finite mass. Within the Weinberg-Salam theory, however, the Majorana mass term is necessarily non-renormalisable, and is taken as an effective interaction [23] that arises from more fundamental interactions at a higher energy scale. In turn, this would provide an explanation as to why the mass of the neutrino is so small compared with the mass of other charged particles.

In fact, the problem of how to understand the smallness of the mass of neutrinos is one of the central issues of this review. The simplest example is the so-called seesaw mechanism [216], in which the right-handed neutrino ν_R has a large Majorana mass M and the left-handed neutrino ν_L is given a mass through leakage of the order of $\sim (m/M)$, with m the Dirac mass which causes mixing between ν_L and ν_R and is taken to be of the order of the other charged particle mass. Such a mechanism, indeed, plays a crucial role in grand unification, which usually requires that the neutrino Dirac mass and the charge $-2/3$ quark mass be related.

The feature of the Majorana field, that the particle is not distinguished from the antiparticle, makes the quantisation somewhat non-trivial. A discussion concerning quantisation is important in order to calculate the process involving Majorana fields. This problem is discussed in full in the next chapter. The phenomenological consequences of the massive neutrinos are discussed mostly in Chapter IV.

10 Dirac matrices

In order to show the features that are characteristic of the Dirac, Weyl and Majorana fields, it is convenient to introduce different Dirac matrix representations which are related by unitary transformations to each other. We take the convention

$$\begin{aligned} \{\gamma^\mu, \gamma^\nu\} &= 2g^{\mu\nu} , \\ g^{\mu\nu} &= \mathrm{diag}(1,-1,-1,-1) . \end{aligned} \tag{3}$$

<u>For Dirac field</u> The representation is what is most conventionally taken in the literature (Feynman representation):

$$\begin{aligned} \gamma^0 &= \begin{pmatrix} 1 & 0 \\ 0 & -1 \end{pmatrix}, \gamma^i = \begin{pmatrix} 0 & \sigma^i \\ -\sigma^i & 0 \end{pmatrix} , \\ \gamma^5 &= i\gamma^0\gamma^1\gamma^2\gamma^3 = \begin{pmatrix} 0 & 1 \\ 1 & 0 \end{pmatrix} , \\ \sigma^{\mu\nu} &= \frac{i}{2}[\gamma^\mu, \gamma^\nu] , \end{aligned} \tag{4}$$

where all elements stand for 2×2 matrices and σ^i is the Pauli matrix.
<u>For Weyl field</u> We take

$$\gamma^0 = \begin{pmatrix} 0 & -1 \\ -1 & 0 \end{pmatrix}, \gamma^i = \begin{pmatrix} 0 & \sigma^i \\ -\sigma^i & 0 \end{pmatrix} ,$$

$$\gamma^5 = \begin{pmatrix} 1 & 0 \\ 0 & -1 \end{pmatrix}, \tag{5}$$

so that

$$\frac{1+\gamma_5}{2} = \begin{pmatrix} 1 & 0 \\ 0 & 0 \end{pmatrix}, \quad \frac{1-\gamma_5}{2} = \begin{pmatrix} 0 & 0 \\ 0 & 1 \end{pmatrix} \tag{6}$$

are projections to the upper (right-handed) and lower (left-handed) components, respectively. These matrices are related to (4) by a unitary transformation:

$$\gamma_\mu(\text{Weyl}) = \frac{1}{\sqrt{2}}(1-\gamma_5\gamma_0)\gamma_\mu\frac{1}{\sqrt{2}}(1+\gamma_5\gamma_0), \tag{7}$$

where the γ's on the right-hand side are those of (4).
For Majorana field Because the field is real in nature, it is convenient to adopt the representation

$$\text{Re}[\gamma^\mu] = 0 \quad , \tag{8}$$

so that all components of the Dirac equation is also real. The representation is expressed [24] as,

$$\begin{aligned}\gamma^0 &= \begin{pmatrix} 0 & \sigma_2 \\ \sigma_2 & 0 \end{pmatrix}, \gamma^1 = -i\begin{pmatrix} \sigma_3 & 0 \\ 0 & \sigma_3 \end{pmatrix}, \gamma^2 = \begin{pmatrix} 0 & \sigma^2 \\ -\sigma_2 & 0 \end{pmatrix}, \gamma^3 = i\begin{pmatrix} \sigma_1 & 0 \\ 0 & \sigma_1 \end{pmatrix} \\ \gamma^5 &= \begin{pmatrix} 0 & \sigma_1 \\ \sigma_1 & 0 \end{pmatrix}\end{aligned} \tag{9}$$

The unitary transformation between (9) and (4) is given by

$$\gamma_\mu(\text{Majorana}) = \frac{1}{\sqrt{2}}(1-\gamma_2)\gamma_\mu\frac{1}{\sqrt{2}}(1+\gamma_2) \quad . \tag{10}$$

11 Massless neutrinos

We define here the Weyl and Majorana fields and show their equivalence. A more rigorous argument will be found in Sect.17. We now work with the Weyl representation. The two-component Weyl spinors are defined by

$$\psi_L \equiv \frac{1-\gamma_5}{2}\psi = \begin{pmatrix} 0 \\ \eta \end{pmatrix} \tag{11}$$

and

$$\psi_R \equiv \frac{1+\gamma_5}{2}\psi = \begin{pmatrix} \xi \\ 0 \end{pmatrix}, \tag{12}$$

where ψ is the Dirac spinor

$$\psi = \begin{pmatrix} \xi \\ \eta \end{pmatrix}. \tag{13}$$

Charge conjugation is given by

$$\psi^c = C^\dagger\psi C = C\bar{\psi}^T \ . \tag{14}$$

With the choice of $C = i\gamma^2\gamma^0$, we have

$$\psi^c = i\gamma^2\psi^* \ . \tag{15}$$

Then, charge conjugation of ψ_L and ψ_R is

$$(\psi_L)^c = \begin{pmatrix} i\sigma_2\eta^* \\ 0 \end{pmatrix} = \frac{1+\gamma_5}{2}\begin{pmatrix} i\sigma_2\eta^* \\ 0 \end{pmatrix} , \tag{16}$$
$$(\psi_R)^c = \begin{pmatrix} 0 \\ -i\sigma_2\xi^* \end{pmatrix} = \frac{1-\gamma_5}{2}\begin{pmatrix} 0 \\ -i\sigma_2\xi^* \end{pmatrix} .$$

The kinetic term is written

$$\mathcal{L} = i\bar{\psi}_R \not{\partial}\psi_R + i\bar{\psi}_L \not{\partial}\psi_L \ . \tag{17}$$

The Majorana field is defined by the Majorana condition that

$$\chi^c = \pm\chi \tag{18}$$

imposed on a four-component spinor χ. Let us define the two fields by [217]

$$\chi \equiv \psi_L + \psi_L^c , \tag{19}$$
$$\omega \equiv \psi_R - \psi_R^c . \tag{20}$$

These two fields obviously satisfy the Majorana condition, and are taken to be Majorana fields. Conversely, we have

$$\frac{1-\gamma_5}{2}\chi = \psi_L , \ \frac{1+\gamma_5}{2}\chi = \psi_L^c ,$$
$$\frac{1-\gamma_5}{2}\omega = -\psi_R^c , \ \frac{1+\gamma_5}{2}\omega = \psi_R . \tag{21}$$

Substituting (19) and (20) into (17), the kinetic term is written

$$\mathcal{L} = \frac{i}{2}(\bar{\chi} \not{\partial}\chi + \bar{\omega} \not{\partial}\omega), \tag{22}$$

the two terms corresponding to left- and right-handed fields.

The advantage of (17) over (22) is that introducing the gauge coupling is simply made by replacing ∂ with $D = \partial - igT^aA^a$ in (17), while the procedure appears to be more complicated with the kinetic term (22), unless the representation of the gauge group is real.

12 Massive Majorana neutrinos

The Lagrangian with a mass term for the Majorana field is given by

$$\mathcal{L} = \frac{i}{2}\bar{\chi} \not{\partial}\chi - \frac{1}{2}m\bar{\chi}\chi \ , \tag{23}$$

by omitting the term for the ω field in (22). We assume here that m is real. As a generalisation of (19), if we define

$$\chi = (e^{i\alpha}\psi_L + e^{i\beta}\psi_L^c), \tag{24}$$

χ satisfies the Majorana condition in an extended sense,

$$\chi^c = e^{-i(\alpha+\beta)}\chi. \tag{25}$$

With (24), (23) reads

$$\mathcal{L}' = i\bar{\psi}_L \,\partial\!\!\!/\psi_L - \frac{m}{2}(e^{-i(\alpha-\beta)}\bar{\psi}_L\psi_L^c + e^{i(\alpha-\beta)}\bar{\psi}_L^c\psi_L). \tag{26}$$

Writing

$$M = \frac{1}{2}me^{i(\alpha-\beta)}, \tag{27}$$

we obtain

$$\mathcal{L}' = i\bar{\psi}_L \,\partial\!\!\!/\psi_L - M\bar{\psi}_L^c\psi_L + h.c. \tag{28}$$

Here, the mass M is complex, but its phase can be absorbed into the phase of the ψ_L field. (The phase in (24), in fact, accounts for the phase of ψ_L.) The second term obviously breaks the lepton number carried by the ψ_L field.

13 Massive Dirac neutrinos

If there are two Weyl fields $\psi_L^{(1)}$ and $\psi_L^{(2)}$, we can construct a mass term as

$$\mathcal{L} = -\frac{m_{ij}}{2}\overline{\psi_L^{(i)c}}\psi_L^{(j)} + h.c. \ , \tag{29}$$

as a generalisation of (28). If $m_{ii} = 0$, the "lepton number" $L_i - L_j$ is conserved. Hence, if we define two fields ψ_L and ψ_R by

$$\psi_L = \psi_L^{(i)}, \ \psi_R = [\psi_L^{(j)}]^c \ , \tag{30}$$

(29) gives a conventional Dirac mass term,

$$\mathcal{L}_{\text{mass}} = -m\bar{\psi}\psi = -m(\bar{\psi}_L\psi_R + \bar{\psi}_R\psi_L) \ , \tag{31}$$

for the Dirac field,

$$\psi = \begin{pmatrix} \psi_R \\ \psi_L \end{pmatrix}. \tag{32}$$

The kinetic term is given by

$$\mathcal{L}_{\text{kin}} = i\bar{\psi}\,\partial\!\!\!/\psi = i(\bar{\psi}_L\,\partial\!\!\!/\psi_L + \bar{\psi}_R\,\partial\!\!\!/\psi_R) \ , \tag{33}$$

as usual.

If we assign $\psi_L^{(i)}$ and $\psi_L^{(j)c}$ to fields with different flavours (e.g., $\psi_L^{(1)} = \nu_{eL}$, $\psi_L^{(2)} = \nu_{\mu L}$), we have a Dirac mass term which is off-diagonal in the flavour degrees of freedom. This possibility was originally discussed by Zeldovich, Konopinski and Mahmoud [218], and re-presented in a modern form by Wolfenstein [219]. This case is often called a pseudo-Dirac mass term.

14 Massive neutrinos in the Weinberg-Salam theory

From the argument given above, we can introduce a Dirac mass term if there exists ν_R in addition to ν_L, i.e.,

$$\mathcal{L}_{\text{mass}} = m\bar{\nu}_R\nu_L \quad , \tag{34}$$

which is induced by giving the Higgs field ϕ^0 in eq.(II.32) a vacuum expectation value through the Yukawa coupling,

$$\mathcal{L}_{\text{mass}} = f_\nu \bar{\nu}_R \phi l_L \quad . \tag{35}$$

One can take this term in an equal footing way, as is done for the other mass terms in (II.32).

If there is no ν_R, however, the Majorana mass term is the only mass term that gives the neutrino a mass. Since $\bar{\nu}_L^c\nu_L$ is an SU(2) triplet state, one may introduce a triplet Higgs particle [220, 221] to make the Lagrangian SU(2) invariant.[8] The existence of such a particle, however, is excluded by the Z^0 width experiment, and the simplest possible mass term is of the form

$$\mathcal{L} = G\bar{l}_L^c l_L \frac{\phi\phi}{M} \quad , \tag{36}$$

and the neutrino mass is given by $m_\nu = G\langle\phi_0^2\rangle/M$. This Lagrangian is non-renormalisable and M is an effective mass. The form seen in (36) gives a hint as to how the smallness of the neutrino mass is realised, i.e., $m_\nu \to 0$ as $M \to \infty$. On the contrary, for the Dirac mass term (35) we must impose that $f_\nu \ll f_e$ by many orders of magnitude by hand.

As a final remark, we consider $\nu_L\phi \to \nu_L^c\phi$ scattering with interaction (36). The cross section is calculated to be

$$\sigma = \frac{2}{\pi}\frac{G^2}{M^2} \quad . \tag{37}$$

On the other hand, the ℓ-th partial wave cross section is bounded by unitarity to be

$$\sigma_\ell \leq \frac{4\pi(2\ell+1)}{2}\frac{1}{k^2} \quad . \tag{38}$$

Since the considered process takes place with the s-wave, we must have

$$\frac{2}{\pi}\left(\frac{G}{M}\right)^2 \leq \frac{8\pi}{s} \quad . \tag{39}$$

The requirement that this holds up to $s = M_{\text{pl}}$ leads to the bound $G/M \leq 2\pi/M_{\text{pl}}$, or

[8] The possibility was conjectured that there exists the coupling $\mathcal{L} = h\bar{\nu}_L^c\nu_L\chi$ with χ an SU(2) triplet. When χ develops a small vacuum expectation value, lepton number is violated and ν_L acquires a small Majorana mass (triplet Majoron model) [220]. However, physical χ particles, which should be massless, couple to the Z boson [221], and such a model is excluded by the Z^0 decay width experiment. For a more complicated model to avoid this constraint, see [222].

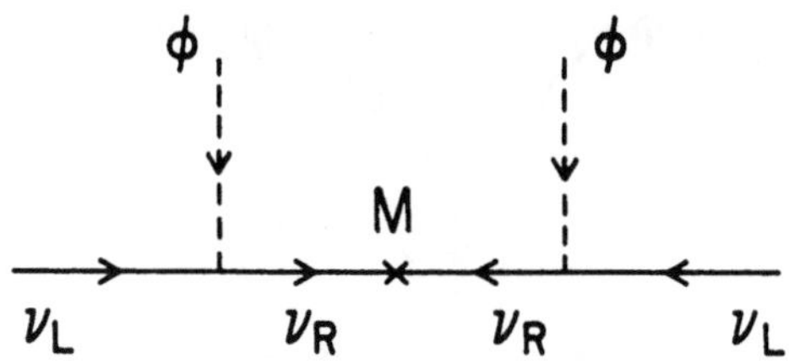

Figure 9: Seesaw mechanism that gives the left-handed neutrino a small Majorana mass.

$$m_\nu < 1.5 \times 10^{-5}\text{eV} \quad , \tag{40}$$

for $\langle\phi\rangle = 250/\sqrt{2}$ GeV. If m_ν turns out not to satisfy (40), there must be a new physical scale below the Planck mass.

15 Seesaw mechanism

The seesaw mechanism [216] is perhaps the simplest model which leads to an effective interaction of form (36) within a renormalisable class of interactions. Let us assume that the mass term is given by

$$\mathcal{L} = f\overline{\nu_R}\nu_L\langle\phi^0\rangle + M\overline{\nu_R^c}\nu_R + h.c. \tag{41}$$

Then, the diagram shown in Fig.9, when the heavy field ($M \gg \langle\phi\rangle$) is integrated out, yields

$$\mathcal{L}_{\text{eff}} = \frac{f^2}{M}\phi^0\phi^0\overline{\nu_L^c}\nu_L \quad , \tag{42}$$

the same form as in (36). Equivalently, this is also obtained by diagonalising the mass matrix,

$$\begin{matrix} & \nu_L \quad \nu_R & \\ & \begin{pmatrix} 0 & m \\ m & M \end{pmatrix} & \begin{matrix} \nu_L \\ \nu_R \end{matrix} \end{matrix} \quad , \tag{43}$$

where two rows (columns) refer to left- and right-handed neutrinos, and the Dirac mass $m = f\langle\phi\rangle$ in (41) induces mixing between the two sectors. We then have

$$m_{\nu_L} \simeq \frac{m^2}{M} \tag{44}$$

in agreement with (42). We note that we have reversed the sign of m_{ν_L} using the degree of freedom for the phase factor (27). The attractive feature of this model is that the smallness of the neutrino mass can be understood in terms of a large-mass scale M, which often appears in higher unification theories.

An alternative idea to give an effective form of (36) is to introduce an SU(2) triplet scalar particle χ, the mass of which is sufficiently large ($M_\chi \gtrsim 45$GeV) that

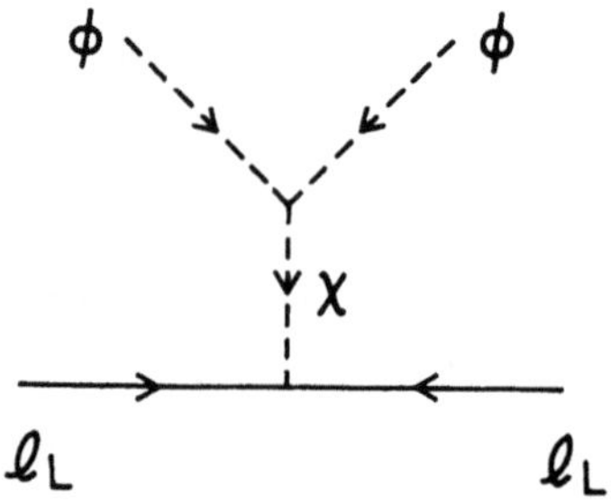

Figure 10: Another seesaw mechanism without a heavy fermion.

it is not visible in the decay of Z^0. Another constraint is that χ can not have a large vacuum expectation value (typically $\langle\chi\rangle \lesssim$ 5GeV) from the ρ parameter, which is very close to unity. A possible mechanism giving rise to the effective term (36) is therefore to consider the process depicted in Fig.10 which is induced from the Lagrangian

$$\mathcal{L} = h\overline{l_L^c} l_L \chi + \lambda M \phi\phi\chi + M^2|\chi|^2 \ + \ h.c. \ . \tag{45}$$

In fact, integration of the χ field in this diagram leads to

$$\mathcal{L}_{\rm eff} = \frac{h\lambda}{M}\phi\phi\overline{l}_L^c l_L \ , \tag{46}$$

which is also in agreement with (36).

The use of the triplet χ is similar to the triplet Majoron model. However, there are two advantages with (45). First, there is no massless Majoron, since the lepton number symmetry is explicitly broken by the $\lambda M\phi\phi\chi$ term and, hence, is experimentally viable. Second, the small vacuum expectation value, $\chi = \lambda\langle\phi\rangle^2/M$, or the small neutrino mass is natural because of the large M.

IV. Quantisation of the Majorana field

16 Necessity of quantisation

A proper formulation of the quantisation is essential for developing a quantum field theory with Majorana neutrinos. This is, however, not so obvious a problem, since there exists a constraint which identifies a particle with an antiparticle. Because of this constraint, which makes the field real, the Majorana field has both positive- and negative-frequency parts. In order to define the energy eigenstate, separating the positive frequency part from the other, quantisation plays a crucial role.

In this chapter we develop a quantum field theory of Majorana particles following Case [223]. For this purpose we first provide a full proof for the unitary equivalence of the Majorana field and the Weyl field, a subject we have already discussed in Sect.11. We then discuss quantisation of the massive Weyl field, which is much easier to handle due to the absence of a constraint. A few elementary examples are given for calculating Feynman diagrams involving Majorana fields. Examples include neutrinoless double beta decay, which is practically the most important application that involves a propagator of the Majorana field.

17 Unitary equivalence of the Majorana and the Weyl fields

We start with the four-component Dirac spinor $\psi(x)$, which is defined by the transformation property under the Lorentz group, as

$$\psi(x) \to \psi'(x') = D(\alpha)\psi(x) \equiv \exp\left(-\frac{\alpha_{\mu\nu}}{2}\sigma^{\mu\nu}\right)\psi \ , \tag{1}$$

$$x'^{\mu} = \Lambda^{\mu}{}_{\nu}x^{\nu} \ , \tag{2}$$

where $\alpha_{\mu\nu}$ are six real parameters corresponding to the Lorentz-transformation $\Lambda^{\mu}{}_{\nu}$, and $\sigma^{\mu\nu}$ in the spinor representation $D(\alpha)$ are defined by the commutator of the Dirac γ matrices (III.4), which satisfy the Cliford algebra (III.3).

If we take the Weyl representation of the γ matrices (III.5), we can show straight-forwardly that $D(\alpha)$ is reduced into 2×2 diagonal matrices,

$$D(\alpha) = \begin{bmatrix} \exp[-i(\frac{i}{2}\alpha_{0i}\sigma_i + \frac{\alpha_{ij}}{2}\epsilon_{ijk}\sigma_k)] & 0 \\ 0 & \exp[-i(-\frac{i}{2}\alpha_{0i}\sigma_i + \frac{\alpha_{ij}}{2}\epsilon_{ijk}\sigma_k)] \end{bmatrix} . \tag{3}$$

There are two convenient ways to decompose a four-component Dirac spinor into two irreducible representations of the Lorentz group. One of them is decomposition into two Weyl spinors,

$$\psi = \psi_L + \psi_R = \begin{pmatrix} \xi \\ \eta \end{pmatrix} , \tag{4}$$

where ψ_L and ψ_R are defined in (III.11) and (III.12). From (3) it is obvious that η and ξ form independent irreducible vector spaces, and transform under the Lorentz group as

$$\begin{aligned}\eta(x) \to \eta'(x') &= M\eta(x) = \exp[-i(\frac{\alpha_{ij}}{2}\epsilon_{ijk}\sigma_k - i\frac{\alpha_{0j}}{2}\sigma_j)]\eta(x) \ , \quad (5)\\ \xi(x) \to \xi'(x') &= \tilde{M}\xi(x) = \exp[-i(\frac{\alpha_{ij}}{2}\epsilon_{ijk}\sigma_k + \frac{\alpha_{0j}}{2}\sigma_j)]\xi(x) \ . \quad (6)\end{aligned}$$

Here M and $\tilde{M}$ satisfy

$$\det M = \det\tilde{M} = 1 \ , \quad (7)$$

and are thus independent two-dimensional representations of SL(2,C). This proves that the Weyl representation of the Lorentz group is a 2×2 irreducible representation of SL(2,C) [45].

Another reduction of the Dirac spinor is given by [24]

$$\chi_\pm = \frac{\psi \pm \psi^c}{\sqrt{2}}, \quad (8)$$

where ψ^c is the charge conjugation of ψ defined by (III.14). Since ψ^c has the same transformation property as that for ψ under the Lorentz group, $\chi_\pm$ are also Lorentz spinors. $\chi_\pm$ obviously satisfy

$$\chi_\pm^c = \pm\chi_\pm. \quad (9)$$

These $\chi_\pm$ are the fields which were introduced by Majorana. By taking the Majorana representation of γ matrices (III.9) it is readily shown that $\chi_\pm$ are four-component real fields. In what follows we show that these $\chi_\pm$ and (χ, ω) defined in (III.19–21) are unitary equivalent [224]–[227], [217], [228]–[229].

Let us now construct a unitary operator U that connects the Majorana spinor with a Weyl spinor,

$$\begin{aligned}U\chi_+U^{-1} &= \psi_L + \psi_L^c \equiv \chi \ ,\\ U\chi_-U^{-1} &= \psi_R - \psi_R^c \equiv \omega \ ,\end{aligned} \quad (10)$$

or equivalently,

$$\begin{aligned}U^{-1}\psi_L U &= \frac{1-\gamma_5}{2}\chi_+ \ , \quad U^{-1}\psi_R U = \frac{1+\gamma_5}{2}\chi_- \ ,\\ U^{-1}\psi_L^c U &= \frac{1+\gamma_5}{2}\chi_+ \ , \quad U^{-1}\psi_R^c U = -\frac{1-\gamma_5}{2}\chi_- \ ,\end{aligned} \quad (11)$$

so that the two sets of spinors are unitary equivalent.

For this purpose, we first expand the Dirac field $\psi(x)$ in terms of wave functions $u_r(\vec{p})$ and $v_r(\vec{p})$ of a massless fermion as

$$\psi(x) = \frac{1}{(2\pi)^{3/2}} \sum_{r=1}^{2} \int d^3p[a_r(\vec{p})u_r(\vec{p})e^{ipx} + b_r^\dagger(\vec{p})v_r(-\vec{p})e^{-ipx}] \ , \tag{12}$$

where u and v satisfy

$$\not{p}u_r(\vec{p}) = 0, \quad v_r(-\vec{p}) = Cu_r^*(\vec{p}) \ , \tag{13}$$

$$\gamma_5 u_1(\vec{p}) = u_1(\vec{p}), \quad \gamma_5 u_2(\vec{p}) = -u_2(\vec{p}) \ . \tag{14}$$

The $\psi_L, \psi_L^c, \psi_R, \psi_R^c$ and $\chi_\pm$ fields are expanded similarly:

$$\psi_L(x) = \frac{1}{(2\pi)^{3/2}} \int d^3p[a_1(\vec{p})u_1(\vec{p})e^{ipx} + b_2^\dagger(\vec{p})v_2(-\vec{p})e^{-ipx}] \ , \tag{15}$$

$$\psi_L^c(x) = \frac{1}{(2\pi)^{3/2}} \int d^3p \left[b_2(\vec{p})u_2(\vec{p})e^{ipx} + a_1^\dagger(p)v_1(-\vec{p})e^{-ipx}\right] \ , \tag{16}$$

$$\psi_R(x) = \frac{1}{(2\pi)^{3/2}} \int d^3p[a_2(\vec{p})u_2(\vec{p})e^{ipx} + b_1^\dagger(\vec{p})v_1(-\vec{p})e^{-ipx}] \ , \tag{17}$$

$$\psi_R^c(x) = \frac{1}{(2\pi)^{3/2}} \int d^3p[b_1(\vec{p})u_1(\vec{p})e^{ipx} + a_2^\dagger(\vec{p})v_2(-\vec{p})e^{-ipx}] \ , \tag{18}$$

$$\begin{aligned} \chi_+(x) \ &= \ \frac{1}{(2\pi)^{3/2}} \sum_{r=1}^{2} \int d^3p \frac{1}{\sqrt{2}}[(a_r(\vec{p}) + b_r(\vec{p}))u_r(\vec{p})e^{ipx} \\ &\quad + (a_r^\dagger(\vec{p}) + b_r^\dagger(\vec{p}))v_r(-\vec{p})e^{-ipx}] \ , \end{aligned} \tag{19}$$

$$\begin{aligned} \chi_-(x) \ &= \ \frac{1}{(2\pi)^{3/2}} \sum_{r=1}^{2} \int d^3p \frac{1}{\sqrt{2}}[(a_r(\vec{p}) - b_r(\vec{p}))v_r(\vec{p})e^{ipx} \\ &\quad + (-a_r^\dagger(\vec{p}) + b_r^\dagger(\vec{p}))v_r(-\vec{p})e^{-ipx}] \ . \end{aligned} \tag{20}$$

Let us write U in the form

$$U = e^{\theta(b^\dagger a - a^\dagger b)}. \tag{21}$$

For a real θ, U satisfies $U^{-1} = U^\dagger$.
Now,

$$\begin{aligned} U^{-1}aU \ &= \ e^{-\theta(b^\dagger a - a^\dagger b)} a e^{\theta(b^\dagger a - a^\dagger b)} \\ &= \ a + \theta[(b^\dagger a - a^\dagger b), a] \\ &\quad + \frac{\theta^2}{2!}[(b^\dagger a - a^\dagger b), [(b^\dagger a - a^\dagger b), a]] \\ &\quad + \dots \\ &= \ a\cos\theta + b\sin\theta \ , \end{aligned} \tag{22}$$

where the use is made of $[(b^\dagger a - a^\dagger b), a] = b$ from the second to the third line. Similarly,

$$U^{-1} b U = a\cos\theta - b\sin\theta \ . \tag{23}$$

With the choice of $\theta = \pi/4$, i.e.,

$$U = \exp[\frac{\pi}{4}(b_1^\dagger a_1 - a_1^\dagger b_1 - b_2^\dagger a_2 + a_2^\dagger b_2)] \ , \tag{24}$$

we obtain

$$\begin{aligned} \frac{1}{\sqrt{2}}(a_1(\vec{p}) + b_1(\vec{p})) &= U^{-1} a_1(\vec{p}) U \ , \\ \frac{1}{\sqrt{2}}(b_2(\vec{p}) + a_2(\vec{p})) &= U^{-1} b_2(\vec{p}) U \ , \\ \frac{1}{\sqrt{2}}(a_2(\vec{p}) - b_2(\vec{p})) &= U^{-1} a_2(\vec{p}) U \ , \\ \frac{1}{\sqrt{2}}(b_1(\vec{p}) - a_1(\vec{p})) &= U^{-1} b_1(\vec{p}) U \ , \end{aligned} \tag{25}$$

which is equivalent to (10), i.e., our end [229]. We note that U is unitary only in the 4-component Dirac spinor space. If one applies it to the Dirac spinor ψ, one obtains

$$U^\dagger \psi U = \frac{1}{\sqrt{2}}(\psi - \gamma_5 \psi^c). \tag{26}$$

This is the Pauli-Gürsey transformation [226, 227].

18 Quantisation of the Majorana fermions [223]

Let us consider the Lagrangian for the free massive Weyl field $\psi_L(x)$,

$$\mathcal{L} = \bar{\psi}_L i \partial\!\!\!/ \psi_L - \frac{1}{2} m \bar{\psi}_L^c \psi_L + h.c \ . \tag{27}$$

This Lagrangian is written in terms of the Weyl spinor field η of (III.11) to be

$$\mathcal{L} = i\eta^\dagger(\sigma^\mu \partial_\mu)\eta - \frac{i}{2} m \eta^T \sigma_2 \eta + \frac{i}{2} m \eta^\dagger \sigma_2 \eta^* \ , \tag{28}$$

where

$$\sigma^\mu \equiv (1, -\sigma_i) \ . \tag{29}$$

We also introduce

$$\tilde{\sigma}^\mu = (1, \sigma_i) \tag{30}$$

for later convenience [230]. Eq.(28) is invariant under the SL(2,C) transformation of eq.(5). The Euler-Lagrange equation of motion is derived with the aid of the variational principle to be

$$\begin{aligned} \sigma^\mu \partial_\mu \eta + m\sigma_2 \eta^* &= 0 \ , & (31) \\ \tilde{\sigma}^\mu \sigma_2 \partial_\mu \eta^* - m\eta &= 0 \ . & (32) \end{aligned}$$

Multiplying $\tilde{\sigma}^\mu \partial_\mu$ on the left of eq.(31), we obtain

$$(\partial_0^2 - \nabla^2 + m^2)\eta = 0 \ . \tag{33}$$

Procedures of quantisation are now standard: We can carry out quantisation by imposing the equal-time anticommutation relations on $\eta(x)$ and its canonical conjugate field $\pi(x)$,

$$\begin{aligned} \{\eta(\vec{x},t), \pi(\vec{x}',t)\} &= i\delta(\vec{x}-\vec{x}') \ , \\ \{\eta(\vec{x},t), \eta(\vec{x}',t)\} &= \{\eta^*(\vec{x},t), \eta^*(\vec{x}',t)\} = 0 \ . \end{aligned} \tag{34}$$

We expend $\eta(x)$ as

$$\begin{aligned} \eta(x) &= \frac{1}{(2\pi)^{3/2}} \int d^3p \sqrt{\frac{p_0 + |\vec{p}|}{2p_0}} [\{a_-(\vec{p})\beta(\vec{p}) + b_+(\vec{p})\alpha(\vec{p})\} e^{-ipx} \\ &\quad + \{c_-(\vec{p})\beta(\vec{p}) + d_+(\vec{p})\alpha(\vec{p})\} e^{ipx}] \ , & (35) \\ \eta^*(x) &= \frac{1}{(2\pi)^{3/2}} \int d^3p \sqrt{\frac{p_0 + |\vec{p}|}{2p_0}} [\{c_-(\vec{p})^\dagger \beta^*(\vec{p}) + d_+^\dagger(\vec{p})\alpha^*(\vec{p})\} e^{-ipx} \\ &\quad + \{a_-^\dagger(\vec{p})\beta^*(\vec{p}) + b_+^\dagger(\vec{p})\alpha^*(\vec{p})\} e^{ipx}] \ , & (36) \end{aligned}$$

where $\alpha(\vec{p})$ and $\beta(\vec{p})$ are helicity eigenspinors that satisfy

$$\begin{aligned} \vec{\sigma}\vec{p}\alpha(\vec{p}) &= |\vec{p}|\alpha(\vec{p}) \ , \\ \vec{\sigma}\vec{p}\beta(\vec{p}) &= -|\vec{p}|\beta(\vec{p}) \ , \end{aligned} \tag{37}$$

and the orthonormal condition,

$$\begin{aligned} \alpha^\dagger(\vec{p})\alpha(\vec{p}) &= \beta^\dagger(\vec{p})\beta(\vec{p}) = 1 \ , & (38) \\ \alpha^\dagger(\vec{p})\beta(\vec{p}) &= \beta^\dagger(\vec{p})\alpha(\vec{p}) = 0 \ . & (39) \end{aligned}$$

We take the convention for their relative phase that

$$\begin{aligned} \sigma_2 \alpha^*(\vec{p}) &= +i\beta(\vec{p}) \ , \\ \sigma_2 \beta^*(\vec{p}) &= -i\alpha(\vec{p}) \ . \end{aligned} \tag{40}$$

The equation of motion (31–32) determines the relation between the expansion coefficients,

$$
\begin{aligned}
d_+ &= \frac{m}{p_0+|p|} a_-^\dagger \ , \\
b_+ &= -\frac{m}{p_0+|p|} c_-^\dagger \ ,
\end{aligned} \tag{41}
$$

leaving the two independent coefficients (operators) a_- and c_-. Upon writing $c_-^\dagger \to a_+$, $\eta(x)$ is written

$$
\begin{aligned}
\eta(x) &= \frac{1}{(2\pi)^{3/2}} \int d^3p \sqrt{\frac{p_0+|\vec{p}|}{2p_0}} [\{a_-(\vec{p})\beta(\vec{p}) - \frac{m}{p_0+|\vec{p}|} a_+(\vec{p})\alpha(\vec{p})\} e^{-ipx} \\
&+ \{a_+^\dagger(\vec{p})\beta(\vec{p}) + \frac{m}{p_0+|\vec{p}|} a_-^\dagger(\vec{p})\alpha(\vec{p})\} e^{ipx}\}] \ .
\end{aligned} \tag{42}
$$

The anticommutation relations of (34) yield

$$
\begin{aligned}
\{a_\pm(\vec{p}), a_\pm^\dagger(\vec{p}')\} &= \delta(\vec{p}-\vec{p}') \ , \\
\text{or otherwise} &= 0 \ .
\end{aligned} \tag{43}
$$

The stress energy-momentum tensor is defined by

$$
\begin{aligned}
T^{\mu\nu} &= \frac{\partial \mathcal{L}}{\partial(\partial_\mu \eta)} \frac{\partial \eta}{\partial x^\nu} - g^{\mu\nu}\mathcal{L} \\
&= i\eta^\dagger \sigma^\mu \partial^\nu \eta - g^{\mu\nu}\mathcal{L} \ .
\end{aligned} \tag{44}
$$

With the aid of the equation of motion (31–32) the energy-momentum four vector

$$
P^\mu \equiv \int d^3x T^{0\mu} \ , \tag{45}
$$

is given by

$$
P^\mu = \int d^3x [i\eta^* \partial^\mu \eta - g^{0\mu} \frac{i}{2} \partial_0(\eta^*\eta)] \ , \tag{46}
$$

or

$$
E = \frac{i}{2} \int d^3x \ (\eta^\dagger \frac{\partial \eta}{\partial t} - \frac{\partial \eta^\dagger}{\partial t} \eta) \ , \tag{47}
$$

$$
\vec{P} = -\frac{i}{2} \int d^3x \ (\eta^\dagger \vec{\nabla}\eta - (\vec{\nabla}\eta^\dagger)\eta) \ . \tag{48}
$$

In terms of the expansion coefficients, they are written

$$
E = \int d^3p \ p_0 \ [a_-^\dagger(\vec{p})a_-(\vec{p}) + a_+^\dagger(\vec{p})a_+(\vec{p}) \ ] \ , \tag{49}
$$

$$\vec{P} = \int d^3p\, \vec{p}\, [a_-^\dagger(\vec{p})a_-(\vec{p}) + a_+^\dagger(\vec{p})a_+(\vec{p})\,] \ , \tag{50}$$

where the zero-point energy is omitted. This allows the interpretation that $a_\pm^\dagger(\vec{P})$ and $a_\pm(p)$ are, respectively, the creation and annihilation operators of a particle with a definite energy and a momentum, E and $\vec{P}$. The Hilbert space can be constructed in the standard manner by operating $a_\pm^\dagger$ onto the vacuum $|0\rangle$. We note that $\eta(x)$ contains creation operators for helicity + states and annihilation operators for helicity − states. This interpretation, however, holds only when $m = 0$, and becomes meaningless for $m \neq 0$, since helicity is not a conserved quantity in the presence of a mass term.

We conclude this section by calculating the Feynman propagator for the free massive Weyl spinor field $\eta(x)$. Using the properties satisfied by the spinors $\alpha(\vec{p})$ and $\beta(\vec{p})$,

$$\begin{aligned} \beta_a\beta_b^* &= \frac{1}{2|\vec{p}|}\,(|\vec{p}| - \sigma_i p_i)_{ab} \ , \\ \alpha_a\alpha_b^* &= \frac{1}{2|\vec{p}|}\,(|\vec{p}| + \sigma_i p_i)_{ab} \ , \end{aligned} \tag{51}$$

we have

$$\{\eta_a(x),\ \eta_b^*\} = i(\tilde{\sigma}^\mu\partial_\mu)_{ab}D(x - x') \ , \tag{52}$$

and

$$\{\eta_a(x),\ \eta_b(x')\} = im(\sigma_2)_{ab}D(x - x') \tag{53}$$

with $D(x)$ defined by

$$D(x) = \frac{1}{(2\pi)^3}\int \frac{d^3p}{2p_0}\,[e^{ipx} - e^{-ipx}] \ . \tag{54}$$

The Feynman propagators are defined by

$$\begin{aligned} \langle 0|T[\eta(x)\eta^\dagger(x)]\ |0\rangle &= \theta(x_0 - x_0')\langle 0|\eta(x)\eta^\dagger(x')|0\rangle \\ &\quad + \theta(x_0' - x_0)\langle 0|\eta^\dagger(x')\eta(x)|0\rangle \ , \end{aligned} \tag{55}$$

$$\begin{aligned} \langle 0|T[\eta(x)\eta^T(x)]\ |0\rangle &= \theta(x_0 - x_0')\langle 0|\eta(x)\eta^T(x')|0\rangle \\ &\quad + \theta(x_0' - x_0)\langle 0|\eta(x')\eta^T(x)|0\rangle \ . \end{aligned} \tag{56}$$

With the aid of the formula

$$\theta(x_0) = \frac{1}{2\pi i}\int \frac{d\lambda}{\lambda - i\varepsilon}\, e^{i\lambda x_0} \tag{57}$$

we obtain

$$\langle 0|T[\eta(x)\eta^\dagger(x')]\ |0\rangle = \frac{1}{(2\pi)^4}\int d^4p \frac{\tilde{\sigma}^\mu p_\mu}{p^2 - m^2}\ e^{-ip(x-x')} \tag{58}$$

and

$$\langle 0|T[\eta(x)\eta^T(x')]\ |0\rangle = \frac{1}{(2\pi)^4}\int d^4p \frac{im\sigma_2}{p^2 - m^2}\ e^{-ip(x-x')}\ . \tag{59}$$

The non-vanishing right-hand side of eq.(59) indicates that lepton number conservation is violated by the mass (Majorana mass) term.

19 Path integral derivation of Feynman propagators

The Feynman propagators of eqs.(58) and (59) can be readily obtained by the path-integral quantisation method [231]. By adding to the Lagrangian (27) the source term $\bar{J}_R\psi_L + \psi_L J_R$, we write the path integral

$$Z[J] = \int d\psi_L d\bar{\psi}_L e^{-i\int d^4x \mathcal{L}_J}\ , \tag{60}$$

where $\mathcal{L}_J$ is written in the form

$$\mathcal{L}_J = \frac{1}{2}(\bar{\Phi}O\Phi + \bar{J}\Phi + \bar{\Phi}J) \tag{61}$$

with

$$\Phi = \psi_L + \psi_L^c\ , \tag{62}$$

$$J = J_R + J_R^c\ , \tag{63}$$

and

$$O = i\partial\!\!\!/ - m\ . \tag{64}$$

Writing

$$J_R = \begin{pmatrix} j \\ 0 \end{pmatrix}\ , \tag{65}$$

we have

$$\begin{aligned} \bar{J}_R\, p\!\!\!/ J_R &= j^\dagger p_\mu \tilde{\sigma}^\mu j\ , \\ \bar{J}_R^c m J_R &= -imj^T\sigma_2 j\ . \end{aligned} \tag{66}$$

Integration of (60) over the ψ_L and $\bar{\psi}_L$ fields yields

$$Z[J] = e^{i\int d^4x(-\frac{1}{2}JO^{-1}J)}\ , \tag{67}$$

with

$$
\begin{aligned}
-\frac{1}{2}\bar{J}O^{-1}J &= -j^{\dagger}[\frac{1}{(2\pi)^4}\int d^4p\ \frac{p_\mu\tilde{\sigma}^\mu}{p^2-m^2}\ e^{-ipx}]\ j \\
&\quad -\frac{1}{2}j^T[\frac{1}{(2\pi)^4}\int d^4p\ \frac{-im\sigma_2}{p^2-m^2}\ e^{-ipx}]\ j \\
&\quad + h.c., \qquad (68)
\end{aligned}
$$

where the expressions in the square brackets are the Feynman propagators in agreement with (58) and (59).

20 Simple examples of applications

As examples, we show elementary calculations for two processes involving Majorana particles, the decay of a Majorana neutrino N into e_L^- and a scalar particle φ, and the annihilation process $N + N \to e^+ + e^-$.

20.1 $N \to e_L^- + \varphi^{(+)}$

Let us suppose that the interaction Lagrangian is given by

$$\mathcal{L}_{\rm int} = g(\bar{e}_L N_L^c \varphi^\dagger + \bar{N}_L^c e_L \varphi) \quad . \qquad (69)$$

With the use of the Weyl spinors, η for N_L and ζ for e_L^-, (69) is written

$$\mathcal{L}_{\rm int} = ig(\zeta^*\sigma_2\eta^*\varphi^\dagger - \eta\sigma_2\zeta\varphi) \quad . \qquad (70)$$

Assigning the momentum as in Fig.11, we calculate

$$\langle q,\ k \mid i\mathcal{L}_{\rm int} \mid p\rangle = -g\int d^4x \langle k|\varphi^\dagger|0\rangle\langle q|\xi^*|0\rangle\sigma_2\langle 0|\eta^*|p\rangle \quad . \qquad (71)$$

For notational convenience we write (42) in the form

$$
\begin{aligned}
\eta(x) &= \frac{1}{(2\pi)^{3/2}}\sum_s\int d^3p\sqrt{\frac{p_0+|\vec{p}|}{2p_0}}\Big[a_s(\vec{p})U^{(s)}(\vec{p})e^{-ipx} \\
&\quad + a_s^\dagger(\vec{p})V^{(s)}(\vec{p})e^{+ipx}\Big] \ , \qquad (72) \\
\eta^*(x) &= \frac{1}{(2\pi)^{3/2}}\sum_s\int d^3p\sqrt{\frac{p_0+|\vec{p}|}{2p_0}}\Big[a_s^\dagger(\vec{p})U^{(s)*}(\vec{p})e^{+ipx} \\
&\quad + a_s(\vec{p})V^{(s)*}(\vec{p})e^{-ipx}\Big] \ , \qquad (73)
\end{aligned}
$$

where

$$
\begin{aligned}
U^{(-)}(\vec{p}) &= u(\vec{p}) \ , \quad U^{(+)}(\vec{p}) = -\frac{M_N}{p_0+|\vec{p}|}v(\vec{p}) \ , \\
V^{(-)}(\vec{p}) &= \frac{M_N}{p_0+|\vec{p}|}v(\vec{p}) \ , \quad V^{(+)}(\vec{p}) = u(p) \ . \qquad (74)
\end{aligned}
$$

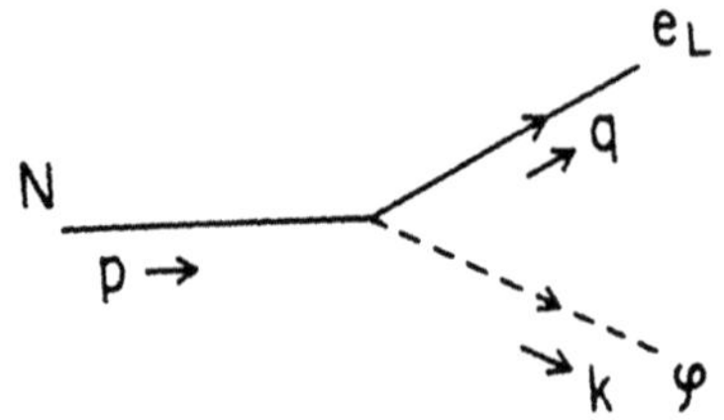

Figure 11: Feynman diagram for decay of a Majorana particle: $N \to e_L^+ + \varphi$. Momentum flows are represented by p, q and k.

The scattering amplitude (71) is then

$$\begin{aligned}&\langle q, k|i\mathcal{L}_{\rm int}|p\rangle \\ &= -(2\pi)^4\delta^4(p-k-q)g\,\frac{1}{\sqrt{2k_0}}\sqrt{\frac{p_0+|\vec{p}|}{2p_0}}\sqrt{\frac{q_0+|\vec{q}|}{2q_0}}\,U^{(s)*}(\vec{q})\sigma_2V^{(s')*}(\vec{p})\ .\end{aligned} \tag{75}$$

Since

$$\frac{1}{2}\sum_{s,s'}|U^{(s)*}(\vec{q})\sigma_2V^{(s')*}(\vec{p})|^2 = \frac{1}{2}T_r(\sigma_2V^*V\sigma_2UU^*) = \frac{1}{2}\frac{2(pq)}{(p_0+|\vec{p}|)(q_0+|\vec{q}|)}\ , \tag{76}$$

we obtain the decay rate

$$\begin{aligned}\Gamma(N\to e^- + \varphi^\dagger) &= \int \frac{d^3q}{(2\pi)^3}\,\frac{d^3k}{(2\pi)^3}\,(2\pi)^4(p-q-k)\,\frac{g^2}{2k_0} \\ &\quad \frac{p_0+|\vec{p}|}{2p_0}\,\frac{q_0+|\vec{q}|}{2q_0}\,\frac{1}{2}\sum_{ss'}|U^*\sigma_2V^*|^2 \\ &= \frac{1}{(2\pi)^2}\int d^3q\,d^3k\,\delta(p-q-k)\,\frac{(pq)}{2p_0 2q_0}\frac{1}{2k_0} \\ &= \frac{g^2}{32\pi}M_N\ ,\end{aligned} \tag{77}$$

where we have assumed $m_e = 0$ and $m_\varphi = 0$ for simplicity. Similarly, we have

$$\Gamma(N\to e^+ + \varphi^-) = \frac{g^2}{32\pi}M_N\ . \tag{78}$$

Then,

$$\Gamma_{\rm total}(N) = \frac{g^2}{16\pi}M_N\ . \tag{79}$$

In case when N is a Dirac fermion,

$$\mathcal{L}_{\rm int} = g(\bar{e}_L N_R \varphi^\dagger + \bar{N}_R e_L \varphi) \tag{80}$$

instead of (69). Therefore, $N \to e_L + \varphi^\dagger$ is allowed, but $N \to e_L^+ + \varphi^-$ is forbidden. Namely,

$$\Gamma_{\rm total}(N) = \frac{g^2}{32\pi} M_N \ , \tag{81}$$

which is half the rate for the Majorana case.

20.2 $N + N \to e_L^- + e_L^+$

Let us consider the following interaction Lagrangian:

$$\mathcal{L}_{\rm int} = g(\bar{e}_L N_R \varphi^\dagger + \overline{N}_R \ e_L \ \varphi). \tag{82}$$

Instead of N_L^c in (69), we use here N_R to define the Majorana field, $N = N_R - N_R^c$. The annihilation process of Majorana N_R, $N + N \to e_L^- + e_L^+$, is given by the φ exchange. With $m_\varphi \gg m_N, m_e$, the effective Lagrangian is

$$\begin{aligned} \mathcal{L}_{\rm eff} &= \frac{g^2}{m_\varphi^2}(\overline{N}_R e_L)(\bar{e}_L N_R) \\ &= -\frac{g^2}{2m_\varphi^2}(\overline{N}_R \gamma_\mu N_R)(\bar{e}_L \gamma_\mu e_L) \ . \end{aligned} \tag{83}$$

Substitution of the Weyl spinors ξ for N_R and ζ for e_L^- yields

$$\mathcal{L}_{\rm eff} = -\frac{g^2}{2m_\varphi^2}(\xi^* \tilde{\sigma}_\mu \xi)(\zeta^* \sigma_\mu \zeta) \ , \tag{84}$$

where σ_μ and $\tilde{\sigma}_\mu$ are given in (29) and (30). With the momentum assignment given in Fig.12, we calculate

$$\langle p', q' | i\mathcal{L}_{\rm eff} | p, q \rangle = +i\frac{g^2}{2m_\varphi^2} \int dx \langle p', q' | \zeta^*(x) \sigma_\mu \zeta(x) | 0 \rangle \langle 0 | \xi^*(x) \tilde{\sigma}_\mu \xi(x) | p, q \rangle \ . \tag{85}$$

The Weyl spinors, $\xi(x)$ and $\xi^*(x)$, are expanded in a way similar to eq.(72) and (73) with

$$\begin{aligned} U^{(+)}(\vec{p}) &= u(\vec{p}), \ \ U^{(-)}(\vec{p}) = -\frac{M_N}{p_0 + |\vec{p}|} v(\vec{p}) \ , \\ V^{(-)}(\vec{p}) &= \frac{M_N}{p_0 + |\vec{p}|} v(\vec{p}), \ \ V^{(+)}(\vec{p}) = u(\vec{p}) \ , \end{aligned} \tag{86}$$

instead of (74). The scattering amplitude (85) is then

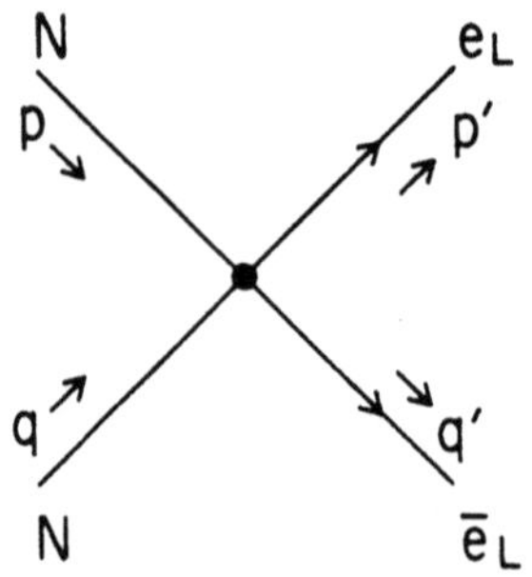

Figure 12: Feynman diagram for scattering of two Majorana particles: $N+N \to e_L^- + e_L^+$.

$$\begin{aligned}\langle p', q'|s|p, q\rangle &= i\frac{g^2}{2m_\varphi^2}\frac{1}{(2\pi)^2}\delta^{(4)}(p+q-p'-q')\sqrt{\frac{p_0+|\vec{p}|}{2p_0}}\sqrt{\frac{q_0+|\vec{q}|}{2q_0}} \\ &\sqrt{\frac{p'_0+|\vec{p'}|}{2p'_0}}\sqrt{\frac{q'_0+|\vec{q'}|}{2q'_0}}\left[u_e^*(\vec{p'})\sigma_\mu v_e(\vec{q'})\right] \\ &\times\left[V^{*(s)}(\vec{p})\tilde{\sigma}_\mu U^{(s')}(\vec{q})+V^{*(s')}(\vec{q})\tilde{\sigma}_\mu U^{(s)}(\vec{p})\right].\end{aligned} \tag{87}$$

It is now straightforward to obtain the annihilation cross section σ,

$$\begin{aligned}\sigma &= \frac{1}{(2\pi)^2}\frac{1}{v_{\rm rel}}\left(\frac{g^4}{4m_\varphi^4}\right)\int d^3p'd^3q'\delta^{(4)}(p+q-p'-q') \\ &\times\frac{1}{2}\frac{p_0+|\vec{p}|}{2p_0}\frac{q_0+|\vec{q}|}{2q_0}\frac{2p'q'}{2q_0'2p_0'}\frac{2}{(q_0+|\vec{q}|)(p_0+|\vec{p}|)}\left(qp+\frac{\vec{q}\cdot\vec{p}}{|\vec{q}||\vec{p}|}M_N^2\right) \\ &\simeq\left(\frac{g^2}{4m_\varphi^4}\right)\frac{M_N^2}{8\pi}\beta\ ,\end{aligned} \tag{88}$$

with $\beta = |\vec{p}|/m$ ($m_e = 0$ is assumed). In this calculation we used

$$\begin{aligned}\frac{1}{4}\sum_{s,s'}|V^{*(s')}(\vec{q})\tilde{\sigma}_\mu U^{(s)}(\vec{p})|^2 &= \frac{1}{4}\frac{2qp}{(q_0+|\vec{q}|)(p_0+\vec{p}|)}\ , \\ \frac{1}{4}\sum_{s,s'}\left(V^{*(s)}(\vec{p})\tilde{\sigma}_\mu U^{(s')}(\vec{q})V^{(s')}(\vec{q})\tilde{\sigma}^\mu U^{*(s)}(\vec{p})\right) &= \frac{1}{4}\frac{2\vec{q}\cdot\vec{p}}{qp}\frac{1}{q_0+|\vec{q}|}\frac{1}{p_0+|\vec{p}|}M_N^2.\end{aligned} \tag{89}$$

We note that the annihilation cross section σ is proportional to β. The $\overline{N}_R\gamma_\mu N_R$ part in eq.(83) is written in terms of the Majorana field N:

$$\overline{N}_R\gamma_\mu N_R \to \frac{1}{2}\overline{N}\gamma_\mu\gamma_5 N\ .$$

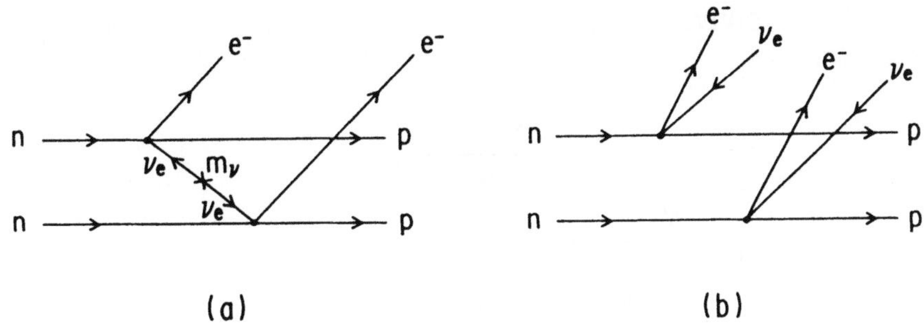

Figure 13: Feynman diagrams giving rise to neutrinoless double beta decay (a) and two neutrino double beta decay (b). In the first diagram m_ν stands for the Majorana mass of the neutrino.

This means that N has only axial-vector coupling, and the annihilation process near the threshold is thus dominated by the p-wave [232]. This significantly suppresses the low-energy cross section, as compared with the annihilation of Dirac fermions, $f + \bar{f} \rightarrow e^+ + e^-$, which proceeds with the s-wave.

21 Double beta decay

The most conspicuous role of the propagator (59) characteristic of the Majorana neutrino is that it induces neutrinoless double beta decay ${}^A_Z\Box \rightarrow {}^A_{Z+2}\Box + e^- + e^-$($0\nu 2\beta$ in brevity), shown in Fig.13 [32]. Because of a strong pairing force between like nucleons, it sometimes happens that even-even nuclei have large binding energies and ordinary β decay ${}^A_Z\Box \rightarrow {}^A_{Z+1}\Box + e^- + \overline{\nu}_e$ is energetically forbidden, which leaves double beta decay as the only allowed decay mode (see Fig.14). There is also another double beta-decay mode ${}^A_Z\Box \rightarrow {}^A_{Z+2}\Box + e^- + e^- + \bar{\nu}_e + \bar{\nu}_e$ ($2\nu 2\beta$ in short), which takes place irrespective of whether the neutrino is of the Dirac or the Majorana type [31]. Here, we are mostly concerned with double beta decay of the first type ($0\nu 2\beta$). As the published date of the original literature [31, 32] shows, this is a classical subject and much work has been done over many years (see [233] for a classical review). More recently, interest has been revived from the point of view that the neutrino naturally appears as a Majorana particle in most of the unified theories of particle interactions (see e.g., [234]). This is, however, a rather involved subject and we must therefore defer any description of details to dedicated review articles [235]–[237] (for experimental status see [238]). Here, we confine ourselves to an elementary account of the subject.

<u>$0\nu 2\beta$ decay rate</u>

The $0\nu 2\beta$ amplitude shown in Fig.13a is calculated in the following manner [233]–[237], [239, 240]. With the charged-current Hamiltonian (142) the effective

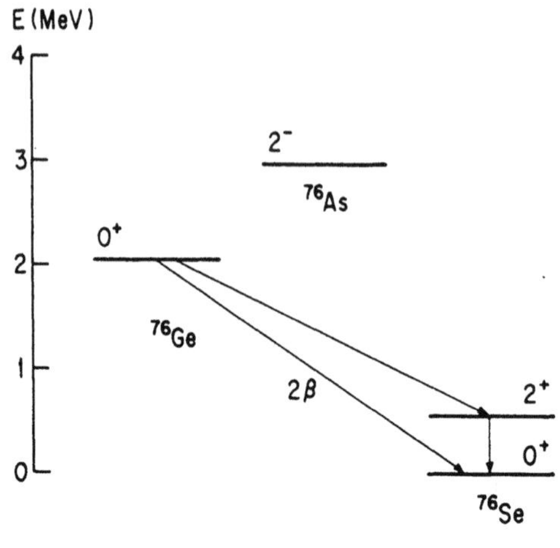

Figure 14: Energy level scheme of the ^{76}Ge$-^{76}$As$-^{76}$Se system.

operator for $0\nu 2\beta$ decay is

$$\begin{aligned} H_{\rm eff} &= (\sqrt{2}G_F\cos\theta_c)^2(\bar{e}_L\gamma_\mu\nu_L)(-\overline{\nu_L^c}\gamma_\nu e_L^c) \\ & [\bar{p}\gamma^\mu(1-g_A\gamma_5)n][\bar{p}\gamma^\nu(1-g_A\gamma_5)n]. \end{aligned} \tag{90}$$

For the $\nu_L\overline{\nu_L^c}$ contraction we obtain from (59),

$$\langle T[\psi_\nu(x)\overline{\psi_\nu^c}(y)]\rangle = \frac{1}{(2\pi)^4}\int d^4k\frac{m_\nu}{k^2-m_\nu^2}e^{-ik(x-y)} \; . \tag{91}$$

Hence, the $0\nu 2\beta$ amplitude reads

$$\begin{aligned} M &= -2G_F^2\cos^2\theta_c\int\frac{d^4k_\nu}{(2\pi)^4}\bar{u}(p_1)\gamma_\mu\frac{m_\nu}{k_\nu^2-m_\nu^2}\frac{1-\gamma_5}{2}\gamma_\nu v(p_2)e^{-ik_\nu r} \\ & \langle N_1'|J^\mu|N_1\rangle\langle N_2'|J^\nu|N_2\rangle 2\pi\delta(k^0-...) \\ &= -2G_F^2\cos^2\theta_c L_{\mu\nu}H^{\mu\nu} \;, \end{aligned} \tag{92}$$

where L and H refer to the leptonic and hadronic parts, $\vec{r}$ is the relative distance of the two nucleons N_1 and N_2, and k_ν is the neutrino four-momentum. The leptonic part is

$$L_{\mu\nu} = \int\frac{d^3k_\nu}{(2\pi)^3}\bar{u}(p_1)\gamma_\mu\gamma_\nu\frac{1+\gamma_5}{2}v(p_2)\frac{m_\nu}{k_\nu^2-m_\nu^2}e^{i\vec{k}_\nu\vec{r}} \; , \tag{93}$$

where $1/(k^2-m^2)$ is written

$$\frac{1}{k_\nu^2-m^2} = \frac{1}{2\epsilon_\nu}\frac{1}{(E_{1i}-E_{1n}-\epsilon_1)^2-\epsilon_\nu^2}$$

$$= \frac{-1}{2\epsilon_\nu}\left(\frac{1}{E_{1i}-E_{1n}-\epsilon_1-\epsilon_\nu}+\frac{1}{E_{2i}-E_{2n}-\epsilon_2-\epsilon_\nu}\right). \tag{94}$$

Here, E_{1i} and E_{2i} are the energies of the two nucleons in the initial state, E_{1n} and E_{2n} are those in the intermediate state, ϵ_1, and ϵ_2 are those of the emitted two electrons $(1,2)$, and $\epsilon_\nu = \sqrt{m_\nu^2+\vec{k}_\nu^2}$. We used in (94) the energy conservation $E_{1i}+E_{2i}=E_{1n}+E_{2n}+\epsilon_1+\epsilon_2$. The two terms in (94) correspond to the diagrams emitting electron 1 first and electron 2 first. We replace $E_{1i}+E_{2i}\to E_i$ and $E_{1n}+E_{2n}\to E_n$ in (94). For the practically most important $0^+\to 0^+$ transitions, two electrons are emitted in the s-wave. For this case, it is sufficient to retain only $g_{\mu\nu}$ of $\gamma_\mu\gamma_\nu = g_{\mu\nu}+\frac{1}{2}[\gamma_\mu,\gamma_\nu]$. We then have

$$L_{\mu\nu} = g_{\mu\nu} m_\nu \bar{u}(p_1)\frac{1+\gamma_5}{2}v(p_2)\int\frac{d^3k}{(2\pi)^3}\sum_\kappa\frac{-1}{2\epsilon_\nu}\frac{1}{E_i-E_n-\epsilon_\kappa-\epsilon_\nu}e^{i\vec{k}_\nu\cdot\vec{r}} \quad . \tag{95}$$

and after squaring $L_{\mu\nu}$,

$$|M|^2 = 4G_F^2\cos^2\theta_c H_\mu^\mu H_\rho^\rho m_\nu^2 2(p_1p_2)[\frac{1}{4\pi}F(r)]^2, \tag{96}$$

where

$$F(r) = \frac{1}{4\pi^2}\sum_{\kappa=1}^{2}\int d^3k_\nu\frac{e^{i\vec{k}_\nu\vec{r}}}{(-\epsilon_\nu)(E_i-E_n-\epsilon_\kappa-\epsilon_\nu)} \tag{97}$$

and

$$H_\mu^\mu = \langle f|\tau_+(1)\tau_+(2)(1-g_A^2\vec{\sigma}(1)\vec{\sigma}(2))|i\rangle,$$

using the non-relativistic expansion for the nuclear matrix elements; p_1 and p_2 are the momenta of the two electrons.

The decay rate is given by

$$\begin{aligned}\Gamma &= \frac{1}{2}\int\frac{d^3p_1}{2\epsilon_1(2\pi)^3}\int\frac{d^3p_2}{2\epsilon_2(2\pi)^3}(2\pi)\delta(\epsilon_1+\epsilon_2-Q)|M|^2\\ &= \frac{G_F^4\cos^4\theta_c}{64\pi^5}\int d\epsilon_1 d\cos\theta\; p_1p_2(\epsilon_1\epsilon_2-\vec{p}_1\vec{p}_2)m_\nu^2\\ &\quad \left|\langle f|\tau_+(1)\tau_+(2)[1-g_A^2\vec{\sigma}(1)\vec{\sigma}(2)]F(r)|i\rangle\right|^2.\end{aligned} \tag{98}$$

When we consider a nucleus we must sum over all nucleons in the nuclei, $(1,2)\to\Sigma(i,j)$. In (97) we should also sum over the intermediate excited states n as,

$$\sum_n\frac{|n\rangle\langle n|}{E_i-E_n-\epsilon_\kappa-\epsilon_\nu}. \tag{99}$$

For the latter, a closure approximation is commonly used to evaluate the matrix element. Namely, (99) is approximated as

$$\frac{1}{E_i - \langle E_n \rangle - \epsilon_\kappa - \epsilon_\nu}. \tag{100}$$

For $0\nu 2\beta$ decay this is expected to be a good approximation, since the neutrino energy is much greater than the typical energy of the nuclear levels. This process takes place only for $|f\rangle = |0^+\rangle$ if $|i\rangle = |0^+\rangle$.

Finally, the differential rate is written

$$\begin{aligned}\frac{d\Gamma}{d\cos\theta d\epsilon_1} &= \frac{G_F^4 \cos^4\theta_c}{16\pi^5}\epsilon_1\epsilon_2 p_1 p_2 (1 - \frac{\vec{p}_1}{\epsilon_1}\frac{\vec{p}_2}{\epsilon_2}) m_\nu^2 \\ & \left|\langle 0_f^+ \| \frac{1}{2}\tau_+(i)\tau_+(j)[1 - g_A^2 \vec{\sigma}(i)\vec{\sigma}(j)]F(r_{ij})\|0_i^+\rangle\right|^2, \end{aligned} \tag{101}$$

where

$$F(r) = \frac{1}{4\pi^2}\sum_{k=1}^{2}\int d^3k_\nu \frac{e^{i\vec{k_\nu}\vec{r}}}{(-\epsilon_\nu)(E_i - \langle E_n\rangle - \epsilon_k - \epsilon_\nu)} \quad . \tag{102}$$

We expect that since the typical neutrino momentum is of the order of

$$\langle k_\nu \rangle \sim \frac{1}{\langle \gamma_{ij} \rangle} \sim 100\text{MeV} \tag{103}$$

for $\langle r_{ij}\rangle \sim 2$fm, $\epsilon_\nu \gg |\langle E_n\rangle + \epsilon_k - E_i| \approx 10$MeV. Eq.(102) then simplifies to

$$\begin{aligned} F(r) &= \frac{1}{2\pi^2}\int d^3k_\nu \frac{e^{i\vec{k_\nu}\vec{r}}}{\epsilon_\nu^2} \\ &= \frac{e^{-m_\nu r}}{r}. \end{aligned} \tag{104}$$

If $m_\nu \ll \langle k_\nu \rangle \sim (2\text{fm})^{-1} \simeq 100$MeV,

$$F(r) \simeq 1/r \quad , \tag{105}$$

i.e., the matrix element becomes independent of the neutrino mass, and m_ν is factored out from the amplitude. If the energy denominator $\langle E_n\rangle + \epsilon_k - E_i \equiv \Delta$ is retained, (102) is

$$F(r) \simeq \frac{2}{\pi r}[\sin(r\cdot\Delta)\text{ci}(r\cdot\Delta) + \cos(r\cdot\Delta)\text{si}(r\cdot\Delta)] \quad , \tag{106}$$

where

$$\text{si}(x) = \int_x^\infty dt\frac{\sin\ t}{t}, \quad \text{ci}(x) = \int_x^\infty dt\frac{\cos\ t}{t} \tag{107}$$

and $m_\nu \ll \Delta$ is assumed. The nuclear matrix element has often been evaluated using (96). Hence, such an evaluation applies only to $m_\nu \ll 1$MeV.

The differential rate (101) is integrated to give

$$\Gamma = (G_F\cos\theta)^4 G_{0\nu}(\frac{m_\nu}{m_e})^2 |\langle 0_f^+ \| \frac{1}{2}\Sigma\tau_\pm(i)\tau_\pm(j)[1 - g_A^2\vec{\sigma}(i)\vec{\sigma}(j)]\|0_i^+\rangle|^2, \quad (108)$$

where

$$\begin{aligned} G_{0\nu} &= \frac{m_e^2}{8\pi^5}\int_{m_e}^{T_0+m_e} p_1p_2\epsilon_1\epsilon_2 d\epsilon_1 \\ &= \frac{m_e^7}{8\pi^5}\left(\frac{t_0^5}{30}+\frac{t_0^4}{3}+\frac{4}{3}t_0^3+2t_0^2+t_0\right) \end{aligned} \quad (109)$$

with $t_0 = T_0/m_e$, the maximum kinetic energy of the electron normalised by its mass. A correction for the Coulomb distortion of the wave function $F(\epsilon_1, Z)F(\epsilon_2, Z)$ must be inserted into the integrand of (109) for practical use. A more elaborate treatment of the Coulomb correction is necessary for such heavy nuclei that often appear in double beta decay [241]. The Coulomb correction enhances the decay rate by a factor of ~ 5 for tellurium. The sum of the electron energies is monochromatic.

Nuclear matrix elements

The most unclear part of the analysis of double beta decay is the evaluation of nuclear matrix elements. The lightest candidate nucleus for double beta decay is ^{48}Ca, and the nuclei often used for experimental studies are ^{76}Ge,^{82}Se,^{100}Mo,^{128}Te, ^{130}Te, ^{136}Xe etc. Even for the lightest nucleus, ^{48}Ca, we need an fp shell $\{2p_{1/2}, 2p_{3/2}, 1f_{5/2}, 1f_{7/2}\}$ wave function, and it already requires a large model space. In fact, ^{48}Ca $\rightarrow$ ^{48}Ti is the only case for which a full shell model calculation has been made [235, 242]. For more familiar candidate nuclei of ^{76}Ge and ^{82}Se we need a shell model space for magic numbers of 28–50, i.e., $\{2p_{1/2}, 2p_{3/2}, 1f_{5/2}, 1g_{9/2}\}$. The diagonalisation of such a large model space is extremely tedious; also, there is no guarantee that such a calculation gives a reliable answer. For heavier nuclei, sophisticated approximations based on the shell model are employed in ref.[235] to estimate the matrix elements.

For heavy nuclei, a quasi-particle random-phase approximation (QRPA [243]) model is more often used for the transition from the initial 0^+ state, taken to be a "vacuum", to 1 particle-1 hole states in the intermediate state [244]–[247]. The validity of QRPA has been tested against transitions for Gamow-Teller resonances to a qualitatively reasonable accuracy. Unfortunately, it turned out that the prediction is very sensitive, at least as far as the matrix element relevant for $2\nu2\beta$ decay is concerned, to particle-particle interactions g^{pp}, which, however, are unimportant for the transition to giant Gamow-Teller resonances [248]. With the inclusion of g^{pp} an exact cancellation occurs among the $2\nu2\beta$ matrix elements close to the value which is thought to be realistic; this would modify the prediction of the $2\beta2\nu$ lifetime by 3 orders of magnitude, or even more. It is not clearly understood as to whether such a cancellation takes place for the $0\nu0\beta$ matrix element;

Table 15: Examples of $\beta\beta$ decay candidates and the experimental work.

Transition	Q value (MeV)	Natural abundance (%)	$2\beta 2\nu$	$2\beta 0\nu$
$^{48}Ca \rightarrow {}^{48}Ti$	4.272	0.187	[256]	[256]
$^{76}Ge \rightarrow {}^{76}Se$	2.045	7.8	[257]–[262]	[263]
$^{82}Se \rightarrow {}^{82}Kr$	3.005	9.2	[264]	[268]
			[265]–[266]*	
			[267]**	
$^{100}Mo \rightarrow {}^{100}Ru$	3.033	9.6	[269]–[271]	
$^{128}Te \rightarrow {}^{128}Xe$	0.869	31.7	} [265]–[266]*	
$^{130}Te \rightarrow {}^{130}Xe$	2.533	34.5		
$^{136}Xe \rightarrow {}^{136}Ba$	2.481	8.9	[272]	[272]

$*$ = geochemical measurement, $**$ = measurement with a meteorite.

the estimated matrix element squares differ by 10–100 among different calculations (e.g., compare [249] with [250]).

On the other hand, there is a belief among nuclear physicists that calculations of the $0\nu 2\beta$ matrix element are much more reliable than that for $2\nu 2\beta$ (e.g., [251]). For the time being, however, we are not able to judge which estimate is more conceivable, and therefore, we list in the tables below uncritical compilations of the results obtained by nuclear physicists.

Experimental studies

35 $\beta^-\beta^-$ candidates and 6 $\beta^+\beta^+$ candidates have been identified for naturally occurring isotopes in ref. [235]. Although a number of processes have been experimentally investigated in the past, recent efforts have rather concentrated on the several processes listed in Table 15 (for a more complete list see [233] and [235]). We include here $^{48}Ca \rightarrow^{48} Ti$ which is practically the lightest candidate for $\beta\beta$ decay, though no experimental effort has been made for the last two decades for this process.

Together with direct counting in the laboratory [252], a geochemical method, which measures the content of rare gas produced from $\beta\beta$ decay and has accumulated over a geological time scale, has been used to explore $\beta\beta$ processes [253]–[255]. The crucial points regarding this method are: (i) whether one can know the geological time during which the daughter element has been accumulated, (ii) the rock has not received metamorphosis, and (iii) the rock had been placed deep enough so that the daughter nuclei were not significantly produced by cosmic rays. For ^{82}Se all geologically measured halflives [265]–[267] (including those using meteorite) show a good agreement with the more modern laboratory value [264]. On the other hand, two geological measurements for ^{128}Te and ^{130}Te [265, 266] give different answers.

With laboratory experiments, the best limit on the Majorana neutrino mass ($0\nu 2\beta$) is obtained from ^{76}Ge $\rightarrow ^{76}$Se [257]–[262]. The technical advantage of ^{76}Ge is that germanium, itself, provides a good solid-state detector with an excellent energy resolution, and that germanium can be purified against radioactive nucleids to a high degree. However, a relatively small Q value (2MeV) partially offsets the advantage.

$2\nu 2\beta$ decay

While this process is not our direct interest as well as it does not give direct information on the matrix element for $0\nu 2\beta$ decay, the prediction for the $2\nu 2\beta$ lifetime could be used as diagnostics for the nuclear model used to calculate the $0\nu 2\beta$ matrix elements. The calculation for the $2\nu 2\beta$ decay rate is straightforward, and the result is

$$d\Gamma = \frac{(G_F \cos\theta)^4}{16\pi^2} p_1^2 p_2^2 dp_1 dp_2 k_1^2 k_2^2 dk_1 \Big[(K^2 + L^2 - KL)|M_F|^2 \tag{110}$$

$$+ \frac{g_A^4}{3}(K^2 + L^2 + KL)|M_{GT}|^2 - 2g_A^2 KL \mathrm{Re}(M_F M_{GT}^*)\Big] \ , \tag{110}$$

where

$$\begin{aligned} K &= \frac{1}{E_i - \epsilon_1 - k_1 - \langle E_n \rangle} + \frac{1}{E_i - \epsilon_2 - k_2 - \langle E_n \rangle} \ , \\ L &= \frac{1}{E_i - \epsilon_1 - k_2 - \langle E_n \rangle} + \frac{1}{E_i - \epsilon_2 - k_1 - \langle E_n \rangle} \ , \end{aligned} \tag{111}$$

$$\begin{aligned} M_F &= \langle 0_f^+ \| \frac{1}{2} \sum \tau_\pm(i) \tau_\pm(j) \| 0_i^+ \rangle \ , \\ M_{GT} &= \langle 0_f^+ \| \frac{1}{2} \sum \vec{\sigma}(i) \tau_\pm(i) \vec{\sigma}(j) \tau_\pm(j) \| 0_i^+ \rangle \end{aligned} \tag{112}$$

and k_i is the neutrino momentum. For the case relevant to the double beta decay that concerns us, the Fermi-type transition contributes only through mixing of high-lying isobaric analogue states and, generally, is very small. We then have [273]

$$\Gamma = (G_F \cos\theta)^4 g_A^4 G_{2\nu} |M_{GT}|^2 \ , \tag{113}$$

with

$$\begin{aligned} G_{2\nu} &= \frac{1}{24\pi^7} \int_{m_e}^{W_0 - m_e} p_1 \epsilon_1 d\epsilon_1 \int_{m_e}^{W_0 - \epsilon_1} p_2 \epsilon_2 d\epsilon_2 \int_{m_e}^{W_0 - \epsilon_1 - \epsilon_2} k_1^2 k_2^2 dk_1 [K^2 + L^2 + KL] \\ &= \frac{m_e^{10}}{1800\pi^7} \Big[\frac{1}{E_i - \langle E_n \rangle - \epsilon_0 / 2} \Big]^2 t\, (t_0 - t)^5 (t^4 + 10\, t^3 + 40\, t^2 + 60\, t + 30), \end{aligned} \tag{114}$$

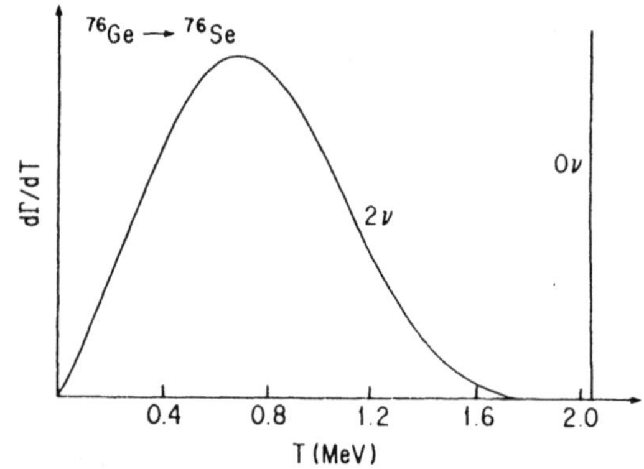

Figure 15: Kinetic energy spectrum of the two electrons emitted from double beta decay of ^{76}Ge. T is the sum of the kinetic energies of the two electrons.

the phase-space volume multiplied by the averaged energy denominator (the Fermi functions $F(\epsilon_1, Z)F(\epsilon_2, Z)$ should be inserted in the integrand for actual evaluations; see also the comment after eq.(109)); ϵ_0 is the maximum electron energy and $t = T/m_e$ is the sum of the kinetic energy of the two electrons. The electron energy spectrum is shown in Fig.15, together with that of $0\nu 2\beta$ decay.

Table 16 compares the model predictions with the experiment. We note that (1) and (9) are calculations based on a shell model, and the others are basically QRPA type calculations. The QRPA calculations (2)-(5) do not include g^{pp} and those of (6)-(8) are with $g^{pp} \neq 0$ after the suggestion of ref. [248].

Experimentally, the best established is the lifetime for ^{82}Se $\rightarrow ^{82}$Kr (and also ^{100}Mo$\rightarrow$ ^{100}Ru more recently). The QRPA predictions with $g^{pp} = 0$ gives a lifetime a factor of 4–25 shorter. One can obtain a value close to the experiment by adjusting g^{pp}. Let us note, however, that the result is very sensitive to the value of g^{pp}, and that the signs of the matrix element are also different between Engel et al. [249] and Tomoda-Faessler [250] in spite of their agreement on the lifetime. A factor > 10 disagreement for ^{128}Te is brought into reasonable agreement after $g^{pp} \neq 0$ is adequately chosen. For ^{100}Mo, all of the calculations are still an order of magnitude off, even with $g^{pp} \neq 0$. On the other hand, the predicted lifetime for ^{76}Ge is longer than is observed by a factor of 5–10, though the authors of the experiment have made a rather weak statement concerning their discovery of the finite lifetime listed in this table. We may conclude that nuclear physics calculations are usually correct, allowing for a factor $10^{\pm 1}$ in $T_{1/2}$.

$0\nu 2\beta$ decay and the limit on the Majorana neutrino mass

An uncritical compilation of nuclear matrix element calculations is given in Table 17, where the numbers shown are halflives for an assumed Majorana neutrino mass of 1eV. The experimental limit is given in the last column. The best limit

Table 16. Half-life of two-neutrino double beta decay (in units of years). Nuclear model calculations are compared with the expriment.

	Haxton 84	Vogel-Fisher 85	Grotz-Klapdor 85		Tomoda et al. 86	Engel et al. 88	Muto-Klapdor 88	Engel et al. 89	Muto et al. 89	Zhao et al. 90	experiment
	(1)	(2)	(3a)	(3b)	(4)	(5)	(6)	(7)	(8)	(9)	
^{48}Ca	2.9×10^{19}									1.9×10^{19}	$> 3.6 \times 10^{19}$ (80%) [256]
^{76}Ge	0.41×10^{21}	1.8×10^{21}	0.22×10^{21}	0.11×10^{21}	0.21×10^{21}	$1.3 - 0.15 \times 10^{21}$	5.5×10^{21}	13.2×10^{21}	3.0×10^{21}		$1.1 ^{+.6}_{-.3} \times 10^{21}$ (95%) [263]
^{82}Se	0.26×10^{20}	0.33×10^{20}	0.15×10^{20}	0.045×10^{20}		$1.2 - 0.17 \times 10^{20}$	1.9×10^{20}	5.1×10^{20}	1.1×10^{20}		$1.08^{+.26}_{-.06} \times 10^{20}$ (68%) [264]
^{100}Mo		0.21×10^{19}		0.18×10^{19}		$0.6 \times 10^{19} - \cdots$			0.11×10^{18}		$1.25 ^{+.32}_{-.21} \times 10^{19}$ (68%) [270]
^{128}Te	0.088×10^{24}	0.11×10^{24}	0.57×10^{24}	0.12×10^{24}		$0.55 - 0.092 \times 10^{24}$	2×10^{24}	11×10^{24}	2.6×10^{24}		$> 5 \times 10^{24}$ (95%) [265]
^{130}Te	0.017×10^{21}	0.023×10^{21}	0.12×10^{21}	0.019×10^{21}		$0.22 - 0.046 \times 10^{21}$	2×10^{21}	4.8×10^{21}	1.8×10^{21}		$1.5 - 2.75 \times 10^{21}$ (95%) [265]

Note— (1) shell model calculation [235]; (2), (3a) QRPA [245, 246]; (3b) PBCS [246]; (4) QRPA [247]; (5), (6) QRPA with $g_{pp} \neq 0$ [249, 274]; (7) generalised seniority-based truncation [251]; (8) QRPA [274]; (9) shell model [242].

Table 17. Half-life of neutrinoless double beta decay (in units of years) for an assumed Majorana neutrino mass $\langle m_\nu \rangle = 1\text{eV}$. The last column shows the experimental limit.

	Haxton 84	Grotz-Klapdor 85		Tomoda et al. 86	Tomoda -Faessler 87	Engel et al. 88	Engel et al. 89	Muto et al. 89	experiment
	(1)	(2a)	(2b)	(3)	(4)	(5)	(6)	(7)	
^{48}Ca	3.2×10^{24}								$> 2.6 \times 10^{21}$ (80%) [256]
^{76}Ge	1.7×10^{24}	0.26×10^{23}		1.1×10^{24}	2.9×10^{24}	$27 - 140 \times 10^{24}$	9.2×10^{24}	3.2×10^{24}	$> 1.9 \times 10^{24}$ (90%) [262]
^{82}Se	5.8×10^{23}	0.95×10^{23}			7.4×10^{23}	$110 - 450 \times 10^{23}$	73×10^{23}	7.2×10^{23}	$> 1.1 \times 10^{22}$ (68%) [268]
^{100}Mo			0.33×10^{23}			19×10^{23}		15×10^{24}	$> 5.1 \times 10^{21}$ (68%) [270]
^{128}Te	3.9×10^{24}	0.98×10^{24}			18×10^{24}	$25 - 45 \times 10^{24}$	18×10^{24}	14×10^{24}	$> 5 \times 10^{24}$ (95%) [265]
^{130}Te	1.6×10^{23}	0.46×10^{23}			6.6×10^{23}	$13 - 23 \times 10^{23}$	11×10^{23}	11×10^{23}	$> 1.5 - 2.75 \times 10^{21}$ (95%) [265]

Note— (1) shell model calculation [235]; (2a) QRPA and (2b) PBCS [246]; (3) QRPA [247]; (4), (5) QRPA with $g_{pp} \neq 0$ [250, 249]; (6) generalised seniority-based truncation [251]; (7) QRPA [274].

Table 18: Limit on $\langle m_\nu \rangle$(eV) from $T_{1/2}(^{76}\text{Ge} \to {}^{76}\text{Se}) > 1.9 \times 10^{24}$yr.

Haxton 84 (1)	Grotz-Klapdor 85 (2)	Tomoda et al. 86 (3)	Tomoda-Faessler 87 (4)	Engel et al. 88 (5)	Engel et al. 89 (6)	Muto et al. 89 (7)
< 0.95	< 0.37	< 0.78	< 1.3	$< 3.7 - 8.6$	< 2.2	< 1.3

on the Majorana neutrino mass is obtained from the ^{76}Ge result by the LBL-UCSB group: $T_{1/2} > 1.2 \times 10^{24}$yr (90% C.L.) [259], and most recently by the Heidelberg-Moscow experiment using enriched (86%) ^{76}Ge: $T_{1/2} > 1.9 \times 10^{24}$yr (90% C.L.) [262]. With the exception of Engel et al. [249], who claim that a matrix element cancellation due to $g^{pp} \neq 0$ is also important in $0\nu 0\beta$, the limits on m_ν derived from various nuclear physics calculations are in agreement up to a factor of 3 (see Table 18, where limits on m_ν are given for $T_{1/2} > 1.9 \times 10^{24}$yr). The result by Engel et al., however, gives a bound one order of magnitude looser. A limit $< 1-3$eV is a value agreed upon by most of the authors. The limits on m_ν from other processes are generally an order of magnitude weaker.

^{128}Te/^{130}Te ratio

Because of the small Q value of $^{130}\text{Te} \rightarrow ^{130}\text{Xe}$ (0.87MeV), $2\nu 2\beta$ decay which involves a 5-body phase-space volume is strongly suppressed. By simply accounting for the difference in the phase-space volume, we expect that $T^{2\nu}_{1/2}(^{128}\text{Te})/T^{2\nu}_{1/2}(^{130}\text{Te}) = 5700$ and $T^{0\nu}_{1/2}(^{128}\text{Te})/T^{0\nu}_{1/2}(^{130}\text{Te}) = 24.5$. We expect that much of the experimental uncertainty in geological measurements cancels in the ratio of two isotopes, as well as a part of the uncertainties in nuclear physics calculations. Therefore, $T_{1/2}(^{128}\text{Te})/T_{1/2}(^{130}\text{Te})$ may be used as a good indicator for the importance of the $0\nu 2\beta$ process and, hence, of m_ν [275] (see also [234]). Table 19 presents a compilation for the predicted ratios from various nuclear physics calculations. The experiment by Kirsten et al. [265] gives $T_{1/2}(^{128}\text{Te})/T_{1/2}(^{130}\text{Te}) > 3040$ (at 95% C.L.), and that by Manuel et al. [266] gives a finite value $T_{1/2}(^{128}\text{Te})/T_{1/2}(^{130}\text{Te}) = 2000$. In the theory side, Muto and Klapdor [274] give a very small value for $T^{2\nu}_{1/2}(^{128}\text{Te})/T^{2\nu}_{1/2}(^{130}\text{Te})$, while all other authors agree within a factor of 2.5 on this ratio. Allowing for this nuclear physics uncertainty (even omitting the calculation by Muto and Klapdor), the Misouri result for a finite value of $T_{1/2}(^{128}\text{Te})/T_{1/2}(^{130}\text{Te})$, which once was taken as evidence that $m_\nu \neq 0$ [234], does not compulsory lead to $m_\nu \neq 0$. The result by Kirsten et al. is consistent with $m_\nu = 0$.

Several Remarks

(i) In the presence of generation mixing, m_ν discussed above is not the Majorana mass of the electron neutrino, but receives a leakage contribution from other neutrinos, e.g., $\pm m_{\nu_\mu} \sin^2\theta_{\nu_e \nu_\mu}$ (in this sense the effective mass is often denoted as $\langle m_\nu \rangle$). A full discussion on this point is given in Sect.27.3. If there would exist a mediumly heavy neutrino (<1MeV, however), the limit on $\langle m_\nu \rangle$ would give a very strong constraint on its mixing with ν_e.

(ii) If the mass of the neutrino is $\gtrsim$1 MeV, (106) should be modified (the neutrino mass cannot be factored out), and thus the resulting limit on the neutrino mass [276]. For even heavier "neutrinos", $m_\nu \gtrsim$1GeV, the effect of the nucleon size becomes important [240]. For $m_\nu \rightarrow \infty$, $F(r)$ is calculated to be [235]

Table 19: $T_{1/2}(^{128}\mathrm{Te})/T_{1/2}(^{130}\mathrm{Te})$. Predictions for $2\nu 0\beta$ and $0\nu 0\beta$ decays by various authors. Experiments indicate $T_{1/2}(^{128}\mathrm{Te})/T_{1/2}(^{130}\mathrm{Te}) >$ 3040 at 95% C.L. [265] or = 2000 [266].

	$T^{2\nu}_{1/2}(^{128}\mathrm{Te})/T^{2\nu}_{1/2}(^{130}\mathrm{Te})$	$T^{0\nu}_{1/2}(^{128}\mathrm{Te})/T^{0\nu}_{1/2}(^{130}\mathrm{Te})$
phase space	5685	24.5
Haxton 84	5130	25
Grotz-Klapdor 85	4750	21.3
Engel et al. $g^{pp} = 0$	2820	18.5
$g^{pp} \neq 0$	2500 - 2000	19.2 - 19.6
Tomoda-Faessler		27.3
Muto-Klapdor 89 $g^{pp} = 0$	2290	
$g^{pp} \neq 0$	1000	
Engel et al. 89	2290	16.4
Muto et al. 89	1430	23.3

$$F(r) \sim \frac{1}{48}\frac{1}{r}\left(\frac{M_A}{m_\nu}\right)^2 e^{-M_A r}[(M_A r)^3 + 3(M_A r)^2 + 3(M_A r)] \ , \qquad (115)$$

rather than (104), where a dipole form factor is assumed for the nucleon weak vertex: $\sim 1/(1 - k_\nu^2/M_A^2)^2$.

(iii) In addition to the diagram shown in Figs.13(a)(b) (two nucleon mechanism), there are other diagrams that cause double beta decays through isobar excitation in nuclei, or a contribution from the exchange current (Fig. 16) [277]. We expect, however, that they give, at most, only corrections to the two-nucleon mechanism, because the probability of isobar excitation or a contribution form exchange current, themselves, are small.

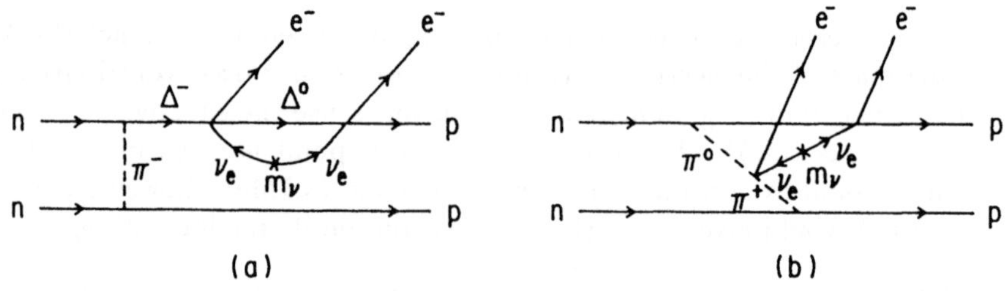

Figure 16: (a) Example of neutrinoless double beta decay induced by Δ (1232) excitation. (b) Example of neutrinoless double beta decay induced by pion exchange current.

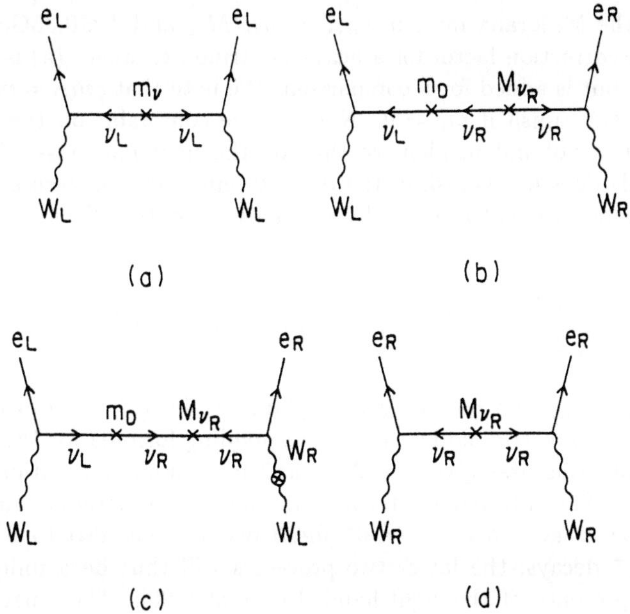

Figure 17: Feynman diagrams leading to neutrinoless double beta decay in the presence of the right-handed current. The nucleon part is omitted in the figure. (a) is the standard process, and (b)-(d) are those with the right-handed current; see text.

(iv) If there exists right-handed symmetry at high energy, we expect a suppressed contribution at low energy to the right-handed current from the exchange of the right-handed weak boson W_R. In this case, the double beta decay takes place even if $m_\nu = 0$[278]. The Feynman diagrams that give rise to $0\nu 2\beta$ decay are depicted in Fig. 17. The corresponding amplitudes are

$$\text{(a)} \quad \sim \quad G_F^2 \frac{m_\nu}{k_\nu} \ , \tag{116}$$

$$\text{(b)} \quad \sim \quad G_F \frac{g^2}{m_{W_R}^2} \frac{m_D M_{\nu R}}{M_{\nu R}^2} \sim G_F^2 \gamma^2 \xi \ , \tag{117}$$

$$\text{(c)} \quad \sim \quad G_F^2 \frac{m_D M_{\nu R}}{M_{\nu R}^2} \sim G_F^2 \epsilon \xi \ , \tag{118}$$

$$\text{(d)} \quad \sim \quad \left(\frac{g^2}{m_{W_R}^2} \right)^2 \frac{f}{M_{\nu R}} \sim G_F^2 \gamma^4 f / M_{\nu R} \ , \tag{119}$$

where $\gamma \equiv m_{W_1}/m_{W_2} \simeq m_{W_L}/m_{W_R}$, ε is the W_R component in the mass eigenstate W_1, which is

$$\begin{aligned} W_1 &= W_L + \varepsilon W_R \\ W_2 &= -\varepsilon W_L + W_R, \end{aligned} \tag{120}$$

M_{ν_R} is the Majorana mass of $\nu_R, \xi = m_D/M_{\nu_R}$ and $f \simeq$0.35GeV [279] is an effective correction factor for a heavy neutrino exchange. (a) is the standard process, but is added for a comparison. We note that $\xi m_D = m_\nu$, and hence (b) and (c) vanish if $m_\nu = 0$. Namely, the constraint on the right-handed coupling is not independent of that on the neutrino mass. On the other hand, (d) does not vanish in the $m_\nu = 0$ limit. The processes involving the right-handed current is generally summarised by the effective interaction,

$$H_{\text{eff}} = \frac{G_F \cos\theta_c}{\sqrt{2}}[J_L^{(\ell)\dagger} J_L^{(h)} + \eta_{LR} J_L^{(\ell)\dagger} J_R^{(h)} + \eta_{RL} J_R^{(\ell)\dagger} J_L^{(h)} + \eta_{RR} J_R^{(\ell)\dagger} J_R^{(h)}], \quad (121)$$

where η_{LR}, η_{RL} and η_{RR} are given by amplitudes (b)–(d) above if we take a left-right symmetric model. The corresponding formula for the double beta-decay rate was first given in [233], and in a modern notation using (121) in [280]. The interference term from the left- and right-handed currents gives rise to not only $0^+ \rightarrow 0^+$ $0\nu2\beta$ decays, but also to $0^+ \rightarrow 2^+$ and $0^+ \rightarrow 1^+$ decays; the latter two processes will thus be a unique signature for the presence of the right-handed current [280]. The current limits are $\eta_{LR} < 0.2 - 2 \times 10^{-7}$ and $\eta_{RR} < 2 \times 10^{-6}$. The combined limits on m_{W_R} and m_ν are discussed in ref.[281].

(v) Majoron emission in association with double beta decay has been discussed in the literature. The Majoron that couples to $\overline{\nu^c}\nu$ is necessarily an $SU(2)_L$ triplet, and it is excluded by the Z^0 width measurement, as we mentioned in Sect.14. Therefore, we have not discussed it in this section.

(vi) Although there have been some discussions concerning the $0\nu2\beta$ process induced by SUSY particle exchanges in the literature [282], such a process occurs only when R parity is broken, which usually induces a number of unwanted effects (fast proton decay etc.).

V. Generation mixing of neutrinos

22 Generation mixing – the case of Dirac neutrinos

Let us generalise the mass term (III.34) so as to incorporate generation

$$\mathcal{L}_{\text{mass}} = -\bar{\nu}_R^\alpha m_{\alpha\beta} \nu_L^\beta + h.c. \,, \tag{1}$$

where $\nu^\alpha = (\nu^e, \nu^\mu, \nu^\tau)$ are the eigenstates of the weak interaction specified by

$$\mathcal{L}_{\text{int}} = \frac{g}{\sqrt{2}} W_\mu^+ [\bar{\nu}_L^e \gamma^\mu e_L + \bar{\nu}_L^\mu \gamma^\mu \mu_L + \bar{\nu}_L^\tau \gamma^\mu \tau_L] + h.c. \tag{2}$$

Taking the basis in which the charged lepton mass term is diagonal, we further diagonalise the neutrino mass term (see remarks made after eq.(II.53)) by

$$\nu_L^\alpha = U_{\alpha i} \nu_L^{'i}, \tag{3}$$

$$\nu_R^\alpha = V_{\alpha i} \nu_R^{'i} \,, \tag{4}$$

viz.,

$$\begin{aligned} \mathcal{L}_{\text{mass}} &= -\bar{\nu}_R^{'i} V_{i\alpha}^\dagger m_{\alpha\beta} U_{\beta j} \nu_L^{'j} + h.c. \\ &= -\bar{\nu}_R^{'i} m_i \nu_L^{'i} + h.c. \end{aligned} \tag{5}$$

i.e.,

$$V^\dagger m U = m_{\text{diag}} \,. \tag{6}$$

Here, we choose, without loss of generality, the phases of $\nu_{L,R}^{'i}$ so that m_i is real and semi-positive definite. Using the basis of the mass eigenstate $\nu^{'i}$ of eq.(3), we rewrite the interaction Lagrangian (2) as,

$$\mathcal{L}_{\text{int}} = \frac{g}{\sqrt{2}} W_\mu^+ (\bar{\nu}^{'1}, \bar{\nu}^{'2}, \bar{\nu}^{'3})_L U^\dagger \gamma_\mu \begin{pmatrix} e \\ \mu \\ \tau \end{pmatrix}_L + h.c. \,, \tag{7}$$

where $U^\dagger$ is the " CKM matrix " for leptons. We note that neutrinos which appear most often in laboratory experiments are the eigenstate of the weak interaction, ν^α. The mass eigenstate of the neutrinos $\nu^{'i}$ appears only in the neutrino mass experiment. (This is in contrast to the case of quarks, for which we usually deal with the mass eigenstate.)

If neutrinos have finite masses, the lepton numbers (e-, μ-number etc.) are not conserved separately within the standard electroweak theory, unless the off-diagonal elements of the mass matrix vanish. The effect of the lepton-number violation caused by generation mixing, however, is greatly suppressed in the charged lepton sector, because such amplitudes should necessarily involve a small neutrino mass. For instance, $l_\alpha \to l_\beta \gamma$ is given by [283]

$$\Gamma(l_\alpha \to l_\beta\gamma) = \frac{1}{2}\alpha G_F^2(\frac{1}{32\pi^2})^2 m_{l_\alpha}^5 \mid \sum_i U_{\alpha i}^* U_{\beta i} \frac{m_{\nu_i}^2}{M_W^2} \mid^2 . \tag{8}$$

The branching ratio is thus

$$B(l_\alpha \to l_\beta\gamma) = (3\alpha/32\pi) \mid \sum_i U_{\alpha i}^* U_{\beta i}(m_{\nu_i}^2/M_W^2) \mid^2 \lesssim 10^{-17}, \tag{9}$$

even with the upper-limit value of the ν_τ mass (II.119). The experimental limit is far too large compared with (9):

$$\begin{aligned} B(\mu \to e\gamma) &< 4.9 \times 10^{-11} , \\ B(\tau \to e\gamma) &< 2.0 \times 10^{-4} , \\ B(\tau \to \mu\gamma) &< 5.5 \times 10^{-4} . \end{aligned} \tag{10}$$

We expect a similar suppression for $\mu \to ee\bar{e}$ [283]:

$$B(\mu \to ee\bar{e}) = \frac{3\alpha^2}{16\pi^2} \mid \sum_i U_{\mu i}^* U_{ei} \frac{m_{\nu_i}^2}{M_W^2} \ln \frac{m_{\nu_i}^2}{M_W^2} \mid^2 \lesssim 10^{-17} , \tag{11}$$

which is compared with the experimental limit,

$$B(\mu \to ee\bar{e}) < 1.0 \times 10^{-12} . \tag{12}$$

This means that the present laboratory limits on flavour-changing leptonic decays do not give any constraints on the lepton-number violation arising from generation mixing.

23 Generation mixing – the case of Majorana neutrinos

The mass term for Majorana neutrinos is written

$$\mathcal{L}_{\text{mass}} = -\frac{1}{2}\nu_L^{\alpha T} C^{-1} m_{\alpha\beta} \nu_L^\beta + h.c. \tag{13}$$

Taking again the basis in which the charged lepton-mass term is diagonal, we diagonalise this mass term by

$$\nu_L^\alpha = U_{\alpha i}\nu_L^{\prime i}, \tag{14}$$

so that

$$\begin{aligned} \mathcal{L}_{\text{mass}} &= -\frac{1}{2}\nu_L^{\prime i} C^{-1} U_{i\alpha} m_{\alpha\beta} U_{\beta j} \nu_L^{\prime j} + h.c. \\ &= -\frac{1}{2}\nu_L^{\prime i} C^{-1} m_i \nu_L^{\prime i} + h.c., \end{aligned} \tag{15}$$

i.e.,

$$U^T m U = m_{\mathbf{diag}} \,, \tag{16}$$

or

$$m = U^* m_{\mathbf{diag}} U^\dagger \,. \tag{17}$$

If the elements of the mass matrix (13) are real, $\mathcal{L}_{\mathbf{mass}}$ is CP invariant and the Majorana neutrinos can be classified by the CP property. This was first pointed out by Yang and Tiomno (1950) [284] and later re-discovered by Wolfenstein [219]. Namely, in this case the unitary matrix U of (14) becomes an orthogonal matrix O and (16) reads

$$O_{k\alpha} m_{\alpha\beta} O^T_{\beta\ell} = \eta_k m_k \delta_{k\ell}, \tag{18}$$

where m_k is real positive and $\eta_k = +1$ or -1. Defining a Majorana field by

$$\chi_k = \sum_\alpha O_{k\alpha} \nu^\alpha_L + \eta_k O_{k\alpha} (\nu^\alpha_L)^c \tag{19}$$

with $\overline{(\nu^\alpha_L)^c} = \nu^{\alpha T}_L C^{-1}$, we can write $\mathcal{L}_{\mathbf{mass}}$

$$\mathcal{L}_{\mathbf{mass}} = \sum m_k \overline{\chi_k} \chi_k \,. \tag{20}$$

This χ_k is nominally the eigenstate of C, since

$$C \chi_k C^{-1} = \eta_k \chi_k \,. \tag{21}$$

Since the neutrino interaction does not respect a charge-conjugation invariance, however, the eigenstate of C is not physically a useful concept. Now let us discuss the relation between χ_k and the CP eigenstate. Under a parity transformation,

$$P \chi_k P^{-1} = \xi_P \gamma^0 \chi_k \,. \tag{22}$$

then,

$$(CP) \chi_k (CP)^{-1} = \xi_P \eta_k \gamma^0 \chi_k \,. \tag{23}$$

We take $\xi_P = +1$ by convention. χ_k is not an eigenstate of CP, but the relative sign of η_k is still meaningful, since the interaction Lagrangian is invariant under the CP transformation.

For example, let us consider a tensor-type transition between two Majorana neutrinos χ_1 and χ_2 (see Sect.41 for detail). If $\eta_1/\eta_2 = +1$, only the term $\bar{\chi}_1 \gamma_5 \sigma_{\mu\nu} \chi_2$ is allowed. If $\eta_1/\eta_2 = -1$, on the other hand, the allowed term is $\bar{\chi}_1 \sigma_{\mu\nu} \chi_2$ instead. The two terms coexist if CP is violated [285].

Pseudo-Dirac neutrinos

If the Majorana mass matrix m_{ij} takes the form

$$m_{ij} = \begin{pmatrix} 0 & * \\ * & 0 \end{pmatrix}, \tag{24}$$

there are two neutrinos with $\eta = +1$ and $\eta = -1$ degenerate in mass. In this case the relative lepton number of ν_1 and ν_2 is conserved and, thus, the set of (ν_1, ν_2) is called a pseudo-Dirac neutrino [219] (see also the argument at the end of Sect. 13 above).

24 Phases of the mixing matrices

The 3×3 mixing matrix U in (3) can be represented by

$$U = e^{i\alpha} e^{i\beta\lambda_3} e^{i\gamma\lambda_8} u e^{i\beta'\lambda_3} e^{i\gamma'\lambda_8}, \tag{25}$$

where λ_3 and λ_8 are Gell-Mann's λ matrices for the SU(3) group, and u stands for the conventional KM representation of a 3×3 matrix with 4 parameters. For the case of Dirac neutrinos the phases β' and γ' are absorbed into the phase rotation of the wave function of ν', which is always multiplied on U from the right (see (3)). Then,

$$U = e^{i\alpha} e^{i\beta\lambda_3} e^{i\gamma\lambda_8} u \,. \tag{26}$$

The interaction Lagrangian (7) is written

$$\mathcal{L}_{\text{int}} = \frac{g}{\sqrt{2}} W_\mu^\dagger (\bar{\nu}'^1 \bar{\nu}'^2 \bar{\nu}'^3) u e^{-i\alpha} e^{-i\beta\lambda_3} e^{-i\gamma\lambda_8} \gamma_\mu \begin{pmatrix} e \\ \mu \\ \tau \end{pmatrix}. \tag{27}$$

The three phases α,β,γ are now absorbed into the phase of the charged leptons, and we are left with

$$\mathcal{L}_{\text{int}} = \frac{g}{\sqrt{2}} W_\mu^\dagger \bar{\nu}' u \gamma_\mu \ell + h.c., \tag{28}$$

where $\ell = (e, \mu, \tau)^T$. Only the phases of a matrix which cannot be factored out as in (25) are physical, and would give rise to CP violation. For n flavours the number of such physical phases is easily calculated to be

$$n^2 - \frac{n(n-1)}{2} - (n-1) - n = \frac{(n-1)(n-2)}{2} \tag{29}$$

for an $n \times n$ matrix. The last two terms on the left-hand side correspond to phases which we factored out and hence absorbed into the wave function.

For the Majorana case, the two phases β' and γ' cannot be absorbed into the wave function of ν', because the mass term takes the form $\nu'^T U^T m U \nu'$, where the same U appears twice, rather than $V^\dagger$ and U for the Dirac case. The other three phases α, β and γ are absorbed into the charged-lepton wave function, as in (27), in the weak-interaction Lagrangian. Therefore, we are left with

$$\mathcal{L}_{\rm int} = \frac{g}{\sqrt{2}} W_\mu^\dagger \bar{\nu}_L' e^{-i\beta'\lambda_3} e^{-i\gamma'\lambda_8} u \gamma_\mu \ell + h.c. \tag{30}$$

instead of (28). The number of physical phases for the n generation case is

$$n^2 - \frac{n(n-1)}{2} - n = \frac{n(n-1)}{2} \,. \tag{31}$$

Namely, we may see effects of CP violation even for $n = 2$ for Majorana neutrinos [286, 287].

If one absorbs the phase in (30) into the neutrino wave function, as $\nu_L'' = e^{i\gamma'\lambda_8} e^{i\beta'\lambda_3} \nu_L'$, the phase appears in the mass matrix,

$$\mathcal{L}_{\rm mass} = -\frac{1}{2} \overline{\nu_L''^c} e^{-i\gamma'\lambda_8} e^{-i\beta'\lambda_8} m_{\rm diag} e^{-i\beta'\lambda_3} e^{-i\gamma'\lambda_8} \nu_L'' \,. \tag{32}$$

This means that the effect of CP violation arising from this extra Majorana phase always appears together with the neutrino mass and hence the effect is suppressed by $(m_\nu/E)^2$ [288].

Let us exemplify the above argument more explicitly for $n = 2$. The neutrino mass term,

$$\mathcal{L}_{\nu\,\rm mass} = \bar{\nu}_L^c \begin{pmatrix} m_{11} & m_{12} \\ m_{21} & m_{22} \end{pmatrix} \nu_L + h.c. \,, \tag{33}$$

has 6 parameters ($m_{12} = m_{21}$). We may write (25) as

$$U = e^{i\alpha} e^{i\beta\tau_3} \begin{pmatrix} \cos\theta & \sin\theta \\ -\sin\theta & \cos\theta \end{pmatrix} e^{i\beta'\tau_3}. \tag{34}$$

The diagonalisation of the mass matrix is now

$$U^* \begin{pmatrix} m_1 & \\ & m_2 \end{pmatrix} U^\dagger = \begin{pmatrix} m_{11} & m_{12} \\ m_{21} & m_{22} \end{pmatrix} \,, \tag{35}$$

corresponding to (14)–(17). We note that $m_1, m_2, \alpha, \beta, \theta, \beta'$ are determined by the 6 parameters of m_{ij}. With (34), (35) reads

$$e^{-2i\alpha} e^{-i\beta\tau_3} \begin{pmatrix} \cos\theta & \sin\theta \\ -\sin\theta & \cos\theta \end{pmatrix} \begin{pmatrix} m_1 e^{-2i\beta'} & 0 \\ 0 & m_2 e^{2i\beta'} \end{pmatrix} \begin{pmatrix} \cos\theta & -\sin\theta \\ \sin\theta & \cos\theta \end{pmatrix} e^{-i\beta\tau_3}$$

$$= \begin{pmatrix} m_{11} & m_{12} \\ m_{21} & m_{22} \end{pmatrix} \,. \tag{36}$$

The weak current is now

$$\begin{aligned} J_\mu &= \bar{\ell}_L \gamma_\mu \nu_L \\ &= \bar{\ell}_L \gamma_\mu U \nu_L' \\ &= \bar{\ell}_L \gamma_\mu e^{i\alpha} e^{i\beta\tau_3} \begin{pmatrix} \cos\theta & \sin\theta \\ -\sin\theta & \cos\theta \end{pmatrix} e^{i\beta'\tau_3} \nu_L'. \end{aligned} \tag{37}$$

The two phases α and β are absorbed into the phase of $\bar{\ell}_L$ upon the redefinition

$$\bar{\ell}'_L = \bar{\ell}_L e^{i\alpha} e^{i\beta\tau_3} \,. \tag{38}$$

On the other hand, β' cannot be absorbed into the phase of ν'_L, and remains physical. If we define

$$\nu''_L = e^{i\beta'\tau_3} \nu'_L \,, \tag{39}$$

we have

$$\mathcal{L}_{\nu\ \mathrm{mass}} = \overline{\nu^{c''}_L} \begin{pmatrix} m_1\ e^{-2i\beta'} & \\ & m_2\ e^{2i\beta'} \end{pmatrix} \nu''_L \,. \tag{40}$$

The phase β' generally causes CP violation.

There are a few special cases in which CP is not violated. If $\beta' = 0$ or $\beta' = \pi/2$ the mass terms are real, and CP is conserved. In this case the relative CP, as discussed in (23) above, is $+1$. The other case is $\beta' = \pi/4$, for which we have

$$\begin{pmatrix} m_1\ e^{-2i\alpha - \frac{\pi}{2}i} & 0 \\ 0 & m_2\ e^{-2i\alpha + \frac{\pi}{2}i} \end{pmatrix} = \begin{pmatrix} -m_1 & 0 \\ 0 & m_2 \end{pmatrix}, \tag{41}$$

using a freedom to choose α. This corresponds to a relative CP of -1.

The case of the pseudo Dirac neutrino (24) is written

$$U^\dagger m U = U^\dagger \begin{pmatrix} 0 & m_{12} \\ m_{21} & 0 \end{pmatrix} U = \begin{pmatrix} -m_1 & 0 \\ 0 & m_1 \end{pmatrix}, \tag{42}$$

which is a special case of $\beta' = \pi/4$.

25 Neutrino oscillation in vacuum

25.1 Formulation

In the presence of off-diagonal components of the neutrino mixing matrix a neutrino of the weak interaction eigenstate may change into another kind while propagating through the vacuum [34],[289]–[291]. A neutrino of the generation α after a time interval of t is given by

$$|\nu^\alpha\rangle_t = \sum_i U_{\alpha i} e^{-E_i t} |\nu'^i\rangle \,, \tag{43}$$

and the transition amplitude to the state $|\nu^\beta\rangle$ is

$$\langle \nu^\beta | \nu^\alpha \rangle_t = \sum U_{\alpha i} (U^\dagger)_{i\beta} e^{-iE_i t}. \tag{44}$$

For $p = |\vec{p}| \gg m_i$ we expand $E_i = \sqrt{p^2 + m_i^2} \simeq p + m_i^2/2p \simeq p + m_i^2/2E$, and using (3) we write (43) as

$$\begin{aligned} |\nu^\alpha\rangle_t &\simeq e^{-ipt} U \begin{bmatrix} e^{-im_1^2 t/2E} & & \\ & e^{-im_2^2 t/2E} & \\ & & \ddots \end{bmatrix} U^\dagger |\nu^\beta\rangle \\ &= e^{-ipt} U \begin{bmatrix} 1 - i\frac{m_1^2}{2E}t + \cdots & & \\ & 1 - i\frac{m_2^2}{2E} + \cdots & \\ & & \ddots \end{bmatrix} U^\dagger |\nu^\beta\rangle \,. \end{aligned} \tag{45}$$

Since

$$U^\dagger m^\dagger m U = m^2_{\text{diag}} = \begin{pmatrix} m_1^2 & & 0 \\ & m_2^2 & \\ 0 & & \ddots \end{pmatrix}, \tag{46}$$

we may write

$$|\nu^\alpha\rangle_t \simeq e^{-ipt}[e^{-i\frac{m^\dagger m}{2E}t}]_{\alpha\beta}|\nu^\beta\rangle, \tag{47}$$

i.e., $|\nu^\alpha\rangle_t$ is a solution of the Schrödinger equation,

$$i\frac{d}{dt}|\nu^\alpha\rangle_t \simeq \frac{m^\dagger m}{2E}|\nu^\alpha\rangle \,, \tag{48}$$

up to the phase factor e^{-ipt}. If we restrict ourselves to the case of mixing between the two generations, we can generally parametrise U to be

$$U = \begin{pmatrix} e^{i\alpha} & \\ & e^{-i\alpha} \end{pmatrix} \begin{pmatrix} \cos\theta & \sin\theta \\ -\sin\theta & \cos\theta \end{pmatrix} \begin{pmatrix} e^{i\beta} & \\ & e^{-i\beta} \end{pmatrix}, \tag{49}$$

where the phase factors $e^{i\tau_3\beta}$ do not appear in $\langle\nu^\beta|\nu^\alpha\rangle_t$ and $e^{i\tau_3\alpha}$ can be absorbed into the wave functions of $|\nu^\alpha\rangle_t$. We may then take

$$U = \begin{pmatrix} \cos\theta & \sin\theta \\ -\sin\theta & \cos\theta \end{pmatrix} \tag{50}$$

without loss of generality. The situation does not differ between the Dirac case and the Majorana case, since the mass matrix related to neutrino oscillation is always $m^\dagger m = U m^2_{\text{diag}} U^\dagger$, which appears in (46), and not $m^T m$. With (50) we have

$$\begin{aligned} m^\dagger m &= U m^2_{\text{diag}} U^\dagger = \begin{pmatrix} m_1^2\cos^2\theta + m_2^2\sin^2\theta & \frac{1}{2}(m_2^2 - m_1^2)\sin 2\theta \\ \frac{1}{2}(m_2^2 - m_1^2)\sin 2\theta & m_1^2\sin^2\theta + m_2^2\cos^2\theta \end{pmatrix} \\ &= \frac{m_1^2 + m_2^2}{2} + \frac{\Delta m^2}{2}\begin{pmatrix} -\cos 2\theta & \sin 2\theta \\ \sin 2\theta & \cos 2\theta \end{pmatrix}, \end{aligned} \tag{51}$$

where $\Delta m^2 = m_2^2 - m_1^2$. Hence, (47) becomes

$$|\nu^\alpha\rangle_t = e^{-i(p+\frac{m_1^2+m_2^2}{4E})t}\left[e^{-i\frac{\Delta m^2}{4E}\tau^a r^a t}\right]_{\alpha\beta}|\nu^\beta\rangle \tag{52}$$

with $r^a = (\sin 2\theta, 0, -\cos 2\theta)$, so that

$$|\nu^\alpha\rangle_t = \begin{pmatrix} \cos\frac{\Delta m^2}{4E}t - i\sin\frac{\Delta m^2}{4E}t\cos 2\theta & -i\sin\frac{\Delta m^2}{4E}t\sin 2\theta \\ -i\sin\frac{\Delta m^2}{4E}t\sin 2\theta & \cos\frac{\Delta m^2}{4E}t + i\sin\frac{\Delta m^2}{4E}t\cos 2\theta \end{pmatrix}|\nu^\beta\rangle, \tag{53}$$

omitting the unimportant common phase factors. The transition from $|\nu^e\rangle = (1,0)^T$ to $|\nu^e\rangle_t$ is then

$$\langle\nu^e|\nu^e\rangle_t = \cos\frac{\Delta m^2}{4E}t - i\sin\frac{\Delta m^2}{4E}t\cos 2\theta \ , \tag{54}$$

and its transition probability reads

$$\begin{aligned} P_{\nu_e\to\nu_e} &= |\langle\nu^e|\nu^e\rangle_t|^2 \\ &= 1 - \sin^2 2\theta\sin^2\frac{\Delta m^2}{4E}t. \end{aligned} \tag{55}$$

This shows oscillation of the neutrino beam propagating through the vacuum. For $\nu_e \to \nu_\mu, \nu_\mu = (0,1)^T$, and we have

$$\begin{aligned} P_{\nu_e\to\nu_\mu} &= |\langle\nu^\mu|\nu^e\rangle|^2 \\ &= \sin^2 2\theta\sin^2\frac{\Delta m^2}{4E}t, \end{aligned} \tag{56}$$

and the same expression also for $\nu_\mu \to \nu_e, \bar\nu_e \to \bar\nu_\mu, \bar\nu_\mu \to \bar\nu_e$. We call l_0, which satisfies

$$\frac{\Delta m^2}{4E}(\ell_0/c) = \pi \tag{57}$$

($c = 1$, light velocity), the oscillation length. If we observe the neutrino flux at a distance L from the source, we have from (55)

$$P_{\nu_e\to\nu_e} = 1 - \sin^2 2\theta\sin^2(\pi\frac{L}{\ell_0}) \ . \tag{58}$$

For $L \gg \ell_0$ or

$$\Delta m^2 \gg 4\pi E/L = \Delta m^2_{\rm min}, \tag{59}$$

the second sine factor oscillates rapidly, and we expect that

$$P_{\nu_e\to\nu_e} = 1 - \frac{1}{2}\sin^2 2\theta, \tag{60}$$

which stands for a reduction of the ν_e flux averaged over a sufficiently long time. (In practice this averaging is caused by the momentum spread of the neutrino beam). This is often called "time-averaged oscillation". However, it does not represent an actual oscillation phenomenon any longer; P measures merely the neutrino mixings (see the next subsection). When (59) is satisfied, we generally expect that

$$P_{\nu_e \to \nu_e} \geq 1/2. \tag{61}$$

For the case of N generations (55) is modified to

$$P_{\nu_e \to \nu_e} = 1 - \sum_{i,j} 4 \mid U_{ei} \mid^2 \mid U_{ej} \mid^2 \sin^2 \frac{\Delta m_{ij}^2}{4E} t, \tag{62}$$

and then

$$P_{\nu_e \to \nu_e} \geq 1/N, \tag{63}$$

when $\Delta m_{ij}^2 \gg 4\pi E/\ell_0$ for all (i,j) pairs.

We emphasize again that the same argument applies to both Dirac [290] and Majorana [289] neutrinos; hence, the experiment does not distinguish between the two cases.

25.2 Neutrino oscillation and uncertainty principle

If the energy and the momentum of the neutrino are measured sufficiently accurately that the error δm_ν of $m_\nu = \sqrt{E_\nu^2 - p_\nu^2}$ is smaller than the mass difference of two neutrinos, the virtual transition into the other neutrino state cannot take place. This means, for instance, that neutrino oscillation disappears if both π and μ momenta are measured with extreme precision in a neutrino oscillation experiment using the beam from $\pi \to \mu\nu$ decays. Although this statement sounds paradoxical in view of what we have argued in Sect.25.1, this "puzzle" is resolved if one considers the uncertainty principle of quantum mechanics. In fact, if the momentum is measured with extreme precision, the information concerning the position, as to where the ν particle is produced, will be lost.

In order to keep the error of $m_\nu^2 (= \delta m^2)$ smaller than Δm^2, one must measure p_ν with a precision of δp_ν that satisfies

$$\Delta m^2 \gg \delta m^2 = (\partial m_\nu^2/\partial p_\nu)\delta p_\nu = 2p_\nu \delta p_\nu. \tag{64}$$

This means that the position becomes undetermined to the order of $\delta x \delta p_\nu > \hbar$ (we take $\hbar = 1$), i.e.,

$$\delta x > \frac{2p_\nu}{\Delta m^2} = \frac{1}{2\pi}\ell_0. \tag{65}$$

The positional precision is lost, at least to the order of the oscillation length, and the oscillation pattern is washed out by this uncertainty. The result then reduces to time-averaged oscillation (60), where the deviation of $P_{\nu_e \to \nu_e}$ from unity is a consequence of the fact that either of the mass eigenstates chosen by the observation is the sum of the two weak eigenstates: namely,

$$P_{\nu_e\to\nu_e} = \cos^4\theta + \sin^4\theta, \tag{66}$$

where the first term is a contribution when ν_1 is chosen in the intermediate state and the second is that when ν_2 is chosen, and they are added incoherently. The probability (66), of course, agrees with (60).

A similar effect is seen in the following situation: if the neutrino beam is originally localised, the neutrino clouds corresponding to the two different mass eigenstates move with a different velocity and will be separated by an amount of $(\Delta v)ct \approx (\Delta m^2/E^2)ct$ after time t. If this separation is larger than the neutrino coherence length, the two clouds do not overlap any more, and hence do not interfere. This means that neutrino oscillation eventually disappears, and that $P_{\nu_e\to\nu_e}$ reduces to (60) again [292]. The condition for oscillation can also be obtained by applying the uncertainty principle.

Wave-packet formalism for neutrino oscillation

In order to discuss the problems related to the uncertainty principle in a more formalistic manner, we now develop a wave-packet formalism [293, 294]. We write a normalised wave packet for the mass eigenstate to be

$$\psi_i(x,t) = \frac{1}{\sqrt{2\pi}\sigma_x}\exp\left[i\left(\overline{p_i}x - \overline{E_i}t\right) - \frac{(x-v_it)^2}{4\sigma_x^2}\right], \tag{67}$$

where $\overline{p_i}$ is the mean momentum, $\overline{E_i}$ the mean energy $\overline{E_i} = \sqrt{\overline{p_i}^2 + m_i^2}, v_i$ the group velocity $v_i = \partial\overline{E_i}/\partial\overline{p_i} = \overline{p_i}/\overline{E_i}$ and σ_x the Gaussian width in the coordinate space. In momentum space (67) is written

$$\psi_i(p) = \frac{1}{\sqrt{2\pi}\sigma_p}\exp\left[-\frac{(p-\overline{p_i})^2}{4\sigma_p^2}\right], \tag{68}$$

where σ_p satisfies

$$\sigma_x\sigma_p = 1/2. \tag{69}$$

Let us consider the two-neutrino case and a neutrino of flavour α created at $x = t = 0$ by a weak interaction process. The probability to find a neutrino of flavour β at x and t is given by

$$P_{\alpha\to\beta}(x,t) = |\sum_i U_{\beta i}^*\psi_i(x,t)U_{\alpha i}|^2. \tag{70}$$

Using (34) for U we obtain for α=1, β=2,

$$\begin{aligned} P_{1\to 2} &= \frac{1}{2\pi}\frac{1}{\sigma_x^2}\sum_{i,j} U_{2i}^*U_{1i}U_{2j}U_{1j}^* \\ &\quad\times\exp[i(\overline{p_i}-\overline{p_j})x - i(\overline{E_i}-\overline{E_j})t] \end{aligned}$$

$$\times \exp\left[-\frac{(x-v_1t)^2}{4\sigma_x^2}-\frac{(x-v_2t)^2}{4\sigma_x^2}\right]$$

$$= \frac{1}{2\pi\sigma_x^2}2\sin^2\theta\cos^2\theta\{1-\cos[(\overline{p_1}-\overline{p_2})x-(\overline{E_1}-\overline{E_2})t]$$

$$\times \exp\left[-\frac{(x-v_1t)^2}{4\sigma_x^2}-\frac{(x-v_2t)^2}{4\sigma_x^2}\right]. \tag{71}$$

For the relativistic limit (71) becomes

$$P_{1\to2}=2\sin^2\theta\cos^2\theta\left\{1-\cos\left(\frac{\Delta m^2}{2\overline{p}}x\right)\exp\left(-\frac{x^2}{8\sigma_x^2}\left(\frac{\Delta m^2}{2\overline{p}}\right)^2-(1+\kappa)\frac{(\Delta m^2)^2}{32\sigma_p^2\overline{p}^2}\right)\right\}, \tag{72}$$

where

$$\kappa=\frac{\overline{p_1}^2-\overline{p_2}^2}{\Delta m^2}. \tag{73}$$

For

$$x \ll \frac{\sigma_x}{\Delta m^2/2\overline{p}^2}\simeq\frac{\sigma_x}{|v_1-v_2|} \tag{74}$$

we obtain

$$P_{1\to2}(x)\simeq\frac{1}{2}\sin^2 2\theta\left(1-\cos\left[\frac{\Delta m^2}{2\overline{p}}x\right]\right)=\sin^2 2\theta\sin^2\left(\frac{\Delta m^2}{4E}t\right), \tag{75}$$

in agreement with (56). The second factor in the exponential of (72) ensures energy conservation within the uncertainty of σ_p [295].

For $x>\sigma_x/|v_1-v_2|$, on the other hand, the damping factor in (72) works, and we obtain

$$P_{1\to2}=2\sin^2\theta\cos^2\theta \tag{76}$$

in agreement with the formula for time-averaged oscillation. This represents an incoherent sum of $\nu_1\to\nu_1'\to\nu_2$ and $\nu_1\to\nu_2'\to\nu_2$.

Examination of the experimentally relevant case

In practically interesting cases the oscillation pattern is erased by the momentum spread of the neutrino beam. The condition that neutrino oscillation can actually be observed is

$$\frac{2\pi E}{L}\lesssim\Delta m^2\lesssim\frac{2\pi E}{L}(\frac{\delta E}{E})^{-1}, \tag{77}$$

where δE is the energy (momentum) spread of the beam, in addition to (74), $\Delta m^2\lesssim(2\pi E/L)(E\sigma_x/\pi)$. Since the second inequality of (77) is usually much stronger than (74), the above consideration concerning the uncertainty principle is rather a matter of academic interest. For example, for laboratory experiments with a neutrino beam from $\pi \to \mu\nu$ the coherence length is $\sigma_x \sim \tau_\pi c$, and thus $E\sigma_x/\pi \sim 10^{16}(E/1\text{ GeV})$, which is a huge factor compared with $(\delta E/E)^{-1} \sim 1-3$.

The only case for which consideration may be worthwhile is for solar neutrinos [292]. For solar neutrino emission the coherence length is prescribed by the mean-free time of thermal-nuclear collisions in the core. For $\rho \sim 100\text{g/cm}^3$ and $kT \sim 1\text{keV}$, $\sigma_x = ct_{\text{collision}} = 1.5 \times 10^{-6}\text{cm}$, or $\delta E \sim 10\text{eV}$ in the effective energy width (see ref.[296] for a more elaborate estimate, which gives the result virtually the same as this), and hence

$$\Delta m^2 < 2 \times 10^{-7}\text{eV}^2 \left(\frac{E}{1\text{MeV}}\right)^2 . \tag{78}$$

On the other hand, solar neutrinos from e capture of ^{7}Be are almost monoenergetic ($E_\nu = 0.86\text{MeV}$). The momentum spread of such a neutrino beam is of the order of the thermal Doppler broadening, $\delta E_\nu \sim 1.3\text{keV}$, in the centre of the Sun; thus, $\delta E_\nu/E_\nu \sim 700$. The oscillation effect then persists up to $\Delta m^2 \approx (2\pi E_\nu/L) \times 700$; that is, oscillation may be visible for the range $\Delta m^2 \approx 0.5 \times 10^{-11}$ to $6 \times 10^{-9}\text{eV}^2$. The oscillation condition from the uncertainty principle (78) is satisfied for the range of our interest, at least with an allowance for one order of magnitude [292].

For neutrinos from a supernova the oscillation condition is stringent [297]. This consideration, however, is physically uninteresting, since oscillation does not play any sensible role, due to a large energy spread of the neutrino flux.

25.3 Experimental test for neutrino oscillation

The detection of $P_{\nu_e \to \nu_e} \neq 1$ or $P_{\nu_e \to \nu_\mu} \neq 0$ could provide important information on a small neutrino mass which otherwise would be too small to detect directly. In physical units the oscillation length given by (57) is

$$\ell_0 = 4\pi E/\Delta m^2 = 2.5 \times 10^2\text{cm}\frac{E/\text{MeV}}{\Delta m^2/\text{eV}^2}. \tag{79}$$

This means that we need a long neutrino beam line, $L \gtrsim l_0/2$ say, and/or a small enough energy so as to obtain information on a small-mass neutrino. For example, a neutrino beam from an accelerator can explore only down to $\sqrt{\Delta m^2} \gtrsim 1\text{eV}$ $(E/1\text{GeV})^{1/2}(L/2.5\text{km})^{1/2}$. One cannot go much beyond this with reactor neutrinos: $\sqrt{\Delta m^2} \gtrsim 0.3\text{eV}$ $(E/2\text{MeV})^{1/2}$ $(L/20\text{m})^{1/2}$. In order to explore a mass difference of the order of 10^{-2}eV, we need more than 250km, even for a 10MeV energy neutrino beam. One way to investigate such a small mass difference is to use "atmospheric neutrinos" generated from reactions of cosmic rays with the atmosphere, and to compare the fluxes produced at both sides of Earth ($L \sim 10^4$km). With a typical energy of 1GeV, we can detect effects down to $\sqrt{\Delta m^2} \gtrsim 0.02\text{eV}$. Long base-line experiments ($L \sim 100-1000$km) are also being proposed to explore a small neutrino mass.

The current experimental status is summarised for $\nu_\mu \to \nu_e$ [298, 299] and $\nu_\mu \to \nu_\tau$ [300, 301] in Figs.18 and 19, where constraints are plotted on the

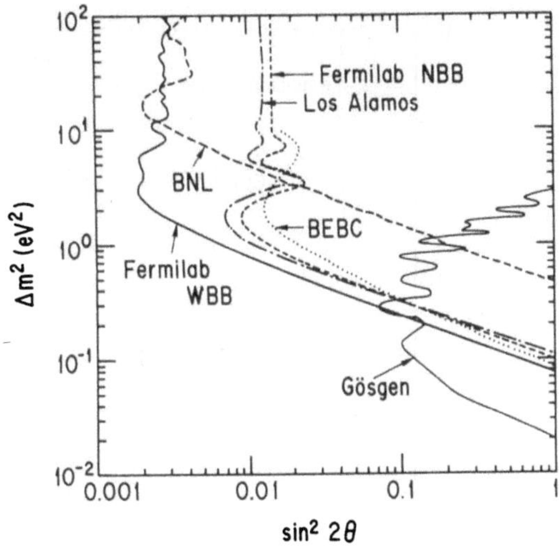

Figure 18: Limit on $\nu_\mu - \nu_e$ (or $\bar{\nu}_\mu - \bar{\nu}_e$) oscillation from Ref. [298]: Fermilab - wide-band beam (Borodowsky et al.); Fermilab - narrow-band beam (Blumenfeld et al.); Los Alamos (Durkin et al.); CERN/BEBC (Angelini et al.); BNL (Ahrens et al.). The limit from a disappearance experiment (Gösgen, Zacek et al.) [299] on $\bar{\nu}_e - \bar{\nu}_x$ ($x = \mu$ and τ) is also plotted.

$(\Delta m^2, \sin^2 2\theta)$ plane. The boundaries of the constraints are set by the detectability of $\nu_\mu \to \nu_e(\nu_\tau)$ (appearance experiment) or of a decrease in the ν_μ (ν_e) neutrino flux (disappearance experiment). The boundaries usually consist of two segments: a vertical line corresponding to $\sin^2 2\theta \simeq 2(1 - P^{\rm lim}_{\nu_\mu \to \nu_e})$ for $\Delta m^2 > 4E/L$ and an oblique line segment corresponding to $\sin^2 2\theta(\Delta m^2/4E)L \approx P^{\rm lim}_{\nu_\mu \to \nu_e}$ for $\Delta m^2 \sim 4E/L$. For $\nu_\mu \to \nu_e$ one can in principle explore oscillation down to $P_{\nu_e \to \nu_\mu} > 10^{-4}$, the limitation basically set due to confusion with ν_e from $\pi \to e\nu_e$, which has a branching ratio of 1×10^{-4}. Current accelerator experiments [298] reach $P^{\rm lim} \gtrsim 1.5 \times 10^{-3}$. For $\nu_\mu \to \nu_\tau$ the experiment with emulsions [301] gives a bound far better than that from disappearance ($\nu_\mu \not\to \nu_\mu$) experiments [300]. An interesting point is that $\sin^2 \theta_{\nu_\tau \nu_\mu} < 0.03$ [301] is already around the lower end of the allowed value for U_{13} of the quark mixing [see (II.89)], if $m_{\nu_\tau} >$5eV. If the bound can be improved in future experiments, this would exclude the interesting possibility that U_{ij}(lepton)$\simeq U_{ij}$(quark), which is expected, for instance, in simple SO(10) models (see [302]), and at the same time, ν_τ has a cosmologically interesting value ($\nu_\tau >$ a few eV).

For the case $m_{\nu_\tau} > 1$MeV, a stronger limit is available on $|U_{\tau i}|^2$ ($i = e, \mu$) from experiments of different types (see Sect.28).

Solar-neutrino problem and neutrino oscillation

The solar-neutrino experiment with ^{37}Cl has continuously given, since the first

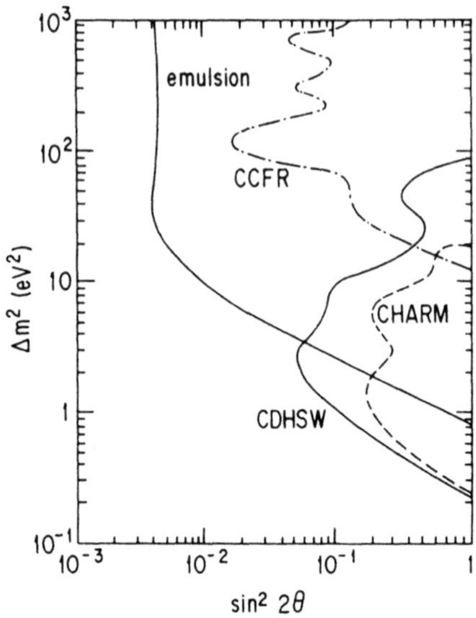

Figure 19: Limit on oscillation for ν_τ. The results of CCFR (Stockdale et al.), CHARM (Bergsma et al.) and CDHSW (Dydak et al.) are from disappearance experiment for $\nu_\mu \to \nu_x$ [300]. The emulsion experiment of Ref.[301] (Fermilab, Ushida et al.) is a search for $\nu_\mu \to \nu_\tau$.

report in 1968 [37], a neutrino capture rate 3–4 times smaller than expected in the standard solar model [303], and this problem has motivated many studies which are looking for an explanation, either by changing the astrophysical model of the Sun or by introducing non-standard properties for the neutrino [38, 39]. From 1987 the Kamiokande group (Kamiokande II, Kam-II in brevity) started measurements of the solar-neutrino flux, which also resulted in a deficit of the neutrino flux by a factor of 2 [40]. Some deficit in the capture rate has also been seen, though with a large statistical error in the more recently started ^{71}Ga experiments [41, 42].

An important point concerning these experiments is that they are sensitive to neutrinos with different energy ranges and hence yield spectroscopic information on neutrino-flux suppression (see Fig.20). The pattern of suppression is quite curious: we observe a stronger suppression, R =(observed)/(expected)= 0.29 ± 0.03 in the ^{37}Cl experiment (2.28 ± 0.23SNU), which is most sensitive to ^{8}B neutrinos (75–80% of total capture) but its sensitivity range extends to a lower energy region.[9] On the

[9]Here we take the calculation by Bahcall and Ulrich [304] as a standard solar model prediction. The calculation gives $7.9(1 \pm 0.33)$ SNU for $^{37}\mathrm{Cl}(\nu_e, e^-)^{37}\mathrm{Al}$, $132(1^{+0.15}_{-0.13})$ SNU for $^{71}\mathrm{Ga}(\nu_e, e^-)^{71}\mathrm{Ge}$, and $\phi(^8\mathrm{B}) = 5.8 \times 10^6(1 \pm 0.37)$ cm^{-2}s^{-1} for the flux from ^{8}B beta decay. Another standard model calculation that has often been referred to in the literature is that by Turck-Chièze et al. [305], which gives a substantially smaller ^{8}B neutrino flux ($\phi(^8\mathrm{B}) = 3.8 \times 10^6(1 \pm 0.29)$ cm^{-2}s^{-1}) and hence a smaller capture rate on ^{37}Cl (5.8 ± 1.3 SNU). Most of the discrepancy between the two values arises from the difference in input parameters of the $^7\mathrm{Be} + p$ reaction cross section, the $^3\mathrm{He}+^3\mathrm{He}$ reaction cross section and opacity (if the same

other hand, the suppression is rather modest ($R = 0.48 \pm 0.08$) at Kam-II, which measures only a higher energy tail of ^{8}B neutrinos ($E > 7.5$MeV), and also at the gallium experiments ($R = 0.63 \pm 0.16$),[10] where the weight of the sensitivity lies in an energy range substantially lower than the other two experiments. The results of the ^{37}Cl experiment and of Kam-II already make the astrophysics explanations unlikely[307]; all such explanations take the view that the temperature at the centre of the Sun is lowered by some non-standard effects, so that the generation of ^{8}B neutrinos is strongly reduced, whereas little is modified for the neutrino flux of other origins (for example if we decrease the temperature by 4%, we correctly obtain the reaction rate at Kam-II, but we expect at Homestake 4.2 SNU , 7σ away from the experiment.[11]) This is contrary to what we observe in the experiments. We also remark that a neutrino flux other than that of ^{8}B origin receives only small uncertainties ($\lesssim 5\%$). The observed suppression pattern is explained best, if the ^{7}Be neutrino flux is strongly suppressed, whereas high-energy ^{8}B neutrinos are suppressed moderately (by a factor of 2–2.5) and low-energy pp neutrinos are reduced only slightly or not at all. It would be difficult to imagine any reasonable astrophysical or nuclear-reaction mechanism which would reduce the ^{7}Be neutrino flux more significantly than the ^{8}B neutrino flux. By using neutrino oscillation the suppression pattern described here may be accounted for, if the neutrino mass difference takes the value which just causes oscillation of neutrinos at the Sun-Earth distance for ^{7}Be neutrinos.

The idea to use neutrino oscillation to explain the deficit in solar-neutrino captures is quite old. The possibility that the solar-neutrino flux would be affected by oscillation had been discussed by Gribov [308] before the result of the ^{37}Cl experiment was reported. Since then, considerations have been made occasionally to explain the low capture rate in the ^{37}Cl experiment (e.g. [292], [309]–[312]). The simplest possibility was to suppose that three generations of neutrinos are maximally mixed ($|U_{ei}|^2 = 1/3$, $i = e, \mu, \tau$) and that all of the mass differences Δm_{ij}^2 are substantially larger than $2\pi E/L = 0.7 \times 10^{-10}\text{eV}^2(E/8\text{MeV})$, with L the Sun-Earth distance. In such a case all solar-neutrino fluxes are reduced energy-independently to 1/3. Interest has also been emphasised, however, in a more accidental possibility that one of the neutrino mass differences takes a value just so that the oscillation length (or its integer multiple) coincides with the Sun-Earth distance [309]–[312]. In this case the neutrino capture rates can fall below 1/3, and the suppression would depend strongly on the neutrino energy. Indeed, this is what we need to explain the suppression pattern mentioned above.

A few detailed calculations were carried out in order to explore this possibility

---

input parameters are adopted, the agreement of the two calculations are expected to be good). See Bahcall and Pinsonneault for discussion concerning this point [306]. Since the reasoning of the choice of the parameters used by Turck-Chièze et al. is not clear to us, we preferentially choose the calculation by Bahcall and Ulrich in this article.

[10] We take the value reported by the GALLEX collaboration $83 \pm 19 \pm 8$ SNU. While the earlier report by the SAGE collaboration [41] gave a substantially smaller capture rate of $20^{+15}_{-20} \pm 32$ SNU, their statistics is 1/7 that of GALLEX, if the efficiency of detection is taken into account (Note that GALLEX measures both K and L electrons, while SAGE measures only K electrons form ^{71}Ge). Therefore, the result would change little even if we combine the two results.

[11] The discrepancy becomes worse, if we use the neutrino flux of Turck-Chièze et al.

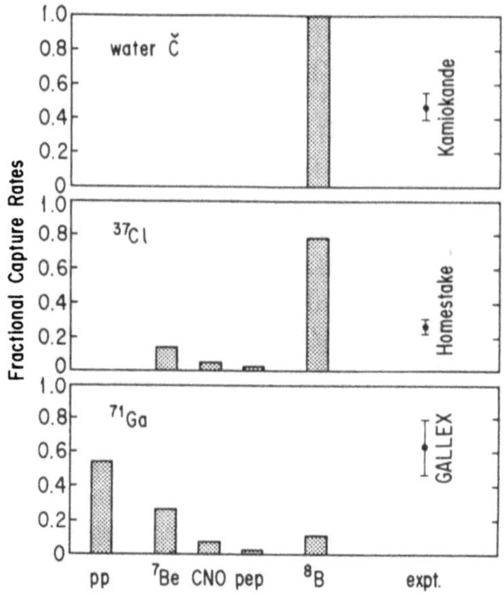

Figure 20: Fractional capture (reaction) rates for water Čerenkov, ^{37}Cl and ^{71}Ga detectors (the total rate normalised to unity). The solar-neutrino flux is taken from the solar model of Bahcall and Ulrich [39]. The current experimental values [303, 40, 42] are also indicated.

(see Fig. 21) [311, 312]; it was shown that the observed pattern of suppression can be explained if Δm^2 and $\sin^2 2\theta$ take specific sets of values between

$$\begin{aligned} \Delta m^2 &= (0.6-1.0)\times 10^{-10}\mathrm{eV}^2, \\ \sin^2 2\theta &= 0.8-1.0. \end{aligned} \tag{80}$$

It has also been pointed out that there should be a semiannual variation of the neutrino capture rate related to the eccentricity of Earth's orbit, which varies by 3.4%. This would be caused by the fact that the oscillation pattern persists up to $\Delta m^2 \sim 6\times 10^{-9}\mathrm{eV}^2$ for ^{7}Be neutrinos; around $\Delta m^2 \sim 1\times 10^{-10}\mathrm{eV}^2$ we would expect a 40% semiannual variation in the ^{7}Be flux and hence a 6% variation in the neutrino capture rate with a ^{37}Cl detector or a 12% variation with a ^{71}Ga detector [309]–[313].

An inherent prejudice against neutrino-oscillation explanations of the solar neutrino problem lies in the maximal (or very large) flavour mixing angle they always require and, for the "just so" case, an accidental coincidence of the neutrino mass with the Sun-Earth distance in addition. However, it may be interesting to recall that $m_\nu \approx (\Delta m^2)^{1/2}$ of eq.(80) is very close to the value from the unitarity bound for Majorana neutrinos (III.40).

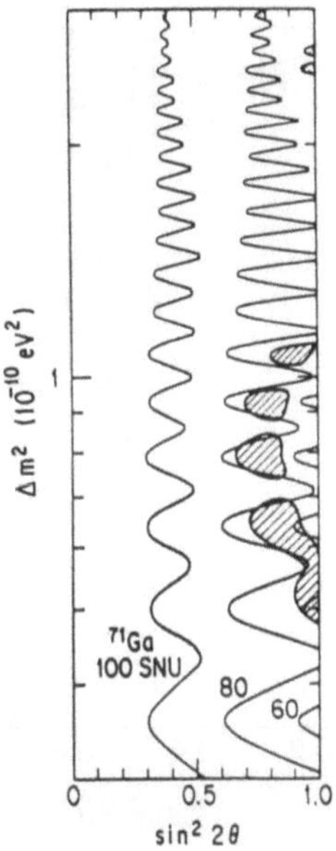

Figure 21: Possibility of the long wavelength neutrino-oscillation solution to the solar neutrino problem. The solid curves denote iso-SNU contours of the predicted ^{71}Ga capture rate, and the hatched part indicates regions allowed by the three solar neutrino experiments. Note that the adopted experimental values, $R(^{37}\mathrm{Cl})=0.26 \pm 0.05$ and $R(\mathrm{Kam})=0.60^{+0.15}_{-0.13}$, are slightly different from those taken in the text. The parameters indicated by P and Q are used in Fig.31 below. (The figure is taken from Ref. [311].)

Neutrino oscillation with atmospheric neutrinos

Neutrinos are copiously produced from decays of π and μ which are produced by the interactions of cosmic-ray protons with the atmosphere [314, 315]. Because of a long beam line (10km-10^4km) the detection of atmospheric neutrinos offers a unique possibility to explore a small neutrino mass, down to $(\Delta m^2)^{1/2} \sim 10^{-2}$eV [316]–[318].

The Kamiokande group [319] has claimed that the ν_μ flux is substantially smaller than is expected from the calculation, while the ν_e flux agrees with the expectation. Considering the fact that calculations of the absolute flux are less certain and depend more on models, the Kamiokande authors claimed the double ratio $R = [(\nu_\mu + \overline{\nu}_\mu)/(\nu_e + \overline{\nu}_e)_{\mathrm{obs}}/(\nu_\mu + \overline{\nu}_\mu)/(\nu_e + \overline{\nu}_e)_{\mathrm{calc}}] = 0.60 \pm 0.09 \neq 1$ be taken to be evidence for neutrino oscillation, more likely to be between ν_μ and ν_τ (see also [320]). The favoured region is displayed in Fig. 22. The Kamiokande authors noted that this "anomaly" is seen independent of the zenith angle of the neutrino flux. The IMB group [321] also sees a similar deficit $R = 0.72 \pm 0.06$, but the authors seem to be reluctant to interpret it as being a result of neutrino oscillation. A general doubt concerns the problem of the efficiency of μ/e separation (which the Kamiokande group claimed to be very good) and the detection

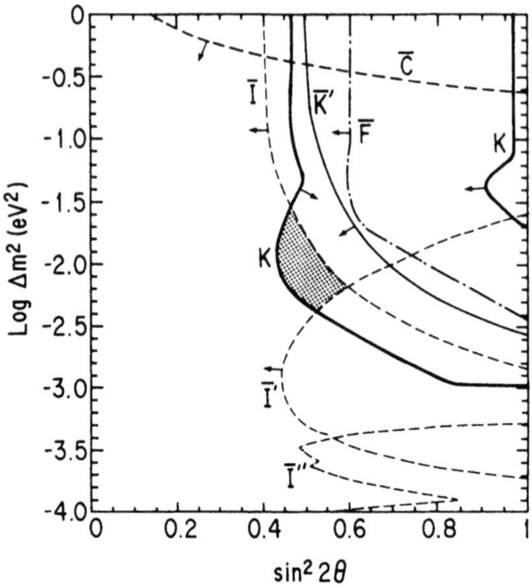

Figure 22: Suggested region for $\nu_\mu - \nu_\tau$ oscillation from the ν_μ deficit in atmospheric neutrinos in the Kamiokande detector (K), and the limits from other atmospheric neutrino experiments: $\bar{I}$ (IMB, upward going muons [321]) $\bar{I}', \bar{I}''$ (IMB, upward going muons stopped in the detector [322]), $\bar{F}$ (Frejus [323]) and $\bar{K}'$ (Kamiokande, upward going muons; preliminary result), $\bar{C}$ stands for a laboratory limit from the CDHSW collaboration [300]. Shade indicates an allowed region. The figure is by courtesy of A. Suzuki.

efficiency of the ν_μ event. These results are based on the neutrino events contained in the detector ($E_\nu \lesssim 1$GeV). On the other hand, no anomaly has been detected for the upward-going muon flux which passes through the detector ($E_\nu \gtrsim 100$GeV) and that stops in the detector ($E_\nu \approx 10$GeV), which sets significant limits on the oscillation scenario [322] (see Fig. 22).

If one would take the deficit as a result of neutrino oscillation, rather than as some instrumental effects, the allowed range would be limited to around (see Fig. 22)

$$\Delta m^2_{\nu_\mu \nu_\tau} \approx 10^{-2\pm 0.5}\text{eV}^2, \qquad \sin^2 2\theta \approx 0.4 - 0.6. \tag{81}$$

Let us finally mention a calculation of the atmospheric neutrino flux [315], [324]–[325]. Most naively, we expect that $(\nu_\mu + \overline{\nu}_\mu)/(\nu_e + \overline{\nu}_e) = 2$ for $\pi \to \mu\nu \to (e\nu\overline{\nu}) + \nu$. Realistic calculations take account of:

(i) The flux and chemical compositions of primary cosmic rays; (a) cosmic ray spectrum of p, He and heavy elements ($Z \geq 3$); (b) effect of solar modulations which affect the low-energy cosmic rays ($\lesssim 1$GeV) and hence the low-energy neutrino flux; (c) low-energy cutoff by geomagnetic fields [326].

(ii) Interactions of cosmic rays with air; (a) A Monte-Carlo code (e.g., LUND, NUCRIN etc.) is usually used to calculate the production of π and K mesons; (b) absorption of μ^- and π^- in nuclei; (c) secondary scattering effects.

(iii) Energy loss of muons in the air and in Earth.

(iv) Muon polarisation effect [327].

(v) Density and compositions of atmosphere.

The most recent calculations are given in [325]. Most of the calculations are made while assuming the one-dimensional propagation of particles (thus ignoring momentum conservation). The calculated absolute flux differs by 30% among the authors. The ratio ν_μ/ν_e, however, converges well within $\lesssim$5–10%; it seems unlikely that the calculation is incorrect for this ratio.

25.4 $\nu_L - \nu_R$ oscillation: oscillation to the opposite helicity state

A left-handed neutrino may also oscillate into a right-handed component on the order of $(m/E)^2$. Let us consider the Dirac equation

$$(i\gamma^\mu\partial_\mu - m)\psi = 0. \tag{82}$$

By multiplying γ^0 from the left, we write the equation in momentum space to be

$$\left[\begin{pmatrix} i\partial_0 + \vec{\sigma}\vec{p} & 0 \\ 0 & i\partial_0 - \vec{\sigma}\,\vec{p} \end{pmatrix} - m\begin{pmatrix} 0 & 1 \\ 1 & 0 \end{pmatrix}\right]\psi(t,\vec{p}) = 0. \tag{83}$$

Choosing $\vec{p} = (0,0,p)$ and writing $\psi = (\psi_{L1}, \psi_{L2}, \psi_{R1}, \psi_{R2})^T$, we have

$$\begin{pmatrix} i\partial_0 + p & 0 & -m & 0 \\ 0 & i\partial_0 - p & 0 & -m \\ -m & 0 & i\partial_0 - p & 0 \\ 0 & -m & 0 & i\partial_0 + p \end{pmatrix}\begin{bmatrix} \psi_{L1} \\ \psi_{L2} \\ \psi_{R1} \\ \psi_{R2} \end{bmatrix} = 0, \tag{84}$$

that is,

$$\left[i\partial_0 + \begin{pmatrix} p & -m \\ -m & -p \end{pmatrix}\right]\begin{pmatrix} \psi_{L1} \\ \psi_{R1} \end{pmatrix} = 0, \tag{85}$$

and a similar equation for $(\psi_{L2}, \psi_{R2})^T$. The matrix

$$\begin{pmatrix} p & -m \\ -m & -p \end{pmatrix} \tag{86}$$

has eigenvalues $\pm\sqrt{p^2+m^2}$. If we denote the energy eigenstates by ψ_+ and ψ_-, then

$$\begin{pmatrix} \psi_{L1} \\ \psi_{R1} \end{pmatrix} = \begin{pmatrix} \cos\theta & -\sin\theta \\ \sin\theta & \cos\theta \end{pmatrix}\begin{pmatrix} \psi_+ \\ \psi_- \end{pmatrix}, \tag{87}$$

$$\mathrm{tg}2\theta = -\frac{m}{p}. \tag{88}$$

The solution of (85) is given by [328]

$$\begin{aligned}\psi_{L1}(t) &= \cos\theta\psi_+(t) - \sin\theta\psi_-(t)\\ &= \cos\theta e^{i\sqrt{p^2+m^2}t}\psi_+(0) - \sin\theta e^{-i\sqrt{p^2+m^2}t}\psi_-(0)\ ,\\ \psi_{R1(t)} &= \sin\theta e^{i\sqrt{p^2+m^2}t}\psi_+(0) + \cos\theta e^{-i\sqrt{p^2+m^2}t}\psi_-(0)\ ,\end{aligned} \tag{89}$$

and so,

$$\begin{aligned}P_{L\to R} &= |\langle\psi_{R_1}(0)|\psi_{L_1}(t)\rangle|^2\\ &= 4\cos^2\theta\sin^2\theta\ \sin^2\sqrt{p^2+m^2}t\\ &\simeq (\frac{m}{E})^2\sin^2 Et,\end{aligned} \tag{90}$$

where $m \ll E$ and $\sin\theta \sim -m/2E$ is used. The transition $\nu_L \to \nu_R$ shows an oscillation pattern. For $Et \gg \pi$, the average transition probability is

$$\langle P_{L\to R}\rangle \simeq \frac{1}{2}(\frac{m}{E})^2. \tag{91}$$

Here we have discussed the case for the Dirac neutrino. An identical result is obtained for the case of the Majorana neutrino, with ν_R replaced by ν_L^c.

If there exist more than two families, the Dirac equation is

$$(i\gamma^\mu\partial_\mu - m_i)\psi_i = 0. \tag{92}$$

We define the weak interaction eigenstates in accordance with (3) and (4) by

$$\begin{aligned}\psi_{\alpha L} &= U_{\alpha i}\psi_{iL}\ ,\\ \psi_{\alpha R} &= V_{\alpha i}\psi_{iR}\ .\end{aligned} \tag{93}$$

By repeating the procedure from (83) to (91), we obtain

$$P_{\alpha L\to\beta R} = \frac{1}{2}|\sum_i V_{\beta i}\frac{-m_i}{E}U^\dagger_{i\alpha}e^{-iE_it}|^2 \tag{94}$$

for the Dirac neutrino.

For the Majorana case since $U^*_{\beta i}$ appears in place of $V_{\beta i}$,

$$P(\nu_{\alpha L}\to\nu^c_{\beta L}) = \frac{1}{2}|\sum_i U^*_{\beta i}\frac{-m_i}{E}U^*_{\alpha i}e^{-iE_it}|^2\ . \tag{95}$$

Note that (95) differs from (94), and that the phase characteristic of the Majorana neutrino appears in Eq.(95). This means that a Majorana neutrino can in principle be distinguished from a Dirac neutrino [288]. Unfortunately, this is experimentally not feasible due to the large chiral suppression factor $(m/E)^2$. Therefore, $\nu_L \to \nu_R(\nu_L^c)$ oscillation is not of practical interest.

25.5 $\nu_L \to \nu_R^c$ oscillation

If the right-handed neutrino exists and if its mass is as small as that of the left-handed neutrino, ν_L may oscillate into ν_R^c (the same helicity as ν_L) without a suppression factor [329, 330]. The general Lagrangian of the neutrino mass is

$$\begin{aligned}\mathcal{L}_{\rm mass} &= \frac{1}{2}(\overline{\nu^c}_{\alpha L} m^L_{\alpha\beta}\nu_{\beta L} + \overline{\nu^c}_{\alpha R} m^R_{\alpha\beta}\nu_{\beta R}) \\ &\quad + \overline{\nu}_{\alpha R} m_{\alpha\beta}\nu_{\beta L} + h.c. \end{aligned} \tag{96}$$

Both m^L and m^R are symmetric due to Fermi statistics. The Dirac neutrino mixing corresponds to a special case of $m^L = m^R = 0$, and the Majorana neutrino mixing corresponds to the case $m^R = m = 0$. If $m^L \sim m^R \sim m$, mass diagonalisation mixes ν_L with ν_R^c; the weak eigenstates $(\nu_L^\alpha, \nu_R^{c\alpha})$ are related to the mass eigenstates (ν_L^i, ν_R^{ci}) by four $N \times N$ matrices $U^{(a)} (a = 1 \sim 4)$, to be

$$\begin{pmatrix} \nu_{\alpha L} \\ \nu^c_{\alpha R} \end{pmatrix} = \begin{pmatrix} U^{(1)} & U^{(2)} \\ U^{(3)} & U^{(4)} \end{pmatrix} \begin{pmatrix} \nu_{iL} \\ \nu^c_{iR} \end{pmatrix}, \quad (\alpha, i = 1 \sim N) \tag{97}$$

with N the number of families. Using the mass eigenstates, (96) reads

$$\mathcal{L}_{\rm mass} = \frac{1}{2}(\overline{\nu^c}_{iL} m^L_{{\rm diag}\,i}\nu_{iL} + \overline{\nu^c}_{iR} m^R_{{\rm diag}\,i}\nu_{iR}) + h.c. \tag{98}$$

The diagonalised mass matrix is given by

$$\begin{pmatrix} m^L_{\rm diag} & \\ & m^{\dagger R}_{\rm diag} \end{pmatrix} = \begin{pmatrix} U^{(1)T} & U^{(3)T} \\ U^{(2)T} & U^{(4)T} \end{pmatrix} \begin{pmatrix} m^L & m^T \\ m & m^{R\dagger} \end{pmatrix} \begin{pmatrix} U^{(1)} & U^{(2)} \\ U^{(3)} & U^{(4)} \end{pmatrix}, \tag{99}$$

where $U^{(i)}$ satisfies the unitarity condition,

$$\begin{aligned} U &\equiv \begin{pmatrix} U^{(1)} & U^{(2)} \\ U^{(3)} & U^{(4)} \end{pmatrix}, \\ U^\dagger U &= UU^\dagger = \mathbf{1}. \end{aligned} \tag{100}$$

To see the consequence of this mixing scheme, let us consider a single family: the left-handed lepton doublet $(\nu_e, e)_L$ and the right-handed lepton singlets e_R and ν_{eR}. The weak eigenstates, ν_{eL} and ν^c_{eR}, are given by the mass eigenstates, ν_{1L} and ν^c_{1R}, as

$$\begin{aligned} \nu_{eL} &= \cos\theta \nu_{1L} + \sin\theta \nu^c_{1R}, \\ \nu^c_{eR} &= -\sin\theta \nu_{1L} + \cos\theta \nu^c_{1R}. \end{aligned} \tag{101}$$

If both m_{1L} and m_{1R} are small enough ($<$ 1MeV, say) and $m_1^L \neq m_1^R, \nu_{eL} \to \nu^c_{eR}$ oscillation takes place. The probability of finding ν_{eL} at a distance L from a source of ν_{eL} is

$$P(\nu_{eL} \to \nu_{eL}) = 1 - (\sin 2\theta)^2 \left(\sin \frac{\Delta m^2 L}{4E}\right)^2 . \tag{102}$$

Since ν_{eR} is a gauge singlet, it does not interact with detectors. Therefore, unlike in the case of neutrino mixing among different flavours, neutral-current induced reactions are also affected [329, 330]. (For the single-family case the ratio of charged- to neutral-current induced reactions is unaffected). We remark that the possibility discussed here is rather unnatural, and we find no sensible models that would lead to such a small m^R.

26 Neutrino oscillation in matter

26.1 Neutrino conversion

When neutrinos propagate through matter ν_e and ν_μ(or ν_τ) feel different potentials, because ν_e scatters off electrons via both neutral and charged currents, whereas $\nu_\mu(\nu_\tau)$ scatters only via the neutral current (see Fig. 23) [331]. This induces a coherent effect in which maximal conversion of ν_e into ν_μ takes place even for a rather small intrinsic mixing angle in the vacuum, when the phase arising from the potential difference between the two neutrinos cancels the phase caused by the mass difference in the vacuum as pointed out by Mikheyev and Smirnov [332].[13] Currently, this mechanism (Mikheyev-Smirnov-Wolfenstein; MSW mechanism) is supposed to be the most attractive solution to the solar-neutrino problem [332, 334].

In the presence of the effective interaction

$$H_{\text{eff}} = \frac{G_F}{\sqrt{2}} \, \bar{\nu}_e \gamma_\mu (1-\gamma_5)\nu_e \; \bar{e}\gamma_\mu(1-\gamma_5)e \; , \tag{103}$$

the electron neutrino energy in (43) receives an extra contribution of $\sqrt{2}G_F n_e$, where n_e is the electron number density in the matter. [This is $V(\nu_e, e) - V(\nu_\mu, e)$ in (II.167).][14] Therefore, the time variation of neutrino wave function is given by

$$i\frac{d}{dt}\begin{pmatrix} \nu_e \\ \nu_\mu \end{pmatrix} = \begin{pmatrix} -\frac{\Delta m^2}{4E}\cos 2\theta + \sqrt{2}G_F n_e & \frac{\Delta m^2}{4E}\sin 2\theta \\ \frac{\Delta m^2}{4E}\sin 2\theta & \frac{\Delta m^2}{4E}\cos 2\theta \end{pmatrix} \begin{pmatrix} \nu_e \\ \nu_\mu \end{pmatrix} \tag{104}$$

instead of (48) and (51).

It is quite obvious that (104) is valid for both Dirac and Majorana cases, because only the left-handed neutrino interacts with matter and the theory is not concerned with whether the right-handed counter part is a neutrino (the Dirac case) or an antineutrino (the Majorana case) (e.g., ref. [335]).

---

[13] This effect was also noted, though less explicitly, by Barger et al. [333] for atmospheric neutrinos propagating through Earth. Unfortunately, its importance in the solar-neutrino problem was dismissed in their work.

[14] The original paper by Mikheyev and Smirnov [332] contained an error of sign in (103). This error was corrected in a paper by Bethe [334].

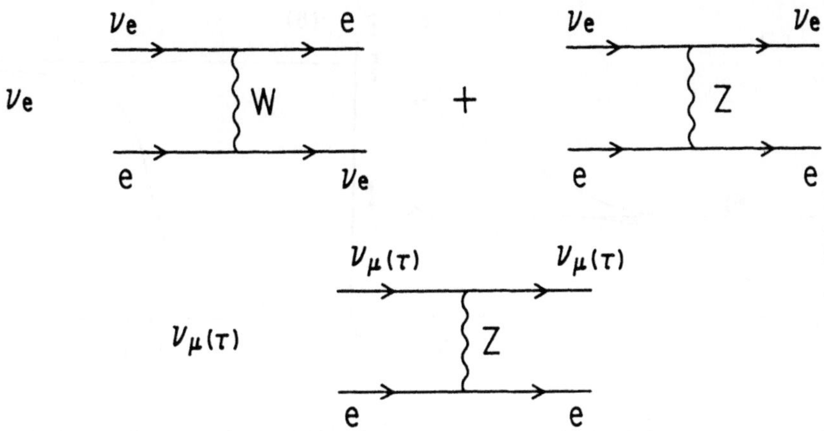

Figure 23: Coherent scattering of ν_e and $\nu_\mu(\nu_\tau)$ in matter.

We diagonalise the energy matrix of (104) by

$$\begin{pmatrix} \nu_e \\ \nu_\mu \end{pmatrix} = U \begin{pmatrix} \tilde{\nu}_1 \\ \tilde{\nu}_2 \end{pmatrix} = \begin{pmatrix} \cos\tilde{\theta} & \sin\tilde{\theta} \\ -\sin\tilde{\theta} & \cos\tilde{\theta} \end{pmatrix} \begin{pmatrix} \tilde{\nu}_1 \\ \tilde{\nu}_2 \end{pmatrix}, \tag{105}$$

where $\tilde{\nu}$ is the energy eigenstate in matter, and $\tilde{\theta}$ is given by

$$\cos 2\tilde{\theta} = \frac{-A/\Delta m^2 + \cos 2\theta}{\sqrt{(A/\Delta m^2 - \cos 2\theta)^2 + \sin^2 2\theta}} \tag{106}$$

and

$$\sin 2\tilde{\theta} = \frac{\sin 2\theta}{\sqrt{(A/\Delta m^2 - \cos 2\theta)^2 + \sin^2 2\theta}} \tag{107}$$

with

$$A = 2\sqrt{2} E G_F n_e \ . \tag{108}$$

The energy eigenvalues are

$$\tilde{m}^2_{\substack{1\\2}} = \frac{A}{2} \mp \frac{1}{2}\sqrt{(A - \Delta m^2 \cos 2\theta)^2 + (\Delta m^2)^2 \sin^2 2\theta} \ . \tag{109}$$

We plot $\tilde{m}^2$ and $\tilde{\theta}$ as a function of n_e in Figs. 24.

This is a level crossing phenomenon of the two-level problem of quantum mechanics, and its importance in the solar neutrino problem was first realised by Mikheyev and Smirnov [332] (see also Bethe [334]). The level crossing ("resonance") occurs at

$$n_e = n_{e,\text{crit}} \equiv \frac{1}{2\sqrt{2} G_F} \frac{\Delta m^2}{E} \cos 2\theta \ . \tag{110}$$

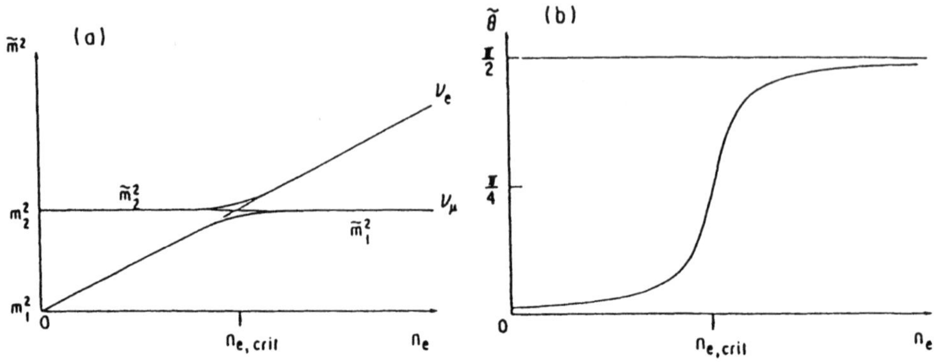

Figure 24: (a) Effective neutrino mass squared in the medium with electron density n_e. $n_{e,crit}$ is the crossing point defined by eq. (110). This figure corresponds to $\sin^2 2\theta = 0.01$. (b) Mixing angle $\tilde{\theta}$ in the medium as a function of n_c, corresponding to the case shown in (a).

As can be seen in Fig. 24, if $n_e \ll n_{e,\text{crit}}$, $\tilde{\theta} \simeq \theta$, and neutrinos oscillate with a mixing length in matter of $l_0 = 4\pi E/\Delta m^2$, as in the vacuum. At $n_e = n_{e,\text{crit}}, \tilde{\theta} = \pi/4$. For $n_e \gg n_{e,\text{crit}}$, $\tilde{\theta}$ approaches $\pi/2$. The mixing length in matter

$$\tilde{l} = \frac{4\pi E}{\tilde{m}_2^2 - \tilde{m}_1^2} = \frac{4\pi E}{\Delta m^2} \frac{1}{[(A/\Delta m^2 - \cos 2\theta)^2 + \sin^2 2\theta]^{1/2}} \tag{111}$$

is much shorter than l_0 for a large n_e.

If ν_e is produced in the region $n_e > n_{e,\text{crit}}$ and propagates into the region $n_e < n_{e,\text{crit}}$, the state follows the upper branch given in Fig. 24; then, ν_e fully converts into ν_μ, provided that the density gradient, $d(\ln\, n_e)/dr$, is sufficiently small that neutrino conversion occurs adiabatically. This adiabatic condition may be derived from

$$(\text{energy gap}) \times (\text{transition time}) \gg \hbar \ , \tag{112}$$

where the energy gap $\delta E = (\tilde{m}_+^2 - \tilde{m}_-^2)_{\text{min}}/2E$ is

$$\delta E = \frac{1}{2E}\,(\Delta m^2)\sin 2\theta \tag{113}$$

from (109), and the transition time in the level crossing region is

$$\begin{aligned} \delta t &= \delta r/c = \left(\frac{1}{n_e}\frac{dn_e}{dr}\right)^{-1} \frac{\delta n_e}{n_e} \\ &= \left(\frac{1}{n_e}\frac{dn_e}{dr}\right)^{-1} \frac{\delta A}{A} \ . \end{aligned} \tag{114}$$

Since "resonance" occurs at $A = \Delta m^2 \cos 2\theta$ and its width is $\delta A \sim \Delta m^2 \sin 2\theta$ from (107), we find that (112) is written

$$\frac{1}{n_e}\frac{dn_e}{dr} \ll \frac{\Delta m^2}{2E}\frac{\sin^2 2\theta}{\cos 2\theta} \ . \tag{115}$$

The probability for $\nu_e \to \nu_e$ is easily calculated in a manner similar to (66) by averaging out the time-varying part, viz.,

$$\begin{aligned} P(\nu_e \to \nu_e) &= \sin^2\theta \sin^2\tilde{\theta} + \cos^2\theta\cos^2\tilde{\theta} \\ &= \frac{1}{2}[1+\cos 2\theta \cos 2\tilde{\theta}], \end{aligned} \tag{116}$$

where $\tilde{\theta}$ is the mixing angle at the initial point.

If the adiabatic condition (115) is not satisfied, the state of the upper branch undergoes a transition to the lower branch while passing through the crossing point with the probability given by [336, 337]

$$P_f = \exp(-\frac{\pi}{2}\gamma), \tag{117}$$

where

$$\gamma = \frac{\Delta m^2 \sin^2 2\theta}{2E\cos 2\theta (1/n_e)(dn_e/dr)} \tag{118}$$

is the ratio of the right-hand side to the left-hand side of inequality (115). This is a straightforward application of the well-known Landau-Zener formula for level crossings [338], and its derivation is recapitulated in Sect.26.3.

With this P_f we write the probability for $\nu_e \to \nu_e$ for a general case [336]:

$$\begin{aligned} P(\nu_e \to \nu_e) &= (\sin^2\theta\sin^2\tilde{\theta} + \cos^2\theta\cos^2\tilde{\theta})(1-P_f) \\ &\quad + (\sin^2\theta\cos^2\tilde{\theta} + \cos^2\theta\sin^2\tilde{\theta})P_f \\ &= \frac{1}{2} + (\frac{1}{2} - P_f)\cos 2\theta\cos 2\tilde{\theta}. \end{aligned} \tag{119}$$

For $n_e \to \infty$, $\cos 2\tilde{\theta} \to -1$, and this reduces to

$$P_{\nu_e\to\nu_e} = \sin^2\theta + \cos 2\theta \ P_f \ . \tag{120}$$

In summary, a virtually complete conversion of ν_e to ν_μ takes place when the three conditions are satisfied [336]: (i) $n_e > n_{e,\rm crit}$ (see (110)), which gives $\Delta m^2 \lesssim$ const (resonance condition) [332, 334]; (ii) $\Delta m^2 \sin^2 2\theta \gtrsim$ const (115) (for a small θ; adiabatic condition) [332, 339, 340]; and (iii) $\sin^2\theta < 1$ (from (119)). These three conditions complete a rectangular triangle (which we refer to as the MSW triangle) in the $(\Delta m^2, \sin^2 2\theta)$ plane (see Fig. 25). Let us remark here concerning the spectroscopic characteristics of neutrino conversion. The conversion is neutrino energy dependent: for the continuum energy spectrum of neutrinos, only the high-energy component that satisfies (110) is cut off by conversion into ν_μ or ν_τ (it is

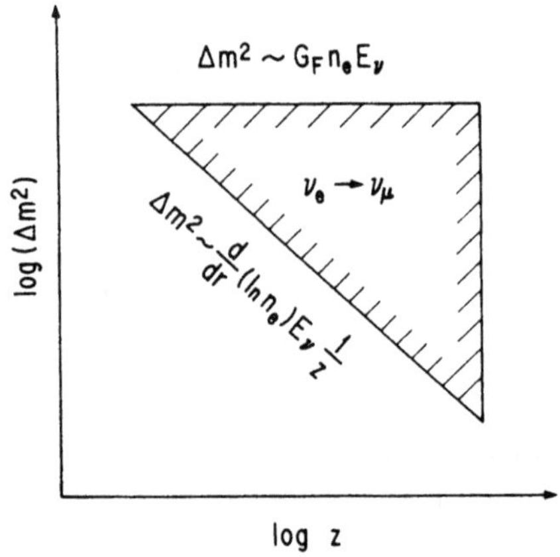

Figure 25: Schematic figure of the MSW triangle. The abscissa is the logarithm of $z = \sin^2 2\theta / \cos 2\theta$. Complete conversion of $\nu_e \to \nu_\mu$ takes place inside the triangle.

called adiabatic conversion). If $\sin^2\theta$ is small this cut-off takes place suddenly for a narrow region, $\delta E/E(\sim \delta\Delta m^2/\Delta m^2) \sim \sin\theta$. On the other hand, the energy dependence of (115) indicates that the lower-energy component is suppressed. For this case, however, an energy dependence appears in an exponential and is rather gradual. For the vertical branch the reduction of the ν_e flux is basically energy independent; the suppression is simply $\sim \sin^2\theta$. In Fig. 26 we show a typical $P(\nu_e \to \nu_e)$ calculated with (117) and (119) with two choices of the parameters: (A) $\sin^2 2\theta = 0.01$ and (B) $\sin^2 2\theta = 0.85$. For other parameters we take $\rho_c = 100$g H/cm^3 and $d(\ln n_e)/dR = 10.5/R_\odot$. The horizontal axis,

$$x \equiv \frac{5.4 \times 10^{-6}\mathrm{eV}^2}{\Delta m^2}\left(\frac{E}{1\mathrm{MeV}}\right) , \tag{121}$$

corresponds to going down along a line of constant $\sin^2 2\theta$ in Fig. 25; the suppressed region of P corresponds to the inside of the triangle. [$x \simeq 1$ is the resonance condition eq.(110).] The features discussed above are apparent in these examples.

The case for the three generations are discussed in ref. [341]. For reviews of the MSW effect see ref. [342, 343].

26.2 Derivation of the transition probability

Let us first discuss the propagation of neutrinos in more detail and derive formula (119) in a more rigorous manner [336]. We assume that ν_e is produced at t_i, passes through the resonance at t_r, and is detected at $t_f \to \infty$ in the vacuum. At $t = t_i$,

$$\psi_{\nu_e}(t_i) = \cos\tilde\theta\ |\tilde\nu_1,\ t_i\rangle\ + \sin\tilde\theta\ |\tilde\nu_2,\ t_i\ \rangle\ , \tag{122}$$

according to (105). At $t = t_r - \epsilon$,

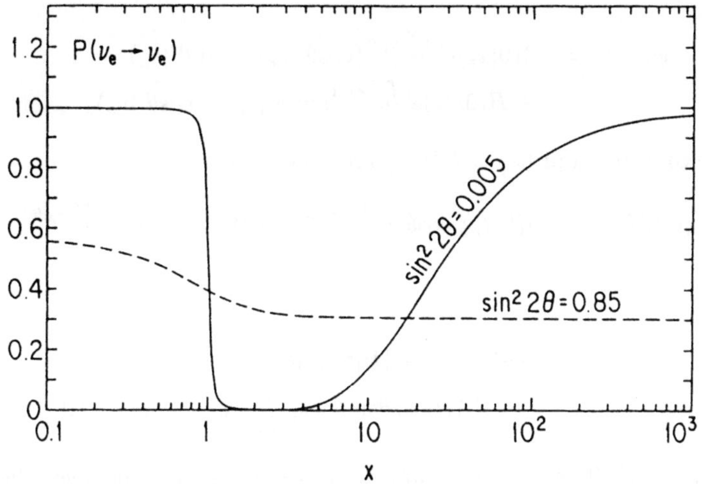

Figure 26: MSW filter function $P(\nu_e \to \nu_e)$ as a function of $x = 2(E/\Delta m^2)\sqrt{2}G_F n_e$ $= (E/1\text{MeV})(1.52 \times 10^{-5}\ \text{eV}^2/\Delta m^2)^{-1}$ (ρ = 100gHcm^{-3} is adopted). The two curves show a small angle case (solid curve, $\sin^2 2\theta = 0.005$) and a large angle case (dashed curve, $\sin^2 2\theta = 0.85$). The Landau-Zener formula is used to calculate these curves. The parameters are chosen so that they are close to the two solutions (A, B) described in the text below.

$$\psi(t_r - \epsilon) = \cos\tilde{\theta}\ e^{i\int_{t_i}^{t_r}\varepsilon_1 dt}\ |\nu_1,\ t_r\rangle + \sin\tilde{\theta}\ e^{i\int_{t_i}^{t_r}\varepsilon_2 dt}\ |\nu_2,\ t_r\rangle\ , \tag{123}$$

where $\varepsilon_{1,2} = \tilde{m}_{1,2}^2/2E$. At $t \simeq t_r$ a flip to the other state may take place, and the state is written

$$\begin{aligned} |\tilde{\nu}_1\rangle &\to \alpha|\tilde{\nu}_1\rangle + \beta|\tilde{\nu}_2\rangle \\ |\tilde{\nu}_2\rangle &\to -\beta^*|\tilde{\nu}_1\rangle + \alpha^*|\tilde{\nu}_2\rangle \end{aligned} \tag{124}$$

with $|\alpha|^2 + |\beta|^2 = 1$. Hence, after the resonance point

$$\begin{aligned} \psi(t_r + \epsilon) &= \cos\tilde{\theta} e^{i\int_{t_i}^{t_r}\varepsilon_1 dt}[\alpha\ |\tilde{\nu}_1,\ t_r\rangle\ + \beta\ |\tilde{\nu}_2,\ t_r\rangle\ ] \\ &\quad + \sin\tilde{\theta} e^{i\int_{t_i}^{t_r}\varepsilon_2 dt}[-\beta^*|\tilde{\nu}_1,\ t_r\rangle\ + \alpha^*|\tilde{\nu}_2,\ t_r\rangle\ ] \\ &\equiv A(0;t_r)\ |\tilde{\nu}_1,\ t_r\rangle + B(0;t_r)\ |\tilde{\nu}_2,\ t_r\rangle\ . \end{aligned} \tag{125}$$

For $t > t_r$ it propagates as

$$\psi(t) = A(0;t_r)e^{i\int_{t_r}^{t}\varepsilon_1 dt}|\tilde{\nu}_1,\ t\rangle + B(0;t_r)e^{i\int_{t_r}^{t}\varepsilon_2 dt}|\tilde{\nu}_2,\ t\rangle\ . \tag{126}$$

At $t_f = \infty$

$$\begin{aligned}\psi(\infty) &= A(0;t_r)e^{i\int_{t_r}^{\infty}\varepsilon_1 dt}(\cos\theta\ |\nu_e\rangle - \sin\theta\ |\nu_\mu\rangle) \\ &+ B(0;t_r)e^{i\int_{t_r}^{\infty}\varepsilon_2 dt}(\sin\theta\ |\nu_e\rangle + \cos\theta\ |\nu_\mu\rangle)\ ,\end{aligned} \tag{127}$$

where $|\tilde{\nu}_i,\ \infty\rangle = |\nu_i\rangle$ and $|\nu_i\rangle = U^\dagger|\nu^\alpha\rangle$ are used. Then,

$$\langle\nu_e|\psi(\infty)\rangle = A(0;t_r)\ \cos\theta\ e^{i\int_{t_r}^{\infty}\varepsilon_1 dt} + B(0;t_r)\ \sin\theta\ e^{i\int_{t_r}^{\infty}\varepsilon_2 dt} \tag{128}$$

and

$$\begin{aligned}P_{\nu_e\to\nu_e} &= |A|^2\cos^2\theta + |B|^2\sin^2\theta \\ &+ 2|AB|\cos\theta\sin\theta\cos[\int_{t_r}^{\infty}(\varepsilon_1-\varepsilon_2)dt + \Omega]\end{aligned} \tag{129}$$

with $\Omega = \arg A^*B$. The last term vanishes if we take an average over the detector position. Inserting expressions of $A(0;t_r)$ and $B(0;t_r)$ [see eq.(125)], we obtain

$$\begin{aligned}\langle P_{\nu_e\to\nu_e}\rangle_\infty &= |A|^2\cos^2\theta + |B|^2\sin^2\theta \\ &= \frac{1}{2} + \frac{1}{2}(|\alpha|^2 - |\beta|^2)\cos 2\tilde{\theta}\cos 2\theta \\ &- |\alpha\beta|\sin 2\tilde{\theta}\cos 2\theta\cos[\int_0^{t_r}(\varepsilon_1-\varepsilon_2)dt + \omega]\ ,\end{aligned} \tag{130}$$

where $\omega = \arg \alpha^*\beta$. The last term also vanishes if we take an average over the position of neutrino production. Identifying

$$|\beta|^2 = P_f\ , \tag{131}$$

we obtain (119), i.e.,

$$\langle P_{\nu_e\to\nu_e}\rangle_{i,\ \infty} = \frac{1}{2} + \left(\frac{1}{2} - P_f\right)\cos 2\theta\cos 2\tilde{\theta}\ . \tag{132}$$

In particular, for $n_e \to \infty$ $(\cos 2\tilde{\theta} \to -1)$,

$$\langle P_{\nu_e\to\nu_e}\rangle_{i,\ \infty} = \sin^2\theta + P_f\cos 2\theta\ , \tag{133}$$

in agreement with (120).

26.3 Derivation of Landau-Zener formula [338]

We recapitulate the derivation of P_f following Zener [338]. Let us write (104) in the form

$$H\psi = \begin{pmatrix}\epsilon_1 & \epsilon_{12} \\ \epsilon_{12} & \epsilon_2\end{pmatrix}\psi, \tag{134}$$

where

$$\begin{aligned}\epsilon_1 &= [a(t) - \cos 2\theta]\Delta m^2/4E \ , \\ \epsilon_2 &= \cos 2\theta \ \Delta m^2/4E \ , \\ \epsilon_{12} &= \sin 2\theta \ \Delta m^2/4E \ , \end{aligned} \tag{135}$$

while ignoring the unimportant constant terms. Here, $a(t) = \sqrt{2}G_F n_e(t)$ is a function of time. With an orthonormal basis (c_1, c_2), the Schrödinger equation $(H - i\partial/\partial t)\psi = 0$ is written

$$\left(H - i\frac{d}{dt}\right)[c_1(t)e^{-i\int \epsilon_1 dt} + c_2(t)e^{-i\int \epsilon_2 dt}] = 0. \tag{136}$$

Using (134), (136) becomes

$$\begin{aligned} i\frac{dc_1}{dt} &= \epsilon_{12}e^{-i\int(\epsilon_2-\epsilon_1)dt}c_2 \ , \\ i\frac{dc_2}{dt} &= \epsilon_{12}e^{-i\int(\epsilon_1-\epsilon_2)dt}c_1 \ . \end{aligned} \tag{137}$$

The boundary condition is that ψ is in the eigenstate of $|\nu_e>=|c_1>$, that is, on the upper branch in Fig. 23 at $t \to \infty$ $(n_e \to \infty)$,

$$\begin{aligned} |c_1(-\infty)| &= 1 \ , \\ c_2(-\infty) &= 0 \ . \end{aligned} \tag{138}$$

The problem is to find the probability that ψ jumps from the upper to the lower branch in the resonance region, i.e., that ψ is in the state $|\nu_e>=|c_1>$ at $t \to +\infty$,

$$P = |c_1(\infty)|^2 = 1 - |c_2(\infty)|^2. \tag{139}$$

We now eliminate c_1 from eq. (137),

$$\frac{d^2c_2}{dt^2} + [i(\epsilon_1 - \epsilon_2) - \frac{\dot\epsilon_{12}}{\epsilon_{12}}]\frac{dc_2}{dt} + \epsilon_{12}^2 c_2 = 0. \tag{140}$$

We assume that $n_e(t)$ varies linearly in t in the resonance region, which is taken to be $t = 0$,

$$n_e(t) = n_e^0 + \dot n_e^0 t, \tag{141}$$

where n_e^0 satisfies the resonance condition, $\sqrt{2}G_F n_e^0 = (\Delta m^2/2E)\cos 2\theta$. Writing

$$\epsilon_1 - \epsilon_2 = \sqrt{2}G_F \dot n_e^0 t \equiv \alpha t, \tag{142}$$

and putting

$$c_2(t) = e^{-\frac{i}{2}\int(\epsilon_2-\epsilon_1)dt}U(t), \tag{143}$$

(140) reads

$$\frac{d^2U}{dt^2} + \left[\epsilon_{12}^2 - \frac{i}{2}\alpha + \frac{1}{4}(\alpha t)^2\right] U = 0. \tag{144}$$

Upon changing variables,

$$z = \sqrt{\alpha}\, e^{-i\pi/4} t \tag{145}$$

and

$$n = i\epsilon_{12}^2/\alpha\ , \tag{146}$$

(144) becomes the Weber equation [344],

$$\frac{d^2U}{dz^2} + \left(n + \frac{1}{2} - \frac{z^2}{4}\right) U = 0\ . \tag{147}$$

The solution is called the Weber function, $D_{-n-1}(\pm iz)$. Since $D_{-n-1}(iz)$ has an asymptotic behaviour, $D_{-n-1}(iz) \to 0$ as $z \to \infty e^{-\frac{3}{4}\pi i}$ or $\infty e^{-\frac{\pi}{4}i}$,

$$U = A_+ D_{-n-1}(-iz) \tag{148}$$

satisfies the required boundary condition, $c_2 \sim U \to 0$ as $t \to -\infty$. Here, A_+ is the normalisation factor which is determined by $|c_1(-\infty)| = 1$. Using the second equation of (137) and the asymptotic behaviour of $D_{-n-1}(iRe^{-\pi i/4})$ as $R \to \infty$,

$$R_{-n-1}(iRe^{-\pi i/4}) \sim e^{-\frac{\pi}{4}(n+1)i} e^{-iR^2/4} R^{-n-1}, \tag{149}$$

we have

$$c_1(\infty) = \frac{\sqrt{\alpha}}{\epsilon_{12}} A_+ e^{-\frac{\pi}{4}\nu} e^{-\frac{3}{4}\pi i - i\nu \ln(-\sqrt{\alpha}t)}. \tag{150}$$

The normalisation condition is then given by

$$|A_+| = \sqrt{\nu} e^{-\frac{\pi}{4}\nu}, \tag{151}$$

where

$$\nu = -in = \epsilon_{12}^2/\alpha. \tag{152}$$

For $t \to +\infty$ we pick up the leading term of the asymptotic expansion of $D_{-n-1}(iRe^{3\pi i/4})$ as $R \to \infty$,

$$D_{-n-1}(iRe^{3\pi i/4}) \sim \frac{\sqrt{2\pi}}{\Gamma(n+1)} e^{\frac{\pi}{4}in} e^{iR^2/4} R^n + e^{\frac{3}{4}\pi i(n+1)} e^{-iR^2/4} R^{-n-1}\ . \tag{153}$$

Then,

$$\begin{aligned}|c_2(\infty)|^2 &= \nu e^{-\frac{\pi}{2}\nu}\frac{\sqrt{2\pi}}{[\Gamma(i\nu+1)\Gamma(-i\nu+1)]^{1/2}}e^{-\frac{\pi}{2}\nu} \\ &= 2\sinh(\pi\nu)e^{-\pi\nu} \\ &= 1-e^{-2\pi\nu}, \end{aligned} \tag{154}$$

where

$$\nu = \frac{1}{4}\frac{\Delta m^2}{2E}\frac{\sin^2 2\theta}{\cos 2\theta}\frac{1}{d\ln n_e/dr} \equiv \frac{1}{4}\gamma \tag{155}$$

from (146), (135), (142) and the resonance condition. We therefore obtain

$$P_f = \exp(-\frac{\pi}{2}\gamma),$$

the desired result (117). The Landau-Zener formula applies when the variation in the density is linear in r, and is exact when the mixing angle is small and γ is not too small. Practically, however, this gives a reasonably accurate result for most of the cases relevant to the solar-neutrino problem.

A number of different derivations have been discussed in the literature, especially for the case $n_e \propto \exp(ct)$ etc. Exact solutions for a general case are known for this exponential case [345] and also for the case $n_e \propto 1/r$ [346]. On the other hand, for $n_e \propto r$ an exact solution that applies to the general case is not known. See ref. [343] for a review.

26.4 Slab model

It would be instructive to consider the case in which $\sin\theta$ is small as a typical example of a non-adiabatic neutrino conversion [339]. In this case, neutrino conversion takes place in a very narrow region around the position of the resonance; outside this region a neutrino propagates without oscillation, since $\sin 2\tilde{\theta}$ is also very small, as seen in eq.(107).

Writing the energy matrix of (104) in the form

$$\begin{pmatrix} A & B \\ B & C \end{pmatrix} \tag{156}$$

and eliminating the ν_μ component, eq.(104) is written in the form

$$\ddot{P} - i(A-C)\dot{P} + B^2P^2 = 0, \tag{157}$$

where $P(t)$ is defined by

$$\nu_e(t) = P(t)\exp[-i\int_0^t A(t')dt']. \tag{158}$$

Here, we assume that ν_e is created at $t=0$. The probability of finding ν_e at time t is given by

$$P(\nu_e \to \nu_e; t) = |P(t)|^2. \tag{159}$$

Inside the resonance region $(A = C), P(t)$ obeys

$$\ddot{P} + B^2 P = 0. \tag{160}$$

Assuming $P = e^{i\alpha}$ when ν_e enters the resonance region, we find

$$P(t) = e^{i\alpha}\cos[B(t - t_0 + \delta t_0)], \quad t_0 - \delta t_0 \le t \le t_0 + \delta t_0, \tag{161}$$

where t_0 corresponds to the time when the neutrino arrives at the position of the resonance and $2\delta t_0$ is the time interval for a neutrino propagating through the resonance region. The time interval $\delta t_0 = \delta R_0$ was calculated in (114) to be $[n_e/(dn_e/dr)]\mathrm{tg}2\theta$. The probability of finding ν_e at $t(> t_0 + \delta t_0)$ is then

$$\begin{aligned} P(\nu_e \to \nu_e;\ t > t_0) &= \cos^2(2B\delta t_0) \\ &= \cos^2\gamma, \end{aligned} \tag{162}$$

where γ was defined by eq.(118). We note that the distance $2\delta r_0$ over which maximal mixing $\sin^2 2\tilde{\theta} = 1$ occurs is assumed to be small compared with the oscillation length at the resonance point. This condition is

$$2\delta r_0 \ll \tilde{\ell}_{\rm osc} \equiv \frac{4\pi E}{\Delta m^2 \sin 2\theta} = \frac{\pi}{B}, \tag{163}$$

i.e.,

$$\frac{\Delta m^2}{2E}\frac{\sin^2 2\theta}{\cos 2\theta}\left(\frac{n_e}{dn_e/dr}\right)_{\rm res} \equiv \gamma \ll \pi. \tag{164}$$

Eq.(162) is compared with the result of the Landau-Zener approximation (for $\sin^2\theta \sim 0$) (117), i.e., $P \simeq \exp(-\pi\gamma/2)$. We see that (162) agrees with (117) within 30%, as long as $0 \le \gamma \lesssim 1.2$, or equivalently, $1 \ge P_f \gtrsim 0.15$.

26.5 The MSW effect in the solar-neutrino problem

The MSW effect was proposed in order to solve the solar-neutrino problem. The density at the centre of the Sun is equivalent to $\rho_c \simeq 100\mathrm{gH\ cm^{-3}}$ for electrons. The position dependence of the electron density is close to exponential, and is described well by $\rho_e(R) \propto \exp[-10.5(R/R_\odot)]$ for $0.2 \lesssim R/R_\odot \lesssim 1$. Numerically, the resonance condition (110) is

$$\Delta m^2 = 1.52 \times 10^{-5}\mathrm{eV}^2(E/1\mathrm{MeV})(\rho/100\mathrm{gH\ cm^{-3}})\frac{1}{\cos 2\theta}. \tag{165}$$

The adiabatic condition (115) is

$$\Delta m^2 \sin^2 2\theta = 5.7 \times 10^{-10}\mathrm{eV}^2(E/1\mathrm{MeV})\cos 2\theta(\frac{d\ln\rho}{dr})R_\odot. \tag{166}$$

Hence, the MSW effect would affect the propagation of neutrinos produced in the core of the Sun if $\Delta m^2 \lesssim 1.5\times 10^{-4}\mathrm{eV}^2$ and $\Delta m^2 \sin^2 2\theta \gtrsim 6\times 10^{-9}\mathrm{eV}^2$. The experiments suggest that ν_e produced in the Sun are only partially converted into ν_μ (or ν_τ): in terms of the MSW triangle, the parameters (Δm^2, $\sin^2 2\theta$) should be placed near to the boundaries of the triangle. The attractive feature of this explanation is that it does not require any particular fine tuning of the parameters, and that a large suppression of neutrino captures results from a reasonably small mixing angle between the two neutrinos. It can naturally explain the energy-dependent suppression patterns observed in the three experiments (Table 20 above). For its attractiveness, a large number of calculations have been made for detailed analyses of the effect [332, 334, 336, 337, 339, 347, 348]. The calculations have yielded basically a well convergent answer, which does not depend much on the method or approximations used; semi-realistic calculations agree within ~10% with each other.

We remark that the Landau-Zener formula with the above parameters gives a reasonably accurate approximation to a more realistic solution of the neutrino propagation in the Sun. (Fig. 26 above has been drawn with these parameters.) In fact, the LZ formula is known to give a solution which agrees with that obtained by numerically integrating (104) within a few percent for $\sin^2 2\theta \lesssim 0.5$.

We present in Fig.27 one such semi-realistic calculation for solar-neutrino captures (reactions) with ^{37}Cl, $\nu e^- \to \nu e^-$ and ^{71}Ga detectors. (This figure virtually agree with those given by Krastev and Petcov and Bludman et al. [349].) The region encircled by two thick curves is that allowed by the Homestake ^{37}Cl experiment for the 90% C.L.; that by dashed curves is for the Kamiokande II $\nu e^- \to \nu e^-$ experiment (90% C.L.). Some wiggles which can be seen in the upper part of the triangle for ^{37}Cl are due to an adiabatic conversion of neutrinos from ^{7}Be, which contribute by a non-negligible fraction (~ 15%). The capture rate with ^{71}Ga detectors is shown by equi-SNU contours. For the solution close to the adiabatic branch for ^{37}Cl, the capture rate with ^{71}Ga is hardly modified, as expected from the fact that only high-energy neutrinos are cut-off with this solution. Along the non-adiabatic branch we see a significantly varying suppression of the neutrino capture rate with the ^{71}Ga detector.

The adiabatic branch (horizontal branch) allowed for the ^{37}Cl experiment does not overlap with that for the Kam-II experiment. This can be easily understood by considering the fact that the Kamiokande experiment, which measures a higher energy component, sees less suppressions of the neutrino flux, whereas adiabatic conversion suppresses the higher energy component, opposite to what is needed to explain the two experiments. Namely, in order to have a consistent explanation for the two experiments, low-energy neutrinos should be more strongly suppressed. This requirement is satisfied by the non-adiabatic branch (oblique branch) solution, and the regions allowed by the two experiments indeed overlap in this branch. The fact that the adiabatic branch solution is unfavourable can also be seen from the energy spectrum of neutrinos in the Kam-II detector; the energy spectrum for $E > 7.5$MeV is what is expected for the decay of ^{8}B, and this excludes a substantial part of the adiabatic branch [40].

An overlap of the region allowed by the two experiments also exists along the

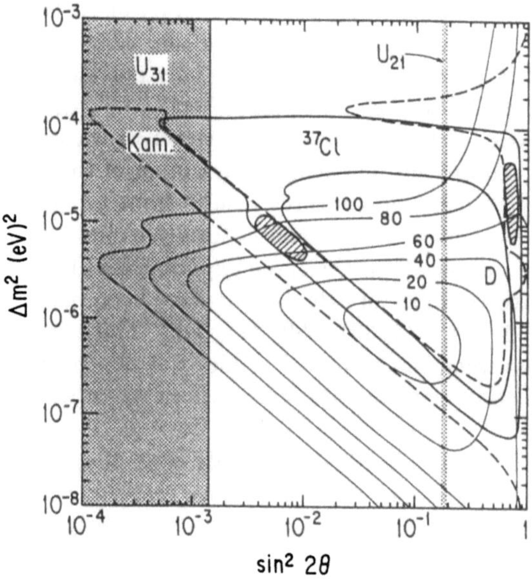

Figure 27: The MSW triangle for solar neutrinos. The region enclosed by thick solid lines is 90%C.L. allowed regions from the Homestake ^{37}Cl experiment, and that by dashed lines is from the Kamiokande experiment (90%C.L.). Equi-SNU contours are shown by thin curves for ^{71}Ga experiment. The hatched parts are the regions allowed by the three experiments (combined 90%; D is the region excluded by the absence of diurnal effect). The mixing angles for the quark sector corresponding to U_{31} and U_{21} are also indicated for comparison.

vertical branch. With this branch the suppression is almost energy independent (except for very low-energy neutrinos which do not satisfy the resonance condition). The existence of this solution is in virtue of large errors, which we adopt to draw this figure (i.e., 90% C.L.); if we take the values at the 68% C.L. for both ^{37}Cl experiment and Kam-II, the overlap disappears.

If we overlay the result of the ^{71}Ga experiments, 83$\pm$21 SNU [42], two solutions are allowed [349] (here 2σ value is taken), which are shaded in Fig. 27[14]:

(A) small-angle solution,

$$\sin^2 2\theta = 0.003 - 0.012, \ \Delta m^2 = (0.3 - 1) \times 10^{-5}\mathrm{eV}^2, \tag{167}$$

and

(B) large-angle solution,

$$\sin^2 2\theta = 0.6 - 0.9, \ \Delta m^2 = (0.4 - 3) \times 10^{-5}\mathrm{eV}^2. \tag{168}$$

[14]Solutions A and B given here agree with those by Krastev and Petcov and by Bludman et al. [349]. The large angle solution given by Anselmann et al. [349], however, shows the mixing angle somewhat smaller than the value given in (168).

The mixing angle of solution A, $\sin\theta = 0.03 - 0.055$, is $< 1/4$ of the quark mixing angle U_{12}, and even the possibility of $U(\text{lepton})_{12} \simeq (1/3)U(\text{quark})_{12}$ (see Sect.31) is disfavoured more than at a 2σ level. (We note that this is a rather model-independent feature, which arises from the fact that both *pp* neutrinos and the higher energy tail of the ^{8}B neutrinos should not be suppressed too strongly.) On the other hand, the angle is far too large compared with $U(\text{quark})_{13}$. The unnaturalness of solution B is the very large mixing angle, as is the case for all solutions with oscillation in the vacuum. We remark that the lower part of the large-angle solution lies in the range excluded due to the absence of the day-night effect, although this is at a 1σ level (see below).

In order to illustrate the suppression pattern, we present in Fig.28 the result of a simple calculation using (117–119). For the A-type solution, ^{7}Be neutrinos (and *pep* neutrinos) are very strongly suppressed to a level of 2%, ^{8}B neutrinos are suppressed to 1/2-1/3 (the suppression is a little stronger for those captured by ^{37}Cl), and *pp* neutrinos do not receive any reduction of the flux. For solution B, the flux reduction is almost uniform, 30–50% for all neutrino components. This is similar to the case with time-averaged oscillation in the vacuum, as discussed in Sect.25.3, but a slightly larger flux of the *pp* neutrino for the present case makes the agreement with the gallium results better than for the latter case.

The most notable feature which distinguishes between the two solutions are the marked energy dependence of the filter $P(\nu_e \to \nu_e)$ for A, and in particular the extremely suppressed capture rate for ^{7}Be neutrinos. Let us recall that the Kam-II experiment is also sensitive to the neutral current, the cross section of which is 1/7 that for the charged current. [$\Delta R \simeq 0.08$ out of 0.48 is a contribution form $\nu_i e$ scattering ($i \neq e$).]

MSW effect in Earth

The MSW effect may also take place for a neutrino propagating through Earth. In a certain range of neutrino parameters the Earth effect flips ν_μ(or ν_τ), converted from ν_e in the Sun, back to ν_e. Neutrinos propagate through Earth only during the night, and this conversion is observed as a day-night effect [350].

We see that the resonance condition is satisfied if

$$\Delta m^2 \approx 4 \times 10^{-6}\text{eV}^2(\rho/5Y_e\text{g cm}^{-3}) \,, \tag{169}$$

with $\rho = 3.5 - 13\text{g cm}^{-3}$ [351] for neutrinos with $E \simeq 10\text{MeV}$ ($Y_e = 1/2$ the electron number per atomic number). The neutrino oscillation length on the resonance $\tilde{\ell}|_{\text{res}} \simeq 6.8 \times 10^8\text{cm}/\sin 2\theta$, however, is of the order of the diameter of Earth (1.3×10^9cm), even for a large $\sin 2\theta$; therefore, the effect is cut at

$$\tilde{\ell}|_{\text{res}} = \frac{4\pi E}{\Delta m^2}\frac{1}{\sin 2\theta} < 2\pi R_\oplus . \tag{170}$$

Namely, we expect a large conversion in the region sandwiched by conditions (169) with $\rho \simeq 13\text{g cm}^{-3}$and (170). We note that the slope of (170) on the $\Delta m^2 - \sin^2 2\theta$ plane is half that of the nonadiabatic conversion (115).

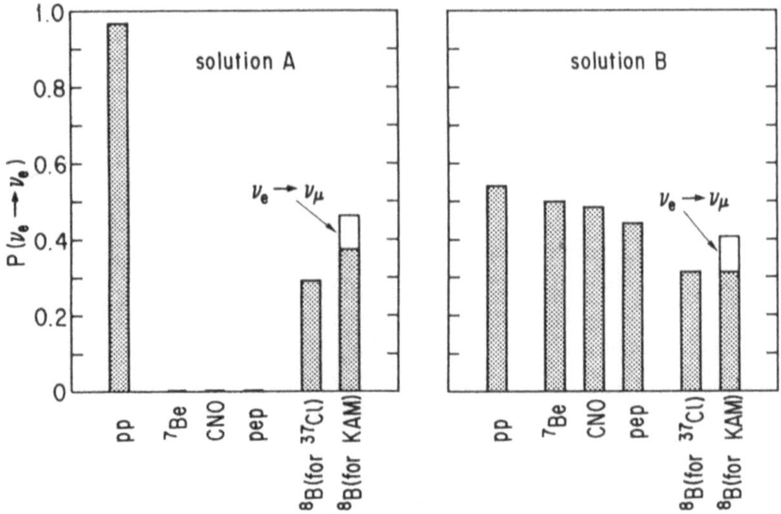

Figure 28: Suppression patterns (filter functions, $P(\nu_e \to \nu_e)$) for each solar neutrino component for the two MSW solutions A and B. The reaction rate in the detector is obtained by multiplying the filter function with the detector response function given in Fig. 20. The contribution from ν_μ for the Kamiokande detector, $P(\nu_e \to \nu_\mu) \times (1/7)$, is also indicated.

In Fig. 29 we show the result from a numerical computation of the reduction factor R of the ^{8}B solar ν flux. Here $E \geq 10$MeV is taken to fit the condition to the earlier Kamiokande experiment. The peninsula region around $\Delta m^2 = 10^{-5} - 10^{-6}$eV in Fig. 29 represents the Earth effect. We also plot in Fig. 30 the day-night asymmetry $r =$ [(night flux) $-$ (day flux)]/[(night flux) + (day flux)]. We note that part of the interesting parameter region, (A, B of eqs.(167) and (168)) could be explored by studying the day-night asymmetry at the $r = 0.05 - 0.10$ level. This would be an interesting task for the Superkamiokande project [352].

While a rough feature of Fig. 29 can be understood by (169) and (170), it may be instructive to develop another approximation to solve neutrino propagation through Earth in an analytic way. We assume here a constant density for Earth. Let us write the initial condition at the surface of Earth by a mixture of $|\nu_e>$ and $|\nu_\mu>$, i.e.,

$$\psi(0) = \alpha\psi_1(0) + \beta\psi_2(0) \quad (\alpha^2 + \beta^2 = 1), \tag{171}$$

where

$$\begin{aligned} \psi_1(0) &= |\nu_e>= \begin{pmatrix} 1 \\ 0 \end{pmatrix}, \\ \psi_2(0) &= |\nu_\mu>= \begin{pmatrix} 0 \\ 1 \end{pmatrix}. \end{aligned} \tag{172}$$

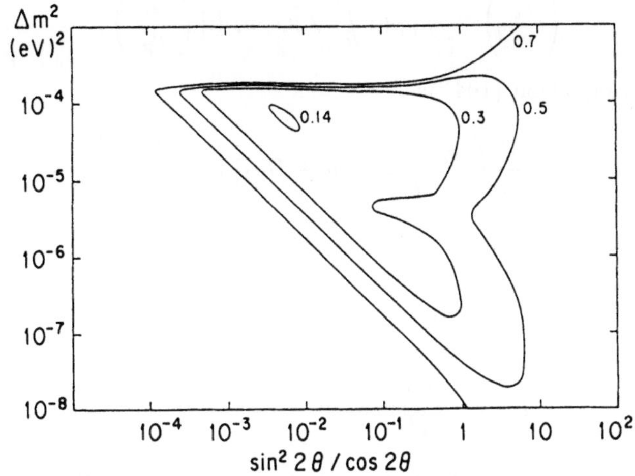

Figure 29: Deformation of the MSW triangle due to the Earth effect expected at the KAMIOKANDE water detector. The contours show fractions of the reaction rate compared to the original solar model value. The threshold energy is taken to be $E_\nu = 10\text{MeV}$. The figure is taken from Hiroi et al. [350].

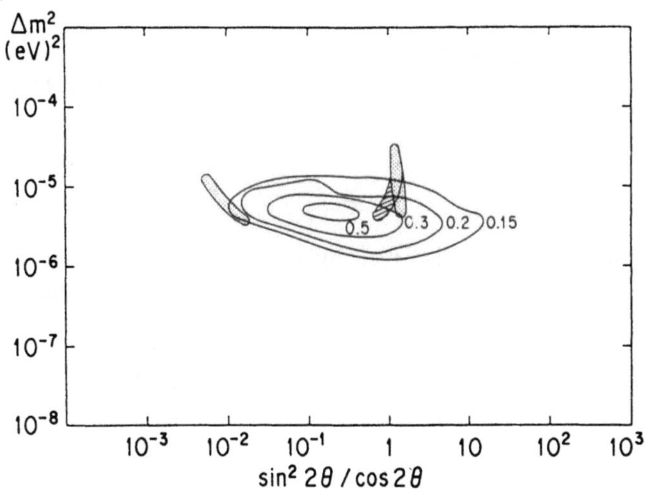

Figure 30: Contours of the diurnal asymmetry r =[(night)-(day)]/[(night)+(day)] expected at the Kamiokande detector. Shaded regions are two solutions A [eq. (167)] and B [eq. (168)] (hatched region is a large angle solution of Anselmann et al. [349]). The figure is taken from Hiroi et al. [350].

If we now write

$$\begin{pmatrix} c_e \\ c_\mu \end{pmatrix} \equiv \exp[+i\int_0^t \sqrt{2}G_F n_e dt] \begin{pmatrix} \nu_e \\ \nu_\mu \end{pmatrix}, \tag{173}$$

the Schrödinger equation (104) is

$$i\frac{d}{dt}\begin{pmatrix} c_e \\ c_\mu \end{pmatrix} = [A\sigma_3 + B\sigma_1]\begin{pmatrix} c_e \\ c_\mu \end{pmatrix}, \tag{174}$$

with

$$\begin{aligned} A &= \frac{1}{2}(\sqrt{2}G_F n_e - \frac{\Delta m^2}{2E}\cos 2\theta), \\ B &= \frac{\Delta m^2}{4E}\sin 2\theta. \end{aligned}$$

It is easy to solve this equation for a constant density. If we impose the boundary conditions $c_e(0) = 1$ and $c_\mu(0) = 0$, the solution is

$$\begin{aligned} c_e(t_f) &= \cos\left(\frac{\pi t_f}{\tilde{\ell}}\right) - i\left(\frac{\tilde{\ell}}{\ell_0}\right)\sin\left(\frac{\pi t_f}{\tilde{\ell}}\right)\left(\frac{\ell_0}{\ell_N} - \cos 2\theta\right), \\ c_\mu(t_f) &= -i\left(\frac{\tilde{\ell}}{\ell_0}\right)\sin\left(\frac{\pi t_f}{\tilde{\ell}}\right)\sin 2\theta, \end{aligned} \tag{175}$$

where t_f is the epoch when neutrinos come out of Earth, ℓ_0 and $\tilde{\ell}$ are the oscillation lengths in the vacuum and in matter, defined by eqs.(57) and (111), respectively, and

$$\ell_N \equiv \frac{\sqrt{2}\pi}{n_e G_F}. \tag{176}$$

We note that if one solution is given by $\psi = (c_e, c_\mu)^T$, another solution which is orthogonal to ψ is given by $(-c_\mu^*, c_e^*)^T$. This is seen by taking the complex conjugate of (174),

$$-i\frac{d}{dt}\begin{pmatrix} c_e^* \\ c_\mu^* \end{pmatrix} = (A\sigma_3 + B\sigma_1)\begin{pmatrix} c_e^* \\ c_\mu^* \end{pmatrix}, \tag{177}$$

and by multiplying σ_2 from the left, i.e.,

$$+i\frac{d}{dt}\begin{pmatrix} -c_\mu^* \\ c_e^* \end{pmatrix} = (A\sigma_3 + B\sigma_1)\begin{pmatrix} -c_\mu^* \\ c_e^* \end{pmatrix}, \tag{178}$$

which has the same form as (174). We then take the two solutions to be ψ_1 and ψ_2.

The probability to find ν_e after propagating through Earth is

$$\begin{aligned} P_{\nu_e} &= |\alpha\langle\nu_e|\psi_1\rangle|^2 + |\beta\langle\nu_e|\psi_2\rangle|^2 \\ &\quad + \alpha\beta^*\langle\nu_e|\psi_1\rangle\langle\nu_e|\psi_2\rangle^* + \alpha^*\beta\langle\nu_e|\psi_1\rangle^*\langle\nu_e|\psi_2\rangle \\ &= |\alpha|^2|c_e(t_f)|^2 + |\beta|^2|c_\mu(t_f)|^2 - \alpha\beta^* c_e(t_f)c_\mu(t_f) - \alpha^*\beta c_e^*(t_f)c_\mu^*(t_f). \end{aligned} \tag{179}$$

The neutrino incident on Earth is generally written as

$$\psi = \exp(iE_1 t + \phi)\cos\alpha|\nu_1\rangle + \exp(iE_2 t)\sin\alpha|\nu_2\rangle. \tag{180}$$

In the parameter region that concerns us, however, solar neutrinos that leave the Sun are almost in the mass eigenstate $|\nu_2>$. For this case,

$$\begin{aligned} \alpha &= +\sin\theta\exp(iE_2 t), \\ \beta &= +\cos\theta\exp(iE_2 t). \end{aligned} \tag{181}$$

Substitution of (175) into eq.(179) then yields

$$\begin{aligned} P_{\nu_e} &= \sin^2\theta\ |c_e(t_f)|^2 + \cos^2\theta|c_\mu(t_f)|^2 \\ &\quad - \sin\theta\cos\theta\{c_e(t_f)c_\mu(t_f) + h.c.\}. \end{aligned} \tag{182}$$

When the resonance condition is approximately satisfied, we have

$$c_e(t_f) \simeq \cos\left(\frac{\sin 2\theta}{\ell_0}\pi t_f\right), \qquad c_\mu(t_f) = \sin\left(\frac{\sin 2\theta}{\ell_0}\pi t_f\right), \tag{183}$$

and then

$$P_{\nu_e} = \frac{1}{2} + \frac{1}{2}\cos 2\theta\cos\left(\frac{2\sin 2\theta}{\ell_0}\pi t_f\right). \tag{184}$$

This probability almost agrees with what is plotted in Fig. 29.

Experimentally, a null result has been reported from the KAM-II collaboration for the day-night effect [353], r =[(night)−(day)]/[(night)+(day)]=0.08 ± 0.11 ± 0.03, which excludes the region where we expect the day-night effect $r > 0.2$ (see Fig. 30). This includes the lower half of the large-angle solution. For the region allowed by the MSW solutions (both two solutions) we expect that $r \gtrsim 0.05 - 0.1$.

<u>Some remarks.</u>

It should be emphasised that the necessity of a particle-physics explanation for the solar-neutrino problem rests on the very small capture rate of the ^{37}Cl experiment at Homestake; so far as this result remains true, it is unlikely to explain the problem by modifying the astrophysics model of the Sun [307].

A decisive test for the hypothesis that the deficit of neutrino captures originates from neutrino oscillation in the vacuum or in matter will be made by the detection of excess neutral-current induced reactions, which are not affected by neutrino oscillation, compared with charged-current ones. One such attempt is being planned by the experiment at the Sudbury Neutrino Observatory [354], which consists of 1000tons of heavy water and aims to observe, in addition to $\nu e \to \nu e$, the neutrino deuterium reactions $\nu_e d \to ppe^+$ and $\nu d \to pn\nu$, the latter being neutral current reactions. Another interesting experiment being proposed is BOREX, which will measure the nuclear excitation of ^{11}B through the neutral current, together with charged-current reactions [189].

Perhaps an equally important experiment is to measure the ^{7}Be neutrino flux. The combined result of the Kam-II and Homestake experiments lead us to suppose that the ^{7}Be neutrino flux is strongly suppressed, as is the case with the most attractive solution A of the MSW scenario as well as some solutions of neutrino oscillation in the vacuum. If such suppression is observed, the oscillation scenario for the solar-neutrino problem would be confirmed. An attempt to detect low-energy neutrinos is being made as the BOREXINO experiment, which aims to measure $\nu_e e \to \nu_e e$ due to the ^{7}Be neutrino [355].

For the discrimination between oscillation in the vacuum and that in matter, a detection of either semiannual or diurnal variation, if observed, would provide the most clear signatures for the respective scenarios. The energy spectrum also provides an important means to discriminate among various solutions. It can clearly distinguish between at least the two solutions which belong to the same class (large- vs. small-angle solutions for MSW; three neutrino time-averaged oscillation vs. long-wavelength oscillation in vacuum). Examples of the filter function $P(\nu_e \to \nu_e)$ are shown in Fig.31. This figure shows that all of the cases shown (except for one case) can be discriminated if the energy spectrum is measured over a substantial energy range (10–5MeV, say) with reasonably good accuracy. This will be a challenging task for Superkamiokande. The discrimination between the three neutrino time-averaged oscillation and the large-angle solution for MSW is difficult. They differ only in the very low-energy region ($E_\nu < 1$MeV), and the relative flux of ^{8}B neutrinos compared with pp neutrinos suffers from a significant uncertainty from astrophysics and nuclear physics. The detection of ^{7}Be neutrinos will also be very helpful for a spectroscopic discrimination among the various solutions, as discussed above.

26.6 Other applications of the MSW effect

Atmospheric neutrino

The matter effect may also be important for atmospheric neutrinos [333, 317]. For a neutrino flux with an energy of $\approx$1GeV the resonance condition (110) is satisfied if $\Delta m^2 \simeq 10^{-3} - 10^{-4}$eV^2. For this mass range a large matter effect is expected if $\sin^2 2\theta \gtrsim 3 \times 10^{-2}$[356], and is detected as the zenith-angle dependence of the ν_e/ν_μ ratio. The flat zenith-angle distribution of ν_e/ν_μ from Kamiokande [319] (within statistical accuracy of 30%), however, excludes a significant part of the region relevant to the matter effect.

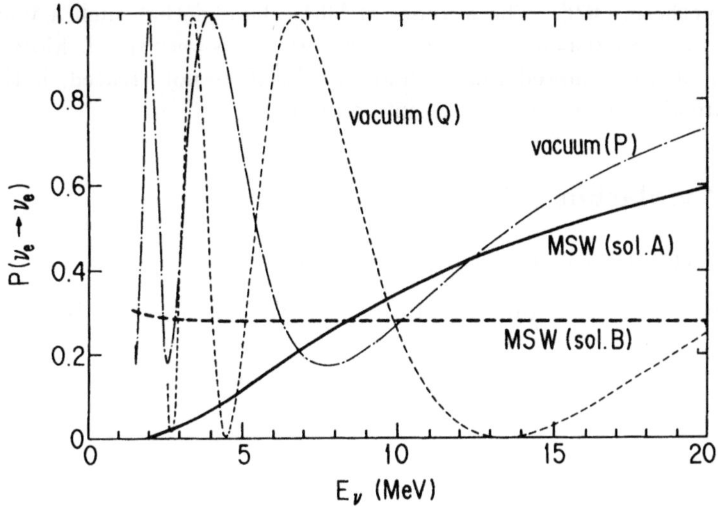

Figure 31: Detailed energy dependence of the filter function $P(\nu_e \to \nu_e)$ for the two MSW solutions (A, B) and for the two cases of long-wavelength neutrino oscillation in the vacuum (for the parameters indicated by P and Q in Fig. 21.) The figure is taken from Ref. [311].

Supernova neutrino

The energetically dominant part of supernova neutrinos are those from pair creation. For these neutrinos almost equally abundant ν_e and ν_μ interchange under the matter effect, and the observable net effect is expected to be small [357]. (Note that dominantly observed in the detector is $\bar{\nu}_e$, which does not receive the resonance effect.) On the other hand, neutrinos from the initial burst (dileptonisation neutrinos) are entirely ν_e, and are most likely to undergo complete conversion to other types of neutrinos, if the MSW effect is ever important in the Sun (the MSW triangle for the supernova progenitor include that for the Sun almost completely). For neutrino detection from SN1987A, the authors of [358] interpreted the first event of Kamiokande, which points towards the direction of SN1987A, as this dileptonisation neutrino, and considered that most of the region acceptable for the solar neutrino problem was "ruled out". This conclusion, however, is based on one event (which can also be ascribed to $\bar{\nu}_e$ accidentally scattered to the forward direction), and is not statistically conclusive.

The matter effect was also studied for $\bar{\nu}_e - \bar{\nu}_\mu$ oscillation in the progenitor, and some constraint was set on oscillation with a large mixing angle [359].

Neutrinos in the early universe

Since the MSW effect works for either ν_e or $\bar{\nu}_e$, it may create asymmetry in the abundance between ν_e and $\bar{\nu}_e$, which would disturb the standard calculation of nucleosynthesis [360]. The authors of [361] claimed that such asymmetry is indeed created, but that its effect on nucleosynthesis is very small. More recently, the author of [214] showed that asymmetry, itself, is not created, if the finite-temperature effect is properly taken into account.

27 CP violation

27.1 CP violation in neutrino oscillation

The transition amplitude from $|\nu^\alpha\rangle$ to $|\nu^\beta\rangle$ after a time interval t is given by

$$\langle\nu^\beta|\nu^\alpha\rangle_t = \sum_i U_{\alpha i}\, e^{i\frac{m_i^2}{2E}t} U^\dagger_{i\beta} \tag{185}$$

(see eq.(45)). Hence,

$$\begin{aligned} P_{\alpha\to\beta} &= |\langle\nu^\beta|\nu^\alpha\rangle_t|^2 \\ &= \sum_{i,j} U_{\alpha i}U^*_{\beta i}U^*_{\alpha j}U_{\beta j}\, e^{i\frac{m_i^2-m_j^2}{2E}t} \\ &= \sum_{i,j} |U_{\alpha i}U_{\beta i}U_{\alpha j}U_{\beta j}|\, e^{i\phi_{\alpha\beta ij}}(e^{i\Delta_{ij}t}-1)+\delta_{\alpha\beta}, \end{aligned} \tag{186}$$

where

$$\phi_{\alpha\beta ij} = \arg\,(U_{\alpha i}U^*_{\beta i}U^*_{\alpha j}U_{\beta j}), \tag{187}$$

and $\Delta_{ij} = (m_i^2 - m_j^2)/2E$. Since $\phi_{\alpha\beta ij} = -\phi_{\alpha\beta ji}$, we have

$$P_{\alpha\to\beta} = \delta_{\alpha\beta} + 2\sum_{j>i} |U_{\alpha i}U_{\beta i}U_{\alpha j}U_{\beta j}|\,[\cos(\Delta_{ij}t+\phi_{\alpha\beta ij}) - \cos\phi_{\alpha\beta ij})]. \tag{188}$$

Using expression (25) for $U, \phi_{\alpha\beta ij}$ is written [329]

$$\phi_{\alpha\beta ij} = \arg\,(u_{\alpha i}u^*_{\beta i}u^*_{\alpha j}u_{\beta j})\ . \tag{189}$$

We note that phases α, β, γ, β', γ' all cancel in this combination. This is another explicit demonstration that the properties of neutrino oscillation do not depend on whether the neutrinos are of the Dirac type or the Majorana type.

It is easy to show that

$$P_{\alpha\to\beta} = P_{\bar{\beta}\to\bar{\alpha}}. \tag{190}$$

This is a consequence of CPT invariance [362], which always holds in local field theory. This relation is shown by noting that

$$\begin{aligned}\langle\nu_\beta|\nu_\alpha\rangle_t &= U_{\alpha i}\, e^{i\frac{m_i^2}{2E}t}U^\dagger_{i\beta} \\ &= \exp[i\,\frac{1}{2E}\,U_{\alpha i}m_i^2U^\dagger_{i\beta}] = [\exp\,i\frac{m^\dagger m}{2E}\,t]_{\alpha\beta}\ ,\end{aligned} \tag{191}$$

and similarly that

$$\langle\bar\nu_\beta|\bar\nu_\alpha\rangle_t = U^*_{\alpha i}e^{i\frac{m_i^2}{2E}t}U^T_{i\beta} = [\exp\,i\frac{m^\dagger m}{2E}t]^T_{\alpha\beta}\ , \tag{192}$$

Hence,

$$\langle\bar\nu_\alpha|\bar\nu_\beta\rangle_t = \langle\nu_\beta|\nu_\alpha\rangle_t\ . \tag{193}$$

If CP invariance holds, we should have [362]

$$\langle\nu_\beta|\nu_\alpha\rangle_t = \langle\bar\nu_\beta|\bar\nu_\alpha\rangle_t = \langle\nu_\alpha|\nu_\beta\rangle_t\ , \tag{194}$$

i.e.,

$$P_{\alpha\to\beta} = P_{\bar\alpha\to\bar\beta} = P_{\beta\to\alpha}\ . \tag{195}$$

In terms of the mass matrix, this condition is represented by

$$(m^\dagger m)^T = m^\dagger m\ , \tag{196}$$

as is obvious from (191) and (192).

By inspecting (188), we see that (195) is satisfied if $\phi_{\alpha\beta ij} = 0$. Eq. (189) shows that CP violation in neutrino oscillation may take place only when $N \geq 3$ through the phase of the KM matrix; other phases, such as those characteristic of Majorana neutrinos, do not contribute to CP violation. Furthermore, it is easy to show that CP violation does not take place if $m_{\nu_e} = m_{\nu_\mu} = 0$; namely, the CP-violation effect is suppressed to the order of $m^2_{\nu_\mu}L/2E$. Taking $i = 1$ to be ν_τ, the arguments of (191) and (192) are $U_{\alpha 1}U^*_{\beta 1}m^2_{\nu_\tau}/2E$ and $U^*_{\alpha 1}U_{\beta 1}m^2_{\nu_\tau}/2E$, respectively. If we take the conventional KM representation of a 3×3 matrix the $(i,\ 1)$ elements are $(\cos\theta_1,\ \sin\theta_1\cos\theta_2,\ \sin\theta_1\sin\theta_2)$ for $i = 1,\ 2,\ 3$, which do not contain a complex phase. This proves $\langle\nu_\beta|\nu_\alpha\rangle_t = \langle\bar\nu_\beta|\bar\nu_\alpha\rangle_t$, i.e., CP invariance. This subject is detailed in the review of [363].

27.2 CP violation in small-component oscillation

If neutrinos are of the Majorana type, CP violation arising from the Majorana phase appears in $\nu_i \to \nu_j^c$, as discussed in Sect.25.4 [288, 364]. Let us suppose the 2×2 case and consider $\nu_\mu \to \nu_e^c$ and its conjugate $\nu_\mu^c \to \nu_e$. The matrix of (37) is

$$U = \begin{pmatrix} \cos\theta & \sin\theta \\ -\sin\theta & \cos\theta \end{pmatrix}\begin{pmatrix} e^{i\beta'} & 0 \\ 0 & e^{-i\beta'} \end{pmatrix}, \tag{197}$$

where the phases α and β in (37) are absorbed into the charged lepton wave functions. From (95)

$$P(\nu_{\mu L} \to \nu_{eL}^c) = \cos^2\theta \sin^2\theta | -m_1 e^{-2i\beta' - iE_1 t} + m_2 e^{2i\beta' - iE_2 t}|^2$$
$$= \cos^2\theta \sin^2\theta \left\{ (m_1 - m_2)^2 + 4m_1 m_2 \sin^2\left[-2\beta' + \frac{E_1 - E_2}{2} t\right]\right\} \frac{1}{E^2}. \quad (198)$$

For the conjugate process, we have

$$P(\nu_{\mu L}^c \to \nu_{eL}) = \cos^2\theta \sin^2\theta \left\{ (m_1 - m_2)^2 + 4m_1 m_2 \sin^2\left[2\beta' + \frac{E_1 - E_2}{2} t\right]\right\} \frac{1}{E^2}. \quad (199)$$

Namely, $P(\nu_{\mu L} \to \nu_{eL}^c) \neq P(\nu_{\mu L}^c \to \nu_{eL})$ for $\beta' \neq 0$. This appears as a disparity between the yield of e^+ produced from the neutrino (ν_μ) generated in $\pi^+ \to \mu^+ \nu_\mu$ and the e^- produced from $\pi^- \to \mu^- + \nu_\mu^c$. Unfortunately, these processes are suppressed to the order of m_ν^2/E^2.

27.3 *CP* violation characteristic of the Majorana phase

The CP-violating phases β', γ' characteristic of the Majorana neutrino appear only in the effect associated with the neutrino mass term. Let us define the quantity [365]

$$\psi_{\alpha\beta ij} = m_i m_j U_{\alpha i} U_{\beta j}^* (U_{\alpha j} U_{\beta i}^*)^* . \quad (200)$$

It is easily shown that $\psi_{\alpha\beta ij}$ is invariant under the redefinition of phases of neutrino wave functions. This combination appears when the process involves the Majorana phase. In fact, the phase which appears in (95) takes this form (Sect.27.2).

As an example for CP-violation effects due to the Majorana phase, we consider the decay $\mu \to e\nu_i\nu_j$. There are two Feynman diagrams which contribute to this decay, as shown in Fig. 32, since the neutrino and the antineutrino are identical. The interference of the two diagrams leads to a term proportional to

$$\sum_{i,j} (U^\dagger)_{j\mu} m_j (U^\dagger)_{je} \times U_{ei} m_i U_{\mu i} = \sum_{i,j} m_i m_j (U_{ei} U_{\mu j}^*)(U_{ej} U_{\mu i}^*)^*$$
$$= \sum_{i,j} \psi_{e\mu ij} . \quad (201)$$

The imaginary part of eq.(201) gives rise to CP-violating effects. This CP-violating term, however, is suppressed again by a factor of $m_1 m_2/m_\mu^2 < O(10^{-10})$, as compared with the CP-conserving part of the $\mu \to e\nu_i\nu_j$ decay.

$|\Delta L| = 2$ processes may also include the Majorana phase β' and γ'. The probability is proportional to

$$\zeta = |\sum_i m_i U_{\alpha i} U_{\beta i}|^2, \quad (202)$$

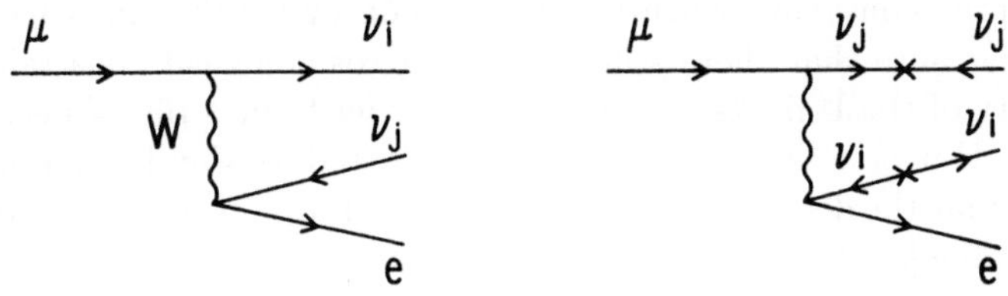

Figure 32: Feynman diagrams for the decay $\mu \to e + \nu_i + \bar{\nu}_j$. The cross in the second diagram stands for the Majorana mass insertion.

which is invariant under the phase transformation of neutrinos. Although the quantity ζ is real and has nothing to do with CP-violating phenomena, it depends on the Majorana phase β' and γ' [366]. A well-known example of this kind is the neutrinoless double-β decay when all neutrino masses are small. This process measures the average neutrino mass

$$\langle m_{\nu_e} \rangle = |\sum_i m_i U_{ei} U_{ei}| \,. \tag{203}$$

To see the dependence on the Majorana phases, we consider the two-family case. Using an explicit representation of U in (197) we obtain

$$\langle m_{\nu_e} \rangle = |m_1 \cos^2\theta e^{2i\beta'} + m_2 \sin^2\theta e^{-2i\beta'}| \,. \tag{204}$$

If $\beta' = \pi/4$ (relative $CP = -1$), the two contributions interfere destructively in the double β-decay amplitude, and it may happen that $\langle m_{\nu_e} \rangle < m_1$, where m_1 is the mass measured, for instance, in the tritium beta decay. For a more special case, $m_1 = m_2$ and $\theta = \beta' = \frac{\pi}{4}$, $\langle m_{\nu_e} \rangle$ vanishes and hence double β decay does not take place [219, 366]. This is the case for the pseudo-Dirac neutrino.

For the CP-conserving case (204) is generalised for N generations to

$$\langle m_{\nu_e} \rangle = \sum_i m_i \eta_i |U_{ei}|^2 \,, \tag{205}$$

where η_i denotes the relative CP of i-th Majorana neutrino [see (23)].

28 Heavy neutrinos and neutrino decays

In the presence of generation mixing a massive neutrino may decay into a light one by emitting a photon or two light fermions. The wide belief that all neutrinos probably have light masses smaller than a few tens of eV comes from the cosmological mass-density argument, which seems to be very solid. The constraint does not allow the neutrino to have a mass between $100h^2$eV (II.126) and $3h^{-1}$GeV, above which the relic neutrino density ($n_\nu \propto m_\nu^{-3}$, if $m_\nu > 10$MeV) is sufficiently small that the mass density carried by the neutrinos, $m_\nu n_\nu$, is also small [367, 368]. (The latter limit, however, already exceeds the laboratory limit from the Z^0 decay, and the only sensible case is $m_\nu > 45$GeV.) This argument assumes the stability of the neutrino in the cosmological time scale. Here, we discuss whether one can avoid this strong constraint when the neutrino decay is taken into account.

Let us first note that there still is a strong cosmological constraint from the energy density of the Universe. If neutrinos heavier than ~100eV decay into light particles, the Universe becomes radiation dominated by the decay products. The constraint is that the decay products be redshifted away so that they do not exceed the closure value [368]:

$$\frac{m_\nu n_{\nu+\bar\nu}}{\rho_{\rm crit}}\left(\frac{\tau_\nu}{t_0}\right)^{1/2} \leq 1 \,, \tag{206}$$

or

$$m_\nu^2 \tau_\nu \lesssim 3.8 \times 10^{21} h^4 {\rm s\ eV}^2 \,, \tag{207}$$

where the age of Universe t_0 is taken to be 15Gyr.[15] (The figure showing this constraint is given in Chap. VII; see Fig. 49 below.)

The most natural possibility for unstable neutrinos is radiative decay. This process takes place through an electromagnetic transition moment of the neutrino, which, in the standard electroweak theory, arises from radiative corrections. The radiative decays, however, generally receive strong constraints from cosmology and astrophysics, unless the lifetime is much longer than the age of the Universe (in case when $m_\nu \lesssim 100$eV), or shorter than 10–100 sec (in case when $m_\nu \gtrsim 1$MeV). We discuss details concerning radiative decays and the constraints on them in Chap. VII.

A left-handed neutrino cannot decay into three neutrinos in the leading order of the weak interaction, because of the cancellation of the flavour-changing neutral current. In higher order of the weak interaction we have $\Gamma(\nu_L \to \nu\nu\bar\nu) \sim G_F^2 (m_\tau/M_W)^4 m_\nu^5$, which gives too long a lifetime to avoid the constraint from the cosmological mass density.

If the mass of the neutrino is heavier than 1MeV (we must take it to be ν_τ), it decays into $\nu_{e,\mu} e^+ e^-$ through the Z exchange. The lifetime is [369]

$$\begin{aligned}\tau(\nu_\tau \to \nu_e + e^+ + e^-) &= \tau_\mu (m_\mu/m_{\nu_\tau})^5 [f(m_\mu/m_{\nu_\tau})(\cos\theta \sin\theta)^2]^{-1} \\ &= 2.9 \times 10^4 {\rm s} (m_{\nu_\tau}/1{\rm MeV})^{-5} [f(m_\mu/m_{\nu_\tau})(\cos\theta\sin\theta)^2]^{-1} \,,\end{aligned} \tag{208}$$

where

$$\begin{aligned} f(x) &= (1 - 14x^2 - 2x^4 - 12x^6)(1-4x^2)^{1/2} \\ &\quad + 24x^4(1-x^4)\log[\{1 - 2x^2 + (1-4x^2)^{1/2}\}/2x^2] \end{aligned} \tag{209}$$

[15] This limit, however, is a conservative one; in order to have successful cosmic structure formation, the Universe must have been matter dominated essentially all the time after hydrogen recombination at redshift $z \simeq 1500$.

and θ is the mixing angle between ν_τ and ν_e. This e^+e^- decay also receives cosmological constraints almost as strong as those for the radiative decay. For example, if neutrino lifetime is longer than 100 sec, the range between 0.5MeV (0.3 MeV, if Majorana) and 35MeV is excluded from successful nucleosynthesis ($N_\nu < 3.6$) [370]. The advantage of the e^+e^- decay is that (208) is quite fast, and the cosmological constraints can be cleared if $m_\nu \gtrsim 10$MeV.

For a massive neutrino with m_ν = 100eV–1MeV the only possibility to avoid the cosmological mass-density constraint is to consider decay into a light neutrino and a hypothetical zero mass particle, the best known of which is a familon. This case is discussed in Sect.34. Some cosmology scenarios with such decaying neutrinos are discussed in ref.[371].

If the massive neutrino is of the Dirac type, there is also a possible constraint on the neutrino mass from the neutrino observation from supernova 1987A. In ref.[372] the authors argue that the duration of neutrino burst would be shortened by 40% if $m_\nu \gtrsim 30$keV. The significance as to whether this gives a bound on the neutrino mass or not is not clear, partly because of statistics of the Kamiokande observation, and more importantly because the model of supernova (e.g. equation of state, neutrino transport) is not well understood. It is also discussed that a massive Dirac neutrino, if it decays, would yield a high-energy $\bar{\nu}_e$; this, however, was not observed at Kamiokande, thus giving a constraint on a combination of m_ν and τ_ν [373]. This argument also depends on the temperature profile and neutrino transport, and the reliability is not clear, however.

Much experimental effort has been made to look for any anomaly due to a heavy neutrino emission in the charged-particle spectrum for a number of processes involving neutrinos: $\pi \to \mu(e)\nu$, $K \to \mu(e)\nu$, $\nu' \to \nu\gamma$, $\nu' \to \nu e^+e^-$, and nuclear beta decay. The search for heavy neutrino emission [374] in $\pi \to \mu\nu$ gives a limit on $|U_{\mu\tau}|^2$ for $m_\nu \gtrsim 1$MeV stronger than is obtained from a neutrino oscillation experiment; for 3MeV$\lesssim m_\nu \lesssim$30MeV, the limit is $\lesssim 3 \times 10^{-5}$ [375]. A similar search in $\pi \to e\nu$ gives a strong limit on $|U_{e\tau}|^2$ only for $m_\nu \gtrsim 15$MeV [376]. For this mixing angle a search for $\nu' \to \nu e^+e^-$ gives a limit valid to a lower mass. The limit is stronger than the oscillation experiment for $m_\nu \gtrsim 1.5$MeV, and a representative value is $|U_{e\tau}|^2 < 10^{-3}$ at $m_\nu \approx 3$MeV [377].

One interesting case which had been given attention was the possibility of a neutrino with $m_\nu \simeq 17$keV "found" in beta decay of ^{3}H with $|U_{ex}|^2 \approx 0.01$ [378]. In spite of repeated experiments which indicated negative evidence [379], this possibility was not completely ruled out until recently [380], and motivated much theoretical speculation [381, 434, 428]; the most unusual feature of this "neutrino" was that it is placed deep inside the cosmologically forbidden region, and it must decay much faster than is allowed by interactions in the standard theory. Such a "neutrino" would also receive strong constraints from neutrino oscillation, and especially from double beta decay (see (204)). A few recent experiments have now conclusively excluded this possibility, however [382]. The limit now available is $|U_{ex}|^2 \approx 0.73 \times 10^{-3}$ (95% C.L.) for $m_\nu \simeq 17$keV and $(1.5 - 2) \times 10^{-3}$ (95% C.L.) for $m_\nu \simeq 10 - 45$keV.

Here, we do not discuss a number of experiments which are searching for "neutrinos" heavier than 30MeV, since the Z^0 width experiment does not allow (other than the three neutrinos) any particles which carry an $SU(2)_L$ quantum number for $m < 45$GeV. No limit is available for $m_\nu > 45$GeV, which is the only possibility to have a fourth generation.

If there exists a modestly light right-handed neutrino, the decay process would also allow $\nu_{jR} \to \nu_{jL} + \nu_{iL} + \bar{\nu}_{iL}$ in the lowest order of the weak interaction. The decay rate is given by [383]

$$\begin{aligned} \tau(\nu_{\tau R} \to \nu_{\tau L} + \nu_{iL} + \bar{\nu}_{iL}) &= \tau_\mu (m_\mu / m_{\nu_\tau})^5 \left[\frac{\sin\theta\cos\theta}{2} \times 2 \times 3 \right]^{-2} \\ &= 1.9 \times 10^4 \mathrm{s} (m_{\nu_\tau}/1\mathrm{MeV})^{-5} (\sin\theta\cos\theta)^{-2} \,. \quad (210) \end{aligned}$$

For $m_{\nu_R} > 1$MeV, e^+e^- decay channel also opens with the lifetime given in (208). The number density of cosmologically produced right-handed neutrinos may not be as many as that of left-handed neutrinos if the interaction of right-handed neutrinos are weak enough so that they decouple from thermal equilibrium before the QCD phase transition. In such a case the cosmological constraint (II.126) is weakened to be $m_\nu \lesssim 500\mathrm{eV}h^2$, or (207) is weakened by a factor of five.

VI. Models for massive neutrinos

29 Classification of models

An interesting view is to suppose that some physical reasons exist for the smallness of the neutrino mass. In the case of the Majorana neutrino, two classes of models are so far known which lead naturally to a small neutrino mass: (i) models in which the seesaw mechanism works and (ii) those in which the neutrino mass is induced by a radiative correction. The central idea of the former class of models is to suppose some higher symmetry (global or local) which is broken at an energy scale much higher than is explored in high-energy experiments. If this symmetry breaking takes place so that it allows the right-handed neutrino to have a mass, this acts as a large M needed in eq.(III.44) and, hence, a small mass is induced for the left-handed neutrino. The smallness of the left-handed neutrino mass is explained by new physics at a very large energy scale; this is why it is often claimed that neutrino mass searches to explore a high-energy frontier. Typical models which belong to this class are:

	symmetry	energy scale
left-right symmetry	local	> 300 GeV
SO(10) model	local	$10^{13} - 10^{16}$ GeV
horizontal symmetry	local	$> 10^6$ GeV
	global	$> 10^9$ GeV
Peccei-Quinn symmetry	global	$10^{10} - 10^{12}$GeV
lepton number	global	no bound

If we take MSW solution A of (V.167) for the solar-neutrino problem, the mass of the right-handed neutrino is

$$M_{\nu_R} = 1.1 \times 10^9 - 3.3 \times 10^{11}\text{GeV} \tag{1}$$

if the partner of the ν_e mixing is ν_μ, and

$$M_{\nu_R} = 3.2 \times 10^{11} - 2.9 \times 10^{15}\text{GeV} \tag{2}$$

if the partner is ν_τ. If we take the large-angle solution B of eq.(V.168), the lower ends of (1, 2) extend to an energy scale that is a factor 3 lower. The large uncertainty reflects the model dependence as to whether the Dirac mass is taken to be the quark mass or the lepton mass. In most of the unified models, the Dirac mass for the neutrino is identified with the mass of charge 2/3 quarks. Among the known models the simplest version of the SO(10) model and the horizontal symmetry model predict specific mixing angles. Their predictions are U(lepton) $= U$(quark) or U_{ij}(lepton)$=\frac{1}{3}U_{ij}$(quark) for some i, j; neither of them, however, agree with the mixing angles (V.167), (V.168) inferred from the solar-neutrino experiment. This does not mean, however, that the basic model of SO(10) or that with horizontal symmetry is ruled out.

We should also remark that there is no known model which predicts the generation-dependence of the neutrino mass, unless the model is further supplemented by a strong assumption on the right-handed neutrino mass. The mass formula $m_{\nu_e} : m_{\nu_\mu} : m_{\nu_\tau} = m_e^2 : m_\mu^2 : m_\tau^2$ or $m_u^2 : m_c^2 : m_t^2$, often mentioned in the literature, is obtained only when one assumes $M_{\nu_e R} = M_{\nu_\mu R} = M_{\nu_\tau R}$, which is almost equivalent to assuming the left-handed neutrino masses, themselves.

In the other classes of models (ii) one introduces a scalar particle (with a mass of the order of the electroweak energy scale) which breaks the lepton number in the Higgs potential. A left-handed neutrino mass is then induced by a radiative correction from a scalar particle loop, and the neutrino mass is typically $m_\nu \sim f^2 m_\ell$, where f is the Yukawa coupling and m_ℓ is the charged lepton mass. This model requires some new physics only at the electroweak energy scale.

In the case of the Dirac neutrino a small mass should be put by hand, blindly assuming a large hierarchy typically $\gtrsim 10^6$ between the Yukawa couplings for charged leptons and neutrinos. While this seems to be a somewhat unnatural procedure to many people, it is perhaps a matter of taste at this level. This view, however, encounters some difficulty once one attempts to understand the particle mass spectrum within grand unification [384, 385], where quarks and leptons are always treated in an equal-footing manner. This is the reason why theorists generally prefer the Majorana neutrino as a more natural possibility to the Dirac neutrino. For the sake of completeness, we should remark that it is not impossible to think of models in which a seesaw-like mechanism suppresses the Dirac neutrino mass.

We should emphasise that no models are known so far which have a predictive power for the neutrino mass matrix which is in agreement with any of experimental indications.

30 The left-right symmetry model

The prototype model in which the seesaw mechanism [216] is most naturally embedded is the left-right symmetric model [386, 387]. This model is one of the simplest extensions of the Weinberg-Salam theory incorporating the right-handed neutrino, and this symmetry often appears in various grand unified models.

Let us suppose that there is $\mathrm{SU}(2)_L \times \mathrm{SU}(2)_R \times \mathrm{U}(1)$ gauge symmetry, and that fermions belong to

$$\begin{aligned}
\ell_L &\equiv \begin{pmatrix} \nu \\ e \end{pmatrix}_L, \quad (\mathbf{2}, \mathbf{1}, -1); \\
\ell_R &\equiv \begin{pmatrix} \nu \\ e \end{pmatrix}_R, \quad (\mathbf{1}, \mathbf{2}, -1); \\
q_L &\equiv \begin{pmatrix} u \\ d \end{pmatrix}_L, \quad (\mathbf{2}, \mathbf{1}, 1/3); \\
q_R &\equiv \begin{pmatrix} u \\ d \end{pmatrix}_R, \quad (\mathbf{1}, \mathbf{2}, 1/3),
\end{aligned} \tag{3}$$

where the representations refer to $(\mathrm{SU}(2)_L, \mathrm{SU}(2)_R, \mathrm{U}(1))$. This is the most natural extension of Weinberg-Salam's assignment of eqs.(II.7) and (II.31). We introduce two kinds of Higgs scalars, ξ and ϕ, which break

$$\mathrm{SU}(2)_L \times \mathrm{SU}(2)_R \times \mathrm{U}(1) \to \mathrm{SU}(2)_L \times \mathrm{U}(1)_Y \quad \text{with } \langle\xi\rangle \neq 0 \tag{4}$$

and

$$\mathrm{SU}(2)_L \times \mathrm{U}(1)_Y \to \mathrm{U}(1)_{\mathrm{em}} \quad \text{with } \langle\phi\rangle \neq 0 \tag{5}$$

We assume that ξ belongs to a triplet representation of $\mathrm{SU}(2)_R$, i.e., (**1**, **3**, +2) [387]. The doublet Higgs ϕ should give masses to fermions and we must take it to be (**2**, **2***, 0). This ϕ transforms as

$$\phi \to u_L\ \phi\ u_R^\dagger \tag{6}$$

under an $\mathrm{SU}(2)_L \times \mathrm{SU}(2)_R$ transformation. For ξ we write it in a 2×2 matrix form,

$$\xi \equiv \frac{1}{\sqrt{2}}\tau_i\xi_i, \tag{7}$$

so that ξ transforms under $\mathrm{SU}(2)_R$ as

$$\xi \to u_R\ \xi\ u_R^\dagger\ . \tag{8}$$

To specify the symmetry breaking of (4), we define the $\mathrm{U}(1)_Y$ charge by

$$Y = I_R^3 + \frac{I}{2}, \tag{9}$$

where I_R^3 is the third component of $\mathrm{SU}(2)_R$ and I is the U(1) charge. The $\mathrm{U}(1)_Y$ charge content of ξ is $(\xi^{++}, \xi^{+}, \xi^{0})$. This can be seen by applying a $\mathrm{U}(1)_Y$ rotation; under the rotation, ξ transforms as

$$\begin{aligned} \xi &\to e^{-i\epsilon}\, e^{-i\epsilon\frac{\tau_3}{2}}\ \xi\ e^{+i\epsilon\frac{\tau_3}{2}} \\ &= \xi - i\epsilon(\xi + [\frac{\tau_3}{2}, \xi]) + \cdots \\ &= \xi - i\frac{\epsilon}{2}\begin{pmatrix} \xi_3/\sqrt{2} & 2(\xi_1 + i\xi_2)/\sqrt{2} \\ 0 & \xi_3/\sqrt{2} \end{pmatrix} + \cdots \end{aligned} \tag{10}$$

Therefore, the $\mathrm{U}(1)_Y$ charge content of ξ is

$$\xi = \begin{pmatrix} \xi^{+}/\sqrt{2} & \xi^{++} \\ \xi^{0} & \xi^{+}/\sqrt{2} \end{pmatrix} \tag{11}$$

The electromagnetic charge is given by

$$Q_{\mathrm{em}} = I_L^3 + Y = I_L^3 + I_R^3 + \frac{I}{2}. \tag{12}$$

For the SU(2) singlet ξ the $\mathrm{U}(1)_Y$ charge and the electromagnetic charge are the same. When ξ^0 develops a vacuum expectation value V,

$$\langle\xi\rangle = \begin{pmatrix} 0 & 0 \\ V & 0 \end{pmatrix} \tag{13}$$

$SU(2)_L \times SU(2)_R \times U(1)$ is broken into $SU(2)_L \times U(1)_Y$.

The electromagnetic charge assignment of ϕ is obvious:

$$\phi = \begin{pmatrix} \phi_1^0 & \phi_1^+ \\ \phi_2^- & \phi_2^0 \end{pmatrix}. \tag{14}$$

If ϕ develops a vacuum expectation value,

$$\langle\phi\rangle = \begin{pmatrix} v_1 & 0 \\ 0 & v_2 \end{pmatrix} \tag{15}$$

$SU(2)_Y \times$ U(1) breaks down to $U(1)_{em}$, where $v_1^2 + v_2^2 = v^2/2$.

We now discuss the Yukawa interaction, which may be generally written in the form

$$\begin{aligned} \mathcal{L} &= g\overline{\ell}_R\phi\ell_L + g'\overline{\ell}_R\tilde{\phi}\ell_L \\ &\quad + f\overline{q}_R\phi q_L + f'\overline{q}_R\tilde{\phi}q_L \\ &\quad + h\tilde{\ell}_R^c\xi\ell_R + \text{h.c.}, \end{aligned} \tag{16}$$

where

$$\tilde{\phi} \equiv \tau_2\phi^*\tau_2 = \begin{pmatrix} \phi_2^{0*} & -\phi_2^+ \\ -\phi_2^- & \phi_1^{0*} \end{pmatrix} \tag{17}$$

$$\tilde{\ell}_R^c \equiv \overline{\ell_R^c}\, i\tau_2 = (\overline{e^c},\ -\overline{\nu^c})_R. \tag{18}$$

The non-zero vacuum expectation value $\langle\xi\rangle \neq 0$ induces a Majorana mass of ν_R,

$$\begin{aligned} \mathcal{L}_{\text{mass}}^M &= -h(\overline{e^c}, -\overline{\nu^c})_R \begin{pmatrix} 0 & 0 \\ V & 0 \end{pmatrix} \begin{pmatrix} \nu \\ e \end{pmatrix}_R + \text{h.c.} \\ &= hV(\overline{\nu_R^c}\nu_R + \text{h.c.})\ . \end{aligned} \tag{19}$$

Because of (15) the neutrino has a Dirac mass of the form

$$\begin{aligned} \mathcal{L}_{\text{mass}}^D &= gv_1\overline{\nu}_R\nu_L + g'v_2^*\overline{\nu}_R\nu_L + \text{h.c.} \\ &= (gv_1 + g'v_2^*)\overline{\nu}_R\nu_L + \text{h.c.} \end{aligned} \tag{20}$$

In sum, the mass term which refers to neutrinos is

$$\mathcal{L}_{\text{mass}} = \frac{1}{2}(\overline{\nu_R^c}\ \ \overline{\nu}_L)\begin{pmatrix} A & B \\ B & 0 \end{pmatrix}\begin{pmatrix} \nu_R \\ \nu_L \end{pmatrix} + \text{h.c.} \tag{21}$$

with

$$\begin{aligned} A &= 2hV \\ B &= (gv_1 + g'v_2^*) \end{aligned} \tag{22}$$

Writing

$$\begin{aligned} N &= \nu_R + \nu_R^c \\ \nu &= \nu_L + \nu_L^c\ , \end{aligned} \tag{23}$$

we have

$$\mathcal{L}_{\rm mass} = \frac{1}{2}(\overline{N}\ \overline{\nu})\begin{pmatrix} A & B \\ B & 0 \end{pmatrix}\begin{pmatrix} N \\ \nu \end{pmatrix}, \tag{24}$$

where $N^c = N$ and $\nu^c = \nu$. The diagonalisation with

$$\begin{aligned} \nu' &= \nu\cos\theta + N\sin\theta \\ N' &= -\nu\sin\theta + N\cos\theta \end{aligned} \tag{25}$$

yields, when $A \gg B$ $(V \gg v_1, v_2)$,

$$\begin{aligned} m_{N'} &\simeq A \\ m_{\nu'} &\simeq -B^2/A\,, \end{aligned} \tag{26}$$

and the mixing tg $2\theta = 2B/A$, in agreement with eq.(III.44). The minus sign in $m_{\nu'}$ is absorbed into the phase of the wave function.

31 Left-right symmetry in SO(10)

The simplest grand unification with ν_R automatically included is the SO(10) model [388]. In this scheme all left-handed fermions are incorporated in the spinor representation of **16**:

$$\mathbf{16}_L = \left[\begin{pmatrix} \nu \\ e \end{pmatrix}, \begin{pmatrix} u \\ d \end{pmatrix}, \begin{pmatrix} \nu^c \\ e^c \end{pmatrix}, \begin{pmatrix} u^c \\ d^c \end{pmatrix}\right]_L \tag{27}$$

This representation contains ν_L^c, i.e., the antiparticle of the right-handed neutrino. A desirable feature of this representation is that the model is automatically triangle-anomaly free [390].

The SO(10) unification includes left-right symmetry, as described in the preceding section, and the scale of left-right symmetry breaking is determined by low-energy phenomenology, thus having a predictive power for the neutrino mass. There are many ways to break SO(10) down to low-energy symmetry of SU(3) $\times$ SU(2) $\times$ U(1). We defer any full description of the model construction and its phenomenology to the following sections, with further mathematical details given in Appendix C (the irreducible representation of SO(10) is summarised in Tables 20 and 21 [391] for convenience). In this section we confine ourselves to the case

$$\mathrm{SO}(10) \to \mathrm{SU}(3) \times \mathrm{SU}(2)_L \times \mathrm{SU}(2)_R \times \mathrm{U}(1) \to \mathrm{SU}(3) \times \mathrm{SU}(2)_L \times \mathrm{U}(1)_Y$$

and discuss only the aspect related to the neutrino mass. In order to break SO(10) to SU(3) $\times$ SU(2)$_L\times$SU(2)$_R\times$ U(1) we introduce adjoint Higgs H of **45**. Further symmetry breaking to SU(3)$\times$SU(2)$_L\times$U(1)$_Y$ is caused by **126**, which includes $\xi(\mathbf{1},\mathbf{3},2)$ of the previous section, and by a vector **10**, which includes $\phi(\mathbf{2},\mathbf{2}^*,0)$, down to SU(3)$\times$ U(1)$_{\rm em}$.

The Yukawa interaction is written

$$\mathcal{L}_{\rm Yukawa} = \frac{g}{2}\mathbf{16}_L\cdot\mathbf{16}_L\cdot\mathbf{10} + \frac{h}{2}\mathbf{16}_L\cdot\mathbf{16}_L\cdot\mathbf{126} + \text{h.c.} \tag{28}$$

Table 20: Irreducible representation (IR) of SO(10).

dimensionality	IR	indices
10	vector	Γ_{μ}
16	spinor	ψ
45	adjoint	$\Sigma_{\mu\nu}$
54	symmetric second-rank tensor	$\Gamma_{\{\mu\nu\}}$
120	antisymmetric third-rank tensor	$\Gamma_{[\mu\nu\lambda]}$
126	antisymmetric fifth-rank tensor	$\Gamma_{[\mu\nu\lambda\rho\sigma]}$
210	antisymmetric fourth-rank tensor	$\Gamma_{[\mu\nu\lambda\kappa]}$

Table 21: SU(5) and $SU(2)_L \times SU(2)_R \times SU(4)$ contens of SO(10) IR's.

SO(10)	SU(5)	$SU(2)_L \times SU(2)_R \times SU(4)$
10	**5+5***	**(2,2,1)+(1,1,6)**
16	**1+5*+10**	**(2,1,4)+(1,2,4*)**
45	**1+10+10*+24**	**(3,1,1)+(1,3,1)+(1,1,15)+(2,2,6)**
54	**15+15*+24**	**(1,1,1)+(3,3,1) +(1,1,20)+(2,2,6)**
126	**1+5*+10+15*+45+50***	**(1,1,6)+(3,1,10*)+(1,3,10)+(2,2,15)**
210	**1+5+5*+10+10*+24**	**(1,1,1)+(1,1,15)+(2,2,6)+(3,1,15)**
	+40+40*+75	**+(1,3,15)+ (2,2,10)+(2,2,10*)**

The first term gives a Dirac mass to the fermions. The Lagrangian of this part reads [see (C.18)]

$$\mathcal{L}^{(10)}_{\text{Yukawa}} = \frac{g}{2}\overline{\psi^c_L}\,\beta\Gamma_\mu\psi_L\phi^\mu + \text{h.c.}, \tag{29}$$

where ψ_L is a spinor in the **16** representation, and ϕ^μ is a vector of **10**. The Γ^μ matrices ($\mu = 1 - 10$) of SO(10) satisfy the Cliford algebra,

$$\{\Gamma_\mu, \Gamma_\nu\} = 2\delta_{\mu\nu}, \tag{30}$$

and the charge conjugation matrix β is written

$$\beta = \Gamma_2\Gamma_4\Gamma_5\Gamma_8\Gamma_9. \tag{31}$$

Noting that ϕ is a real field, and consulting with the explicit representation given in Appendix C (see (C.14)), (29) is

$$\mathcal{L}^{(10)}_{\text{Yukawa}} = g(\overline{u}_R u_L + \overline{\nu}_R\nu_L)(\phi^9 - i\phi^{10}) + g^*(\overline{d}_L d_R + \overline{e}_L e_R)(\phi^9 - i\phi^{10}) + \cdots \tag{32}$$

The quarks and leptons acquire masses when ϕ develops a vacuum expectation value,

$$<\phi^\mu> = \begin{bmatrix} 0 \\ 0 \\ \vdots \\ v \end{bmatrix} \begin{matrix} 1 \\ 2 \\ \vdots \\ 10 \end{matrix} \tag{33}$$

This, however, leads to the mass pattern

$$m_u = m_\nu^D = m_d = m_e \;, \tag{34}$$

which is not consistent with the experiment. To leave the freedom which makes the prediction consistent with the experiment, we need two **10**'s [392] which have the vacuum expectation values

$$\langle\phi^{10}\rangle = v_1, \quad \langle\phi'^9 - i\phi'^{10}\rangle = v_2, \tag{35}$$

with v_1 and v_2 real numbers, or

$$\langle\phi^\mu\rangle = \begin{bmatrix} 0 \\ 0 \\ \vdots \\ v_1 \end{bmatrix}, \quad \langle\phi'^\mu\rangle = \begin{bmatrix} 0 \\ 0 \\ \vdots \\ \text{Re}\; v_2 \\ \text{Im}\; v_2 \end{bmatrix}, \tag{36}$$

so that

$$\begin{aligned} m_u = m_\nu^D &= gv_1 + g'v_2 \\ m_d = m_e &= gv_1 + g'v_2^*. \end{aligned} \tag{37}$$

This mass formula does not change even if we introduce more **10**'s. The right-handed neutrino acquires a Majorana mass when $SU(2)_R$ is broken by $\langle \mathbf{126} \rangle \neq 0$. The Yukawa term in (28) gives

$$M_{\nu_R} = h\langle \mathbf{126} \rangle . \tag{38}$$

With the seesaw mechanism, we have

$$\begin{aligned} m_\nu &= m_{\nu D}^2 / M_{\nu_R} \\ &= m_u^2 / M_{\nu_R} \end{aligned} \tag{39}$$

or

$$\begin{aligned} m_\nu = & \ m_{\nu D}^T \ M_{\nu_R}^{-1} m_{\nu D} \\ = & \ m_u^T \ M_{\nu_R}^{-1} m_u \ , \end{aligned} \tag{40}$$

writing the mass in a matrix form taking the generation into account.

If we take the basis in which m_u is diagonal, m_d and m_e are expressed as

$$m_d = m_\ell = V^\dagger \begin{pmatrix} m_e & & \\ & m_\mu & \\ & & m_\tau \end{pmatrix} U \tag{41}$$

with two unitary matrices V and U. Since the mass matrices of charge $-1/3$ quarks and charged leptons are given by a vector **10** in our SO(10) model, they are symmetric, i.e., $m_d^T = m_d$ and $m_\ell^T = m_\ell$, and, hence, $V^\dagger$ can be identified with U^T. A unitary matrix is generally expressed as

$$U = K U_{KM}^{-1} L \ , \tag{42}$$

where K and L are the diagonal (phase) matrix and U_{KM} is the Kobayashi-Maskawa matrix. Since L is absorbed into the phase of the state [see (V.27)], we write

$$m_d = m_e = U_{KM}^{-1T} K \begin{pmatrix} m_e & & 0 \\ & m_\mu & \\ 0 & & m_\tau \end{pmatrix} K U_{KM}^{-1} . \tag{43}$$

On the other hand, the neutrino mass matrix is

$$m_\nu = \begin{pmatrix} m_u & & 0 \\ & m_c & \\ 0 & & m_t \end{pmatrix} M_{\nu R}^{-1} \begin{pmatrix} m_u & & 0 \\ & m_c & \\ 0 & & m_t \end{pmatrix} . \tag{44}$$

Because of the hierarchy $m_u \ll m_c \ll m_t$, we expect m_ν to be already close to diagonal, unless the matrix elements of M_{ν_R} has the strong hierarchy that compensates the hierarchy in the quark mass matrix. Then, an extra mixing angle which enters in the ν sector is very small, and we expect [393]

$$U_{\text{lepton}} \simeq U_{\text{KM}} , \tag{45}$$

i.e.,

$$\theta_{e\mu} \simeq \theta_{ds}, \ \ \theta_{e\tau} \simeq \theta_{db}, \cdots \tag{46}$$

This is shown in the following way: in the $m_u = 0$ limit the neutrino mass matrix is given by

$$m_\nu = \begin{pmatrix} 0 & 0 & 0 \\ 0 & A & B \\ 0 & B & C \end{pmatrix}, \tag{47}$$

and it is diagonalised by

$$U_\nu = \begin{pmatrix} 1 & 0 \\ 0 & V \end{pmatrix}, \tag{48}$$

with

$$V \begin{pmatrix} A & B \\ B & C \end{pmatrix} V^T = \begin{pmatrix} m_2 & 0 \\ 0 & m_3 \end{pmatrix}. \tag{49}$$

Expression (48) means that the $\nu_e - \nu_\mu$ mixing angle $\theta_{e\mu} = \theta_{\mathsf{Cabibbo}}$. Repeating this procedure for the (2,3) elements, we can show (45) in the limit of $m_u/m_t \to 0$ and $m_c/m_t \to 0$.

The generation dependence of the neutrino mass depends explicitly on the assumption on the spectrum of M_{ν_R}. It is often assumed to have the simplest form $M_{\nu_R} = M \cdot \mathbf{1}$ ($\mathbf{1}$ is the unit matrix). With this assumption we have

$$m_{\nu_e} : m_{\nu_\mu} : m_{\nu_\tau} = m_u^2 : m_c^2 : m_t^2 \ . \tag{50}$$

We note that the quark masses receive a large radiative correction to compare them with the experiment [394]. In fact, relation (50) holds only at the unification scale, and the up, charm and top quark masses at the renormalisation point μ=1GeV, $2m_c$ and $2m_t$ are related with the values at the unification scale (which are to be identified with the Dirac masses in eq.(50)) as [395, 396]

$$\frac{m_{\nu_e}(\text{Dirac})}{m_u(1\text{GeV})} \simeq 0.22 \ , \tag{51}$$

$$\frac{m_{\nu_\mu}(\text{Dirac})}{m_c(3\text{GeV})} \simeq 0.26 \ , \tag{52}$$

$$\frac{m_{\nu_\tau}(\text{Dirac})}{m_t(300\text{GeV})} \simeq 0.42 \ . \tag{53}$$

Therefore, (50) is

$$m_{\nu_e} \ : \ m_{\nu_\mu} \ : \ m_{\nu_\tau} = 1.6 \times 10^{-5} \ : \ 1 \ : \ 2.6 \times 10^4 \tag{54}$$

for $m_t = 150$GeV.

We note that the relations (45) and (46) do not receive QCD radiative corrections at the one loop level, since they cancel in the mass ratios.

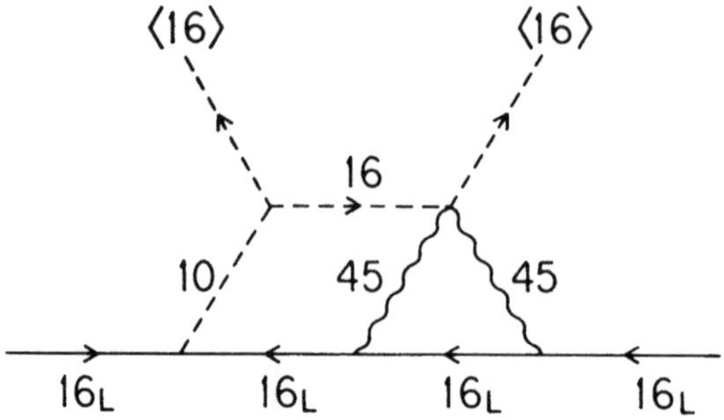

Figure 33: Radiative correction involving vacuum expectation value of **16** giving the right-handed neutrino a Majorana mass.

Neutrino mass without **126**

If intermediate symmetry is broken by a Higgs scalar $\Phi(\mathbf{16})$ instead of $\xi(\mathbf{126})$, the right-handed neutrino remains massless at the tree level, since $\Phi(\mathbf{16})$ has no Yukawa coupling to the fermions. In this case, however, we expect the Majorana masses of the right-handed neutrino from the two loop diagrams depicted in Fig.33 [397]. This gives

$$M_{\nu_R} \sim \frac{\alpha_G^2}{\pi^2}(g\tilde{g} + g\tilde{g}')\frac{\langle\Phi(\mathbf{16})\rangle^2}{M_G}, \tag{55}$$

where $\tilde{g}$ and $\tilde{g}'$ are trilinear couplings defined by

$$\mathcal{L} = \tilde{g}M_{\rm G}\Phi(\mathbf{16})\Phi(\mathbf{16})\phi(\mathbf{10}) + \tilde{g}'M_{\rm G}\Phi(\mathbf{16})\Phi(\mathbf{16})\phi(\mathbf{10}')\ . \tag{56}$$

In (55) g and g' are Yukawa couplings between $\phi(\mathbf{10}), \phi(\mathbf{10}')$ and matter fields (see eq.(37)), and M_G is the mass scale of grand unification. Since g and g' are proportional to m_u and m_d, it is likely that the masses of the light neutrinos are proportional to a single power of the u quark mass. If we put approximate energy scales of $\langle\Phi(\mathbf{16})\rangle$ and $M_{\rm G}$ resulting from a phenomenological analysis, $\langle\Phi(\mathbf{16})\rangle \simeq 10^{12}$GeV, $M_{\rm G} = 10^{16}$GeV (see Sect.32), however, M_{ν_R} turns out to be too small to take it seriously. This model gives a reasonable M_{ν_R} only when $\langle\Phi(\mathbf{16})\rangle$ is taken to be as large as $M_{\rm G}$ [397]. The idea of effective interactions of the form $\langle\Phi(\mathbf{16})\rangle^2/M$ is extensively developed in the next section.

32 General discussion of SO(10) grand unification

32.1 Symmetry-breaking pattern and the neutrino mass

The prototype of grand unification is based on the SU(5) group [385], and in its prototype model the neutrino is massless. As an extension, it is possible to introduce ν_R as an SU(5) singlet and to invoke the seesaw mechanism to give a small Majorana mass to ν_L. However, since the singlet ν_R is completely independent from other multiplets, no prediction is made for the neutrino mass. The simplest

Table 22: Breaking of SO(10) GUT and Higgs particles.

intermediate symmetry(G_I)	Higgs: SO(10) $\to G_I$	Higgs: $G_I \to G_0$	Higgs: $G_0 \to U(1)_{em}$
SU(5)	**16** or **126**	**45**,**54** or **210**	**10**
SU(5) × U(1)	**45**	**126** ⊕**45**	**10**
		or **16**⊕**45**	**10**
SU(4) × SU(2)$_L$× SU(2)$_R$	**54** or **210**	**16** or **126**	**10**
SU(4) × SU(2)$_L$× U(1)$_R$	**45**	**16** or **126**	**10**
SU(3) × SU(2)$_L$× SU(2)$_R$× U(1)$_{B-L}$	**45**	**16** or **126**	**10**
SU(3) × SU(2)$_L$× U(1)$_R$× U(1)	**45**	**16** or **126**	**10**

grand unification in which ν_R is automatically incorporated and, hence, prediction is given for the neutrino mass is that with SO(10).

There are many ways to break SO(10) down to $G_0 = SU(3)\times SU(2)\times U(1)$ (e.g., see [398]). Here, we restrict our consideration to those cases in which symmetry is broken to G_0 by one or two steps:

$$\begin{aligned}
&\text{(i)} && SO(10) \xrightarrow{M_G} G_0 \\
&\text{(ii)} && SO(10) \xrightarrow{M_G} SU(5) \xrightarrow{M_I} G_0 \\
&\text{(iii)} && SO(10) \longrightarrow SU(5)\times U(1) \longrightarrow G_0 \\
&\text{(iv)} && SO(10) \longrightarrow SU(4)\times SU(2)_L \times SU(2)_R \longrightarrow G_0 \\
&\text{(v)} && SO(10) \longrightarrow SU(4)\times SU(2)_L \times U(1) \longrightarrow G_0 \\
&\text{(vi)} && SO(10) \longrightarrow SU(3)\times SU(2)_L \times SU(2)_R \times U(1) \longrightarrow G_0 \\
&\text{(vii)} && SO(10) \longrightarrow SU(3)\times SU(2)_L \times U(1)_R \times U(1) \longrightarrow G_0
\end{aligned} \tag{57}$$

In (i) and (ii), SO(10) (rank=5) is reduced to rank 4 symmetry at the GUT scale M_G. In the remainder of the cases rank 5 is maintained at M_G, and reduces to rank-4 symmetry at the intermediate scale M_I. In general, ν_R acquires a mass when the symmetry is reduced to rank 4; ν_R becomes massive at $M = M_G$ in (i) and (ii), and at $M = M_I$ in (iii)-(vii).

Generally speaking, there exist so many choices of the Higgs scalars to cause a breaking of (57). We consider only cases with Higgs particles of low dimentionalities. We note that (i) is regarded as the special case $M_I = M_G$ of the others. The Higgs particles needed to cause symmetry breaking of (57) are summarised in Table 22. We discuss the possible consequence on the neutrino mass in the following:

Case (ii). The first symmetry breaking around the grand-unified mass scale is caused by either **16** or **126**. If $\langle \mathbf{126} \rangle \neq 0$ breaks the symmetry, the mass of ν_R is expected to be of the order of M_G, since the SU(2)$_R$ triplet ξ is contained in **126** and ν_R obtains a Majorana mass through $\langle \mathbf{126} \rangle \neq 0$. We expect a somewhat different scenario if $\langle \mathbf{16} \rangle \neq 0$ breaks SO(10) symmetry: since **16** does not couple to ν_R, it does not acquire a Majorana mass. However, we may have an effective interaction, such as

$$\frac{1}{M_{\text{Planck}}} \nu_R \nu_R \langle \mathbf{16} \rangle \langle \mathbf{16} \rangle \tag{58}$$

from the physics at the Planck mass scale, which leads to a Majorana mass of ν_R of the order of M_G^2/M_{Planck}. The breaking of SU(5) to SU(3) × SU(2) × U(1) is caused by **24** of SU(5), i.e. $\langle \mathbf{45} \rangle \neq 0$, $\langle \mathbf{54} \rangle \neq 0$ or $\langle \mathbf{210} \rangle \neq 0$ in terms of the SO(10) language.

Case (iii). The first breaking is caused by $\langle \mathbf{45} \rangle \neq 0$. We need two Higgs' $\mathbf{126} \oplus \mathbf{45}$ or $\mathbf{16} \oplus \mathbf{45}$ to break the intermediate symmetry. In the former case ν_R receives a mass of the order of M_I. In the latter case the Majorana mass is perhaps of the order of M_I^2/M_{Planck} from an argument similar to that given for case (ii).

Case (iv). The first breaking is caused by **54** or **210**, and the second by **16** or **126**. The argument for the ν_R mass is the same as (iii).

Case (v)–(vii). SO(10) is broken by **45**, and the intermediate symmetry by **16** or **126**. The argument for the ν_R mass is the same as (iii) and (iv).

In all cases SU(3) × SU(2)$_L$× U(1) is broken to SU(3) × U(1)$_{\text{em}}$ by a Higgs of **10**, which contains a doublet needed in the Weinberg-Salam theory. With one **10**, however, we are led to an undesirable relation, $m_u = m_d = m_e$, as noted in the previous section. In order to relax this relation we need two or more **10**'s. In this case the relation still holds that

$$m_d = m_e, \quad m_u = m_\nu(\text{Dirac}) \ . \tag{59}$$

The argument for the left-handed neutrino mass and its generation mixing is the same as we described in eqs.(39)–(50) above.

32.2 Prediction of the mass scale

The unification mass scales M_G and M_I can be predicted from low-energy coupling constants with the use of the renormalisation group equation,

$$\mu \frac{d\alpha_i}{d\mu} = -2b_i\alpha_i^2, \tag{60}$$

by requiring that all or a part of the running couplings $\alpha_i(\mu)$'s agree at some energy scales [399]. Here, μ is the energy scale of renormalisation and b_i are one-loop beta function coefficients. In the case of SU(5) grand unification, all three couplings associated with SU(3), SU(2) and U(1) should agree at one scale M_G, leading to a unique prediction of $\sin^2\theta_W$ [399]. The same is also true for case (i)–(iii). For (iv)–(vii) the three low-energy couplings are used to predict both M_G and M_I, i.e., the intermediate mass scale is predicted by the experimentally determined

$\sin^2\theta_W$ as an input parameter [400]–[402]. In the following argument we assume Higgs particles of the minimal extent needed to break symmetries. In the presence of a left-right symmetry, however, we assume the Higgs particles in a left-right symmetric way.

Case of SU(5). Let us first remember the case of SU(5) GUT [399] . From (60), the evolution of the three running coupling constants $\alpha_3(\mu), \alpha_2(\mu)$ and $\alpha_1(\mu)$ from $\mu = M_G$ to m_{Z^0} is given by

$$\alpha_i^{-1}(\mu) = \alpha_i^{-1}(M_G) - 2b_i \log(M_G/\mu) \ , \tag{61}$$

with

$$\begin{aligned} b_3 &= \frac{1}{4\pi}(11 - \frac{4}{3}N_f), \\ b_2 &= \frac{1}{4\pi}(\frac{22}{3} - \frac{4}{3}N_f - \frac{1}{6}N_H), \\ b_1 &= -\frac{1}{4\pi}(\frac{4}{3}N_f + \frac{1}{10}N_H), \end{aligned} \tag{62}$$

where N_H is the number of Higgs doublets. We note that g_1 associated with U(1) in SU(5) is related to g' of $U(1)_Y$ by $g_1 = (5/3)^{1/2}g'$. By requiring unification, that $\alpha_3(M_G) = \alpha_2(M_G) = \alpha_1(M_G)$ (this gives $\sin^2\theta_W \equiv g'^2/(g^2+g'^2) = 3/8$ at $\mu = M_G$), we obtain for $N_H = 1$,

$$\log\left(\frac{M_G}{\mu}\right) = \frac{\alpha_{em}^{-1}(\mu) - (\frac{8}{3})\alpha_3^{-1}(\mu)}{2[(\frac{8}{3})b_3 - b_2 - \frac{5}{3}b_1]} \tag{63}$$

$$\sin^2\theta_W(\mu) = \frac{b_3 - b_2 + \frac{5}{3}(b_2 - b_1)\dfrac{\alpha_{em}(\mu)}{\alpha_3(\mu)}}{\frac{8}{3}b_3 - b_2 - \frac{5}{3}b_1}. \tag{64}$$

These expressions are independent of N_f. With the low-energy parameters [403],

$$\alpha_{em}^{-1}(M_Z) = 127.9 \pm 0.2 \tag{65}$$

$$\alpha_3(M_Z) = 0.12 \pm 0.012, \tag{66}$$

we obtain

$$M_G = 7.6^{+6.6}_{-4.0} \times 10^{14}\text{GeV}, \tag{67}$$

$$\sin^2\theta_W(M_Z) = 0.207^{+0.004}_{-0.003}. \tag{68}$$

The predictions are significantly modified if we include two-loop corrections [404],

$$M_G = 4.5^{+3.4}_{-2.2} \times 10^{14}\text{GeV}, \tag{69}$$

$$\sin^2\theta_W(M_Z) = 0.2102^{+0.0031}_{-0.0037}. \tag{70}$$

We note that neither values agree with the experiment, in particular with

$$\sin^2\theta_W(M_Z)_{\rm exp} = 0.2334 \pm 0.008 \ . \tag{71}$$

See Sect.33 below for more details. The same also applies to case (i)–(iii) of SO(10).

Case (iv) Pati-Salam symmetry SU(4) $\times$ SU(2)$_L\times$ SU(2)$_R$ [384] appears in the intermediate mass scale. Evolution of the three running couplings $\alpha_i(\mu)$, which associate with SU(4), SU(2)$_L$ and SU(2)$_R$, from $\mu = M_{\rm G}$ to M_I obeys

$$\alpha_i^{-1}(\mu) = \alpha_{\rm G}^{-1} - 2b_i \ \log(M_{\rm G}/\mu) \ , \tag{72}$$

where

$$\begin{aligned} b_4 &= \frac{1}{4\pi}(\frac{44}{3} - \frac{4}{3}N_f) + b_4(H), \\ b_{2L} &= \frac{1}{4\pi}(\frac{22}{3} - \frac{4}{3}N_f) + b_{2L}(H), \\ b_{2R} &= \frac{1}{4\pi}(\frac{22}{3} - \frac{4}{3}N_f) + b_{2R}(H) \end{aligned} \tag{73}$$

and $b_i(H)$ is the contribution from Higgs particles that depend further on the model. To calculate $b_i(H)$ we may consider two models as we discussed above:

(a) $\chi(\mathbf{16})$ and $\phi(\mathbf{10})$. The SU(4), SU(2)$_L$ and SU(2)$_R$ contents of χ and ϕ are

$$\begin{aligned} \chi_L &= (\mathbf{4},\ \mathbf{2},\ \mathbf{1}), \\ \chi_R &= (\mathbf{4},\ \mathbf{1},\ \mathbf{2}), \\ \phi &= (\mathbf{1},\ \mathbf{2},\ \mathbf{2^*}). \end{aligned} \tag{74}$$

We thus have

$$b_4(H) = \frac{1}{4\pi}(-\frac{2}{3}), \quad b_{2L}(H) = \frac{1}{4\pi}(-1), \quad b_{2R}(H) = \frac{1}{4\pi}(-1). \tag{75}$$

(b) $\xi(\mathbf{126})$ and $\phi(\mathbf{10})$. In this case we have

$$\begin{aligned} \xi_L &= (\mathbf{10},\ \mathbf{3},\ \mathbf{1}), \\ \xi_R &= (\mathbf{10},\ \mathbf{1},\ \mathbf{3}), \\ \phi &= (\mathbf{1},\ \mathbf{2},\ \mathbf{2^*}) \end{aligned} \tag{76}$$

as well as

$$b_4(H) = \frac{1}{4\pi}(-6), \ b_{2L}(H) = \frac{1}{4\pi}(-7), \ b_{2R}(H) = \frac{1}{4\pi}(-7). \tag{77}$$

For the energy scale from $\mu = M_I$ to M_W the Weinberg-Salam doublet in $\phi(\mathbf{10})$ is the only Higgs particle that contributes to the beta function, and the evolution of the couplings for SU(3), SU(2)$_L$ and U(1) is governed by

$$\alpha_i^{-1}(\mu) = \alpha_i^{-1}(M_I) - 2b_i \ \log(M_I/\mu), \tag{78}$$

with

$$
\begin{aligned}
b_3 &= \frac{1}{4\pi}(\frac{33}{3} - \frac{4}{3}N_f), \\
b_2 &= \frac{1}{4\pi}(\frac{22}{3} - \frac{4}{3}N_f - \frac{1}{6}), \\
b_1 &= \frac{1}{4\pi}(-\frac{4}{3}N_f + \frac{1}{10}).
\end{aligned} \tag{79}
$$

We note that the doublet has the SU(3), $SU(2)_L$, and U(1) content:

$$\phi = (\mathbf{1},\ \mathbf{2},\ 1/2)\ . \tag{80}$$

We then impose a matching condition on (72) and (78) at $\mu = M_I$, that

$$
\begin{aligned}
\alpha_3^{-1}(M_I) &= \alpha_4^{-1}(M_I), \\
\alpha_2^{-1}(M_I) &= \alpha_{2L}^{-1}(M_I), \\
\alpha_1^{-1}(M_I) &= \frac{1}{5}[2\alpha_4^{-1}(M_I) + 3\alpha_{2R}^{-1}(M_I)].
\end{aligned} \tag{81}
$$

With the low-energy parameters of (65), (66) and (71) we obtain for (a)

$$
\begin{aligned}
M_G &= 10^{15.5} - 10^{14.8}\mathrm{GeV}, \\
M_I &= 10^{13.6} - 10^{13.5}\mathrm{GeV}
\end{aligned} \tag{82}
$$

and for (b)

$$
\begin{aligned}
M_G &= 10^{15.4} - 10^{14.7}\mathrm{GeV}, \\
M_I &= 10^{13.7} - 10^{13.5}\mathrm{GeV}.
\end{aligned} \tag{83}
$$

(See Fig. 34a). For the ν_R Majorana mass we then find

$$M_{\nu_R} \simeq \alpha_Y M_I^2/M_{\mathrm{Planck}} \simeq 10^8\alpha_Y \mathrm{GeV} \tag{84}$$

for (a), and

$$M_{\nu_R} \simeq \alpha_Y M_I \simeq 10^{13}\alpha_Y \mathrm{GeV} \tag{85}$$

for (b).

Case (v). Similar to case (iv), we have two models depending upon whether intermediate symmetry SU(4) × $SU(2)_L$ × $U(1)_R$ is broken either by (a) **16** or (b) **126**. The beta function coefficient for α_{1R} is now

$$b_{1R} = \frac{1}{4\pi}(-\frac{4}{3}N_f) + b_{1R}(H). \tag{86}$$

The coefficients b_4 and b_{2L} are the same as in (73) up to $b_i(H)$. (a) $\chi(\mathbf{16})$ and $\phi(\mathbf{10})$ are given by

$$
\begin{aligned}
\chi &= (\mathbf{4},\ \mathbf{1},\ -1/2) \\
\phi &= (\mathbf{1},\ \mathbf{2},\ -1/2),
\end{aligned} \tag{87}
$$

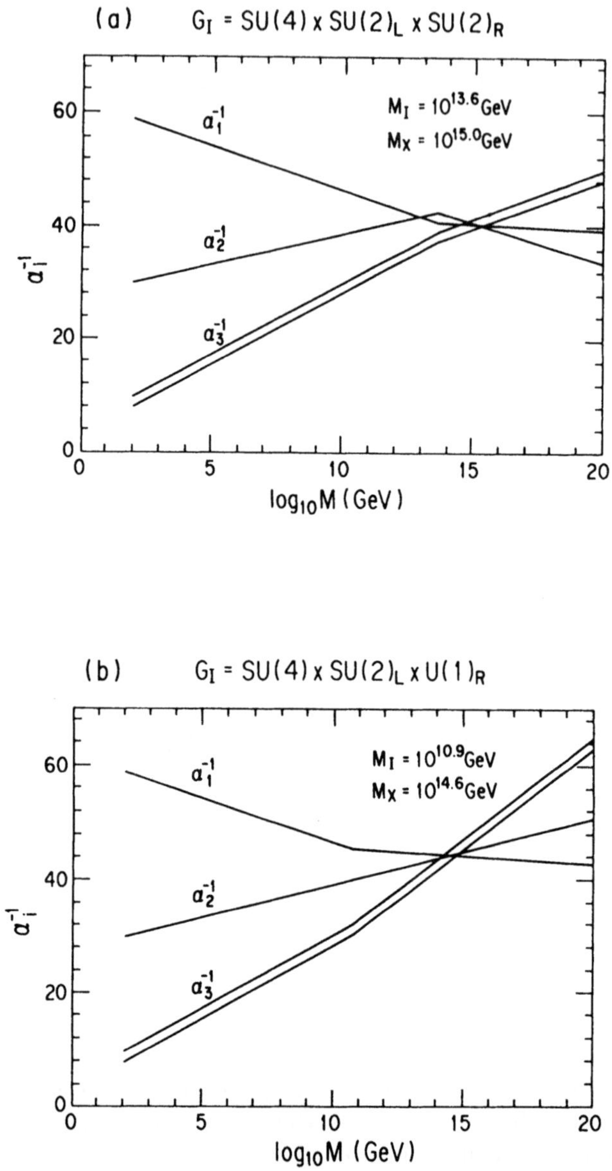

Figure 34: Matching of the three coupling constants in the SO(10) model. Examples are shown for intermediate symmetry (a) SU(4) $\times$ SU(2)$_L\times$ SU(2)$_R$ and (b) SU(4) $\times$ SU(2)$_L\times$ U(1)$_R$. The figures show one-loop calculations.

where the last number is the $U(1)_R$ charge. The contribution to the Higgs loop is

$$b_4(H) = \frac{1}{4\pi}(-\frac{1}{6}),\ \ b_{2L}(H) = \frac{1}{4\pi}(-\frac{1}{6}),\ \ b_{1R}(H) = \frac{1}{4\pi}(-\frac{1}{2}). \tag{88}$$

(b) $\xi(\mathbf{126})$ and $\phi(\mathbf{10})$ are

$$\begin{aligned} \xi &= (\mathbf{10},\ \mathbf{1},\ -1) \\ \phi &= (\mathbf{1},\ \mathbf{2},\ 1/2). \end{aligned} \tag{89}$$

Then,

$$b_4(H) = \frac{1}{4\pi}(-1),\ \ b_{2L}(H) = \frac{1}{4\pi}(-\frac{1}{6}),\ \ b_{1R}(H) = \frac{1}{4\pi}(-\frac{7}{3}). \tag{90}$$

Evolution from M_I to M_W is the same as in case (iv). With the matching condition

$$\alpha_3^{-1} = \alpha_4^{-1},\ \alpha_3^{-1} = \alpha_2^{-1},\ \alpha_1^{-1} = \frac{1}{5}(2\alpha_4^{-1} + 3\alpha_{1R}^{-1}) \tag{91}$$

at $\mu = M_I$, we obtain (Fig. 34b)

$$\begin{aligned} (a)\ \ M_G &= 10^{15.1} - 10^{14.6}\text{GeV}, \\ M_I &= 10^{12.1} - 10^{12.3}\text{GeV} \qquad (92) \\ (b)\ \ M_G &= 10^{14.9} - 10^{14.4}\text{GeV} \\ M_I &= 10^{10.6} - 10^{11.1}\text{GeV}. \qquad (93) \end{aligned}$$

We see that M_I is substantially smaller than that in (iv). This gives for the ν_R mass

$$(a)\ \ M_{\nu_R} \simeq \alpha_Y \frac{(10^{12})^2}{M_{\text{Planck}}} \simeq 10^5 \alpha_Y \text{GeV} \tag{94}$$

$$(b)\ \ M_{\nu_R} \simeq 10^{11}\alpha_Y \text{GeV} \tag{95}$$

Case (vi). Intermediate symmetry is $SU(3) \times SU(2)_L \times SU(2)_R \times U(1)_{B-L}$. The renormalisation group coefficients are

$$\begin{aligned} b_3 &= \frac{1}{4\pi}(\frac{33}{3} - \frac{4}{3}N_F), \\ b_{2L} &= b_{2R} = \frac{1}{4\pi}(\frac{22}{3} - \frac{4}{3}N_F) + b_{2L,R}(H), \\ b_{B-L} &= \frac{1}{4\pi}(-\frac{4}{3}N_F) + b_{R-L}(H). \end{aligned} \tag{96}$$

For the Higgs content we have again two cases:
(a) $\chi(\mathbf{16})$ and $\phi(\mathbf{10})$.

$$\begin{aligned} \chi_L &= (\mathbf{1},\ \mathbf{2},\ \mathbf{1},\ 1)\ \in \chi(\mathbf{16}), \\ \chi_R &= (\mathbf{1},\ \mathbf{1},\ \mathbf{2},\ 1)\ \in \chi(\mathbf{16}), \\ \phi &= (\mathbf{1},\ \mathbf{2},\ \mathbf{2}^*,\ 0)\ \in \phi(\mathbf{10}), \end{aligned} \tag{97}$$

yielding

$$b_3(H) = 0, \quad b_{2L,R} = \frac{1}{4\pi}(-\frac{1}{2}), \quad b_{B-L} = \frac{1}{4\pi}(-\frac{1}{2}). \tag{98}$$

(b) $\xi(\mathbf{126})$ and $\phi(\mathbf{10})$.

$$\begin{aligned} \xi_L &= (\mathbf{1},\ \mathbf{3},\ \mathbf{1},\ +2) \ \in \xi(\mathbf{126}), \\ \xi_R &= (\mathbf{1},\ \mathbf{1},\ \mathbf{3},\ +2) \ \in \xi(\mathbf{126}), \\ \phi &= (\mathbf{1},\ \mathbf{2},\ \mathbf{2}^*,\ 0) \ \in \xi(\mathbf{10}). \end{aligned} \tag{99}$$

For $b_i(H)$,

$$b_3(H) = 0, \ b_{2L,R} = \frac{1}{4\pi}(-1), \ b_{B-L} = \frac{1}{4\pi}(-3). \tag{100}$$

Imposing the matching condition,

$$\alpha_3^{-1} = \alpha_3^{-1}, \ \alpha_{2L}^{-1} = \alpha_{2R}^{-1}, \ \alpha_1^{-1} = \frac{1}{5}(2\alpha_{B-L}^{-1} + 3\alpha_{2R}^{-1}) \tag{101}$$

we obtain from (72) and (96) for $M_I \le \mu \le M_G$ and (78) and (79) for $M_W \le \mu \le M_I$,

$$\begin{aligned} (a) \ M_G &= 10^{16.9} - 10^{15.8} \text{ GeV}, \\ M_I &= 10^{10.2} - 10^{10.9} \text{ GeV} \end{aligned} \tag{102}$$

$$\begin{aligned} (b) \ M_G &= 10^{16.2} - 10^{15.2} \text{ GeV}, \\ M_I &= 10^{9.6} - 10^{10.5} \text{ GeV}. \end{aligned} \tag{103}$$

Hence,

$$(a) \ M_{\nu_R} = 10^3 \alpha_Y \text{ GeV} \tag{104}$$

$$(b) \ M_{\nu_R} = 10^{10} \alpha_Y \text{ GeV}. \tag{105}$$

Case (vii). With intermediate symmetry SU(3) × $SU(2)_L$× $U(1)_R$× U(1), $\sin^2\theta_W$ is obviously smaller than the value predicted by SU(5) grand unification. This case does not agree with the experiments.

33 Phenomenology of grand unified theories

The two basic tests of grand-unification models are the lifetime of proton decays and the Weinberg angle. The minimal SU(5) model predicts $\tau_p = 10^{29\pm2}$yr and $\sin^2\theta_W \simeq 0.210 \pm 0.004$ [395]. By 1985 the minimal version of SU(5) model had already become disfavoured due to the null result that $\tau(p \to e^+\pi^0) > 3 \times 10^{32}$yr at the 90% C.L. (see, e.g. [405]; the current lower limit is $> 5.5 \times 10^{32}$yr [406]). It also turned out that minimal SU(5) gives a $\sin^2\theta_W$ that is too small compared with the value determined from neutral-current phenomenology (e.g. [407]). After the precision determination of $\sin^2\theta_W$ became available from the LEP experiments [69], the SU(5) prediction of $\sin^2\theta_W$ becomes off from the experiment at the $7-8\sigma$ level (cf. (70) with (71)). Equivalently, if the experimentally determined low-energy

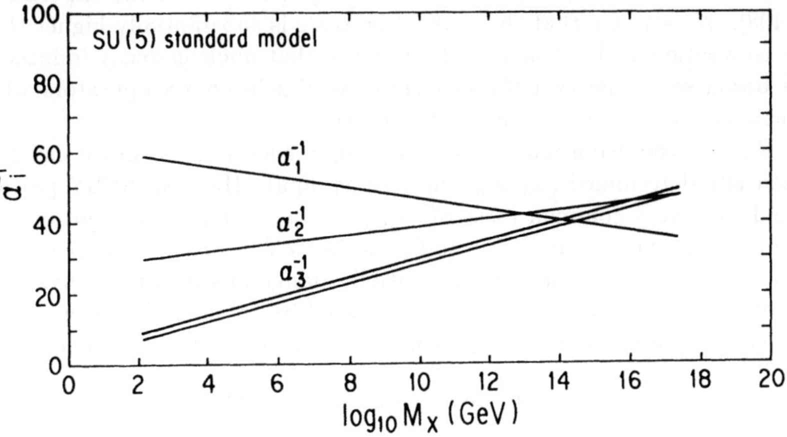

Figure 35: Matching of the three running couplings in the SU(5) standard model of grand unification using two-loop calculations.

parameters ($\alpha_3, \alpha_{em}, \sin^2\theta_W$) are used as input, the three running couplings, α_1, α_2 and α_3, do not match at a single unification energy scale (Fig. 35) [403, 408].

There are two directions for eliminating this difficulty: one is to consider a supersymmetric extension of SU(5), so that $\alpha_i(\mu)$ is affected by the contribution from SUSY particles; the other is to take SO(10) models in which the three couplings are unified in two steps (as shown above). In the following we review the phenomenology of those two viable GUT models.

33.1 SU(5) SUSY GUT

In SUSY GUT supersymmetric particles contribute to the coefficients of beta functions for α_i by the amount [409]

$$\begin{aligned}
\Delta b_3 &= \frac{1}{4\pi}(-2 - \frac{2}{3}N_f), \\
\Delta b_2 &= \frac{1}{4\pi}(-\frac{4}{3} - \frac{2}{3}N_f - \frac{1}{3}N_H), \\
\Delta b_1 &= \frac{1}{4\pi}(-\frac{2}{3}N_f - \frac{1}{5}N_H)\,,
\end{aligned} \tag{106}$$

where $N_H = 2$ for the minimal SUSY GUT. With these modifications the evolution of $\alpha_3(\mu)$ slows down relative to $\alpha_2(\mu)$ and $\alpha_1(\mu)$, and then the three gauge couplings α_3, α_2 and α_1 meet at a single energy scale, which is higher than that of the SU(5) GUT. In parallel to (67), if α_3 and α_{em} are used, we obtain $M_G = 1.4^{+1.7}_{-1.0} \times 10^{16}$GeV and $\sin^2\theta_W = 0.2370^{+0.0031}_{-0.0027}$ from (63)–(64). With two-loop corrections the minimal SU(5) GUT predicts

$$\sin^2\theta_W(M_Z) = 0.2334^{+0.0031}_{-0.0027}, \tag{107}$$

$$M_G = 1.9^{+1.7}_{-1.0} \times 10^{16}\text{GeV}. \tag{108}$$

The predicted value of (107) is in excellent agreement with the experiment (71) [403, 408]. We also see that the unification scale is substantially higher than that of the conventional SU(5) model. This means that nucleon decay induced by the gauge boson exchange (via the conventional dimension six operators which are proportional to M_G^{-2}) is sufficiently suppressed.

In Fig. 36 the evolution of the three gauge couplings is shown using the experimentally determined $(\alpha_3, \alpha_{\rm em}, \sin^2\theta_W)$ as input. Here, all SUSY particles are assumed to have a common mass of $m_{\rm SUSY} = m_{Z^0}$. The agreement of the three couplings would not be spoiled in so far as $100{\rm GeV} \le m_{\rm SUSY} \le 10{\rm TeV}$.

For nucleon decays there are potentially dangerous dimension-five operators (suppressed only by M_G^{-1}), which are generated by colour-triplet Higgs (H_c) multiplets [410]. The mass of this H_c, however, is suggested to lie between [412]

$$2.4 \times 10^{13}{\rm GeV} \le m_{H_c} \le 2.3 \times 10^{17}{\rm GeV} \tag{109}$$

from the threshold effect in a renormalisation group analysis by requiring that unification of the gauge couplings at a single mass scale takes place for the given range of $m_{\rm SUSY}$ =0.1–10TeV. Allowing for other uncertainties (e.g., Yukawa coupling, SUSY breaking parameters etc.), it is shown that a substantially wide parameter range exists, with which the minimal SUSY SU(5) model (with the SUSY particle lighter than 1TeV) gives a nucleon decay lifetime that is consistent with the present experimental limit [412, 413].

In the framework of SUSY SU(5) no predictions are made on the neutrino mass, since the right-handed neutrino is a singlet of SU(5). However, SUSY SU(5) GUT may be embedded into SUSY SO(10) GUT with successful features of SU(5) SUSY GUT being retained. In this case, there are two possibilities for M_{ν_R}, depending on the Higgs representation that breaks SO(10). For $\chi(\mathbf{16})$ we expect

$$M_{\nu_R} \simeq \alpha_Y \frac{\langle\chi\rangle^2}{M_{\rm Planck}} \simeq 10^{13}\alpha_Y {\rm GeV}, \tag{110}$$

and for $\xi(\mathbf{126})$

$$M_{\nu_R} \simeq \alpha'_Y \langle\xi\rangle \simeq 10^{16}\alpha'_Y {\rm GeV}. \tag{111}$$

33.2 SO(10) GUT

The two difficulties with the minimal SU(5) GUT can be avoided in SO(10). The correct Weinberg angle can be derived by assuming a reasonable intermediate energy scale M_I. The unification scale M_G is significantly higher than that with SU(5), which makes the lifetime of nucleon decay consistent with the experimental limit [400]–[402] (see also [414]).

Unfortunately, precise predictions of $\sin^2\theta_W$ and M_G cannot be made with SO(10) GUT. The reason is that there are a large number of scalar bosons with masses on the order of the unification scales. Their masses may lie above or below the unification scales by an order of magnitude or so, and these scalar particles contribute to the beta function differently. Although the threshold effect from a

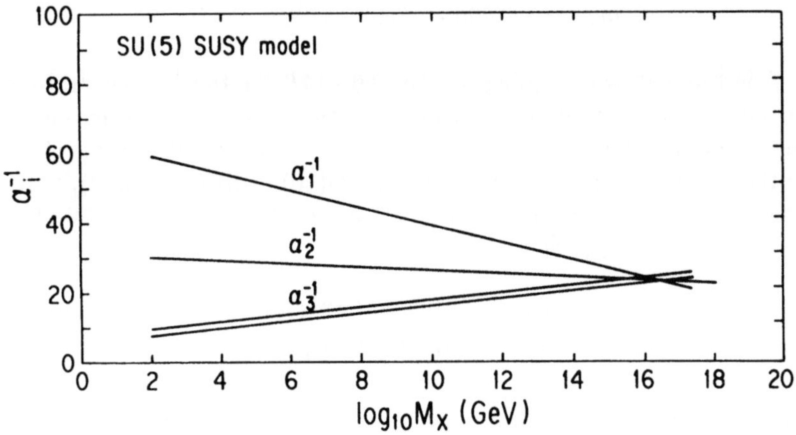

Figure 36: Matching of the three running couplings in the minimal SUSY SU(5) model (two-loop result). The masses of SUSY particles are all assumed to be M_Z.

single scalar boson is very small, the contribution from hundreds of such bosons in SO(10) causes large uncertainties, also leading to large uncertainties in the low-energy predictions [402, 415]. Indeed, for the symmetry-breaking cases discussed in Sect.32 the uncertainties are estimated to be as large as 4 orders of magnitude in nucleon lifetime and ± 0.06 in $\sin^2\theta_W$ [415], assuming that the scalar bosons lie at energies that are $10^{\pm 1}$ times the symmetry-breaking scale. These threshold uncertainties cancel out for a special class of intermediate symmetry group, such as $SU(4)\times SU(2)_L \times SU(2)_R \times P$ [416], with P a Z_2 symmetry corresponding to an interchange of $SU(2)_L$ and $SU(2)_R$ [417]. Such a cancellation, however, does not occur for more general left-right symmetric cases.

A more detailed analysis has been made for cases (iv) and (vi) of Sect.32 [418, 420]. For case (iv), $SO(10) \to SU(4) \times SU(2)_R \times SU(2)_L \to G_0$, Higgs multiplets of **210**, **126** and **10** are assumed. It is also assumed that a submultiplet of **126**, $\Delta(\mathbf{10},\mathbf{1},\mathbf{3})$, and $\phi(\mathbf{1},\mathbf{2},\mathbf{2}^*)$ of **10** lie near to the intermediate scale $M_I \times 10^{\pm 1}$, and that other components of these multiplets are around M_G. A submultiplet $\varphi(\mathbf{1},\mathbf{2})$ of $\phi(\mathbf{1},\mathbf{2},\mathbf{2}^*)$ is taken to have a mass of the order of the electroweak energy scale. With the two-loop renormalisation effect [421] taken into account, the authors of ref. [420] obtained

$$\begin{aligned} M_G &= 10^{15.8\pm 0.2^{+0.8}_{-0.7}}\mathrm{GeV} \\ M_I &= 10^{11.5\pm 0.03^{+2.8}_{-1.5}}\mathrm{GeV}, \end{aligned} \tag{112}$$

where the first errors are from the uncertainty in the low-energy parameters and the second are due to the threshold effects of Higgs particles. This gives a proton decay lifetime of $\tau_p \simeq 1.6 \times 10^{35\pm 0.9^{+3.2}_{-2.8}}$ years. If the see-saw mechanism is invoked, the intermediate-scale M_I leads to a neutrino mass spectrum of

$$m_{\nu_e} = 6 \times 10^{-12} - 2 \times 10^{-7}\mathrm{eV} \cdot f_e^{-1},$$

$$
\begin{aligned}
m_{\nu_\mu} &= 4\times10^{-7} - 10^{-2}\mathrm{eV}\cdot f_\mu^{-1}, \\
m_{\nu_\tau} &= 0.006 - 180\mathrm{eV}\cdot f_\tau^{-1},
\end{aligned}
\tag{113}
$$

where f_i is the Yukawa coupling for **16** · **16** · **126** for the i-th generation right-handed neutrinos. This ν_μ mass appears in the right range to account for the solar-neutrino problem with the MSW scenario (see eqs.(V.167, 168)) [419, 420].

For the symmetry-breaking chain of (vi), SO(10) $\to$ SU(3)$\times$SU(2)$_L\times$SU(2)$_R\times$ U(1), the Higgs scalar multiplets are taken to be **45**, **54**, **126** and **10** of SO(10). A similar analysis [420] yields

$$
\begin{aligned}
M_G &= 10^{15.8\pm0.25\pm0.2}\mathrm{GeV}, \\
M_I &= 10^{9\pm0.18{}^{+0.3}_{-0.0}}\mathrm{GeV},
\end{aligned}
\tag{114}
$$

which predicts the proton lifetime to be $1.6\times10^{35\pm1\pm0.8}$ years. The neutrino masses are

$$
\begin{aligned}
m_{\nu_e} &= 10^{-7} - 10^{-6}\mathrm{eV} f_e^{-1}, \\
m_{\nu_\mu} &= 10^{-2} - 10^{-1}\mathrm{eV} f_\mu^{-1}, \\
m_{\nu_\tau} &= 1 - 10^{3}\mathrm{eV} f_\tau^{-1}.
\end{aligned}
\tag{115}
$$

The ν_μ mass, thus predicted, is consistent with the large-angle solution (V.168) of the MSW scenario, but not with the small-angle solution as long as we take $f_\mu \lesssim 1$.

As we emphasised before, these simple models naturally lead to $\theta_{e\mu} \simeq \theta_{\mathrm{Cabibbo}}$, which is not supported by experiments (V.167, 168). If one introduces **126** Higgs scalars, which couple only to the (2,2) component of the fermion mass matrix [422], we have the mass relation [423],

$$
\begin{aligned}
m_d &\simeq 3m_e \\
m_s &\simeq \frac{1}{3}m_\mu,
\end{aligned}
\tag{116}
$$

which is favoured from phenomenology, and the mixing angle [422]:

$$
\theta_{e\mu} \simeq \frac{1}{3}\theta_{\mathrm{Cabibbo}}. \tag{117}
$$

Unfortunately, this mixing angle is still too large to account for the solar-neutrino experiments.

It is still possible to obtain the correct lepton mixing angles by either tuning f_i's so that $M_{\nu_{Ri}}$ compensates the hierarchy of the charged fermion mass matrix or by assuming an artificial mixture of Higgs' **10** + **126**, which give the fermions a mass term. However, this means that the predictive power of the simple SO(10) models is lost. The bottom line is that simple SO(10) models with a predictive power fail to account for the lepton mixing pattern that is inferred from the MSW solution to the solar neutrino problem.

34 Seesaw mechanism with other symmetries

34.1 Model with U(1) symmetry

It is possible to identify the symmetry which protects the right-handed neutrino from having a mass with some global or local symmetry proposed in other contexts.

For instance, we may take Peccei–Quinn (PQ) U(1) [424] as protecting symmetry. We usually assign the PQ charge to be

$$\begin{array}{lclccc} q_L & = & \begin{pmatrix} u \\ d \end{pmatrix}_L & 1 & \begin{array}{c} u_R \\ d_R \end{array} & \begin{array}{c} -1 \\ -1 \end{array} \\ \ell_L & = & \begin{pmatrix} \nu \\ e \end{pmatrix}_L & 1 & e_R & -1. \end{array} \tag{118}$$

We may add the right-handed neutrino

$$\nu_R \quad -1. \tag{119}$$

If we introduce a Higgs χ with a PQ charge of +2 to break PQ symmetry, we expect to have a Yukawa coupling of the form

$$\mathcal{L}_Y = h\overline{\nu_R^c}\nu_R\chi. \tag{120}$$

Thus, ν_R acquires the mass

$$M_{\nu_R} = h\langle\chi\rangle \tag{121}$$

as χ develops a vacuum expectation value. An interesting point regarding this model is that it has predictive power for the neutrino mass: the vacuum expectation value of χ is constrained to lie between 10^9 and 10^{12}GeV from astrophysics and cosmology arguments (see Sect.36.3); therefore,

$$M_{\nu_R} \simeq h \times (10^9 - 10^{12})\text{GeV}. \tag{122}$$

This value is close to what is required from the MSW solution to the solar-neutrino problem. In practice, however, there are some subtleties in constructing a model; they will be discussed in detail in Sect. 36. If U(1) symmetry is identified with that of lepton number, the model is essentially the singlet Majoron model [425].

34.2 Model with family symmetry

Family (horizontal) symmetry [426, 427] $\mathrm{SU}(3)_F$ may also be used as a protecting symmetry. We usually take the assignment

$$\begin{array}{lcl} q_L = \begin{pmatrix} u \\ d \end{pmatrix}_L & \mathbf{3} & \begin{array}{cccc} u_R & \mathbf{3} & \text{or} & \mathbf{3}^* \\ d_R & \mathbf{3} & \text{or} & \mathbf{3}^* \end{array} \\ \ell_L = \begin{pmatrix} \nu \\ e^- \end{pmatrix}_L & \mathbf{3} & \begin{array}{cccc} \nu_R & \mathbf{3} & \text{or} & \mathbf{3}^* \\ e_R & \mathbf{3} & \text{or} & \mathbf{3}^* \end{array} \end{array} \tag{123}$$

The Majorana mass term $M\nu_R\nu_R$ is forbidden by $\mathrm{SU}(3)_F$. In order to break $\mathrm{SU}(3)_F$ symmetry completely we should prepare Higgs particles of (i) one **6**, (ii) two **8**'s, or (iii) two $\mathbf{3}^*(\mathbf{3})$'s. In case (i) or (iii) the Higgs particle gives ν_R a Majorana mass. If we introduce χ (**6** or $\mathbf{6}^*$) to break family symmetry, a Majorana mass arises from $h\nu_R^i\nu_R^j\chi_{ij}$, and it is

$$[M_{\nu_R}]_{ij} = h\langle\chi\rangle_{ij}. \tag{124}$$

The case with symmetry breaking by triplet Higgs' (iii) leads to a qualitatively different scenario [428]; it will be discussed separately in Sect.35. The Dirac mass is given by two $\mathrm{SU}(2)_L$ doublet Higgs', either $(\mathbf{1},\mathbf{2}) \oplus (\mathbf{8},\mathbf{2})$ or $(\mathbf{6},\mathbf{2}) \oplus (\mathbf{6},\mathbf{2})$ of $(\mathrm{SU}(3)_F, \mathrm{SU}(2)_L)$. Two Higgs' are necessary to give the correct fermion mass spectrum.

An interesting feature with the choice $(\mathbf{6},\mathbf{2}) \oplus (\mathbf{6},\mathbf{2})$ is that it leads to $U_{KM} \simeq U_{\text{lepton}}$. Since these two Higgs' have $Y = 1/2$ and $Y = -1/2$, respectively, the Yukawa couplings are

$$\begin{aligned}\mathcal{L} &= f_d\bar{q}_L\phi_1 d_R + f_u\bar{q}_L\phi_2 u_R \\ &\quad + f_e\bar{\ell}_L\phi_1 e_R + f_\nu\bar{\ell}_L\phi_2\nu_R + h.c. \,.\end{aligned} \tag{125}$$

Both d-type quarks and leptons acquire masses from the same Higgs $\langle\phi_1\rangle$, and u-type quarks and neutrinos (Dirac mass) from $\langle\phi_2\rangle$, so that the quark and lepton mass matrices are simultaneously diagonalised.

Complete symmetry breaking generally produces 8 Nambu-Goldstone bosons (φ), which are called familons [427]. Such bosons couple to flavour-changing neutral currents as

$$\begin{aligned}\mathcal{L} &= \frac{1}{v_{e\mu}}\bar{e}\gamma^\mu(a+b\gamma_5)\mu\partial_\mu\varphi \\ &\quad + \frac{1}{v_{ds}}\bar{d}\gamma^\mu(a'+b'\gamma_5)s\partial_\mu\varphi + \ldots + h.c. \,, \\ &\quad (a^2+b^2 = a'^2+b'^2 = 1),\end{aligned} \tag{126}$$

which cause μ or K to decay

$$\mu \rightarrow e + \varphi \,, \tag{127}$$

$$K \rightarrow \pi + \varphi \,, \tag{128}$$

where $v_{ij} = \langle\chi\rangle$ is the relevant symmetry-breaking scale. The coupling of φ to quarks and leptons is suppressed by $1/v_{ij}$ due to the Goldstone theorem; the existence of such zero mass particles is thus harmless, provided that v_{ij} is sufficiently large. The decay rates are given by

$$\Gamma(\mu^+ \rightarrow e^+ + \varphi) = m_\mu^3(|a^2| + |b^2|)/16\pi v_{e\mu}^2 \,, \tag{129}$$

$$\Gamma(K^+ \rightarrow \pi^+ + \varphi) = m_K^3|a'|^2/16\pi v_{ds}^2 \,. \tag{130}$$

From the experimental limits,

$$Br(\mu^+ \rightarrow e^+ + \varphi) < 2.6 \times 10^{-6}[429], \tag{131}$$

$$Br(K^+ \rightarrow e^+ + \varphi) < 6.4 \times 10^{-9}[430], \tag{132}$$

we obtain

$$v_{e\mu} > 1.0 \times 10^{10}\text{GeV} \tag{133}$$
$$v_{ds} > 7.5 \times 10^{10}\delta\text{GeV}, \tag{134}$$

where $\delta = A/(|A|^2 + |B|^2)^{1/2}$, which depends further on the details of the model [431]. These values of the lower limits roughly agree with the scale presumed for the PQ symmetry breaking. For this scale the mass of the left-handed neutrino is $m_{\nu_\tau} \simeq m_\tau{}^2/M_{\nu_R} \lesssim h^{-1} \times (1\text{eV})$ if the Dirac mass is taken to be the charged lepton mass. If the Dirac mass is identified with the quark mass, e.g., as in SO(10), then $m_{\nu_\tau} \lesssim h^{-1} \times (1\text{keV})$.

An important feature with this model is that the presence of φ makes massive neutrinos unstable [427]. For effective coupling of the form

$$\mathcal{L} = \frac{1}{v_{ij}} \overline{\nu_i} \gamma^\mu (A + B\gamma_5) \nu_j \partial_\mu \varphi, \quad |A|^2 + |B|^2 = 1 \tag{135}$$

the decay rate of ν_j is given by

$$\begin{aligned} \Gamma &= m_{\nu_j}^3 / 16\pi v_{ij}^2 \\ &= (3.5 \times 10^{15}\text{s})^{-1} (m_{\nu_j}/1\text{keV})^3 (v_{ij}/10^{10}\text{GeV})^{-2}. \end{aligned} \tag{136}$$

This decay process could be much faster than $\nu' \to \nu + \gamma$ (Chap. VII) and sufficiently fast to avoid the cosmological constraint, if $\langle v \rangle$ is not too large. The condition that the energy density carried by decay products of unstable neutrino does not exceed the closure value (V.206) leads to

$$v \leq 1.0 \times 10^{10} (m_\nu/1\text{keV})^{1/2} h^2 \text{GeV}. \tag{137}$$

The cosmological constraint may be avoided if $m_\nu >$1keV, yet consistently with the laboratory limits.

We may also use gauge symmetry (horizontal gauge symmetry) [432] instead of global symmetry. Such a model differs from the one with global symmetry in two respects: the theory has a triangle anomaly and the existence of ν_R is inevitably required from internal consistency of the theory [433]. [We must choose **3** for right-handed fermions in (123).] The second point is that the symmetry-breaking scale does not receive a constraint as strong as in the case with global symmetry, because the Nambu-Goldstone bosons that would cause the decay (126), (127) or excess stellar cooling (Sect.36.3) do not exist. The strongest bounds on this model are derived from the $K^0 - \overline{K^0}$ mass difference.

Extensions of the idea presented here can be carried out straightforwardly to models with other symmetries.

35 Model for the fourth generation neutrino

If it happens that M_{ν_R} vanishes in Eq.(III.43) for the n-th generation, the corresponding left-handed neutrino appears as a Dirac neutrino with $m_\nu = m$ [428, 434].

Such a case is naturally realised if family $\mathrm{SU}(n)_F$ symmetry (either global or local) is broken by Higgs scalars in the fundamental representation [428].

Let us assume that the left-handed leptons and quarks ℓ_L transform as the $\boldsymbol{n}$ representation of $\mathrm{SU}(n)_F$ and the right-handed ν_R, e_R, u_R and d_R as either $\boldsymbol{n}$ or $\boldsymbol{n}^*$ of $\mathrm{SU}(n)_F$. (For local $\mathrm{SU}(n)_F$ we must take $\boldsymbol{n}$ for the right-handed fermions to cancel the gauge anomalies, however). A complete breaking of $\mathrm{SU}(n)_F$ is caused by $n-1$ Higgs scalars $\xi^{(a)}$ in the fundamental $\boldsymbol{n}$ representation of $\mathrm{SU}(n)_F$ (they are assumed to be singlets of $\mathrm{SU}(2)_L$). The vacuum expectation values of $\xi^{(a)}(a = 1 \sim n-1)$ can generally be written

$$\langle\xi^{(1)}\rangle = \begin{pmatrix} v_1^{(1)} \\ 0 \\ \vdots \\ 0 \end{pmatrix}, \ \ldots, \ \langle\xi^{(n-1)}\rangle = \begin{pmatrix} v_1^{(n-1)} \\ \vdots \\ v_{n-1}^{(n-1)} \\ 0 \end{pmatrix} . \tag{138}$$

The Higgs $\xi^{(a)}$ do not couple directly to ν_R and, hence, it is natural to suppose effective interactions, the lowest-dimensional form of which are written as

$$\frac{f_{ab}}{M}(\xi_i^{(a)}\nu_R^i)(\xi_j^{(b)}\nu_R^j), \tag{139}$$

where i and j denote the family indices, and M is the effective mass scale. Thus, after the $\mathrm{SU}(n)_F$ breaking, ν_R acquires the Majorana mass term

$$M_{\nu_R} = \frac{f_{ab}}{M}\langle\xi^{(a)}\rangle\langle\xi^{(b)}\rangle \ . \tag{140}$$

This $n \times n$ matrix has vanishing matrix elements for either $i = n$ or $j = n$ and, therefore, one of the eigenvalues vanishes.

The Dirac mass terms for quarks and leptons are given by the Yukawa couplings of $\mathrm{SU}(2)_L$ doubled Higgs scalars ϕ's which break $\mathrm{SU}(2)_L\times$ U(1) down to $\mathrm{U(1)_{em}}$. The transformation property of ϕ under $\mathrm{SU}(n)_F$ depends upon whether $\boldsymbol{n}$ or $\boldsymbol{n}^*$ is chosen for the right-handed fermions ψ_R. If ψ_R is taken to be $\boldsymbol{n}$, we should prepare an $\mathrm{SU}(n)_F$ adjoint ϕ^i_j and a singlet ϕ. The Yukawa couplings in the lepton sector are given by

$$\overline{e_R}^j(f_e^{(1)}\tilde{\phi}^i_j + f_e^{(2)}\tilde{\phi}\delta^i_j)\ell_{Li} + \overline{\nu_R}^j(f_\nu^{(1)}\phi^i_j + f_\nu^{(2)}\phi\delta^i_j)\ell_{Li} + h.c. \tag{141}$$

For $\psi_R = \boldsymbol{n}^*$, we need two $\mathrm{SU}(n)_F$ symmetric $(n(n+1)/2)$ Higgs ϕ^{ij} and ϕ'^{ij}. The Yukawa couplings are

$$\overline{e}_{Ri}(f'_e\phi'^{ij})\ell_{Lj} + \overline{\nu}_{Ri}(f'_\nu\phi^{ij})\ell_{Lj} + h.c. \tag{142}$$

There is a critical difference in the symmetry structure of the Higgs sector between these two choices. For $\psi_R = \boldsymbol{n}^*$, there remains an unwanted $\mathrm{U}(n)_F^\xi \times \mathrm{U}(n)_F^\phi$ symmetry within the class of renormalisable interactions, where the two $\mathrm{U}(n)_F$'s correspond to rotations of ξ_i^a and ϕ^{ij} (and ϕ'^{ij}) with respect to family indices. Since $\mathrm{U}(n)_F^\phi$ is spontaneously broken by the condensation of ϕ^{ij} and

ϕ'^{ij}, there appear n^2 Nambu-Goldstone bosons, which, however, are excluded by experiments, because their coupling strengths are set by the weak scale. However, we may get rid of unwanted $\mathrm{U}(n)_F^\phi$ symmetry by supposing an effective operator,

$$\frac{1}{M^2}\phi^{ij}\phi'^{k\ell}\xi_i^{(a)}\xi_j^{(b)}\xi_k^{(c)}\xi_\ell^{(d)}. \tag{143}$$

With this interaction the n^2 Nambu-Goldstone bosons acquire masses of the order of $\langle\xi\rangle^2/M$, which is usually large enough.

For $\psi_R = \boldsymbol{n}$, on the other hand, unwanted $\mathrm{U}(n)_F^\phi$ is broken by the renormalisable interaction,

$$\phi_j^i\phi^*\xi_i^{(a)}\xi_{(b)}^{*j}. \tag{144}$$

In either of the two models, the seesaw mechanism works only for the family $i = 1 \sim n-1$ to give small Majorana masses to left-handed neutrinos,

$$m_\nu \simeq m_D M_{\nu_R}^{-1} m_D^T\ , \tag{145}$$

with

$$m_D = f_\nu^{(1)}\langle\phi_j^i\rangle + f_\nu^{(2)}\langle\phi\rangle \qquad (\text{for} \quad \psi_R = \boldsymbol{n}) \tag{146}$$

or

$$m_D = f'_\nu\langle\phi^{ij}\rangle \qquad (\text{for} \quad \psi_R = \boldsymbol{n}^*)\ . \tag{147}$$

For the neutrino of the n-th generation, however, the seesaw mechanism does not work and its mass is given by the Dirac mass of the n-th generation ($n = 4$), which, if exists at all, is of the order of a few $\times$ 100 GeV, the mass supposed for the fourth generation heavy lepton or quark ("partially broken" seesaw mechanism).

If we suppose that the family symmetry is broken at $\sim 10^{15}$GeV and $M \sim M_{\mathrm{planck}}$, we have $M_{\nu_R} \sim 10^{11}$GeV. If we take the Dirac masses of the neutrinos to be those of the charge 2/3 quarks, we expect $m_{\nu_e} \sim 10^{-8}$eV, $m_{\nu_\mu} \sim 10^{-4}$eV and $m_{\nu_\tau} \sim 1$–10eV while for the fourth generation $m_{\nu'_\tau} \gtrsim 100$ GeV.

36 Peccei–Quinn Symmetry and the Neutrino Mass

36.1 Strong CP problem and Peccei–Quinn Symmetry

The presence of the instanton in SU(3) gauge theory leads us to suppose that QCD vacuum has a non-trivial topological structure, with the true strong interaction Lagrangian

$$\mathcal{L} = \mathcal{L}_{\mathrm{QCD}} + \frac{1}{32\pi^2}\theta F_{\mu\nu}^a\tilde{F}^{a\mu\nu}, \tag{148}$$

where a stands for the colour index of the gluon field [435, 436], and $\tilde{F}^{\mu\nu} = \frac{1}{2}\varepsilon^{\mu\nu\rho\sigma}F_{\mu\nu}$. The second term is a total divergence of the axialvector current. Diagonalisation of the quark mass matrix, $V^{-1}m_qU = m_q^{\mathrm{diag}}$, is also a chiral transformation that rotates the phase of the fermion fields by $\psi \to \exp(-\frac{i}{2}\gamma_5 \arg\det m_q)\psi$, which also contributes to θ. Hence θ is modified to

$$\theta_{\mathrm{eff}} = \theta + \arg\ \det \parallel m_d m_u \parallel, \tag{149}$$

taking account both $Q = 2/3$ and $Q = -1/3$ quark matrices. $\theta_{\rm eff}$ labels the QCD vacuum, and the vacuum of any $\theta_{\rm eff}$ must be equally acceptable.

On the other hand, the second term of (148) is $\propto \vec{E} \cdot \vec{H}$ that violates P and CP (P : $\vec{E} \to -\vec{E}, \vec{H} \to \vec{H}$, and CP : $\vec{E} \to \vec{E}, \vec{H} \to -\vec{H}$). Hence, the $\theta_{\rm eff}$ must be very small to make the theory consistent with the experiment. In fact, the parameter $\theta_{\rm eff}$ is bounded by the absence of the electric dipole moment of the neutron ($d < 1.2 \times 10^{-25} e$ cm [437]) to be [438]

$$|\theta_{\rm eff}| < 10^{-10} - 10^{-9}. \tag{150}$$

The reason for this smallness is unknown, and this is often referred to as the strong CP problem [435]. ($|\theta_{\rm eff}|$ induced from the CP-violating KM matrix is of the order of 10^{-14}–10^{-15}.)

The most well accepted solution to this problem is the one proposed by Peccei and Quinn [424]. Let us consider the standard electroweak model extended to include two Higgs doublets ϕ_1 and ϕ_2 . Here ϕ_1 is supposed to give a mass to the electron and the d quark, and ϕ_2 to the u quark instead of ϕ_1^* [see (II.32)]. This extension allows us to have a new chiral U(1) symmetry. In parallel to (II.32) we write the Yukawa coupling

$$\begin{aligned} \mathcal{L}_Y &= -f_e \bar{\ell}_L e_R \phi_1 \\ &\quad - f_d \bar{q}_L d_R \phi_1 - f_u \bar{q}_L u_R \phi_2 + h.c. \end{aligned} \tag{151}$$

This Lagrangian is invariant under the chiral rotation $\psi \to e^{i\alpha\gamma_5}\psi$, i.e.,

$$\begin{aligned} \ell_L &\longrightarrow e^{-i\alpha}\ell_L, \ e_R \longrightarrow e^{+i\alpha}e_R, \ q_L \longrightarrow e^{-i\alpha}q_L, \\ u_R &\longrightarrow e^{+i\alpha}u_R, \ d_R \longrightarrow e^{+i\alpha}d_R, \\ \phi_1 &\longrightarrow e^{-2i\alpha}\phi_1, \ \phi_2 \longrightarrow e^{-2i\alpha}\phi_2 \end{aligned} \tag{152}$$

This contrasts to the case with the minimal standard model where the third term of (151) violates the invariance by $\phi^* \to e^{+2i\alpha}\phi^*$.

With this chiral rotation the QCD Lagrangian transforms as

$$\delta \int d^4x \mathcal{L}_{\rm QCD} = -2\alpha \frac{g^2}{32\pi^2} \int d^4x F^a_{\mu\nu} \tilde{F}^{a\mu\nu} \tag{153}$$

and the strong CP-violating term in (148) is now

$$\mathcal{L}_{\not{CP}} = \frac{g^2}{32\pi^2}(\theta_{\rm eff} - 2\alpha) F^a_{\mu\nu} \tilde{F}^{a\mu\nu}. \tag{154}$$

This means that the QCD vacuum angle $\theta_{\rm eff}$ is physically meaningless and can be rotated away [424].

$\mathrm{U(1)_{PQ}}$, however, is broken, when ϕ_1 and ϕ_2 obtain vacuum expectation values at the electroweak scale v. This gives rise to a Nambu-Goldstone (NG) boson, and the argument given above cannot be valid as it is. Furthermore, this $\mathrm{U(1)_{PQ}}$ symmetry is also explicitly broken by a chiral anomaly at a low energy (QCD energy scale), as can be seen in eq.(153). Hence, $\mathrm{U(1)_{PQ}}$ remains only as approximate symmetry, and the Nambu-Goldstone boson acquires a small mass [439] of the order of

$$m_A \sim \frac{f_\pi m_\pi}{v} \sim 100\text{keV}. \tag{155}$$

This pseudo NG boson is called an axion [439]. The mass and coupling property of axion are prescribed by the Goldstone theorem, and searches for such a particle have already excluded its existence [440].

It was shown, however, that this undesirable feature of the model can be avoided if $\mathrm{U}(1)_{\mathrm{PQ}}$ is broken at a scale much higher than v, so that the mass of the pseudo NG boson is $m_A \sim f_\pi m_\pi / v_{\mathrm{PQ}}$ [441]–[443]. For this purpose one introduces an $\mathrm{SU}(2)_L\times$ U(1) singlet Higgs χ, which carries a $\mathrm{U}(1)_{\mathrm{PQ}}$ charge +2, and develops a vacuum expectation value $v_{\mathrm{PQ}} \gg v$. This χ transforms as

$$\chi \longrightarrow e^{+2i\alpha}\chi. \tag{156}$$

By virtue of the Goldstone theorem, this axion not only has a very small mass, but its coupling to matter is suppressed by m_f/v_{PQ} with m_f the mass of the fermion. This is the model of invisible axions [441]–[443].

We now return to the issue of the strong CP problem. The current of Peccei–Quinn symmetry has an anomaly that contributes to the Lagrangian

$$\mathcal{L} = \frac{1}{32\pi^2} A(x) F^a_{\mu\nu} \tilde{F}^{a\mu\nu}, \tag{157}$$

which should be added to (148). On the other hand, axions induce a mass term, through the QCD instanton effect, of the form [444]

$$\mathcal{L}_{\text{mass}} = m_A^2 (A(x) + \bar{\theta}_{\text{eff}})^2, \tag{158}$$

so that the vacuum takes the expectation value

$$\langle A(x) \rangle = -\bar{\theta}_{\text{eff}}. \tag{159}$$

The anomaly term (148) + (157) thus vanishes, restoring the CP invariance.

36.2 Model for the neutrino mass

We now introduce the right-handed neutrino [445, 447], which transforms as

$$\nu_R \longrightarrow e^{+i\alpha}\nu_R. \tag{160}$$

The mass term which respects $\mathrm{U}(1)_{\mathrm{PQ}}$ is then

$$\mathcal{L}_Y = -f_\nu \overline{\ell_L} \nu_R \phi_2 + h.c. \tag{161}$$

We may also have a mass term of the form

$$\mathcal{L} = h \overline{\nu_R^c} \nu_R \chi + h.c. \tag{162}$$

The introduction of ν_R substantially changes the symmetry structure of the model. In this model there are four U(1) symmetries, baryon number, lepton number,

$U(1)_{PQ}$ and $U(1)_Y$, of which the latter three are eventually broken. When χ develops the vacuum expectation value, both lepton number $U(1)_L$ and $U(1)_{PQ}$ appear to be broken. However, what is actually broken is the linear contribution of two U(1) symmetries; another U(1) remains unbroken. An analysis shows that the current corresponding to the remaining U(1) symmetry has a U(1) anomaly; this U(1) is identified as a new PQ symmetry, meaning that symmetry broken by $\langle\chi\rangle \neq 0$ corresponds to the lepton number. This new PQ symmetry is broken only after the doublet ϕ_i develops a vacuum expectation value. Accordingly, the mass of the axion is suppressed only as $f_\pi m_\pi / v^2$, contrary to our desire. This is essentially the case of the standard axion, which is already excluded by the experiment.

To save the model from this problem the authors of ref. [447] introduced an explicit symmetry-breaking term in the Higgs potential, so that the lepton number conservation is explicitly violated. They introduced a soft breaking term[16],

$$\mathcal{L}_{\Delta L} = \lambda \phi_1 \phi_2 \chi + h.c. \tag{163}$$

Then, $\langle\chi\rangle \neq 0$ in eq.(162) leads to $U(1)_{PQ}$ violation, and the mass of axion is given by $m_A \sim m_\pi f_\pi / \langle\chi\rangle$, as desired.

In this model $\langle\chi\rangle \neq 0$ at the same time gives the right-handed neutrino a large Majorana mass,

$$M_{\nu_R} = h\langle\chi\rangle. \tag{164}$$

The argument of Sect.34 now applies to this model. We should remark, however, that PQ symmetry must be re-defined in the presence of the symmetry-breaking term (163); namely, under this $U(1)_{PQ}$ the fermion fields should transform as:

$$\begin{aligned}
q_L &\to e^{-i\alpha} q_L, \quad u_R \to e^{+i\alpha} u_R, \quad d_R \to e^{+i\alpha} d_R, \\
\ell_L &\to \ell_L, \quad \nu_R \to e^{+2i\alpha} \nu_R, \quad e_R \to e^{+2i\alpha} e_R, \\
\phi_1 &\to e^{-2i\alpha} \phi_2, \quad \phi_2 \to e^{-2i\alpha} \phi_2, \\
\chi &\to e^{-4i\alpha} \chi
\end{aligned} \tag{165}$$

to make the Lagrangian invariant.

36.3 Astrophysical and cosmological constraints on the symmetry - breaking scale

A lower limit can be derived on the symmetry-breaking scale v_{PQ} from the argument for stellar evolution [448]–[449]. If axions do exist, they would be copiously produced in a high-temperature environment of stellar plasma, and would carry out stellar energy produced in nuclear reactions. In order to obtain the observed luminosity of stars, more nuclear fuels are then necessary to compensate for the energy loss by axion emission. Since stellar evolution is controlled by the chemical

---

[16] If one introduces a symmetry-breaking term of dimension four, $\mathcal{L}_{\Delta L} = g\phi_1\phi_2\chi\chi$, radiative corrections induce new quadratically divergent terms in the Higgs potential that destroy the desired hierarchy $\langle\chi\rangle \gg \langle\phi_i\rangle$. This problem is avoided by introducing a soft symmetry-breaking term, since it does not give rise to a quadratic divergence; a hierarchy $\langle\chi\rangle \gg \langle\phi_i\rangle$, once satisfied at the tree level, also holds after radiative corrections, if $\lambda\langle\chi\rangle \lesssim \langle{\phi_i}^2\rangle$ is satisfied.

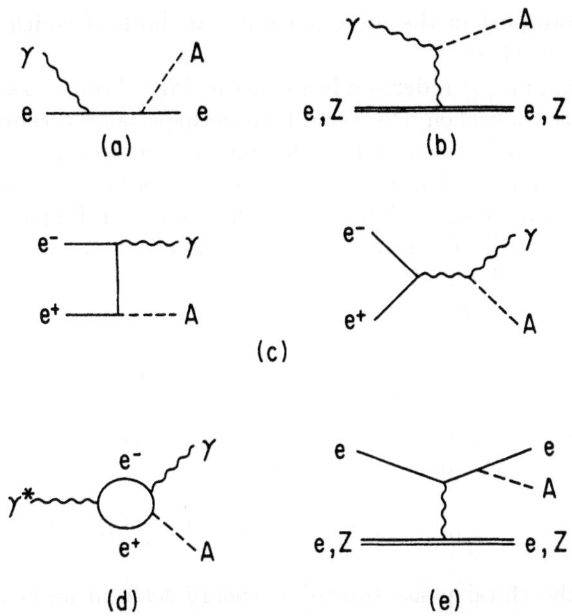

Figure 37: Feynman diagrams for the axion emission processes in stars. (a) Compton process; (b) Primakoff process; (c) e^+e^- pair annihilation; (d) plasmon decay; (e) bremsstrahlung axion emission.

composition of the core, extra nuclear reactions shorten the lifetime of stars at each stage, especially that at the initial helium-burning stage (which can easily be identified in the colour-magnitude diagram of stars).

The processes that are important in stars are shown in Fig.37: Compton process $\nu+e \to A+e$, Primakoff process $\gamma+\gamma_v$ (virtual) $\to A$, plasmon decay γ_ω (plasmon) $\to A + \gamma_{\omega'}(\omega' < \omega)$, e^+e^- pair annihilation $e^+e^- \to A + \gamma$, and bremsstrahlung e + nucleus $\to e$ + nucleus + A. The importance of each process depends on the temperature and density of the stellar cores. In helium-burning stars, for instance, the first three dominate over the others. Since the coupling of χ with a fermion is suppressed by $m_f/\langle\chi\rangle$, all of the processes are essentially suppressed by the factor $\langle\chi\rangle^{-2}$. When the axion emissivity becomes of the order of the optical luminosity of stars, stellar evolution is affected (to derive a limit quantitatively one should compare the change which would be caused by axion emission with observation; whether the change is detectable is the crucial point in deriving a constraint [449]). The most reliable, yet most stringent, limit is the bound derived from a star at the helium-burning stage [449]; it is

$$v_{\rm PQ} = \langle\chi\rangle > 10^9 {\rm GeV}. \tag{166}$$

Limits as strong as $v_{\rm PQ} > 10^9 - 10^{11}$ are inferred from a consideration of supernova SN1987A [450]. These limits, however, cannot be taken too literally, since the argument seems strongly model dependent, especially on the temperature profile and neutrino transport in the protoneutron star, both of which are subject to a large uncertainty[17] [451].

An upper limit on $\langle\chi\rangle$ is derived from a cosmological consideration [452]. When the PQ symmetry is broken, the axion field perhaps takes a random phase which varies form horizon to horizon. When the temperature decreases to the QCD scale, the axion field acquires a finite mass [453], and it starts oscillating as the Hubble expansion time scale becomes longer than the oscillation frequency ($\sim m_A$). The oscillation is eventually damped by the viscous effect of the Hubble expansion and settles to the value of (159). This dynamics of the free axion field $\phi(x,t)$ ($= A(x,t)$) is described by

$$\ddot{\phi} + 3\frac{\dot{a}}{a}\dot{\phi} + (m_A^2 - V^2/a^2)\phi = 0, \tag{167}$$

where V is the Higgs potential and a is the scale factor of the Universe. This oscillating coherent field gives a mass density of

$$\rho = \frac{1}{2}m_A^2\phi^2 \approx 2\times 10^{-30}\mathrm{gcm}^{-3}\Lambda_{200}^{-3/4}{v_{\mathrm{PQ}}}^{7/16}, \tag{168}$$

where Λ_{200} is the chiral phase-transition energy scale in units of 200MeV. The constraint that oscillation is sufficiently damped and ρ then be smaller than the critical density of the Universe, ρ_{crit} of (II.125), gives an upper limit on v_{PQ} [452],

$$v_{\mathrm{PQ}} < 7\times 10^{12}\mathrm{GeV}h^{12/7}\Lambda_{200}^{9/14}. \tag{169}$$

When (169) is satisfied the oscillation amplitude today correspondes to $\theta = \phi/v_{\mathrm{PQ}} < 10^{-21}$, which is sufficiently small to satisfy (150).

In this way the symmetry-breaking scale v_{PQ} is constrained to be in the narrow range. Corresponding to this v_{PQ}, the mass of axions is

$$10^{-5}\mathrm{eV} < m_A < 10^{-3}\mathrm{eV}. \tag{170}$$

36.4 Domain wall problem

In (154) we have seen that $\mathrm{U(1)_{PQ}}$ is not symmetry of the full Lagrangian. In the presence of N_f families, (154) generalises to

$$\mathcal{L}_{\mathrm{eff}} = \frac{1}{32\pi^2}(\theta_{\mathrm{eff}} - 4\alpha N_f)F^a_{\mu\nu}\tilde{F}^{a\mu\nu}. \tag{171}$$

Since θ_{eff} is an angle of mod 2π, we see that the Lagrangian is invariant if we replace

$$\alpha \to \alpha + \frac{2\pi}{4N_f}n \quad (n = \text{integer}), \tag{172}$$

[17] Axion emission depends on the temperature with a high power $T^8 \sim T^{10}$. Therefore, the argument obtained by setting the axion luminosity equal to the neutrino luminosity is incorrect. A slight decrease in the temperature deep inside the core greatly reduces axion emissivity. Although some authors take into account its feedback effect on the neutrino emissivity, the result is still largely model dependent. Let us recall that we do not yet understand the mechanism of supernova explosion.

i.e., we still have Z_{4N_f} symmetry [454]. The condensation of χ then causes a spontaneous breakdown of this discrete symmetry, which gives rise to a domain wall problem[18].

We can avoid this problem if we embed the Z_N discrete symmetry into the centre of some non-Abelian group [456]; in such a case we have a path which connects the Z_N vacua with a zero energy gradient. For this to be realised one has to extend the standard model. One of the simplest cases is to consider SO(10) $\times$ $\mathrm{SU}(3)_F$ [457]. The matter field spinor $\mathbf{16}_L$ of SO(10) is decomposed into $\psi_L(\mathbf{4},\mathbf{2},\mathbf{1})$ and $\bar{\psi}_L(\mathbf{4}^*,\mathbf{1},\mathbf{2})$ under its maximal subgroup $\mathrm{SU}(4)\times\mathrm{SU}(2)_L\times\mathrm{SU}(2)_R$. Under the transformation corresponding to the centre of $\mathrm{SU}(4)\times\mathrm{SU}(2)_L\times\mathrm{SU}(2)_R$, these spinors transform as

$$\begin{aligned} \psi_L &\rightarrow e^{-i\frac{2\pi}{2}\alpha}e^{-i\frac{2\pi}{4}\gamma}\psi_L\,, \\ \bar{\psi}_L &\rightarrow e^{-i\frac{2\pi}{2}\beta}e^{+i\frac{2\pi}{4}\gamma}\psi_L\,. \end{aligned} \tag{173}$$

Taking $\beta=\alpha+\gamma$, we see that both ψ_L and $\bar{\psi}_L$ transform under Z_4 as

$$\begin{pmatrix}\psi_L\\ \bar{\psi}_L\end{pmatrix}\rightarrow e^{-i\frac{2\pi}{4}\ell}\begin{pmatrix}\psi_L\\ \bar{\psi}_L\end{pmatrix} \tag{174}$$

with $\ell=2\alpha+\gamma$. Therefore, under $\mathrm{SO}(10)\times\mathrm{SU}(3)_F$ $\mathbf{16}_L$ transforms as

$$\begin{aligned} \begin{pmatrix}\psi_L\\ \bar{\psi}_L\end{pmatrix} &\rightarrow e^{-i\frac{2\pi}{4}\ell}e^{-i\frac{2\pi}{3}m}\begin{pmatrix}\psi_L\\ \bar{\psi}_L\end{pmatrix} \\ &= e^{-i\frac{2\pi}{12}n}\begin{pmatrix}\psi_L\\ \bar{\psi}_L\end{pmatrix}, \end{aligned} \tag{175}$$

where $n=3\ell+4m$. We take here the assignment that left-handed fermions belong to $\mathbf{3}$ of $\mathrm{SU}(3)_F$ and right-handed fermions to $\mathbf{3}^*$, rather than $\mathbf{3}$. Transformation (175) is identified with the remaining Peccei–Quinn Z_{4N_f} symmetry for $N_f=3$. The argument here shows that starting with Peccei–Quinn U(1) symmetry we are led to $\mathrm{U}(1)_{\mathrm{PQ}}\times\mathrm{SU}(3)_F\times\mathrm{SO}(10)$ as the simplest possibility to make the model fully acceptable.

37 Model with horizontal gauge symmetry

Let us consider the gauge group $\mathrm{SU}(2)_L\times\mathrm{U}(1)\times\mathrm{SU}(3)_H$, where the last $\mathrm{SU}(3)_H$ stands for the horizontal gauge group [431]. We assume that q_L and ℓ_L belong to $(\mathbf{2},\mathbf{3})$ of $\mathrm{SU}(2)_L\times\mathrm{SU}(3)_H$ and u_R, d_R, e_R, to $(\mathbf{1},\mathbf{3})$. It is obvious from mismatched left-right symmetry for the neutrino that this model has an $\mathrm{SU}(3)_H$ gauge anomaly. This anomaly can be canceled, however, if one adds $\nu_R(\mathbf{1},\mathbf{3})$.

[18]When discrete symmetry is broken in an early universe, walls are formed between two regions with different vacua. Such domain walls store a very large energy, and it greatly exceeds the critical mass density of the Universe, unless the symmetry-breaking energy scale is smaller than <100MeV [455].

In order to explore the energy scale of this model, let us specify the Higgs sector. The simplest choice may be

$$\begin{aligned} \phi &= (\mathbf{2}, \mathbf{8}+\mathbf{1}), \\ \chi &= (\mathbf{1}, \mathbf{6}). \end{aligned} \tag{176}$$

$(\mathbf{8}+\mathbf{1})$ ϕ are necessary to give the correct mass spectrum to the quarks and leptons. The horizontal gauge fields $S^i_\mu (i = 1-8)$ acquire their masses by $\langle\chi\rangle \neq 0$.

The exchange of S^i_μ gives rise to flavour-changing neutrino currents, and one has to assume that $\langle\chi\rangle \gg \langle\phi\rangle$ in order to suppress unwanted effects. The most stringent limit is derived from the $K^0 - \overline{K^0}$ mass difference, leading to $\langle\chi\rangle/\langle\phi\rangle \gtrsim 10^5$ [458]. Therefore, the right-handed neutrino mass induced by

$$\mathcal{L} = h\nu_R^T \nu_R \chi + h.c. \tag{177}$$

is

$$M_{\nu_R} \gtrsim h\langle\chi\rangle \gtrsim h \times 10^7\text{GeV}. \tag{178}$$

If we take $m_D = m_\ell$ and $\langle\chi\rangle = V\delta_{ij}$, we have

$$\begin{aligned} m_{\nu_e} &\lesssim h^{-1} \times 3 \times 10^{-5}\text{eV}, \\ m_{\nu_\mu} &\lesssim h^{-1} \times 10\text{eV}, \\ m_{\nu_\tau} &\lesssim h^{-1} \times 400\text{eV}. \end{aligned} \tag{179}$$

In this model relatively large neutrino masses can be obtained compared with those in SO(10) or in the global symmetry models.

38 Neutrino mass induced by radiative corrections

A small Majorana neutrino mass may be induced from radiative corrections if one introduces a scalar particle that breaks lepton number in the Higgs potential [459]. As an example, let us suppose h^-, which transforms as $(\mathbf{1}, -1)$ under $SU(2)_L\times U(1)_Y$. The Yukawa coupling of h^- is

$$\mathcal{L}_{\text{Yukawa}} = h^{ab}\epsilon_{ij}(\overline{\ell^a_{Li}})^c\ \ell^b_{Lj}\ h^\dagger + \text{h.c.}, \tag{180}$$

where indices i, j refer to $SU(2)_L$ and a, b to generation. We necessarily have

$$h^{ab} = -h^{ba} \tag{181}$$

from the Fermi statistics. If there exist two (or more than two) Higgs doublets, we have a cubic coupling term of the form

$$\mathcal{L}_{\phi\phi h} = M_{\alpha\beta}\ \epsilon_{ij}\ \phi^\alpha_i\ \phi^\beta_j\ h + \text{h.c.}, \tag{182}$$

where

$$M_{\alpha\beta} = -M_{\beta\alpha}. \tag{183}$$

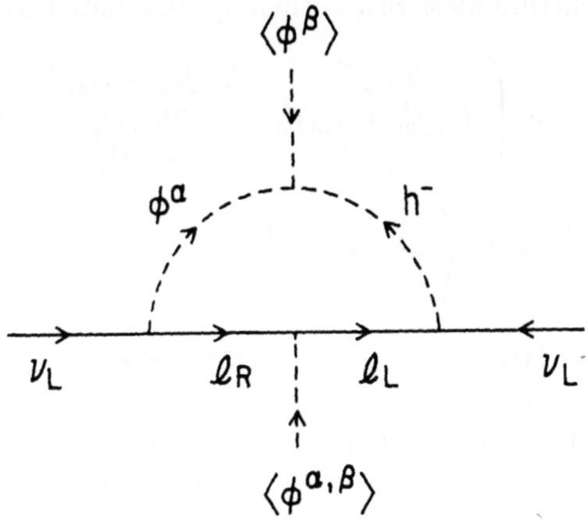

Figure 38: One-loop radiative correction giving rise to a neutrino mass (Majorana mass). h^- is an $SU(2)_L$-singlet charged scalar particle.

Having (180) and (182) together, lepton number is broken by two units, and a Majorana mass is induced from the diagram of Fig.38, which yields the effective mass term $m_{ab}(\overline{\nu_L^a})^c\ \nu_L^b$:

$$m_{ab} \simeq \{h^{ac} m_c (f_\beta^\dagger)^{cb} + (a \leftrightarrow b)\}\ \frac{v_\alpha M_{\alpha\beta}}{m_h^2}\ [\log(m_\phi^2/m_h^2)]_{\beta\alpha}, \tag{184}$$

where m_c is the charged lepton mass of generation c, f_β^{ab} is the usual Yukawa coupling of ϕ^β to $\overline{\ell_L^a} e_R^b$, and v_α is the vacuum expectation value of $< \phi_0^\alpha >$. If $M_{\alpha\beta}$ is of the order of the h^- mass m_h and $v_\alpha \sim v_\beta \sim v$, we find

$$m_{ab} \sim \{h^{ac}\ \frac{m_c}{m_h}\ f^{\dagger cb} + (a \leftrightarrow b)\}\ v. \tag{185}$$

The model has too many free parameters to predict any definite phenomenological effects. Motivated by the empirical hierarchy in the KM matrix, however, we may assume that f and h take the forms

$$f_\alpha \simeq f_\beta \simeq C \begin{pmatrix} \varepsilon^3 & \varepsilon^2 & \varepsilon^2 \\ \varepsilon^2 & \varepsilon & \varepsilon \\ \varepsilon^2 & \varepsilon & 1 \end{pmatrix} \tag{186}$$

and

$$h \simeq D \begin{pmatrix} 0 & \varepsilon & \varepsilon \\ -\varepsilon & 0 & 1 \\ -\varepsilon & -1 & 0 \end{pmatrix}. \tag{187}$$

In this case the neutrino mass matrix in eq.(185) is dominated by the τ-exchange contribution. The neutrino mass m_{ab} is then approximated by

$$m_{ab} \sim \frac{m_\tau}{m_h} v \begin{pmatrix} 2h_{13}f^\dagger_{31} & h_{13}f^\dagger_{32} + h_{23}f^\dagger_{31} & h_{13}f^\dagger_{33} \\ h_{13}f^\dagger_{32} + h_{23}f^\dagger_{31} & 2h_{23}f^\dagger_{32} & h_{23}f^\dagger_{33} \\ h_{13}f^\dagger_{33} & h_{23}f^\dagger_{33} & 0 \end{pmatrix} \tag{188}$$

$$\sim \frac{m_\tau}{m_h} vCD \begin{pmatrix} \varepsilon^3 & \varepsilon^2 & \varepsilon \\ \varepsilon^2 & \varepsilon & 1 \\ \varepsilon & 1 & 0 \end{pmatrix}, \tag{189}$$

and we then expect a large mixing, $\theta_{23} \sim 45°$, between ν_μ and ν_τ [460]. This may be taken as the most natural consequence of this model. If the ν_μ deficit (as discussed in Sect.25.3) could be ascribed to neutrino oscillation between ν_μ and ν_τ, this feature becomes an interesting prediction. The experimental constraints on the Yukawa couplings f_{ab} and h_{ab} are discussed in Sect.43 in more detail.

39 Seesaw models for Dirac neutrinos

For Dirac neutrinos, the small mass is usually put by hand. In this section we show that it is not impossible to make a model in which a small Dirac mass is ascribed to the seesaw mechanism.

To demonstrate this, we consider a specific model employing chiral U(1) symmetry. Let us assume that $q_L, \bar{u}_R, \bar{d}_R, \ell_L$ and $\bar{e}_R$ have U(1) charge $Q = 1$. On the other hand, we assume a charge of -1 for $\overline{\nu_R}$. (The assignment differs from (118) and (119) only for $\overline{\nu_R}$). If we have two Higgs doublets ϕ_1 and ϕ_2 which have $Q = -2$, the possible Yukawa couplings are

$$f_1 \bar{u}_R q_L \phi_1 \ + \ f_2 \bar{d}_R q_L \phi_2 \ + \ f_3 \bar{e}_R \ell_L \phi_2 + h.c. \tag{190}$$

The above U(1) charge assignment for ν_R does not allow the neutrino to have a Dirac mass term.

We now introduce a pair of heavy singlet fermions $\overline{N}_R$ and N_L with the U(1) charge of $+1$ and -1 , respectively , and assume the Dirac mass term,

$$M\overline{N}_R N_L + h.c. \tag{191}$$

We may also have the coupling

$$g\overline{N}_R \ell_L \phi_1. \tag{192}$$

We introduce a Higgs χ with a U(1) charge of $+2$ in order to break the U(1) symmetry. We then have a Yukawa coupling of the form

$$\mathcal{L} = h\overline{\nu}_R N_L \chi \ . \tag{193}$$

Since there are no $\overline{\nu}_R\overline{\nu}_R\chi$ or $N_L N_L \chi$ terms, $\langle\chi\rangle \neq 0$ does not break lepton number. The full Dirac mass matrix for $\nu_{L,R}$ and $N_{L,R}$ is then

$$(\overline{\nu}_R \ \overline{N}_R) \begin{pmatrix} 0 & h\langle\chi\rangle \\ g\langle\phi_1\rangle & M \end{pmatrix} \begin{pmatrix} \nu_L \\ N_L \end{pmatrix}. \tag{194}$$

By diagonalising this matrix, we obtain a light Dirac neutrino mass,

$$m_\nu^D \simeq hg\frac{\langle\chi\rangle\langle\phi_1\rangle}{M} \,. \tag{195}$$

If we take $M \simeq M_{pl} = 10^{19}$GeV and $\langle\chi\rangle \simeq 10^{10}$GeV, we have

$$m_\nu^D \simeq 10^{-9} hg\langle\phi_1\rangle \sim 10^{-9}(hg/f)m_\ell \,, \tag{196}$$

where m_ℓ is the mass of a charged lepton. The U(1) symmetry discussed here might be identified with $U(1)_{PQ}$ in Sect.34. The possibility of the seesaw mechanism for the Dirac neutrino was first discussed in ref.[461] in the context of a superstring "inspired" model.

40 Model with a Fritzsch mass matrix – a phenomenological description

Since none of the concrete models so far discussed have been successful in describing the neutrino mass and mixing inferred from experiments, let us try a more phenomenological approach. We try to use the quark mass matrix as a guide line in our consideration.

For the quark it is known that one can successfully describe the masses and mixings by assuming the matrix form proposed by Fritzsch [462]. The Fritzsch mass matrices extended to the 3×3 case are

$$m_u = \begin{pmatrix} 0 & |a_u|e^{i\varphi_{12}} & 0 \\ |a_u|e^{i\varphi_{21}} & 0 & |b_u|e^{i\varphi_{23}} \\ 0 & |b_u|e^{i\varphi_{32}} & |c_u|e^{i\varphi_{33}} \end{pmatrix} \tag{197}$$

for the charge 2/3 quarks and

$$m_d = \begin{pmatrix} 0 & |a_d|e^{i\varphi'_{12}} & 0 \\ |a_d|e^{i\varphi'_{21}} & 0 & |b_d|e^{i\varphi'_{23}} \\ 0 & |b_d|e^{i\varphi'_{32}} & |c_d|e^{i\varphi'_{33}} \end{pmatrix} \tag{198}$$

for the charge $-1/3$ quarks. For the mass matrices given in (197) and (198) the CKM matrix is given by

$$U = R_u^T \begin{pmatrix} 1 & & 0 \\ & e^{i\sigma} & \\ 0 & & e^{i\tau} \end{pmatrix} R_d, \tag{199}$$

where R_u and R_d are the orthogonal matrices which diagonalise

$$m_u^* = \begin{pmatrix} 0 & |a_u| & 0 \\ |a_u| & 0 & |b_u| \\ 0 & |b_u| & |c_u| \end{pmatrix} \,, \; m_d^* = \begin{pmatrix} 0 & |a_d| & 0 \\ |a_d| & 0 & |b_d| \\ 0 & |b_d| & |c_d| \end{pmatrix} \,, \tag{200}$$

respectively, viz.,

$$R_u^T m_u^* R_u = \begin{pmatrix} m_u & & \\ & -m_c & \\ & & m_t \end{pmatrix}, \quad (201)$$

$$R_d^T m_d^* R_d = \begin{pmatrix} m_d & & \\ & -m_s & \\ & & m_b \end{pmatrix}. \quad (202)$$

For $m_u \ll m_c \ll m_t$ and $m_d \ll m_s \ll m_b$, one obtains

$$\begin{aligned} U_{12} &\simeq -\sqrt{\frac{m_d}{m_s}} + e^{i\sigma}\sqrt{\frac{m_u}{m_c}}, & U_{21} &\simeq -\sqrt{\frac{m_u}{m_c}} + e^{i\sigma}\sqrt{\frac{m_d}{m_s}} \\ U_{23} &\simeq e^{i\sigma}\sqrt{\frac{m_s}{m_b}} - e^{i\tau}\sqrt{\frac{m_c}{m_t}}, & U_{32} &\simeq e^{i\sigma}\sqrt{\frac{m_c}{m_t}} - e^{i\tau}\sqrt{\frac{m_s}{m_b}} \\ U_{13} &\simeq \sqrt{\frac{m_u}{m_c}} U_{23}, & U_{31} &\simeq \sqrt{\frac{m_d}{m_s}} U_{32}. \end{aligned} \quad (203)$$

It has been discussed that $U_{12}(\simeq U_{21})$, $U_{23}(\simeq U_{32})$, U_{13} and U_{31} can be predicted within the range allowed by the experiment including CP phenomenology with the preferred choice of $\sigma \simeq \tau \simeq \pm\pi/2$, in so far as m_t in eq.(203) is set to be $\lesssim$200GeV [463, 464].

We now suppose that the Fritzsch matrix also applies to the lepton sector [465]. We assume

$$m_{\nu D} = \begin{pmatrix} 0 & |A_\nu|e^{i\psi_{12}} & 0 \\ |A_\nu|e^{i\psi_{21}} & 0 & |B_\nu|e^{i\psi_{23}} \\ 0 & |B_\nu|e^{i\psi_{32}} & |C_\nu|e^{i\psi_{33}} \end{pmatrix}, \quad (204)$$

$$m_e = \begin{pmatrix} 0 & |A_l|e^{i\psi'_{12}} & 0 \\ |A_l|e^{i\psi'_{21}} & 0 & |B_l|e^{i\psi'_{23}} \\ 0 & |B_l|e^{i\psi'_{32}} & |C_l|e^{i\psi'_{33}} \end{pmatrix}, \quad (205)$$

where $m_{\nu D}$ is the Dirac mass matrix of the neutrino and m_e is the charged lepton mass matrix. If the smallness of the neutrino mass is to be in virtue of the seesaw mechanism, the mass of the left-handed neutrino is given by

$$m_\nu = m_{\nu D}^T M_{\nu_R}^{-1} m_{\nu D} . \quad (206)$$

The lepton CKM matrix may generally differ from the form given in (199), since extra phases may come in from the ν_R sector. In order to keep the simplicity of the model, however, let us take M_{ν_R} as being diagonal and proportional to the unit matrix in the basis where $m_{\nu D}$ is diagonal. In such a case, the CKM matrix is given by form (199) with R_u and R_d replaced by $R_{\nu D}$ and R_e, defined in a way similar to (201) and (202) (we use η and ρ for leptons in place of σ and τ). We remark that this conclusion receives only little modification, in so far as the eigenvalues of $m_{\nu D}$ have a hierarchical structure $m_1 \ll m_2 \ll m_3$ for a reasonable variety of M_{ν_R}.

The argument may then be developed in a way parallel to that for the quark sector, with m_u, m_c, m_t replaced by the eigenvalues of $m_{\nu D}$, i.e. m_1, m_2, m_3, and m_d, m_s, m_b by the charged lepton masses m_e, m_μ, m_τ. The neutrino masses are given by

$$m_{\nu_e} = m_1^2/M_{\nu R},\ m_{\nu_\mu} = m_2^2/M_{\nu R},\ m_{\nu_\tau} = m_3^2/M_{\nu R}\ . \qquad (207)$$

With $m_e \ll m_\mu \ll m_\tau$ and $m_{\nu_e} \ll m_{\nu_\mu} \ll m_{\nu_\tau}$ we obtain

$$\begin{aligned}
U_{12} &= -\left(\frac{m_e}{m_\mu}\right)^{1/2} + e^{i\eta}\left(\frac{m_{\nu_e}}{m_{\nu_\mu}}\right)^{1/4}, \\
U_{23} &= e^{i\eta}\left(\frac{m_\mu}{m_\tau}\right)^{1/2} - e^{i\rho}\left(\frac{m_{\nu_\mu}}{m_{\nu_\tau}}\right)^{1/4}, \\
U_{13} &= \left(\frac{m_{\nu_e}}{m_{\nu_\mu}}\right)^{1/4} U_{23}, \\
U_{31} &= -\left(\frac{m_e}{m_\mu}\right)^{1/2}\left(e^{i\rho}\left(\frac{m_\mu}{m_\tau}\right)^{1/2} - e^{i\eta}\left(\frac{m_{\nu_\mu}}{m_{\nu_\tau}}\right)^{1/4}\right).
\end{aligned} \qquad (208)$$

If $m_{\nu_e} = 0$ we would obtain $\sin^2 2\theta_{12} = 4m_e/m_\mu \simeq 0.020$. This value is virtually the same as that of (117) obtained in a modified SO(10) model. We have already discussed that this mixing lies beyond the experimentally allowed range (V.167). This means that the second term should be crucial in eq.(208). Due to the quartic root nature of the second term, however, only a tiny value of $m_{\nu_e}/m_{\nu_\mu} \simeq 5\times10^{-9} - 2\times10^{-6}$ will suffice to bring $\sin^2 2\theta_{12}$ into the allowed range, provided that $\eta \simeq 0$ [465].

Let us now consider U_{23}. If we take $m^2_{\nu_\tau} \simeq 10^{(-2\pm0.5)}\mathrm{eV}^2$, as inferred from the neutrino oscillation hypothesis for the atmospheric ν_μ deficit [see (V.81)], and the value of m_{ν_μ} from the small angle MSW solution (V.167), U_{23} of eq.(208) becomes

$$U_{23} \simeq 0.24e^{i\eta} - (0.32 - 0.50)e^{i\rho}\ . \qquad (209)$$

This means that

$$\sin^2 2\theta_{23} = 0.5 - 1, \qquad (210)$$

unless the two terms of (209) interfere destructively. This large $\sin^2 2\theta_{23}$ is the mixing needed to explain the ν_μ deficit in the Kamiokande and IMB experiments, as discussed in Sect.25.3.

If the free parameters are fixed in this way, the model also predicts

$$\theta_{13}/\theta_{23} = (m_{\nu_e}/m_{\nu_\mu})^{1/4} \simeq 0.01 - 0.04\ , \qquad (211)$$

and

$$\theta_{31}/\theta_{23} = (m_e/m_\mu)^{1/2} \simeq 0.070 \qquad (212)$$

if the phases in (208) are taken so that the two terms add up either constructively or incoherently. Hence, this model suggests that $\sin^2 2\theta_{12} \simeq 0.01, \sin^2 2\theta_{23} \simeq 0.6, \sin^2 2\theta_{13} \simeq 3\times10^{-4\pm1}$ and $\sin^2 2\theta_{31} \simeq 5\times10^{-3}$ as a preferred solution.

We may conclude that a simple Fritzsch form of the lepton mass matrix may describe the mixing pattern fairly well, as inferred from the solar neutrino experiments and the ν_μ deficit in the atmospheric neutrino flux, although the solution is by no means unique.

VII. Magnetic moment of the neutrino

41 Neutrino magnetic moment in the Weinberg - Salam model

From the general argument (see Sect.5) only the Dirac neutrino can have a magnetic moment. The transition moment, which is relevant to $\nu_i \to \nu_j$, may exist, however, for both Dirac and Majorana neutrinos. A straightforward calculation of the one-loop diagram [106, 285] (see also [466]) depicted in Fig.39 gives for the Dirac neutrino

$$T = -i\varepsilon^{\mu} q^{\nu} \bar{\nu}_j(p')\sigma_{\mu\nu}(A + B\gamma_5)\nu_i(p), \tag{1}$$

where $q = p' - p$ and

$$A = -\frac{eG_F}{8\sqrt{2}\pi^2}(m_{\nu_i} + m_{\nu_j})\sum_{\ell} U^{\dagger}_{j\alpha}U_{\alpha i}F([m_{\ell_\alpha}/m_W]^2) \tag{2}$$

$$B = -\frac{eG_F}{8\sqrt{2}\pi^2}(m_{\nu_i} - m_{\nu_j})\sum_{\ell} U^{\dagger}_{j\alpha}U_{\alpha i}F([m_{\ell_\alpha}/m_W]^2) \tag{3}$$

with

$$F(x) = \frac{1}{(1-x)^2}[-\frac{3}{4}(2 - 5x + x^2) + \frac{3}{2}\frac{x^2 \log x}{1-x}] \tag{4}$$

$$\simeq -\frac{3}{2} + \frac{3}{4}x. \quad (x \ll 1) \tag{5}$$

Using the general expression (II.94)

$$T = -i\varepsilon^{\mu}\bar{\nu}(p')[F_2(q^2) - G_2(q^2)\gamma_5]\sigma_{\mu\nu}q^{\nu}\nu(p), \tag{6}$$

the magnetic dipole moment is defined by

$$\mu = F_2(0) = A, \tag{7}$$

and similarly the electric dipole moment is

$$d = G_2(0) = B. \tag{8}$$

Comparing (6) to (1) with $i = j$, we obtain [106, 107, 108]

$$\mu_{\nu_i} = \frac{3eG_F}{8\pi^2\sqrt{2}}m_{\nu_i}\left[1 - \frac{1}{2}\sum_{\ell=1}^{3}(\frac{m_\ell}{m_W})^2|U_{\ell i}|^2\right], \tag{9}$$

where the second term is negligible. We note that the chirality flip, which is necessary to induce the magnetic moment, arises only from the neutrino mass; for this reason (9) is a very small quantity,

$$\mu_{\nu_i} = 3.20 \times 10^{-19}\mu_B(\frac{m_{\nu_i}}{1\text{eV}}), \tag{10}$$

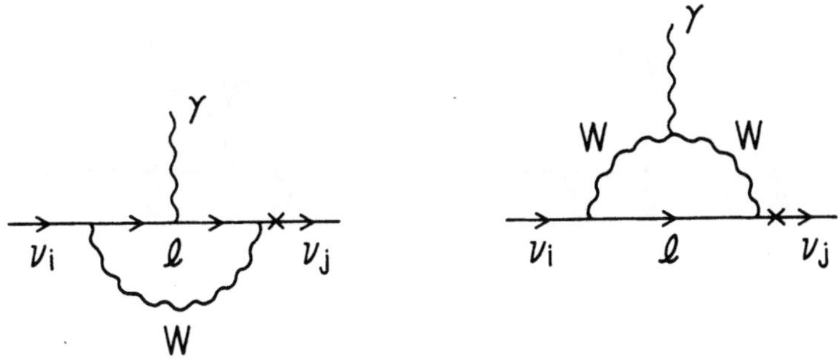

Figure 39: Radiative correction giving rise to a magnetic moment of the Dirac neutrino within the Weinberg-Salam theory. The cross is a Dirac mass insertion.

where $\mu_B = e/2m_e$ is the Bohr magneton (II.101). The electric dipole moment always vanishes because of the factor $m_{\nu_i} - m_{\nu_j}$ in eq.(3), even if U includes CP-violating phases.

For the off-diagonal transition moment, the first term of (5) vanishes in the summation over ℓ due to the orthogonality condition of U (GIM mechanism [50]); thus,

$$\mu_{\bar{\nu}_j \nu_i} = \frac{3eG_F}{32\sqrt{2}\pi^2}(m_{\nu_i} + m_{\nu_j}) \sum_{\alpha=1}^{3} U^{\dagger}_{j\alpha} U_{\alpha i} (\frac{m_{\ell_\alpha}}{m_W})^2, \tag{11}$$

$$d_{\bar{\nu}_j \nu_i} = \frac{3eG_F}{32\sqrt{2}\pi^2}(m_{\nu_i} - m_{\nu_j}) \sum_{\alpha=1}^{3} U^{\dagger}_{j\alpha} U_{\alpha i} (\frac{m_{\ell_\alpha}}{m_W})^2. \tag{12}$$

Hence the decay width of $\nu_i \to \nu_j + \gamma$ is given by [467]

$$\Gamma(\nu_i \to \nu_j + \gamma) = \frac{m^3_{\nu_i}}{8\pi}(1 - \frac{m^2_{\nu_j}}{m^2_{\nu_i}})^3 [|\mu_{\bar{\nu}_j \nu_i}|^2 + |d_{\bar{\nu}_j \nu_i}|^2] \tag{13}$$

$$\simeq \frac{\alpha m^5_{\nu_i}}{128\pi^4} G_F^2 \frac{9}{16} (\frac{m_\tau}{m_W})^4 |U^{\dagger}_{j3} U_{3i}|^2 \tag{14}$$

Here we used the fact that $m_\tau \gg m_\mu, m_e$ and also assumed $m_{\nu_i} \gg m_{\nu_j}$. Numerically,

$$\Gamma(\nu_i \to \nu_j + \gamma) = (10^{29}\text{yr})^{-1} (\frac{m_{\nu_i}}{30\text{eV}})^5 |U^{\dagger}_{j3} U_{3i}|^3, \tag{15}$$

i.e.,

$$\tau(\nu_i \to \nu_j + \gamma) \geq 4 \times 10^{29}\text{yr}(m_{\nu_i}/30\text{eV})^{-5}, \tag{16}$$

where the use is made of $|U^{\dagger}_{j3} U_{3i}|^2 \leq |\sin\theta\cos\theta|^2 \leq 1/4$ by unitarity. We remark that the transition moment is very small, typically $\mu \sim 10^{-22}\mu_B(m_{\nu i}/1\text{eV})$, as a result of the chiral suppression and the GIM cancellation.

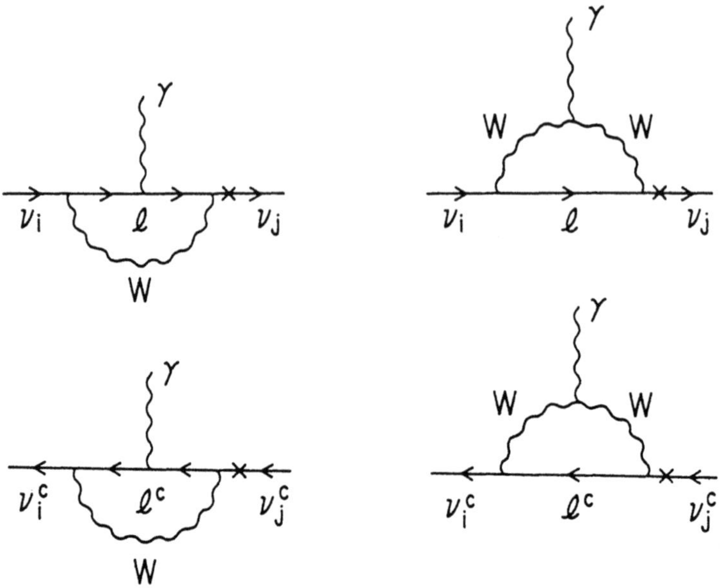

Figure 40: Same as Fig. 39 for a Majornana neutrino. The cross is a Majorana mass insertion.

The case for Majorana neutrinos

The Feynman diagrams for the Majorana neutrino case are given in Fig.40. We have both a charged lepton and an antilepton propagating in the internal loop. Correspondingly, we have the transition matrix

$$\begin{aligned} T_{ji} &= -i\varepsilon^{\mu}\bar{\nu}_j(p')[\{F_2(q^2)_{ji} - F_2(q^2)_{ij}\} - \{G_2(q^2)_{ji} - G_2(q^2)_{ij}\}\gamma_5]\sigma_{\mu\nu}q^{\nu}\nu_i(p) \\ &= -i\varepsilon^{\mu}\bar{\nu}_j(p')[2i\ \mathrm{Im}F_2(q^2)_{ji} - 2\ \mathrm{Re}\ G_2(q^2)_{ji}\gamma_5]\sigma_{\mu\nu}q^{\nu}\nu_i(p). \end{aligned} \tag{17}$$

The relative minus sign in each bracket comes from an exchange of the two fermions corresponding to arrows in the opposite direction. Eq.(17) shows explicitly (e.g. [466]) that

$$\mu_{\nu_i} = d_{\nu_i} = 0. \tag{18}$$

For the transition moment, only one of the two terms of (17) is non-vanishing if the interaction respects the CP invariance: the first term vanishes if the relative CP of ν_i and ν_j is even, and the second term vanishes if odd [468, 285, 469, 466].

In the standard model we obtain from (2) and (3)

$$\begin{aligned} d_{\nu_i\nu_j} &= 2\mathrm{Re}\ G_2(0)_{ji} \\ &= \frac{3eG_F}{16\pi^2\sqrt{2}}(m_{\nu_i} - m_{\nu_j})\sum_{\ell=1}^{3}\mathrm{Re}\ [U_{j\alpha}^{\dagger}U_{\alpha i}(\frac{m_{\ell_\alpha}}{m_W})^2], \qquad (19) \\ \mu_{\nu_i\nu_j} &= 2i\ \mathrm{Im}\ F_2(q^2) \end{aligned}$$

$$= \frac{3eG_F}{16\pi^2\sqrt{2}}(m_{\nu_i}+m_{\nu_j})\sum_{\ell=1}^{3} i \ \mathrm{Im}\ [U^\dagger_{j\alpha}U_{\alpha i}(\frac{m_{\ell_\alpha}}{m_W})^2], \tag{20}$$

allowing for violation of the CP invariance. The decay rate of $\nu_i \to \nu_j + \gamma$ is calculated by (13). If CP is conserved $U^\dagger_{j\alpha}U_{\alpha i}$ is either real (when relative CP of ν_i and ν_j are even) or pure imaginary (when it is odd), and only one of (19) and (20) remains non-vanishing, respectively. Therefore, the decay rate of $\nu_i \to \nu_j + \gamma$ for the Majorana neutrino is just twice that for the Dirac case (14) [285].

42 Neutrino magnetic moment in the left-right symmetric model

If there is a right-handed current which mixes with the left-handed current, the chiral suppression does not work and the magnetic moment is greatly enhanced.

Let us suppose the $SU(2)_L \times SU(2)_R \times U(1)$ model with the interaction Lagrangian

$$\begin{aligned}\mathcal{L}_{\mathrm{int}} &= \frac{g_L}{\sqrt{2}}W^\dagger_{L\mu}(\bar\nu_{eL}\gamma_\mu e_L + \bar\nu_{\mu L}\gamma_\mu\mu_L + \bar\nu_{\tau L}\gamma_\mu\tau_L) + h.c. \\ &+ \frac{g_R}{\sqrt{2}}W^\dagger_{R\mu}(\bar\nu_{eR}\gamma_\mu e_R + \bar\nu_{\mu R}\gamma_\mu\mu_R + \bar\nu_{\tau R}\gamma_\mu\tau_R) + h.c.\end{aligned} \tag{21}$$

The mass term is given by

$$\begin{aligned}\mathcal{L}_{\mathrm{mass}} &= (\bar e, \bar\mu, \bar\tau)_R \begin{pmatrix} m_e & & 0 \\ & m_\mu & \\ 0 & & m_\tau \end{pmatrix} \begin{pmatrix} e \\ \mu \\ \tau \end{pmatrix}_L \\ &+ (\bar\nu_1, \bar\nu_2, \bar\nu_3)_R \begin{pmatrix} m_1 & & 0 \\ & m_2 & \\ 0 & & m_3 \end{pmatrix} \begin{pmatrix} \nu_1 \\ \nu_2 \\ \nu_3 \end{pmatrix}_L,\end{aligned} \tag{22}$$

where the unitary transformations

$$\begin{aligned}\nu_{iL} &= U^\dagger_{i\alpha}\nu_{\alpha L}, \\ \nu_{iR} &= V^\dagger_{i\alpha}\nu_{\alpha R}\end{aligned} \tag{23}$$

are applied for the mass matrix for ν_e, ν_μ and ν_τ to diagonalise the neutrino mass term. With (23) the interaction term (21) is written

$$\begin{aligned}\mathcal{L}_{\mathrm{int}} &= \frac{g_L}{\sqrt{2}}W^\dagger_{L\mu}\bar\nu_{iL}U^\dagger_{i\alpha}\gamma_\mu\ell_{\alpha L} + h.c. \\ &+ \frac{g_R}{\sqrt{2}}W^\dagger_{R\mu}\bar\nu_{iR}V^\dagger_{i\alpha}\gamma_\mu\ell_{\alpha R} + h.c.\end{aligned} \tag{24}$$

After symmetry breaking of both $SU(2)_L$ and $SU(2)_R$ we may write the mass term for the gauge bosons to be

$$\begin{aligned}\mathcal{L}_{\mathrm{mass}} &= \frac{1}{2}(W^+_L \ W^+_R)\begin{pmatrix} m^2_{W_L} & \delta \\ \delta^* & m^2_{W_R}\end{pmatrix}\begin{pmatrix} W^-_L \\ W^-_R\end{pmatrix} \\ &= \frac{1}{2}(W^+_1 \ W^+_2)\begin{pmatrix} m^2_{W_1} & 0 \\ 0 & m^2_{W_2}\end{pmatrix}\begin{pmatrix} W^-_1 \\ W^-_2\end{pmatrix},\end{aligned} \tag{25}$$

where the second line is obtained by

$$\begin{aligned} W_1 &= W_L \cos\phi - W_R \sin\phi e^{i\psi}, \\ W_2 &= W_L \sin\phi e^{-i\psi} + W_R \cos\phi \end{aligned} \tag{26}$$

and

$$\mathrm{tg}2\phi = \frac{2|\delta|}{m_{W_R}^2 - m_{W_L}^2}. \tag{27}$$

We note that the off-diagonal element of the mass matrix (25) can generally be complex allowing for CP violation.

The diagrams giving radiative transitions are depicted in Fig.41. A calculation then yields for the magnetic moment [470, 466]

$$\begin{aligned} \mu_{\nu_i} &= \frac{eg_L^2 m_{\nu_i}}{32\pi^2}\left[\frac{3}{2}\frac{\cos^2\phi}{m_{W_1}^2}(1-\frac{1}{2}\sum_{\alpha=1}^{3}|U_{\alpha i}|^2\mathcal{E}_\alpha^{(1)}) + \frac{3}{2}\frac{\sin^2\phi}{m_{W_2}^2}(1-\frac{1}{2}\sum_{\alpha=1}^{3}|U_{\alpha i}|^2\mathcal{E}_\alpha^{(2)})\right] \\ &+ \frac{eg_R^2 m_{\nu_i}}{32\pi^2}\left[\frac{3}{2}\frac{\sin^2\phi}{m_{W_1}^2}(1-\frac{1}{2}\sum_{\alpha=1}^{3}|V_{\alpha i}|^2\mathcal{E}_\alpha^{(1)}) + \frac{3}{2}\frac{\cos^2\phi}{m_{W_2}^2}(1-\frac{1}{2}\sum_{\alpha=1}^{3}|V_{\alpha i}|^2\mathcal{E}_\alpha^{(2)})\right] \\ &+ \frac{eg_L g_R}{32\pi^2}\sin\phi\cos\phi\sum_{\alpha=1}^{3} m_{\ell_\alpha}\mathrm{Re}(e^{-i\psi}V_{i\alpha}^\dagger U_{\alpha i}) \\ &\times \sum_{k=1}^{2}(-1)^k\frac{4}{m_{W_k}^2}\left[-1+\mathcal{E}_\alpha^{(k)}(\ln\frac{1}{\mathcal{E}_\alpha^k}-\frac{9}{8})\right], \end{aligned} \tag{28}$$

where $\mathcal{E}_\alpha^k = m_{\ell_\alpha}^2/m_{W_k}^2$. We note that the last term does not vanish, even if $m_{\nu_i} = 0$. This gives an example, where the neutrino magnetic moment is not proportional to the mass of the neutrino.

In this model the electric dipole moment is calculated to be

$$\begin{aligned} d_{\nu_i} &= \frac{eg_L g_R}{8\pi^2}\sin\phi\cos\phi\sum_{\alpha=1}^{3} m_{\ell_\alpha}\mathrm{Im}(e^{i\psi}U_{i\alpha}^\dagger V_{\alpha i}) \\ &\times \sum_{k=1}^{2}(-1)^k\frac{1}{m_{W_k}^2}\left[-1+\mathcal{E}_\alpha^{(k)}(\ln\frac{1}{\mathcal{E}_\alpha^{(k)}}-\frac{9}{8})\right], \end{aligned} \tag{29}$$

which is non-vanishing only if CP is violated, i.e., $\mathrm{Im}(e^{i\psi}...) \neq 0$.

The first four terms of Eq.(28) gives only small contributions and are negligible. Assuming that $m_{W_2} \gg m_{W_1} \simeq m_W$ and $g_L = g_R = g$, we have

$$\begin{aligned} \mu_{\nu_i} &\simeq -\frac{eg^2}{8\pi^2}\sin\phi\cos\phi\sum_\alpha m_{\ell_\alpha}\mathrm{Re}(e^{-i\psi}V_{i\alpha}^\dagger U_{\alpha i}) \\ &\times \frac{1}{m_W^2}[-1+\frac{m_{\ell_\alpha}^2}{m_W^2}(\ln\frac{m_W^2}{m_{\ell_\alpha}^2}-\frac{9}{8})] \\ &\simeq \frac{eG_F}{\sqrt{2}\pi^2}\sin\phi\cos\phi\sum m_{\ell_\alpha}\mathrm{Re}(e^{-i\psi}V_{i\alpha}^\dagger U_{\alpha i}) \\ &= \frac{\sqrt{2}G_F}{\pi^2}\sin\phi\cos\phi m_e\sum_\alpha m_{\ell_\alpha}\mathrm{Re}(e^{-i\psi}V_{i\alpha}^\dagger U_{i\alpha}^\dagger)\mu_B \end{aligned} \tag{30}$$

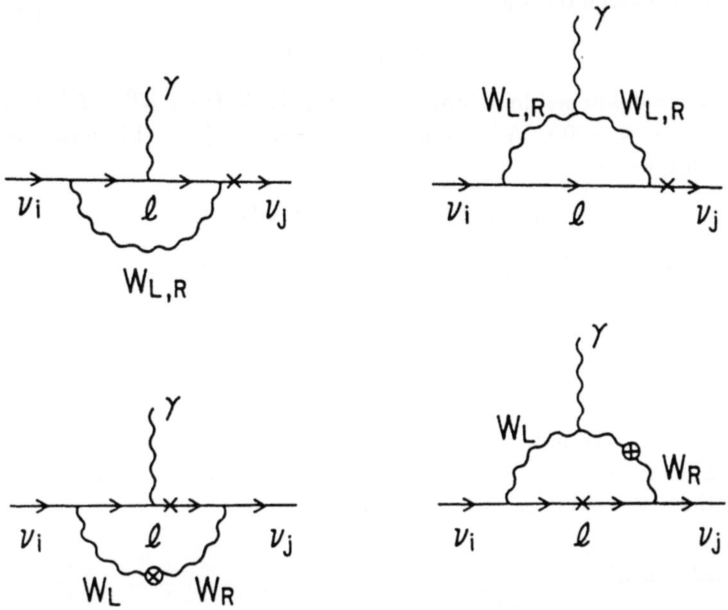

Figure 41: Radiative correction giving rise to a magnetic moment in the left-right symmetric model. The circled cross shows $W_L - W_R$ mixing.

$$\begin{aligned} = \; & \sin\phi\cos\phi[1.3\times 10^{-9}\ \mathrm{Re}(e^{-i\psi}V^{\dagger}_{i\tau}U_{\tau i}) \\ & + 1.0\times 10^{-10}\ \mathrm{Re}(e^{-i\psi}V^{\dagger}_{i\mu}U_{\mu i}) \\ & + 0.2\times 10^{-14}\ \mathrm{Re}(e^{-i\psi}V^{\dagger}_{ie}U_{ei})]\mu_B. \end{aligned} \tag{31}$$

Which term dominates depends on the mixing angle.

A constraint on the $W_L - W_R$ mixing is derived from a measurement of the Michel parameter of μ decay [471], which yields

$$|\sin\phi| \leq 0.05 \quad (90\%\ \mathrm{C.L.}). \tag{32}$$

The endpoint energy spectrum of e^+ in stopped polarised muon decay [429] gives

$$|\sin\phi| \leq 0.035 \qquad \text{for } m_{W_R}\to\infty\ , \tag{33}$$

and

$$m_{W_R} \geq 475\mathrm{GeV} \qquad \text{for a small } \phi. \tag{34}$$

A slightly stronger constraint on $|\sin\phi|$ may be derived in the following way. If there is a $W_L - W_R$ mixing, the mass of W_L shifts by the amount

$$\begin{aligned} \delta m^2_{W_L} & \simeq -\delta^2/m^2_{W_R} \\ & = \sin^2\phi\ m^2_{W_R}. \end{aligned} \tag{35}$$

By requiring that the Weinberg-Salam prediction for m_W should not be disturbed by more than 1.6 standard deviations (90%) (0.78GeV), we obtain with the aid of the bound (34) given on m_{W_R},

$$|\sin\phi| \leq 0.024\ . \tag{36}$$

If we use a model-dependent limit $m_{W_R} \geq 1.6$TeV from $K^0 - \bar{K}^0$ mixing [472], we have $|\sin\phi| \leq 0.006$. Adopting the conservative (model-independent) bound of (36) we are led to

$$\mu_{\nu_i} \lesssim 0.3 \times 10^{-10} \mathrm{Re}(e^{-i\psi} V^{\dagger}_{i\tau} U_{\tau i}) \mu_B \tag{37}$$

from the τ-loop contribution. If $V_{i\tau}$ and $U_{\tau i} (i \neq \tau)$ are very small, we still have

$$\mu_{\nu_e} \lesssim 0.5 \times 10^{-15} \mu_B \tag{38}$$

from the $e-$loop and

$$\mu_{\nu_\mu} \lesssim 2.4 \times 10^{-12} \mu_B \tag{39}$$

from the μ-loop. We note a large magnetic moment compared with that in the standard model due to an absence of the chirality suppression in the present model, provided that $W_L - W_R$ mixing is substantial.

Transition moment

The transition moment is calculated to give

$$\begin{aligned}\mu_{\nu_j\nu_i} = & \frac{3e(m_{\nu_i}+m_{\nu_j})}{256\pi^2}\sum_{\alpha=1}^{3}\Big[-g_L^2\Big(\frac{\cos^2\phi}{m_{W_1}^2}\mathcal{E}_\alpha^{(1)}+\frac{\sin^2\phi}{m_{W_2}^2}\mathcal{E}_\alpha^{(2)}\Big)U^{\dagger}_{j\alpha}U_{\alpha i} \\ & -g_R^2\Big(\frac{\sin^2\phi}{m_{W_1}^2}\mathcal{E}_\alpha^{(1)}+\frac{\cos^2\phi}{m_{W_2}^2}\mathcal{E}_\alpha^{(2)}\Big)V^{\dagger}_{j\alpha}V_{\alpha i}\Big] \\ & +\frac{eg_Lg_R}{16\pi^2}\sin\phi\cos\phi\sum_{\alpha=1}^{3}m_{\ell_\alpha}\Big[e^{i\psi}U^{\dagger}_{j\alpha}V_{\alpha i}+e^{-i\psi}V^{\dagger}_{j\alpha}U_{\alpha i}\Big] \\ & \times\Big[\sum_{k=1}^{2}\frac{(-1)^k}{m_{W_k}^2}\{-1+\mathcal{E}_\alpha^{(k)}(\ln\frac{1}{\mathcal{E}_\alpha^{(k)}}-\frac{9}{8})\}\Big]\end{aligned} \tag{40}$$

and

$$\begin{aligned}d_{\nu_j\nu_i} = & \frac{3e(m_{\nu_i}-m_{\nu_j})}{256\pi^2}\sum_{\alpha=1}^{3}\Big[-g_L^2\Big(\frac{\cos^2\phi}{m_{W_1}^2}\mathcal{E}_\alpha^{(1)}+\frac{\sin^2\phi}{m_{W_2}^2}\mathcal{E}_\alpha^{(2)}\Big)U^{\dagger}_{j\alpha}U_{\alpha i} \\ & +g_R^2\Big(\frac{\sin^2\phi}{m_{W_1}^2}\mathcal{E}_\alpha^{(1)}+\frac{\cos^2\phi}{m_{W_2}^2}\mathcal{E}_\alpha^{(2)}\Big)V^{\dagger}_{j\alpha}V_{\alpha i}\Big] \\ & +\frac{eg_Lg_R}{16\pi^2}\sin\phi\cos\phi\sum_{\alpha=1}^{3}m_{\ell_\alpha}\Big[e^{i\psi}U^{\dagger}_{j\alpha}V_{\alpha_i}-e^{-i\psi}V_{j\alpha}U_{\alpha i}\Big] \\ & \times\Big[\sum_{k=1}^{2}\frac{(-1)^k}{m_{W_k}^2}\{-1+\mathcal{E}_\alpha^{(k)}(\ln\frac{1}{\mathcal{E}_\alpha^{(k)}}-\frac{9}{8})\Big]\ .\end{aligned} \tag{41}$$

For $m_\nu \ll m_\ell$, $m_{W_2} \gg m_{W_1} = m_W$ and $g_L = g_R (= g)$, (40) and (41) are

$$\mu_{\nu_j\nu_i} \simeq \frac{eG_F}{2\sqrt{2}\pi^2}\sin\phi\cos\phi\ \ m_\tau[e^{i\psi}U^{\dagger}_{j\tau}V_{\tau i}+e^{-i\psi}V^{\dagger}_{j\tau}U_{\tau i}] \tag{42}$$

and

$$d_{\nu_j\nu_i} \simeq \frac{eG_F}{2\sqrt{2}\pi^2} \sin\phi\cos\phi \;\; m_\tau[e^{i\psi}U^\dagger_{j\tau}V_{\tau i} - e^{-i\psi}V^\dagger_{j\tau}U_{\tau i}], \tag{43}$$

where we retained only the contribution from the τ-loop. The decay rate of $\nu_i \to \nu_j + \gamma$ calculated using (13) is

$$\begin{aligned}\Gamma(\nu_i \to \nu_j + \gamma) &= \frac{\alpha}{8\pi^4} G_F^2 \sin^2\phi\cos^2\phi \;\; m_\tau^2 m_{\nu_i}^3 \left(1 - \frac{m_{\nu_j}^2}{m_{\nu_i}^2}\right)^3 \\ &\quad \times (|U^\dagger_{j\tau}V_{\tau i}|^2 + |V^\dagger_{j\tau}U_{\tau i}|^2). \end{aligned} \tag{44}$$

Numerically,

$$\begin{aligned}&\Gamma(\nu_i \to \nu_j + \gamma) \\ &\quad = (0.22 \times 10^{15}\mathrm{sec})^{-1} \sin^2\phi\cos^2\phi(|U^\dagger_{j\tau}V_{\tau i}|^2 + |V^\dagger_{j\tau}U_{\tau i}|^2)(m_{\nu_i}/10\mathrm{eV})^3 \\ &\quad \lesssim (0.38 \times 10^{18}\mathrm{sec})^{-1}(|U^\dagger_{j\tau}V_{\tau i}|^2 + |V^\dagger_{j\tau}U_{\tau i}|^2)(m_{\nu_i}/10\mathrm{eV})^3, \end{aligned} \tag{45}$$

using constraint (36). If $V_{\tau 1} \sim 0.01$ and $V_{\tau 2} \sim 0.1$, the ν_τ decay can be as fast as

$$\tau(\nu_3 \to \nu_{2,1} + \gamma) \gtrsim 10^{12}\mathrm{yr}(m_{\nu_3}/10\mathrm{eV})^{-3} . \tag{46}$$

The fact that there is no chirality suppression appears in the m_ν^3 dependence of (46), rather than m_ν^5 in (16). The transition amplitude does not receive the GIM suppression in the left-right symmetric model.

43 Models with a charged scalar particle

Another interesting case in which the neutrino magnetic moment does not receive a chirality suppression is models with a charged scalar particle. In such models one may expect a large magnetic moment [473]. Although the scalar particle can be either an $SU(2)_L$ singlet or a doublet, the former is the simplest choice [459, 473]. This is because an $SU(2)_L$ singlet scalar does not couple to quarks and hence we do not need to worry about the flavour-changing neutral current induced by a scalar particle in the hadron sector, which otherwise would lead to a very strong constraint on such an extra particle. The singlet scalar carries a unit electric charge.

The allowed Yukawa coupling of the charged singlet scalar η^+ is given by

$$\mathcal{L}_{\mathrm{int}} = g_{ij}\, \overline{\ell^{ci}_L}\, \ell^j_L\, \eta^+ + f_{ij}\, \overline{\nu^{ci}_R}\, e^j_R\, \eta + \mathrm{h.c.}\,, \tag{47}$$

where

$$g_{ij} = -g_{ji} \tag{48}$$

from the $SU(2)_L$ property and the Fermi statistics (we ignore here the neutrino mass). We may also impose $f_{ij} = -f_{ji}$ for the right-handed coupling only for simplicity of the argument.

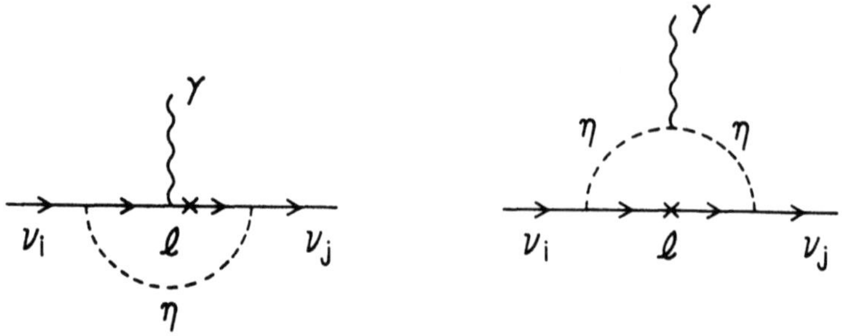

Figure 42: Dominant diagrams of radiative corrections giving rise to a magnetic moment in a model with a charged scalar particle η.

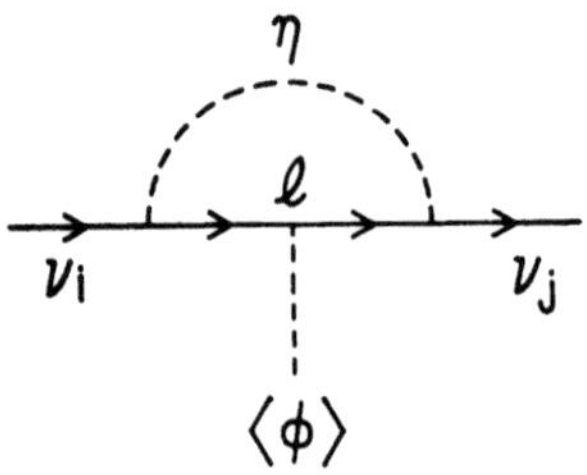

Figure 43: Diagram for neutrino mass induced by radiative corrections in the η model.

A calculation of diagram depicted in Fig.42 yields

$$\mu_{\nu_i} = e \sum_{j=1}^{3} \frac{f_{ij} g_{ji}^{\dagger} + g_{ij} f_{ji}^{\dagger}}{32\pi^2} \frac{m_{\ell_j}}{m_\eta^2} \left[\ln \frac{m_\eta^2}{m_{\ell_j}^2} - 1 \right]. \tag{49}$$

If τ dominates in the intermediate state,

$$\mu_{\nu_e} = e \frac{f_{13} g_{31}^{\dagger} + g_{13} f_{31}^{\dagger}}{32\pi^2} \frac{m_\tau}{m_\eta^2} \left[\ln \frac{m_\eta^2}{m_\tau^2} - 1 \right]. \tag{50}$$

The η loop may also contribute to the neutrino mass (Fig.43). This diagram, however, is logarithmically divergent, and it must be canceled by a counter term; i.e., the mass of the neutrino is arbitrary and is not calculable.

One of the motivations to consider such a model is a possible requirement for a large magnetic moment for $\nu_e, \mu_{\nu_e} = (10^{-10} - 10^{-11})\mu_B$, to explain the solar neutrino problem, as we detail in Sect.45. In order to obtain $\mu_{\nu_e} = 10^{-11}\mu_B$, we must have

$$\frac{g_{13} f_{31}^{\dagger} + f_{13} g_{31}^{\dagger}}{2} \frac{1}{m_\eta^2} \left[\ln \frac{m_\eta^2}{m_\tau^2} - 1 \right] \simeq 10^{-6} \mathrm{GeV}^{-2}. \tag{51}$$

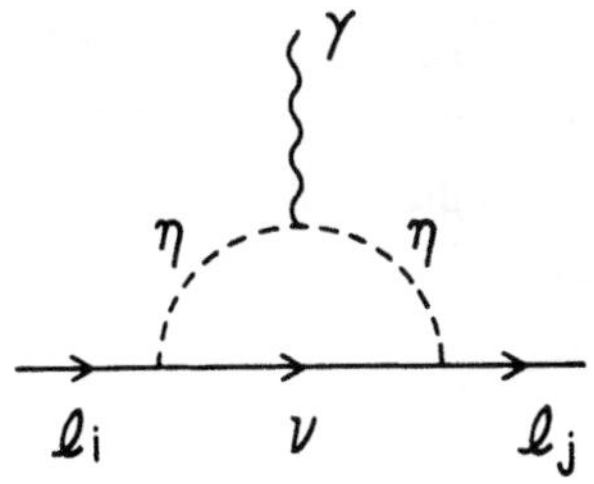

Figure 44: Contribution to the lepton anomalous magnetic moment from radiative corrections involving the η particle.

Constraints on the model

A number of constraints are derived on this non-standard model from varieties of experiments, and the condition (51) must be checked against these constraints [473]. For the consideration given here we assume

$$f_{ij} \simeq g_{ij} \tag{52}$$

in order to reduce the number of parameters, so that we are left with four parameters: m_η, g_{12}, g_{13} and g_{23}.
(i) Anomalous magnetic moment $(g-2)$ of leptons. Radiative corrections by η gives rise to an extra contribution to the anomalous magnetic moment of leptons. From a calculation of Fig.44 we have

$$\mu_{\ell\alpha} = -\frac{em_{\ell\alpha}}{16\pi^2}\frac{1}{12m_\eta^2}[(gg^\dagger)_{\alpha\alpha} + (ff^\dagger)_{\alpha\alpha}] \tag{53}$$

or

$$\delta(g_\alpha - 2) = \frac{2\mu_{\ell\alpha}}{\mu_B} = -\frac{1}{24\pi^2}\frac{m_{\ell\alpha}^2}{m_\eta^2}(g^\dagger g)_{\alpha\alpha} \tag{54}$$

with the use of (52). From $\delta(g_e - 2)/2 < 2.9 \times 10^{-10}$ we obtain

$$\frac{(g^\dagger g)_{11}}{m_\eta^2} < 2.7 \times 10^{-1}\mathrm{GeV}^{-2}. \tag{55}$$

For the muon we have $\delta(g_\mu - 2)/2 < 1.3 \times 10^{-8}$, which leads to

$$\frac{(g^\dagger g)_{22}}{m_\eta^2} < 0.83 \times 10^{-3}\mathrm{GeV}^{-2}. \tag{56}$$

(ii) Cabibbo and μe universalities. The η exchange induces an anomalous μ decay (Fig.45a). In particular, the first diagram interferes with the standard W exchange, and an effective G_μ^{eff} obtained from μ decay is modified to [474]

$$(G_\mu^{\mathrm{eff}})^2 = G_F^2\left[1 + \frac{1}{4}\frac{|g_{12}|^2/m_\eta^2}{G_F/\sqrt{2}}\right]. \tag{57}$$

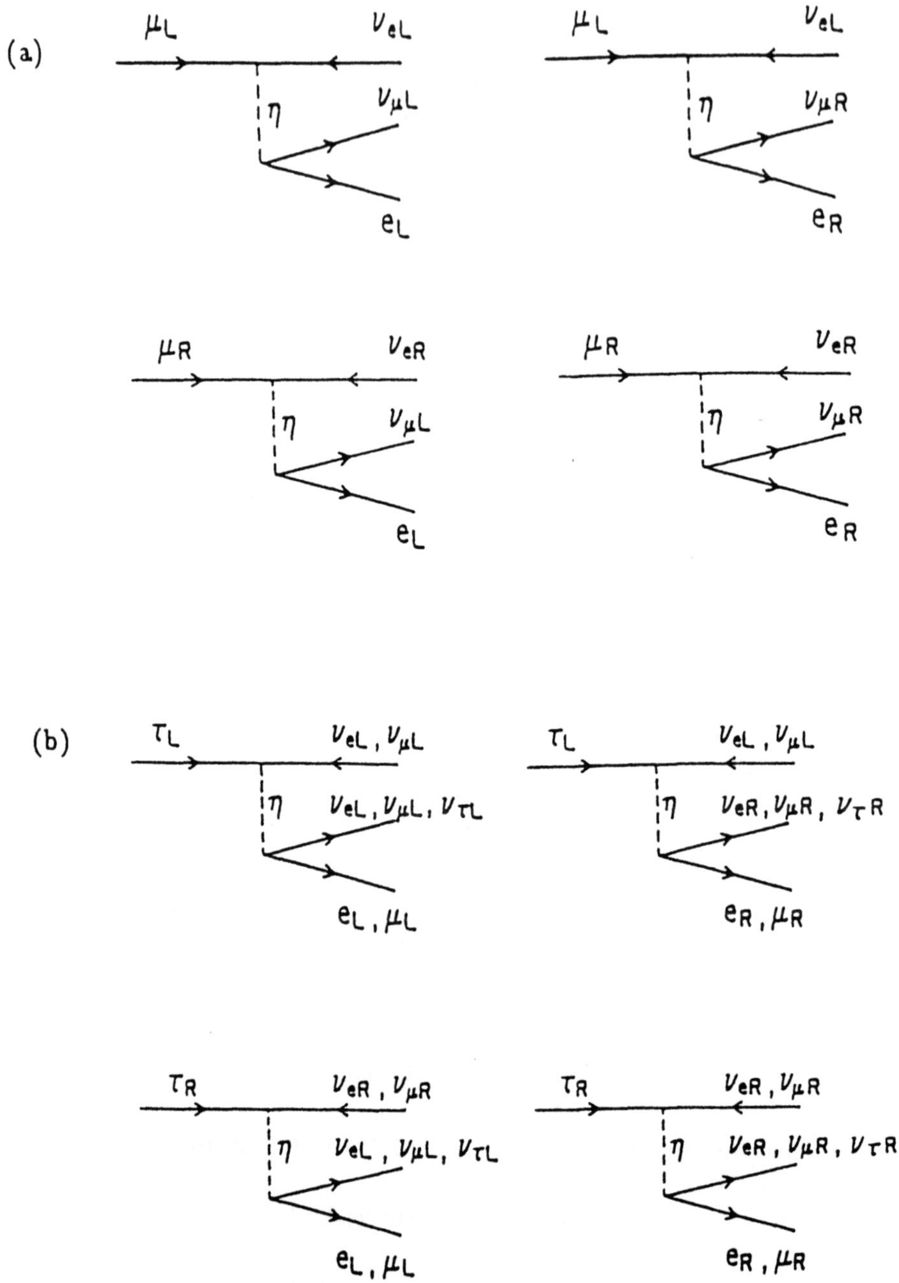

Figure 45: (a) Types of anomalous muon decay induced by the η particle. (b) Same for tau decay.

From the unitarity test (II.91) we get

$$\frac{|g_{12}|^2}{m_\eta^2} < 1.6 \times 10^{-7}\mathrm{GeV}^{-2}. \tag{58}$$

The η exchange also disturbs μe universality in the τ decay through a diagram of Fig.45b. We expect

$$\frac{B_e}{B_\mu} = 1 + \frac{\sqrt{2}}{4G_F}\frac{|g_{13}|^2 - |g_{23}|^2}{m_\eta^2}\ . \tag{59}$$

The experiment $B_e/B_\mu = 1.020 \pm 0.031$ means that

$$\frac{|g_{13}|^2}{m_\eta^2} < 0.9 \times 10^{-6}\mathrm{GeV}^{-2}, \tag{60}$$

allowing for a 2σ error, where we have used $|g_{23}| \ll |g_{13}|$, which is derived below in eq.(63). Note that this process directly constrains g_{13} that appears in (51).
(iii) $\mu \to e + \gamma$. The charged scalar η induces a flavour-changing neutral current in the lepton sector. In particular, the diagram shown in Fig.46 causes $\mu \to e\gamma$ with the amplitude

$$T(\mu \to e\gamma) = \frac{e}{32\pi^2}\frac{m_e + m_\mu}{12m_\eta^2}[(gg^\dagger)_{12} + (ff^\dagger)_{12}]. \tag{61}$$

The experimental limit on this process (V.10) yields

$$\frac{(g^\dagger g)_{12}}{m_\eta^2} \simeq \frac{g_{13}g_{32}}{m_\eta^2} < 3 \times 10^{-8}\mathrm{GeV}^{-2}. \tag{62}$$

This is a strong constraint, and means that

$$\frac{g_{32}}{g_{13}} < 3 \times 10^{-2}\left(\frac{g_{13}^2/m_\eta^2}{10^{-6}\mathrm{GeV}^2}\right)^{-1}\ . \tag{63}$$

In conclusion it seems marginally possible to have parameters that satisfy (51), but we need disparities in the Yukawa couplings, such that

$$|\frac{g_{12}}{g_{13}}| < 0.4, \quad |\frac{g_{23}}{g_{13}}| < 0.03 \tag{64}$$

for $g_{13}^2/m_\eta^2 \sim 10^{-6}\mathrm{GeV}^{-2}$.

Transition moment

The expression for the transition moment is essentially the same as (49),

$$\mu_{\nu_i\nu_j} = e\sum_k \frac{f_{ik}g_{kj}^\dagger + g_{ik}f_{kj}^\dagger}{32\pi^2}\frac{m_{\ell_k}}{m_\eta^2}\left[\ln\frac{m_\eta^2}{m_{\ell_k}^2} - 1\right], \tag{65}$$

$$d_{\nu_i\nu_j} = \mu_{\nu_i\nu_j}. \tag{66}$$

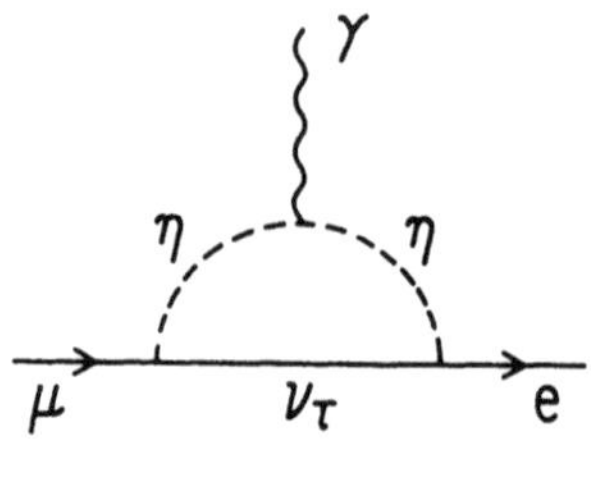

Figure 46: Diagram that induces $\mu \to e\gamma$.

The decay rate for $\nu_i \to \nu_j + \gamma$ is given by (13).

For $\nu_\mu \to \nu_e + \gamma$ (65) is

$$\begin{aligned} \mu_{\nu_\mu \nu_e} &\simeq e\frac{g_{23}g_{31}}{16\pi^2}\frac{m_\tau}{m_\eta^2}\left[\ln\frac{m_\eta^2}{m_\tau^2} - 1\right] \\ &\lesssim 3.4 \times 10^{-10} e\left[\ln\frac{m_\eta^2}{m_\tau^2} - 1\right]\ \mathrm{GeV}^{-1} \end{aligned} \tag{67}$$

upon the use of assumption (52) and the constraint given in (62). Hence,

$$\tau(\nu_\mu \to \nu_e + \gamma) \gtrsim 0.2 \times 10^{14} \mathrm{yr}(10\mathrm{eV}/m_{\nu_\mu})^3 \tag{68}$$

for $m_{\nu_\mu} \gg m_{\nu_e}$. For $\nu_\tau \to \nu_e + \gamma$ the intermediate state is μ and the Yukawa coupling $g_{32}g_{21}$ receives constraints from (58) and (62). The lower limit on the lifetime is

$$\tau(\nu_\tau \to \nu_\mu + \gamma) \gtrsim 1.3 \times 10^{14} \mathrm{yr}(10\mathrm{eV}/m_{\nu_\tau})^3. \tag{69}$$

This is similar to the lifetime that we obtained for the left-right symmetric model.

<u>η-induced ν oscillation in matter</u>

Another interesting possibility with the η model is that the η particle may induce neutrino oscillation in matter, even when the neutrinos are massless [475]. As discussed in Sect.8.5, the neutrinos acquire an effective mass in matter as a result of a refractive interaction with electrons. This is compared with the standard weak interaction which generates only diagonal effective masses for neutrinos; neutrino oscillation does not occur in the $m_{\nu_i} = 0$ limit. The refractive effect of the η exchange interaction, however, also induces an off-diagonal effective mass, which causes neutrino oscillation, even if neutrinos have vanishing intrinsic masses, as discussed first by Wolfenstein [331, 476].

Let us consider only the left-handed neutrinos and assume that they are massless. In this case η has the Yukawa interactions given by the first term of eq.(47). The neutrinos acquire effective masses in matter through the η exchange diagram

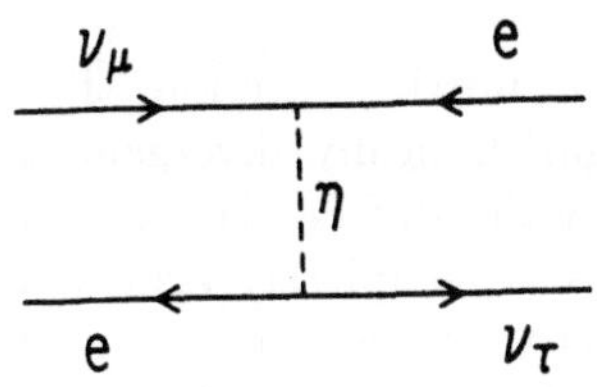

Figure 47: $\nu_\mu - \nu_\tau$ exchange interaction induced by the η particle.

shown in Fig.47. The effective Hamiltonian is then given by

$$H = n_e \begin{pmatrix} \sqrt{2}G_F & 0 & 0 \\ 0 & \frac{g_{12}^2}{4m_\eta^2} & \frac{-g_{12}g_{13}}{4m_\eta^2} \\ 0 & -\frac{g_{12}g_{13}}{4m_\eta^2} & \frac{g_{13}^2}{4m_\eta^2} \end{pmatrix} \tag{70}$$

where the (1,1) element represents a contribution from the W exchange, and the standard neutral current interactions from Z^0 exchange are omitted, since they contribute to all neutrino species in the same way. With (70) the mixing angle ψ between ν_μ and ν_τ is given by

$$\tan 2\psi = \frac{2g_{12}g_{13}}{g_{12}^2 - g_{13}^2}. \tag{71}$$

Thus, the maximal mixing $\psi = \pi/4$ occurs if $g_{12} = g_{13}$, even if they are small. The oscillation length is given by

$$L_{\rm osc} = \frac{2\pi}{(g_{12}^2 + g_{13}^2)n_e/4m_\eta^2}, \tag{72}$$

which is independent of the neutrino energy.

We may apply the above result to the atmospheric ν_μ flux propagating through Earth (so called upward-going neutrinos). Using the density profile of Earth, $L_{\rm osc}$ is estimated to be

$$L_{\rm osc} \simeq 37 R_\oplus G_6^{-1}[1 + (g_{13}^2/g_{12}^2)]^{-1} , \tag{73}$$

where $G_6 \simeq g_{12}^2/m_\eta^2 \times 10^6$ and $R_\oplus$ is the radius of Earth. The deficit of the up-going muon neutrino flux is given by [475]

$$P(\nu_\mu \to \nu_\tau) = \frac{1}{2}\sin^2 2\psi[1 - \cos(4\pi R_\oplus/L_{\rm osc})]. \tag{74}$$

Using the present experimental constraints on g_{12}^2/m_η^2 and g_{13}^2/m_η^2, we find the deficit is at most 10% for the allowed parameters of g_{ij}.

44 Variations of the scalar model

In the model described in the previous section, the η loop radiative correction for the neutrino mass is logarithmically divergent and, hence, a counter term must be introduced to cancel the divergence. Although this model is logically consistent, the neutrino mass is completely arbitrary and it does not give any insight concerning the smallness of the neutrino mass. Many extensions of the η model have been considered in order to obtain a large magnetic moment of the neutrino while keeping a small neutrino mass. In this subsection we briefly summarise these attempts.

44.1 Voloshin's symmetry

The most general mass term for neutrinos is written

$$\mathcal{L} = -\frac{1}{2} m_{ij} \overline{\psi_L^{ic}} \psi_L^j + h.c., \tag{75}$$

where ψ is a Weyl spinor and indices $i, j = 1, ..., 2N$ with N the number of neutrino species; $\psi_1, \psi_3, ...$ are $\mathrm{SU}(2)_L$ doublets and $\psi_2, \psi_4, ...$ are singlets (for instance, a Dirac spinor is constructed from a pair of ψ and the Dirac mass term is given by $m_{11} = 0$ and $m_{12} = m_{21}$ etc.). Due to Fermi statistics we should have

$$m_{ij} = m_{ji} \ . \tag{76}$$

On the other hand, the interaction with the magnetic moment is written

$$\mathcal{L}_{\text{mag}} = \mu_{ij} \overline{\psi_L^{ic}} \sigma_{\mu\nu} \psi_L^j F^{\mu\nu} \ , \tag{77}$$

with antisymmetric μ_{ij},

$$\mu_{ij} = -\mu_{ji} \ . \tag{78}$$

If we take two neutrino spinors ψ_L^1 and ψ_L^2 and assume that they form a doublet of some SU(2) symmetry, the mass term transforms like a triplet and the magnetic moment transforms like a singlet. Thus, if this SU(2) (referred to as $\mathrm{SU}(2)_\nu$) is imposed on the model to be exact symmetry, the mass term is forbidden while the neutrino has a non-vanishing magnetic moment [477].

Unfortunately, we cannot impose this $\mathrm{SU}(2)_\nu$ symmetry in a straightforward manner, because ψ_L^1 and ψ_L^2 belong to different multiplets of $\mathrm{SU}(2)_L$. To implement this idea consistently with the standard electroweak model, various possibilities have been explored. Examples are:

(i) Extension of the electroweak gauge group.

Electroweak $\mathrm{SU}(2)_L \times$ U(1) symmetry is enlarged to $\mathrm{SU}(3)_L \times$ U(1), so that it may include Voloshin's $\mathrm{SU}(2)_\nu$ [478]. ψ_1 and ψ_2 are embedded in a multiplet as (ν_L^c, ν_L, e^-).

(ii) Horizontal symmetry

One may use some symmetry which commutes with the electroweak $\mathrm{SU}(2)_L \times$ U(1) group. An example is horizontal (family) $\mathrm{SU}(2)_F$, where ν_e and ν_μ from a doublet of $\mathrm{SU}(2)_F = \mathrm{SU}(2)_\nu$ [479]. This resembles the case of a pseudo Dirac neutrino.

Other examples include a model with discrete abelian [480] or nonabelian symmetry [481].

The problem inherent in these models is that it is not easy to keep the $SU(2)_\nu$ breaking scale sufficiently low without a fine-tuning of the parameters in the Higgs potential; once $SU(2)_\nu$ is broken, a large mass is induced. If we characterise the $SU(2)_\nu$ breaking by $\delta m_\phi^2 = m_{\phi_1}^2 - m_{\phi_2}^2$, the neutrino mass is written [477, 478]

$$m_{\nu_e} \sim \frac{1}{e}\mu_{\nu_e}\frac{\delta m_\phi^2}{\ln(m_\phi^2/m_\tau^2)}\,, \tag{79}$$

where the scalar bosons ϕ_i are $SU(2)_\nu$ doublets, which induce a neutrino magnetic moment. If we require $m_{\nu_e} < 10\text{eV}, \mu_{\nu_e} > 10^{-11}\mu_B$, we must have

$$\frac{\delta m_\phi^2}{m_\phi^2} < 3 \times 10^{-3}\,. \tag{80}$$

All of the models proposed so far need a fine tuning of this kind to give a large magnetic moment of $\mu_{\nu_e} = 10^{-10} - 10^{-11}\mu_B$ with the neutrino mass kept small enough.

44.2 The helicity suppression mechanism

Let us suppose that the Feynman diagram in Fig.48a gives the dominant contribution to the magnetic moment [482]. The "blob" represents an effective vertex which converts a spin-zero boson η^+ into a weak boson by emitting a photon. To generate a transition moment a chirality flip is needed, which is provided by η's Yukawa coupling to the lepton.

On the other hand, the mass term is generated by the diagram in Fig.48b. The "blob" here converts a spin-zero field into a vector field, which takes place only when the vector boson is longitudinally polarised. Now, the longitudinal component of the weak boson W^+ is an unphysical Higgs whose Yukawa couplings to fermions (f) always involves m_f/m_W. This factor suppresses the neutrino mass term. Furthermore, since both vertices on the fermion line of Fig.48b flip chirality, another lepton mass insertion is necessary to obtain a net chirality flip of the mass term. Hence, the suppression factor in the mass diagram amounts to m_f^2/m_W^2.

A concrete model is given by Barr et al. [482], who obtained that

$$\mu_{ij} \sim \frac{e}{(16\pi^2)^2}\frac{g_{ij}}{M}\,, \tag{81}$$

while the mass is suppressed as

$$m_{ij} \simeq \frac{1}{(16\pi^2)^2}\frac{g_{ij}(m_{\ell i}^2 - m_{\ell j}^2)}{M}. \tag{82}$$

Here $g_{ij} = -g_{ji}$ is the antisymmetric Yukawa coupling of η given in eq.(47).

At this level it appears possible to obtain a large magnetic moment, $\mu \simeq (10^{-10} - 10^{-11})\mu_B$, while keeping a neutrino mass of $m_{\nu_e\nu_\mu} < 10^{-2}\text{eV}$. In this

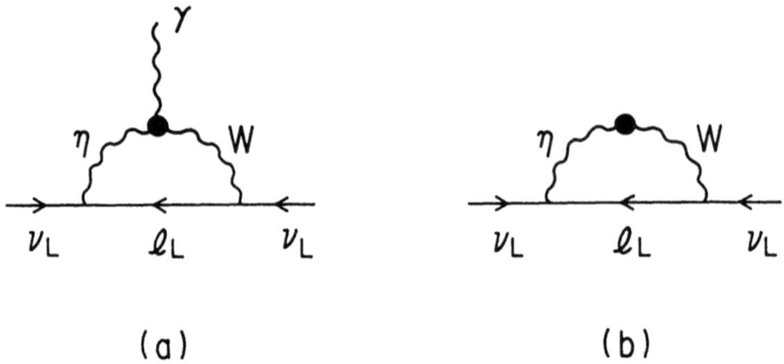

Figure 48: (a) A variant η model giving a magnetic moment to the neutrino. (b) Diagram that induces neutrino mass.

model, however, another one-loop diagram of Fig.38 should contribute to the mass term; this contribution is not necessarily suppressed. Thus, one must impose some extra discrete symmetry to remove this contribution in order to make the scenario viable.

45 Phenomenological significance of the neutrino magnetic moment

In the standard model the magnetic moment of neutrinos is a very small quantity due to chirality suppression. There are many orders of magnitude between the standard model prediction (10) and the present empirical limit, $\mu_{\nu_e} < 10^{-10} - 10^{-9}\mu_B$, from laboratory experiments or $\mu_{\nu_i} < 10^{-11}\mu_B$ from astrophysics. If the magnetic moment would be found to have a value beyond (10), this would indicate the presence of interactions which violate chirality conservation beyond the standard model. In this section we review the problems where a large magnetic moment would be experimentally relevant.

45.1 Solar neutrino problem

An interesting indication on the presence of a large magnetic moment is given from the solar-neutrino experiment. In 1986 Davis reported [303] that the solar-neutrino capture rate at the Homestake ^{37}Cl detector varies with time, and it anticorrelates (see also [483] for an earlier suggestion) with solar activity characterised by sunspot numbers. It is later shown that this anticorrelation persists to the next solar cycle (solar cycle 21, starting from the autumn of 1986) [303]. While the possibility is not excluded that the observed anticorrelation is merely due to a statistical fluke (for statistical tests, see Ref.[484]), the effect is quite intriguing, if it really exists, since it is difficult to imagine any astrophysical mechanisms which correlate the neutrino flux produced in the core of the Sun with an outer-layer phenomenon, such as the solar activity.

Voloshin, Vysotsky and Okun (VVO) [485] have proposed a fascinating solution in which this anticorrelation could be explained if the neutrino has a magnetic moment as large as

$$\mu_{\nu_e} \sim 1 \times 10^{-10} \mu_B, \tag{83}$$

just below the limit set by laboratory experiments. The mechanism is that the left-handed neutrino (Dirac neutrino) precesses into the right-handed component under the influence of the magnetic field expected in the convective layer of the Sun [486]. For a relativistic neutrino the probability of the $\nu_L \to \nu_R$ flip is given by

$$P(\nu_L \to \nu_R) = \sin^2(\mu B L \sin\phi) , \tag{84}$$

where ϕ is the angle between the direction of the neutrino beam and the magnetic field ($\phi \sim 90°$ for the case of the solar neutrinos). The solar magnetic field is expected to be stronger when the Sun is more active. If $\mu BL \sim \pi/2$ is satisfied for the solar maximum, we anticipate a complete suppression of neutrino captures in the ^{37}Cl detector, since the right-handed neutrino is sterile to the detector. For $L \sim 0.3R_\odot$ (the length of the convective layer) and $B_{\rm max} \sim$ 2kGauss, (83) gives $\mu BL \sim 1.2$, about the desired condition, and the observed anticorrelation is accounted for. This is the only explanation so far known to explain the anticorrelation, if it exists at all.

The neutrino may also be of the Majorana type. In this case, however, the transition is $\nu_{eL} \to \nu_{\mu L}^c$ (or $\nu_{\tau L}^c$) and the mass difference between ν_e and $\nu_\mu(\nu_\tau)$ must be much smaller than $\Delta m^2 \sim \mu B_{\rm max} E_\nu \sim 1\times10^{-7}{\rm eV}^2$ in the original scenario.

VVO noted that in a realistic case ν_L receives a refractive effect in matter from scattering via the neutral current, while ν_R does not. This causes a difference in the potential between ν_L and ν_R, i.e., the Hamiltonian for the neutrino propagation is

$$H = \begin{pmatrix} \sqrt{2}G_F(n_e - \frac{1}{2}n_n) & \mu_\nu B \\ \mu_\nu B & 0 \end{pmatrix} \tag{85}$$

Here the (1,1) component is $V(\nu_e, e) + V(\nu_e, p) + V(\nu_e, n) = \sqrt{2}G(n_e - n_n/2) \simeq 5.5 \times 10^{-15}{\rm eV}(\rho/0.1g{\rm cm}^{-3})$ [see (II.167, 168); $n_e = n_p$ is assumed] and the off-diagonal components are $\mu_\nu B = 5.8 \times 10^{-16}{\rm eV}(\mu_\nu/10^{-10}\mu_B)(B/1{\rm kGauss})$. For a typical density in the convective layer of the Sun, $\rho = 0.2$g cm^{-3} at $0.7R_\odot$ and 0.02g cm^{-3} at $0.9R_\odot$, it is conceived that the precession by the off-diagonal element is suppressed unless $\mu_\nu BL \gtrsim 10$. An explicit calculation [487] shows that one needs

$$\mu BL \sim 3 \times (\pi/2) \tag{86}$$

to bring the flux suppressed to the level of 1/4 of the standard solar-model value. This means that $B_{\rm max} \sim$ 10kGauss is necessary for the magnetic moment[19] of (83).

Barbieri and Fiorentini [487] also discussed that, if the magnetic moment is as larger as $\gtrsim 1\times10^{-10}\mu_B$, the interference effect between the process $\nu_L \to [\nu_L + e^- \to$

[19]The authors of [487] assume that a large magnetic field is expected between $r = 0.8R_\odot$ and $0.9R_\odot$, and take $L \sim 0.1R_\odot$. The required value is thus about a factor 3 larger.

$\nu_L + e^-]$ and $\nu_L \to \nu_R \to [\nu_R + e^- \to \nu_L + e^-]$ (the latter through the magnetic moment) may be sizable in the azimuthal asymmetry with respect to the solar magnetic field axis. They calculated that it is about 10% (15%) for the electron energy cut-off at 7.5(5)MeV for $\mu_\nu = 1 \times 10^{-10}\mu_B$.

Let us note that the value of (83) is already one order of magnitude larger than the limit from stellar cooling (II.107), and such a value would require a drastic alteration of the theory of helium-burning stars and also of cooling white dwarfs. On the other hand, if $\mu \sim 10^{-11}\mu_B$, one needs $B_{\max} \sim 40$kGauss. Although this is a large value, the present understanding of the solar magnetic field does not exclude this case [488].

A more contriving possibility has been proposed in ref.[489], that the $\nu_L \to \nu_R$ flip and the MSW-type resonance can occur at once, which alleviates the suppression from the refractive effect in matter.

Let us consider the case of two Dirac neutrinos,

$$\psi = \begin{pmatrix} \nu_{eL} \\ \nu_{\mu L} \\ \nu_{eR} \\ \nu_{\mu R} \end{pmatrix} \tag{87}$$

The evolution equation is given by

$$i\frac{d}{dt}\psi = \mathcal{H}\psi \ , \tag{88}$$

where

$$\mathcal{H} = \begin{pmatrix} -\frac{\delta m^2}{4E}\cos 2\theta + V_e, & \frac{\delta m^2}{4E}\sin 2\theta, & \mu_{ee}B, & \mu_{e\mu}B \\ \frac{\delta m^2}{4E}\sin 2\theta, & \frac{\delta m^2}{4E}\cos 2\theta + V_\mu, & \mu_{\mu e}B, & \mu_{\mu\mu}B \\ \mu^*_{ee}B, & \mu^*_{\mu e}B, & -\frac{\delta m^2}{4E}, & 0 \\ \mu^*_{\mu\mu}B, & \mu^*_{\mu\mu}B, & 0, & \frac{\delta m^2}{4E}, \end{pmatrix} . \tag{89}$$

Here, $\delta m^2 = m^2_{\nu_\mu} - m^2_{\nu_e}$, and V_e and V_μ are the potentials felt by the left-handed ν_e and ν_μ which are given by $V_e = \sqrt{2}G_F(n_e - \frac{1}{2}n_n)$ and $V_\mu = -\frac{1}{\sqrt{2}}G_F n_n$. From expression (89) the $\nu_{eL} - \nu_{\mu R}$ resonance occurs when

$$V_e = \frac{\delta m^2}{4E}(1 + \cos 2\theta) \tag{90}$$

or

$$(n_e - \frac{1}{2}n_n)_{\mathrm{Res}} = \frac{\delta m^2 \cos^2\theta}{2\sqrt{2}G_F E} \ . \tag{91}$$

This is similar to the condition for the MSW resonance (V.110). While an analysis is complicated for a general case, the 4×4 matrix of (89) reduces into two blocks of 2×2 matrices in the limit of θ=0 and $\mu_{ee} = \mu_{\mu\mu} = 0$. The block that involves ν_{eL} represents the $\nu_{eL} \to \nu_{\mu R}$ transition, which is not suppressed by matter effects [489]. One can thus proceed with an analysis in a way similar to the MSW case, finding that a sufficiently large spin precession, $\nu_{eL} \to \nu_{\mu R}$, occurs under a magnetic field of 3kGauss for Δm^2 as large as 10^{-4}eV^2 [490].

If neutrinos are of the Majorana type, a possibility exists for a net $\nu_e \to \bar{\nu}_e$ conversion [491]; $\nu_e \to \nu_\mu$ through the MSW effect and subsequently $\nu_\mu \to \bar{\nu}_\mu$ by the helicity-flavour flip, or $\nu_e \to \bar{\nu}_\mu$ by helicity-flavour flip followed by $\bar{\nu}_\mu \to \bar{\nu}_e$ oscillation in vacuum. This is an interesting possibility, since the reaction rate of $\bar{\nu}_e$ is one order of magnitude larger than of ν_e in a water detector; however, the directionality information cannot be used to reduce the background.

The most important problem for the VVO scenario and its extensions is that the observation by KAMIOKANDE collaboration does not see any time variation for the most recent solar cycle (cycle 21) [40, 353]. The KAM-II observation was started in February 1987, and extended to April 1990 [40]. This time span corresponds to the period during which Davis et al. recorded the maximum capture rate of 4.2SNU and the minimum rate of 1.0SNU. The flux observed in KAM-II, however, stays at a constant value within a statistical accuracy of 20% and excludes the large time variation as observed in the Homestake detector at the 99% C.L.

This leaves the interpretation that the anticorrelation observed in the Homestake detector is merely a statistical fluke as the most likely possibility. The result of KAM-II has higher run-to-run statistics, and its detection seems to be solid since it observes the directionality of the neutrino flux.

It is not impossible, nevertheless, to think of particle models which can reconcile the time variation at the Homestake detector with the time invariability of the KAM-II result; we can explain the two experiments simultaneously if we would have some interactions of right-handed neutrinos with electrons with a strength almost equal to that of the standard weak interaction. One of the two proposed possibilities is that ν_R scatters off electrons in a water detector by a large magnetic moment with a strength comparable to weak interaction cross sections [492]. This, however, requires μ_ν as large as $6 \times 10^{-10}\mu_B$, which exceeds the limit from helium stars by two orders of magnitude. Another possibility is that ν_R interacts with electrons via a scalar particle exchange [493]. In fact, such a possibility is included in the model discussed in Sect.43; if f_{11} is non-vanishing and is as large as $f_{11}^2/m_\eta^2 \approx 10^{-5}\text{GeV}^{-2}$, ν_R scatters off electrons through the η exchange, which fills that part of the reaction rate missed by the $\nu_L \to \nu_R$ flip. The η particle, of course, does not couple to nucleons, and the capture rate with ^{37}Cl remains unchanged.

Attempts were also made to understand this experiment-dependent time variation by a resonant helicity-flavour flip mechanism using the energy dependence of the helicity-flavour flip [494]. The result of these analyses shows that it can explain the time variation of the Homestake experiment, allowing, however, for a 30% time variation at Kam-II during the period between the solar maximum and minimum. The capture rate with ^{71}Ga depends largely on the assumption on the magnetic field in the radiative zone ($R < 0.7R_\odot$).

45.2 Neutrino radiative decay and cosmology

The place where we expect interesting physical effects from neutrino radiative decay is cosmology, since the number of cosmologically produced neutrinos is very large and the energy stored in massive neutrinos is enormous compared with any astrophysically produced energies in a later epoch. Fig.49 shows in the $\tau_\nu - m_\nu$ plane the region where we expect cosmological (and astrophysical) effects from radiatively decaying neutrinos.

If neutrinos decay with a lifetime $\gtrsim 10^5$yr, the photons produced in the decay would remain unthermalised and would contribute to a diffuse uv, x-ray, or γ-ray background radiation [367, 495, 107, 496, 497]. The observed background radiation sets a limit on the lifetime, that it should be longer than $\approx 10^{23}$sec (denoted by U and X in Fig.49). If the lifetime is 10^5sec$< \tau_\nu < 10^5$yr, the emitted photons will be thermalised by Compton scattering, while the total number of photons of cosmic microwave background (CMB) is kept fixed; hence, they would distort the CMB spectrum [107, 496, 498]. (The number of CMB photons freezes at $t \sim 10^5$s, where inverse bremsstrahlung and double Compton scattering decouple from expansion, and since then the energy injected into the cosmic plasma is used to distort the black-body spectrum, see e.g. [498].) The constraint on the CMB spectral distortion is very strong, typically of the order of $\Delta\epsilon/\epsilon < 10^{-5}$ [499], where $\Delta\epsilon/\epsilon$ is the fraction of the energy injected into CMB, and the decay within the above lifetime interval is essentially forbidden (C). If the lifetime is shorter, high-energy photons are consumed to increase the CMB temperature. If this occurred after big-bang nucleosynthesis ($\sim 10^2$sec), it would disturb the agreement of nucleosynthesis calculation of n_B/n_γ with the observed elemental abundance (N). The photons emitted during the nucleosynthesis epoch also directly disturb element synthesis by fission (F) [500]. If the decay takes place before nucleosynthesis, it is harmless.

Neutrinos are also emitted copiously from supernovae and white dwarfs, much more than emission of photons. Therefore, if these neutrinos would decay with a lifetime shorter than the age of the universe, they would substantially increase the diffuse background photons (S_1, W) [501, 502]. If more than 1% of the neutrinos emitted from stellar collapse would decay before they reach the envelope of the progenitor star, they would drastically alter the energetics of type II supernovae (S_2) [502]. Constraints are also available from the absence of excess γ-rays emitted from SN1987A (S_3) [503]. The constraint against excess cooling of helium stars gives a limit displayed as He in the figure. We also plotted the constraints from the mass density of the Universe, (II.109; V.207) etc. as M.

Neutrinos with a very short lifetime do not receive constraints from astrophysics and cosmology, but for ν_μ such a case is excluded by laboratory experiments searching for $\nu_\mu \to \nu + \gamma$ ($\tau/m_{\nu_\mu} >$15.4 s$\cdot$ eV^{-1} at the 90% C.L.) [504].

The regions expected from (16) (standard model)and (46) are indicated in Fig. 49. The line corresponding to (46) is practically the limit for the fastest radiative decay for which we can make a particle physics model consistently with other laboratory experiments. From this figure, we understand that the most interesting case is long-lifetime ($\tau \sim 10^{23}$sec) neutrino decays with $m_\nu \simeq 0.1$eV-100eV, which are observed as a slight excess of the uv photon flux. The short-lifetime ($\tau \lesssim 10$sec) decay of heavy ($m_\nu \gtrsim 1$MeV) neutrinos (ν_τ) is also an allowed case. For a neutrino with mass $\gtrsim 1$ MeV, $\nu' \to \nu + e^+ + e^-$ is generally faster than $\nu' \to \nu + \gamma$.

In this way radiative decays of the neutrino receive very tight constraints from cosmology and astrophysics. The only way to avoid such strong constraints is to

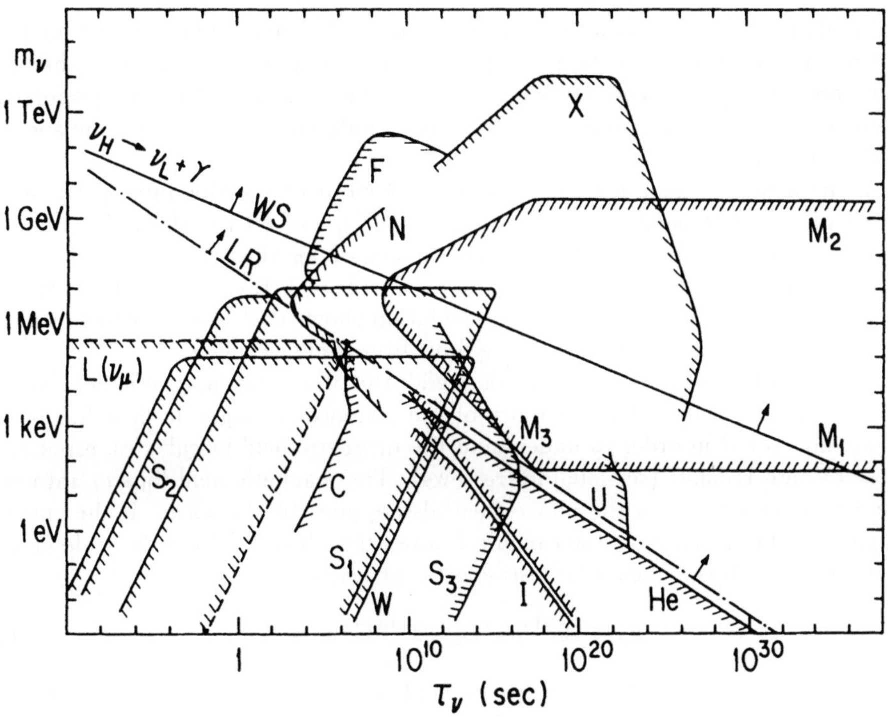

Figure 49: Constraints on radiatively decaying neutrinos. M: limit from cosmological mass density [36,367,368]. This limit applies to any unstable (or stable) neutrinos; X: limit from diffuse X-(γ-) ray radiation [107]; U: limit from uv radiation; C: constraint against distortion of the cosmic microwave background radiation [496]; I: allowed window for U due to interstellar infrared emission [496]; N: constraint from nucleosynthesis [500]; F: limit from photofission of light elements [500]; S_1: limit from diffuse X-ray from supernovae [502]; S_2: limit from supernova energetics [502]; S_3: limit from the absence of γ-ray from SN1987A [503]; W: limit from diffuse X-ray from white dwarfs [501]; He: limit against excessive cooling of He star (corresponding to $\mu_\nu = 10^{-11}\mu_B$) [123–125]. L stands for a laboratory limit from a search for $\nu_\mu \to \nu_e + \gamma$ [504] (the limit applies only to ν_μ). The two oblique lines represent maximally allowed decay rate in the Weinberg-Salam model [eq.(16)] (WS) and that in the left-right symmetric model [eq.(46)] (LR).

invoke competing decays to electromagnetically non-interacting particles, as we discussed in Sect.34.

45.3 Neutrino magnetic moment and the early Universe

Calculations of primordial nucleosysthesis leads to the constraint that the number of neutrino species be less than 3.6 [162]. This also applies to the right-handed neutrino; i.e. right-handed neutrinos of any generation should not be populated more than 0.6 times the ordinary left-handed neutrino abundance for one species at the time of decoupling of beta equilibrium ($T \sim 0.7$MeV).

If the neutrino has a magnetic (transition) moment, ν_L flips into ν_R via scattering off charged particles [505] (see also [123]); see (II.102,103). The nucleosythesis constraint means that the abundance of ν_R thus produced before decoupling of the neutrino from other radiations ($T \sim 4$MeV) be smaller than 0.6 times the equilibrium abundance. A calculation shows that this is satisfied unless $\mu_\nu > 0.6 \times 10^{-10} \mu_B$ which violates the limit from steller cooling.

A more interesting result can be derived if the primordial magnetic field would exist in the Universe. The existence of the primordial magnetic field has often been conjectured in order to understand the magnetic field in galaxies, especially those at high redshift (see [506] for reviews). The magnetic field flips ν_L into ν_R; the argument similar to the above then also applies to this case. If the cosmic magnetic field decreases as the canonical power law, $B \approx a^{-2}$ (a is the scale factor of the Universe), the nucleosysthesis constraint leads to

$$\mu_\nu < 10^{-15} \mu_B (B_0/10^{-10}\,\text{Gauss})^{-1}, \qquad (92)$$

where B_0 is the intergalactic magnetic field today. If μ_ν would be as large as $10^{-11}\mu_B$, (92) means $B_0 < 10^{-14}$ Gauss, which is too small for the primordial magnetic field to have any significance in cosmology.

VIII. Baryon numbers in the early Universe and neutrinos

46 Introduction

The modern view ascribes the origin of the baryon number in the Universe to the action of baryon number violating particle interactions at some early epoch of the Universe. Sakhalov has pointed out that the Universe's baryon asymmetry can be explained if three conditions cooperate: (i) baryon number is violated in a fundamental process, (ii) C and CP invariance is violated at the same time, and (iii) there is a deviation from thermal equilibrium [507]. After the advent of grand unification people realised that the first condition, which is the most difficult one in the standard theory, is naturally satisfied [508, 509]. The first attempts, where the scattering process $qq \rightarrow \bar{q}\ell$ was considered, however, almost dismissed the third condition, which made baryon numbers generated in this scenario very small [510]. The most conventional view today is that baryon asymmetry was generated during the epoch of grand unification by out-of-equilibrium, delayed decays of heavy coloured Higgs'[510] or Majorana neutrinos [511]. If grand unification were to be confirmed, this would certainly be the most promising view to explain the presence of baryon numbers.

More recently, Kuzmin, Rubakov and Shaposhnikov (KRS) [512] have pointed out that baryon-number violation takes place at temperatures at around the electroweak phase transition by a non-perturbative effect which makes possible the transition between different vacua classified with a topological property. This is actually a finite-temperature generalisation of what was discovered by 't Hooft [435] more than a decade ago: that baryon number is violated in the Weinberg-Salam model due to the presence of an anomaly in the baryon current. The 't Hooft's effect, however, is strongly suppressed, and it had been thought that it is not relevant to nature. KRS found that such a suppression does not exist at a high temperature comparable to that of the electroweak phase transition. While this KRS effect (sphaleron-induced baryon number violation), perhaps, has not yet been established beyond doubt, many analyses made to date indicate that such an effect is indeed likely to take place.

In this chapter we consider two cases in which the neutrino mass has an important interplay with baryon-number generation. In the next subsection we discuss the case with SO(10) grand unification in which the mass of the neutrino is related to the size of baryon asymmetry. In the subsequent subsection we describe the case of baryon number nonconservation with the KRS effect, which will turn out to have an interplay with the physics of massive neutrinos.

The presence of this baryon-number violating process does not necessarily mean that baryon number was generated during this epoch. Although the KRS effect erases very efficiently any pre-existing baryon numbers if they were generated by $B-L$ conserving interactions, the generation of a baryon-number excess by this mechanism is a non-trivial issue. An interplay with the physics of the neutrino mass is seen in the following way. If lepton-number violation processes are strong enough during the epoch when the KRS mechanism works, we are left with no baryon numbers nor lepton numbers, irrespective of whether baryon numbers are generated in $B - L$ conserving processes during an earlier epoch or not. The requirement that such a case should not happen leads to a new class of constraints on Majorana-type interactions in unified theories. On the other side, it is possible to use such a $\Delta L \neq 0$ process to generate the ΔL excess, which is eventually converted into the ΔB excess under the action of the KRS effect, if appropriate conditions are satisfied.

We refer to Ref.[44] for GUT baryogenesis and Ref.[513] for a list of ideas concerning non-GUT baryogenesis.

47 Baryon number and the neutrino mass in SO(10) GUT

Let us show that the amount of the baryon number generated during the early Universe is likely to be related to the neutrino mass in the SO(10) grand-unification models [514]. This arises from the fact that the loop diagram that contributes to the baryon asymmetry is counterbalanced by a corresponding diagram in which all fermions of definite chiralities in the original diagram are replaced by those of opposite chiralities: $\nu_L \leftrightarrow \nu_R, q_L \leftrightarrow q_R$ etc.; (see Fig.50 below). This pair of contributions exactly cancels each other if the phase-space volumes for these two diagrams are equal. This means that baryon number is generated from the disparity in the mass of ν_L and ν_R. The mass of ν_R must be roughly as large as the GUT scale in order to produce a sizable baryon asymmetry. This in turn leads to a constraint on the light neutrino mass through the seesaw formula.

Let us consider the baryon-number generation mechanism from out-of-equilibrium decays of the colour-triplet Higgs [510]. We assume the presence of two colour-triplet Higgs multiplets which belong to the vector representation **10**. As discussed in Sect.31, the quark masses are given by two Yukawa couplings or, conversely, Yukawa couplings g_1 and g_2 are expressed in terms of the two quark mass matrices m_u and m_d as

$$g_1 = (2\sqrt{2}G_F)^{1/2}(\eta_1 m_u + \eta_1^* m_d), \qquad (1)$$

$$g_2 = (2\sqrt{2}G_F)^{1/2}(\eta_2 m_u + \eta_2^* m_d), \qquad (2)$$

where $\eta_1/\eta_2 = -(v_2/v_1)^*$ with v_i the vacuum expectation values of the SU(2) doublet Higgs $H^{(i)}$.

We now consider the baryon asymmetry produced in the decay of coloured Higgs particles $H_c^{(i)}$. In the presence of CP violation there is a disparity in the partial width for $H_c \to \ell q$ (and $\bar{q}\bar{q}$) and $\bar{H}_c \to \bar{\ell}\bar{q}$ (and qq). This causes a baryon-number excess if other necessarily conditions are satisfied. The net baryon number produced by the pair decay of Higgs bosons H_c and its antiboson $\overline{H}_c$ is given by

$$\epsilon = \frac{\frac{1}{3}\Gamma(H_c \to \ell q) - \frac{2}{3}\Gamma(H_c \to \bar{q}\bar{q})}{\Gamma(H_c \to \ell q) + \Gamma(H_c \to \bar{q}\bar{q})} + \frac{-\frac{1}{3}\Gamma(\overline{H}_c \to \bar{\ell}\bar{q}) + \frac{2}{3}\Gamma(\overline{H}_c \to qq)}{\Gamma(\overline{H}_c \to \bar{\ell}\bar{q}) + \Gamma(\overline{H}_c \to qq)}. \qquad (3)$$

In order to generate baryon asymmetry left behind the GUT epoch during the early Universe, the baryon-number producing process must take place out of equilibrium. It has been shown that such an out-of-equilibrium condition is satisfied when the decay proceeds under the condition that the particle mass is larger than the temperature of the Universe, i.e., under the the circumstance that its inverse process is sufficiently suppressed by the Boltzmann factor [510]. (Note, however, that baryon-number generation is not a direct result of an imbalance between the decay and inverse decay; if the decay of H_c contributes to $\epsilon > 0$, its inverse decay also results in $\epsilon > 0$ as a result of CPT invariance (see [515])). Roughly speaking, the condition is

$$\Gamma/\gamma_{\rm exp} \lesssim 1 \ , \tag{4}$$

where Γ is the decay rate and

$$\gamma_{\rm exp} = \dot{a}/a = \left(\frac{16\pi^3 g_*}{45}\right)^{1/2} \frac{T^2}{M_{pl}} \tag{5}$$

is the expansion rate of the Universe. Writing $\Gamma = \alpha_H m_H$, the condition is

$$m_H > \alpha_H \times 10^{17}{\rm GeV} \ , \tag{6}$$

where $T \simeq m_H$ and the number of relativistic degrees of freedom at the GUT epoch, $g_* \simeq 160$, are put into $\gamma_{\rm exp}$. This is often quoted as the delayed decay scenario.

To quantitatively estimate the Universe's baryon excess we must solve the Boltzmann equation for the baryon-number density and the Higgs-particle density. Fry et al. [516] have given a convenient numerical formula for the final baryon number by fitting the solution of the Boltzmann equation; for the ratio of final baryon numbers to specific entropy,

$$\frac{kn_B}{s} = 0.5 \times 10^{-2} \epsilon \frac{1}{1 + (3K)^{1.2}} \tag{7}$$

for Higgs decays, where K is the ratio of the decay rate to the expansion rate,

$$K \equiv \frac{1}{2} \frac{\Gamma}{\gamma_{\rm exp}}\bigg|_{T=m_H} \tag{8}$$

$$\simeq 2.9 \times 10^{17}{\rm GeV} \alpha_H \frac{1}{m_H} \left(\frac{160}{g_*}\right)^{1/2} . \tag{9}$$

From a successful nucleosynthesis calculation [162] the left-hand side of (7) must be $kn_B/s = 10^{-10.30\pm0.15}$.

The net baryon number ϵ of (3) is computed from the one-loop diagrams in Fig.50 [517]. We note that the complex phases of g_i alone do not cause the required disparity in the partial widths; it arises from the interference term between a tree diagram and a one-loop rescattered diagram whose discontinuity (imaginary part) is non-vanishing. Therefore, Fig.50 is the simplest diagram which leads to the

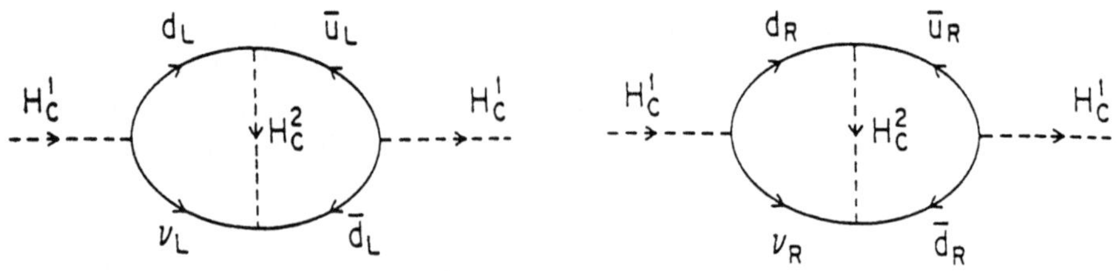

Figure 50: A pair of the simplest diagrams giving rise to baryon number generation by coloured Higgs decay in the SO(10) model. The decay width is given by the discontinuity (imaginary part) of these diagrams.

CP violation effect in the branching ratio and, hence, to baryon asymmetry. A computation [517] gives

$$\begin{aligned}\epsilon^{(1)} &= \frac{1}{8\pi}[\mathrm{tr}(f_1^\dagger f_1)]^{-1} \\ &\quad \times \mathrm{Im}\ \mathrm{tr}(f_2^\dagger f_1 f_2^\dagger f_1 \rho_i^\dagger \rho_i) F(m_2^2/m_1^2, \mu_i^2/m_1^2) \qquad (10)\end{aligned}$$

for the decay of $H_c^{(1)}$, and similarly for $H_c^{(2)}$. Here, m_i is the mass of $H_c^{(i)}$, f_i is the Yukawa coupling matrix of $H_c^{(i)}$ to fermions, $\rho_i\mu_i$ is the Majorana mass matrix of M_{ν_R} of ν_R (ρ_i is the phase factor) and

$$\begin{aligned}F(x,y) &\simeq y + x\ln[(1+x-y)/(1+x)] \ \text{ if } y<1 \\ &\simeq 1 - x\ln\left[\frac{(1+x)}{x}\right] \ \text{ if } y>1. \qquad (11)\end{aligned}$$

For the limit $y \ll 1, F(x,y) \sim y/(1+x)$. This explicitly demonstrates the cancellation between the two contributions, one containing ν_L and the other containing ν_R in the intermediate state; the degree of cancellation depends on the mass ratio $y = (M_{\nu_R}/m_{H_c})^2$.

For $M_{\nu_{Ri}} \ll m_{H_c^{(i)}} (y \ll 1)$ the dominant term of eq.(10) is

$$\epsilon^{(1)} \simeq \left(\frac{3^6}{\sqrt{2}\pi}\right)^{-1} G_F|\eta_2|^2 \sin\left(\arg\left(\frac{\eta_1}{\eta_2}\right)^2\right) m_t^6/[m_\nu^2(m_1^2+m_2^2)]\ , \qquad (12)$$

where we assume that the Majorana mass matrix for ν_R is a linear combination of m_u and m_d, as motivated in the Witten's SO(10) model (see Sect.31) to simplify the general expression and use the seesaw mass formula of eq.(VI.40). Upon adding the contribution from $H_c^{(2)}$ decay we obtain

$$\begin{aligned}\epsilon &= \epsilon^{(1)} + \epsilon^{(2)} \\ &= \left(\frac{3^6}{\sqrt{2}\pi}\right)^{-1} G_F(|\eta_1|^2 - |\eta_2|^2) \\ &\quad \times \sin\left[\arg\left(\frac{\eta_1}{\eta_2}\right)^2\right] \frac{m_t^6}{[m_\nu^2(m_1^2+m_2^2)]}\ . \qquad (13)\end{aligned}$$

If we use the constraint on m_{H_c} derived from proton-decay experiments,

$$m_{H_c} > 3 \times 10^{10}\text{GeV}, \tag{14}$$

and assume that $\eta_1/\eta_2 = O(1)$, we obtain from the empirical value of kn_B/s

$$(0.1 - 0.5)\text{eV} < m_\nu^{\max} < (10 - 50)\text{eV} \tag{15}$$

for $m_t = 100 - 200\text{GeV}$. This constraint applies to the heaviest neutrino mass.

48 Sphaleron-induced baryon-number violation

48.1 Instanton-induced baryon-number nonconservation

If we define the baryon- and lepton-number currents by

$$j_B^\mu = \frac{1}{3}\sum(\bar{q}_L\gamma^\mu q_L + \bar{u}_R\gamma^\mu u_R + \bar{d}_R\gamma^\mu d_R) \tag{16}$$

and

$$j_L^\mu = \sum(\bar{\ell}_L\gamma^\mu \ell_L + \bar{e}_R\gamma^\mu e_R + [\bar{\nu}_R\gamma^\mu \nu_R])\ , \tag{17}$$

where the first summation is made over colour and family and the second over family, we can show that the currents are conserved:

$$\partial_\mu j_B^\mu = 0; \qquad \partial_\mu j_L^\mu = 0\ . \tag{18}$$

Thus, the charges

$$B = \int d^3x\ j_B^0 \quad \text{and} \quad L = \int d^3x\ j_L^0, \tag{19}$$

are also conserved at the classical level.

These conservations, however, are violated by a quantum effect called the U(1) anomaly arising from the triangle diagram [435] depicted in Fig.51. Namely, in the presence of an anomaly (18) is modified to

$$\begin{aligned} \partial_\mu j_B^\mu &= \frac{N_f}{32\pi^2}g^2 F^a_{\mu\nu}\tilde{F}^a_{\mu\nu}\ , \\ \partial_\mu j_L^\mu &= \frac{N_f}{32\pi^2}g^2 F^a_{\mu\nu}\tilde{F}^a_{\mu\nu}\ , \end{aligned} \tag{20}$$

where

$$\begin{aligned} F^a_{\mu\nu} &= \partial_\mu W^a_\nu - \partial_\nu W^a_\mu + g\epsilon^{abc}W^b_\mu W^c_\nu\ , \\ \tilde{F}^a_{\mu\nu} &= \epsilon_{\mu\nu\lambda\sigma}F^{a\lambda\sigma}\ , \end{aligned} \tag{21}$$

and N_f is the number of generations. W^a_μ in eq.(21) is the SU(2) gauge field ($a = 1 - 3$). Eq.(20) means that

$$\partial_\mu(j_B^\mu - j_L^\mu) = 0 \tag{22}$$

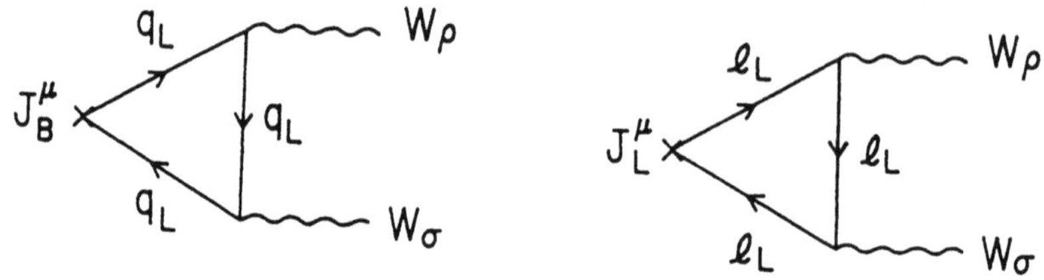

Figure 51: Triangle diagrams giving anomaly to the baryonic and leptonic currents.

and

$$\partial_\mu(j_B^\mu + j_L^\mu) = \frac{N_f}{16\pi^2} g^2 F^a_{\mu\nu} \tilde{F}^a_{\mu\nu} \tag{23}$$

The Euclidean integral of (23) over d^4x gives a change of $B+L$ between $\tau = -\infty$ and $\tau = +\infty$ (here $\tau = it$), i.e.,

$$\begin{aligned} \int d^4x \; \partial_\mu \; j^\mu_{B+L} &= \int d^3x \; j^0_{B+L} \Big|_{\tau=+\infty} - \int d^3x \; j^0_{B+L} \Big|_{\tau=-\infty} \\ &= (B+L)\Big|_{\tau=+\infty} - (B+L)\Big|_{\tau=-\infty} \\ &= \Delta(B+L) \; . \end{aligned} \tag{24}$$

On the other hand, the right-hand side of (23) is

$$\begin{aligned} \int d^4x \partial_\mu j^\mu_{B+L} &= 2N_f \frac{g^2}{32\pi^2} \int d^4x F^a_{\mu\nu} \tilde{F}^{a\mu\nu} \\ &= 2N_f n, \end{aligned} \tag{25}$$

where n is an integer and is called the winding number. Therefore,

$$\Delta(B+L) = 2N_f n \; . \tag{26}$$

The change in $B+L$ is caused by a tunneling effect due to instantons [435, 518]; the tunneling rate is

$$\begin{aligned} \Gamma &\sim |e^{-S(\text{instanton})}|^2 \\ &= e^{-16\pi^2/g^2} \; . \end{aligned} \tag{27}$$

This value, however, is very small (10^{-170}) and had been considered not to be physically relevant.

48.2 *B* violation in the Minkowski space-time

We consider the SU(2) model, omitting the U(1) part which does not play any essential role in the argument. We take the Lagrangian

$$\mathcal{L} = -\frac{1}{4} F^a_{\mu\nu} F^{a\mu\nu} + |D_\mu \phi|^2 - V(\phi) \; , \tag{28}$$

where

$$D_\mu \phi = (\partial_\mu - \frac{ig}{2}\tau^a W^a_\mu)\phi \; , \tag{29}$$

and

$$V(\phi) = \lambda(\phi^\dagger\phi - \frac{1}{2}v^2)^2 \, , \qquad \lambda > 0, \tag{30}$$

with ϕ an SU(2) doublet. By adopting the gauge for which $W^a_0 = 0$, (28) is written

$$\mathcal{L} = \frac{1}{2}(\dot{W}^a_i)^2 - \frac{1}{4}F^a_{ij}F^a_{ij} + |\dot{\phi}|^2 - |D_i\phi|^2 - V(\phi). \tag{31}$$

With the use of the canonical variables,

$$\begin{aligned} \pi_{Wi} &= \frac{\delta\mathcal{L}}{\delta\dot{W}_i} = \dot{W}_i \; , \\ \pi_\phi = \dot{\phi}^*, &\qquad \pi_{\phi^*} = \dot{\phi} \; , \end{aligned} \tag{32}$$

the Hamiltonian density is given by

$$\begin{aligned} \mathcal{H} &= \pi_{W_i}\dot{W}_i + \pi_\phi\dot{\phi} + \pi_{\phi^*}\dot{\phi}^* - \mathcal{L} \\ &= \frac{1}{2}(\dot{W}_i)^2 + |\dot{\phi}|^2 + \frac{1}{4}F^a_{ij}F^a_{ij} + |D_i\phi|^2 + V(\phi) \; . \end{aligned} \tag{33}$$

The vacuum solution is given by

$$\begin{aligned} \dot{W}^a_i &= 0, \qquad \dot{\phi} = 0, \\ F_{ij} &= 0, \qquad D_i\phi = 0, \qquad V(\phi) = 0 \; . \end{aligned} \tag{34}$$

$F_{ij} = 0$ means that $W^a_i(\vec{x})$ is given by a pure gauge function, i.e., the solution is

$$\begin{aligned} \widehat{W}_i(\vec{x}) \equiv W^a_i\tau^a &= -\frac{2i}{g}\partial_i g(\vec{x}) \cdot g^{-1}(\vec{x}) \\ \phi(\vec{x}) &= \frac{v}{\sqrt{2}}g(\vec{x})\begin{pmatrix}0\\1\end{pmatrix} , \end{aligned} \tag{35}$$

where $g(\vec{x})$ is an element of the SU(2) group and represents a mapping from the three-dimensional sphere S^3 to SU(2). Since

$$\pi_3(\mathrm{SU}(2)) = Z, \tag{36}$$

the function $g(\vec{x})$ is classified by a topological number [519, 518]. This means that the vacuum is classified into topologically distinct classes, labeled by the Chern-Simons charge,

$$Q = \frac{1}{24\pi^2}\int d^3x \; \epsilon^{ijk} Tr[\partial_i g \cdot g^{-1}\partial_j g \cdot g^{-1}\partial_k g \cdot g^{-1}] \; , \tag{37}$$

which takes an integer number.

In the presence of fermions the zero energy level changes when Q changes; this causes a baryon-number violation [520]. Let us suppose that all levels with

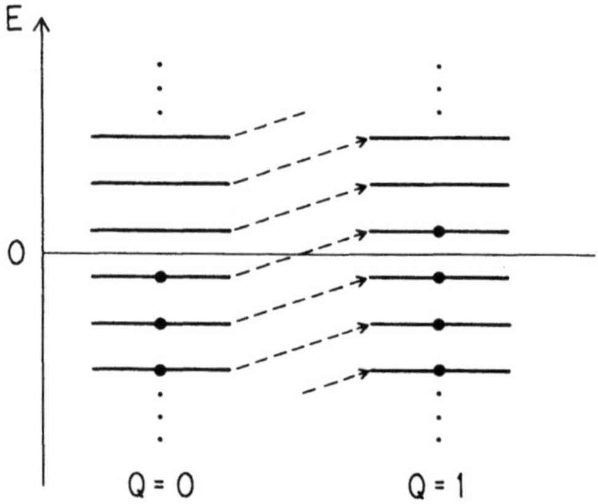

Figure 52: Fermion energy levels for the two states with different topological numbers, showing that the state with $Q = 1$ has unit baryon number.

$E < 0$ $(k = -1, -2, ...)$ are filled (Dirac sea) for $Q = 0$. This may be defined to be a $B = 0$ state. The change into $Q = 1$ causes the zero energy level to come to between $k = -1$ and $k = -2$ in the original level scheme. This means that the state with $Q = 1$ has $B = 1$ (see Fig.52).

We can see that this level crossing is associated with the "creation" of each fermion doublet, either an up or down component, that is associated with the "creation" of $3N_f$ quarks and N_f leptons. We therefore have a selection rule that $\Delta B = \Delta L = N_f \Delta Q$. This agrees with (20) derived from the U(1) anomaly argument.

48.3 Sphalerons

Klinkhammer and Manton [521] found a static, classical, saddle-point solution which connects the $Q = 0$ state with the $Q = 1$ state. The field equations derived from (31) are

$$\begin{aligned} (D_i F_{ij})^a &= \frac{i}{2} g \{\phi^\dagger \tau^a (D_j \phi) - (D_j \phi)^\dagger \tau^a \phi\}, \\ D_i^2 \phi &= 2\lambda(\phi^\dagger \phi - \frac{1}{2} v^2)\phi. \end{aligned} \tag{38}$$

The solution was found under the assumption that W_i^a and ϕ take the form

$$\begin{aligned} W_i^a &= \frac{2i}{g} f(r) \left(i\varepsilon_{iaj} x_j \frac{1}{r^2} \right), \\ \phi &= \frac{v}{\sqrt{2}} h(r)\, i \frac{\vec{\tau}\vec{x}}{r} \begin{pmatrix} 0 \\ 1 \end{pmatrix}. \end{aligned} \tag{39}$$

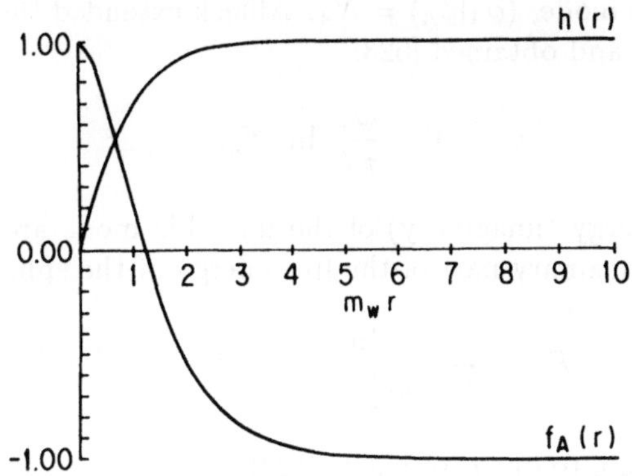

Figure 53: The functions $f_A(r)$ and $h(r)$ for the sphaleron.

The behaviour of W and ϕ is displayed in Fig.53. This solution is appropriately called a "sphaleron". With (33) the mass of the sphaleron is calculated to be

$$M_{\rm sph} = c\, m_W \frac{8\pi}{g^2}, \tag{40}$$

with c varying from 1.6 (for $\lambda \simeq 0$) to 2.7 (for $\lambda \to \infty$). The solution has a topological charge of 1/2.

At a finite temperature the transition from the $Q = 0$ state to the $Q = 1$ state takes place across the potential barrier by thermal fluctuations. The sphaleron corresponds to the solution corresponding to the top of the barrier. The transition crossing over this barrier is not suppressed at sufficiently high temperatures [512].

48.4 Calculation for the rate of the B-violating process

The transition probability can be calculated with the aid of Langer's theory [522] for the transition between two degenerate ground states. According to this theory, the transition amplitude is given by

$$\Gamma = [Nv]T\ e^{-\Delta F/T}\ e^{-M_{\rm sph}/T}, \tag{41}$$

where ΔF is the free-energy contribution from excitations around the sphaleron solution, ΔF=(free energy at sphaleron) − (free energy at vacuum), Nv is the product of the volume and the normalisation factor arising from the zero mode in the sphaleron background, and is given by

$$Nv = \Pi_A \left[\frac{N_A T}{2\pi}\right]^{1/2} \Omega_A\ . \tag{42}$$

Here, Ω_A is the volume for the zero mode (e.g., $\Omega_A = V$ for the translational mode, $8\pi^2$ for the rotational mode) and N_A is the adopted normalisation for the wave function of the zero mode, $\langle \phi_A | \phi_A \rangle = N_A$. Affleck extended the theory to the case of quantum theory, and obtained [523]

$$\Gamma = \frac{\omega_-}{\pi T} \text{ Im } \mathcal{F}_{\text{sp}} , \tag{43}$$

where ω_- is the energy (imaginary) of the unstable mode around the sphaleron, and Im$\mathcal{F}_{\text{sp}}$ is the imaginary part of the free energy of the sphaleron contribution,

$$\text{Im } \mathcal{F}_{\text{sp}} = \frac{Nv}{4\beta \sin(\frac{1}{2}\omega_-/T)} e^{-\Delta F/T} \; e^{-M_{\text{sph}}/T} . \tag{44}$$

Applying this formula to our case yields [524]

$$\Gamma = \frac{T\omega_-}{m_W(T)} \left(\frac{\alpha_w}{4\pi}\right)^4 (Nv) \left[\frac{2m_W(T)}{\alpha_w T}\right]^7 \kappa e^{-M_{\text{sph}}(T)/T} , \tag{45}$$

where $\alpha_w = g^2/4\pi$,

$$(Nv) = (Nv)_{\text{trans}} \cdot (Nv)_{\text{rot}} \approx 5 \times 10^4 T^3 V , \tag{46}$$

and $\kappa \sim$ O(1) is the determinant of the non-zero modes near to the sphaleron [525], and ω_- is a known quantity.

We note that the most important part in (45) is the Boltzmann factor and that $M_{\text{sph}}(T)$ and $m_W(T)$ are T-dependent. Since $m_W(T)$, and hence $M_{\text{sph}}(T)$, vanish towards the critical temperature T_c of SU$(2)_L$ breaking as $\sim [1-(T/T_c)^2]^{1/2}$, the suppression is removed as $T \to T_c$. A numerical evaluation shows that at a temperature T of 100GeV $\lesssim T \lesssim T_c$, Γ becomes faster than the expansion rate $\dot{a}/a$ of the universe, and that this baryon violating process becomes equilibrium.

For $T > T_c$ it is not clear what takes places, since the sphaleron configuration does not exist and the semiclassical treatment fails. On the other hand, there is no apparent reason for the $B+L$ violation to be suppressed [526]. In fact, the result of computer simulations on the lattice using a microcanonical technique [527, 528] suggest that the $B+L$ violating transition also takes place for this case at a rate given by [528]

$$\Gamma \sim (0.1-1)\alpha_w^4 T . \tag{47}$$

This rate is larger than the expansion rate of the Universe, $\gamma_{\text{exp}} \equiv \dot{a}/a \sim 30T^2/M_{pl}$, up to

$$T \lesssim \frac{\alpha_w^4}{30} M_{pl} \simeq 10^{12}\text{GeV}. \tag{48}$$

We remark that the result described here does not conflict with the conclusion by Gross et al. [453], who claimed that the instanton effect is suppressed at finite temperatures. What causes the effect here is classical thermal fluctuations, and not quantum tunneling, as described by instantons.

The important conclusion is that neither baryon number nor lepton number is conserved. The only conserved quantity is $B-L$. Starting from an arbitrary initial state, what is left after the action of sphalerons can be seen in the following way. When the sphaleron action is in equilibrium, we must have

$$N_f(\mu_{u_L} + 2\mu_{d_L}) + \sum_{i=1}^{N_f} \mu_{\nu_L^i} = 0 \ , \tag{49}$$

where μ_i denotes the chemical potential of particle i. In addition, above the critical temperature of the electroweak phase transition, both the sum of electroweak charge (Q) and that of the third component of weak isospin (T_3) must vanish. Namely,

$$\begin{aligned} Q &= 2N_f(\mu_{u_L} + \mu_{u_R}) - N_f(\mu_{d_L} + \mu_{d_R}) \\ &\quad - \sum_i(\mu_{e_L^i} + \mu_{e_R^i}) - 4\mu_W - 2N_H\mu_L = 0 \end{aligned} \tag{50}$$

$$\begin{aligned} T_3 &= \frac{3}{2}N_f(\mu_{u_L} + \mu_{d_L}) + \frac{1}{2}\sum_i(\mu_{\nu_L^i} + \mu_{e_L^i}) \\ &\quad - 4\mu_W - N_H(\mu_{H^0} - \mu_{H^-}) = 0 \ . \end{aligned} \tag{51}$$

The equilibrium relations also demand that:

$$\begin{aligned} \mu_W &= \mu_{H^-} + \mu_{H^0} \ , \\ \mu_{d_L} &= \mu_{u_L} + \mu_W \ , \\ \mu_{\nu_L^i} &= \mu_{e_L^i} + \mu_W \ , \\ \mu_{u_R} &= \mu_{H^0} + \mu_{u_L} \ , \\ \mu_{d_R} &= -\mu_{H^0} + \mu_{d_L} \ , \\ \mu_{e_R^i} &= -\mu_{H^0} + \mu_{e_L^i} \ . \end{aligned} \tag{52}$$

From (49)–(52) we obtain [526] (see also [529])

$$\Delta B_0 = \frac{8N_f + 4N_H}{22N_f + 13N_H}\Delta(B-L)_i \ , \tag{53}$$

where $\Delta(B-L)_i$ is the initial value of $B-L$, ΔB_0 is the final baryon number, and N_H is the number of Higgs doublets. This means that if $B-L=0$ initially, no baryons remain after the action of sphalerons; for instance, since SU(5) GUT respects $B-L$, the baryon numbers generated by SU(5) GUT are all erased by the sphaleron effect above the electroweak energy scale.[20]

See Ref.[531] for a general review of the sphaleron-induced baryon number non-conservation.

48.5 Constraints on the $B-L$ violating interactions

If some interactions that violate $B-L$ become thermal equilibrium at the temperature where the sphaleron action is effective, the combined effects erase all

[20]It is argued that a small amount of the baryon number excess survives the sphaleron-induced processes even if $B-L=0$ for all temperatures, if one takes account of the charged lepton mass [530]. This effect, however, is very small ($< 10^{-6}$), and is negligible in any case.

pre-existing baryon- and lepton-number excess in the Universe. The presence of the baryon asymmetry today requires that such erasure should not have taken place. This requirement leads to constraints on the $B-L$ violating interactions [532].

The simplest example is given by the dimension-five operator discussed in Sect.14,

$$L = \frac{1}{M}\ell_L\ell_L\phi\phi \, , \tag{54}$$

which violates L by two units. The interaction (54) causes $\ell_L + \phi \to \bar{\ell}_L + \bar{\phi}$ scattering with a cross section

$$\sigma(\ell_L + \phi \to \bar{\ell}_L + \bar{\phi}) \simeq \frac{1}{\pi}\left[\frac{1}{M}\right]^2 \, , \tag{55}$$

where the masses of particles are neglected. The rate of this lepton-number violating process is

$$\begin{aligned} \Gamma_{\Delta L\neq 0} &\simeq \ < \sigma n_\phi > \\ &\simeq \ 0.12 \times \frac{1}{\pi}\frac{T^3}{M^2} \, , \end{aligned} \tag{56}$$

where $n_\phi = [\zeta(3)/\pi^2]T^3$ is the number density of ϕ particles in the Universe at a temperature $T \gg m_\phi$. The condition that this process does not become thermal equilibrium at temperature T_0 is

$$\Gamma_{\Delta L\neq 0}(T_0) < \gamma_{\rm exp} \simeq 17T_0^2/M_{pl} \, , \tag{57}$$

which leads to

$$M > 10^{14}\left(\frac{T_0}{10^{12}\text{GeV}}\right)^{1/2} \text{GeV} \, . \tag{58}$$

In Ref.[532] T_0 was taken to be the electroweak energy scale. On the other hand, the authors of [529, 533, 534] extended it to 10^{12}GeV, the temperature below which the $B+L$ violating transition becomes faster than the expansion rate $\gamma_{\rm exp}$ of the Universe, as discussed in the previous subsection. Since the neutrino mass is given by $m_\nu = \langle\phi\rangle^2/M$, eq.(58) also gives a constraint on m_ν as [532, 529, 533, 534]

$$m_\nu < 1\text{eV}\left(\frac{10^{12}\text{GeV}}{T_0}\right)^{1/2} \, . \tag{59}$$

If there are no mechanisms which generate baryon asymmetry below $T \simeq 10^{12}$GeV, T_0 is taken to be 10^{12}GeV and (59) gives a very strong constraint. On the other hand, some $B-L$ violation process may give rise to a $B-L$ number excess, once the out-of-equilibrium condition happens to be satisfied. In such a case, T_0 must be taken to be that temperature, and the constraint becomes weaker[21]. In

[21]Some other reasons to truncate T_0 at a smaller energy scale are also discussed. For instance, the authors of [535] argued in the context of SUSY theories that chiral gaugino charges inhibit erasure of the baryon/lepton asymmetry above the effective SUSY breaking scale, 10^8GeV, say.

the next subsection we discuss the possibility of baryon asymmetry generation at $T \gtrsim 400$GeV. In this case the constraint is weakened to be $m_\nu < 50$keV. If a baryon asymmetry is generated at the electroweak transition energy, the constraint is further weaker; it becomes the original constraint in [532].

The argument here can be extended to the case which involves any $B-L$ violating operators [533]. Let us consider the interaction with a non-renormalisable effective operator of dimension $D = 4 + n$,

$$\mathcal{L}_D = \frac{O_D}{M^n} . \tag{60}$$

The rate of interactions induced by (60) is

$$\Gamma_D(T) \sim \frac{T^{2n+1}}{M^{2n}} . \tag{61}$$

If we impose

$$\Gamma_D(T) \lesssim \gamma_{\rm exp}(T) \tag{62}$$

to avoid thermal equilibrium of $B-L$ violating processes, we have the condition that

$$M^{2n} \gtrsim 0.05 M_{pl} T^{2n-1} . \tag{63}$$

If we set $T \approx 10^{12}$GeV again, we obtain

$$M \gtrsim 10^{12+3/n}{\rm GeV} . \tag{64}$$

If we apply this constraint to the $\Delta B = 2$ interaction that causes $n-\bar{n}$ oscillation ($D = 9$) [536], we have

$$M \gtrsim 10^{12.6}{\rm GeV} . \tag{65}$$

This is compared with the present experimental limit from the absence of $n-\bar{n}$ oscillation ($\tau_{n\bar{n}} \gtrsim 1 \times 10^7$sec from a free neutron experiment[537], or $> 1.2 \times 10^8$sec from bound neutron [^{16}O, ^{56}Fe] experiments using a calculation for the nuclear suppression factor [538]):

$$(M)_{\rm exper.} \gtrsim 2 \times 10^6{\rm GeV} . \tag{66}$$

Constraint (65) means that the detection of $n-\bar{n}$ oscillation in the laboratory is hopeless. However, if the $B-L$ asymmetry is produced at $T \sim 400$GeV, constraint (62) can be weakened to be

$$M \gtrsim 10^2{\rm GeV} , \tag{67}$$

which is still compatible with the present experimental limit.

48.6 Model for baryogenesis

Two types of scenarios have been discussed for baryon asymmetry generation using sphaleron-induced baryon nonconservation. They are:

(i) The asymmetry is generated at the time of the electroweak phase transition. For this to happen, this transition must be strong first order so that the $B+L$

violation rate is already sufficiently small after the transition in order to have generated asymmetry left intact [539, 540, 541].
(ii) The baryon number asymmetry is the result of a primordial L (or $B-L$) asymmetry generated during some epoch earlier than the electroweak phase transition. This L asymmetry is converted into a B asymmetry with the action of sphalerons [542].

Scenario (i) has the advantage that we do not need to suppose any new physics beyond the standard model. It is, in principle, possible to satisfy Sakharov's condition if the phase transition at the electroweak energy scale is of first order, since C and CP are violated by the electroweak interaction including the Yukawa couplings. For this scenario to be realised, the $B+L$ violation rate must be small enough after the phase transition. This condition is satisfied if the mass of the Higgs scalar is $m_\phi \leq 37$GeV according to a one-loop calculation [539, 543], which, however, conflicts with the LEP experiment ($m_\phi > 58.4$GeV) [75]. To make the scenario viable one must add some extra scalar particles in order to relax the constraint[544]. Another problem with this scenario is that the mechanism of baryon asymmetry production is not clear. Shaposhnikov and others [539, 540] ascribe its origin to non-trivial behaviours of Higgs and gauge fields during the electroweak phase transition, for instance, breaking of the degeneracy of the Chern-Simon vacua caused by CP violation and the expansion of the Universe. Another class of authors [541] argue that a CP violating interaction of the fermions with the bubble of the true vacuum formed during the phase transition generates a baryon-number excess. This argument, however, depends heavily on the dynamics of first-order phase transitions, which has not been well understood yet. With either of the two scenarios, it seems difficult to carry out a convincing calculation of the amount of the baryon-number excess.

In this section we present a simple model for generating a $B-L$ asymmetry for case (ii) [542]. For this purpose we add to the standard model three families of right-handed Majorana neutrinos ν_R^i ($i = 1-3$). The Lagrangian which includes ν_R^i is

$$\mathcal{L}_{N_R} = \overline{\nu^i}_R \, \partial\!\!\!/ \nu_R^i + M_i \overline{\nu^{ic}}_R \nu_R^i + h_{ij} \overline{\nu^i}_R \ell_L^j \phi^\dagger + h.c. \tag{68}$$

We take a basis in which the Yukawa coupling $f_{ij}\bar{e}_R^i \ell_L^j \phi$ is diagonal ($f_{ij} = f_i \delta_{ij}$). We assume that the Yukawa coupling h_{ij} involves a CP-violating phase, as in the Kobayashi-Maskawa matrix.

The Majorana neutrino, $N = \nu_R + \nu_R^c$, has two decay channels,

$$\begin{aligned} N &\rightarrow \ell_L + \bar{\phi} \,, \\ &\rightarrow \bar{\ell}_L + \phi \,. \end{aligned} \tag{69}$$

If CP is violated, a difference emerges between the branching ratios for the two decay channels in (69), through the interference term between the tree diagram and the one-loop diagram of Higgs decay (Fig.54). This process yields a net lepton-number (equivalently $B-L$) generation. For simplicity, we assume that $M_1 < M_2 < M_3$ and that the baryon-number generation is dominantly due to the decay of the lightest particle N_R^1. Then, the magnitude of net lepton-number

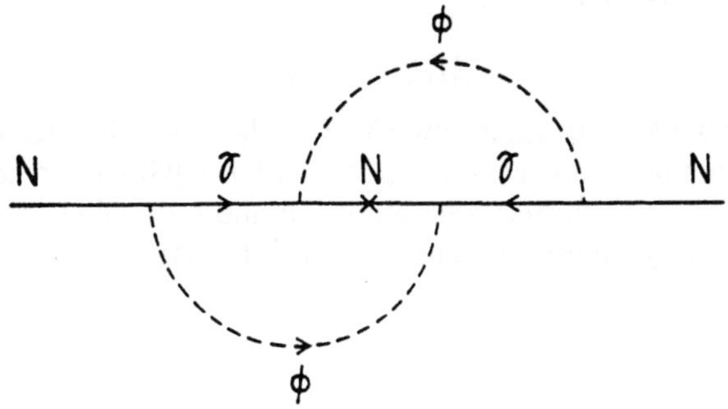

Figure 54: Diagram leading to lepton-number generation by heavy Majorana neutrino decay. The decay width is given by the discontinuity (imaginary part) of the diagram.

production by the decay of one N_R^1 is calculated to be

$$\epsilon = \frac{\Gamma(N^1 \to \ell_L + \bar{\phi}) - \Gamma(N^1 \to \bar{\ell}_L + \phi)}{\Gamma(N^1 \to \ell_L + \bar{\phi}) + \Gamma(N^1 \to \bar{\ell}_L + \phi)} \tag{70}$$

$$\simeq \left(\frac{1}{4\pi}\right) \mathrm{Im}(h_{1i}h^\dagger_{ij}h^\dagger_{jk}h_{k1}) I(M_j^2/M_1^2) \Big/ (hh^\dagger)_{11} \tag{71}$$

with

$$I(x) = x^{1/2}\{1 + (1+x)\ln(x/(1+x))\} . \tag{72}$$

If we assume h_{33} to be the largest entry of h_{ij} and $M_3 \gg M_1$, we obtain

$$\varepsilon \simeq \left(\frac{1}{8\pi}\right) |h_{33}|^2 \left(\frac{M_1}{M_3}\right) \delta , \tag{73}$$

where δ is the CP violating phase in h_{ij}.

The argument is now parallel to the case for baryon asymmetry generation discussed in Sect.47. The out-of-equilibrium condition is satisfied by a delayed decay scenario. The lepton excess left behind is expected to be given by (7)

$$\frac{kn_L}{s} \sim 10^{-3}\epsilon K^{-1.2} \tag{74}$$

for $K \ll 1$ with K defined by (8). An explicit Boltzmann equation calculation was also carried out for this leptogenesis case, which confirmed the result using (74) [545]. This lepton excess is immediately converted into a baryon excess,

$$n_B = -\frac{28}{79} n_L , \tag{75}$$

from (53) for $N_f = 3$ and $N_H = 1$ by the action of sphalerons.

If we assume a maximal CP violation δ=1, $\epsilon \simeq 10^{-2}(M_1/M_3)$ for an assumed Dirac mass of 100GeV for the third-generation neutrino ($m_3 = h_{33}v/\sqrt{2}$). It therefore seems reasonable to suppose that $\epsilon \approx 10^{-3}$. In order to have the required baryon excess, $kn_B/s \simeq 10^{-10.3}$, $K < 2 \times 10^{-3}$ is necessary. This leads to

$$M_1 \gtrsim 3.6 \times 10^{12} (hh^\dagger)_{11} \mathrm{GeV}, \tag{76}$$

using $\Gamma(N_\nu^1) = (hh^\dagger)_{11}M_1/16\pi$. If we take $(hh^\dagger)_{11}^{1/2} \sim 10^{-5}$ (i.e. $m_1 \sim 1\text{MeV}$), we obtain

$$M_1 \gtrsim 400\text{GeV} . \tag{77}$$

This M_1 corresponds to $m_{\nu_1} \lesssim$ a few eV. On the other hand, the condition for the survival of this baryon excess is $m_\nu < 50(M_1/400\text{GeV})^{-1/2}$ keV for the heaviest neutrino, as derived in Sect.48.5 above. A more detailed analysis is given in Ref.[546]. For supersymmetric extensions, see [547, 548].

Acknowledgment

We would like to thank Jiro Arafune, Ken-ichi Hikasa, Chong-Sa Lim, Volodya Miransky, Hitoshi Murayama, Alexei Smirnov, Atsuto Suzuki and Eiichi Takasugi for their valuable comments during the various stages of the completion of this article.

Appendix A
Triangle anomaly and anomaly cancellation

When the gauge theory contains chiral fermions, the triangle diagram contains a divergence which cannot be absorbed into renormalisation of the coupling constants; this divergence is called an anomaly. Since this anomaly, if it exists, spoils renormalisability of the gauge theory, it must be canceled among contributions from various fermion loops to make the theory internally consistent.

The triangle diagram of a left-handed fermion ψ_L is proportional to [390]

$$\mathrm{Tr}[T_a\{T_b, T_c\}] \ , \tag{A1}$$

where T_a is the generator of the Lie algebra of the gauge group G, and the coupling of the gauge field A^a_μ is defined by

$$g\overline{\psi_L} T_a \gamma_\mu A^a_\mu \psi_L \ . \tag{A2}$$

To cancel the anomaly, we should have

$$\sum_i \mathrm{Tr}[T^i_a\{T^i_b, T^i_c\}] = 0 \ , \tag{A3}$$

where T^i_a is the coupling matrix of the gauge field A^a_μ to the i-th fermion ψ^i_L.

We now consider the case for the standard model based on SU(3) $\times$ SU(2) $\times$ U(1)$_Y$, and take the U(1) hypercharge Y for the quark, lepton and Higgs fields to be

$$\begin{array}{ccc} & & Y \\ \ell_L & : & A \\ e_R & : & B \\ q_L & : & C \\ u_R & : & D \\ d_R & : & E \end{array} \tag{A4}$$

and

$$\phi \ : \ F \ .$$

Without loss of generality, we choose $F = 1$ for the normalisation of hypercharge. By requiring that a cancellation of the gauge anomalies takes place within one family, we obtain

$$\begin{array}{rll} 2C = D + E & \text{for the} & [\mathrm{SU}(3)]^2 \cdot \mathrm{U}(1) \ \text{anomaly} \ , \\ 3C + A = 0 & \text{for the} & [\mathrm{SU}(2)]^2 \cdot \mathrm{U}(1) \ \text{anomaly} \ , \\ 2A^3 + 6C^3 = B^3 + 3(D^3 + E^3) & \text{for the} & [\mathrm{U}(1)]^3 \ \text{anomaly} \ . \end{array} \tag{A5}$$

Other combinations, such as $[\mathrm{SU}(2)]^3$, automatically vanish by a trace operation.

If we impose that all fermions, except for the neutrino, obtain masses through the Yukawa coupling to ϕ, as in Eq.(II.29), we should have

$$\begin{aligned} A - B - 1 &= 0 , \\ C - E - 1 &= 0 , \\ C - D + 1 &= 0 . \end{aligned} \tag{A6}$$

Note that the first equation of (A5) is not independent from (A6). Eqs.(A5) and (A6) have a unique solution:

$$\begin{aligned} A &= -1 , \\ B &= -2 , \\ C &= 1/3 , \\ D &= 4/3 , \\ E &= -2/3 , \end{aligned} \tag{A7}$$

in agreement with the $\mathrm{U}(1)_Y$ charge assignment of the standard electroweak theory. This means that electromagnetic charges are quantised, and in particular $Q_\nu = 0$.

We now discuss the anomaly-free condition when the right-handed neutrino ν_R is introduced. We take the U(1) hypercharge $Y = G$ for ν_R. The last equation of (A5) is modified to be

$$2A^3 + 6C^3 = B^3 + 3(E^3 + D^3) + G^3 , \tag{A8}$$

which, when combined with other equations of (A5) and (A6), gives

$$G = -3C + 1 . \tag{A9}$$

This means that we lose the quantisation of the electromagnetic charges.

If we write the charge of neutrino to be $G = \varepsilon$, we obtain

$$\begin{aligned} A &= -1 + \varepsilon , \\ B &= -2 + \varepsilon , \\ C &= (1 - \varepsilon)/3 , \\ D &= (4 - \varepsilon)/3 , \\ E &= -(2 + \varepsilon)/3 , \end{aligned} \tag{A10}$$

which gives the condition

$$Q_n + Q_\nu = Q_p + Q_e = 0 . \tag{A11}$$

This guarantees the neutrality of the hydrogen atom. This is what we discussed in Sect.5.1.

Appendix B
Beta decay

A brief summary is given for the elementary account for the theory of nuclear beta decays. More details can be found in the standard text books, e.g., Bohr and Mottelson [549], Morita [550] and Blin-Stoyle [551].

The weak-interaction Hamiltonian is given by (1), i.e.,

$$H = \frac{G_V}{\sqrt{2}} \bar{e}\gamma^\mu(1-\gamma_5)\nu \; \bar{p}\gamma_\mu(1-g_A\gamma_5)n \; , \tag{B1}$$

where $G_V = G_\mu |U_{ud}| = G_\mu \cos\theta_c$ up to the radiative corrections. With a non-relativistic reduction (Pauli-reduction) of the nuclear part, the matrix-element square becomes

$$\begin{aligned} |-i\langle f|H|i\rangle|^2 &= \frac{G_V^2}{2} 8E_e E_\nu [(1+\beta\cos\theta)\langle 1\rangle^2 \\ &\quad + g_A^2(1-\frac{\beta}{3}\cos\theta)\langle\vec{\sigma}\rangle^2] \; , \end{aligned} \tag{B2}$$

where E_e and E_ν are the electron and neutrino energies $\beta = p_e/E_e, \theta$ is the angle between e and $\bar{\nu}_e$, and

$$\begin{aligned} \langle 1\rangle &= \langle f|\tau_+|i\rangle \\ \langle\vec{\sigma}\rangle &= \langle f|\tau_+\vec{\sigma}|i\rangle \end{aligned} \tag{B3}$$

are the Fermi and Gamow-Teller matrix elements, which comes from γ_μ ($\mu = 0$) and $\gamma_\mu\gamma_5$ ($\mu = i$), respectively.

The decay rate is given by

$$\begin{aligned} \Gamma &= \int \frac{d^3p_e}{2E_e(2\pi)^3} \frac{d^3p_\nu}{2E_\nu(2\pi)^3} (2\pi)\delta(Q-E_e-E_\nu)|\langle f|H|i\rangle|^2 F(E_e, Z) \\ &= \frac{G_V^2}{2\pi^3} \int_{m_e}^{E_0} dE_e E_e p_e E_\nu p_\nu \; [\langle 1\rangle^2 + g_A^2\langle\vec{\sigma}\rangle^2] F(E_e, Z), \end{aligned} \tag{B4}$$

where $E_0 = Q - m_\nu$. Writing

$$f(E_0) = \int_1^{w_0} w\sqrt{w^2-1}(w_0-w)^2 \; F(w,z)dw \tag{B5}$$

for $m_\nu = 0$ ($w_0 = E_0/m_e$),

$$\Gamma = \frac{G_V^2}{2\pi^3} m_e^5 \; f(E_0)[\langle 1\rangle^2 + g_A^2\langle\vec{\sigma}\rangle^2]. \tag{B6}$$

Using the half-life, $t = \ln 2/\Gamma$, we have

$$\begin{aligned} |M_\beta|^2 \equiv \langle 1\rangle^2 + g_A^2\langle\vec{\sigma}\rangle^2 &= \frac{2\pi^3 \ln 2}{(ft)m_e^5 G_V^2} \\ &= \frac{6147}{(ft)} \mathrm{s} \end{aligned} \tag{B7}$$

with the ft value by Hardy et al. [82] (often used in the literature is the Wilkinson's recommended value 6166±2s in the numerator of (B7) [86]).

The Fermi matrix element $\langle 1\rangle$ is non-vanishing if $\Delta T = 0$ and $\Delta J = 0$, and

$$\langle 1\rangle^2 = T(T+1) - T_{3i}T_{3f} \; . \tag{B8}$$

Such beta decays are called superallowed transitions, and $\log ft$ is smaller than 3.8, corresponding to $|M_\beta|^2 \gtrsim 1$. For the Gamow-Teller transition the selection rule is $\Delta T = 0, \pm 1$ and $\Delta J = 0, \pm 1 (J = 0 \to J = 0$ is forbidden). Although a typical $\log ft$ is around 5, it may vary from 4 to 9. For $n \to p + e^- + \nu_e$, $\langle 1 \rangle^2 = 1$ and $\langle \vec{\sigma} \rangle^2 = 3$. In eqs.(B4)–(B5) $F(E_e, Z)$ is the Fermi function for the correction of the Coulomb wave function distortion given by

$$F(E,Z) = \frac{1+\gamma}{2} 4(pR)^{2\gamma-2} e^{\pi\nu} \frac{|\Gamma(\gamma+i\nu)|^2}{[\Gamma(2\gamma+1)]^2} \tag{B9}$$

$$\simeq \frac{2\pi\nu}{1-e^{-2\pi\nu}}, \tag{B10}$$

where $\gamma = \sqrt{1-\alpha^2 Z^2}$, $\nu = \alpha Z E/p$ and the last expression is for $\gamma \approx 1$. For β^+ decay the sign of ν in (B9) is reversed, and we have

$$F(E,Z) \simeq \frac{2\pi\nu}{e^{2\pi\nu}-1} \tag{B11}$$

for $\gamma \approx 1$. Roughly speaking, the Coulomb correction factor is $|\psi_{\rm Coulomb}(0)|^2 / |\psi_{\rm plane\ wave}(0)|^2$.

When the argument is made to an accuracy of 1–2%, QED radiative corrections should be taken into account [79, 80]. This is made by replacing $|M_\beta|^2 d^3p$ with

$$|M_\beta|^2 d^3p \left[1 + \frac{\alpha}{2\pi} g(E, E_0) + \delta_{\rm in}(\beta) - \delta_{\rm in}(\mu)\right], \tag{B12}$$

where $\delta_{\rm in}(\beta) - \delta_{\rm in}(\mu)$ is the universal part given by eq.(II.80) and

$$\begin{aligned} g(E,E_0) &= 3\ln\left(\frac{m_p}{m_e}\right) - \frac{3}{4} + \frac{4}{\beta} L\left(\frac{2\beta}{1+\beta}\right) \\ &+ 4(\mathrm{th}^{-1}\beta/\beta - 1)\left[\frac{E_0 - E}{3E} - \frac{3}{2} + \ln\frac{2(E_0-E)}{m_e}\right] \\ &+ \frac{1}{\beta}\mathrm{th}^{-1}\beta\left[2(1+\beta^2) + \frac{(E_0-E)^2}{6E^2} - 4\mathrm{th}^{-1}\beta\right] \end{aligned} \tag{B13}$$

$$L(x) = \int_0^x \frac{dt}{t} \ln(1-t). \tag{B14}$$

This gives $(\alpha/2\pi) g(E, E_0) = \delta_{\rm out}$ in eq.(I.79).

For an accurate calculation the integral of f in (B5) is also replaced by

$$f(E_0) = \int w\sqrt{w^2-1}(w_0 - w)^2 F(w,z) C(w) dw\,, \tag{B15}$$

with C the spectral shape factor, which takes account of (i) the screening effects of atomic electrons, (ii) the dependence of the nuclear matrix element on E_e, and (iii) the correction from second-forbidden matrix elements.

Inverse β decay (ν–nucleus scattering) is given by simply replacing the integral of (B4) with

$$\frac{1}{2E_\nu} \int \frac{d^3p_e}{2E_e(2\pi)^3}$$

i.e.,

$$\sigma(\nu_e A \to e^- A') = \frac{1}{\pi} G_V^2 E_e p_e [\langle 1 \rangle^2 + g_A^2 \langle \vec{\sigma} \rangle^2] F(E_e, Z) \ . \tag{B16}$$

The coefficient of the angular dependence is obtained straightforwardly from (B2).

Appendix C
Representation of SO (10) group

1. SO(N) group

The SO(N) group has $N(N-1)/2$ generators

$$\Sigma_{ij} = -\Sigma_{ji} \quad (i, j = 1 - N) \ , \tag{C1}$$

which satisfy the Lie algebra,

$$[\Sigma_{ij}, \Sigma_{\ell m}] = i\{\delta_{i\ell}\Sigma_{jm} + \delta_{jm}\Sigma_{i\ell} - \delta_{im}\Sigma_{j\ell} - \delta_{j\ell}\Sigma_{im}\} \ . \tag{C2}$$

2. Spinor representations of SO($N = 2n$)

The spinor representations of SO($2n$) are easily constructed by introducing γ matrices $\Gamma_i (i = 1 \sim 2n)$ of SO($2n$), which satisfy Clifford algebra,

$$\{\Gamma_i, \Gamma_j\} = 2\delta_{ij} \ . \tag{C3}$$

These Γ_i are represented by $2^n \times 2^n$ matrices. The spinor representation of SO($2n$) Lie algebra is defined by

$$\Sigma_{ij} = \frac{1}{2i}[\Gamma_i, \Gamma_j] \ . \tag{C4}$$

The spinor field ξ, which consists of 2^n components, transforms under SO(10) as

$$\xi \to e^{i\omega_{k\ell}\Sigma_{k\ell}}\xi \ . \tag{C5}$$

For SO(N) with N even there exists a chiral projection operator,

$$\Gamma_\chi \equiv -i\Gamma_1 \ldots \Gamma_{2n} \ . \tag{C6}$$

We define chiral spinor fields, ξ_+ and ξ_-, by

$$\xi_\pm \equiv \frac{1}{2}(1 \pm \Gamma_\chi)\xi, \tag{C7}$$

each of which has 2^{n-1} components. Since all Σ_{ij} commute with Γ_χ, the two spinors ξ_+ and ξ_- are independent and transform as

$$\xi_\pm \to e^{i\omega_{k\ell}\Sigma_{k\ell}}\xi_\pm \ . \tag{C8}$$

The Γ matrices of SO($2n$) can be constructed from those of SO($2n-2$) by induction, as

$$\begin{aligned}
\Gamma_i &= \begin{pmatrix} 0 & -i\tilde{\Gamma}_i \\ i\tilde{\Gamma}_i & 0 \end{pmatrix}, \qquad i = 1 \sim 2n-2, \\
\Gamma_{2n-1} &= \begin{pmatrix} 0 & -i\tau_3 \\ i\tau_3 & 0 \end{pmatrix}, \\
\Gamma_{2n} &= \begin{pmatrix} 0 & \mathbf{1} \\ \mathbf{1} & 0 \end{pmatrix},
\end{aligned} \tag{C9}$$

where $\tilde{\Gamma}_i (i = 1 \sim 2n-2)$ are the Γ matrices of SO($2n-2$). The induction starts from $n = 1$ with $\Gamma_1 = \tau_1$ and $\Gamma_2 = \tau_2$. In this basis Γ_χ is given by

$$\Gamma_\chi = \begin{pmatrix} \mathbf{1} & 0 \\ 0 & -\mathbf{1} \end{pmatrix}. \tag{C10}$$

and the $2^n \times 2^n$ generators Σ_{ij} are

$$\Sigma_{ij} = \begin{pmatrix} \sigma_{ij} & 0 \\ 0 & -\sigma_{ij} \end{pmatrix}.$$

It is clear that the $2^n \times 2^n$ matrix Σ_{ij} are reduced to two irreducible $2^{n-1} \times 2^{n-1}$ matrices σ_{ij} and $-\sigma_{ij}$.

3. Representation of SO(10)

The SO(10) Γ matrices are written

$$\Gamma_i = \left(\begin{array}{cccccccccccccccc}
&&&&&&&&&&&&&&& \tau_i \\
&&&&&&&&&&&&&& -\tau_i & \\
&&&&&&&&&&&&& -\tau_i && \\
&&&&&&&&&&&& \tau_i &&& \\
&&&&&&&&&&& -\tau_i &&&& \\
&&&&&&&&&& \tau_i &&&&& \\
&&&&&&&&& \tau_i &&&&&& \\
&&&&&&&& -\tau_i &&&&&&& \\
&&&&&&& -\tau_i &&&&&&&& \\
&&&&&& \tau_i &&&&&&&&& \\
&&&&& \tau_i &&&&&&&&&& \\
&&&& -\tau_i &&&&&&&&&&& \\
&&& \tau_i &&&&&&&&&&&& \\
&& -\tau_i &&&&&&&&&&&&& \\
& -\tau_i &&&&&&&&&&&&&& \\
\tau_i &&&&&&&&&&&&&&&
\end{array}\right), i = 1-3,$$

$$
\Gamma_4 = \begin{pmatrix}
 & & & & & & & i\tau_1^{(4)} \\
 & & & & & & -i\tau_1^{(4)} & \\
 & & & & & -i\tau_1^{(4)} & & \\
 & & & & i\tau_1^{(4)} & & & \\
 & & & -i\tau_1^{(4)} & & & & \\
 & & i\tau_1^{(4)} & & & & & \\
 & i\tau_1^{(4)} & & & & & & \\
 -i\tau_1^{(4)} & & & & & & &
\end{pmatrix},
$$

$$
\Gamma_5 = \begin{pmatrix}
 & & & & & & & i\tau_3^{(4)} \\
 & & & & & & -i\tau_3^{(4)} & \\
 & & & & & -i\tau_3^{(4)} & & \\
 & & & & i\tau_3^{(4)} & & & \\
 & & & -i\tau_3^{(4)} & & & & \\
 & & i\tau_3^{(4)} & & & & & \\
 & i\tau_3^{(4)} & & & & & & \\
 -i\tau_3^{(4)} & & & & & & &
\end{pmatrix},
\tag{C12}
$$

$$
\Gamma_6 = \begin{pmatrix}
 & & 0 & -\tau_1^{(8)} \\
 & 0 & \tau_1^{(8)} & 0 \\
 0 & \tau_1^{(8)} & 0 & \\
 -\tau_1^{(8)} & 0 & &
\end{pmatrix},
$$

$$
\Gamma_7 = \begin{pmatrix}
 & & 0 & -\tau_3^{(8)} \\
 & 0 & \tau_3^{(8)} & 0 \\
 0 & \tau_3^{(8)} & 0 & \\
 -\tau_3^{(8)} & 0 & &
\end{pmatrix},
$$

$$
\Gamma_8 = \begin{pmatrix} 0 & -i\tau_1^{(16)} \\ i\tau_1^{(16)} & 0 \end{pmatrix},
$$

$$
\Gamma_9 = \begin{pmatrix} 0 & -i\tau_3^{(16)} \\ i\tau_3^{(16)} & 0 \end{pmatrix},
$$

$$
\Gamma_{10} = \begin{pmatrix} 0 & \mathbf{1}^{(16)} \\ \mathbf{1}^{(16)} & 0 \end{pmatrix},
$$

where $\tau_i^{(n)}$ is an $n \times n$ matrix extension of the Pauli matrix τ_i and hence $\tau_i^{(2)} = \tau_i$,

$$
\tau_1^{(4)} = \begin{pmatrix} 0 & \mathbf{1}^{(2)} \\ \mathbf{1}^{(2)} & 0 \end{pmatrix}, \tau_2^{(4)} = \begin{pmatrix} 0 & -i\mathbf{1}^{(2)} \\ i\mathbf{1}^{(2)} & 0 \end{pmatrix}, \tau_3^{(4)} = \begin{pmatrix} \mathbf{1}^{(2)} & 0 \\ 0 & -\mathbf{1}^{(2)} \end{pmatrix},
$$

and so on. From these Γ matrices we can easily construct 16×16 irreducible matrices, σ_{ij}. It is more convenient, however, to use a new σ_{ij} obtained from the unitary transformation,

$$\sigma_{ij} \to U^{-1}\sigma_{ij}U \ ,$$

$$U = \begin{pmatrix} \begin{matrix} \mathbf{1}^{(4)} & 0 \\ 0 & u \end{matrix} & 0 \\ 0 & \begin{matrix} \mathbf{1}^{(4)} & 0 \\ 0 & u \end{matrix} \end{pmatrix} \text{ with } u = \begin{pmatrix} \begin{matrix} 0 & 1 \\ 1 & 0 \end{matrix} & 0 \\ 0 & \begin{matrix} 0 & 1 \\ 1 & 0 \end{matrix} \end{pmatrix} . \tag{C13}$$

With the new basis the **16**-dimensional spinor field $\xi_+(\mathbf{16})$ is

$$\xi_+(\mathbf{16}) = (u\ u\ u\ \nu\ \bar{u}\ \bar{u}\ \bar{u}\ \bar{\nu}\ d\ d\ d\ e\ \bar{d}\ \bar{d}\ \bar{d}\ \bar{e})_L^T \ . \tag{C14}$$

The gauge interactions for $\xi_+(\mathbf{16})$ are given by

$$\mathcal{L} = \overline{\psi}(\partial\!\!\!/ - \frac{ig}{2\sqrt{2}}\sigma_{jk}A\!\!\!/_{jk})\psi \ , \tag{C15}$$

where $\psi \equiv \xi_+(\mathbf{16})$ and A^{μ}_{jk} are 45-dimensional gauge fields (real field). The factor $1/2\sqrt{2}$ is inserted to normalise the coupling in agreement with the conventional definition. To write the Yukawa couplings to ψ fields, we must introduce a new spinor $\tilde{\psi}$ to represent the charge conjugate field ψ^*:

$$\tilde{\psi} \equiv \beta\psi^* \tag{C16}$$

with

$$\beta \equiv \Gamma_2\Gamma_4\Gamma_5\Gamma_8\Gamma_9 \ .$$

Since $\beta\sigma^*_{ij}\beta^{-1} = -\sigma_{ij}$, the thus-defined $\tilde{\psi}$ transforms as a spinor field of SO(10). The possible forms of the Yukawa interactions for ψ are then

$$\begin{aligned} &\tilde{\psi}^\dagger\Gamma_\mu\psi\phi^\mu \ , \qquad \tilde{\psi}_{[}\Gamma_\mu\Gamma_\nu\Gamma_{\rho]}\psi\phi^{\mu\nu\rho} \\ &\tilde{\psi}^\dagger{}_{[}\Gamma_\mu\Gamma_\nu\Gamma_\rho\Gamma_\sigma\Gamma_{\eta]}\psi\phi^{\mu\nu\rho\sigma\eta} \qquad \text{etc.} \end{aligned} \tag{C17}$$

Here, ϕ^μ is a vector, $\phi^{\mu\nu\rho}$ is an antisymmetric third-rank tensor, and $\phi^{\mu\nu\rho\sigma\eta}$ is an antisymmetric fifth-rank tensor, whose dimensions are **10**, **120** and **126**, respectively.

The Yukawa couplings of $\phi^\mu(\mathbf{10})$ which yield the fermion mass matrix are given explicitly by

$$\begin{aligned} \mathcal{L}_Y &= f\tilde{\psi}^\dagger\Gamma_\mu\phi_\mu\psi + h.c. \\ &= 2f(\bar{u}_R u_L + \bar{\nu}_R\nu_L)(\phi_9 - i\phi_{10}) \\ &\quad + 2f^\dagger(\bar{d}_L d_R + \bar{e}_L e_R)(\phi_9 - i\phi_{10}) + \dots \ . \end{aligned} \tag{C18}$$

This is eq.(32) of Sect. VI.

Representations in terms of SU(4)$\times$SU(2)$_L\times$SU(2)$_R$ subalgebra can be constructed by noting that the five Cartan generators are given by:

$$\lambda_3 = \tfrac{1}{2}(\sigma_{12} - \sigma_{34})$$

$$\lambda_8 = -\tfrac{1}{2\sqrt{3}}(\sigma_{12} + \sigma_{34} + 2\sigma_{56}) \tag{C19}$$

$$\lambda_{15} = \tfrac{1}{\sqrt{6}}(\sigma_{12} + \sigma_{34} - \sigma_{56}),$$

for the SU(4) part,

$$\tau_{L3} = -\frac{1}{2}(\sigma_{78} + \sigma_{910}), \tag{C20}$$

for $SU(2)_L$, and

$$\tau_{R3} = -\frac{1}{2}(\sigma_{78} - \sigma_{910}), \tag{C21}$$

for $SU(2)_R$, λ_i being the SU(4) generalisation of Gell-Mann's λ matrices.

With a straightforward calculation we can write the gauge interaction of (C15) explicitly in terms of the $SU(3)_c \times SU(2)_L \times SU(2)_R$ components. The interaction part reads,

$$\begin{aligned}
\mathcal{L}_{int} = & -\bar{\psi}^{ia}_{L,R}\frac{ig}{2}[\lambda^\alpha]_{ab}\gamma_\mu\psi^{ib}_{L,R}G^\alpha_\mu \\
& -\bar{\psi}^{ia}_L\frac{ig}{2}[\tau^\beta]_{ij}\gamma_\mu\psi^{ja}_L W^\beta_{L\mu} \\
& -\bar{\psi}^{ia}_R\frac{ig}{2}[\tau^\beta]_{ij}\gamma_\mu\psi^{ja}_R W^\beta_{R\mu} \\
& +\frac{q}{\sqrt{2}}[\bar{q}^{ia}_L\gamma_\mu(\ell^i_R)^c + \bar{\ell}^j_L\gamma_\mu(q^{ja}_R)^c + \epsilon_{abc}\overline{(q^{jb}_R)^c}\gamma_\mu q^{ic}_L]X^{ij}_{\mu a} \\
& + h.c.,
\end{aligned} \tag{C22}$$

where ψ in (C15) is

$$\psi_L = (\psi_L, \psi^c_R), \tag{C23}$$

and

$$\begin{aligned}
\psi_{L,R} &= (q,\ell)_{L,R} \\
q_{L,R} &= \begin{pmatrix} u_1 & u_2 & u_3 \\ d_1 & d_2 & d_3 \end{pmatrix}_{L,R} \\
\ell_{L,R} &= \begin{pmatrix} \nu \\ e \end{pmatrix}_{L,R}
\end{aligned} \tag{C24}$$

The indices α run 1–15 and β run 1–3, i, j take 1, 2 and a, b take 1–3. The gauge fields are

$$G^\alpha = (\mathbf{8} + \mathbf{3} + \mathbf{3}^* + \mathbf{1}) \tag{C25}$$

of $SU(3)_c$,

$$W_{L,R} = \mathbf{3} \tag{C26}$$

of $SU(2)_{L,R}$, and

$$X_a^{ij} = (\mathbf{3}, \mathbf{2}, \mathbf{2}) \tag{C27}$$

of $SU(3)_c \times SU(2)_L \times SU(2)_R$. We note that G and $W_{L,R}$ are real, and X are complex fields.

The decomposition of the irreducible representations of SO(10) into those of subgroup components is given in [391]. It is often convenient to construct representations on the basis of those of SU(5), see e.g. [552].

References

[1] W. Pauli, Letter to L. Meitner and her colleagues (letter open to the participants of the conference in Tübingen) (1930), recorded in W. Pauli, *Wissenschaftlicher Briefwechsel*, Band II (Springer, Berlin, 1985) S. 39; the first publication was given in W. Pauli, Septieme Conseil de Physique Solvay 1933: "*Noyaux Atomiques*", Paris 1934, p.324f.

[2] J. Chadwick, Verh. d. deutschen phys. Ges. **16**, 383 (1914).

[3] E. Fermi, Ricercha Scient. 2, No 12 (1933); Z. Phys. **88**, 161 (1934).

[4] F. Perrin, Compt. Rendues **197**, 1625 (1933).

[5] M. Fierz, Z. Phys. **104**, 553 (1937).

[6] B. Stech and J. H. D. Jensen, Z. Phys. **141**, 175 (1955).

[7] L. Landau, Nucl. Phys. **3**, 127 (1957).

[8] E. C. G. Sudarshan and R. E. Marshak, in *Proc. Padua Conference on Mesons and Recently Discovered Particles* (1957); R. P. Feynman and M. Gell-Mann, Phys. Rev. **109**, 193 (1958); E. C. G. Sudarshan and R. E. Marshak, Phys. Rev. **109**, 1860(L) (1958).

[9] T. D. Lee and C. N. Yang, Phys. Rev. **104**, 254 (1956).

[10] F. Reines and C. L. Cowan, Jr., Nature **178**, 446 (1956); C. L. Cowan et al., Science **124**, 103 (1956).

[11] W. B. Herrmannsfeldt, D. R. Maxson, P. Stähelin and J. S. Allen, Phys. Rev. **107**, 641(L) (1957).

[12] M. Goldhaber, L. Grodzins and A. W. Sunyar, Phys. Rev. **109**, 1015(L) (1958).

[13] W. Pauli, in *Aufsätze und Vorträge über Physik und Erkenntnistheorie* (F. Vieweg und Sohn, Braunschweig, 1961) S. 156.

[14] H. Yukawa, Proc. Phys. Math. Soc. Japan **17**, 48 (1935).

[15] J. Schwinger, Ann. Phys. (NY) **2**, 407 (1957).

[16] S. L. Glashow, Nucl. Phys. **22**, 579 (1961).

[17] S. Weinberg, Phys. Rev. Lett. **19**, 1264 (1967).

[18] A. Salam, in *Elementary Particle Theory, Proc. of the 8th Nobel Symposium*, Aspenäsgården, 1968, ed. by N. Svartholm (Almqvist and Wiksell, Stockholm, 1968) p.367.

[19] G. 't Hooft, Nucl. Phys. **B35**, 167 (1971).

[20] F. J. Hasert et al., Phys. Lett. **46B**, 138 (1973).

[21] S. Weinberg, Rev. Mod. Phys. **46**, 255 (1974); in *Proc. of the 19th International Conference on High Energy Physics*, Tokyo, 1978, ed. by S. Homma, M. Kawaguchi and H. Miyazawa (Physical Society of Japan, Tokyo, 1979) p.907.

[22] G. Arnison et al., Phys. Lett. **122B**, 103 (1983); **126B**, 398 (1983); M. Banner et al., Phys. Lett. **122B**, 476 (1983).

[23] S. Weinberg, in *Proc. of the First Workshop on Grand Unification*, ed. by P. H. Frampton, S. L. Glashow and A. Yildiz (Math Sci Press, 1980) p.347; in *Proc. of 23rd International Conference on High Energy Physics*, Berkeley 1986, ed. by S. C. Loken (World Scientific, Singapore, 1987), vol. I, p.271.

[24] E. Majorana, Nuovo Cim. **14**, 171 (1937).

[25] E. M. Lyman, Phys. Rev. **55**, 234(A) (1939).

[26] R. O. Haxby, W. E. Shoupp, W. E. Stephens and W. H. Wells, Phys. Rev. **58**, 1035 (1940).

[27] E. J. Konopinski, Phys. Rev. **72**, 518(L) (1947).

[28] L. M. Langer and R. J. D. Moffat, Phys. Rev. **88**, 689 (1952).

[29] K.-E. Bergkvist, Nucl. Phys. **B39**, 317; 371 (1972).

[30] V. A. Lubimov et al., Phys. Lett. **94**, 266 (1980).

[31] M. Goeppert-Mayer, Phys. Rev. **48**, 512 (1935).

[32] W. H. Furry, Phys. Rev. **56**, 1184 (1939).

[33] B. Pontecorvo, Zh. Eksp. Teor. Fiz. **33**, 549 (1957) [JETP **6**, 429 (1958)]

[34] Z. Maki, M. Nakagawa, S. Sakata, Prog. Theor. Phys. **28**, 870 (1962).

[35] S. S. Gershtein and Ya. B. Zeldovich, Pis'ma Zh. Eksp. Teor. Fiz. **4**, 174 (1966) [JETP Lett. **4**, 120 (1966).]

[36] R. Cowsik and J. McClelland, Phys. Rev. Lett. **29**, 669 (1972).

[37] R. Davis, Jr., D. S. Harmer and K. C. Hoffman, Phys. Rev. Lett. **20**, 1205 (1968).

[38] J. N. Bahcall and R. Davis, Jr., in *Essays in Nuclear Astrophysics*, ed. by C. A. Barnes, D. D. Clayton and D. Schramm (Cambrige University Press, Cambridge, 1982) p.243.

[39] J. N. Bahcall, *Neutrino Astrophysics* (Cambridge University Press, Cambridge, 1989).

[40] K. S. Hirata et al., Phys. Rev. Lett. **63**, 16 (1989); **65**, 1297 (1990).

[41] A. I. Abazov et al., Phys. Rev. Lett. **67**, 3332 (1991).

[42] P. Anselmann et al., Phys. Lett. **B285**, 376; 390 (1992).

[43] A. D. Dolgov and Ya. B. Zeldovich, Rev. Mod. Phys. **53**, 1 (1981).

[44] E. W. Kolb and M. S. Turner, *The Early Universe* (Addison-Wesley, Redwood, U. S. A., 1990).

[45] H. Weyl, Z. f. Phys. **56**, 330 (1929). More discussion was made by W. Pauli, Handbuch d. Phys. **24.1**, 83 (1933) [especially p.226].

[46] E. S. Abers and B. W. Lee, Phys. Rep. C**9**, 1 (1973); J. C. Taylor, *Gauge Theories of Weak Interactions* (Cambridge University Press, Cambridge, 1976).

[47] C. N. Yang and R. L. Mills, Phys. Rev. **96**, 191 (1954); R. Utiyama, Phys. Rev. **101**, 1597 (1956).

[48] Y. Nambu, G. Jona-Lasinio, Phys. Rev. **122**, 345 (1961); **124**, 246 (1961); J. Goldstone, Nuovo Cim. Serie X, **19**, 154 (1961).

[49] P. W. Higgs, Phys. Lett. **12**, 132 (1964); Phys. Rev. Lett. **13**, 508 (1964); F. Englert and R. Brout, Phys. Rev. Lett. **13**, 321 (1964); G. S. Guralnik, C. R. Hagen and T. W. B. Kibble, Phys. Rev. Lett. **13**, 585 (1964); T. W. B. Kibble, Phys. Rev. **155**, 1554 (1967).

[50] S. L. Glashow, J. Iliopoulos and L. Maiani, Phys. Rev. D**2**, 1285 (1970).

[51] M. Gell-mann and M. Lévy, Nuovo Cim. **10**, 705 (1960).

[52] N. Cabibbo, Phys. Rev. Lett. **10**, 531 (1963) .

[53] M. Kobayashi and T. Maskawa, Prog. Theor. Phys. **49**, 652 (1973).

[54] R. E. Behrends, R. J. Finkelstein and A. Sirlin, Phys. Rev. **101**, 866 (1956).

[55] Particle Data Group, Phys. Rev. **D45**, Part II, 1 (1992).

[56] B. Adeva et al. (L3 collaboration), Z. Phys. C**51**, 179 (1991).

[57] G. Alexander et al. (OPAL collaboration), Z. Phys. C**52**, 175 (1991).

[58] P. Abreu et al. (Delphi collaboration), Nucl. Phys. **B367**, 511 (1991).

[59] D. Decamp et al. (Aleph collaboration), Z. Phys. C**53**, 1 (1992).

[60] The LEP collaborations: ALEPH, DELPHI, L3 and OPAL, Phys. Lett. B**276**, 247 (1992).

[61] J. Alitti et al. (UA2 collaboration), Phys. Lett. B**276**, 354 (1992).

[62] F. Abe et al. (CDF collaboration), Phys. Rev. Lett. **65**, 2243 (1990).

[63] H. Abramowicz et al. (CDHS collaboration), Phys. Rev. Lett. **57**, 298 (1986).

[64] J. V. Allaby et al. (CHARM Collaboration) Z. Phys. **C36**, 611 (1987).

[65] A. Sirlin, Phys. Lett. B**232**, 123 (1989); S. Fanchiotti and A. Sirlin, Phys. Rev. **D41**, 319 (1990).

[66] A. Sirlin, Phys. Rev. **D22**, 971 (1980); D**29**, 89 (1984).

[67] K.-I. Aoki, Z. Hioki, R. Kawabe, M. Konuma and T. Muta, Prog. Theor. Phys. Suppl. **73**, 1 (1982).

[68] W. F. L. Hollik, Fortsch. Phys. **38**, 165 (1990).

[69] P. Langacker, in *Electroweak Physics Beyond the Standard Model*, ed. J. W. F. Valle and J. Velasco (World Scientific, Singapore, 1991), p.75; Phys. Rev. Lett. **63**, 1920 (1989); P. Langacker and M. Luo, Phys. Rev. D**44**, 817 (1991).

[70] J. Ellis and G. L. Fogli, Phys. Lett. B**249**, 543 (1990).

[71] L. F. Abbott and R. M. Barnett, Phys. Rev. **D18**, 3214 (1978); J. E. Kim, P. Langacker, M. Levine and H. H. Williams, Rev. Mod. Phys. **53**, 211 (1981).

[72] U. Amaldi et al., Phys. Rev. **D36**, 1385 (1987); G. L. Fogli and D. Haidt, Z. Phys. **C40**, 379 (1988); G. Costa et al., Nucl. Phys. B**297**, 244 (1988); J. Ellis and G. L. Fogli, Phys. Lett. B**232**, 139 (1989).

[73] D. Decamp et al., Phys. Lett. B**236**, 233 (1990); **245**, 289 (1990); **246**, 306 (1990).

[74] S. Coleman and E. Weinberg, Phys. Rev. D**7**, 1888 (1973).

[75] D. Buskulic et. al., Phys. Lett. B**313**, 299 (1993).

[76] M. Veltman, Acta. Phys. Pol. B**8**, 475 (1977).

[77] B. W. Lee, C. Quigg and H. B. Thacker, Phys. Rev. D**16**, 1519 (1977).

[78] J. Kuti, L. Lin and Y. Shen, Phys. Rev. Lett. **61**, 678 (1988); A. Hasenfratz et al., Nucl. Phys. B**317**, 81 (1989); G. Bhanot, K. Bitar, U. M. Heller and H. Neuberger, Nucl. Phys. B**353**, 551 (1991); U. M. Heller, H. Neuberger and P. Vranas, Florida preprint FSU-SCRI-91-94 (1991); the first proposal was made in R. Dashen and H. Neuberger, Phys. Rev. Lett. **50**, 1897 (1983).

[79] T. Kinoshita and A. Sirlin, Phys. Rev. **113**, 1652 (1959); G. Källén, Nucl. Phys. **B1**, 225 (1967); A. Sirlin, Phys. Rev. **164**, 1767 (1967).

[80] A. Sirlin, Nucl. Phys. **B71**, 29 (1974).

[81] W. J. Marciano and A. Sirlin, Phys. Rev. Lett. **56**, 22 (1986); W. J. Marciano, Ann. Rev. Nucl. Part. Sci. **41**, 469 (1992).

[82] J. C. Hardy et al., Nucl. Phys. **A509**, 429 (1990); I. S. Towner and J. C. Hardy, Nucl. Phys. **A205**, 33 (1973)

[83] A. Sirlin, Phys. Rev. **D35**, 3423 (1987); see also W. Jaus and G. Rasche, Nucl. Phys. **A143**, 202 (1970); A. Sirlin and R. Zucchini, Phys. Rev. Lett. **57**, 1994 (1986); W. Jaus and G. Rasche, Phys. Rev. **D35**, 3420 (1987) for history of the radiative correction of $O(\alpha^2)$.

[84] P. Bopp et al., Phys. Rev. Lett. **56**, 919 (1986).

[85] W. Mampe et al., Phys. Rev. Lett. **63**, 593 (1989); J. Byrne et al., Phys. Rev. Lett. **65**, 289 (1990).

[86] D. H. Wilkinson, Nucl. Phys. **A377**, 474 (1982).

[87] M. Ademollo and R. Gatto, Phys. Lett. **10**, 339 (1964).

[88] H. Leutwyler and M. Roos, Z. Phys. **C25**, 91 (1984).

[89] J. F. Donoghue, B. R. Holstein and S. W. Klimt, Phys. Rev. **D35**, 934 (1987).

[90] M. Bourquin et al., Z. Phys. C**21**, 27 (1983).

[91] D. I. Britton et al., Phys. Rev. Lett. **68**, 3000 (1992); G. Czapek et al., Phys. Rev. Lett. **70**, 17 (1993).

[92] F. Abe et al., Phys. Rev. Lett. **69**, 28 (1992).

[93] J. Alitti et al., Z. Phys. C**52**, 209 (1991).

[94] D. A. Dryman, preprint Triumf TRI-PP-92-3 (1992).

[95] H. Albrecht et al., Phys. Lett. B**292**, 221 (1992).

[96] J. Z. Bai, et. al., Phys. Rev. Lett. **69**, 3021 (1992).

[97] D. Buskulic et al., Phys. Lett B**297**, 432 (1992).

[98] J. Bernstein, M. Ruderman and G. Feinberg, Phys. Rev. **132**, 1227 (1963).

[99] S. L. Adler, Phys. Rev. **177**, 2426 (1969); J. S. Bell and R. Jackiw, Nuovo Cim. **60A**, 47 (1969).

[100] N. G. Deshpande, Oregon preprint OITS-107 (1979); C. Q. Geng and R. E. Marshak, Phys. Rev. D**39**, 693 (1989); R. Foot, G. C. Joshi, H. Lew, and R. R. Volkas, Mod. Phys. Lett. A**5**, 95 (1990); K. S. Babu and R. N. Mohapatra, Phys. Rev. Lett. **63**, 938 (1989).

[101] K. S. Babu and R. N. Mohapatra, Phys. Rev. D**42**, 3866 (1990).

[102] L. Davis, Jr., A. S. Goldhaber and M. M. Nieto, Phys. Rev. Lett. **35**, 1402 (1975).

[103] H. F. Dylla and J. G. King, Phys. Rev. **A7**, 1224 (1973); see also J. C. Zorn, G. E. Chamberlain and V. W. Hughes, Phys. Rev. **129**, 2566 (1963).

[104] J. Baumann, R. Gähler, J. Kalus and W. Mampe, Phys. Rev. **D37**, 3107 (1988); R. Gähler, J. Kalus and W. Mampe, Phys. Rev. **D25**, 2887 (1982).

[105] G. Barbiellini and G. Cocconi, Nature **329**, 21 (1987).

[106] W. J. Marciano and A. I. Sanda, Phys. Lett. **67**B, 303 (1977); B. W. Lee and R. E. Shrock, Phys. Rev. D**16**. 1444 (1977).

[107] J. E. Gunn, B. W. Lee, I. Lerche, D. N. Schramm and G. Steigman, Astrophys. J. **223**, 1015 (1978).

[108] K. Fujikawa and R. E. Shrock, Phys. Rev. Lett. **45**, 963 (1980).

[109] H. A. Bethe, Proc. Cam. Phil. Soc. **31**, 108 (1935).

[110] C. L. Cowan, Jr. and F. Reines, Phys. Rev. **107**, 528 (1957).

[111] A. V. Kyuldjiev, Nucl. Phys. **B243**, 387 (1984).

[112] F. Reines, H. S. Gurr and H. W. Sobel, Phys. Rev. Lett. **37**, 315 (1976).

[113] W. J. Marciano, cited in Particle Data Group, Phys. Lett. B**239**, 1 (1990).

[114] P. Vogel and J. Engel, Phys. Rev. D**39**, 3378 (1989).

[115] D. A. Krakauer et al., Phys. Lett. B**252**, 177 (1990).

[116] L. A. Ahrens et al., Phys. Rev. **D41**, 3297 (1990).

[117] J. Dorenbosch et al., Z. Phys. **C41**, 567 (1989).

[118] H. Grotch and R. W. Robinett, Z. Phys. **C39**, 553 (1988).

[119] G. Beaudet, V. Petrosian and E. E. Salpeter, Astrophys. J. **150**, 979 (1967); see also J. B. Adams, M. A. Ruderman and C.-H. Woo, Phys. Rev. **129**, 1383 (1963); V. N. Tsytovich, Zh. Eksp. Teor. Fiz. **45**, 1183 (1964) [Sov. Phys. JETP **18**, 816 (1964)].

[120] G. V. Domogatskiĭ and D. K. Nadezhin, Yad. Fiz. **12**, 1233 (1970); [Sov. J. Nucl. Phys. **12**, 678 (1971)].

[121] P. Sutherland et al., Phys. Rev. **D13**, 2700 (1976).

[122] M. A. B. Bég, W. J. Marciano and M. Ruderman, Phys. Rev. **D17**, 1395 (1978).

[123] M. Fukugita and S. Yazaki, Phys. Rev. **D36**, 3817 (1987).

[124] S. I. Blinnikov, Moscow preprint ITEP-88-19 (1988).

[125] G. G. Raffelt, Phys. Rev. Lett. **64**, 2856 (1990).

[126] J. M. Lattimer and J. Cooperstein, Phys. Rev. Lett. **61**, 23 (1988); R. Barbieri and R. N. Mohapatra, Phys. Rev. Lett. **61**, 27 (1988); D. Nötzold, Phys. Rev. **D38**, 1658 (1988).

[127] M. Fukugita, N. Iwamoto, L. Qin and S. Tsuruta, Toledo preprint (1993).

[128] W. A. Bardeen, R. Gastmans and B. Lautrup, Nucl. Phys. **B46**, 319 (1972); A. D. Dolgov and Ya. B. Zeldovich, Ref.[43]; G. Auriemma, Y. Srivastava and A. Widom, Phys. Lett. **B195**, 254 (1987); J. L. Lucio, A. Rosado and A. Zepeda, Phys. Rev. **D31**, 1091 (1985); for early discussion of form factors see J. Bernstein and T. D. Lee, Phys. Rev. Lett. **11**, 512 (1963).

[129] G. Degrassi, A. Sirlin and W. J. Marciano, Phys. Rev. **D39**, 287 (1989).

[130] M. Fukugita and K. Kubodera, Z. Phys. **C9**, 365 (1981).

[131] J. Law, Phys. Lett. **102B**, 371 (1981).

[132] O. Fackler et al., Phys. Rev. Lett. **55**, 1388 (1985).

[133] R. L. Martin and J. S. Cohen, Phys. Lett. **110A**, 95 (1985).

[134] H. Ågren and V. Carravetta, Phys. Rev. **A38**, 2707 (1988).

[135] I. G. Kaplan, V. N. Smutny and G. V. Smelov, Phys. Lett. **112B**, 417 (1982); **161**B, 389 (1985); Zh. Eksp. Teor. Fiz. **84**, 833 (1983) [Sov. Phys. JETP **57**, 483 (1983)]; I. Kaplan et al., Dokl. Akad. Nauk. SSSR **279**, 1110 (1984) [Sov. Phys. Doklady **29**, 1023 (1984)].

[136] J. Arafune and T. Watanabe, Phys. Rev. **C34**, 336 (1986); J. Arafune, N. Koga, K. Morokuma and T. Watanabe, J. Phys. Soc. Japan **55**, 3806; 3813 (1986).

[137] S. Boris et al., Phys. Lett. 159B, 217 (1985); Phys. Rev. Lett. **58**, 2019 (1987).

[138] M. Fritschi et al., Phys. Lett. B**173**, 485 (1986).

[139] H. Kawakami et al., Phys. Lett. **B187**, 198 (1987).

[140] J. F. Wilkerson et al., Phys. Rev. Lett. **58**, 2023 (1987).

[141] E. Holzschuh, M. Fritschi and W. Kündig, Phys. Lett. **B287**, 381 (1992).

[142] H. Kawakami et al., Phys. Lett. **B256**, 105 (1991).

[143] R. G. H. Robertson et al., Phys. Rev. Lett. **67**, 957 (1991).

[144] H. Backe et al., in *Proc. Neutrino 92*, Granada, 1992, ed. by A. Morales, Nucl. Phys. B (Proc. Suppl.) **31**, 46 (1993).

[145] IBEC was first studied by E.G.C. Stueckelberg [Nature **137**, 1070 (1936); Helv. Phys. Acta. **9**, 533 (1936)] and C. Møller [Phys. Zeits d. Sowjetunion, **11**, 9 (1937)]. The process was first observed by H. Bradt et al. [Helv. Phys. Acta **19**, 222 (1946)]. The theory was further developed by R. J. Glauber and P. C. Martin, Phys. Rev. **104**, 158 (1956); P. C. Martin and R. J. Glauber **109**, 1307 (1958), A. de Rújula (Ref.[146]) and others. For a review, see W. Bambynek et al., Rev. Mod. Phys. **49**, 77 (1977).

[146] A. de Rújula, Nucl. Phys. **B188**, 414 (1981); see also A. de Rújula and M. Lusignoli, Nucl. Phys. **B219**, 277 (1983).

[147] J. U. Andersen et al., Phys. Lett. **113B**, 72 (1982).

[148] S. Yasumi et al., Phys. Lett. B**181**, 169 (1986).

[149] P. T. Springer, C. L. Bennett and P. A. Baisden, Phys. Rev. **A35**, 679 (1987).

[150] C. L. Bennett et al., Phys. Lett. **107B**, 19 (1981).

[151] R. Abela et al., Phys. Lett. **146B**, 431 (1984).

[152] B. Jeckelmann et al., Phys. Rev. Lett. **56**, 1444 (1986).

[153] M. Daum et al., Phys. Lett. **B265**, 425 (1991).

[154] H. B. Anderhub et al., Phys. Lett. **114B**, 76 (1982).

[155] H. Albrecht et al., Phys. Lett., B**202**, 149 (1988).

[156] D. Cinabro et al., Phys. Rev. Lett. **70**, 3700 (1993).

[157] A partial list for this work is: J. Arafune and M. Fukugita, Phys. Rev. Lett. **59**, 367 (1987); K. Sato and H. Suzuki, Phys. Rev. Lett. **58**, 2722 (1987); J. N. Bahcall and S. L. Glashow, Nature (London) **326**, 476 (1987); W. D. Arnett and J. L. Rosner, Phys. Rev. Lett. **58**, 1906 (1987); E. W. Kolb, A. J. Stebbins and M. S. Turner, Phys. Rev. **D35**, 3598 (1987); A. Burrows and J. M. Lattimer, Astrophys. J. Lett. **318**, L63 (1987); K. Hikasa, M. Kobayashi and M. Yoshimura, Prog. Theor. Phys. (Lett.) **78**, 180 (1987); S. H. Kahana, J. Cooperstein and E. Baron, Phys. Lett. **B196**, 259 (1987).

[158] E. Ma and J. Okada, Phys. Rev. Lett. **41**, 287 (1978).

[159] M. Z. Akrawy et al., Z. Phys. C **50**, 373 (1991).

[160] V. F. Shvartsman, Zh. Eksp. Teor. Fiz. Pisma **9**, 315 (1969) [JETP Lett. **9**, 184 (1969)]. The dependence of the helium abundance on the number of neutrino species was originally noted by Hoyle and Tayler [F. Hoyle and R. J. Tayler, Nature **203**, 1108 (1964)]. See also, P. J. E. Peebles, Astrophys. J. **146**, 542 (1966), R. V. Wagoner, W. A. Fowler and F. Hoyle, Astrophys. J. **148**, 3 (1967) and H. Sato, Prog. Theor. Phys. **38**, 1083 (1967) for pioneering attempts for light element nucleosynthesis.

[161] G. Steigman, D. N. Schramm and J. E. Gunn, Phys. Lett. **66B**, 202 (1977).

[162] T. P. Walker et al., Astrophys. J. **376**, 51 (1991).

[163] M. Green and M. Veltman, Nucl. Phys. **B169**, 137 (1980).

[164] K.-I. Aoki et al., Prog. Theor. Phys. **65**, 1001 (1981); K.-I. Aoki and Z. Hioki, Prog. Theor. Phys. **66**, 2234 (1981).

[165] S. Sarantakos, A. Sirlin and W. J. Marciano, Nucl. Phys. **B217**, 84 (1983).

[166] W. J. Marciano and A. Sirlin, Phys. Rev. **D22**, 2695 (1980).

[167] P. Vilain et al., Phys. Lett. B**281**, 159 (1992); D. Geiregat et al., Phys. Lett. B**259**, 499 (1991).

[168] N. Cabibbo and R. Gatto, Nuovo Cim. Serie X, **15**, 304 (1960); Y. Yamaguchi, Prog. Theor. Phys. **23**, 1117 (1960); M. Igarashi and S. Nakamura, Prog. Theor. Phys. **40**, 576 (1968).

[169] R. P. Feynman, *Photon-Hadron Interactions* (W. A. Benjamin, Reading, U. S. A., 1972).

[170] T. W. Donnelly, Phys. Lett. **43**B, 93 (1973).

[171] J. N. Bahcall and S. C. Frautschi, Phys. Rev. **136**, B1547 (1964); N. Itoh and Y. Kohyama, Nucl. Phys. A**306**, 527 (1978).

[172] J. N. Bahcall and C. A. Barnes, Phys. Lett. **12**, 48 (1964).

[173] J. C. Hardy and R. I. Verrall, Phys. Rev. Lett. **13**, 764 (1964); P. L. Reeder, A. M. Poskanzer and R. A. Esterlund, Phys. Rev. Lett. **13**, 767 (1964).

[174] R. G. Sextro, R. A. Gough and J. Cerny, Nucl. Phys. A**234**, 130 (1974); the most modern data are given by A. Garcia et al., Phys. Rev. Lett. **67**, 3654 (1991).

[175] M. Fukugita, T. Kuramoto and K. Shimizu, to be published.

[176] S. Cohen and D. Kurath, Nucl. Phys. **73**, 1 (1965).

[177] For a review, A. Arima, K. Shimizu, W. Bentz and H. Hyuga, Adv. Nucl. Phys. **18**, 1 (1987).

[178] B. H. Wildenthal, Prog. Part. Nucl. Phys. **11**, 5 (1984); B. A. Brown and B. H. Wildenthal, At. Data and Nucl. Data. Tables **33**, 347 (1985).

[179] C. D. Goodman et al., Phys. Rev. Lett. **44**, 1755 (1980); F. Petrovich, W. G. Love and R. J. McCarthy, Phys. Rev. **C21**, 1718 (1980); T. N. Taddeucci et al., Phys. Rev. C**25**, 1094 (1982); J. Rapaport et al., Phys. Rev. Lett. **47**, 1518 (1981); Nucl. Phys. **A410**, 371 (1983); D. Krofcheck et al., Phys. Rev. Lett., **55**, 1051 (1985).

[180] T. N. Taddeucci et al., Nucl. Phys. **A469**, 125 (1987).

[181] B. Pontecorvo, Chalk River report PD-205 (1946); L. W. Alvarez, University of California Radiation Laboratory report UCRL-328 (1949); R. Davis, Jr., Ref.[215].

[182] V. A. Kuzmin, Zh. Eksp. Theor. Fiz. **49**, 1532 (1965) [Sov. Phys. JETP **22**, 1051 (1966)].

[183] A. J. Baltz, J. Weneser, B. A. Brown and J. Rapaport, Phys. Rev. Lett. **53**, 2078 (1984); D. Krofcheck et al. Ref.[179]; G. J. Mathews et al. S. D. Bloom, G. M. Fuller and J. N. Bahcall, Phys. Rev. C**32**, 796 (1985); K. Grotz et al., Astron. Astrophys. **154**, L1 (1986); for the ground-to-ground transition, W. Hampel and L. P. Remsberg, Phys. Rev. C**31**, 666 (1985).

[184] J. N. Bahcall, Rev. Mod. Phys. **50**, 881 (1978); see also Ref.[39].

[185] S. S. Gershteĭn, N. van Hieu and R. A. Éramzhyan, Zh. Eksp. Teor. Fiz. **43**, 1554 (1962) [Sov. Phys. JETP **16**, 1097 (1963)].

[186] H. C. Lee., Nucl. Phys. **A294**, 473 (1978).

[187] J. Bernabéu and P. Pascual, Nucl. Phys. A**324**, (1979).

[188] T. W. Donnelly and R. D. Peccei, Phys. Rep. **50**, 1 (1979).

[189] R. S. Raghavan, S. Pakvasa and B. A. Brown, Phys. Rev. Lett., **57**, 1801 (1986); R. S. Raghavan and S. Pakvasa, Phys. Rev. **D37**, 849 (1988).

[190] M. Fukugita, Y. Kohyama and K. Kubodera, Phys. Lett. **B212**, 139 (1988).

[191] M. Pourkaviani and S. L. Minz, J. Phys. G **16**, 569 (1990).

[192] M. Aglietta et al., Astroparticle Phys. **1**, 1 (1992).

[193] J. Arafune, M. Fukugita, Y. Kohyama and K. Kubodera, Phys. Lett. **B217**, 186 (1989)

[194] F. J. Kelly and H. Überall, Phys. Rev. Lett. **16**, 145 (1966).

[195] S. D. Ellis and J. N. Bahcall, Nucl. Phys. **A114**, 636 (1968).

[196] A. Ali and C. A. Domingnez, Phys. Rev. **D12**, 3673 (1975)

[197] S. Nozawa, Y. Kohyama, T. Kanata and K. Kubodera, J. Phys. Soc. Japan, **55**, 263 (1986); J. N. Bahcall, K. Kubodera and S. Nozawa, Phys. Rev. **D38**, 1030 (1988).

[198] W. C. Haxton, Phys. Rev. **D36**, 2283 (1987).

[199] J. B. Langworthy, B. A. Lamers and H. Überall, Nucl. Phys. **A280**, 351 (1977).

[200] E. V. Bugaev, G. S. Bisnovatyi-Kogan, M. A. Rudzsky and Z. F. Seidov, Nucl. Phys. **A324**, 350 (1979).

[201] T. K. Gaisser and J. S. O'Connell, Phys. Rev. **D34**, 822 (1986).

[202] T. Kuramoto, M. Fukugita, Y. Kohyama and K. Kubodera, Nucl. Phys. **A512**, 711 (1990).

[203] J. Engel, E. Kolbe, K. Langanke and P. Vogel, preprint BA-93-20 (1993).

[204] T. W. Donnelly and W. C. Haxton, Nucl. Phys. A**287**, 506 (1977).

[205] R. C. Allen et al., Phys. Rev. Lett. **64**, 1871 (1990); D. A. Krakauer et al. Phys. Rev. C**45**, 2450 (1992).

[206] B. Bodmann et al., Phys. Lett. B**267**, 321 (1991); in *Neutrino 92*, Granada, 1992, ed. by A. Morales, Nucl. Phys. B (Proc. Suppl.) **31**, 301 (1993).

[207] B. Bodmann et al., Phys. Lett. B**280**, 198 (1992).

[208] S. L. Minz and M. Pourkaviani, Phys. Rev. C**40**, 2458 (1989).

[209] D. Z. Freedman, Phys. Rev. D**9**, 1389 (1974); see H. A. Bethe, in *Proc. International School of Physics, 'Enrico Fermi'*, Course 91, Varenna 1984 (North Holland, Amsterdam, 1986), for a review of the stellar collapse theory.

[210] A. Drukier and L. Stodolsky, Phys. Rev. D**30**, 2295 (1984).

[211] M. W. Goodman and E. Witten, Phys. Rev. D**31**, 3059 (1985).

[212] M. Fukugita, D. Nötzold, G. Raffelt and J. Silk, Phys. Rev. Lett. **60**, 879 (1988); D. Nötzold and G. Raffelt, Nucl. Phys. **B307**, 924 (1988).

[213] F. J. Botella, C. S. Lim and W. J. Marciano, Phys. Rev. **D35**, 896 (1987).

[214] K. Enqvist, K. Kainulainen and J. Maalampi, Nucl. Phys. B**349**, 754 (1991).

[215] R. Davis, Jr., Phys. Rev. **97**, 766 (1955)

[216] T. Yanagida, in *Proc. of the Workshop on the Unified Theory and Baryon Number in the Universe*, ed. by O. Sawada and A. Sugamoto (KEK report 79-18, 1979), p.95; M. Gell-Mann, P. Ramond and R. Slansky, in *Supergravity*, ed. by P. van Nieuwenhuizen and D. Z. Freedman (North Holland, Amsterdam, 1979), p.315.

[217] J. A. McLennan, Jr., Phys. Rev. **106**, 821 (1957).

[218] Ya. B. Zeldovich, Dokl. Akad. Nauk. SSSR **86**, 505 (1952); E. J. Konopinski and H. M. Mahmoud, Phys. Rev. **92**, 1045 (1953).

[219] L. Wolfenstein, Phys. Lett. **107B**, 77 (1981).

[220] G. B. Gelmini and M. Roncadelli, Phys. Lett. **99B**, 411 (1981).

[221] H. M. Georgi, S. L. Glashow and S. Nussinov, Nucl. Phys. **B193**, 297 (1981).

[222] K. Choi, C. W. Kim, W. P. Lam and A. Santamaria, Carnegie-Melon preprint CMU-HEP 89-22 (1989).

[223] K. M. Case, Phys. Rev. **107**, 307 (1957); see also P. D. Mannhein, Int. J. Theor. Phys. **23**, 643 (1984).

[224] B. Touschek, Z. Phys. **125**, 108 (1949).

[225] J. Serpe, Physica **18**, 295 (1952).

[226] W. Pauli, Nuovo Cim. Serie X, **6**, 204 (1957).

[227] F. Gürsey, Nuovo Cim. Serie X, **7**, 411 (1958).

[228] L. A. Radicati and B. Touschek, Nuovo Cim. **5**, 1693 (1957).

[229] C. Ryan and S. Okubo, Nuovo. Cim. Suppl. **2**, 234 (1964).

[230] H. Jehle, Phys. Rev. **75**, 1609(L) (1949).

[231] M. Fukugita and T. Yanagida, unpublished note (1991).

[232] H. Goldberg, Phys. Rev. Lett. **50**, 1419 (1983).

[233] e.g., H. Primakoff and S. P. Rosen, Rep. Prog. Phys. **22**, 121 (1959); Ann. Rev. Nucl. Part. Sci. **31**, 145 (1981).

[234] M. Doi, T. Kotani, H. Nishiura, K. Okuda and E. Takasugi, Phys. Lett. **103B**, 219 (1981).

[235] W. C. Haxton and G. J. Stephenson, Jr., Prog. Part. Nucl. Phys. **12**, 409 (1984).

[236] M. Doi, T. Kotani and E. Takasugi, Prog. Theor. Phys. Suppl. **83**, 1 (1985).

[237] J. D. Vergados, Phys. Rep. C**133**, 1 (1986).

[238] D. O. Caldwell, in *Proc. of the XXIII International Conference on High Energy Physics*, Berkeley, 1986, ed. by S. C. Loken (World Scientific, Singapore, 1987) vol. II, p.951.

[239] A. Molina and P. Pascual, Nuovo. Cim. **41**A, 756 (1977).

[240] J. D. Vergados, Phys. Rev. C**24**, 640 (1981).

[241] W. C. Haxton, G. J. Stephenson, Jr. and D. Strottman, Phys. Rev. D**25**, 2360 (1982); M. Doi, T, Kotani, H. Nishiura and E. Takasugi, Prog. Theor. Phys. **70** 1353 (1984); Relativistic corrections to beta decay is given in, H. Behrens and J. Jännecke, *Numerical Tables for Beta-Decay and Electron Capture* (J. Springer, Berlin, 1969).

[242] L. Zhao, B. A. Brown and W. A. Richter, Phys. Rev. C**42**, 1120 (1990).

[243] J. A. Halbleib and R. A. Sorensen, Nucl. Phys. **A98**, 542 (1967).

[244] A. H. Huffman, Phys. Rev. C**2**, 742 (1970).

[245] P. Vogel and P. Fisher, Phys. Rev. C**32**, 1362 (1985).

[246] K. Grotz and H. V. Klapdor, Phys. Lett. **153B**, 1 (1985); **157B**, 242 (1985); Nucl. Phys. A**460**, 395 (1986).

[247] T. Tomoda, A. Faessler, K. W. Schmid and F. Grümmer, Nucl. Phys. A**452**, 591 (1986).

[248] P. Vogel and M. R. Zirnbauer, Phys. Rev. Lett. **57**, 3148 (1986).

[249] J. Engel, P. Vogel and M. R. Zirnbauer, Phys. Rev. C**37**, 731 (1988).

[250] O. Civitarese, A. Faessler and T. Tomoda, Phys. Lett. **B194**, 11 (1987), T. Tomoda and A. Faessler, Phys. Lett. **199**B, 475 (1987).

[251] J. Engel, P. Vogel, X. Ji and S. Pittel, Phys. Lett. **225B**, 5 (1989).

[252] For earlier attempts, E. L. Fireman, Phys. Rev. **74**, 1238(A) (1948); **75**, 323(L) (1949); J. S. Lawson, Phys. Rev. **81**, 299(A) (1951); M. I. Kalkstein and W. F. Libby, Phys. Rev. **85**, 368(L) (1952).

[253] M. G. Inghram and J. H. Reynolds, Phys. Rev. **76**, 1265(L) (1949); **78**, 822(L) (1950).

[254] N. Takaoka and K. Ogata, Z. Naturforsch. **21a**, 84 (1966).

[255] T. Kirsten, W. Gentner and O. Müller, Z. Naturforsch. **22a**, 1783 (1967); T. Kirsten, W. Gentner and O. A. Schaeffer, Z. Phys. **202**, 273 (1967).

[256] R. K. Bardin, P. J. Gollon, J. D. Ullman and C. S. Wu, Nucl. Phys. A**158**, 337 (1970).

[257] E. Bellotti et al., Nuovo Cim. **95A**, 1 (1986); E. Fiorini et al., Phys. Lett. B**25**, 602 (1967) used a ^{76}Ge detector at the first time.

[258] F. T. Avignone III, H. S. Miley, R. L. Brodzinski and J. H. Reeves, Phys. Rev. D**35**, 1713 (1987).

[259] D. O. Caldwell et al., Phys. Rev. Lett. **59**, 419 (1987); ibid **54**, 281 (1985).

[260] J. J. Simpson et al., Phys. Rev. Lett. **53**, 141 (1984).

[261] H. Ejiri et al., Nucl. Phys. A**448**, 271 (1986).

[262] H. V. Klapdor-Kleingrothaus, in *Neutrino 92*, Granada, 1992, ed. by A. Morales, Nucl. Phys. B (Proc. Suppl.) **31**, 72 (1993); A. Balysh et al., Phys. Lett. B**283**, 32 (1992).

[263] H. S. Miley et al., Phys. Rev. Lett. **65**, 3092 (1990).

[264] S. R. Elliott, A. A. Hahn and M. K. Moe, Phys. Rev. Lett. **59**, 2020 (1987); S. R. Elliott et al. Phys. Rev. C**46**, 1535 (1992).

[265] T. Kirsten et al., in *Nuclear Beta Decays and Neutrino*, Osaka, 1986, ed. by T. Kotani, H. Ejiri and E. Takasugi (World Scientific, Singapore, 1986) p.81; T. Kirsten, H. Richter and E. K. Jessberger, Phys. Rev. Lett. **50**, 474 (1983); Z. Phys. C**16**, 189 (1983).

[266] O. Manuel, in *Nuclear Beta Decays and Neutrino*, Osaka, 1986, ed. by T. Kotani, H. Ejiri and E. Takasugi (World Scientific, Singapore, 1986) p.71; E. W. Hennecke, O. K. Manuel and D. D. Sabu, Phys. Rev. C**11**, 1378 (1975).

[267] K. Marti and S. V. S. Murty, Phys. Lett. **163**B, 71 (1985).

[268] S. R. Elliott, A. A. Hahn and M. K. Moe, in *Weak and Electromagnetic Interactions in Nuclei*, Heidelberg, 1986, ed. by H. V. Klapdor (Springer Verlag, Berlin, 1987) p.692.

[269] M. Alston-Garnjost et al., Phys. Rev. Lett. **63**, 1671 (1989).

[270] H. Ejiri et al., Phys. Lett. B**258**, 17 (1991).

[271] S. R. Elliott et al., J. Phys. G**17**, S145 (1991).

[272] A. S. Barabash et al., Phys. Lett. B**223**, 273 (1989).

[273] E. J. Konopinski, *The Theory of Beta Radioactivity* (Clarendon Press, Oxford, 1966).

[274] K. Muto and H. V. Klapdor, Phys. Lett. B**201**, 420 (1988); K. Muto, E. Bender and H. V. Klapdor, Z. Phys. A**334**, 177; 187 (1989).

[275] B. Pontecorvo, Phys. Lett. **26**B, 630 (1968).

[276] A. Halprin, S. T. Petcov and S. P. Rosen, Phys. Lett. **125B**, 335 (1983).

[277] H. Primakoff and S. P. Rosen, Phys. Rev. **184**, 1925 (1969).

[278] C. P. Enz, Nuovo Cim. **6**, 250 (1957).

[279] R. N. Mohapatra and G. Senjanović, Phys. Rev. D**23**, 165 (1981).

[280] M. Doi, T. Kotani, H. Nishiura and E. Takasugi, Prog. Theor. Phys. **69**, 602 (1983).

[281] R. N. Mohapatra, Phys. Rev. D**34**, 909 (1986).

[282] G. K. Leontaris, K. Tamvakis and J. D. Vergados, Phys. Lett. **171**B, 412 (1986); J. D. Vergados, Phys. Lett. B**184**, 55 (1987); R. N. Mohapatra, Phys. Rev. D**34**, 3457 (1986).

[283] T. P. Cheng and L. F. Li, Phys. Rev. Lett. **38**, 381 (1977); Phys. Rev. D**16**, 1425 (1977)* ; B. W. Lee, S. Pakvasa, R. E. Shrock and H. Sugawara, Phys. Rev. Lett. **38**, 937 (1977)*; H. Terazawa, Prog. Theor. Phys. **57**, 1808 (1977); B. W. Lee and R. E. Shrock, Phys Rev. D**16**, 1444 (1977)* ; J. D. Bjorken, K. Lane and S. Weinberg, Phys. Rev. D**16**, 1474 (1977); S. T. Petcov, Yad. Phys. **25**, 641 (1977)* [Sov. J. Nucl. Phys. **25**, 340 (1977)]. (* presents also a calculation of $\mu \to eee$.)

[284] C. N. Yang and J. Tiomno, Phys. Rev. **79**, 495 (1950).

[285] P. B. Pal and L. Wolfenstein, Phys. Rev. **D25**, 766 (1982).

[286] J. Schechter and J. W. F. Valle, Phys. Rev. **D22**, 2227 (1980).

[287] S. M. Bilenky, J. Hošek and S. T. Petcov, Phys. Lett. **94B**, 495 (1980), M. Doi et al., Phys. Lett. **102B**, 323 (1981).

[288] J. Schechter and J. W. F. Valle, Phys. Rev. D**23**, 1666 (1981).

[289] V. Gribov and B. Pontecorvo, Phys. Lett. **28B**, 493 (1969).

[290] S. Eliezer and D. A. Ross, Phys. Rev. D**10**, 3088 (1974); S. Eliezer, A. R. Swift, Nucl. Phys. B**105**, 45 (1976).

[291] Earlier studies include: J. N. Bahcall and S. Frautschi, Phys. Lett. **29**B, 623 (1969); S. M. Bilenky and B. Pontecorvo, Phys. Lett. **61B**, 248 (1976); H. Fritzsch, P. Minkowski, Phys. Lett. **62B**, 72 (1976); A. K. Mann and H. Primakoff, Phys. Rev. D**15**, 655 (1977); for a review, S. M. Bilenky and B. Pontecorvo, Phys. Rep. **41**C, 225 (1978).

[292] S. Nussinov, Phys. Lett. **63B**, 201 (1976).

[293] B. Kayser, Phys. Rev. D**24**, 110 (1981).

[294] C. Giunti, C. W. Kim and U. W. Lee, Phys. Rev. D**44**, 3635 (1991).

[295] R. G. Winter, Lett. Nuovo Cim. **30**, 101 (1981).

[296] A. Loeb, Phys. Rev. D**39**, 1009 (1989).

[297] H. Anada and H. Nishimura, Phys. Rev. D**37**, 552 (1988).

[298] L. Borodovsky et al., Phys. Rev. Lett. **68**, 274 (1992); B. Blumenfeld et al., Phys. Rev. Lett. **62**, 2237 (1989); L. S. Durkin et al., Phys. Rev. Lett. **61**, 1811 (1988); C. Angelini et al., Phys. Lett. B**179**, 307 (1986); L. A. Ahrens et al., Phys. Rev. **D31**, 2732 (1985).

[299] G. Zacek et al., Phys. Rev. **D34**, 2621 (1986).

[300] F. Bergsma et al., Phys. Lett. **142B**, 103 (1984); F. Dydak et al., Phys. Lett. **134B**, 281 (1984); I. E. Stockdale et al., Z. Phys. C**27**, 53 (1985).

[301] N. Ushida et al., Phys. Rev. Lett. **57**, 2897 (1986).

[302] H. Harari, in *Weak Interactions and Neutrinos – 1989, Proc. of the 12th International Workshop on Weak Interactions and Neutrinos*, Sea of Galilee, 1989, ed. by P. Singer and G. Eilam, Nucl. Phys. B (Proc. Suppl.) **13**, 103 (1990).

[303] The representative reports from the ^{37}Cl experiment at Homestake include: R. Davis, Jr. in *Proc. of International Symposium on Neutrino Astrophysics*, Takayama 1992 (to be published); in *Proc. of Seventh Workshop on Grand Unification, ICOBAN 86*, ed. by J. Arafune (World Scientific, Singapore, 1987) p.237; J. K. Rowley, B. T. Cleveland and R. Davis, Jr., in *Solar Neutrinos and Neutrino Astronomy*, Homestake, 1984, ed. by M. L. Cherry, K. Lande and W. A. Fowler (AIP Conf. Proc. No.126, 1985) p.1; R. Davis Jr., in *Proc. Informal Conference on Status and Future of Solar Neutrino Research*, BNL, 1978, BNL report 50879 ed. G. Friedlander (1978) p.1.

[304] J. N. Bahcall and R. K. Ulrich, Rev. Mod. Phys. **60**, 297 (1988).

[305] S. Turck-Chièze, S. Cahen, M, Cassé and C. Doom, Astrophys. J. **335**, 415 (1988).

[306] J. N. Bahcall and M. H. Pinsonneault, Rev. Mod. Phys. **64**, 885 (1992).

[307] J. N. Bahcall, in *Neutrino 90, Proc of the 14th International Conference on Neutrino Physics and Astrophysics*, CERN, 1990, ed. by J. Panman and K. Winter, Nucl. Phys. B (Proc. Suppl.) **19**, 94 (1991); M. Fukugita and T. Yanagida, Mod. Phys. Lett. A**6**, 645 (1991); J. N. Bahcall and H. A. Bethe, Phys. Rev. Lett. **65**, 2233 (1990); Phys. Rev. D**47**, 1298 (1993); S. Bludman, N. Hata and P. Langacker, Pennsylvania preprint UPR-0572T (1993).

[308] B. Pontecorvo, Zh. Eksp. Teor. Fiz. **53**, 1717 (1968). [Sov. Phys. JETP **26**, 984 (1968)].

[309] V. Barger, K. Whisnant and R. J. N. Phillips, Phys. Rev. D**24**, 538 (1981).

[310] S. L. Glashow and L. Kraus, Phys. Lett. **190B**, 179 (1987).

[311] V. Barger, R. J. N. Phillips and K. Whisnant, Phys. Rev. Lett. **65**, 3084 (1990); **69**, 3135 (1992).

[312] A. Acker, S. Pakvasa and J. Pantaleone, Phys. Rev. D**43**, R1754 (1991).

[313] S. Pakvasa and J. Pantaleone, Phys. Rev. Lett. **65**, 2479 (1990).

[314] F. Reines, Ann. Rev. Nucl. Sci. **10**, 1 (1969); M. A. Markov, in *Proc. of the 1960 Annual International Conference on High Energy Physics at Rochester*, 1960, ed. by E. C. G. Sudershan, J. H. Tinlot and A. C. Melissinos (Rochester University, New York, 1960) p.578; M. A. Markov and I. M. Zheleznykh, Nucl. Phys. **27**, 385 (1961).

[315] G. T. Zatsepin and V. A. Kuzmin, Zh. Eksp. Teor. Fiz. **41**, 1818 (1961) [Sov. Phys. JETP **14**, 1294 (1962)].

[316] P. H. Frampton and S. L. Glashow, Phys. Rev. D**25**, 1982 (1982).

[317] D. S. Ayres et al., Phys. Rev. D**29**, 902 (1984); G. V. Dass and K. V. L. Sarma, Phys. Rev. **D30**, 80 (1984).

[318] J. M. LoSecco et al., Phys. Rev. Lett. **54**, 2299 (1985).

[319] K. S. Hirata et al., Phys. Lett. B**280**, 146 (1992); see also, ibid B**205**, 416 (1988).

[320] J. G. Learned, S. Pakvasa and T. J. Weiler, Phys. Lett. B**207**, 79 (1988); V. Barger and K. Whisnant, Phys. Lett. B**209**, 365 (1988); K. Hidaka, M. Honda and S. Midorikawa, Phys. Rev. Lett. **61**, 1537 (1988);

[321] R. Becker-Szendy et al., Phys. Rev. D**46**, 3720 (1992); see also, D. Casper et al., Phys. Rev. Lett. **66**, 2561 (1991).

[322] R. Becker-Szendy et al., Phys. Rev. Lett. **69**, 1010 (1992); see also LoSecco et al., Ref.[318].

[323] C. Berger et al., Phys. Lett. B**245**, 305 (1990).

[324] E. C. M. Young, in *Cosmic Rays at Ground Level*, ed by A. W. Wolfendale (Institute of Physics, London, 1973) p.105 ; L. V. Volkova, Yad. Fiz. **31**, 1510 (1980) [Sov. J. Nucl. Phys. **31**, 784 (1980)]; T. K. Gaisser, T. Stanev, S. A. Bludman and H. Lee, Phys. Rev. Lett. **51**, 223 (1983); A. Dar, Phys. Rev. Lett. **51**, 227 (1983).

[325] T. K. Gaisser, T Stanev and G. Barr, Phys. Rev. D**38**, 85 (1988) T. K. Gaisser and T. Stanev, Phys. Rev. D**39**, 3532 (1989); M. Honda et al., Phys. Lett. B**248**, 193 (1990); E. V. Bugaev and V. A. Naumov, Phys. Lett. B**232**, 391 (1989); H. Lee and Y. S. Koh, Nuovo Cim. **105**B, 883 (1990); M. Kawasaki and S. Mizuta, Phys. Rev. D**43**, 2900 (1991).

[326] C. Störmer, Astrophysics **1**, 237 (1930); D. J. Cooke, Phys. Rev. Lett. **51**, 320 (1983).

[327] L. V. Volkova, unpublished (1988) (quoted in G. Barr et al., Ref.[325].)

[328] M. Fukugita, H. Murayama and T. Yanagida, unpublished (1992).

[329] T. Yanagida and M. Yoshimura, Prog. Theor. Phys. **64**, 1870 (L) (1980).

[330] V. Barger, P. Langacker, J. P. Leveille and S. Pakvasa, Phys. Rev. Lett. **45**, 692 (1980).

[331] L. Wolfenstein, Phys. Rev. **D17**, 2369 (1978).

[332] S. P. Mikheyev and A. Yu. Smirnov, Yad. Fiz. **42**, 1441 (1985) [Sov. J. Nucl. Phys. **42**, 913 (1985)]; Nuovo Cim. **9**C, 17 (1986).

[333] V. Barger, K. Whisnant, S. Pakvasa and R. J. N. Phillips, Phys. Rev. D**22**, 2718 (1980).

[334] H. A. Bethe, Phys. Rev. Lett. **56**, 1305 (1986).

[335] P. D. Mannheim, Phys. Rev. D**37**, 1935 (1988).

[336] S. J. Parke, Phys. Rev. Lett. **57**, 1275 (1986).

[337] W. C. Haxton, Phys. Rev. Lett. **57**, 1271 (1986).

[338] L. Landau, Phys. Zeits. d. Sowjetunion, **2**, 46 (1932); C. Zener, Proc. Roy. Soc. London A**137**, 696 (1932); E. C. G. Stueckelberg, Helv. Phys. Acta **5**, 369 (1932).

[339] S. P. Rosen and J. M. Gelb, Phys. Rev. D**34**, 969 (1986).

[340] E. W. Kolb, M. S. Turner and T. P. Walker, Phys. Lett. B**175**, 478 (1986).

[341] T. K. Kuo and J. Pantaleone, Phys. Rev. Lett. **57**, 1805 (1986).

[342] S. P. Mikheev and A. Yu. Smirnov, Usp. Fiz. Nauk **153**, 3 (1987) [Sov. Phys. Usp. **30**, 759 (1987)].

[343] T. K. Kuo and J. Pantaleone, Rev. Mod. Phys. **61**, 937 (1989).

[344] E. T. Whittaker and G. N. Watson, Modern Analysis (Cambridge University Press, Cambridge 1927).

[345] S. Toshev, Phys. Lett. **B196**, 170 (1987); T. Kaneko, Prog. Theor. Phys. **78**, 532 (1987).

[346] T. K. Kuo and J. Pantaleone, Phys. Rev. D**39**, 1930 (1989).

[347] V. Barger, R. J. N. Phillips and K. Whisnant, Phys. Rev. D**34**, 980 (1986); J. Bouchez et al., Z. Phys. **C32**, 499 (1986); S. J. Parke and T. P. Walker, Phys. Rev. Lett. **57**, 2322 (1986); W. C. Haxton, Phys. Rev. **D35**, 2352 (1987); J. N. Bahcall, J. M. Gelb and S. P. Rosen, Phys. Rev. **D35**, 2976 (1987); A. Dar, A. Mann, Y. Melina and D. Zajfman, Phys. Rev. **D35**, 3607 (1987); P. I. Krastev and S. T. Petcov, Phys. Lett. **207B**, 64 (1988); J. N.

Bahcall and W. C. Haxton, Phys. Rev. **D40**, 931 (1989); K. S. Hirata et al., Phys. Rev. Lett. **65**, 1301 (1990); P. I. Krastev, S. P. Mikheev and A. Yu. Smirnov, Inst. Nucl. Research (Moscow) preprint, P-695 (1991).

[348] Analyses with three neutrinos are carried out by: H. W. Zaglauer and K. H. Schwarzer, Phys. Lett. B 198, 556 (1987); C. W. Kim and W. K. Sze, Phys. Rev. D**35**, 1404 (1987); T. K. Kuo and J. Pantaleone, Phys. Rev. D**35**, 3432 (1987); S. T. Petcov, Phys. Lett. B**214**, 259 (1988); S. P. Mikheyev and A. Yu. Smirnov, Phys. Lett. B**200**, 560 (1988).

[349] More recent analyses are given in: P. Anselmann, Phys. Lett. B**285**, 390 (1992); X. Shi, D. N. Schramm and J. N. Bahcall, Phys. Rev. Lett. **69**, 717 (1992); P. I. Krastev and S. T. Petcov, Phys. Lett. B**299**, 99 (1993); S. A. Bludman, N. Hata, D. C. Kennedy and P. G. Langacker, Phys. Rev. D**47**, 2220 (1993); M. White, L. Krauss and E. Gates, Phys. Rev. Lett. **70**, 375 (1993).

[350] S. P. Mikheyev and A. Yu. Smirnov, in *'86 Massive Neutrinos in Astrophysics and in Particle Physics, Proc. of the 6th Moriond Workshop*, Tignes, 1986, ed. by O. Fackler and J. Tran Thanh Van (Editions Frontières, Gif-sur-Yvette, 1986) p.355; J. Bouchez et al., Z. Phys. C**32**, 499 (1986); A. Dar and A. Mann, Nature (London) **325**, 790 (1987); A. J. Baltz and J. Weneser, Phys. Rev. D**35**, 528 (1987); M. Cribier, W. Hampel, J. Rich and D. Vignaud, Phys. Lett. B**182**, 89 (1986); M. L. Cherry and K. Lande, Phys. Rev. D**36**, 3571 (1987); S. Hiroi, H. Sakuma, T. Yanagida and M. Yoshimura, Phys. Lett. **B198**, 403 (1987): Prog. Theor. Phys. **78**, 1428 (1987); A. J. Baltz and J. Weneser, Phys. Rev. **D37**, 3364 (1988).

[351] F. D. Stacey, *Physics of the Earth* (Wiley, New York, 1969).

[352] M. Koshiba, in *Proc. of Workshop on Grand Unified Theories and Cosmology*, Tsukuba, 1983 ed. by K. Odaka and A. Sugamoto, KEK report 84-12 (1984) p.24; Y. Totsuka, in *Seventh Workshop on Grand Unification / ICOBAN'86*, Toyama, 1986, ed. by J. Arafune (World Scientific, Singapore, 1987) p.118.

[353] K. S. Hirata et al., Phys. Rev. Lett. **66**, 9 (1991).

[354] G. Sinclair et al. Nuovo Cim. **9**C, 308 (1986); G. Aardsma, Phys. Lett. **194**B, 321 (1987).

[355] C. Arpesella et al. (BOREXINO Collaboration) *BOREXINO at Gran Sasso, Proposal for a Real Time Detector for Low Energy Neutrinos* (Gran Sasso Lab. 1991) vol. I and II.

[356] E. D. Carlson, Phys. Rev. D**34**, 1454 (1986); J. M. LoSecco, Phys. Rev. Lett. **57**, 652 (1986); J. M. LoSecco et al., Phys. Lett. B**184**, 305 (1987).

[357] P. O. Lagage, M. Cribier, J. Rich and D. Vignaud, Phys. Lett. **B193**, 127 (1987); L. Wolfenstein, Phys. Lett. B**194**, 197 (1987).

[358] J. Arafune and M. Fukugita, Ref.[157]; J. Arafune, M. Fukugita, T. Yanagida and M. Yoshimura, Phys. Rev. Lett. **59** 1864 (1987); Phys. Lett. B**194**, 477 (1987); H. Minakata, H. Nunokawa, K. Shiraishi and H. Suzuki, Mod. Phys. Lett. A**2**, 827 (1987); S. P. Mikheyev and A. Yu. Smirnov, Pisma Zh. Eksp. Teor. Fiz. **46**, 11 (1987) [JETP Lett. **46**, 10 (1987)].

[359] J. N. Bahcall, A. Yu. Smirnov and D. N. Spergel, Princeton preprint IASSNS-93-15 (1983).

[360] S. P. Mikeyev and A. Yu. Smirnov, Ref.[350]

[361] P. Langacker, S. T. Petcov. G. Steigman and S. Toshev, Nucl. Phys. B**282**, 589 (1987).

[362] N. Cabibbo, Phys. Lett. **72B**, 333 (1978).

[363] S. M. Bilenky and S. T. Petcov, Rev. Mod. Phys. **59**, 671 (1987).

[364] L. F. Li and F. Wilczek, Phys. Rev. D**25**, 143 (1982).

[365] H. -Y. Chen, Phys. Rev. D**34**, 2794 (1986).

[366] M. Doi et al., Phys. Lett. **102B**, 323 (1981).

[367] B. W. Lee and S. Weinberg, Phys. Rev. Lett. **39**, 165 (1977).

[368] D. A. Dicus, E. W. Kolb and V. L. Teplitz, Phys. Rev. Lett. **39**, 168 (1977); Astrophys. J. **221**, 327 (1978).

[369] E. W. Kolb and T. Goldman, Phys. Rev. Lett. **43**, 897 (1979).

[370] E. W. Kolb, M. S. Turner, A. Chakravorty and D. N. Schramm, Phys. Rev. Lett. **67**, 533 (1991); A. D. Dolgov, I. Z. Rothstein, Phys. Rev. Lett. **71**, 476 (1993); see also E. W. Kolb and R. J. Scherrer, Phys. Rev. **D25**, 1481 (1982).

[371] M. S. Turner, G. Steigman and L. M. Krauss, Phys. Rev. Lett. **52**, 2090 (1984); M. Fukugita nad T. Yanagida, Phys. Lett. **144**B, 386 (1984).

[372] R. Gandhi and A. Burrows, Phys. Lett. **B246**, 149 (1990); A. Burrows, R. Gandhi and M. S. Turner, Phys. Rev. Lett. **68**, 3834 (1992).

[373] S. Dodelson, J. A. Frieman and M. S. Turner, Phys. Rev. Lett. **68**, 2572 (1992).

[374] R. E. Shrock, Phys. Rev. D**24**, 1232 (1981).

[375] M. Daum et al., Phys. Rev. D**36**, 2624 (1987); R. Abela et al., Phys. Lett. **105B**, 263 (1981); see also R. C. Minehart et al., Phys. Rev. Lett. **52**, 804 (1984).

[376] D. A. Bryman et al., Phys. Rev. Lett. **50**, 1546 (1983); N. De Leener-Rosier et al., Phys. Rev. D**43**, 3611 (1991).

[377] L. Oberauer, F. von Feilitzsch and R. L. Mössbauer, Phys. Lett. B**198**, 113 (1987).

[378] J. J. Simpson, Phys. Rev. Lett. **54**, 1891 (1985).

[379] A. M. Apalikov et al., Pis'ma Zh. Eksp. Teor. Fiz. **42**, 233 (1985) [JETP Lett. **42**, 289 (1985)]; J. Markey and F. Boehm, Phys. Rev. C**32**, 2215 (1985); T. Altzitzoglou et al., Phys. Rev. Lett. **55**, 799 (1985); T. Ohi et al., Phys. Lett. **160B**, 322 (1985); V. M. Dater et al., Nature **318**, 547 (1985).

[380] J. J. Simpson, Phys. Lett. **B174**, 113 (1986); J. J. Simpson and A. Hime, Phys. Rev. D**39**, 1825; 1837 (1989); B. Sur et al., Phys. Rev. Lett. **66**, 2444 (1991); A. Hime and N. A. Jelley, Phys. Lett. B**257**, 441 (1991).

[381] M. J. Dugan, G. B. Gelmini, H. Georgi and L. J. Hall, Phys. Rev. Lett. **54**, 2302 (1985); S. L. Glashow and A. Manohar, Phys. Rev. Lett. **54**, 2306 (1985); J. F. Valle, Phys. Lett. **B159**, 49 (1985); M. J. Dugan, A. Manohar and A. E. Nelson, Phys. Rev. Lett. **55**, 170 (1985).

[382] T. Ohshima et al., Phys. Rev. D**47**, 4840 (1993); M Chen et al., Phys. Rev. Lett. **69**, 3151 (1992); J. Mortara et al., Phys. Rev. Lett. **70**, 394 (1993); F. Wietfeldt et al., Phys. Rev. Lett. **70**, 1759 (1993); A. Hime, Phys. Lett. B**299**, 165 (1993).

[383] Y. Hosotani, Nucl. Phys. **B191**, 411 (1981).

[384] J. C. Pati and A. Salam, Phys. Rev. Lett. **31**, 661 (1973).

[385] H. Georgi and S. L. Glashow, Phys. Rev. Lett. **32**, 438 (1974).

[386] J. C. Pati and A. Salam, Phys. Rev. **D10**, 275 (1974); R. N. Mohapatra and J. C. Pati, Phys. Rev. **D11**, 2558 (1975); G. Senjanovic and R. N. Mohapatra, Phys. Rev. **D12**, 1502 (1975)

[387] R. N. Mohapatra and G. Senjanovic, Phys. Rev. Lett. **44**, 912 (1980); Phys. Rev. D**23**, 165 (1981).

[388] H. Fritzsch and P. Minkowski, Ann. Phys. (NY) **93**, 193 (1975).

[389] H. Georgi, in *Particles and Fields, APS/PPF Williamsburg*, ed. C. E. Carlson (AIP, New York, 1975).

[390] H. Georgi and S. L. Glashow, Phys. Rev. D**6**, 429 (1972); D. J. Gross and R. Jackiw, Phys. Rev. **D6**, 477 (1972).

[391] R. Slansky, Phys. Rep. C**79**, 1 (1981).

[392] M. S. Chanowitz, J. Ellis and M. K. Gaillard, Nucl. Phys. **B128**, 506 (1977).

[393] G. Branco and A. Masiero, Phys. Lett. B**97**, 95 (1980); T. Yanagida and M. Yoshimura, Phys. Lett. **B97**, 99 (1980).

[394] A. J. Buras, J. Ellis, M. K. Gaillard and D. V. Nanopoulos, Nucl. Phys. **B135**, 66 (1978).

[395] P. Langacker, Phys. Rep. **C72**, 185 (1981).

[396] S. F. Tuan, Mod. Phys. Lett. A**7**, 641 (1992).

[397] E. Witten, Phys. Lett. **91B**, 81 (1980).

[398] S. Rajpoot, Phys. Rev. **D22**, 2244 (1980).

[399] H. Georgi¡ H. R. Quinn and S. Weinberg, Phys. Rev. Lett. **33**, 451 (1974).

[400] H. Georgi and D. V. Nanopoulos, Nucl. Phys. B**159**, 16 (1979); Q. Shafi, M. Sondermann and C. Wetterich, Phys. Lett. **92**B, 304 (1980); T. G. Rizzo and G. Senjanovic, Phys. Rev. Lett. **46**, 1315 (1981).

[401] W. E. Caswell, J. Milutinović and G. Senjanović, Phys. Rev. D**26**, 161 (1982); R. W. Robinett and J. L. Rosner, Phys. Rev. D**25**, 3036 (1982).

[402] Y. Tosa, G. C. Branco and R. E. Marshak, Phys. Rev. **D28**, 1713 (1983).

[403] P. Langacher and M. X. Luo, Ref.[69].

[404] T. J. Goldman and D. A. Ross, Phys. Lett. **84B**, 208 (1979); Nucl. Phys. **B171**, 273 (1980); M. B. Einhorn and D. R. T. Jones, Nucl. Phys. B**196**, 475 (1982).

[405] For example, Y. Totsuka, in *Proc of the 1985 International Symposium on Lepton and Photon Interaction at High Energies*, Kyoto, 1985 ed. by M. Konuma and K. Takahashi (Yukawa Hall, Kyoto 1986), p.120.

[406] R. Becker-Szendy et al., Phys. Rev. D**42**, 2974 (1990).

[407] For example W. J. Marciano, in *Proceedings of Eighth Workshop on Grand Unification*, ed. C. Wali (World Scientific, Singapore, 1987) p.185.

[408] U. Amaldi, W. de Boer and H. Fürstenau, Phys. Lett. B**260**, 447 (1991); W. J. Marciano, Ref.[81]; J. Ellis, S. Kelly and D. V. Nanopoulos, Phys. Lett. B**260**, 131 (1991).

[409] S. Dimopoulos, S. Raby and F. Wilceek, Phys. Rev. **D24**, 1681 (1981).

[410] N. Sakai and T. Yanagida, Nucl. Phys. **B197**, 533 (1982).

[411] S. Weinberg, Phys. Rev. **D26**, 287 (1982).

[412] J. Hisano, H. Murayama and T. Yanagida, Phys. Rev. Lett. **69**, 1014 (1992).

[413] P. Nath and R. Arnowitt, NUB-TH-3056/92 (1992); J. L. Lopez, D. V. Nanopoulos and A. Zichichi, CERN preprint TH 6934/93 (1993).

[414] M. Machacek, Nucl. Phys. **B159**, 37 (1979).

[415] V. V. Dixit and M. Sher, Phys. Rev. **D40**, 3765 (1989).

[416] M. K. Parida and P. K. Patra, Phys. Rev. Lett. **66**, 858 (1991).

[417] D. Chang, R. N. Mohapatra and M. K. Parida, Phys. Rev. Lett. **52**, 1072.

[418] N. G. Deshpande, E. Keith and P. B. Pal, Phys. Rev. D**46**, 2261 (1992); N. T. Shaban and W. J. Stirling, Phys. Lett. B**291**, 281 (1992); M. K. Parida, P. K. Patra and C. C. Hazra, Phys. Rev. D**43**, 2351 (1991); G. M. Asatryan and S. G. Matinyan, Yad. Fiz. **55**, 479 (1992).

[419] M. Fukugita and T. Yanagida, Ref. [307].

[420] R. N. Mohapatra and M. K. Parida, Phys. Rev. D**47**, 264 (1993).

[421] D. Chang, R. N. Mohapatra, J. M. Gipson, R. E. Marshak and M. K. Parida, Phys. Rev. **D31**, 1718 (1985).

[422] J. A. Harvey, D. B. Reiss and P. Ramond, Nucl. Phys. **B199**, 223 (1982).

[423] H. Georgi and C. Jarlskog, Phys. Lett. **86B**, 297 (1979).

[424] R. D. Peccei and H. R. Quinn, Phys. Rev. Lett. **38**, 1440 (1977); Phys. Rev. **D16**, 1791 (1977).

[425] Y. Chikashige, R. M. Mohapatra and R. N. Peccei, Phys. Rev. Lett. **45**, 1926 (1980).

[426] D. B. Reiss, Phys. Lett. **B115**, 217 (1982).

[427] F. Wilczek, Phys. Rev. Lett. **49**, 1549 (1982); G. B. Gelmini, S. Nussinov and T. Yanagida, Nucl. Phys. **B219**, 31 (1983).

[428] M. Fukugita and T. Yanagida, Phys. Rev. Lett. **66**, 2705 (1991).

[429] A. Jodidio et al., Phys. Rev. **D34**, 1967 (1986).

[430] M. S. Atiya et al., Phys. Rev. Lett. **64**, 21 (1990).

[431] M. Fukugita and T. Yanagida, Phys. Rev. Lett. **55**, 2645 (1985).

[432] T. Maehara and T. Yanagita, Nuovo. Cim. Lett. **19**, 424 (1977); Prog. Theor. Phys. **60**, 822 (1978); K. Akama and H. Terazawa, INS-Report-257 (1977); F. Wilczek and A. Zee, Phys. Rev. Lett. **42**, 421 (1979).

[433] T. Yanagida, Phys. Rev. D**20**, 2986 (1979).

[434] S. L. Glashow, Phys. Lett. B**256**, 255 (1991).

[435] G. t'Hooft, Phys. Rev. Lett. **37**, 8 (1976); Phys. Rev. **D14**, 3432 (1976).

[436] C. G. Callan, Jr., R. F. Dashen and D. J. Gross, Phys. Lett. B**63**, 334 (1976); R. Jackiw and C. Rebbi, Phys, Rev. Lett. **37**, 172 (1976).

[437] M. F. Smith, Phys. Lett. B**234**, 191 (1990).

[438] V. Baluni, Phys. Rev. **D19**, 2227 (1979); R. J. Crewther, P. DiVecchia, G. Veneziano and E. Witten, Phys. Lett. **88B**, 123 (1979).

[439] S. Weinberg, Phys. Rev. Lett. **40**, 223 (1978); F. Wilczek, Phys. Rev. Lett. **40**, 279 (1978).

[440] For example, see R. D. Peccei, in Proceedings of Grand Unified Theories and Related Topics, Kyoto, 1981, ed. by M. Konuma and T. Maskawa (World Scientific, Singapore, 1981) p.347.

[441] J. E. Kim, Phys. Rev. Lett. **43**, 103 (1979).

[442] M. Dine, W. Fischler and M. Srednicki, Phys. Lett. **104B**, 199 (1981).

[443] M. A. Shifman, A. I. Vainshtein and V. I. Zakharov, Nucl. Phys. **B166**, 493 (1980).

[444] C. Vafa and E. Witten, Phys. Rev. Lett. **53**, 535 (1984); Nucl. Phys. B**234**, 173 (1984).

[445] R. N. Mohapatra and G. Senjanović, Z. Phys. **C17**, 53 (1983).

[446] R. Holman, G. Lazarides and Q. Shafi, Phys. Rev. **D27**, 995 (1983).

[447] P. Langacker, R. D. Peccei and T. Yanagida, Mod. Phys. Lett. **A1**, 541 (1986).

[448] D. A. Dicus, E. W. Kolb, V. L. Teplitz and R. V. Wagoner, Phys. Rev. D**18**, 1829 (1978); see also M. I. Vysotskii, Ya. B. Zeldovich, M. Yu. Khlopov and V. M. Chechetkin, Pisma Zh. Eksp. Teor. Fiz. **27**, 533 (1978) [JETP Lett. **27**, 502 (1978)]; the argument for the constraint on exotic scalar particle emission from stars was first made by K. Sato and H. Sato, Prog. Theor. Phys. **54**, 1564 (1975).

[449] M. Fukugita, S. Watamura and M. Yoshimura, Phys. Rev. Lett. **48**, 1522 (1982); Phys. Rev. D**26**, 1840 (1982); D. S. P. Dearborn, D. N. Schramm and G. Steigman, Phys. Rev. Lett. **56**, 26 (1986); G. G. Raffelt, Phys. Rev. D**33**, 897 (1986); D**41**, 1324 (1990); constraints from neutron star cooling was discussed in N. Iwamoto, Phys. Rev. Lett. **53**, 1198 (1984).

[450] R. Mayle et al., Phys. Lett. B**203**, 188 (1988); B**219**, 515 (1989); A. Burrows, M. S. Turner and R. P. Brinkmann, Phys. Rev. D**39**, 1020 (1989).

[451] M. Fukugita, in *Weak Interactions and Neutrinos – 1989, Proc. of the 12th International Workshop on Weak Interactions and Neutrinos*, Sea of Galilee, 1989, ed. by P. Singer and G. Eilam, Nucl. Phys. B (Proc. Suppl.) **13**, 401 (1990)

[452] J. Preskill, M. B. Wise and F. Wilczek, Phys. Lett **120**B, 127 (1983); L. F. Abbott and P. Sikivie, Phys. Lett **120**B, 133 (1983); M. Dine and W. Fishler, Phys. Lett **120**B, 137 (1983).

[453] D. J. Gross, R. D. Pisaraki and L. G. Yaffe, Rev. Mod. Phys. **53**, 43 (1981).

[454] P. Sikivie, Phys. Rev. Lett. **48**, 1156 (1982).

[455] T. W. B. Kibble, J. Phys. **A9**, 1387 (1976).

[456] G. Lazarides and Q. Shafi, Phys. Lett. **115B**, 21 (1982).

[457] S. M. Barr, D. B. Reiss and A. Zee, Phys. Lett. **116**B, 227 (1982).

[458] T. Maehara and T. Yanagida, Prog. Theor. Phys. **61**, 1434 (1979); R. N. Cahn and H. Harari, Nucl. Phys. B**176**, 135 (1980).

[459] A. Zee, Phys. Lett. **93B**, 389 (1980).

[460] L. Wolfenstein, Nucl. Phys. **B175**, 93 (1980).

[461] R. N. Mohapatra and J. W. F. Valle, Phys. Rev. **D34**, 1642 (1986).

[462] H. Fritzsch, Phys. Lett. **70B**, 436 (1977); **73B**, 317 (1978); Nucl. Phys. **B155**, 189 (1979).

[463] M. Shin, Phys. Lett. **B145**, 285 (1984); H. Harari and Y. Nir, Phys. Lett. **B195**, 586 (1987); C. H. Albright, Phys. Lett. **B227**, 171 (1989).

[464] S. Dimopoulos, L. J. Hall and S. Raby, Phys. Rev. Lett. **68**, 1984 (1992); V. Barger, M. S. Berger, T. Han and M. Zralek, Phys. Rev. Lett. **68**, 3394 (1992).

[465] M. Fukugita, M. Tanimoto and T. Yanagida, Prog. Theor. Phys. **89**, 263 (1993).

[466] R. E. Shrock, Nucl. Phys. **B206**, 359 (1982).

[467] S. T. Petcov, Yad, Fiz. **25**, 641 (1977) [Sov. J. Nucl. Phys. **25**, 340 (1977)]; T. Goldman and G. J. Stephenson, Jr., Phys. Rev. **D16**, 2256 (1977); K. Sato and M. Kobayashi, Prog. Theor. Phys. **58**, 1775 (1977).

[468] J. Schechter and J. W. F. Valle, Phys. Rev. D**24**, 1883 (1981).

[469] B. Kayser, Phys. Rev. **D26**, 1662 (1982).

[470] J. E. Kim, Phys. Rev. D**14**, 3000 (1976); M. A. Bég, W. J. Marciano and M. Ruderman, Phys. Rev. **D17**, 1395 (1978).

[471] D. P. Stoker et al., Phys. Rev. Lett. **54**, 1887 (1985).

[472] G. Beall, M. Bander and A. Soni, Phys. Rev. Lett. **48**, 848 (1982).

[473] M. Fukugita and T. Yanagida, Phys. Rev. Lett. **58**, 1807 (1987).

[474] A. Zee, Phys. Lett. **161**B, 141 (1985).

[475] M. Fukugita and T. Yanagida, Phys. Lett. **206B**, 93 (1988).

[476] J. W. F. Valle, Phys. Lett. **199**B, 432 (1987).

[477] M. B. Voloshin, Yad. Fiz. **48**, 804 (1988) [Sov. J. Nucl. Phys. **48**, 512 (1988)].

[478] R. Barbieri and R. N. Mohapatra, Phys. Lett. **218B**, 225 (1989); J. Liu, Phys. Lett. **225**B, 148 (1989).

[479] M. Leurer and N. Marcus, Phys. Lett. **B237**, 81 (1990); K. S. Babu and R. N. Mohapatra, Phys. Rev. Lett. **63**, 228 (1989); Phys. Rev. **D42**, 3778 (1990).

[480] H. Georgi and L. Randall, Phys. Lett. **244B**, 196 (1990).

[481] K. S. Babu and R. N. Mohapatra, Phys. Rev. Lett. **64**, 1705 (1990); G. Ecker, W. Grimus and H. Neufeld, Phys. Lett. **B232**, 217 (1990); D. Chang, W. Y. Keung and G. Senjanović, Phys. Rev. **D42**, 1599 (1990); D. Chang, W. Y. Keung, S. Lipovaca and G. Senjanović, CERN-TH 6027-91 (1991).

[482] S. M. Barr, E. M. Freire and A. Zee, Phys. Rev. Lett. **65**, 2626 (1990).

[483] G. A. Bazilevskaya, Yu. I. Stozhkov and T. N. Charakhchyan, Pis'ma Zh. Eksp. Teor. Fiz. **35**, 273 (1982) [JETP Lett. **35**, 341 (1982)]; G. A. Bazilevskaya et al., Yad. Fiz. **39**, 856 (1984) [Soviet J. Nucl. Phys. **39**, 543 (1984)].

[484] J. N. Bahcall, G. B. Field and W. H. Press, Astrophys. J. **320**, L69 (1987); J. N. Bahcall and W. H. Press, Astrophys. J. **370**, 730 (1991); B. W. Filippone and P. Vogel, Phys. Lett. **B246**, 546 (1990); H. Nunokawa and H. Minakata, Int. J. Mod. Phys. A**6**, 2347 (1991); J. W. Bieber, D. Seckel, T. Stanev and G. Steigman, Nature **348**, 407 (1990).

[485] M. B. Voloshin and M. I. Vysotskiĭ, Yad. Fiz. **44**, 845 (1986) [Sov. J. Nucl. Phys. **44**, 544 (1986)]; L. B. Okun, Yad. Fiz. **44**, 847 (1986) [Sov. J. Nucl. Phys. **44**, 546 (1986)]; M. B. Voloshin, M. I. Vysotskĭ and L. B. Okun, Yad. Fiz. **44**, 677 (1986) [Sov. J. Nucl. Phys. **44**, 440 (1986)]; Zh. Eksp. Teor. Fiz. **91**, 754 (1986) [Sov. Phys. JETP **64**, 446 (1986)].

[486] A. Cisneros, Astrophys. Sp. Sci. **10**, 87 (1971).

[487] R. Barbieri and G. Fiorentini, Nucl. Phys. **B304**, 909 (1988).

[488] W. A. Dziembowski and P. R. Goode, Astrophys. J. **347**, 540 (1989).

[489] C. S. Lim and W. J. Marciano, Phys. Rev. **D37**, 1368 (1988); E. Kh. Akhmedov, Phys. Lett. B**213**. 64 (1988).

[490] H. Minakata and H. Nunokawa, Phys. Rev. Lett. **63**, 121 (1989); J. Pulido, Phys. Lett. **B251**, 305 (1990).

[491] C. S. Lim, M. Mori, Y. Oyama and A. Suzuki, Phys. Lett. B**243**, 389 (1990).

[492] A. Suzuki, M. Mori, K. Numata and Y. Oyama, Phys. Rev. D**43**, 3557 (1991).

[493] M. Fukugita and T. Yanagida, Phys. Rev. Lett. **65**, 1975 (1990).

[494] K. S. Babu, R. N. Mohapatra and I. Z. Rothstein, Phys. Rev. D**44**, 2265 (1991); Y. Ono and D. Suematsu, Phys. Lett. B**271**, 165 (1991); E. Kh. Akhmedov, A. Lanza and S. T. Petcov, Phys. Lett. B**303**, 85 (1993).

[495] K. Sato and M. Kobayashi, Ref.[467]; A. De Rujúla and S. Glashow, Phys. Rev. Lett. **45**, 942 (1980).

[496] J. Silk and A. Stebbins, Astrophys. J. **269**, 1 (1983); M. Kawasaki and K. Sato, Phys. Lett. **169**B, 280 (1986).

[497] J. Maalampi, K. Mursula and M. Roos, Phys. Rev. Lett. **56**, 1031 (1986); J. Murthy and R. C. Henry, Phys. Rev. Lett. **58**, 1581 (1987).

[498] M. Fukugita and M. Kawasaki, Astrophys. J. **353**, 384 (1990).

[499] J. C. Mather et al., COBE preprint 93-01 (9993).

[500] P. Hut, Phys. Lett. **69**B, 85 (1977); D. A. Dicus, E. W. Kolb, V. L. Teplitz and R. V. Wagoner, Phys. Rev. D**17**, 1529 (1978); S. Miyama and K. Sato, Prog. Theor. Phys. **60**, 1703 (1978); E. W. Kolb and R. J. Scherrer Phys. Rev. D**25**, 1481 (1982); M. Kawasaki, N. Terasawa and K. Sato, Phys. Lett. **178**, 71 (1986).

[501] R. Cowsik, Phys. Rev. Lett. **39**, 784 (1977).

[502] S. W. Falk and D. N. Schramm, Phys. Lett. **79**B, 511 (1978).

[503] F. von Feilitzsch and L. Oberauer, Phys. Lett. **200**B, 580 (1988); E. L. Chupp, W. T. Vestrand and C. Reppin, Phys. Rev. Lett. **62**, 505 (1989); E. W. Kolb and M. S. Turner, Phys. Rev. Lett. **62**, 509 (1989).

[504] D. A. Krakauer et al., Phys. Rev. D**44**, R6 (1991); J. S. Frank et al., Phys. Rev. D**24**, 2001 (1981).

[505] J. Morgan, Phys. Lett. **102**B, 247 (1981); Mon. Not. Roy. Astr. Soc. **195**, 173 (1981).

[506] E. Asseo and H. Sol, Phys. Rep. **148**C, 307 (1987); see also, P. Coles, Comm. Astrophys. **16**, 45 (1992) for cosmological implication of the primordial magnetic field.

[507] A. D. Sakharov, Pis'ma Zh. Eksp. Teor. Fiz. **5**, 32 (1967) [JETP Lett. **5**, 24 (1967)].

[508] A. Yu. Ignatiev, N. V. Krasnikov, V. A. Kuzmin and A. N. Tavkhelidze, in *Neutrino 77, Proc. of the International Conference on Neutrino Physics and Neutrino Astrophysics*, Baksan Valley, 1977, ed. by M. A. Markov et al. (Nauka, Moscow, 1978) vol. II, p.293; Phys. Lett. **76B**, 436 (1978).

[509] M. Yoshimura, Phys. Rev. Lett. **41**, 281 (1978).

[510] D. Toussaint, S. B. Treiman, F. Wilczek and A. Zee, Phys. Rev. D**19**, 1036 (1979); S. Weinberg, Phys. Rev. Lett. **42**, 850 (1979); M. Yoshimura, Phys. Lett. **88B**, 294 (1979).

[511] T. Yanagida and M. Yoshimura, Phys. Rev. Lett. **45**, 71 (1980).

[512] V. A. Kuzmin, V. A. Rubakov and M. E. Shaposhnikov, Phys. Lett. **155B**, 36 (1985): see also S. Dimopoulos and L. Susskind, Phys. Rev. D**18**, 4500 (1978).

[513] A. D. Dolgov, Phys. Rep. **222**, 309 (1992).

[514] M. Fukugita, T. Yanagida and M. Yoshimura, Phys. Lett. **106**B, 183 (1981).

[515] A. D. Dolgov, Yad. Fiz. **32**, 1606 (1980) [Sov. J. Nucl. Phys. **32**, 831 (1980)].

[516] J. N. Fry, K. A. Olive and M. S. Turner, Phys. Rev. Lett. **45**, 2074 (1980); Phys. Rev. D**22**, 2953; 2977 (1980).

[517] T. Yanagida and M. Yoshimura, Phys. Rev. D**23**, 2048 (1981).

[518] For a review, S. Coleman, in *The Whys of Subnuclear Physics*, Erice, 1977, ed. by A. Zichichi (Plenum Press, New York 1979) p.805.

[519] J. Arafune, P. G. O. Freund and C. Goebel, J. Math. Phys. **16**, 433 (1975).

[520] N. H. Christ, Phys. Rev. **D21**, 1591 (1980).

[521] F. R. Klinkhamer and N. S. Manton, Phys. Rev. **D30**, 2212 (1984); see also N. S. Manton, Phys. Rev. D**28**, 2019 (1983).

[522] J. S. Langer, Ann. Phys. (NY) **54**, 258 (1969).

[523] I. Affleck, Phys. Rev. Lett. **46**, 388 (1981).

[524] P. Arnold and L. McLerran, Phys. Rev. D**36**, 581 (1987); A. Ringwald, Phys. Lett. B**201**, 510 (1988); D. Yu. Grigoriev and V. A. Rubakov, Nucl. Phys. **B299**, 67 (1988); S. Yu. Khlebnikov and M. E. Shaposhnikov, Nucl. Phys. B**308**, 885 (1988).

[525] T. Akiba, H. Kikuchi and T. Yanagida, Phys. Rev. D**40**, 588 (1989).

[526] S. Yu. Khlebnikov and M. E. Shaposhnikov, Nucl. Phys. **B308**, 885 (1988).

[527] D. Yu. Grigoriev, V. A. Rubakov and M. E. Shaposhnikov, Phys. Lett. **216B**, 172 (1989); Nucl. Phys. **B326**, 737 (1989).

[528] J. Ambjϕrn, T. Askgaard, H. Porter and M. E. Shaposhnikov, Nucl. Phys. B**353**, 346 (1991).

[529] J. A. Harvey and M. S. Turner, Phys. Rev. **D42**, 3344 (1990).

[530] V. A. Kuzmin, V. A. Rubakov and M. E. Shaposhnikov, Phys. Lett. **191B**, 171 (1987); H. Dreiner and G. G. Ross, OUTP-92-08P (1992).

[531] M. Shaposhnikov, in *Lattice 91, Proc. of the International Symposium on Lattice Field Theory*, Tsukuba, 1991, ed. by M. Fukugita et al., Nucl. Phys. B (Proc. Suppl.) **26**, 78 (1992).

[532] M. Fukugita and T. Yanagida, Phys. Rev. **D42**, 1285 (1990).

[533] B. A. Campbell, S. Davidson, J. Ellis and K. A. Olive, Phys. Lett. **B256**, 457 (1991); Astroparticle Phys. **1**, 77 (1992).

[534] W. Fischler, G. F. Giudice, R. G. Leigh and S. Paban, Phys. Lett. **B258**, 45 (1991); A. E. Nelson and S. M. Barr, Phys. Lett. **B246**, 141 (1990).

[535] L. E. Ibáñez and F. Quevedo, Phys. Lett. B**283**, 261 (1992).

[536] V. A. Kuzmin, Pisma Zh. Eksp. Teor. Fiz. **12**, 335 (1970) [JETP Lett **12**, 228 (1970)]; R. N. Mohapatra and R. E. Marshak, Phys. Rev. Lett. **44**, 1316 (1980).

[537] M. Baldo-Ceolin et al. Phys. Lett. **236**, 95 (1990).

[538] M. Takita et al., Phys. Rev. D**34**, 902 (1986); C. Berger et al., Phys. Lett. B**240**, 237 (1990); see C. B. Dover, A. Gal and J. M. Richard, Phys. Rev. D**27**, 1090 (1983) for a nuclear physic for the binding effect.

[539] M. E. Shaposhnikov, Nucl. Phys. **B287**, 757 (1987); **B299**, 797 (1988).

[540] J. Ambjϕrn, M. L. Laursen and M. E. Shaposhnikov, Nucl. Phys. B**316**, 483 (1989); L. McLerran, Phys. Rev. Lett. **62**, 1075 (1989); N. Turok and J. Zadrozny, Phys. Rev. Lett. **65**, 2331 (1990); L. McLerran, M. Shaposhnikov, N. Turok and M. Voloshin, Phys. Lett. **B256**, 451 (1991); M. Dine, P. Huet and R. Singleton, Jr., Nucl. Phys. B**375**, 625 (1992).

[541] A. G. Cohen, D. B. Kaplan and A. E. Nelson, Nucl. Phys. **B349**, 727 (1991); A. E. Nelson, D. B. Kaplan and A. G. Cohen, Nucl. Phys. **B373**, 453 (1992).

[542] M. Fukugita and T. Yanagida, Phys. Lett. **B174**, 45 (1986).

[543] M. Dine, R. G. Leigh, P. Huet, A. Linde and D. Linde, Phys. Rev. D**46**, 550 (1992).

[544] A. I. Bochkarev, S. V. Kuzmin and M. E. Shaposhnikov, Phys. Lett. **B244**, 275 (1990); J. R. Espinosa, M. Quiros and F. Zwirner, CERN preprint TH6833/93 (1993).

[545] M. A. Luty, Phys. Rev. **D45**, 455 (1992).

[546] B. A. Campbell, S. Davidson, J. Ellis and K. A. Olive, Phys. Lett. **B257**, 457 (1991).

[547] H. Murayama, H. Suzuki, T. Yanagida and J. Yokoyama, Phys. Rev. Lett. **70**, 1912 (1993).

[548] B. A. Campbell, S. Davidson, and K. Olive, Nucl. Phys. **B399**, 111 (1993).

[549] A. Bohr and B. R. Mottelson, *Nuclear Structure*, (W. A. Benjamin, NY, 1969) vol.1.

[550] M. Morita, *Beta Decay and Muon Capture* (W. A. Benjamin, Reading (U.S.A.), 1973).

[551] R. J. Blin-Stoyle, *Fundamental Interactions and the Nucleus* (North-Holland, Amsterdam, 1973).

[552] R. N. Mohapatra and B. Sakita, Phys. Rev. D**21**, 1062 (1980); F. Wilczek and A. Zee, Phys. Rev. D**25**, 553 (1982).

Kamiokande

Kenzo Nakamura, Takaaki Kajita, Masayuki Nakahata

Institute for Cosmic Ray Research, University of Tokyo

Atsuto Suzuki

Faculty of Science, Tohoku University

1. Introduction

Kamiokande[†] is an imaging water Cherenkov detector with 3,000 tons of pure water in the main detector tank, and an additional 1,500 tons of water forming, itself, an active water Cherenkov anticounter layer. The unique feature of this experiment is the use of the world's largest 20" diameter photomultiplier tubes (PMTs), which provide a large coverage of the photosensitive area over the inner detector surface. This feature is essential for making the detector sensitive to neutrinos with energies as low as $\lesssim 10$ MeV.

The original motivation which stimulated the construction of the Kamiokande detector was the search for nucleon decay. As yet, it has not been discovered, with only the lower limits of the nucleon partial lifetimes having been determined [1]. However, it was realized at the very beginning of the detector operation that the energy spectrum of decay electrons from the stopping muons could be measured down to a kinetic energy of $\sim$12 MeV. The trigger rate below this energy rapidly increased because of radioactive impurities dissolved in the water. Therefore, if they could be reduced, the Kamiokande detector could be converted to a real-time solar neutrino detector.

As is well known, solar neutrinos have been observed by Davis *et al.* [2] with the use of a chlorine detector since $\sim$ 1970, and the deficit of the solar-neutrino flux, known as "the solar-neutrino problem", has been reported. However, this experiment is based on a radiochemical technique which provides no information concerning the energy, direction, or timing of the observed neutrinos. It is therefore difficult for this experiment to prove unambiguously that the Sun is really emitting neutrinos.

In a water Cherenkov detector νe scattering is utilized to detect solar neutrinos. This reaction offers a nice feature of the directional correlation of the recoil electron

[†]Kamiokande is an acronym of <u>Kamioka</u> <u>n</u>ucleon <u>d</u>ecay <u>e</u>xperiment.

with the incoming neutrino, thus providing an opportunity to obtain clear evidence that the observed neutrinos are coming from the Sun. Also, the measurement of recoil electron energies provides the sensitivity to the energy spectrum of solar neutrinos. The real-time nature of the detector allows us to measure possible short-term variations such as the day-night difference in the solar-neutrino flux.

As the results of many efforts to remove uranium, radium and radon contaminated in the detector water, it was actually achieved to lower the trigger threshold down to 7.6 MeV (50% efficiency) at the beginning of 1986. In February 23, 1987, a historic supernova explosion occurred in the Large Magellanic Cloud, ~ 50 kpc away from the solar system, and a neutrino burst was detected by Kamiokande for the first time [3, 4]. Also, the observation of solar neutrinos has been successfully continued since January, 1987, and a number of interesting results have already been published [5, 6, 7, 8].

In addition to the Sun and type-II supernovae, there are a number of other neutrino sources, both of astrophysical and terrestrial origin. The fluxes and energies of neutrinos from these sources extend over many orders of magnitude. Figure 1 shows the expected spectra of neutrinos, relevant to the underground experiments, from some of these sources.

As a neutrino detector, Kamiokande is characterized by its capability to provide access to a remarkably wide range of energies. Medium-energy neutrinos (a few hundred MeV $\sim$ a few GeV) are detected as fully contained events, i.e., the same event sample as that used in a search for nucleon decay. High-energy neutrinos ($\gtrsim$ a few GeV) are detected with the use of upward-going muons. This feature allows a wide variety of topics to be investigated with Kamiokande. Figure 2 summarizes these topics and the respective energy regions of the search.

In the medium-energy region, the investigation of the atmospheric neutrinos is an important subject of the Kamiokande experiment, not only because they constitute the background to the search for nucleon decay as well as to the measurement of astrophysical neutrinos, but also because they deserve special attention in their own right. Namely, an anomalously small ν_μ/ν_e ratio was found [9, 10] with respect to the theoretical expectation; this might indicate neutrino oscillations. A search for high-energy neutrinos from the Sun or from the center of the Earth produced by annihilation of dark-matter particles is another important subject, though no positive signal has yet been observed.

This paper is organized as follows. The Kamioka Underground Laboratory and the Kamiokande detector are described in Sect. 2. The principle of neutrino detection in various energy regions is discussed in Sect. 3. Section 4 gives a description of detector simulation. The physics results obtained with Kamiokande are presented in the following sections. First of all, the results on measurements of atmospheric neutrinos and their implications are dealt with in Sect. 5, since the background from atmospheric neutrinos must be taken into consideration to develop a discussion concerning astrophysical neutrinos and nucleon decay. Section 6 is then devoted to a presentation of the results concerning versatile topics related to astrophysical neutrinos. The results concerning a search for baryon-number violating processes are given in Sect. 7. Finally, Sect. 8 focuses on the next-generation experiment at Kamioka, Superkamiokande.

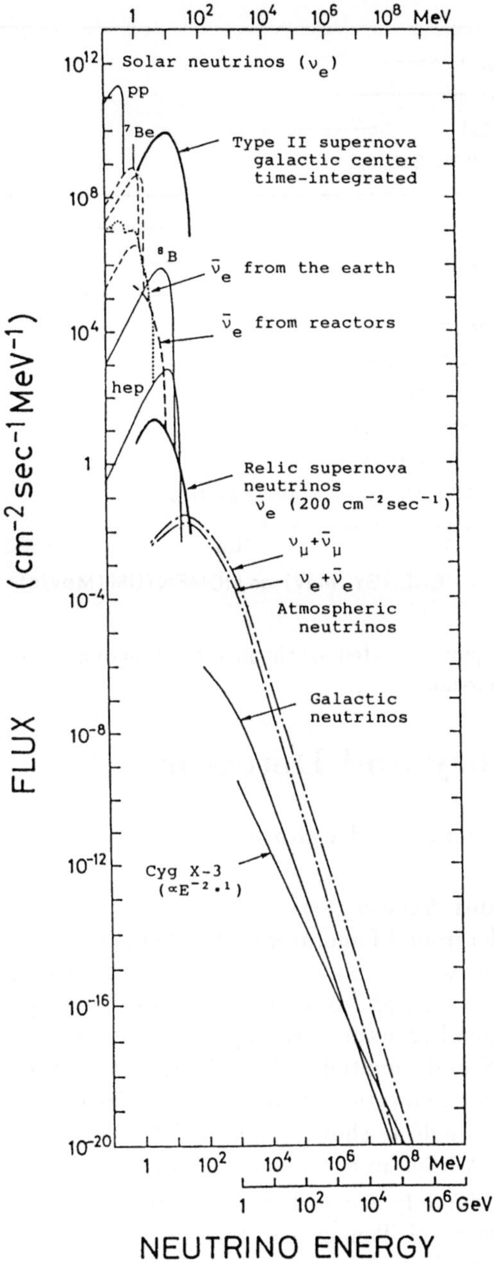

Figure 1: Expected neutrino energy spectra from various sources.

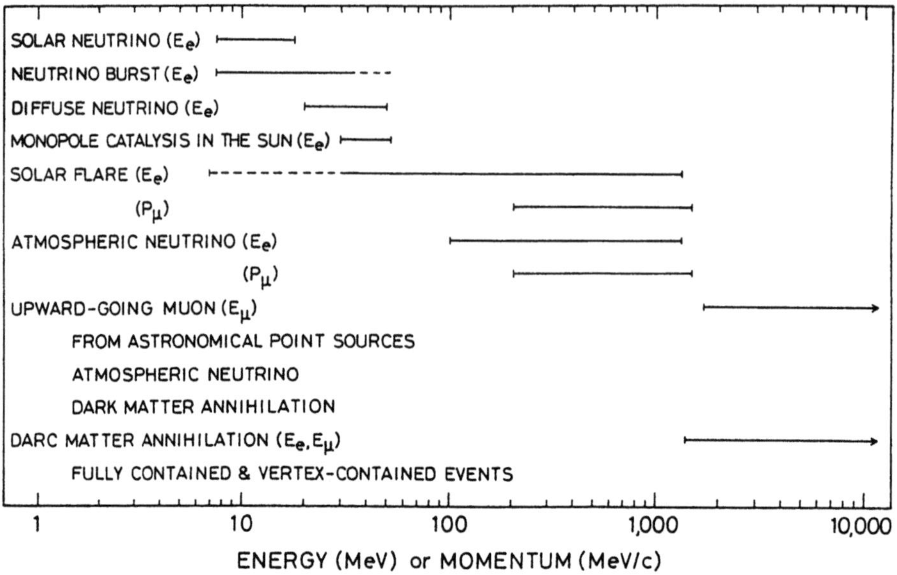

Figure 2: Energy regions selected for the investigation of astrophysical and atmospheric neutrinos with Kamiokande.

2. Laboratory and Detector

2.1 Kamioka Underground Laboratory

2.1.1 Location and Access

The Kamioka Underground Laboratory is located in the Mozumi Mine in Kamioka Town, which is operated by Kamioka Mining and Smelting Company. Kamioka Town, about 250 km west of Tokyo, lies in the northern region of Gifu Prefecture, bordering on Toyama Prefecture (see Fig. 3). The exact location of the laboratory is latitude 36.42°N and longitude 137.31°E; the geomagnetic latitude is 25.8°N. The laboratory has an elevation of 350 m, but it lies nearly under the top of Mt. Ikenoyama, 1,350 m high, as shown in Fig. 4. Therefore, the laboratory has a rock overburden of about 1,000 m in the vertical direction.

Until recently, the only access to the laboratory has been provided by trolley car through the Mozumi Tunnel (see Fig. 4), which runs horizontally about 3 km from the Mozumi Entrance into the mine. This tunnel has a width of 2.6 m and a height of 2.5 m. When the Kamiokande detector was constructed, all of the construction materials were carried into the laboratory through this tuunel. Toward the end of 1991, Kamioka Mining and Smelting Company completed the Atotsu Tunnel, and provided it as an alternative route to the laboratory. It is about 2 km long, and is a horizontal trackless tunnel with a width of 4.5 m and a height of 4.2 m. It has much larger capacity for transportation than does the

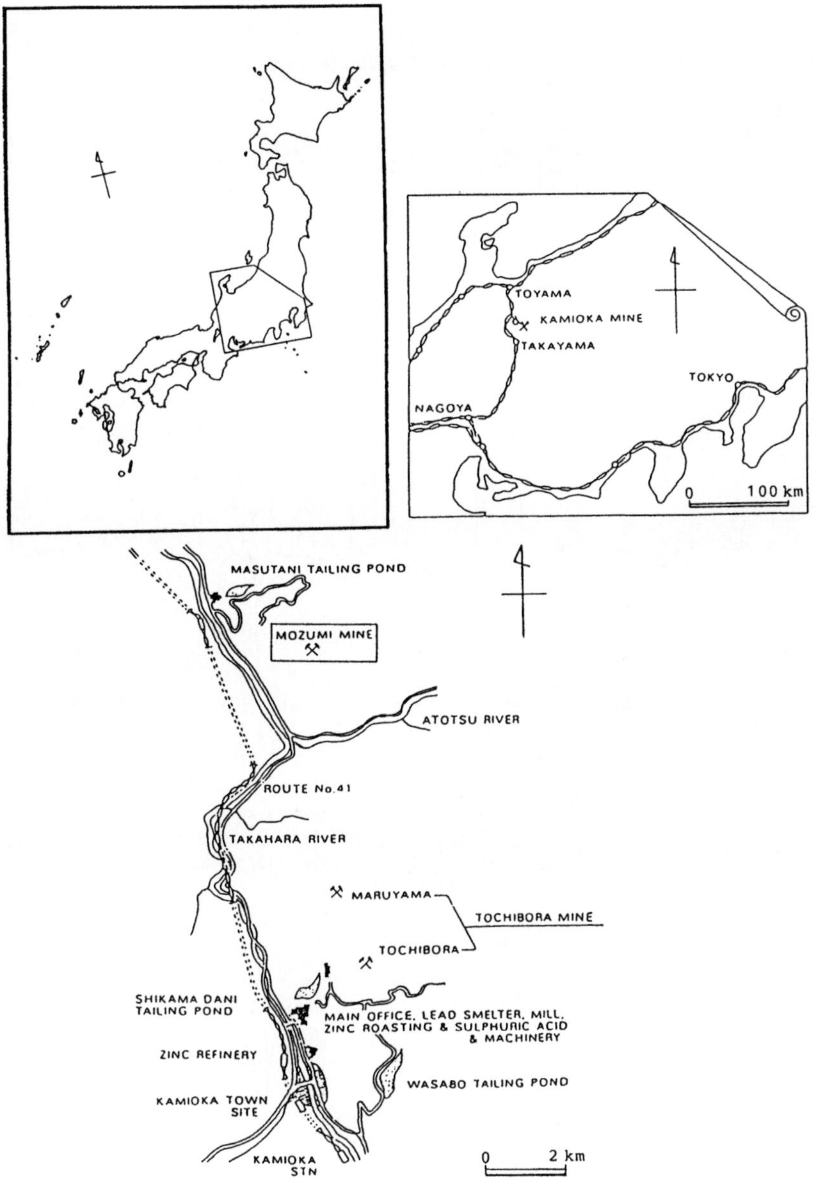

Figure 3: Location of the Mozumi Mine.

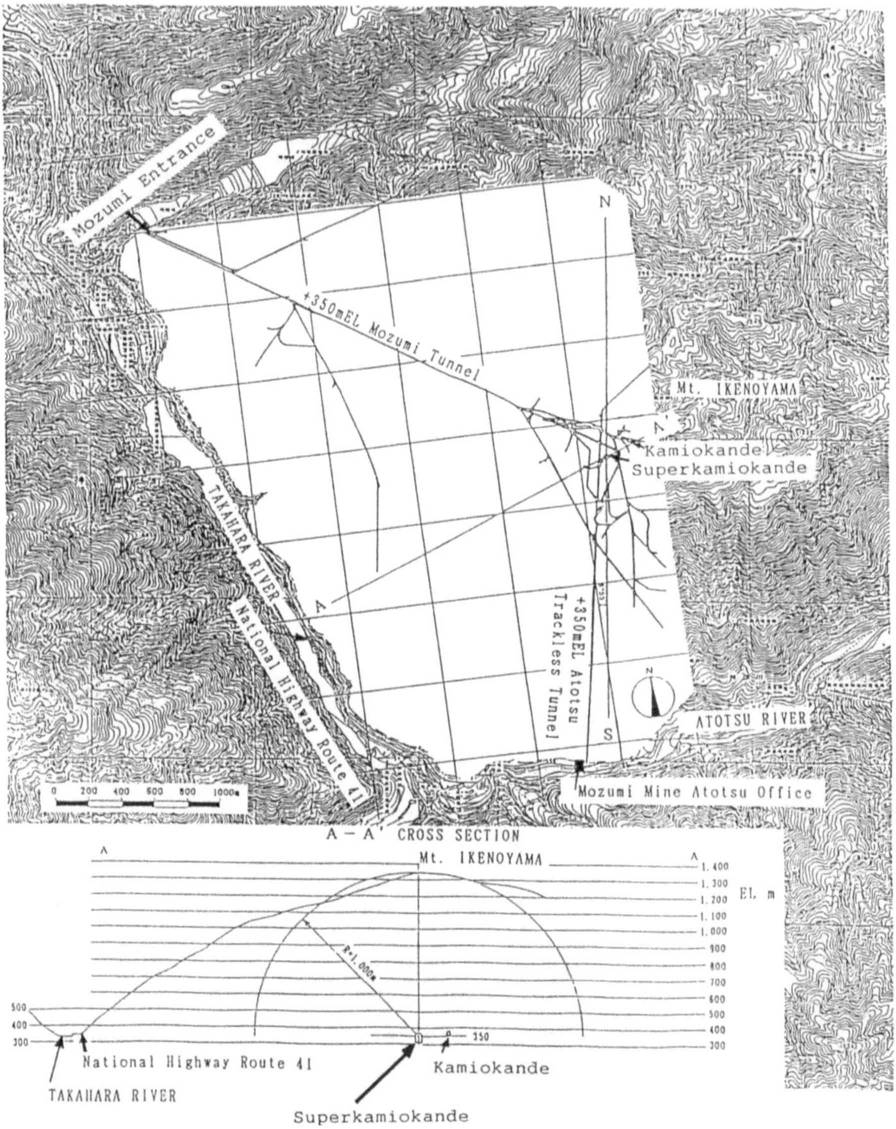

Figure 4: Location of Kamiokande and Superkamiokande inside the Mozumi Mine.

Table 1: Some characteristics of test pieces of a gneiss sampled at the Superkamiokande site.

Density	2.7 g/cm^3
Uniaxial compressive strength	1,520 kgf/cm^2
Tensile strength	99.1 kgf/cm^2
Static modulus of elasticity	5.09 $\times 10^5$ kgf/cm^2
Dynamic modulus of elasticity	6.36 $\times 10^5$ kgf/cm^2

Mozumi Tunnel. The construction of Superkamiokande is being carried out with heavy use of this tunnel.

The access into the mine is controlled by the company. It is prohibited to enter the mine when it is closed. Also, researchers are normally not allowed to stay overnight inside the mine. These constraints regulate the mode of detector operation and data-taking, as described in subsect. 2.3.

2.1.2 Geology around the Laboratory

The region covering from southern Toyama Prefecture to northern Gifu Prefecture is called the Hida region. It forms a highland comprising the Hida plateau, 1,000 to 1,500 m above sea level, and the Hida mountains with more than 3,000 m high peaks along its eastern margin. Kamioka lies in the central part of this region and is chiefly underlain by gneissic masses called Hida gneiss and early Mesozoic granitoids, as shown in Fig. 5. The Hida gneisses have long been suggested as the oldest rock in Japan, i.e., the Precambrian basement. However, they were mostly metamorphosed during the late Paleozoic, and extensively reheated by the intrusion of early Mesozoic granitoids. The record indicating the Precambrian age of the gneisses is therefore rare and indefinite, though not always deniable [11].

The mineralization in the Kamioka ore deposits are mainly of metasomatic replacement embedded in intercalated marble beds within the Hida gneisses [11]. The major constituent ore minerals are sphalerite (ZnS) and galena (PbS).

The cavity for the Kamiokande detetor was excavated in stable gneissic rock which had undergone no mineralization. The location of the cavity for the Superkamiokande detector was also determined to be in stable gneissic rock, about 200 m west-southwest of the Kamiokande. Table 1 lists some characteristics of test pieces of a gneiss sampled at the Superkamiokande site.

2.1.3 Outline of the Laboratory

Figure 6 shows a schematic of the site for Kamiokande. The plan for Superkamiokande is given in Sect. 8. The temperature at the laboratory site is nearly constant ($\sim$ 14°C) throughout the year. The cosmic-ray muon flux at the Laboratory is shown in Fig. 7 (a), where the contours of the vertical muon flux, defined by $\cos\theta \, I(\theta,\phi)$, are plotted. Here, θ is the zenith angle and $I(\theta,\phi)$ is the muon flux at the laboratory from the direction with zenith θ and azimuth ϕ. In this figure, a constant-radius circle with respect to the origin corresponds to a constant zenith

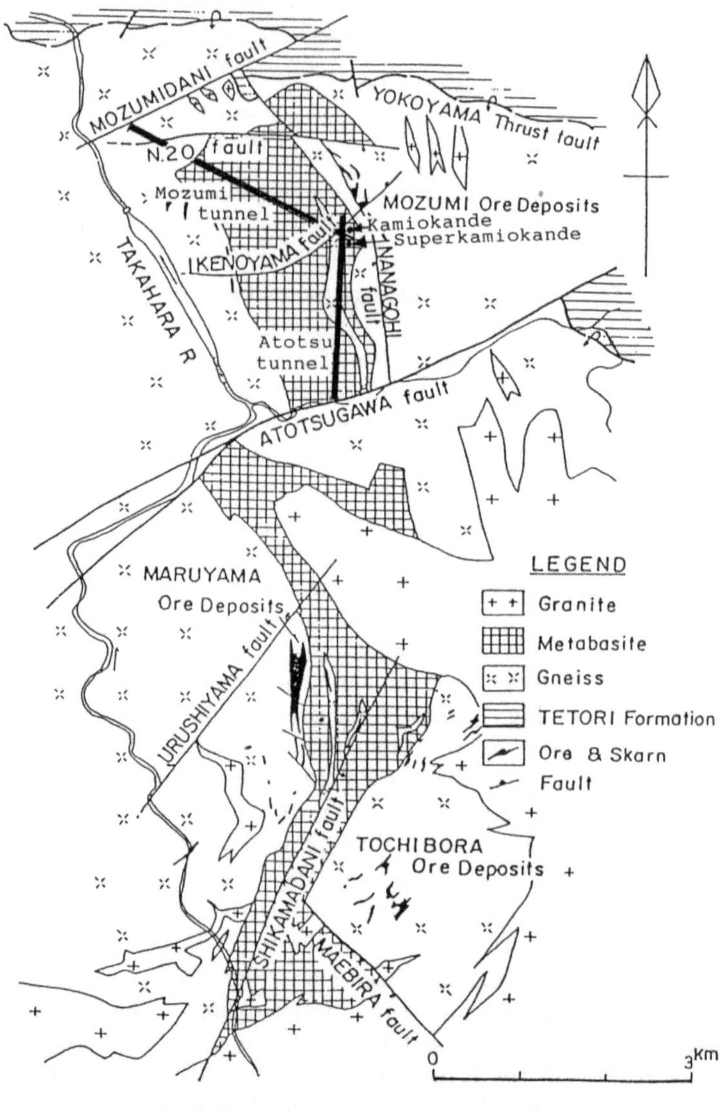

Figure 5: Geology of the Kamioka Mine [12]. Although "Metabasite" generally means metamorphosed basalt, the major body of the metabasite in the Kamioka area is made up mainly of coarse-grained diorites including gabbroic rocks, more or less metamorphosed and having gneissic features [11]. Note that this map represents the geologic sketch of the surface, and does not represent the geology at the level of the Kamioka Underground Laboratory. The cavities for the Kamiokande and Superkamiokande detectors are located in gneiss at their level.

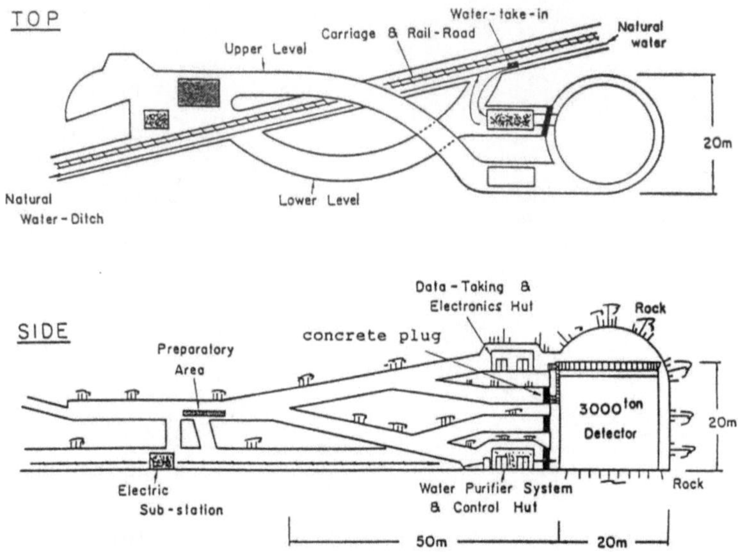

Figure 6: Schematic representation of the Laboratory site for Kamiokande.

angle. This plot may be compared with Fig. 7 (b), which shows contours of the slant depth, which is proportional to the amount of matter in the direction (θ, ϕ).

The cavity for the Kamiokande detector has a diameter of 20 m and a height of 24 m. On the surface of the cavity, a system of current network was installed in order to nearly cancel the geomagnetic field inside the cavity (see Fig. 8), because the 20″ PMT is very sensitive to magnetic fields.

A water-purification system is located at the bottom level of the cavity. Plenty of clean natural water with a temperature of $\sim 10°C$ comes from crevices about 500 m away from the laboratory site and flows through the ditch shown in Fig. 6. When the emptied tank is filled with pure water again, this natural water is taken in and purified with the water-purification system.

There is an electronics hut at the same level as the top of the tank. Cooling of this hut is provided by heat exchange with water pumped-up from the ditch. Starting at the beginning of 1990, an online computer in the electronics hut has been connected with another computer placed outside of the Mozumi Entrance via an optical-fiber link; the latter computer receives the data transmitted by the online computer and stores them on magnetic tapes. Until then, the online computer could be reached from outside of the mine only via a modem; although this allowed the control and monitoring of runs, a large amount of data could not be transmitted.

At the preparatory area, a small experiment which is searching for double β-decay and other rare nuclear processes is being conducted by a group from Osaka University. A detector called ELEGANT V is now being operated [13].

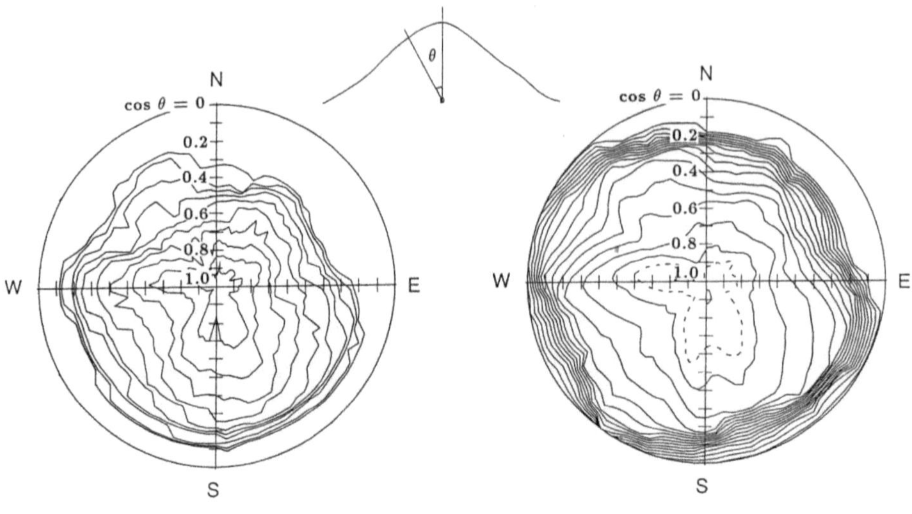

(a) (b)

Figure 7: (a) Contour plot of the vertical muon flux at the Kamiokande site. A constant-radius circle with respect to the origin corresponds to a constant zenith angle (see the insert). The innermost contour corresponds to $10^{-9.8}$ cm^{-2} s^{-1} sr^{-1}. The contours are plotted at intervals of $10^{0.2}$ cm^{-2} s^{-1} sr^{-1}, so that the outermost contour corresponds to 10^{-8} cm^{-2} s^{-1} sr^{-1}. (b) Contour plot of the slant depth. The innermost two separated contours correspond to 900 m. The contour represented by the dashed curve corresponds to 1,000 m. The contours represented by the solid curves are plotted at intervals of 200 m.

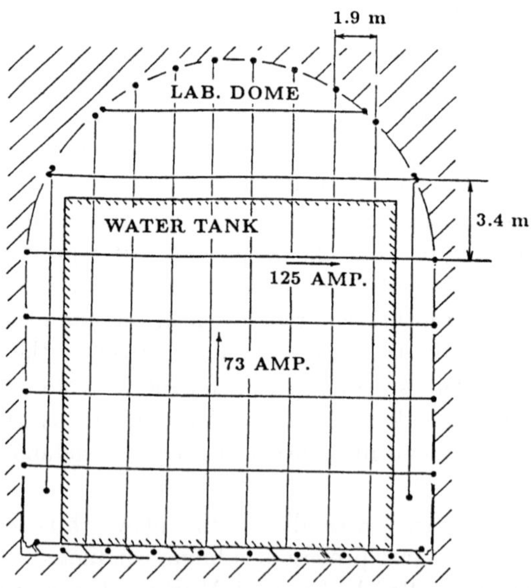

Figure 8: Current network on the surface of the cavity.

2.2 The Kamiokande Detector

2.2.1 Brief History

The Kamiokande detector has been continually improved. The first phase of the experiment (July 1983 - November 1985) is called Kamiokande-I. It was initially operated with no anticounter layer, which was constructed later in October - December 1984 (top and bottom sections, both inside the main detector tank) and in October 1985 (barrel section, between the rock cavity and the main detector tank).

The second phase, Kamiokande-II, was aimed at observing solar neutrinos, and has been operated since November 1985 with a new electronics system capable of recording the arrival time of each PMT signal in addition to its charge. The continued effort to purify water and, in particular, that of removing radioactive contamination from water, should be specially noted, since it has been essential for the successful observation of low-energy neutrinos.

In June 1988 the detector performance of Kamiokande-II was improved by doubling the PMT gain. The effects of the PMT gain increase are: (i) improved energy resolution for low-energy electrons and (ii) improvement in the event vertex reconsruction. These improvements were achieved owing to the increase ($\sim$20%) of hit PMTs corresponding to the unit deposited energy by electrons. This led to a significant reduction of background events.

The second phase was completed in April, 1990. Up to then, since more than 10% of the PMTs had been out of order due to various reasons, the detector was emptied and dead PMTs were replaced. At this occasion, the entire electronics were replaced with a more compact, low-power system. A reflector was also introduced to each PMT in the inner detector to collect more light. Observation was resumed in December, 1990; this phase is called Kamiokande-III. The following description of the detector is mainly that of Kamiokande-II, since this phase was most thoroughly studied. Table 2 gives some selected parameters at the final stage of Kamiokande-II.

2.2.2 The Detector

The Kamiokande-II (and also -III) detector is schematically shown in Fig. 9. Pure water with a mass of 3,000 tons is contained in a 16-m high, 15.6-m diameter cylindrical steel tank, 2,140 tons of which defines the inner detector (14.4 m in diameter and 13.1 m in height) and is viewed by 948 PMTs covering 20% of the surface. The top and bottom surfaces of the inner detector contain, respectively, 164 and 160 PMTs located on square lattice points with a spacing of 1 m. The barrel part of the inner detector contains 624 PMTs; vertically, 13 PMTs are arranged in a column with a spacing of 1 m, and the barrel wall comprises 48 such columns with a spacing of 0.945 m. A view of the emptied inner detector is shown in Fig. 10.

A 4π water-Cherenkov anticounter layer, viewed by 123 PMTs, surrounds the inner detector. The typical thicknesses of water in the anticounter are 0.7 m, 1.2 m and 1.5 m for the top, bottom, and barrel sections, respectively. The top and bottom sections are within the steel tank, and are optically separated from

Table 2: Parameters of the Kamiokande-II detector after the PMT gain change.

Depth	1,000 m (2,700 m water equivalent)
Dimensions of main tank	15.6 m in diameter × 16 m in height
Total volume	4,500 tons
Fiducial volume	2,140 tons (for supernova neutrinos)
	680 tons (for solar neutrinos)
Anticounter thickness	~ 1.5 m
Number of PMTs	948 (inner detector)
	123 (anticounter)
Photo-sensitive area	20% of the inner detector surface
Trigger efficiency for electrons	50% at 6.1 MeV, 90% at 9.0 MeV
Energy resolution	$3.6\% / \sqrt{E_e/1\,\mathrm{GeV}}$ for electrons
	4% for muons with $\lesssim$ 1 GeV
	$20\% / \sqrt{E_e/10\,\mathrm{MeV}}$ for electrons
	with $\lesssim$ 20 MeV
Vertex resolution	±1.1 m for 10 MeV electrons
Angular resolution	±28° for 10 MeV electrons
Uncertainty in the energy scale	< ±3 %

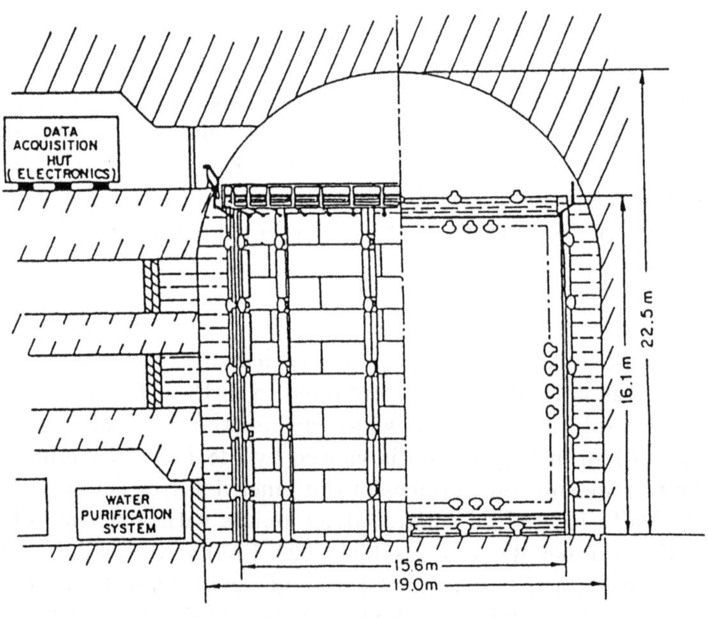

Figure 9: Schematic drawing of the Kamiokande-II detector showing the location of the water-purification system, anticounter, and electronics hut. The fiducial volume for the solar-neutrino observations is also shown. The appearance of the Kamiokande-III detector is the same.

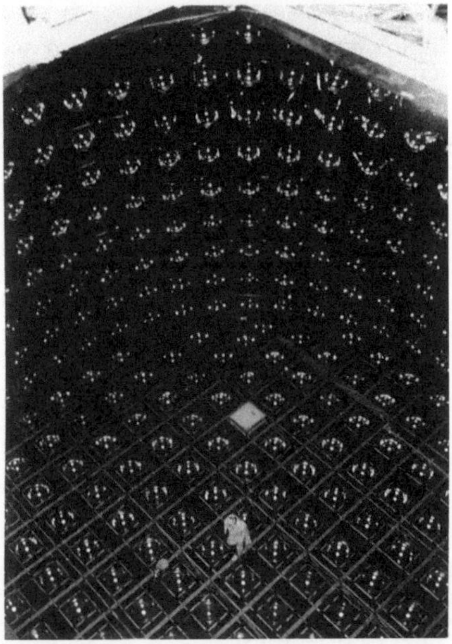

Figure 10: Photograph of the emptied inner detector of Kamiokande-II.

the inner detector by black plastic sheets. The barrel section was constructed by water-sealing the cavity and filling it with natural water. The anticounter serves to detect any signals outside the inner detector volume, and to shield against γ-rays and neutrons from outside the detector.

The 20″ PMT is a product (R1449) of Hamamatsu Photonics K. K. Its structure is schematically shown in Fig. 11 (a). Figure 12 shows a typical pulse-height distribution of a single photoelectron signal. In spite of its large size, a timing resolution of $\sim$ 5 ns was obtained † for a single photoelectron at a gain of 10^7. Further details concerning its performance can be seen in Ref. [14]. When the Kamiokande experiment started, the gain of each PMT was adjusted to 3×10^6. During the phase of Kamiokande-II, it was doubled to 6×10^6. The PMTs are protected against residual geomagnetic fields inside the current network (see subsect. 2.1.3) by a hemispherical mesh for the front part and a cone for the rear part, both made of μ-metal, as shown in Fig. 11 (b). With this magnetic shield, the residual magnetic field inside the PMT is less than 30 mG.

2.2.3 Electronics and Event Trigger

In the phase of Kamiokande-I, the electronics mainly consisted of (Analog to Digital Converter), linear adders, and transient digitizers which were later replaced

---

†Recently, the quality of the 20″ PMT has been greatly improved aiming at the production for Superkamiokande [15]. The timing resolution for a single photoelectron of $\sim$ 3 ns has been obtained at a gain of 10^7, and a clear single-photoelectron peak can be seen in the pulse-height distribution.

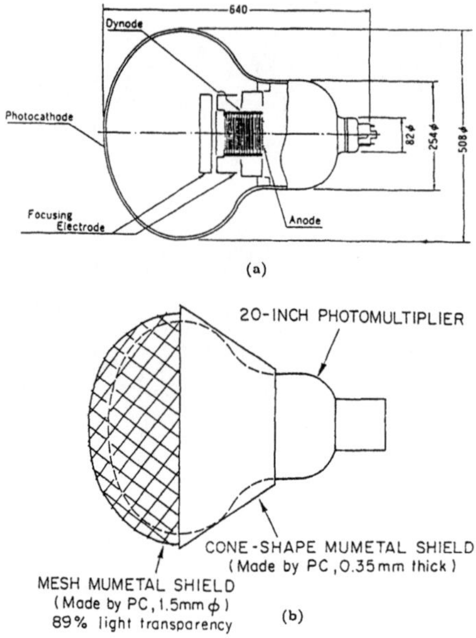

Figure 11: (a) Structure and (b) magnetic shield of the 20″ PMT.

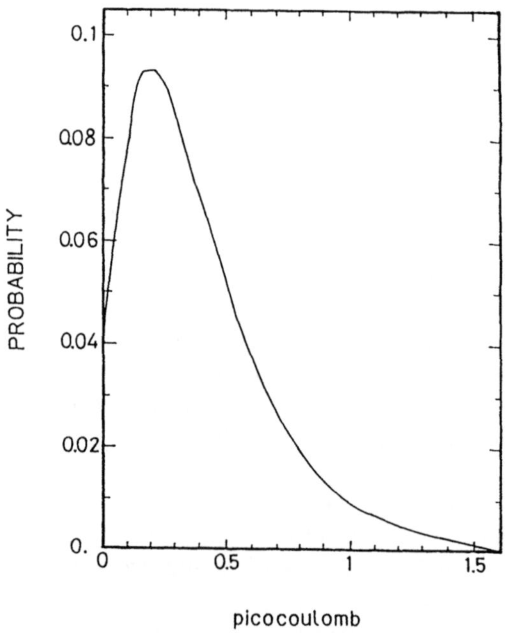

Figure 12: Typical pulse-height distribution of a single photoelectron signal from a 20″ PMT.

by flash ADCs. The event trigger basically required the summed PMT signals to be greater than 30 MeV for electrons. The transient digitizers and flash ADCs were used to record the $\mu \rightarrow e$ decay signals. This electronics system has been conventinally called the "Tokyo electronics".

In November, 1985, a new electronics system called "Pennsylvania electronics" was installed for the purpose of detecting solar neutrinos, and the new phase, Kamiokande-II, was started. This system was a charge and timing measuring system with multi-depth analog buffers, with the dead time reduced to nearly the limits imposed by the PMT's recovery times (less than 50 ns). This system allowed new trigger configurations to be set for a variety of physical events. The Tokyo electronics, supplemented with ADCs for the barrel anticounter, was also used in parallel. Each PMT signal was split and fed to both electronics systems.

The Kamiokande-II trigger system was designed to collect all events above ~ 6 MeV with almost no dead time. If the PMT signal was above the computer-controled threshold of ~ 0.35 photoelectron, a rectangular discriminator output was produced. An event trigger was generated when the analog sum of the discriminator outputs from all the inner-detector PMTs exceeded the adjustable trigger threshold.

In the phase of Kamiokande-III, the electronics were completely replaced with a high-density, low-power system. The front end of this system comprises semi-custom analog integrated circuits [16], each corresponding to one PMT and containing two time-to-amplitude converters and two current integrators to accommodate successive pulses, such as those from $\mu \rightarrow e$ decay with a reduced dead time. The scheme of the event trigger is the same as that of the Kamiokande-II electronics.

The trigger threshold was determined to keep the trigger rate at $1 \sim$ a few Hz. The trigger rate due to cosmic-ray muons is 0.37 Hz, and the remaining triggers are almost entirely caused by low-energy background events. Consequently, as the result of the effort to reduce the amount of radon dissolved in water (see the next subsection), the trigger threshold could be lowered. The trigger threshold at the beginning of Kamiokande-II was 7.6 MeV (50% efficiency), which was lowered to 6.7 MeV after October, 1987. The trigger threshold was further lowered to 6.1 MeV after an increase in the PMT gain in June, 1988, 5.5 MeV at the beginning of Kamiokande-III, and 5 MeV in June, 1992.

2.2.4 Water-Purification System

The water in the inner detector must have a high transparency (attenuation length $\gtrsim 35$ m) in order to minimize the loss of Cherenkov photons. For this purpose, a water-purification system is operated. The water in the tank is continuously circulated at a rate of ~ 4.5 tons/hr through the water-purification system. It consists of (a) 5 membrane filters, (b) a degasifier, (c) an ultra-violet sterilizer to destroy bacteria, (d) ion-exchange columns for removing uranium (CR-55), (e) a mixed-bed-type deionization system, (f) pumps, and (g) a radon-free air system; for a schematic diagram, see Fig. 13. The most important task of the water-purification system in Kamiokande-II and -III has been to remove any radioactive elements dissolved in water. Continued efforts have been made to improve this

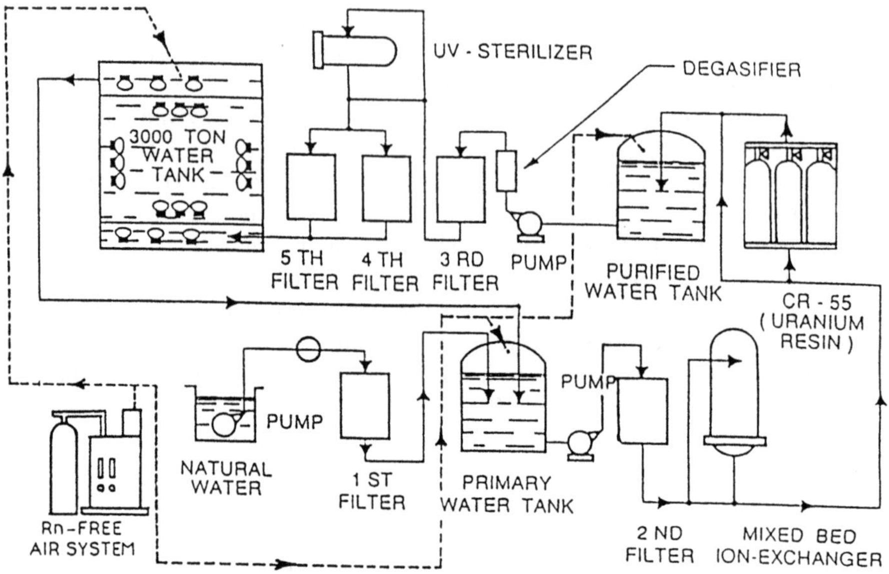

Figure 13: Block diagram of the Kamiokande-II water-purification system.

system.

At low energies, there are many background sources which prevent the trigger threshold from being lowered. Among them, γ-rays and neutrons from the surrounding rock are shielded by the anticounter layer. Hence, the most important low-energy background sources remaining are radioactive elements dissolved in the water: ^{238}U, ^{226}Ra, and ^{222}Rn. Spontaneous fission of ^{238}U produces prompt γ-rays with a summed energy of up to $\sim$8 MeV. ^{226}Ra and ^{222}Rn produce ^{214}Bi in the decay chain $^{226}\text{Ra} \to {}^{222}\text{Rn} \to \cdots \to {}^{214}\text{Bi} \to {}^{214}\text{Po}$. The maximum electron energy from the β-decay of ^{214}Bi is 3.26 MeV. Though this is below the trigger threshold, a fraction (estimated to be between 10^{-5} and 10^{-4}) of β-decay electrons cause triggering due to the energy resolution of the detector.

^{226}Ra was efficiently removed by a mixed-bed-type deionization system. The ^{226}Ra content in the primary water was 1.5×10^{-2} pCi/ℓ. It was reduced to about 10^{-4} pCi/ℓ. ^{238}U is selectively absorbed by CR-55 chelate resin. Initially, the uranium content in the detector water was 0.17 pCi/ℓ. It became $\lesssim 2.4 \times 10^{-4}$ pCi/ℓ after installation of a deionization column for ^{238}U. Fig. 14 shows the history of removing ^{236}Ra and ^{238}U in the circulating water.

The sources of the ^{222}Rn are as follows: (1) It is primarily contained in the mine water with a concentration of 200 $\sim$ 300 pCi/ℓ. The presence of ^{222}Rn was confirmed by the observation of a sudden rise in the trigger rate when fresh water was added to the tank, followed by a reduction of the trigger rate, consistent with the 3.8-day half-life (see Fig. 15). (2) It is dissolved in the detector water from the air. The concentration of radon in the air inside the Kamioka mine is $\sim$50 pCi/ℓ,

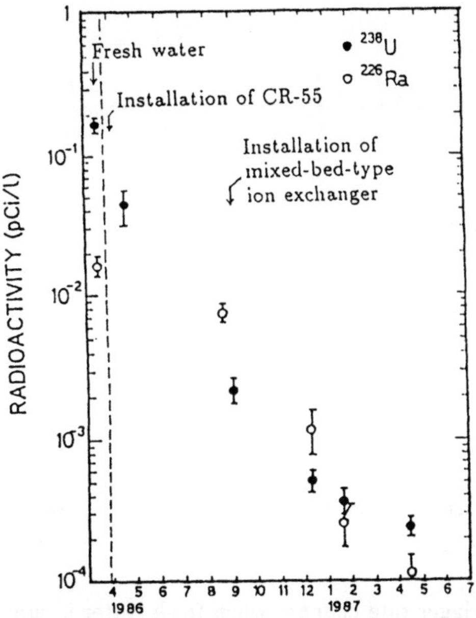

Figure 14: History of removing ^{238}U and ^{226}Ra in the circulating water. The times when the CR-55 chelate resin for removing ^{238}U and the mixed-bed-type ion exchanger were installed are indicated.

which is at least 100-times higher than that of the air outside the mine. (3) It is also produced from the decay of ^{226}Ra in the detector water.

To prevent radon in the mine air from dissolving into water, the top of the detector tank was sealed and the water-purification system was made air-tight in Spring of 1987. In addition, radon-free air has been supplied to the free space at the top of the main detector tank as well as the buffer tanks in the purification system since the middle of 1987. A degasification system was also installed to remove air (radon) from the circulating water. The concentration of ^{222}Rn in the detector water is typically 0.01 pCi/ℓ at the time of writing this manuscript.

To develop methods to remove radon from the water, it is essentially important to measure the amount of radon in any place within the closed system of the main detector tank and water-purification system, either in the gas phase or in the liquid phase. Recently, real-time radon detectors developed by Tasaka [17] have been installed at a number of locations both inside and outside of the detector and water-purification system, and have proved to be very useful. This detector utilizes a PIN silicon photodiode exposed to a gas volume which includes the radon to be measured. Positively charged daughters of ^{222}Rn, decayed inside the mesh electrode (see Fig. 16), are collected at the surface of the diode by a bias potential; the energies of α-rays from RaA and RaC$'$ are measured with an accuracy of $\sim$ 40 keV (FWHM). To measure the amount of radon dissolved in water, the radon detector volume is separated from water with a 25-μm thick polypropylene microporous film (see Fig. 16). The sensitivity of this radon monitor is typically 0.1

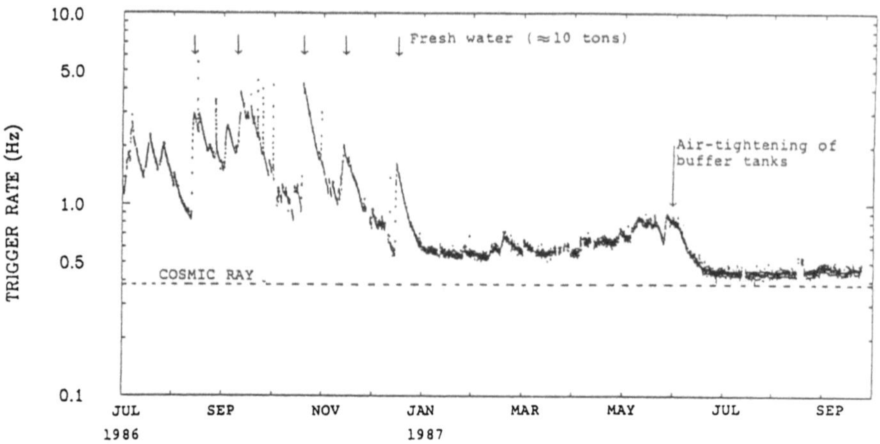

Figure 15: How the trigger rate changes when fresh water is supplied to the tank. The dashed line shows the trigger rate due to cosmic-ray muons (0.37 Hz). The decrease in the trigger rate since May, 1987, is due to the air-tightning of the buffer tanks in the water-purification system.

Bq/m^3 or 0.003 pCi/ℓ, but improves with the integrated time of the measurement.

The trigger rate has been reduced with time by these efforts to eliminate radioactivity in the detector water. As a result, the trigger threshold has been lowered by several times (see subsect. 2.2.3).

2.3 Data-Taking and Monitoring

Until January, 1990, when the online computer in the electronics hut was linked with another computer placed in a room outside of the mine, the data were written on magnetic tapes in the underground laboratory. Consequently, on long weekends and holidays, during which the laboratory was inaccessible, a different data-taking mode from that of week days was adopted so as to reduce the recorded data. In this case, part or all of the downward-going cosmic-ray muon data were discarded, depending on the length of no-access days, except for those muons with a total pulse height $\geq 20,000$ photoelectrons (these were needed to make the spallation cut in the analysis of low-energy neutrinos; see subsect. 6.1.1). This "holiday data-taking mode" has no longer been necessary since January, 1990.

The events recorded on magnetic tapes include pedestal calibration events and dummy events triggered by 1-Hz pulses from a high-precision crystal clock. The pedestal calibration is performed at the beginning of each run. In addition, calibration was performed every one or two hours during data-taking in Kamiokande-II; in Kamiokande-III, one pedestal calibration event is recorded every 30 seconds. A

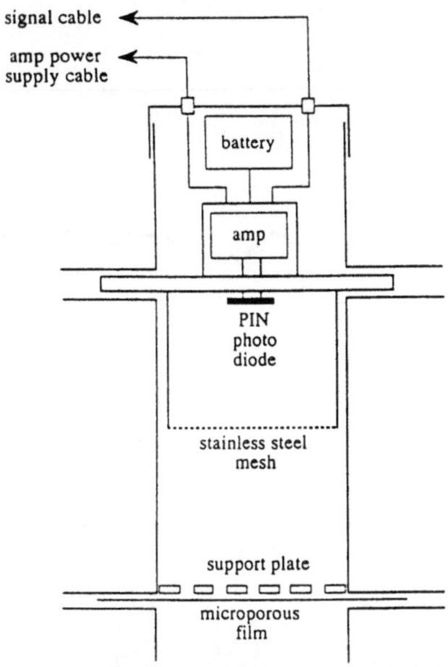

Figure 16: Schematic drawing of the real-time radon detector. This figure shows the detector configuration for measuring radon dissolved in water. When the detector is used for measuring radon in a gas volume, it is directly connected to this volume without a polypropylene microporous film and a supporting plate.

high-precision crystal clock was introduced to the system only after the event of SN1987A. This clock is calibrated by a LORAN-C receiver.

The status of the data-taking is monitored by the online computer. The trigger rate as a function of time, frequencies of each PMT hit, and summed pulse-height distributions are particularly important, and are monitored routinely. With a recent increase in the computing power due to the installation of additional computers and work stations at Kamioka, online monitoring tasks are rapidly improving. For example, a semi-realtime supernova watch is now at work, by which it is possible to detect a neutrino burst within 20 minutes of its occurence.

Nevertheless, for a full-scale analysis of the data, recorded magnetic tapes must be sent to Tokyo. In the future, a large computer facility will be available at Kamioka; the data from Superkamiokande will be fully analyzed there.

2.4 Calibrations and Detector Performance

2.4.1 Gain Calibration

Before installation, the high voltage applied to each PMT was adjusted so as to eqalize the gain within 6%. In order to keep the PMT gain equal, it is necessary to monitor the long-term stability of the individual PMT gain. For this purpose, a "scintillator ball" made of acrylic resin, including 50 ppm of BBOT for wavelength shifter and 500 ppm of MgO powder for light diffuser, is suspended at the center

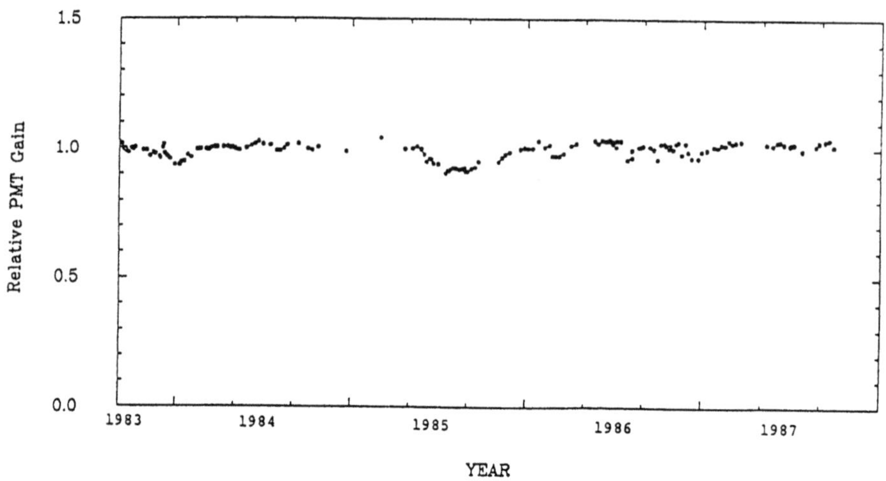

Figure 17: Long-term stability of 20″ PMTs in the detector water measured with the scintillator-ball system. Each point represents the mean gain of the PMTs relative to the initial value.

of the detector, and ultraviolet light produced by a spark gap in Ar gas is guided to the scintillator ball. It absorbs ultraviolet light and emits wavelength-shifted light isotropically. Figure 17 shows the mean of the PMT gains over 4 years; it was stable within 4%. The rms value of the gain drift of all PMTs was less than 6% over the same period.

The gain of the detector as a whole depends on the PMT gains and the attenuation length of water in the detector. Though the PMT gain is quite stable, the attenuation length of water occasionally shows a large variation, especially when the top of the detector tank is not closed (before spring 1987). The gain variation of the detector has to be monitored continuously and corrected for. The data of penetrated muons are used for this purpose and a certain average of the relative PMT gains and the attenuation length of water can be determined separately; for details, see Ref. [18]. Figure 18 shows the variation of the attenuation length of the detector water determined by using this method. After the detector tank was completely sealed in the Spring of 1987, the attenuation length of the detector water has been stable, and reached ~ 60 m.

2.4.2 Timing Calibration

Because of the difference in the applied high voltage, the transit time differs from PMT to PMT. Also, the time-slew of a discriminator causes its timing response to be dependent on the input pulse height. Therefore, the timing response of each PMT must be calibrated as a function of the pulse height. In Kamiokande-II and -III, a nitrogen laser calibration system is used to calibrate the timing response

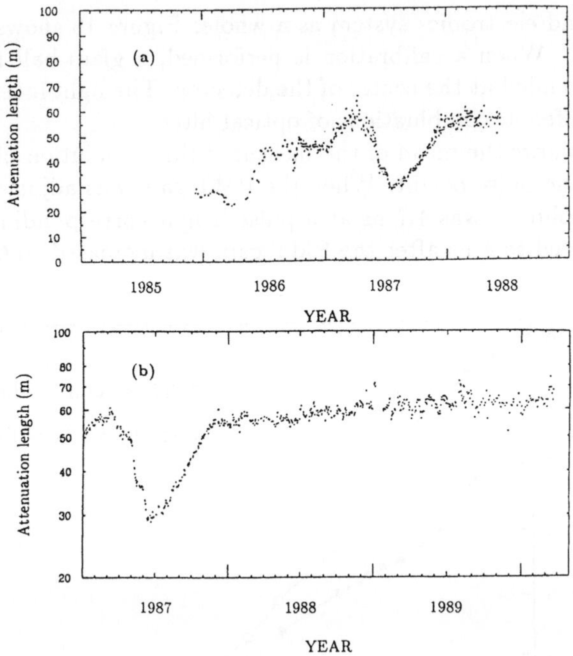

Figure 18: Variation of the attenuation length in the detector water. (a) December 1985 – June 1988, (b) January 1987 – June 1990.

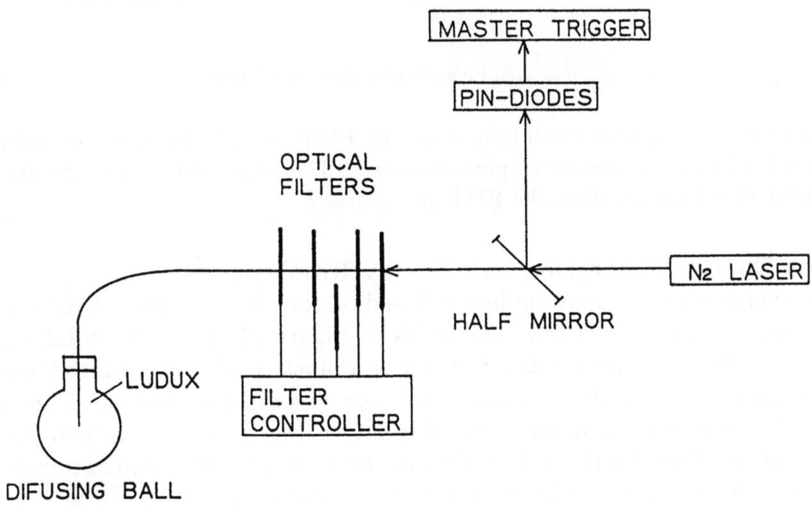

Figure 19: Block diagram of the laser calibration system.

of the PMT and electronics system as a whole. Figure 19 shows a block diagram of this system. When a calibration is performed, a glass ball filled with a light diffuser is suspended at the center of the detector. The light intensity is controlled by selecting different combinations of optical filters.

Figure 20 shows the mean of the measured time resolution of all the PMTs as a function of the pulse height. When the PMT gain was adjusted to 3×10^6, the mean time resolution was 4.7 ns at a pulse height corresponding to 1 photoelectron. It improved to 4 ns after the PMT gain was increased to 6×10^6.

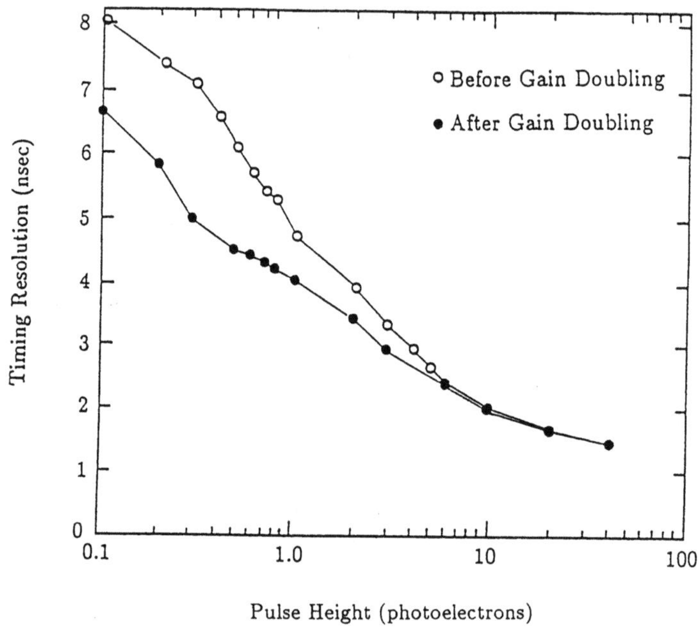

Figure 20: Mean time resolution of all the PMTs as a function of the pulse height, which is given by a number of photoelectrons. The open circles show the data before, and the closed circles after, the PMT gain increase.

2.4.3 Energy Calibration and Energy Resolution

The energy-measurement method is based on the observed pulse height at a total electron energy of $\gtrsim$ a few tens of MeV. Figure 21 shows the total number of photoelectrons observed in the detector as a function of momentum. These results were obtained by a Monte-Carlo simulation described in Sect. 4. The relatively low photoelectron yield for charged pions is due to nuclear interactions. The fluctuation of the total number of photoelectrons gives the momentum resolution of the detector. The resultant momentum resolution is shown in Fig. 22.

For solar neutrino events, the recoil electron energy is $\lesssim 15$ MeV. In this case the typical PMT response corresponds to the detection of one photoelectron. However, the pulse-height distribution of a one-photoelectron signal is very broad, as can be seen from Fig. 12. Therefore, it is anticipated that the use of the total pho-

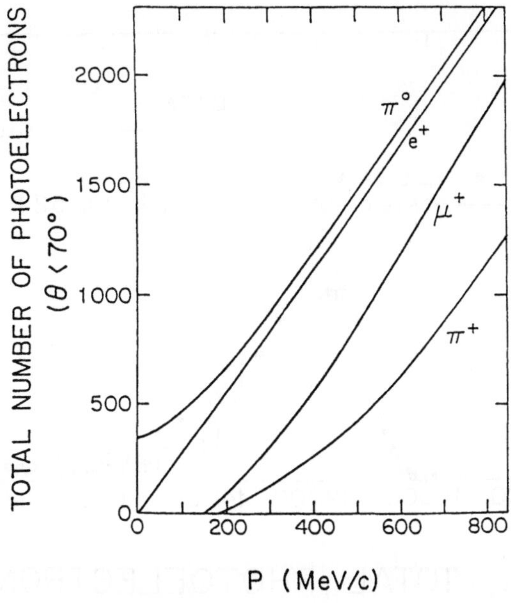

Figure 21: Total number of photoelectrons as a function of momentum for various particles. Contributions from PMTs located inside a cone with a half opening angle of 70° with respect to the vertex point are counted.

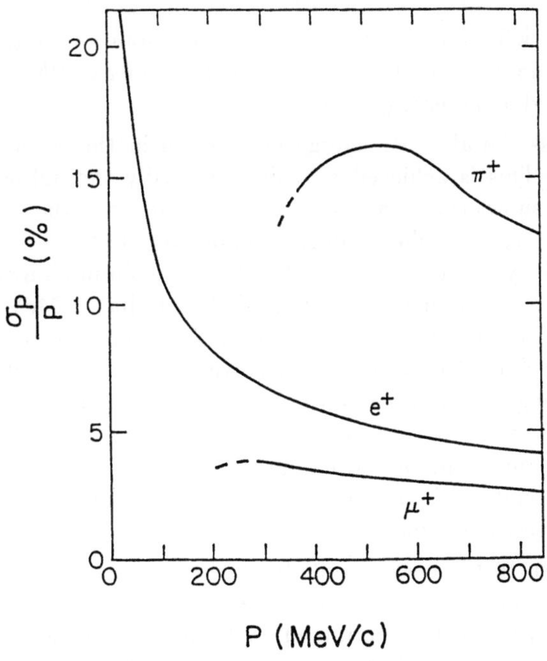

Figure 22: Momentum resolution for various particles.

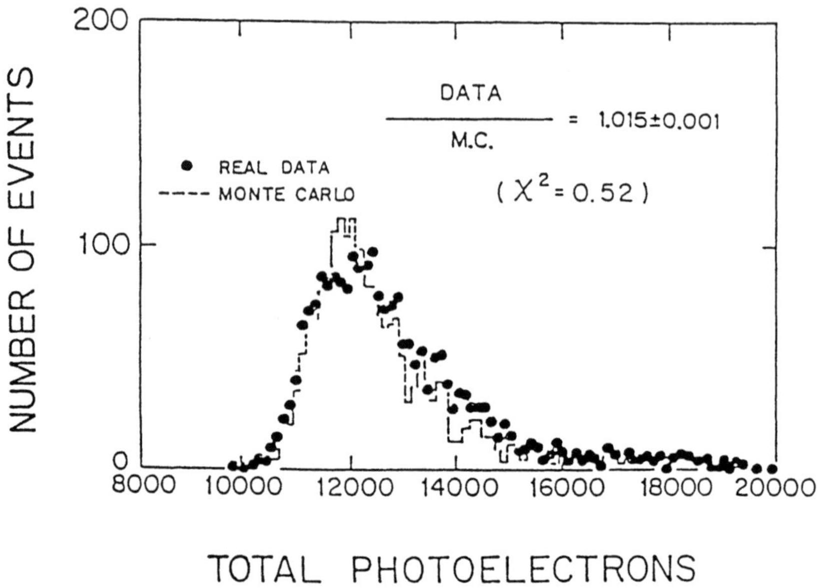

Figure 23: Distribution of total photoelectron number for vertically through-going muons. The filled circles represent the data and the dashed histogram shows the result of a Monte Carlo simulation.

toelectron number will give a rather poor energy resolution at the energy region of the solar neutrino measurement. Instead, the number of PMTs which respond to the signal is used as an energy scale.

We now discuss the absolute energy calibration in the energy region of $\gtrsim$ a few tens of MeV. This is achieved by using vertically through-going cosmic-ray muons. Figure 23 shows the total photoelectron number of these events. The tail in the high-energy region is due mainly to knock-on electrons. This distribution is well reproduced by the result of the Monte-Carlo simulation indicated by the dashed histogram. From this result, it is deduced that 1 MeV corresponds to 3.4 photoelectrons when the attenuation length of water is normalized to 35 m in Kamiokande-II. In Kamiokande-III, 1 MeV corresponds to 4.7 photoelectrons when the attenuation length of water is normalized to 54 m. This energy calibration is further checked in the lower energy region by using $\mu \to e$ decay electrons and relatively low-energy muons which go through the detector, but stop before entering the anticounter. Good agreement has been obtained between the real data and the Monte-Carlo simulation.

Let us now turn to the energy scale for low-energy events. Single low-energy electrons or positrons produced inside the detector cause the emission of Cherenkov light with the total intensity determined by their total energy. The broken line in Fig. 24 shows the number of photoelectrons as a function of the total electron energy (E_e), as obtained from a simulation of the Kamiokande-II detector. The

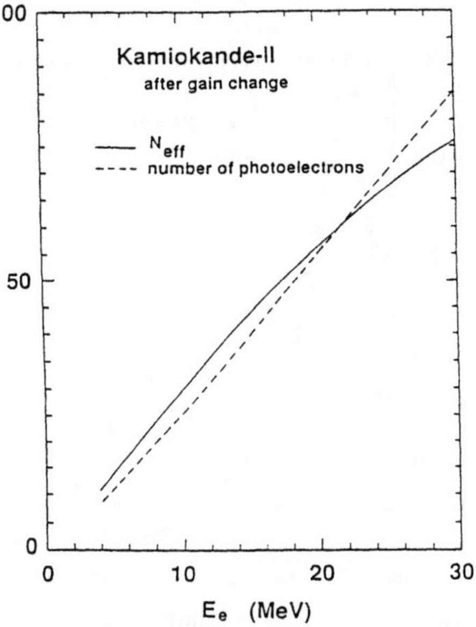

Figure 24: Number of photoelectrons (dashed curve) and N_{eff} (solid curve) as a function of the total electron energy (E_e). This figure corresponds to Kamiokande-II after the PMT gain increase.

actual response of the detector to Cherenkov light depends on the attenuation of Cherenkov light in the water and on the geometrical effects arising from event location and the location of the responding PMTs. The number of hit PMTs corrected for these effects is expressed by the effective number of hit PMTs (N_{eff}). The relation between N_{eff} and E_e is also shown in Fig. 24. To a good approximation, N_{eff} is proportional to E_e in the energy region above the analysis threshold of 7.5 MeV and below $\sim$ 15 MeV. The proportionality constant to convert N_{eff} to the total electron energy (E_e; in units of MeV) depends on the PMT gain if all other factors remain stable. In Kamiokande-II, an electron of E_e = 10 MeV gave an average N_{eff} of 26 and 30, respectively, before and after the PMT gain increase. In Kamiokande-III, the corresponding number is 39.

The accuracy of energy measurements is vital for an accurate measurement of the solar neutrino flux. An energy calibration of the detector at low energies is provided by utilizing γ-rays from a thermal-neutron capture reaction, Ni(n, γ)Ni*. The γ-rays are produced in transitions with total emitted energies of 9.0, 7.8, and 6.8 MeV with relative intensities of 0.71, 0.16, and 0.13, respectively. As a neutron source, ^{252}Cf is used. It is enclosed in a water-filled nickel can whose dimensions are 9.5 cm in diameter, 16 cm in hight and 1 mm in thickness.

Figure 25 shows the N_{eff} distribution of γ-rays from ^{252}Cf enclosed in the nickel can. Since ^{252}Cf, itself, emits γ-rays, the N_{eff} spectrum of γ-rays with ^{252}Cf only (without a nickel can) was also measured. The N_{eff} spectrum of γ-rays from the neutron capture reaction in Ni is obtained by subtraction, and is shown in Fig. 26.

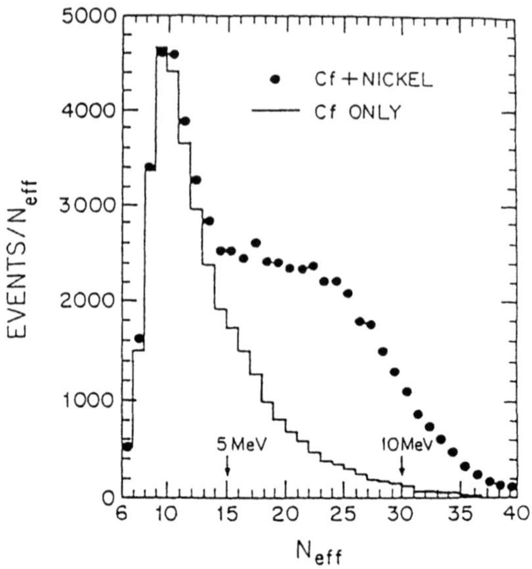

Figure 25: Histogram showing the measured γ-ray spectrum from a ^{252}Cf source. The data points show that from the same ^{252}Cf source enclosed in a water-filled nickel can.

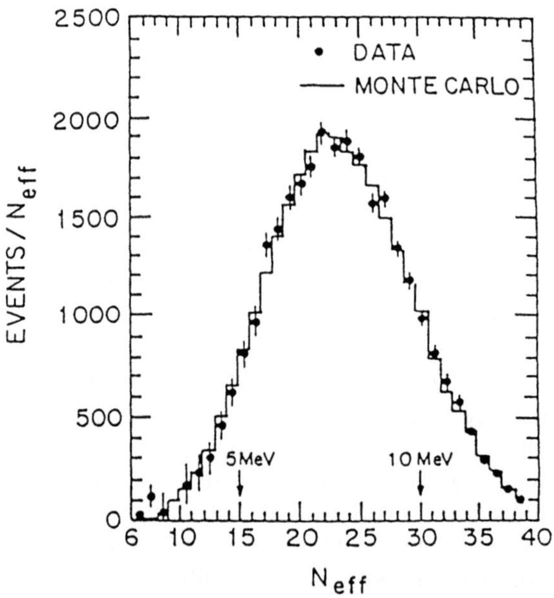

Figure 26: Data points showing the γ-ray spectrum from the reaction Ni(n,γ)Ni* obtained by subtracting the Cf spectrum from the Cf+Ni spectrum in Fig. 25. The histogram shows the results of a Monte-Carlo calculation adjusted so as to reproduce the peak of the measured spectrum.

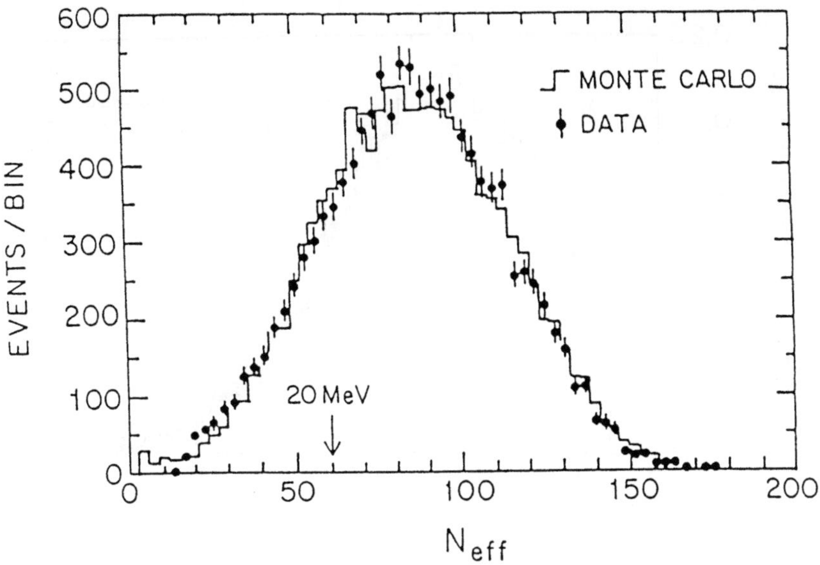

Figure 27: Data points showing the N_{eff} distribution of electrons from stopping muons. It is compared with the results of a Monte-Carlo calculation (histogram).

It should be noted that we actually wish to know the energy calibration for electrons. At a given total energy of E, where $E \lesssim 15$ MeV, the responses of photons and electrons are different in a water Cherenkov detector. At these energies, although photons lose energy through both Compton scattering and pair creation, the Compton scattering has a larger cross section than pair creation in water. On the average, several electrons receive recoil; these electrons emit Cherenkov light only when their kinetic energies exceed the Cherenkov threshold (0.24 MeV). Therefore, photons of energy E correspond to a smaller N_{eff} than that of electrons of the same energy.

It is a Monte-Carlo simulation program that relates the calibration for photons with that for electrons. This program simulates the Kamiokande detector as well as the physical processes in it using EGS4. It was tuned so as to reproduce the peak of the N_{eff} distribution of the ^{252}Cf/Ni calibration data within $\pm 1\%$. The shape of the observed spectrum in Fig. 26, especially the width, which represents the energy resolution, is well reproduced by this simulation.

The thus-obtained energy calibration can be tested by the following two methods:
(i) Electrons from stopping-muon decays are sampled and compared with a Monte-Carlo simulation of the known muon decay spectrum. As shown in Fig. 27, the simulated spectrum well reproduces the data. (Note, however, that the horizontal scale of Fig. 27 is N_{eff}. At energies $\gtrsim 15$ MeV, N_{eff} is no longer proportional to the electron total energy, as shown in Fig. 24).
(ii) A Monte-Carlo simulation is performed of the β decay of the short-lived isotopes produced in spallation events induced by cosmic-ray muons. The results are

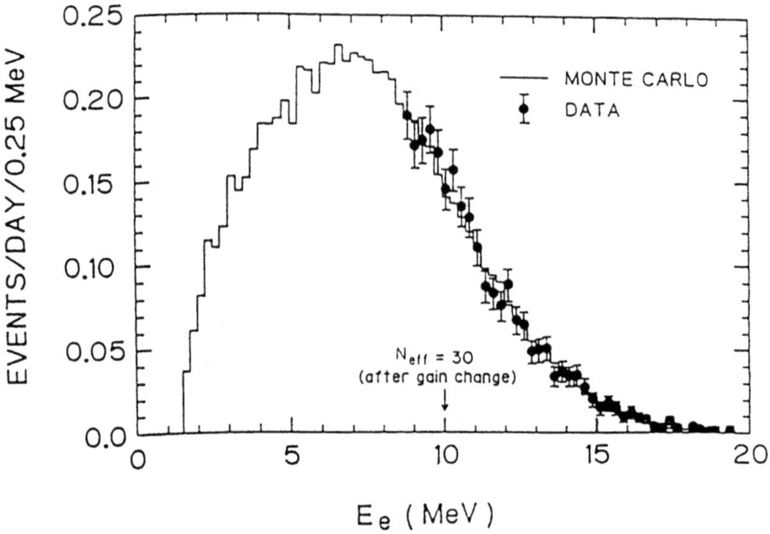

Figure 28: Energy distribution of the β decays from the short-lived spallation products compared with the histogram obtained by a Monte-Carlo simulation of β decays from a mixture of ^{12}B ($\tau_{1/2}$ = 20.4 ms) and ^{12}N ($\tau_{1/2}$ = 11.0 ms) with a relative ratio of 0.85 to 0.15. This mixture reproduces the lifetime distribution (event rate against time from the preceding muon event) of short-lived spallation products.

compared with the data in Fig. 28.

From the ^{252}Cf/Ni calibration as well as two consistency tests, the uncertainty in the absolute energy calibration is estimated to be less than 3%. The energy resolution can be estimated by the Monte-Carlo simulation program, and the result is shown in Fig. 29. At the end of Kamiokande-II (after the gain increase), it is approximately given by

$$\frac{\sigma(E_e)}{E_e} = \frac{0.63}{\sqrt{E_e}}$$

for $E_e < 12$ MeV, where E_e is given in units of MeV.

2.4.4 Vertex Resolution

The vertex position and direction of a track are reconstructed using the timing information and geometrical pattern of hit PMTs. The vertex is determined, in principle, by searching for the position which minimizes the width of the residual time distribution of hit PMTs with a constraint of the expected Cherenkov pattern of a relativistic particle in water. The residual time of a PMT is given by the measured signal time minus the time needed for light emitted from the vertex to reach that PMT. The first approximation of the vertex point obtained in this way must be refined by taking into account the finite track length (the tracks for low-energy electrons are safely regarded as being points, because the track length of a 10 MeV electron in water is only ~ 5 cm). The actual procedure of vertex

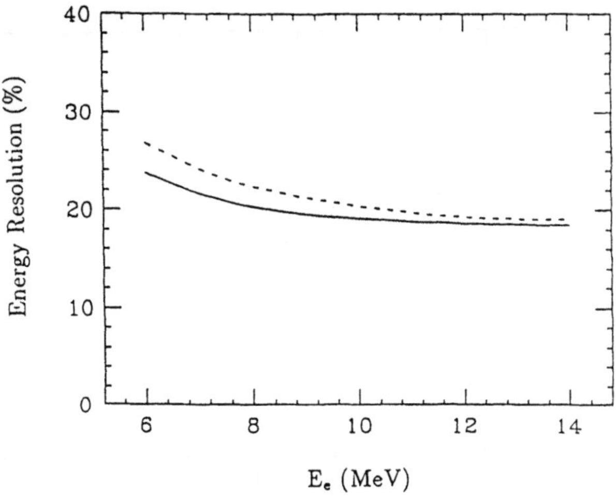

Figure 29: Energy resolution as a function of E_e for low-energy electrons indicated by the dashed curve for Kamiokande-II (after the PMT gain increase) and by the solid curve for Kamiokande-III.

reconstruction and directional reconstruction is more sophisticated, and is fully described in Ref. [19, 20, 21, 22].

The vertex resolutions for low-energy electrons estimated from Monte-Carlo simulations are shown in Fig. 30 for Kamiokande-II (after the PMT gain increase) and Kamiokande-III. Figure 31 shows the vertex resolutions for higher energy electrons and muons estimated from Monte-Carlo simulations in which the reconstruction is performed either using ADCs (for the case of Kamiokande-I only, see Ref. [19]) or TDCs. There is no appreciable difference between the vertex resolutions obtained by using the TDCs for Kamiokande-II before and after the PMT gain change and for Kamiokande-III.

2.4.5 Angular Resolution

The accuracy in measuring the direction of the electron from νe scattering is very important in both solar neutrino observations and determining the neutrino-burst source. This accuracy is determined by multiple Coulomb scattering in water. Figure 32 shows the superimposed trajectories of 10 MeV electrons in water. EGS4 was used for the simulation. The angular resolution obtained from this simulation is shown in Fig. 33 as a function of E_e. To confirm the validity of this simulation, a calibration was performed using collimated γ-rays from the Ni(n,γ)Ni* reaction which were incident on the water from the top of the detector. The measured directional distribution of these γ-rays was satisfactorily reproduced. From this result, it is concluded that the uncertainty of the angular resolution shown in Fig. 33 is less than 20%.

The angular resolutions for higher energy electrons and muons were estimated from Monte-Carlo simulations in which the reconstruction was performed either using ADCs (Kamiokande-I only, see Ref. [19]) or TDCs. Figure 34 shows the results.

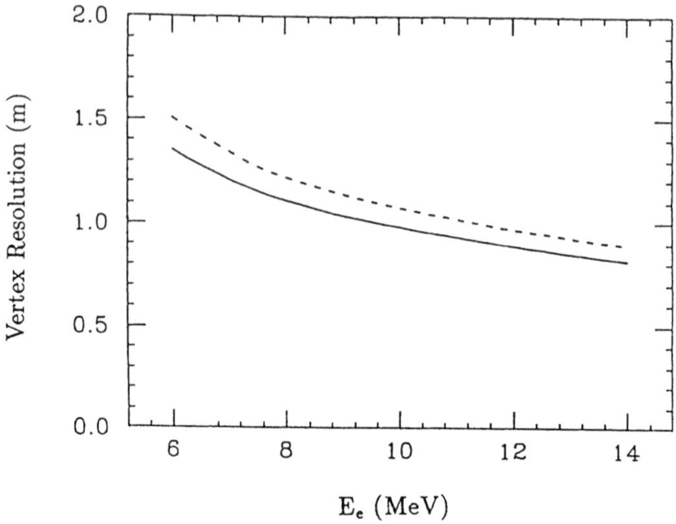

Figure 30: Vertex resolution as a function of E_e for low-energy electrons indicated by the dashed curve for Kamiokande-II (after the PMT gain increase) and by the solid curve for Kamiokande-III.

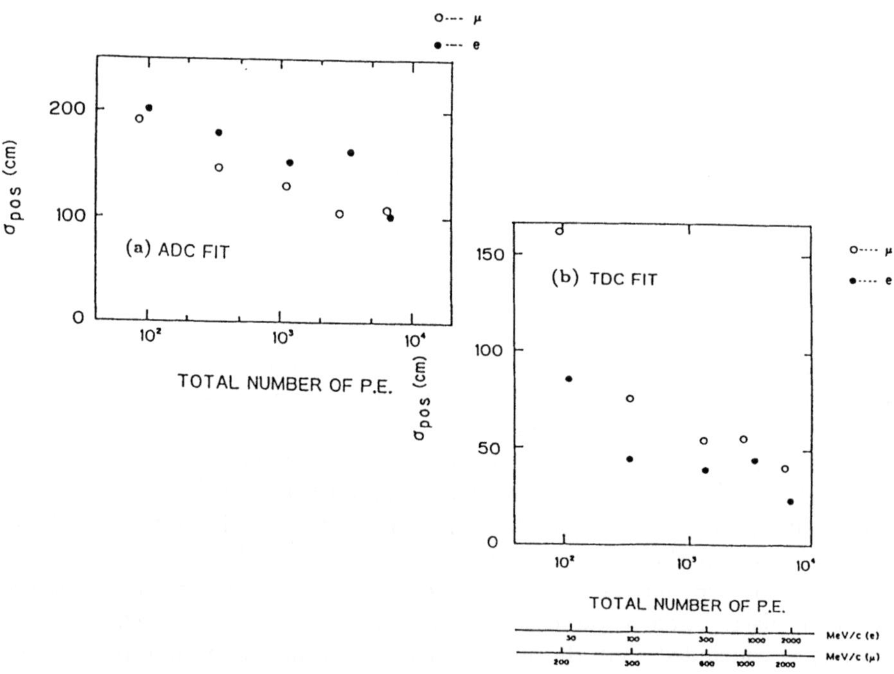

Figure 31: Vertex resolution as a function of the number of photoelectrons with the reconstruction performed using (a) ADCs and (b) TDCs.

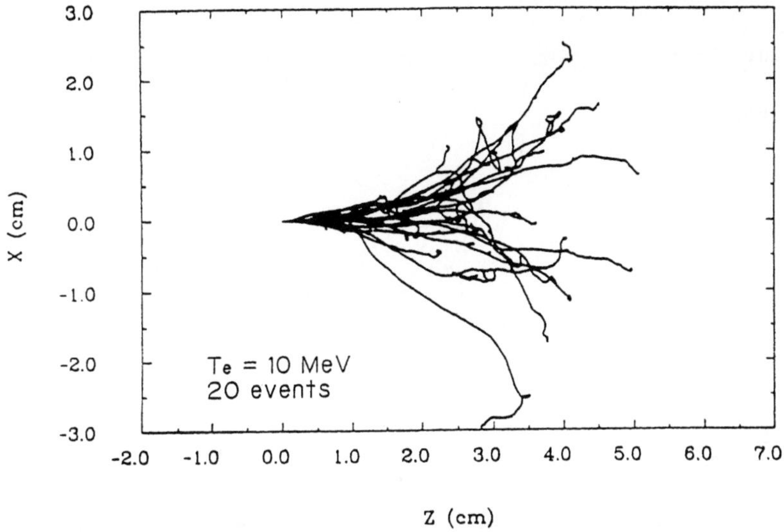

Figure 32: Trajectories of electrons in water, projected onto the x-z plane. Twenty electrons were Monte-Carlo generated with the same kinetic energy (10 MeV) and the same initial direction (parallel to the z-axis).

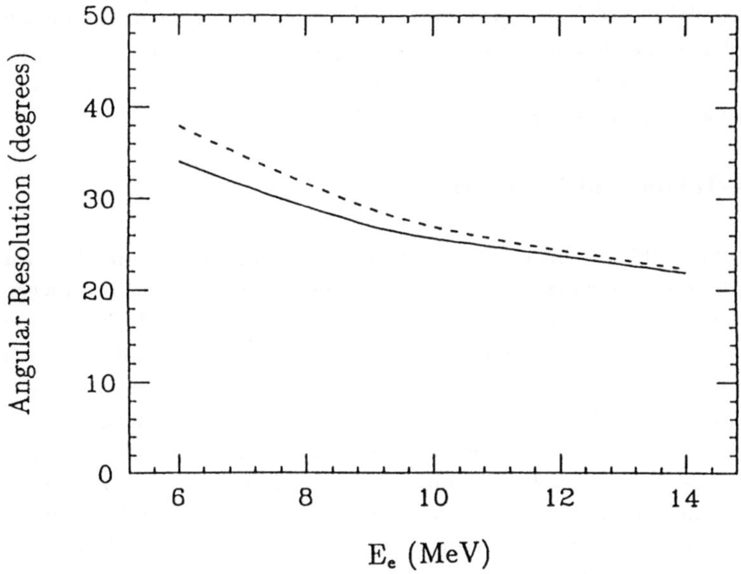

Figure 33: Angular resolution as a function of E_e for low-energy electrons indicated by the dashed curve for Kamiokande-II (after the PMT gain increase) and by the solid curve for Kamiokande-III.

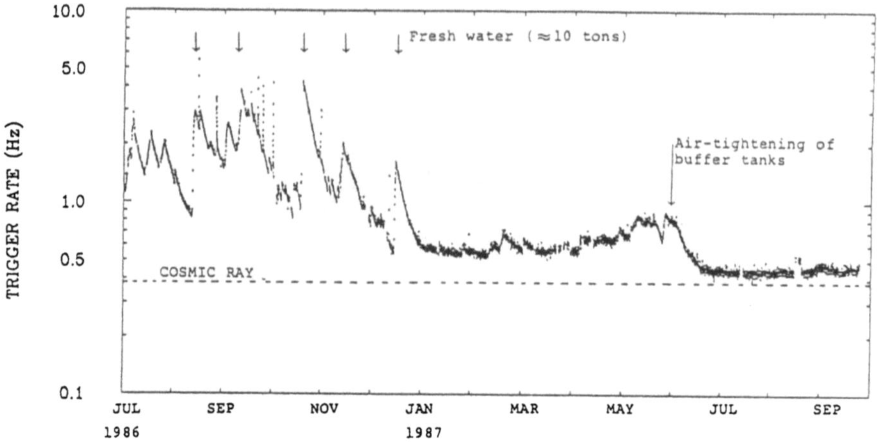

Figure 15: How the trigger rate changes when fresh water is supplied to the tank. The dashed line shows the trigger rate due to cosmic-ray muons (0.37 Hz). The decrease in the trigger rate since May, 1987, is due to the air-tightning of the buffer tanks in the water-purification system.

Bq/m^3 or 0.003 pCi/ℓ, but improves with the integrated time of the measurement.

The trigger rate has been reduced with time by these efforts to eliminate radioactivity in the detector water. As a result, the trigger threshold has been lowered by several times (see subsect. 2.2.3).

2.3 Data-Taking and Monitoring

Until January, 1990, when the online computer in the electronics hut was linked with another computer placed in a room outside of the mine, the data were written on magnetic tapes in the underground laboratory. Consequently, on long weekends and holidays, during which the laboratory was inaccessible, a different data-taking mode from that of week days was adopted so as to reduce the recorded data. In this case, part or all of the downward-going cosmic-ray muon data were discarded, depending on the length of no-access days, except for those muons with a total pulse height $\geq 20,000$ photoelectrons (these were needed to make the spallation cut in the analysis of low-energy neutrinos; see subsect. 6.1.1). This "holiday data-taking mode" has no longer been necessary since January, 1990.

The events recorded on magnetic tapes include pedestal calibration events and dummy events triggered by 1-Hz pulses from a high-precision crystal clock. The pedestal calibration is performed at the beginning of each run. In addition, calibration was performed every one or two hours during data-taking in Kamiokande-II; in Kamiokande-III, one pedestal calibration event is recorded every 30 seconds. A

where

$$g_L = \frac{1}{2} + \sin^2\theta_W \qquad \text{for} \quad \nu_e\, e \quad \text{and} \quad \bar{\nu}_e\, e \tag{8}$$

$$= -\frac{1}{2} + \sin^2\theta_W \qquad \text{for} \quad \nu_{\mu,\tau}\, e \quad \text{and} \quad \bar{\nu}_{\mu,\tau}\, e \tag{9}$$

and

$$g_R = \sin^2\theta_W, \tag{10}$$

with G_F, E_ν, m, T, and θ_W being the Fermi coupling constant, incident neutrino energy, electron mass, electron kinetic energy, and the Weinberg angle, respectively.

The total cross sections of reactions (1)–(5) are given by integrating Eq. (6) or (7) from $T = 0$ to $T_{max} = E_\nu/(1 + m/2E_\nu)$. For $E_\nu \gg m$, $\sigma_{\nu e}$ and $\sigma_{\bar{\nu} e}$ are proportional to E_ν. With $\sin^2\theta_W = 0.23$, the numerical representation of the total cross sections of these reactions are given in Table 3. Roughly, $\sigma(\bar{\nu}_e\, e^-) = \frac{1}{2.5}\,\sigma(\nu_e\, e^-)$, $\sigma(\nu_{\mu,\tau}\, e^-) = \frac{1}{6}\,\sigma(\nu_e\, e^-)$, and $\sigma(\bar{\nu}_{\mu,\tau}\, e^-) = \frac{1}{7}\,\sigma(\nu_e\, e^-)$.

Table 3: Total cross sections of low-energy neutrino interactions in water.

Reaction	Total cross section (cm^2)
$\nu_e e^- \rightarrow \nu_e e^-$	$9.3 \times 10^{-44}\,(E_\nu/10\text{MeV})$
$\bar{\nu}_e\, e^- \rightarrow \bar{\nu}_e\, e^-$	$3.9 \times 10^{-44}\,(E_\nu/10\text{MeV})$
$\nu_{\mu,\tau}\, e^- \rightarrow \nu_{\mu,\tau}\, e^-$	$1.5 \times 10^{-44}\,(E_\nu/10\text{MeV})$
$\bar{\nu}_{\mu,\tau}\, e^- \rightarrow \bar{\nu}_{\mu,\tau}\, e^-$	$1.3 \times 10^{-44}\,(E_\nu/10\text{MeV})$
$\bar{\nu}_e\, p \rightarrow e^+ n$	$8.9 \times 10^{-42}\,(E_e/10\text{MeV})^2$
	$[E_\nu \simeq E_e + m_n - m_p = E_e + 1.3\ \text{MeV}]$
$\nu_e\, {}^{16}\text{O} \rightarrow e^-\, {}^{16}\text{F}$	$1.1 \times 10^{-44}\,[E_\nu(\text{MeV}) - 13]^2$

A very characteristic feature of νe scattering is the directional correlation of the recoil electron with the incident neutrino for $E_\nu \gg m$. Denoting the laboratory scattering angle of the electron with respect to the direction of the incident neutrino by θ, we obtain the following kinematical relation:

$$\cos\theta = \frac{1 + \frac{m}{E_\nu}}{\sqrt{1 + \frac{2m}{T}}}\,. \tag{11}$$

For $T \gg m$, angle θ is obviously small. (Note that in a νe scattering experiment there is a threshold for T. It is 7.5 MeV in the present Kamiokande experiment.) By taking advantage of this directional correlation in νe scattering, we can unambiguously observe the neutrino signal from a point source. This fact plays an important role in solar neutrino observations.

Reaction (5) is an inverse β decay which mainly contributes in the detection of supernova neutrinos. The differential cross section is approximately given as [24]

$$\frac{d\sigma}{d\cos\theta} = \frac{G_F^2}{\pi}\cos^2\theta_C\, p_e\, E_e\left\{\left(1 + \frac{g_A^2}{g_V^2}\right) + \frac{1}{2}\left(\frac{g_A^2}{g_V^2} - 1\right)\left(1 - \frac{p_e}{E_e}\cos\theta\right)\right\} \tag{12}$$

by neglecting terms of order E_e/M, where θ is the angle between the positron and the incident neutrino, p_e and E_e are the momentum and energy of the positron, M is the proton mass, $g_A/g_V \simeq -1.26$ is the ratio of vector and axial-vector couplings of the nucleon and θ_C is the Cabbibo angle, $\cos\theta_C \sim 0.975$. From Eq. (12),

$$\frac{\mathrm{d}\sigma}{\mathrm{d}\cos\theta} \propto 1 + 0.11\,(1 - \cos\theta) \tag{13}$$

for $E_e \sim 10$ MeV. Therefore, the angular distribution is nearly isotropic with a slight backward enhancement. Neglecting the term proportional to $\cos\theta$, the total cross section is expressed as $\sigma(\bar{\nu}_e p \to e^+ n) = \frac{G_F^2}{\pi}\cos^2\theta_C\, p_e\, E_e\,(1 + 3\,\frac{g_A^2}{g_V^2})$. A better numerical representaion [24], however, is given in Table 3.

For the detection of supernova neutrinos with relatively higher energies, the reaction

$$\nu_e\,{}^{16}\mathrm{O} \;\rightarrow\; e^-\,{}^{16}\mathrm{F} \tag{14}$$

should be also considered. The total cross section of this reaction [25] is also given in Table 3.

3.1.2 Measurements of $e^{\pm}$ from Neutrino Interactions in Kamiokande

To detect the low-energy neutrinos of astrophysical origin in a water Cherenkov detector, a recoil electron from νe scattering or the positron from reaction (5) is measured. The relevant quantities are the energy and direction of the electron or positron and the position of the interaction vertex. The measurements of these quantities and their calibrations have already been discussed in subsect. 2.4.

The recoil-electron energy spectrum in $\nu e \to \nu e$ scattering is given by

$$\frac{dN}{dT} = N_e \times \int \phi(E_\nu)\frac{d\sigma(E_\nu)}{dT}dE_\nu, \tag{15}$$

where $\phi(E_\nu)$ is the neutrino energy spectrum, $\frac{d\sigma(E_\nu)}{dT}$ is the differential cross section given by Eq. (6) or Eq. (7), and N_e is the number of electrons contained in the detector fiducial volume. As an example, the recoil-electron energy spectrum is shown in Fig. 35(b) for the ^{8}B solar neutrino spectrum of $\phi_{\mathrm{B}}(E_\nu)$ which is also shown in Fig. 35(a).

Electron antineutrinos are mainly detected through the $\bar{\nu}_e\, p \to e^+ n$ reaction. The energy of the produced positron is given by $E_{\bar{\nu}_e} - 1.3$ MeV. Hence, the positron energy spectrum is a product of a given neutrino energy spectrum, $\phi(E_{\bar{\nu}_e})$, and the reaction cross section presented in Table 3.

3.1.3 Background Considerations

From the viewpoint of the background, the detection of a neutrino burst due to stellar collapse has the cleanest signal, since many neutrinos are detected within a short duration (~10 sec). Two background sources are important for the observation of solar neutrinos. One is low-energy background from radioactive nuclei in the detector material and in the surrounding rock. ^{222}Rn dissolved in the detector water becomes particularly serious if the trigger threshold is further lowered

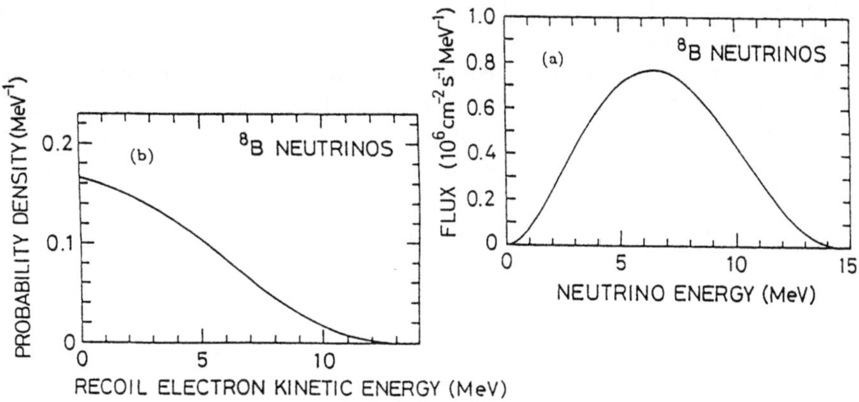

Figure 35: (a) Energy spectrum of the ^{8}B solar neutrinos. The total flux is normalized to 5.8×10^6 cm^{-2} s^{-1}, predicted by the standard solar model of Bahcall and Ulrich [26]. (b) Recoil-electron energy spectrum calculated by using the energy spectrum of the ^{8}B solar neutrinos.

from the present level. The other is the β decay of radioactive nuclei produced by cosmic-ray muons.

The cosmic-ray muons directly cause fragmentation of oxygen nuclei, or create negative pions which are captured by oxygen nuclei, thereby producing radioactive nuclides. Also, stopped negative muons are captured by oxygen nuclei, which occasionally leads to the production of radioactive ^{16}N nuclei, which have long half life (7.13 sec). Most of the produced nuclei are temporarily and are spatially correlated with the preceding muon track. Therefore, most of the background caused by cosmic-ray muons can be eliminated by exploiting these correlations. For example, events which have a spatial correlation within 3 m and a timing correlation within 20 sec with stopping muons are rejected so as to eliminate the background caused by ^{16}N.

The γ-ray and neutron background from the surrounding rock are absorbed in the anticounter layer. A further reduction in the external γ-ray background is necessary for solar neutrino observations; this is achieved by limiting the fiducial volume of the solar neutrino observation to the inner 680 tons. With this fiducial-volume cut and the elimination of muon-induced background, the solar neutrino signal can be seen in the Kamiokande-II experiment. However, in order to further eliminate the remaining external γ-ray background, a "γ-ray cut" is applied, which rejects events whose direction is nearly perpendicular to, and whose vertex lies within 1 m from, the boundary of the fiducial volume.

Figure 36 shows how the background can be reduced by these methods. The open circles show the energy spectrum of the events in the total detector volume (2140 tons), while the closed circles show that in the 680-ton fiducial volume. Notice that the fiducial-volume cut reduces the number of events by a factor of ~10, though the ratio of the total detector volume to the fiducial volume is ~ 3. The solid line represents the recoil electron energy spectrum expected from the standard solar model of Bahcall and Ulrich [26]. Clearly, the background events must be reduced further. This is achieved by eliminating muon-induced background events, which leads to the energy spectrum shown by the open diamonds. Finally, the closed diamonds show the result of the γ-ray cut. The event rate afer all of these cuts is still higher than that expected from the standard solar model [26]. However, the directional correlation of νe scattering helps to obtain a clean solar-neutrino signal at this level, as described in subsect. 6.1.

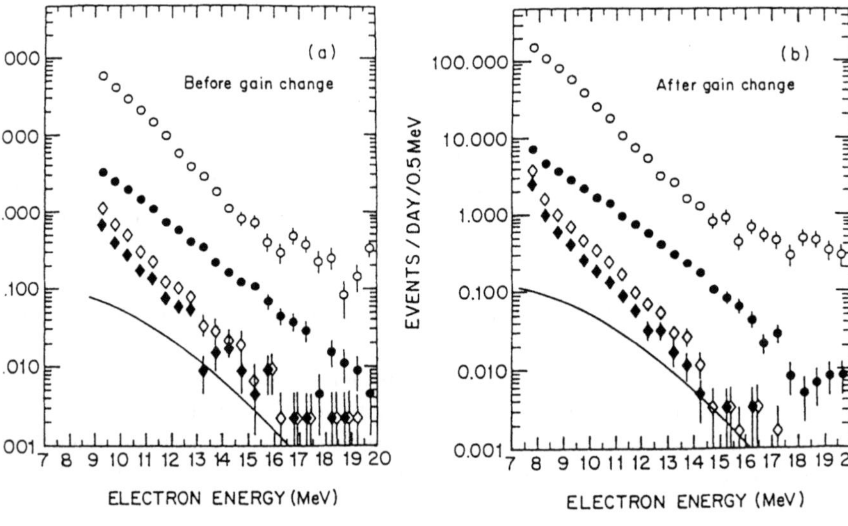

Figure 36: Event rates as a function of E_e (a) before the PMT gain change and (b) after the PMT gain change. o all events; • events in the fiducial volume; ◇ fiducial volume events after the spallation cut and ◇ after the γ-ray cut. The curve shows the expected event rate for the ^{8}B solar neutrinos calculated using Bahcall and Ulrich's standard solar model.

3.2 Measurement of Medium-Energy Neutrinos

3.2.1 Interactions of Medium-Energy Neutrinos

The medium-energy (~ a few GeV) neutrinos observed in Kamiokande are almost entirely atmospheric neutrinos. For these neutrinos, neutrino interactions can be sub-divided into the following processes:

(A)	Charged-current (CC) elastic scattering	$\nu N \rightarrow l^{\pm} N'$
(B)	CC Δ resonance	$\nu N \rightarrow l^{\pm} N' \pi$
(C)	CC multipion production	$\nu N \rightarrow l^{\pm} N' \pi ...$
(D)	Neutral-current (NC) elastic scattering	$\nu N \rightarrow \nu N$
(E)	NC Δ resonance	$\nu N \rightarrow \nu N' \pi$
(F)	NC multipion production	$\nu N \rightarrow \nu N' \pi ...$

The cross sections and kinematics of the CC elastic scattering can be calculated using a standard Q^2 distribution with nuclear dipole form factors measured by neutrino accelerator experiments [27], where $-Q^2$ is the square of the 4-momentum transfer between the incident neutrino and the outgoing lepton. Single-pion productions are mainly via a Δ resonance. The cross sections of the CC single-pion production were measured in neutrino experiments [28, 29] and theoretical calculations are available [30]. The total cross sections of the CC interactions were measured by Baker *et al.* [31], from which the cross sections of the CC multipion production can be deduced.

The cross sections of the NC single-pion production were measured by several experiments [32, 33]. These cross sections are consistent with the theoretically calculated cross sections by Rein and Sehgal [34] within 20%. The ratios of various inclusive cross sections of NC to CC interactions were measured [35]. The total cross sections of NC interactions can be obtained from these data.

3.2.2 Event Types

Neutrino events in the Kamiokande detector are classified into the two categories; fully contained events and vertex-contained events. The fully contained events are required to have their vertex position inside the detector fiducial volume and to have no visible energy deposited in the anticounter. The vertex-contained events are those events which have at least one prong that penetrates the inner detector. Obviously, the total visible energy can be measured for the fully contained events, but not for the vertex-contained events. For studying atmospheric neutrinos, fully contained events are exclusively selected. For some special purposes, however, vertex-contained events are also studied; see subsect. 6.7.

Neutrino events are also classified according to the number of Cherenkov rings. Table 4 gives the decomposition of fully contained single-ring events into various neutrino reaction modes, as derived from a Monte-Carlo simulation of the atmospheric neutrino interactions (see Sect. 4).

From this table, it can be seen that approximately 95% of the fully contained single-ring events are produced by leptons which remember the flavor of the parent neutrinos; the contribution from neutral-current reactions is only about 5%. The contamination of events other than the charged-current elastic scattering in the fully contained single-ring events results from the detector response and nuclear effects in ^{16}O nuclei. Many of the produced π^+ undergo nuclear interactions in ^{16}O nuclei and lose their energies down below the Cherenkov threshold; then they become invisible in the detector.

For the multi-ring events, the contamination of neutral-current interactions is significant (~30%). Also, the production cross sections of the multi-ring events have larger ambiguities than those of the single-ring events. Furthermore, lepton identification (see the following subsection) in a multi-ring event is more difficult than in a single-ring event. For these reasons, the single-ring events are more relevant for the study of the atmospheric-neutrino fluxes than the multi-ring events.

Table 4: Fractional contributions from various neutrino reaction modes to the fully contained single-ring events, as given by a Monte-Carlo simulation of the atmospheric neutrino interactions (third column). For comparison, decomposition of all the neutrino events before any event selection is also given in the second column.

Reaction mode	All events (%)	Fully contained single-ring events (%)
$\nu N \to l^{\pm} N'$	45	75
$\nu N \to l^{\pm} N' \pi$	21	18
$\nu N \to l^{\pm} N' \pi ...$	22	2
$\nu N \to \nu N' \pi$	6	3
$\nu N \to \nu N' \pi ...$	6	2

3.2.3 Particle Identification

Single-ring events are further classified into electron(e)-like ($e^{\pm}$, γ) and muon(μ)-like ($\mu^{\pm}$, $\pi^{\pm}$) events, paying attention to the diffuseness of their Cherenkov-ring pattern. A ring produced by an e-like event exhibits more diffuse pattern than that produced by a μ-like event, since an electron suffers considerable multiple Coulomb scattering in water. The identification of the particle type in the fully contained single-ring events is very important and, therefore, a rather detailed description is given below.

The particle momentum can be estimated from the observed total photoelectron (p.e.) number, PE_{TOT}. Then the expected p.e. number of each PMT can be calculated by assuming the particle type, as described below. A comparison between the expected and observed p.e. numbers for each PMT indicates the most probable assumption of the particle type.

To calculate the expected p.e. number distribution for e-like events, a method based on Monte-Carlo generated events is adopted since an analytic approach is next to impossible due to the fluctuation of electromagnetic showers. The expected p.e. number of a given PMT is expressed as:

$$p.e._{exp'd}(e) = C \times Table_{p.e.}(PE_{TOT}, \theta) \times \frac{1}{l^2} \times \exp(-\frac{l}{l_{abs}}) \times S(\Theta),$$

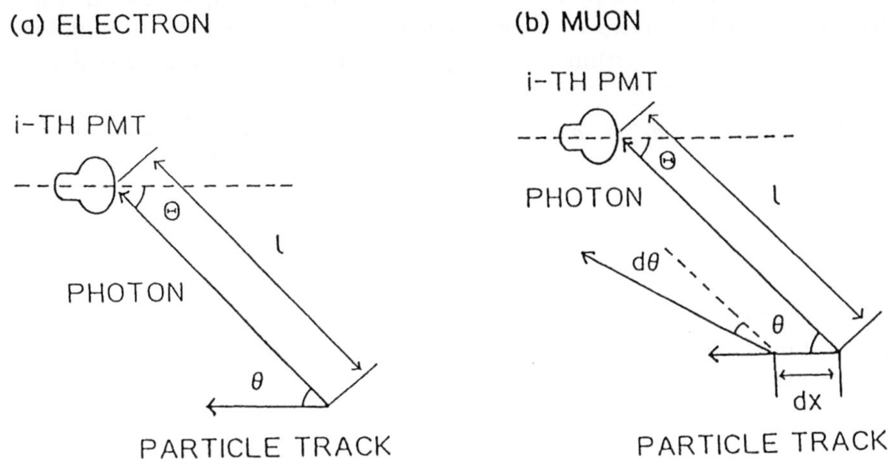

Figure 37: Definition of the quantities used in the particle identification in Kamiokande for (a) e-like and (b) μ-like events.

where C is a normalization factor to conserve PE_{TOT}, $Table_{p.e.}(PE_{TOT}, \theta)$ is a table of the expected p.e. number for e-like events, which is constructed by using the Monte-Carlo generated events, θ is an angle between the momentum vector and the direction of the PMT (see Fig. 37 (a)), and l is a distance between the vertex point and the PMT. An l^{-2} dependence on l originates from the diffuseness of the Cherenkov ring. The fourth factor in the right-hand side is a correction factor to take account of light attenuation in the water with an attenuation length of l_{abs}. $S(\Theta)$ is a factor related to an effective area of the PMT for those Cherenkov photons entering at an angle Θ on the PMT photocathode.

The propagation of a muon in water is much simpler than that of an electron. Consequently, for μ-like events the expected p.e. number for a PMT which receives Cherenkov radiation can be calculated without relying on the Monte Carlo method, and is expressed as:

$$p.e._{exp'd}(\mu) = C \times \frac{1}{l(\sin\theta + l \cdot d\theta/dx)} \times \sin^2\theta \times \exp(-\frac{l}{l_{abs}}) \times S(\Theta).$$

Here, the definition of l and θ are different from that for e-like events; θ is a Cherenkov angle, and l is a distance between a given point on the particle track and a PMT, which is located at an angle θ with respect to the vector direction of the particle (see Fig. 37 (b)). The second factor in the right-hand side originates from energy loss ($\mathrm{d}E/\mathrm{d}x$) of a μ-like particle in water; if the energy loss is negligible, this factor simply becomes $1/(l\sin\theta)$. The third factor represents the relative intensity of Cherenkov light.

In estimating $p.e._{exp'd}$ for both assumptions of e-like and μ-like events, photons scattered in the water or at the detector wall must also be properly taken into account.

Let us now define the probability, $Prob_i(p.e._{exp'd}, p.e._{obs})$, for the ith PMT to observe $p.e._{obs}$ when an expected p.e. number is $p.e._{exp'd}$. For $p.e._{exp'd} > 10$, the p.e. number distribution is assumed to be approximated by a gaussian distribution.

$$Prob_i\left(p.e._{exp'd}(\mathrm{e}\ or\ \mu),\ p.e._{obs}\right) = \frac{1}{\sqrt{2\pi}\sigma}\exp\left(\frac{-(p.e._{obs} - p.e._{exp'd})^2}{2\sigma^2}\right),$$

with

$$\sigma^2 = 1.2^2 \times p.e._{exp'd} + (0.1 \times p.e._{exp'd})^2,$$

where a factor 1.2 results from the PMT pulse height resolution compared with the ideal one, and the second term in the right-hand side takes account of a 10% uncertainty in the PMT relative calibration.

For $p.e._{exp'd} < 10$, the expression for $Prob_i(p.e._{exp'd}, p.e._{obs})$ is more complicated because one must take into account the Poisson statistics and the pulse-height response of the PMT for a single p.e. Therefore, a table of *Prob* was constructed by taking these effects into account, and the particle identification program refers to this table to obtain $Prob_i(p.e._{exp'd}, p.e._{obs})$.

A likelihood function is then defined as:

$$L(\mathrm{e}\ or\ \mu) = \prod_{\theta_i < 1.5 \times \theta_{Cer}} Prob_i\left(p.e._{exp'd}(\mathrm{e}\ or\ \mu),\ p.e._{obs}\right),$$

where θ_{Cer} is the observed half opening angle of the Cherenkov ring. The PMTs located inside a cone with a half opening angle of $1.5\times\theta_{Cer}$ from the particle's direction are used in calculating this likelihood function. An event is more e-like than μ-like, if $\log_{10} L(\mu) < \log_{10} L(e)$, and *vice versa.*

Figures 38 and 39 show $\log_{10} L(\mu) - \log_{10} L(e)$ distributions for the Monte-Carlo generated and real cosmic-ray stopping muons. From these figures, it can be concluded that the misidentification probability of the particle type for stopping muons is estimated to be about 2%.

Figure 40 shows the same plot for the Monte-Carlo generated single-ring atmospheric-neutrino events. The e-like and μ-like events are separated well. Figure 41 shows the same plot for the observed fully contained single-ring atmospheric neutrino events. It can be seen that these two distributions look similar.

Figure 42 shows the particle mis-identification probability as a function of the observed total p.e. number as estimated by using the Monte-Carlo generated single-ring atmos-pheric neutrino events. It can be seen that the mis-identification probability varies from 7% (below 300 p.e.) to 0.5% (above 2000 p.e.). From this figure, it can be estimated that the particle identification efficiency averaged over the total p.e. number in the range of atmospheric-neutrino analysis ($110 < PE_{TOT} < 4500$, see Sect. 5) is 98%.

3.3 Measurement of High-Energy Neutrinos

3.3.1 Interactions of High-Energy Muon Neutrinos

The fluxes of high-energy muon neutrinos can be measured with underground

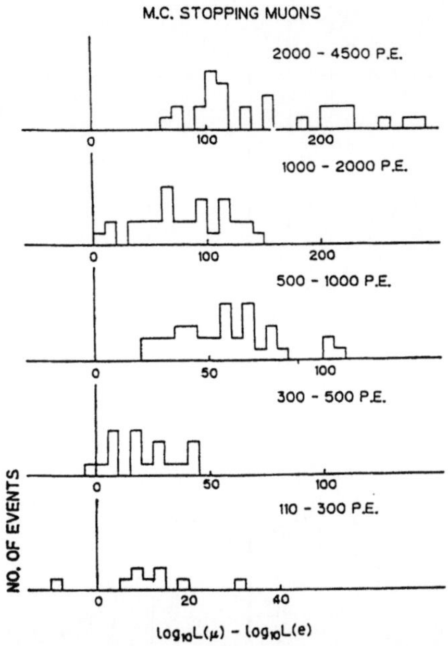

Figure 38: $\log_{10} L(\mu) - \log_{10} L(e)$ distribution, as defined in the text, for cosmic-ray stopping muons generated by a Monte Carlo simulation in various regions of total p.e. number. An event is classified as e-like (μ-like) by the particle identification program if $\log_{10} L(\mu) - \log_{10} L(e)$ is negative (positive).

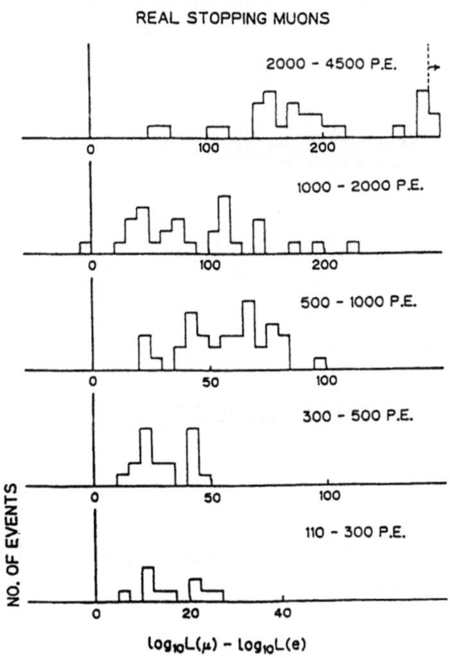

Figure 39: Same as Fig. 38, but for real cosmic-ray stopping muons.

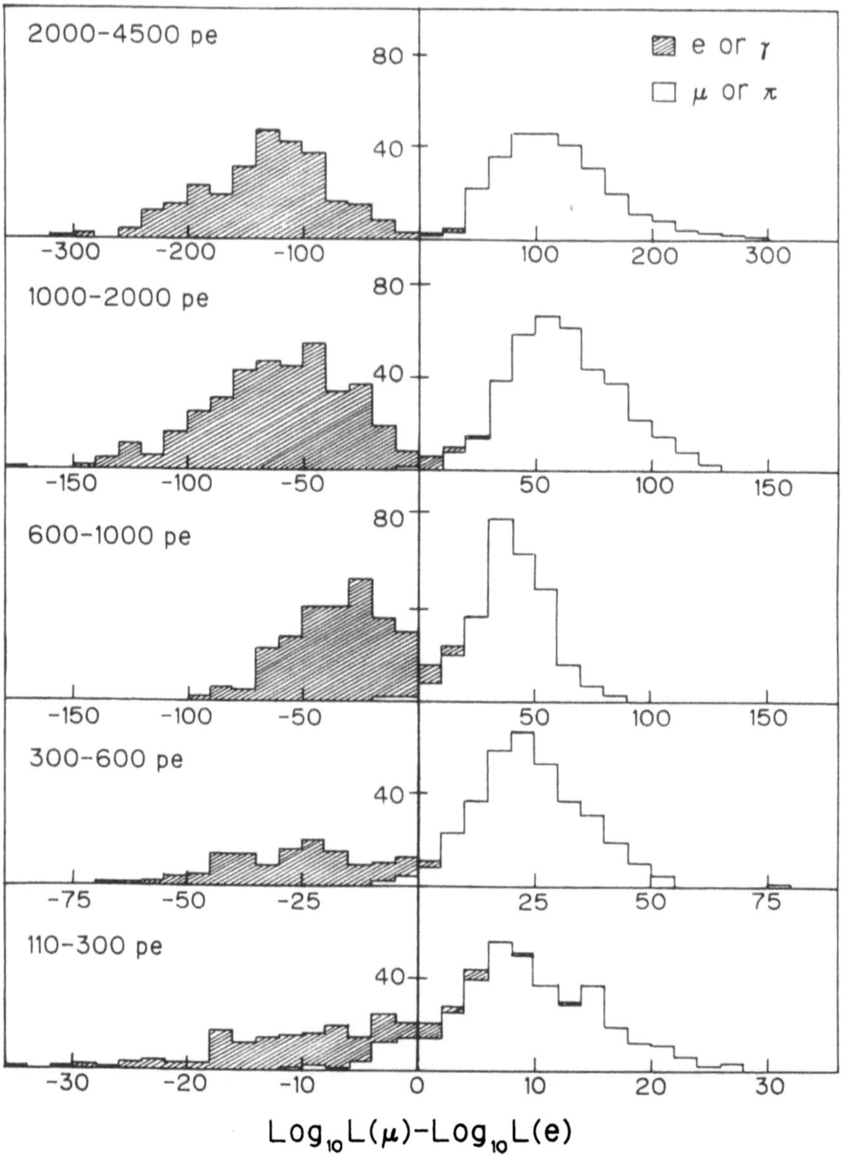

Figure 40: Same as Fig. 38, but for single-ring atmospheric-neutrino interactions generated by a Monte-Carlo simulation.

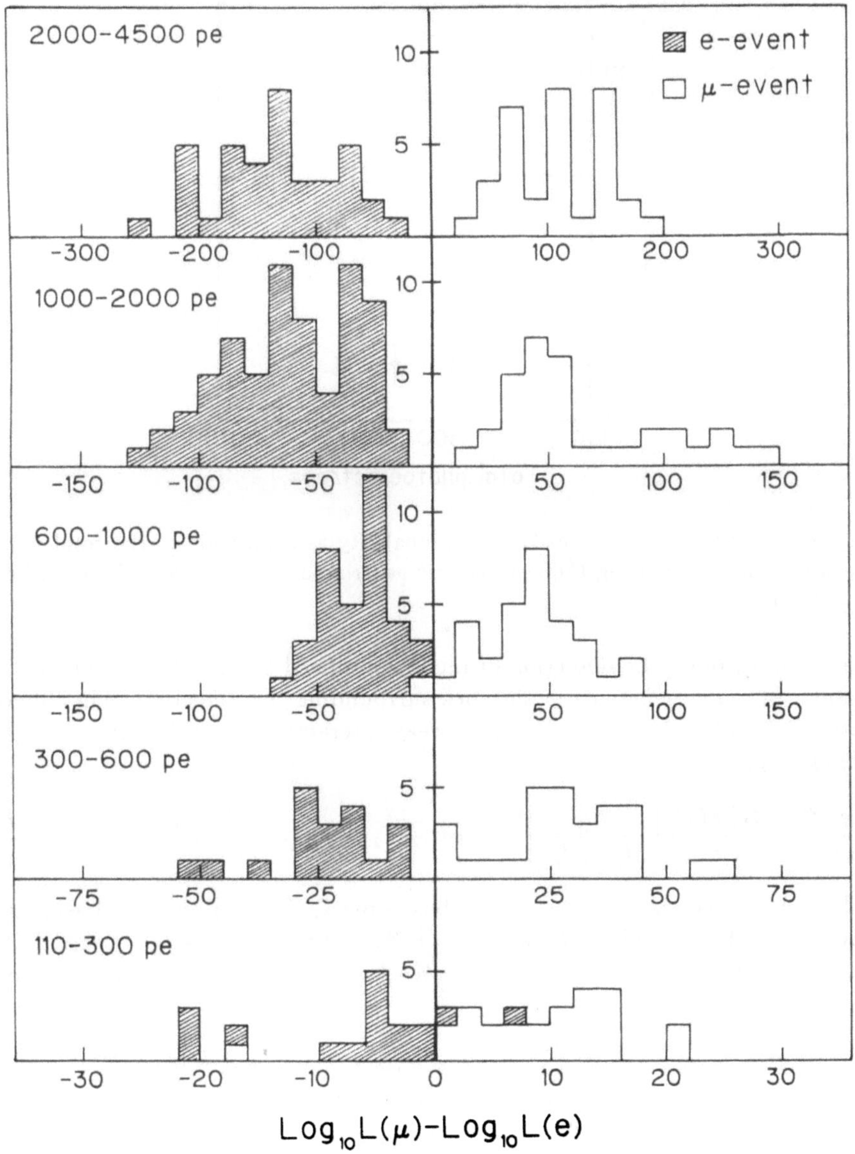

Figure 41: Same as Fig. 38, but for the single-ring atmospheric-neutrino events observed in Kamiokande.

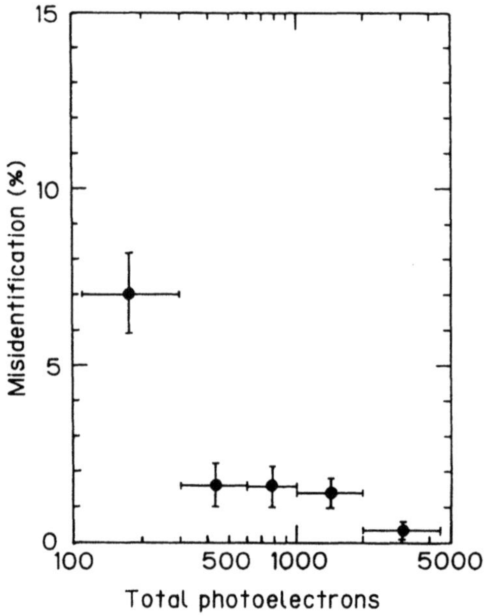

Figure 42: Particle mis-identification probability as a function of the total p.e. number as estimated by using the atmospheric-neutrino events generated in a Monte-Carlo simulation.

detectors through the detection of muons produced by the charged-current interactions of muon neutrinos in the rock surrounding the detectors. The differential cross section for the charged-current weak interaction ν_μ $(\bar{\nu}_\mu) + N \rightarrow \mu^\mp$ + *anything* is given by

$$\frac{d\sigma^{\nu(\bar{\nu})}}{dx\,dy} = \frac{G_F^2 M E_\nu}{\pi}\left[\frac{M_W^2}{M_W^2+Q^2}\right]^2\left[F_1 xy^2 + F_2\left(1-y-\frac{M}{2E_\nu}xy\right) \pm F_3 xy\left(1-\frac{y}{2}\right)\right],$$

where M_W is the intermediate weak-boson mass. The scaling variables (x and y) are defined by $x = Q^2/2M\nu$ and $y = \nu/E_\nu$ with $\nu = E_\nu - E_\mu$, where E_μ is the energy of the produced muon. The structure functions on isoscalar targets are given by:

$$\begin{aligned} F_2^\nu &= x(u+d+2s+2b+\bar{u}+\bar{d}+2\bar{c}+2\bar{t}),\\ xF_3^\nu &= x(u+d+2s+2b-\bar{u}-\bar{d}-2\bar{c}-2\bar{t}),\\ F_2^{\bar{\nu}} &= x(u+d+2c+2t+\bar{u}+\bar{d}+2\bar{s}+2\bar{b}),\\ xF_3^{\bar{\nu}} &= x(u+d+2c+2t-\bar{u}-\bar{d}-2\bar{s}-2\bar{b}),\end{aligned}$$

and

$$2xF_1^{\nu(\bar{\nu})} = F_2^{\nu(\bar{\nu})},$$

where the structure functions and parton distribution functions all depend on x and Q^2.

Below, in this subsection, neutrinos and antineutrinos are not distinguished, because the Kamiokande detector cannot identify the charge state of the muon.

At the detector, the flux of upward-going muons produced by atmospheric neutrinos, integrated over the energy region above the threshold E_{th}, is given as

$$\begin{aligned}\frac{d\phi_\mu(E_{th},\cos\theta)}{d\Omega} &= \int_{E_{th}}^{\infty} dE_\nu \int_0^1 dy \int_0^1 dx\, \frac{d\sigma}{dx\,dy}\, \frac{d\phi_\nu(E_\nu,\cos\theta)}{dE_\nu\, d\Omega}\, R(E_\mu, E_{th}) \\ &= \int_{E_{th}}^{\infty} \left[\int_0^1 \left[\int_0^1 \frac{d\sigma}{dx\,dy}\, dx\right] R(E_\mu, E_{th})\, dy\right] \frac{d\phi_\nu(E_\nu,\cos\theta)}{dE_\nu\, d\Omega}\, dE_\nu \\ &= \int_{E_{th}}^{\infty} P(E_\nu, E_{th})\, \frac{d\phi_\nu(E_\nu,\cos\theta)}{dE_\nu\, d\Omega}\, dE_\nu,\end{aligned}$$

where $\frac{d\phi_\nu(E_\nu,\cos\theta)}{dE_\nu\, d\Omega}$ is the differential spectrum of the parent neutrinos as a function of E_ν and the zenith angle θ, $R(E_\mu, E_{th})$ is the distance that the muon travels while its energy decreases from E_μ to E_{th}, and $P(E_\nu, E_{th})$ is defined as

$$P(E_\nu, E_{th}) = \int_0^1 \left[\int_0^1 \frac{d\sigma}{dx\,dy}\, dx\right] R(E_\mu, E_{th})\, dy.$$

The function $P(E_\nu, E_{th})$ is interpreted as being the probability that neutrinos with energy E_ν are converted and observed as muons at the detector with an energy larger than E_{th}.

For studying upward-going muons produced by atmospheric neutrinos, two alternative approaches were examined for the structure functions (Field and Feynmann parton model [36] and QCD evolution given by Eichten *et al.* [37]), the atmospheric neutrino flux (Volkova [38] and Mitsui *et al.* [39]), and the muon energy loss (linear approximation $dE_\mu/dX = -\alpha - \beta E$ with $\alpha = 2$ MeV/(g/cm^2) and $\beta = 3.9 \times 10^{-6}$ (g/cm^2)$^{-1}$ and the calculation by Lohman *et al.* [40], which takes into account the muon energy loss by ionization, Bremsstrahlung, direct pair production and nuclear interactions).

The upward-going muon flux produced by extraterrestrial neutrinos from an astronomical point source is given by

$$\Phi_\mu = \int_{E_{th}}^{E_c} P(E) f(E) \frac{d\Phi_\nu}{dE}\, dE, \tag{16}$$

where E_c is the cutoff energy and $\frac{d\Phi_\nu}{dE}$ is the parent neutrino spectrum; $f(E)$ is defined as the probability that a neutrino with energy E from the point source survives after having passed through the Earth, and is given as

$$f(E) = \Big\langle \exp\left[-N_A \rho \sigma_\nu(E)\right] \Big\rangle_t ,$$

where N_A is the Avogadro number, ρ is the column density of the Earth in the direction of the point source, $\sigma_\nu(E)$ is the total cross section of the charged-current interaction, and $< >_t$ means the time average because the column density (ρ) in the direction of the point source changes with time due to the Earth's rotation. This effect is ignored in the study of the atmospheric neutrinos, but should be taken into account when much-higher-energy astrophysical neutrinos are searched for. In the latter case, the function $P(E)$ is almost independent of the energy threshold (E_{th}) and, therefore, the calculated results given in Ref. [41] are used.

If the upward-going muon flux is determined experimentally and a power-law spectrum is assumed for the parent neutrino spectrum,

$$\frac{\mathrm{d}\Phi_\nu}{\mathrm{d}E} = A\,E^{-\gamma}, \tag{17}$$

the numerical factor A is deduced from Eq.(16), and the integrated neutrino flux above E_{th} is given by

$$\Phi_\nu = \int_{E_{th}}^{E_c} A\,E^{-\gamma}\,\mathrm{d}E. \tag{18}$$

The neutrino luminosity of the point source at a distance R from the Earth is then given by

$$L = 4\pi R^2 \int_{E_{th}}^{E_c} A\,E^{-\gamma}\,\mathrm{d}E. \tag{19}$$

3.3.2 Measurement of Upward-Going Muons

The data sample considered were taken from July, 1983, through April, 1990 [42] with different modes of online cuts. The "weekday mode" was 100% efficient for upward-going muons; the total live time corresponding to this mode is 1479 days. The "holiday mode" was about 50% efficient and the corresponding live time was 441 days. The missing upward-going muons in the holiday-mode data were those entering the detector nearly horizontally. A total of 2.27×10^8 events were recorded and 120,684 events were software-selected (97% efficient) as upward-going (zenith angle larger than 90°) muon candidates which went through the detector. Most of these events were either nearly horizontal, but actually downward-going, muon events or multiple muon events. Either a further selection of the upward-going muons was made by a visual scan, requiring a path length of more than 7 m in the inner detector volume. With this requirement, the nominal detection area is about 165 m^2 for Kamiokande-I and 150 m^2 for Kamiokande-II. A total of 252 upward-going muons were finally selected. Figure 43 shows how an upward-going muon event looks in an exploded view of the Kamiokande detector.

Although the minimum path length of 7 m corresponds to the minimum muon energy of 1.7 GeV, the effective threshold (E_{th}) is higher than this value. This is because only through-going muons were selected by the off-line analysis. With this condition, the calculated relative acceptance for the upward-going muons is shown in Fig. 44 as a function of the muon energy. Figure 45 shows the path-length distribution of the upward-going muons, which was calculated by assuming the adopted online and off-line cuts and the spectrum of atmospheric neutrinos [38, 39]; it is the distribution expected for the finally selected data sample. An average path length of 12.2 m corresponds to an effective threshold of $E_{th} = 3.0$ GeV [43].

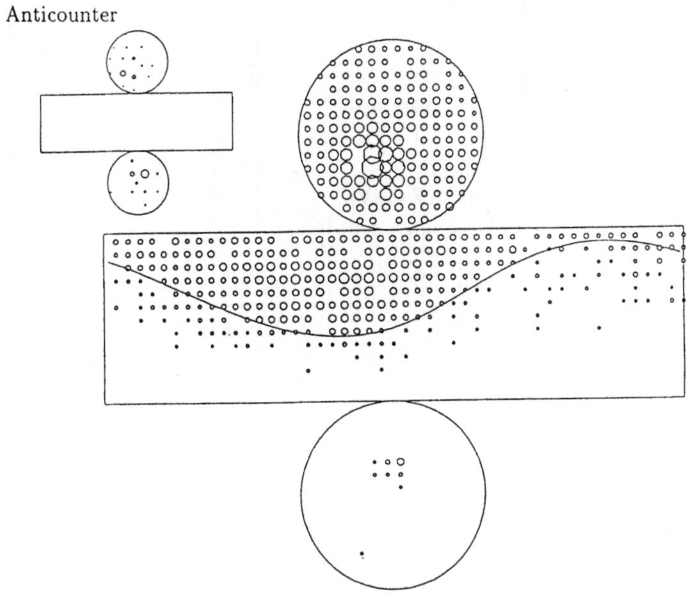

Figure 43: The typical upward-going muon event in an exploded view of Kamiokande. Each small circle represents a struck PMT with a diameter proportional to the charge output of the PMT.

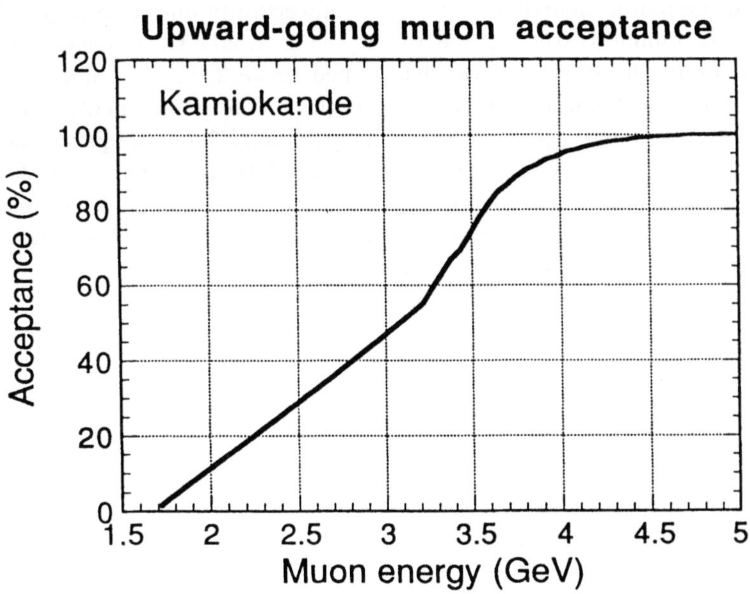

Figure 44: Relative acceptance for the upward-going muons as a function of the muon energy, corresponding to the selection criteria described in the text.

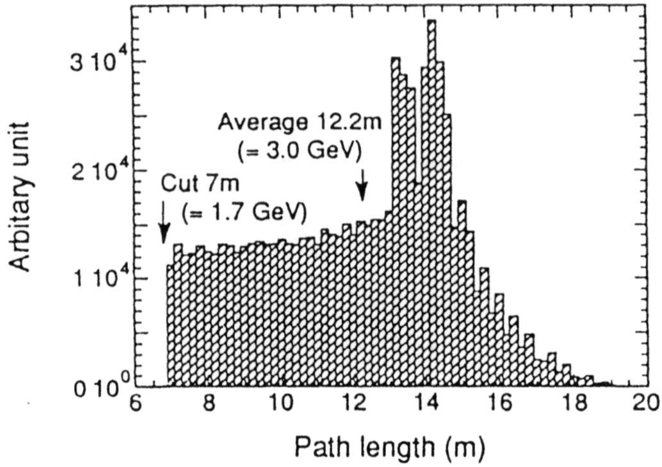

Figure 45: Expected path-length distribution of upward-going muons which went through the Kamiokande detector.

In order to discuss the observation of high-energy neutrinos from astronomical point sources, the angular resolution for the parent neutrinos must be considered. The angular difference between the reconstructed muon and the parent neutrino directions ($\Delta\theta$) is given by $\Delta\theta = (\Delta\theta_1^2 + \Delta\theta_2^2 + \Delta\theta_3^2)^{1/2}$, where $\Delta\theta_1$ is the error due to manual reconstruction, $\Delta\theta_2$ the scattering angle between the produced muon and the parent neutrino, and $\Delta\theta_3$ the uncertainty due to multiple Coulomb scattering of muons traversing the rock. Using Monte Carlo-generated events, $\Delta\theta_1$ (one standard deviation) was determined to be 2.1°. $\Delta\theta_2$ and $\Delta\theta_3$ depend on the energy spectrum and the cutoff energies of the parent neutrinos. Assuming a power-law energy spectrum where $d\Phi_\nu/dE \propto E^{-\gamma}$ for the parent neutrinos, $(\Delta\theta_2^2 + \Delta\theta_3^2)^{1/2}$ was calculated by a computer simulation for various cutoff energies ($E_c = 10^{12} - 10^{15}$ eV) and spectral indices ($\gamma = 2.1 - 2.7$). The resulting one-standard-deviation angular interval for $\Delta\theta$ is 2.2° for $E_c = 10^{15}$ eV and $\gamma = 2.1$, increasing to 3.5° for $E_c = 10^{12}$ eV and $\gamma = 2.7$.

4. Detector Simulation

4.1 General

The response of the detector to the neutrino interactions is simulated by a Monte-Carlo method. The simulation is extensively used for estimating the detector performance and comparing the observed data with theoretical expectations.

The detector simulation is performed in three steps. The first step simulates production of energetic particles from neutrino interactions, where the kinematics of the neutrino interactions is taken into account. The second step simulates transportation of the produced particles through the detector. Interactions of the particles in water are simulated in this step. In the last step, photons from charged particles are generated and a Cherenkov ring pattern is simulated.

In this section, essential features of the Kamiokande's detector simulation are briefly described. Further details are referred to Refs. [19, 44].

4.2 Particle Generation

Low-energy neutrinos are detected through $\nu e \rightarrow \nu e$ and $\bar{\nu}_e p \rightarrow e^+ n$ interactions as described in subsect. 3.1. These interactions generate energetic electrons or positrons, which can be detected by observing Cherenkov radiation. The production of these particles is simulated according to the kinematics given in subsect. 3.1.1.

For the simulation of medium-energy neutrino interactions, the cross sections described in subsect. 3.2.1 are used. When the kinematics of the single-pion production is simulated, the mass spectrum of the Δ resonance is taken into account. The kinematics of the lepton side in CC single- and multi-pion productions is simulated according to the following Bjorken scaling:

$$\frac{\mathrm{d}^2\sigma^{\nu N,\bar{\nu}N}}{\mathrm{d}x'\mathrm{d}y} = \frac{G_F^2 M E_\nu}{\pi} \left\{ \left[1 - y + \frac{1}{2}y^2 \right] F_2(x') \pm y(1 - \frac{1}{2}y) \left[x' F_3(x') \right] \right\} ,$$

where x' is defined as

$$x' = \frac{Q^2}{2M(E_\nu - E_l) + M^2} ,$$

where E_l is the scattered lepton energy. The mean multiplicity of pions in the CC multi-pion productions is determined by using a ln W^2 scaling observed in neutrino beam experiments [45], where W is the hadronic invariant mass. The pion multiplicity for individual events is simulated by using the KNO scaling [46]. The kinematics of the NC reactions is simulated essentially in the same way as that of the CC reactions.

The Kamiokande's detector simulation also takes into account the nuclear effects for mesons produced in a ^{16}O nucleus due to neutrino interactions or nucleon decay. If a neutrino interacts with an ^{16}O nucleus, Fermi motion of the target nucleon and the Pauli exclusion principle are taken into account. Due to the latter effect, the momentum of the recoil nucleon is required to exceed the Fermi surface momentum (250 MeV/c) on the basis of the Fermi gas model. Detailed description of these meson nuclear effects are referred to Refs. [19, 44].

As described in subsect. 3.3, high energy neutrinos are observed by means of detecting upward-going muons. The energy spectrum of the upward-going muons generated by the parent neutrinos with a given energy spectrum can be calculated according to the prescription given in subsect. 3.3.1.

4.3 Track Generation

Tracks of the particles produced by the neutrino interactions are simulated by taking into account the interactions of the particles with water. The particles are classified according to their interaction types as (1) electromagnetic particles ($e^\pm$ and γ), (2) muons ($\mu^\pm$) and (3) pions ($\pi^\pm$).

4.3.1 Electromagnetic Particles

Low energy electrons and positrons mainly loose energy by ionization loss and their tracks are deflected by multiple Coulomb scattering in water. Medium energy electrons, positrons, and γ-rays cause electromagnetic cascade showers by the physical processes of bremsstrahlung and pair creations. These processes are simulated by a standard simulation program EGS4 [47]. An example of a track of a low energy electron is shown in Fig. 32.

4.3.2 Muons

For $\mu^\pm$, physical processes of the ionization loss, knock-on electrons and Coulomb multiple scattering are taken into account. The knock-on electrons are simulated by the EGS4 program. Other processes such as the pair creation, bremsstrahlung and nuclear interactions are negligible in the energy range investigated in Kamiokande.

4.3.3 Pions

For the generation of $\pi^\pm$ tracks, nuclear interactions are taken into account in addition to the ionization loss, knock-on electrons and Coulomb multiple scattering. For the total cross sections of $\pi - {}^{16}\mathrm{O}$ and $\pi - p$ interactions are taken from Refs. [48] and [49], respectively. The nuclear interactions considered are: (1) elastic scattering, (2) inelastic scattering, (3) charge exchange and (4) nuclear absorption. The relative strengths of these interactions are determined on the basis of the $\pi^+ - \mathrm{Al}$ scattering data [50]. The angle and energy of the scattered pion are determined according to the $\pi - {}^{12}\mathrm{C}$ data [51] and the $\pi^- p \to \pi^0 n$ data.

4.4 Cherenkov-Light Generation

A charged particle traversing water with a velocity larger than c/n emits Cherenkov photons, where n is the index of refraction of water (1.344) and c is the light velocity in vacuum. The photons are emitted at an angle θ with respect to the direction of the charged particle. The Cherenkov angle θ is given by $\cos\theta = \frac{1}{n\beta}$, where β is

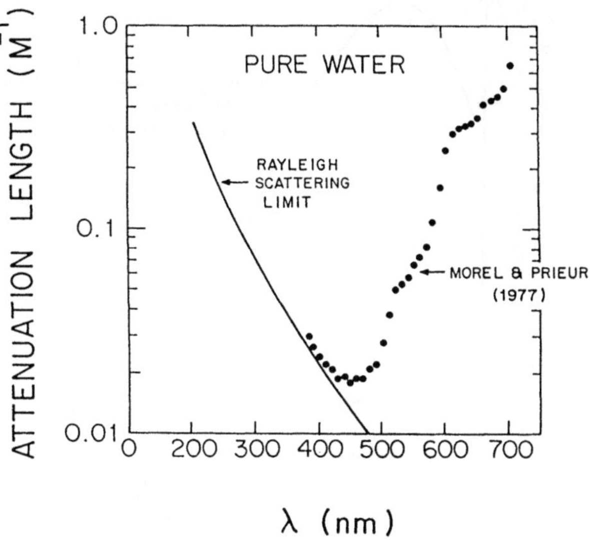

Figure 46: Attenuation length of light in pure water from 390 nm to 700 nm [52]. The Rayleigh scattering limit is also shown for short wavelengths.

the particle velocity relative to c. In water, $\theta = 42°$ for $\beta = 1$.

The number of photons emitted per unit path length per unit wavelength is given by

$$\frac{d^2N}{dx\,d\lambda} = 2\pi\alpha(1 - \frac{1}{n^2\beta^2})\frac{1}{\lambda^2},$$

where λ is the wavelength and α is the fine structure constant. This equation shows that in the relativistic limit ($\beta = 1$) a particle emits ~340 photons per 1 cm of track length in the wavelength range of 300 - 600 nm (sensitive range of PMTs).

In the simulation, the generated Cherenkov photons are traced in the detector until they are detected by the PMTs or absorbed by the detector materials. Scattering of photons in water and that by the inner surface of the detector are also taken into account. The wavelength dependence of the photon attenuation length used in the simulation is shown in Fig. 46. The simulation of the generation of photoelectrons by Cherenkov photons which hit PMTs is performed by using the quantum efficiency shown in Fig. 47. Output signals from PMTs are simulated on the basis of the measured PMT response. Figure 12 shows the PMT response to a single-photoelectron signal.

4.5 Detector Response

An example of the simulation of a 10 MeV electron is shown in Fig. 48. Clear Cherenkov-ring patterns are seen even for such low-energy particles. Various resolutions for low-energy electrons are obtained by the simulation. The results are

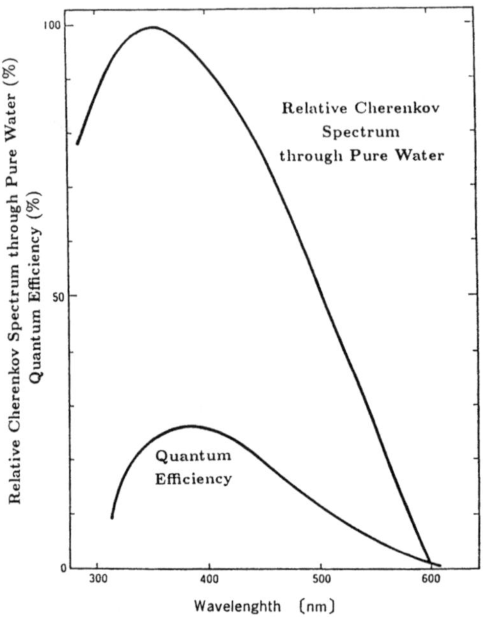

Figure 47: Spectral dependence of the quantum efficiency of a 20″ PMT. Also shown is the relative Cherenkov radiation spectrum transmitted through 15 m of pure water.

shown in Fig. 29 for the energy resolution, in Fig. 33 for the angular resolution, and in Fig. 30 for the vertex resolution.

The simulated results concerning total number of photoelectrons observed in the detector for medium-energy particles is shown in Fig. 21 as a function of momentum, and the momentum resolution is shown in Fig. 22. Figures 49 (a) and (b) show examples of medium-energy electron and muon events generated by a Monte Carlo simulation.

5. Investigation of Atmospheric Neutrinos

5.1 Fully Contained Events

The interactions in massive underground detectors of atmospheric neutrinos provide a means of studying the intrinsic properties of neutrinos. Specifically, sensitive searches for neutrino oscillations are made possible by the well-understood ratio of muon-type neutrinos (ν_μ and $\overline{\nu}_\mu$) to electron-type neutrinos (ν_e and $\overline{\nu}_e$), both constituting the atmospheric neutrino flux, and by the large range of distances traveled by these neutrinos ($\sim$10 to 1.3×10^4 km) to reach a detector on earth.

Atmospheric neutrinos are produced by the decay of π and K mesons produced in the nuclear interactions of the primary component of cosmic rays in the atmosphere. Since pions are dominant and they decay according to $\pi^\pm \rightarrow \mu^\pm + \nu_\mu\,(\bar{\nu}_\mu)$, $\mu^\pm \rightarrow e^\pm + \nu_e\,(\bar{\nu}_e) + \bar{\nu}_\mu\,(\nu_\mu)$, roughly speaking $\nu_\mu : \nu_e \approx 2 : 1$. (Hereafter in this section ν represents $\nu + \bar{\nu}$.) More elaborate calculations of the

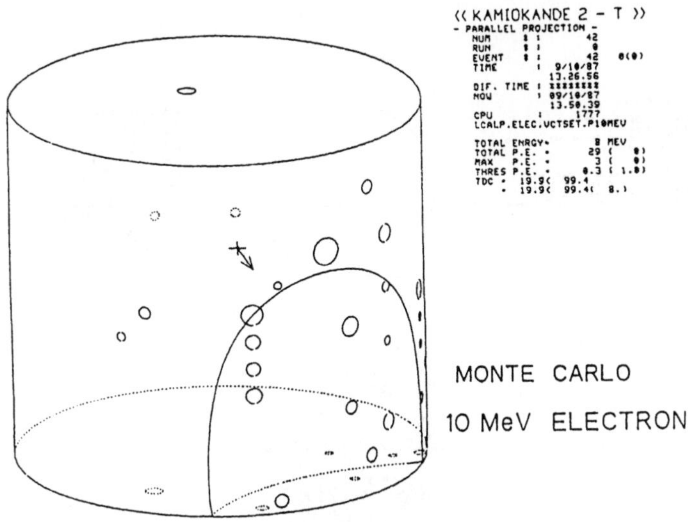

Figure 48: Three-dimensional representation of a Monte-Carlo generated 10 MeV electron event. The vertex position of the Cherenkov ring is indicated by a + mark. The arrow starting from this mark indicates the direction of the generated electron. Each small circle represents a struck PMT with the diameter proportional to the charge output of the PMT. The large circle shows the outline of the Cherenkov ring, which is drawn as a cone with an opening angle of 42° with respect to the vertex position.

atmospheric neutrino fluxes [53, 54] were performed with a typical uncertainty of $\sim 20\%$. However, the ratio of the fluxes ν_μ/ν_e has much less uncertainty ($\sim 5\%$ at neutrino energies around 1 GeV) mainly attributed to the uncertainty of the K/π production ratio in the atmosphere [9].

In Kamiokande, the energy range of the analysis is 110 - 4500 total photoelectrons, corresponding to 30 - 1330 MeV/c for electrons and 205 - 1500 MeV/c for muons. The fiducial volume is defined to be 2 m (Kamiokande-I) or 1.5 m (Kamiokande-II) inside from the PMT planes. The corresponding fiducial masses are 880 tons (780 tons after 1985) for Kamiokande-I and 1040 tons for Kamiokande-II. During 4.92 kt·yr of the detector exposure, 475 fully contained events were observed within the fiducial volume. Of these, 329 events had single Cherenkov rings and they were classified as either μ- or e-like. The probability of misidentifying the particle species of these single-ring events is estimated to be 2% (see subsect. 3.2.3). As can be seen from Table 4, a Monte-Carlo simulation tells us that 75% of the the single-ring events are caused by the charged-current (CC) elastic scattering and contamination from neutral-current interactions is small. Although the contribution from the CC pion-production processes to the single-ring events is 20%, the particles observable in these processes are mostly μ or e. Therefore, it is clear that ν_μ/ν_e ratio can be observed by studying the single-ring μ-like/e-like ratio. Consequently, hereafter we discuss single-ring events only.

In 1988, the Kamiokande Collaboration reported [9], using 277 fully contained events observed during 2.87 kt·yr of the detector exposure, that the number of

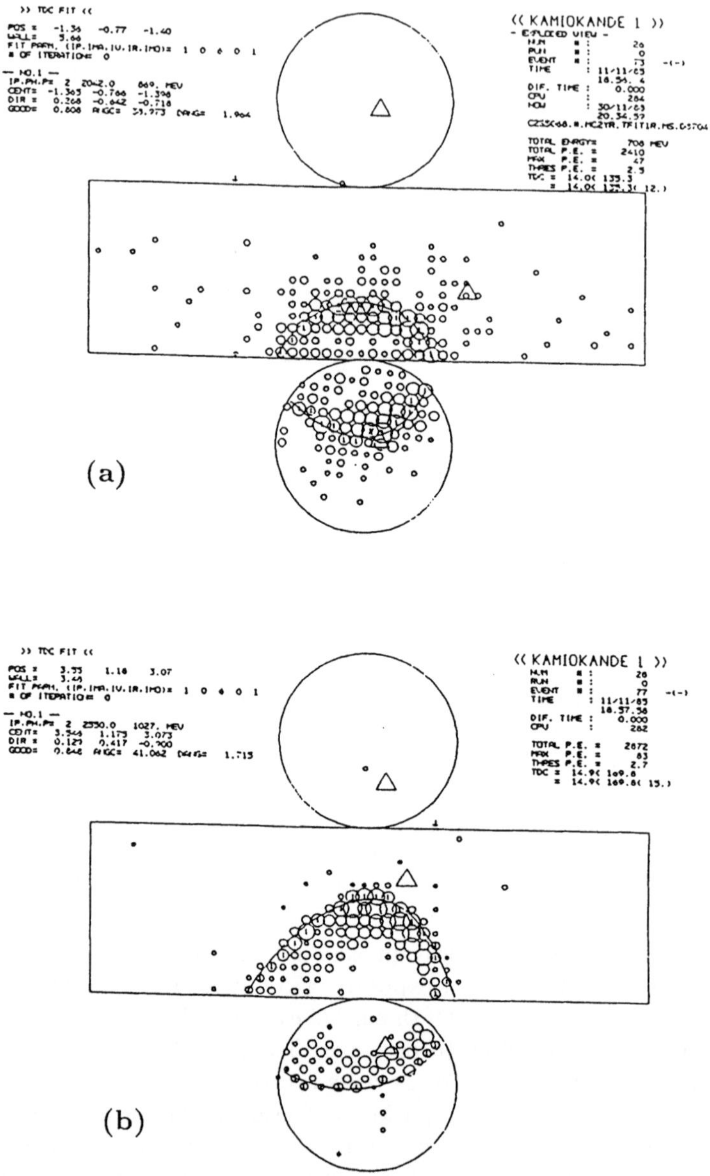

Figure 49: Single-ring, fully contained (a) e-like and (b) μ-like events generated by a Monte-Carlo simulation of atmospheric-neutrino interactions.

Table 5: Summary of the fully contained single-ring events observed during 4.92 kt·yr of the detector exposure. Predictions of Monte-Carlo calculations with the atmospheric neutrino fluxes given by Barr *et al.* [56], Honda *et al.* [57], and Lee and Koh [58] are are also shown. $\mu \to e$ in the third and fifth columns represents the number of events accompanied by the $\mu \to e$ decay signal. The numbers in parentheses in the fourth row show events with $p_e > 100$ MeV/c.

	Data		Monte Carlo			
			Barr *et al.*		Honda *et al.*	Lee & Koh
	total ev.	$\mu \to e$	total ev.	$\mu \to e$	total ev.	total ev.
Single ring	329	112	452.2	206.8	405.9	352.4
	(310)		(425.5)		(380.2)	(332.8)
μ-like	151	99	260.6	190.4	234.2	205.2
e-like	159	13	164.9	16.4	146.0	127.7
(>100MeV/c)						
$(\mu/e)_{Data}/(\mu/e)_{MC}$			$0.60^{+0.07}_{-0.06}$		$0.59^{+0.07}_{-0.06}$	$0.59^{+0.07}_{-0.06}$

μ-like single-ring events had a significant deficit compared with the Monte-Carlo prediction which was based on the atmospheric neutrino fluxes calculated for the Kamioka site, and for mean solar activity [53, 54], while the number of e-like single-ring events was in good agreement with the expected number. Defining $\mu/e \equiv$ (μ-like)/(e-like, with electron momentum $p_e > 100$ MeV/c), $(\mu/e)_{\rm Data}/(\mu/e)_{\rm MC}$ was $(85/93)/(144.0/88.5) = 0.56^{+0.09}_{-0.08}$, where the associated errors are statistical ones. It is this observation that stimulated the current interest in atmospheric neutrinos.

Soon thereafter, it was pointed out [55] that the effect of polarization must be taken into account in the flux calculation, since the muons from pion decay are polarized. This effect, however, was turned out to be less than a 15% effect [56] on the μ/e ratio.

The Kamiokande Collaboration then published an updated result [10] based on the data sample from 4.92 kt·yr of detector exposure and on the recent flux calculations. Table 5 summarizes the data and compares them with the Monte-Carlo prediction. One sees that there is as much as 25% difference in the predictions of the total number of events. This table therefore suggests that the ambiguity in the absolute flux value is as large as 25%. Also there is an ambiguity of about 10% in the cross sections of neutrino interactions at an energy region around 1 GeV. Consequently, the total systematic error in the absolute event number predicted by the Monte-Carlo calculations is estimated to be about 30%. Because of this, the disagreement in the total number of events between the data and the Monte-Carlo predictions, as seen in Table 5, is not a serious problem.

The momentum spectra of the events are shown in Fig. 50. The shape and area of the spectrum of the e-like events in Fig. 50 (a) are well reproduced by the Monte-Carlo simulation using the flux calculation of Ref. [56], while the use of Refs. [57] and [58] respectively suggests about a 9% and 25% excess of e-like events relative to the Monte-Carlo prediction. Figure 50 (b) indicates a deficit of μ-like events of 42%, 35% and 24%, respectively, relative to the calculated values using Refs. [56],

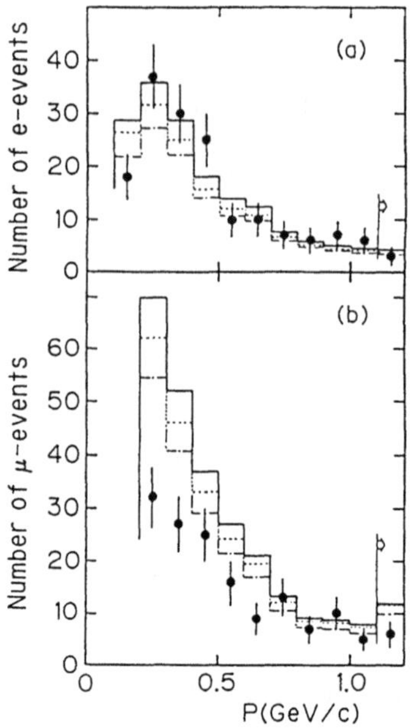

Figure 50: Momentum distributions of (a) the e-like and (b) the μ-like events. The histograms show the Monte-Carlo predictions based on Ref. [56] (solid line), Ref. [57] (dotted line), and Ref. [58] (dot-dashed line).

[57], and [58]. It is emphasized, however, that the ratio $(\mu/e)_{Data}/(\mu/e)_{MC}$ is essentially independent of the calculated absolute values of the ν_μ and ν_e fluxes, as shown in Table 1. Because of this fact, hereafter we refer to the flux calculation of Ref. [56] for simplicity. We then obtain:

$$R(\mu/e)_{\mathrm{Data/MC}} \equiv \frac{(\mu/e)_{Data}}{(\mu/e)_{MC}} = 0.60^{+0.07}_{-0.06} \pm 0.05,$$

where the first error is statistical and the second error is systematic. This result indicates more than 4σ deviation (statistically) from unity.

The momentum and the zenith-angle dependences of $R(\mu/e)_{\mathrm{Data/MC}}$ are shown in Fig. 51. One finds that the small μ/e ratio does not show any strong momentum and zenith-angle dependence.

The relative deficit of the μ-like events of the data can also be tested by studying the number of events accompanied by a $\mu - e$ decay signal. A large fraction of CC ν_μ-induced single-ring events are accompanied by μ-decay electrons, while most of CC ν_e-induced single-ring events are not. Therefore, a ratio $R_{decay} \equiv$ (single-ring events with μ decay) /(single-ring events without μ decay) is closely related to the ν_μ/ν_e ratio. The ratio Data/MC of R_{decay} is:

$$R_{decay}(\mathrm{Data})/\mathrm{R_{decay}}(\mathrm{MC}) = 0.61 \pm 0.07,$$

where the associated error is statistical. This result is in good agreement with the value obtained from the particle identification.

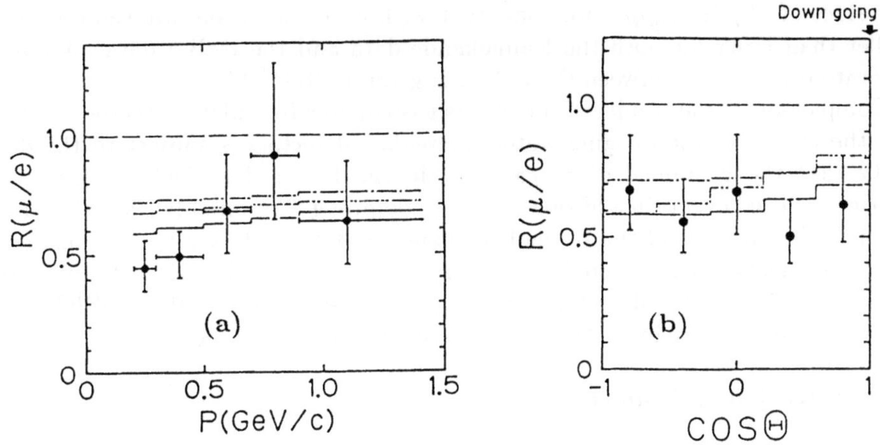

Figure 51: (a) The momentum and (b) the zenith-angle dependences of $R(\mu/e)_{\rm Data/MC}$. The data are shown by the circles with vertical bars which indicate a $\pm 1\sigma$ statistical error. The solid, dot-dashed, and double-dot-dashed lines show the expected values of $R(\mu/e)_{\rm Data/MC}$ for the neutrino-oscillation parameters ($\Delta m^2_{\nu_\mu \nu_\tau}$ in eV^2, $\sin^2 2\theta_{\nu_\mu \nu_\tau}$) = ($0.8 \times 10^{-2}$, 0.87) which is the best fit parameter set, (0.8×10^{-2}, 0.43), and (1.1×10^{-3}, 0.87).

Let us now examine the systematic errors involved in the μ/e ratio of the Monte-Carlo calculations and the data. Most of the theoretical uncertainties almost cancel in $R(\mu/e)_{\rm Data/MC}$. However, there are still several sources of systematic errors in $R(\mu/e)_{\rm Data/MC}$. This ratio suffers an uncertainty of 2 times 2% due to the particle identification error. The other sources of errors in $R(\mu/e)_{\rm Data/MC}$ are: 2% from an uncertainty in the separation of single-ring and multi-ring events, <1% from the difference in vertex fits for μ-like and e-like events, <1% from an uncertainty in the absolute energy calibration, <5% from the incident ν_μ/ν_e flux ratio, 4% from a statistical fluctuation of the Monte-Carlo events, 2% from an uncertainty in the $\overline{\nu}_e/\nu_e$ flux ratio, 3% from an uncertainty in the ν_μ/ν_e ratio of the charged-current interactions including the nuclear effects, 2% from an uncertainty in the neutral-current interactions and ~1% from the influence on particle identification of γ-rays emitted by an excited target nucleus. Adding all these errors in quadrature, the systematic error in the value of $R(\mu/e)_{\rm Data/MC}$ is estimated to be 9%. One finds that the possible sources of the systematic errors are well controlled and too small to account for the small μ/e ratio.

The Kamiokande result is compared with the results from the other underground detectors [10, 59]. The IMB Collaboration accomplished, with about 90% confidence in IMB-3, differentiation between a showering track and a nonshowering track. Therefore, comparison between the Kamiokande and the IMB-3 results is made through the values of $R(\mu/e)_{\rm Data/MC}$. The result from IMB-3, combining statistical and systematic errors, is $R(\mu/e)_{\rm Data/MC} = 0.67 \pm 0.17$ [60].[†] Clearly, the values of $R(\mu/e)_{\rm Data/MC}$ are consistent with each other, and are significantly smaller than unity for both the Kamiokande data and the IMB-3 data. A more elaborate comparison between these data is given in Ref. [59].

Comparison of the results of the Fréjus gas counter-iron plate detector [62, 63] with the results of the imaging water Cherenkov detectors is subject to the difficulty that the contained event category in the Fréjus data includes inelastic neutrino-induced events in addition to the single-prong (presumably elastic) events on which the analyses of the Kamiokande and IMB-3 results are based. Furthermore, the energy range covered by the Fréjus data is slightly higher than those of the water Cherenkov detectors. The lower statistics data of the NUSEX [64] neither support nor exclude the Kamiokande and IMB-3 results.

5.2 Upward-Going Muons

The upward-going muons observed in Kamiokande are almost entirely due to atmospheric neutrino interactions. From the observed 252 events (see subsect. 3.3.2) an average upward-going muon flux with an effective threshold of 3 GeV[†] is found to be

$$\phi_\mu^{average} = (2.04 \pm 0.13) \times 10^{-13}\,\mathrm{cm^{-2}\,s^{-1}\,sr^{-1}}.$$

This number is compared with a theoretical expectation of upward-going muon flux which is derived from the atmospheric neutrino flux [38], the cross section of charged-current weak interactions with the quark distributions given in Ref. [37], and the muon energy loss in the rock [40]:

$$\phi_\mu^{expected} = 1.92 \times 10^{-13}\,\mathrm{cm^{-2}\,s^{-1}\,sr^{-1}}.$$

It can be concluded that the experimental result agrees well with the theoretical calculation.

The differential flux as a function of the cosine of the zenith angle, $(\frac{d\phi_\mu}{d\Omega})$, is also obtained. The experimental result and the theoretical expectation are plotted in Fig. 52. They are also in good agreement.

[†]The recent result from IMB-3 [61] on the μ/e ratio, $R(\mu/e)_{\rm Data/MC} = 0.54 \pm 0.05 \pm 0.12$, is also in good agreement with the Kamiokande result.

[†]As described in subsect. 3.3.2, a minimum path length of 7 m was required in the event selection of upward-going muons. This corresponds to a threshold of 1.7 GeV. However, considering the acceptance as a function of muon energy (see Fig. 44) upward-going muons are observed with full efficiency, effectively above 3 GeV.

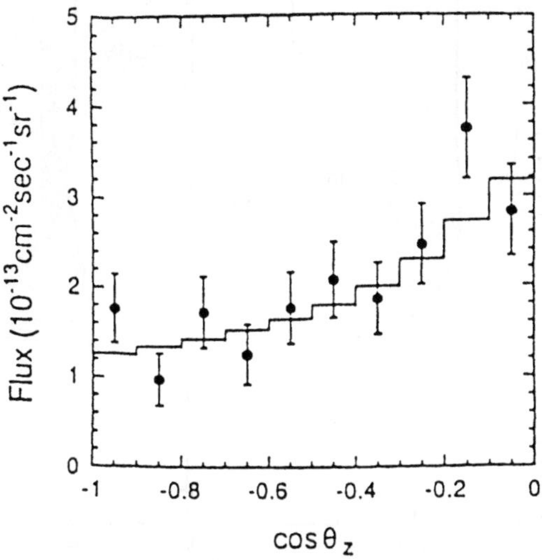

Figure 52: Zenith-angle distribution of 252 upward-going muons observed in Kamiokande. The histogram shows the expected distribution calculated from the atmospheric-neutrino flux of Ref. [38], the cross section of charged-current weak interactions with the quark distributions given in Ref. [37], and the muon energy loss in the rock given in Ref. [40].

5.3 Implications for Neutrino Oscillations

A possible explanation of the deviation from unity of the value of $(\mu/e)_{Data}/(\mu/e)_{MC}$ for the fully contained events may be sought in neutrino oscillations. Here we examine the two possible channels $\nu_\mu \leftrightarrow \nu_e$ and $\nu_\mu \leftrightarrow \nu_\tau$. There also is the possibility of ν_μ oscillating to a sterile, *i.e.*, right-handed, neutrino of unknown mass and flavor, which would be consistent with the observed small value of $(\mu/e)_{Data}/(\mu/e)_{MC}$. The consequences of this oscillation channel need to be explored elsewhere.

To test neutrino oscillations, the e-like and μ-like events are respectively mapped on zenith angle ($\cos\Theta$) and momentum (p) planes, where ($\cos\Theta$, p) is meshed into 10×11 cells. Then a χ^2 is defined to draw contours of allowed regions on the (Δm^2, $\sin^2 2\theta$) plane:

$$\chi^2 = \mathrm{Min}(\alpha,\beta)\left(L(\alpha,\beta) + \frac{\alpha^2}{\sigma_\alpha^2} + \frac{\beta^2}{\sigma_\beta^2}\right),$$

$$L(\alpha,\beta) = -2\times\sum_{i,j}\left(\ln\left\{\frac{(X_{ij}(e))^{N_{ij}(e)}\cdot\exp\left(-X_{ij}(e)\right)}{N_{ij}(e)!}\right\} + \ln\left\{\frac{(X_{ij}(\mu))^{N_{ij}(\mu)}\cdot\exp\left(-X_{ij}(\mu)\right)}{N_{ij}(\mu)!}\right\}\right),$$

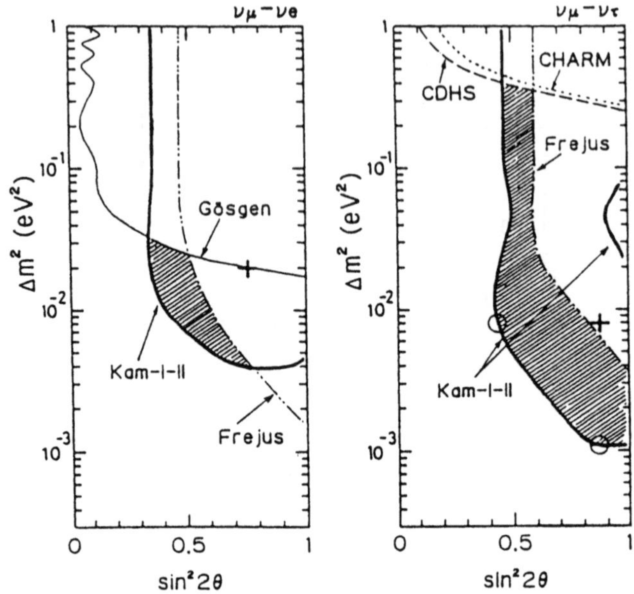

Figure 53: The 90% CL allowed regions on neutrino-oscillation parameters by the atmospheric-neutrino data from Kamiokande. The best-fit parameter sets are shown by the crosses. The two open circles in the figure for $\nu_\mu \leftrightarrow \nu_\tau$ indicate the points which are used in Fig. 51. The results from other experiments [65] show excluded regions at 90% CL.

$$X_{ij}(e) = (1+\alpha)(1-\beta/2)M_{ij}(e),$$

$$X_{ij}(\mu) = (1+\alpha)(1+\beta/2)M_{ij}(\mu),$$

where $N_{ij}(e \text{ or } \mu)$ is the number of observed e-like or μ-like events in the $(\cos\Theta(\mathrm{i}), \mathrm{p(j)})$ cell, $M_{ij}(e \text{ or } \mu)$ is that of a Monte-Carlo sample with given oscillation parameters, α is a factor relevant to the absolute normalization with (systematic) error σ_α=30%, and β to the μ/e ratio with (systematic) error σ_β=9%. In calculating $M_{ij}(e \text{ or } \mu)$ of $\nu_\mu \leftrightarrow \nu_e$, the matter effect is taken into account. A set of (α, β) which minimizes the above χ^2 is calculated for each combination of the oscillation parameters. More details of the method of calculation are given in Ref. [19].

Figure 53 shows the constraint on neutrino-oscillation parameters obtained from the Kamiokande data, where the crosses indicate the best-fit parameters and the shaded regions give the 90% CL allowed regions. For the Monte-Carlo prediction, the flux calculation of Ref. [56] was used. Concerning this result, it should be noted that the allowed region for the $\nu_\mu \leftrightarrow \nu_e$ channel conflicts with the MSW solution of the solar neutrino problem, as can be seen in Fig. 60. The 90% CL allowed region for $\nu_\mu \leftrightarrow \nu_\tau$, on the other hand, is not in conflict with any other experimental results.

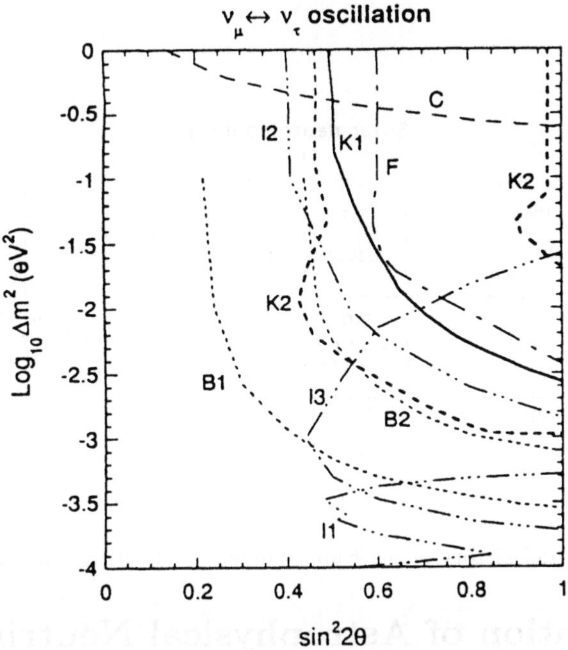

Figure 54: The 90% CL excluded regions on neutrino-oscillation parameters by the upward-going muon data from Kamiokande for the channel $\nu_\mu \leftrightarrow \nu_\tau$ (K1). The region surrounded by the K2 curves show the 90% CL allowed region by the contained event data from Kamiokande. The results from other experiments (C, Ref. [70]; F, Ref. [63]; I1, I2, and I3, Ref. [67]; B1 and B2, Ref. [68]) show excluded regions at 90% CL.

Neutrino oscillations may also be manifested in the momentum and the zenith-angle distributions of the single-ring events which would reflect a difference in neutrino energy and in path length or matter traversed. Shown in Fig. 51 are the expectations from the Monte-Carlo simulations for the best-fit parameter set and some other parameter sets for $\nu_\mu \leftrightarrow \nu_\tau$, indicated in Fig. 53. One sees that the Monte-Carlo prediction which includes neutrino oscillations reproduces the data quite well.

Comparison of the measured upward-going muon flux with the theoretical expectation also yields a constraint on neutrino-oscillation parameters [66]. The result for the channel $\nu_\mu \leftrightarrow \nu_\tau$ is shown in Fig. 54. Since the measured flux is in good agreement with the theoretical prediction, the 90% CL excluded region (the right-hand side of the curve) indicated by K1 in Fig. 54 is obtained. In the same figure, the constraints from the IMB [67] and Baksan [68] results on upward-going muon fluxes are also shown. If these constraints are taken seriously, the allowed region by the Kamiokande atmospheric neutrino result of contained events is ruled out at 90% CL. However, various ambiguities involved in calculating these constraints from upward-going muon flux measurements [69] make it difficult to draw a definite conclusion.

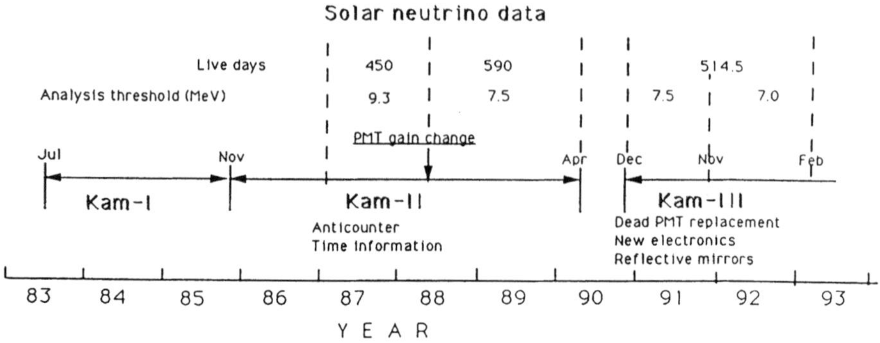

Figure 55: Brief history of the Kamiokande solar-neutrino observations.

6. Investigation of Astrophysical Neutrinos

6.1 Observation of Solar Neutrinos

6.1.1 The Data Samples

The history of the Kamiokande's solar-neutrino observations is schematically shown in Fig. 55. Through the entire observation period, the fiducial volume for the final data sample has been limited to the inner 680 tons.

In the Kamiokande-II solar-neutrino observations [5, 6, 8], the data corresponding to the detector live time of 1040 days were taken in the period January 1987 - April 1990. The first 450 days of the data sample corresponding to the data-taking period up to May, 1988, were analyzed at $E_e \geq 9.3$ MeV (for the trigger threshold, see subsect. 2.2.3). In June, 1988, the PMT gain was increased by a factor of 2, which led to an improved single-photoelectron response of the detector, improved energy resolution, and an improved accuracy of spatial reconstruction. Consequently, the remaining 590 days of the data were analyzed at $E_e \geq 7.5$ MeV.

The data-taking of the new phase, Kamiokande-III, then started in December, 1990. At the time of writing this manuscript the data up to February, 1993, have been analyzed, the corresponding live time being 514.5 days [71]. During this period, the amount of radon dissolved in water had been decreasing further. Because of this, it became possible to lower the analysis threshold from 7.5 MeV to 7 MeV for the data taken after November, 1991. A study of the Kamiokande-III detector performance, however, is still under way. Consequently, the Kamiokande-III data are all preliminary and may be subject to changes when the detector performance is completely understood.

With this rather high threshold energy of the Kamiokande-II and -III experiments, the detectable solar neutrinos are those from the decay of ^{8}B nuclei, which

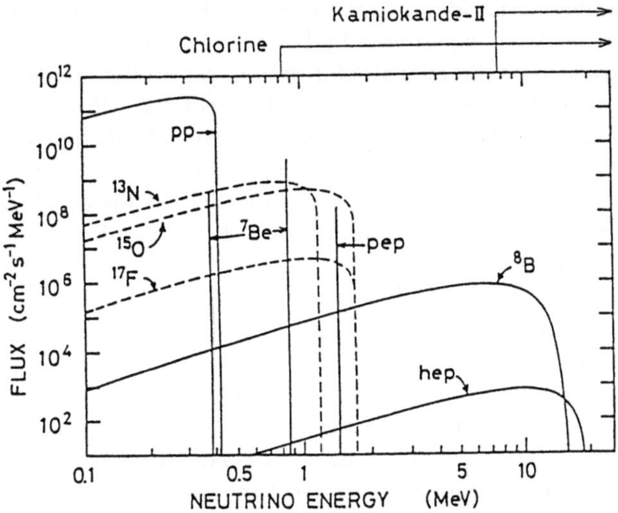

Figure 56: Solid curves showing the energy spectra of solar neutrinos from the pp chain and the dashed curves showing those from the CNO cycle. The fluxes were calculated by Bahcall and Ulrich [26]. For labels other than pp and ^{8}B, see Ref. [26]. Also indicated are the energy regions studied by the Kamiokande-II and chlorine experiments.

are created in the chain of nuclear fusion reactions generating the solar energy. This can be clearly seen in Fig. 56, where the energy spectrum of solar neutrinos from various processes as predicted by the standard solar model calculations by Bahcall and Ulrich [26] is shown. The predicted ^{8}B solar neutrino flux is 5.8×10^6 cm^{-2}s^{-1} with 37% of the theoretical uncertainty (effectively 3 standard deviations), while the flux of the most abundant pp solar neutrinos is 6.0×10^{10} cm^{-2}s^{-1}.

6.1.2 The Solar-Neutrino Signal

Figure 57 (a) shows the $\cos\theta_\odot$ distribution of the candidate solar neutrino events with a recoil electron energy $E_e \geq 9.3$ MeV, where $\theta_\odot$ is an angle between the direction of a recoil electron and that from the Sun to the Earth. As is evident from the figure, there is a prominent signal peak correlated with the direction of the Sun, which is superimposed on the flat background. As a check, the cos $\theta_\odot$ distribution was examined by subdividing the data sample, and by changing the analysis threshold. Figs. 57 (b) and (c) show the angular distributions for the subdivided data corresponding to the data-taking periods before and after the PMT gain change with the same analysis threshold of 9.3 MeV. Fig. 57 (d) shows the angular distribution with an analysis threshold of 7.5 MeV for the data after the PMT gain change. In each of these figures a peak which is directionally correlated with the Sun can be clearly seen.

The existence of the forward peak in the $\cos\theta_\odot$ distribution means that the Sun is "seen" from underground through use of a neutrino. In this sense, the

Kamiokande detector is literally a "neutrino" telescope. It should be noted that this is the unambiguous evidence that the Sun really emits neutrinos, since the first successful solar neutrino experiment by Davis and his collaborators using the ^{37}Cl detector [2] provided no directional information concerning the detected neutrinos.

The Kamiokande observation of the solar neutrinos directly implies that ^{8}B nuclei are produced in the Sun, i.e., the pp-III branch in the pp chain is at work in the Sun. This statement is valid without invoking the energy distribution of the recoil electrons, because solar neutrinos with energies higher than 7.5 MeV and with a sufficiently large flux to be observed in the Kamiokande detector are only produced in the decay of ^{8}B nuclei.

6.1.3 Average Solar-Neutrino Flux

In order to calculate the flux of ^{8}B solar neutrinos, the $\cos\theta_{\odot}$ distribution is fitted by a maximum-likelihood method assuming an isotropic background and a signal which has the shape expected from the β-decay of ^{8}B and convoluted by the detector resolution. This shape of the signal angular distribution is calculated using a Monte-Carlo method for a given analysis threshold of the recoil electron energy. This method implicitly assumes that the observed recoil-electron energy spectrum agrees with that given by the β-decay of ^{8}B. As will be shown in the following subsection, the Kamiokande-II results are consistent with this assumption [5, 6]. The histogram in each of Figs. 57 (a) – (d) represents the shape of the signal expected from a Monte-Carlo calculation in which the ^{8}B solar-neutrino flux predicted by the standard solar model (SSM) calculation by Bahcall and Ulrich [26] is assumed. The flat background level obtained by the fit is also shown.

The solar-neutrino flux deduced from the Kamiokande experiment is usually represented as a value relative to the SSM calculations, such as that given by Bahcall and Ulrich [26]. This representation method is adopted because only the recoil-electron energy spectrum above the analysis threshold is observed. Nothing can be said about the recoil-electron energy spectrum below the threshold. If one assumes, however, that the shape of the recoil-electron energy spectrum is consistent with that given by the β-decay of ^{8}B all the way down to 0 MeV, one can convert the relative flux value to the absolute value.

Assuming a constant ^{8}B solar neutrino flux, the Kamiokande-II result [5, 6] relative to the center value of the Bahcall and Ulrich's SSM prediction [26] is

$$\frac{\text{Data}}{\text{SSM}} = 0.46 \pm 0.05 \pm 0.06,$$

where the first error is statistical and the second error is systematic. The possible sources of the systematic error are uncertainties in the absolute energy calibration, in angular resolution, and in the calculated dead time of various cuts. An uncertainty of 3% in the absolute energy calibration, as described in subsect. 2.4.3, corresponds to 12% of the obtained flux. Also, an uncertainty of 20% in the measurement of the angular resolution, as described in subsect. 2.4.5 corresponds to 7% of the obtained flux. The uncertainty in the dead time of various event-selection criteria is estimated to be 4% of the measured flux. The quadratic sum of these uncertainties is 14% of the obtained flux.

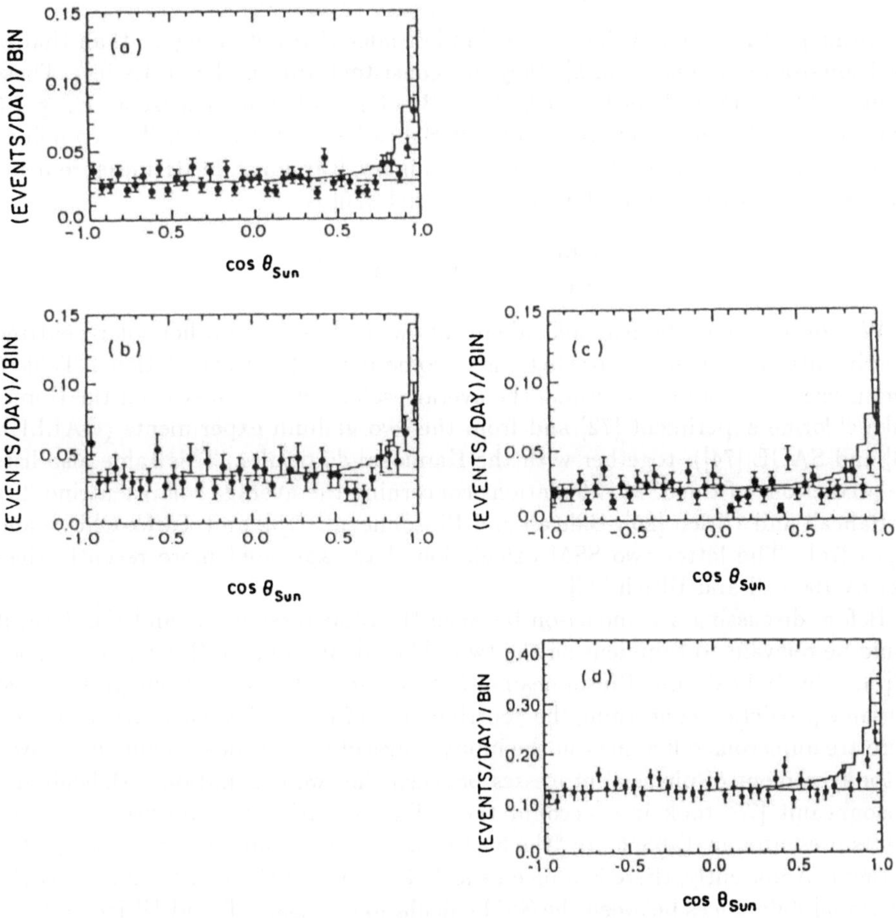

Figure 57: (a) Angular distribution of the finally selected event sample with $E_e \geq 9.3$ MeV in the entire Kamiokande-II observation period, January 1987 through April 1990. Here, $\theta_\odot$ is the angle between the electron direction and the direction from the Sun to the Earth. The histogram shows the angular distribution of the solar neutrino signal expected from a Monte-Carlo simulation in which the solar neutrino flux calculated with the standard solar model of Bahcall and Ulrich [26] is assumed and the angular and energy resolutions of the detector are taken into account. The isotropic background level is shown by the thin solid line. (b) Same as (a) but for the period January 1987 through May 1988, a total of 450 days of live time. (c) Same as (a) but for the period June 1988 through April 1990, a total of 590 days of live time. (d) Same as (c) but for $E_e \geq 7.5$ MeV.

Also, the preliminary Kamiokande-III result from 514.5 days of live time [71], relative to the prediction by Bahcall and Ulrich [26], is

$$\frac{\text{Data}}{\text{SSM}} = 0.56 \pm 0.06 \pm 0.06.$$

Although the central value of the Kamiokande-III result is higher than that of the Kamiokande-II result [5, 6], they are consistent within the statistics. These results indicate that there is a deficit of ^{8}B solar neutrinos at least at $E_e \geq 7.5$ MeV, if one believes the predicted flux of Bahcall and Ulrich [26]. The combined result of the Kamiokande-II and the preliminary Kamiokande-III measurements relative to the prediction by Bahcall and Ulrich [26] is

$$\frac{\text{Data}}{\text{SSM}} = 0.50 \pm 0.04 \pm 0.06.$$

We now compare the Kamiokande results with those from other solar-neutrino experiments, as well as the predictions of some recent SSM calculations. Table 6 summarizes the results concerning the average solar-neutrino flux from the Homestake chlorine experiment [72] and from the two gallium experiments (GALLEX [73] and SAGE [74]), together with the Kamiokande results. This table also lists the predictions of the SSM calculations concerning the average solar-neutrino flux by Bahcall and Ulrich [26], Bahcall and Pinsonneault [75], and Turck-Chièze and Lopes [76]. The latter two SSM calculations have appeared more recently than that by Bahcall and Ulrich [26].

Before discussing a comparison between the SSM predictions and the data, it would be relevant to comment on the two SSM calculations by Bahcall and Ulrich [26] and by Bahcall and Pinsonneault [75]. These SSM calculations give almost the same prediction concerning the neutrino fluxes from varios reactions. However, there are numerous differences in the input values of the physical quantities as well as the treatment of physical processes between the two calculations. Bahcall and Pinsonneault [75] took into account the helium diffusion for the first time; the results presented in Table 6 are "the best results with helium diffusion", as quoted by them. Apparently, there is a nice cancellation among the effects caused by the individual differences between the SSM calculation by Bahcall and Ulrich [26] and that by Bahcall and Pinsonneault [75]. If helium diffusion is not considered, their best standard solar model yields a ^{37}Cl capture rate of 7.2 ± 2.7 SNU, a ^{71}Ga capture rate of 127.5^{+19}_{-16} SNU, and a ^{8}B solar-neutrino flux of 5.06 $\times 10^6$ cm^{-2}s^{-1} [75], where SNU means "solar neutrino unit" defined by 1 SNU = 10^{-36} ν captures per detector atom per second.

Figs. 58 and 59 show a comparison of the experimental data with the SSM calculation by Bahcall and Ulrich [26] and that by Turck-Chièze and Lopes [76], respectively. In these figures, the experimental data relative to the central value of the SSM predictions are plotted. A box associated with each datum point shows the systematic uncertainty; the bars at both ends of the box represent the 1σ statistical error. This plot was made with the following fact in mind: the experimentally determined central value may be shifted over the full range of systematic uncertainty. This plot therefore represents the maximal uncertainty involved in

Table 6: Predictions of three representative standard solar model calculations for the capture rates of ^{37}Cl and ^{71}Ga radiochemical solar neutrino experiments and for the ^{8}B solar neutrino flux. The experimental results are also listed with units of SNU for ^{37}Cl and ^{71}Ga, and $cm^{-2}s^{-1}$ for Kamiokande. The uncertainties in the SSM predictions by Bahcall and Ulrich [26] and by Bahcall and Pinsonneault [75] represent "total theoretical ranges" or "effective 3σ" as the authors quote. They used 3σ errors for measured quantities. On the other hand, Turck-Chièze and Lopes [76] used 1σ error for measured quantities in estimating the uncertainties in their SSM calculations.

		Standard solar models		
		Bahcall & Ulrich	Bahcall & Pinsonneault	Turck-Chièze & Lopes
^{37}Cl	pep	0.2	0.2	0.22
	^{7}Be	1.1	1.2	1.10
	^{8}B	6.1	6.2	4.63
	^{13}N	0.1	0.1	0.063
	^{15}O	0.3	0.3	0.21
	total	7.9 ± 2.6	8.0 ± 3.0	6.4 ± 1.4
	Homestake (1970.8 – 1991.5)		2.28 ± 0.23	
	Homestake (1986.7 – 1991.7)		2.95 ± 0.34	
^{71}Ga	pp	70.8	70.8	71.1
	pep	3.0	3.1	2.99
	^{7}Be	34.3	35.8	30.9
	^{8}B	14.0	13.8	10.77
	^{13}N	3.8	3.0	2.36
	^{15}O	6.1	4.5	3.66
	total	132^{+20}_{-17}	131.5^{+21}_{-17}	123 ± 7
	GALLEX*		$83 \pm 19 \pm 8$	
	SAGE (1990)		$20^{+15}_{-20} \pm 32$	
	SAGE (1991)		$85^{+22}_{-32} \pm 20$	
	SAGE (1990&1991)		$58^{+17}_{-24} \pm 14$	
$\phi(^8B)$	SSM	$(5.8 \pm 2.1) \times 10^6$	$(5.7 \pm 2.5) \times 10^6$	$(4.4 \pm 1.1) \times 10^6$
	Kam-II/SSM	$0.46 \pm 0.05 \pm 0.06$	$0.47 \pm 0.05 \pm 0.06$	$0.61 \pm 0.07 \pm 0.08$
	Kam-III/SSM	$0.56 \pm 0.06 \pm 0.06$	$0.57 \pm 0.06 \pm 0.06$	$0.74 \pm 0.08 \pm 0.08$

*GALLEX Collaboration recently reported [77] that the final result from the 15 runs in the first period (GALLEX-I) of the data taking (May 1991 - April 1992) is $81 \pm 17 \pm 9$ SNU (1σ) and the preliminary result from the first 6 runs of GALLEX-II (August 1992 - February 1993) is $97 \pm 23 \pm 7$ SNU (1σ). The joint result for all 21 runs is also reported to be $87 \pm 14 \pm 7$ SNU (1σ).

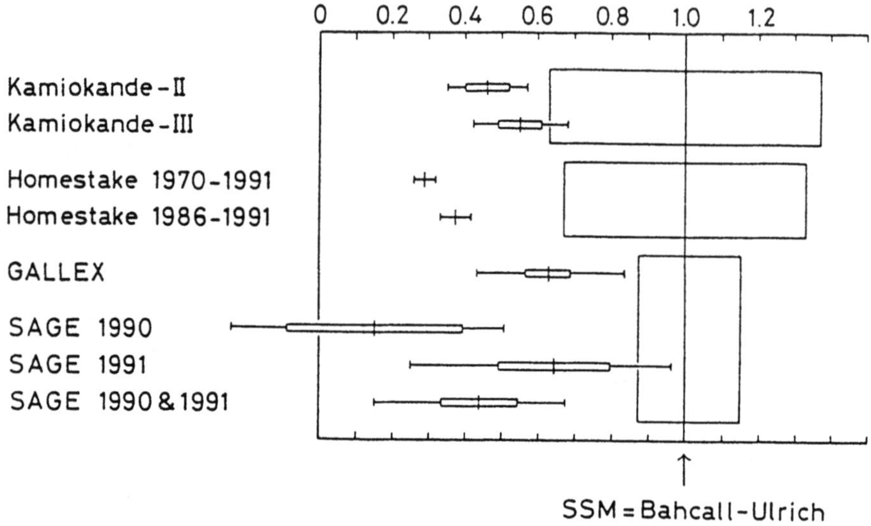

Figure 58: A comparison of the solar-neutrino data with the SSM calculation by Bahcall and Ulrich [26]. The experimental data relative to the central value of the SSM predictions are plotted. The box associated with each datum point shows the systematic uncertainty, and the bars at both ends of the box represent the 1σ statistical error. The uncertainties associated with the theoretical calculation are shown by large uncertainty boxes around 1.0.

the data. Also, the uncertainties associated with the SSM calculations are shown by large uncertainty boxes at around 1.0. The uncertainty boxes in Fig. 58 represent the "total theoretical range" or "effective 3σ" as Bahcall and Ulrich quote [26]. For the measured quantities they used 3σ errors. The uncertainty boxes in Fig. 59 represent the uncertainties which Turck-Chièze and Lopes quote [76]. They used 1σ errors for the measured quantities. In their calculation, they consider any theoretical error as a minimal error (absence of error on the assumptions of the calculation) and are only indicative of the present status and of the occurring improvements. This is quite different from the case of Bahcall and Ulrich. In any case, since the uncertainties in the theoretical quantities do not have a statistical nature, it may be dangerous to interpret the total uncertainty as being 1σ or 3σ.

It is quite a distinctive feature that all of the results concerning the average flux from the present solar-neutrino observations are less than the SSM predictions. Nevertheless, it seems that the average solar-neutrino fluxes observed by Kamiokande, GALLEX, and SAGE (1991 or 1990 & 1991 combined) are all not very significantly far away from the SSM predictions, considering the rather large experimental and theoretical uncertainties. Only the Homestake result seems to be very difficult to reconcile with the SSM prediction, if their possible systematic error is not very large.

The average solar neutrino capture rate observed by the chlorine experiment over the period 1970 $-$ 1988 is 2.3 ± 0.3 SNU. Referring to the SSM calculations

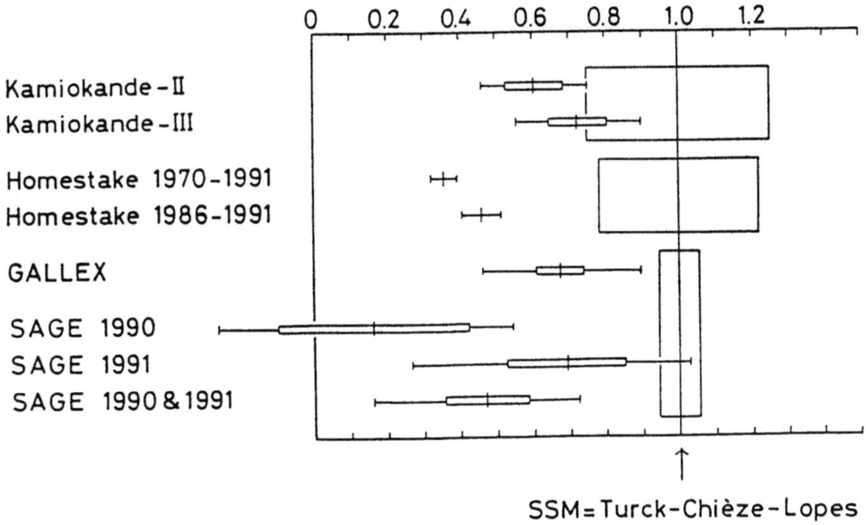

Figure 59: Same as in Fig. 58, but the data are compared with the SSM calculation by Turck-Chièze and Lopes [76].

by Bahcall and Ulrich [26], this result is expressed as

$$\frac{\text{Cl Data}}{\text{SSM}} = 0.29 \pm 0.04.$$

The chlorine experiment has a lower detection threshold than does the Kamiokande νe scattreing experiment (see Fig. 56), but to the chlorine experimentthe, dominant contribution comes from the ^{8}B solar neutrinos (see Table 6). Therefore, by taking the face values, the Homestake and Kamiokande results are mutually inconsistent if one assumes standard neutrinos. Namely, with the reduction factor of the ^{8}B solar-neutrino flux as determined from the Kamiokande result, the Homestake ^{37}Cl capture rate would be oversaturated, and there would be no room to accommodate the ^{7}Be solar neutrinos. This makes it extremely difficult to consistently explain all of the solar-neutrino observations in the context of solar physics.

In this regard, the MSW mechanism in the Sun is very attractive for explaining all of the experimental data concerning the average solar-neutrino flux consistently. Figure 60 is taken from Ref. [78] and shows the 90% CL allowed region in the neutrino-oscillation parameter space for the Homestake result (average capture rate of 2.1 ± 0.3 SNU is assumed), for the combined Kamiokande-II and -III results, and for the combined GALLEX and SAGE 1990 and 1991 results (the authors use an average ^{71}Ga capture rate of 71 ± 15 SNU), in the two-flavor mixing scheme. The reference solar neutrino flux is taken to be the central value of the Bahcall and Pinsonneault SSM calculation [75]. The shaded regions in this figure are allowed at the 90% CL in the combined MSW fit. There are two allowed regions, one being $\Delta m^2 = (0.3 - 1.2) \times 10^{-5}$ eV^2 and $\sin^2 2\theta = (0.4 - 1.5) \times 10^{-2}$ (non-adiabatic solution) and the other being $\Delta m^2 = (0.3 - 3) \times 10^{-5}$ eV^2 and

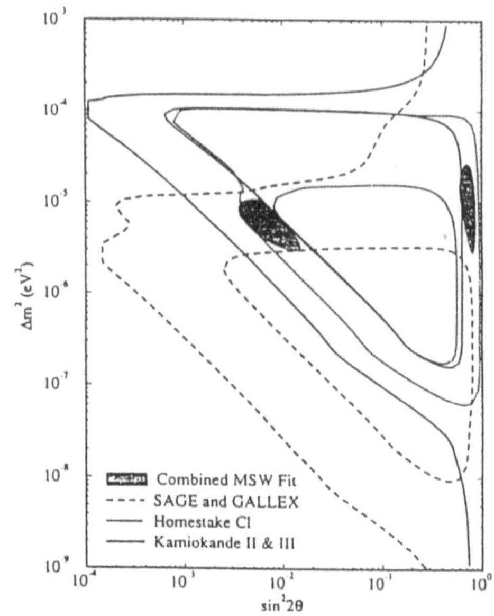

Figure 60: The 90% CL allowed regions for the Homestake, Kamiokande, and the combined GALLEX and SAGE results in the MSW scheme. The SSM prediction for the solar-neutrino spectra are taken from the work by Bahcall and Pinsonneault [75]. The shaded regions are the solutions at the 90% CL from a combined fit of the four experiments.

$\sin^2 2\theta = (0.6 - 0.9)$ (large mixing angle solution).

6.1.4 Recoil-Electron Energy Spectrum

The energy spectrum of the recoil electrons from νe scattering at $E_e \geq 7.5$ MeV [5, 6] is shown in Fig. 61. This spectrum was obtained by subdividing the data into 0.5 MeV bins according to the value of E_e and performing a maximum-likelihood fit in each E_e bin independently. The histogram in this figure shows the energy spectrum expected from a Monte-Carlo calculation in which the SSM prediction by Bahcall and Ulrich [26] is assumed. The shape of this spectrum is determined by the β-decay spectrum of ^{8}B, the cross section of the νe scattering, and the detector response. The ratio of the observed-to-the-expected event rate is shown in Fig. 61(b). Evidently, the shapes of the observed and expected recoil-electron energy spectra are in agreement within the statistical uncertainties.

The consistency of the shape of the recoil-electron energy spectrum with those expected from ^{8}B decay constitutes further evidence for the pp-III branch at work in the Sun. It also implies that there is no appreciable distortion in the energy spectrum of the ^{8}B solar neutrinos at $E_e \geq 7.5$ MeV.

In the MSW scheme, the solar-neutrino energy spectrum receives a distortion which depends on the values of Δm^2 and $\sin^2 2\theta$. In particular, a high-energy side of the ^{8}B solar-neutrino spectrum, to which the Kamiokande experiment is

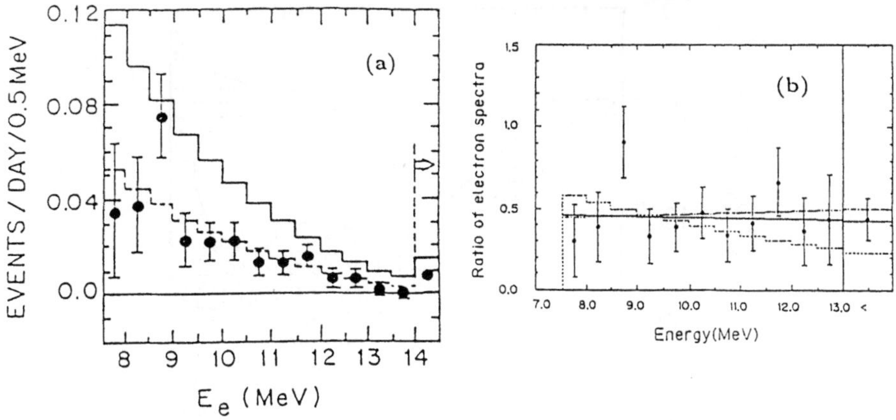

Figure 61: (a) Differential energy distribution of the recoil electrons from νe elestic scattering. The last bin corresponds to E_e = 14 MeV. The solid histogram shows the area and shape of the distribution predicted by the SSM [26]. The dashed histogram is the best fit (0.46 × SSM) to the data. (b) The points show the ratio of the Kamiokande-II differential recoil-electron energy spectrum to the spectrum expected from SSM [26]. The lines are the distorted electron spectra due to neutrino oscillations with representative parameters ($\sin^2 2\theta$, Δm^2): (6.3×10^{-1}, 10^{-4}) for the solid line, (10^{-2}, 3.2×10^{-6}) for the dot-dashed line, and (2×10^{-3}, 1.4×10^{-4}) for the dashed line.

sensitive, is suppressed if the parameters lie in the region of the "adiabatic solution" of the Homestake result, *i.e.*, $\Delta m^2 \sim 10^{-4}$ eV^2 and $\sin^2 2\theta \gtrsim 10^{-4}$. It is therefore expected that the Kamiokande result on the average solar-neutrino flux together with that on the recoil-electron energy spectrum can constrain the allowed region in the neutrino-oscillation parameter space at around the adiabatic solution, without invoking the results from other experiments.

In Fig. 61 (b) the distorted recoil-electron energy spectra due to the MSW effect are shown relative to the expectation of the SSM calculation by Bahcall and Ulrich [26] for three representative pairs of parameters ($\sin^2 2\theta$, Δm^2). Figure 62 shows the 90% CL allowed regions of the MSW solutions obtained by using both the average solar-neutrino flux and the recoil-electron energy spectrum [7, 8]. Here, the regeneratin of ν_e by the Earth is taken into account. As expected, the adiabatic region, $\Delta m^2 \sim 1.3 \times 10^{-4}$ eV^2 and $\sin^2 2\theta = 7.2 \times 10^{-4} \sim 6.3 \times 10^{-3}$ is disfavored at the 90% CL.

6.1.5 Search for Day-Night and Semiannual Flux Variations

If the neutrino-oscillation parameters lie in the relevant regions, a regeneration of ν_e by the Earth through the MSW effect would cause an observable difference between the daytime and night-time solar neutrino fluxes, as well as a small seasonal variation in the observed flux. In addition, if the neutrino has a magnetic moment (either direct or transition) of the order of $10^{-10}\mu_B$, where μ_B is the Bohr magneton, the solar neutrino flux might vary semiannually.

To test these ideas, the Kamiokande-II data were divided into: (i) daytime and night-time samples, (ii) Spring (4 February – 5 May), Summer (6 May – 4

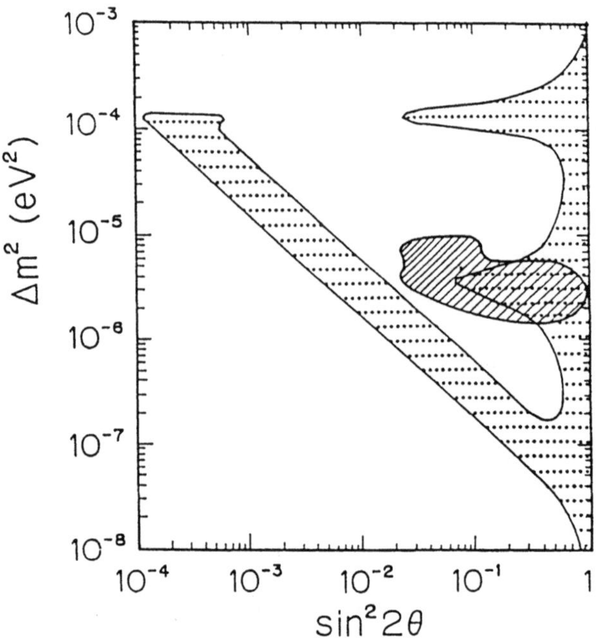

Figure 62: Dotted region showing the 90% CL allowed region in the Δm^2 - $\sin^2 2\theta$ space obtained from the total flux and the recoil-electron energy spectrum, measured in the Kamiokande-II detector. The crosshatched region shows the 90% CL excluded region by the null result of the day-night flux difference.

August), Fall (5 August – 3 November), and Winter (4 November – 3 February) samples, and (iii) period I (22 April – 21 July and 21 October – 20 January) and period II (21 January – 21 April and 22 July – 20 October) samples. The flux was determined for each data sample. The daytime is defined as the period during which the Sun is above the horizon, and *vice versa.* Period I corresponds to the time interval in which the Earth is near to the intersection of the ecliptic plane with the solar equatorial plane. At these times the Kamiokande detector views the core of the Sun through the solar equator, where the magnetic field is expected to be weaker than at a higher solar latitude. Therefore, for solar neutrinos pointing to the Earth, the strength of the interaction of the neutrino magnetic moment with the solar magnetic field would be less at those times. Period II, on the other hand, corresponds to the time interval in which the Earth is far from the intersection of the ecliptic plane with the solar equatorial plane.

The results [8] are shown in Fig. 63. In obtaining these results, a correction for the small variation ($\lesssim$ 6%) of the flux induced by the eccentricity of the Earth's orbit was made when it was necessary. These results indicate no statistically significant flux variation in any of the studied data samples. Specifically, the daytime flux was 0.91 ± 0.15 and the nighttime flux was 1.07 ± 0.16, both relative to the the average value. The indicated errors are only statistical. The relative

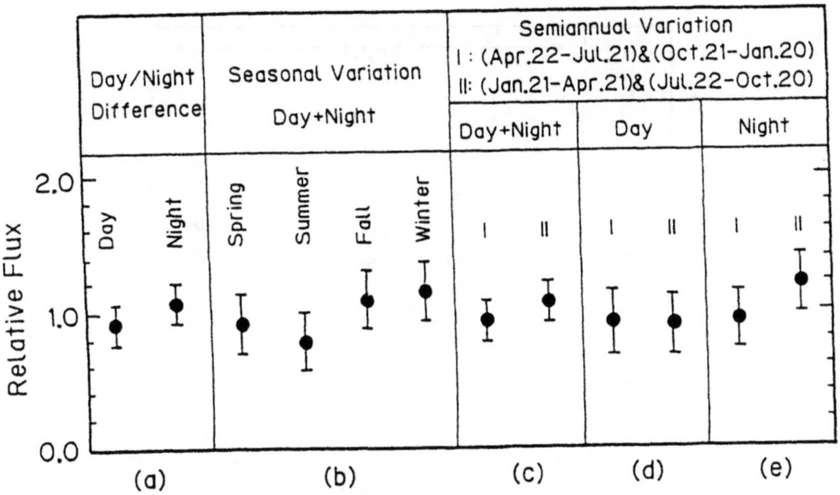

Figure 63: Measured solar-neutrino fluxes relative to the averaged value: (a) daytime, night-time; (b) spring, summer, fall, winter; (c) periods I and II; (d) periods I and II, daytime; (e) periods I and II, night-time.

difference between the daytime and night-time fluxes is erxpressed by

$$\frac{\text{day} - \text{night}}{\text{day} + \text{night}} = -0.08 \pm 0.11 \pm 0.03,$$

where the first error is statistical and the second error is systematic. Also, the relative difference between period I and period II is expressed by

$$\frac{\text{period I} - \text{period II}}{\text{period I} + \text{period II}} = -0.06 \pm 0.11 \pm 0.02.$$

The data concerning the day-night difference can be used to constrain the neutrino-oscillation parameters. In doing so, more information can be extracted from the data by dividing the total sample into subsamples with different path lengths and density profiles through the Earth. For this purpose, the data were divided into six subsamples according to the values of $\cos \delta_{\text{Sun}}$. Here, the angle δ_{Sun} is that of the Sun relative to the detector coordinate system with the definition that $\delta_{\text{Sun}} = 0$ corresponds to the direction in which the Sun is just below the detector. Thus, the daytime sample has $\cos \delta_{\text{Sun}} < 0$. Figure 64 shows the relative fluxes obtained from subsamples of the data [8]. A comparison of the data and calculations with the MSW effect was performed based on the data plotted in this figure; the calculated fluxes with some neutrino-oscillation parameters are indicated for illustrative purposes. Since the data show no significant day-night variation, this analysis yielded the excluded parameter region without any assumption on the absolute value of the expected solar neutrino flux. The result [8] is shown as the crosshatched region in Fig. 62, which is excluded at the 90% CL.

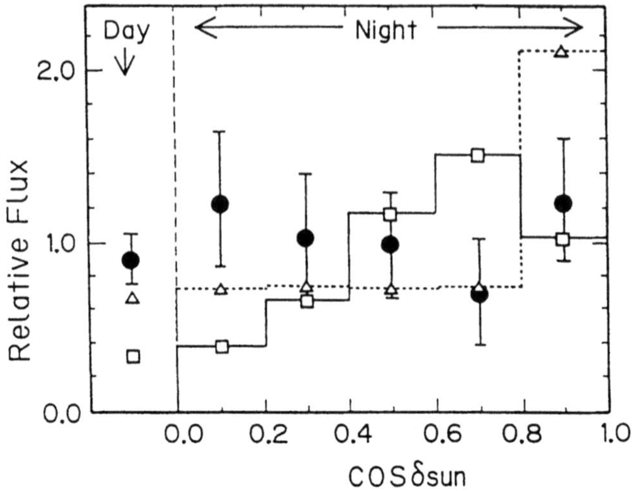

Figure 64: Measured solar-neutrino fluxes of daytime and subdivided night-time relative to the averaged value (solid circles). The horizontal axis ($\cos\delta_{\mathrm{Sun}}$) is the cosine of the zenith angle of the Sun relative to the z axis of the detector. $\delta_{\mathrm{Sun}} = 0$ corresponds to the direction in which the Sun is just below the detector. The solid histogram with open squares is the flux calculated for $\Delta m^2 = 3.5 \times 10^{-6}$ eV^2 and $\sin^2 2\theta = 0.11$; the dashed histogram with open triangles is for $\Delta m^2 = 7.9 \times 10^{-6}$ eV^2 and $\sin^2 2\theta = 0.05$.

6.1.6 Is the Solar-Neutrino Flux Related with the Sunspot Numbers?

It is often claimed that the solar neutrino flux observed at the Homestake chlorine experiment [2] shows a time variation that is anticorrelated with the sunspot numbers. The only natural explanation for such an effect is provided by the unusually large neutrino's magnetic moment [79, 80, 81]. Should this be the case, a semiannual flux time variation would also be expected, as already noted in the preceding subsection.

Figure 65 shows the observed solar neutrino flux as a function of time [82]. Both the Kamiokande-II and -III results are plotted. Each point corresponds to about 200 days of live time. The Kamiokande results do not show any statistically significant flux time variation over the recent six years, during which period the sunspot numbers increased from a minimum to a maximum, as shown in Fig. 66 [83]. Specifically, the Kamiokande results do not show any evidence for a statistically significant correlation or anticorrelation between the solar-neutrino flux and the sunspot numbers, as is evident from Fig. 67, where the relative flux value is plotted against the sunspot numbers. Indeed, a fit with a linear relation between the relative flux value and the sunspot numbers is quite consistent with the constant flux. At the 90% CL

$$\frac{|\,\mathrm{flux(SSN}=0) - \mathrm{flux(SSN}=160)\,|}{\mathrm{flux(SSN}=0)} < 39.5\%,$$

as shown by the dashed lines in Fig. 67.

It should be noted, however, that the Kamiokande results do not preclude a

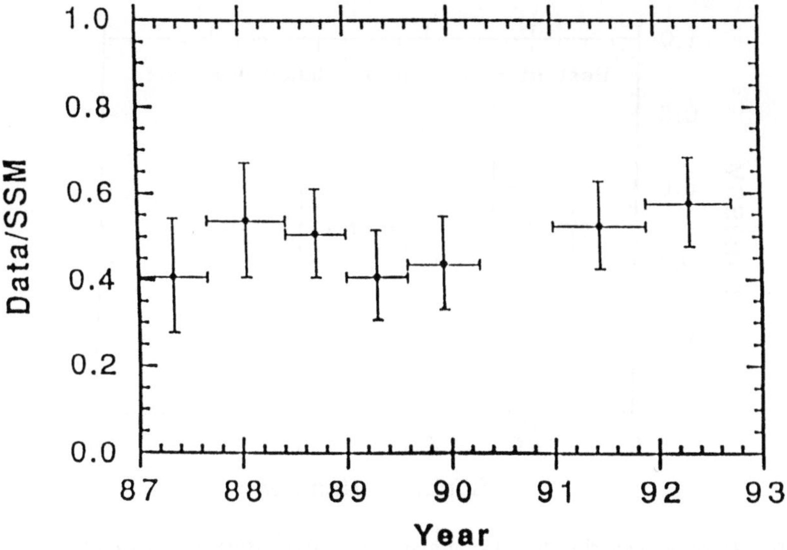

Figure 65: Time dependence of the solar neutrino flux relative to the standard solar model prediction. Both the Kamiokande-II and -III results are plotted.

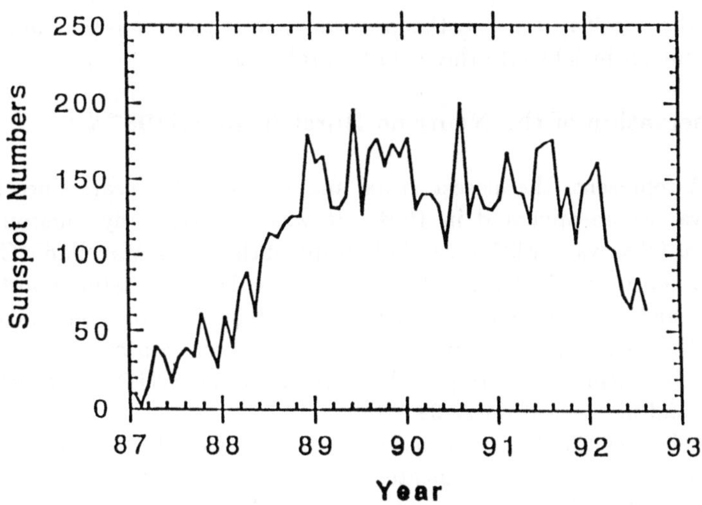

Figure 66: Sunspot numbers as a function of time.

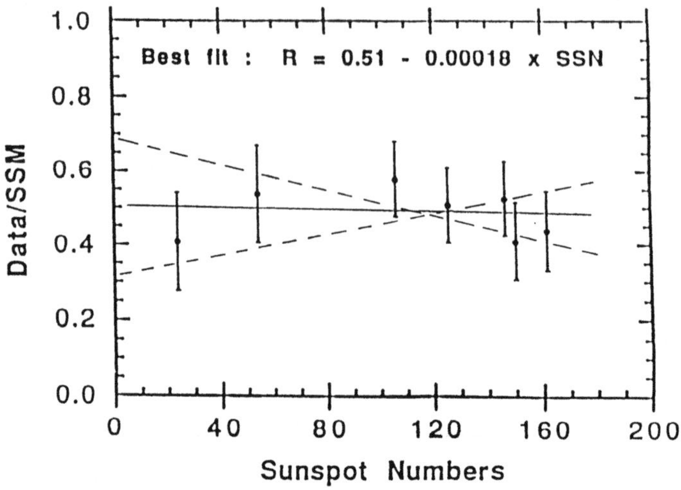

Figure 67: Solar neutrino flux vs. sunspot numbers. Both the Kamiokande-II and -III results are plotted. The solid line shows the best-fit result when a linear relation is assumed between the flux and the sunspot numbers. The dashed lines show the time variation allowed at the 90% CL.

flux time variation that is either correlated or anticorrelated with the sunspot numbers within the limit quoted above. Also note that the Kamiokande results do not constrain the periodic flux time variations with a period that is much shorter than 200 days, if any. Such a possibility might arise from twisting toroidal magnetic fields in the Sun [84]. Only forthcoming high-statistics solar-neutrino experiments will provide an insight into these subtle problems.

6.2 Observation of the Neutrino Burst from SN1987A

SN1987A represents the closest visual supernova in 383 years since the Kepler's supernova, which appeared in 1604. It was discovered by Shelton at the Las Campanas Observatory [85] on a photograph of the Large Magellanic Cloud (LMC) taken on February 24.23 UT, 1987. McNaught, however, later reported [86] that SN1987A had already been there at magnitude 6.5 in a photograph taken on 23.443 UT. Further useful information was provided by an amateur astronomer, Jones, through his routine monitoring of the LMC; he found SN1987A on February 24.37 UT, but had not seen it on 23.39 UT.

On February 25, the presence of hydrogen was observed in the spectra of SN1987A [87]. This feature indicates that SN1987A is a type-II supernova caused by the gravitational collapse of a massive star. According to supernova theory, this means the birth of a neutron star which is accompanied by the emission of $\sim$ 99% of the released gravitational energy in the form of neutrinos within a few seconds. Thus, the neutrino burst signals the time of core collapse, which is earlier than the time of brightning by 1 $\sim$ 10 hours, depending on the size of the progenitor.

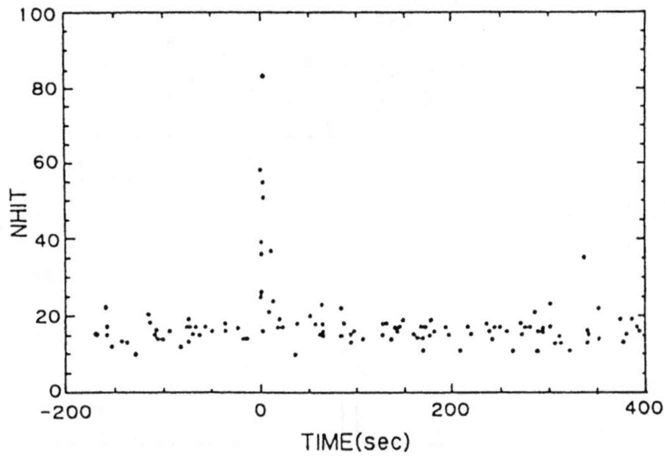

Figure 68: Scatter plot of N_{hit} against time relative to 7:35:35 UT, February 23, 1987, for low-energy electron events observed in Kamiokande-II during a 600-sec interval. A burst of events is evident at 7:35:35 UT.

After having been informed of the optical observation of SN1987A, a search was made for a burst of low-energy electron events in the data tapes recorded by the Kamiokande-II detector.† The fiducial volume was taken to be 2,140 tons throughout this analysis. By inspecting the scatter plot of N_{hit} (the number of hit PMTs) against time, a burst of events was found on February 23 at 7:35:35 UT (±1 min), as shown in Fig. 68‡

The time sequence of events in a 45-sec interval centered at the burst time is shown in Fig. 69, where muon events preceding the burst of events are also shown. A total of 11 electron events were observed during 12.5 sec.† The probability that any of these preceding muons caused spallation, which led to the multiple low-energy events, is negligibly low because the estimated probability for an incident muon to cause low-energy electron events with multiplicity ≥ 3 is $\sim 3 \times 10^{-3}$. No other burst candidates of low-energy events were observed over a longer period including February 23. In particular, there was no evidence of a neutrino burst on February

†At that time, no attempt was made for a real-time analysis of the data, and the recorded data tapes were analyzed only after having been brought to Tokyo. It should be mentioned, however, that an online semi-realtime analysis has been implemented at present to detect a burst of events within ~20 min from its occurrence.

‡The uncertainty of ±1 min associated with the absolute time of the beginning of the neutrino burst in Kamiokande arose from the absence of an absolute time calibration source in the Kamiokande equipment at that time. Though a precise relative time was assigned to each event by the clock in the event-time circuit, a power failure which took place in the Kamioka mine on February 25 precluded an absolute calibration of the clock after the detection of the neutrino burst in the off-line analysis.

†Event 6 in Figs. 69 and 72 and Table 7 has an energy less than the analysis threshold of 7.5 MeV, and is consistent with background. Therefore, it is disregarded.

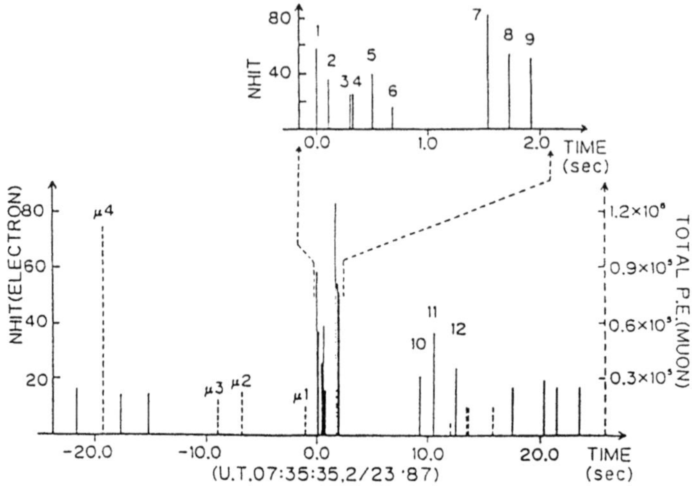

Figure 69: Time sequence of events in a 45-sec interval centered on 7:35:35 UT, February 23, 1987. The vertical height of each line represents the relative energy of the event. The solid lines represent low-energy electron events in units of N_{hit} (left-hand scale). The dashed lines represent muon events in units of the number of photoelectrons (right-hand scale). Events $\mu 1$ - $\mu 4$ are muon events which preceded the burst of electron events at time zero. The upper-right figure is the $0 \sim 2$-sec time interval on an expanded scale.

23 around 2:52 when the LSD liquid scintillation detector at Mont Blanc observed 5 events in 7 sec [88].

Having been informed of Kamiokande's burst time, the IMB Collaboration [89] also found a burst of 8 events within 6 sec at essentially the same time as that of Kamioka, but it did not find any burst of events in coincidence with the Mont Blanc events. The early optical observations of SN1987A shown in Fig. 70 is consistent with theoretical models for the supernova explosion of the progenitor, Sk -69202, having a known radius of $(3 \pm 1) \times 10^{12}$ cm, with the time of core collapse defined by the Kamiokande observation of the neutrino burst. The time of the Mont Blanc event is inconsistent with the observed light curve.

The properties of the 11 events observed in Kamiokande are summarized in Table 7, where an additional one event (Event 6) is also included. Two independent reconstruction programs were used to determine these properties. Within the known error in reconstruction, the results obtained from the two programs are in excellent agreement, except for the electron angle in Event 2 with respect to the LMC. Though one reconstruction gives $\theta_{LMC} = 15° \pm 27°$ [3], the other gives $\theta_{LMC} = 40° \pm 27°$ [4], where θ_{LMC} is the angle between the measured electron direction and the direction from the LMC to the Earth. This result lessons the likelihood that Event 2 originated from νe scattering. With this exception, the results listed in Table 7 are from the first reconstruction program [3]. Figure 71 shows two examples of the observed patterns of hit PMT.

Figure 72 shows a scatter plot of E_e *vs.* $\cos\theta_{LMC}$ and a projection on each axis

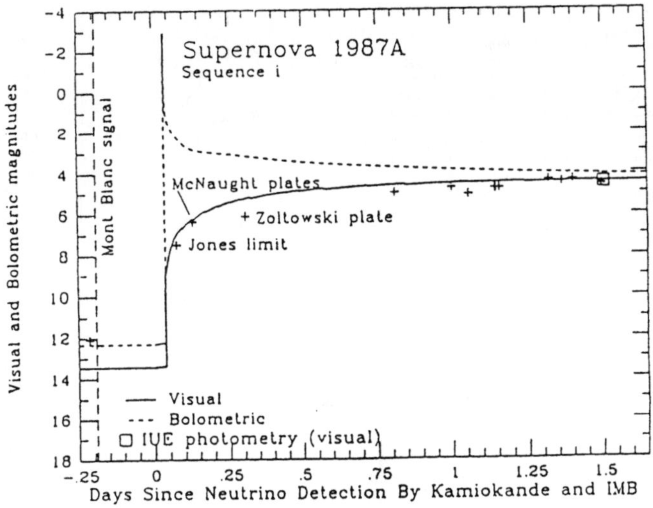

Figure 70: Visual-magnitude light curve of SN1987A during its first two days, with the time of core collapse defined by the neutrino signal from Kamiokande and IMB (February 23.316 UT). The time of the Mont Blanc event is indicated by the vertical dashed line. The curves show the results of theoretical calculations by Arnett. The solid curve, the visual magnitude; and the dashed curve, the bolometric luminosity. This figure is taken from Ref. [90]. For the references to the data points and theoretical curves, see Ref. [90].

(for Event 2, θ_{LMC} is taken to be 15°). The angular distribution is not inconsistent with the isotropic distribution of the electrons relative to the LMC. This is an expected feature of the supernova neutrino detection by a water-Cherenkov detetor, since the dominant contribution results from the reaction $\bar{\nu}_e + p \to e^+ + n$ with positrons emitted almost isotropically. The electron angle of the first event, $\theta_{LMC} = 18° \pm 18°$, should be particularly noted. For this event, the second reconstruction program yields $\theta_{LMC} = 10° \pm 18°$ [4]. Therefore, the first event is consistent with νe scattering. Note that the total νe scattering events expected in Kamiokande for 10 $\bar{\nu}_e$ absorption is $\lesssim 0.5$ [25][91]. However, it is unlikely that this first event is caused by the neutronization burst, since more than a one-order-of-magnitude smaller number of νe scattering events is expected from the neutronization burst.

The time structure of the observed 11 events is noteworthy. Three clusters of events can be recognized: five events within the first $\sim$ 0.5 sec, the next three events between 1.5 and 2 sec, and the last three events after 9 sec with respect to the detection time of the first event. Whether such a time structure bears any significance causes some discussion; however, the number of the observed events

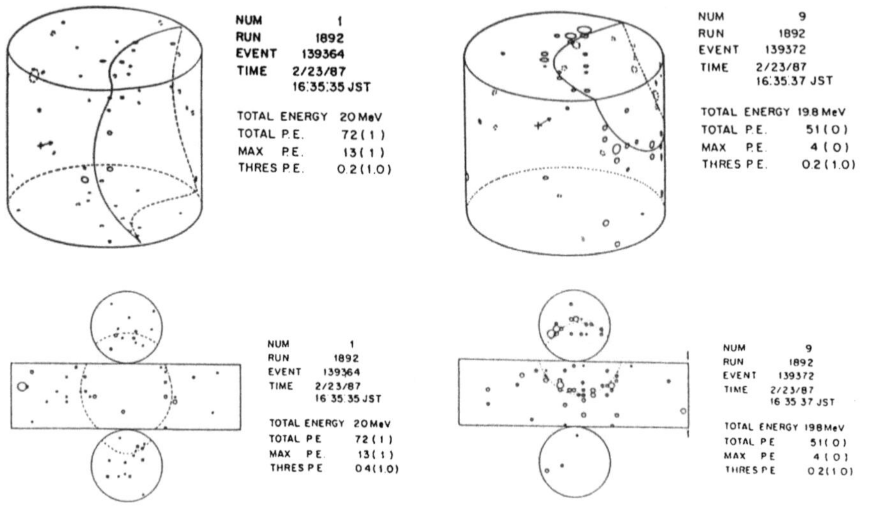

(a) (b)

Figure 71: (a) Above: three-dimensional reproduction of Event 1 listed in Table 7. Bottom: exploded view of Event 1. Each small circle represents a struck PMT with the diameter proportional to the charge output of the PMT. The numbers in parentheses refer to the quantities measured in the anticounter. (b) Same as (a) but for Event 9.

Table 7: Measured properties of the 11 electron events above a threshold of 7.5 MeV and one additional event below 7.5 MeV detected in the neutrino burst. The electron angle in the last column is relative to the direction of SN1987A. The errors quoted are one standard deviation Gaussian errors.

Event number	Event time (sec)	Number of PMT ($N_{\rm hit}$)	Electron energy (MeV)	Electron angle (deg)
1	0	58	20.0 ± 2.9	18 ± 18
2	0.107	36	13.5 ± 3.2	15 ± 27[a]
				40 ± 27[b]
3	0.303	25	7.5 ± 2.0	108 ± 32
4	0.324	26	9.2 ± 2.7	70 ± 30
5	0.507	39	12.8 ± 2.9	135 ± 23
6	0.686	16	6.3 ± 1.7	68 ± 77
7	1.541	83	35.4 ± 8.0	32 ± 16
8	1.728	54	21.0 ± 4.2	30 ± 18
9	1.915	51	19.8 ± 3.2	38 ± 22
10	9.219	21	8.6 ± 2.7	122 ± 30
11	10.433	37	13.0 ± 2.6	49 ± 26
12	12.439	24	8.9 ± 1.9	91 ± 39

[a] Ref. [3], seetext. [b] Ref. [4], see text.

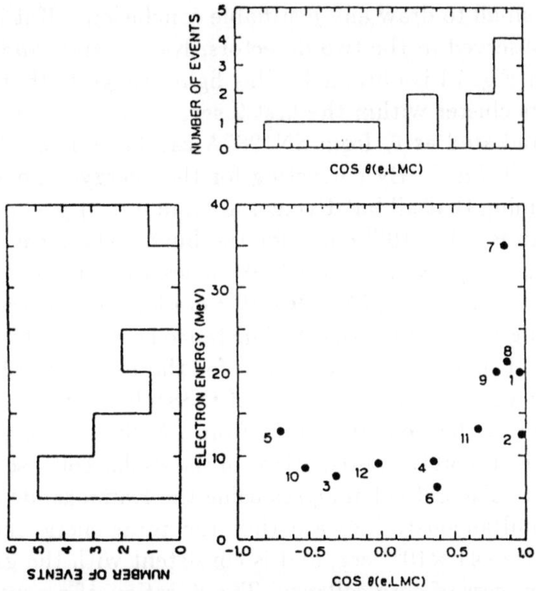

Figure 72: (a) Scatter plot of the detected electron energy and cosine of the angle between the measured electron direction and the direction from the Large Magellanic Cloud to the Earth. The number associated with each entry corresponds to the time-sequential event number shown in Table 7. (b) and (c) Projections of the scatter plot.

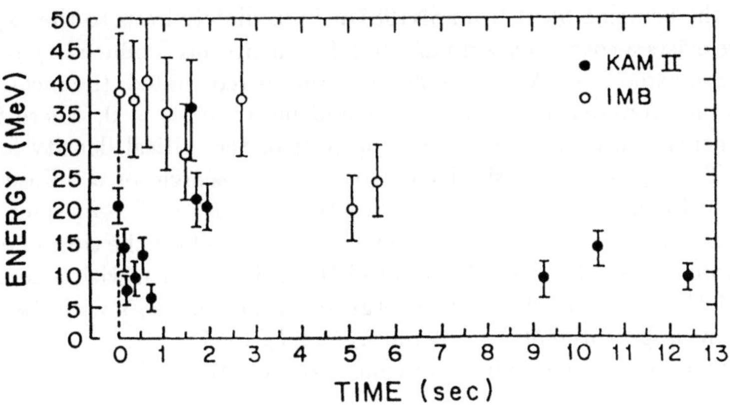

Figure 73: Scatter plot of the energy and time of the 12 events (including Event 6 in Table 7) in the burst sample observed in Kamiokande-II, and the 8 events in the burst sample observed in the IMB detector. The earliest event in the sample of each detector has, arbitrarily, but not unreasonably, been assigned $t = 0$.

seems to be too small to draw any justifiable conclusions. If it is assumed that the earliest events observed in the two detectors, Kamiokande and IMB, coincided in time, the plot in Fig. 73 is obtained. This figure suggests that 13 events out of a total of 19 events cluster within the first 2 sec.

The energy radiated as $\bar{\nu}_e$ from SN1987A can be estimated using the electron energies given in Table 7. By correcting for the energy-dependent detection efficiency, and assuming that all but Event 1 are due to $\bar{\nu}_e + p \rightarrow e^+ + n$, the resultant integrated $\bar{\nu}_e$ flux is 1.1×10^{10} cm^{-2} for the burst, where the $\bar{\nu}_e$ energy given by $E_{\bar{\nu}_e} = E_{e^+} + (m_n - m_p) = E_{e^+} + 1.3$ MeV, is greater than 8.8 MeV. This in turn leads to a $\bar{\nu}_e$ output for SN1987A of 9×10^{52} erg for an observed average energy of 15 MeV, assuming that the Earth-LMC distance is 55 kpc. This enormous energy is both direct and unambiguous evidence for the explosion of SN1987A to have been induced by a gravitational collapse of the stellar core.

The detection of the neutrino burst from SN1987A brought about profound consequences. The theoretical predictions of the stellar collapse has been proven to be largely correct. Theoretical analyses using the Kamiokande and IMB neutrino-burst events simultaneously indicate that the total energy emitted by the hot neutron star is $(2 \sim 4) \times 10^{53}$ erg, and is consistent with the gravitational energy released in the process of core collapse. The duration of a neutrino burst of ~ 10 sec and an average antineutrino energy of $< E_{\bar{\nu}_e} > \sim 13$ MeV are consistent with the cooling of a proto-neutron star. Significant results were also obtained concerning the neutrino properties: for example, $m_{\bar{\nu}_e} \lesssim 15 \sim 20$ eV and $\gamma\tau_{\bar{\nu}_e} \gtrsim 1.6 \times 10^5$ yr, where $\tau_{\bar{\nu}_e}$ is the lifetime of $\bar{\nu}_e$ and $\gamma = 1/\sqrt{1-\beta^2}$. For more about the information obtained by the observation of the neutrino burst from SN1987A, see Ref. [90].

6.3 Search for Diffuse Neutrinos

The correctness of the theory for a collapse-driven supernovae, proved by the observation of the neutrino burst from SN1987A, immediately leads to the expectation that there exists a cosmic background of diffuse neutrinos produced by past Type-II supernova explosions. Although the spectrum is red-shifted, the mean neutrino energy is only lowered by 0.6 if $\Omega_0 = 1$ and by 0.5 if $\Omega_0 = 0$, where Ω_0 is the present energy density of the universe in units of the critical density for closure [92, 93]. This opens the possibility of a meaningful search for a diffuse neutrino background by Kamiokande using the reaction $\bar{\nu}_e + p \rightarrow e^+ + n$, since even the red-shifted $\bar{\nu}_e$ spectrum has a reasonable high-energy tail (at the time of explosion, the spectrum is well described by a Fermi-Dirac distribution with a temperature $T_F \sim 4$ MeV). Figure 74 shows the energy spectra of relic $\bar{\nu}_e$'s with $T_F = 4$ MeV and a chemical potential of $\mu_F = 0$ and $\mu_F = 2T_F$ for $\Omega_0 = 1$.

The total flux of these diffuse neutrinos can be estimated by

$$F_{\bar{\nu}_e} = n_G(t_0)\, R_{SN}\, c(t_0 - t_s) \frac{L_{\bar{\nu}_e}}{< E_{\bar{\nu}_e} >}, \qquad (20)$$

where t_0 is the present time; t_s the initial time of the occurrence of the stellar collapse; $n_G(t_0)$ the present number density of spiral galaxies; R_{SN} the average core-collapse supernova rate per year per galaxy, which is assumed to be constant

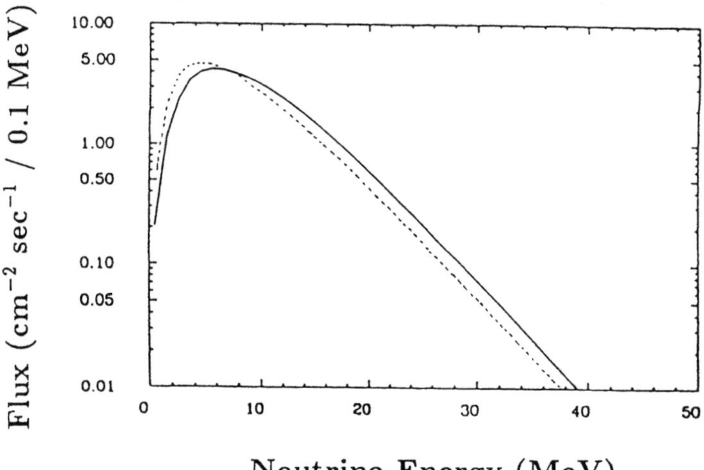

Figure 74: Red-shifted energy spectra of relic $\bar{\nu}_e$'s with $T_F = 4$ MeV and chemical potential $\mu_F = 0$ and $\mu_F = 2T_F$ for $\Omega_0 = 1$. The total flux is normalized to 50 cm^{-2} s^{-1}.

in time; $L_{\bar{\nu}_e}$ the total energy; and $< E_{\bar{\nu}_e} >$ the average energy of the emitted $\bar{\nu}_e$'s. Although R_{SN} is highly uncertain, we take it to be 1/(20 yr). Putting into Eq. (20) $t_0 - t_s = 10^{10}$ yr, $n_G(t_0) = 1/(50\text{Mpc}^3)$ [94], $L_{\bar{\nu}_e} = \frac{1}{6} \times 3 \times 10^{53}$ erg, and $< E_{\bar{\nu}_e} > = 13$ MeV, we obtain $F_{\bar{\nu}_e} \sim 24\,\text{cm}^{-2}\,\text{s}^{-1}$. A more optimistic estimate of $F_{\bar{\nu}_e} \sim 250\,\text{cm}^{-2}\,\text{s}^{-1}$ is given by the argument of massive star formation during the era of galaxy formation [93]. An unambiguous detection of diffuse neutrinos, in turn, would provide valuable insight into the early evolution of the universe.

For the detection of these diffuse neutrinos, the background sources are solar neutrinos and atmospheric neutrinos. There is a relatively narrow window above the energy region of the solar neutrinos, in which the atmospheric neutrino flux is expected to be less than the signal flux. A search for diffuse neutrinos should be performed within this energy window, $20 \lesssim E_e \lesssim 50$ MeV [21].† In the 680-ton fiducial volume, 2 events were found in this window before increasing the PMT gain (see Fig. 75). After the PMT gain change, no candidate events were found.

Considering the detection efficiency (see Fig. 76), which is estimated from a Monte-Carlo simulation of the detector response to the relic $\bar{\nu}_e$'s with red-shifted energy spectrum, these 2 events place the 90% CL upper limit of the total relic $\bar{\nu}_e$ flux [21],

$$F_{\bar{\nu}_e} < 1200\,\text{cm}^{-2}\,\text{s}^{-1}\ ,$$

for $T_F = 4$ MeV, $\mu_F = 0$, and $\Omega_0 = 1$. It is expected, however, that there are background events from "invisible muons" which are produced by charged-current

†In the early analysis of the Kamiokande-II data, an energy interval 19 − 35 MeV was investigated [95] and a model-independent 90% CL upper limit of the relic $\bar{\nu}_e$ flux in this energy interval of 226 cm^{-2} s^{-1} was obtained.

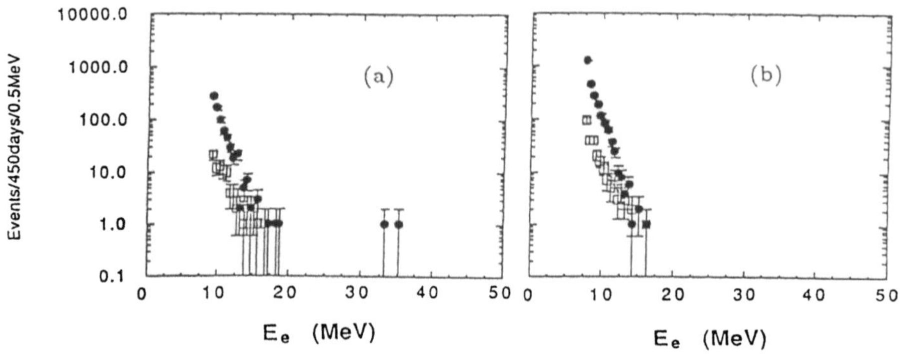

Figure 75: (a) Energy spectrum of the finally remaining events before the PMT gain change while searching for diffuse neutrinos in Kamiokande-II. The closed circles represent the events with cos $\theta_{\odot} < 0.88$ and the open squares represent the events with cos $\theta_{\odot} \geq 0.88$. (b) Same as (a) but after the PMT gain change.

interactions of atmospheric ν_μ's and $\bar{\nu}_\mu$'s in the detector. The invisible muons are those produced with a momentum less than the Cherenkov threshold (120 MeV/c), or those missed by the trigger because the trigger efficiency for a single muon is less than 100% in the 120 − 190 MeV/c momentum range. The estimated number of background events was 2.9 throughout the Kamiokande-II experiment. From this, the background-subtracted 90% CL upper limit of the total relic $\bar{\nu}_e$ flux was obtained as [21]

$$F_{\bar{\nu}_e} < 780\,\mathrm{cm}^{-2}\,\mathrm{s}^{-1}$$

for $T_F = 4$ MeV, $\mu_F = 0$, and $\Omega_0 = 1$. If it is assumed that the chemical potential μ_F is given by $2T_F$, the corresponding limit becomes [21]

$$F_{\bar{\nu}_e} < 540\,\mathrm{cm}^{-2}\,\mathrm{s}^{-1}.$$

Figure 77 shows the background-subtracted flux upper limit at the 90% CL as a function of T_F for $\mu_F = 0$ and $\Omega_0 = 1$.

In passing, it should be noted that the sensitivity of the 50,000-ton water Cherenkov detector Superkamiokande, which is under construction, to the diffuse neutrino background would be $F_{\bar{\nu}_e} \sim 20\,\mathrm{cm}^{-2}\,\mathrm{s}^{-1}$ during 5 years of operation [96]. It is therefore hoped that Superkamiokande will be able to positively detect a new source of astrophysical neutrinos for the first time.

6.4 Search for Magnetic Monopoles Based on Nucleon Decay Catalysis

The production of superheavy magnetic monopoles during the very early Universe is predicted by Grand Unified Theories combined with the standard Big-Bang cosmology. Among the methods used to search for magnetic monopoles is the use of

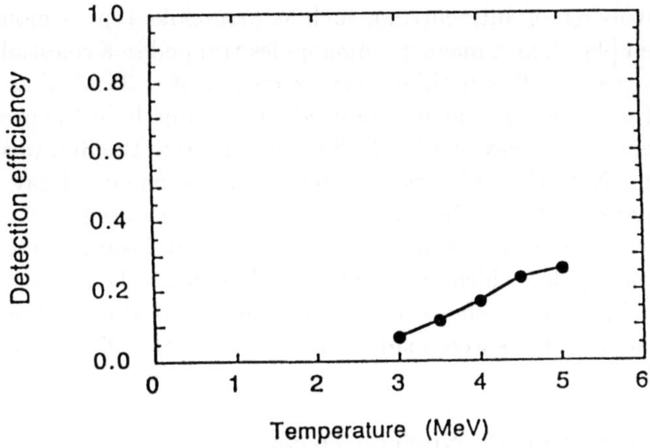

Figure 76: Detection efficiency as a function of T_F, estimated from a Monte-Carlo simulation of the detector response to the relic $\bar{\nu}_e$'s with red-shifted energy spectrum. This result was obtained for $\Omega_0 = 1.0$ corresponding to the detector condition after the PMT gain increase. For $\Omega_0 = 0.1$ and/or the detector condition before the PMT gain increase, however, the calculated detection efficiency shows no significant difference from the result shown in this figure.

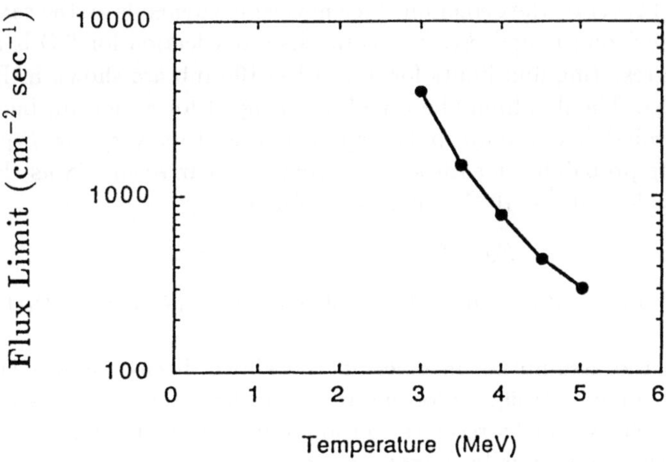

Figure 77: Background-subtracted upper limit of the diffuse $\bar{\nu}_e$ flux at the 90% CL as a function of the temperature (T_F), where Ω_0 is assumed to be unity.

the Rubakov effect [97], namely, nucleon decay catalized by magnetic monopoles with cross sections typical of strong interactions. With SU(5) magnetic monopoles, the process involving an antineutrino, such as monopole + p $\rightarrow$ monopole + $\bar{\nu}$ + π^+, is forbidden [98]. Thus, magnetic monopoles trapped in a celestial object, such as the Sun, causes a ν_e flux with an average energy of $\sim$ 35 MeV as the result of the β decay of μ^+ or K^0, μ^+ being produced either directly in the proton decay or indirectly through the decay of π^+ [99]. This ν_e flux from the Sun was searched for in Kamiokande. Note that π^- mesons produced as secondary decay products are absorbed by hydrogen before decay, so that no $\bar{\nu}_e$ flux is anticipated. A search was also performed for multiple nucleon decay in the Kamiokande detector induced by magnetic monopoles which traversed it. Though the latter has no relevance to astrophysical neutrinos, the result of the search will also be described in this subsection. These analyses were made using the data taken during the early phase of Kamiokande-I.

6.4.1 Search for Multiple Nucleon Decay

Candidate events were selected by demanding the existence of at least one delayed pulse larger than 330 p.e. (110 MeV for e) within the maximum range (88 μs) of Transient Digitizers, which were triggered by the prompt event and recorded delayed pulses. A further analysis of the selected events (93 events with only one delayed pulse) showed that they could be well explained by random coincidences of cosmic-ray muons and PMT dark pulses, and that no candidate events remained.

The 90% CL upper limits for a monopole flux with velocity β_{M} are given by

$$F_{\mathrm{M}} < 2.3\,/\,[S\Omega T\,\varepsilon(\sigma,\beta_{\mathrm{M}})],$$

where $S\Omega$ is the effective aperture (3,550 m^{2}sr), T the live time (335 days in this analysis), and $\varepsilon(\sigma,\beta_{\mathrm{M}})$ the detection efficiency, which depends on the catalysis cross section per nucleon (σ) and β_{M}. The catalysis cross section for ^{16}O is assumed to be 16σ. The resulting flux limits for $\sigma = 0.1 - 100$ mb are shown in Fig. 78 as a function of β_{M}. The flux limits become less stringent for a high β_{M} because of the decreasing probability of multi-pulse separation, and for very low β_{M} because of the decreasing probability for the second event to occur within 88 μs. For $\sigma = 100$ mb and $\beta_{\mathrm{M}} = 5\times10^{-5} - 10^{-3}$, the upper limit is

$$F_{\mathrm{M}} < 2.5\times10^{-15}\,\mathrm{cm^{-2}s^{-1}sr^{-1}}.$$

For $\sigma > 100$ mb and $\beta_{\mathrm{M}}\sigma < 0.1$ mb, the flux limit is the same as that mentioned above.

However, the assumption involved in the above discussion was criticized by Arafune and Fukugita [100]. According to them, the catalysis cross section of slow monopoles for heavy nuclei receives considerable suppression due to a long-range force originating from an extra angular momentum carried by a monopole-electric charge system, and σ for ^{16}O is negligible for $\beta_{\mathrm{M}} < 3\times10^{-3}$ compared to that for hydrogen nuclei, which can be written as $\sigma = \sigma_0/\beta_{\mathrm{M}}^2$. σ_0 is typically of the order of strong interactions. The 90% CL monopole flux limit is given in Fig. 78 for $\sigma_0 = 10^{-4}$ mb.

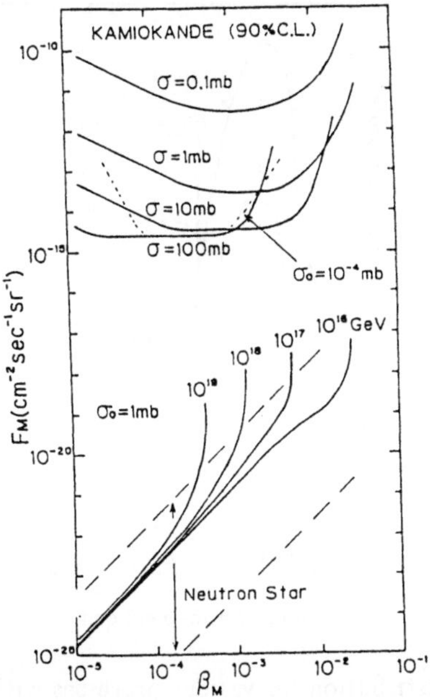

Figure 78: Upper solid curves: 90% CL upper limits for the monopole flux for various catalysis cross section (σ), obtained from a search for multiple nucleon decay in the Kamiokande detector. Also shown by the dotted curve is the upper limit for $\sigma_0 = 10^{-4}$ mb, when $\sigma = \sigma_0/\beta_{\rm M}^2$ is assumed for hydrogen, while the catalysis cross section for ^{16}O is neglected. Lower solid curves: 90% CL upper limits for the monopole flux for various monopole masses, derived from the intensity limit of low-energy neutrinos from the Sun. The flux limit derived from the old neutron stars [101] is also shown.

6.4.2 Search for Characteristic ν_e Flux from the Sun

In the Kamiokande detector, ν_e's in the energy region relevant to the monopole-catalyzed nucleon decay are detected through the reactions $\nu_e\,^{16}\mathrm{O} \to e^-\,^{16}\mathrm{F}$ and $\nu_e\, e^- \to \nu_e\, e^-$. There are also small contributions from $\nu_\mu\, e^- \to \nu_\mu\, e^-$ and $\bar{\nu}_\mu\, e^- \to \bar{\nu}_\mu\, e^-$. Figure 79 shows the total p.e. distributions for these processes. Shower-type contained events with a momentum between 30 MeV/c (trigger threshold in Kamiokande-I) and 52 MeV/c and with no $\mu \to e$ decay signal were searched in the limited fiducial volume of 450 tons (the event vertices must be at least 3 m inside from the PMT surface), in order to reject decay electrons from low-energy cosmic-ray muons entering the detector invisibly (note that no anticounter layer existed in Kamiokande-I). During 274 days of live time, 3 events were found. On the other hand, 2.5 background events were estimated to come from decay electrons from invisible low-energy muons produced in the fiducial volume by atmospheric ν_μ and $\bar{\nu}_\mu$. The background due to low-energy atmospheric ν_e and $\bar{\nu}_e$ is much less (< 0.1 event). Consequently, the 3 observed events are consistent with the background from atmospheric neutrino interactions.

Without subtracting the background, these 3 events give the following 90% CL upper limit on the electron neutrino flux, considering the detection efficiency for

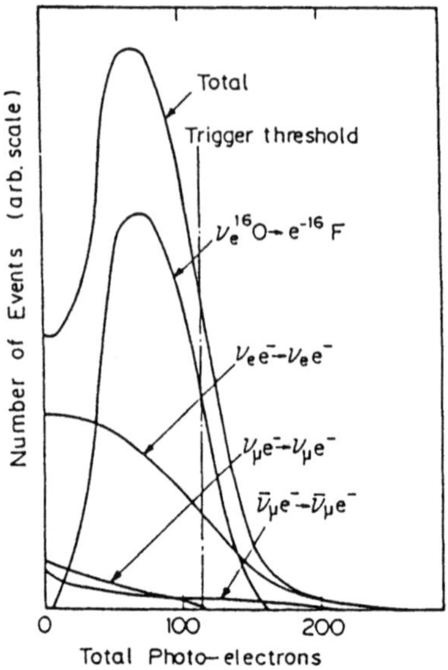

Figure 79: Total p.e. distribution for various processes with which neutrinos produced by the Rubakov effect in the Sun are detected in Kamiokande-I. The trigger threshold is also shown.

neutrinos produced by the Rubakov effect in the Sun (11%, mainly determined by the trigger efficiency of the detector; see Fig. 79):

$$I_{\nu_e} < 1.9 \times 10^4 \,\mathrm{cm}^{-2}\mathrm{s}^{-1}.$$

The contributions from ν_μ and $\bar{\nu}_\mu$ are neglected. Following the argument given by Arafune and Fukugita [100] (i.e., helium gives a negligible contribution), and assuming the branching ratio of proton decay into π^+ + anything to be 0.57, as given by the standard SU(5) model [102] with the branching ratios for μ^+ + mesons and $\bar{\nu}$ + mesons being set to be zero, the above limit on the electron neutrino flux leads to the following 90% CL monopole flux limit:

$$F_\mathrm{M} \cdot (\sigma/1\mathrm{mb}) < 6.0 \times 10^{-10} \beta_\mathrm{M}^2 \,\mathrm{cm}^{-2}\mathrm{s}^{-1}\mathrm{sr}^{-1}.$$

Writing $\sigma = \sigma_0/\beta_\mathrm{rel}^2$ with $\sigma_0 \sim 1$ mb, where $\beta_\mathrm{rel} = 1.7 \times 10^{-3}$ is the relative velocity between the monopole and hydrogen in the Sun, the monopole flux limit is given as

$$F_\mathrm{M} \cdot (\sigma_0/1\mathrm{mb}) < 1.6 \times 10^{-21} (\beta_\mathrm{M}/10^{-3})^2 \,\mathrm{cm}^{-2}\mathrm{s}^{-1}\mathrm{sr}^{-1}.$$

If the monopole mass is heavier than 10^{17} GeV, a correction is needed to take into account the fact that the probability for the Sun to trap monopoles decreases with the increasing monopole mass [103]. The lower solid curves in Fig. 78 show the resulting limits on the monopole flux for various masses. For a comparison, the monopole flux limit obtained from the X-ray excess of old neutron stars is also shown. It is noted that the present limits depend on less ambiguous astrophysical assumptions than that derived from old neutron stars.

6.5 Search for Neutrino Events in Correlation with Solar Flares

In addition to the famous solar-neutrino deficit, Davis and collaborators suggested another puzzling phenomena in their chlorine experiment, *i.e.*, the possible observation of neutrino events associated with large solar flares [104, 105]. Three runs (Run Nos. 27, 51, and 71) with an unusually high neutrino-capture rate (20 $\sim$ 250 excess neutrino captures) were recorded during the time in which large solar flares occurred. Additionally, $\sim$ 40 excess neutrinos were estimated to have been captured in Run No. 86, which was not obviously correlated with a solar flare, but was in coincidence with a strong γ-ray burst observed by the Solar Maximum Mission satellite on August 5, 1984, in the constellation Hydra [106].

If the suggested result from the ^{37}Cl experiment [105] is true, the existing underground neutrino detectors can observe copious neutrino events with an energy of E_ν in the 100 MeV region in correlation with large solar flares [107, 108], provided that these neutrinos originate from the decay of mesons produced by the collision of accelerated particles at the surface of the Sun.

In fact, the sensitivity of the Kamiokande detector (for a precise definition, see Ref. [109]) is $\sim$ 1,000-times greater than that of the ^{37}Cl detector for E_ν = 100 MeV, as shown in Fig. 80, where the time-integrated ν_e flux on Earth corresponding to one neutrino event in the detector is plotted for two assumptions concerning the neutrino flux: ν_e only and $\nu_e : \bar{\nu}_e : \nu_\mu : \bar{\nu}_\mu = 1:1:2:2$ (the latter assumption is the same as a naive expectation for atmospheric neutrinos). The reason why only the ν_e flux is plotted in Fig. 80 is to directly compare the Kamiokande result with the neutrino flux suggested by the ^{37}Cl experiment [105], which is also shown based on the assumption that the excess neutrino captures in the ^{37}Cl detector are due to solar-flare neutrinos. The curves in Fig. 80 were calculated using the neutrino reaction cross sections on ^{16}O for $E_\nu \gtrsim 50$ MeV and on ^{37}Cl for $E_\nu \gtrsim 10$ MeV given by Kuramoto *et al.* [110], and that on ^{16}O for $E_\nu \lesssim 50$ MeV given by Haxton [111].

Searches for neutrino events in correlation with solar flares have so far been performed by selecting two data-taking periods: one from July 1983 to July 1988 [112], and the other from February 1 to April 30, 1989 [109]. During the latter period, in March 1989, particularly intense solar flares occurred [113]. The solar activity during this period was among the strongest since record-taking began. In particular, it was the strongest since the Kamiokande detector started observations. Two importance 4, seven importance 3, and 35 importance 2 flares were observed throughout this period.† For a comparison, during August 2 – 7, 1972, during

†The optical importance is a measure of flare size and brilliance with five ranks (S; 1, 2, 3, and 4 in increasing order). In Ref. [112], the quoted optical importance is the maximum value if its observation was reported by more than one observatory. The optical importance defined in this way is biased toward higher values and, in fact, the corresponding average optical importance given in Solar Geophysical Data (comprehensive reports) are less than 3; no importance 4 flare occurred during the period July 1983 - July 1988. In Ref. [109], the average optical importance is quoted from Preliminary Report and Forecast of Solar Geophysical Data [113]. For this reason,

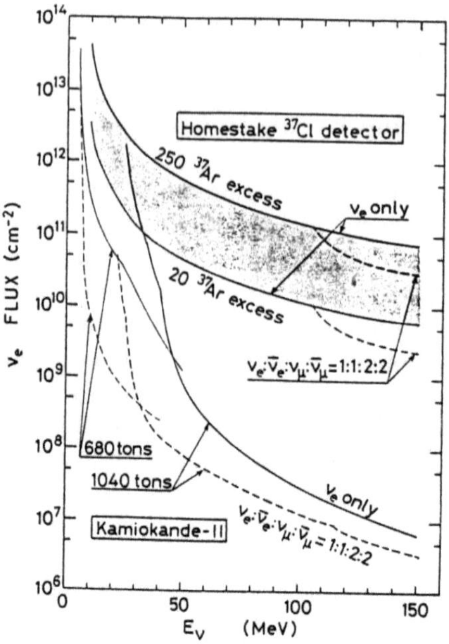

Figure 80: Sensitivity of the Kamiokande-II detector to the solar-flare ν_e flux plotted as a function of the ν_e energy, corresponding to one neutrino event in the detector. The shaded region indicates the solar-flare ν_e flux estimeted from the ^{37}Cl experiment.

Run No. 27 of the ^{37}Cl experiment, in which the largest excess neutrino captures (250 ± 130) were reported, two importance 3 and one 2 flares were observed as well as several importance 1 flares [114].

By examining the contained events in the 880 tons of fiducial volume in Kamiokande-I and in the 1040 tons of fiducial volume in Kamiokande-II, recorded during the interval of (start of flare − 1h, end of flare + 1h), the following results were obtained:

(i) Although one ν_μ type event was found during the search period between July 1983 and July 1988, this was statistically consistent with the background level of 0.35 event expected from the random distribution of atmospheric neutrino events during the flare periods.

(ii) No events were observed associated with the large flares which occurred during the period between February 1 and April 30, 1989. Based on the null result, the 90% CL upper limits were determined for the time-integrated ν_e flux per flare on Earth in the following way. First, we assumed that the solar-flare neutrinos are monochromatic with $E_\nu = 100$ MeV. If no neutrino event is observed during a large solar flare, a conservative 90% CL upper limit for the time-integrated ν_e flux is given by $2.3\,S(100\text{ MeV}) = 7 \times 10^7\text{ cm}^{-2}$, where $S(100\text{ MeV})$ is the sensitivity at $E_\nu = 100$ MeV corresponding to the assumption of ν_e flux only. This is also a conservative limit for the solar-flare neutrinos having a continuous energy spectrum with $E_\nu \geq 100$ MeV. If the solar-flare neutrinos with $E_\nu \geq 50$ MeV are considered,

---

the neutrino flux limits of the solar-flare origin are quoted from Ref. [109].

a conservative 90% CL upper limit is 2.3 $S(50\,\mathrm{MeV}) = 1.9 \times 10^9\,\mathrm{cm}^{-2}$. Now, during the period February 1 – April 30, 1989, importance 4 solar flares occurred twice, and the Kamiokande detector was at work at these two flare times. Therefore, the 90% CL upper limit for the time-integrated ν_e flux per importance 4 flare at E_ν=100 (50) MeV is 3.5×10^7 (9.4×10^8) cm^{-2}. Likewise, six importance 3 and twenty-six importance 2 flares occurred when the Kamiokande detector was at work during the same time period, and therefore the limits for importance 3 and 2 flares at E_ν=100 (50) MeV are 1.4×10^7 (3.7×10^8) and 2.4×10^6 (6.5×10^7) cm^{-2}, respectively. Also, the 90% CL upper limits for the time-averaged ν_e flux during the flare times at E_ν=100 (50) MeV were obtained to be 6.0×10^3 (1.6×10^5), 2.0×10^3 (5.4×10^4), and 6.2×10^2 (1.7×10^4) cm^{-2}s^{-1} for the importance 4, 3, and 2 flares, respectively. These results indicate that even the upper limit for the importance 4 flare is far below the ν_e flux at the corresponding energy, which is required to ascribe the excess neutrino captures reported in the ^{37}Cl experiment to large solar flares.
(iii) Considering the possibility that the strong neutrino flux arises from flares on the side of the Sun opposite to the Earth, which is not visible by optical observations, a correlation between the number of flares and neutrino events summed over each month was examined during the period July 1983 – July 1988 with the result that no correlation between the two event types was found.
(iv) No events were recorded in correlation with a γ-ray burst observed by the Solar Maximum Mission satellite on August 5, 1984. It can be seen from the sensitivity shown in Fig. 80 (scaled to 880 tons of fiducial volume of Kamiokande-I), corresponding to the 40 excess events in the ^{37}Cl detector, $\gtrsim 600$ (40) events should have been observed in the Kamiokande-I detector if $E_\nu \geq 100$ (50) MeV. Therefore, the possible neutrino flux associated with this γ-ray burst is at most $\frac{1}{600}$ times ($\frac{1}{40}$ times) that estimated from the ^{37}Cl data, provided that $E_\nu \geq 100$ MeV (50 MeV).

A conceivable neutrino source related to the energetic flare particles that reach the Earth is the production of $\beta^\pm$ emitters in the atmosphere of the Earth. The flare that occurred on 1972 August 4 is known to have produced the largest energetic proton event during solar cycle 20, with a maximum proton flux (≥ 10 MeV) of 2.5×10^4 p.f.u.[†] [115]. However, the estimated proton fluence (≥ 10 MeV) of $\sim 2 \times 10^{10}$ cm^{-2} [116], can only produce many orders of magnitude less neutrino flux from positron emitters in the Earth's atmosphere than is needed to account for the excess neutrino captures in the ^{37}Cl detector [117].

Nevertheless, low-energy events with a recoil electron (or positron) energy of $E_e \gtrsim 10$ MeV were examined in the reduced fiducial volume of 680 tons in order to directly search for low-energy neutrinos in the Earth's atmosphere, produced by solar particles of flare origin. The sensitivity of the Kamiokande-II detector to low-energy neutrinos is also shown in Fig. 80. However, no significant difference was found between the event rate during the flare times and that off the flare times. These results indicate that the study of low-energy events is background-limited, thus giving an upper limit for the possible flare-associated low-energy event rate

[†]p.f.u. means "particle flux unit" and is defined as: number of particles cm^{-2}s^{-1}sr^{-1}.

of $\sim 10^{-5}$ s^{-1}. If these events are due to ν_e's and $\bar{\nu}_e$'s from $\beta^{\pm}$ emitters in the atmosphere of the Earth, $\bar{\nu}_e$'s mainly contribute. Then, the upper limit for a flare-associated low-energy $\bar{\nu}_e$ flux of $\sim 10^5$ $cm^{-2}s^{-1}$ is obtained from the sensitivity curve corresponding to the assumption of $\nu_e : \bar{\nu}_e : \nu_\mu : \bar{\nu}_\mu = 1:1:2:2$ in Fig. 80 if these $\bar{\nu}_e$'s have an energy of 10 MeV. (Although with this assumption the ν_e flux is the same as the $\bar{\nu}_e$ flux, this assumption is not justified for cosmogenic radioactivity. In order that the flux limit for $\bar{\nu}_e$'s be used to limit the flare-associated low-energy ν_e flux, a realistic energy spectra for both the ν_e's and $\bar{\nu}_e$'s from cosmogenic radioactivity in the atmosphere should be given.)

In summary, the results presented above do not support the conjecture that large solar flares producing energetic neutrinos from meson production and decay caused the more than 20 excess neutrino captures in the ^{37}Cl detector.

6.6 Search for High-Energy Neutrinos from Point Sources

The search for point sources of high-energy neutrinos has an important bearing on investigations of the acceleration mechanisms of cosmic-ray particles. If matter exists near to a compact source of cosmic-ray acceleration, collisions between the accelerated protons and the matter produce charged and neutral pions, the dacay of which gives rise to the production of high-energy neutrinos and γ-rays, respectively. Such a production mechanism implies that high-energy neutrinos are accompanied by high-energy γ-rays with the same power-law spectral index [118], provided that high-energy γ-rays are not absorbed during propagation.

Among the promising sources are a newly born neutron star with expanding ejecta [119, 120] and an X-ray-emitting accreting binary system, in which a neutron star is accompanied by a companion star [121]. Accordingly, high-energy neutrinos from the remnant of SN1987A were searched for in Kamiokande for about 1 year since its explosion [122, 123]. A search has also been made for high-energy neutrinos from active X-ray binaries (Cyg X-3, Vela X-1, LMC X-4, and Her X-1) as well as other potential sources using the data from both Kamiokande-I and -II [124].

6.6.1 Search for High-Energy Neutrinos from SN1987A

Upward-going muons from the direction of SN1987A were searched for using the data taken during the time period February 23, 1987 – March 14, 1988 (385th day from the explosion) [123]. The corresponding live time was 338 days and the exposure in the direction of SN1987A was 132.0 m^2 yr. A total of 59 upward-going muons were recorded during this period. To pick up the signal, the angular window around the direction of SN1987A was defined in such a way that 95% of the signal would be accepted. It is dependent on the assumed energy spectrum of the neutrinos [see Eq. 17 in subsect. 3.3.1, and also see subsect. 3.3.2]; for example, a 4.2° half-angle cone for the spectral index γ=2.1 and cutoff energy $E_c = 10^{15}$ eV, and a 7.0° half-angle cone for γ=2.7 and $E_c = 10^{12}$ eV.

One muon event was observed 6° away from the direction of SN1987A on October 19, 1987, 11:39 UT. No other event was observed within a 7.0° half-angle cone around SN1987A. Whether or not this one event is significant depends on the

Table 8: 90% CL upper limits for the muon-neutrino flux from the direction of SN1987A (top, in $cm^{-2}\,s^{-1}$) and the corresponding muon-neutrino luminosity (bottom, in erg s^{-1}) during the observation period between 23 February 1987 and 14 March 1988. The flux and luminosity limits for $\gamma = 2.7$ and $E_c = 10^{12}$ eV are determined based on one observed event, but those for other γ and E_c are determined based on zero observed event.

E_c (eV)	Spectral index (γ)		
	2.1	2.4	2.7
10^{12}	1.1×10^{-4}	3.8×10^{-4}	2.1×10^{-3}
	4.5×10^{41}	1.0×10^{42}	4.0×10^{42}
10^{13}	2.4×10^{-5}	1.5×10^{-4}	7.3×10^{-4}
	1.3×10^{41}	4.0×10^{41}	1.4×10^{42}
10^{14}	1.4×10^{-5}	1.1×10^{-4}	6.6×10^{-4}
	8.6×10^{40}	3.2×10^{41}	1.3×10^{42}
10^{15}	1.3×10^{-5}	1.1×10^{-4}	6.5×10^{-4}
	8.2×10^{40}	3.1×10^{41}	1.3×10^{42}

atmospheric neutrino background. Based on the observed 59 events, the average flux of the upward-going muons was $(2.31 \pm 0.32) \times 10^{-13}$ $cm^{-2}s^{-1}sr^{-1}$. This flux value is consistent with the expected atmospheric neutrino background of $(2.04 \pm 0.13) \times 10^{-13}$ $cm^{-2}s^{-1}sr^{-1}$ [42], and the expected background in a 7° half-angle cone around SN1987A is 0.4 event. This indicates that the observed one event from the direction near SN1987A is consistent with the background.

The 90% CL upper limit of the muon-neutrino flux from the direction of SN1987A averaged over the observation period, was calculated to be 6.1×10^{-14} $cm^{-2}\,s^{-1}$ or 10.4×10^{-14} $cm^{-2}\,s^{-1}$, depending on whether the upper limit was calculated based on zero events or one event. Table 8 shows the 90% CL upper limits of the muon flux from SN1987A with an energy greater than 1.7 GeV for spectral indices $\gamma = 2.1$, 2,4, and 2.7, and cutoff energies of $E_c = 10^{12}$ to 10^{15} eV. The distance between the Earth and SN1987A was assumed to be 50 kpc. The upper limit for γ=2.7 and $E_c = 10^{12}$ eV was given on the basis of one observed event, because it occurred within the angular window. The upper limits for all other combinations of γ and E_c were given on the basis of zero observed events.

There are some theoretical models which relate the flux of upward-going muons or high-energy neutrinos with the luminosity of accerelated protons. For example, according to Gaisser and Stanev [120] a total of 20 $\sim$ 40 upward-going muons will be detected with a $100m^2$ detector up to $\sim$200 days after a type-II supernova explosion if the accelerated proton luminosity amounts to 10^{43} erg s^{-1}. Through these models, the Kamiokande result constrains the luminosity of protons having been accerelated in the remnant of SN1987A during the early days since the explosion.

6.6.2 Search for High-Energy Neutrinos from γ-Ray Sources

A strong interest in high-energy neutrinos from binary systems was prompted by the observations of UHE (ultra-high-energy) γ-rays from Cygnus X-3 [125], which is one of the most luminous X-ray sources in our galaxy. Observations of UHE γ-rays were also reported from Vela X-1, LMC X-4 and that of TeV γ-rays from Her X-1; these are all active X-ray binaries. Furthermore, there are some other reported UHE and VHE (very high-energy) γ-ray observations from various point sources. Unfortunately, none of these observations have been confirmed to date.† Nevertheless, it seems worthwhile to search for high-energy neutrinos from these objects, since they are all known to be active sources of radio waves, X-rays, and γ-rays up at least to $\sim$ 100 MeV.

The results presented here are based on the same data sample of 252 upward-going muon events (as discussed in subsect. 3.3.2) and update the already published results [124]. The angular window which accepts 68% (one standard deviation) of the upward-going muons from a point source is discussed in subsect. 3.3.2. Based on this discussion, it is assumed in the following that 68% of the upward-going muons come from an angular window of a 2.5° half-angle cone.

In Fig. 81, the arrival directions of these 252 upward-going muons in the celestial coordinates are plotted together with eight astronomical objects: Cyg X-3, Her X-1, Crab nebula, Geminga, SS433, Galactic center, Vela X-1 and LMC X-4. Since the Kamiokande detector is located in the northern hemisphere, objects in the southern hemisphere sky are observable with high efficiency. On the other hand, objects with a larger declination ($\geq$ 53.6°N, i.e., 90° $-$ Kamioka latitude) are not observable at all, because they are always seen above the horizon.

The upward-going muons shown in Fig. 81 seem to be uniformly distributed in proportion to the exposure corresponding to each celestial position. No candidate event was observed from the angular window around 6 possible neutrino sources; one upward-going muon came from the direction of Her X-1 and one from the Galactic center. The celestial position of the considered astronomical objects, distance from the earth, exposure in units of m^2 yr, number of candidates from the adopted window, and number of expected background events from that window are listed in Table 9. The number of observed events is consistent with that of the expected background events, and no significant signal is obtained. Assuming that 68% of the upward-going muons from a point source be found within the adopted angular window, the 90% CL upper limits for the upward-going muon flux were calculated for the 8 astronomical objects and are listed in Table 9 together with the corresponding 90% CL upper limits for the muon-neutrino flux and the muon-neutrino luminosity of each object based on the assumption that $E_c = 10^{15}$ eV and $\gamma = 2.1$.

The Kamiokande results should be compared with the published results from IMB [128] and Fréjus [129] Collborations. The upper limits for the muon flux or high-energy neutrino flux from the studied astronomical objects are better than

---

†Recently the Whipple collaboration reported the observation of TeV γ-rays from the Crab nebula with a high statistical significance [126]. Also, the observation of TeV γ-rays from the active galaxy Markarian 421 was reported by the Whipple Collaboration[127].

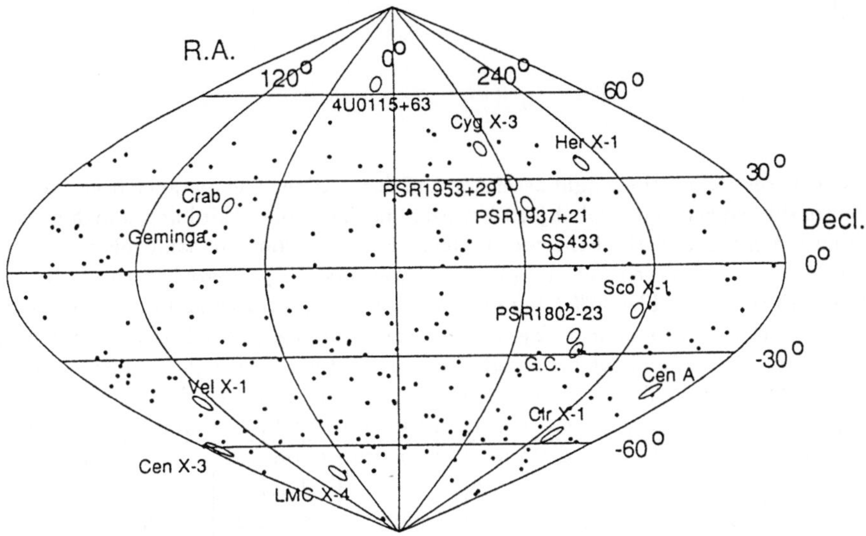

Figure 81: Arrival direction of the 252 upward-going muon events in the celestial coordinates. 2.5° windows around possible sources are also shown.

the upper limits given by the other two experiments if the same spectral index and cutoff energy are used. Moreover, the observation in Kamiokande can be continued with a small background, while the IMB and Fréjus observations are already background-limited due to the following reasons. The difference in the expected background between the Kamiokande and IMB observations mainly stems from the angular resolution of the track reconstruction, which is 2.1° (68% acceptance) in Kamiokande and 7.0° (90% acceptance) in IMB. The Fréjus experiment lacks the sense of the track direction; this causes background from downward-going muons.

Finally, it should be mentioned that three upward-going muons were observed within a 1° half-angle cone centered at the celestial position of 353° right ascension and 19° declination, where no known VHE γ-ray source is located. Moreover, these events occurred within a year: July 22, 1985, February 6, 1986, and July 7, 1986. The chance probability that three muons are observed within an angular window of a 2.5° half-angle cone is calculated as follows. The expected number of upward-going muons within this window and during the period July 22, 1985 – July 7, 1986 (the corresponding exposure is 28.4 m^2 yr) is calculated to be 0.011. The probability that 3 events are observed when the expected number is 0.011 is

$$\frac{e^{-0.011}(0.011)^3}{3!} = 2.19 \times 10^{-7}.$$

On the other hand, the number of 2.5° half-angle cones within a 4π solid angle is given by

$$\frac{4\pi}{\pi(2.5 \times \frac{\pi}{180})^2} = 2.1 \times 10^3.$$

Table 9: Celestial position (right ascension and declination) and distance from the Earth of the considered astronomical objects are listed in the second, third and fourth columns. Exposure in m^2 yr and the expected number of muons from the atmospheric neutrino background, and observed upward-going muon events are presented in the fifth, sixth, and seventh columns. The 90% CL upper limit on the upward-going muon flux from the direction of the 8 astronomical objects, the corresponding 90% CL upper limit on the muon-neutrino fluxes and the luminosities are also listed in the eighth, ninth, and tenth columns. The threshold energy of upward-going muons is 1.7 GeV. The cutoff energy and spectral index of parent neutrinos are assumed to be 10^{15} eV and 2.1, respectively.

source	R.A. (deg)	Decl. (deg)	distance (kpc)	exposure (m^2 yr)	expected back-ground events	observed events
Cyg X-3	307.6	40.8	12	185.94	0.10	0
Her X-1	254.0	35.4	5	214.26	0.10	1
Crab	82.9	22.0	2	272.80	0.11	0
Geminga	97.7	17.8		283.33	0.11	0
SS433	285.1	4.8	3	316.92	0.11	0
Gal.Cen.	265.5	−28.6	10	428.02	0.14	1
Vel X-1	128.4	−45.0	2	524.33	0.18	0
LMC X-4	81.0	−69.5	50	669.90	0.22	0

	90% C.L. muon flux limit ($cm^{-2}s^{-1}$)	90% C.L. neutrino flux limit ($cm^{-2}s^{-1}$)	90% C.L. neutrino luminosity limit ($erg\, s^{-1}$)
Cyg X-3	5.8×10^{-14}	9.9×10^{-6}	3.7×10^{39}
Her X-1	8.5×10^{-14}	14.4×10^{-6}	9.7×10^{38}
Crab	3.9×10^{-14}	7.3×10^{-6}	7.6×10^{37}
Geminga	3.8×10^{-14}	7.2×10^{-6}	
SS433	3.4×10^{-14}	5.8×10^{-6}	1.3×10^{38}
Gal.Cen.	4.2×10^{-14}	7.3×10^{-6}	1.9×10^{39}
Vel X-1	2.0×10^{-14}	4.8×10^{-6}	2.5×10^{37}
LMC X-4	1.6×10^{-14}	3.3×10^{-6}	2.1×10^{40}

Also, the ratio of the total exposure (277.0 m^2 yr) to the exposure during the period between July 22, 1985 and July 7, 1986, is 9.75. The probability that such a coincidence is observed in any celestial position by chance is found to be

$$(2.19 \times 10^{-7}) \times (2.1 \times 10^{3}) \times 9.75 = 4.5 \times 10^{-3}.$$

To further clarify this problem, a longer exposure as well as an independent observation in other experiments is needed. However, if the activity of the possible source† of these events does not repeat, it would be impossible to confirm it.

6.6.3 Evidence against High-Energy Muons from Cyg X-3

In 1985, two groups of underground experiments, Soudan-I [130] and NUSEX [131], reported evidence for a 4.8-h time-modulated muon flux from Cyg X-3. The Soudan-I observation [130] extended over the period from September 1981 to November 1983, and the signal was observed in the phase interval 0.65 – 0.90. Later, muon events correlated with the large radio outburst of Cyg X-3 during September – October 1985 were also observed in the phase interval 0.725 – 0.750 [132]. The NUSEX [131] observed an enhancement of the muon flux between June 1982 and February 1985 at the phase interval 0.7 – 0.8. Also, IMB Collaboration reported [133] a possible observation of muons from the direction of Cyg X-3 correlated with the strong radio outburst in October, 1983, though the muon-flux enhancement was at a phase of 0.5 – 0.6. If confirmed, these observations imply the possibility of a new type of neutral primary particle and/or unexpected interactions of photons or neutrinos, because the observed muon flux is not consistent with the VHE and UHE γ-ray fluxes reported at that time.

No excess muon flux from Cyg X-3, however, was observed in Kamiokande between July 1983 and September 1984 [134] ($< 3.7\times10^{-12}$ cm^{-2} sec^{-1} at a minimum rock overburden of 2400 mwe for the phase interval 0.65 – 0.90 and $< 1.7 \times 10^{-12}$ cm^{-2} sec^{-1} at a minimum slant depth of 5000 mwe for the phase interval 0.7 – 0.8). Furthermore, no burst of muon flux was observed in correlation with the Cyg X-3 radio outburst in October, 1983 ($< 1.9 \times 10^{-11}$ cm^{-2} sec^{-1} at a minimum rock overburden of 2400 mwe for the phase interval 0.5 - 0.6) nor with the outburst in September – October 1985 ($< 2.1 \times 10^{-11}$ cm^{-2} sec^{-1} at a minimum rock overburden of 2400 mwe for the phase interval 0.725 - 0.750) [135]. Other large underground detectors at Baksan [136] and at Fréjus [137] did not observe any excess muon fluxes from Cyg X-3, either. Figure 82 summarizes both the positive and negative results.

6.7 Search for Dark Matter

Suppose that dark matter in the Universe [139] consists of unknown weakly interacting, massive elementary particles. Such matter would be captured by massive astronomical objects, such as the Sun or Earth, and would accumulate at the center, where it would annihilate into hadrons and leptons and produce energetic

---

†It should be called Pegasus ν-1, if confirmed.

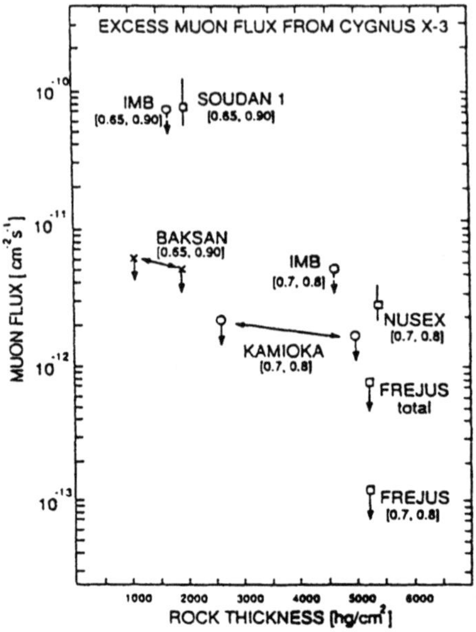

Figure 82: Positive and negative (90% CL upper limits are indicated) results for excess muon flux from Cyg X-3 as a function of depth. This figure is taken from Ref. [138].

neutrinos both directly and/or indirectly [140]. Therefore, huge underground neutrino detectors provide a means to search for massive dark-matter particles by observing the flux of energetic neutrinos from the Sun or from the center of the Earth. For the case of the Earth, dark-matter particles would be efficiently trapped if their masses were to be close to those of constituent nuclei of the Earth ($30 \sim 80$ GeV [141]). There are obviously background events caused by atmospheric neutrinos, which, however, can be reduced by using information concerning the incident neutrino direction and energy. In the following, a search in the Kamiokande is described for heavy Dirac/Majorana neutrinos [142, 143] and neutralinos [42] as potential dark-matter candidates.

6.7.1 Search for Candidate Events

For a dark-matter search we classify the observed events related to high-energy neutrino interactions into: (1) fully contained events, (2) vertex-contained events, and (3) upward-going muons. Type-(1) events are those in which all the charged secondary particles are absorbed in the detector, while type-(2) events are those in which one or more energetic charged particles go out of the detector. Only the data taken in Kamiokande-II were used for the analysis of type-(1) and type-(2) events, because the information from the anticounter simplifies the event reduction from the raw data and plays a crucial role in selecting these events. For an analysis of type-(3) events, both Kamiokande-I and Kamiokande-II data were used. Note that most of the $\nu_\mu(\bar{\nu_\mu})$-induced events with an energy larger than a few GeV are

Table 10: Number of events and total exposure for each type of neutrino event observed in Kamiokande. For fully contained events, those comprising a single Cherenkov ring or multiple Cherenkov rings are classified as "single" and "multi", respectively.

	Type (1) (Fully contained)	Type (2) (Vertex contained)	Type (3) (Upgoing muon)
Total exposure	4.4 kt·yr	1.6 kt·yr	770 m^2yr
No. of events	single 64		
	multi 54	31	252
Analysis threshold	3000 p.e.	1500 p.e.	1.7 GeV
Data period	Nov. '85 ~ Jul. '88	Dec. '85 ~ Jun. '88	Jul. '83 ~ Apr. '90
Fiducial volume	2140 tons	1040 tons	

not contained in the detector; they are detected as type-(2) and -(3) events. Also, in the sample of vertex-contained events, the percentage of $\nu_e(\bar{\nu}_e)$-induced events increases with increasing neutrino energy.

Although the selection criteria for type-(1) (fully contained) events are similar to those employed in an analysis of atmospheric neutrino events, for this study the total p.e. number in the inner counter was required to be > 3000, corresponding to a ν_e energy of about 1.4 GeV, in order to obtain good angular resolution. Also, a larger fiducial volume (2140 tons) was defined for type-(1) events. Type-(2) (vertex-contained) events were selected after having eliminated through-going and stopping cosmic-ray muons, with the criteria that the total p.e. number in the inner counter be > 1500, and that the vertex be seperated by more than 1.5 m from the PMT wall. The fiducial volume for type-(2) events was defined to be 1040 tons. The selection criteria for type-(3) events (upward-going muons) were the same as those already described in the preceding subsection. Table 10 lists the number of events and the total exposure for each type of selected events.

Figure 83 shows two-dimensional plots of $\cos\theta_\odot$ *vs* the total p.e. number for fully contained and vertex-contained events, where $\theta_\odot$ is the angle between the direction from the Sun and the reconstructed direction. These results indicate that no statistically significant excess is observed in the direction of the Sun. The zenith-angle distribution of upward-going muons is shown in Fig. 52. It is consistent with the expected flux of the atmospheric neutrinos, and no enhancement toward the nadir (or the Earth core) is observed. Figure 84 shows the distribution of upward-going muons against $\cos\theta_\odot$. Here again, no extra flux of upward-going muons from the direction of the Sun can be recognized.

Thus, no evidence was found for an excess of neutrino events which might be attributed to dark-matter annihilation; the Kamiokande data also yield constraints on the parameters describing the dark-matter candidates. For this purpose, the flux upper limits of energetic neutrinos from the Sun or from the center of the Earth was calculated based on the observed number of events and expected atmospheric neutrino background in a relevant angular window, which was chosen so as to enhance the signal/background ratio. In calculating the flux limit, type-(3) events within angular windows for both the Earth and Sun were considered as possible

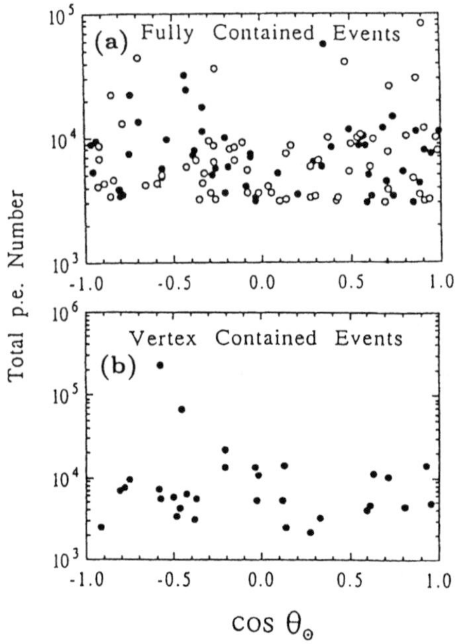

Figure 83: Two-dimensional plot of $\cos\theta_\odot$ *vs* the total p.e. number for (a) fully contained events (4.4 kt·yr) and (b) vertex-contained events (1.6 kt·yr). The angle $\theta_\odot$ is that between the direction from the Sun and that of events. The solar direction is $\cos\theta_\odot = 1$. For fully contained events, the open and closed circles correspond to single-ring and multiring events, respectively.

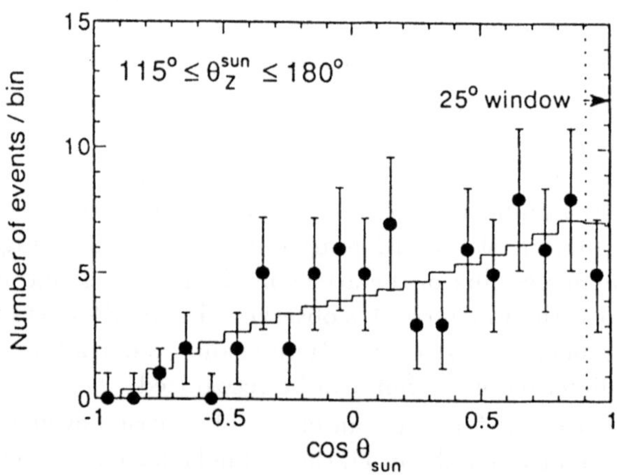

Figure 84: Correlation of upward-going muon events with the direction of the Sun. The angle θ_{sun} is the angle between the upward-going muon direction and a radius vector of the sun. Events which occurred when the zenith angle of the sun (θ_z^{sun}) was greater than 115° were taken into account. The histogram shows the expected distributions of atmospheric neutrino events.

candidates from dark-matter annihilation, while type-(1) and -(2) events from only the direction of the Sun were considered as being such. This is because the sensitivity of the type-(1) and -(2) events to the incoming neutrino energy does not extend to the region relevant to a search for dark matter trapped in the Earth ($30 \sim 80$ GeV).

The size of the angular window depends on the deposited energy (for type-(1) and -(2) events) and on the assumed dark-matter particle, and was determined from the results of a Monte-Carlo simulation of the expected angular spread of the signals. The latter dependence results from the fact that the neutrino spectrum from dark-matter annihilation is specific to the particle assumed to be dark matter. For type-(1) and -(2) events, the adopted angular window is $\cos\theta_{\odot} > 0.98$ for the total p.e. number > 20,000, but becomes wider as the total p.e. number decreases. For type-(3) events, it is a $15° \sim 30°$ half-angle cone around the source (Earth/Sun) direction, depending on the mass and particle type for Dirac and Majorana neutrinos, whereas it is fixed at 30° for neutralinos.

6.7.2 Constraints on Massive Dirac and Majorana Neutrinos

To obtain the constraints on the mass of Dirac and Majorana neutrinos as dark-matter candidates, the neutrino fluxes expected to be observed must be calculated for a given particle type and a mass, and compared with the experimentally obtained flux limits. They are mostly calculated according to the prescription presented in Ref. [140]; a brief description is given below.

The capture rate of dark matter by a massive astronomical object is proportional to the product of the mass density (ρ_{DM}), the inverse of the root-mean-square velocity (v_{DM}), and the elastic scattering cross section with nuclei of dark-matter particles. Here, ρ_{DM} and v_{DM} are assumed to be those of the halo of our galaxy (0.3 GeV/cm^3 and 300 km/s). The cross section is the sum of the spin-independent and spin-dependent terms. Although for Dirac neutrinos both terms contribute, for Majorana neutrinos only the latter term is relevant. Thus, the Earth is not very effective in capturing Majorana neutrinos, since it almost comprises spin-zero nuclei. The expression for the capture rate also includes the kinematical factor which takes into account the following effects, pointed out by Gould [141]: a resonant enhancement when the dark-matter mass is near to the masses of constituent nuclei, and a suppression factor for Dirac neutrinos (and neutralinos which are discussed in subsect. 6.7.3) due to a lack of coherence in elastic scattering. In calculating these, the density profile of the Earth is taken from Ref. [144] and that for the Sun from Ref. [26].

The annihilation cross sections of massive neutrinos into fermion pairs are given in Ref. [140]. In this analysis the top-quark mass (m_t) is arbitrarily assumed to be 150 GeV. For neutrinos heavier than the W-boson mass, a $\nu\overset{(-)}{\nu} \rightarrow W^+W^-$ channel opens, and for neutrinos heavier than the Z-boson mass a $\nu\overset{(-)}{\nu} \rightarrow Z^0Z^0$ channel opens in addition [145]. These cross sections are calculated in the framework of the standard model. The reaction $\nu\overset{(-)}{\nu} \rightarrow W^+W^-$ proceeds through a Z-boson, neutral Higgs, and heavy lepton (which is associated with the heavy neutrino) exchange, and the reaction $\nu \overset{(-)}{\nu} \to Z^0 Z^0$ through a heavy neutrino exchange. Decays of gauge bosons are dealt with by using the JETSET Monte-Carlo program [147]; the resulting neutrino spectra are derived. For particles produced through dark-matter annihilation in the Sun, the energy loss of hadrons in the solar medium, and muon-neutrino production from the decay of τ-leptons generated in collisions of ν_τ's with solar medium may be significant in the high-energy region. These effects are taken into account, following Ref. [146] and Ref. [148].

The neutrino interactions in the detector are calculated as described in subsect. 3.2.1, but the ν_τ and $\bar{\nu}_\tau$ interactions are supplemented, since they constitute a significant fraction of dark-matter signals (for details, see Ref. [142]). The neutrino conversion process into muons via a charged-current interaction in the rock surrounding the detector is calculated as in subsect. 3.3.1.

As a result of all these calculations, the fully contained and vertex-contained event rates and upward-going muon flux are estimated and compared with the upper limits derived from the Kamiokande data. Figure 85 shows the background-subtracted upper limits for the rates of fully contained and vertex-contained events from the Sun for Dirac neutrinos (ν_D) and Majorana neutrinos (ν_M) at the 90% CL as a function of the mass, where the expected rates are also shown. For Majorana neutrinos, the expected event rates are calculated both with and without the EMC effect [149] for the spin-dependent cross section taken into account. Figure 86 shows the upper limits and expected flux of upward-going muons from annihilation of massive neutrinos captured by the Earth and the Sun as a function of the mass. Here, the expected flux for Majorana neutrinos was calculated by taking account of the EMC effect [149] for the spin-dependent cross section. The sharp peaks near to masses of 26 and 52 GeV of the expected fluxes from the Earth correspond to the resonant scattering of massive neutrinos off ^{28}Si and ^{56}Fe nuclei inside the Earth. The bump seen in the Majorana neutrino case reflects the opening of a $\nu \overset{(-)}{\nu} \to Z^0 Z^0$ channel (sudden increase near to a mass of 90 GeV) and the opening of a $\nu \overset{(-)}{\nu} \to t\bar{t}$ channel [sudden drop near mass 150 GeV ($= m_t$)]. Here, the associated heavy lepton mass and the Higgs mass are assumed to be $m_\nu + 100$ GeV and 1 TeV, respectively; however, the dependence on these values is weak.

Combining the results given in Figs. 85 and 86, Dirac neutrinos with a mass of $m_{\nu_D} \geq 3$ GeV and Majorana neutrinos with a mass of $m_{\nu_M} \geq 24$ GeV are excluded from the candidates for halo dark matter, based on the assumption that the galactic halo consists of a single component. However, if the cosmological relic abundance of dark matter is smaller than the density parameter of the universe (Ω_0; $0.05 \leq \Omega_0 \leq 4$ [150]), ρ_{DM} might be smaller than the halo density and the excluded mass region claimed above would be reduced. This effect is model-dependent and a detailed discussion is given in Refs. [140, 143].

6.7.3 Constraints on Neutralinos

In the analysis presented below, neutralinos in the minimal supersymmetric standard model (MSSM) [151] are considered. Although the method of analysis for neutralino dark matter is mostly in parallel with that for massive Dirac and Majorana neutrinos, there are more parameters for the MSSM neutralinos. In MSSM,

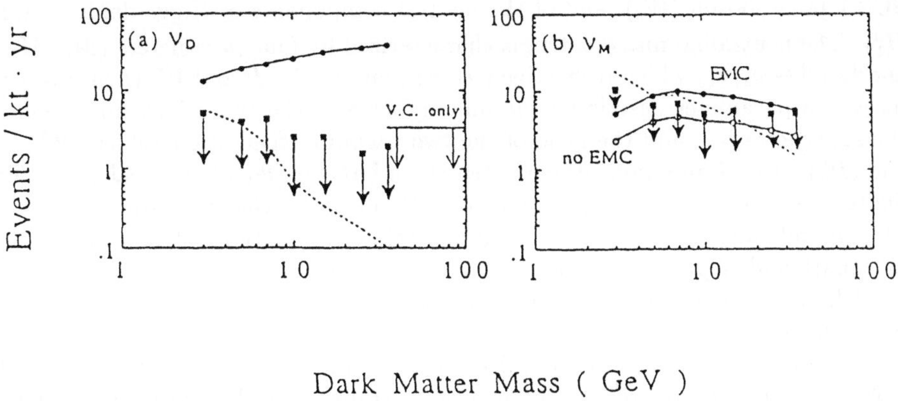

Figure 85: Upper limits on the rate of contained events from the Sun for (a) Dirac neutrinos (ν_D) and (b) Majorana neutrinos (ν_M) at 90% CL. The circles show the expected neutrino event rates. In (a), an upper limit obtained from vertex-contained (V.C.) events only is also shown. In (b), the closed circles show the expected event rate calculated with the spin-dependent cross section, which includes the EMC effect. The open circles indicate the expected event rate calculated with the spin coupling derived from a naive quark model (no EMC effect). The dashed lines indicate the atmospheric neutrino background.

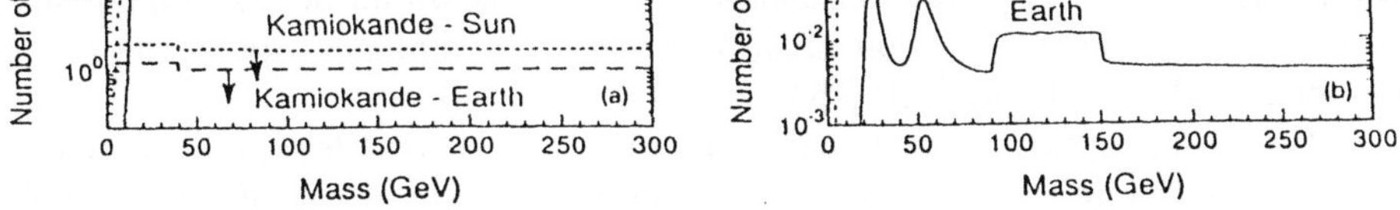

Figure 86: Number of muons expected from massive neutrino dark-matter annihilation in the Earth or Sun (for $\rho_{DM} = 0.3$ GeV/cm^3 and $v_{DM} = 300$ km/s; see text) and the limits from upward-going muons observed in Kamiokande, plotted as a function of the assumed neutrino mass: (a) for Dirac neutrinos, (b) for Majorana neutrinos. The associated heavy lepton mass and the Higgs mass are assumed to be $m_\nu + 100$ GeV and 1 TeV, respectively.

there are four neutralino mass eigenstates which can be expressed as linear combinations of superpartners of the hypercharge gauge boson ($\tilde{B}$), of the neutral SU(2) gauge boson ($\tilde{W}_3$), and of the neutral components of Higgs doublets ($\tilde{H}_1^0$, $\tilde{H}_2^0$). The neutralino mass matrix is characterized by four parameters (M_1, M_2, μ and $\tan\beta \equiv v_2/v_1$) which respectively correspond to the $\tilde{B}$ and $\tilde{W}_3$ (non-physical) masses, a parameter describing the mixing between the two Higgs superfields in the superpotential, and the ratio of the two vacuum expectation values $\langle H_1^0 \rangle = v_1$ and $\langle H_2^0 \rangle = v_2$. Grand unification relates M_1 to $\frac{5}{3} M_2 \tan^2\theta_W$ at the weak scale. The lightest neutralino mass (m_χ) is consequently given by the three free parameters (M_2, μ and $\tan\beta$). In the analysis of neutralino interactions, it is assumed that all squark and slepton masses are so heavy that they do not show up explicitly in the following discussion. The lightest scalar Higgs mass (M_{H_2}) is also taken as a free parameter.

Neutralinos are conjectured to have strong coherent interactions with nuclei [152, 148]; they therefore may be efficiently trapped at the center of the Earth if their masses are close to those of the constituent nuclei of the Earth (30 – 80 GeV). Here, the analysis is limited to models which give the neutralino mass as being below the W-boson mass, and therefore neutralino pair annihilation produces directly or indirectly (via Higgs decay) fermion-antifermion pairs as the final products. For this reason, only the Earth is considered as a source of energetic neutrinos produced in the annihilation of accumulated neutralinos. For a calculation of the expected upward-going muon flux, the annihilation cross section based on MSSM [152, 148] is used; the resulting secondary particles are dealt with by JETSET.

The constraints for neutralinos are presented as regions excluded in (μ,M_2)-space for some combinations of parameters $\tan\beta$ and M_{H_2} under the assumptions of $\rho_\chi = 0.3$ GeV/cm^3 and $v_\chi = 300$ km/s. Figure 87 shows the results for the 90% CL excluded regions in (μ,M_2)-space based on the upper limits for annihilation of massive neutralinos captured by the Earth: (a) for $\tan\beta = 2$ and $M_{H_2} = 30$, 40, and 50 GeV, (b) for $\tan\beta = 8$ and $M_{H_2} = 50$, 70, and 80 GeV, and (c) for $\tan\beta = 20$ and $M_{H_2} = 50$, 80 and 90 GeV. The mass range of the lightest scalar Higgs (H_2) is restricted by the ($\tau\bar{\tau}$ + hadrons)-jet analysis at LEP ($M_{H_2} > 23$ (38) GeV for $\tan\beta = 2$ (8) [153]), and by the MSSM prediction of $M_{H_2} < m_Z$ at the tree-level. In Figs. 87 (a) – (c), neutralino isomass curves are shown for 30, 52 and 66 GeV. Notice that the excluded region in (μ,M_2)-space is complimentary to the region excluded by the results at LEP [153].

If the cosmological relic abundance of neutralinos is smaller than the density parameter of the universe, the same comment as given at the end of subsect. 6.7.2 applys. For detailed discussion, see Refs. [152, 148, 42].

The analysis has also been extended to a neutralino mass range above the W-boson mass [154]. In this case, neutralino pair annihilation proceeds through $\chi\chi \to f\bar{f}$ (f stands for a fermion), W^+W^-, Z^0Z^0, $H_2^0H_3^0$, $H_1^0H_3^0$, $Z^0H_1^0$, $Z^0H_2^0$, and $W^\pm H^\mp$ [148, 155], where H_1^0, H_2^0, H_3^0, and $H^\pm$ are three neutral and two charged, physical Higgs bosons, and fermions are also produced by subsequent decay of gauge/Higgs bosons. Possible signal from neutralino dark-matter annihilations in the Sun or the Earth has been searched for, and within the framework of

MSSM a large portion of the parameter space is excluded up to a neutralino mass range of ~ 1 TeV, as shown in Fig. 88 if the neutralino is the major component of the dark matter in our galaxy. This mass range has never been analyzed in previous indirect searches for dark matter.

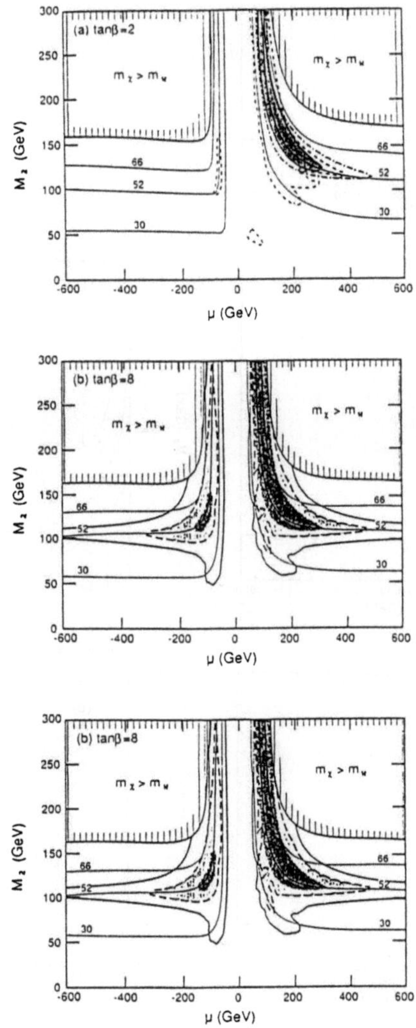

Figure 87: The 90% CL excluded regions in (μ, M_2)-space: (a) for $\tan\beta = 2$ and $M_{H_2} = 30$ GeV (bounded by dashed line), 40 GeV (dot-dashed line), and 50 GeV (solid line), (b) for $\tan\beta = 8$ and $M_{H_2} = 50$ GeV (solid line), 70 GeV (dashed line), and 80 GeV (dot-dashed line), and (c) for $\tan\beta = 20$ and $M_{H_2} = 50$ GeV (solid line), 80 GeV (dot-dashed line), and 90 GeV (dotted line), based on the Kamiokande results with the assumptions of $\rho_\chi = 0.3$ GeV/cm^3 and $v_\chi = 300$ km/s. The thin solid lines represent neutralino isomass curves corresponding to $m_\chi = 30$, 52 and 66 GeV. The areas $(m_\chi > m_W)$ in the upper-left and upper-right corners are not considered in this analysis.

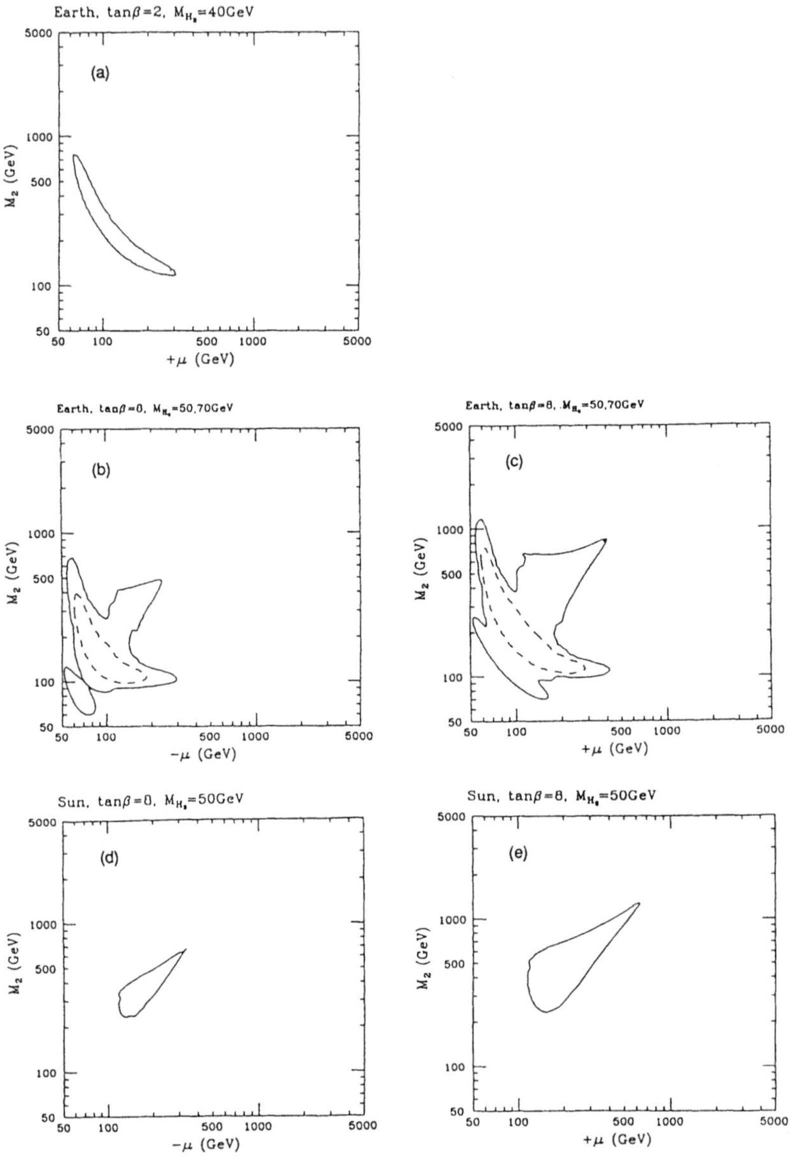

Figure 88: The 90% CL excluded regions in (μ, M_2)-space: (a) for $\tan\beta = 2$, (b) - (e) for $\tan\beta = 8$, and (f) - (i) for $\tan\beta = 20$; (a) - (c), (f), and (g) for the Earth, (d), (e), (h), and (i) for the Sun. The regions surrounded by the lines are excluded when M_{H_2} is (a) 40 GeV (solid line), (b) - (i) 50 GeV (solid line), 70 GeV (dashed line), and 90 GeV (dotted line). In these calculations, it is assumed that all squark and slepton masses are infinitely heavy and no radiative correction is applied. (In Ref. [152], however, the results of calculations are also presented, in which all squark and slepton masses are assumed to be $m_{\tilde{q}} = \max(2.5\, m_\chi,\ m_{\tilde{q},min} = 126 \text{ GeV})$ and radiative corrections are applied.)

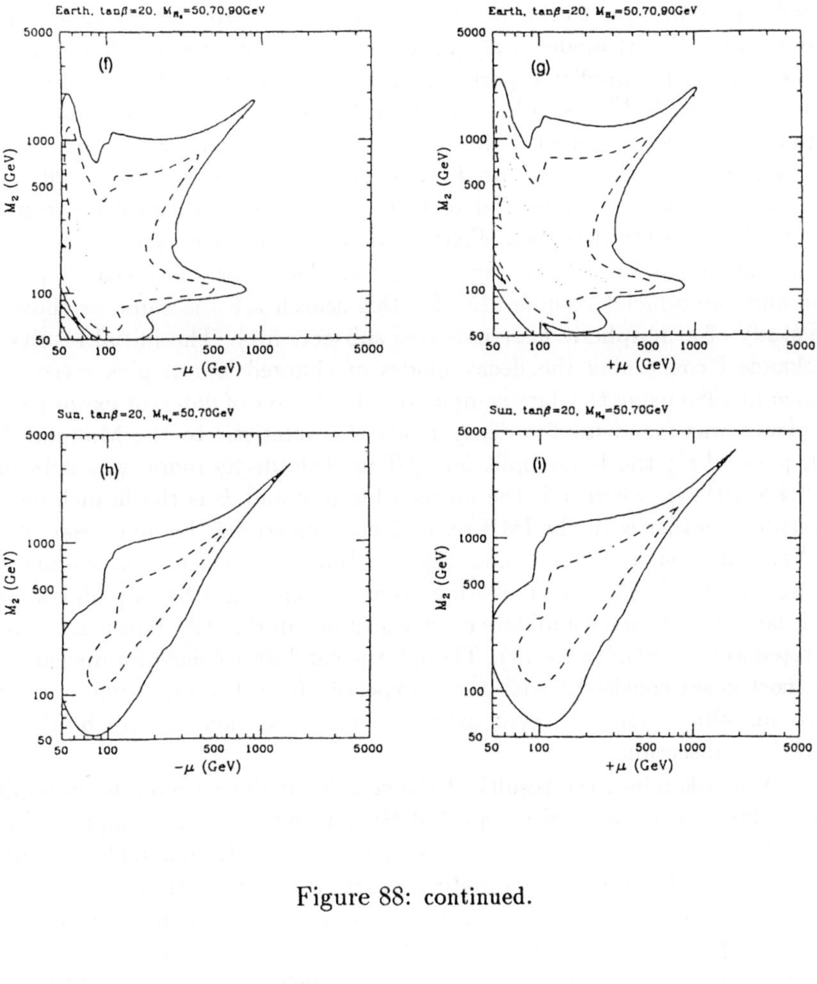

Figure 88: continued.

7. Search for Baryon-Number Violating Processes

7.1 Search for Nucleon Decay

It was a prediction of the Grand Unified Theories (GUT) [156] that stimulated the experimental search for nucleon decay with dedicated underground detectors. In the early 1980's, there was some optimistic hope that nucleon decay would soon be observed.† This was due to the following reason. The main decay mode predicted by the Minimal SU(5) model [158], which was admired very much at that time, was $p \rightarrow e^+\pi^0$ with a predicted partial lifetime of $4 \times 10^{29\pm0.7}\,(M_X/2 \times 10^{14}\ \mathrm{GeV})^4$ yr, where $M_X = (2.0^{+2.1}_{-1.0}) \times 10^{14}$ GeV. This decay mode is among those that are experimentally easy to detect. Thus, if the lifetime of the proton were just the central value of the theoretical prediction, an experiment watching 100 kton of protons would have been able to detect at least $150 \times \varepsilon$ proton decay signals in one year. Here ε is the detection efficiency of $p \rightarrow e^+\pi^0$; it is typically $0.3 \sim 0.5$.

Kamiokande has been searching for nucleon decay for many years. The data sample and the fiducial volume used for this search are the same as those used in the study of atmospheric neutrinos (see subsect. 5.1). The initial results from Kamiokande-I concerning the decay modes of charged lepton plus mesons were published in 1985 using the data sample from 0.83 kt·yr of detector exposure [159]. No evidence was found for the decay mode characteristic to the Minimal SU(5) model, $p \rightarrow e^+\pi^0$; the lower limit on τ/B for this decay mode was determined to be 6.4×10^{31} yr, where τ is the nucleon lifetime and B is the branching ratio. This result, together with the IMB result [160] concerning the non-observation of $p \rightarrow e^+\pi^0$, ruled out the simple and elegant Minimal SU(5) model. Kamiokande, however, reported [159] four interesting events which might be candidates for nucleon decay. One of these candidate events is shown in Fig. 89. This event could be interpreted as $p \rightarrow \mu^+\eta^0$ ($\eta \rightarrow \gamma\gamma$). Though the candidate events for nucleon decay are in most cases consistent with those expected from the atmospheric neutrino background, this particular event calls for some attention, as will be discussed later in this subsection.

From Kamiokande-I, the results of the search for the nucleon decay modes of neutrino plus mesons were also reported [161] using the data sample from 1.11 kt·yr of the detector exposure. In these decay modes Kamiokande did not see any excess of nucleon-decay candidates, either, above the atmospheric neutrino background. The 90% CL lower limits on the partial lifetime of nucleons were obtained for $p \rightarrow \bar{\nu}K^+$ and $n \rightarrow \bar{\nu}K^0$ to be 4.2×10^{31} yr and 3.2×10^{31} yr, respectively. These decay modes are important, since the Supersymmetric Grand Unified Theories (SUSY-GUTs) [162] which appeared in the early 1980's predicted that they are dominant in nucleon decay.

After these publications of the initial results, the performance of the Kamiokande detector was upgraded from Kamiokande-I to Kamiokande-II. This upgrade im-

---

†A pioneering experiment at the Kolar gold fields (KGF) indeed reported [157] a few candidate events for proton decay.

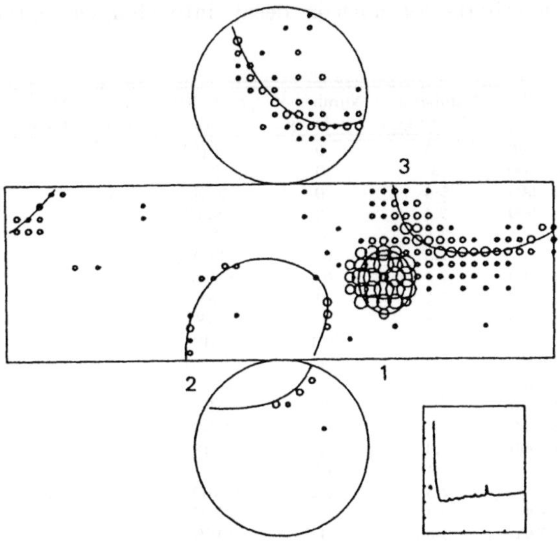

Figure 89: Candidate of nucleon decay observed in Kamiokande-I. A possible decay mode is $p \to \mu^+ \eta^0 (\eta \to \gamma\gamma)$. The Cherenkov ring indicated by 2 is assigned as μ and the other two rings (1 and 3) are assigned as γ's.

proved the detector performance not only for the detection of low-energy neutrinos, but also for the search for nucleon decay. The analysis programs for nucleon decay were also improved. Consequently, less neutrino background events were expected in Kamiokande-II than those in Kaniokande-I. In the rest of this subsection, the analysis adopted in the Kamiokande-II nucleon decay search [1] is descibed. The final results, however, will be given by combining the Kamiokande-I and -II searches.

In Kamiokande-II, the selection of nucleon decay candidates for each decay mode was carried out based on the following information: the total p.e. number, number of Cherenkov rings, number of observed μ-decay signals, particle identification, momentum imbalance, total invariant mass, meson momentum, and meson mass. The criteria for charged lepton + meson and antineutrino + meson modes are summarized in Tables 11 and 12, respectively. Here, a non-interacting neutral, massless particle is referred to as an "antineutrino". The cuts were determined from Monte-Carlo studies of nucleon-decay events.

These selection criteria are quite similar to those adopted in Kamiokande-I [159, 161], except for the improvements in kinematical reconstruction, i.e., the cuts on the momentum imbalance, total invariant mass, and meson mass are tighter than in Kamiokande-I for most of the decay modes. Also, particle identification was applied for all Cherenkov rings after spatial reconstruction. The particle type of each ring (if successfully identified) was required to be not incompatible with those shown in the fifth column of Tables 11 and 12. If particle species were not determined uniquely for any ring, all possible particle species for that ring were

Table 11: Selection criteria for nucleon decay into charged lepton + mesons in Kamiokande-II.

Modes	Total[(a)] p.e.	Number of Ch. rings	Number of μ-decay	particle[(b)] id.	ΔP & $M_{tot}^{(c)}$ tight	loose	$M_{meson}^{(d)}$	
$e^+\pi^0$	2900-3500	2,3	0	SS(S)	o		90-180	(e)
$e^+\eta$ $(\gamma\gamma)$	2900-3500	3	0	SSS	o		480-620	
$(3\pi^0)$	2900-3500	$\geq$4	0	SSS..	o		450-650	
$e^+\omega$ $(\gamma\pi^0)$	2900-3500	3,4	0	SSS(S)	o		700-870	(e)
(3π)	–	3-5	0,1	SMS..		o	600-1000	(f)
$e^+\rho^0$	–	2,3	0,1	SM(M)		o	600-900	(g)
e^+K^0 $(\pi^+\pi^-)$	1100-1700	2,3	0,1	SM(M)	o		440-550	(g)
$(2\pi^0)$	2900-3500	3-5	0	SSS..	o		400-590	
e^+K^{*0} $(3\pi^0)$	2900-3500	$\geq$4	0	SSS..	o		700-1000	
$(\pi^+\pi^-\pi^0)$	–	3-5	0,1	SMS..		o	600-1000	(f)
$(\pi^-\mu^+\nu)$	200-800	3	1	SMM	–	–	–	(h)
$\mu^+\pi^0$	2200-2900	2,3	1	MS(S)	o		90-180	(e)
$\mu^+\eta$ $(\gamma\gamma)$	2200-2900	3	1	MSS	o		480-620	
$(3\pi^0)$	2200-2900	$\geq$4	1	MSS..	o		450-650	
$\mu^+\omega$ $(\gamma\pi^0)$	2400-3000	2,3	1	SS(S)	–	–	700-870	(e,i)
(3π)	–	2-4	1,2	MS..	–	–	600-1000	(f,j)
$\mu^+\rho^0$	–	2,3	1,2	MM(M)		o	600-900	(g)
μ^+K^0 $(\pi^+\pi^-)$	400-900	2,3	1,2	MM(M)	o		440-550	(g)
$(2\pi^0)$	2200-2900	3-5	1	MSS..	o		400-590	
$e^+\pi^-$	1500-2500	2	0	SM	o		–	
$e^+\rho^-$	–	2-4	0	S...		o	600-900	(f,g)
$\mu^+\pi^-$	800-1800	2	1	MM	o		–	
$\mu^+\rho^-$	–	2-4	1	M...		o	600-900	(f,g)

(a) Total p.e. corresponds to electron energy by a relation 3.4 p.e. = 1 MeV.
(b) M = (μ or π-like), and S = (e or γ-like).
(c) For decay modes indicated by 'tight', total invariant mass, M_{tot}, and momentum imbalance, ΔP, must satisfy, 830 - 1040 MeV/c^2 and < 250 MeV/c, respectively. For decay modes indicated by 'loose', M_{tot} and ΔP must satisfy 800 - 1100 MeV/c^2 and < 350 MeV/c, respectively.
(d) Requirement on meson mass (MeV/c^2).
(e) π^0 mass (90 - 180 MeV/c^2) is required for 3-ring events ($e\pi^0$, $\mu\pi^0$ and $\mu\omega$ $(\gamma\pi^0)$) or for 4-ring events ($e\omega$ $(\gamma\pi^0)$).
(f) If $\gamma\gamma$ are assigned to any of 2 rings, then π^0 mass (90 - 180 MeV/c^2) is required.
(g) If the observed number of Cherenkov rings is 2 and the opening angle between the 2 rings is larger than 130°, then one missing particle (μ or π) is assumed, and the momentum of the missing particle is assigned to minimize Eq. (21). In this case, M_{tot} and ΔP is calculated including the missing particle. After this procedure the criteria listed in the table are imposed to the event with the assumed one invisible particle.
(h) P_μ = 215 - 255 MeV/c is additionally required.
(i) Meson momentum, P_{meson}, < 250 MeV/c is additionally required.
(j) P_{meson} < 350 MeV/c is additionally required.

tried in order to test the various kinematical quantities, such as the momentum imbalance.

The selection criteria for the decay mode $p \to \bar{\nu}K^+$ require a special explanation, because most K^+ stop before they decay, and only the decay products of K^+ are observed. For example, μ^+ from $K^+ \to \mu^+\nu$ has a monochromatic momentum of 236 MeV/c. Therefore, if this decay mode is to be observable, an accumulation of μ-like single-ring events is expected at this momentum. Figure 90 shows the momentum distribution of μ-like single ring events associated with one μ-decay signal. Here, both the Kamiokande-I and -II data are combined. Clearly, no significant accumulation of real events can be seen within the signal region. Contrary to $K^+ \to \mu^+\nu$, there is no significant background in another K^+ decay channel, $K^+ \to \pi^+\pi^0$.

The analysis then goes as follows. Three kinds of data sets (the real data,

Table 12: Selection criteria for nucleon decay into antineutrino + mesons in Kamiokande-II.

Modes	Total p.e.	Number of Ch. rings	Number of μ-decay	particle[(a)] id.	$M^{(b)}_{meson}$	$P^{(c)}_{meson}$	
$\bar{\nu}\pi^+$	–	1	0,1	M	–	340-580	
$\bar{\nu}\rho^+$	1060-2230	2,3	0,1	MS(S)	600-900	40-400	(d)
$\bar{\nu}K^+$ $(\mu^+\nu)$	–	1	1	M	–	–	(e)
$(\pi^+\pi^0)$	740-1040	2,3		SS(M)	440-550	0-50	(d,f)
$\bar{\nu}K^{*+}$ $(\pi^0\pi^0\pi^+)$	1660-2440	3-5	0,1	MSS..	600-1000	0-400	
$(\pi^+\pi^+\pi^-)$	110-580	2,3	0-2	MM(M)	600-1000	0-400	
$(\pi^0\mu^+\nu)$	920-1560	2,3	1	MS(S)	–	–	(d,g)
$\bar{\nu}\pi^0$	1330-2000	2	0	SS	90-180	340-580	
$\bar{\nu}\eta$ $(\gamma\gamma)$	1850-2510	2	0	SS	480-620	200-430	
$(3\pi^0)$	1850-2510	≥ 3	0	SSS..	450-650	200-430	
$\bar{\nu}\omega$ $(\gamma\pi^0)$	2450-3010	2,3	0	SS(S)	700-870	60-300	(h)
(3π)	680-1440	2-4	0,1	MS..	600-1000	0-400	(d)
$\bar{\nu}\rho^0$	110-990	2	0,1	MM	600-900	40-400	
$\bar{\nu}K^0$ $(\pi^+\pi^-)$	110-510	2	0,1	MM	440-550	240-450	
$(2\pi^0)$	1740-2430	3,4	0	SSS(S)	400-590	240-450	
$\bar{\nu}K^{*0}$ $(3\pi^0)$	2570-3460	≥ 3	0	SSS..	700-1000	0-300	
$(\pi^+\pi^-\pi^0)$	860-1660	2-4	0,1	MS..	600-1000	0-400	(d)
$(\pi^-\mu^+\nu)$	130-470	2	1	MM	–	–	(g)

(a) M = (μ or π-like), and S = (e or γ-like).
(b) Requirement on meson mass (MeV/c^2). 'Meson mass' corresponds to the total invariant mass of an event in these modes.
(c) Requirement on meson momentum (MeV/c). 'Meson momentum' corresponds to the total momentum imbalance of an event in these modes.
(d) If $\gamma\gamma$ are assigned to any of 2 rings, then π^0 mass (90 - 180 MeV/c^2) is required.
(e) P_μ = 215 - 255 MeV/c is additionally required.
(f) For 2 ring events, $M_{\gamma\gamma}$ = 90 - 180 MeV/c^2 and 1 μ-decay are required and M_{meson} and P_{meson} are not required. For 3 ring events, the requirement on μ-decay is 0 or 1.
(g) P_μ = 215 - 255 MeV/c and $\theta_{\mu\pi} > 90°$ are additionally required.
(h) π^0 mass (90 - 180 MeV/c^2) is required for 3-ring events.

the simulated nucleon decay events, and the simulated atmospheric neutrino interactions) were passed through the same selection criteria in order to obtain the number of nucleon decay candidates, the detection efficiency and the neutrino background rate. In calculating the detection efficiency, various nuclear effects, such as the Fermi motion of nucleons in the oxygen nuclei and the subsequent nuclear interactions of mesons in the oxygen nuclei and in the water, are considered as detailed in Ref. [44] (see Sect. 4). It should be noted that in a water detector 20% of the proton-decay events are caused by free protons, and are free from these nuclear effects in the nuclei. However, the remaining nucleon-decay events are caused by bound nucleons. Consequently, the dominant systematic error of the detection efficiency, which depends on the produced mesons and their momentum, results from these nuclear effects in the oxygen nuclei. For pions, a systematic error of ±30% is estimated for their interaction cross sections in the nuclei. This error translates into an uncertainty in the quoted detection efficiency of typically ±10% (*e.g.*, for $p \to e^+\pi^0$) or maximally ±35% (for $n \to \bar{\nu}\rho^0$). For K mesons with strangeness = 1, on the other hand, the uncertainty is estimated to be ±1% because of their small interaction probability (3%) in the oxygen nuclei. There is also a statistical error in the detection efficiency coming from a number of analyzed nucleon-decay events. For each decay mode, the detection efficiency is estimated

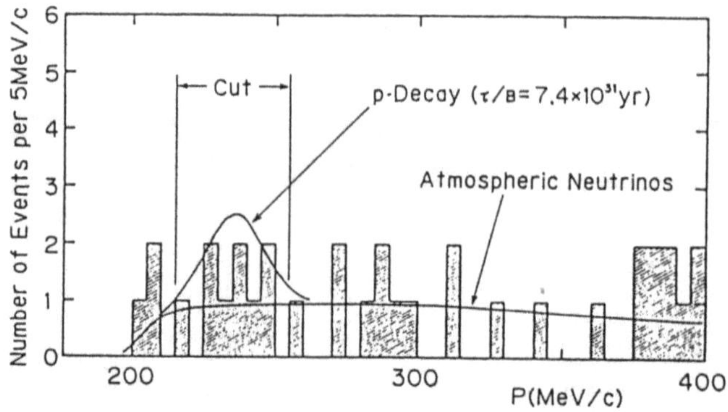

Figure 90: Momentum distribution of μ-like single-ring events associated with one μ-decay signal. Also shown are the expected background distribution normalized to the total observed number of μ-like events, the adopted cuts for nucleon decay search, and the expected signal for $\tau/B = 0.74 \times 10^{32}$ yr which corresponds to the 90% CL lower limit for this channel.

with the use of $30 \sim 140$ simulated events. Therefore, the overall uncertainty in the efficiency ranges from ±4% (3.7% of statistical fluctuation of the number of events which satisfy the selection criteria and 1% of systematic error) of the quoted efficiency for $p \to \bar{\nu}K^+$ $(\mu^+\nu)$ to ±50% (35% of statistical, and 35% of systematic) for $n \to \bar{\nu}\rho^0$.

The simulated events of atmospheric neutrino interactions in the detector are generated by a Monte-Carlo program [44], as described in Sect. 4. Various (ν_e/ν_μ independent) quantities, such as the distributions of the total visible energy and the number of Cherenkov rings, are reasonably reproduced. To illustrate this, Fig. 91 shows as an example a two-dimensional plot of momentum imbalance (ΔP) *vs* the total invariant mass (M_{tot}) for multi-ring events: (a) the real data, (b) the simulated neutrino interactions† and (c) the simulated nucleon decay, $p \to e^+\pi^0$. These kinematical quantities are not always uniquely determined, due to the ambiguity in the particle identification. In such cases, a point which gives the minimum of the expression

$$(M_{tot} - 938\,\mathrm{MeV/c^2})^2 + (\Delta P - 150\,\mathrm{MeV/c})^2 \cdot \theta(\Delta P - 150\,\mathrm{MeV/c}), \qquad (21)$$

where θ is the step function, is selected and plotted in (a) and (b). Here, the last factor, $\theta(\Delta P - 150$ MeV/c), is included, considering the Fermi motion of

†One may argue, based on Fig. 91 (b), that the background already dominates at 11.4 kt·yr. It should be noted, however, that the events in the allowed region in ΔP and M_{tot} are further subject to various selection criteria, such as meson mass, number of μ-decay, etc. They reduce the number of remaining events approximately by a factor of 2. The remaining events are backgrounds of $e^+\rho^0$, $e^+\rho^-$, $e^+\omega$ $(\pi^+\pi^0\pi^-)$, e^+K^{*0} $(\pi^+\pi^0\pi^-)$, $\mu^+\rho^0$ and $\mu^+\rho^-$. It should be noted also that the decay modes which have high background rates ($\geq$ 1 event per 11.4 kt·yr) involve ρ, ω or K^*. For decay modes involving π or K, the background rates are much lower. See Tables 13 and 14.

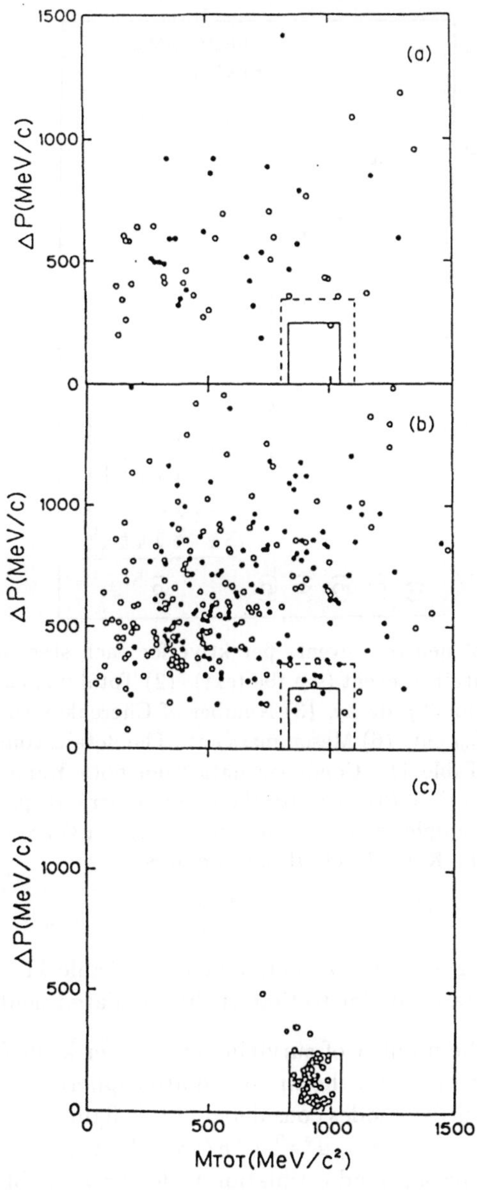

Figure 91: ΔP *vs* M_{tot} for multi-ring events after reconstruction: (a) the Kamiokande-II data, (b) the simulated neutrino interactions of 11.4 kt·yr exposure equivalent and (c) the simulated nucleon decay into $e^+\pi^0$. ∘ and • represent events which have no and at least 1 observed μ-decay, respectively. In plotting points in (c), various nuclear effects such as Fermi motion of nucleons and pion nuclear interactions in oxygen nucleus are included. Events shown in (c) are those selected after having imposed the criteria of the total number of p.e., number of Cherenkov rings, number of μ-decay, and particle identification cuts. Therefore, in (c), most of the events whose π^0 undergo nuclear interaction are removed before plotting the points. The broken and solid lines show the cuts on the momentum imbalance and total invariant mass; see footnote (c) to Table 11.

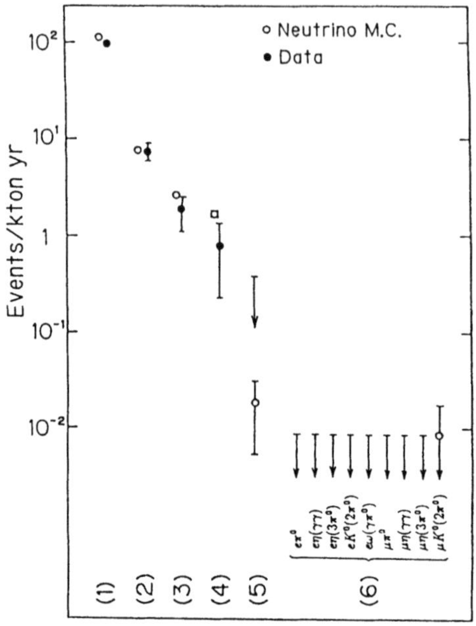

Figure 92: Number of neutrino events per kt·yr for each step of the nucleon decay selection: (1) Fully contained event (see the text). (2) Total p.e. cut; 2200 – 2900 for 1 μ-decay. 2900 – 3500 for 0 μ-decay. (3) Number of Cherenkov rings ≥ 2. (4) Particle id. cut. (5) ΔP and M_{tot} cut. (6) Meson mass cut. The details concerning the selection criteria are shown in Table 11. Combined data from both Kamiokande-I and -II are also shown for criteria (1) $\sim$ (3), because there are no serious quantitative differences between the two data samples as far as the criteria (1) $\sim$ (3) are concerned. After the fourth criterion, only the Kamiokande-II data are used.

the nucleons and the ΔP and M_{tot} cuts defined in Table 11. It is seen that the distribution of the data is similar to that of the simulated neutrino interactions.

Figure 92 shows the number of surviving events per kt·yr for each step of the reduction for both the data and the simulated atmospheric neutrino events of 105 kt·yr equivalent. The decay modes considered in this figure are $p \to e^+\pi^0$, $e^+\eta\,(\gamma\gamma,\ \pi^0\pi^0\pi^0)$, $e^+K^0\,(\pi^0\pi^0)$, $e^+\omega\,(\gamma\pi^0)$, $\mu^+\pi^0$, $\mu^+\eta\,(\gamma\gamma,\ \pi^0\pi^0\pi^0)$, and $\mu^+K^0\,(\pi^0\pi^0)$. This figure shows that the background estimation is done reasonably well.

Possible systematic errors in the background estimation include the uncertainties in the absolute atmospheric neutrino fluxes at $\sim$ 1 GeV ($\pm$20%), in absolute neutrino cross sections for the final states with multi Cherenkov rings ($\sim \pm$20%) and in the cross sections of pion interactions in oxygen nuclei ($\sim \pm$30%). The uncertainty of $\pm$30% in the pion cross section changes the number of multi-ring neutrino events by $\pm$10%. From these considerations, the systematic uncertainty in the background is estimated to be $\sim \pm$30%. The corresponding uncertainty for $p \to \bar{\nu}\pi^+$ and $p \to \bar{\nu}K^+\,(\mu^+\nu)$ is estimated to be $\pm$10%. This is because the background rate is normalized to the total number of single-ring μ-like events.‡ There is also an error in the background estimation resulting from the statistical fluctuation of the number of analyzed neutrino events. Since $0 \sim 18$ simulated neutrino events survive within the nucleon decay criteria for the decay modes with multi-ring final states, the statistical error of the estimated number of neutrino background ranges from $\pm 24\%$ to $\pm(\gtrsim 100)\%$ depending on the decay mode.

Tables 13 and 14 summarize the detection efficiency multiplied by the meson branching ratio ($\varepsilon \cdot B_m$), the number of candidate events, and the number of the expected neutrino background events in Kamiokande-II. The corresponding numbers in Kamiokande-I are also listed in the same Tables. Because of the good energy resolution of e, γ and μ, the expected backgrounds are small for the decay modes which involve only these particles in the final state. On the other hand, the search is limited for some of the decay modes which involve charged pions in the final state. There are several candidate events, which, however, are in most cases consistent with those expected from the neutrino background.†

From these results, it is concluded that in Kamiokande-II no evidence has been observed for nucleon decay. Therefore, the lower limits on the nucleon partial lifetime (τ/B, where B is the branching ratio of nucleon decay for the relevant decay mode) are calculated at the 90% CL combining Kamiokande-I and -II data and subtracting the atmospheric neutrino background. The 90% CL upper limit for the number of nucleon-decay candidates is calculated by integrating the likelihood function to the 90% probability level:

$$\frac{\int_0^{x_{limit}} \left[\prod_{i=1}^{n} P\left(N_{obs}(i),\ X_{BG}(i) + \frac{\varepsilon B_m(i)\cdot T(i)}{\sum_{j=1}^{n}(\varepsilon B_m(j)\cdot T(j))}\cdot x\right)\right] dx}{\int_0^{\infty} \left[\prod_{i=1}^{n} P\left(N_{obs}(i),\ X_{BG}(i) + \frac{\varepsilon B_m(i)\cdot T(i)}{\sum_{j=1}^{n}(\varepsilon B_m(j)\cdot T(j))}\cdot x\right)\right] dx} = 0.90,$$

where x_{limit} is the number to be obtained (i.e., the upper limit for the number of nucleon-decay candidates), n the number of independent observations (for example, n=4 for the $p \to \bar{\nu}K^+$ mode, because there are two K^+ decay modes, $\mu^+\nu$ and $\pi^+\pi^0$, and two experimental phases, Kamiokande-I and -II), $P(N,x)$ the probability function of the Poisson statistics, $N_{obs}(i)$ the number of observed candidates in the i-th observation, $X_{BG}(i)$ the number of neutrino background events for the i-th observation, and $T(i)$ the detector exposure in the i-th observation in units of

‡Because of a deficit of the number of single ring μ-like events relative to the Monte Carlo calculation (0.60 ± 0.06, see Ref. [10]), the estimated number of background events presented in Fig. 90 and Table 14 for $\bar{\nu}K^+$ ($\mu^+\nu$) and $\bar{\nu}\pi^+$ are $\times 0.6$ of the calculated background without any deficit.

†Event A, which is a candidate of $p \to \mu^+\eta$ ($\gamma\gamma$), observed in Kamiokande-I, might be a possible exception, because the expected background is ≤ 0.02. However, the observed momentum imbalance, 396 ± 71 MeV/c, is too large to conclude that this single event is evidence for nucleon decay.

Table 13: Results of the nucleon decay search for charged lepton + meson modes. The detection efficiency multiplied by the meson branching ratio, $\varepsilon \cdot B_m$, the number of observed candidates (a,b,c for KAM-II and A,B,C,D,E for KAM-I) and the number of expected neutrino backgrounds in Kamiokande-II (KAM-II) and Kamiokande-I (KAM-I) are shown in the 2 ~ 4th and the 5 ~ 7th columns, respectively. The background subtracted limits on τ/B at 90 % CL are shown in the eighth column. Both the KAM-I and -II data are combined in deriving these numbers. Background unsubtracted limits are also shown in the last column.

	KAM-II (2.46 kt·yr)			KAM-I (1.30 kt·yr)			KAM-I+II (3.76 kt·yr)	
Modes	$\varepsilon \cdot B_m$		ν BG	$\varepsilon \cdot B_m$		ν BG	τ/B limit BG subt'd ($\times 10^{31}$ yr)	τ/B limit BG unsubt'd ($\times 10^{31}$ yr)
$e^+\pi^0$	0.45		< 0.02(*)	0.53		< 0.02	26	26
$e^+\eta$ $(\gamma\gamma)$	0.14		< 0.02(*)	0.17		< 0.02	14	14
$(3\pi^0)$	0.10		< 0.02(*)	0.11		0.05		
$e^+\omega$ $(\gamma\pi^0)$	0.03		< 0.02(*)	0.04		0.05	4.5	3.4
(3π)	0.08	b	0.5	0.17	B	0.9		
$e^+\rho^0$	0.24	a,c	1.8	0.18		0.9	7.5	5.2
e^+K^0 $(\pi^+\pi^-)$	0.14		0.1	0.18		< 0.1	15	15
$(2\pi^0)$	0.11		< 0.02(*)	0.13		0.05		
e^+K^{*0} $(3\pi^0)$	0.01		< 0.02(*)	0.02		0.05	5.2	3.1
$(\pi^+\pi^-\pi^0)$	0.03	b	0.5	0.04	B	0.7		
$(\pi^-\mu^+\nu)$	0.09		0.2	0.08		0.1		
$\mu^+\pi^0$	0.43		< 0.02(*)	0.40		0.05	23	23
$\mu^+\eta$ $(\gamma\gamma)$	0.13		< 0.02(*)	0.16	A	< 0.02	6.9	6.9
$(3\pi^0)$	0.08		< 0.02(*)	0.06		0.02		
$\mu^+\omega$ $(\gamma\pi^0)$	0.04		< 0.1	0.04		< 0.1	5.7	3.7
(3π)	0.09		0.7	0.17	C,E	1.2		
$\mu^+\rho^0$	0.22		1.0	0.18		0.7	11	11
μ^+K^0 $(\pi^+\pi^-)$	0.20		0.1	0.20	D	0.3	12	9.8
$(2\pi^0)$	0.10		0.02(*)	0.11		0.02		
$e^+\pi^-$	0.28		< 0.1	0.32		0.1	13	13
$e^+\rho^-$	0.15		1.2	0.10		0.7	5.8	5.8
$\mu^+\pi^-$	0.23		< 0.1	0.25		< 0.1	10	10
$\mu^+\rho^-$	0.06		0.9	0.08	C	0.9	2.3	1.7

(*) Neutrino background rates are estimated from Monte Carlo events equivalent to 105 kt·yr for the decay modes denoted by (*) and to 22 kt·yr for the other decay modes (KAM-II).

kt·yr. Then, the 90% CL lower limit on τ/B is obtained as

$$\tau/B > \frac{\left[\sum_{j=1}^{n}\left(\varepsilon B_m(j)\cdot T(j)\right)\right]\cdot N_N}{x_{limit}},$$

where N_N is the number of protons or neutrons in 1 kton of water. The results obtained with a total detector exposure of 3.76 kt·yr (Kamiokade-I and -II combined) are shown in the last two columns of Tables 13 and 14 for charged lepton + meson and for antineutrino + meson decay modes, respectively. Figure 93 summarizes the experimental limits on the nucleon lifetime for various two-body decay modes.

To summarize, Kamiokande has not observed any evidence for nucleon decay. The background-subtracted limits on τ/B range from 0.1 to 2.6×10^{32} yr depending on the decay modes. These limits are the most stringent ones for most of the decay modes. For $p \to e^+\pi^0$, the limit on τ/B is 2.6×10^{32} yr. The limits on

Table 14: Same as Table 13 for antineutrino + meson modes.

Modes	KAM-II (2.46 kt·yr) $\varepsilon \cdot B_m$	No. of Ev.	ν BG	KAM-I (1.30 kt·yr) $\varepsilon \cdot B_m$	No. of Ev.	ν BG	KAM-I+II (3.76 kt·yr) τ/B limit BG subt'd ($\times 10^{31}$yr)	τ/B limit BG unsubt'd ($\times 10^{31}$yr)
$\bar{\nu}\pi^+$	0.24	22	23.5	0.14	10	9.3	2.5	0.6
$\bar{\nu}\rho^+$	0.15	1	0.5	0.16	4	1.0	2.7	2.1
$\bar{\nu}K^+$ $(\mu^+\nu)$	0.51	5	4.5	0.41	4	1.9	10	5.1
$(\pi^+\pi^0)$	0.10		0.5	0.10		0.4		
$\bar{\nu}K^{*+}$ $(\pi^0\pi^0\pi^+)$	0.03	1	0.3	0.07	2	1.3	2.0	1.3
$(\pi^+\pi^+\pi^-)$	0.02		0.5	0.03	2	0.3		
$(\pi^0\mu^+\nu)$	0.02		0.2	0.05		0.4		
$\bar{\nu}\pi^0$	0.28	1	2.0	0.31		1.0	10	7.5
$\bar{\nu}\eta(\gamma\gamma)$	0.15	1	0.2	0.15		0.1	5.4	4.6
$(3\pi^0)$	0.10	1	0.3	0.08		0.3		
$\bar{\nu}\omega(\gamma\pi^0)$	0.03		<0.1	0.05		<0.1	4.3	2.7
(3π)	0.11	1	0.6	0.21	2	2.1		
$\bar{\nu}\rho^0$	0.06	2	1.5	0.08	2	2.1	1.3	0.8
$\bar{\nu}K^0$ $(\pi^+\pi^-)$	0.07		0.2	0.12		0.9	8.6	8.6
$(2\pi^0)$	0.10		0.6	0.13		0.7		
$\bar{\nu}K^{*0}(3\pi^0)$	0.01		0.1	0.02		0.1	2.1	1.1
$(\pi^+\pi^-\pi^0)$	0.01	1	0.6	0.03	2	1.6		
$(\pi^-\mu^+\nu)$	0.06		0.1	0.06	1	0.1		

τ/B are 1.0×10^{32} yr and 0.9×10^{32} yr for $p \to \bar{\nu}K^+$ and $n \to \bar{\nu}K^0$, respectively. These results impose very significant constraints upon the Grand Unified Theories.

7.2 Search for $n - \bar{n}$ Oscillations

Some class of GUTs predict $\Delta B = 2$ and $\Delta L = 0$ baryon-number-violating processes, such as $n - \bar{n}$ oscillations. In view of the fact that there has so far been no experimental evidence for nucleon decay, and that the Minimal SU(5) model is excluded by the non-observation of nucleon decay (and also by the precise measurements of $\sin^2\theta_W$ by the LEP experiments [165]), it is important to search for $\Delta B = 2$ processes. Here, the search for $n - \bar{n}$ oscillations in Kamiokande [166] is described.

An $\bar{n}$ produced by $n - \bar{n}$ oscillations in a ^{16}O nucleus annihilates with a nucleon and produces a burst of pions in the residual nucleus. In order to estimate the detection efficiency of possible $n - \bar{n}$ oscillations, a Monte-Carlo program was developed. The pion multiplicity distribution was estimated from the data of hydrogen and deuterium bubble-chamber experiments with $\bar{p}$ at rest. The total and charged mean multiplicities of these pions were estimated to be 5.1 and 3.2, respectively. Ideally, $n - \bar{n}$ oscillation events should be characterized by $M_{tot} = 2 \cdot M_N$ and $\Delta P \sim P_F$, where M_N and P_F are nucleon mass and Fermi momentum, respectively. However, since the produced pions undergo strong nuclear interactions, these characteristics are somewhat loosened. Also, the energy resolution for charged pions

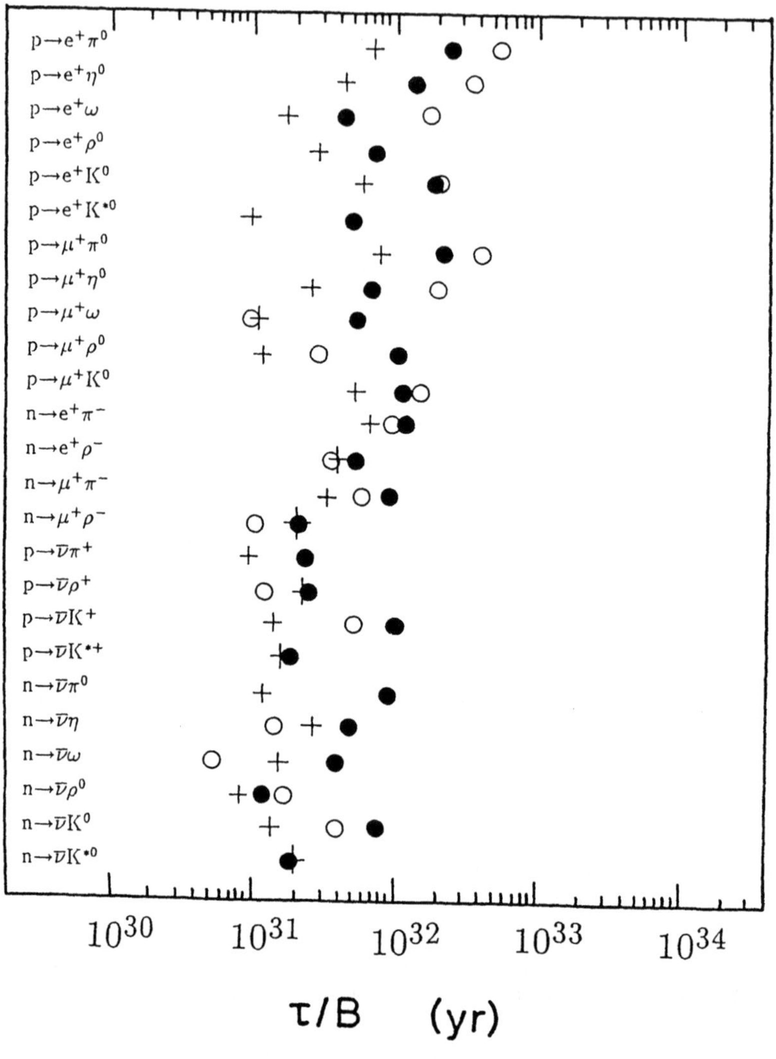

Figure 93: Nucleon partial lifetime limits at the 90% CL from Kamiokande (•), IMB [163] (∘) and Fréjus [164] (+).

are not good compared with those for electrons and muons. The search for $n - \bar{n}$ oscillations has to be performed with these facts in mind.

The data sample used for this study is the same as that used for the nucleon decay search. The selection criteria for $n - \bar{n}$ events are as follows:

(i) Ring multiplicity ≥ 3.

(ii) Kinematical constraint: By assigning possible combination of particle species ($\pi^{\pm}$ or γ) to each ring, the total invariant mass $M_{tot} = [(\Sigma_i p_i)^2]^{1/2}$ and the total residual momentum $\Delta P = | \Sigma_i p_i |$ are calculated. $1200 < M_{tot} < 2200$ MeV/c^2 and $\Delta P < 600$ MeV/c are then imposed, with particle identification whenever possible.

(iii) At least 1 μ-decay for (total p.e. number) < 2500.

Figure 94 shows a M_{tot} - ΔP plot for (a) Monte-Carlo $n - \bar{n}$ oscillation events, (b) the data, and (c) Monte-Carlo atmospheric neutrino events. The detection efficiency was estimated to be 33%. The expected number of background events was 0.9 events/yr, and no real events survived the criteria during 1.11 kt·yr of the detector exposure. The limit on the lifetime ($T_{n-\bar{n}}$) for neutrons in a ^{16}O nucleus was obtained as

$$T_{n-\bar{n}} > 4.3 \times 10^{31}\ \text{yr}\ \ (90\%\ \text{CL}).$$

This limit can be translated to the free-neutron oscillation time ($\tau_{n-\bar{n}}$) by theoretical relations [167],

$$\tau_{n-\bar{n}} = (T_{n-\bar{n}}/T_R)^{1/2}$$

and

$$T_R = 1 \times 10^{23}\ \text{s}^{-1},$$

for a ^{16}O nucleus. Therefore, the Kamiokande result implies that

$$\tau_{n-\bar{n}} = 1.2 \times 10^{8}\ \text{s}\ \ (90\%\ \text{CL}).$$

The estimated error of $\tau_{n-\bar{n}}$ is less than 30% [168]. This result should be compared with $\tau_{n-\bar{n}}$ from various experiments on $n - \bar{n}$ oscillations [169, 170, 171]. The results are summarized in Table 15.

8. Superkamiokande

8.1 Outline of the Project and the Expected Detector Performance

The Superkamiokande project utilizing a 50,000 ton water Cherenkov detector was proposed in 1984, the year after the commissioning of the Kamiokande. The budget request to the Ministry of Education, Science and Culture started in Japanese fiscal year (JFY) 1988, and the entire project was approved in JFY 1991. The ground

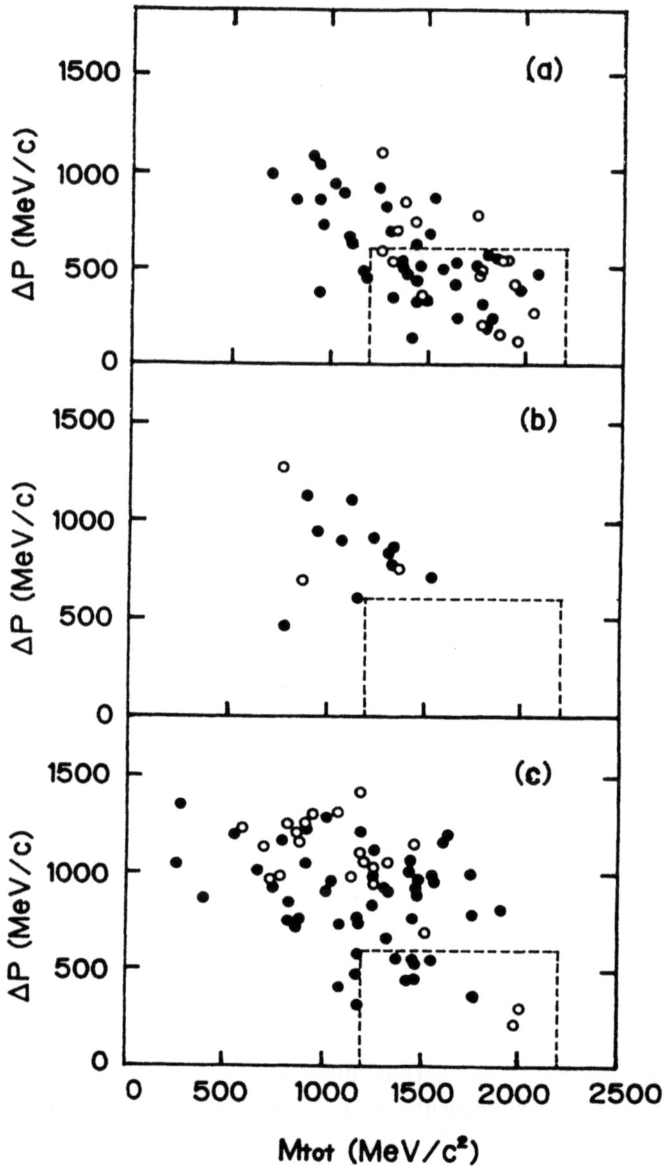

Figure 94: M_{tot} - ΔP plots for (a) Monte-Carlo $n - \bar{n}$ oscillation events (100 events generated), (b) real data (1.11 kt·yr), and (c) Monte-Carlo atmospheric neutrino events (8.8 kt·yr equivalent) after the ring multiplicity cut ($\geq$3). The dotted lines indicate the region of the kinematical cut (see the text). The filled and open circles indicate events with and without μ-decay, respectively.

Table 15: 90% CL lower limits on the $n - \bar{n}$ oscillation time.

Experiment	$T_{n-\bar{n}}$ (nucleus) (yr)	$\tau_{n-\bar{n}}$ (sec)	Comment
ILL	–	1.0×10^{7}	free neutron
IMB	2.4×10^{31}	8.8×10^{7}	H_2O
Kamiokande	4.3×10^{31}	1.2×10^{8}	H_2O
Fréjus	6.5×10^{31}	1.2×10^{8}	Fe

breaking was celebrated at the predetermined site for Superkamiokande in the Mozumi mine on December 6, 1991. The target date for completing the detector construction was aimed at the end of JFY 1995, *i.e.*, the end of March, 1996, with the excavation of the cavity to be completed in August 1994. The total budget foreseen in JFY 1993 is about 10 billion yen including everything (excavation of the cavity, construction of the water tank, PMTs, electronics, etc.).

The location of Superkamiokande in the Mozumi Mine is shown in Fig. 4. Figure 95 shows the excavation design of the cavity for Superkamiokande and associated facilities (the shaded regions). An artist's impression of the completed detector is shown in Fig. 96. At the end of JFY 1992, the excavation of the dome of the cavity was finished and a 2-ton rotating crane was constructed (Fig. 97).

Table 16 lists some of the parameters of Superkamiokande, which can be compared with those in Table 2 for Kamiokande-II after gain change. Superkamiokande will have not only a larger mass, but also a better light collection efficiency than the present Kamiokande detector. Moreover, the quality of the 20" PMT to be used in Superkamiokande has been improved as noted in subsect. 2.2.2. Two examples of simulated events are shown in Fig. 98 for a 10 MeV electron and in Fig. 99 for a 1 GeV muon.

In the following subsection, prospects of Superkamiokande [96, 172, 173, 174] will be described for the selected topics; observation of solar neutrinos, detection of a neutrino burst, measurement of atmospheric neutrinos, and search for nucleon decay. Prospects of a search for relic supernova neutrinos in Superkamiokande has already been mentioned briefly in subsect. 6.3.

8.2 Prospects of the Superkamiokande Experiment

8.2.1 Solar-Neutrino Observation

Among the most important physics objectives of Superkamiokande is to make a significant contribution toward the solution of the solar-neutrino problem. The high counting rate of ν-e scattering due to the solar neutrinos and the capability of measuring the recoil-electron energy spectrum constitute the main points of the Superkamiokande's approach to the solar-neutrino problem.

The expected solar-neutrino event rates in Superkamiokande are summarized in Table 17 for several analysis thresholds and compared with the rate observed in Kamiokande-II. Note that these rates are given for three different conditions: (a)

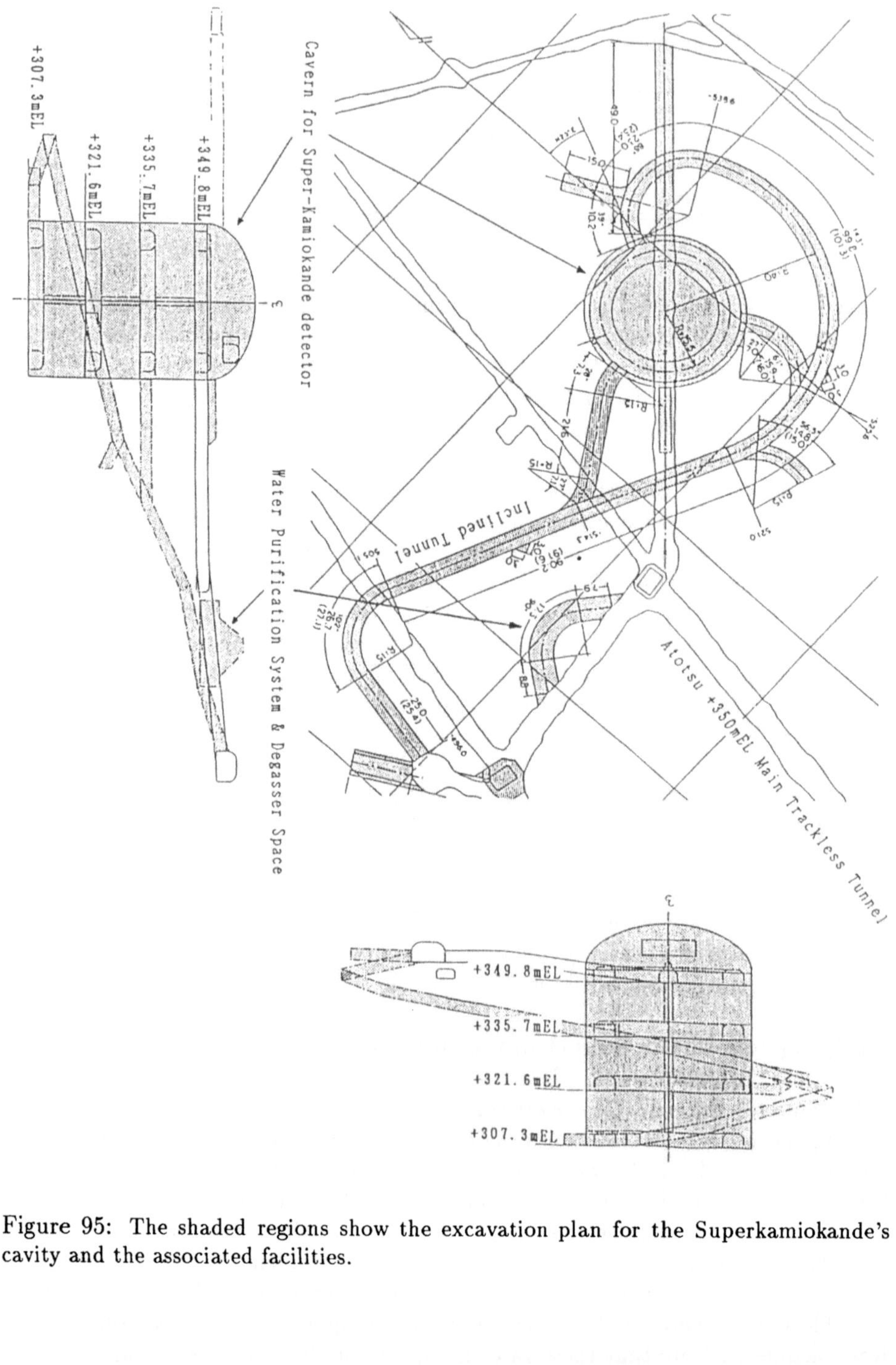

Figure 95: The shaded regions show the excavation plan for the Superkamiokande's cavity and the associated facilities.

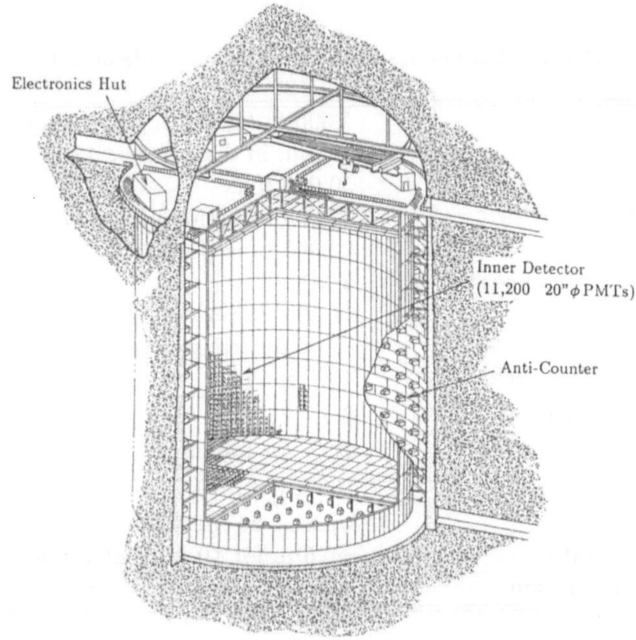

Figure 96: An artist's impression of the Superkamiokande detector.

Figure 97: Photograph of the excavation of the Superkamiokande's cavity. At the end of JFY 1992, the dome of the cavity was completed and a rotating crane was constructed. The crane and its supporting stucture in this picture are protected against the subsequent blasting to excavate the lower level.

Table 16: Parameters of the Superkamiokande detector.

Depth	1,000 m (2,700 m water equivalent)
Dimensions of main tank	39.3 m in diameter × 42 m in height
Total volume	50,000 tons
Fiducial volume	32,000 tons (for supernova neutrinos)
	22,000 tons (for solar neutrinos)
Anticounter thickness	2 m
Number of PMTs	11,200 (inner detector)
	to be determined (anticounter)
Photo-sensitive area	40% of the inner detector surface
Energy resolution	2.6% $/\sqrt{E_e/1\,\mathrm{GeV}}$ for electrons
	2.5% for muons with $\lesssim$ 1 GeV
	16% $/\sqrt{E_e/10\,\mathrm{MeV}}$ for electrons
	with $\lesssim$ 20 MeV
Vertex resolution	±0.5 m for 10 MeV electrons
Angular resolution	±28° for 10 MeV electrons

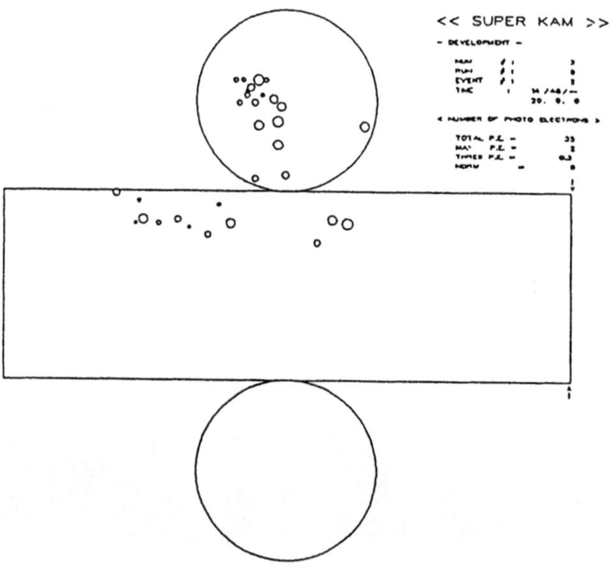

Figure 98: Unfolded view of a simulation event (10 MeV electron) in Superkamiokande.

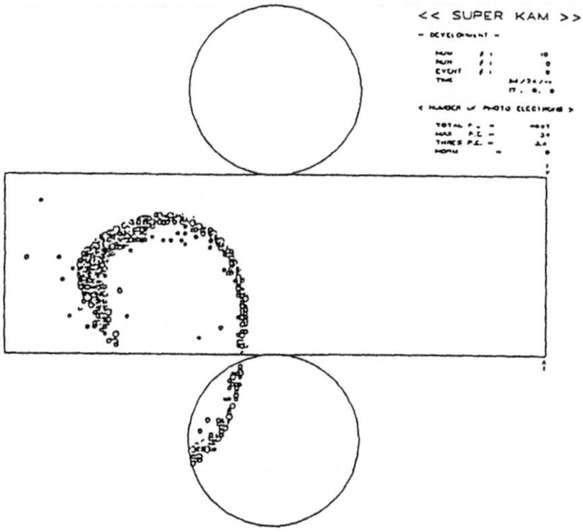

Figure 99: Unfolded view of a simulation event (1 GeV muon) in Superkamiokande.

the ^{8}B solar-neutrino flux as predicted by the SSM calculations by Bahcall and Ulrich [26] is assumed, (b) the ^{8}B solar-neutrino flux is assumed to be 46% of the SSM calculations by Bahcall and Ulrich [26] as observed in Kamiokande-II, and in addition, (c) dead time of 28% is considered due to the cuts imposed in the analysis as in Kamiokande-II. Obviously, the actual event rate in Kamiokande-II is 0.27 per day.

Table 17: Expected solar-neutrino event rates in Superkamiokande for several analysis thresholds. The solar-neutrino event rate observed in Kamiokande-II is also shown in the last column.

	Superkamiokande			Kamiokande-II
Fiducial volume (tons)	22,000			680
Analysis threshold (MeV)	5.0	6.5	7.5	7.5
Counting rate (events/day)				
SSM (Bahcall & Ulrich)	71	42	26	0.82
SSM×0.46	33	19	12	0.38
SSM×0.46×0.72	23	14	8.6	0.27

To what extent the analysis threshold can be lowered depends on the amount of low-energy background due to ^{222}Rn. Even if the amount of ^{222}Rn dissolved in a unit volume of water is the same as the typical value of the Kamiokande-III experiment (~0.01 pCi/ℓ), the improved energy resolution of Superkamiokande will lead to a lower analysis threshold (~6.5 MeV) than that of the present Kamiokande

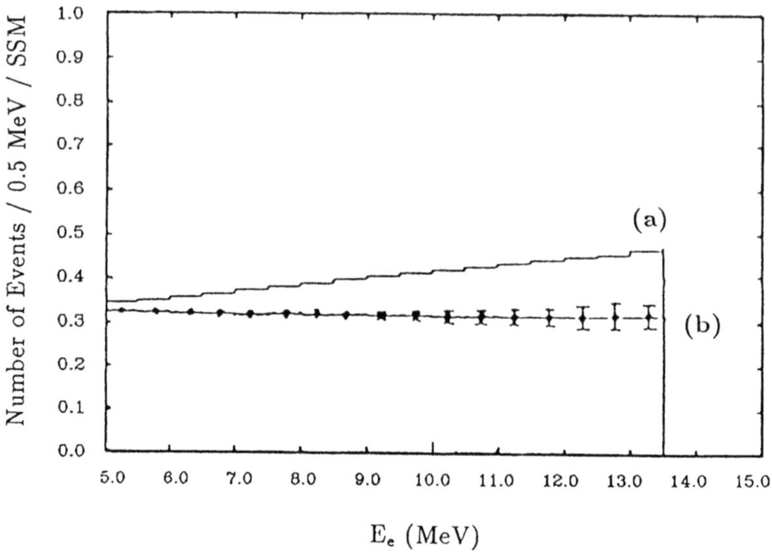

Figure 100: This figure shows the recoil-electron energy spectra relative to the SSM prediction, expected from ^{8}B solar neutrino-electron scattering for the two MSW solutions: (a) $\Delta m^2 = 10^{-5.22}$ eV^2 and $\sin^2 2\theta = 10^{-2.15}$ (a typical non-adiabatic solution) and (b) $\Delta m^2 = 10^{-5.0}$ eV^2 and $\sin^2 2\theta = 10^{-0.22}$ (a typical large mixing-angle solution). The data points indicate the statistical accuracy gained by Superkamiokande with a live time of three years.

experiment while keeping the S/N ratio to be the same. According to a Monte Carlo simulation, reduction of the amount of ^{222}Rn dissolved in water by 1/100 is needed to lower the analysis threshold down to 5 $\sim$ 5.5 MeV [174].

Owing to the high event rate in Superkamiokande, the average ^{8}B solar neutrino flux will be measured yearly with a statistical accuracy of $\pm(1 \sim 2)\%$. With a detector live time of one year, the day/night and seasonal variations of the flux will be measured to an accuracy of $\pm(2 \sim 3)\%$. The eccentricity of the Earth's orbit causes a time variation of the solar neutrino flux of $\sim 7\%$. This trivial time variation will be clearly demonstrated in Superkamiokande, and this will provide calibration for possible flux time variation of different origin, if any.

In the MSW approach to the solar-neutrino problem, there are two regions of possible solution in the $\sin^2 2\theta$ - Δm^2 plane as described in subsect. 6.1.3: one being $\Delta m^2 = (0.3 - 1.2) \times 10^{-5}$ eV^2 and $\sin^2 2\theta = (0.4 - 1.5) \times 10^{-2}$ (non-adiabatic solution) and the other $\Delta m^2 = (0.3 - 3) \times 10^{-5}$ eV^2 and $\sin^2 2\theta = (0.6 - 0.9)$ (large mixing-angle solution). Figure 100 shows the recoil-electron energy spectra corresponding to these two regions. The data points indicate the expected statistical accuracy gained by Superkamiokande with a live time of three years. Clearly, Superkamiokande will be able to distinguish these two MSW solutions.

8.2.2 Observation of a Neutrino Burst

If a type-II supernova explosion occurs at the center of our galaxy (10 kpc away from the Earth), a total of $\gtrsim$4000 neutrino events will be detected in Superkamiokande (see Table 18), assuming the same energy threshold of 7.5 MeV as the Kamiokande-II observation of the neutrino burst from SN1987A. With these large statistics, the detailed information on the explosion mechanism of type-II supernovae and the formation process of a neutron star will be obtained from measurement of the precise time profile as well as measurement of the time evolution of the positron energy spectrum for $\bar{\nu}_e p \to e^+ n$ events.

From the ν-e elastic scattering events, the direction of the supernova will be determined without optical information with a moderate accuracy of $\sim 4°$. This is estimated as follows [174]. With ~300 ν-e elastic scattering events pointing back to the supernova direction with an angular resolution (1σ) of $\sim 30°$, a statistical accuracy of the supernova direction in the presence of isotropic background from $\bar{\nu}_e p \to e^+ n$ may be given by

$$30° \times \frac{\sqrt{300 \times 0.68 + 2 \times 4000 \times 0.067}}{300 \times 0.68} \sim 4°,$$

where 300×0.68 and 4000×0.067 in the above expression are the numbers of ν-e elastic scattering events and $\bar{\nu}_e p \to e^+ n$ events, respectively, contained within the $\sim 30°$ angular window around the supernova direction.

Table 18: Expected neutrino events to be detected in Superkamiokande when a type-II supernova explosion occurs at a distance of 10 kpc from the Earth. The same energy threshold (7.5 MeV) as the Kamiokande-II observation of the neutrino burst from SN1987A is assumed. For comparison, those detected in Kamiokande-II are also shown. The number of events listed in this table are estimated from the observation of SN1987A and Ref. [91].

	Superkamiokande		Kamiokande-II	
	$\nu e \to \nu e$	$\bar{\nu}_e p \to e^+ n$	$\nu e \to \nu e$	$\bar{\nu}_e p \to e^+ n$
$\bar{\nu}_e$	44	4000	3	270
ν_e	130		9	
$\nu_\mu(\bar{\nu}_\mu) + \nu_\tau(\bar{\nu}_\tau)$	80		6	
neutronization ν_e	13		0.9	
total	~4300		~300	

The mass limit for ν_e will be improved down to $1 \sim 2$ eV using the ν_e-e events from the initial ~ 10 ms of neutronization burst. As for the tau-neutrino mass, Krauss *et al.* [175] have shown by a Monte Carlo analysis of a galactic supernova that Superkamiokande will have a sensitivity in the range of at least $\sim 50 - 300$ eV, and possibly down to a cosmologically interesting mass of ~ 25 eV depending upon the distance of the supernova.

These expectations clearly demonstrate the great capabilities of Superkamiokande in supernova physics as well as in neutrino physics, once a type-II supernova explosion occurs in our galaxy. There is a rather broad range of predictions for the type-II supernova rate in our galaxy, i.e., 1/(10~100 yr) [176]. Nevertheless,

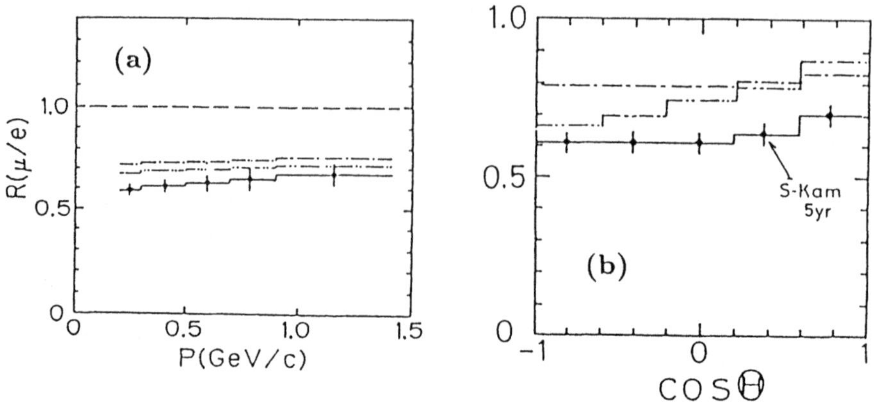

Figure 101: $R(\mu/e)_{\text{Data/MC}}$ as a function of (a) momentum and (b) zenith angle ($\cos\Theta = 1$ corresponds to downward) for some typical neutrino oscillation parameters allowed by Kamiokande-I & -II; $\Delta m^2 = 0.8 \times 10^{-2}$ eV^2 and $\sin^2 2\theta = 0.85$ (solid line), $\Delta m^2 = 1 \times 10^{-3}$ eV^2 and $\sin^2 2\theta = 0.85$ (double-dot-dashed line), and $\Delta m^2 = 0.8 \times 10^{-2}$ eV^2 and $\sin^2 2\theta = 0.45$ (dot-dashed line). Only ν_μ-ν_τ oscillations are considered. The data points indicate the statistical accuracy gained by Superkamiokande with a live time of five years.

there will be a good chance for Superkamiokande to observe a type-II supernova burst in our galaxy during its expected life of $\gtrsim$20 years.

8.2.3 Measurement of Atmospheric Neutrinos

With a live time of five years (110 kt·yr) Superkamiokande will accumulate 3,500 electron-like, 3,400 muon-like, and 3,300 multi-ring events. These represent 22 times higher statistics than Kamiokande-I & -II [10]. Figure 101 shows the momentum and angular distributions of $R(\mu/e)_{\text{Data/MC}}$ for some neutrino oscillation parameters allowed by Kamiokande-I & -II, where only two-flavor ν_μ-ν_τ oscillations are considered. The data points represent the statistical accuracy obtained by Superkamiokande with an exposure of 110 kt·yr. Provided that the anomaly in the value of $R(\mu/e)_{\text{Data/MC}}$ is caused by neutrino oscillations, Superkamiokande will further restrict the allowed region of the oscillation parameters.

8.2.4 Search for Nucleon Decay

Whether the next-generation detectors for the nucleon-decay search such as Superkamiokande can make significant contribution depends on the capability of rejecting background due to atmospheric neutrinos. In this regard, each of the Kamiokande, IMB, and Fréjus groups has investigated the atmospheric-neutrino background level in its detector by a Monte Carlo simulation. The 90% CL results for the decay mode $p \to e^+\pi^\circ$ are:

Kamiokande-II [1]	simulated 105 kt·yr,	background < 0.02 ev/kt·yr
IMB-III [177]	simulated 40 kt·yr,	background < 0.06 ev/kt·yr
Fréjus [177]	simulated 75 kt·yr,	background < 0.03 ev/kt·yr.

Since Superkamiokande is planned to have better light-collection efficiency (40%) than Kamiokande-II, the lower limit of $\tau/B(p \to e^+\pi^\circ)$ would be easily pushed up to $\sim 10^{34}$ yr. (For such a decay mode in which no candidate (background) is observed, the τ/B limit is directly proportional to the detector exposure.) It should also be mentioned that with the Kamiokande-II simulation of 105 kt·yr [1], there are several other decay modes which are still background free (<0.02 ev/kt·yr at 90% CL), i.e., $p \to e\eta\,(\gamma\gamma)$, $e\eta\,(3\pi^\circ)$, $eK^\circ\,(2\pi^\circ)$, $e\omega\,(\gamma\pi^\circ)$, $\mu\pi^\circ$, $\mu\eta\,(\gamma\gamma)$ and $\mu\eta\,(3\pi^\circ)$.

On the other hand, the decay mode $p \to K^+\bar{\nu}$ $(K^+ \to \mu^+\nu_\mu)$, is dominated by the atmospheric-neutrino background as shown in Fig. 90. Another K^+ decay channel, $p \to K^+\bar{\nu}$ $(K^+ \to \pi^+\pi^0)$ will also be dominated by the background for an exposure of 105 kt·yr as inferred from Table 14. For such a case of background dominance, the τ/B limit only improves as the square root of exposure. Therefore, scaling from the present Kamiokande result [1], $\tau/B(p \to K^+\bar{\nu})$ limit will be improved by Superkamiokande to $\sim 5 \times 10^{32}$ years in a live time of 5 years. However, since SUSY-GUTs predict $\tau/B(p \to K^+\bar{\nu}) \sim 10^{29\pm4}$ yr [178], it is very important to examine the τ/B limit for this decay mode up to 10^{33} years. For this to be possible, a clever way to reduce the background rate may be needed. The decay of a proton in the $1p_{3/2}$ shell of a ^{16}O nucleus will cause a ~ 6 MeV prompt γ-ray through the reaction $^{16}\mathrm{O} \to {}^{15}\mathrm{N}^* + (K^+\bar{\nu})$, $^{15}\mathrm{N}^* \to {}^{15}\mathrm{N} + \gamma$. This will give the characteristic $p \to K^+\bar{\nu}$ signal together with the delayed K^+ decay product, μ^+ or $\pi^+\pi^0$, and may serve the τ/B limit to be pushed up to $\sim 10^{33}$ years [96, 172, 174].

The expected lower limits on the partial lifetime for various nucleon decay modes to be obtained in five years of the Superkamiokande operation (110 kt·yr) are shown in Fig. 102. It is seen that a partial lifetime limit of 10^{33-34} years is expected for several decay modes. If the nucleon has a lifetime of less than 10^{34} years, one has a good chance with Superkamiokande to observe its decay. Otherwise, one still obtains strong constraints on the Grand Unified Theories.

9. Conclusion

realtime detection of astrophysical neutrinos. It has an excellent advantage of directionality through ν-e scattering at low energies and Lorentz boost of the produced charged particle(s) at high energies. For the detection of low-energy neutrinos, water is easily purified to remove radioactive contamination. Water is inexpensive and has a long optical attenuation length. These features allow the construction of a large water-Cherenkov detector with reasonable cost. The sensitivity of the detector primarily depends on the density of photosensitive devices. At present, Kamiokande is the most sensitive detector with the trigger threshold being 5 MeV. Superkamiokande, which is under construction and is expected to start operation in 1996, will have higher sensitivity because the density of PMTs will be doubled compared to the present Kamiokande.

With its outstanding achievements of the first observation of stellar-collapse

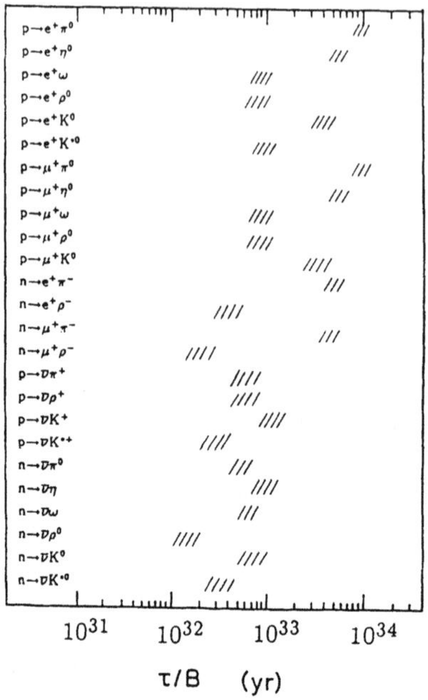

Figure 102: Expected sensitivity of Superkamiokande for various nucleon decay modes. Lower limits (90% CL) of partial lifetime to be obtained with a live time of five years are shown.

neutrinos and directional measurement of solar neutrinos, Kamiokande has opened a new era of neutrino astronomy. The knowledge obtained by Kamiokande will be further expanded by the next-generation water-Cherenkov detector Superkamiokande. In particular, it is hoped that the diffuse neutrino background originated from past supernovae will be observed by Superkamiokande as the M'third" source of astrophysical neutrinos ever detected by the mankind.

Acknowledgments

The authors would like to thank all members of the Kamiokande Collaboration for their efforts in construction, upgrading, and operation of the Kamiokande detector, and in analyzing the data, to which this report owes almost everything. Our special thanks go to Professor S. Tasaka, Dr. M. Takita, Dr. Y. Oyama, and Dr. M. Mori for useful discussions and comments.

References

[1] K.S. Hirata *et al.*, Phys. Lett. B 220 (1989) 308.

[2] R. Davis, Jr. *et al.*, Phys. Rev. Lett. 20 (1968) 1205; J.K. Rowley, B.T. Cleveland, and R. Davis, Jr., *Solar Neutrinos and Neutrino Astronomy (AIP Conference Proceedings No. 126)*, edited by H.C. Wolfe (American Institute of Physics, New York, 1985), p. 1.

[3] K. Hirata *et al.*, Phys. Rev. Lett. 58 (1987) 1490.

[4] K.S. Hirata *et al.*, Phys. Rev. D 38 (1988) 448.

[5] K.S. Hirata *et al.*, Phys. Rev. Lett. 65 (1990) 1297.

[6] K.S. Hirata *et al.*, Phys. Rev. D 44 (1991) 2241, and 45 (1992) 2170 (E).

[7] K.S. Hirata *et al.*, Phys. Rev. Lett. 65 (1990) 1301.

[8] K.S. Hirata *et al.*, Phys. Rev. Lett. 66 (1991) 9.

[9] K.S. Hirata *et al.*, Phys. Lett. B 205 (1988) 416.

[10] K.S. Hirata *et al.*, Phys. Lett. B 280 (1992) 146.

[11] T. Kano and M. Shimizu, *Mineral Deposits and Magnetism in the Hida and Hida Marginal Belts, Central Japan*, a guide book for Field Trip C20 at the 29th Int. Geological Congress, Kyoto, 1992.

[12] Taken from *The Outline of Kamioka Mine* prepared by Kamioka Mining & Smelting Co., Ltd. for the tour to the Kamioka Underground Laboratory at the Int. Symp. on Neutrino Astrophysics, Takayama/Kamioka, 1992.

[13] H. Ejiri, in this book.

[14] H. Kume *et al.*, Nucl. Instr. and Meth. 205 (1983) 443.

[15] A. Suzuki *et al.*, Nucl. Instr. and Meth. A329 (1993) 299.

[16] T.Tanimori *et al.*, IEEE Trans. on Nucl. Sci. NS-36 (1989) 497.

[17] S. Tasaka, private communication.

[18] Y. Oyama, thesis submitted to Univ. of Tokyo, 1989 (ICR-Report-193-89-10).

[19] M. Takita, thesis submitted to Univ. of Tokyo, 1989 (ICR-Report-186-89-3).

[20] M. Nakahata, thesis submitted to Univ. of Tokyo, 1988 (UT-ICEPP-88-01).

[21] K.S. Hirata, thesis submitted to Univ. of Tokyo, 1991 (ICRR-Report-239-91-8).

[22] K. Kihara, thesis submitted to Univ. of Tokyo, 1992 (ICRR-Report-274-92-12).

[23] L.B. Okun, *Leptons and Quarks*, (North-Holland, Amsterdam, 1982).

[24] E.W. Kolb, A.J. Stebbins, and M.S. Turner, Phys. Rev. D 35 (1987) 3598.

[25] J. Arafune and M. Fukugita, Phys. Rev. Lett. 59 (1987) 367.

[26] J.N. Bahcall and R.K. Ulrich, Rev. Mod. Phys. 60 (1988) 297. 72) 261.

[27] K.L. Miller *et al.*, Phys. Rev. D 26 (1982) 537 and references therein.

[28] S.J. Barish *et al.*, Phys. Rev. D 19 (1979) 2521.

[29] G.M. Radecky *et al.*, Phys. Rev. D 25 (1982) 1161.

[30] G.L. Fogli and G. Nardulli, Nucl. Phys. B160 (1979) 116.

[31] N.J. Baker *et al.*, Phys. Rev. D 25 (1982) 617.

[32] S.J. Barish *et al.*, Phys. Rev. Lett. 33 (1974) 448; M. Derrick *et al.*, Phys. Rev. D 23 (1981) 569.

[33] W. Krenz *et al.*, Nucl. Phys. B135 (1978) 45.

[34] D. Rein and L.M. Sehgal, Ann. Phys. 133 (1981) 79.

[35] J.E. Kim *et al.*, Rev. Mod. Phys. 53 (1981) 211 and references therein.

[36] R. Field and R.P. Feynmann, Phys. Rev. D 15 (1977) 1590.

[37] E. Eichten, I. Hinchliffe, K. Lane, and C. Quigg, Rev. Mod. Phys. 56 (1984) 579; 58 (1986) 1065(E).

[38] L.V. Volkova, Yad. Fiz. 31 (1980) 1510 [Sov. J. Nucl. Phys. 31 (1980) 784].

[39] K. Mitsui, Y. Minorikawa, and H. Komori, Nuovo Cimento 9C (1986) 995

[40] W. Lohman, R. Kopp, and R. Voss, CERN 85-03, 1985.

[41] C. Quigg, M.H. Reno and T.P. Walker, Phys. Rev. Lett. 57 (1986) 774; M.H. Reno and C. Quigg, Phys. Rev. D 37 (1988) 657.

[42] M. Mori *et al.*, Phys. Lett. B 270 (1991) 89.

[43] M. Mori, *High Energy Neutrino Astrophysics*, edited by V.J. Stenger *et al.* (World Scientific, Singapore, 1992), p. 221.

[44] M. Nakahata *et al.*, J. Phys. Soc. Jpn. 55 (1986) 3786.

[45] S.J. Barish *et al.*, Phys. Rev. D 17 (1978) 1.

[46] H. Saarikko, *Proc. of the Int. Conf. on Neutrinos, Weak Interactions and Cosmology*, Bergen, 1979, edited by A. Haatuft and C. Jarlskog, Vol. 2, p. 507.

[47] W.R. Nelson, H. Hirayama, and D.W.O. Rogers, *The EGS4 Code System*, SLAC-265, 1985.

[48] A.S. Clough *et al.*, Nucl. Phys. B76 (1974) 15.

[49] E. Bracci *et al.*, *π Induced Reaction*, CERN/HERA 72-1, 1972.

[50] R. Eichler, JADE note No.65, 1980 (unpublished).

[51] F. Binon *et al.*, Nucl. Phys. B17 (1970) 168.

[52] A. Morel and L. Prieur, Limnology and Oceanography 22 (1977) 709.

[53] T.K. Gaisser, T. Stanev, S.A. Bludman, and H. Lee, Phys. Rev. Lett. 51 (1983) 223.

[54] T.K. Gaisser, T. Stanev, and G. Barr, Phys. Rev. D 38 (1988) 85.

[55] L.V. Volkova, Ref. [38] and private communication.

[56] G. Barr, T.K. Gaisser, and T. Stanev, Phys. Rev. D 39 (1989) 3532.

[57] M. Honda, K. Kasahara, K. Hidaka and S. Midorikawa, Phys. Lett. B 248 (1990) 193.

[58] H. Lee and Y.S. Koh, Nuovo Cimento 105B (1990) 883.

[59] E.W. Beier *et al.*, Phys. Lett. B 283 (1992) 446.

[60] D. Casper *et al.*, Phys. Rev. Lett. 66 (1991) 2561; D. Casper, thesis submitted to Univ. of Michigan, 1990.

[61] R. Becker-Szendy *et al.*, Phys. Rev. D. 46 (1992) 3720.

[62] Ch. Berger *et al.*, Phys. Lett. B 227 (1989) 489.

[63] Ch. Berger *et al.*, Phys. Lett. B 245 (1990) 305.

[64] M. Aglietta *et al.*, Europhys. Lett. 8 (1989) 611.

[65] K. Gabathuler *et al.*, Phys. Lett. B 138 (1984) 449; F. Dydak *et al.*, Phys. Lett. B 134 (1984) 281; F. Bergsma *et al.*, Phys. Lett. B 142 (1984) 103; and Ref.[63].

[66] Y. Oyama, *Progress in Atomic Physics Neutrinos and Gravitation* (*Proc. of the 27th Rencontre de Moriond*, Les Arcs, 1992), edited by G. Chardin, O. Fackler, and J. Trân Thanh Vân (Editions Frontiers, Gif-sur-Yvette, 1992), p. 59.

[67] R. Becker-Szendy *et al.*, Phys. Rev. Lett. 69 (1992) 1010.

[68] M.M. Boliev *et al.*, (*Proc. of the Third International Workshop on Neutrino Telescopes*, Venice, 1991), edited by M. Baldo-Ceolin (INFN, Padova), p. 235.

[69] T.K. Gaisser, *Frontiers of Neutrino Astrophysics* (*Proc. of the Int. Symp. on Neutrino Astrophysics*, Takayama, 1992), edited by Y. Suzuki and K. Nakamura (Universal Academy Press, Tokyo, 1993), p. 309.

[70] F. Dydak *et al.*, Phys. Lett. B 207 (1988) 79.

[71] K. Inoue, talk presented at the 28th Rencontres de Moriond, "Electroweak Interactions and Unified Theories", Les Arcs, 1993.

[72] R. Davis, Jr., *Frontiers of Neutrino Astrophysics* (*Proc. of the Int. Symp. on Neutrino Astrophysics*, Takayama, 1992), edited by Y. Suzuki and K. Nakamura (Universal Academy Press, Tokyo, 1993), p. 47.

[73] GALLEX Collaboration, P. Anselmann *et al.*, Phys. Lett. B 285 (1992) 376.

[74] V.N. Gavrin *et al.*, *Proc. of the 26th Int. Conf. on High Energy Physics*, Dallas, 1992, edited by J.R. Sanford (AIP, 1993), Vol. 1, p. 1101.

[75] J.N. Bahcall and M.H. Pinsonneault, Rev. Mod. Phys. 64 (1992) 885.

[76] Turck-Chièze and I. Lopes, Astrophys. J. 408 (1993) 347.

[77] GALLEX Collaboration, P. Anselmann *et al.*, sumbitted to Phys. Lett. B.

[78] S.A. Bludman, N. Hata, D.C. Kennedy, and P.G. Langacker, Phys. Rev. D 47 (1993) 2220.

[79] M.B. Voloshin and M.I. Vysotskii, Sov. J. Nucl. Phys. 44 (1986) 544; M.B. Voloshin, M.I. Vysotskii, and L.B. Okun, Sov. J. Nucl. Phys. 44 (1986) 440; M.B. Voloshin, M.I. Vysotskii, and L.B. Okun, Sov. Phys. JETP 64 (1986) 446.

[80] C.S. Lim and W.J. Marciano, Phys. Rev. D 37 (1988) 1368.

[81] E. Kh. Akhmedov, Phys. Lett. B 213 (1988) 64; E. Kh. Akhmedov, Sov. Phys. JETP 68 (1989) 690.

[82] Y. Suzuki, *Frontiers of Neutrino Astrophysics* (*Proc. of the Int. Symp. on Neutrino Astrophysics*, Takayama, 1992), edited by Y. Suzuki and K. Nakamura (Universal Academy Press, Tokyo, 1993), p. 61. Also, see K. Nakamura, Nucl. Phys. B (Proc. Suppl.) 31 (1993) 105.

[83] Solar Geophysical Data (Boulder, Colorado, NOAA-USAF Space Environmental Information Services Center).

[84] E.Kh. Akhemedov. P.I. Krastev, and A.Yu. Smirnov, Phys. Lett. B 260 (1991) 161; T. Kubota, T. Kurimoto, M. Ogura, and E. Takasugi, Phys. Lett. B 292 (1992) 195.

[85] I. Shelton, IAU Circular No. 4316, 1987.

[86] R.M. McNaught, IAU Circular No. 4316, 1987.

[87] IAU Circular No. 4317, 1987.

[88] M. Aglietta *et al.*, Europhys. Lett. 3 (1987) 1315.

[89] R.M. Bionta *et al.*, Phys. Rev. Lett. 58 (1987) 1494.

[90] W.D. Arnett, J.N. Bahcall, R.P. Kirshner, and S.E. Woosley, Ann. Rev. Astron. Astrophys. 27 (1989) 629.

[91] K. Sato and H. Suzuki, Phys. Rev. Lett. 58 (1987) 2722.

[92] L.M. Krauss, S.L. Glashow, D.N. Schramm, Nature 310 (1984) 191.

[93] S.E. Woosley, J.R. Wilson, and R. Mayle, Astrophys. J. 302 (1986) 19.

[94] D.N. Schramm, R. Mayle, and J.R. Wilson, Nuovo Cimento 9C (1986) 443.

[95] W. Zhang *et al.*, Phys. Rev. Lett. 61 (1988) 385.

[96] Y. Totsuka, *Proc. of the Int. Symp. on Underground Physics Experiments*, Tokyo, 1990, edited by K. Nakamura (Institute for Cosmic Ray Research, Univ. of Tokyo, 1990), p. 129.

[97] V.A. Rubakov, Pis'ma Zh. Eksp. Teor. Fiz. 33 (1981) 658 [JETP Lett. 33 (1981) 644]; Nucl. Phys. B203 (1982) 311; C.G. Callan, Phys. Rev. D25 (1982) 2141; D26 (1982) 2058.

[98] J. Ellis, D.V. Nanopoulos, and K.A. Olive, Phys. Lett. 116B (1982) 127; F.A. Bais, J. Ellis, D.V. Nanopoulos, and K.A. Olive, Nucl. Phys. B219 (1983) 189.

[99] J. Arafune and M. Fukugita, Phys. Lett. 133B (1983) 380.

[100] J. Arafune and M. Fukugita, Phys. Rev. Lett. 50 (1983) 1901.

[101] E.W. Kolb and M.S. Turner, Astrophys. J. 286 (1984) 702 and references therein.

[102] For a review of Grand Unified Theories, see P. Langacker, Phys. Rep. 72 (1981) 185 and references therein.

[103] A.J.S. Hamilton and C.L. Sarazin, Astrophys. J. 274 (1983) 399; N. Meyer-Vernet, Astrophys. J. 290 (1985) 21.

[104] R. Davis, Jr. and J.C. Evans, *Proc. of the 13th Int. Cosmic Ray Conf.*, Denver, 1973, Vol. 3, p. 2001.

[105] R. Davis, Jr. *Proc. of 7th Workshop on Grand Unification/ICOBAN'86*, Toyama, Japan, 1986, edited by J. Arafune (World Scientific, Singapore, 1987), p. 237.

[106] G.H. Share *et al.*, Adv. Space Res. 6 (1986) 15.

[107] M. Fukugita, Y. Kohyama, K. Kubodera, and T. Kuramoto, Astrophys. J. 337 (1989) L59.

[108] J.N. Bahcall, Phys. Rev. Lett. 61 (1988) 2650.

[109] K.S.Hirata *et al.*, Astrophys. J. 359 (1990) 574.

[110] T. Kuramoto, M. Fukugita, Y. Kohyama, and K. Kubodera Nucl. Phys. A512 (1990) 711.

[111] W.C. Haxton, Phys. Rev. D 36 (1987) 2283.

[112] K.S. Hirata *et al.*, Phys. Rev. Lett. 61 (1988) 2653.

[113] *Preliminary Report and Forecast of Solar Geophysical Data* (Boulder, Colorado, NOAA-USAF Space Environmental Services Center), SESC PRF 706-710, 1989.

[114] *Solar Geophysical Data*, Comprehensive Reports (Boulder, Colorado, NOAA Environmental Data and Information Service), No. 342, Part 2, 1973.

[115] R.P. Lin, and H.S. Hudson, Solar Phys. 50 (1976) 153.

[116] R.E. Lingenfelter and H.S. Hudson, *Proc. of the Conf. on the Ancient Sun: Fossil Record in the Earth, Moon, and Meteorites*, Boulder, Colorado, 1979, edited by R.O. Peppin *et al.* (Pergamon, New York, 1980), p. 69.

[117] K. O'Brien and A. de la Zerda, Trans. Am. Nucl. Soc. 46 (1984) 641; A. de la Zerda Lerner and K. O'Brien, Nature 330 (1987) 353.

[118] See, for example, T. Stanev, Nucl. Phys. B (Proc. Suppl.) 14A (1990) 17 and references therein.

[119] H. Sato, Prog. Theor. Phys. 58 (1977) 549; V.S. Berezinski and J.P. Ostriker, Astron. Astrophys. 66 (1978) 325; V.S. Berezinsky and O.F. Prilutsky, Astron. Astrophys. 66 (1987) 325; H. Sato, Mod. Phys. Lett. A2 (1987) 801; T. Nakamura, Y. Yamada, and H. Sato, Prog. Theor. Phys. 78 (1987) 1065.

[120] T.K. Gaisser and T. Stanev, Phys. Rev. Lett. 58 (1987) 1695 and 59 (1987) 844(E).

[121] D. Eichler, Nature 275 (1978) 725; V.S. Berezinski, C. Castagnoli, and P. Galotti, Nuovo Cimento 8C (1985) 185.

[122] Y. Oyama *et al.*, Phys. Rev. Lett. 59 (1987) 2604.

[123] Y. Oyama *et al.*, *Proc. of the Yamada Conference XX on Big Bang, Active Galactic Nuclei and Supernovae*, Tokyo, 1988, edited by S. Hayakawa and K. Sato (Universal Academy Press, Tokyo, 1989), p. 557.

[124] Y. Oyama *et al.*, Phys. Rev. D 39 (1989) 1481.

[125] See, for example, T. C. Weekes, Phys. Reports 160 (1988) 1.

[126] T.C. Weekes *et al.*, Astrophys. J. 342 (1989) 379; G. Vacanti *et al.*, Astrophys. J. 377 (1991) 467.

[127] M. Punch *et al.*, Nature 358 (1992) 477.

[128] R. Svoboda *et al.*, Astrophys. J. 315 (1987) 420.

[129] Ch. Berger *et al.*, Z. Phys. C 48 (1990) 221.

[130] M.L. Marshak *et al.*, Phys. Rev. Lett. 54 (1985) 2079; Phys. Rev. Lett. 55 (1985) 1965.

[131] G. Battistoni *et al.*, Phys. Lett. 155B (1985) 465.

[132] M.L. Marshak *et al.*, talk presented at the 23rd Int. Conf. on High Energy Phys. Berkeley, 1986.

[133] R.M. Bionta *et al.*, Phys. Rev. D 36 (1987) 30.

[134] Y. Oyama *et al.*, Phys. Rev. Lett. 56 (1986) 991.

[135] Y. Oyama *et al.*, Phys. Rev. D 36 (1987) 3537.

[136] Y.M. Andreyev *et al.*, Sov. Phys. JETP 44 (1987) 401.

[137] Ch. Berger *et al.*, Phys. Lett. B 174 (1986) 118.

[138] G. Chardin and G. Gerbier, Astron. Astrophys. 210 (1989) 52.

[139] For a recent review of dark matter, see, for example, M.S. Turner, Fermilab preprint FERMILAB-Conf-90/230-A, 1990; J. Primack, D. Seckel, and B. Sadoulet, Ann. Rev. Nucl. Part. Sci. 38 (1989) 751; D.O. Caldwell, Mod. Phys. Lett. A5 (1990) 1543.

[140] J. Silk, K. Olive, and M. Srednicki, Phys. Rev. Lett. 55 (1985) 257; L.M. Krauss, M. Srednicki, and F. Wilczek, Phys. Rev. D 33 (1986) 2079; T.K. Gaisser, G. Steigman, and S. Tilav, Phys. Rev. D 34 (1986) 2206; J. Hagelin, K.-W. Ng, and K.A. Olive, Phys. Lett. B180 (1986) 375; M. Srednicki, K.A. Olive, and J. Silk, Nucl. Phys. B279 (1987) 804.

[141] A. Gould, Astrophys. J. 321 (1987) 571.

[142] N. Sato *et al.*, Phys. Rev. D 44 (1991) 2220.

[143] M. Mori *et al.*, Phys. Lett. B 289 (1992) 463.

[144] F. D. Stacey, *Physics of the Earth*, 2nd ed. (Wiley, New York, 1977).

[145] For the case of Dirac neutrinos, see K. Enqvist, K. Kainulainen and J. Maalampi, Nucl. Phys. B317 (1989) 647. The formulation for the case of Majorana neutrinos was given by K. Hikasa and M.M. Nojiri (unpublished).

[146] S. Ritz and D. Seckel, Nucl. Phys. B304 (1988) 877.

[147] T. Stöstrand and M. Bengtsson, Comput. Phys. Commun. 43 (1987) 367.

[148] M. Kamionkowski, Phys. Rev. D 44 (1991) 3021.

[149] J. Ellis and R.A. Flores, Nucl. Phys. B307 (1988) 885.

[150] K. Hikasa *et al.* (Particle Data Group), Phys. Rev. D 45 (1992) III.2.

[151] J.F. Gunion and H.E. Haber, Nucl. Phys. B272 (1986) 1.

[152] G.B. Gelmini *et al.*, Nucl. Phys. B351 (1991) 623; A. Bottino *et al.*, Phys. Lett. B 265 (1991) 57.

[153] D. Decamp *et al.* (ALEPH Collaboration), Phys. Lett. B 241 (1990) 141.

[154] M. Mori *et al.*, to be published in Phys. Rev. D.

[155] M. Drees and M.M. Nojiri, Phys. Rev. D 45 (1993) 376.

[156] For a review, see, for example, P. Langacker, Phys. Rep. 72 (1981) 185.

[157] M.R. Krishnaswamy *et al.*, Phys. Lett. B 115 (1982) 349.

[158] H. Georgi and S.L. Glashow, Phys. Rev. Lett. 32 (1974) 438.

[159] K. Arisaka *et al.*, J. Phys. Soc. Jpn. 54 (1985) 3213.

[160] G. Blewitt *et al.*, Phys. Rev. Lett. 55 (1985) 2114.

[161] T.Kajita *et al.*, J. Phys. Soc. Jpn. 55 (1986) 711.

[162] S. Dimopoulos, S. Raby, and F. Wilczek, Phys. Rev. D 24 (1981) 1681; S. Dimopoulos and H. Georgi, Nucl. Phys. 193B (1981) 150; N. Sakai, Z. Phys. C 11 (1981) 153.

[163] T.J. Haines *et al.*, Phys. Rev. Lett. 57 (1986) 1986; S. Seidel *et al.*, Phys. Rev. Lett. 61 (1988) 2522; R. Becker-Szendy *et al.*, Phys. Rev. D 42 (1990) 2974.

[164] Ch. Berger *et al.*, Nucl. Phys. B313 (1989) 509: Ch. Berger *et al.*, Z. Phys. C 50 (1991) 385.

[165] See, for example, L. Rolandi, *Proc. of the 26th Int. Conf. on High Energy Physics*, Dallas, 1992, edited by J.R. Sanford (AIP, 1993), Vol. 1, p. 56; R. Tanaka, *ibid.*, p. 681.

[166] M. Takita *et al.*, Phys. Rev. D 34 (1986) 902.

[167] C.B. Dover, A. Gal, and J.M. Richard, Phys. Rev. D 27 (1983) 1090.

[168] C.B. Dover, A. Gal, and J.M. Richard, Phys. Rev. C 31 (1985) 1423.

[169] M. Baldo-Ceolin *et al.*, Phys. Lett. B 236 (1990) 95.

[170] T.W. Jones *et al.*, Phys. Rev. Lett. 52 (1984) 720.

[171] Ch. Berger *et al.*, Phys. Lett. B 240 (1990) 237.

[172] Y. Totsuka, *Proc. of the 7th Workshop on Grand Unification/ICOBAN'86*, Toyama, Japan, 1986, edited by J. Arafune (World Scientific, Singapore, 1987), p. 118.

[173] T. Kajita, *Proc. of the Third Workshop on Elementary-Particle Picture of the Universe*, Fujiyoshida, Japan, 1988, edited by C.-S. Lim *et al.* (KEK, 1989), p. 163.

[174] M. Takita, *Frontiers of Neutrino Astrophysics* (*Proc. of the Int. Symp. on Neutrino Astrophysics*, Takayama, 1992), edited by Y. Suzuki and K. Nakamura (Universal Academy Press, Tokyo, 1993), p. 135.

[175] L.M. Krauss *et al.*, Nucl. Phys. B380 (1992) 507.

[176] S. van den Bergh and G.A. Tammann, Annu. Rev. Astron. Astrophys. 29 (1991) 363.

[177] Cited from R. Barloutaud, *Proc. of the 13th Int. Conf. on Neutrino Physics and Astrophysics*, Boston, 1988, edited by J. Schneps *et al.* (World Scientific, Singapore, 1989), p. 540.

[178] P. Langacker, talk presented at the Franklin Symposium in Celebration of the Discovery of the Neutrino, Philadelphia, May, 1992.

Kamiokande : Historical Account

Atsuto Suzuki

Faculty of Science, Tohoku University

1 Idea of Kamiokande

The initial concept of Kamiokande was proposed by Koshiba in 1979. This idea was presented by one of his collaborators at a workshop on the unified theory and the baryon number in the universe [1]. The detector considered is shown in Fig.1. It comprises water Čerenkov counters A and B, and is located deep underground. Counter B is used to detect nucleon decay which occurs in water within the detector. To be specific, the size is $20 \times 20 \times 5$ m^3, containing 2000 tons of water. The surface is covered with 7500 5-inch photomultiplier tubes (PMT's), one in every 40×40 cm^2. The inside of the box is painted black so as to avoid any reflection of the Čerenkov light: the direction and, in some cases, energy of charged particles and γ-rays can then be measured. Counter A has a size of $25 \times 25 \times 0.1$ m^3, and is placed above counter B while serving as an anticounter to veto atmospheric muons that penetrate. The inside of this counter is painted white, and the water is dissolved with a wavelength shifter so as to increase the light yield, which reduces the number of necessary phototubes: one hundred tubes are hopefully more than enough. These are pasted onto the top surface of the counter. A steel plate having a thickness 25 cm between counters A and B is used to stop any charged particles produced by nucleon decays in counter B. Fig. 2 illustrates how the $p \rightarrow \mu^+ + \gamma$ decay mode would be detected in this detector.

However, this detector requires a huge amount of funding, since it employs, tens of thousand of PMT's. Another approach to develop new PMT's having a large photocathode was taken instead. Thanks to the generous cooperation of Hamamatsu TV Co., Ltd. (the present name: Hamamatsu Photonics Co., Ltd.). a new large tube, the 20-inch diameter PMT, became a reality in 1981 [2]. Fig. 3 is a photograph of the 20-inch PMT, compared with 1/2-inch, 5-inch and 8-inch tubes.

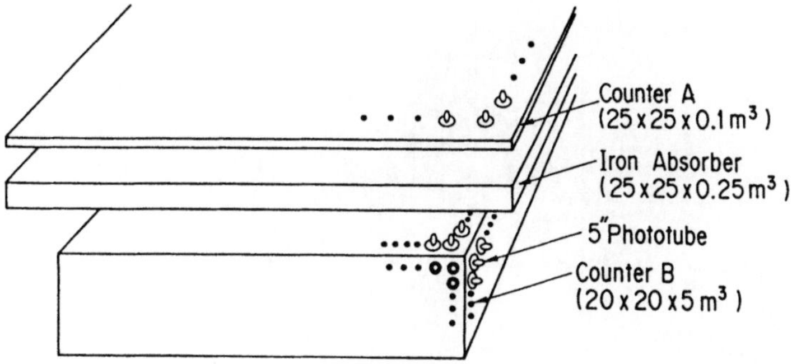

Figure 1: Proposed detector for nucleon decay observations.

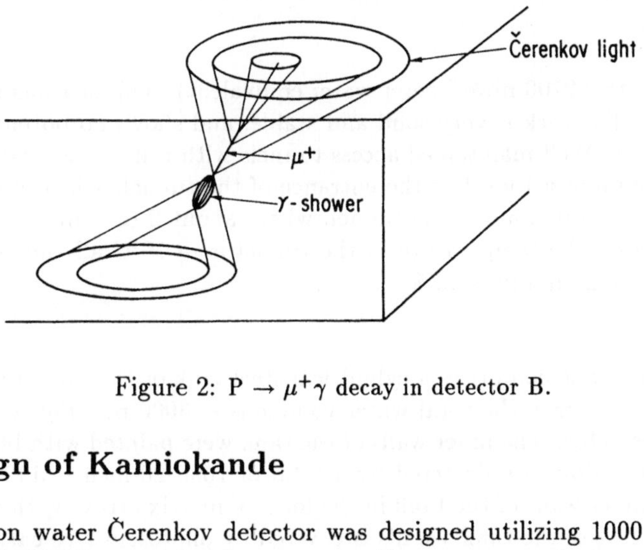

Figure 2: P $\rightarrow \mu^+\gamma$ decay in detector B.

2 Design of Kamiokande

The 3000-ton water Čerenkov detector was designed utilizing 1000 newly developed 20-inch PMT's, resulting in an outstanding energy resolution and capability for identifying various decay modes of proton decay. A cubic detector which is entirely surrounded by a 1 m depth of anticounters was proposed at the early stage of designing (see Fig. 4). However, in terms of the stability of the rock condition, a cylindrical-shape cavity was excavated and, consequently, the detector shape was changed to cylindrical one. The experimental site was determined to be at the Kamioka mine located in Central Japan, 300 km west of Tokyo. This mine has been actively producing zinc and lead. The underground laboratory is situated at a depth of 1000 m directly below the top of a mountain (see Fig. 5). The average rock density was measured to be $\rho \simeq 2.7$ g/cm^3; the average rock overburden is

Figure 3: Photograph of the 20-inch PMT; 1/2-, 5- and 8-inch PMT's are also shown for a comparison.

thus approximately 2700 mwe (meter water equivalent). This site has many desirable functions. The rock is very solid and stable, and shows no potential hazards with water or air. Well-maintained access tunnels with rail-tracks lead to the site, which is at the same sea level as the entrance of the tunnel. Clean natural water is available from a ditch along the tunnel, where about 2 ton/min of water (9°C, PH = 7.8) flows. The temperature in the tunnel is $\sim$ 16°C all the year around, while the relative humidity is 93%.

A pure-water tank comprises a cylindrical steel tank of 15.5 m in diameter and 16 m in height, of which the total water volume is ~ 3000 m^3. Fig. 6 depicts the Kamiokande detector. The inner walls of the tank were painted with black-colored epoxy. Čerenkov rings are detected by a total of 1000 20-inch PMT's which are placed on the inner walls of the tank in the form of matrix arrays with a density of $\sim$ 1PMT/m^2; 20% of the tank's walls is covered by photosensitive cathodes. This offers the possibility of good energy resolution, say $\sim$ 4%, for a 1 GeV electron shower. A good background rejection, and consequently a high detection efficiency, and good identification of various modes of nucleon decay are therefore expected. Fig. 7(a) shows a typical example of simulated Čerenkov-ring patterns appearing on the inner walls of the tank for the decay $p \rightarrow e^+\pi^0$. The response of the phototube array for this type of event is shown in Fig. 7(b). The characteristic Čerenkov-ring patterns can be clearly recognized when we consider the circled tubes with a signal of more than $\sim$ 4 photoelectrons (see Fig. 7(b)). The design and preparation status of the 3000-ton water Čerenkov detector was first presented at Neutrino '81 [3]. Other design-reports are referred to in[4][5][6][7].

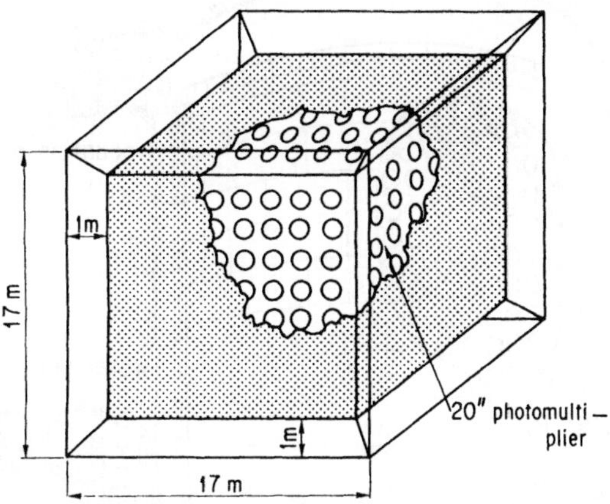

Figure 4: First designed detector of Kamiokande.

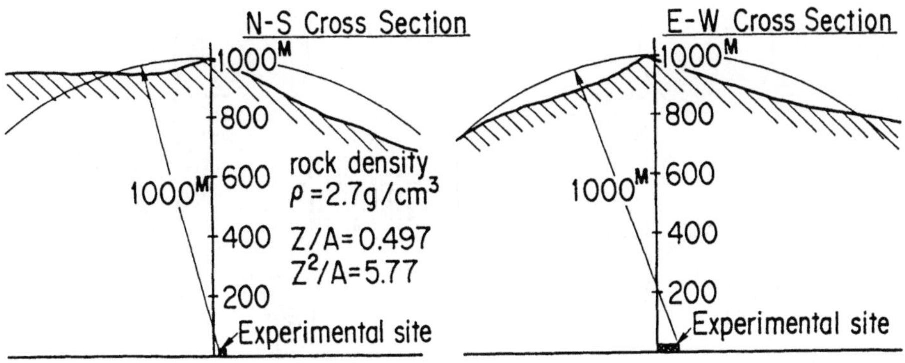

Figure 5: Cross sectional view of the experimental site.

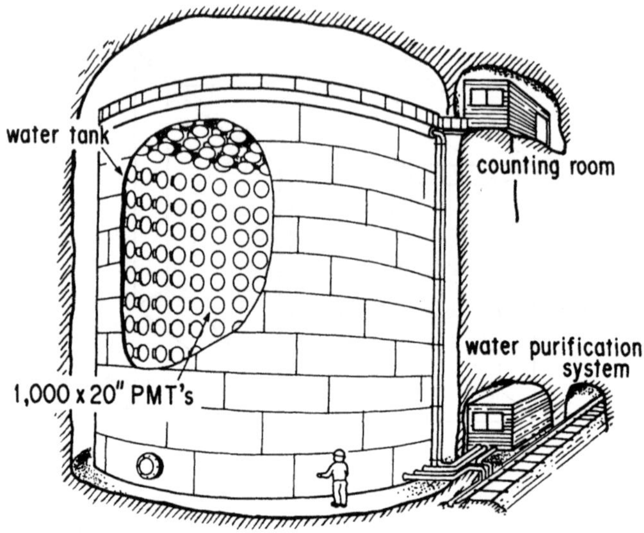

Figure 6: Sketch of the 3000 m^3 water Čerenkov detector.

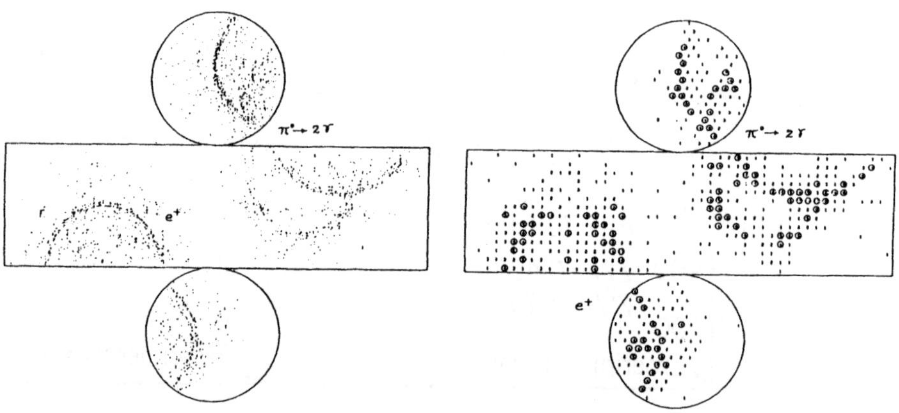

Figure 7: (a) Simulated Čerenkov light rings of $p \rightarrow e^+\pi^0$. One dot corresponds to 10 photons. (b) Phototube response for the event shown in (a), where the number of photoelectrons produced in each PMT is written at the PMT position. PMT's with more than 4 photoelectrons are indicated by a circle.

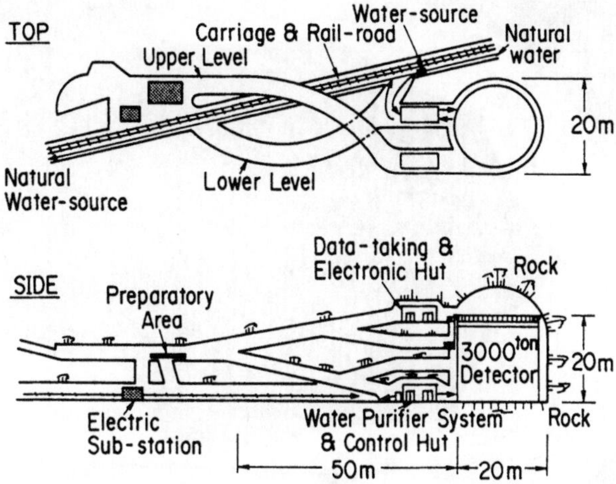

Figure 8: Sketch of the Kamiokande underground laboratory

3 Construction of the Detector

Excavation of the large-volume cavity (19 m in diameter, 23 m high at the center and 20 m high at the edges) started at the beginning of 1982. It took about 10 months. A schematic view of the entire experimental area is shown in Fig. 8. The flat spaces of the tunnels near to the cavity were used for the data-taking electronics room at the top and for the water-purifier system at the bottom. Construction of the 3000-ton water tank continued until March, 1983. The steel was 12 mm thick at the bottom and 4.5 mm thick at the top. The inner walls of the tank were carefully painted black to a thickness of 250 μm with a specially prepared epoxy, a suitable paint for pure water. The construction of the water-purification system and the data-taking electronics system began in early 1983. All of the necessary instruments were delivered to the mine by the end of March, 1983.

The 20-inch PMT's were already ascertained to have sufficiently good characteristics as the detector component of the proton-decay experiment, and the mass-production was established [8]. During detector construction, 1000 20-inch PMT's were delivered to the mine side from Hamamatsu Photonics Co. The installation of 1000 PMT's into the tank began in April, 1983. We utilized three methods of PMT installation. In the bottom plane and the lowest two-barrel rows, we attached PMT's to the support frame by hand. Water was then poured into the tank and installation on the barrel wall was carried out, floating rubber boats. The water was leveled up after completing every 48 tubes in one row on the barrel. Finally, the top plane PMTs, which comprised a 5-PMT module were installed using a crane. Two photographs taken during the detector construction are shown

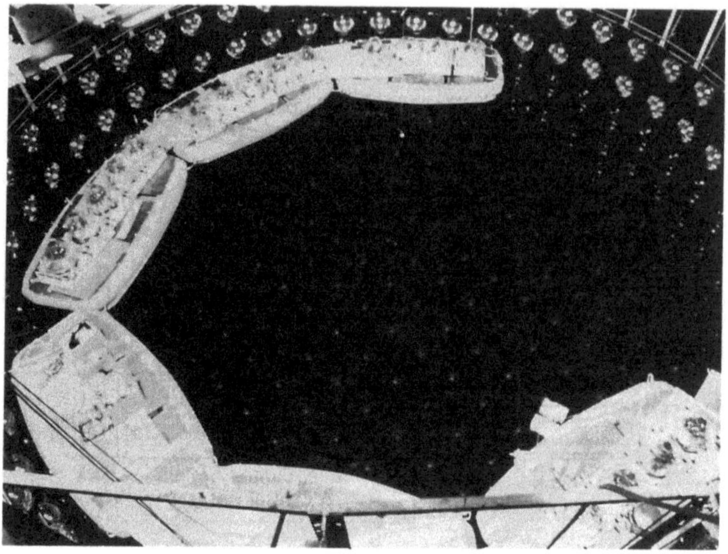

Figure 9: Installation of the barrel PMT's.

here: PMT installation on the barrel wall in Fig. 9 and an inner view of the tank just before completing PMT installation in Fig. 10. The detector was completely ready by July, 1983. This experiment was named Kamiokande after the Kamioka nucleon decay experiment.

4 Kamiokande Ready

The Kamiokande experiment started on July 6, 1983. In July and August we took much time to examine the detector performance and to make analysis programs for nucleon-decay events, atmospheric-neutrino events and so on. Data tapes were regularly sent from Kamioka to the Univ. of Tokyo. The ICEPP (International Center for Elementary Particle Physics) computer was fully utilized for the data analysis. On Sept. 9, 1983, the first candidate event of nucleon decay was observed. Fig. 11 shows an exploded view of this event, in which the PMT walls are opened at the top and bottom as well as at the side. The area of each circle is proportional to the number of photoelectrons detected by the PMT. Three well-separated Čherenkov rings can be clearly seen. The transit digitizer records a small second pulse of the $\mu \rightarrow e$ decay signal shown in the lower-right of Fig. 11. The curves in this figure are the results of a fit which assumed a common vertex for the rings. This event can be interpreted as a nucleon decay in the modes of $p \rightarrow \mu^+\eta(\rightarrow \gamma\gamma)$, $p \rightarrow \mu^+K^0(\rightarrow \pi^0\pi^0)$ and $n \rightarrow e^+\rho^-$. Since this event was so clear, it had a stimulating impact on nucleon-decay searches.

The first result from Kamiokande was presented by Koshiba at ICOBAN '84

Figure 10: Just after completing PMT installation.

(Utah, Jan. 1984) [9] and by Totsuka at the Moriond Conf. (Feb. 1984) [10]. During July 6, 1983 - April 5, 1984, 485-ton year of data were accumulated and two events survived tighter kinematical cuts, while the expected atmospheric neutrino background was small. The result of lower limits on the nucleon decay lifetime (τ/B) are presented in Table 1[10]. Papers concerning the Kamioka nucleon decay results were summarized in three issues. We decided to send these memorable papers of Kamiokande to the Journal of the Physical Society of Japan [11][12][13].

Another search for nucleon decay induced by a magnetic monopole [14] was carried out by detecting nucleon decays by two independent methods: (1) A search for multiple nucleon decays in the detector; (2) A search for low-energy neutrinos with $< E_\nu > = 35$ MeV from the catalyzing decay of solar hydrogen by magnetic monopoles captured in the core of the Sun [15]. These experiments found no evidence for monopole-catalyzed nucleon decay, and gave limits for the monoploe flux (see ref. [16]).

Soon after the Kamiokande experiment had started, the possibility to detect ^{8}B solar neutrinos was pointed out in a careful analysis of cosmic-ray muons which were stopped in the detector and were accompanied by $\mu \rightarrow$ e decay electrons. Fig. 12 shows the decaying electron energy spectrum. One can see that the lower end of the energy spectrum reaches down to $\sim$ 15 MeV. Below this energy, background events rise up rapidly. This figure implies that it is possible to detect ^{8}B solar neutrinos with a maximum energy of 14 MeV, with the aid of reasonable reduction of background events. This was the motivation for upgrading the Kamiokande detector.

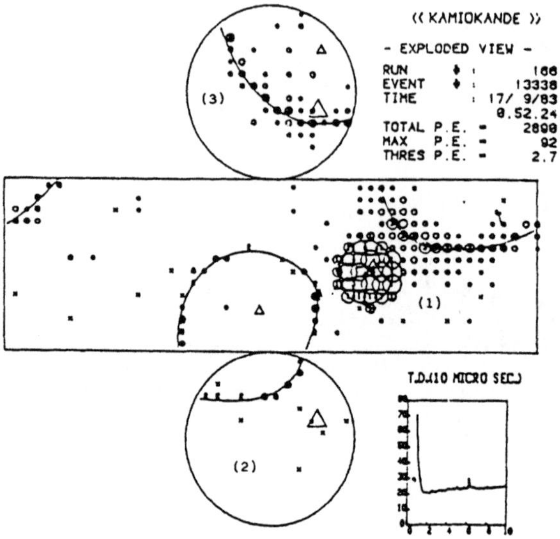

Figure 11: Unfolded view of the first nucleon decay candidate. The particle assignments for the p $\to \mu^+\eta(\gamma\gamma)$ mode are ring(1) = γ, ring(2) = μ and ring(3) = γ.

Table 1: τ/B lower limit (90 % C.L.) for 450-ton·yr data (1984).

Preliminary

T=201.3days WT=485 ton·years

Decay Modes	ε.Bm	ν-B.G.	Candidates	τ/B(90%CL) ×10^{31} years
$P\to e^+\pi^0$	0.45	0	0	3.2
$e^0\eta$	0.38	0	0	2.7
$e^+\omega$	0.29	0.6	1	1.2
$e^+\rho^0$	0.30	0.3	0	2.1
e^+K^0 $(2\pi^0)$	0.16	0	0	
$(\pi^+\pi^-)$	0.23	0.3	0	2.8
e^+K^*	0.20	0	0	1.4
$P\to\mu^+\pi^0$	0.39	0	0	2.8
$\mu^+\eta_0(2\gamma)$	0.27	0	1	1.1
$\mu^+\rho_0$	0.31	0	0	2.2
μ^+K^0 $(2\pi^0)$	0.11	0	1	
$(\pi^+\pi^-)$	0.16	0	0	1.1
$n\to e^+\pi^-$	0.25	0.1	0	1.4
$e^+\rho^-$	0.20	0	2	0.4
$n\to\mu+\pi-$	0.28	0	0	1.6
$\mu+\rho-$	0.14	0.3	0	0.8

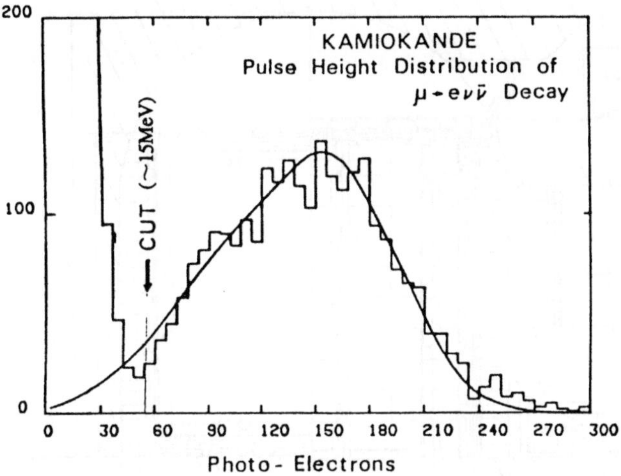

Figure 12: Energy spectrum of observed electrons with the energy <∼ 80 MeV. Distribution above 60 photoelectrons is well fitted to the electron spectrum expected from cosmic-ray $\mu \rightarrow e\nu\bar{\nu}$ decays. Events below 60 photoelectrons are background events coming from the inside and outside the detector. 110 photoelectrons correspond to ∼ 30 MeV.

5 Construction of Kamiokande-II

At ICOBAN'84 (1984) Koshiba also gave a talk concerning the possibility to detect ^{8}B solar neutrinos in addition to proton decay. He asked people there to join in upgrading the Kamiokande detector for this purpose. Soon after this workshop, A.K. Mann of the Univ. of Pennsylvania expressed a desire for collaboration between Japan and the United States, and to prepare new signal discriminators and timing electronics for each PMT. The first collaboration meeting was held in Tokyo in March, 1984. The U.S. party included A.K. Mann, E.W. Beier and R. Van Berg from Penn, B. Cortez from Caltech, and D.H. White from Brookhaven.

The detector design of Kamiokande-II was investigated until July, 1984. Kamiokande-II is an upgraded version of the Kamiokande nucleon decay detector, which not only improved the sensitivity to nucleon decay, but also lowered the energy threshold, thereby allowing it to search for solar neutrinos. Fig. 13 shows a schematic outline of the Kamiokande-II detector. 1071 20-inch PMT's (948 for the inner part and 123 for the anti-part) were mounted on the detector walls. The mean thickness of the top, barrel and bottom antilayers is 1.5 m. These active water layers shield against entering photons and neutrons in addition to vetoing entering cosmic-ray muons. In order to construct the bottom anticounter, after 48 inner-barrel PMT's had been removed, all of the bottom PMT's were lifted up by 1.2 m (see Fig. 14). There were 3 access tunnels to the tank, which should be closed for the construction of the barrel anticounter. The outer wall of the tank and the cavity wall facing each other were sprayed with rubber-asphalt to make

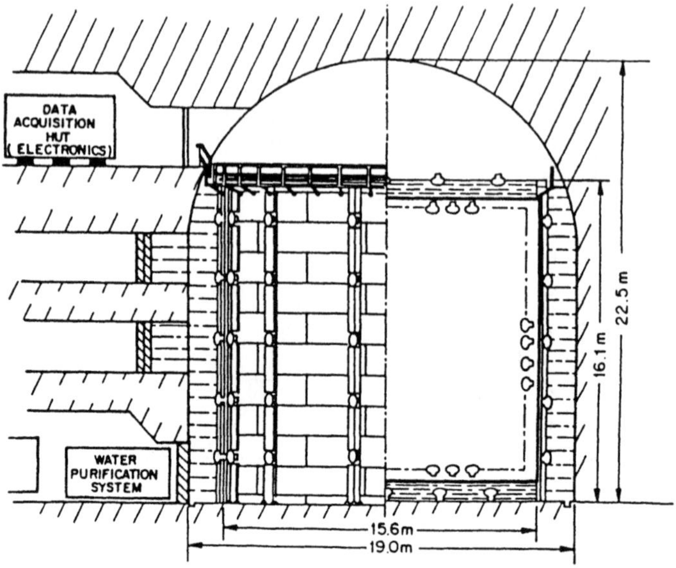

Figure 13: Schematic view of the Kamiokande-II detector.

the cavity water-tight. The entire surface of the anticounter sections was covered by aluminum polyetheylene-coated sheets.

The anticounter construction began in Sept., 1984. Fig. 15 shows the construction of the wall closing of one of the access tunnels for the barrel anticounter. 30 dead PMT's during runs of Kamiokande-I (called for the first stage of the Kamiokande experiment) were replaced by new ones while removing detector water (see Fig 16). This was the first time that we could see the inside of the detector when the tank was empty. The construction of the top and bottom antilayers was completed by the middle of December, 1984. Water was again filled into the tank. Subsequently, filling water into the barrel anticounter started. Soon after filling, we found a water leak from the entire rock wall. This was caused from many regions of unevenness on the rock wall, thus making a tight coating difficult. Although spraying rubber-asphalt was retried again and again, we finally gave up complete water-tightness, and instead filled water by a high-power pump so as to overcome the water leak. Due to this problem, completion of the barrel anticounter was delayed to March, 1985. However data-taking had been carried out only during the night time since December, 1984, without any barrel anticounter data. Fig. 17 shows the PMT's and the supporting frame of the barrel anticounter.

In Kamiokande-I we supplied fresh water into the tank which flows continuously in the mine ditch. However, we found that natural water contained too much ^{222}Rn (200 pci(pico curie)/l), which was one of the major background sources of solar-neutrino detection. Hence, from the beginning of April, 1985, the detector water was recirculated after passing through a water purification system. Improvements of the water purifier in Kamiokande-II are described in the next section.

Figure 14: Construction of the bottom anticounter

Figure 15: Construction of the barrel anticounter wall. The 3000-ton detector stands backwards the wall.

Figure 16: Replacing dead PMT's by new ones.

Figure 17: Barrel anticounter viewed from the bottom.

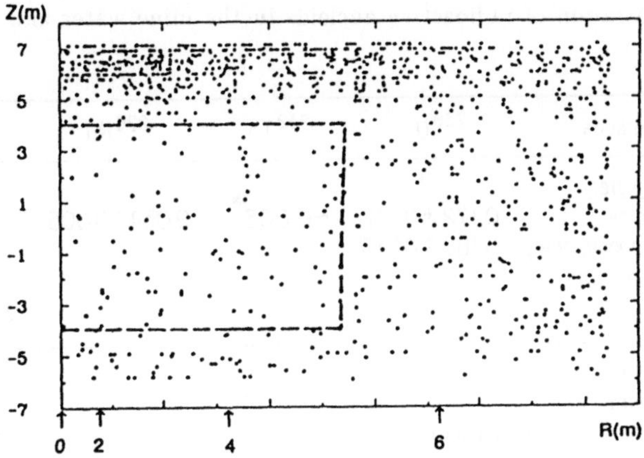

Figure 18: Vertex position distribution for events with $E_e > 10$ MeV. Note that the horizontal axis is scaled by R^2 for volume normalization.

The electronics system was modified in Kamiokande-II. Signals from the PMT's were connected to the Pennsylvania electronics. Each signal was split inside the electronics. One signal goes through a buffer amplifier and then into the Kamiokande-I electronics. The other signal goes into the Pennsylvania electronics. The Pennsylvania electronics became available in November, 1985. They include charge and timing measurement systems with multi-depth analog buffers. The timing information provided a remarkable improvement in the vertex reconstruction, especially for low-energy events. For example, the accuracies (1σ) of the position and direction for a 10 MeV electron were 1.7 m and 28°, respectively. Kamiokande-II data-taking started in December, 1985.

6 Toward Low-Background Detector

6.1 Major Background Sources

To observe ^{8}B solar neutrinos it is necessary to lower the detection threshold energy, at least down to less than $\sim$ 10 MeV. However, there are various kinds of background sources which produce low-energy electrons and γ-rays in the detector. The environmental background events of γ-rays and neutrons from rock walls are attenuated, owing to the anticounter water layer. Surviving events passing through the anticounter are almost rejected by applying a fiducial volume cut of event selection. Fig. 18 shows the vertex point distribution of the reconstructed events in the $R - Z$ plane, where R and Z stand for the radial and height coordinates in the cylindrical detector. The concentration of events can be seen near to the edges (top, barrel and bottom) of the detector. The fiducial volume for the solar neutrino analysis was set in the region of - 3.98 m $\leq Z \leq$ 4.0 m and $R \leq$ 5.2 m, which corresponds to a total fiducial volume of 680 m^3.

Other serious background sources come from interactions of high-energy cosmic-ray muons in the detector, where muons break up ^{16}O and produce various radioac-

Table 2: Concentration of radioactive nucleids in the mine water and air (unit: pico curie per litter).

Measuring methods	^{238}U	^{232}Th	^{226}Ra	^{222}Rn
in the mine water				
α-spectrometry	0.172 ±0.012	~0.003*	0.054±0.005	
Atomic spectrophotometry	0.180 ±0.013			
Scitillation counting				270 ± 14
				300 ± 0.1
in the mine air				
γ-spectrometry				50 ± 0.5
(Rn adsorption by cooled charcoal)				68 ± 2

*estimated from ^{222}Rn concentration

tive nuclei. Pions produced in the cosmic-ray muon interactions also contribute to the production of unstable nuclei through interactions with ^{16}O. These cosmic-ray muon-induced events are eliminated only by off-line data analysis, using the spatial correlation between the muon traversing path and the emitted electron or γ vertex points, the time interval correlation between the preceding muons and the electrons or γ's, and the deposit energy correlation of muons.

The fission products of ^{238}U and ^{232}Th, as well as β-rays and γ-rays from radioactive nuclei in the decay series of ^{238}U and ^{232}Th are also dominant background sources. In the spontaneous fission of uranium, prompt energy is released in the form of γ-rays with an energy of $\sim$ 7 MeV, and the mean number of γ-rays is about 7.4. β-rays with $E_{max} = 3.26$ MeV emitted from ^{214}Bi in the ^{226}Ra $\rightarrow$ ^{222}Rn $\rightarrow \cdots \rightarrow$ ^{214}Bi $\rightarrow$ ^{214}Po decay chain are a serious background source, since ^{226}Ra and ^{222}Rn are highly contaminated in the detector water. Moreover, the content of ^{222}Rn in the mine air is at least 10^3-times higher than that of air outside of the mine. The mine air was always in contact with the detector water at the top surface of the 3000 m^3 tank and two buffer tanks in the water purification system. The improvement in the water-purification system was directed so as to remove the radioactive materials discussed here. This improvement has been successfully carried out since the Kamiokande-II data-taking. Concentrations of ^{238}U, ^{232}Th, ^{226}Ra and ^{222}Rn in the mine water, in the mine air, and in the detector water were measured. The results are listed in Tables 2 and 3.

6.2 Improvement in the Water-Purification System

A normal ion-exchanger is effective for absorbing all types of ions desolved in water. However, the efficiency strongly depends on the concentration of other ions. Amidoxime group chelete resins have the property to selectively adsorb uranium in water. The efficiency was measured to be above 90% for the Kamioka mine water. The ion-exchanger with amidoxime resins was installed in March, 1986. ^{232}Th and

Table 3: Concentration of radioactive nucleids in the detector water (unit: pico curie per litter).

Measuring methods	^{238}U	^{232}Th	^{226}Ra	^{222}Rn
in the detector water				
α-spectrometry (April, 1987)	$(2.43\pm0.04)\times10^{-4}$		$(1.13\pm0.38)\times10^{-4}$	
ICP*-mass spectrometry (Nov., 1987)	$<5\times10^{-5}$	$<5\times10^{-5}$		
Scintillation counting (Sep., 1988)				0.89 ± 0.24
in the Rn-free air				
γ-spectrometry (Sep., 1988)				0.19 ± 0.25

* ICP : Inductively Coupled Source Plasma

^{226}Ra are expected to be removed by the usual type mixed-bed (anion+cation) ion-exchanger. A new powerful mixed-bed ion-exchanger was installed in Sep., 1986. The change in the contents of ^{238} and ^{226}Ra after operating the new ion-exchangers is shown in Fig. 19.

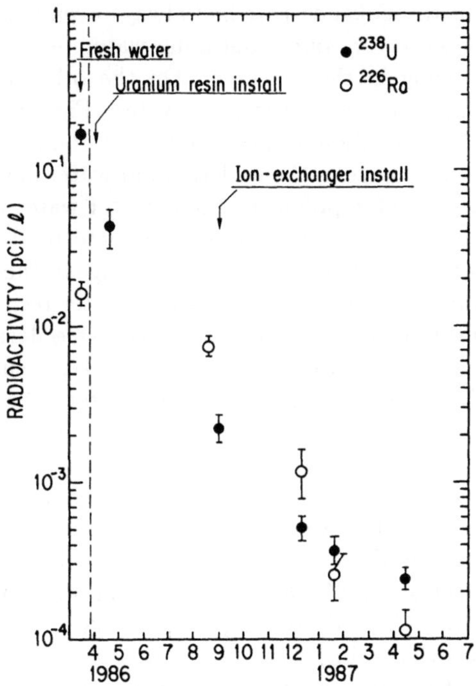

Figure 19: Reduction of the radioactivity (^{238}U and ^{226}Ra) in the detector water since 1986.

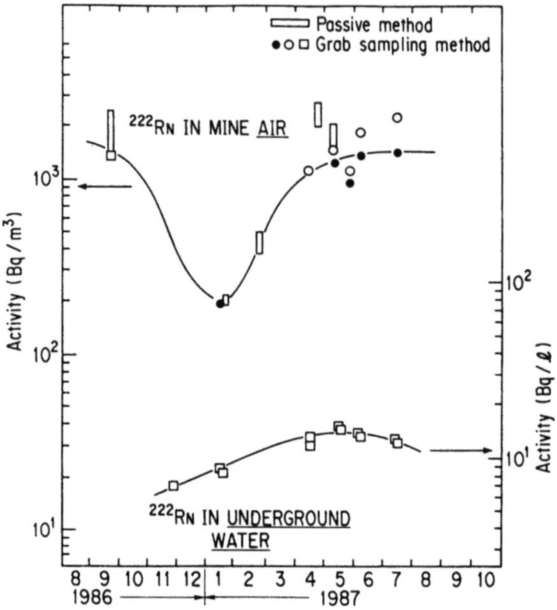

Figure 20: Annual variation of the ^{222}Rn concentration in the mine air and water.

As written above, in Kamiokande-I we had always been supplying fresh water from the mine into the detector, since cool mine water ($\sim$ 9° C) is effective for depressing bacteria growing in the detector. On the other hand, the detector suffered from radon pollution due to the fresh water. Fig. 20 shows the annual variation in the ^{222}Rn concentration in the mine air and the mine water. The clear variation correlates to the change in the underground water flow due to a thaw on the mountain. To stop the radon pollution, the detector water was recirculated in the closed system of the water purification and the detector tank, since the half life of ^{222}Rn is short (3.8 days). Fig. 21 shows the change in the trigger rate under a 8.5 MeV detection threshold energy after starting water recirculation. One can find that the trigger rate decreases rapidly with the same decay time as that of ^{222}Rn, and also big spikes of the trigger rate. These spikes resulted from troubles in the water-purification system, mostly water leaks from the apparatus. Fresh water therefore had to be supplied, resulting in a sudden jump in the trigger rate, as shown in Fig. 21.

The next improvement is to make isolation structures from the mine air both in the 3000-ton tank and in the water-purification system. An air-tight ceiling structure on the top surface of the detector tank was made in March, 1987. This was just after the detection of SN1987A. The sealing of two buffer tanks was completed in May, 1987. Clear decreasing of the trigger rate is shown in Fig. 22 after finishing the air-tightening (May 1987). However, perfect air-tightness is difficult, because a negative pressure state exists upon changing the water level in the tank. Therefore, ^{222}Rn-free air, which is made by flowing the mine air through

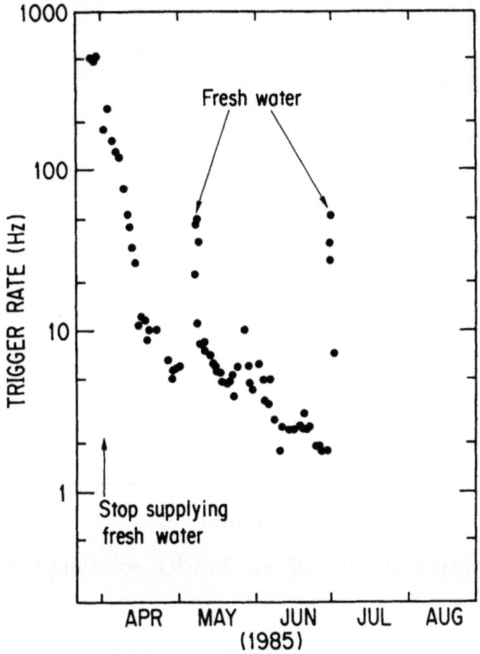

Figure 21: Change in the trigger rate after stopping the fresh-water supply. The threshold energy of the event trigger is 8.5 MeV.

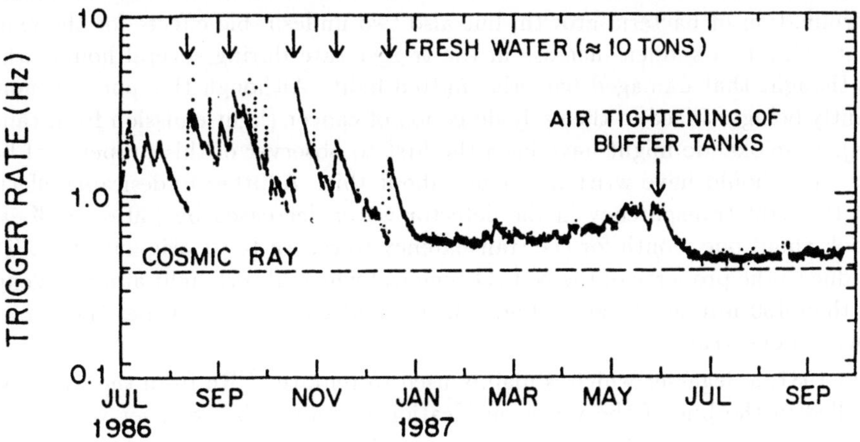

Figure 22: Change in the trigger rate since 1986

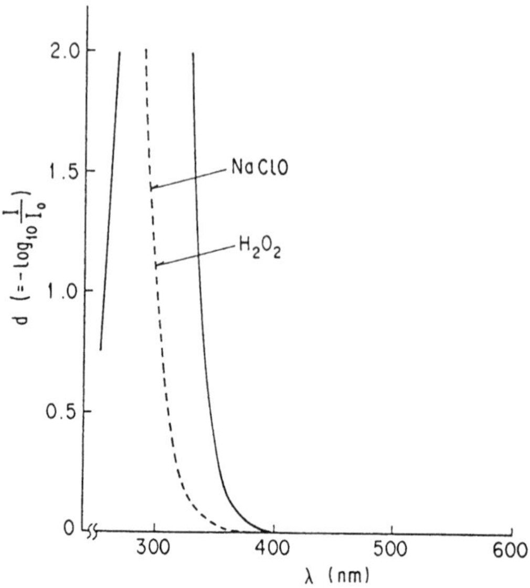

Figure 23: Light-absorption spectra of an NaClO solution(0.1%) and an H_2O_2 solution(3%)

dry chacoal filters with an efficiency of 99%, was constantly supplied into the top space of the detector and buffer tanks.

During the course of improving the water-purification system, we received a big bonus of finding how to stop the propagation of bacteria. Since the recirculation of the detector water started, we have been annoyed about a large amount of bacteria growth. Whenever it occurred, a NaClO solution used as a sterilization chemical was put into the detector tank. This chemical resulted in not only an efficient extermination of bacteria growth, but also two undesirable effects for the experiment. One is a sudden increase in the trigger rate during several hours, which was thought that damaged bacteria emitted light. Although this phenomenon is recently being utilized in the early detection of cancer (light-emission from cancer cells), Kamiokande might have been the first to observer of this property of bacteria. We should have written a paper about this. Another undesirable effect is that the light transparency in the detector water decreased by $(30 \sim 40)\%$, and it took about one month for the transparency to return to its original value. This was due to the property of the NaClO chemical which absorb light at a wavelength less than 400 nm, as shown in Fig. 23. We had two or three times the bacteria growth in one year.

In 1987, a degasser which is mainly used to remove ^{222}Rn from the water, was installed in the line of the water-purification system. The degasser was efficient to also remove oxygen desolved in the water. Since the operation of the degasser, unexpectedly, the bacteria growth vanished. We found that deoxidizing was the best way to destroy bacteria without adding any contamination to the detector water.

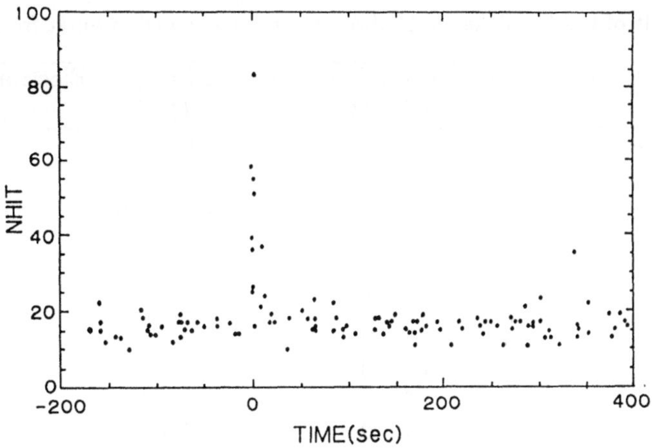

Figure 24: Observed data of supernova neutrinos. The horizontal axis stands for time (sec.) and the vertical axis for energy. The 11 events starting at t = 0 sec. were caused by a supernova neutrino burst. Events with NHIT $<\sim$ 20 distributed contineously were due to background events, where NHIT is defined in the text.

7 Supernova SN1987A

Troubles in the water-purification system continued during 1986. Consequently, a big jump-up of the trigger rate occurred frequently, as shown in Fig. 22. These problems were fortunately fixed by the beginning of 1987. The trigger rate has calmed down since then. Supernova SN1987A exploded during the period of a somewhat stable detector condition. Our painful work toward realizing solar-neutrino detection was paid back dramatically.

On February 25, 1987, a telefax from the U.S. informed us of the appearance of a supernova in the Large Magellanic Clouds. News of this supernova was quickly sent to the shift persons at the Kamioka mine, and the data tapes until February 25 were immediately dispatched to Tokyo. Since the mine was closed on February 22 and 23, data were continuously taken during these days. We were lucky, because from the end of February, we were preparing to construct a steel roof on the top of the detector tank in order to protect from the radon invading from the mine air. However, the company postponed this job due to difficulty of arranging workers; 6 $\sim$ 7 Kamiokande members were thus standing by at the mine side for construction to start. The company's judgement regarding this delay was quite in perspectives and offered a historical observation.

The data tapes arrived at Tokyo on February 27 . The data between 16:09 February 21 and 07:31 February 24 in Japanese Standard Time were analyzed quickly. Analysis programs for low-energy events (space reconstruction, energy determination, and elimination of spallation events induced by cosmic-ray muons) had been already ready for the solar-neutrino search. A two-dimensional plot of NHIT (the number of hit photomultiplier tubes, which is proportional to the energy; NHIT = 2.28 E (MeV)) vs. the event time was made for the entire period. After scanning a few hundred pages of its print-out, a burst of events was

Table 4: Result of the Kamiokande-II observation for a neutrino burst from SN1987A.

Experiment	Event No.	Time (UT) (Feb. 23)	Electron Energy (MeV)	Electron Angle (degree)
Kamiokande	1	$7^h35^m35^s$	20.0 ± 2.9	18 ± 18
	2	$35^s.11$	13.5 ± 3.2	15 ± 27
	3	$35^s.30$	7.5 ± 2.0	108 ± 32
	4	$35^s.32$	9.2 ± 2.7	70 ± 30
	5	$35^s.51$	12.8 ± 2.9	135 ± 23
	*	$35^s.69$	6.3 ± 1.7	68 ± 77
	6	$36^s.54$	35.4 ± 8.0	32 ± 16
	7	$36^s.73$	21.0 ± 4.2	30 ± 18
	8	$36^s.92$	19.8 ± 3.2	38 ± 22
	9	$44^s.22$	8.6 ± 2.7	122 ± 30
	10	$45^s.43$	13.0 ± 2.6	49 ± 26
	11	$47^s.44$	8.9 ± 1.9	91 ± 39

immediately found. Fig.24 shows this scatter-plot. The burst occurred at 16:35:35, 23 February, 1987, JST, or at 7:35:35, 23 February, 1987, UT (± 1 minute) during a time interval of 13 seconds. The signal consisted of 11 electron events of energy 7.5 to 36 MeV, of which the first 2 events pointed back to the Large Magellanic Clouds with angles of $18° \pm 18°$ and $15° \pm 27°$. All of the information for the 11 events are listed in Table 4. A paper was ready by March 6. On March 7, after having sent the paper to a post office, we found a serious mistake in the definition of both the right ascension and declination. We asked the post office to keep the paper aside until sending a new one. First of all, we worried that the major part of the paper had to be rewritten. However as it sometimes happens, double mistakes mostly made the mistake cancel out. The numerical modification gave only a slight change in the table and figures, but no change in the manuscript. A rewritten paper was resubmitted on March 9 (Monday) [17]. A press release was given on the same day. The story concerning how things happened in the Kamiokande group was written in [18] in detail.

The time difference of the burst observation between Kamiokande and IMB resulted from a large uncertainty of the absolute timing at Kamiokande. We experimenced a short power outrage in the mine on February 26. We therefore lost a chance to recalibrate our computer clock. The error estimate of ± 1 minute is based on a check of the data-taking condition on February 21 through a terminal from Tokyo.

A group at the Mont Blanc Underground Neutrino Observatory (LSD) observed a cluster of 5 events in 7 seconds at 2:52:36, UT on February 23. This detection time was about 5-hours earlier than that of Kamiokande. The LSD event data predict that a total of 27 ± 12 events should have been detected in Kamiokande at that time, considering the different detector size, detector material and detection efficiency in both experiments. However, the Kamiokande data around 2:52, UT

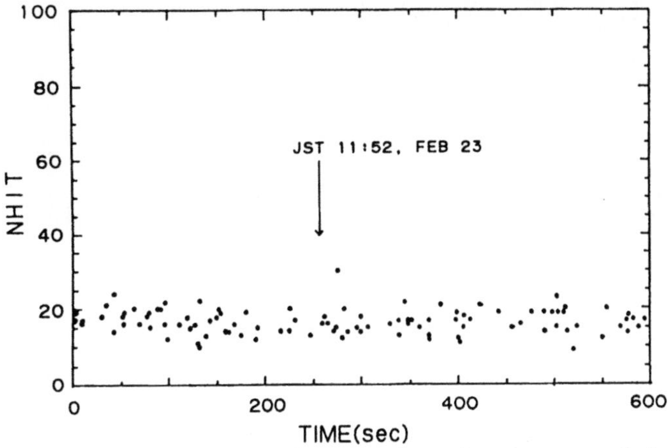

Figure 25: Scatter plot of NHIT vs. the event-detection time around the time of burst detection in LSD.

(11:52, JST) showed no event enhancement (Fig. 25). Instead, one can see only low NHIT background events with an average frequency of about 0.2 Hz.

A neutrino burst from supernova SN1987A was observed at Kamiokande, escaping several accidental problems and happenings. We learned much about the difficulty to continue data-taking under the good detector conditions which are a requisite at the observatory. Fig. 26 shows another example of accidents. The supernova burst came ~ 2 minutes ahead of a data-taking blind period due to an electronics gain check. Gain checks of the Kamiokande electronics, generating test pulses into all of the PMT channels, were repeated every 2 hours. It took about 3 minutes, during which data-taking was blind. Based on this experience, gain checks are now carried out channel by channel instead of all channels simultaneously. Thus, there has been no dead time for the gain check with a victim of one rest channel.

A search for high-energy neutrinos from the direction of SN1987A was carried out by observing upward-going muons produced in the nearby rock by neutrinos[15]. No upward-going muons were found in the data sample on 23 Feb. to 1 Sep., 1987; the 90% confidence-level flux limit is 1.2 $\times 10^{-13}$ cm^{-2} s^{-1} for an energy threshold of 1.7 GeV.

8 Solar Neutrino Detection

Kamiokande-II has permitted a study of low-energy contained electrons produced through the neutrino elastic-scattering reaction $\nu + e^- \rightarrow \nu + e^-$. The detection of neutrinos emitted in the reaction $^8B \rightarrow {}^8Be + e^+ + \nu_e$ occurring deep in the Sun is the principal motivation for this study. This was the first experiment to confirm the results of the radiochemical ^{37}Cl solar neutrino experiment of R. Davis, Jr., and collaborators [20][21]. In Kamiokande-II the initial position and vector momentum of the recoiling electron were measured. The directionality of the electron detection

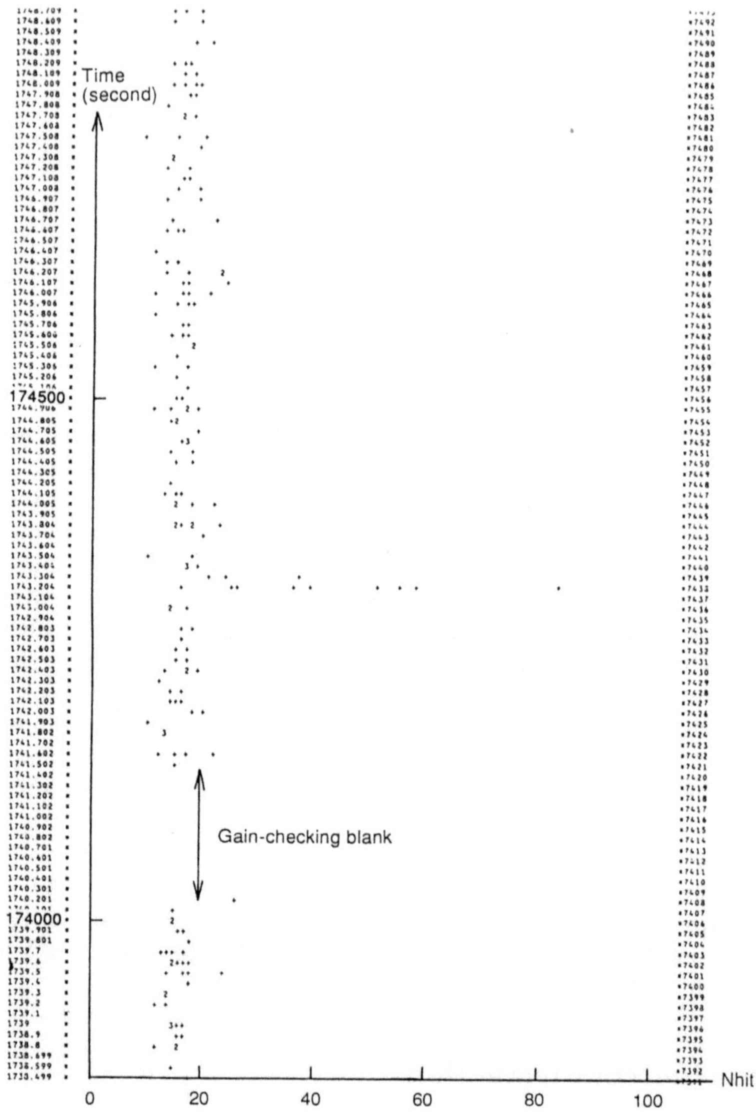

Figure 26: Computer print-out of the scatter plot; NHIT vs. the event detection time. (From ref. [18])

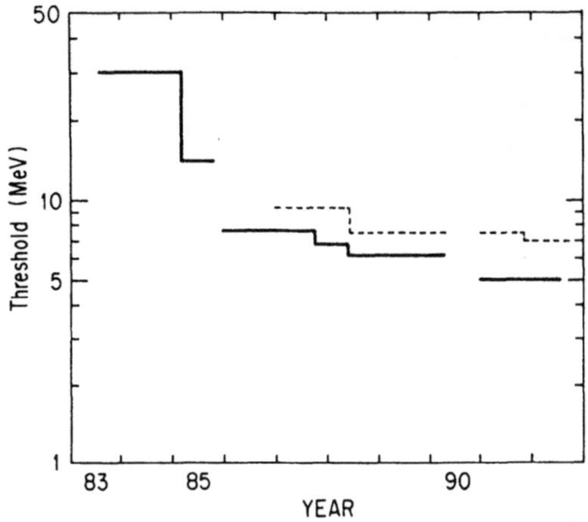

Figure 27: History of the trigger (solid) and analysis (dot) threshold energies.

was a consequence of the kinematic limit for neutrino-electron elastic scattering ($\Theta_e^2 \leq 2M_e/E_e$), independent of E_ν for $E_\nu \gg M_e$. Here, Θ_e, E_e, and M_e are the angle, energy, and mass of the electron, and E_ν is the neutrino energy. For 10 MeV electrons, Θ_e must be less than about 3°. The detector has an angular resolution of 28° at 10 MeV, and is mainly limited by multiple scattering of electrons in water. This therefore makes it possible to point the incident neutrinos back to the Sun. In addition, the observed electrons yield information concerning the energy spectrum of the incident ν_e. This provides confident evidence of ^{8}B solar neutrinos which comes from the fusion processes occurring inside the Sun.

The detector performance of Kamiokande-II was first presented at the 7th Workshop on Grand Unification/ICOBAN'86[22]. The first result concerning the search for solar ^{8}B neutrinos was reported at the 12th Int. Conf. on Neutrino Physics and Astrophysics, Neutrino '86, based on 48.5 days of live data[23]. The finite-flux value of solar ^{8}B neutrinos was reported in 1989[24] and in 1990[25]. The ^{8}B neutrino flux obtained in a Kamiokande-II 1040-day data sample was 0.46±0.13(stat)±0.08(sys) of the value predicted by the standard solar model[26]. Thus, Kamiokande-II reconfirmed the solar neutrino deficit which had been observed in the ^{37}Cl experiment for almost 20 years [27]. The recent result for a 1554-day data sample of combined Kamiokande-II and -III is 0.50±0.04(stat)±0.06(sys) of the standard solar model prediction[28].

Analyses of the Mikheyev-Smirnov-Wolfenstein effect[29][30] inside the Sun and the Earth were reported in refs. [31][32], using 1040 days of data. Both results provided constraints on the neutrino-oscillation parameters.

Thanks to a continuous upgrading the detector, the threshold energies of the trigger and the analysis for the solar neutrino events have been gradually decreasing down to 5 MeV and 7 MeV, as shown in Fig. 27.

In June, 1988, the detector performance was improved by increasing the gain of the PMT's by a factor of two, thereby decreasing the background level due to the following reasons: (1) an improvement of the energy resolution suppressed the

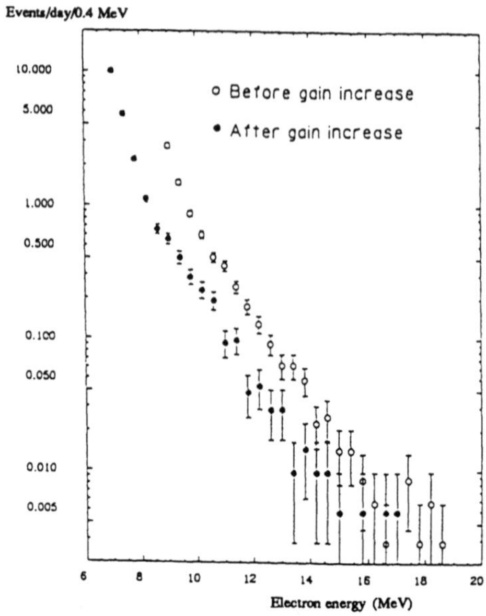

Figure 28: Differential energy distributions of low-energy events after applying all possible cuts to the data before and after the gain increase.

contribution from the smeared low-energy background spectrum due to radioactivity; (2) an increase of the number hit PMT in an event led to an improvement in the vertex reconstruction, resulting in a reduction of poorly reconstructed events in the fiducial volume. The energy spectrum for the events before and after a gain increase survived through all event-selection cuts is shown in Fig 28. The event rate in the low-energy region was especially reduced after a gain increase.

The changes in the distribution of $\cos\theta_{SUN}$ for different live day data samples are summarized in Fig. 29, where $\cos\theta_{SUN}$ is the cosine of the angle between the trajectory of an electron and the direction of the Sun at the time of an event. A clear enhancement at $\cos\theta_{SUN} = 1$ above the isotropic background can be seen in data samples with more than 450 live-day. The 450 live-day data sample was taken from January 1987 through May 1988 with a relatively low background (see Fig. 22). Therefore, the events taken in 1986 were not included in the further solar-neutrino search for more than 450-day data samples.

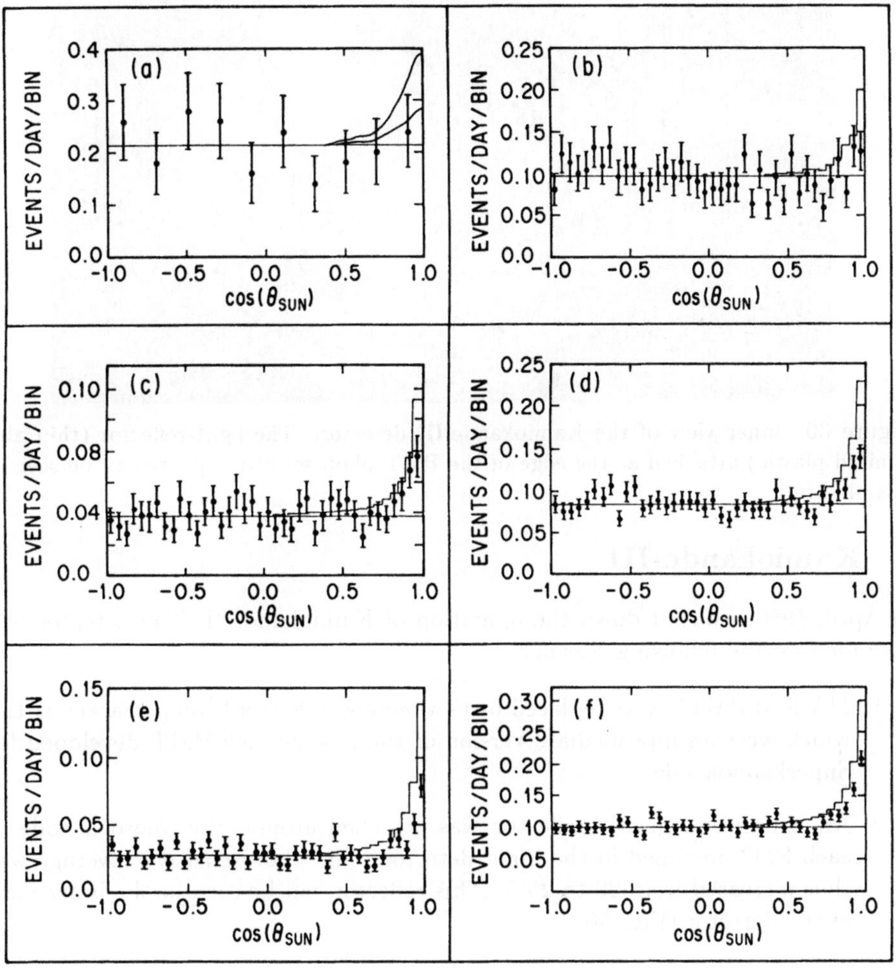

Figure 29: Cos θ_{SUN} distributions: (a) 48.5 live-day data with $E_e \geq 9$ MeV; (b) 220 live-day data with $E_e \geq 9.6$ MeV; (c) 450 live-day data with $E_e \geq 10.1$ MeV; (d) 738 live-day data with $E_e \geq 10.1$ MeV for 450 days and $E_e \geq 7.5$ MeV for 288 days; (e) 1040 live-day data with $E_e \geq 9.3$ MeV for 450 days and $E_e \geq 7.5$ MeV for 590 days; and (f) 1557 live-day data with $E_e \geq 9.3$ MeV for 449 days, $E_e \geq 7.5$ MeV for 794 days and $E_e \geq 7.0$ MeV for 314 days.

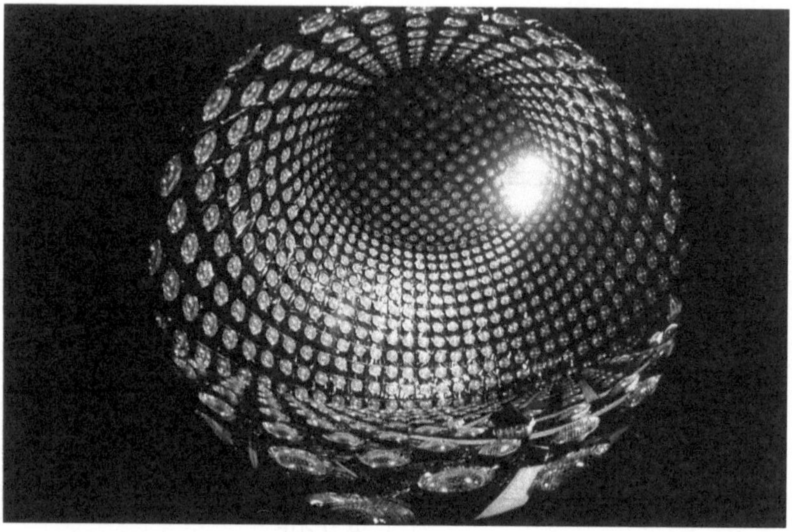

Figure 30: Inner view of the Kamiokande-III detector. The light-reflector (thin aluminized plastic) attached at the edge of the PMT photosensitive sphere can be seen in this picture.

9 Kamiokande-III

In April, 1990, we shut down the operation of Kamiokande-II. The detector was upgraded for the following reasons:

1. 110 dead PMT's were replaced by new ones with better timing characteristics which were an intermediate version of the new 20-inch PMT developed for Superkamiokande.

2. An aluminized plastic reflector was attached around the photocathode of each PMT installed in the inner detector. The photosensitive coverage was thus increased from 20 to 25%. This reflector can be seen in the inner view of the detector (Fig. 30).

3. The electronics were replaced by a new system which was developed as a prototype for Superkamiokande.

Due to the above-mentioned improvements, better detector performance was expected, especially for low-energy events. Kamiokande entered a new phase, called "Kamiokande-III". Kamiokande-III operation started in November, 1990.

10 Superkamiokande

Large-volume underground detectors have opened new research fields in elementary particle physics and astrophysics. As a future major experiment at the Kamioka

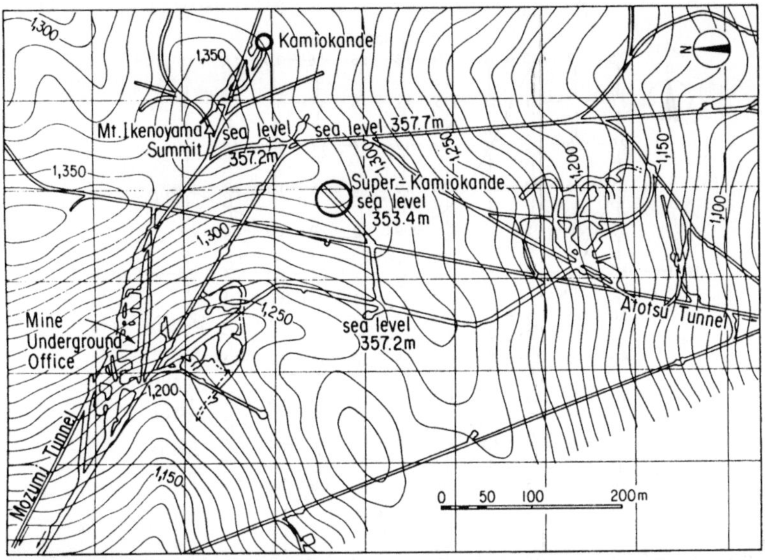

Figure 31: Topographical map of the sites of the Kamiokande and Superkamiokande detectors.

underground facility, a large water Cerenkov detector with a sensitive mass of 32000 tons, named JACK (Japan-America Collaboration at Kamioka), was considered in 1983 [33]. The name Superkamiokande was given to this detector in the following year (1984) [34]. The detector performance of Superkamiokande was presented in detail at the 7th Workshop on Grand Unification/ICOBAN '86 [35].

As an experimental site, the Kamioka mine is still the best suited, due to the tight rock condition and many available mine services: electric power, rock maintenance, transportation and so on. However, it is the first experience in our country to excavate such a large cavity ($\sim 10^5$ m^3) with a dome shape deep underground. Different kinds of tests necessary to investigate the rock mechanical characteristics were carried out by the Kamioka mine company, and also by a project team of civil-engineering companies and universities during 1986 to 1990. The most suitable site for Superkamiokande was decided, based not only on the obtained test results of the rock condition, but also on a consideration of acquiring a rock overburden that is as large as possible. Fig. 31 shows the positions of Kamiokande and Superkamiokande on a topographical map. The Superkamiokande site is about 200 m south-west from the present Kamiokande. The expected cosmic-ray muon rate is 2.9 Hz for the Superkamiokande size detector. This is almost the same as that at the Kamiokande site, as shown in Fig. 32. There are no locations with a rock condition which is adequate for constructing a large cavity in the northern part of the present Kamiokande site, where the cosmic-ray muon rate is relatively small (< 2 Hz).

The size of a water Cerenkov detector is strongly limited by the transparency of Cerenkov light in water. Since the attenuation length is ~ 50 m for 450 nm light

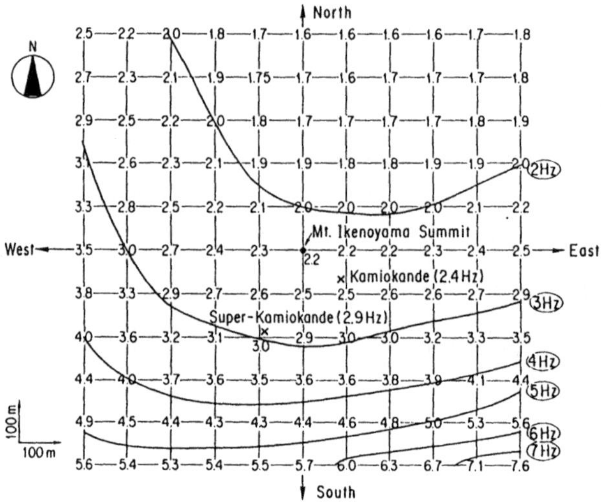

Figure 32: Expected cosmic-ray muon rate for the detector with the same size of Superkamiokande at the Kamiokande and Superkamiokande sites.

in pure water, a 50000 − 60000 m^3 total volume is the maximum size for an efficiently available volume as the Kamiokande-type detector. The final design of the Superkamiokande detector was determined to have a 50000 ton cylindrical tank 42 m high and 39 m in diameter. The barrel and bottom walls of the cavity are laid by thin stainless steel. This container is used as a water tank by itself. 11200 20-inch PMT's are placed uniformly over the entire inner surface, which gives a 40% photo-sensitive coverage. This is 2-times larger than that of Kamiokande. The support structure of the 20-inch PMT's is positioned at 2.7 m inside from the tank walls (top, barrel and bottom). The total mass of water inside the PMT surface is 32000 tons. The outer volume of the detector is used for anticounter layers. Black polyethylene sheets are set just behind the inner 20-inch PMT's and optically separate the inner and outer volumes. The detector parameters of Superkamiokande are listed in Table 5, and are compared with those of Kamiokande. The 22000-ton fiducial mass is 20-times larger for proton-decay detection and 32-times lager for solar-neutrino detection, than those of Kamiokande.

The quality of the 20-inch PMTs has been improved over the required specifications for those used in Superkamiokande [36]. In particular, the resolution of the timing response for single-photoelectron light (light produces one electron from the photosensitive surface of PMT) has been reduced to ~ 3 ns (1σ width), and a clear single-photoelectron peak can be seen in the pulse-height distribution. The typical relative transit-time distribution and pulse-height distribution taken by fully diffused illumination of single-photoelectron light with a wavelength of 410 nm are shown in Figs. 33 and 34, respectively. Such a good timing response allows a precise determination of the event-vertex point and the direction of emitted Cerenkov

Table 5: Detector performance of Superkamiokande and Kamiokande-II.

Parameters	Super-Kamiokande	Kamiokande-II	Remarks
Total size	41mh×39mϕ	16mh×19mϕ	
Total mass	50000t	4500t	
Fiducial mass	32000t	2140t	supernova ν burst
	22000t	1040t	proton decay
	22000t	680t	solar ν
Thickness of anti-counter	2m	1.2m∼1.5m	
Number of PMTs	11200	948	
Photosensitive coverage	40%	20%	
PMT timing resolution	2.5nsec	4nsec	at 1 P.E.
Energy	$2.6\%/\sqrt{E}$	$3.6\%/\sqrt{E}$	e of E(GeV)
	2.5%	4%	$\mu(\lesssim 1$ GeV)
resolution	$16\%/\sqrt{E(10MeV}$	$20\%/\sqrt{E(10MeV)}$	e($\lesssim$20 MeV)
Position	50 cm	110 cm	10 MeV e
resolution	∼10 cm	15 cm	$p \to e^+\pi^0$
Angular	28°	28°	10 MeV e
resolution	∼1°	2.7°	Thru-going μ
E_{th}(trigger)	4∼5 MeV	5.2 MeV	
E_{th}(analysis)	5 MeV	7.5 MeV	Solar ν
e/μ separation	99 %	98±1 %	$0.03<p_e<1.33$GeV/c
			$0.2<p_\mu<1.5$GeV/c
$\epsilon_{\mu^+\to e^+\nu\nu}$	95 %	87±1 %	

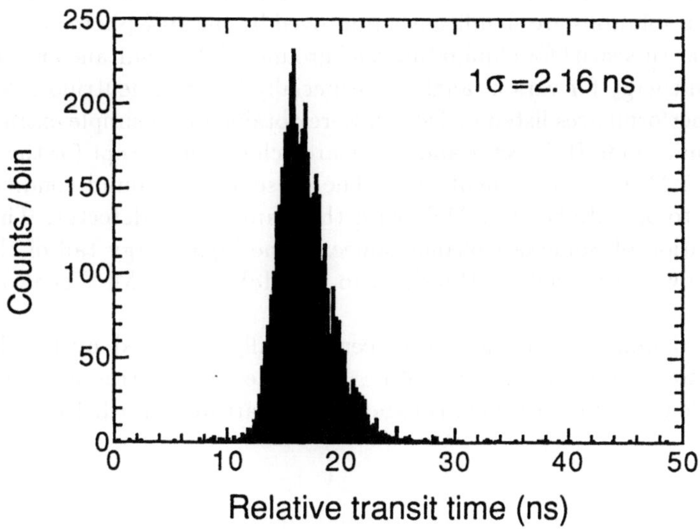

Figure 33: Relative transit time distribution of a new 20-inch PMT for a single-photoelectron light.

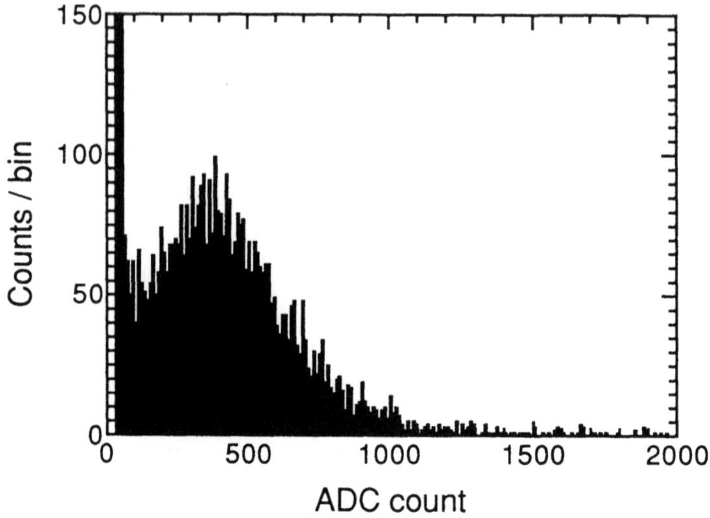

Figure 34: Pulse-height distribution of Single-photoelectron in a new 20-inch PMT.

light. The well-separated pulse-height distribution for single-photoelectron light makes it possible to eliminate γ- and β-rays produced from the decay of ^{238}U, ^{232}Th, ^{222}Rn and ^{40}K contaminated in the detector materials. This is because the lower energy resolution of the detector causes a misidentification of the energies of the γ- and β-rays as being more energetic.

The expected detector performances are summarized in Table 5. The energy and position resolutions are significantly improved by the 40% photo-sensitive coverage. They are essential for eliminating background while simultaneously lowering the minimum energy for various analyses, especially for solar neutrinos. Note that some of the performances listed in Table 6 were obtained by a simple extrapolation from the Kamiokande-II detector and, thus, are achievable, except for the analysis threshold of 5 MeV for solar neutrinos. The present background condition does not allow us to operate below 7 MeV with the Kamiokande detector. The better energy resolution of Superkamiokande squeezes the high-energy tail of the background and lowers the analysis threshold to 6.5 MeV. The 5 MeV goal, however, is not trivial.

The Superkamiokande project has been officially approved by the Japanese government to start excavation in 1991. The detector construction schedule is given in Table 6. The data taking is expected to start at 1 April, 1996.

Table 6: Superkamiokande time schedule. (From ref. [37])

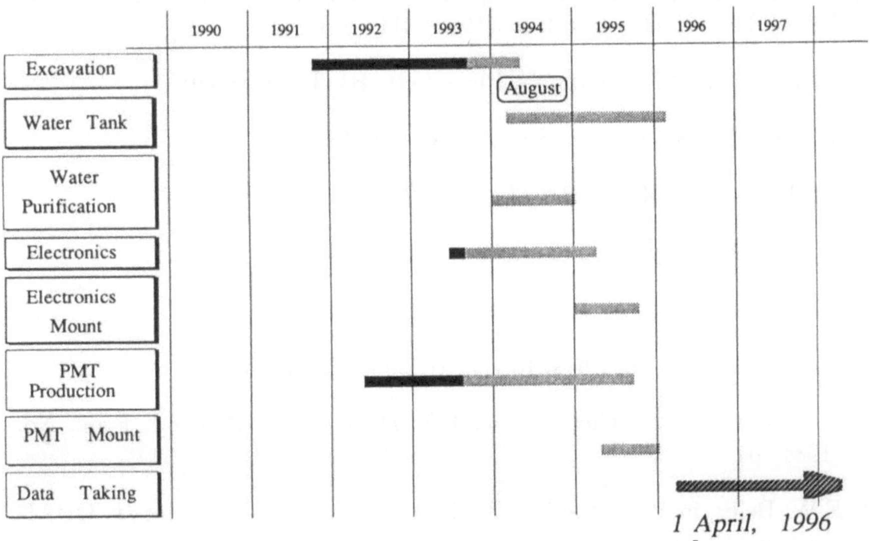

References

[1] Y. Watanabe, Proc. Workshop on the unified theory and the baryon number in the universe, KEK Report, KEK-79-18 (1979).

[2] T. Hayashi, Proc. INS Int. Symp. on nuclear radiation detectors, Tokyo (1981).

[3] T. Suda, Proc. Neutrino '81, Hawaii (1981).

[4] K. Takahashi, Proc. Third workshop on grand unification, North Carolina (1982).

[5] M. Koshiba, Proc. 21th High energy Int. conf., Paris (1982); Proc. Topical symposium on high energy physics, Tokyo (1982).

[6] K. Arisaka, Proc. Topical symposium on high energy physics, Tokyo (1982).

[7] A. Suzuki, Proc. Topical symposium on high energy physics, Tokyo (1982).

[8] H. Kume et al., Nucl. Instr. of Meth. 205 (1983), 443.

[9] M. Koshiba, Proc. ICOBAN '84, Utah (1984).

[10] Y. Totsuka, Proc. XIXth Moriond Conf. (1984)

[11] K. Arisaka et al., J. Phys. Soc. Jpn. 54 (1985) 3213.

[12] T. Kajita et al., J. Phys. Soc. Jpn. 54 (1985) 4065.

[13] T. Kajita et al., J. Phys. Soc. Jpn. 55 (1986) 711.

[14] V.A. Rubakov, Pis'ma Zh. Eksp. & Teor. Fiz. 33 (1981) 658 (JETP Lett. 33 (1981) 644); Nucl. Phys. B203(1982) 311.

[15] J. Arafune and M. Fukugita, Phys. Lett. B133 (1983) 380.

[16] T. Kajita et al., J. Phys. Soc. Jpn. 54 (1985) 4065.

[17] K. Hirata et al., Phys. Rev. Lett. 58 (1987) 1490.

[18] M. Koshiba, Phys. Rep. 220 (1992) 229.

[19] Y. Oyama et al., Phys. Rev. Lett. 59 (1987) 2604.

[20] R. Davis, Jr., et al., Phys. Rev. Lett. 20 (1968) 1205.

[21] J.K. Rowley, B.T. Cleveland and R. Davis, Jr., AIP Conf. Proc. No. 126 (1984) p1.

[22] E.W. Beier, in Proc. the 7th Workshop on Grand Unification/ICOBAN '86, Toyama (1986).

[23] A. Suzuki, in Proc. the 12th Int. Conf. on Neutrino Physics and Astrophsics, Neutrino '86, Sendai (1986).

[24] K.S. Hirata et al., Phys. Rev. Lett. 63 (1989) 16.

[25] K. S. Hirata et al., Phys. Rev. Lett. 65 (1990) 1297.

[26] J.N. Bahcall and P.K. Ulrich, Rev. Mod. Phys. 60(1988)297.

[27] R. Davis, Jr., Proc. 13th Int. Conf. on Neutrino Physics and Astrophysics, Neutrino '88, Boston (1988).

[28] K. Inoue, Proc. XXVIIIth Recontre de Moriond, Electroweak Interactions and Unified Theories (1993).

[29] S.P. Mikheyev and A.Yu Smirnov, Yad. Fiz. 42(1985)1441 (Sov. J. Nucl. Phys. 42(1985)913).

[30] L. Wolfenstein, Phys. Rev. D17 (1978) 2369; 20 (1979) 2634.

[31] K. S. Hirata et al., Phys. Rev. Lett. 65 (1990) 1301.

[32] K.S. Hirata et al., Phys. Rev. Lett. 66 (1991) 9.

[33] M. Koshiba, Proc. of Workshop on Grand Unified Theories and Cosmology, KEK Report 84-12 (1983).

[34] M. Koshiba and Y. Totsuka, Proc. of Workshop: Toward Unification and its Verification, KEK Report (1984).

[35] Y. Totsuka, Proc. of 7th Workshop on Grand Unification/ICOBAN '86, Toyama (1986); UT-ICEPP-86-06; in Proc. of Int. Symposium on Undergraound Physics Experiments, Tokyo (1990).

[36] A. Suzuki et al., Nucl. Instr. and Meth. (1993).

[37] M. Takita, in Proc. Frontiers of Neutrino Astrophysics, Takayama (1992).

List of Publications

11-NOV-1993 **Kamiokande collaboration**

1. **20-Inch Diameter Photomultiplier**

   Nucl. Inst. Meth. **205** (1983) 443–449

   H. Kume, S. Sawaki, M. Ito, K. Arisaka, T. Kajita, A. Nishimura and A. Suzuki

2. **Search for Nucleon Decay into Charged Lepton + Mesons**

   J. Phys. Soc. Jpn. **54** (1985) 3213–3216

   K. Arisaka, T. Kajita, M. Koshiba, M. Nakahata, Y. Oyama, A. Suzuki, M. Takita, Y. Totsuka, T. Kifune, T. Suda, K. Takahashi and K. Miyano

3. **Search for Nucleon Decays Catalyzed by Magnetic Monopoles**

   J. Phys. Soc. Jpn. **54** (1985) 4065–4068

   T. Kajita, K. Arisaka, M. Koshiba, M. Nakahata, Y. Oyama, A. Suzuki, M. Takita, Y. Totsuka, T. Kifune, T. Suda, K. Takahashi and K. Miyano

4. **Search for Nucleon Decays into Anti-Neutrino + Mesons**

   J. Phys. Soc. Jpn. **55** (1986) 711–714

   T. Kajita, K. Arisaka, M. Koshiba, M. Nakahata, Y. Oyama, A. Suzuki, M. Takita, Y. Totsuka, T. Kifune, T. Suda, N. Sato, K. Takahashi and K. Miyano

5. **Search for High-Energy Muons from Cygnus X-3**

   Phys. Rev. Lett. **56** (1986) 991–994

   Y. Oyama, K. Arisaka, T. Kajita, M. Koshiba, M. Nakahata, A. Suzuki, M. Takita, Y. Totsuka, T. Kifune, T. Suda, N. Sato, K. Takahashi and K. Miyano

6. **Atmospheric Neutrino Background and Pion Nuclear Effect for KAMIOKA Nucleon Decay Experiment**

   J. Phys. Soc. Jpn. **55** (1986) 3786–3805

   M. Nakahata, K. Arisaka, T. Kajita, M. Koshiba, Y. Oyama, A. Suzuki, M. Takita, Y. Totsuka, T. Kifune, T. Suda, N. Sato, K. Takahashi and K. Miyano

7. **Search for neutron-antineutron oscillation in ^{16}O nuclei**

Phys. Rev. D **34** (1986) 902–904

M. Takita, K. Arisaka, T. Kajita, T. Kifune, M. Koshiba, K. Miyano, M. Nakahata, Y. Oyama, N. Sato, T. Suda, A. Suzuki, K. Takahashi and Y. Totsuka

8. Observation of a Neutrino Burst from the Supernova SN1987A

Phys. Rev. Lett. **58** (1987) 1490–1493

K. Hirata, T. Kajita, M. Koshiba, M. Nakahata, Y. Oyama, N. Sato, A. Suzuki, M. Takita, Y. Totsuka, T. Kifune, T. Suda, K. Takahashi, T. Tanimori, K. Miyamo, M. Yamada, E.W. Beier, L.R. Feldscher, S.B. Kim, A.K. Mann, F.M. Newcomer, R. Van Berg, W. Zhang and B.G. Cortez

9. Search for High-Energy Neutrinos from SN1987A: First Six Months

Phys. Rev. Lett. **59** (1987) 2604–2606

Y. Oyama, K. Hirata, T. Kajita, M. Koshiba, M. Nakahata, N. Sato, A. Suzuki, M. Takita, Y. Totsuka, T. Kifune, T. Suda, K. Nakamura, K. Takahashi, T. Tanimori, K. Miyamo, M. Yamada, E.W. Beier, L.R. Feldscher, S.B. Kim, A.K. Mann, F.M. Newcomer, R. Van Berg, W. Zhang and B.G. Cortez

10. Search for high-energy muons from Cygnus X-3 during the radio outbursts in 1983 and 1985

Phys. Rev. D **36** (1987) 3537–3540

Y. Oyama, K. Hirata, T. Kajita, M. Koshiba, M. Nakahata, N. Sato, A. Suzuki, M. Takita, Y. Totsuka, T. Kifune, T. Suda, K. Takahashi and K. Miyano

11. Experimental Study of the Atmospheric Neutrino Flux

Phys. Lett. B **205** (1988) 416–420

K.S. Hirata, T. Kajita, M. Koshiba, M. Nakahata, S. Ohara, Y. Oyama, N. Sato, A. Suzuki, M. Takita, Y. Totsuka, T. Kifune, T. Suda, K. Nakamura, K. Takahashi, T. Tanimori, K. Miyamo, M. Yamada, E.W. Beier, L.R. Feldscher, E.D. Frank, W. Frati, S.B. Kim, A.K. Mann, F.M. Newcomer, R. Van Berg, W. Zhang and B.G. Cortez

12. Observation in the Kamiokande-II detector of the neutrino burst from supernova SN1987A

Phys.Rev. D **38** (1988) 448–458

K.S. Hirata, T. Kajita, M. Koshiba, M. Nakahata, Y. Oyama, N. Sato, A. Suzuki, M. Takita, Y. Totsuka, T. Kifune, T. Suda, K. Takahashi, T. Tanimori, K. Miyamo, M. Yamada, E.W. Beier, L.R. Feldscher, W. Frati, S.B. Kim, A.K. Mann, F.M. Newcomer, R. Van Berg, W. Zhang and B.G. Cortez

13. Experimental Limit on the Flux of Relic Antineutrinos from Past Supernovae

Phys. Rev. Lett. **61** (1988) 385–388

W. Zhang, E.W. Beier, L.R. Feldscher, E.D. Frank, W.Frati, S.B. Kim, A.K. Mann, F.M. Newcomer, R. Van Berg, K.S. Hirata, T. Kajita, M. Koshiba, M. Nakahata, S. Ohara, Y. Oyama, N. Sato, A. Suzuki, M. Takita, Y. Totsuka, T. Kifune, T. Suda, K. Nakamura, K. Takahashi, T. Tanimori, K. Miyamo, M. Yamada and B.G. Cortez

14. Search for Correlations of Neutrino Events with Solar Flares in Kamiokande

Phys. Rev. Lett. **61** (1988) 2653–2656

K.S. Hirata, T. Kajita, T. Kifune, K. Kihara, M. Nakahata, K. Nakamura, S. Ohara, Y. Oyama, N. Sato, M. Takita, Y. Totsuka, Y. Yaginuma, M. Mori, A. Suzuki, K. Takahashi, T. Tanimori, M. Yamada, M. Koshiba, T. Suda, K. Miyano, H. Miyata, H. Takei, K. Kaneyuki, Y. Nagashima, Y. Suzuki, E. W.Beier, L.R. Feldscher, E.D. Frank, W. Frati, S.B. Kim, A.K. Mann, F.M. Newcomer, R. Van Berg, W. Zhang and M. Fukugita

15. Design and Performance of Semi-custom Analog IC Including Two TACs and Two Current Integrators for "Superkamiokande"

IEEE Trans. Nucl. Sci. **36** (1989) 497–501

T. Tanimori, H. Ikeda, M. Mori, K. Kihara, H. Kitagawa and Y. Haren

16. Experimental study of upward-going muons in Kamiokande

Phys. Rev. D **39** (1989) 1481–1491

Y. Oyama, K.S. Hirata, T. Kajita, T. Kifune, K. Kihara, M. Nakahata, K. Nakamura, S. Ohara, N. Sato, M. Takita, Y. Totsuka, Y. Yaginuma, M. Mori, A. Suzuki, K. Takahashi, T. Tanimori, M. Yamada, M. Koshiba, T. Suda, K. Miyano, H. Miyata, H. Takei, K. Kaneyuki, Y. Nagashima, Y. Suzuki, E.W. Beier, L.R. Feldscher, E.D. Frank, W. Frati, S.B. Kim, A.K. Mann, F.M. Newcomer, R. Van Berg and W. Zhang

17. Experimental Limits on Nucleon Lifetime for Lepton + Meson Decay Modes

Phys. Lett. B **220** (1989) 308–316

K.S. Hirata, T. Kajita, T. Kifune, K. Kihara, M. Nakahata, K. Nakamura, S. Ohara, Y. Oyama, N. Sato, M. Takita, Y. Totsuka, Y. Yaginuma, M. Mori, A. Suzuki, K. Takahashi, T. Tanimori, M. Yamada, M. Koshiba, T. Suda, K. Miyano, H. Miyata, H. Takei, K. Kaneyuki, Y. Nagashima, Y. Suzuki, E.W. Beier, L.R. Feldscher, E.D. Frank, W. Frati, S.B. Kim, A.K. Mann, F.M. Newcomer, R. Van Berg and W. Zhang

18. Observation of ^{8}B Solar Neutrinos in the Kamiokande-II Detector

Phys. Rev. Lett. **63** (1989) 16–19

K.S. Hirata, T. Kajita, T. Kifune, K. Kihara, M. Nakahata, K. Nakamura, S. Ohara, Y. Oyama, N. Sato, M. Takita, Y. Totsuka, Y. Yaginuma, M. Mori, A. Suzuki, K. Takahashi, T. Tanimori, M. Yamada, M. Koshiba, T. Suda, K. Miyano, H. Miyata, H. Takei, K. Kaneyuki, Y. Nagashima, Y. Suzuki, E.W. Beier, L.R. Feldscher, E.D. Frank, W. Frati, S.B. Kim, A.K. Mann, F.M. Newcomer, R.Van Berg and W. Zhang

19. Search for Neutrino Events in the Kamiokande II Detector in Correlation with the Solar Flare Activity in 1989 March

Astrophys. J. **359** (1990) 574–577

K.S. Hirata, T. Kajita, T. Kifune, K. Kihara, M. Nakahata, K. Nakamura, S. Ohara, N. Sato, Y. Suzuki, Y. Totsuka, Y. Yaginuma, M. Mori, Y. Oyama, A. Suzuki, K. Takahashi, H. Takei, T. Tanimori, M. Koshiba, T. Suda, K. Miyano, H. Miyata, M. Yamada, Y. Fukuda, K. Kaneyuki, Y. Nagashima, M. Takita, E.W. Beier, L.R. Feldscher, E.D. Frank, W. Frati, S.B. Kim, A.K. Mann, F.M. Newcomer, R.Van Berg and W. Zhang

20. Results from One Thousand Days of Rea-Time, Directional Solar-Neutrino Data

Phys. Rev. Lett. **65** (1990) 1297–1300

K.S. Hirata, K. Inoue, T. Kajita, T. Kifune, K. Kihara, M. Nakahata, K. Nakamura, S. Ohara, N. Sato, Y. Suzuki, Y. Totsuka, Y. Yaginuma, M. Mori, Y. Oyama, A. Suzuki, K. Takahashi, M. Yamada, M. Koshiba, T. Suda, T. Tajima, K. Miyano, H. Miyata, H. Takei, Y. Fukuda, E. Kodera, Y. Nagashima, M. Takita, K. Kaneyuki, T. Tanimori, E.W. Beier, L.R. Feldscher, E.D. Frank, W. Frati, S.B. Kim, A.K. Mann, F.M. Newcomer, R.Van Berg and W. Zhang

21. Constraints on Neutrino-Oscillation Parameters from the Kamiokande-II Solar-Neutrino Data

Phys. Rev. Lett. **65** (1990) 1301–1304

K.S. Hirata, K. Inoue, T. Kajita, T. Kifune, K. Kihara, M. Nakahata, K. Nakamura, S. Ohara, N. Sato, Y. Suzuki, Y. Totsuka, Y. Yaginuma, M. Mori, Y. Oyama, A. Suzuki, K. Takahashi, M. Yamada, M. Koshiba, T. Suda, T. Tajima, K. Miyano, H. Miyata, H. Takei, Y. Fukuda, E. Kodera, Y. Nagashima, M. Takita, K. Kaneyuki, T. Tanimori, E.W. Beier, L.R. Feldscher, E.D. Frank, W. Frati, S.B. Kim, A.K. Mann, F.M. Newcomer, R.Van Berg and W. Zhang

22. Search for Day-Night and Semiannual Variations in the Solar Neutrino Flux Observed in the Kamiokande-II Detector

Phys. Rev. Lett. **66** (1991) 9–12

K.S. Hirata, K. Inoue, T. Kajita, K. Kihara, M. Nakahata, K. Nakamura, S. Ohara, N. Sato, Y. Suzuki, Y. Totsuka, Y. Yaginuma, M. Mori, Y. Oyama, A. Suzuki, K. Takahashi, M. Yamada, M. Koshiba, K. Nishijima, T. Suda, T. Tajima, K. Miyano, H. Miyata, H. Takei, Y. Fukuda, E. Kodera, Y. Nagashima, M. Takita, K. Kaneyuki, T. Tanimori, E.W. Beier, L.R. Feldscher, E.D. Frank, W. Frati, S.B. Kim, A.K. Mann, F.M. Newcomer, R.Van Berg and W. Zhang

23. Search for Fractionally Charged Particles in Kamiokande-II

Phys. Rev. D **43** (1991) 2843–2846

M. Mori, Y. Oyama, A. Suzuki, K. Takahashi, M. Yamada, K. Miyano, H. Miyata, H. Takei, K.S. Hirata, T. Kajita, K. Kihara, M. Nakahata, K. Nakamura, S. Ohara, N. Sato, Y. Suzuki, Y. Totsuka, Y. Yaginuma, M. Koshiba, T. Suda, T. Tajima, Y. Fukuda, Y. Nagashima, M. Takita, K. Kaneyuki, T. Tanimori, E.W. Beier, E.D. Frank, W. Frati, S.B. Kim, A.K. Mann, F.M. Newcomer, R.Van Berg and W. Zhang

24. Measurements of the Charge Ratio and Polarization of 1.2-TeV/c Cosmic-ray Muons with the Kamiokande II Detector

Phys. Rev. D **44** (1991) 617–621

M. Yamada, K. Miyano, H. Miyata, H. Takei, M. Mori, Y. Oyama, A. Suzuki, K. Takahashi, K.S. Hirata, T. Kajita, T. Kifune, K. Kihara, M. Nakahata, K. Nakamura, S. Ohara, N. Sato, Y. Suzuki, Y. Totsuka, Y. Yaginuma, M. Koshiba, T. Suda, T. Tajima, Y. Fukuda, Y. Nagashima, M. Takita, K. Kaneyuki, T. Tanimori, E.W. Beier, E.D. Frank, W. Frati, S.B. Kim, A.K. Mann, F.M. Newcomer, R.Van Berg and W. Zhang

25. Time Variation of the Cosmic Ray Muon Flux in Underground Detectors and Correlation with Atmospheric Temperature

J. Phys. Soc. Jpn. **60** (1991) 2808–2811

K. Munakata, T. Tsuchida, S. Sagisaka, M. Shimada, Yikihiko Suzuki, M. Furuhata, S. Yasue, S. Mori, M. Mori, Y. Oyama, A. Suzuki, K. Takahashi, M. Yamada, K. Miyano, H. Miyata, H. Takei, K.S. Hirata, T. Kajita, K. Kihara, M. Nakahata, K. Nakamura, S. Ohara, N. Sato, Y. Suzuki, Y. Totsuka, Y. Yaginuma, M. Koshiba, T. Suda, T. Tajima, Y. Fukuda, Y. Nagashima, M. Takita, K. Kaneyuki, T. Tanimori, E.W. Beier, E.D. Frank, W. Frati, S.B. Kim, A.K. Mann, F.M. Newcomer, R.Van Berg and W. Zhang

26. Mass Limits for Dark-matter Particles Derived from High-energy Neutrinos from the Sun

Phys. Rev. D **44** (1991) 2220–2240

N. Sato, K.S. Hirata, T. Kajita, T. Kifune, K. Kihara, M. Nakahata, K. Nakamura, S. Ohara, Y. Suzuki, Y. Totsuka, Y. Yaginuma, M. Mori, Y. Oyama, A. Suzuki, K. Takahashi, H. Takei, T. Tanimori, M. Koshiba, T. Suda, T. Tajima, K. Miyano, H. Miyata, M. Yamada, Y. Fukuda, K. Kaneyuki, Y. Nagashima, M. Takita, E.W. Beier, L.R. Feldscher,E.D. Frank, W. Frati, S.B. Kim, A.K. Mann, F.M. Newcomer, R.Van Berg and W. Zhang

27. Real-time, Directional Measurement of ^{8}B Solar Neutrinos in the Kamiokande-II Detector

Phys. Rev. D **44** (1991) 2241–2260

K.S. Hirata, K. Inoue, T. Ishida, T. Kajita, K. Kihara, M. Nakahata, K. Nakamura, S. Ohara, N. Sato, Y. Suzuki, Y. Totsuka, Y. Yaginuma, M. Mori, Y. Oyama, A. Suzuki, K. Takahashi, M. Yamada, M. Koshiba, K. Nishijima, T. Suda, T. Tajima, K. Miyano, H. Miyata, H. Takei, Y. Fukuda, E. Kodera, Y. Nagashima, M. Takita, K. Kaneyuki, T. Tanimori, E.W. Beier, L.R. Feldscher, E.D. Frank, W. Frati, S.B. Kim, A.K. Mann, F.M. Newcomer, R.Van Berg and W. Zhang

28. Search for Neutralino Dark Matter in Kamiokande

Phys. Lett. B **270** (1991) 89–96

M. Mori, M.M. Nojiri, Y. Oyama, A. Suzuki, K. Takahashi, M. Yamada, H. Takei, K. Miyano, H. Miyata, K.S. Hirata, K. Inoue, T. Ishida, T. Kajita, K. Kihara, M. Nakahata, K. Nakamura, N. Sato, Y. Suzuki, Y. Totsuka, Y. Yaginuma, M. Koshiba, K. Nishijima, T. Suda, T. Tajima, Y. Fukuda, E. Kodera, Y. Nagashima, M. Takita, K. Kaneyuki, T. Tanimori, E.W. Beier, E.D. Frank, W. Frati, S.B. Kim, A.K. Mann, F.M. Newcomer, R.Van Berg and W. Zhang

29. A Limit on Spontaneous R Parity Breaking from Kamiokande

Phys. Lett. B **278** (1992) 217–220

M. Mori, M.M. Nojiri, Y. Oyama, A. Suzuki, K. Takahashi, M. Yamada, H. Takei, K. Miyano, H. Miyata, K.S. Hirata, K. Inoue, T. Ishida, T. Kajita, K. Kihara, M. Nakahata, K. Nakamura, N. Sato, Y. Suzuki, Y. Totsuka, Y. Yaginuma, M. Koshiba, K. Nishijima, T. Suda, T. Tajima, Y. Fukuda, E. Kodera, Y. Nagashima, M. Takita, K. Kaneyuki, and T. Tanimori

30. Observation of a Small ν_μ/ν_e Ratio in Kamiokande

Phys. Lett. B **280** (1992) 146–152

K.S. Hirata, K. Inoue, T. Ishida, T. Kajita, K. Kihara, M. Nakahata, K. Nakamura, S. Ohara, A. Sakai, N. Sato, Y. Suzuki, Y. Totsuka, Y. Yaginuma, M. Mori, Y. Oyama, A. Suzuki, K. Takahashi, M. Yamada, M. Koshiba, K. Nishijima, T. Kajimura, T. Suda, T. Tajima, K. Miyano, H. Miyata, H. Takei, Y. Fukuda, E. Kodera, Y. Nagashima, M. Takita, H. Yokoyama, K. Kaneyuki, Y. Takeuchi, T. Tanimori, E.W. Beier, E.D. Frank, W. Frati, S.B. Kim, A.K. Mann, F.M. Newcomer, R.Van Berg and W. Zhang

31. Search for Neutrino-induced Low-energy-electron-event Clusters in Kamiokande-II

Phys. Rev. D **45** (1992) 3355–3360

K.S. Hirata, K. Inoue, T. Ishida, T. Kajita, K. Kihara, M. Nakahata, K. Nakamura, S. Ohara, N. Sato, Y. Suzuki, Y. Totsuka, Y. Yaginuma, M. Mori, Y. Oyama, A. Suzuki, K. Takahashi, M. Yamada, T. Suda, T. Tajima, K. Miyano, H. Miyata, H. Takei, Y. Fukuda, E. Kodera, Y. Nagashima, M. Takita, K. Kaneyuki, T. Tanimori, E.W. Beier, E.D. Frank, W. Frati, S.B. Kim, A.K. Mann, F.M. Newcomer, R.Van Berg and W. Zhang

32. Survey of Atmospheric Neutrino Data and Implications for Neutrino Mass and Mixing

Phys. Lett. B **283** (1992) 446–453

E.W. Beier, E.D. Frank, W. Frati, S.B. Kim, A.K. Mann, F.M. Newcomer, R.Van Berg, W. Zhang, K.S. Hirata, K. Inoue, T. Ishida, T. Kajita, K. Kihara, M. Nakahata, K. Nakamura, S. Ohara, A. Sakai, N. Sato, Y. Suzuki, Y. Totsuka, Y. Yaginuma, M. Mori, Y. Oyama, A. Suzuki, K. Takahashi, M. Yamada, M. Koshiba, K. Nishijima, T. Kajimura, T. Suda, T. Tajima, K. Miyano, H. Miyata, H. Takei, Y. Fukuda, E. Kodera, Y. Nagashima, M. Takita, H. Yokoyama, K. Kaneyuki, Y. Takeuchi and T. Tanimori

33. Front-end Hybrid Circuit for SuperKAMIOKANDE

Nucl. Inst. Meth. in Phys. Res. **A320** (1992) 310–316

H. Ikeda, M. Mori, T. Tanimori, K. Kihara, Y. Suzuki and Y. Haren

34. A Limit on Massive Neutrino Dark Matter from Kamiokande

Phys. Lett. B **289** (1992) 463–469

M. Mori, K. Hikasa, M.M. Nojiri, Y. Oyama, A. Suzuki, K. Takahashi, M. Yamada, H. Takei, K. Miyano, H. Miyata, K.S. Hirata, K. Inoue, T. Ishida, T. Kajita, K. Kihara, M. Nakahata, K. Nakamura, A. Sakai, N. Sato, Y. Suzuki, Y. Totsuka, M. Koshiba, K. Nishijima, T. Kajimura, T. Suda, Y. Fukuda, E. Kodera, Y. Nagashima, M. Takita, H. Yokoyama, K. Kaneyuki, Y. Takeuchi, T. Tanimori, E.W. Beier, E.D. Frank, W. Frati, S.B. Kim, A.K. Mann, F.M. Newcomer, R.Van Berg and W. Zhang

35. Improvement of 20 in. Diameter Photomultiplier Tubes

Nucl. Inst. Meth. in Phys. Res., **A329** (1993) 299–313

A. Suzuki, M. Mori, K. Kaneyuki, T. Tanimori, J. Takeuchi, H. Kyushima and Y. Ohashi

36. Study of Invisible Nucleon Decay, $n \to \nu\nu\bar{\nu}$, and a Forbidden Nuclear Transition in the Kamiokande Detector

Phys. Lett. B **311** (1993) 357–361

Y. Suzuki, Y. Fukuda, K. S. Hirata, K. Inoue, T. Ishida, T. Kajita, K. Kihara, M. Nakahata, K. Nakamura, A. Sakai, N. Sato, Y. Totsuka, M. Mori, Y. Oyama, A. Suzuki, K. Takahashi, M. Yamada, M. Koshiba, K. Nishijima, T. Kajimura, T. Suda, A. T. Suzuki, K. Miyano, H. Miyata, H. Takei, E. Kodera, Y. Nagashima, M. Takita, H. Yokoyama, K. Kaneyuki, Y. Takeuchi, T. Tanimori, and H. Ejiri

37. Search for Neutralino Dark Matter in Kamiokande

To appear in *Phys. Rev.* D (1993)

M. Mori, M. M. Nojiri, K. S. Hirata, K. Kihara, Y. Oyama, A. Suzuki, K. Takahashi, M. Yamada, H. Takei, M. Koga, K. Miyano, H. Miyata, Y. Fukuda, T. Hayakawa, K. Inoue, T. Ishida, T. Kajita, Y. Koshio, M. Nakahata, K. Nakamura, N. Sato, A. Sakai, N. Sato, M. Shiozawa, J. Suzuki, Y. Suzuki, Y. Totsuka, M. Koshiba, K. Nishijima, T. Kajimura, T. Suda, A. T. Suzuki T. Hara, Y. Nagashima, M. Takita, H. Yokoyama, A. Yoshimoto, K. Kaneyuki, Y. Takeuchi, and T. Tanimori

List of Presentations at International Conferences

24-NOV-1993 Kamiokande collaboration

1. **Kamioka Proton Decay Experiment**
T. Suda
Talk at the 1981 International Conference on Neutrino Physics and Astrophysics, Maui, Hawaii, July 1–8, 1981;
Published in Proceedings of the 1981 International Conference on Neutrino Physics and Astrophysics, ed. R.J. Cence et al., Univ. of Hawaii, 1981, Vol.1, pp.224–231.

2. **KEK-(UT)^2 Experiment on Proton Decay**
M. Koshiba
Talk at the 17th International Conference on Cosmic Rays, Paris, France, July 13–25, 1981;
Published in Proceedings of the 17th International Conference on Cosmic Rays, 1981, Vol.7, p.185.

3. **The Kamioka Proton Decay Experiment – Present Status –**
K. Takahashi
Talk at the Third Workshop on Grand Unification, University of North Carolina, Chapel Hill, U.S.A., April 15–17, 1982;
Published in "Third Workshop on Grand Unification", ed. P.H. Frampton, S.L. Glashow and H. van Dam, Birkhäuser (Boston), 1982, pp.56–71.

4. **Particle Physics at LICEPP – Present and Future**
M. Koshiba
Talk at Topical Symposium on High Energy Physics, Tokyo, Japan, September 7–11, 1982;
Published in Proceedings of Topical Symposium on High Energy Physics, ed. T. Eguchi and Y. Yamaguchi, World Scientific (Singapore), 1983, pp.245–297.

5. **The Kamioka Nucleon decay Experiment (Kamiokande) – Status Report –**
A. Suzuki
Talk at Topical Symposium on High Energy Physics, Tokyo, Japan, September 7–11, 1982.

6. **The Kamioka Nucleon decay Experiment (Kamiokande) – Expected Performance –**
   K. Arisaka
   Talk at Topical Symposium on High Energy Physics, Tokyo, Japan, September 7–11, 1982.

7. **?**
   M. Koshiba
   Talk at the International Colloquium on Baryon Nonconservation (ICOBAN'84), Park City, Utah, U.S.A., January 4–7, 1984;
   Published in Proceedings of the International Colloquium on Baryon Nonconservation, University of Wisconsin Madison, 1984.

8. **Results from the KAMIOKA Nucleon Decay Experiment**
   Y. Totsuka
   Talk at the XIXth Rencontre de Moriond, La Plague, Savoie, France, February 26 – March 4, 1984;
   Published in "Phenomenology of Gauge Theories", ed. J. Tran Thanh Van, Editions Frontieres, 1984, pp.625–634.

9. **Results from the KAMIOKA Nucleon Decay Experiment**
   M. Koshiba
   Talk at the Fifth Workshop on Grand Unification, Providence, U.S.A., April 12–14, 1984;
   Published in Proceedings of the Fifth Workshop on Grand Unification, ed. K. Kang et al., World Scientific (Singapore), 1984, pp.13–30.

10. **Recent Results on Kamiokande (Atmospheric Neutrinos and Neutrinos from Cyg X-3)**
    A. Suzuki
    Talk at the Aspen Summer Conference on Neutrino Physics, Aspen, Colorado, U.S.A., July, 1984.

11. **Results from KAMIOKANDE**
    M. Koshiba
    Talk at the XXII International Conference on High Energy Physics, Leipzig, July 19–25, 1984;
    Published in Proceedings of the XXII International Conference on High Energy Physics, ed. A. Meyer and E. Wieczorek, Akademie der Wissenschaften der DDR, 1984, Vol.1, pp.250–251.

12. **Experiments on Baryon Number Non-Conservation**
    M. Koshiba
    Talk at the XXII International Conference on High Energy Physics, Leipzig, July 19–25, 1984;
    Published in Proceedings of the XXII International Conference on High Energy Physics, ed. A. Meyer and E. Wieczorek, Akademie der Wissenschaften der DDR, 1984, Vol.2, pp.67–82.

13. **Proton Decay**
    M. Koshiba
    Talk at the Tenth International Conference on Particles and Nuclei, Heidelberg, Fed. Rep. of Germany, July 30 – August 3, 1984;
    Published in Nucl. Phys. **A434** (1985) 431c–449c.

14. **Proton Decay**
    M. Koshiba
    Lecture at the 22nd International School of Subnuclear Physics, Erice, Sisily, Italy, August 5–15, 1984;
    Published in "Quarks, Leptons and their Constituents", ed. A. Zichichi, Plenum Press, New York and London, 1988, pp.349–390.

15. **KAMIOKA Nucleon Decay Experiment**
    M. Koshiba
    Talk at the Sixth Workshop on Grand Unification, Minneapollis, Minnesota, U.S.A., April 18–20, 1985;
    Published in Proceedings of the Sixth Workshop on Grand Unification, ed. S. Rudaz and T.F. Walsh, World Scientific (Singapore), 1985, pp.65–88.

16. **Baryon Number Non-Conservation and Related Topics**
    Y. Totsuka
    Talk at the 1985 International Symposium on Lepton and Photon Interactions at High Energies, Kyoto, August 19–24, 1985;
    Published in Proceedings of the 1985 International Symposium on Lepton and Photon Interactions at High Energies, ed. M. Konuma and K. Takahashi, Kyoto Univ., 1985, pp.120–160.

17. **On the Oscillation of Solar ^{8}B Neutrinos**
    M. Koshiba
    Talk at the Symposium in honor of Prof. Y. Nambu, Chicago, January 1986;
    Published in Prog. Theor. Phys. Supplement **86** (1986) 357–360.

18. **Recent Results from Kamioka Nuceon Decay Experiment**
T. Suda
Talk at the 7th Workshop on Grand Unification / ICOBAN'86, Toyama, Japan, April 16–18, 1986;
Published in "Seventh Workshop on Grand Unification / ICOBAN'86", ed. J. Arafune, World Scientific (Singapore), 1987, pp.70–78.

19. **The Kamiokande II Detector**
E. Beier
Talk at the 7th Workshop on Grand Unification / ICOBAN'86, Toyama, Japan, April 16–18, 1986;
Published in "Seventh Workshop on Grand Unification / ICOBAN'86", ed. J. Arafune, World Scientific (Singapore), 1987, pp.79–91.

20. **SUPER-KAMIOKANDE**
Y. Totsuka
Talk at the 7th Workshop on Grand Unification / ICOBAN'86, Toyama, Japan, April 16–18, 1986;
Published in "Seventh Workshop on Grand Unification / ICOBAN'86", ed. J. Arafune, World Scientific (Singapore), 1987, pp.118–136.

21. **KAMIOKA Nucleon Decay Experiment**
M. Koshiba
Published in Nuovo Cimento, **9c** (1986) 141–158.

22. **The Search for Solar ^{8}B Neutrinos at KAMIOKANDE II**
A. Suzuki
Talk at the 12th International Conference on Neutrino Physics and Astrophysics, Sendai, Japan, June 3–8, 1986;
Published in Proceedings of the 12th International Conference on Neutrino Physics and Astrophysics, ed. T. Kitagaki and H. Yuta, World Scientific (Singapore), 1986, pp.306–316.

23. **The Status of the Search for Solar Neutrinos at Kamiokande-II**
E. W. Beier
Talk at the 1986 Summer Study on the Particle Physics of the Superconducting Supercollider, June 23 – July 11, 1986;
Published in the Proceedings of the 1986 Summer Study on the Particle Physics of the Superconducting Supercollider, ed. R. Donaldson and J. Marx, 1986.

24. **Search for Solar ^{8}B Neutrinos at KAMIOKANDE II**
T. Kajita
Talk at the Leptonic Session of the Twenty-Second Rencontre de Moriond, March 8–15, 1987;
Published in "The Standard Model, The Supernova 1987A", ed. J. Tranh Than Vanh, Editions Frontieres, 1987, pp.689–700.

25. **Observation of a Neutrino Burst from Supernova SN1987A**
T. Kajita
Talk at the Leptonic Session of the Twenty-Second Rencontre de Moriond, March 8–15, 1987;
Published in "The Standard Model, The Supernova 1987A", ed. J. Tranh Than Vanh, Editions Frontieres, 1987, Vol.1, pp.727–734.

26. **Observation of a Neutrino Burst from the Supernova Sn1987A**
E. W. Beier
Talk at the Telemark IV, Neutrino Masses and Neutrino Astrophysics, Ashland, Wisconsin, U.S.A., March 16–18, 1987;
Published in "Neutrino Masses and Neutrino Astrophysics", ed. V. Barger et al., World Scientific (Singapore), 1987, pp.51–62.

27. **Observation of a Neutrino Burst from the Supernova SN1987A**
M. Nakahata
Talk at the Eighth Workshop on Grand Unification, Syracuse, New York, U.S.A., April 16–18, 1987;
Published in Proceedings of the Eighth Workshop on Grand Unification, ed. K.C. Wali, World Scientific (Singapore), 1987, pp.425–434.

28. **Observation of a Neutrino Burst from Supernova SN1987A**
Y. Totsuka
Talk at the XIth International Conference on Particles and Nuclei "PANIC-87", Kyoto, April 20–24, 1987;
Published in Nucl. Phys. **A478** (1988) 189c–198c.

29. **Observational Particle Physics**
M. Koshiba
Talk at the XIth International Conference on Particles and Nuclei "PANIC-87", Kyoto, April 20–24, 1987;
Published in Nucl. Phys. **A478** (1988) 355c–364c.

30. **Toward Observational Neutrino Astrophysics**
M. Koshiba

Talk at the International Workshop on SN1987A, Minneapolice, U.S.A., June 5–6, 1987;
Published in Proceedings of the International Workshop on SN1987A, ed. K. Olive and T. Walsh, Independence Press (Minneapolice), 1988.

31. **Observation of Neutrinos from SN1987A by KAMIOKANDE-II**
M. Koshiba
Talk at the ESO Workshop on the SN1987A, Garching bei München, July 6–8, 1987;
Published in Proceedings of ESO Workshop on the SN1987A, ed. I.J. Danziger, 1987, p.219.

32. **Supernova Neutrino Signal and KAMIOKANDE-II**
M. Koshiba
Talk at the 13th International Symposium on Lepton and Photon Interactions at High Energies, Hamburg, July 27–31, 1987;
Published in Nucl. Phys. B (Proc. Suppl.) 3 (1988) 441–452.

33. **Search for ^{8}B Solar Neutrinos at KAMIOKANDE-II**
Contributed paper to the 1987 International Symposium on Lepton and Photon Interactions at High Energies, Hamburg, July 27–31, 1987.

34. **Observation of a Neutrino Burst from the Supernova 1987A**
Published in Proceedings of IAU Colloquium, 1987.

35. **Kamiokande Result on a Neutrino Burst from the Supernova SN1987a**
Published in Proceedings of the 20th International Cosmic Ray Conference, Moscow, August 2–15, 1987, Vol.9, p.53.

36. **Toward Observational Neutrino Astrophysics**
M. Koshiba
Lecture at ESO Summer School, Alpbach, Australia, August 1987.

37. **Toward Observational Neutrino Astrophysics**
M. Koshiba
Lecture at the 15th SLAC Summer Institute on Particle Physics, Stanford, U.S.A., August 10–21, 1987.
Published in "Looking Beyond the Z", ed. E.C. Brennan, SLAC Report No.328, 1987.

38. **Toward Observational Neutrino Astrophysics**
M. Koshiba
Talk at the 4th George Mason Astrophysics Workshop, Fairfax, Virginia, U.S.A., 1987;
Published in Proceedings of the 4th George Mason Astrophysics Workshop, 1987, p.130.

39. **Underground Experiments in Japan**
M. Koshiba
Talk at the Symposium in Honor of Marchello Conversi's 70th Birthday, Rome, November 3–4, 1987; Published in Proceedings of the Symposium in Honor of Marchello Conversi's 70th Birthday, ed. G. Baroni, L. Maiani and G. Salvini, 1987, p.87.

40. **Kamiokande Results on Neutrino Burst and UHE Neutrinos from Supernova 1987A**
A. Suzuki
Talk at Aspen Winter Conference on Elementary Particle Physics 'CP-Violation', Aspen, Colorado, U.S.A., January 10–17, 1988.

41. **Recent ν-Results from KAMIOKANDE-II and Future**
M. Koshiba
Talk at the XXIIIrd Rencontre de Moriond, Les Arcs, Savoie, France, January 23–30, 1988;
Published in "5th Force – Neutrino Physics", ed. O. Fackler and J. Tranh Thanh Van, Editions Frontieres, 1988, pp.215–222.

42. **Recent ν-Results from KAMIOKANDE-II and Future**
M. Koshiba
Talk at Les Rencontre de Physique de la Vallee d'Aoste, La Thule, Aoste Valley, France, February 26 – March 5, 1988;
Published in Proceedings of Les Rencontre de Physique de la Vallee d'Aoste, 1988, p.51.

43. **Experimental Study of the Atmospheric Neutrino Flux**
M. Takita
Talk at the Leptonic Session of the XXIIIrd Rencontre de Moriond, Les Arcs, Savoie, France, March 6–13, 1988;
Published in "'88 Electroweak Interactions and Unified Theories", ed. J. Tran Thanh Van, Editions Frontieres, 1988, pp.447–452.

44. **Search for High Energy Neutrinos from Dark Matter Annihilations in the Sun**
M. Takita
Talk at the Leptonic Session of the XXIIIrd Rencontre de Moriond, Les Arcs, Savoie, France, March 6–13, 1988;
Published in "'88 Electroweak Interactions and Unified Theories", ed. J. Tran Thanh Van, Editions Frontieres, 1988, pp.459–466.

45. **Search for Solar Neutrinos at Kamiokande-II**
T. Suda
Talk at the XVI INS International Symposium in Neutrino Mass and Related Topics, Tokyo, March 16–18, 1988;
Published in "Neutrino Mass and Related Topics", ed. S. Kato and T. Ohshima, World Scientific, Singapore, 1988, pp.226–238.

46. **Present Status on Solar Neutrinos ar Kamiokande-II**
Y. Totsuka
Talk at the YAMADA Conference XX on Big Bang, Active Galactic Nuclei and Supernovae, Tokyo, Japan, March 28 – April 1, 1988;
Published in "Big Bang, Active Galactic Nuclei and Supernovae", ed. S. Hayakawa and K. Sato, Universal Academy Press (Tokyo), 1989, pp.385–397.

47. **Neutrino Burst from Supernova and Neutrino Astronomy**
M. Koshiba
Talk at the YAMADA Conference XX on Big Bang, Active Galactic Nuclei and Supernovae, Tokyo, Japan, March 28 – April 1, 1988;
Published in "Big Bang, Active Galactic Nuclei and Supernovae", ed. S. Hayakawa and K. Sato, Universal Academy Press (Tokyo), 1989, pp.407–425.

48. **Search for High-Energy Neutrinos from SN1987A**
K. Nakamura
Talk at the YAMADA Conference XX on Big Bang, Active Galactic Nuclei and Supernovae, Tokyo, Japan, March 28 – April 1, 1988;
Published in "Big Bang, Active Galactic Nuclei and Supernovae", ed. S. Hayakawa and K. Sato, Universal Academy Press (Tokyo), 1989, pp.557–562.

49. **Neutrino Astrophysics: Its Birth and Future (as seen by KAMIOKANDE)**
M. Koshiba
Talk at the Third ESO-CERN Symposium, Bologna, Italy, May 16–20, 1988;
Published in "Astronomy, Cosmology and Fundamental Physics", ed. M. Caffo et al., Kluwer Academic Publishers, Dordrecht, 1989, p.317–326.

50. **Latest Results on Nucleon Decay Search from Kamiokande**
Y. Totsuka
Talk at the Ninth Workshop on Grand Unification, Aix-Les-Bains, France, April 28–30, 1988;
Published in Proceedings of the Ninth Workshop on Grand Unification, ed. R. Barloutaud, World Scientific (Singapore), 1988, pp.29–41.

51. **Anomaly in the Atmospheric Neutrino Flux observed at Kamiokande**
Y. Totsuka
Talk at the Ninth Workshop on Grand Unification, Aix-Les-Bains, France, April 28–30, 1988;
Published in Proceedings of the Ninth Workshop on Grand Unification, ed. R. Barloutaud, World Scientific (Singapore), 1988, pp.145–158.

52. **Present Status on Solar Neutrino Observation ar Kamiokande-II**
Y. Totsuka
Talk at the Ninth Workshop on Grand Unification, Aix-Les-Bains, France, April 28–30, 1988;
Published in Proceedings of the Ninth Workshop on Grand Unification, ed. R. Barloutaud, World Scientific (Singapore), 1988, pp.242–255.

53. **Review of Atmospheric Neutrino Phenomena in Underground Nucleon Decay Detectors**
A. Suzuki
Talk at the 13th International Conference on Neutrino Physics and Astrophysics, Boston, U.S.A., June 5–11, 1988;
Published in "Neutrino '88", ed. J. Schneps et al., World Scientific, Singapore, 1989, pp.631–643.

54. **Non-accelerator Particle Physics**
Y. Totsuka
Lecture at Summer School in High Energy Physics and Cosmology, Trieste, Italy, June 27 – August 5, 1988;
Published in "Superstrings, Unified Theories and Cosmology", World Scientific, ed. G. Ellis et al., 1988, pp.612–631.

55. **Non-Accelerator Particle Physics**
Y. Totsuka
Talk at the XXIV International Conference on High Energy Physics, Munich, Fed. Rep. of Germany, August 4–10, 1988;
Published in Proceedings of the XXIV International Conference on High Energy Physics, ed. R. Kotthaus and J.H. Kühn, Springer Verlag (Berlin),

1988, pp.282–308.

56. **Solar Neutrinos**
Y. Totsuka
Talk at the XXIV International Conference on High Energy Physics, Munich, Fed. Rep. of Germany, August 4–10, 1988;
Published in Proceedings of the XXIV International Conference on High Energy Physics, ed. R. Kotthaus and J.H. Kühn, Springer Verlag (Berlin), 1988, pp.1313–1320.

57. **Results in Non-Accelerator Particle Physics**
K. Nakamura
Talk at the 8th International Conference on Physics in Collision, Capari, 1988;
Published in "Physics in Collision 8", ed. P. Strolin, Editions Frontieres, Gif-sur-Yvette, 1989, pp.63–93.

58. **High Enrgy Neutrinos form the Sun: Kamiokande Results**
Y. Totsuka
Talk at the Particle Astrophysics Workshop: Forefront Experimental Issues, Berkley, U.S.A., December 8–10, 1988;
Published in the Proceedings of the Particle Astrophysics Workshop: Forefront Experimental Issues, ed. E.B. Norman, World Scientific (Singapore) 1989.

59. **Search for WIMPs**
Y. Totsuka
Talk at the Third Nishinomiya-Yukawa Memorial Symposium, Nishinomiya, Japan, November 10–11, 1988;
Published in "Dark Matter in the Universe", ed. H. Sato and H. Kodama, Springer-Verlag, Berlin, 1990, pp.79–93.

60. **Imaging Water Cerenkov Detector: Why and How?**
M. Koshiba
Talk at the Workshop on "Neutrino Telescopes", Venice, Italy, November 17–19, 1988;
Published in Proceeding of the International Workshop on Neutrino Telescopes, ed. M. Baldo Ceolin, Univ. Padova, 1989, pp.151–169.

61. **Nucleon Decay and Solar Neutrinos – Status and Prospects**
E. W. Beier

Talk at "Beyond the Standard Model", Iowa State University, November 18–20, 1988;
Published in "Beyond the Standard Model", ed. K. Whisnant and Bing-Lin Young, World Scientific (Singapore), 1989.

62. **Status of the Super-Kamiokande**
Published in Proceedings of the Workshop on Particle Astrophysics, 1988, p.316.

63. **Observation of Solar Neutrinos and Search for Solar-Flare Neutrino Events in Kamiokande**
M. Mori
Talk at The IXth Moriond Workshop of the XXIVth Rencontre de Moriond, Les Arcs, Savoie, France, January 21–28, 1989;
Published in "Tests of Fundamental Laws in Physics", ed. O.Fackler and J. Tran Thanh Van, Editions Frontieres, 1989, pp.391–400.

64. **Results from Kamiokande-II Experiment**
Y. Oyama
Talk at the Twelfth International Workshop on Weak Interactions and Neutrinos, Ginosar, Sea of Galilee, Israel, April 9–14, 1989;
Published in "Weak Interactions and Neutrinos", ed. P. Singer and G. Eilam, North-Holland, 1990 (Nucl. Phys. B (Proc. Suppl.) **13** (1990)) pp.362–367.

65. **Kamiokande, 1983-1989-1990's**
Y. Totsuka
Talk at the Tenth and Final Workshop on Ground Unification, North Carolina, April 20–22, 1989;
Published in Proceedings of the Last Workshop on Ground Unification, ed. P.H. Frampton, World Scientific (Singapore), 1989, pp.3–17.

66. **The Kamiokande Solar Neutrino Experiment**
M. Nakahata
Talk at the 121st Colloquium of the International Astronomical Union, Verseilles, France, May 22–26, 1989;
Published in "Inside the Sun", ed. G. Berthomieu and M. Cribier, Kluwer Academic Publishers (Dordrecht), 1990, pp.179–186.

67. **Recent Results and Status on the Kamiokande-Project**
K. Takahashi
Talk at the XII Warsaw Symposium on Elementary Particle Physics, Kazimierz, Poland, May 29 – June 2, 1989;
Published in "Frontiers in Particle Physics", ed. Z. Adduk et al., World

Scientific (Singapore), 1990, pp.82–109.

68. **Solar ^{8}B Neutrino Observation at Kamiokande-II**
A. Suzuki
Talk at the YAMADA Conference XXIII on Nuclear Weak Process and Nuclear Structure, Osaka, June 12–15, 1989;
Published in "Nuclear Weak Process and Nuclear Structure", ed. M. Morita et al., World Scientific (Singapore), 1989, pp.254–263.

69. **Recent Solar Neutrino Data from the Kamiokande-II Detector**
Contributed paper to the XIV International Symposium on Lepton and Photon Interactions, Stanford, U.S.A., August 7–12, 1989.

70. **Solar Neutrino Experiments**
M. Nakahata
Published in Proceedings of Progress of Seismology of the Sun and Stars, 1989.

71. **Underground Low Background Detector (Kamiokande)**
A. Suzuki
Talk at the International Workshop on Ultra Radioactive Background Experiments, Berkely, California, U.S.A., December 1989.

72. **Measurements of Solar Neutrinos in kamiokande-II and Status of Super-Kamiokande**
A. Suzuki
Talk at the International Workshop on Signature of Particle Dark Matter, Berkely, California, U.S.A., December 1989.

73. **Analysis of Atmospheric Neutrinos and Search for dark Matter Using the Kamiokande Detector**
M. Nakahata
Talk at the Xth Moriond Workshop of the XXVIth Rencontre de Moriond, Les Arcs, Savoie, France, January 20–27, 1990;
Published in "New and Exotic Phenomena '90", ed. O. Fackler and J. Tran Thanh Van, Editions Frontieres, 1990, pp.449–454.

74. **Super-Kamiokande**
M. Nakahata
Talk at the Xth Moriond Workshop of the XXVIth Rencontre de Moriond, Les Arcs, Savoie, France, January 20–27, 1990;

Published in "New and Exotic Phenomena '90", ed. O. Fackler and J. Tran Thanh Van, Editions Frontieres, 1990, pp.605–608.

75. **^{8}B Solar Neutrino Observation and Possible Tests on Solar Neutrino Problem in Kamiokande-II**
A. Suzuki
Talk at the Second International Workshop on Neutrino Telescopes, Venezia, Italy, February 13–15, 1990;
Published in "Neutrino Telescopes", ed. M. Baldo Ceolin, Univ. Padova, 1990, pp.15–32.

76. **Present Status and Future Prospects of Underground Physics Experiments**
K. Nakamura
Talk at the International Symposium on Underground Physics Experiments, Tokyo, April 25, 1990; Published in Proceedings of International Symposium on Underground Physics Experiments, ed. K. Nakamura, Institute for Cosmic Ray Research, University of Tokyo, 1990, pp.3–55.

77. **Super-Kamiokande**
Y. Totsuka
Talk at the International Symposium on Underground Physics Experiments, Tokyo, April 25, 1990;
Published in Proceedings of International Symposium on Underground Physics Experiments, ed. K. Nakamura, Institute for Cosmic Ray Research, University of Tokyo, 1990.

78. **Recent Results on Solar Neutrinos from Kamiokande**
Y. Totsuka
Talk at the 14th International Conference on Neutrino Physics and Astrophysics, CERN, Geneva, Switzerland, June 10–15, 1990;
Nucl. Phys. (Proc. Suppl.) 19, ed. J. Panman and K. Winter, North-Holland, 1991, pp.69–76.

79. **Non-Accelerator Particle Physics**
K. Nakamura
Talk at the 25th International Conference on High Energy Physics, Singapore, August 2–8, 1990;
Published in "Proceedings of the 25th International Conference on High Energy Physics", ed. K.K. Phua and Y. Yamaguchi, South East Asia Thoeretical Physics Association , Japan Physical Society of Japan and co-sponsoring asscociations, 1991, Vol.1, pp.281–308.

80. **Results from Kamiokande on Solar and Atmospheric Neutrino**
T. Kajita
Talk at the 25th International Conference on High Energy Physics, Singapore, August 2–8, 1990;
Published in "Proceedings of the 25th International Conference on High Energy Physics", ed. K.K. Phua and Y. Yamaguchi, South East Asia Thoeretical Physics Association , Japan Physical Society of Japan and co-sponsoring asscociations, 1991, pp.685–688.

81. **Kamiokande Atmospheric Neutrino Experiment**
M. Takita
Talk at the 25th International Conference on High Energy Physics, Singapore, August 2–8, 1990.

82. **A Search for Dark Matters in the Kamiokande-II**
Y. Suzuki
Talk at the the IUPAP Conf. on Primordial Nucleosynthesis and Evolution of Early Universe, Tokyo, September 4–8, 1990.

83. **Kamiokande II Results on Solar Neutrinos and Super-Kamiokande Experiment**
Y. Suzuki
Talk at the XIth Moriond Workshop of the XXVIth Rencontre de Moriond, Les Arcs, Savoie, France, January 26 – February 2, 1991;
Published in "Massive Neutrino, Tests of Fundamental Symmetries", ed. O. Fackler and J. Tran Thanh Van, Editions Frontieres, 1991, pp.427–435.

84. **Kamiokande Solar Neutrino Experiment**
T. Tanimori
Talk at the Leptonic Session of the XXVIth Rencontre de Moriond, Les Arcs, Savoie, France, March 11–17, 1991;
Published in "'91 Electroweak Interaction and Unified Theories", ed. J. Tran Thanh Van, Editions Frontieres, 1991, pp.183–192

85. **Solar Neutrinos – Kamiokande**
K. Nakamura
Talk at the Second Annual SSCL Spring Conference "Electroweak and Flavor Physics: Implications for the SSC", SSCL, Dallas, May 20–23, 1991.

86. **Kamiokande Solar Neutrino Experiment**
Y. Totsuka
Talk at the International School of Particle of Cosmology, Backsan, USSR,

May 6–12, 1991;
Published in Proceedings of the International School of Particle of Cosmology, World Scientific (Singapore) 1992.

87. **Solar Neutrinos in Kamiokande and Super-Kamiokande**
A. Suzuki
Talk at the XIIIth International Workshop on Weak Interactions and Neutrino Physics, Gran Sasso, L'Aquilla, Italy, July 8–12,1991.

88. **Atmospheric Neutrino observation in Kamiokande**
A. Suzuki
Talk at XIIIth International Workshop on Weak Interactions and Neutrino Physics, Gran Sasso, L'Aquilla, Italy, July,1991.

89. **Solar and Atmospheric Neutrino Observed at Kamiokande**
Y. Totsuka
Talk at the Second International Workshop on Theoretical and Phenomenological Aspects of Underground Physics, Toledo, Spain, September 9–13, 1991;
Published in Nucl. Phys. B (Proc. Suppl.) **28A** (1992) 67–84.

90. **Observation of a Small Atmospheric ν_μ/ν_e Ratio in Kamiokande and its Implications on Neutrino Oscillations**
T. Kajita
Talk at the Workshop on "The Many Aspects of Neutrino Physics", Fermilab, Batavia, Illinois, U.S.A., November 14–17, 1991;
Published in Proceedings of the Workshop on Long-Baseline Neutrino Oscillation, Fermilab, ed. M. Goodman (1992) pp.145–158.

91. **Solar Neutrino Experiments: Kamiokande II and III Results**
T. Kajita
Talk at the Workshop on "The Many Aspects of Neutrino Physics", Fermilab, Batavia, Illinois, U.S.A., November 14–17, 1991;

92. **Recent results from Kamiokande solar neutrino observations**
K. Nakamura
Talk at the 15th International Conference on Neutrino Physics and Astrophysics, Granada, Spain, June 7–11, 1992;
Published in "Neutrino '92", Nucl. Phys. B (Proc. Suppl.) **31** (1993) 105–110.

93. **Atmospheric Neutrinos – Updated Results –**
Y. Totsuka

Talk at the 15th International Conference on Neutrino Physics and Astrophysics, Granada, Spain, June 7–11, 1992;
Published in "Neutrino '92", Nucl. Phys. B (Proc. Suppl.) **31** (1993) 428–436.

94. **Solar Neutrinos Observed ar Kamiokande**
Y. Totsuka
Talk at the IVth Rencontre de Blois Particle Astrophysics, de Blois, France, June 15–20, 1992.

95. **Recent Results from Kamiokande on Solar and Atmospheric Neutrinos**
T. Kajita
Talk at the 26th International Conference on High Energy Physics, Dallas, U.S.A., August 6–12, 1992.

96. **Kamiokande Solar Neutrino Experiment and Solar Neutrino Problem**
A. Suzuki
Talk at the International Workshop on Electroweak Physics beyond the Standard Model, Valencia, Spain, October 2–5, 1991;
Published in "Electroweak Physics beyond the Standard Model", ed. J.W.F. Valle and J. Valesco, World Scientific (Singapore, 1992), pp169–185.

97. **Recent Results from Kamiokande**
Y. Oyama
Talk at the XIIth Moriond Workshop "Progress in Atomic Physics, Neutrinos and Graviataion", Les Arcs, Savoie, France, January 25 – February 1, 1992;
Published in "Progress in Atomic Physics, Neutrinos and Gravitation", Proceedings of the XXIIth Rencontre de Moriond, ed. G. Chardin, O. Fackler and J. Tran Thanh Van, Editions Frontieres (1992) pp.59–66.

98. **Upward-going muons in Kamiokande**
M. Mori
Talk at the Workshop on High Energy Neutrino Astrophysics, Honolulu, Hawaii, 23–26 March, 1992;
Published in Proceedings of the Workshop on High Energy Neutrino Astrophysics, ed. V.J. Stenger, J.G. Learned, S. Pakvasa and X. Tata, World Scientific (Singapore), 1992, pp.221–234.

99. **Solar Neutrino Problem and Atmospheric Neutrino Problem in Kamiokande**
    A. Suzuki
    Talk at the 5th Asia Pacific Physics Conference, Kuala Lumpur, Malaysia, August 10–15, 1992.

100. **Solar Neutrinos + Supernova Search (Kamiokande Results)**
    Y. Suzuki
    Talk at the International Symposium on Neutrino Astrophysics, Takayama/Kamiok Japan, October 19–22, 1992.

101. **Atmospheric Neutrinos: Results from Kamiokande**
    T. Kajita
    Talk at the International Symposium on Neutrino Astrophysics, Takayama/Kamiok Japan, October 19–22, 1992.

102. **SUPERKAMIOKANDE**
    M. Takita
    Talk at the International Symposium on Neutrino Astrophysics, Takayama/Kamiok Japan, October 19–22, 1992.

103. **Recent Results on Atmospheric Neutrinos from Kamiokande**
    K. Kaneyuki
    Talk at the XXVIIIth Rencontre de Moriond "Perspectives in Neutrinos, Atomic Physics and Gravitaion", Villars sur Ollon, Switzerland, January 30 - February 6, 1993.

104. **Recent Results on Solar Neutrino Observation from Kamiokande**
    K. Kaneyuki
    Talk at the XXVIIIth Rencontre de Moriond "Perspectives in Neutrinos, Atomic Physics and Gravitaion", Villars sur Ollon, Switzerland, January 30 - February 6, 1993.

105. **Recent Results from Kamiokande**
    A. Suzuki
    Talk at the 5th International Workshop on Neutrino Telescopes, Venice, Italy, March 2–4, 1993.

106. **Atmospheric Neutrino flux and ν_μ/ν_e ratio measured by the Kamiokande Detector**
    A. Suzuki
    Talk at the International Workshop on ν_μ/ν_e-problem in Atmospheric Neutrinos, Gran Sasso, L'Aquilla, Italy, March 5–6, 1993;
    Published in Proceedings of the International Workshop on ν_μ/ν_e-problem

in Atmospheric Neutrinos, ed. V. Berezinsky and G. Fiorentini, Laoratori Nazionali del Gran Sasso, L'Aquilla, Italy, 1993, pp.71–99.

107. **Kamiokande-III Solar Neutrino Observation**
K. Inoue
Talk at the XXVIIIth Rencontre de Moriond "Electroweak Interaction and Unified Theories", Les Arcs, Savoie, France, March 13–20, 1993.

108. **The Atmosphereic Neutrino Anomaly: Experimental Results (contained events)**
T. Kajita
Talk at the Atmospheric Neutrino Workshop, Baton Rouge, U.S.A., May 6–8, 1993.

109. **The Atmosphereic Neutrino Anomaly: Experimental Results (upward-going muons)**
A. Suzuki
Talk at the Atmospheric Neutrino Workshop, Baton Rouge, U.S.A., May 6–8, 1993.

110. **SuperKamiokande**
K. Nishikawa
Talk at the Atmospheric Neutrino Workshop, Baton Rouge, U.S.A., May 6–8, 1993.

111. **The Search for Dark Matter in Kamiokande Detector**
M. Mori
Talk at the International Workshop on "The Dark Side of the Universe: Experimental Efforts and Theoretical Framework", Rome, Italy, June 23–25, 1993.

Direct Measurement of Electron Antineutrino Mass

Takayoshi Ohshima

National Laboratory for High Energy Physics (KEK)

Hirokane Kawakami

Institute for Nuclear Study, University of Tokyo

1 Introduction

This article [1] describes the current experimental situation of the electron antineutrino mass measurement from the study of tritium β-decay, $^3H \rightarrow {}^3He + e^- + \nu_e$. This is the most direct method to measure the neutrino mass without relying upon any particular assumptions. Namely, this is compared with the counter examples: the measurement of the neutrino-less double β-decay provides information on the neutrino mass averaged over neutrino species with the weight of the mixing angle squared, under the assumption that the neutrinos are Majorana type particles [1]; study of neutrino oscillation results in the neutrino mass in terms of mass-square difference, $\Delta m^2 = m^2_{\nu_i} - m^2_{\nu_j}$ (i and j are indices for neutrino flavour), only when the mixing exists between different flavours [2]. The method using tritium β-decay is only based upon the kinematics of the decay process. The masses of the muon neutrino, m_{ν_μ}, and tau neutrino, m_{ν_τ}, are also being studied by similar direct methods in the decays of π meson and tau lepton, respectively, and their current respective upper limits are obtained as 270 keV [3] and 31 MeV [4].

Although the neutrino mass has been considered to be either zero or quite small since its historical introduction in 1930, there is no theoretical reason to believe that m_ν=0. Experimental efforts to measure m_ν have thus been continued over 60 years, as listed in Table 1 and also plotted in Fig.1, where only the upper limit has been obtained but the limit gets steadily smaller with time.

The year 1980 was made exciting for particle physicists when an experimental group at ITEP (Institute of Theoretical and Experimental Physics, Moscow)

[1]Both the neutrino and antineutrino are referred to simply as ν in this article unless otherwise specified. Although electron and β-ray (or simply β) are used from place to place, they simply refer to the same object. Mass is denoted as m instead of mc^2.

Table 1: List of experimental results on neutrino mass m_{ν_e}.

Year	m_ν (eV)	Confidence Level (%)	Experiment
1948	<5,000		Cook et al. [5]
1949	<1,700		Curran et al. [6]
1949	< 500		Hanna et al. [7]]
1951	< 250		Hamilton et al. [8]
1952	< 250		Langer et al. [9]
1953	< 150		Hamilton et al. [10]
1958	<1,000		Friedman et al. [11]
1968	<4,100	67	Beck et al. [12]
1969	< 200	80	Salgo et al. [13]
1969	< 75	67	Daris et al. [14]
1972	< 60	90	Bergkvist [15]
1973	< 100		Piel [16]
1976	< 35	90	Tretyakov et al. (ITEP) [17]
1980	14 – 46	99	Lubimov et al. (ITEP) [18]
1981	< 65	95	Simpson [19]
1983	< 50	90	Derbin et al. [20]
1985	20 – 45	90	Boris et al. (ITEP) [21]
1986	< 18	95	Fritschi et al. (Zurich) [22]
1987	17 – 40	90	Boris et al. (ITEP) [23]
1987	< 32	95	Kawakami et al. (INS) [24]
1987	< 27	95	Wilkerson et al. (LANL) [25]
1988	< 29	95	Kawakami et al. (INS) [26]
1991	< 13	95	Kawakami et al. (INS) [27]
1991	< 9.3	95	Robertson et al. (LANL) [28]
1992	< 11	95	Holzschuh et al. (Zurich) [29]
1992	< 8		Decman et al. (LLNL) [30]
1993	< 7.2	95	Weinheimer et al. (Mainz) [31]

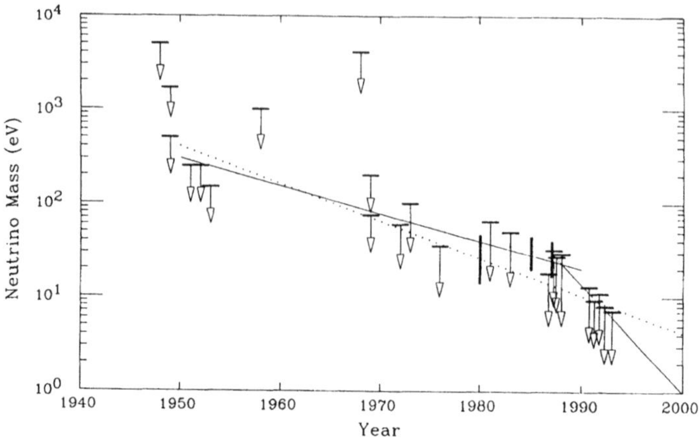

Figure 1: Plot of experimental results on the neutrino mass m_ν, listed in Table 1. The horizontal bar with an arrow corresponds to the upper limit and the thick vertical bar indicates the mass range reported by the ITEP group.

announced a finite mass value of m_ν=14–46 eV for electron neutrino[18]. It was the first experimental indication of a non-zero neutrino mass. Needless to say, the physics community was excited about this unexpected "discovery". A positive indication of neutrino oscillations [32] (which was later denied by other measurements) was also reported in the same year, suggesting the presence of a finite mass of the neutrino. It furthermore triggered studies for the possibility that such a massive neutrino could very well be the solution for the missing mass problem in the universe [33]. Many experiments were then proposed to confirm the finding. If the finding was correct, the confirmation would have been relatively easy. If not, however, conclusively excluding a mass in the indicated range was expected to be a very difficult task, because it was situated in a region difficult to reach with simple extensions of previous experimental techniques. To settle this controversial subject, strict demands were needed not only on the sensitivity on the mass but also on the reliability of the experiments to forestall severe criticisms. With steady efforts of experimentalists over a decade, however, the new upper limit of 10 eV was finally achieved by the three groups: INS (Institute for Nuclear Study, University of Tokyo) [24,26,27], LANL (Los Alamos National Laboratory) [25,28] and Zurich (University of Zurich) [22,29], and the "ITEP discovery" was totally rejected as a result. A new experimental generation has since then been born, aiming at a mass sensitivity of several eV or less. The latest results that have been recently reported are m_ν <8 eV by LLNL group (Lawrence Livermore National Laboratory) [30] and m_ν <7.2 eV by Mainz group (Universität Mainz) [31].

The experimental principle and crucial requirements for the measurement are described in section 2. Section 3 reviews the historical progress of the experimentation, classified into three periods, marked off by the ITEP finding in 1980 and the settlement of the issue in early 1990's. The measurements by ITEP, Zurich and LANL groups are thus described as the second generation experiments. In an attempt to explain in great detail, section 4 is devoted to the description of a series of the INS measurements which are also categorized as the second generation. Some essential issues relating to the mass study such as a deduction method of an upper mass limit, the mass difference between ^{3}H and ^{3}He atoms derived from a variety of methods, are discussed in section 5. Our personal view for future prospects concerning the mass measurement is also given.

2 Experimental Principle and Objectives

2.1 Principle

The principle is simple. It is to detect the mass effect on the electron phase space distribution, which appears in the β-ray energy spectrum in the end-point energy region of nuclear β-decay. Therefore, only a measurement of β-ray momentum or energy spectrum is required; any detection of the neutrino is unnecessary.

The transition probability emitting an electron with momentum p, kinetic energy E, and total energy W in an allowed β-decay, $A \rightarrow A' + e + \nu$, is given as

$$P(E)dE = \frac{G^2}{2\pi^3}|M|^2 F(Z,E)S(E)dE, \tag{1}$$

where the shape factor $S(E)$ represents the phase space volume that dominates the behavior of the electron spectrum,

$$S(E) = pW(E_0 - E)[(E_0 - E)^2 - m_\nu^2]^{1/2}, \tag{2}$$

where G is the Fermi coupling constant and M is the nuclear matrix element of the specific β-decay. $F(Z,E)$ is the Fermi function that takes into account the Coulomb field effect of the daughter nucleus with atomic number Z on the emitted electron:

$$F(\pm Z, E) = 2(1+\gamma)(2pR)^{2\gamma-2} e^{\pm\pi\xi} \frac{|\Gamma(\gamma \pm i\xi)|^2}{[(2\gamma+1)]^2}, \tag{3}$$

where $\gamma = (1-\alpha^2 Z^2)^{1/2}$, $\xi = \alpha Z W/p$ with α the fine structure constant ($\approx 1/137$), and R is the nuclear radius. Upper and lower signs of $\pm$ are for electron and positron emissions, respectively, in the decay. E_0 is the end-point energy which is the maximum electron kinetic energy for the m_ν=0 case. It is simply approximated by neglecting the recoil kinetic energy of the daughter nucleus as $M_A - M_{A'} - m_e$, where M_A and $M_{A'}$ are the masses of parent and daughter nuclei, respectively. In a narrow energy region around E_0, $F(Z,E)$ for ^{3}H β-decay does not show strong energy dependent structure but varies only slightly. Also, radiative corrections [34] change little the spectrum of ^{3}H β-decay.

With a normalization constant A_0, the electron energy spectrum is expressed as

$$N(E) = A_0 F(Z,E)pW(E_0 - E)[(E_0 - E)^2 - m_\nu^2]^{1/2}, \tag{4}$$

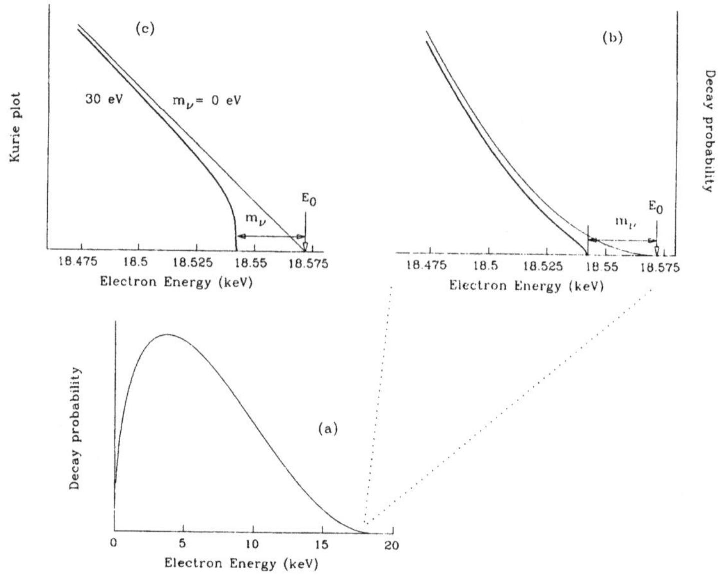

Figure 2: Ideal ^{3}H β-spectrum and its Kurie-plot. (a) is the β-spectrum with m_ν=0. (b) is the enlarged spectrum close to the end-point energy, E_0, for the m_ν=0 and 30 eV cases, shown by thin and thick curves, respectively. (c) is the same with (b) but shown in the form of Kurie plot. E_0 is here set as 18,572.1 eV.

(For the energy region concerned here, the spectrum can be approximated, away from the very end-point region, as being proportional to the square of the energy-difference from E_0, by neglecting the neutrino mass:

$$N(E) \propto \mathcal{E}^2, \tag{5}$$

$$\mathcal{E} \equiv (E_0 - E). \tag{6}$$

An integrated spectrum over $\mathcal{E}$ is then proportional to $\mathcal{E}^3$: $\int_{E_0-E}^{E_0} N(E')dE' \propto \mathcal{E}^3$.) The spectrum is often presented in the form of the Kurie-plot [35] defined by

$$K(E) \equiv \sqrt{\frac{N(E)}{F(Z,E)pW}}. \tag{7}$$

Fig.2 illustrates the ^{3}H β-spectrum and the Kurie-plot for m_ν=0 and 30 eV. For a massless neutrino, $K(E)$ is linear in E (thin line) and crosses the energy axis at E_0: $K(E) \propto (E_0 - E)$. If the neutrino has a finite mass, the maximum electron energy gets lower by the amount of m_ν: $E_0 - m_\nu$, and $K(E)$ tangentially falls to the point. Therefore, the mass measurement is, in other words, a detection of the departure size of the Kurie plot spectrum from the straight line of the $m_\nu = 0$ case.

2.2 Crucial Requirements

In spite of the above simple principle, a precise and reliable measurement of the spectral shape is, in practice, quite difficult from both the statistics and systematics viewpoints.

Statistics

The difficulty arises from the fact that the dominant mass effect appears over only an energy interval of the order of m_ν from E_0. Therefore, a good energy resolution of $\Delta E \sim m_\nu$ is required. In addition, because of the extremely small transition probability in this end-point energy region, as illustrated in Fig.2, a high luminosity is required. Simultaneous demands of high resolving power and high luminosity are, in general, contradictory in usual optical systems. The choice of ^{3}H, as a β-emitter of a superallowed decay, helps reduce this discrepancy:

- An optical property of magnetic spectrometers concerning the resolving power is represented by the relative momentum resolution $\Delta p/p$. The moderately low end-point energy of $E_0(^3\mathrm{H}) = 18.6$ keV makes it possible to obtain a reasonably good energy resolution of ΔE=10 - 20 eV with an achievable resolution, say, of $\Delta p/p = (2\text{–}5)\times 10^{-4}$.[2]

- The half-life of $\tau_{1/2}(^3\mathrm{H})$=12.3 years is moderate enough to produce a highly radioactive β-source. Due to its small E_0 and short $\tau_{1/2}$, an experimental counting rate in the mass sensitive region can be achieved that is relatively high compared to other nuclei, even though the transition rate is still 3×10^{-7} in a 100 eV interval below E_0.

The measured spectral shape with a small signal rate, needless to say, suffers seriously from background contributions. Backgrounds are cosmic-rays, β-rays from ^{3}H atoms evaporated from a β-source, etc, whose dominant components depend upon individual experiments. Not only, does the signal-to-background ratio limit the mass sensitivity, but an incorrect knowledge of the background's spectral shape also substantially affects the electron spectrum and therefore the final result. As described in sections 3 and 4, various technical methods are exploited regarding the spectrometer, β-source, and β-detector by individual experiments to overcome these difficulties.

Systematics

Even though a high statistics spectrum is obtained, the mass can not be reliably extracted without the correct and precise knowledge of the response function and the final state spectrum because both smear the ideal or theoretical spectral shape, as described below.

---

[2]The relation $\Delta E/E \cong 2 \times \Delta p/p$ is valid for all cases discussed in this article.

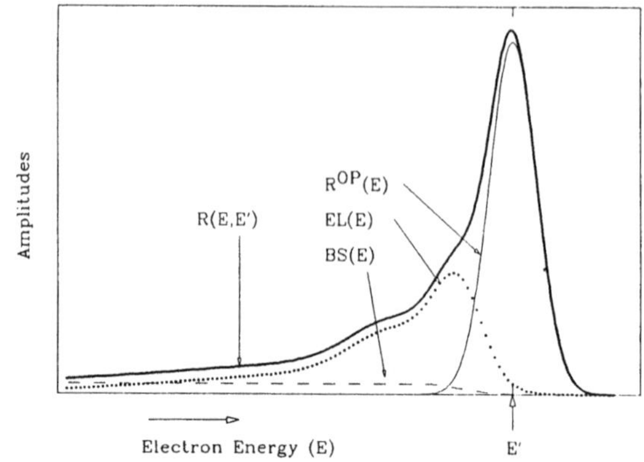

Figure 3: Illustration of the composition of the response function, $R(E, E')$. The optical response, $R^{OP}(E)$, energy-loss, EL, and backscattering, BS, are indicated by the thin solid, dotted and dashed curves, respectively.

(a) Response function

The ideal spectral shape, $N(E)^{th}$, is deformed or smeared by a finite resolving power, $R(E, E')$, so that the observed spectrum, $N(E)^{obs}$, can be expressed as

$$N(E)^{obs} = \int N(E')^{th} \cdot R(E, E')dE'. \tag{8}$$

$R(E, E')$ [3] is called the response function of the system. It is equivalent to an energy-spectrum observed for monoenergetic β-rays with an energy of E'. $R(E)$ consists of three components: the intrinsic optical response of the spectrometer $R^{OP}(E)$, energy-losses of β-rays in a β-source EL, and backscattering of β-rays by a source backing-plate BS.

$$R(E) = R^{OP}(E) \otimes EL \otimes BS, \tag{9}$$

where $\otimes$ denotes the convolution procedure. Fig.3 illustrates these three contributions.

The EL spectrum observed by the INS experiment is described in section 4. BS exhibits approximately a flat energy distribution with small amplitude, δ_{tail} $(\text{eV})^{-1}$, continuing down to zero energy. This long low-energy tail yields an important contribution to the spectrum $N(E)^{obs}$ through eq.(8). The fractional size of the tail contribution to the observed spectrum can be estimated as a function of $\mathcal{E}$,

$$f_{\text{tail}} = \frac{\Delta N(E)_{\text{tail effect}}}{N(E)} = \frac{\int_E^{E_0} \delta_{tail} \cdot N(E')dE'}{N(E)} \sim \frac{\delta_{tail}}{3}\mathcal{E}. \tag{10}$$

Although the importance of backscattering depends on the experiment, roughly speaking, it gives a 20–30% contribution to $R(E)$ and then $\delta_{\text{tail}} \sim (20\text{–}30\%)/(18\,\text{keV})$

[3]For simplicity, $R(E, E')$ is hereafter denoted as $R(E)$.

is on the order of 10^{-5} (eV)$^{-1}$. Therefore, the cumulative effect of eq.(10) results in $\sim 1\%$ of $N(E)$ for $\mathcal{E} \sim$ (a few keV). The fraction of backscattering shows a linear $\mathcal{E}$ dependence and the effect appears as an excess in a Kurie plot; the departure from the ideal curve of $N(E)$ increases as $\mathcal{E}$. In spite of this importance, experimental understanding of this effect is quite limited due to its smallness. For this reason the so-called shape correction term is often introduced in data analysis to absorb the effect by the following form:

$$S(E) \rightarrow S(E) \times (1 + \alpha_1 \mathcal{E} + \alpha_2 \mathcal{E}^2), \tag{11}$$

where α_1 and α_2 are free parameters. Some of the experimental groups assume only linear or quadrature term. It is a historical lesson that the ignorance of these EL and BS effects resulted in the discrepancy in interpreting observed spectra which motivated theorists to revise the Fermi theory in the 1930's (see [36]). It took over 15–20 years to clearly understand what had produced the excess [37]. The recent controversial issue of the 17 keV heavy neutrino [36,38] is likely to be ascribed to a poor understanding of the response function .

$R(E)$ is mostly obtained by experimental observation: individual or whole components are evaluated with the use of a monoenergetic β-ray source produced in electron capture, internal conversion, or Auger transitions. It is to be noted, however, that the primary electron arising from these processes is not monoenergetic but rather has a structure tailing to low energy [39]; the sudden change of atomic Coulomb field associated with electron emission causes excitation or ionization of other orbital electrons, and a part of the released energy is consumed for this effect. This is called shake up or shake off effect. Also, the monoenergetic β-ray, irrespective to its production process, acquires an intrinsic energy spread due to the natural line width, Γ. In order to evaluate the relevant components, $R^{OP}(E)$, EL and BS, from measured spectra of monoenergetic β-rays, these two effects, shake up/off effect and natural line width, must be taken into consideration.

(b) Final state spectrum

Unless a free ^{3}H nucleus is used, a β-emitter is in some bound state. In a sudden change of nuclear electromagnetic field in β-decay, orbital atomic electrons in the final state are excited. The final states of daughter atoms or molecules are then distributed over many different states. The electron spectrum $N(E)$ is therefore a superposition of various spectra $N^i(E)$ with different end-point energy, $E_0^i = E_0 - E_{ex}^i$, where E_{ex}^i is an excitation energy of a state i relative to the ground state and w^i is an excitation probability:

$$S(E) \rightarrow S(E) = A_o p W \sum_i w^i (E_0^i - E)[(E_0^i - E)^2 - m_\nu^2]^{1/2}. \tag{12}$$

In order to estimate the final state spectrum, FSS, we have to resort to a theoretical calculation. For free ^{3}H molecules, being the simplest β-emitters used so far, FSS can be reliably calculated [40]. But, in most of the experiments ^{3}H atoms are bound in complicated compounds, and the theoretical calculations suffer from the uncertainties that limit the achievable mass sensitivity.

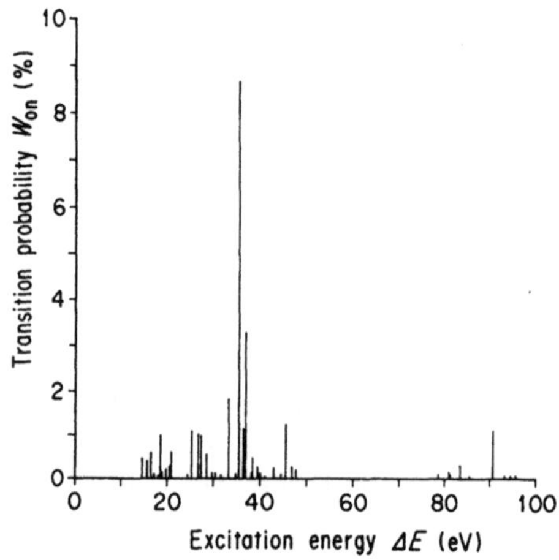

Figure 4: Final state spectrum, *FSS*, calculated by Arafune et al. [42] for propane molecule (cited from ref. [42]). Plotted is the probability ω^i_{ex} of a one electron excitation. The probability for the ground state, not plotted, is 59.80%.

Kaplan, Smutny and Smelov [41] calculated *FSS* for various molecules including a valine acid which was used in the ITEP experiment. Taking the electron correlation into account in the transition probability, they carried out the calculation using the configuration interaction method. Resulting probabilities for the ground state and discrete excitation levels were 61.2% and 28.2%, respectively. Ionization probability was taken as the rest of the transition to be 10.6%, and its spectrum was represented by a single state with an energy of 46.5 eV above the ground state as the mean energy of the transitions. They concluded that *FSS* are most seriously affected by the nearest neighbors of the tritium atom; *FSS* must be almost the same for all hydrocarbon molecules where tritium atom is bound by a carbon atom. Arafune, Koga, Morokuma, and Watanabe [42] carried out a more elaborate calculation with the self-consistent field molecular orbital method for methane (CH_3T, where T denotes a tritium atom) [4] and propane (CH_3-CHT-CH_3). The validity of the adopted basis set of the wave functions was also investigated by examining the saturation of the unitarity and the sum rules for the excitation energy. The probabilities of the ground state, one electron excitations, and two electron excitations were obtained as 59.8%, 34.0%, and 4.8%, respectively; the unitarity was saturated to 98.6%. Resulting *FSS* for propane molecule shows two broad peaks at ~20 eV and 35 eV as seen in Fig.4, being rather similar to the one for valine acid obtained by Kaplan et al. [41]. Their calculation for propane is employed for the *FSS* correction for the INS experiment, which actually used icosanoic acid.

The end-point energy E_0, which is obtained along with m_ν^2 from the observed β-spectrum, is related to the mass difference, ΔM, between ^3H and ^3He atoms, which is also measured by other types of experiments using, for instance, the

[4] For designation of chemical structure, tritium atom is denoted by T instead of ^3H.

radiofrequency-mass-spectrometer-doublet [43] [5] or the ion-cyclotron-resonance methods.[6] Therefore, ΔM that results from the neutrino mass experiment is a good physical quantity for evaluating the reliability of the experiment. ΔM for a free ^{3}H atom is written

$$\Delta M = \Delta \tilde{M} + \Delta E_{B(free)}, \tag{13}$$

where $\Delta \tilde{M}$ is the nuclear mass difference, and $\Delta E_{B(free)}$ is the difference of atomic electron binding energies between the initial and final ground states, 65.4 eV. In the case of ^{3}H in a molecule, $\Delta \tilde{M}$ can be obtained as

$$\Delta \tilde{M} = E_0 - \Delta E_{B(molecule)} + E_{rec}, \tag{14}$$

where the second term on the right-hand side is again the binding energy difference between the two ground states for the relevant molecule: for example it is 48.3 eV for valine [41] and 51.0 eV for propane [42]. The recoil energy E_{rec} is at most 3 eV and is ignored in most cases.

3 Review of the Experimental Progress

3.1 First period: late 1940's – 1980

Electron spectroscopy has an old and rich history, especially regarding the spectroscopic technology [45] including the ^{3}H β-spectroscopy. Technical approaches used in the second period were already embodied in this era, although their precisions were limited by those day's technological levels. Experiments in this period were devoted mostly to achieving better energy resolution and higher statistical accuracy. The systematics problems described in subsection 2.2. had not limited the sensitivity of the experiments until the late 1970's when Bergkvist [15] and Simpson [19] started their experiments.

Prior to experimental investigations, Konopinski [46] made a theoretical estimate on the mass assuming that $|M|^2(ft_{1/2})$ is the same for all allowed nuclear transitions, which is inversely proportional to the Fermi coupling squared, G^2 ($t_{1/2}$ is a half-life). The mass term is included in the integrated Fermi function $f = \int F(Z,E)S(E)dE$. With $|M|^2(ft_{1/2})$=5,760 sec evaluated from ^{6}He decay, he estimated from the decay of ^{3}H that m_ν=11–17 keV. Pruett [47] later pointed out, however, that the influence of the Coulomb field on the electron and a relativistic correction factor had to be included and then no finite neutrino mass would need to be assumed if the end-point energy of ^{3}H is as high as E_0 ~20 keV.

[5]The accuracy of the mass determination is better than 10^{-7}. The details of the instrument is found in ref. [44].

[6]In a homogeneous magnetic field B, an ion with mass m and charge e takes a circular motion with an angular velocity of $\omega_c = eB/m$. Under an additional external radiofrequency (RF) of $f_c = \omega_c/2\pi$ vertical to B, the ion resonantly absorbs the RF energy and is accelerated. The mass of ion can be therefore measured within an accuracy of 10^{-8} from the resonant frequency f_c.

Cook, Langer and Price [5] measured the β-spectrum of ^{35}S instead of ^{3}H by a shaped magnetic field spectrometer of a 40 cm mean radius. Because of an end-point energy of $E_0(^{35}S)$=167 keV that is higher than that of ^{3}H, it was less suitable for the neutrino mass measurement; the achieved upper limit was 5 keV.

The tritium β-spectrum was measured by Langer and Mofatt [9] with $\Delta p/p$=0.7% using the same spectrometer and a small GM-counter as a β-detector. The source was prepared by thermally evaporating tritiated succinic acid onto a thin support in vacuum. The source thus prepared was quite stable at room temperature and had a specific activity of 0.9 mCi/μg. The observed spectrum was consistent with a zero-mass neutrino; the resulting upper limit was 250 eV and E_0 was 17.95$\pm$0.10 keV. This is the original work for the present style of magnetic spectrometer experiments.

Meanwhile, Curran, Angus and Cockroft [6] and Hanna and Pontecorvo [7] independently observed the ^{3}H β-spectrum with the use of a proportional counter by mixing a small amount of tritium in the gas in the chamber. The energy spectrum was in this case observed in terms of the pulse height distribution. Although this method could avoid electron energy-loss in the source, the achievable energy resolution was limited and moreover deteriorated at the counter ends. The mass was found to be less than 1.7 keV in the former experiment and 0.5 keV in the latter.

The first use of an electrostatic β-spectrometer can be found in the ^{3}H β-spectrum measurement by Hamilton, Alford and Gross [8]. A retarding electric potential was applied to electrons emitted from a source at the center of a spherically symmetric spectrometer. Only electrons which had more than a certain energy were therefore detected by the collector. This method thus provided an integral spectrum instead of a differential one. They achieved an energy resolution of 120 eV in the end-point energy region and set an upper limit of 150 eV.

A monograph by Allen (1958) [48] is an excellent text book to learn about neutrino physics in those days. Many experimental efforts were going on at that time as listed in Table 1, but the first period was considered to come to an end by Bergkvist's superb measurement [15], whose pioneering ideas were faithfully followed in the second generation experiments. In his experiment, a high luminosity and high energy resolution were both achieved by the use of a $\pi\sqrt{2}$ type magnetic spectrometer with electrostatic devices. A ^{3}H source was prepared by implanting tritium ions into a thin layer of aluminum with a surface area of 10$\times$20 cm^2 which was divided into many narrow strips. An appropriate electric potential was applied to individual strips to correct the source extent by accelerating and/or decelerating emitted electrons, and an electrostatic corrector was also prepared to reduce the optical aberrations. With these devices, the luminosity was increased by three orders of magnitude in comparison to the experiments using the conventional spectrometer. The energy resolution achieved was 40 eV (FWHM). He was also the first who took into account the atomic effects, *FSS*, in ^{3}H β-decay: he noted that the effect of *FSS* which smears the spectrum could not be ignored compared to the instrumental energy resolution. An upper limit was set to be 60 eV at 90% confidence level.

Simpson carried out a unique experiment with the use of a Si(Li) detector into which tritium atoms were deeply implanted [19]. Since all energy released in both β-decay and de-excitation of final atomic state was measured by this calorimetric device, the uncertainty related to FSS was essentially eliminated. A beautiful linear spectrum on the Kurie-plot was obtained over the whole energy region [7]. However, the achievable energy resolution is intrinsically limited by a solid state detector to 217 eV (FWHM), and the mass limit was set as $m_\nu < 65$ eV with 95% confidence level.

3.2 Second period: 1980 – 1992

Along with technical progress, physics demands finer considerations in experiments such as the final state spectrum and the precise response function including the energy-loss and the backscattering effects.

The second period began in 1980 by the ITEP "discovery" [18] as mentioned in section 1. To clearly settle this surprising affair, the prime experimental goal was set for a mass sensitivity down to the 10 eV level, and twenty or more experiments were proposed, as listed in Table 2. Because of the various difficulties, mainly in the preparation of a high quality ^3H source, only Zurich, INS and LANL could, having spent about 10 years, successfully arrive at the final goal with continuous and steady improvements.

ITEP experiment

Since the first report in 1976 with $m_\nu <35$ eV [17], the group repeated the measurements which eventually resulted in a finite mass with $14< m_\nu <46$ eV (99% Confidence Level) in 1980 [18], $20< m_\nu <45$ eV in 1985 [21], and $17< m_\nu <40$ eV in 1987 [23].

The source was a tritiated valine ($C_5H_9T_2NO_2$), an amino acid, with an 18% enrichment in tritium (about two ^3H atoms per acid molecule), and was prepared by means of the vacuum evaporation method. The source thickness was $2\,\mu g/cm^2$. The doubt about whether the chemical form of the evaporated source would preserve the initial compound structure or not was treated as a systematic error.

The momentum analyzer was an iron-free [8] Tretyakov type toroidal magnetic spectrometer [62]. The principle of the toroidal magnetic field system was proposed by Vladimirskii, Tarasov, and Trebuhkovskii in 1956 [63]: Charged particles

[7] In a series of this measurement, Simpson "discovered" a heavy neutrino having a mass of 17 keV with an admixture of 1% in electron neutrinos in 1985 [38].

[8] "Iron-free", meaning no magnetic material was used, helps avoid undesirable hysteresis phenomena so that the field distribution can be ideally controlled and a better resolution can be achieved than that of a spectrometer with iron. Furthermore, the high precision for determining the absolute electron momentum can be achieved, provided the coil current is highly stabilized. On the other hand, because there is no iron-york, it is essentially influenced by external magnetic field disturbances such as the variation of the earth's magnetic field and influences of cars or trains passing nearby. A compensation system against these disturbances is therefore indispensable in order to maintain its high performance. The spectrometer is usually installed house well isolated from other facilities that generate a strong magnetic field like an accelerator.

Table 2: Lists of experiments proposed. This is an updated version of lists presented by Shavitz [50] and Wilkerson [49]. 'Status' indicates the experiment either is completed (C), published results (R), preliminary results (P), being tested (T), or under construction (U). The experiments marked by * are classified as first generation.

Experiment	Location	Spectrometer	Source §	Status
Lubimov et al. [18,21,23]	(ITEP)	Toroidal Mag.	Valine-T	C
Kündig et al. [22,29]	(Zurich)	Toroidal Mag.	C-T, OTS-T(LB-film)	C
Ohshima et al. [24,26,27]	(INS)	$\pi\sqrt{2}$ Mag.	Icosanoate-T(LB-film)	C
Robertson et al. [25,28]	(LANL)	Toroidal Mag.	T_2 gas	C
Daniel et al. [51]	München	$\pi\sqrt{13/2}$ Mag.	HfO_2-T	P
Fackler et al. [52]	Livermore	Retarding E-S	Al_2O_3, T_2 solid	T
Boyd et al. [53]	Ohio	Retarding E-S	T_2 solid	U
Clark et al. [54]	IBM	Retarding E-S	?	T
Jelley et al. [55]	Oxford	Cylind. Mirror	Cd Palmitate-T(LB-film)	T
Graham et al. [56]	Chark River	$\pi\sqrt{2}$ Mag.	Ti-T	
Bergkvist [57]	Stockholm	$\pi\sqrt{2}$ Mag.	Al-T	
Heller et al. [50]	Berkeley	SSD	Si-T	
Wellenstein et al. [58]	Brandeis	Cylind. Mirror	T_2 gas	U
Sun et al. [59]	Beijing	Iron-core Mag.	PAD organic T	P
Shang et al.* [60]	Beijing	SSD	Si-T	P
Simpson* [19]	Guelph	SSD	Si-T	C
Derbin et al.* [20]	Leningrad	SSD	Si-T	C
Stoeffl et al. [30]	(LLNL)	Toroidal Mag.	T_2 gas	R
Bonn et al. (Mainz) [31]	(Mainz)	Retarding E-S	Cold T_2	R
Lobashov et al. [61]	Moscow	Retarding E-S	T_2 gas	T

§ Tritium atom is denoted by T instead of ^3H.

emitted perpendicularly to the rotation axis are focused at every 180° turn under a magnetic field with an approximate $1/r$ distribution. A characteristic of this toroidal spectrometer is a large solid angle of $\Omega/4\pi = 0.3\%$. Fig.5 is a cross sectional view of the ITEP spectrometer. Looking from the top of the spectrometer, the magnetic field was produced in the circumferential region of a doughnut shape by 72 rectangular current loops. Electrons emitted in the radial direction are focused four times during a 4π rotation so that, while keeping the solid angle large, off-momentum electrons are well discriminated by the windows. The resolution obtained with the finite source size and detector window was $\Delta p/p = 0.12\%$ or ΔE=45 eV at the energy E_0.

Either a GM counter, a proportional counter, or a six channel proportional counter was placed on the focal point as a β-detector in various measurements. Pulse height of the proportional counters provided energy information on an incident β-ray with ~20% resolution and was made use of for background rejection.

An electrostatic element was set up in front of the source in order to accelerate or decelerate emitted electrons. The β-spectrum was thus measured by changing the applied electric voltage while keeping the magnetic field strength constant: it is the so-called electrostatic scan. That any electron impinging on the detector had the same energy irrespective of the initial value, made the detection efficiency independent of the electron energy at the source. The signal-to-background ratio in the end-point energy region was as high as 16 : 1. 100K events were accumulated

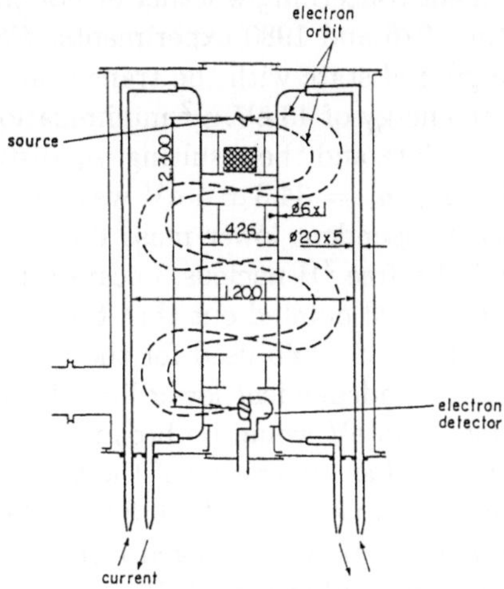

Figure 5: Schematic view of the ITEP toroidal spectrometer [62].

in a 100eV range below E_0.

The individual components of $R(E)$ were measured as follows. Conversion electrons from a ^{169}Yb source, the M1 line (18.42 keV) of the 20.74 keV nuclear transition, were used as monoenergetic β-rays to evaluate $R^{OP}(E)$. By greatly decelerating the electrons with the electrostatic element, $R^{OP}(E)$ got narrower by an order of magnitude, being approximated as a δ-function, while the effect of EL remained unchanged. The EL effect, together with the shake up/off effect and the natural line width effect, were deconvoluted away from the measured conversion spectrum. EL in the ^{3}H source was evaluated from the conversion spectra of the ^{169}Yb source on which the valine layer was additionally mounted with various different thicknesses.

A χ^2 fit to the data was carried out with the FSS for a valine molecule calculated in ref. [41] as:

$$N^{th} = A_0 \int dE' R(E, E')[1 + \alpha_1(E' - E_0)][1 + \alpha_2(E' - E_0)^2] \sum w^i f_i(E', m_\nu, E_0), \tag{15}$$

$$f_i(E', m_\nu, E_0) = pWF(Z, E')(E_0 - E^i_{ex} - E')\sqrt{(E_0 - E^i_{ex} - E')^2 - m^2_\nu}, \tag{16}$$

where α_1 and α_2 are the shape correction coefficients. The parameters in the fit were m^2_ν, E_0, α_1, and α_2, and the normalization constant A_0 was determined from the number of observed total events rather than treated it as a fitting parameter. An analysis was also performed including the FSS for the ^{3}H atom and nucleus, the latter of which was used to evaluate a "model independent" lower mass limit.

Basic instrumental aspects were already embodied in the ITEP 1976 experiment. Although a small background rate was one of the superb aspects, the rate continuously increased with time due to ^{3}H evaporation from the source: it increased by a factor of 4 during a one month period [17,18]. Consequently, it raised a doubt about the experiment concerning whether or not the background had any energy dependence. In the 1976 and 1980 experiments, FSS was represented by only two transitions: the ground state with the transition probability of 70% and an excitation state with an energy of 43 eV. χ^2 minimization was performed independently for 16 samples of data and the resulting m_ν distribution was examined by a Monte Carlo simulation; $m_\nu = 34.3 \pm 4$ eV was then obtained as the most probable value. A model independent lower mass limit was set as $m_\nu > 14$ eV (99% C.L.) with the FSS of a free ^{3}H nucleus, because it should always give the smallest mass value. Simpson [64] pointed out that the neglect of the natural line width of the M conversion lines of ^{169}Yb used for the response function measurement made the claimed model-independent lower limit invalid. He estimated that the instrumental resolution of 45 eV would be broadened by a (3.5–10) eV and a mass of 29 eV would be resulted as an artefact if the true resolution was 56 eV in place of 45 eV for the zero-mass neutrino. In 1983 the ITEP group re-analyzed the 1980 data: the mean mass value was changed from 34 eV to (28 ± 4) eV and the lower limit of 14 eV was reduced to 0 eV [65].

Between 1983 – 1985, the results of a new measurement with the improved apparatus were reported [21,65,66]. A Kurie plot of the ^{3}H β-spectrum is seen in Fig.6(a). Bergkvist [67] indicated the importance of the low-energy tail of the response function consisted of the energy-losse and backscattering, while the ITEP 1983 analysis [65] treated the response function as being symmetrical between higher and lower energies. He also criticized the insufficient allowance for shake-up effects in the conversion line for determining the response function in the 1984 ITEP analysis [66]. In 1985, the most probable mass was obtained as 34.9 ± 1.9 eV and a realistic mass estimate was $20 < m_\nu < 45$ eV [21]. A model-independent lower limit of $m_\nu > 9$ eV was discarded in this estimate, since an application of a free ^{3}H nuclear FSS was quite unrealistic.

In their latest measurement published in 1987 [23], the response function was obtained by the procedure described above in response to Bergkvist's criticism. By combining newly accumulated data with previous ones, χ^2 fits were performed in different energy intervals in order to examine the stability of the resultant parameters: $m_\nu^2 = 966 \pm 50$ (eV)^2 was obtained with χ^2/(degree-of-freedom, DOF)= 429/407 for a fitted energy interval of 1680 eV. Fig.6(b) shows the result of the χ^2 fit in terms of deviations from the best fitted curve. With systematic uncertainties in evaluating $R(E)$ and FSS, the result was

$$m_\nu = 30.3^{+2}_{-8} \ eV. \tag{17}$$

$\Delta M = (18,600.3 \pm 4)$ eV, derived from the resulting $E_0 = (18,580.8 \pm 0.4)$ eV, agreed excellently with $\Delta M = (18,599 \pm 2)$ eV measured by Lippmaa et al. [69] with an ion-cyclotron-resonance spectrometer. By restricting E_0 within an allowable range of these two ΔM values, the ITEP group claimed the model independent mass interval to be

$$17 < m_\nu < 40 \ \text{eV}. \tag{18}$$

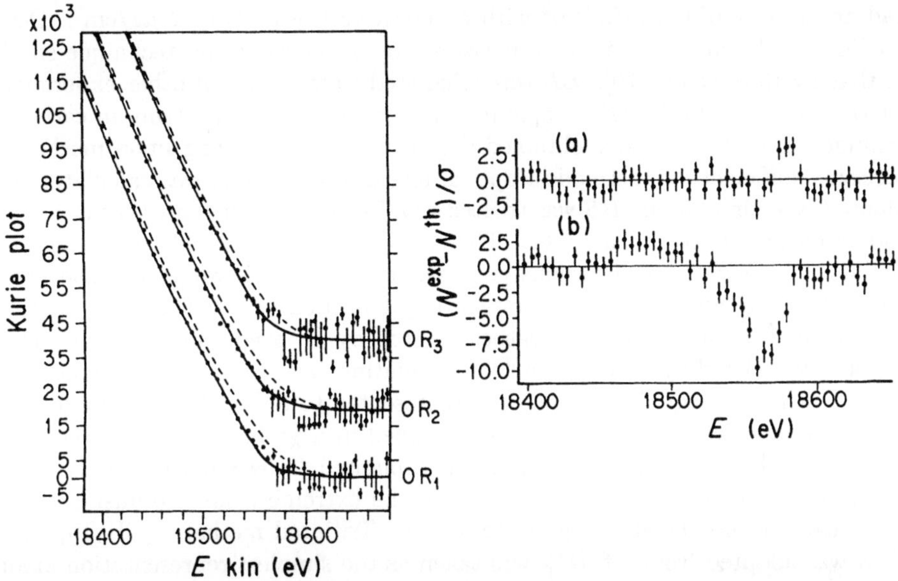

Figure 6: (a) Kurie plot of ITEP 1985 data (cited from ref. [21]). (b) Result of χ^2-fit in the ITEP 1987 experiment (from ref. [23]).

Zurich experiment

The application of an electrostatic retarding potential on a magnetic spectrometer of moderate intrinsic resolution made it possible to realize a good energy resolution. The Zurich experiment was also characterized by its high statistics measurement. The first result appeared in 1986 with $m_\nu < 18$ eV (95% C.L.) [22] and the second in 1992 with $m_\nu < 11$ eV (95% C.L.) [29].

The spectrometer was the Tretyakov type, the same as ITEP's, but a little simpler. An electrostatic potential was formed between the ^{3}H source (applied positive high voltage) and a cylindrical grid (kept near the ground potential) surrounding the source. Its purpose was to greatly retard the electron energy, from 18 keV to 2 keV, as well as to scan the spectrum. Electrons entering the magnetic spectrometer had energies an order of magnitude smaller than their initial values, so that a ten times better absolute momentum resolution Δp could in principle be obtained compared with the case without retardation, provided that the same optical resolving power $\Delta p/p$ was given in both cases. In other words, a ten times poorer optical resolving power would result in the same absolute momentum resolution as that in a case of no potential. Electrons were finally detected at the focus where re-accelerating high voltage was applied to the detector. Characteristics of the system were $\Omega/4\pi = 0.6\%$, dispersion of 1.1 eV/mm, and $\Delta E = 27$ eV (FWHM) in the E_0 region. On the contrary, the retardation method might suffer seriously from background contributions, especially if the spectrometer was polluted even slightly by ^{3}H. No description on the background and signal-to-background ratio was found in their literature.

In their first experiment, the ^{3}H source was implanted into a graphite bulk. It had an activity of 0.3 mCi/cm^2 with an effective thickness of 2 μg/cm^2. Depth profiles of ^{3}H concentration were measured by a nuclear recoil technique similar to that by Ross et al. [70]. EL was calculated from values of other experiments for the mean free path and a single interaction energy-loss spectrum in carbon in conjunction with the above obtained depth profile. By relying upon mechanical precision sufficient for the modest optical resolution, $R^{OP}(E)$ was calculated by a Monte Carlo simulation. BS was treated as a free parameter representing a shape correction factor (α_1) in a χ^2 fit. The calculated $EL + R^{OP}(E)$ was examined against an observation of various conversion spectra. However, the conversion line source was not prepared by implanting it into carbon at the same depth as that of the ^{3}H source, but was rather covered by an evaporated carbon. There might have existed some uncertainty in evaluating BS of the observed conversion spectra, whose energy dependence was represented to be flat from a Monte Carlo simulation and whose intensity was determined in a χ^2 fit analysis for the ^{3}H β-spectrum. The chemical structure of the implanted ^{3}H was not definitely known but supposed to be in C$-$T bonds due to the extremely small diffusion rate of ^{3}H in carbon and the stability of the source. Tritiated methane (CH_3T), whose FSS was adopted from ref. [41], was taken as the simplest representation among a variety of molecules with C$-$T bonds. In their data analysis, FSS was actually represented with only three transition levels, the ground state and two excited states [71]. Data were taken spectrographically by a position sensitive detector and a total of 100K events were accumulated in a 100 eV interval below E_0 for four data samples. The best fit was $m_\nu^2 = (-11 \pm 63(\text{stat.}) \pm 204(\text{sys.}))(\text{eV})^2$ with $\chi^2 = 1,370.2$ for DOF=1,369. Setting this m_ν^2 to be exactly zero, an upper limit of $m_\nu < 18$ eV was obtained [22].

The second measurement was reported in 1992 [29] which reduced systematic uncertainties. The most substantial improvement is the use of a tritiated monomolecular layer of the source. As is described in the next section, certain kinds of molecules with long hydrocarbon chains can easily form a monolayer of 25 Å in thickness by means of the Langmuir-Blodgett technique [72], which was first used by the INS group for preparing their highly radioactive and large tritium sources on tritiated icosanoate acid. The Zurich group employed tritiated octadecyltrichlorosilane: there are 18 carbons in a chain among which three double bonds are hydrogenized to give six ^{3}H atoms. EL was greatly reduced and therefore its uncertainty became a small portion of the total systematic error. While a delicate treatment of the source substrate was made to suppress the BS effect, it was again treated as a shape correction factor in a χ^2 fit; α_1 was 2×10^{-5} (eV)$^{-1}$ in the previous measurement and 1.7×10^{-5} (eV)$^{-1}$ at the present time. The FSS calculated by Claxton, Schafroth, and Meier [73] for propane molecules was employed. The measured ^{3}H β-spectrum is seen in Fig.7 with the result of the χ^2 analysis. A global fit for 6 data samples resulted in the best fit with m_ν^2 as

$$m_\nu^2 = -24 \pm 48(stat.) \pm 61(sys.) \quad (eV)^2, \tag{19}$$

with χ^2/DOF=304.9/271. The systematic error was dominated by the uncertainty of the FSS. With 95% C.L. an upper limit was derived as $m_\nu < 11$ eV. The re-

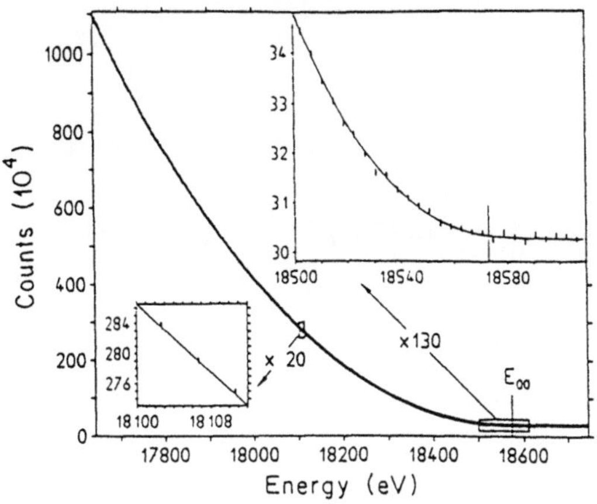

Figure 7: Spectrum obtained by the Zurich 1992 experiment (from ref. [29]).

sulting $E_0 = (18,573.32 \pm 0.18(\text{stat.}))$ eV was in very good agreement with the values of $(18,572.1 \pm 0.4(\text{stat.}) \pm 3.0(\text{sys.}))$ eV and $(18,570.5 \pm 2.0)$ eV obtained, respectively, by the INS [27] and LANL [28] experiments.

LANL experiment

The most distinguished feature of this experiment was the use of a pure gaseous tritium source: the *FSS* was well understood and reliably calculable. Therefore, systematic uncertainty related to *FSS*, the dominant source for other experiments using complex molecules, could be reduced to an insignificant level. It also had other advantages: a small *EL* and no *BS*. On the other hand, the large disadvantage was the difficulty of accumulating sufficient statistics due to the low counting rate. In 1987, the group reported a model independent upper limit of $m_\nu < 27$ eV (95% C.L.) [25] and in 1991 $m_\nu < 9.3$ eV (95% C.L.) [28].

The experimental apparatus is shown in Fig.8 [74]. $^3H_2$ gas was injected into a long tube (3.86 m long by 3.8 cm ϕ) at the midpoint and recirculated by pumping at both ends. Under the solenoidal magnetic field (1.5 KG) the electrons from ^{3}H-decay spiralled toward the ends along the field lines. The tube was uniformly biased to -8 keV in order to accelerate the electrons between the tube and pumping restriction regions. It was designed to discriminate against the background arising from ^{3}H-decay which occurred somewhere other than in the source, and to reduce the beam size at the entrance to the detector. Contrary to the retarding-potential method, this acceleration imposed heavier demand on the spectrometer to have a higher resolving power. To prevent the spectrometer from being contaminated, a differential pumping system was prepared next to the source region, where the accelerated electrons were focused non-adiabatically using a rapidly falling magnetic field to form an image on a collimator (1 cm ϕ) at the entrance to the spectrometer. The spectrometer was a 5 m focal-length toroidal type, similar to the Tretyakov's.

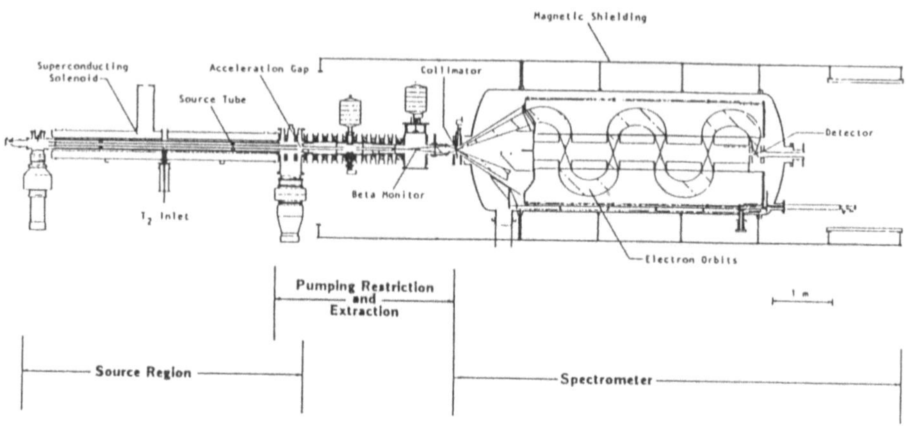

Figure 8: Cross sectional view of the LANL spectrometer (cited from ref. [74]).

1% of the electrons from a 2.2 cm^2 area in the source tube were detected by a position sensitive proportional counter at the focus. Energy resolution of $\Delta E = 36$ eV (FWHM) was achieved. The ^{3}H β-spectrum was scanned by changing the voltage applied to the tube while keeping the magnetic field strength of the spectrometer constant.

Introducing a gaseous ^{83m}Kr source into the apparatus, $R^{OP}(E)$ was determined from an observed K conversion line whose shake up/off effect was estimated from experimental and theoretical data. $R(E)$ was obtained by convoluting $R^{OP}(E)$ with EL calculated by a Monte Carlo simulation. In the 1987 experiment, only 650 counts over 140 background events were accumulated for 4 data samples in a 100 eV interval below E_0. With the use of the FSS for the ^{3}H molecule calculated by Martin and Cohen [40], a fit was performed with five parameters: A_0, E_0, m_ν^2, the background, and α_2, using a maximum likelihood procedure with Poisson statistics instead of a χ^2 fit, due to low statistics data. The last parameter α_2 was introduced to correct for any energy dependent efficiency of the apparatus which was best tuned for the energy E_0; it was found that $\alpha_2 = -1 \times 10^{-8}$ (eV)$^{-2}$. The resulting value was $m_\nu^2 = (-57 \pm 453(\text{stat.}) \pm 118(\text{sys.}))$ (eV)^2, the error of which was dominated by the statistical uncertainty. They obtained $m_\nu < 26.8$ eV as an upper limit with 95% C.L.

The group continued the experiment by employing improved instruments in many aspects, among which the replacement of the proportional counter by an octagonal array of Si microstrip detectors produced the most significant outcome. In 1991 they accumulated 20K events in a 100 eV range below E_0 for 3 data samples, with the signal-to-background ratio twice as good as the 1987 measurement. Systematic uncertainties related to the R^{OP}, EL, and FSS were examined in more detail than in the previous experiments. The energy dependence of the system efficiency was represented again by putting either linear or quadratic shape correction term; the coefficient was determined by a χ^2 fit as $\alpha_1 \sim -2 \times 10^{-5}$ (eV)$^{-1}$ or $\alpha_2 \sim -0.7 \times 10^{-8}$ (eV)$^{-2}$. An average of m_ν^2 values obtained with either α_1 or α_2 put in the fits was taken to be the final result:

$$m_\nu^2 = -147 \pm 68(\text{stat.}) \pm 41(\text{sys.}) \quad (\text{eV})^2. \tag{20}$$

The uncertainty arising from the shape correction term was $\Delta m_\nu^2 = \pm 32$ (eV)^2, which contributed predominantly to the systematic error. The group derived an upper limit of $m_\nu < 9.3$ eV (95% C.L.) as the most stringent limit in their second generation experiment. However, the derived m_ν^2 value was nearly 2 standard deviations below zero: this occurrs only 3% probability if $m_\nu = 0$.

The important features and the results of the four experiments, ITEP, Zurich, LANL, and INS (described in next section), are summarized in Table 3. The long-standing controversial issues was now settled as follows: the finite mass of $\sim$ 30 eV was completely excluded by the Zurich, LANL, and INS experiments, while the reason why the ITEP group had obtained a finite value was not clarified. We know, however, that it was probably not due to the complexity of the *FSS* of the compound source nor the statistical fluctuations. Some unknown systematics problems probably arose from an evaluation of $R(E)$ or background.

3.3 Third period: 1992 –

We may classify the experiments aimed at reaching a mass sensitivity of several eV as those in the third period. Among the many proposals listed in Table 2, the two groups of Mainz and LLNL recently reported their first results, as $m_\nu < 7.2$ eV [31] and $m_\nu < 8$ eV [30], respectively.

Mainz experiment

The source was a frozen ^{3}H molecule of 2.7 mCi, equivalent to 40 monolayers. As was the case with the LANL experiment, the *FSS* of the ^{3}H molecule was reliably calculable and the structure of *EL* was simple due to the small Z of the source material. The apparatus, an electrostatic spectrometer with guiding magnetic field [75] (Fig.9), basically consisted of two superconducting solenoids (B_0 = 2.4 T) separated by a system of ring electrodes (4 m in length). Most of the transverse energy of the electrons was adiabatically transferred into the longitudinal component during the flight from the source with high magnetic field (B_0) to the center of the apparatus (weak field of $B_1 = 2 \times 10^{-4} B_0$). In the left-half stage of the spectrometer, a retarding potential (U_0) decelerated electrons and discriminated against those with energies lower than eU_0. All electrons passing through this filtering are accelerated in the right-hand side of the apparatus and were detected by a silicon device placed in the solenoid coil. An integrated number,

Table 3: Important properties and results of the second generation experiments. The S/N ratio is a ratio between the integrated signal and background counts over a 100 eV interval below E_0, and the statistics in the next row are the integrated signal counts. The experimental results listed on a lower half are the latest reported values of the individual measurements. See the text and the respective references for details.

	ITEP [18]	ITEP [23]	Zurich [22]	Zurich [29]	LANL [25]	LANL [28]	INS [24,26]	INS [27]
Spectrometer	Toroidal mag.		Toroidal mag. retarding potential		Toroidal mag.		$\pi\sqrt{2}$ mag.	
$\Omega/4\pi$	0.3%		0.6%		25%		0.06%	
ΔE (eV)	45	20	27	17	36	23	8-19	30
Source	valine ($C_5H_9T_2NO_2$)	vacuum-evaporation	T_2-C implantation	OTS† ($C_{18}H_xT_6Y$) LB-film	T_2 gas		icosanoate ($(C_{20}H_{32}T_8O_2)_2Cd$) LB-film	
thickness($\mu g/cm^2$)	2		2	0.3			0.6	
intensity(mCi)			47		0.5		8	40
reference source	^{169}Yb			^{195}Au	^{83m}Kr		^{109}Cd	
β − detector	GM/pro. ctr.	6ch pro. ctr.	pro. ctr. (charge div.)		pro. ctr.	12 SSD	40ch pro. ctr.	7ch pro. ctr. (charge div.)
S/N ratio	16:1		0.1:1‡		5:1	(5-9):1	1:1	
Statistics	75K	55K	100K		0.6K	20K	5K,9K	150K
Result of Analysis								
m_ν^2 (eV)^2	970		-24		−147		-65	
Δm_ν^2 (eV)^2	±168		±77.6		±79.4		±107	
statistical	±50		±48		±68		±85	
systematic	±160		±61		±41		±65	
m_ν (eV)	17 – 40		<11		<9.3		<13	
Deviation from m_ν =17 (eV)			4.0σ		5.5σ		3.3σ	
E_0 (eV)	18,580.8		18,573.32		18,570.5		18,572.1	
ΔE_0 (eV)	±0.4		±0.18(stat.)		±2.0		±0.4(stat.)±3.0(sys.)	

† OTS = Octadecyltrichorosilane.

‡ This S/N ratio is inferred from Fig.7.

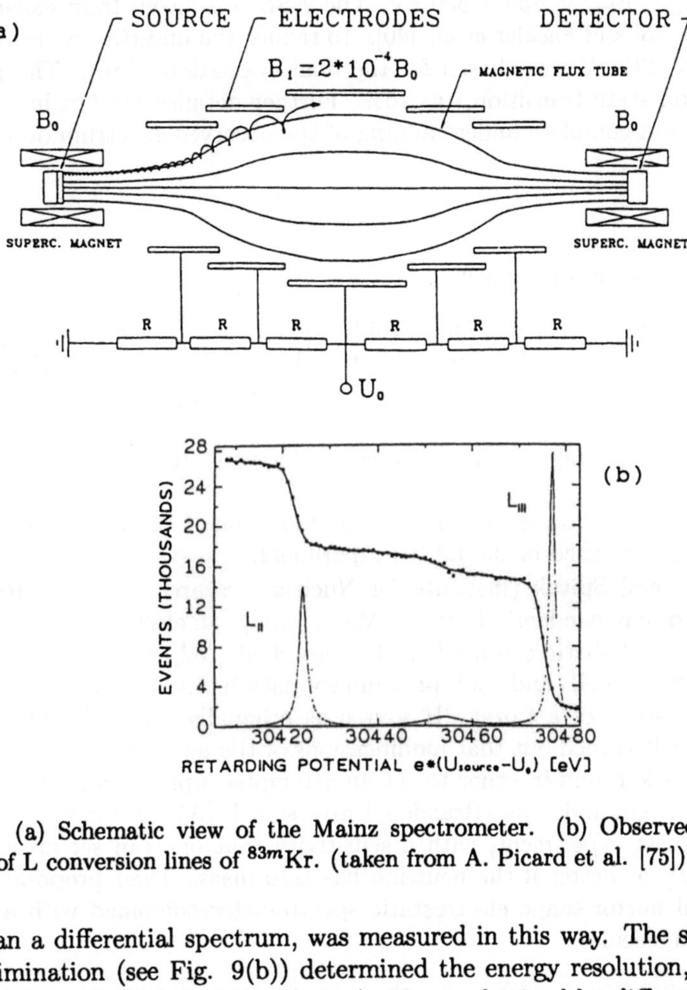

Figure 9: (a) Schematic view of the Mainz spectrometer. (b) Observed integrated spectrum of L conversion lines of ^{83m}Kr. (taken from A. Picard et al. [75])

rather than a differential spectrum, was measured in this way. The sharpness of this discrimination (see Fig. 9(b)) determined the energy resolution, which was given by $\Delta E = (B_1/B_0)E$; $\Delta E =$ (a few) eV was obtained by differentiating the observed spectra of ^{83m}Kr conversion lines. The spectrometer provided a large solid angle of $\Omega/4\pi = 40\%$ because only the solenoidal coil is a limiting factor and the acceptable polar angle is $\leq 78^o$. The signal-to-noise ratio was 1 : 1 at 30 eV below E_0.

The first result has been published in 1993 as

$$m_\nu^2 = -39 \pm 34(\text{stat.}) \pm 15(\text{sys.}) \quad (\text{eV})^2, \tag{21}$$

or $m_\nu < 7.2$ eV with 95% C.L, and $E_0 = (18,574.8 \pm 0.6)$ eV. A χ^2 fit was carried out by changing the lower limit (E_{low}) of the fitting energy interval with five free parameters, A_0, E_0, m_ν^2 and two parameters representing the background, and also with four free parameters by fixing α_1 to be the pre-determined value of $(6.7 \pm 1.7) \times 10^{-5}$ (eV)$^{-1}$. A significant dependence of m_ν^2 and E_0 on E_{low} was found: m_ν^2 were distributed around -100 (eV)^2 with statistical uncertainty of ± 20

$(\mathrm{eV})^2$ for $E_0 - E_{low} = 150 - 500$ eV. The FSS was taken from calculations by Kolos et al. [76] and Fackler et al. [40]. To reduce the uncertainty from the FSS, $E_0 - E_{low} = 137$ eV was chosen for the final evaluation of m_ν^2. The percentage of the ground state transition was 76%. Further detailed studies have yet to be performed for a complete understanding of the observed spectrum over the whole region.

Other modern experiments

The spectrometer by Decman and Stöffl [77] was a large Tretyakov type, similar to that of LANL. A first brief report is found in ref.[30], although the details are not yet known to the authors at the time this manuscript is written:

$$m_\nu^2 = -72 \pm 41(\mathrm{stat.}) \pm 30(\mathrm{sys.}) \quad (\mathrm{eV})^2, \tag{22}$$

or m_ν <8 eV was obtained with the energy resolution of 12 eV. The ^{3}H source was in gaseous form as used in the LANL experiment.

Lobashev and Spivak (Institute for Nuclear Research, Moscow, Russia) [61] had proposed, independently from the Mainz group, an electrostatic integral spectrometer with adiabatic collimation. Fackler et al. [52,78] built an electrostatic spectrometer at LLNL and took preliminary data in 1988 with an ion-implanted ^{3}H source, instead of a frozen ^{3}H source as originally planned; ΔE=6 eV was achieved, but it turned out that modifications of the apparatus were necessary to reduce the background in order to set an attempted upper limit on the neutrino mass. Wellenstein and Fink (Brandeis University, USA) [58] made a proposal for a neutrino mass experiment, with a sensitivity that one can set an upper limit of $m_\nu < 2$ eV or better if the neutrino has zero mass. Their proposal employed a cyrindrical mirror-shape electrostatic spectrometer combined with a source of gaseous ^{3}H molecules.

4 INS Experiment

The INS group intended to reliably determine $R(E)$ by an experimental measurement and not by invoking a Monte Carlo simulation. A unique approach for preparing both the ^{3}H and monoenergetic ^{109}Cd sources of the same quality and same chemical structure made this possible. The whole response function was obtained by a single measurement, not separately evaluating the three components, $R^{OP}(E)$, EL, and BS as done in other experiments. It also eliminated unknown uncertainties mingled in the process of the evaluation of $R(E)$ by convoluting $R^{OP}(E)$ with EL and BS. A series of three measurements was performed and resulted in $m_\nu < 32$ eV (95% C.L.) in 1987 [24], $m_\nu < 29$ eV (95% C.L.) in 1988 [26], and finally $m_\nu < 13$ eV in 1991 [27].

4.1 Experimental Devices

^{3}H and ^{109}Cd sources

Langmuir and Blodgett (LB) [72] invented an excellent method for making a thin organic film. They found that certain kinds of fatty acids spread over the water surface like soap, and under a fine compression form perfectly lining up molecules in a closely packed monomolecular layer with the hydrophobic side of the molecule up and the hydrophilic side down (see Fig.10). A single layer is then transferred onto a substrate by means of a chemical affinity between the molecule and the substrate surface, and successive transferences to this layer form a multi-layer [79]. The thickness and uniformity of the film can be controlled in principle in units of a monomolecular layer of 25 Å and of a molecular size of 20 Å^2, respectively. This is superior to other source preparations such as vacuum evaporation and ion implantation techniques; the uncertainties such as non-uniformities of the thickness and depth profile can essentially be eliminated, and there is in principal no limitation on the film size.

Tritiated [5, 6, 8, 9, 11, 12, 14, 15-^{3}H] icosanoic acid was employed due to its high specific activity – 8 tritium atoms per molecule (see Fig. 10(e)). A two-monolayer source of the size of 60×200 mm^2 was prepared. To the author's knowledge this is the largest size LB film ever made [80,81]. An interesting feature was the use of the acid in the form of Cd-salt, $(C_{20}H_{31}T_8O_2)_2{-}Cd$, where cadmium atoms were bound in the middle of two layers. The advantage of the present compound is that one can use 18.5 keV monoenergetic electrons from the KLL Auger lines[9] of ^{109}Cd to estimate the response function. For this purpose we also made the two-monolayer ^{109}Cd source with the use of a non-radioactive icosanoic acid with ^{109}Cd by applying the LB method. The ^{3}H source, on the other hand, was obtained from the tritiated icosanoic acid with a natural cadmium. The two types of the source had the same chemical structure and the same effective thickness of one monolayer, 0.3 μg/cm^2, so that $R^{OP}(E)$, EL and BS should be the same for both β-rays from the two sources. $R(E)$ was therefore obtained from the measured KLL lines by just deconvoluting contributions of shake up/off effect and the natural line width.

Fig.11(a) illustrates a device developed for the large size film deposition, and Fig.11(b) shows the procedure to build up a film. A very shallow trough of 0.7 mm depth for substrate solution kept the radioactive ^{109}Cd concentration high. Horizontal lifting of the substrate plate avoided unnecessary contamination of the opposite side in contrast to the commonly used vertical dipping method. The mechanical generation of a surface pressure instead of a piston oil made the repeated use of the ^{109}Cd solution possible, because even extremely small chemical impurities in the solution led to failure of perfect deposition. The prepared ^{3}H and ^{109}Cd sources for the third measurement had activities of 40 mCi and 200-500 μCi, respectively. Radioactive uniformity of the sources was examined by autoradiography and photometry. A result of photometry for an exposed X-ray film with

[9] Emission of an L-electron, when a vacancy in the K-shell is filled by another electron from the L-shelland.

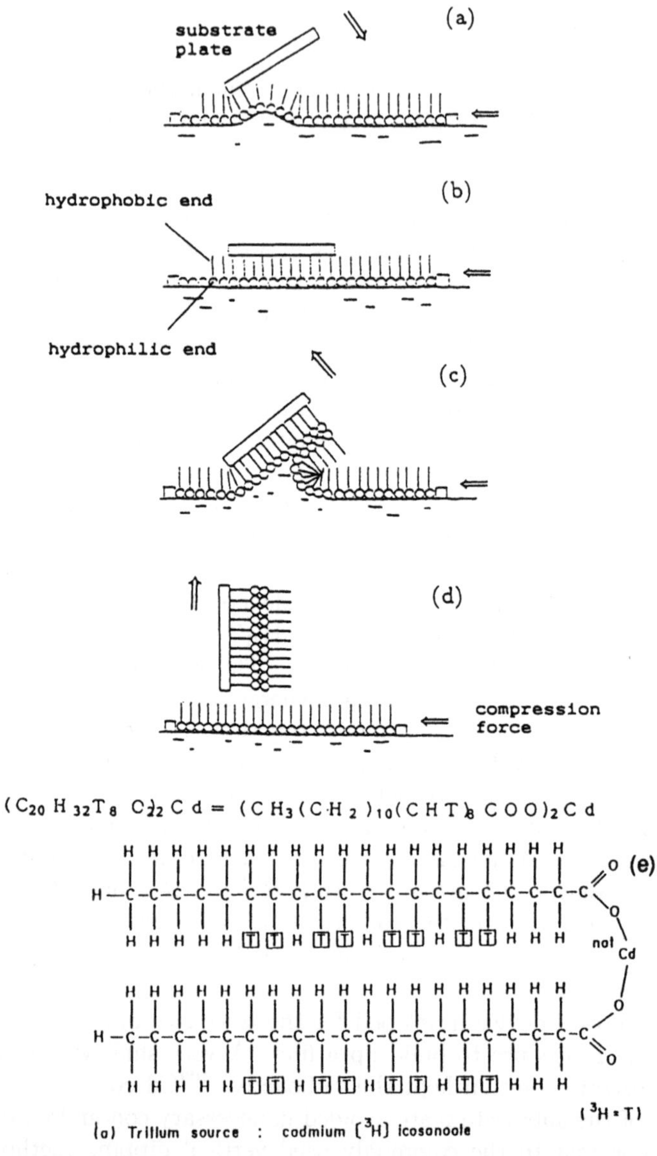

Figure 10: Illustration of building up LB film with horizontal lifting method. A monomolecular layer is formed on the water surface with good ordering as shown in (a) by finely compressing the acid from the right side. Hydrophobic and hydrophilic ends of the acid are illustrated by a solid line and an open circle, respectively. By attaching a chemically treated substrate surface (hydrophobic) onto the acid, a monomolecular layer is built up by means of the chemical affinity between the hydrophobic plate-surface and the acid-end, (b). The next layer is transferred as seen in (c) by the affinity between the hydrophilic ends of the layers. Chemical structure of the tritiated icosanoate used in the INS experiment is illustrated in (e).

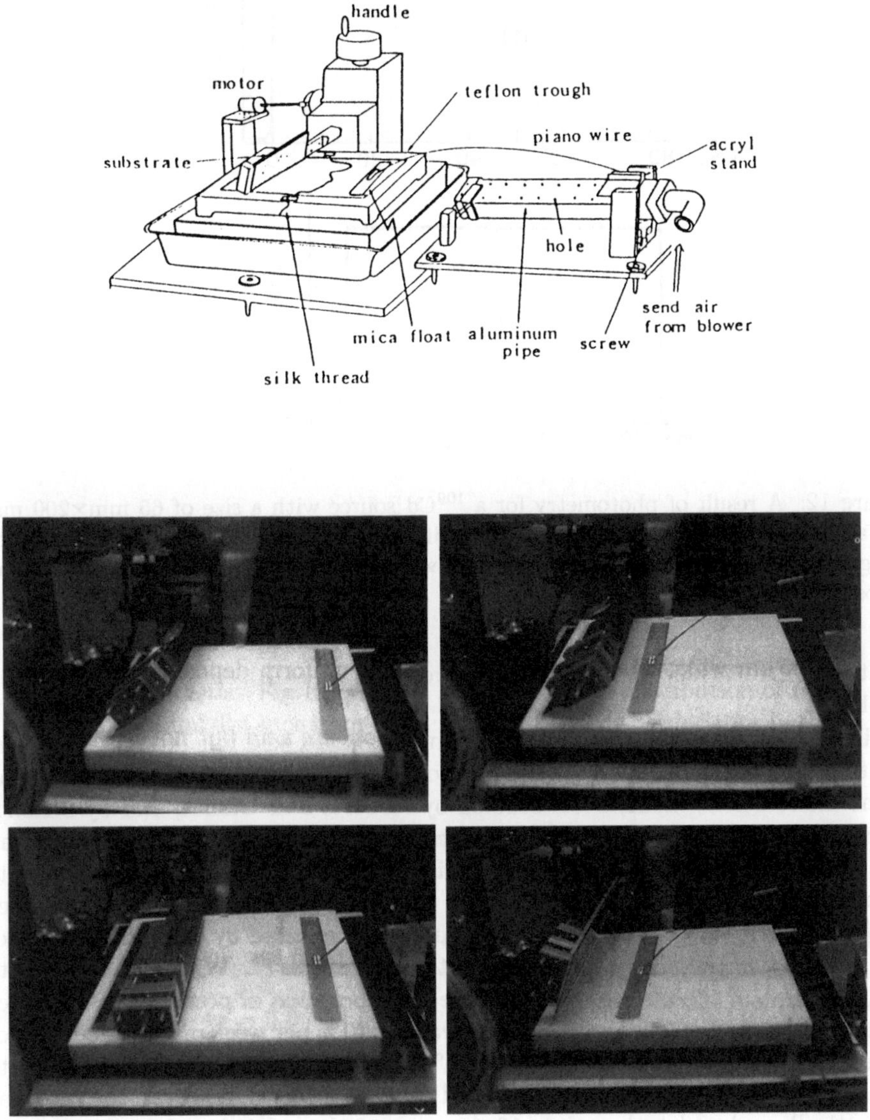

Figure 11: LB film deposition view. The upper-half drawing illustrates the INS device (taken from ref. [80,81]). The actual LB film producing procedure is pictured in the lower half: (1) the top-left picture corresponds to Fig.10(a), (2) the bottom-left to Fig.10(b), (3) the top-right to Fig.10(c), and (4) the bottom-right to Fig.10(d). The movement of the mica float indicates the transformation of the acid layer to the substrate plate.

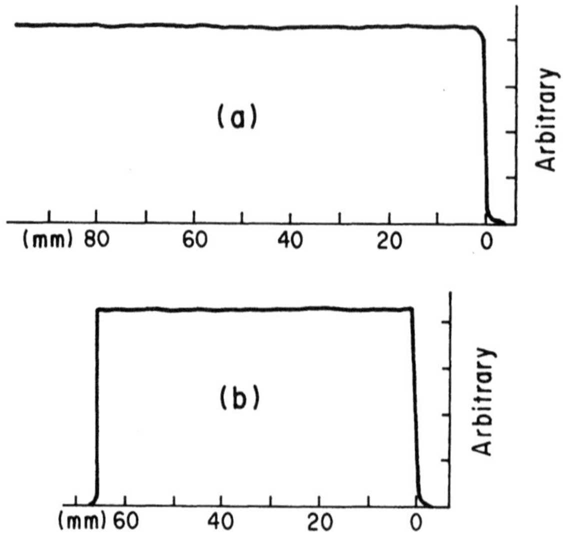

Figure 12: A result of photometry for a ^{109}Cd source with a size of 65 mm×200 mm and activity of 500 μCi. (a) shows radioactivity distribution along the long axis and (b) along the short axis. A slit of 200 μm wide was used in photometry for an X-ray film exposed to the source.

a slit of 200 μm wide, which confirmed excellent uniform deposition, is shown in Fig.12.

The reason for employing the cadmium icosanoate acid but not icosanoic acid was to obtain the monoenergetic electron source as well as to avoid the evaporation of the molecules in vacuum. As studied by Roberts and Gaines, and by Schonohorn, and Spink [82], the LB-film of metal soap is more stable in vacuum than that of the free acid whose molecules easily evaporate in vacuum. The complete formation of the cadmium icosanoate acid was therefore practically important (see ref. [80,81] for details). The evaporation rate was measured by using different specific activities of tritiated acid at different temperatures [10]. With [^{3}H] icosanoate of 5,000 Ci/mole specific activity, the rate of evaporation of prepared test sources each with 2, 4, 6, and 8 layer thicknesses was $\sim$ 0.01% per week under a pressure of 1×10^{-4} Pa and a room temperature of 15°C. In the 200,000 Ci/mole case the rate was $\sim$ 0.01% for 1 day; when it was cooled down to −100°C, however, the rate was dramatically reduced to less than 10^{-4} over a 10 day period and no appreciable time dependence was found [85]. The INS group then decided to cool the sources during the experiments.

Both ^{3}H and ^{109}Cd sources were set vertically in a source chamber and they were interchanged easily with each other.

[10]These studies reported in ref. [83,84] would provide useful data for colloid and interface science, especially because data with such high specific activities are scarce.

Figure 13: Picture of the INS iron-free $\pi\sqrt{2}$ spectrometer. The largest Helmholtz coil pair, above and below a vacuum chamber, can be seen at the center. A part of additional coils to cancel external magnetic field disturbance is also seen at the top and the corners.

$\pi\ \sqrt{2}$ air-core spectrometer

The $\pi\sqrt{2}$ spectrometer (Fig.13) yielded a magnetic field distribution of $B \propto 1/\sqrt{r}$, where r is the radial distance from the center. It focused the electrons both radially and vertically at an angle of $\pi\sqrt{2}$, as illustrated in Fig.14. The INS spectrometer [86] consisted of three sets of Helmholtz coils to form a field with a mean orbital radius of $r_0 = 75$ cm. The main coil current was stabilized to less than 10^{-5}. Three sets of additional Helmholtz coils (east-west, north-south, and up-down) canceled external magnetic fields and their fluctuations down to 0.05 mG, corresponding to 10^{-6} to the field strength of ~ 5 Gauss for an electron at the energy E_0. The optical momentum resolution was determined by a discrimination baffle slit located at 60° from the source. The best achievable resolution was $\Delta p/p = 0.01\%$ (FWHM) with a solid angle of $\Omega/4\pi = 0.09\%$ for a strip source of 250 μm wide. It also provided a wide momentum coverage of $\pm 2\%$ at the focal plane, and then the momentum spectrum could be observed spectrographically with a position sensitive β-detector on the plane.

To overcome the small transmission of the spectrometer, the highly-radioactive and large-sized sources mentioned above were employed with an application of a focusing electric non-equipotential to compensate for optical aberration due to the finite source size. In the case of the combined use of a large source and a position sensitive detector, there were two solutions for the tilt angle relative to the central orbit. One of the solutions allowed a larger size of the source but the momentum resolution varied along the focal plane. The other provided a wider momentum coverage with a fairly uniform resolution. An analytic orbit

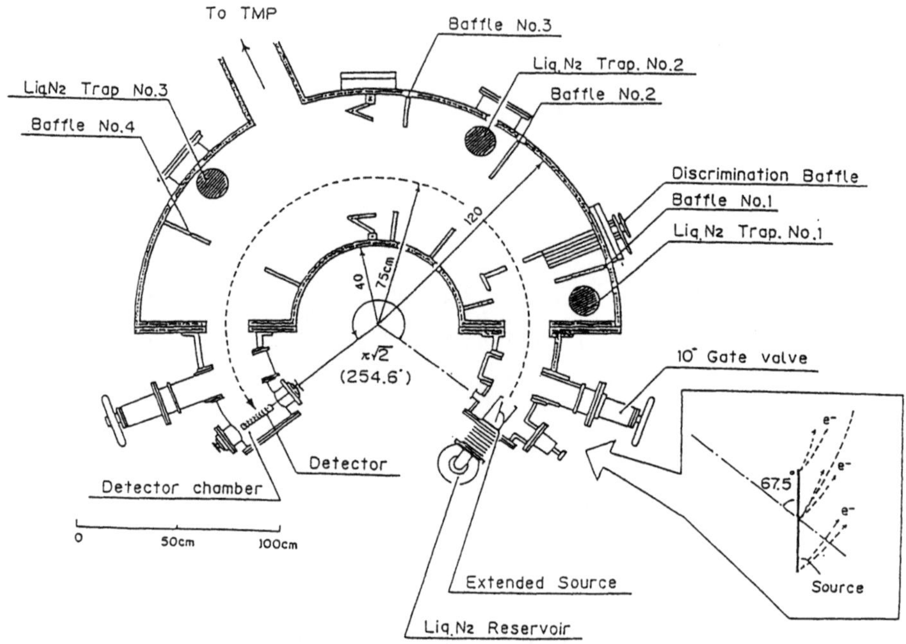

Figure 14: Schematic top view of the INS $\pi\sqrt{2}$ spectrometer.

calculation was performed for the latter case up to 3rd order approximation and the tilt angles were determined to be 67.6° for the source and 4.2° (almost normal incidence) for the detector [87]. The calculation was confirmed by measurements of iso-aberration contours for three different momenta of $\Delta p/p = 0\%$ and $\pm 1.5\%$ with the use of the ^{207}Bi K-conversion line (975 keV). From this study, a baffle slit of $\Delta p/p = 0.01\%$ (FWHM) was designed for the wide extended source and expected line shapes were calculated [88].

The applicability of the non-equipotential method was tested with the use of 5 narrow ^{109}Cd strip sources, each separated 45.5 mm along the source plane with the tilt angle of 67.5°. With the optimized potentials on the individual sources, 5 separated K-line (62 keV) peaks without potential (Fig.15(a)), were well unified into one (Fig.15(b)). The dotted histogram in Fig.15(c) is a sum of the five spectra added by lining up their peak positions and the solid histogram is the spectrum observed with the potentials; no appreciable difference was found. It worked quite satisfactorily with an accuracy of 5×10^{-5} in momentum [89], much better than the required precision.

The non-equipotential was formed by source electrodes and an electrode cage surrounding the extended source as shown in Fig.16, where the exit side of the cage was kept at the ground potential. The 500 μm wide electrodes were vertically printed at every 45 mm interval on a surface of a high purity alumina substrate plate with high-resistive RuO_2 coating . Optimized electric voltage applied on the source electrodes exhibited an approximately parabolic distribution just as predicted by the orbit calculation.

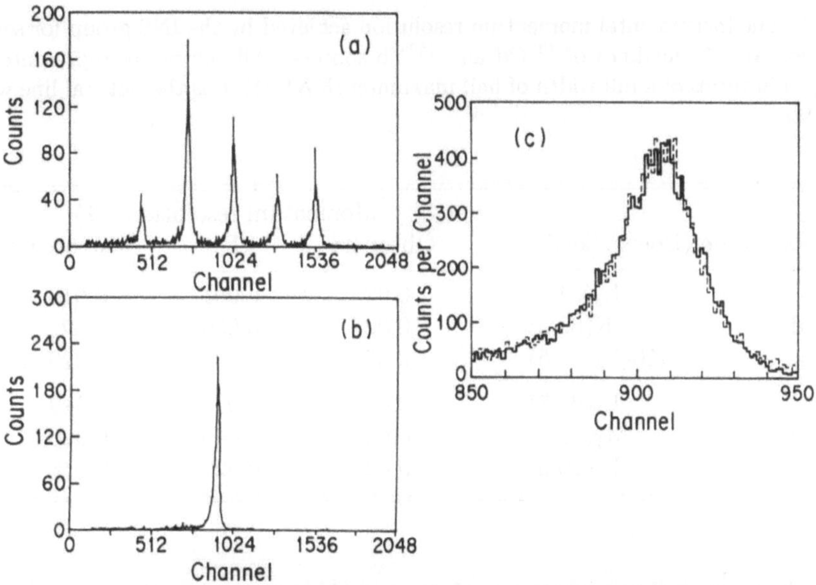

Figure 15: Demonstration of the non-equipotential method applied on five ^{109}Cd sources. (a) and (b) are the observed ^{109}Cd K conversion spectra without and with the non-equipotential on the sources, respectively. The dotted histogram in (c) is the sum of the five spectra in (a) added by lining up individual peak channels, while the solid histogram is the observed spectrum (b). The spectra around the peak are enlarged.

Figure 16: Picture of the source chamber. The central box is the electrode cage, inside which are the two sources (black plates).

Table 4: The instrumental momentum resolution achieved by the INS group for several conversion and Auger lines of ^{109}Cd and ^{169}Yb sources. All values are represented by $\Delta p/p$ (%) in terms of a full width of half maximum (FWHM). Γ is the natural line width (FWHM).

Source	Line (Energy(keV))	Momentum resolution (%) Observed	Γ	Instrumental
^{109}Cd	L_3(84.7)	0.020	0.0012	0.019
	K(62.5)	0.027	0.0067	0.020
	KL_2L_3(18.5)	0.054	0.0340	0.020
^{169}Yb	L_1(99.7)	0.019	0.003	0.019
	M_1(18.4)	0.054	0.040	0.020
	L_1(10.6)	0.038	0.020	0.018

The instrumental momentum resolution was investigated by the measurements of several conversion and Auger lines of ^{109}Cd and ^{169}Yb sources with the application of the non-equipotential method with a 40 cell β-detector as described below. A ^{169}Yb-source of 20×100 mm^2 size was prepared by means of vacuum evaporation; deposition uniformity was less than 8%, the thickness was evaluated to be 0.4 μg/cm^2, and the intensity was 40 μCi [90]. The ^{109}Cd source with two-layer thickness had at the time the same size as the ^{169}Yb source. The results are listed in Table 4. By deconvoluting the respective natural line widths of a Gaussian shape, the instrumental resolution was found to be $\Delta p/p$ = 0.018–0.020% (FWHM), which corresponds to an energy resolution of ΔE = 7–8 eV (FWHM) at $E_0(^3\mathrm{H})$ = 18.6 keV. It agreed very well with the expected value of 0.019% from a Monte Carlo simulation including the effects of source thickness and the finite slit width of the β-detector.

β-detector

A proportional chamber was placed on the focal plane as a position sensitive β-detector for a spectrographic measurement. In the first and second experiments, the detector consisted of 40 isolated cells [2 mm (wide)×50 mm (high)×2 mm (depth)] at 3 mm intervals, each of which had a 500 μm wide slit in front. The chamber gas was isobutane of 6.7 × 10^4 Pa to produce a sufficient number of ionization pairs. A 2 μm thick polyestel film inserted behind the slit plane isolated the gas from the vacuum. In the last experiment, the source size was enlarged to 60 mm×200 mm from 25 mm×100 mm, so that the higher order optical aberration due to the extended source size could not be ignored; this required vertically dividing the detector. A proportional chamber this time consisted of 7 cells [35 mm (wide)×10 mm (high)×4 mm (depth)] placed vertically, and resistive anode wires strung horizontally. The position of incoming β-rays was determined by a charge division readout instead of using slits. The β-rays stopped in isobutane gas at nearly one atmospheric pressure just after passing through an aluminized polyestel film window of 279 μgr/cm^2. The resultant positional resolution was 1.2 mm (FWHM) for 18 keV β-rays. An extra layer of the proportional chamber was used as a veto behind the 7 cells to reject background, which arose mainly from cosmic rays.

4.2 Preliminary Studies

Absolute energy calibration

The KLL Auger line was used to extract $R(E)$ as well as to calibrate the absolute energy and to monitor the system stability in the ^{3}H-spectrum measurement. As the Auger line energies were at the time known with accuracies of only $\pm$60 eV [91], their precise calibration was carried out by the accurately known ^{169}Yb conversion line energies with the momentum resolution of 0.02%. All 9 ^{109}Cd KL$_i$L$_j$ $(i, j = 1-3)$ Auger lines were clearly separated due to the high resolution as seen in Fig.1 in ref. [92]. The absolute energies and relative transition intensities of ^{109}Cd Auger lines were determined within an accuracy of $<$ 1.6 eV and 0.1–0.7%, respectively, which were almost two order of magnitudes more precise than the data from previous measurements. The energy of the KL$_2$L$_3$ line used for $R(E)$ measurement was determined to be $18,511.7 \pm 1.3$ (eV). [11]

Energy-loss spectra

The energy-loss spectrum was studied for a complete understanding of $R(E)$, although it was not really required in this experiment. Due to the source tilting angle of 67°, the average source thickness (0.75 μg/cm^2) was 2.5 times that of a monolayer. KL$_2$L$_3$ spectra were measured for two ^{109}Cd sources: one was the standard type of two-layers and the other had two extra non-active layers. Therefore, EL by the extra two-layers of the cadmium icosanoate was given by the difference between the two spectra. The energy-loss spectrum free from the instrumental resolution, was obtained by deconvoluting the spectrum with two extra layers using the spectrum with standard layers. The resulting spectrum, as shown in Fig.17, exhibited a sharp non-energy-loss peak with a probability of 75%, followed by prominent energy-loss peaks at -23 eV and -43 eV below the original peak and possibly also at -60 eV with 18.9%, 5.9% and 0.7% probabilities, respectively. A simple Possion distribution $(\lambda^n/n!)e^{-\lambda}$ for the probability of multi-interaction of

[11] ^{109}Cd KL$_i$M$_j$ and KL$_i$N$_j$ $(i = 1-3, j = 1-5)$ Auger spectra were also studied over different periods but with the same set-up. Most of the lines in these transitions were well separated and the complex spectral structure was revealed as seen in Fig.1 in ref. [93]. These measurements would provide high quality data for extensive understanding of the complex structure of Auger transitions in atomic physics. The results of measurements suggested the necessity for further theoretical investigations of the relativistic calculation of intermediate couplings with configuration interactions.

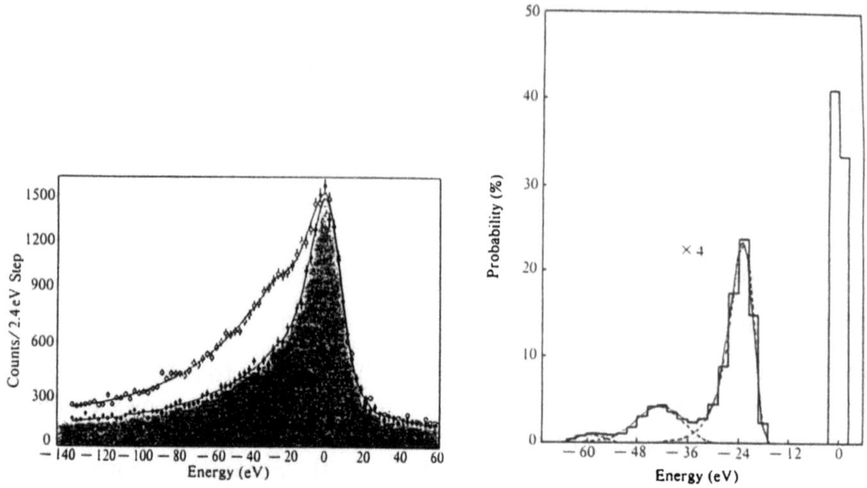

Figure 17: Right: Observed KL_2L_3 spectra for the standard ^{109}Cd source (closed circles) and for the source having an additional non-active two layers (open circles). The spectrum for the latter is normalized to have the same counts as the former at the higher energy side of the peak. The origin of the horizontal axis is set at the peak of the spectra. Left: Energy-loss spectrum. The solid curve is the result of a fit to estimate the probabilities of the energy-loss. The prominent peak at 0 eV is the non-energy-loss component. (taken from ref. [81,94].)

β-rays explained fairly well the observation, where n was the number of interactions and $\lambda = 0.293$ was obtained from the non-energy-loss probability [94]. The energy-loss probability by the effective thickness of 2.5 layers was consequently evaluated to be $\sim 12\%$.

Cooling of the source

In the ^3H-spectrum measurement, the sources were cooled down in order to suppress the evaporation in vacuum, as described in section 4.1. However, the cold sources inevitably adsorbed residual gas onto their surfaces, which consequently caused an origion of another energy-loss. The dependence of the adsorption rate on temperature was measured through changes of the KL_2L_3 spectral shape using the standard two-layer ^{109}Cd source, and the same analysis procedure as explained above was applied employing the observed spectrum at room temperature (24°C) as a reference [95]. With the temperature being reduced, the low-energy tail component of the spectrum increased; the probabilities of the energy-loss were 9, 20 and 48% at −10°C, −35°C, and −60°C, respectively, in the vacuum of 7×10^{-5} Pa. Energy-loss spectra showed peaks at −23 eV, −(40−50) eV and −(60−70) eV below the non-energy-loss peak, which was quite similar to those observed when the extra acid layers were added. It was known from ultraviolet optical measurements of the dielectric constant ε in vacuum [96] that the energy-loss functions $(-\mathrm{Im}(1/\varepsilon))$ of liquid water and ice had a prominent peak at −21 eV. Therefore, it was reasonably assumed that the adsorbed residual gas was mainly of H_2O. It was also observed that the spectral shape responded promptly to a temperature change and was completely reversible with respect to the source temperature. The adsorption could then be due to a physical but not a chemical process. The activation energy, E_d, of desorption was evaluated to be ~16 kcal/mol, which, in fact, was consistent with physical adsorption. As a result of this study, the source temperature was set at −35°C and the electrode cage surrounding the cooled sources at −50°C. The adsorbed gas thickness was equivalent to a 0.7 monolayer thickness of the acid at −35°C.

4.3 Measurement of ^{3}H β-Spectra

Along with preparatory and preliminary work described above, the first experiment was performed in the summer of 1985 [97], almost five years after the proposal in the late 1980 [98]. By accumulating 5K events in a 100 eV region below E_0, an upper limit was set to be $m_\nu < 32$ eV with 95% C.L. [24,99]. Although the energy resolution achieved was superior to those of the other experiments as shown in Fig.18, the mass sensitivity was statistically limited. The second measurement with the same set-up as the first obtained 9K events in the following year. By combining the two data sets, the mass squared of $m_\nu^2 = 223 \pm 244(\text{stat.}) \pm 173(\text{sys.})$ (eV)^2 or an upper limit of $m_\nu < 29$ eV (95% C.L.) was obtained [26,100]. To achieve the statistically allowed maximum mass sensitivity, the source size was then enlarged to 60 mm×200 mm and a 7-cell proportional chamber was used instead of a 40-cell chamber with slits. These improvements increased the counting rate by a factor of ~30 compared with the previous runs, and 150K events were successfully accumulated in 1989. In return for these improvements, however, the energy resolution was deteriorated to $\Delta E = 30$ eV (FWHM) from 10–15 eV in the original set up. This third phase of the experiment then followed.

The electrons from the sources were accelerated by the non-equipotential method with a negative bias voltage, V_{ACC} ~1,500 volts. The ^{3}H β-spectrum was measured with six different settings of V_{ACC}, keeping the magnetic field strength constant: the electron energies at the center of the β-detector were 17.97, 18.24, 18.52, 18.59, 18.69 and 18.79 keV. Data were taken for the period with the ratios of 3 : 9 : 50 : 2 : 2 : 2 corresponding to these six V_{ACC} values with full scans of the spectrum. The energy range at each setting was ~350 eV, and the adjacent settings had overlap regions to check for a smooth continuation. The KL_2L_3 Auger spectrum was also measured daily, with the third V_{ACC} setting, upon completion of the ^{3}H cycle, by replacing the ^{3}H source with the ^{109}Cd source. During the two months of the measurement, no appreciable change was observed in the shape of the spectrum.

The position linearity of the detector was measured by noting the response at various slit settings, and in fact, good linearity was found between the position of the slits and the observed data within an accuracy of 0.1 mm, corresponding to $\Delta p/p = 3 \times 10^{-5}$. By periodically measuring the Auger spectrum with different positions of the β-detector, at 14 values of V_{ACC} (in 20 V steps), the dispersion

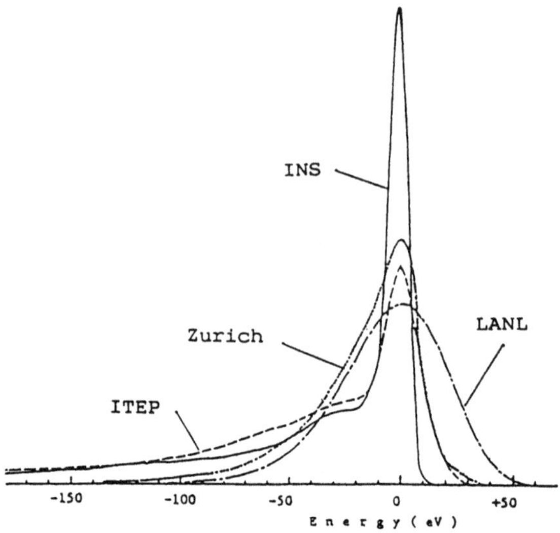

Figure 18: Comparison of the response functions, $R(E)$, among four experiments: INS [24], ITEP [66], Zurich [22], and LANL [25]. The $R(E)$ of Zurich does not include the BS component. The $R(E)$'s are normalized to have the same integral area as each other. (cited from ref. [99].)

relation for the detected position of the electron and its energy was obtained. The efficiency variation on the position of the β-detector was examined by comparing among the Auger peak counts; a slight parabolic behavior of $\leq 2\%$ increase was found for the ends of the each cell and was taken into account in the data analysis.

The background was studied by setting V_{ACC} to the values that corresponded to central β-ray energies of 18.88, 18.98 and 19.38 keV. A small energy dependent component was found, but its rate did not vary during the experiment. By closing the vacuum gate valve just downstream of the source, or upstream of the detector, the origin of the background was shown not to be associated with the β source.

The accumulated number of events, 150,000, are the same or higher statistics compared to those from other experiments [22,23,28]. Data from the first and last cells were not used because of their lower statistics and the data from the remaining five cells were analyzed separately. The background was discriminated from the signal by imposing a cut on the pulse-height distribution. Also, any events that fired the veto counter or several cells were rejected as due to cosmic rays. A small energy dependence of the transmission, the long-term drift of the system, and the dead time of the data acquisition system were corrected for in the analysis. The energy of the ^3H β-spectrum was determined by a comparison with the KL_2L_3 Auger line, and the dispersion was measured to be $\simeq 12.3$ eV/mm. For the final analysis, the β-spectra from the six different V_{ACC} settings for the individual five cells were bound in 5 eV intervals.

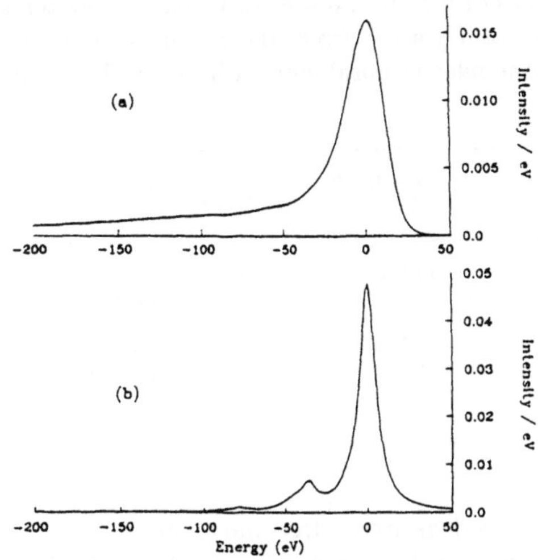

Figure 19: (a) The response function, $R(E)$, of the fourth cell in the 1991 INS experiment [27], which was obtained from the measured KL_2L_3 spectrum via eq.(23). The small but long tail at low energy extended up to -1000 eV. The total intensity was normalized to unity. (b) The shake up/off spectrum, SUO, convoluted with the natural Lorentzian line of width $\Gamma = 11.43$ eV. (taken from ref. [27].)

4.4 Deduction of the Neutrino Mass

The response function $R(E)$ was obtained from the measured KL_2L_3 spectrum as

$$R(E) = (Auger) \otimes \Gamma^{-1} \otimes SUO^{-1}, \tag{23}$$

where -1 denotes the deconvolution procedure [101]. The natural line width Γ was estimated to be 11.4 eV with a $\pm 10\%$ uncertainty from ref. [45], and the shake up/off effect, SUO was taken from a newly refined calculation by Takizawa and Kubodera [102]. The two mechanisms of the Auger transition were considered: one was that associated with a K-vacancy produced in the initial K-capture by ^{109}Cd, and the other was that associated with a K-vacancy produced by a K-conversion emission of the ^{109}Ag metastable nuclear state. The SUO and the resulting $R(E)$ are shown in Fig.19.

The m_ν^2 was deduced by a χ^2 fit of the data to the following formula for each cell:

$$N^j(E) = \eta(\epsilon) A_0 \int R(E, E')[1 + \alpha_2 (E_0 - E')^2] \sum_i \omega^i f_i(E', m_\nu, E_0) dE' + [BG_0 + BG_1(E_0 - E)], \tag{24}$$

$$f_i(E', m_\nu, E_0) = p'W'F(Z, E')(E_0 - E_{ex}^i - E')\sqrt{(E_0 - E_{ex}^i - E')^2 - m_\nu^2}, \tag{25}$$

where $\eta(\epsilon) \equiv 1 + \zeta_1 \epsilon + \zeta_2 \epsilon^2$ with $\epsilon \equiv E - E_C^j$. Instead of using the value obtained from the Auger spectrum, the energy dependent efficiency, $\eta(\epsilon)$, parametrized by

Table 5: Results of the χ^2 fit in the INS 1991 experiment [27]. The second and the third terms of the final values of m_ν^2 and E_0 are statistical and systematic errors, respectively. The systematic error for E_0 arose from uncertainties in the energy calibration of the KL_2L_3 line and from the relative misalignment between the ^3H and ^{109}Cd sources. (taken from ref. [27].)

Cell No.	χ^2/dof	A_0 (10^{-11})	$E_0 - 18500$ (eV)	α_2 $(10^{-8}/eV^2)$	BG_0	BG_1 $(10^{-5}/eV)$	m_ν $(eV)^2$
2	465/345	0.843 ± 0.005	73.1 ± 0.7	2.68 ± 1.16	613.4 ± 9.5	7.6 ± 1.0	-1.1 ± 181.7
3	392/345	1.358 ± 0.009	70.7 ± 0.8	-0.67 ± 1.27	713.6 ± 10.7	6.5 ± 1.0	-37.9 ± 170.1
4	322/345	1.279 ± 0.009	71.4 ± 0.9	-0.96 ±1.30	716.6 ±10.6	8.7 ± 1.0	-59.8 ± 180.6
5	421/345	1.267 ± 0.008	72.9 ±0.8	1.67 ± 1.27	753.3 ±11.3	8.6 ± 1.0	-15.9 ± 165.5
6	334/285	1.291 ± 0.015	71.2 ±1.6	1.75±2.19	718.6 ±10.9	3.7 ±1.1	-765.1 ±380.7

$m_\nu^2 = -65 \pm 85 \pm 65(eV)^2, E_0 = 18572.1 \pm 0.4 \pm 3.0(eV), m_\nu^2 < 166(eV)^2$ or $m_\nu < 13(eV)$

ζ_1 and ζ_2 was introduced as extra free parameters. The superscript j refers to different V_{ACC} settings for ^3H spectra, E_C^j was the central energy of the β-detector at the setting j, and ζ_1 and ζ_2 were common to all six data sets. The quadratic shape correction factor α_2 was required in fitting the data. Free parameters thus consisted of A_0, E_0, α_2, BG_0 and BG_1, ζ_1 and ζ_2, and m_ν^2. There were typically 354 data points used in each cell within the energy range of 17.88 – 18.95 keV.

The final state spectrum was most importantly affected by the nearest neighbors of the tritium atom [41], and, consequently, the acid was expected to have a FSS similar to that calculated for the valine acid by Kaplan et al. [41]. The FSS calculated by Arafune et al. for propane [42] was on the whole very close to those for values in ref. [41]. Since electron correlation effects were not included in the former calculation, the FSS in the latter was considered to be more appropriate and its E_{ex}^i and ω_i were used for the data analysis.

The resulting parameters from the best fit are listed in Table 5, and the corresponding Kurie plots for the individual cells are shown in Fig.20. Fig.21(a,b) and (c) show the typical deviations of the experimental data from the fitted curve, and the relation between χ^2 and m_ν^2 for one of the cells. The resulting η's for individual cells agreed very well with the values deduced from the Auger spectrum. Using the latter η's, the fit with six parameters resulted in consistent values not only for m_ν^2 but also for the other parameters. The sensitivity to m_ν^2 was examined by varying the lowest energy data points to 18.13 keV in the fits; the resulting m_ν^2 values were found to be stable within the statistical accuracy.

Systematic errors in $R(E)$ arising from ambiguities in estimating Γ and SUO were evaluated by changing their values by ±10%. Their effect was to shift the resulting m_ν^2 by an amount of ±34 (eV)^2. The uncertainty from the FSS was difficult to estimate, so that the result was necessarily model dependent. Arafune et al. [42] examined the dependence of the transition probabilities on the basis set by taking a larger number of wave functions for representing the initial and final molecules than those by Kaplan et al. [41], while the latter calculation emphasized the importance of configuration mixing. By referring to the result of ref. [42], ω_i

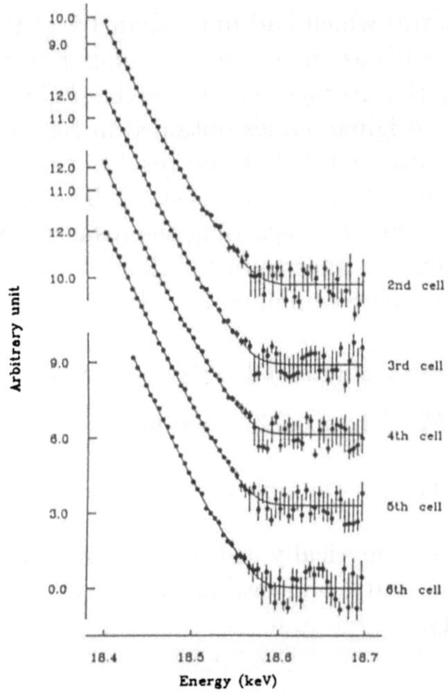

Figure 20: Observed data and the best fit curves are shown for the energy region from 18.4 to 18.7 keV in the form of Kurie plots. The detection efficiency η was corrected for and overlapping data were summed up in the figure. (taken from ref. [27].)

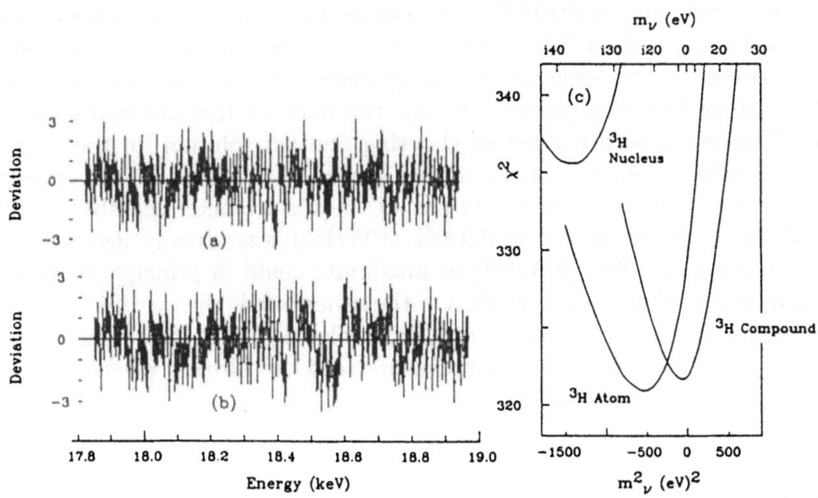

Figure 21: Deviation of the data from the fit divided by the error on each data point, and the relation between χ^2 and m_ν^2 for the case of the fourth cell. (a) shows the best fit case for all eight free parameters including the m_ν^2. (b) The case for the seven free parameters with m_ν^2 fixed at $(30\ \mathrm{eV})^2$. (c) shows χ^2 obtained as a function of m_ν^2 for the full fit to seven free parameters, assuming the *FSS* for the ^{3}H nucleus, ^{3}H atom, and ^{3}H compound as calculated by ref. [41]. (taken from ref. [27].)

of the ground and excited states which had more than 1% of the probability were varied randomly by $\Delta\omega_i = \pm 1\%$ as the systematic uncertainty of the FSS used in the fit. In this process, the average energy of the FSS and the ionization probability were fixed to the original values obtained in ref. [41], namely 18.8 eV and 10.6%, respectively. It was found that the resulting m_ν^2 shifted at most by ± 31 (eV)^2 relative to the value of m_ν^2 for the best fit. The above two systematic uncertainties behaved in the same way regarding the resulting m_ν^2 and were added linearly to yield a total systematic uncertainty of $\Delta m_\nu^2 = \pm 65$ (eV)^2.

The five different data sets were combined by taking a weighted mean average to yield

$$m_\nu^2 = -65 \pm 85(\text{stat.}) \pm 65(\text{sys.}) \quad (\text{eV})^2. \tag{26}$$

The upper limit on m_ν^2 at 95% C.L. was obtained as

$$m_\nu^2 < 166 \ \ (\text{eV})^2 \qquad \text{or} \qquad m_\nu < 13 \ \ \text{eV}, \tag{27}$$

by taking into account only the physically acceptable region of $m_\nu^2 \geq 0$, renormalizing the total probability to unity, and evaluating the m_ν^2 (≥ 0) value yielding 95% of integrated probability.

4.5 Summary

The unique features of the INS sources allowed a reliable evaluation of $R(E)$. The evaporation of ^{3}H source was negligibly suppressed. The superior property of the LB film demonstrated by the INS group was then also employed in the experiments of Zurich [29] and Oxford [55]. In spite of the good optical energy resolution of the INS spectrometer, the large source-tilting angle and the adsorbed gas made the effective source thickness greater and deteriorated the instrumental energy resolution. This could be overcome by choosing another solution for the source and detector configuration, an almost normal source setting and a large detector-tilt angle, and also by improving the vacuum of the spectrometer chamber. While the momentum resolution of $\Delta p/p$=0.020% (FWHM) was already achieved, higher resolution $\Delta p/p$=0.006% (FWHM) at maximum could in principle be attainable according to an orbit analysis [103]. On the other hand, the intrinsic FSS uncertainty arising from the complex compound molecule limited the mass sensitivity. Therefore, we believe that the third measurement had already reached close to its realistic limit on the sensitivity.

5 Discussion

5.1 Definition of the Upper-Limit

The experimentally obtained m_ν^2 is given by a distribution, which may also extend to negative values. If the true neutrino mass square $\tilde{m}_\nu^2$=0, m_ν^2 <0 would happen with a 50% probability [104]. A probability distribution $\mathcal{P}(\tilde{m}_\nu^2)$ obtained as a

Table 6: Upper limits on m_ν^2 (m_ν) at 95% confidence level deduced by different methods discussed in the text. Case-A is the conventional statistical method, case-B is the method recommended by PDG, and case-C is the method described in the text. Total error for the average is obtained by a quadrature sum of the statistical and systematic errors.

Experiment	m_ν^2	Upper-limit on m_ν^2 (m_ν) with 95% confidence level					
		case-A		case-B		case-C	
	$(eV)^2$	$(eV)^2$	(eV)	$(eV)^2$	(eV)	$(eV)^2$	(eV)
INS [27]	-65±85±65	111	(10.5)	166	(13)	176	(13.3)
LANL [28]	-147±68±41	-16.4		87	(9.3)	131	(11.4)
Zurich [29]	-24±48±61	104	(10.2)	137	(11.7)	128	(11.3)
Mainz [31]	-39±34±15	22.1	(4.7)	51.6	(7.2)	61.1	(7.8)
LLNL [30]	-72±41±30	11.6	(3.4)	62.8	(7.9)	83.6	(9.1)
Average	-59.3±25.6	-17.2		24.4	(4.9)	42.1	(6.5)

Gaussian with a mean $\mu = m_\nu^2$ and a variance $\sigma^2 = (\Delta m_\nu^2)^2$ could have a significant probability in a negative unphysical region. No rigorously correct procedure is known to estimate the upper limit on m_ν^2 with a given confidence level. Therefore, the quoted upper limit is biased by a specific method used in the analysis, as discussed by Robertson and Knapp [105], James and Roos [104], and by the Particle Data Group [106]. The most conventional statistical method (referred to as the case-A in Table 6) is to quote an upper limit, for instance, as $m_\nu^2 + 1.645\Delta m_\nu^2$ for 95% confidence level. When m_ν^2 is largely negative, this procedure results in an upper limit smaller than the true experimental uncertainty and moreover provides sometimes an unphysical negative limit. PDG suggests the following approach (case-B): consider only the physical region of the probability distribution $\mathcal{P}(\tilde{m}_\nu^2)$, renormalize it to total of unity, and then find a value of the upper limit on m_ν^2 such that the integral of the distribution from m_ν^2=0 to this limit corresponds to the desired confidence level. This method gives a conservative upper limit in the sense that the true probability of $\tilde{m}_\nu^2$ being in the relevant interval might be larger than that specified by the quoted confidence level. Yet another method often used is to quote a 95% confident upper limit as $m_\nu^2 + 1.645\Delta m_\nu^2$ when m_ν^2 is positive or as $1.645\Delta m_\nu^2$ when m_ν^2 is negative (case-C). The upper limits calculated by these different methods are given in Table 6 for the five recent experiments. In the publications of these recent m_ν experiments [27-31], the method recommended by PDG is used consentaneously.

Because of this uncertainty it is essential to provide the experimental result by the actually observed m_ν^2 value with its error, when the distribution of the variable is approximately Gaussian, as emphasized by James and Roos [104]. This is the unbiased data, and so, different independent results can in principle be combined in terms of the weighed mean average. The average value for the five experiments is

$$m_\nu^2 = -59.3 \pm 25.6 \ (\mathrm{eV})^2, \tag{28}$$

where the statistical and systematic errors are added in quadrature. An upper limit with 95% C.L. is then deduced by the case-B (PDG method) as

$$m_\nu^2 < 24.4 \ (\mathrm{eV})^2 \quad \text{or} \quad m_\nu < 4.9 \ \mathrm{eV}, \tag{29}$$

while the method of case-C provides

$$m_\nu^2 < 42.1 \ (\mathrm{eV})^2 \quad \text{or} \quad m_\nu < 6.5 \ \mathrm{eV}. \tag{30}$$

The central value of eq.(28) is a 2.3 standard deviations away from $m_\nu^2 = 0$ towards the unphysical region. We note that all experiments resulted in negative central values of m_ν^2: -0.6σ by INS, -0.3σ by Zurich, -1.0σ by Mainz, -1.4σ by LLNL and LANL gave the deviation as large as -1.9σ. Probabilistically this is a very unlikely situation. Literally taking these negative values together with a negative mass squared for the muon-neutrino, $m_{\nu_\mu}^2 = -0.097 \pm 0.072 \ (\mathrm{MeV})^2$ [3,107], it was discussed that these particles might have "spacelike energy-momentum dispersion like a tachyon" [108]. The forthcoming third generation experiments should clear up this situation.

5.2 E_0 and ΔM

The end-point energy E_0 is one of the fitting parameters as well as m_ν^2. The correlation between E_0 and m_ν^2 is weak because they appear in the β-spectral shape function in a quite different way. As indicated in eqs.(13) and (14), E_0 is related to the atomic mass difference, ΔM, as

$$\Delta M = E_0 - \kappa, \tag{31}$$

$$\kappa \equiv \Delta E_{B(molecule)} - \Delta E_{B(free)} - E_{rec}, \tag{32}$$

and ΔM is measured also by other methods such as those using the radiofrequency-mass (RFM) spectrometer, the ion-cyclotron-resonance (ICR) spectrometer, and the ^3H implanted Si(Li) detector. Therefore, the resulting ΔM value plays an important role in assessing the reliability of the experiments.

Audi, Graham, and Geiger [109] summarized the data on ΔM and found that the consistent values were obtained within those using the same techniques: the mass doublet RFM [43,110] and ICR [44,69,111] measurements, ^3H β-spectrum measurements by the magnetic spectrometer [15,16,18,66], and those by the Si(Li) detector [19,20,112,113]. When compared among the values with different techniques, however, they show a notable disagreement. Redondo and Robertson [114] pointed out that the discrepancy between the ΔM values measured by the mass doublet method and the Si(Li) detector [113] was reconciled by taking into account the chemical binding effects in the latter. Table 7 and Fig.22 give a summary of ΔM. As is seen in Fig.22 the data scatter more widely than their estimated uncertainties; thus the presence of some systematic errors is suggested. For instance,

Table 7: The ^{3}H - ^{3}He atomic mass difference and E_0.

Year	Experiment	Method	ΔM (eV)	E_0 (eV)
1987	ITEP [21]	β-spect. †1	18,600.3±4	18,580.8±0.4
1991	INS [27]	β-spect. †1	18,592.2±3	18,572.1±3.0
1991	LANL [28]	β-spect. †2	18,590.5 §1	18,570.5±2.0
1992	Zurich [29]	β-spect. †3	18,593.4 §2	18,573.3±0.2 (stat. only)
1992	LLNL [30]	β-spect. †1	18588.5±2.5	
1993	Mainz [31]	β-spect. †2	18,591±3	18,574.8±0.6
1985	Simpson [113]	Si(Li)	18,590±8 §3	
1975	Smith [43]	RFM †3	18,595±7 §4	
1981	Smith [110]	RFM †3	18,579±8 §4	
1984	Lippmaa [44]	ICR †4	18,599±3	
1984	Nikolaev [111]	ICR †4	18,584±4	
1985	Lippmaa [69]	ICR †4	18,599±2	

†1 Iron-type magnetic spectrometer.

†2 Electrostatic spectrometer.

†3 Radiofrequency Mass spectrometer.

†4 Ion Cyclotron Resonance spectrometer.

§1 The ΔM value is not given in ref. [28], thus it was calculated with κ=20.0 eV which was obtained from ref. [115].

§2 The ΔM value is not given in ref. [29], thus it was calculated with κ=20.1 eV which was assumed to be the same as that of the INS measurement.

§3 Value corrected for the chemical shift by ref. [114].

§4 Values re-evaluated with the use of an averaged accurate H_2-D mass difference by ref. [109].

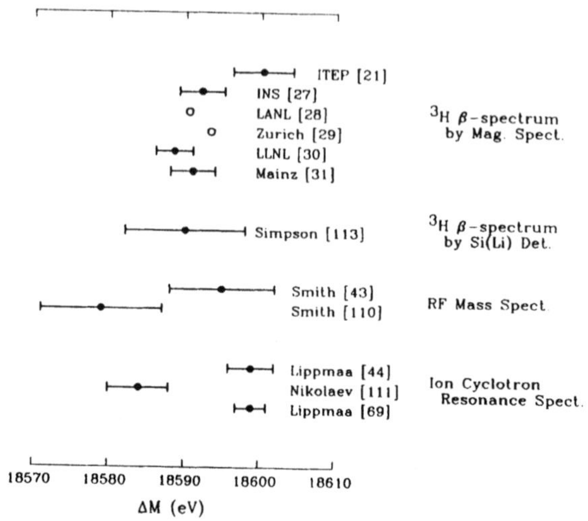

Figure 22: The atomic mass difference, ΔM, obtained by different types of experiments. The open circles for LANL and Zurich experiments are the inferred values indicated in Table 7, since no deduction of ΔM from the measured E_0 value is yet performed.

Audi et al. [109] introduced a "partial consistency factors" of 1.5 and 2.5 to the data of the RFM and ICR measurements, respectively, to estimate more realistic errors. The only obvious statement is that all of the ΔM values from ^{3}H β-spectrum measurements [27-31,113] except for the ITEP data [21] are distributed consistently around 18,590 eV; the ITEP value is 10 eV higher. If $\Delta M \approx 18,590$ eV represents the true value, the "model-independent" lower limit on m_ν reported by the ITEP group might be reduced to approximately 8 eV, instead of 17 eV [21].

5.3 Shape Correction

Although the shape correction term was used by all experimental groups, the meaning imparted to it varied from one group to another in such a way as to absorb the unknown amounts of the energy dependent detection efficiency, or to represent the backscattering amplitude, or to implement the uncertainty of the evaluated low-energy tail component of $R(E)$. It then took either a positive or negative sign, but was mostly of an order of 10^{-5} (eV)$^{-1}$ for α_1 and 10^{-8} (eV)$^{-2}$ for α_2. Its contribution to the β-spectrum amounted to about 1% at an energy of 1 keV below E_0. As it had a smooth energy dependence irrespective of its linear or quadratic form, α correlated only weakly with m_ν^2, but strongly with E_0 in the fit. The influence of the term was made smaller by limiting the fitting region to only around E_0, although the statistical sensitivity on m_ν^2 and E_0 was reduced by this.

Irrespective of the groups' intentions, the shape correction term functioned to absorb any components of the observed spectrum which could not be represented by the ideal formula. Although it is unnatural and unsatisfactory, it is in practice absolutely necessary to introduce the shape correction term to precisely represent the observed high statistics spectrum over the wide energy region without affecting m_ν^2. The size of α reflected the level of understanding by the experimentalists of their own apparatus; the smaller the α was, the better the system was controlled. It would therefore be considered best if the obtained α was consistent with zero within statistical accuracy.

5.4 Background

Not only the signal-to-background ratio restricted the statistical sensitivity of m_ν^2, but also uncertainty in its energy dependence seriously affected m_ν^2. The background in most of the experiments was associated with the presence of the ^{3}H source so that its solo contribution could not be measured. If the spectrometer was polluted with evaporated tritium and efficient filtering elements were not equipped, a strongly energy-dependent background should exist below E_0. There was no way to confirm the correctness of the functional form for the energy-dependent background used in the fit, especially below E_0, since the form must be inferred by extrapolating from above E_0.

5.5 Future Prospect

It seems now obvious to us that the third generation experiment is characterized by two common features: a ^{3}H molecular source in gaseous or frozen form, and an electrostatic spectrometer with or without the retarding potential. The former reduces the uncertainty associated in the FSS and also the importance of the EL component. Furthermore in the gasous case no BS exists. Serious problems in using these types of ^{3}H sources are the necessity of preventing the system from ^{3}H contamination and rejecting the background down to a tolerable level. The use of an electrostatic spectrometer provides a large solid angle with a few eV of energy resolution, which is an essential requirement also from the statistical viewpoint. A control of the systematics, especially the complete understanding of $R(E)$, is required at a more elaborate level than that was made for second generation experiments.

Motivated by the ITEP "discovery" of a finite neutrino mass, much effort has been invested over the last decade on the study of the electron neutrino mass with the result as summarized in Fig. 1. We may hope that an upper limit of several eV will be obtained with high confidence in the near future. However, since the limitation from systematics will be extremely serious problems at the level, it seems almost impossible to achieve a limit as small as 1 eV by the year 2000 as naively anticipated in Fig. 1.

References

[1] Concerning the double β-decay physics, see this book.

[2] Concerning the physics and experiment regarding the neutrino oscillation, see this book.

[3] B. Jeckelmann, T. Nakada, W. Beer, G. de Chambrier, O. Elsenhans, K.L. Giovanetti, P.F.A. Goudsmit, H.J. Leisi, A. Rüetschi, O. Piller, and W. Schwitz, Phys. Rev. Lett. 56, 1444 (1986).

[4] ARGUS Collaboration, Phys. Lett. 292B, 221 (1992).

[5] C.S. Cook, L.M. Langer, and H.C. Price, Jr., Phys. Rev. 73, 548, (1948); ibid. 1395.

[6] S.C. Curran, J. Angus, and A.L. Cockroft, Nature 162, 302 (1948); Phil. Mag. 40, 53 (1949); Phys. Rev. 76, 853 (1949).

[7] G.C. Hanna and B. Pontecorvo, Phys. Rev. 75, 983 (1949).

[8] D.R. Hamilton, W.P. Alford, and L.Gross, Phys. Rev. 83, 215 (1951).

[9] L.M. Langer and R.J.D. Mofatt, Phys. Rev. 88, 689 (1952).

[10] D.R. Hamilton, W.P. Alford, and L.Gross, Phys. Rev. 92, 1521 (1953).

[11] L. Friedman and L. Smith, Phys. Rev. 109, 2214 (1958).

[12] E. Beck and H. Daniel, Zeit. Phys. 216, 229 (1968).

[13] R.C. Salgo and H.H. Staub, Nucl. Phys. A138, 417 (1969).

[14] R. Daris and C. St-Pierre, Nucl. Phys. A138, 545 (1969).

[15] K.-E. Bergkvist, Nucl. Phys. B39, 317 (1972); ibid. 371; Phys. Scripta 4, 23 (1971).

[16] W.F. Piel, Jr., Nucl. Phys. A203, 369 (1973).

[17] E.T. Tretyakov, N.F. Myasoedov, A.M. Apalikov, V.F. Konyaev, and E.G. Novikov, Proceedings of the Int. Neutrino Conference, Aachen, Germany, p.663, June 8-12, 1976, eds. H. Faissner, H. Reithler and P. Zerwas (Vieweg).

[18] V.A. Lubimov, E.G. Novilov, V.Z. Nozik, E.F. Tretyakov and V.S. Kosik, Phys. Lett. B94, 266 (1980); V.A. Lyubimov, E.G. Novilov, V.Z. Nozik, E.F. Tret'yakov, V.S. Kosik, and N.F. Myasoedov, Sov. Phys. JETP 54(4), 616 (1981); V. Lubimov, E. Novikov, V. Nozik, E. Tretyakov, and V. Kosik, Proceedings of Wisconsin Neutrino Mass Mini-Conference, p.7, Telemark, Wisconsin, USA, Oct. 2-4, 1980, eds. V. Barger and D. Cline.

[19] J.J. Simpson, Phys. Rev. D23, 649 (1981).

[20] A.V. Derbin and L.A. Popeko, Yad. Fiz. 38, 1105 (1983) [Sov. J. Nucl. Phys. 38, 665 (1983)].

[21] S. Boris, A. Golutvin, L. Laptin, V. Lubimov, V. Nagovizin, E. Novikov, V. Nozik, V. Soloshenko, I. Tihomirov, and E. Tretjakov, Phys. Lett. 159B, 217 (1985).

[22] M. Fritschi, E. Holzschih, W. Kündig, J.W. Petersen, R.E. Pixley, and H. Stüssi, Phys. Lett. 173B, 485 (1986).

[23] S. Boris, A. Golutvin, L. Laptin, V. Lubimov, V. Nagovizin, V. Nozik, E. Novikov, V. Soloshenko, I. Tihomirov, and E. Tretjakov, Phys. Rev. Lett. 58, 2019 (1987).

[24] H. Kawakami, K. Nishimura, T. Ohshima, S. Shibata, Y. Shoji, I. Sugai, K. Ukai, T. Yasuda, N. Morikawa, N. Nogawa, T. Nagafuchi, F. Naito, T. Suzuki, H. Taketani, M. Iwahashi, K. Hisatake, M. Fujioka, Y. Fukushima, T. Matsuda, and T. Taniguchi, Phys. Lett. 187B, 198 (1987).

[25] J.F. Wilkerson, T.J. Bowles, J.C. Browne, M.P. Maley, R.G.H. Robertson, J.S. Cohen, R.L. Martin, D.A. Knapp, and J.A. Helffrich, Phys. Rev. Lett. 58, 2023 (1987).

[26] H. Kawakami, S. Kato, F. Naito, K. Nishimura, T. Ohshima, S. Shibata, T. Suzuki, K. Ukai, N. Morikawa, N. Nogawa, T. Nagafuchi, H. Taketani, M. Iwahashi, K. Hisatake, Y. Fukushima, T. Matsuda, and T. Taniguchi, J. Phys. Soc. Jpn. 57, 2873 (1988).

[27] H. Kawakami, S. Kato, T. Ohshima, S. Shibata, K. Ukai, N. Morikawa, N. Nogawa, K. Haga, M. Shigeta, Y. Fukushima, and T. Taniguchi, Phys. Lett. 256B, 105 (1991).

[28] R.G.H. Robertson, T.J. Bowles, G.J. Stephenson, Jr., D.L. Wark, J.F. Wilkerson, and D.A. Knapp, Phys. Rev. Lett. 67, 957 (1991).

[29] E. Holzschuh, M. Fritschi, and W. Kündig, Phys. Lett. 287B, 381 (1992).

[30] D.J. Decman and W. Stoeffl, Bull. Am. Phys. Soc. 37, 1286 (1992).

[31] Ch. Weinheimer, M. Przyrembel, H. Backe, H. Barth, J. Bonn, B. Degen, Th. Edling, F. Fischer, L. Fleischmann, J.U. Grooß, R. Haid, A. Hermann, G. Kube, P. Leiderer, Th. Loeken, A. Molz, R.B. Moore, A. Osipowicz, E.W. Otten, A. Picard, M. Schrader, and M. Steininger, Phys. Lett. B300, 210 (1993).

[32] F. Reines, H.W. Sobel, and E. Pasierb, Phys. Rev. Lett. 45, 1307 (1980).

[33] E. Witten, Proceedings of First Workshop on Grand Unification, p.275, New Hampshire, USA, April 10-12, 1980, eds. P.H. Frampton, S.L. Glashow, and A. Yildiz (MATH SCI PRESS); The following proceedings include many discussions on dark matter: Proceedings of XXIII Rencontre de Moriond, Dark Matter, Les Arcs, France, March 8-15, 1988, eds. J. Audouze and J. Tran Thanh Van (Editions Frontieres); Proceedings of XXV Rencontre de Moriond, Particle Astrophysics, The early universe and cosmic structures, Les Arcs, France, March 4-11, 1990, eds. J.-M. Alimi, A. Blanchard, A. Bouquet, F. Martin de Volnay, and J. Tran Thanh Van (Editions Frontieres).

[34] A. Sirlin, Phys. Rev. 164, 1767 (1967); G. Källén, Nucl. Phys. B1, 225 (1967); D.H. Wilkinson and B.E.F. Macefield, Nucl. Phys. A158, 110 (1970).

[35] F.N.D. Kurie, J.R. Richardson and H.C. Paxton, Phys. Rev. 49, 368 (1936).

[36] T. Ohshima, Proceedings of the XXVI Int. Conf. on High Energy Physics, p.1128, Dallas, Texas, USA, Aug. 6-12, 1992, ed. by J.R. Sanford.

[37] Concerning the study on the effects of energy-loss and backscattering in electron spectroscopy in the 1930's and 1940's, see the following references, for instance: C.D. Ellis and W.J. Henderson, Proc. Roy. Soc. A146, 206 (1934); C.S. Cook, L.M. Langer, and H.C. Price, Jr., Phys. Rev. 74, 547 (1948); C.S. Wu and R.D. Albert, Phys. Rev. 75, 315 (1948); L.M. Langer and R.J.D. Moffat, Phys. Rev. 88, 548 (1952). A review on nuclear β-decay in this era can be found in E.J. Konopinski, Rev. Mod. Phys. 15, 209 (1943).

[38] J.J. Simpson claimed in 1985 evidence of a 17 keV heavy neutrino with a 1% admixture in the measurement of ^{3}H β-decay: J.J. Simpson, Phys. Rev. Lett. 54, 1891 (1985). As an explanatory report, for instance; B. Schwarzschild, Phys. Today, 44, 17 (1991). A review on the experimental investigation is also found: D.R.O. Morrison, Proceedings of XXVII Rencontre de Moriond, Progress in Atomic Physics, Neutrinos and Gravitations, p.207, Les Arcs, France, Jan. 25 - Feb. 1, 1992, eds. G. Chardin, O. Fackler, and J. Tran Thanh Van (Editions Frontieres).

[39] For instance, see F.T. Porter, M.S. Freedman, and F. Wagner, Jr. Phys. Rev. C3, 2246 (1971); T.A. Carlson, C.W. Nestor, Jr., T.C. Tucker, and F.B. Malik, Phys. Rev. 169, 27 (1968).

[40] R.L. Martin and J.S. Cohen, Phys. Lett. 110A, 95 (1985); O. Fackler, B. Jeziorski, W. Kolos, H.J. Moukhorst, and K. Szalewicz, Phys. Rev. Lett. 55, 1388 (1985); H. Ågren and V. Carravetta, Phys. Rev. A38, 2707 (1988).

[41] I.G. Kaplan, V.N. Smutny, and G.V. Smelov, Phys, Lett. B112, 417 (1982); Sov. Phys. JETP 57(3), 483 (1983); I.G. Kaplan, G.V. Smelov, and V.N. Smutny, Phys. Lett. B161, 389 (1985).

[42] J. Arafune, N. Koga, K. Morokuma, and T. Watanabe, J. Phys. Soc. Jpn. 55, 3806 (1986); ibid. 3813.

[43] L.G. Smith and A.H. Wapstra, Phys. Rev. C11, 1392 (1975).

[44] É.T. Lippmaa, R.Ĭ. Pikver, É.R. Suurmaa, Ya.O. Past, Yu.Kh. Puskar, I.A. Koppel', and A.A. Tammik, JETP Lett. 39, 646 (1984).

[45] As a textbook, see K.D. Sevier, Low Energy Electron Spectrometry (Wiley-Interscience, New York, 1971).

[46] E.J. Konopinski, Phys. Rev. 72, 518 (1947).

[47] J.R. Pruett, Phys. Rev. 73, 1219 (1948).

[48] J.S. Allen, The Neutrino (Princeton University Press, Princeton, New Jerseu, 1958).

[49] J.F. Wilkerson, Proceedings of the Salt Lake City Meeting, p.339, Utah, USA, Jan. 14-17, 1987, eds. C. Detar and J. Ball (World Scientific, Singapore).

[50] M.H. Shaevitz, Proceedings of the Int. Symp. on Lepton and Photon Interactions at High Energies, p.132, Ithaca, New York, USA, Aug. 4-9, 1983, eds. D.G. Cassel and D.L. Kreinick.

[51] H. Daniel, K.-H. Hiddemann, and O. Schwentker, Proceedings of XVI INS Int. Symp. Neutrino Mass and Related Topics, p.125, Tokyo, Japan, March 16-18, 1988, eds. S. Kato and T. Ohshima (World Scientific, Singapore).

[52] O. Fackler, B. Balke, M. Mugge, and R.M. White, Proceedings of XVI INS Int. Symp. Neutrino Mass and Related Topics, p.143, Tokyo, Japan, March 16-18, 1988, eds. S. Kato and T. Ohshima (World Scientific, Singapore).

[53] B. Sur, R.N. Boyd, M.S. Islam, J. Spizuoco, and P. Koncz, Proceedings of the Telemark IV, Neutrino Masses and Neutrino Astrophysics, p.1, Wisconsin, USA, 1987, eds. V. Barger, F. Halzen, M. Marshak, and K. Olive (World Scientific, Singapore).

[54] M.A. Frisch, G.J. Clark, and P. Chaudhari, Proceedings of the Telemark IV, Neutrino Masses and Neutrino Astrophysics, p.7, Wisconsin, USA, 1987, eds. V. Barger, F. Halzen, M. Marshak, and K. Olive (World Scientific, Singapore).

[55] N.A. Jelley, S.S. Williams, P.A. Holmes, R.Alizadeh, and P.A. Le Bas, Proceedings of the Telemark IV, Neutrino Masses and Neutrino Astrophysics, p.12, Wisconsin, USA, 1987, eds. V. Barger, F. Halzen, M. Marshak, and K. Olive (World Scientific, Singapore).

[56] R.L. Graham, M.A. Lone, H.R. Andrews, J.L. Gallant, J.S. Geiger, G.E. Lee-Whiting, and G. Audi, Proceedings of the 7th Int. Conf. Atomic Masses and Fundamental Constants, AMCO-7, p.646, Darmstadt-Seeheim, Germany, Sep. 3-7, 1984, ed. O. Klepper.

[57] K.E. Bergkvist, Proceedings of the 4th Moriond Workshop, Massive Neutrinos in Astrophysics and in Particle Physics, p.243, Savoie, France, Jan. 15-21, 1984, ed. J. Tran Thanh Van (Editions Frontieres).

[58] M. Fink and H.F. Wellenstein, private communication.

[59] H. Sun, D. liang, S. Chen, G. Si, N. Mao, Q. Sun, H. Du, Z. Li, S. Liang, Z. Hu, Z. Chen, C. Yan, Y. Zhao, C. Yang, Q. Yang, and Z. Mei, Proceedings of the Int. Symp. Nuclear Beta Decays and Neutrino, p.322, Osaka, Japan, June, 1986, eds. T. Kotani, H. Ejiri, and E. Takasugi (World Scientific, Singapore).

[60] R. Shang, T. Zhang, F. Lei, H. He, Y. Zha, Z.Zhao, Z. Xu, S. Xu, and H. Ding, Proceedings of the Int. Symp. on Nuclear Beta Decays and Neutrino, p.329, Osaka, Japan, June, 1986, eds. T. Kotani, H. Ejiri, and E. Takasugi (World Scientific, Singapore).

[61] V.M. Lobashev and P.E. Spivak, Nucl. Instr. and Methods, A240, 305 (1985).

[62] E.F. Tret'yakov, Izv. Akad. Nauk USSR, Ser. Fiz. 39, 583 (1975).

[63] V.V. Vladimirskii, E.K. Tarasov, and Yu.V. Trebuhkovskii, Pribory i Tekhnika Eksperimenta, 1, 13 (1956).

[64] J.J. Simpson, Phys. Rev. D30, 1110 (1984).

[65] S. Boris, A. Golutvin, L. Laptin, V. Lubimov, V. Nagovizin, E. Novikov, V. Nozik, V. Soloshenko, I. Tichomirov, and E. Tretjakov, Proceedings of Int. Europhysics Conf. on High Energy Physics, p.386, Brighton, UK, July, 20-27, 1983, eds. J. Guy and C. Costain.

[66] S. Boris, A. Golutvin, L. Laptin, V. Lubimov, V. Nagovizin, E. Novikov, V. Nozik, V. Soloshenko, I. Tichomirov, and E. Tretjakov, Proceedings of XXII Int. Conf. on High Energy Physics, Vol.I, p.259, Leipzig, Germany, July 19-25, 1984, eds. A. Meyer and E. Wieczorek.

[67] K.E. Bergkvist, Phys. Lett. 154B, 224 (1985).

[68] K.E. Bergkvist, Phys. Lett. 159B, 408 (1985).

[69] E. Lippmaa, R. Pikver, E. Suurmaa, J. Past, J. Puskar, I. Koppel, and A. Tammik, Phys. Rev. Lett. 54, 285 (1985).

[70] G.G. Ross, B. Terreault, B. Gobeil, G. Abel, C. Boucher, and G. Veilleux, J. Nucl. Mat. 128, 730 (1984).

[71] W. Kündig, M. Fritschi, E. Holzschih, J.W. Petersen, R.E. Pixley, and H.Stüssi, Proceedings of the Int. Symp. on Nuclear Beta Decays and Neutrino, p.283, Osaka, Japan, June, 1986, eds. T. Kotani, H. Ejiri, and E. Takasugi (World Scientific, Singapore).

[72] K.B. Blodgett, J. Am. Chem. Soc. 56, 1007 (1935); K.B. Blodgett and I. Langmuir, Phys. Rev. 51, 964 (1939).

[73] T.A. Claxton, S. Schafroth, and P.F. Meier, Phys. Rev. A45, 6209 (1992).

[74] J.F. Wilkerson, T.J. Bowles, D.A. Knapp, M.P. Maley, and R.G.H. Robertson, Proceedings of VIth Moriond Workshop, '86 Massive Neutrinos in Astrophysics and in Particle Physics, p.483, Tignes, France, Jan.25 - Feb.1, 1986, eds. O. Fackler and J. Tran Thanh Van (Editions Frontieres).

[75] H. Backe, J. Bonn, Th. Edling, H. Fischer, A. Hermanni, P. Leiderer, Th. Loeken, R.B. Moore, A. Osipowicz, E.W. Otten, and A. Picard, Physica Scripta. T22, 98 (1988); A. Picard, H. Becke, J. Bonn, B. Degen, R. Haid, A. Hermanni, P. Leiderer, A. Osipowicz, E.W. Otten, M. Przyrembel, M. Schrader, M. Steininger, and Ch. Weinheimer, Z. Phys. A342, 71 (1992); A. Picard, H. Becke, H. Barth, J. Bonn, B. Degen, Th. Edling, R. Haid, A. Hermanni, P. Leiderer, Th. Loeken, A. Molz, R.B. Moore, A. Osipowicz, E.W. Otten, M. Przyrembel, M. Schrader, M. Steininger, and Ch. Weinheimer, Nucl. Instr. and Methods, B63, 345 (1992).

[76] W. Kolos, B. Jeziorski, J. Rychlewski, K. Szalewicz, H.J. Monkhorst, and O. Fackler, Phys. Rev. A37, 2297 (1988).

[77] W. Stoeffl, D. Decman, and J. Engelage, Proceedings of XVI INS Int. Symp. Neutrino Mass and Related Topics, p.135, Tokyo, Japan, March 16-18, 1988, eds. S. Kato and T. Ohshima (World Scientific, Singapore).

[78] O. Fackler, M. Mugge, H. Sticker, R. White, and R. Woerner, Proceedings of Neutrino Mass and Low Energy Weak Interactions, p.48, Telemark, Wisconsin,

USA, 1984, eds. V. Barger and D. Cline (World Scientific, Singapore); O. Fackler, B. Jeziorski, W. Kolos, H. Monkhorst, M. Mugge, H. Sticker, K. Szalewicz, R.M. White, and R. Woerner, Proceedings of Rencontres de Moriond, Perspective of Electroweak Interactions, p.313, Les Arcs, France, March 17-23, 1985 ed. J. Tran Thanh Van (Editions Frontieres); O. Fackler, M. Mugge, H. Sticker, R.M. White, and R. Woerner, Proceedings of '86 Massive Neutrinos in Astrophysics and in Particle Physics, p.529, Tignes, France, Jan.25 - Feb.1, 1986, eds. O. Fackler and J. Tran Thanh Van (Editions Frontieres).

[79] As a textbook regarding to LB film, see for instance; Langmuir-Blodgett Films, ed. G. Roberts (Plenum Press, New York).

[80] S. Shibata and K. Haga, Nippon Kagaku Kaishi, (12), 1978 (1989), (in Japanese).

[81] S. Shibata and K. Haga, Nippon Butsuri Gakkaishi, Vol.45, No.3, p.178 (1990) (in Japanese).

[82] R.W. Roberts and G.L.Jr. Gaines, Trans. Ninth Natl. Vac. Symp., Amer. Vac. Soc., 515 (1962); Nature (London) 197, 787 (1963); H.J. Schonohorn, J. Colloid Sci. 18 (1963) 445; J.A. Spink, J. Colloid Interface Sci. 24, 61 (1967).

[83] M. Iwasaki, T. Yasuda, Y. Shoji, T. Ohshima, H. Kawakami, S. Shibata, K. Ukai, I. Sugai, K. Nishimura, F. Naito, T. Suzuki, T. Nagafuchi, N. Morikawa, N. Nogawa, T. Matsuda, Y. Fukushima, T. Taniguchi, M. Fujioka, and K. Hisatake, J. Colloid Interface Sci., 122, 9 (1988).

[84] M. Iwasaki, F. Naito, N. Watanabe, T. Seimiya, N. Morikawa, N. Nogawa, T. Ohshima, H. Kawakami, K. Ukai, I. Sugai, S. Shibata, T. Yasuda, Y. Shoji, T. Suzuki, T. Nagafuchi, H. Taketani, T. Matsuda, Y. Fukushima, M. Fujioka, and K. Hisatake, Bull. Chem. Soc. Jpn. 58, 2093 (1985).

[85] F.Naito, T.Suzuki, H.Taketani, M.Iwahashi, N.Morikawa, N.Nogawa, Y.Shoji, M.Fujioka, S.Shibata, I.Sugai, H.Kawakami, T.Yasuda, K.Ukai, and T.Ohshima, Genshikaku Kenkyu, Vol.30, No.5, p.5 (1985), (in Japanese).

[86] M. Fujioka and K. Hisatake, preprint, INS-TL-89 (1966).

[87] Y. Shoji, H. Kawakami, K. Hisatake, M. Fujioka and T. Ohshima, preprint, INS-TL-144 (1983) (in Japanese).

[88] Y. Shoji, H. Kawakami and T. Ohshima, Genshikaku Kenkyu, Vol.28, No.1, p.47 (1984), (in Japanese).

[89] T. Suzuki, F. Naito, H. Taketani, Y. Shoji, M. Fujioka, M. Iwahashi, N. Morikawa, K. Hisatake, T. Yakuda, H. Kawakami, S. Shibata, I. Sugai, K. Ukai, and T. Ohshima, Genshikaku Kenkyu, Vol.30, No.5, p.29 (1985), (in Japanese).

[90] H. Kawakami, K. Nishimura, T. Ohshima, S. Shibata, Y. Shoji, I. Sugai, K. Ukai, T. Yasuda, N. Morikawa, N. Nogawa, T. Nagafuchi, F. Naito, T. Suzuki, H. Taketani, M. Iwahashi, K. Hisatake, M. Fujioka, Y. Fukushima, T. Matsuda, and T. Taniguchi, Jpn. J. App. Phys., 27, 1488 (1988); Y. Shoji, M. Fujioka, S. Shibata, I. Sugai, H. Kawakami, T. Yasuda, K. Ukai, T. Ohshima, N. Morikawa, M. Iwahashi, F. Naito, T. Suzuki, H. Taketani, and K. Hisatake, Genshikaku Kenkyu, Vol.30, No.5, p.13 (1985), (in Japanese).

[91] F.A. Johnson and J.S. Foster, Can. J. Phys. 31, 469 (1953).

[92] H. Kawakami, K. Nishimura, T. Ohshima, S. Shibata, Y. Shoji, I. Sugai, K. Ukai, T. Yasuda, N. Morikawa, N. Nogawa, T. Nagafuchi, F. Naito, T. Suzuki, H. Taketani, M. Iwahashi, K. Hisatake, M. Fujioka, Y. Fukushima, T. Matsuda, and T. Taniguchi, Phys. Lett. 121A, 414 (1987).

[93] H. Kawakami, S. Kato, F. Naito, K. Nishimura, T. Ohshima, S. Shibata, T. Suzuki, K. Ukai, N. Morikawa, N. Nogawa, T. Nagafuchi, H. Taketani, M. Iwahashi, K. Hisatake, Y. Fukushima, T. Matsuda, and T. Taniguchi, Phys. Lett. 127A, 175 (1988).

[94] H. Kawakami, K. Nishimura, T. Ohshima, S. Shibata, Y. Shoji, I. Sugai, K. Ukai, T. Yasuda, N. Morikawa, N. Nogawa, T. Nagafuchi, F. Naito, T. Suzuki, H. Taketani, M. Iwahashi, K. Hisatake, M. Fujioka, Y. Fukushima, T. Matsuda, and T. Taniguchi, Nucl. Inst. and Methods, A264, 511 (1988).

[95] H. Kawakami, K. Nishimura, T. Ohshima, S. Shibata, Y. Shoji, I. Sugai, K. Ukai, T. Yasuda, N. Morikawa, N. Nogawa, T. Nagafuchi, F. Naito, T. Suzuki, H. Taketani, M. Iwahashi, K. Hisatake, M. Fujioka, Y. Fukushima, T. Matsuda, and T. Taniguchi, Jpn. J. App. Phys. 28, 883 (1989).

[96] J.M. Heller Jr., R.N. Hamm, R.D. Birkhoff, and L.R. Painter, J. Chem. Phys. 60, 3483 (1974); M.W. Williams, R.N. Hamm, E.T. Arakawa, L.R. Painter, and R.D. Birkoff, Int. J. Radiat. Phys. Chem. 7, 95 (1975).

[97] H. Kawakami, K. Nishimura, T. Ohshima, S. Shibata, Y. Shoji, I. Sugai, K. Ukai, T. Yasuda, N. Morikawa, N. Nogawa, T. Nagafuchi, F. Naito, T. Suzuki, H. Taketani, M. Iwahashi, K. Hisatake, M. Fujioka, Y. Fukushima, and T. Matsuda, Proceedings of 1985 INS Int. Symp. on Composite Models of Quarks And Leptons, p.115, Tokyo, Japan, Aug. 13-15, 1985, eds. H. Terazawa and M. Yasue.

[98] T. Ohshima, Proceedings of the Japan-Italy Symp. on Fundamental Physics, p.244, Tokyo, Japan, Jan. 27-30, 1981, eds. S. Fikui and T. Toyoda; Soryuushiron Kenkyu, Vol.62, No.1, p.B-23 (1980), (in Japanese).

[99] T. Ohshima, Proceedings of Seventh Workshop on Grand Unification / ICOBAN'86, p.187, Toyama, Japan, Apr. 16-18, 1986, ed. J. Arafune (World Scientific, Singapore); T. Ohshima, Proceedings of the 12th Int. Conf. on Neutrino Physics and Astrophysics, p.36, Sendai, Japan, June, 3-8, 1986, eds. T. Kitagaki and H. Yuta (World Scientific, Singapore); T. Ohshima, Proceedings of Telemark IV, Neutrino Masses and Neutrino Astrophysics, p.488, Wisconsin, USA, 1987, eds. V. Barger, F. Halzen, M. Marshak, and K. Olive (World Scientific, Singapore).

[100] S. Kato, H. Kawakami, F. Naito, K. Nishimura, T. Ohshima, S. Shibata, T. Suzuki, K. Ukai, N. Morikawa, N. Nogawa, T. Nagafuchi, H. Taketani, M. Iwahashi, K. Hisatake, Y. Fukushima, T. Matsuda, and T. Taniguchi, Nucl. Phys. A478, 433c (1988).

[101] T.J. Kennett, W.V. Prestwish and A. Robertson, Nucl. Instr. and Methods 151, 414 (1987).

[102] M. Takizawa and K. Kubodera, Proceedings of XVI INS Int. Symp. on Neutrino Mass And Related Topics, p.349, Tokyo, Japan, March 16-18, 1988, eds. S. Kato and T. Ohshima (World Scientific, Singapore).

[103] H. Kawakami, S. Kato, F. Naito, K. Nishimura, T. Ohshima, S. Shibata, T. Suzuki, K. Ukai, N. Morikawa, N. Nogawa, T. Nagafuchi, H. Taketani, M. Iwahashi, K. Hisatake, Y. Fukushima, T. Matsuda, and T. Taniguchi, Proceedings of XVI INS Int. Symp. on Neutrino Mass And Related Topics, p.82, Tokyo, Japan, March 16-18, 1988, eds. S. Kato and T. Ohshima (World Scientific, Singapore).

[104] F. James and M. Ross, Phys. Rev. D44, 299 (1991).

[105] R.G.H. Robertson and D.A. Knapp, Ann. Rev. Nucl. Part. Sci. 38, 185 (1988).

[106] Particle Data Group, Phys. Rev. D45, III.39 (1992).

[107] R. Abela, M. Daum, G.H. Eaton, R. Frosch, B. Jost, P.-R. Kettle, and E. Steiner, Phys. Lett. 146B, 431 (1984).

[108] A. Chodos, A.I. Hauser, and V.A. Kostelecký, Phys. Lett. 150B, 431 (1985); A. Chodos, V.A. Kostelecký, R. Potting, and E. Gates, Mod. Phys. Lett. A7, 467 (1992).

[109] G. Audi, R.L. Graham, and J.S. Geiger, Z. Phys. A321, 533 (1985).

[110] L.G. Smith, E. Koets, and A.H. Wapstra, Phys. Lett. 102B, 114 (1981).

[111] E.N. Nikolaev, Yu.I. Neronov, M.V. Gorshkov, and V.L. Tal'rose, JETP Lett. 39, 535 (1984).

[112] W.R. Dixon, W.F. Davidson, R.S. Storey, and D.M. Rehfield, Phys. Lett. B108, 265 (1982).

[113] J.J. Simpson, W.R. Dixon, and R.S. Storey, Phys. Rev. C31, 1891 (1985).

[114] A. Redondo and R.G.H. Robertson, Phys. Rev. C40, 368 (1989).

[115] S.T. Staggs, R.G.H. Robertson, D.L. Wark, P.P. Nguyen, J.F. Wilkerson, and T.J. Bowles, Phys. Rev. C39, 1503 (1989).

[116] T. Ohshima, Physics Monthly, Vol.7, No.9, p.589 (Kaiyo Shuppan) in Japanese; E. Holzschuh, Rep. Prog. Phys. 55, 1035 (1992).

Double Beta Decays and Neutrinos

Hiroyasu Ejiri

Department of Physics and Research Center for Nuclear Physics,
Osaka University

1 Introduction

Interest in double beta decays ($\beta\beta$) has recently increased in nuclear and particle physics. Within the framework of the minimal $SU(2)_L \times U(1)$ standard theory, double β decay is followed by two neutrino (2ν) emission, and the lepton number (L) is conserved. It is written as

$$A \rightarrow B + \beta + \beta + \nu + \nu. \tag{1}$$

Here, the ν's are the electron anti neutrino and the electron neutrino in the cases of β^- (electron) decay and β^+ (positron) decay, respectively. Double beta decay with two neutrinos is called "two neutrino double beta decay ($2\nu\beta\beta$)".

Double beta decay without neutrinos violates the lepton-number conservation law, and is beyond the standard theory. It is written as

$$A \rightarrow B + \beta + \beta. \tag{2}$$

This is called "neutrino-less double beta decay ($0\nu\beta\beta$)". The $0\nu\beta\beta$ process is indeed a very sensitive probe for studying the properties of the neutrino as well as those of the weak interactions beyond the standard theory.

The $\beta\beta$ decay scheme is illustrated in Fig. 1. Since the single beta-decay rate is much larger than the $\beta\beta$ decay rate, experimental studies of $\beta\beta$ decay are feasible only in the case that a single beta decay is energetically forbidden. In some even-even nuclei of $A(Z, N)$ and $B(Z \pm 2, N \mp 2)$, with Z and N being the proton and neutron numbers, their masses/energies (m_A and m_B) are smaller than the mass (m_C) of the neighboring odd-odd nucleus of $C(Z \pm 1, N \mp 1)$, because the pairing interaction pushes down the energies of the 0^+ ground states in the even-even nuclei (see Fig. 1). Thus, whereas the single beta decay $A \rightarrow C$ is forbidden, the double beta decay $A \rightarrow B$ is allowed.

The half-lives for typical $\beta\beta$ decay neuclei are of the order of or longer than 10^{19}y (years), which is far longer than the age (10^{10}y) of the universe. Thus, $\beta\beta$ isotopes are naturally existing isotopes with finite abundance ratios.

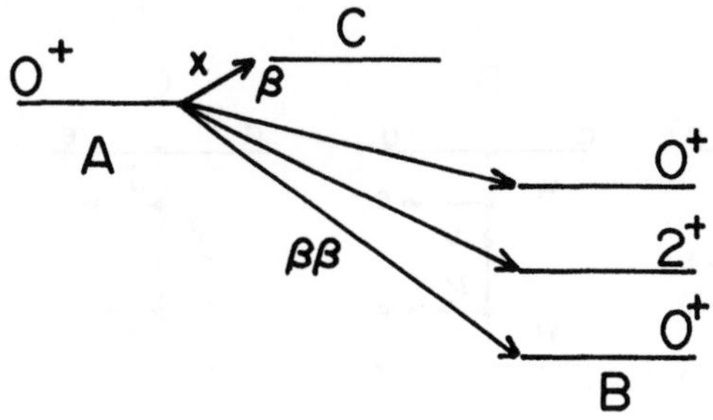

Figure 1: Level scheme of the $\beta\beta$ decay $A \rightarrow B$. The single β decay $A \rightarrow C$ is energetically forbidden.

The theoretical and experimental aspects of $\beta\beta$ decay have been given in several review articles [1]–[7].

In what follows we briefly describe theories concerning $\beta\beta$ decay (2), the nuclear and detector sensitivities for $\beta\beta$ decay (3), experiments concerning $\beta\beta$ decay (4), nuclear matrix elements for $\beta\beta$ decay (5) and concluding remarks (6). A part of the present article is based on a recent review [6].

2 Theories of Double Beta Decay

Decay diagrams for the $2\nu\beta\beta$ and $0\nu\beta\beta$ processes are schematically shown in Fig. 2. Diagram A is the $2\nu\beta\beta$ process in the two-neutron ($2n$) mode, where two-neutrons (n_1 and n_2) in one nucleus decay as $n_1 \rightarrow p_1 + \beta_1^- + \bar{\nu}_1$ and $n_2 \rightarrow p_2 + \beta_2^- + \bar{\nu}_2$. Diagrams B, C and D show the $0\nu\beta\beta$ processes in three different modes. In diagram B, the antineutrino emitted from one neutron (d-quark in one neutron) is reabsorbed by the other neutron (d-quark in the other neutron). This is a neutrino exchange process between two neutrons. Thus, the neutrino cannot be a Dirac particle, but a Majorana particle with a finite mass (m_ν). The helicity-matching condition in this process leads to a finite mass term $\langle m_\nu \rangle$ and/or a finite right-handed current $\langle$RHC$\rangle$ ($\langle V + A \rangle$). In diagram C, one of the super symmetric particles (SUSY) is exchanged between the two neutrons (d-quarks). Here, both lepton-number violating interaction and R-parity violation are required. In diagram D, the mass-less Goldston boson arising from a spontaneous breaking of the lepton number is emitted in the neutrino exchange process. Theoretical discussions of these $0\nu\beta\beta$ processes are given in refs. [8]–[19].

The Hamiltonian for the $0\nu\beta\beta$ process with a ν-exchange between two d-quarks (two neutrons) in one nucleus (see Fig. 2B) is generally written as [1, 2]

$$H \simeq \frac{G}{\sqrt{2}}(j_L J_L + \eta j_R J_L + \lambda j_R J_R) + h.c., \tag{3}$$

where $j_L(J_L)$ and $j_R(J_R)$ are the left- and right-handed lepton (hadron) currents, respectively, and η and λ are the coupling constants relevant to the right-handed currents. The constant λ is given as $\lambda = (M_W^L/M_W^R)^2$, where M_W^L and M_W^R are the

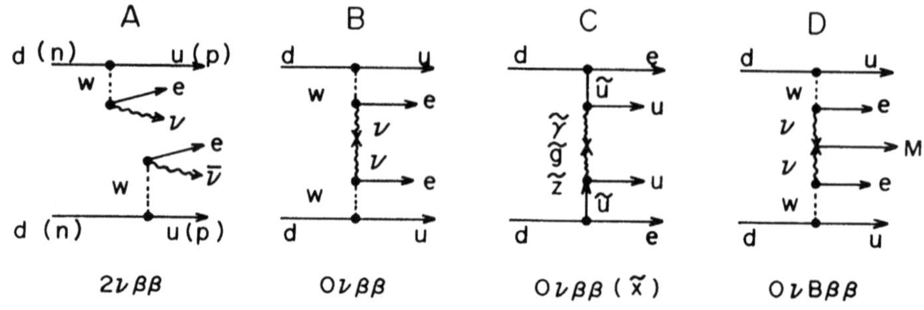

Figure 2: Schematic diagrams of $\beta\beta$ decay processes: A, two-neutrino process; B, neutrino-less process with the Majorana neutrino exchange; C, neutrino-less process in the SUSY model; and D, neutrino-less process followed by the Goldstone boson (Majoron). d and u are d- and u-quarks, respectively, and n and p are neutron and proton, respectively.

masses of the left- and right-handed weak bosons (W^L and W^R), respectively, and η is the mixing ratio of the W^R and W^L bosons.

The transition probability is written as [1]–[7]

$$\begin{aligned} T^{0\nu} &= 2G^{0\nu}|M^{0\nu}|^2[\langle m_\nu\rangle^2 + C_{\lambda\lambda}\langle\lambda\rangle^2 + C_{\eta\eta}\langle\eta\rangle^2 \\ &\quad + C_{m\lambda}\langle m_\nu\rangle\langle\lambda\rangle + C_{m\eta}\langle m_\nu\rangle\langle\eta\rangle + C_{\eta\lambda}\langle\lambda\rangle\langle\eta\rangle]. \end{aligned} \tag{4}$$

Here, $G^{0\nu}$ and $M^{0\nu}$ are the phase-space factor and the nuclear matrix element, respectively, and $\langle m_\nu\rangle$, $\langle\lambda\rangle$ and $\langle\eta\rangle$ are the effective neutrino mass and the relevant RHC coupling terms. They are written as

$$\begin{aligned} \langle m_\nu\rangle &= \sum m_j U_{ej}^2, \\ \langle\lambda\rangle &= \lambda\sum U_{ej}V_{ej} \end{aligned}$$

and

$$\langle\eta\rangle = \eta\sum U_{ej}V_{ej}. \tag{5}$$

Here, U_{ej} and V_{ej} are the coefficients of the mass eigenstates (N_{jL} and N_{jR}) for the left- and the right-handed neutrinos, respectively, as given by

$$V_{eL} = \sum U_{ej}N_{jL} \tag{6}$$

and

$$V_{eR} = \sum V_{ej}N_{jR}. \tag{7}$$

It should be noted that the matrix element $|M^{0\nu}|$ and the coefficients C_{ij} must be either known or evaluated in order to deduce the neutrino mass term $\langle m_\nu\rangle$ and the right-handed current terms $\langle\eta\rangle$ and $\langle\lambda\rangle$ from the measured decay rate (Eq. (4)). The relative values for $\langle m_\nu\rangle$, $\langle\eta\rangle$ and $\langle\lambda\rangle$ are obtained from the energy and angular correlations of two beta rays (β_1, β_2).

Nuclear diagrams associated with the $0\nu\beta\beta$ process are illustrated in Fig. 3. Diagrams A and B concern the two-nucleon ($2n$) modes due to the $\langle m_\nu\rangle$ and $\langle$RHC$\rangle$ terms, respectively, where the neutrino is exchanged between two neutrons in the same neucleus. Diagrams C, D and E are for the isobar (Δ) mode, where the neutrino is exchanged between two d-quarks in the same isobar. The meson modes are shown in diagrams F, G and H, where the exchange current (pion) is involved in the $\beta\beta$ decay.

The $0\nu\beta\beta$ transition rate for the process induced by SUSY particles (Fig. 2C) is given in terms of the SUSY particle masses involved in the process and the relevant interaction constant f. The transition rate for the gluino exchange process is expressed as [12]

$$T_s(\tilde{g}) = G^s f^2 M_{\tilde{g}} A(M_{\tilde{g}})/M_{\tilde{u}}^4, \tag{8}$$

where $M_{\tilde{g}}$ and $M_{\tilde{u}}$ are the masses of the gluino ($\tilde{g}$) and squark ($\tilde{u}$), respectively, and G^s and $A(M_{\tilde{g}})$ are the phase space factor and the nuclear form factor, respectively.

The transition rate for the $0\nu\beta\beta B$ process accompanied by the Goldston boson (Majoron) [14] (Fig. 2D) is written as

$$T = G^B (M^B)^2 \langle g_B\rangle^2, \tag{9}$$

where G^B, M^B and $\langle g_B\rangle$ are the phase-space factor, the matrix element and the effective coupling constant, respectively. The double Majoron emission process of $0\nu\beta\beta BB$ has also been discused [15].

The $2\nu\beta\beta$ transition rate is given in terms of the phase-space factor $G^{2\nu}$ and the nuclear matrix element $M^{2\nu}$ as

$$T^{2\nu} = G^{2\nu} |M^{2\nu}|^2. \tag{10}$$

The $2\nu\beta\beta$ nuclear matrix element $M^{2\nu}$ is directly derived from the measured transition rate. In fact, an evaluation of the nuclear matrix element $M^{2\nu}$ is difficult since the value is very sensitive to the spin and isospin interactions in the nucleus. Thus, a comparison of the observed $M^{2\nu}$ with the calculated value is used to check the calculation of $M^{2\nu}$ and to obtain the appropriate $\tau\sigma$ interactions to be used. They can then be applied for evaluating $M^{0\nu}$.

Some energy spectra of $E = \epsilon_1 + \epsilon_2$ for the two β rays from the ^{100}Mo $\beta\beta$ decay are calculated as shown in Fig. 4. The spectrum for the $0\nu\beta\beta$ process shows a delta-function like peak at the Q-value of $E = Q_{\beta\beta}$. On the other hand, the spectra for the $2\nu\beta\beta$, $0\nu\beta\beta B$ and $0\nu\beta\beta BB$ processes show continuum shapes because large fractions of the decay energy are carried away by 2ν, B and BB, respectively.

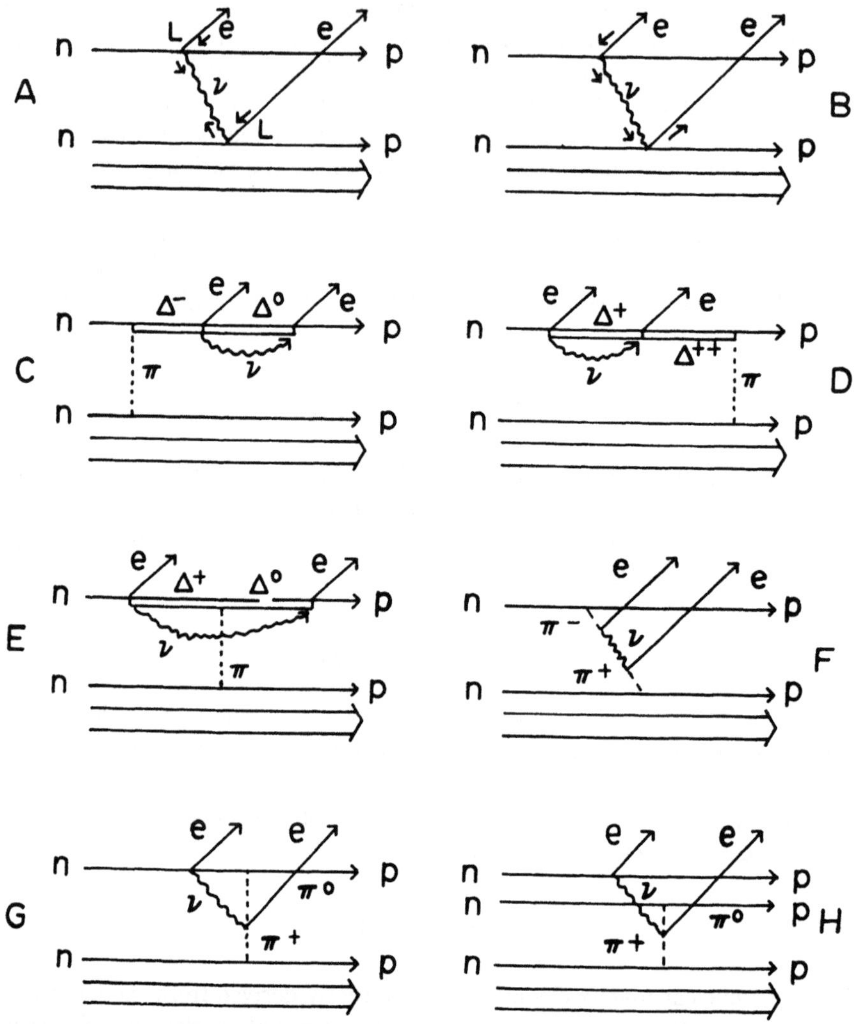

Figure 3: Neutrinoless double beta-decay schemes: A, $2n$ mode with the mass term $\langle m_\nu \rangle$; B, 2n with the right handed current term $\langle \mathrm{RHC} \rangle$; C, Δ^- in ψ_i; D, Δ in ψ_f; E, $\Delta^+\Delta^0$ mode; F, 2n double pion exchange current; G, $2n$ single pion exchange; and H, $3n$ single pion exchange, The arrows in A and B give the relevant helicity (From Refs. [2] and [6]).

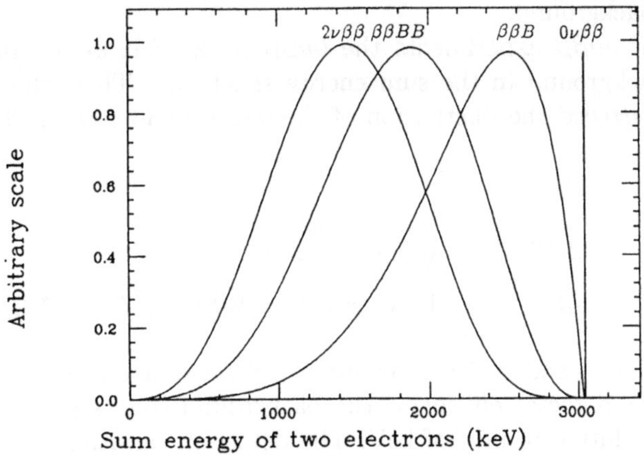

Figure 4: Sum energy spectra for $\beta\beta$ decays from ^{100}Mo(0^+) to the 0^+ ground state of Ru. The solid lines indicate the calculated shapes on an arbitrary scale.

3 Nuclear and Detection Sensitivities for Double Beta Decays

Since $\beta\beta$ decays are second-order weak processes, their decay rates are extremely small. The half-lives for $2\nu\beta\beta$ decays are of the orders of $10^{19} \sim 10^{24}$y, and those for the $0\nu\beta\beta$ type with $\langle m_\nu \rangle \approx 1$eV are of the order of $10^{22} \sim 10^{26}$y, both of which depend greatly on their Q-values ($Q_{\beta\beta}$) as well as on their nuclear matrix elements (M). Therefore, one must choose $\beta\beta$ nuclei with large M and Q (Fig. 5). The transition rate is rewritten by introducing the nuclear sensitivity $S_N^{0\nu}$ as

$$T^{0\nu} = S_N^{0\nu} \langle m_\nu \rangle^2 \tag{11}$$

and

$$S_N^{0\nu} = G^{0\nu} |M^{o\nu}|^2, \tag{12}$$

where $\langle m_\nu \rangle$ is the neutrino mass term in unit of the electron mass (m_e), and the nuclear sensitivity ($S_N^{0\nu}$) is defined as the product of the square of the nuclear matrix element ($|M^{0\nu}|^2$) and the phase-space factor ($G^{0\nu}$). Then, the small quantity $\langle m_\nu \rangle$ to be studied is amplified by the nuclear sensitivity ($S_N^{0\nu}$) which gives a measurable transition rate ($T^{0\nu}$), as in Eq. (12). Thus, the $\beta\beta$ nucleus can be regarded as a microdetector (laboratory) for studying the small quantity of $\langle m_\nu \rangle$. Then, $|M^{0\nu}|^2$ and $G^{2\nu}$ correspond to the intrinsic sensitivity and the acceptance, respectively. Since the phase-space factor $G^{0\nu}$ for the two particles of 2β is much larger than the phase space factor $G^{2\nu}$ for the four particles of $2\nu\beta\beta$, the nuclear sensitivity $S_N^{0\nu}$ becomes larger by a factor $10^{(6\sim8)}$ than the nuclear sensitivity $S_N^{2\nu} = G^{2\nu}|M^{2\nu}|^2$ for the $2\nu\beta\beta$ process. Since single β decays are energetically forbidden in most $\beta\beta$ nuclei which can be studied, huge single-β events are screened off. Therefore, the Majorana mass induced $\beta\beta$ nucleus acts as a sensitive detector (microscope) having a large enhancement factor ($G^{0\nu}/G^{2\nu}$) and a large filtering power against single β-ray background.

In most of $0\nu\beta\beta$ experiments the $0\nu\beta\beta$ peak, if it exists, lies on the large continuum background in the sum energy spectrum. Thus, the number of true events has to exceed the fluctuation of the background counts. This condition is expressed as

$$N_0 tkT^{0\nu} > \sqrt{N_{BG} \cdot \Delta E \cdot t}, \tag{13}$$

$$S_D^{0\nu} T^{0\nu} > 1, \text{ with } S_D^{0\nu} = N_0 k\sqrt{t}/\sqrt{N_{BG} \cdot \Delta E} \tag{14}$$

where N_0, t, k, N_{BG} and ΔE are the number of $\beta\beta$ source nuclei, the measurement (run) time, the detector efficiency, the background counts per year per keV and the energy resolution in unit of keV. The detection sensitivity (S_D) is defined as the inverse of the minimum transition rate which can be detected in a one-year measurement. As shown in Eq. (14), one needs a large volume for the $\beta\beta$ source (large N_0), a high detection efficiency (large k), a low background (small N_{BG}), and a high energy resolution (small ΔE). If the background counts are negligible, the detector sensitivity is defined as

$$S_D^{0\nu} T^{0\nu} > 1 \text{ with } S_D^{0\nu} = N_0 kt. \tag{15}$$

The total sensitivity is then given as

$$\langle m_\nu \rangle \sqrt{S^{0\nu}} > 1, \tag{16}$$

where

$$S^{0\nu} = S_N^{0\nu} \cdot S_D^{0\nu} \cdot \tag{17}$$

In the case of the $2\nu\beta\beta$ process, the transition rate, $T^{2\nu} = G^{2\nu}|M^{2\nu}|^2$, is proportional to the four-particle phase space ($G^{2\nu}$) for $\beta\beta\nu\nu$. Since $G^{2\nu}$ has the leading term of $Q_{\beta\beta}^{11}$, the selection of $\beta\beta$ nuclei with large $Q_{\beta\beta}$ is curcial.

Some of the $\beta\beta$ decay nuclei and their decay schemes are shown in Fig. 5. The double β decays studied so far have been mostly $\beta^-\beta^-$ (electron) decays, since positron ($\beta^+\beta^+$) decays are greatly suppressed by the Coulomb distortion (f-factor).

4 Double Beta-Decay Experiments

The experimental methods for studying $\beta\beta$ decays are classified into two classes. One is an indirect geochemical method; the other is a direct counting method. The geochemical method is an inclusive measurement of the $\beta\beta$ decay product nuclei for a geological long-time interval. The direct counter measurement can identify individual decay processes of the $0\nu\beta\beta, 0\nu\beta\beta B, 0\nu\beta\beta BB$ and $2\nu\beta\beta$ processes, and individual $\beta\beta$ decay branches to the ground (0^+) and excited states. Spectroscopic studies of the energy and/or angular correlations of two β rays in the direct counter

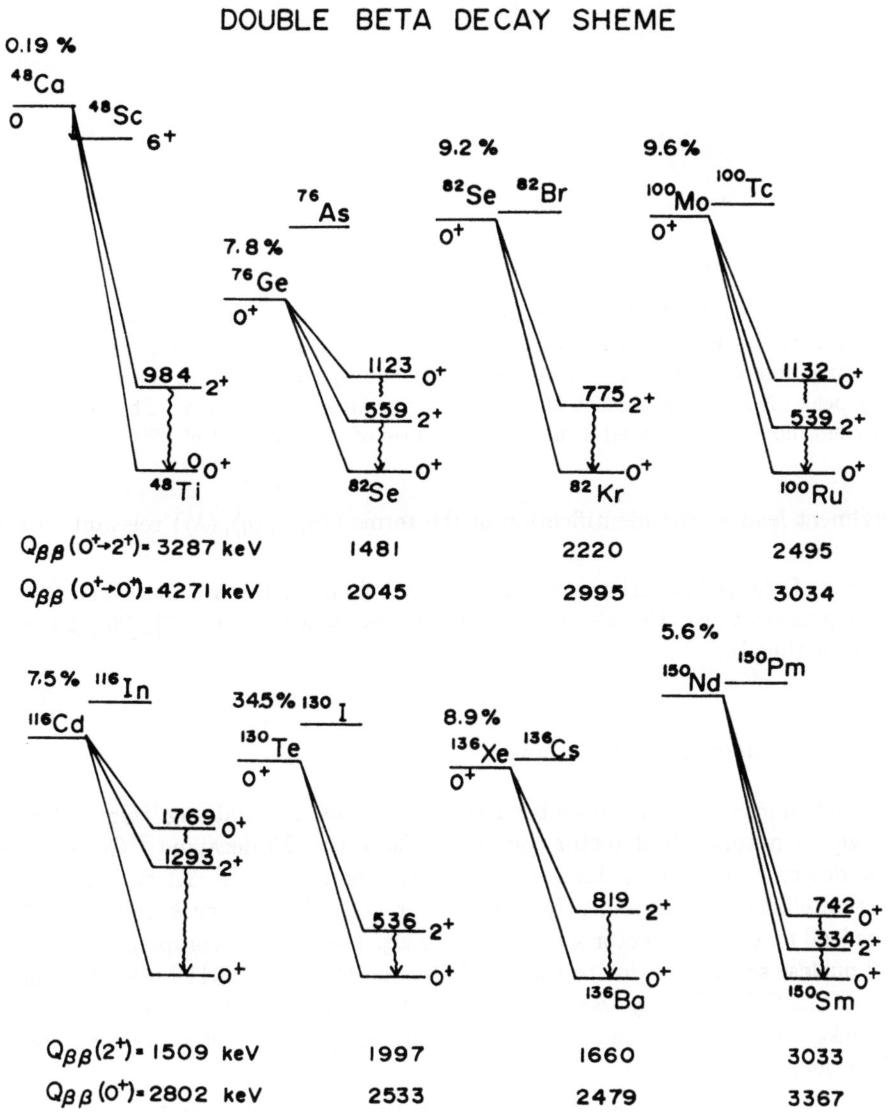

Figure 5: Level schemes of typical double beta decay.

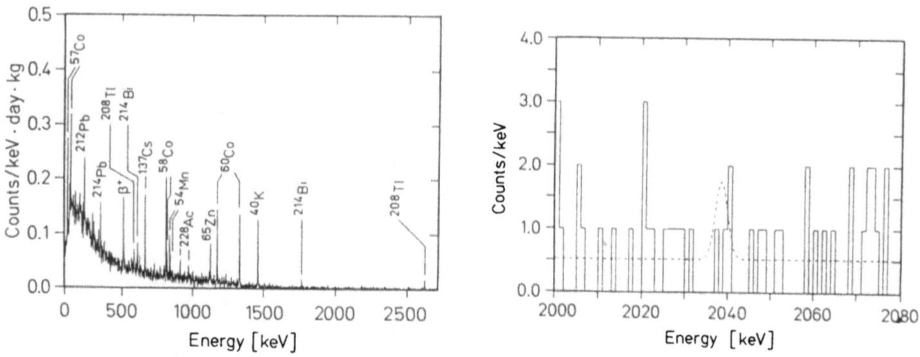

Figure 6: *Left hand side:* Combined background spectrum of the ^{76}Ge detectors. The presence of peaks demonstrates the excellent resolution and stability of the experiment. Measuring time is 1.29 kg y. *Right hand side:* Energy range around the hypothetical $\beta\beta$ 0ν-peak. No peak is present after a measuring time of 1.29 kg y. The dotted line represents the excluded signal at a confidence level of 90% (From Ref. [29]).

experiment lead to the identification of the terms $(\langle m_\nu \rangle, \langle \eta \rangle, \langle \lambda \rangle)$ relevant to the $0\nu\beta\beta$ process.

Some of the $\beta\beta$ experimental methods and their results are described in the following subsections. Details are presented in review articles [1]–[7], [20], [21] and references therein.

4.1 Ge Semiconductor Detector

Ge semiconductor detectors have been extensively used for studying ^{76}Ge$\beta\beta$ decays [22]-[30]. A natural Ge detector contains 7.8% of the $\beta\beta$ decaying ^{76}Ge isotopes in the detector, itself, and the energy resolution is as good as $\Delta E \approx 2$ keV. The detection efficiency is thus nearly 100%, i.e. $k \approx 1$. These large k and small ΔE values lead to a large detector sensitivity (see Eq. (14)), which compensates for the small nuclear sensitivity due to the small Q-value of $Q_{\beta\beta} = 2.041$ MeV. A pioneer study of the ^{76}Ge$\beta\beta$ decay was made by the Milano group [22]. Some groups use a low-background pure Ge detector, while others use pure Ge detectors surrounded by NaI detectors.

The Ge detector gives a sum energy spectrum of $E = \varepsilon_1 + \varepsilon_2$ for two β rays. Due to the good energy resolution, it is very useful to search for a sharp peak arising from the $0\nu\beta\beta$ process.

Recently, Ge detectors made of enriched ^{76}Ge isotopes have been developed [29, 30]. Their sensitivities are indeed higher by one order of magnitude than those of natural-Ge detectors. The energy spectrum is shown in Fig. 6. The half-life limit [29] of $T^{0\nu} > 1.4 \times 10^{24}$y for the $0\nu\beta\beta$ $0^+ \rightarrow 0^+$ process gives the most stringent limit of $\langle m_\nu \rangle \leq 1.5$ eV, $\langle \lambda \rangle \leq 2.3 \times 10^{-6}$ and $\langle \eta \rangle \leq 1.5 \times 10^{-8}$. Recently, the $2\nu\beta\beta$ half-life of around 10^{21}y for ^{76}Ge has been observed [24, 25].

4.2 Time Projection Chamber Method

Exclusive measurements of the $\beta\beta$ trajectories and their momenta are carried out by using a time projection chamber (TPC). The direct observation of the two β-ray tracks is effective for selecting true $\beta\beta$ events from other background events, even though the energy resolution is not very good.

The finite $2\nu\beta\beta$ half-life for ^{82}Se transition has been measured for the first time by using the TPC by a UCL Irvine group [31]. The TPC used by the Irvine group and the sum energy spectrum for the $\beta\beta$ decay are shown in Fig. 7. The obtained half-life of 1.1×10^{20}y is consistent with the values suggested by the geochemical method [32, 33].

The $\beta\beta$ decay of ^{136}Xe has been studied by using the TPC with enriched ^{136}Xe gas [34].

4.3 Detector Complex ELEGANTS

Highly sensitive low-background spectrometers [27], [35]–[38] have been developed at Osaka University for studying double beta decays. They are the ELEGANT III, IV and V (Electron Gamma-ray Neutrino Telescope). EL III is a high-purity Ge detector for $\beta\beta$-rays; it is surrounded by a 4π NaI scintillator for X- and γ-rays [27]. Here, the Ge detector, which includes 7.8% of the ^{76}Ge isotope, is used as the $\beta\beta$ source for the $\beta\beta$ detector (as mentioned in sec. 4.1). The detection efficiency is almost 100%, and a good energy resolution of about 2 keV for β-rays is maintained without suffering from the target (source) thickness. The X- and γ-ray signals from the NaI detector are used to identify $\beta\beta$ rays to the excited state, followed by the deexciting γ-ray, and to reject such background electrons (β-rays, Compton-electons, conversion electrons, etc.) as are accompanied by specific γ- and X-rays. EL IV is same as the EL III, except that the Ge detector is replaced by a set of eleven Si-detector disks [35]. The $\beta\beta$ source foils are interleaved between the Si detector disks for the $\beta\beta$ coincidence measurement. Any sources in the form of a thin foil can be used.

It is noted here that the Si detector array developed by LBL and others was also used for studying the ^{100}Mo $\beta\beta$ decay [39].

EL V consists of two drift chambers (DCA and DCB) for measuring the $\beta\beta$ trajectories, 16 modules of plastic scintillators for measuring the $\beta\beta$ energies and times, and 20 modules of NaI scintillators for X- and γ-rays [37]. The large source (70 cm $\times$ 70 cm) can be set between DCA and DCB. It is schematically shown in Fig. 8.

Recently, the $2\nu\beta\beta$ spectrum for ^{100}Mo was measured [37] by using EL V, as shown in Fig. 9. This was the first direct measurement of the counter experiment for this mode. The observed half-life of $T_{\beta\beta} = 1.2 \times 10^{19}$y leads to a $2\nu\beta\beta$ nuclear matrix element of $M^{2\nu} = 0.09$. The TPC gave a similar $2\nu\beta\beta$ half-life [31]. The small value of the observed $M^{2\nu}$ indicates a severe reduction of the GT strength due to the spin-isospin correlation. It is used to evaluate $M^{0\nu}$ for the $0\nu\beta\beta$ mode in ^{100}Mo. No $0\nu\beta\beta$ peaks have been found in the $\beta\beta$ spectrum observed so far. The upper limit on the possible mass term has been derived as $\langle m_\nu \rangle < 6$ eV by

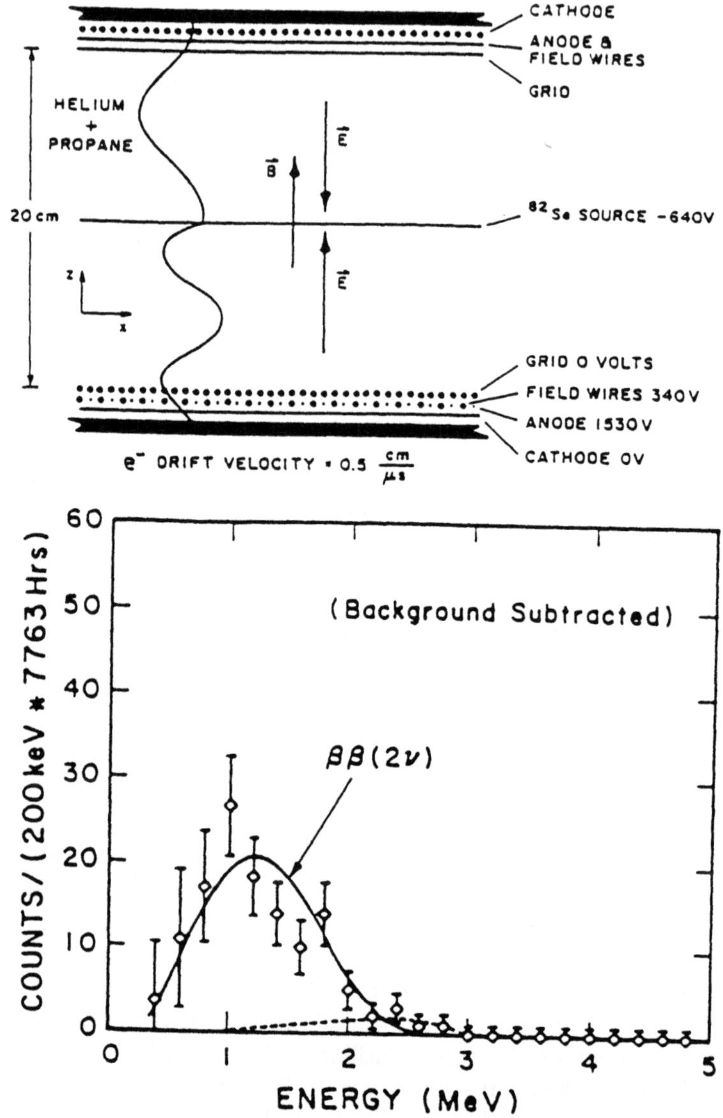

Figure 7: *Top:* Partial section of the TPC used for the ^{82}Se$\beta\beta$ experiment. The configuration is identical for TPC1 and TPC2. (From Ref. [31]). *Bottom:* Sum spectrum of double beta-decay candidates after cuts and subtraction of the known background. The dashed curve is the spectrum for single Majoron emission, arbitrarily normalized. The endpoint energy for ^{82}Se is 3.0 MeV (From Ref. [31]).

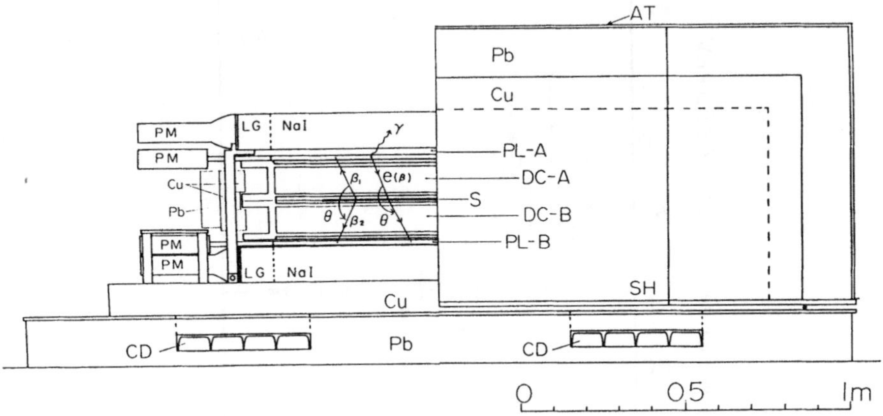

Figure 8: Schematic view of ELEGANTS V. ^{100}Mo, Mo source sheet; DC, drift chamber; NaI, NaI(Tl) crystal; Q, quartz light guide; PL, plastic scintillator; PMT, photomultiplier; Cu, OFHC cupper brick; Pb, lead brick; CD, cable duct (From Ref. [36]).

using the $M^{0\nu}$ estimated on the basis of the observed $M^{2\nu}$.

Limits on the $0\nu\beta\beta B$ and $0\nu\beta\beta BB$ are also obtained for ^{100}Mo and limits on $|\langle g_B\rangle|$ and the SUSY particles are deduced [37].

Double beta decays of ^{100}Mo to the excited states of 2^+ and 0^+ in ^{100}Ru have been studied by coincidence measurement of the $\beta\beta$ rays and γ-rays from the excited states [37]. The $0\nu\beta\beta$ to the 2^+ state is interesting because the process involves only the right-handed interaction $\langle V + A\rangle$. Stringent limits on the half-lifes were obtained as 1.5×10^{20}y and 1.1×10^{21}y, respectively, for the $2\nu\beta\beta$ and $0\nu\beta\beta$ decays to the 2^+ states [37].

Measurements of double beta decays of ^{116}Cd are now in progress [38] by means of the EL V collaboration. The preliminary data indicate a half-life of the order of 10^{19}y for the $2\nu\beta\beta$ $0^+ \to 0^+$ process. An improvement in the EL V by adding a new drift chamber is now under progress in order to search for $\langle m_\nu\rangle$ and $\langle V + A\rangle$ in the region of 1 eV and $10^{-(7\sim9)}$, respectively.

4.4 Geochemical Method

The geochemical method measures the daughter isotopes which are accumulated in the ore containing the parent $\beta\beta$ isotopes for a long time interval t of the order of $10^{(6\sim8)}$y. The daughter isotopes are separated chemically, and are analyzed by a mass separator.

The geochemical method is an inclusive measurement of the total decay rate of all $\beta\beta$ processes ($0\nu\beta\beta, 2\nu\beta\beta\ldots$) feeding the ground and excited states. Actually, the $\beta\beta$ decay branches to excited states are much smaller than the branch to the ground state because of the strong $Q_{\beta\beta}$ dependence of the decay rate. In most $\beta\beta$ nuclei with $Q_{\beta\beta} \geq 2.5$ MeV, the $0\nu\beta\beta$ rates are considered to be smaller than the $2\nu\beta\beta$ rates in the realistic region of $\langle m_\nu\rangle \leq 5$ eV. Consequently, the finite half-life

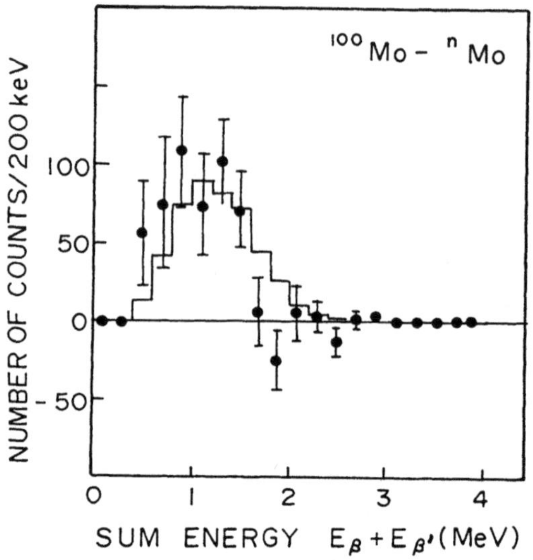

Figure 9: Sum energy ($E_\beta + E'_\beta$) spectrum of ^{100}Mo, being corrected for the spectrum of ^{n}Mo (From Ref.[37]).

derived by the geochemical method may be considered to be the half-life of the $2\nu\beta\beta$ process to the ground state (0^+). On the other hand, the lower limit on the half-life gives the upper limits on all $0\nu\beta\beta, 0\nu\beta\beta B, 2\nu\beta\beta, \ldots$ processes to the ground and excited states.

The $\beta\beta$ nuclei of ^{128}Te with Q = 0.868 MeV, ^{130}Te with Q = 2.533 MeV, and of ^{82}Se with Q = 2.995 MeV have been measured by the geochemical method [32, 33, 41, 42]. The decay products of Te and Se are rare gass Xe and Kr isotopes. The half-lives of $1.5 \sim 2.7 \times 10^{21}$y and 1.0×10^{20}y for ^{130}Te and ^{82}Se may be considered to be the half-lives for the $2\nu\beta\beta$ processes to the ground states of ^{130}Xe and ^{82}Kr, respectively. The lower limit [32] of $5 \sim 8 \times 10^{24}$y on the ^{128}Te $\beta\beta$ decay leads to a stringent limit on the neutrino mass, $\langle m_\nu \rangle \leq 1.5$ eV, depending somewhat on the $M^{0\nu}$ used.

In case that daughter nuclei (decay products) are radioactive, they are counted by measuring the remaining activity. The ^{238}U $\beta\beta$ decay [43] was studied by measuring the α decay of ^{238}Pu.

4.5 Low Background Measurements

Since $\beta\beta$ decay rates are extremely small, low-background measurements are crucial. Since the $\beta\beta$ energy is as small as $1 \sim 2$ MeV, all backgrounds from natural radioactive isotopes have to be reduced. Both $\beta\beta$ source and materials used for detectors have to be purified to the level of the order of 100 ppt (100×10^{-12}) for U and Th chain isotopes. Since air contains Rn gas, space around the $\beta\beta$ source and detector elements has to be filled with dry nitrogen or other pure gas, which are free from Rn gas. In order to avoid room backgrounds, the detector system is shielded by heavy metal bricks. OFHC (Oxygen-free high conductive copper) is quite good for this purpose because it contains less radioactive materials than other metals.

Detectors are covered by active shields made of plastic detectors, NaI detectors, etc. to reject charged particles and γ-rays. They are set at an underground laboratory with 500 $\sim$ 4000 mwe (water equivalent) depth to avoid background associated with cosmic rays. Note that neutrons produced by nuclear interaction of cosmic muons activate detector elements. Therefore $\beta\beta$ sources, detectors and other parts have to be well kept at an underground laboratory.

5 Nuclear Matrix Elements for $\beta\beta$ Decays

Nuclear matrix elements $M^{0\nu}$ and $M^{2\nu}$ for the $0\nu\beta\beta$ and $2\nu\beta\beta$ processes are concerned with the spin-isospin ($\sigma\tau$) correlation (structure) in the nucleus. In the case of the $2\nu\beta\beta$ process, low-energy neutrinos are emitted in real space. Thus, they are mostly s-wave neutrinos. The major component of the $2\nu\beta\beta$ matrix element is the GT (Gamow-Teller) type of $M^{2\nu} \approx \langle\sigma\tau\sigma\tau\rangle$ and the $2\nu\beta\beta$ decay proceeds through 1^+ intermediate states. Here, the Fermi-type matrix element through the 0^+ intermediate states is small, since most of the Fermi-type β strength is concentrated in one isobaric-analogue state. The major component of $M^{2\nu}$ is the GT ($\langle\sigma\tau\sigma\tau\rangle$) term. As is well known, the GT strengths are greatly reduced due to the $\tau\sigma$ correlation in nuclei.

In the case of the $0\nu\beta\beta$ process a virtual neutrino is interchanged between two nucleons in the nucleus. Since the distance between the nucleons is of the order of 1 fm, the momentum of the neutrino can be in the range 0.1 $\sim$ 0.5 GeV/c, and, accordingly, many high J intermediate states with $J^\pi = 0^\pm, 1^\pm, 2^\pm, \cdots, 6^\pm$ are involved in the $0\nu\beta\beta$ process.

The GT β strength for the 1^+ states and the β strengths for the higher J states are reduced in the present low-excitation region because of the strong spin-isospin correlations in nuclei [44]. In other words, these spin-isospin correlations must be properly taken into account for calculations of $M^{0\nu}$ and $M^{2\nu}$.

Shell model calculations for $M^{2\nu}$ and $M^{0\nu}$ have been made by including spin-isospin residual interactions [45]–[48]. QRPA (Quasi-particle Random Phase Approximation) calculations have been made by including pairing, quadrupole and spin-isospin interactions [49]–[52]. Here, these residual interactions are classified into two types: one is the particle-particle interaction and other is the particle-hole interaction. It has been pointed out by Vogel et al. [49] that $M^{2\nu}$ is very sensitive to the particle-particle interactions, $g_{pp}V_{pp}$. Actually, $M^{2\nu}$ with mainly 1^+ intermediate states crosses zero at around the proper value of the particle-particle interaction strength of $g_{pp} \sim 1$, while $M^{0\nu}$ with many high J intermediate states does not, as shown in Fig. 10. Thus, $M^{2\nu}$ can be very small, and is hard to predict. On the other hand, $M^{0\nu}$ is stable and well predicted. Consequently, if one can observe $T^{2\nu}_{1/2}$ for the $2\nu\beta\beta$ process, the value of $M^{2\nu}$(exp) extracted from $T^{2\nu}_{1/2}$ can be used to check the calculation (to obtain proper parameters used for the calculations). Noting that the $M^{0\nu}(1^+)$ component in $M^{0\nu} = M^{0\nu}(1^+) + \sum_J M^{0\nu}(J)$

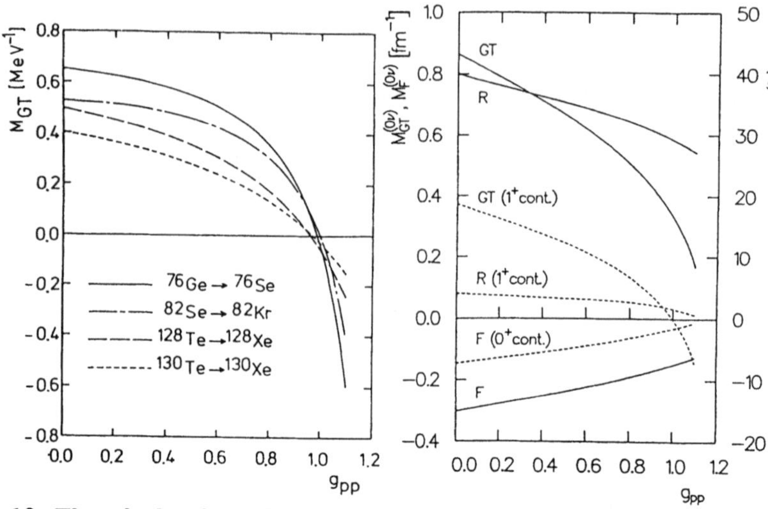

Figure 10: The calculated $2\nu\beta\beta$ decay matrix elements M_{GT} as functions of the strength g_{pp} of the particle-particle interaction (left-hand side), and the calculated nuclear matrix elements $M^{(0\nu)}_{GT,F}$ and $M^{(0\nu)}_{R}$ for the $0\nu\beta\beta$ decay of ^{76}Ge as function of g_{pp} (right-hand side). GT, F and R stand for the GT, Fermi and recoil matrix elements, respectively. (From Ref. [50]).

is hard to evaluate, one can use $M^{2\nu}$ to evaluate $M^{0\nu}(1^+)$.

Calculations of $M^{0\nu}$ and $M^{2\nu}$ have so far been made mainly for the two-nucleon $\beta\beta$ mode, where two nucleons are involved in the two beta decays. As mentioned in sec. 2, the isobars (Δ) and exchange currents ($\pi, \cdots$) in nuclei contribute to the $\beta\beta$ process. In particular, these non-nucleonic modes can be important for the $0\nu\beta\beta$ process involving the large momentum of the virtual neutrino. Some calculations [53, 54] indicate that these modes are not major. They, however, can be important when the matrix element for the $2n$ mode is suppressed.

6 Concluding Remarks

1. The neutrino-less $\beta\beta$ decays ($0\nu\beta\beta$) are very sensitive to the possible violation of the lepton-number conservation law, the possible finite mass of the Majorana neutrino, the possible right-handed weak currents and to other particles and interactions associated with the lepton-number non-conservation. These are beyond the minimal $SU(2)_L \times U(1)$ standard theory.

2. The $0\nu\beta\beta$ process is used to test the possible theories, such as grand unified theories and supersymmetric theories, which are beyond the standard theory.

3. In the $0\nu\beta\beta$ process the Majorana neutrino (or some SUSY particle) is exchanged between two nucleons (quarks) in the nucleus. The nucleus thus acts as a microscope to enhance the $0\nu\beta\beta$ process and to filter the true signal against the single beta background.

4. The two-neutrino double beta decay ($2\nu\beta\beta$) is a process within the standard theory. The nuclear matrix element for the $2\nu\beta\beta$ process is derived from the observed half-life of the $2\nu\beta\beta$ process. It is used to check the nuclear-structure calculation and the spin-isospin interaction. It, in turn, is used to evaluate the nuclear matrix elements ($M^{0\nu}$) for the $0\nu\beta\beta$ process, which are crucial in deducing the terms $\langle m_\nu \rangle$, $\langle \mathrm{RHC} \rangle$, etc from the observed (limit on) $0\nu\beta\beta$ decay rate.

5. The recent development of pure Ge detectors made of the enriched ^{76}Ge isotopes, TPC (time projection chamber) and of the detector complex ELEGANT have made it possible to measure the finite $2\nu\beta\beta$ decays in the region of $T^{2\nu}_{1/2} \sim (10^{19} \sim 10^{22})$y and to search for the $0\nu\beta\beta$ decays in the region of $T^{0\nu}_{1/2} \sim (10^{23} \sim 10^{24})$y.

6. The finite $2\nu\beta\beta$ half-lives for ^{76}Ge, ^{82}Se, ^{100}Mo, and ^{116}Cd gave small values of $M^{2\nu} \approx 0.05 \sim 0.10$. These values indicate severe cancellations due to the spin-isospin correlation in the nucleus. The obtained value for $M^{2\nu}$ and the spin-isospin interaction are used to evaluate the $0\nu\beta\beta$ matrix element.

7. The upper limits obtained so far on the $0\nu\beta\beta$ decay rates lead to upper limits of ($\sim$5) eV on the Majorana neutrino mass $\langle m_\nu \rangle$. Thus, the Majorana electron nuetrino is not a major candidate of dark matter. The most stringent limit on the half-life derived from the direct counter experiment of the ^{76}Ge $0\nu\beta\beta$ decay gives upper limits of $\langle m_\nu \rangle \leq 1.5$ eV, $\langle \lambda \rangle \leq 2.3 \times 10^{-6}$ and $\langle \eta \rangle \leq 1.5 \times 10^{-8}$ on the neutrino mass, the square of the mass ratio of right-to left-handed weak bosons, and on the mixing ratio of the right- and left-handed weak bosons, respectively [29]. Useful limits are also given on the coupling of the Majoron, the masses of relevant SUSY particles, the mass of the heavy right-handed neutrino, and on other related quantities.

8. Further developments and measurements of the $0\nu\beta\beta$ decays for various $\beta\beta$ nuclei with large-volume ^{76}Ge detectors, TPC, ELEGANTS, and others [55] will be used to search for the smaller mass region of $\langle m_\nu \rangle = (0.1 \sim 0.5)$ eV in the near future. New techniques, such as thermal detectors and resonance ionization mass-separators, are also expected to be promising.

References

[1] W.C. Haxton and G. J. Stephenson, Jr., Prog. in Part Nucl. Phys. **12** (1984) 409.

[2] M. Doi, T. Kotani and E. Takasugi, Prog. Theor. Phys. Supp. **No.83** (1985).

[3] D. O. Caldwell, Int. Journ. of Mod. Phys. **A4** (1989) 1851.

[4] F. T. Avignone, III and R. L. Brondzinski in Neutrinos, ed. H. V. Klapdor, Springer-Verlag, 1988, p147.

[5] K. Muto and H. V. Klapeor, in Neutrinos, ed. H. V. Klapdor, Springer-Velag, 1988, p183; K. Grotz and H. V. Klapdor, in The Weak Interaction in Nuclear, Particle and Astrophysics, Adam Hilger, 1990.

[6] H. Ejiri and M. Voigt, in Gamma-ray and Electron Spectroscopy in Nuclear Physics, Oxford Univ. Press, 1989; H. Ejiri, Proc. Workshop on Weak Interactions and Neutrinos, ed. J. Kim, Seoul, 1993.

[7] M. K. Moe, Proc. Neutrino 92, Granada, Nucl. Phys., to be published.

[8] M. Gell-Mann, P. Ramond and R. Slansky, in Supergravity, ed. van Niewwenyhuizen and Freedman, North-Holland, Amsterdam, 1979.

[9] T. Yanagida, Proc. Workshop on Unified Theory and Baryon Number in the Universe, ed. Sawada and Sugamoto, KEK, 1979.

[10] F. J. Gilman and S. H. Rhie, Phys. Rev. **D32** (1985) 324; R. N. Mohapatra, Phys. Rev. **D34** (1986) 909.

[11] P. Langacker and S. Umasankar, Phys. Rev. **D40** (1989) 1569.

[12] R. N. Mohapatra, Phys. Rev. **D34** (1986) 3457.

[13] J. D. Vergados, Phys. Lett. **184B** (1987) 55.

[14] G. B. Gelmini and M. Roncadelli, Phys. Lett. **99B** (1981) 411; H. M. Georgi, S. L. Glashow and S. Nussinov, Nucl. Phys. **B193** (1981) 297; Y. Chikashige, R. N. Mohapatra and R. D. Peccei, Phys. Lett. **98B** (1981) 265.

[15] R. N. Mohapatra and E. Takasugi, Phys. Lett. **B211** (1988) 192.

[16] M. C. Gonzalez-Garcia and Y. Nir, Phys. Lett. **B232** (1988) 383.

[17] Mark II Collaboration, G. S. Abrams et al., Phys. Rev. Lett. **63** (1989) 724. .

[18] E. Takasugi, private communication (1989, Sep.); T. Kotani, KEK Report 90-9 (1990, Jan.).

[19] H. Primakoff and S. P. Rosen, Phys. Rev. **184** (1969) 1925.

[20] H. Ejiri, Proc. Int. Conf. Grand Unification and ICOBAN '86, Toyama, 1985. p. 215; Proc. INS Int. Symposium on Neutrino Mass and Related Topics, March 1989, Tokyo.

[21] A. Morales, Proc. 1st Int. Workshop on Theoretical and Phenomenological Aspects of Underground Physics, TAUP '89 and private communication (1990) Gran Sasso, Italy, Sep, 1989.

[22] E. Fiorimi, A. Pullia, G. Bertolini, F. Cappelani and G. Restelli, Phys. Lett. **25B** (1967) 602, Nuovo Cimento **13A** (1973) 747; E. Bellotti et al., Phys. Lett. **146B** (1984) 450, Nuovo Cimento **95A** (1986) 1.

[23] D. O. Caldwell et al., Phys. Rev. Lett. **54** (1985) 281; Phys. Rev. **D33** (1986) 2737; Phys. Rev. Lett. **59** (1987) 419; Nucl. Phys. **B13** (Proc. Suppl) (1990) 547.

[24] F. T. Avignone, III, et al., Phys. Rev. Lett. **50** (1983) 721; Phys. Rev. **C34** (1986) 666; Prog. Part Nucl. Phys. **21** 99; H. S. Miley et al., Proc. TAUP '89 Workshop Italy; Phys. Rev. Lett. **65** (1990) 3092.

[25] I. V. Kirpichnikov, Double Beta Decay Experiments, to appear in the Proc. of the Conference on Lepton-Photon Interaction, Stanford, 1989; A. A. Vasenko et al., Mod. Phys. Lett. **A5** (1990) 1299.

[26] P. Fisher, et al., Phys. Lett. **B218** (1989) 257; **B192** (1987) 460.

[27] H. Ejiri et al., Proc. Int. Symp. Nucl. Spectroscopy and Nucl. Interactions, Osaka, March 1984, World Scientific Pub., p. 284; H. Ejiri et al., Nucl. Phys. **A448** (1986) 271; N. Kamikubota et al., Nucl. Instr. Methods **A245** (1986) 379; H. Ejiri et al., J. Phys. G: Nucl. Phys. **13** (1987) 839.

[28] P. Dassie, Ph. Hubert et al., Nuovo Cimento **A85** (1985) 19; A. Morales et al., Nuovo Cim. 100A (1988) 525, Ph. Hubert et al., Nucl. Phys. (1990).

[29] A. Balysh et al., Phys. Lett. **B283** (1992) 32; M. Beck et al., Z. Phys. A: Hadrons and Nuclei, **343** (1992) 397.

[30] F. T. Avignone, R. L. Brodzinski, H. S. Miley and J. H. Reeves, Proc. WEIN'89 Conf. Montreal, May '89, World Scientific; R. L. Brodzinsky et al., Proc. Neutrino 92, Granada, Nucl. Phys. B, to be published.

[31] S. R. Elliot, A. A. Hahn and M. K. Moe, Phys. Rev. Lett. **59** (1987) 2020; S. R. Elliot, M. K. Moe, M. A. Nelson and M. A. Vient, J. Phys. G: Nucl. Part. Phys. **17** (1991) S145.

[32] T. Kirsten, E. Heusser, D. Kaether, J. Oehm, E. Pernicka and H. Richter, Proc. Int. Symp. on Nuclear Beta Decays and Neutrino. eds. T. Kotani, H. Ejiri and E. Takasugi, World Scientific Pub. Singapore (1986) 81.

[33] O. K. Manuel, Proc. Int. Symp. on Nuclear Beta Decays and Neutrino. eds. T. Kotani, H. Ejiri and E. Takasugi, World Scientific Pub. Singapore (1986) 71; J. Phys. G: Nucl. Part. Phys. **17** (1991) S221.

[34] M. Z. Igbal et al., Nucl. Instr. Meth. in Phys. Res. **A259** (1987) 459; H.T. Wong et al., Phys. Rev. Lett. **67** (1991) 1218; J.-C. Vuilleumier et al., Proc. Neutrino 92, Granada, Nucl. Phys., to be published.

[35] H. Ejiri et a., Proc. Telemark IV Conf., Ashland, Wisconsin, March 1987, ed. V. Barger and D. Cline, World Scientific, Singapore (1987) 281; H. Ejiri, Proc. of the Yamada Conf. on Nuclear Weak Proceses and Nuclear Structure, Osaka, June 1989 ed. Morita, Ejiri, Ohtusbo and Sato, World Scientific, Singapore (1989) p190; T. Watanabe, Ph. D. Thesis, Osaka Univ., 1988.

[36] K. Okada et al., Nucl. Phys. **A478** (1988) 447C; H. Ejiri, et al., Nucl. Instr. Method **A302** (1991) 304.

[37] H. Ejiri, Nucl. Phys. **A522** (1991) 305C; H. Ejiri et al., J. Phys. G. Nucl. Phys. G: Nucl. Part. Phys. **17** (1991) S155; T. Shima, Ph. D. Thesis, Osaka University 1990; H. Ejiri et al., Phys. Lett. **258B** (1991) 17; J. Tanaka and H. Ejiri, Phys. Lett. (1993), to be published; N. Kudomi, H. Ejiri, K. Nagata, K. Okada, T. Shibata, T. Shima, and J. Tanaka, Phys. Rev. **C46** (1992) R1232.

[38] K. Kume and the Ejiri/ELEGANT group and Zdesenko/Kiev group, private communication (1992).

[39] M. Alston-Garnjost et al., Phys. Rev. Lett. **63** (1989) 1671.

[40] Ke You et al., Phys. Lett. **B265** (1991) 53; A. A. Kliminko et al., INR (Moscow) Preprint 11-89; A. S. Barabash et al., WEIN'89, Montreal (1989) and TAUP'89 Workshop, Italy (1989).

[41] T. Kirsten, H. Richter, and E. Jessberger, Phys. Rev. Lett. **50** (1983) 474.

[42] E. W. Hennecke, O. K. Manuel, and D. D. Sabu, Phys. Rev. **C11** (1975) 1378; T. Bernatowicz et al., Phys. Rev. Lett. **69** (1992) 2341; N. Takaoka and H. Sagawa, 45th Mtg. Phys. Soc. Japan, abstract, p.48.

[43] A. L. Turkevich, T. E. Economou, and G. A. Cowan, Phys. Rev. Lett. **67** (1991) 3211.

[44] J. I. Fujita, S. Fujii, and K. Ikeda, Phys. Rev. **133B** (1964) 549; H. Ejiri, K. Ikeda, and J. I. Fujita, Phys. Rev. **176** (1968) 1277; H. Ejiri and J. I. Fujita, Phys. Report **38C** (1978) 85.

[45] W. C. Haxton, G. J. Stephenson Jr. and D. Strottman, Phys. Rev. **D25** (1982) 2360; W. C. Haxton and G. J. Stephenson Jr. Prog. Part. Nucl. Phys. **12** (1984) 409.

[46] L. Zamick and N. Auerbach, Phys. Rev. **C26** (1982) 2185.

[47] J. D. Vergados, Phys. Rep. **133** (1986) 1 and references threin.

[48] B. A. Brown and L. Zhao, Proc. Int. Yamada Conf. XXIII, Nucl. Weak Process and Nuclear Structure, Osaka 1989, World Scientific, p291; L. Zhao, B. A. Brown, W. A. Richter, preprint MSUCL-714 (1990).

[49] P. Vogel and M. R. Zirnbauer, Phys. Lett. **57** (1986) 3148; J. Engel, P. Vogel and M. R. Zirnbauer, Phys. Rev. **C37** (1988) 731; J. Engel, P. Vogel, O. Civitarese and M. R. Zirnbauer, Phys. Lett. **208B** (1988) 187; J. Engel, P. Vogel, J. I. Xiangdong and S. Pittel, Phys. Lett. **225B** (1989) 5.

[50] O. Civitarese, A. Faessler and T. Tomoda, Phys. Lett. **B194** (1987) 11; T. Tomoda and A. Faessler, Phys. Lett. **B199** (1987) 475; T. Tomoda, Nucl. Phys. **A484** (1988) 635.

[51] K. Muto and H. V. Klapdor, Phys. Lett. **B201** (1988) 420; K. Muto, E. Bender and H. V. Klapdor, Z. Phys. **A334** (1989) 177, 187; A. Staudt, T. T. Kuo and H. V. Klapdor-Kleingrothaus, Phys. Rev. **C46** (1992) 871.

[52] T. T. S. Kuo, Proc. INS Symp. Neutrino Mass and Related Topics, Tokyo 1988.

[53] H. Wu and T. Huang, Nucl. Phys. **A487** (1988) 667; M. Ericson and J. D. Vergados, private communication (1989).

[54] J. D. Vergados, A. Faessler and T. Tomoda, private communication (1989).

[55] A. Alessandrello et al., Phys. Lett. **B285** (1992) 176; A. Morales, Nuovo Cimento **A105** (1992) 535; D. Blum et al., Phys. Lett. **B274** (1992) 506; A. A. Klimenko et al., Nucl. Instr. Meth. in Phys. Res. **B17** (1986) 445.

Neutrino Oscillations in Emulsion Experiment

Kimio Niwa

Department of Physics, Nagoya University

1 Emulsion experiments for high energy physics

1.1 Introduction

Emulsion experiments have a long history. Becquerel discovered radioactivity of uranium in 1896 following the discovery of Roentgen Rays [1]. He observed some activity of Uranium on a photographic emulsion penetrating through several sheets of black papers. Kinoshita then showed in 1910 [2] that α-particles produce a blackening of photographic emulsion so that individual α-tracks could be recognized. An attempt to improve the sensitivity of emulsions by doubling the concentration of AgBr was made by Bloch (Ilford, Ltd.) and Powell, in 1939 [3]. This effort was eventually rewarded by the discovery of "two mesons" in 1947. The first electron-sensitive emulsion (Kodak NT4) was produced by Berriman (1949) [4]. The improvement was made not only in the sensitivity of the emulsion, but also in the increased thickness. An emulsion up to 2 mm thick was successfully processed. We note a number of important discoveries made for new particle physics in cosmic rays in the 1940-1950's using emulsions [5].

The counter and bubble chamber experiments eventually took over the role of emulsions in high energy physics following the development of accelerators. Since then, emulsions have been used mainly by cosmic ray physicists who studied high energy interactions. In 1970, four year before the discovery of J/ψ, one pair of short lived particles, which are now known as charmed particles, were discovered in an emulsion experiment [6]. The discovery had not attracted world-wide attention of the particle physics community, but a few Japanese physicists took the event seriously, which stimulated the discovery of the Kobayashi-Maskawa matrix prior to the discovery of J/ψ. After the J/ψ epoch, accelerator experiments using emulsions have been re-considered much more seriously. A Japanese group centered at Nagoya University have been engaged in high energy experiments using their emulsion techniques to find short-lived charmed and tau particles, the most recent effort being a search for τ particles produced in the neutrino (ν_μ, ν_e) beam.

Table 1: The properties of Fuji ET-7B emulsion

element	symbol	atomic number	mass number	weight fraction*%	mole fraction %
Iodine	I	53	126.9	0.3	0.06
Silver	Ag	47	107.9	45.4	11.2
Bromine	Br	36	79.9	33.4	11.1
Sulfur	S	16	32.1	0.2	0.2
Oxygen	O	8	16.0	6.8	11.3
Nitrogen	N	7	14.0	3.1	5.9
Carbon	C	6	12.0	9.3	20.6
Hydrogen	H	1	1.0	1.5	40.0

Quark contents : 48% up, 52% down
Mean nucleus : 36 protons, 45 neutron
Density = 3.73 g/cm^3
Radiation length = 2.94 cm
Nuclear interaction mean free path = 38 cm
Concentration of AgBr = 45.5 % in volume

* At 68 % relative humidity

An example of the composition of an emulsion currently being used (Fuji ET-7B) is shown in Table 1 [7]. AgBr, whose crystal size of 0.3 μm, takes up 78.8% by weight and 45.5% by volume. A charged particle ionizes a AgBr crystal by the Coulomb force, and this ion acts as an imager. After development this ion changes into a Ag particle with a size of 0.8 μm along the trace of the charged particle like a bubble in the bubble chamber.

Emulsions are very useful, not only for a study of short lifetime particles such as charm/beauty and the tau lepton, but also as a superfine tracking detector. The method of using emulsion has also been advanced drastically from the stand-alone emulsion into an emulsion-counter hybrid detector.

1.2 Pellicle emulsion

The plate of emulsion gel was called pellicle and its typical thickness was 600 μm with a size of around 10 cm square. The thickness was limited by the working distance of the objective lens of the microscope.

Many pellicles are placed together on a surface parallel to incident particles like a pack of cards, which is then called *horizontal* type emulsion. After the exposure, the pellicle is developed using a supporting plate on which pellicle was treated, and a special coordinate system is prepared by printing the location on the surface of pellicles using a photo-mask.

The interaction vertices are identified by general scanning. In some cases incident particle is traced to obtain the primary vertex; this method is called the track-follow method. Detailed characteristics of the pellicle emulsion are described in ref. [5] .

The emulsion measures the ionization energy loss of a penetrating charged particle; this energy loss is proportional to the grain density along the track in an emulsion. The absolute grain density (I) depends on the batch of gel production and processing conditions of development etc., but the ratio (I/I_0) is relatively insensitive to various conditions and is given roughly as

$$I/I_0 = 1/\beta^2 , \tag{1}$$

where β is the velocity divided by the light velocity, and I_0 is the minimum ionization grain density typically of $I_0 = 30$ counts/100 μm. I and I_0 are sensitive to a fading effect which depends on the temperature and humidity during the exposure and the storage. By measuring the range of a track in the emulsion, the energy of a charged particle stopping in the emulsion can be obtained. The relation of energy (E in MeV) to range (R in μm) is obtained experimentally as

$$E = KZ * 2M * 0.42[(\rho/\rho_0)R]^{0.58} \tag{2}$$

where a constant $K = 0.006$, Z and M are the charge and mass (MeV) of the stopping particle, ρ is the density of the emulsion in mass weight, and ρ_0 is the standard density taken to be 3.73 g/cm^3.

In the case of a very short ranged track, especially shorter than 10 μm, the crystal size (0.3 μm) of AgBr that occupies 45.5 % of the volume causes significant fluctuations in the measured range.

The results of experiments using pellicles are summarized in ref. [5].

1.3 Emulsion chamber (ECC)

An emulsion group of Chicago University exposed a big emulsion stack (80 liter pellicle) in a large balloon in 1965. Cosmic ray physicists in Japan noted the limits of pellicle emulsion technology for studying high energy particle interactions and introduced the new idea of emulsion chambers (called *ECC*) using an emulsion plate coated with a thin gel of 50 μm on a plastic base of 1300 μm, instead of a pellicle of 600 μm pure emulsions. Such *ECC* were used in a balloon experiment in 1956. The *ECC* costs 90% cheaper than a pellicle stack of the same weight.

The configuration of the *ECC* is shown in Fig. fig:1. To detect the energy of π^0 instead of charged particles, many emulsion plates were alternated with lead plates and exposed to cosmic rays perpendicular to the surface of the emulsion plate. In *ECC* the emulsion plates are used as a tracking detector of vector type, and additional target layers are placed on the top of the *ECC* to avoid overlapping of particles in a cascade shower. The number of electron-positron tracks was measured by a microscope plate by plate and was compared with a calibrated number. In such a way, the energy of each γ-ray (π^0) was measured separately with an error of about 10 % for a particle with an energy around a TeV.

The *ECC* experiments, placed on the top of a mountain or balloon, produced significant results in the study of high energy interactions by cosmic rays.

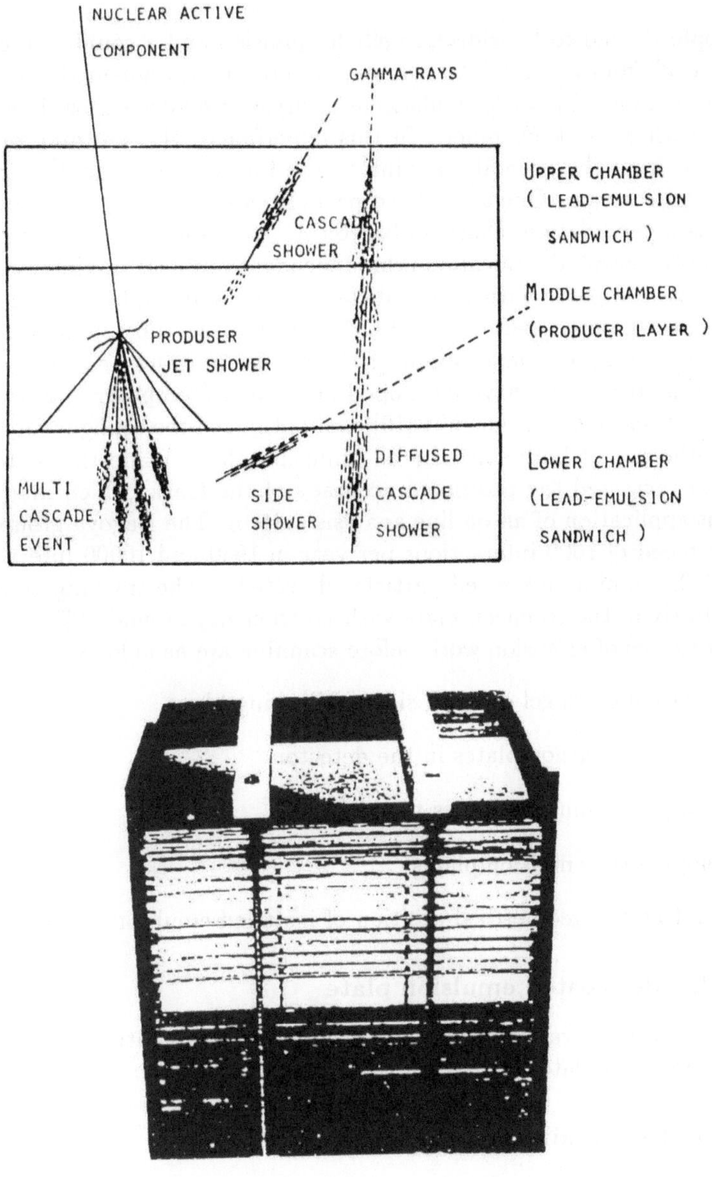

Figure 1: Schematic illustration of the emulsion chamber and photograph.

2 Hybrid emulsion detector

2.1 Introduction

Many people started to hybridize a pellicle emulsion and a counter in cosmic-ray experiment [8] from around 1960. In an accelerator experiment, Burhop and his group started to use a pellicle emulsion as a target to detect a short lived particles combined with a spark chamber. In this experiment, the scanning of emulsion was changed from the general scanning to the limited-area scanning oriented by the counter prediction. One short lived particle was eventually observed [9] in an emulsion as a candidate of charmed baryon Λ_c. It turned out, however, that the efficiency of the event identification is low; the identified neutrino interactions were only 16 events compared with 160 predicted events. This limitation was overcome by the Nagoya group based on the ECC emulsion. A substantial progress was made in both detector configuration of the emulsion target and analysis system of emulsion plates. They have developed an on-line (computer and TV camera) emulsion analysis system from 1974 [10], and optimized the configuration of target emulsion: the use of both-sides-coated emulsion plates made the material hard enough and increased the positional accuracy of the track, which made possible an efficient application of an on-line analysis system. The Nagoya group achieved a location speed of 1000 interactions per year in 1980 and 10000 interactions per year in 1990. Minimum ionized particles detected in the tracking counter were located quickly in the emulsion plate with an efficiency around 99%.

The processes of emulsion work before scanning are as follows:

1. pouring emulsion gel on both sides of backing film.
2. assembling emulsion plates in the detector.
3. exposing the emulsion plates to the beam.
4. developing the emulsion plates.
5. calibrating the geometrical position of installed emulsion plates.

2.2 Both-sides-coated emulsion plate

In 1981 Nagoya University and Fuji-Ltd. started a collaboration to find a configuration of emulsion plates for the emulsion-counter hybrid experiment. [7]

Material of the backing film and basic treating

Both-sides-coated emulsion plate is made by pouring nuclear emulsion gel on both sides of a transparent backing film. Polystyrene film with a $70\,\mu$m thickness is used for the target emulsion. A Lucite with a 200 $\sim$ 800 μm thickness is used to make an emulsion plate for a tracker. Tri-acetate-cellulose film with a $90\,\mu$m thickness is used for the target emulsion. Before coating the gel, it is necessary to coat the surface of backing film with very thin gelatine. This process is called a *basic treating* of the backing film. Quality of this *basic treating* correlates strongly

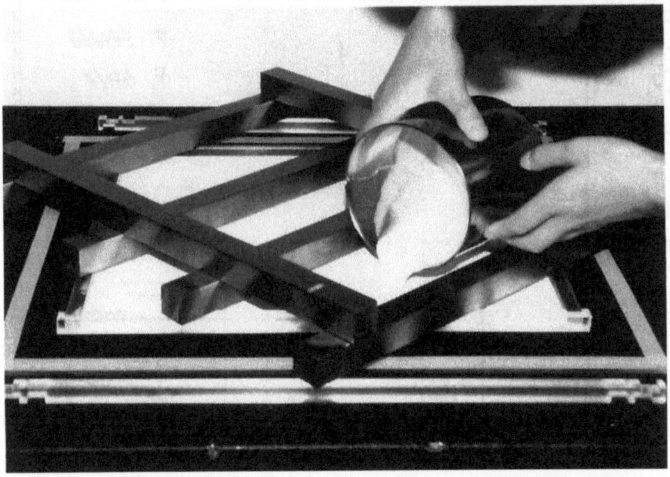

Figure 2: Photograph of puring process on a flat table

with an appearance rate of water bubbles in emulsion during the development process. The way of the basic treating depends on used materials: in the case of Lucite, its surface has to be chemically activated by Primer 80; in the case of Polystyrene, the surface has to be electrically activated by glow discharge; and for Tri-acetate-cellulose, activating is not required.

In selecting the material for the backing film, together with the mechanical strength, the optical characteristics of single reflection must be examined using a transmission light microscope with high magnification.

Pouring on a thin and a thick backing plate

The nuclear emulsion is stored in gel form in a refrigerator at 5°C. A gel is weighed with an accuracy of 0.2% in order to control the thickness of an emulsion plate. A 1 cm^2 dried emulsion needs 11.2 g gel of Fuji ET-7B. Before pouring, gel is melted in a bath, keeping the temperature around 40°C. The surface of working table used in pouring must have a flatness of 30 μm to obtain a uniform emulsion plate. Fig. 2 shows a pouring table which is made of stone.

The treated backing film is put on this flat table and chacked by lowering pressure. In making a thin plate with a thickness of 50 $\sim$ 70 μm, the pouring process requires an elaborate technique that can be carried out only by a few experts using the surface tension of emulsion gel. On the other hand, a thick emulsion with a thickness of 300 $\sim$ 600 μm is easily made in the following way: some volume of emulsion gel is poured in the box which is made of plastic channels or stainless steel square rods; the liquefied gel poured in the box is left for about 20 minutes under the room temperature around 20°C; during this drying process of a emulsion plate, in order to avoid curling of a emulsion plate, the stiff substrata made of Lucite or stainless steel are used in the way of sticking along the surrounding

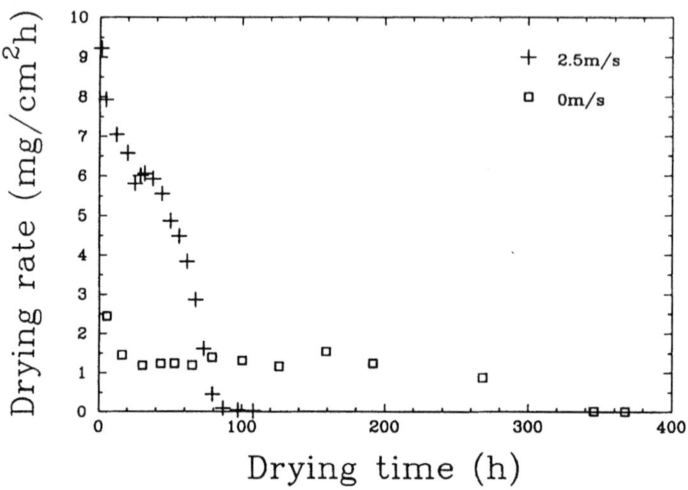

Figure 3: Drying rate vs. drying time for two different blowing speed. A poured emulsion plate, 25 cm × 25 cm with 550 μm thickness, was dried in the condition of 80% humidity at 20 °C.

edges with treating tapes or mechanical clippers; Pouring the emulsion into the other side can be done in the same way. After drying, the emulsion gel shrinks by one eighth of its original thickness before pouring. Irregularity of the thickness of the emulsion plate coming from the non-uniformity of a pouring table is reduced by the same rate (1/8).

Drying and distortion of the emulsion plate

There are two important points in the drying process. One is a drying speed and the other is a distortion of the emulsion. The drying time depends on the thickness of the emulsion, blowing speed and humidity, as shown in Fig. 3. We have to care about hard blowing to avoid cracks made on the surface of the emulsion plate.

The track image of high energy particles in the emulsion plate is deviated from a straight line by a distortion which originates mainly from a lack of uniformity in the drying process of pouring, but less from the development process. This distortion results in a significant reduction of the analysis speed of emulsion plates in the semi-automated and fully automated analysis systems described below.

In principle, near-edges of the emulsion plate drys faster than the central part and this difference of drying speed causes a stress. This effect is observed by measuring the distortion of images of penetrating tracks (see Fig. 4). Making a large size emulsion plate is one of the solutions to alleviate the edge loss. A more useful technique named a *picture frame method* has recently been developed by the Nagoya group: after pouring emulsion gel, fake gel[1] is poured around the edge

[1]The amount of the gelatine is tuned to be 50% and cheap SiO_2 is used instead of expensive AgBr.

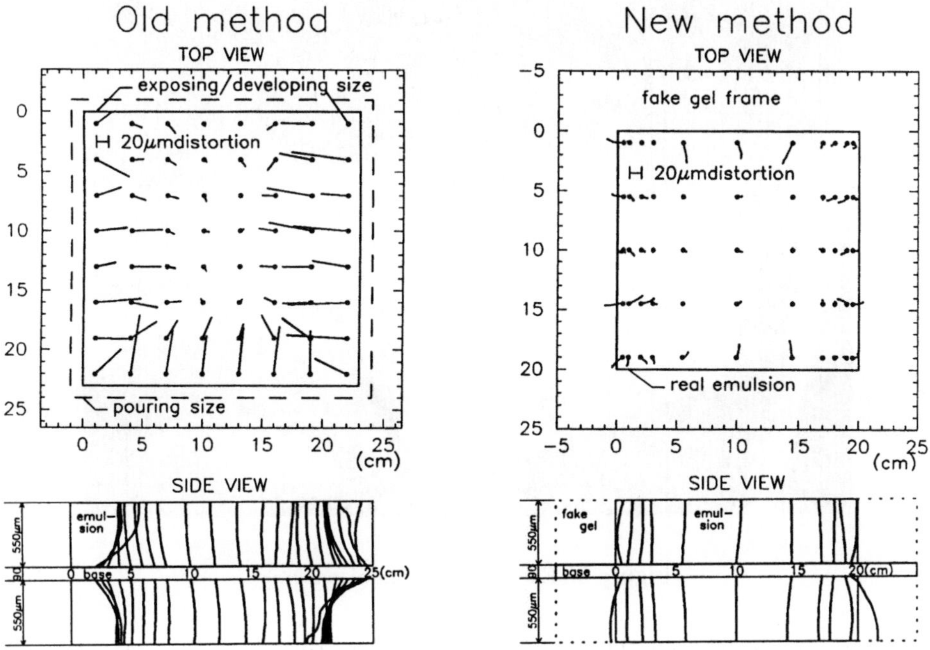

Figure 4: Maps of the distortion measured by penetrating tracks. The distortion near edges is larger than that in the central part. The figure on the right-hand side (denoted by "New method") shows the reduction of the distortion using the picture-frame-pouring method, dried with a vertical air flow.

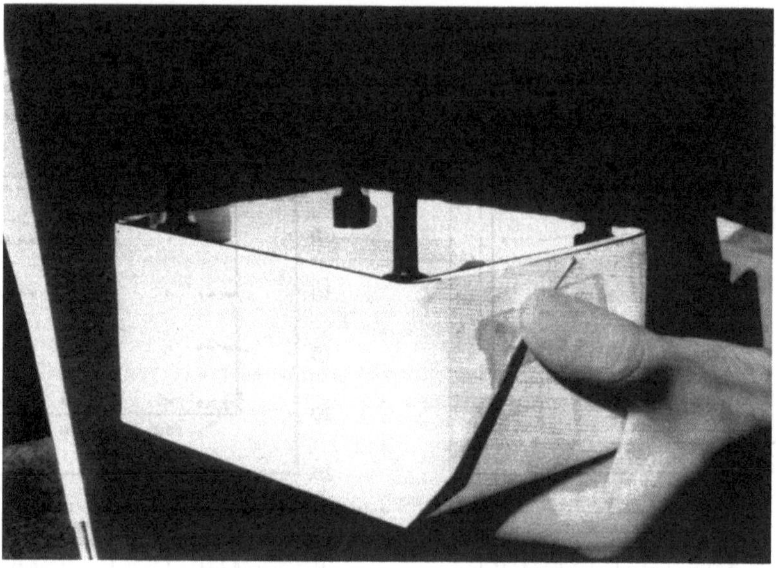

Figure 5: Example of origami packing

part like a picture frame. This new technique is effective as shown in Fig. 3.

Cutting and *origami*-packing

Stacking of the emulsion plates is the final task before the beam exposure. The size of the emulsion plates are arranged to make a precise emulsion chamber by using a specially-ordered cutting machine. In this process, the emulsion plates have to be protected from some kinds of stain and dusts. Well arranged emulsion plates are stacked in flat by putting them in a box of aluminized wrapping papers and by lowering the pressure in the box. This method is called "*origami*" packing[2][11], with which it is also easy to keep humidity to be constant . Fig. 5 is the example of emulsion chamber arranged by origami packing.

Fading of the emulsion plates during a long time period strongly depends on the relative humidity and temperature in the store. Fig. 6 shows this fading characteristics of Fuji ET-7B. From Fig. 6, it is required to keep the relative humidity to be less than 50% or temperature lower than 5°C to avoid substantial fading. Fig. 7 shows the time required to reduce the relative humidity of 85% in emulsion to 49%, 58% or 70%, as in the container.

[2]Origami is traditional Japanese art of folding paper into figure.

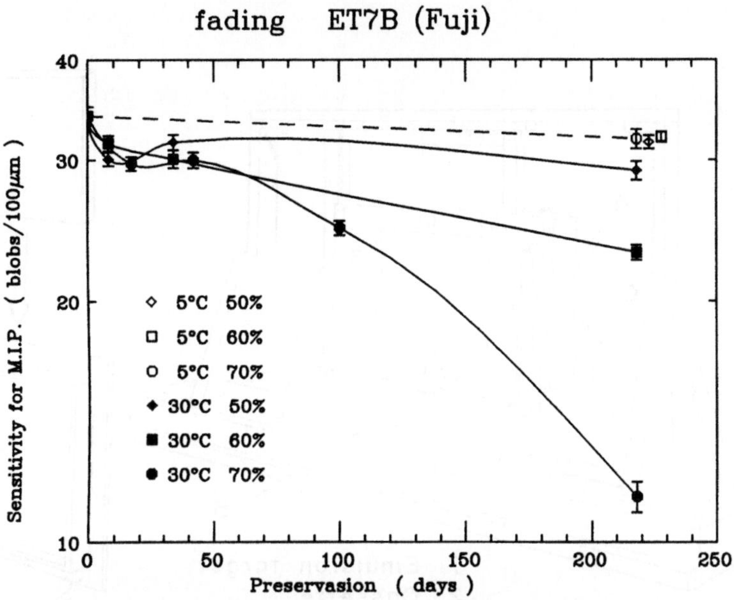

Figure 6: Fading characteristics of Fuji ET-7B

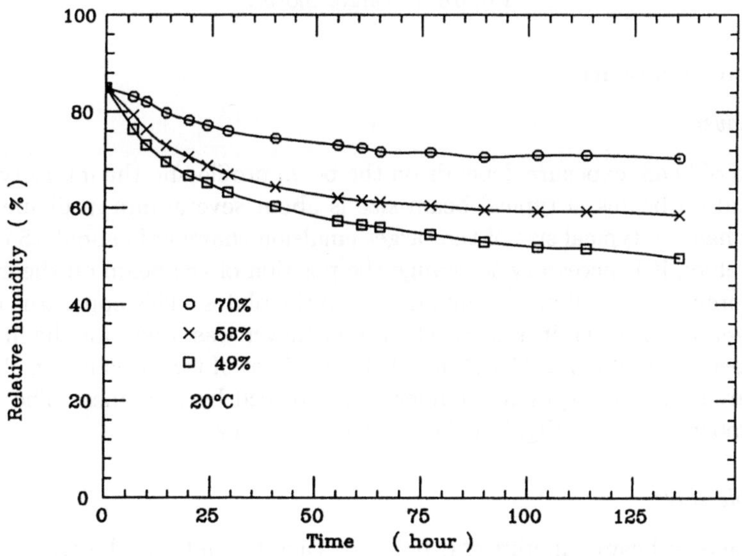

Figure 7: Time dependence of a relative humidity in emulsion

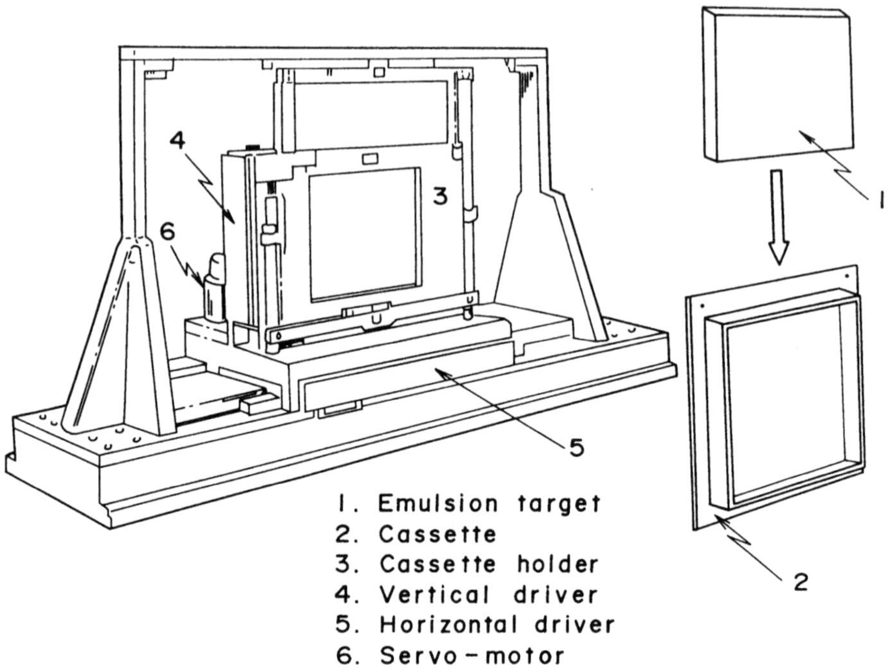

Figure 8: Target mover.

2.3 Beam exposure

Target mover

A method of beam exposure depends on the beam profile and the intensity. In the case of hadron beams, a typical beam size is about several mm in diameter. On the other hand, a typical size of the target emulsion chamber is about 25 cm $\times$ 25 cm. Therefore, it is necessary to change the position of the beam on the emulsion plates in order to get uniform beam density on the plates. This apparatus is called *"target mover"* as shown in Fig. 8. The target mover was used , for the first time, in the experiment WA75 at CERN in 1983 [12]. A lot of target emulsion chambers are prepared and the exposed chambers are replaced by new ones. The cassette type chamber as seen in Fig. 8 makes its handling easy.

Changeable sheet

In high density beam experiments, it is not easy to find out the track which is predicted by tracking counter out of a huge number of background tracks. Nagoya group used a special emulsion plate, which was changed many times during the beam exposure in 1978 (E531) [11]. This changeable sheet with a 75 μm thickness of emulsion coated on both sides of 800 μm Lucite base, was used as an interface

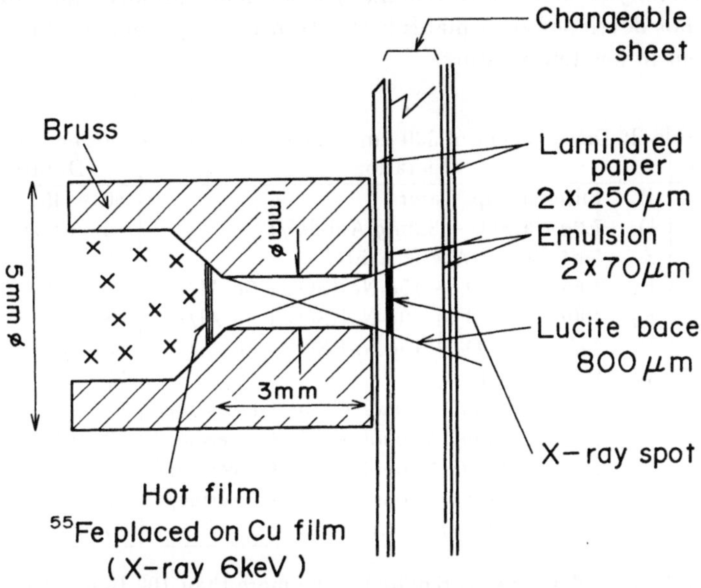

Figure 9: X-ray gun that emits collimated X-rays into the changeable sheet. A source intensity is typically 1 μCi for two week exposure.

detector to connect the tracks reconstructed in the counter to those observed in the target emulsion. The changeable sheet is regarded as a tracking detector of vector type with an angular resolution of 1 mrad and a position resolution of 1 μm. As an extension of this idea, a film type emulsion like a movie film was developed and used in the experiment of E653 at Fermilab. in 1987 [13].

The changeable sheet is placed in the form of origami packing. With this structure, it is not easy to achieve a mechanical alignment inside the emulsion plates with high accuracy. Collimated X-rays are then illuminated to make calibration marks on the emulsion plates. Fig. 9 shows an example of an X-ray collimator used in the experiment E531, where ^{55}Fe isotope with 1 μCi activity was placed in the brass collimator and its intensity was determined to fit the time interval ($10 \sim 20$ days) of replacing the changeable sheet.

2.4 Processing and mini-plate

Processing

Both-sides-coated emulsion plates can be developed by a method similar to that for a pellicle emulsion. Typical development condition is listed in Table 2. Both sides emulsion plate are hung vertically through the developing and fixing processes. Insufficient development lowers the detection sensitivity and over developing also results in the same effect due to too many hogs (grains) created randomly. To keep the transparency of thick emulsion plate high, the development process must be done carefully as well as a careful fixing treatment. Usually short-time development in a hot developer gives plates better transparency, compared to a long-time development in low temperatures.

Table 2: Development of 330 μm double-side coated emulsion sheets.

Stage	Operation		Duration		
1	P	Filtered Tap Water	50	(35)	min.
2	D	$7.2g$ $Na_2SO_3 + 3.2mg$ Amidol/ℓ	6	(3.5)	hour
3	S	$5m\ell$ Glacial Acetic Acid/ℓ	50	(50)	min.
4	F_1	$200g$ $Na_2S_2O_3 + 15g$ $NaHSO_3 + 5g$ Al/ℓ	50	(35)	min.
5	F_2	$400g$ $Na_2S_2O_3 + 30g$ $NaHSO_3 + 5g$ Al/ℓ	48		hour
6	F_3	$200g$ $Na_2S_2O_3 + 15g$ $NaHSO_3 + 5g$ Al/ℓ	50	(35)	min.
7	W	Water Wash	72		hour
8	A	Soak in $200m\ell$ Alcohol $+115g$ Glycerine/ℓ	50	(35)	min.
9	A	Soak in $400m\ell$ Alcohol $+115g$ Glycerine/ℓ	50	(35)	min.
10	A	Soak in $600m\ell$ Alcohol $+115g$ Glycerine/ℓ	50	(35)	min.
11	A	Soak in $800m\ell$ Alcohol $+115g$ Glycerine/ℓ	2		hour
12	D	Dry at 60 to 70 % R.H.	24		hour

1. The volume of developer is required to be more than 100 times that of the dried emulsion volume and that of main fixer is required to be 250 times.

2. Stages 1–11 are processed at 5°C. It is highly desired to keep 5°C for stages 1–6.

3. The numbers in the parentheses denote the case with an operation at 10°C. (Add 10 g Na_2SO_4 to P and 5 g Na_2SO_4 to F in the case of the 10°C operation)

Mini Plate

In scanning tracks from one plate to another, emulsion plates on the measuring stage have to be replaced quickly. For this purpose the *"mini-plate"* method shown in Fig. 10 was introduced by the Nagoya group [12]. The processed emulsion plates are sliced by a slicing machine (shown in Fig. 11) into the mini-plate with the size of 3 cm × 3 cm to 7 cm × 7 cm, and these mini-plates are assembled on a backing plastic sheet. This assembled sheet is called "scanning plate"

The position of each mini-plate on a scanning plate is measured by using special marks printed on the emulsion surface by the grid film technique. By using this mini-plate method, the required positional accuracy is somewhat relaxed and changing the emulsion plates is done by moving the measuring stage with the aid of computer control. For well calibrated mini-plates, the time needed to connect the track from one plate to another is only 1 $\sim$ 2 sec.

2.5 Calibrations

There are three kinds of calibrations; external calibration, internal calibration, and sticking-address calibration of mini-plate emulsions. Before the development

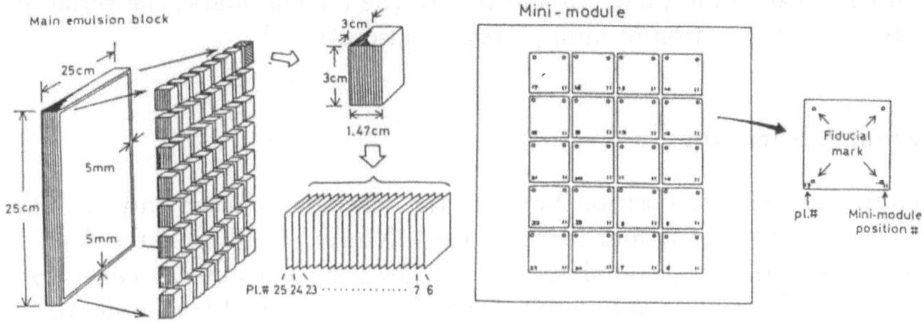

Figure 10: Mini-plate structure

Figure 11: Photograph of the slicing machine and mini-plates

of the exposed emulsion plates, the special marks is printed by using the grid film method on all of emulsion plates to obtain fiducial marks which give the coordinate on the emulsion plate. These fiducial marks, as shown in Fig. 10, are printed at the corner of each mini-plate. In addition to the fiducial mark, the serial number of plates and the position of mini-plates are also printed.

External calibration

To obtain the relative position between the tracking counter and the emulsion plates (changeable sheets and target plates), X-rays are illuminated from the X-ray gun outside the emulsion plates. Fig. 9 shows the 6 keV X-ray generator made of radioactive ^{55}Fe source of typically 1μCi. X-rays with higher energies of 20 $\sim$ 60 keV from a medicine/industrial X-ray generator are also used when one needs a high penetrating power. An attenuation length of X-rays in emulsions is 6 μm, 103 μm, 464 μm for 5 keV, 20 keV and 50 keV X-rays, respectively.

Internal calibration

The relative positional accuracy of each emulsion plate to 100 μm in the chamber is obtained only by aligning the edge of each plate. However, this accuracy is insufficient for connecting a track from one emulsion plate to another. The required alignment accuracy must be better than 20 μm. This is achieved by using high energy particles penetrating through the emulsion chamber; for example cosmic rays and charged particles associated in the incident beam.

Sticking address calibration

After sticking mini-plates on the backing plate, the relative position of each mini-plate is aligned with an accuracy of $\sim$ 10 μm.

2.6 Emulsion analysis system

Nagoya group started to construct an automatic analysis system for emulsion plates (Fig. 12)[10]. First attempt was to construct an on-line system to measure the track position on a TV screen by pointing quickly a position on the screen with a control stick (or mouse). The coordinate of the track position on a surface of an emulsion plate and the track angle were measured by using two grain information of both-sides-coated emulsion like a vector chamber (tracking jet-chamber). Developing the automatic track identification scheme was the next task.

With the use of an on-line analysis system, the scanning work became a routine work from an “art” This system has been operated by well-trained scanning technicians.

Figure 12: Scanning room of Nagoya University

Vertex Search

Under the guidance of the track information measured in the counter system, primary vertices are located on the emulsion target. There are two methods to locate the primary vertex.

One is the *"volume-scanning"* method, where the vertex is searched around the predicted position by a general scanning. The scanning volume depends on the reconstruction accuracy of the counters. Since the uncertainty along the beam direction is not so small, typically several emulsion plates need to be scanned.

The other is the *"scan-back"* method. The predicted track is searched in the emulsion plate at the end of the emulsion target. If a track is found, scanning moves to the next plate until arriving at the primary vertex. By using the mini-plate technology, this scan-back method is very effective to get quickly the vertex.

The scan-back method based on a combination of the mini-plate method and on-line system is a unique way to identify low-energy interactions or neutrino interactions quickly (about 10 min.) without losing efficiency.

Semi-automatic system

The semi-automatic emulsion analysis system is shown schematically in Fig. 13. This system consist of large mechanical stage controlled by a computer. The optics with a TV camera, and a video-image analyzer with a computer are also shown (Fig. 14).

A scanner sets the emulsion plate on the stage and operates the system from a key-board and a stick (or mouse) or a hand-box drawn on a monitor TV screen. Video images of the emulsion are superimposed on the character/graphics and a tracking window controlled by a scanner. The identification of the track is done by eye-scanning; the position is quickly measured by computer commands, and is stored in a storage disk. The mechanical stage can cope with many kinds of emulsion plates. A typical size of an emulsion plate is around 30 cm square and

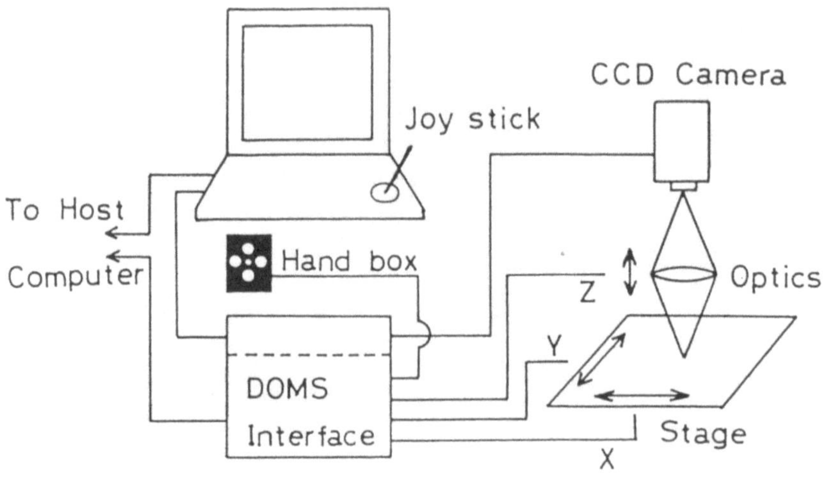

Figure 13: Block diagram of the semi-automated emulsion analysis system

Figure 14: Photograph of a measuring stage and a video processor

the maximum one amounts to 100 cm square.

Optics An objective lens requires a long working distance[3]. This allows one to examine the bottom side of both-sides-emulsion plate with a 1000 μm thickness. A TV camera is mounted on the pipe of micrometer optics, while keeping a constant magnification. The view size of the monitor screen is 150×150 μm^2 with a magnification factor of 50. Digitization is made on the TV screen into a mesh of 256 × 256 in the x and y axes. The image on the objective lens is projected onto CCD directly, without corrections through intermediate optics. The optical distortion of objective lens does not give any significant effect on the position measurement, because the CCD (10 mm in size) covers only the central part of the image (28 mm in diameter).

Stage mechanics The accuracy of stage mechanics is 1 μm with a moving stroke more than 15 cm. On the other hand, the size of the emulsion plate (scanning sheet) is 15cm×15cm to 35cm×35cm, and the size of mini-plate is 3×3 (to 7×7) cm^2. In order to change one plate with the next, it is necessary to drive 3 (to 7) cm quickly. The required speed is more than 2 cm/sec. The stage movement is controlled with an accuracy of 10 μm. Associated with the stage movement, immersion oil is injected between the emulsion and the objective lens (see Fig. 15).

TV camera and system controller In the case of an emulsion plate used in the hybrid experiment, the coordinates of the track passing perpendicularly through the emulsion plate must be measured. By adjusting a focal plane with a depth of the 3 μm at the surface of an emulsion (magnification by a factor of 50), the coordinate of the identified track can be measured as follows: a scanner moves the tracking window (0.1 % of the TV screen) on the track roughly using a stick or mouse; the position of the track is then automatically measured by detecting the darkest point in the tracking window. This system measures the darkness of the track which indicates ionization loss.

To assist a measurement by the scanner, the host computer gives the necessary data (for example, plate position, track slope, momentum, and so on) on the TV screen by graphics or by characters according to the request of the scanner.

Full automated system [10],[14]

Principle of track recognition *"Track selector"* is a type of the video image processor. Its basic operation is shown in Fig. 16. It has 16 frame memories, which can be used separately. By changing the focal plane at regular intervals of the depth inside the emulsion, tomographic images of grains are obtained for slices of the emulsion and they are stored into each frame memory. After reading the images of 16 slices in the emulsion, the track selector recognizes track candidates by examining the coincidence rate of images of 16 layers at every pixel.

---

[3]Long working distance of 1.1 mm (with a magnification factor of 90 (with numerical aperture (NA) = 1.25) and 50 (NA = 0.95)) was made available by Tiyoda Optics Ltd. (The numerical

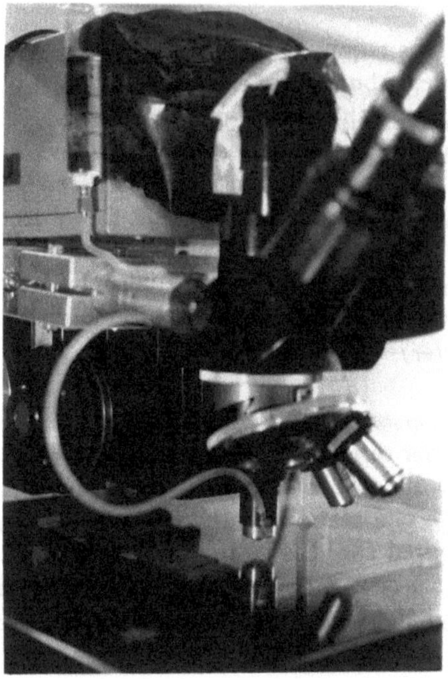

Figure 15: Photograph of an immersion oil injector

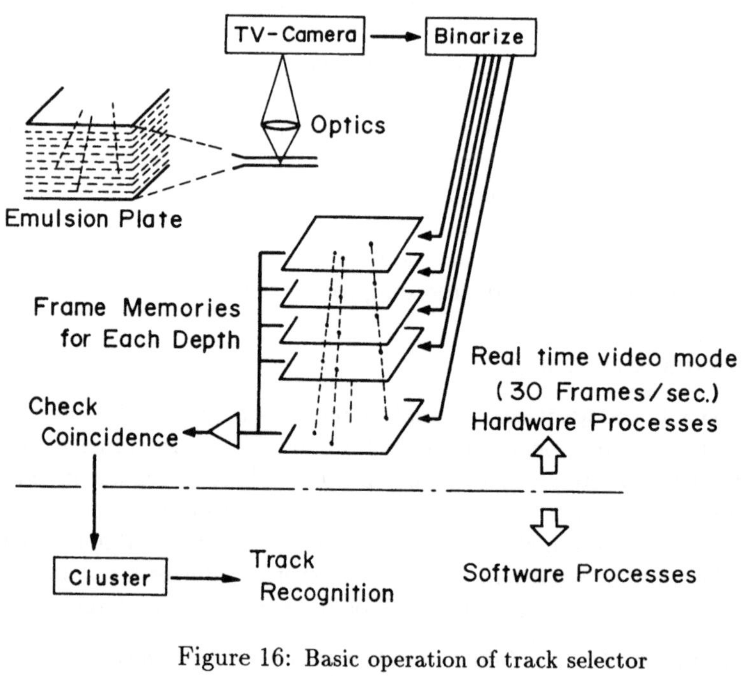

Figure 16: Basic operation of track selector

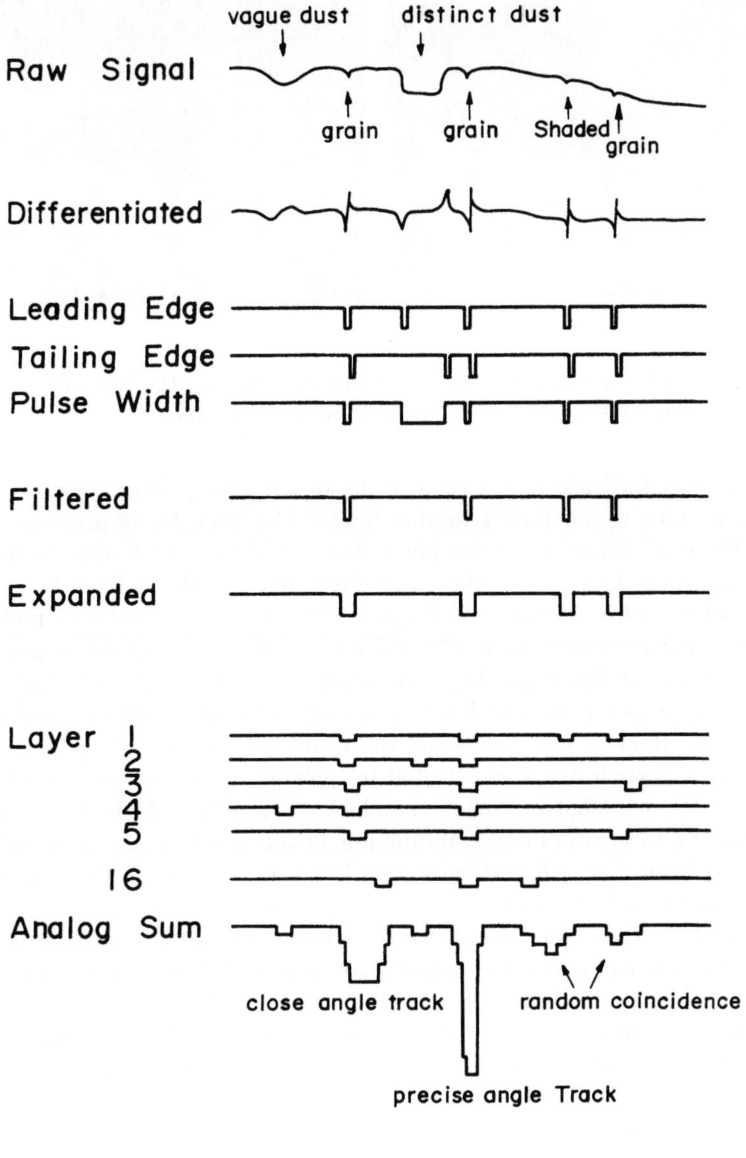

Figure 17: Step by step transfer of a video signal.

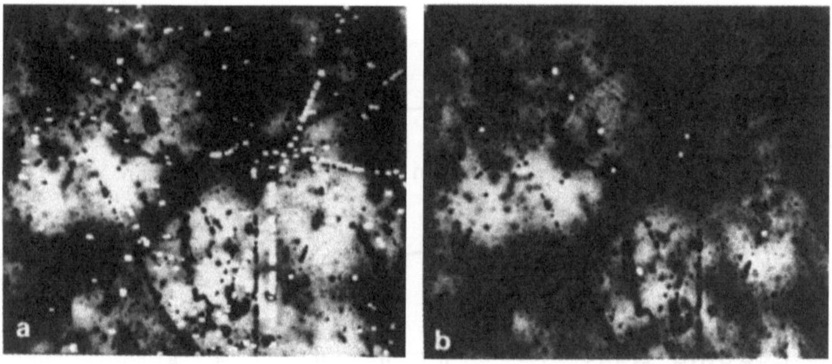

Figure 18: (a) Stored signal of one layer superimposed on the raw video image. (b) Discriminated track signal superimposed on the raw video image.

Fig. 17 schematically shows the transfer of a microscopic image of an emulsion plate to a video signal (one scanning line). This transfer is processed step by step in the track selector. In the raw video signal, in-focus grains corresponding to particle tracks form thin spikes, and dusts and heavily ionizing tracks usually form broader peaks. At first, the signal is differentiated in order to pick up only spike signals of in-focus grains. The differentiation also eliminates vague shadows and unevenness of lightings. Discrimination of the differentiated signal is made by setting a proper threshold level. Leading and trailing edges are detected to recognize the track of particles using the width information of each signal; using the width, a particle track is selected by picking up narrow and sharp signals. The information is digitized using the clock of 12 MHz, which gives digitization corresponding to 1/512 of one scanning line of the video signal. (The video signal has already been digitized vertically with 512 scanning lines.) The video signal is thus digitized into 512×512.

The field of view is 160×160 μm^2 with an objective lens of a magnification factor of 50, and hence one pixel corresponds to about 0.3×0.3 μm^2. A digitized image of a grain occupies typically 3×3 pixels. Fig. 18 shows stored signals superimposed on a raw video image. All of these processes are carried out by hardware, and thereby the video signal is processed in real-time. The process time for each frame is about 33 ms which corresponds to 30 frames per second.

Each frame memory works independently. After storing all the signals of 16 layers, the digital signals in all frame memories are integrated. A sharp peak as shown in the bottom line of Fig. 17 is identified as a penetrating track. In order to detect a slopping track, the frame memories are shifted regularly in proportion to a desired angle before the integration.

A group of coordinates (typically 5 to 7) are recorded for each track, because the size of a pixel is smaller than grains that constitute a track. Finally, the group

aperture is defined by $n \sin\theta$ with n the reflection index and θ the angle of the incident beam.)

of the coordinates are clustered by a software. Track selector sends the central coordinates of the cluster of observed tracks to the host computer.

It takes much shorter time to process the video images than to control the optical system mechanically and to communicate with the host computer, because most of processes are carried out by hardware in a track selector. If we use only software techniques, it would take an enormous amount of time to accomplish this series of video image processes.

3 Application

3.1 Neutrino experiment in E531 [11]

An emulsion hybrid experiment was originally proposed to search for charmed particles at Fermilab in Sept. 1975 (P517) as one large international collaboration of emulsion-counter groups. Before starting the experiment, however, this large collaboration split into three groups: E531 organized by B. Reay and the emulsion group of Japan, E382 organized by L. Hand based on an emulsion-spark chamber, and E564 organized by L. Voyvodic using the 15 feet Bubble chamber.

The apparatus of E531, shown in Fig. 19, consisted of two major parts, a nuclear emulsion target in which short decay paths of charmed particles were accurately measured, and the spectrometer with an analyzing magnet in which decay products of charmed particles were efficiently measured. In this experiment, a single horn wide-band neutrino beam produced by 350(400) GeV protons was used for the following reasons: (1) charm is produced in neutrino interactions at a higher relative rate than in other interactions (about 7% of the total charged-current cross section) and (2) neutrinos leave no trace of their passage through the emulsion target unless they interact. The spectrometer predicted the neutrino interaction vertices in the emulsion target and the exit points from the emulsion stack for individual secondary tracks arising from the neutrino interactions. The exit point prediction was a unique feature of this experiment that made possible to locate neutrino interaction vertices in the target emulsion with a very high efficiency, when combined with a single track follow-back method.

Two runs were made in this experiment. In the first run started on November 18, 1978 and ended on February 7, 1979, 7.2×10^{18} proton of 350 GeV/c were incident on the production target, from which an estimated total flux is 1.7×10^{15} neutrinos/m^2 passed through the emulsion target with the total area of 0.49 m^2. 1257 events were actually found in the emulsion target. The second run started from Dec. 1980 and ended May 1981. 6.8×10^{18} protons with 400 GeV/c were used. The energy spectrum of neutrinos is shown in Fig. 20; the fraction of ν_μ is 99.8% in the first run and 98.5% in the second.

The emulsion target consisted of 22.6 liters in the first run and 30 liters in the second run using Fuji ET-7B nuclear emulsion packaged in 39 modules in each run. Two types of emulsion modules were used in the two runs (see Fig. 21). There were 12 modules in the upper half of the target, each of which consisted of 177 pellicles having a size of 5×14 cm and a thickness of 614 μm, and 27 modules in the lower half of the target, each of which consisted of 78 plates having the size of 9.5×12

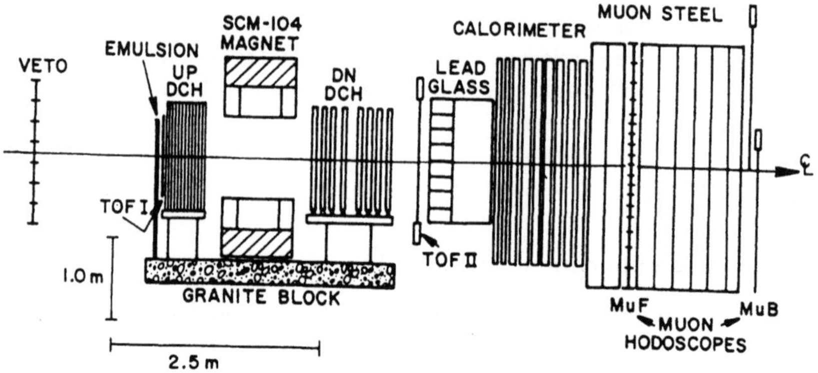

Figure 19: Elevation view of apparatus

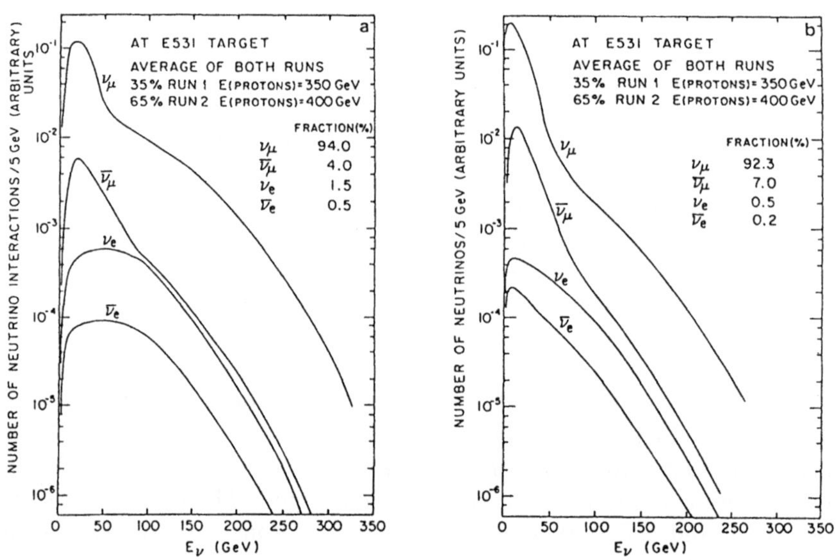

Figure 20: The neutrino energy spectrum (averaged over both runs) predicted by a beam Monte-Carlo program with the parametrization given by Malensek

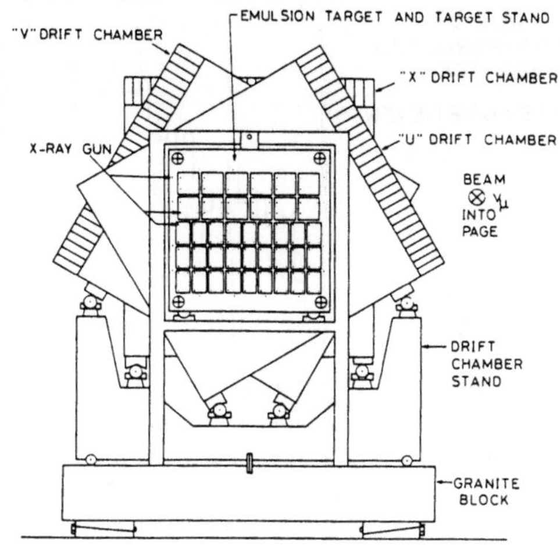

Figure 21: Support stands, emulsion target, and upstream drift chambers.

cm with a 330 μm thick emulsion coated on both sides of polystyrene sheet of 70 μm. Immediately downstream of the emulsion modules, two large sheets of Lucite of 91.5 cm $\times$ 40 cm, 800 μm thick, coated on both sides with 75 μm of emulsion, were attached with the pressure lowered between the modules. These large sheets were made at Fermilab immediately before the exposure, and changed every two or three days to limit the total number of background tracks passing through them. They were replaced 27 times during the run. These changeable sheets were employed in order to follow easily the tracks predicted by the spectrometer into the target emulsion.

Track candidates from neutrino interactions were located on the changeable sheet with an accuracy of 320 μm (1σ) by predictions from the drift chamber; those found on the changeable sheet were then quite easily followed into the emulsion modules with a spatial resolution of 20 μm (1σ). The efficiency for following tracks from the changeable sheet into the emulsion was 95% for all fast tracks in the vertical modules.

The position of the changeable sheet relative to each emulsion module was located to an accuracy of 50 μm through the use of ^{55}Fe X-ray sources (guns) of 1μCi embedded in the four corners of each emulsion module. Sixteen X-ray guns on the Hexcel surrounding the emulsion target also coupled the changeable sheet directly to the coordinate system of the spectrometer with a comparable accuracy. A total of 20 drift chambers, 12 upstream of the analyzing magnet and 8 downstream, were used to locate events and measure the momentum. The upstream drift chamber had 32 wires, 20 μm spaced and 4.0 cm apart, and the downstream drift chamber had the wires spaced 5.08 cm.

The analyzing magnet had the bending power of 6.2 kG m and the error in momentum due to a finite resolution of the drift chamber and scattering is $\sigma_p/p = (0.005p)^2 + (0.01)^2$ with p in units of GeV.

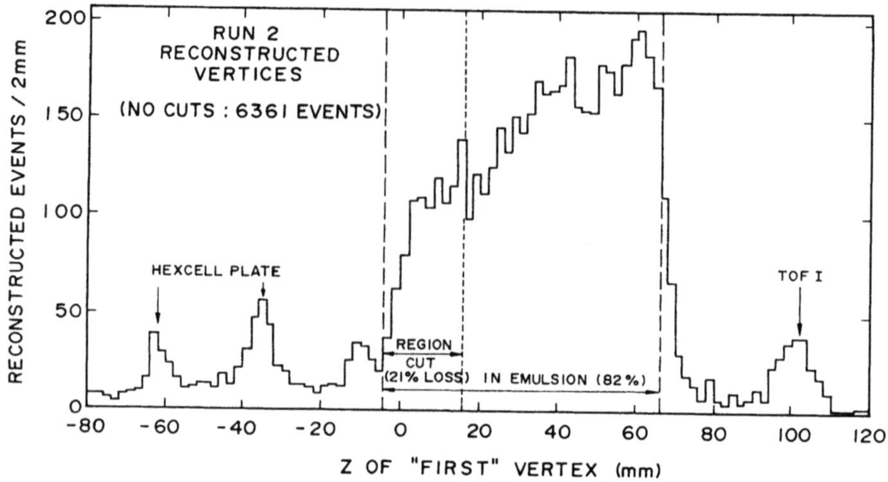

Figure 22: z-distribution (along the beam) of the predicted neutrino interaction vertices for each reconstructed event of the full second run sample; 82% of the events were predicted to lie within the emulsion, while the most upstream events (21%) of those were later discarded. The gradient in the number of events in the emulsion indicates the presence of a problem with scattering of particles in the emulsion.

Emulsion analysis and physics result

The vertices of the neutrino interactions in the target emulsion and also exit points of charged particles produced in neutrino interactions from the emulsion were predicted by reconstructing each event from the information obtained in the downstream spectrometer.

The vertices of neutrino interactions in the emulsion were searched for by following back the tracks, where the track candidates were first located on the changeable sheet and then followed into the emulsion. An area of ± 1 mm^2 on the changeable sheet around the predicted point was scanned. The average time needed to process one predicted event was one hour.

As shown in Fig. 22, the reconstruction efficiency was low in the 2 cm upstream side of the emulsion. 2638 events were located in the emulsion compared to 3220 predicted events. Examples of the charm production are shown in Figs. 23 and 24.

As a byproduct of the charm experiment, a search for direct tau lepton production was done to examine $\nu_\mu/\nu_e \rightarrow \nu_\tau$ oscillation. Since the tau lepton has a lifetime similar to charmed particles, it should be seen in an emulsion target. In total, 3886 (1248 in the first run, 2638 in the second run) neutrino and antineutrino interactions were located in the emulsion and 164 charged particle decay candidates were identified. Because of possible background due to scattering, a kink is not considered as a τ^- decay candidate if the secondary decay track has a momentum perpendicular to the parent direction (P_T) smaller than 125 MeV/c. Finally, 104 charged decay candidates for τ were obtained. Since the beam contained 90%

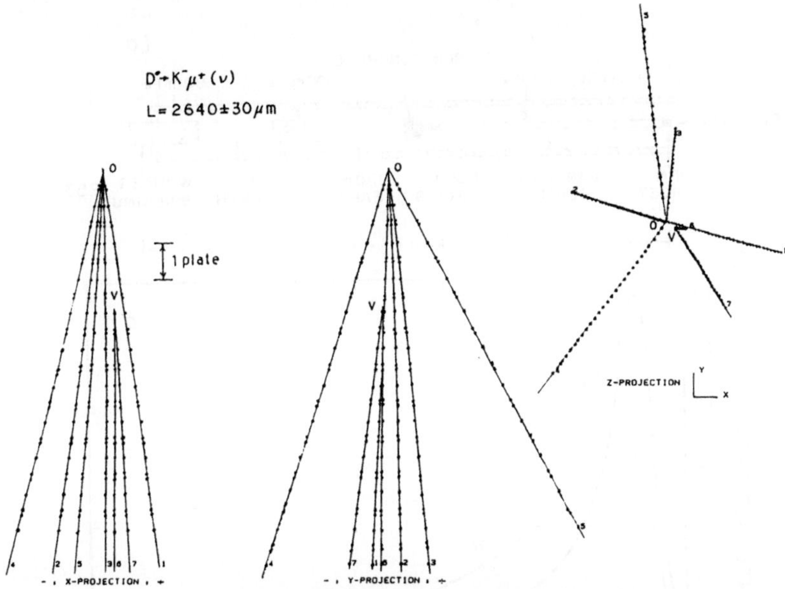

Figure 23: $D^0 \rightarrow K^-\mu^+(\nu)$

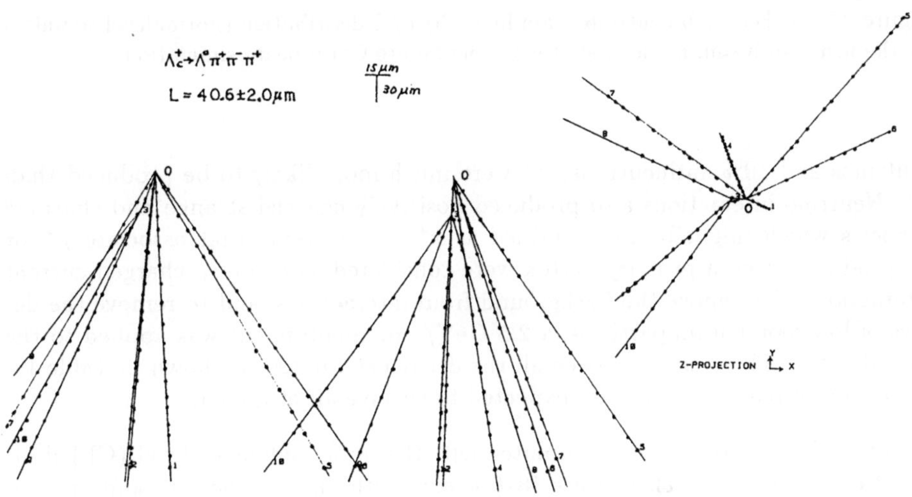

Figure 24: $\Lambda_c^+ \rightarrow \Lambda^0\pi^+\pi^-\pi^+$

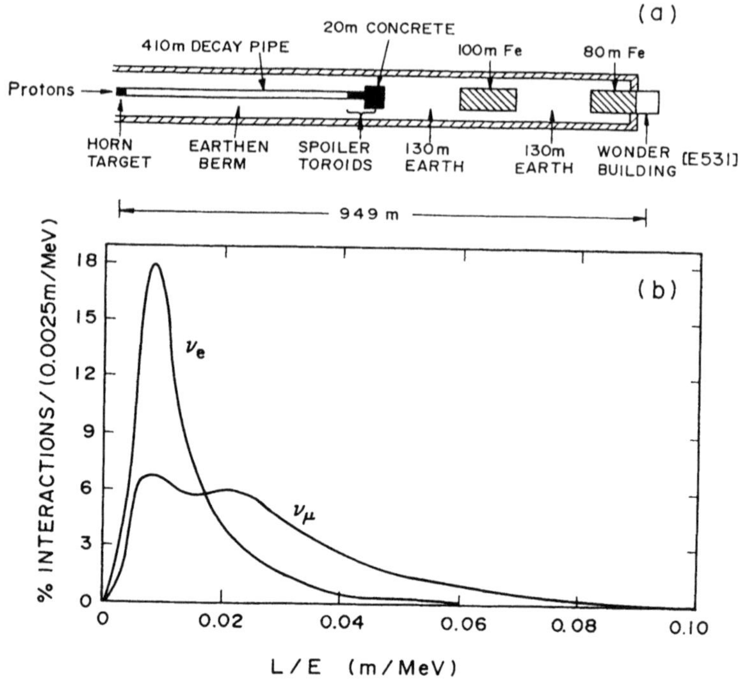

Figure 25: (a) Fermilab neutrino beam line. (b) L/E distribution (normalized to unity) for the neutrino beam, as calculated with our Monte Carlo beam simulation.

neutrinos and 10% antineutrons, τ^- were much more likely to be produced than τ^+. Neutrino interactions also produced positively charged strange and charmed particles which might be misidentified as τ^+. The events that associate μ^+ or μ^- produced from a primary vertex were considered as $\bar{\nu}_\mu$ or ν_μ charged-current interactions. To remove the background from interactions, and to remove the decays of low momentum particles, a 2.5 GeV/c momentum cut was applied to the τ candidates. These cuts removed all the decay candidates, as shown in Table 16, while 95% of real τ^- decays are expected to survive all of the cuts.

Since there were no tau candidates left, the 90% confidence-level (CL) limit of 2.3 events was set. There were 1870 events with an identified μ^- and an estimated 53 e^- events, yielding an uncorrected upper limit of $R(\mu^-) = 2.3/1870 = .0012$(90% CL) and $R(e^-) = 2.3/53 = 0.043$(90% CL), where R is the probability that ν_μ/ν_e oscillates into ν_τ. The correction factors for detection efficiency were analyzed and found to be 1.6 and 1.7, giving upper limits of $R(\mu^-) = 0.002$ and $R(e^-) = 0.073$.

To interpret this limit in terms of $\nu_\mu/\nu_e \rightarrow \nu_\tau$ oscillations, a two-neutrino mixing parametrization is assumed. The probability that a ν_μ/ν_e has oscillated

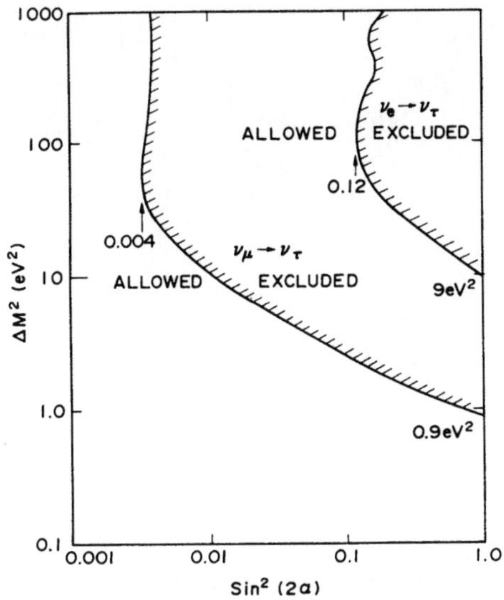

Figure 26: The region excluded by neutrino oscillation experiments E531 on the Δm^2 vs. $\sin^2(2\alpha)$ plane. The curves show the 90% C.L. limits for ν_μ/ν_τ and ν_e/ν_τ oscillations.

into a ν_τ is

$$P(\nu_\mu/\nu_e \to \nu_\tau) = \sin^2(2\alpha) \int \rho(\frac{L}{E}) \cdot \sin^2(1.27 \cdot \Delta m^2 \cdot \frac{L}{E}) \cdot \delta(\frac{L}{E}) \qquad (3)$$

where α is the mixing angle between the neutrinos, Δm^2 is the difference of the squared neutrino masses, L is the neutrino flight path in meters, E is the neutrino energy in MeV, and ρ is the neutrino energy distribution. Fig. 25 shows the distribution of $\rho(L/E)$ for the neutrino beam. Using the limits quoted above, the excluded regions in Δm^2 vs. $\sin^2(2\alpha)$ are shown in Fig. 26 [15][16].

3.2 Short base line neutrino oscillation

The emulsion hybrid experiment (E531) enabled us to obtain a remarkably good upper limit for the $\nu_\mu \to \nu_\tau$ mixing, $\sin^2(2\alpha_{\mu\tau}) < 0.004$, with the mass resolution of $(m^2_{\nu_\tau} - m^2_{\nu_\mu}) > 10\text{eV}^2$. An article by Harari motivated us to propose an improved oscillation experiment of $\nu_\mu \to \nu_\tau$ by the emulsion hybrid experiment [17]. He suggested that among the candidates of cosmological dark matter the light neutrino ν_τ with a mass $m(\nu_\tau) = 15 \sim 65$ eV is the most likely possibility, and that the only practical way to show $m(\nu_\tau)$ to lie in this range is to search for $\nu_\mu \to \nu_\tau$ oscillations down to $\sin^2(2\alpha_{\mu\tau})$ being as low as 4×10^{-4}.

New experiment P803, which was ten times more sensitive compared to the old experiment E531, was proposed to Fermilab in 1988. This proposal, however, was not approved at that time[4], because this required a *Main Injector*, which

[4]P803 was approved in Nov.1993 after the cancellation of SSC.

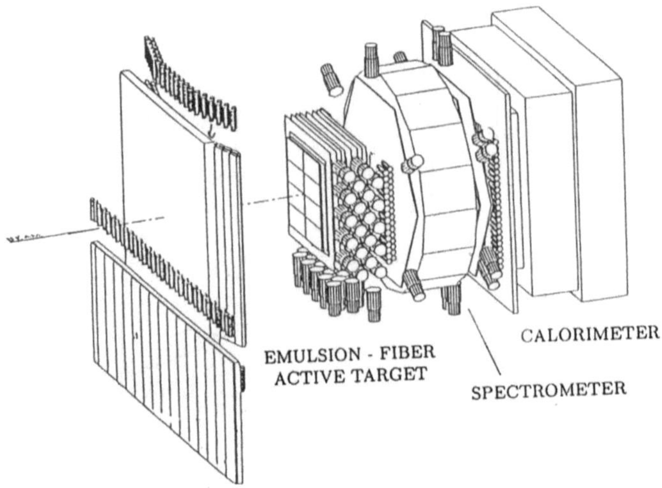

Figure 27: Set up of CHORUS

should give a neutrino flux ten times stronger than that is available using the existing main-ring. Upon the suggestion by S. Hayakawa and Y. Yamaguchi a modified experiment (WA95 [18]) was then proposed to CERN in 1989 using SPS neutrino beam. It was soon approved (nickname is CHORUS[5]). At the same time a counter experiment aiming at a similar physics, NOMAD, was also approved. Both neutrino oscillation experiments are aiming at a search for neutrino oscillation involving ν_τ with a small mixing angle around $\sin^2(2\alpha_{\mu\tau}) < 3 \times 10^{-4}$ for a mass parameter $\Delta m^2 > 1\mathrm{eV}^2$.

CHORUS detector

The CHORUS experiment attempts to detect directly the decay of $\tau^- \to \mu^-\bar{\nu}_\mu\nu_\tau$ (18%) and $\tau^- \to h^-(n\pi^0)\nu_\tau$ (50.3%) and also other decay modes, $\tau^- \to h^-(n\pi^\pm)\nu_\tau$ and $e^-\bar{\nu}_e\nu_\tau$ in the emulsion target. The CHORUS detector, shown in Fig. 27, has a combination of a large volume massive active target and a spectrometer. The configuration of this active target has a sandwich structure of bulk emulsion stacks and fiber trackers. Figure 28 shows the detailed structure of *emulsion-fiber* active target and the schematic view of the procedure of the follow-back location of a neutrino interaction.

The emulsion target has a volume of 230 liters in total and the target mass of about 800 kg is segmented into two units, each unit having 2 stacks of a 1.44 m × 1.44 m target with a thickness of 2.75cm. Each stack is subdivided into 8 sectors of 0.72 m × 0.36 m; one sector comprises 25 emulsion plates. The emulsion plate (2 × 550 μm thick) are made by the film of tri-acetate-cellulose with 90 μm in

[5]Japan (Nagoya-Univ., Gifu-Univ., Aichi-Educ.-Univ., Kinki-Univ., Kobe-Univ., Osaka-City-Univ., Toho-Univ., Utsunomiya-Univ.), Korea (Gyonsan-Univ., Chonnan-Univ., Changwon-Univ.), CERN, Russia (ITEP), Belgium (ULB-VUB, Louvain-la-Neuve), Ankara, Italy (Roma-Univ., Bari-Univ., Napoles-Univ., Salerno-Univ., Ferrara-Univ.), Amsterdam, Münster

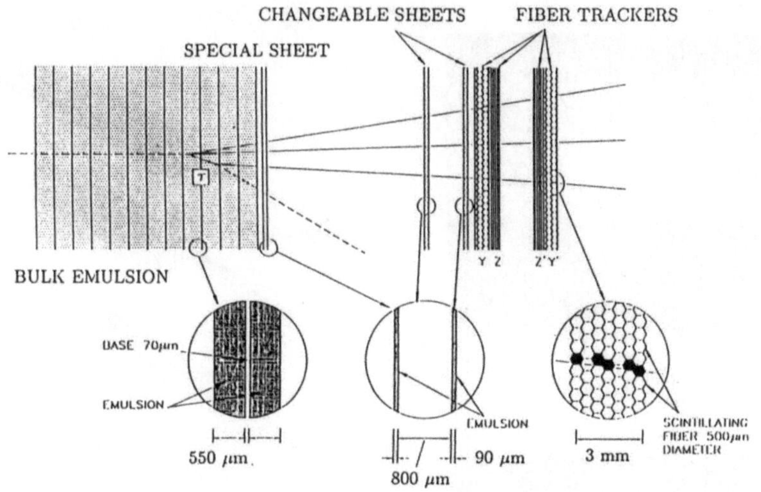

Figure 28: Schematic view of $\nu_\tau N \rightarrow \tau^- X$ event in the emulsion. Nuclear fragments occur at the vertex in $\sim 25\%$ of the events.

thickness coated both sides. At the bottom of the target emulsion, a Lucite of 800 μm in thickness coated both sides with the emulsion of thickness of 70 μm was placed for a convenience to connect the tracks from the changeable sheet to the target emulsion.

A pair of changeable sheet was mounted in front of the fiber trackers (see Fig. 28) to help the track finding in the emulsion from the prediction of fiber trackers. This changeable sheet will be replaced with a new one more than ten times during the run to avoid the integration of muons produced by the neutrino beam and cosmic rays. Pouring of emulsion and development will be done at CERN.

Scintillating fiber trackers predict the track position on the changeable sheet within an error of $(300\ \mu\text{m})^2$. Each scintillating fiber tracker provides 4 projected coordinates y, z, y', z' to determine unambiguously the track; the projections y' and z' are rotated by an angle of 8° with respect to y and z for this purpose. A good resolution for two tracks is required to overcome large particle densities close to the interaction vertex. Each tracker is composed of ribbons of 500 μm diameter scintillating fibers[6] arranged in seven staggered layers (*tawara* structure[7] with a thickness of 3.2mm as shown in Fig. 29(a) and Table 3) to ensure a good efficiency. Each fiber ribbon has three parts; "tracker part" "flexible part" and readout-part shown in Fig. 29(b). These ribbons were fabricated at Nagoya university by a computer controlled ribbonization machine shown in Fig 30 [19]. The fibers having a cladding of 15μm thickness are covered with a TiO_2-based white paint acting as a glue and as an extra-mural absorber. The surface of the end of the ribbon is covered with a sputtered aluminum having a reflectivity better than 80 %. The other end

[6] Fibers SCSF-38 manufactured by Kuraray in Japan
[7] A Japanese straw rice-bag, which is usually stacked as in Fig. 29

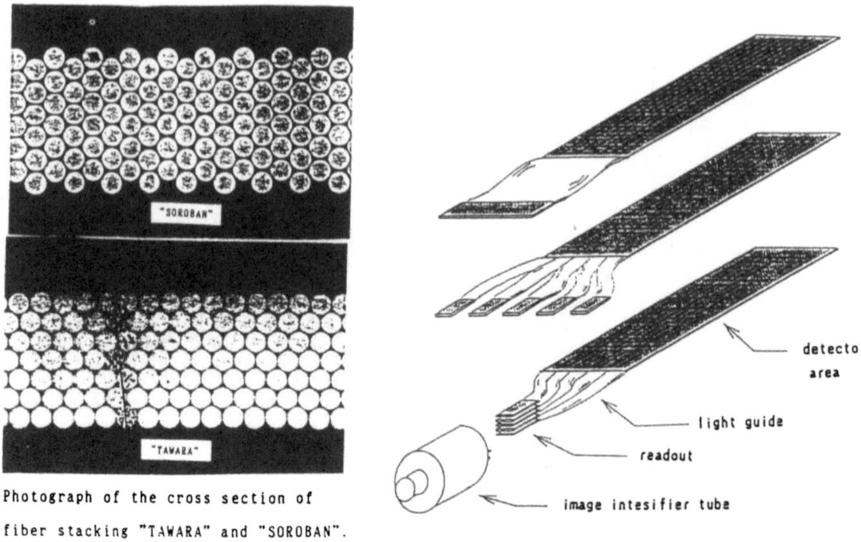

Figure 29: (a) Photograph of the cross section of fiber stacking *tawara'* and *soroban'* (abacus). (b) Fiber tracker and read out.

of the ribbon is bundled together using the flexible part and polished to couple to the photocathode of the first stage image intensifier[8] of 10 cm in diameter. To obtain a sufficient sensitivity to a single photon, 4-stage image-intensifier chains (see Fig. 31) are used to read out scintillating fiber ribbons.

The first stage image intensifier has a fiber optics window to suppress the light divergence on the photocathode. A quantum efficiency of 20-26% including the loss in the fiber optics window are obtained by measuring the cathode current.

Image intensifiers are read out by a CCD chip of 550 $\times$ 288 pixels. Overall

[8]Manufactured by Hamamatsu Ltd.

Number of photon	30 Photons at 50 cm
	10 photons at 240 cm
Diameter of fiber	495 μm
Clading thickness	5% of core = 15μm
Diameter of core	$495 - 2 \times 15 = 465\mu$m
Fablicated Ribbon	7-layers, 43 cm wide (860 fibers)
(The length of tracking part = 150cm)	
(Flexible part without glue $\sim$ 80cm)	
(Reading part = 15cm)	
Number of ribbon	$4 \times 4 \times 8 = 128$ ribbons
Total fiber length	1650 km

Table 3: The summary of fiber tracker of the the active target.

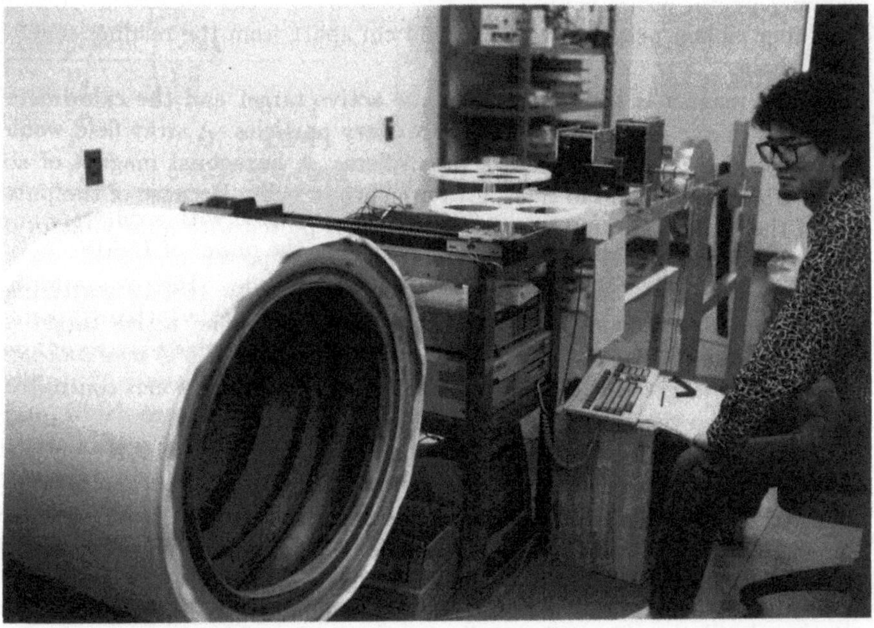

Figure 30: Photograph of the fiber tracker fabrication machine of Nagoya University.

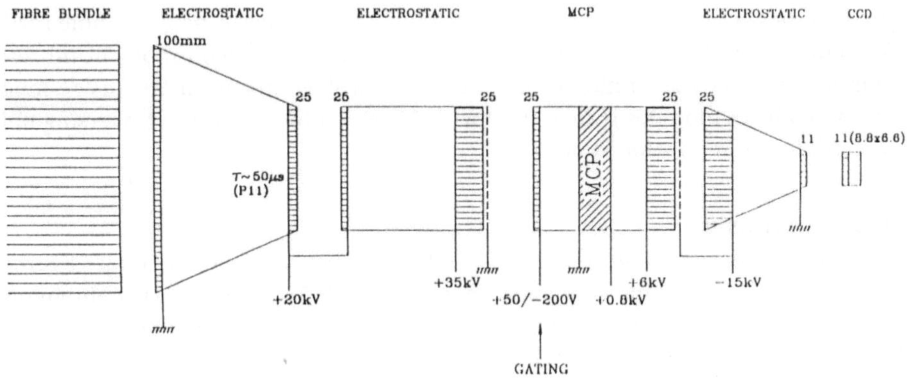

Figure 31: Chain of image intensifiers read out by a CCD. The microchannel plate is gated. The total demagnification is 9.1, providing 6 pixels per fiber.

tracking efficiency of 99% is obtained for a minimum ionizing particle passing throw the fiber ribbon perpendicularly at 250 cm apart from the reading edge of the optical chain.

An air core magnet is located between the active target and the calorimeter to measure the charge and momentum of secondary particles. A stray field would prohibit to use the electrostatic image intensifiers. A hexagonal magnet of six equal-sided triangles with 1.5 m sides, 0.75 m depth is built. Because of the pulse structure of the neutrino beam the magnet-current is pulsed with a duty-factor of $\sim 4 \times 10^{-3}$. A field of 0.12 T is attained by applying the power of 15 kW.

The designed value of a momentum resolution is $(\Delta p/p)_s = 16\%$ by scattering, and $(\Delta p/p)_t = 1.6\% \times p(\mathrm{GeV/c})$ by tracking resolution. The active target of emulsion stacks and fiber trackers are kept at a temperature of 5°C in a cool box covering the whole target area. In this way the fading of the emulsion is controlled, and a lifetime of two years is achieved.

The CHORUS calorimeter is designed to distinguish the reaction $\nu_\tau \mathrm{N} \to \tau^-(\to \mu^-$ or $h^-)\mathrm{X}$ from ν_μ induced charged- and neutral-current reactions. To measure the missing transverse momentum and the angle between the missing particles and a hadron shower, an electromagnetic calorimeter (EM) and two hadron calorimeters (HD1, HD2) are installed.

The EM sector modules are build by piling up extruded layers of grooved lead and plastic scintillating fibers (1mm diameter) positioned in the grooves. The HD1 modulates are built in a similar way, piling up the same fibers and lead. The achieved energy resolution is $\sigma(E_e)/E = 13\%/\sqrt{E}(\mathrm{GeV})$ for electrons and $\sigma(E_h)/E = 30\%/\sqrt{E}(\mathrm{GeV})$ for pions.

The muon spectrometer consists of six circularly magnetized iron modules. Each magnet module consists of twenty iron plates of 2.5 cm thickness; the magnetic field in the iron is close to 2 T. The range of muons can be determined with a sampling step of 12.5 cm of iron (every 5 planes). The muon momentum resolution depends on the number of modules traversed by muons; a 20 GeV/c muon traversing all six modules is measured with 15 % resolution; if it is traversing only one module the resolution is 30 %.

Neutrino exposure

The SPS accelerates protons to 450 GeV with a cycle of 14.4 seconds. Protons are extracted in two 6 ms long spills separated by 2.5 s. The relative abundance and the energy of the neutrino flux are listed in Table 4. The muon flux incident on the detector is $50 \sim 500/\mathrm{m}^2$ for one cycle of 2×10^{13} protons. Effective neutrino interactions in 800 kg emulsion target is estimated as follows: the number of CC-sample (with muon) is 500 K and the number of NC-sample(without muon) is 130 K after the loss caused by selection cuts for reconstruction of the event. An overall neutrino event rate is 0.7 per spill of 10^{13} protons on target area of 2400 kg.

Event location and tau detection

The neutrino beam is nearly perpendicular to the emulsion stacks. The sandwich structure of a emulsion target and a scintillating fiber tracker enable semiau-

Table 4: Neutrino beam components and the mean energies

component	relative abundance	$\langle E_\nu \rangle$ (GeV)	CC events/0.8 ton $\cdot 10^{19}p$
ν_μ	1.0	26.9 ± 0.5	2.1×10^5
$\overline{\nu_\mu}$	0.056 ± 0.004	21.7 ± 0.5	
ν_e	0.007 ± 0.0013	47.9 ± 2.2	
$\overline{\nu_e}$	0.0017 ± 0.0003	35.3 ± 1.7	

tomatic and automatic scanning. The fiber trackers give the position of the track on the changeable sheets within an area of $(300\mu m)^2$ and the angle resolution of $\Delta\theta \sim 3$mrad. An automatic scanning system will easily detect the predicted track on the changeable sheet. Following exposure and development, the target emulsion plates (25 plates each module) will be cut into mini-plates with the size of 7cm $\times$ 7cm, which constitute 25 mini-plates in the tower structure. The assembled mini plates have a size of 35cm $\times$ 35cm. The estimated follow-back time from the bottom plate to the primary vertex is around 10 minutes by using a semiautomatic system. In this collaboration, total of 15 semi-automatic systems are prepared that make possible to locate more than 50,000 neutrino interactions in the emulsion. After the location of the events, all of the charged particles emitted from the neutrino interaction are followed towards the downstream over a distance of 3mm to search for a decay vertex with the finding efficiency of roughly 95% (85% of tau decays show a kink topology).

To perform the analysis in a reasonably short time, some selection criteria will be used: one is to enrich the event with μ^- tracks (CC-sample) to search for $\tau^- \to \mu^- \overline{\nu_\mu} \nu_\tau$, and the other is to search for the event without μ^-(NC-sample). We apply the following selection criteria:

CC-sample selections

1. Identification of a negative charge track reconstructed in the muon spectrometer.

2. P_μ <30 GeV/c is imposed to select a muon with a low momentum associated with the ν_τ reaction; a negative muon from $\nu_\mu N \to \mu^- X$ has a high momentum. This selection reduces the number of candidates to be scanned by a factor (event reduction factor) of 0.75. 92% of the ν_τ events survive this cut (see Fig. 32).

3. A cut of P_T(missing) >0.4 GeV/c is imposed to select the events with an unbalanced (missing) transverse momentum due to undetected neutrinos in $\tau^- \to \mu^- \overline{\nu_\mu} \nu_\tau$, (see Fig. 33). The event reduction factor is 0.45; 87% of the ν_τ events survive the cut.

4. $\Phi(P_T(\text{hadron}), P_T(\text{missing})) > 90°$ is imposed on the relative azimuthal angle. The missing transverse momentum tends to be opposite to that of the hadron jet in ν_τ reactions. In ν_μ reactions the relative azimuthal angle shows a nearly symmetric distribution with some excess in the same direction (see

Fig. 34). The event reduction factor is 0.31. 90% of the ν_τ events survive the cut.

With these selections, the CC reactions are reduced to be 0.067 times the total events, corresponding to 33,500 events to be scanned out of 5×10^5 CC events. From this sample, 23 ν_τ events will be obtained from $\nu_\mu \rightarrow \nu_\tau$ oscillation if $\Delta m^2 > 40$ eV^2 and $\sin^2(2\alpha_{\mu\tau}) = 5 \times 10^{-3}$.

NC-sample selection

In the case of no-muon events, all of the reconstructed events will be scanned in the emulsion. The number of this sample is 13,000.
The estimated yield of ν_τ is approximately 5 times the CC events $\tau^- \rightarrow \mu^-\overline{\nu_\mu}\nu_\tau$. These NC events could be used to confirm the detection of ν_τ, although the quality of ν_τ candidate events from the NC-sample may not be so good because of the more serious background of NC-sample compared to the CC-sample.

Background

The reaction of $\nu_\tau N \rightarrow \tau^-(\rightarrow \mu^-\overline{\nu_\mu}\nu_\tau)X$ could be identified by the direct observation of the tau lepton in the emulsion. The origins of the background are:

1. Single-prong interaction of secondary hadrons ("white kink" This is the most serious source of the background to hadronic tau decays. A negative secondary hadron from a NC interaction may scatter within 3 mm from the vertex without a visible recoil, a blob or an Auger electron. The interaction length λ_{kink} for so-called white kink hadron scattering has been measured in an emulsion exposure at KEK. By selecting events with a decay-$P_T > 0.24$ GeV/c, $\lambda_{\text{kink}} > 80$ m was found. This background is less than one event in ν_τ candidates in our sample.

2. Inefficiency in the detection of muon (μ^+) from primary interaction of $\nu_\mu N \rightarrow \mu^+ D^-(\rightarrow \mu^-\text{neutral})X$. For a positive charm particle production this is not a background, even if the muon coming from the neutrino interaction is missed, because the charge of muon from charm particle is negative. In the case of negative charm particle, however, an undetected primary muon (electron) from $\bar{\nu}_\mu(\bar{\nu})$ charged current interactions becomes a background. This background is expected to be less than 0.5 events in our sample.

3. A missing negative charm particle in pair production through neutral current interactions: D^- decay may give a background if its associated charmed meson is missed. The detection efficiency of the partner of D^- is estimated to be 95 % for D^+, 90 % for Λ_c, and 80 % for D^0, respectively. This background is less than one event.

4. Muonic decay of $K^- \rightarrow \mu^-$ in the target emulsion. Application of a P_T cut of <0.24GeV/c eliminates all hyperon and kaon decays.

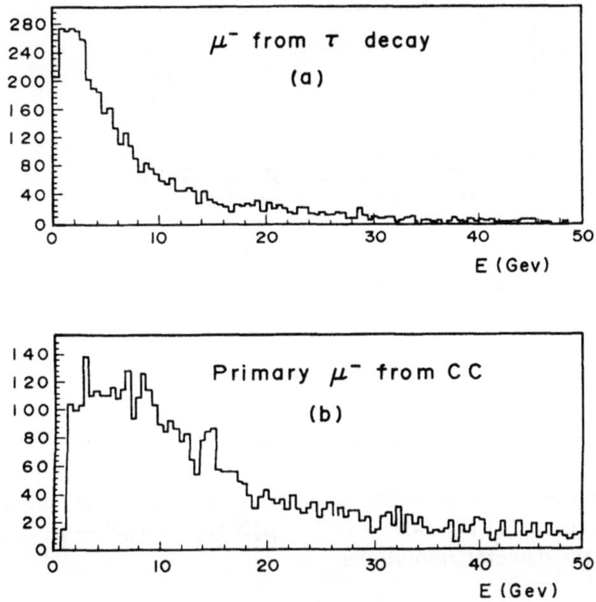

Figure 32: Muon energy distribution in (a) ν_τ and (b) ν_μ induced CC reactions.

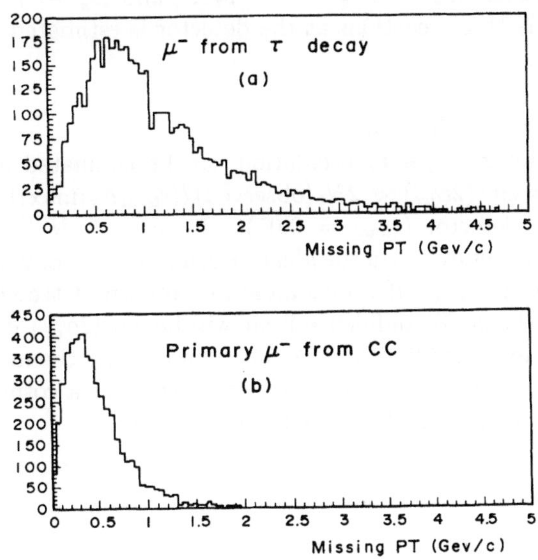

Figure 33: Missing transverse momentum distributions. (a) For reaction $\nu_\tau N \to (\tau^- \to \mu^-)$ and (b) for reaction $\nu_\mu N \to \mu^- X$, including the Fermi motion and the detector resolutions.

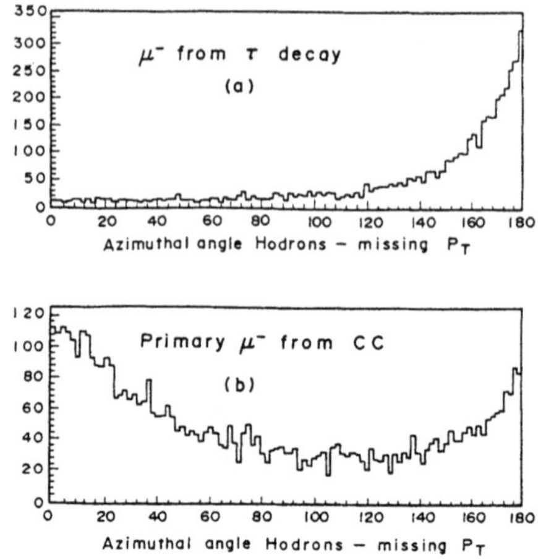

Figure 34: Distribution of the angle between the missing transverse momentum and the hadron shower direction. (a) For $\nu_\tau N \rightarrow (\tau^- \rightarrow \mu)X$ (b) for $\nu_\mu N \rightarrow \mu^- X$, including the Fermi motion and the detector resolutions.

5. Prompt tau neutrinos from the primary proton target and from an interaction in the beam dump tunnel. Direct ν_τ productions may occur through the following reactions: $D_s^+ \rightarrow \tau^+\nu_\tau, \tau^+ \rightarrow \bar{\nu}_\tau X$, and $D_s^- \rightarrow \tau^-\bar{\nu}_\tau, \tau^- \rightarrow \nu_\tau X$. The ν_τ yield via these reactions at the detector is estimated to be less than 0.01 events.

Sensitivity of $\nu_\mu \rightarrow \nu_\tau$ oscillation

The sensitivity for the $\nu_\mu - \nu_\tau$ oscillation can be estimated in terms of the mixing parameter as $\sin^2(2\alpha_{\mu\tau}) < 2N_\tau(\text{observed})/(\epsilon_{total}N_\tau(\text{max}))$, where ϵ_{total} is the efficiency of the detector ($\epsilon_{total} = 0.05$), $N_\tau(\text{max}) = N^{CC}$ is an expected number of charged-current events in the ideal case, and the factor 2 is the correction for the cross section of ν_τ/ν_μ. If no candidate events are detected, the limit on the mixing angle at the 90 % confidence limit will be $\sin^2(2\alpha_{\mu\tau}) < 3.1 \times 10^{-4}$ for $N^{CC} = 5 \times 10^5$. The expected limit on the mixing angle and the mass difference is shown in Fig. 35. This could eliminate the tau neutrino as a cosmic dark matter candidate, unless the mixing angle is unexpectedly small.

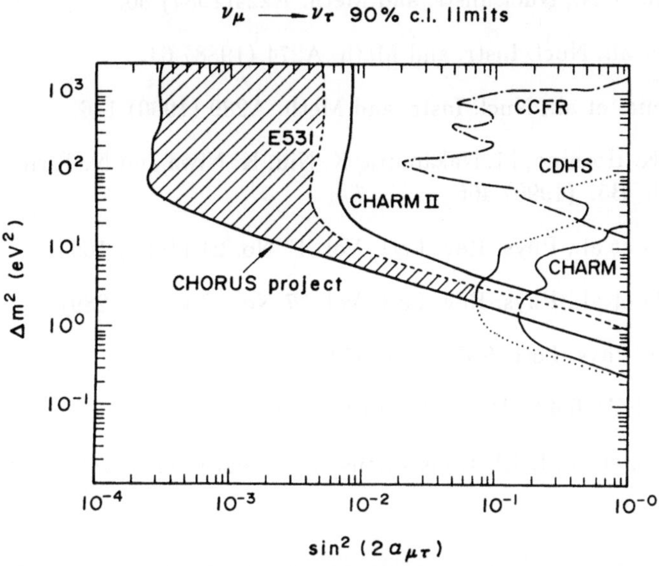

Figure 35: Domains of $(\Delta m)^2$ and $\sin^2 2\alpha_{\mu\tau}$ explored by previous experiments and to be explored by the CHORUS experiment.

References

[1] Becquerer; Comptes Rendus 122, pps. 420, 501 and 1086 (1896)

[2] Kinoshita, Proc. Roy. Soc. (A) 83 (1910) 432.

[3] Powell and Fertel; Nature 144, 115 (1939)

[4] Berriman, Photo. J. 89B (1949) 121.

[5] C. F. Powell, P. H. Fowler and D. H. Perkins, "The study of Elementary Particles by the Photographic Method", Pergamon Press, 1959.

[6] K. Niu et al., Prog. Theor. Phys. 46 (1971) 1644.

[7] K. Hoshino, Ph. D. thesis, Nagoya University, (1989).

[8] S. I. Brikkev et al., Suppl. Nuovo Cim. , 2 (1958) 733; N. B. Mistry et al., Nuovo Ci. , 17 (1960) 429; P. K. F. Gieder et al.,m Suppl. Nuovo Ci. , 26 (1962) 271.

[9] E. H. S. Burhop et al., Phys. Lett. 65B (1976) 299.

[10] K. Niwa, K. Hoshino and K. Niu, Proc. int. Cosmic ray sympo. on High energy Phenomena (Cosmic ray Lab. , Univ. Tokyo, 1974) p. 149. K. Niwa, S. Kaneko, M. Abe and K. Kurasawa, Technical report of television society, Vol. 5 (Jan. 1982) No. 31 p.7.

[11] N. Ushida et al., Nucl. Instr. and Meth. A224(1984) 50.

[12] S. Aoki et al., Nucl. Instr. and Meth. A274 (1988) 64.

[13] K. Kodama, et al., Nucl. Instr. and Meth. A289 (1990) 146.

[14] S. Aoki, K. Hoshino, M. Nakamura, K. Niu, K. Niwa and N. Torii, Nucl. Instr. and Meth. B51 (1990) 466.

[15] N. Ushida et al., Phys. Rev. Lett. Vol. 47 No. 24 (1981) 1694.

[16] N. Ushida et al., Phys. Rev. Lett. Vol. 57 No. 23 (1986) 2897.

[17] H. Harari, Phys. Lett. 6 B (1989) 413.

[18] CERN SPSC/90-42 SPSC p254 15 Dec 1990.

[19] T. Nakano, et al., IEEE transactions on nuclear science Vol. 39, No. 4 (1992) 680.

Observation of Atmospheric Neutrinos

Takaaki Kajita

Institute for Cosmic Ray Research
University of Tokyo

1 Introduction

Atmospheric neutrinos arise from the decay of secondaries (π, K and μ) produced by primary cosmic-ray interactions in the atmosphere. In the energy range below 1GeV, where all secondaries decay, we roughly expect that $(\nu_\mu + \bar{\nu}_\mu)/(\nu_e + \bar{\nu}_e) \simeq 2$. This is because a π-decay produces a ν_μ and a μ; the μ, when it decays, produces another ν_μ and a ν_e. These neutrinos enter into the detector from all directions nearly isotropically. The path length of the neutrinos (from the production point to the detector) ranges from a few km to 13000km, the diameter of the earth. In this article we describe recent measurements of atmospheric neutrinos by large underground detectors. Some of the experiments have observed that the $(\nu_\mu + \bar{\nu}_\mu)/(\nu_e + \bar{\nu}_e)$ ratio is significantly smaller than expected. The interpretation of these interesting results, the future prospect on atmospheric neutrinos and related neutrino physics are also described. In most parts of this article, when mention is made of neutrinos, we imply both neutrinos and antineutrinos.

Experiments aimed at detecting atmospheric neutrinos started in the 1960's. During this phase of experiments, atmospheric neutrinos were detected through the detection of upward (or horizontally) going muons. They enter the detector and run through it in the upward (or horizontal) direction. These muons are produced in the rock surrounding the detector by the interaction of relatively high-energy ($\langle E \rangle \sim$100 GeV), atmospheric $\nu_\mu(\bar{\nu}_\mu)$'s. This technique did not allow the detection of $\nu_e(\bar{\nu}_e)$'s, since most of the electrons produced by $\nu_e(\bar{\nu}_e)$ could not reach the detector.

In the early 1980's, several massive underground detectors were constructed. These detectors were primarily designed for the detection of nucleon decay which was predicted by Grand Unified Theories of particle interactions. One of the characteristics of the nucleon decay signal is "fully-contained", i.e., there should be no entering or no exiting tracks. Since this characteristics is similar to an event due to the interaction of low-energy ($E_\nu \sim 1$ GeV) atmospheric neutrinos in the detector, fully-contained atmospheric neutrino events were also detected in the same detectors. The global features, such as the visible energy distribution, of

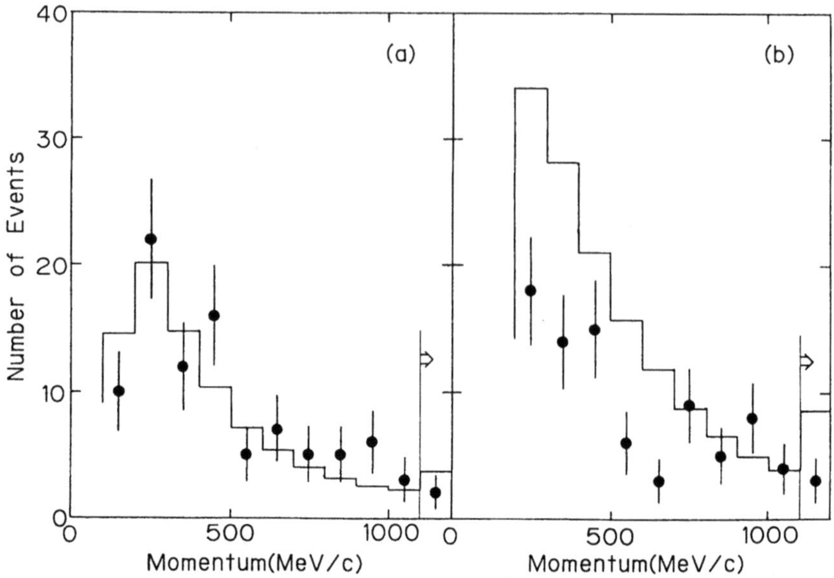

Figure 1: Momentum distributions of e- and μ-like events observed in Kamiokande in 1988 [1]. The histograms are the predicted event rate based on a flux calculation of [4].

the neutrino events were found to be consistent with that expected based on the calculated atmospheric neutrino flux, experimentally determined neutrino cross sections and the detection efficiency of the detector. In spite of experimental progress, interest in atmospheric neutrinos was not very strong until Kamiokande published atmospheric neutrino data.

In 1988, Kamiokande [1] showed a result in which fully-contained single-ring events were separated into muon (μ)- and electron (e)-like events. A single-ring μ-like event is produced by a non-showering straight track and is, in most cases, due either to a μ or a π. A single-ring e-like event is produced by a showering particle and is due either to an electron or a gamma. Kamiokande observed that the number of single-ring μ-like events, which were mostly due to interactions of atmospheric ν_μ and $\bar{\nu}_\mu$ in the detector, showed a significant deficit compared with the Monte Carlo (MC) prediction, while the number of single-ring e-like events was in good agreement with the expected number. Defining $\mu/e \equiv$ (μ-like, 205 $\sim$ 1500 MeV/c)/(e-like, 100 $\sim$ 1330 MeV/c), $(\mu/e)_{Data}/(\mu/e)_{MC}$ was $(85/93)/(144.0/88.5) = 0.56^{+0.09}_{-0.08}$(stat). Fig. 1 shows, as an example, the momentum distribution of the single-ring e-like and μ-like events from [1]. Those events are mostly (~95%) due to charged current (CC) $\nu_e, \bar{\nu}_e, \nu_\mu$ and $\bar{\nu}_\mu$ interactions. One can see that the deficit of μ-like events relative to the MC prediction is quite significant, and does not show any strong momentum dependence. The momentum distribution of e-like events is well explained by the MC prediction. Some authors [2] analyzed the data in terms of neutrino oscillations. This result triggered the current interest in atmospheric neutrinos.

Soon thereafter, it was pointed out [3] that the calculation of the atmospheric neutrino fluxes in [4], which were used in the analysis of the Kamiokande data, did not take into account the effect of the polarization of cosmic-ray muons in their decay. Muons from pion decay are completely polarized. Thus, the neutrinos from these muons should have angular and energy correlations with the direction of the polarization of the muons. However, in the flux calculation carried out before 1988, none (except for [3]) took into account this effect. Since neglecting the μ polarization underestimates the flux of ν_e, the polarization effect accounts for a part of the small μ/e ratio of the data. Following this suggestion, authors of [5] recalculated the atmospheric neutrino fluxes including this effect. It turned out that the change in the $(\nu_\mu + \bar{\nu}_\mu)/(\nu_e + \bar{\nu}_e)$ ratio due to the polarization effect is less than 15 % at $E_\nu \lesssim 1$ GeV.

Successively, three large underground experiments (NUSEX, Frejus and IMB), studied of the flavor composition of atmospheric neutrinos. The observed μ/e ratio by NUSEX [6] and Frejus [7] were in agreement with the MC prediction within the experimental errors. Then, in 1991, IMB-3 [8] published a result on the atmospheric neutrinos, showing a smaller μ/e ratio than expected, in good agreement with the Kamiokande result.

In what follows, we would like to describe the details of these experiments, the present status of atmospheric neutrino observations and future prospects. For details of the flux calculation, see ref. [9].

2 Simulation of the interactions of atmospheric neutrinos in the detector

To understand the characteristics of atmospheric neutrino events, each experimental group developed a Monte Carlo program for simulating the interaction of atmospheric neutrinos in the detector. In what follows, the Monte Carlo program of the Kamiokande experiment [10] is described.

The typical neutrino energy observed by the present underground detectors is $O(1\,\text{GeV})$. In this energy region the following 6 types of reactions have to be taken into account:

1. Charged-current quasi-elastic interaction, $\nu(\bar{\nu})N \rightarrow l^\pm N'$.

2. Charged-current single-pion production through Δ resonance, $\nu(\bar{\nu})N \rightarrow l^\pm \pi N'$.

3. Charged-current multi-pion production, $\nu(\bar{\nu})N \rightarrow l^\pm (m\pi) N'$ $(m \geq 1)$.

4. Neutral-current quasi-elastic interaction, $\nu(\bar{\nu})N \rightarrow \nu(\bar{\nu})N'$.

5. Neutral-current single-pion production through Δ resonance, $\nu(\bar{\nu})N \rightarrow \nu(\bar{\nu})\pi N'$.

6. Neutral-current multi-pion production, $\nu(\bar{\nu})N \rightarrow \nu(\bar{\nu})(m\pi)N'$ $(m \geq 1)$.

Here $l^{\pm}$ and $N(N')$ represent a charged lepton and a nucleon. In a charged-current (CC) quasi-elastic interaction, the energy and momentum of a lepton in the final state are determined from the q^2 ($q = p_\nu - p_{lepton}$; 4-momentum transfer) dependence of the cross section. This cross section is calculated [12] using the standard $V - A$ theory. Neutral-current (NC) quasi-elastic interactions are almost always invisible in water Cerenkov detectors unless a final-state proton recoils with sufficient energy ($p_p \gtrsim 1.5\,\mathrm{GeV/c}$) to emit Cerenkov radiation, or pions are produced by nucleon interactions off water nuclei. Fig. 2 shows the calculated cross sections for CC quasi-elastic interactions off free nucleons. The data from the Argonne 12-foot bubble-chamber experiment are also shown. The calculated cross section seems to be slightly higher than that from the Argonne data. This is because a parameter appearing in the formula of the cross section M_A, called axial-vector mass, measured by the Argonne experiment was $0.95 \pm 0.09\,\mathrm{GeV}$, while the value inputted in our calculation was 1.01 GeV, based on more recent measurements. Very roughly, the cross section of neutrino is proportional to the M_A value.

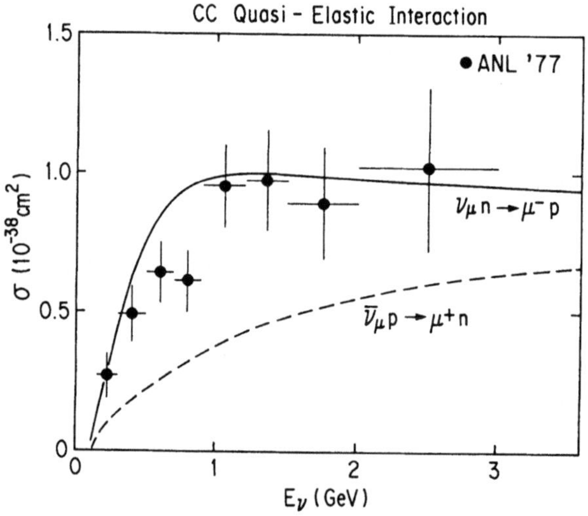

Figure 2: Cross section of charged-current quasi-elastic interactions off a free nucleon. The curves represent the calculation and the black circles with error bars represent the data [13]. The curve tends to have a larger cross section, because the value of M_A used in the simulation (1.01 GeV) is larger than the best-fit value of this experiment ($M_A = 0.95 \pm 0.09\,\mathrm{GeV}$).

CC single-pion production through neutrino interactions off nucleons has been extensively studied, and several satisfactory models have been developed [14],[15]. The agreement of the cross section between these two models is of the order of 20 %. We use the cross section calculated by the models of [14]. Fig. 3 shows some examples of the cross sections of several CC single-pion production reactions off free nucleons as a function of the neutrino energy. In the same figure, data from accelerator experiments are also shown. The agreement between the data and the calculation is satisfactory.

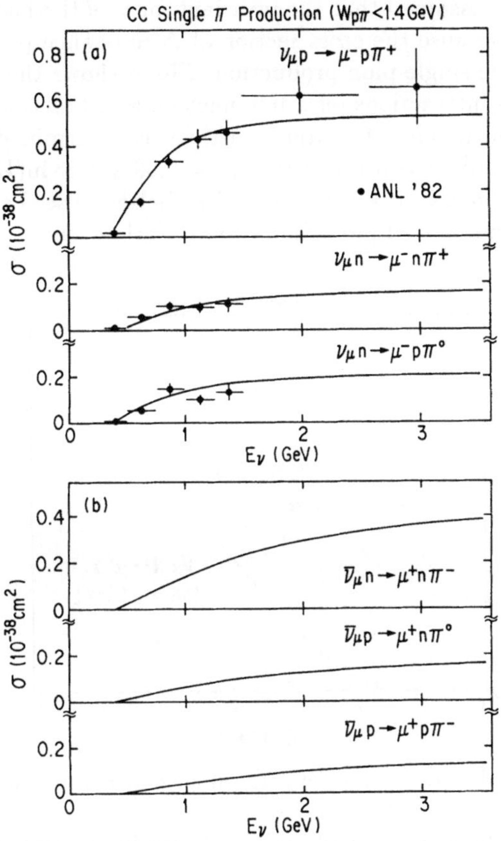

Figure 3: Cross section of charged-current single-pion production at the Δ-resonance off a free nucleon. The curves represent the calculation and the black circles with error bars represent the data [11].

In high-energy neutrino interactions ($E_\nu \gtrsim O(10\,\mathrm{GeV})$), it is known that the cross section and the interaction kinematics can be expressed by Bjorken scaling. We assume that the same scaling can be used in a relatively low-energy region of a few GeV. This assumption could be supported by the Gargamelle data [16], which shows a good agreement with Bjorken scaling, even for a few GeV neutrinos. The total cross section of multi-pion production channels can be calculated by integrating the following equation:

$$\frac{\sigma^{\nu N,\bar{\nu}N}}{dx'dy} = \frac{G_F^2 M E_\nu}{\pi} \times \{(1 - y + \frac{1}{2}y^2)F_2(x') \pm y(1 - \frac{1}{2}y)x'F_3(x')\}, \tag{1}$$

where $x' = -q^2/(2M(E_\nu - E_{lepton}) + M^2)$, $y = (E_\nu - E_{lepton})/E_\nu$, G_F the Fermi coupling constant, M the nucleon mass and $F_i (i = 2, 3)$ the structure functions.

For F_2 and F_3 we use those listed in Review of Particle Properties. In the integration, it is implicitly assumed that the invariant mass of the hadronic system be larger than 1.4 GeV, because the cross section of Δ-resonance production is separately considered in the single-pion production. Fig. 4 shows the cross sections of multi-pion production interactions off a free nucleon as a function of the neutrino energy. The mean multiplicity of charged particles is determined by high-energy neutrino experiments. We use a relation $\langle n_\pi \rangle = 0.09 + 1.83 \ln(W^2)$, where W is the hadronic invariant mass. Then, KNO (Koba-Nielsen-Olsen) scaling is used to determine the actual number of pions in a Monte Carlo event.

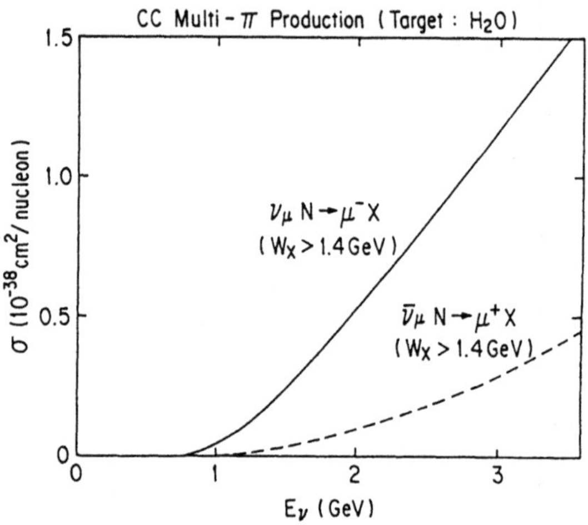

Figure 4: Calculated cross sections of charged-current multi-pion production.

Table 1 compares the event rate of each exclusive interaction channel between the Monte Carlo simulation and the Argonne neutrino data [11]. In this table the Argonnne neutrino fluxes are used as input data for the Monte Carlo simulation. It should also be noticed that the shape of the Argonne neutrino spectrum is similar to those of the atmospheric neutrinos. From these figures and the table, it is clear that the Monte Carlo simulation reproduces the accelerator data well. Fig. 5 shows the charged-current total cross sections. Fig. 6 shows the neutral-current total cross sections. Fig. 7 shows the calculated ratio $\sigma(\nu_\mu\text{: CC} - \text{total})/\sigma(\nu_e\text{: CC} - \text{total})$ as a function of the neutrino energy. One can see, as expected, that the $\sigma(\nu_\mu)/\sigma(\nu_e)$ ratio approaches unity as the neutrino energy increases, because the difference in the cross sections comes from the difference in the mass of the charged leptons.

Table 1: Event rate of each exclusive interaction channel of the Argonne data [11] compared with the results of the Monte Carlo calculation of [10] with the flux of the Argonne neutrino beam. The calculation was normalized to the event rate of the $\nu n \rightarrow \mu p$ channel.

Interaction	Argonne data	Monte Carlo
$\nu p \rightarrow \mu^- p \pi^+$	308 ± 24	301
$\nu p \rightarrow \mu^- p \pi^+ (l\pi^0), l \geq 1$	20 ± 5	22.3
$\nu p \rightarrow \mu^- n \pi^+ \pi^+ (m\pi^0), m \geq 0$	15 ± 5	12.2
$\nu p \rightarrow \mu^- p \pi^+ \pi^+ \pi^-$	10 ± 3	2.5
$\nu p \rightarrow \mu^- p \pi^+ \pi^+ \pi^- \pi^0$	—	1.3
$\nu p \rightarrow \mu^- n \pi^+ \pi^+ \pi^+ \pi^-$	1 ± 1	0.5
$\nu p \rightarrow \mu^- strange\ particle(s)$	1 ± 1	—
$\nu n \rightarrow \mu^- p$	833 ± 41	833
$\nu n \rightarrow \mu^- p \pi^0$	124 ± 14	129
$\nu n \rightarrow \mu^- n \pi^+$	90 ± 11	89.0
$\nu n \rightarrow \mu^- p (k\pi^0), k \geq 2$	31 ± 13	13.1
$\nu n \rightarrow \mu^- n \pi^+ (l\pi^0), l \geq 1$	29 ± 12	24.6
$\nu n \rightarrow \mu^- p \pi^+ \pi^-$	20 ± 5	21.1
$\nu n \rightarrow \mu^- p \pi^+ \pi^- (l\pi^0), l \geq 1$	8 ± 4	9.1
$\nu n \rightarrow \mu^- n \pi^+ \pi^+ \pi^- (m\pi^0), m \geq 0$	3 ± 2	6.1
$\nu n \rightarrow \mu^- p \pi^+ \pi^+ \pi^- \pi^-$	0	1.5
$\nu n \rightarrow \mu^- strange\ particle(s)$	12 ± 4	—

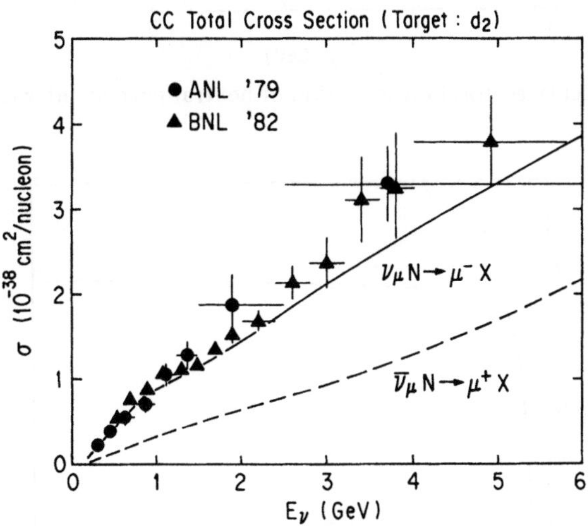

Figure 5: Total cross section of charged-current interactions. The curves represent the calculation and the black circles with error bars represent the data [17].

The actual neutrino interaction cross sections off a target nucleus (H_2O, Fe) are more complicated because nucleons in a nucleus are bounded. For nucleons bounded in a nucleus, the Fermi motion of the target nucleon, i.e., the motion of the nucleons within a nucleus, must be taken into account. For quasi-elastic interactions, the Pauli exclusion principle, i.e., a principle that states that two nucleons in a nucleus cannot share the same state, must be taken into account. The actual treatment of these effects is made by the Fermi-gas model [18], in such a way that the momentum of the recoil nucleon must exceed the Fermi surface momentum (about 250 MeV/c). Fig. 8(a) compares the calculated cross section of CC quasi-elastic interactions off a free nucleon and a bounded nucleon in ^{16}O. Fig. 8(b) shows the calculated cross section of ν_e quasi-elastic interaction off ^{16}O using various models. One can see that the cross section calculated by this method agrees well with the others.

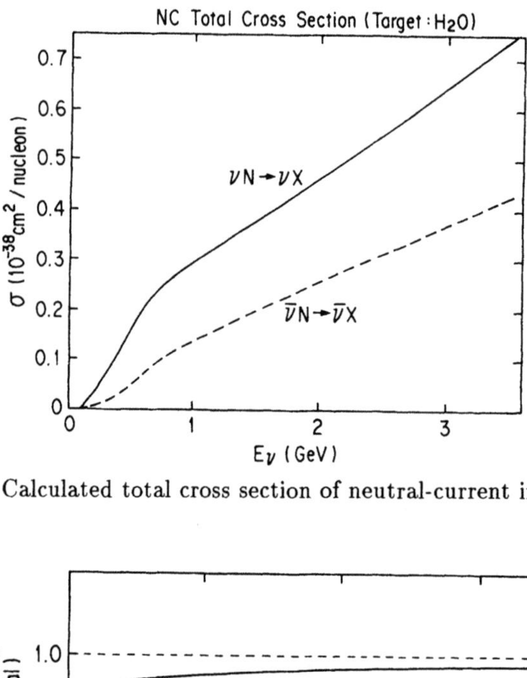

Figure 6: Calculated total cross section of neutral-current interactions.

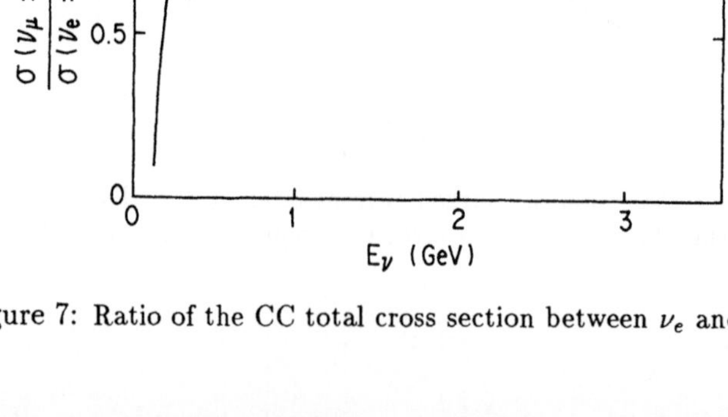

Figure 7: Ratio of the CC total cross section between ν_e and ν_μ.

Pions produced by a neutrino interaction in a nucleus often undergo secondary nuclear interactions before leaving the nucleus. These pions also interact in the detector material. The pion interactions considered in the Monte Carlo simulation are: inelastic interactions ($\pi N \rightarrow \pi N'(m\pi, m \geq 0)$), absorption and charge exchange processes. Fig. 9 shows the probability of various nuclear interactions of pions in a parent nucleus. These effects are also properly taken into account in the Monte Carlo simulation.

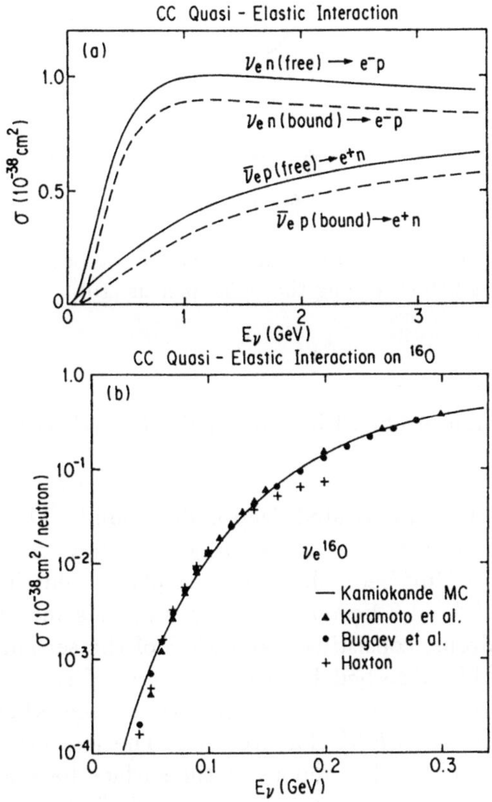

Figure 8: (a) Comparison of the calculated cross section of CC quasi-elastic interactions off a free nucleon and a bounded nucleon in ^{16}O. (b) Calculated cross section of ν_e quasi-elastic interaction off a ^{16}O using various models.

3 Detectors

To date, there have been four experiments which reported results on the atmospheric ν_μ and ν_e composition. Table 2 summarizes the experiments. Two of them, NUSEX (Europe) and Frejus (Europe), are gas-counter iron-plate tracking-calorimeters, which detect the tracks of charged particles. The other two, IMB-3 (USA) and Kamiokande (Japan), are imaging water Cerenkov detectors, which de-

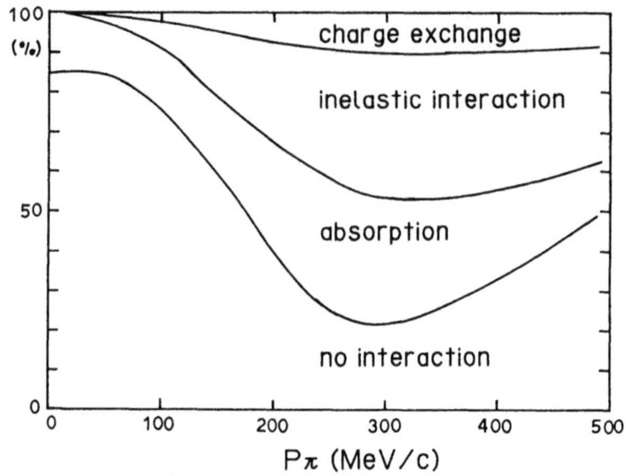

Figure 9: Calculated probability of various nuclear interactions of a π^+ generated randomly in a ^{16}O nucleus before leaving the same nucleus [10].

tect Cerenkov radiation produced by relativistic charged particles in the detector water.

All of these detectors are located deep underground. This is because, at the surface the background events due to cosmic-ray particles (mostly $\mu^{\pm}, p$ and n) are so frequent, ($\sim 2 \times 10^2/\mathrm{m}^2/\mathrm{sec}$), that an atmospheric neutrino experiment with such a low event rate (about 0.1 events per ton per year) is unrealistic using present-day technology. Hadrons, consisting about 25 % of the secondary cosmic rays at the surface, are quickly absorbed by several meters of rock. Muons, consisting the remaining 75 % of the secondary cosmic rays, on the other hand, loose their energy only by ionization (dE/dx for brevity). The loss rate of energy is quite small, about 2 MeV·cm^2/g. The muons at the surface have a mean energy of 2 GeV and a differential energy spectrum falling as E_μ^{-2} at $E_\mu = 10\,\mathrm{GeV}$, steeping smoothly to $E_\mu^{-3.6}$ above 1 TeV [19]. As a consequence, the detector must be located deep underground. Fig. 10 shows the depth of the detectors and the cosmic ray muon rate. Even at a depth of 2700 m.w.e. (meters water equivalent, 1000 m underground), Kamiokande observes 1 muon in 3 seconds. This rate should be compared with the observed atmospheric neutrino rate by the same detector, 1 event in 4 days. The rate of muon events is $\sim 10^5$ times more frequent than that of atmospheric neutrinos, even at the depth of 1000 meters underground.

These detectors are designed to study in detail the characteristics of fully contained events. Also, the detectors are designed to have a good detection efficiency for fully contained events. In the following, we describe each detector in some detail.

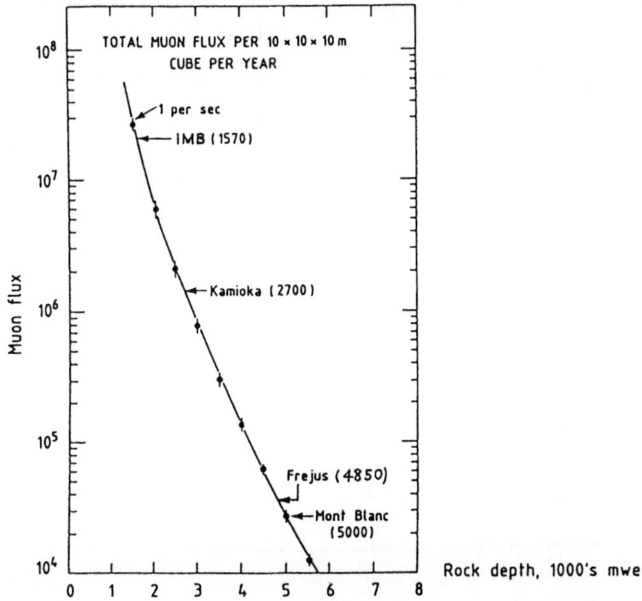

Figure 10: Depth of the detectors mentioned in the text and the cosmic-ray muon rate.

3.1 The NUSEX detector

The NUSEX detector [20] consists of a 3.5×3.5×3.5 m cube having a mass of 150 tons, made by 134 iron plates of 1 cm thickness, interleaved with planes of plastic streamer tubes, 3.5 m long and 1×1 cm in cross section. It is located at a depth of 5000 m.w.e. in the Mont Blanc tunnel. The fiducial volume is about 130 tons. The trigger efficiency reaches 50 % at $\sim$ 300 MeV for both muons and electrons. The trigger rate is about 5 events per hour; most of them are due to cosmic-ray muons. The coordinates on each plane are read bydimensionally by means of X- and Y-strips parallel and orthogonal to the tubes, respectively (see Fig. 11). The read out system enables two-dimensional views of an event. The analysis of an event, such as the determination of the vertex position, the number of charged tracks and the visible energy, is made using information obtained from fired flash tubes. Information of dE/dx is not available. Fig. 12 shows the event pattern of (a) a CC ν_e candidate, and (b) a CC ν_μ candidate observed by this detector. A typical pattern of an electromagnetic shower can be clearly seen for Fig. 12 (a), while a clear single track, as expected by muon passage, is seen for Fig.12 (b). The separation of μ- and e-like events was tested by a beam test, and was claimed to be good.

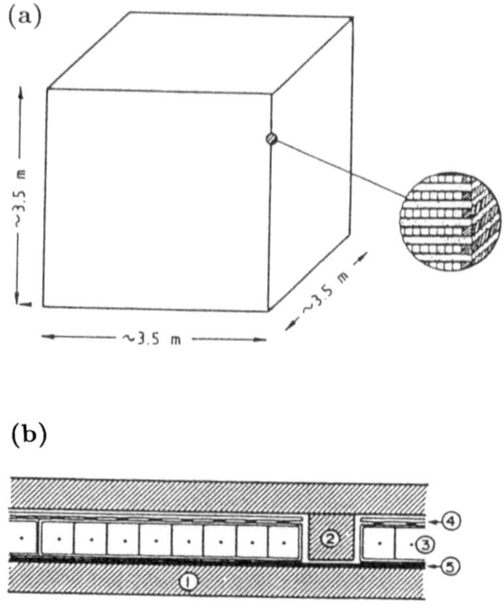

Figure 11: (a) Sketch of the NUSEX detector. (b) A detector plane: (1) iron plate; (2) iron spacer; (3) streamer tubes; (4) X-strips; (5) Y-strips. (From [20]

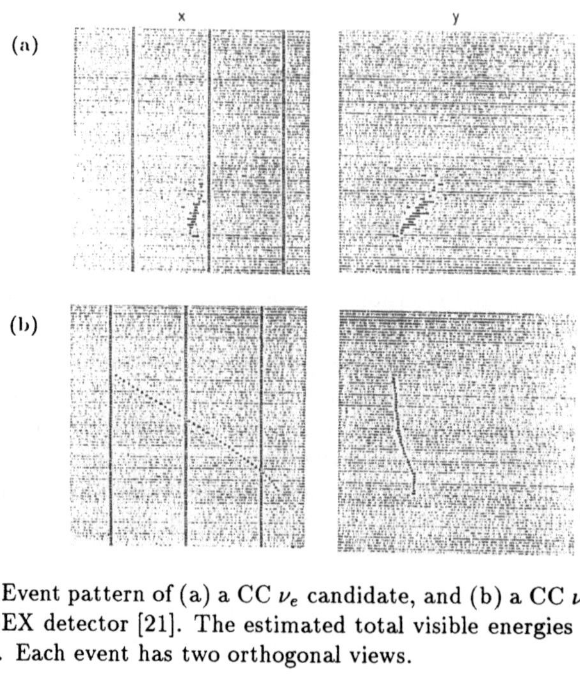

Figure 12: Event pattern of (a) a CC ν_e candidate, and (b) a CC ν_μ candidate observed by the NUSEX detector [21]. The estimated total visible energies are 3.7 and 1.1 GeV, respectively. Each event has two orthogonal views.

3.2 The Frejus detector

The Frejus detector [22], located at a depth of 4850 m.w.e. in the Frejus highway tunnel in France, consists of a 6×6×12.3 m having a mass of 900 tons. The fiducial volume is 554 tons, leaving events with their vertex position located within 50 cm from the surface. It consists of 912 flash-chamber planes and 113 Geiger-tube planes. A flash-chamber plane is made of a sandwich of 3 mm thick iron plate and a 5 mm thick plane of plastic flash tubes. The planes of Geiger tubes, 15×15 mm in cross section, provide the trigger. The trigger efficiency reaches 50 % at ~ 400 MeV for both muons and electrons. The trigger rate is about 45 events per hour; about half of them are due to cosmic-ray muons and the remaining events are mostly due to radioactive contamination in the detector. The detector which consists of 6 m long flash and Geiger tubes, alternatively vertical and horizontal, gives two orthogonal views for each event. The total number of read-out channels is 10^6 for the flash chambers and 4×10^4 for the Geiger tubes. Information of dE/dx is not available. (see Fig. 13). Fig. 14 shows the event pattern of (a) a CC ν_e candidate, and (b) a CC ν_μ candidate observed by this detector. A typical pattern of an electromagnetic shower can be clearly seen in Fig. 14(a). In Fig. 14(b), a muon can be recognized by a long straight track. The other unclear tracks are due to hadrons. The probabilities of correctly identifying the CC ν_μ and ν_e events are 95 and 85 %, respectively.

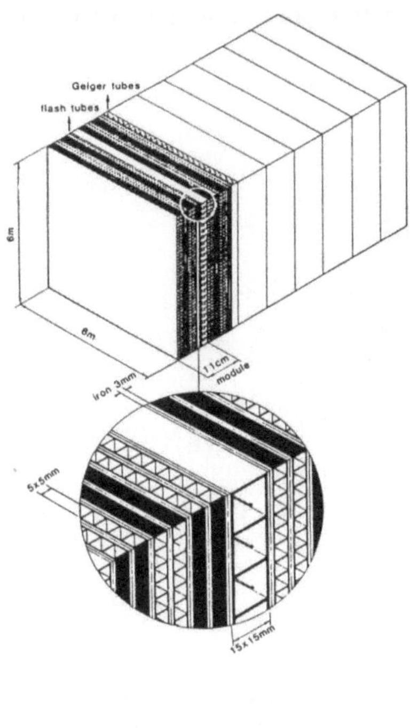

Figure 13: Structure of the Frejus detector [22].

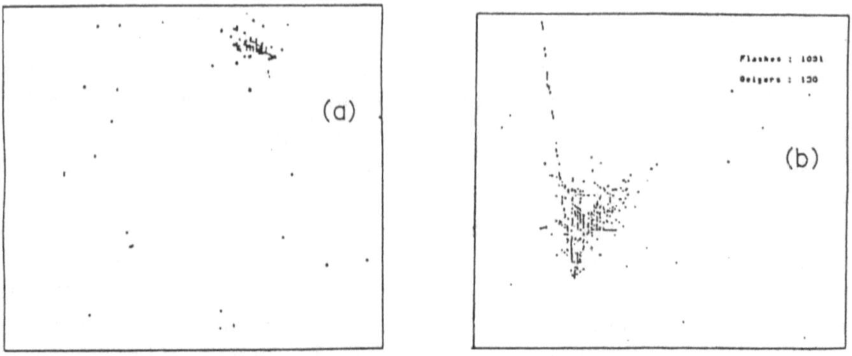

Figure 14: Event pattern of (a) a CC ν_e candidate, and (b) a CC ν_μ candidate observed by the Frejus detector [22].

3.3 The IMB-3 detector

The IMB-3 detector [23] is an 8-kton water Cerenkov detector located at a depth of 1570 m.w.e. in the Morton Salt mine, Ohio, USA. (Although the IMB detector initially started operation in 1982, the early phases of the experiment, (IMB-1 and -2), did not have a sufficient performance to allow an efficient particle identification.) It consists of a tank of water, (18 m$\times$17 m$\times$22.5 m), instrumented on all six faces with 2048 8-inch photomultiplier tubes (PMTs) with a wave shifter plate (see Fig. 15). The wave shifter plate is installed to increase the light correction efficiency.

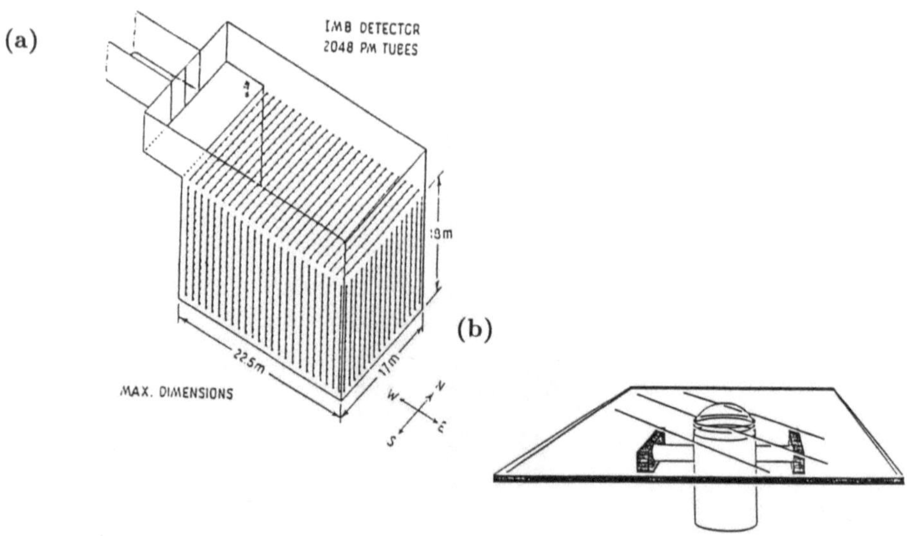

Figure 15: (a) Sketch of the IMB-3 detector. 2048 PMT's are positioned on the surfaces of all six faces with ~1 meter spacing. (b) A 8 inch diameter PMT with a wave length shifter plate. (From [23])

Each charged particle with $\beta > 1/n$ ($n = 1.34$, refraction index of water) emits Cerenkov photons in a circular cone with a half opening angle (θ) to the particle trajectory of $\cos\theta = 1/n\beta$. In water, this θ is 42° for particles with $\beta = 1$. These photons are detected by PMTs. The charge and timing information from each PMT are used for event analysis. The fiducial volume is 3.3 ktons, defined so that the vertex position be at least 2 meters inside from the PMT surface. The trigger threshold is about 15 MeV for electrons. The trigger rate, due mostly to cosmic-ray muons, is about 2.7 events per second. An event due to a charged particle can be recognized by a ring pattern of detected Cerenkov photons, as observed by the PMTs installed on all six surfaces (see Fig. 16). The difference between an electron track and a muon track can be recognized by the characteristic shape of the Cerenkov ring. Figs. 16(a) and 16(b) show a muon and an electron event generated by the Monte Carlo simulation, respectively. One can see a clear outer edge for the muon event, while for the electron event the edge is not clear due to an electromagnetic shower and multiple scattering of low-energy electrons. The probability of correctly identifying a single-ring electron or a muon event is 92 % with negligible bias toward one classification or the other.

A muon which stops inside the detector volume may decay, producing an electron. The spectrum of electrons from muon decay has an end-point energy of 53 MeV. Thus, the detection of a muon decay can be rather easy. A fraction of the μ^-'s which stop in the detector can be absorbed by a nucleus prior to decay, giving no detectable electron. This absorption probability is 18 % in water. (This probability is 90 % in iron). The presence of a muon in the neutrino interaction products can be inferred by searching for a delayed electron signal in the time window between 0.4 and 7.5 μsec after the prompt signal. The detection efficiency of muon decay for cosmic ray stopping muons is 70 %. (The cosmic-ray μ^+/μ^- ratio is about 1.25.)

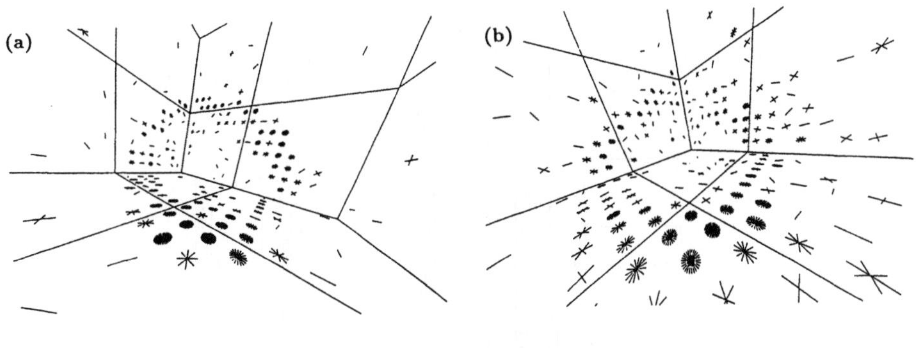

Figure 16: Monte Carlo generated events in the IMB-3 detector; (a) a μ- and (b) an e-like event. The visible energies are 500 MeV for both events [23].

3.4 The Kamiokande detector

Kamiokande (see ref. [24] for a detailed description of the detector and its performances related to this topic) is a water Cerenkov detector located 2700 m.w.e. underground in the Kamioka mine, Japan. The detector was initially a cylindrical volume of water, 15.6 m in diameter and 16 m in height containing 3 kton of pure water (inner-counter). 1000 PMTs are uniformly placed over the entire inner surface of the detector facing inward. The trigger threshold was initially 30 MeV/c for electrons (205 MeV for muons), but has been gradually lowered to 6 MeV/c for electrons and 165 MeV/c for muons after 1988. The trigger rate depended on the trigger condition and the threshold, and ranged from 0.4 to several events per sec. Among the triggered events, 0.37 events/sec are due to cosmic-ray muons and the remaining triggers are mostly due to low-energy radioactivity. After 1986, the detector was improved with an additional 4π water-Cerenkov anticounter surrounding the inner-counter (see Fig. 17). The total weight increased to 4.5 kton, including the anticounter. During the initial phase of operation, only the charge information from each PMT was read out. Then, after 1986, charge and timing information was read out. The first phase of the experiment was called Kamiokande-I (Kam-I), from July 1983, to the end of 1985; the next phase, called Kamiokande-II (Kam-II), from the beginning of 1986 to April 1990. (After the end of 1990, the detector was further upgraded, called Kamiokande-III. However, since this article does not describe the data obtained during this period, we do not describe any details of the new phase of the experiment.) The fiducial volume is defined as 2 m inside from the PMT planes for Kam-I and 1.5 m for Kam-II. The corresponding fiducial masses are 880 tons (780 tons after 1985) and 1040 tons for Kam-I and -II, respectively. Figs. 18(a) and 18(b) show a μ- and an e-like candidate event, respectively. A comparison between Figs. 16 and 18 suggests that the difference in the event pattern between a muon-like event and an electron-like event is clearer in Kamiokande than in IMB-3. This is essentially due to a factor of 3 to 4 better light correction efficiency in Kamiokande. The separation of electron- and muon-like events in Kamiokande is made with an efficiency of 98 %.

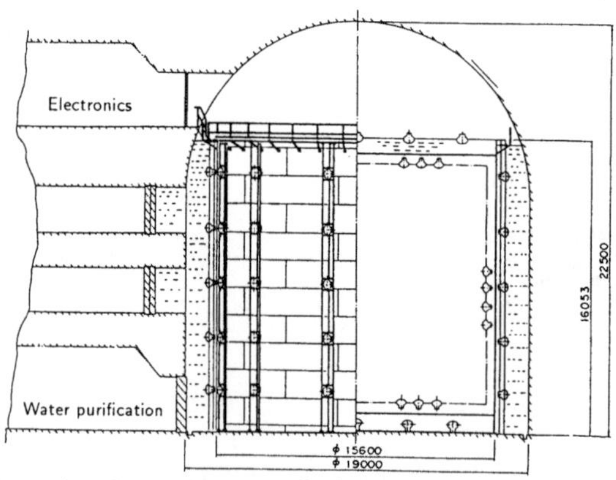

Figure 17: The Kamiokande-II detector [24]. The inner water Cerenkov detector (2140 ton) is surrounded by a 4π water Cerenkov anticounter. Also shown are the location of the water purification system and the electronics hut.

In Kamiokande-I, the detection of μ-decay was made by searching for a delayed pulse in the time window between 0.5 and 7.6 μsec in the total analog-sum signal of the PMTs recorded by a FADC. The detection efficiency of muon decay for cosmic-ray stopping muons was $68 \pm 3\,\%$. In Kamiokande-II, the detection efficiency was improved by additionally searching for a second event (trigger) in the time window between 0.4 and 20 μsec. Fig. 19(b) shows a typical second event. The presence of a delayed signal in the ADC gate (400 nsec) of the prompt event was also searched for. By combining all these methods, the detection efficiency was $80 \pm 1\,\%$ for cosmic ray muons. (see Fig. 20, which shows the detection efficiency of μ-decay in Kam-II). Table 2 summarizes the four atmospheric neutrino experiments.

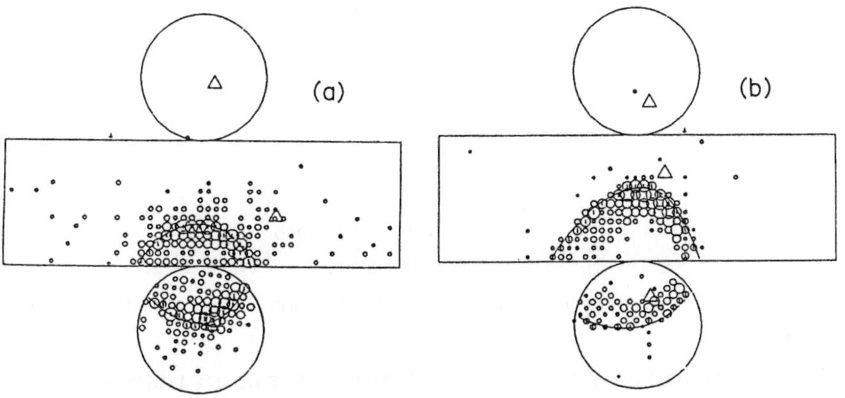

Figure 18: Event pattern of (a) a μ- and (b) an e-like candidate event observed by the Kamiokande detector. Each circle represents a PMT which detected Cerenkov photons; the area of the circle is proportional to the observed photoelectron number.

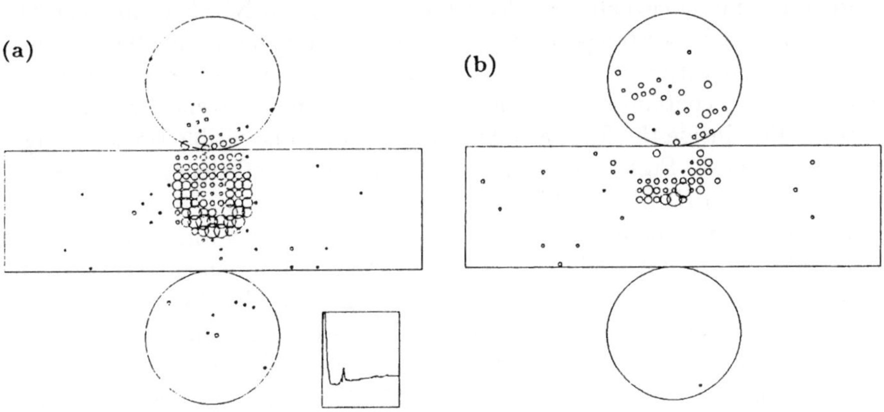

Figure 19: (a) a μ-like event and (b) an electron event from the decay of the proceeding μ-like event observed by the Kamiokande detector. Shown at the lower left of (a) is the total sum signal of 948 inner PMT after the trigger of the prompt event for 7.6 μsec duration. One clearly sees the second pulse at about 2 μsec after the prompt pulse.

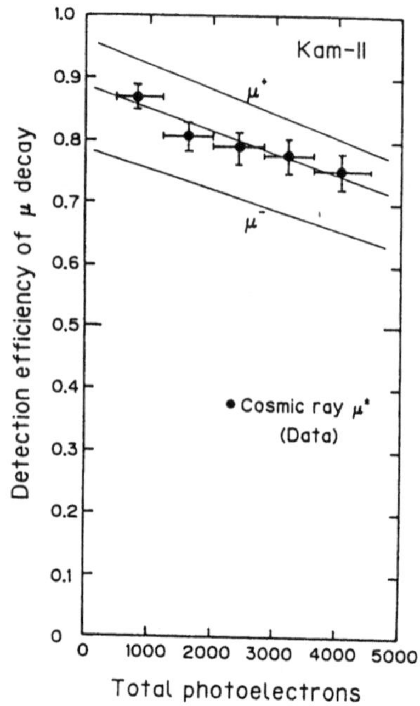

Figure 20: Detection efficiency of μ-decay in Kam-II for cosmic-ray stopping muons.

4 Experimental results on atmospheric neutrinos

4.1 Selection of atmospheric neutrino events

In NUSEX and Frejus, the trigger rates are so low, about 5 and 45 events per hour for NUSEX and Frejus, respectively, that they scan all of the events and the atmospheric neutrino events are selected by scanning. In NUSEX, fully contained events, which have a vertex position inside the fiducial volume and all of the visible tracks contained within the detector, are used to study atmospheric neutrinos. In Frejus both fully contained events and 'partially contained' events, which have the vertex position inside the fiducial volume and at least one visible track exiting the detector, are used. Figs. 21 and 22 show the detection efficiency of atmospheric neutrino events in NUSEX and Frejus, respectively.

In IMB, 10^4 triggers per hour (most of them are cosmic-ray muons) are reduced by two independent data-reduction programs. Data reduction is essentially made by reconstructing the vertex position of an event and the fiducial volume cut. About 10 events per day survive. These events are then scanned by physicists, and any events confirmed to originate inside the 2 m fiducial volume are saved as an atmospheric neutrino sample. Fig. 23 shows the detection efficiency of atmospheric neutrino events in IMB-3.

In Kamiokande-II, several thousands events per hour are subject to the simple data-reduction program. The data reduction is made by a total visible-energy cut and by the requirement of no significant signal in the anticounter. Several tens of events per day survive. These events are then scanned by two groups of

Table 2: Summary of the atmospheric neutrino detectors.

	NUSEX	Frejus	IMB-3	Kamiokande
Detection technique	gas counter-iron plate	gas counter-iron plate	water Cerenkov	water Cerenkov
Fiducial mass	130t	550t	3300t	780 $\sim$ 1040t
Total exposure	0.74kt·yr	1.56	7.7	4.92
Number of ev.	50	$\sim$185	935	476
μ/e identification	good	95% for ν_μ 85% for ν_e	$\sim$92%	98%
μ-decay			70%$^{(1)}$	68%$^{(1)}$ (Kam-I) 80%$^{(1)}$ (Kam-II)
Event selection for ν_μ/ν_e determination	>200 MeV fully-contained	>200 MeV all ev.	100–1500 MeV/c for e 300–1500 MeV/c for μ 1-ring	100–1330 MeV/c for e 205–1500 MeV/c for μ 1-ring

physicists. In this process, background events, mostly due to stopping cosmic-ray muons entering through the less sensitive region of the anticounter (edge of the top-anticounter) and to electric noises, are removed. Then, the vertex position of an event is determined by a reconstruction program using information of timing from each PMT. The final data sample is made after a fiducial volume cut and an empirical cut of maximum photoelectrons per PMT < 200, which further ensures the full containment of an event. Fig. 24 shows the detection efficiency of atmospheric neutrino events in Kam-II. The detection efficiency in Kam-I was similar to that of Kam-II.

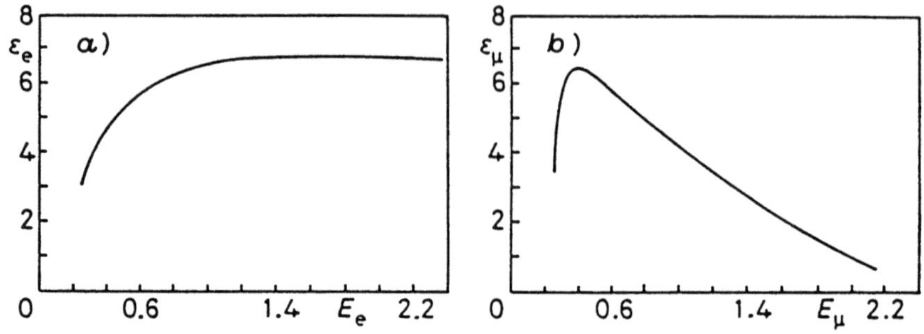

Figure 21: Detection efficiency as a function of lepton energy in the NUSEX detector: (a) for electrons and (b) for muons [6].

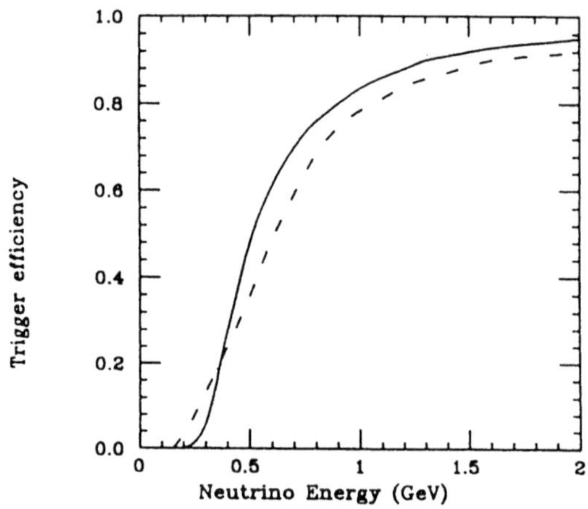

Figure 22: Detection efficiency as a function of neutrino energy for CC interactions in the Frejus detector: the solid line is for CCν_μ and the dashed line for CCν_e [7].

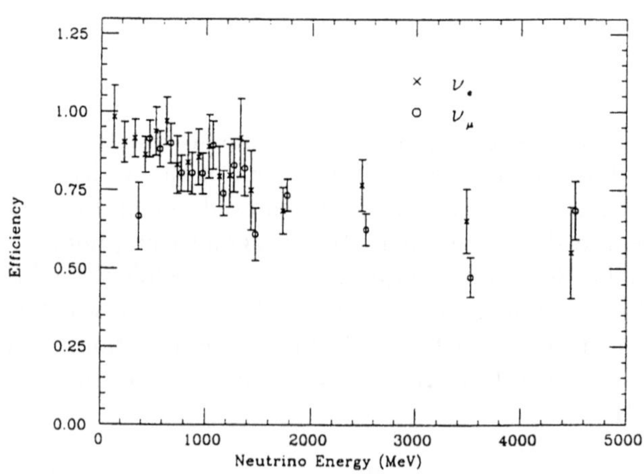

Figure 23: Detection efficiency as a function of neutrino energy for ν_e- and ν_μ-induced events in the IMB-3 detector [8].

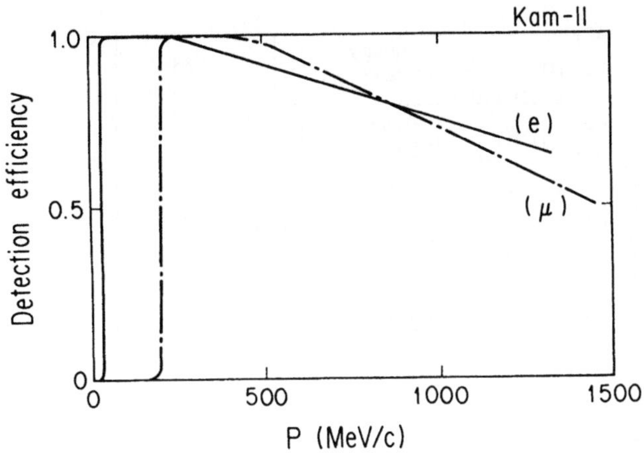

Figure 24: Detection efficiency as a function of lepton momentum for single-ring events in the Kamiokande-II detector.

4.2 Global distributions

Four large underground experiments, (NUSEX, Frejus, IMB-3 and Kamiokande), started taking data in the (early) 1980's. These experiments have already observed a significant number of atmospheric neutrino events (total of ~ 1600), (see Table 2).

Fig. 25 shows the total visible-energy distribution from the 4 experiments. One can see that the shapes of the distributions of the data from each experiment are in reasonable agreement with the corresponding Monte Carlo predictions. The observed event rates are smaller than those of the Monte Carlo predictions. However, one should notice the uncertainty in the total number of predicted events, which is estimated to be about $\pm 30\,\%$, resulting mainly from uncertainties in the flux calculation which is (at least) $\pm 20\,\%$ at $E_\nu = 1\,\mathrm{GeV}$ and in the cross section of low-energy neutrino interactions, about $10\,\%$.

Fig. 26(a) shows the distribution of the number of charged tracks from the Frejus experiment. Fig. 26(b) shows the number of Cerenkov rings from the Kamiokande experiment. In both cases, the shape of distributions are in reasonable agreement between the data and the simulations. Quantatively, for the case of Kamiokande, the fraction of multi-ring events in the total sample is 0.31 ± 0.03 which should be compared with the Monte-Carlo predicted number of 0.27. For the case of Frejus, the fraction of multi-prong events is 0.63 ± 0.05 while the Monte-Carlo predicted number is 0.67. The difference between Frejus and Kamiokande could be explained by the differences in the energy range (the Frejus data are higher in energy range, see Fig. 25) and of the detection technique (Frejus can detect recoiled protons).

One can see that the Monte Carlo simulations of the atmospheric neutrino interactions reasonably reproduce these global distributions of the data.

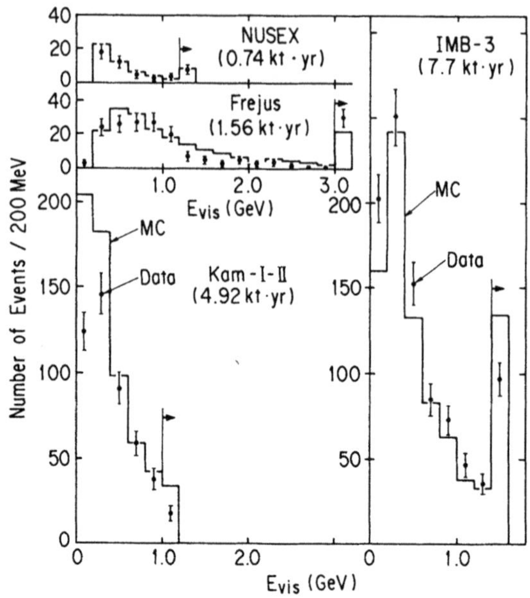

Figure 25: Total visible-energy distribution of the atmospheric neutrino events from NUSEX, Frejus, IMB-3 and Kamiokande.

4.3 Study of ν_μ and ν_e induced events by particle identification

To study the flavor of atmospheric neutrino events, it is necessary to separate the ν_μ interactions from the ν_e interactions. This is achieved by means of particle identification. The efficiency of ν_μ/ν_e or μ/e separation, i.e., the probability that an $e(\mu)$ track is identified as 'e'('μ'), is shown in Table 2 for each experiment. One can see that the quoted efficiencies are quite high, 85 ~ 98 %. In tracking calorimeters, the event pattern is recorded in a 2-dimensional view, and μ/e separation can be carried out clearly. Fig. 14 shows examples of (a) a ν_e and (b) a ν_μ candidate event observed by the Frejus detector. An event due to a CC ν_μ interaction can be clearly recognized by a long straight track, which is supposed to be due to a muon. An event due to a CC ν_e interaction, on the other hand, has an electromagnetic shower linked to the vertex point.

In water Cerenkov detectors, the difference between events due to CC ν_e interactions and to ν_μ interactions can be recognized by the characteristic shape of the Cerenkov ring. Figs. 18 (a) and 18 (b) show a μ- and an e-like candidate event, respectively. One can see a clear outer edge for the μ-like event, while for the e-like event the edge is not clear due to an electromagnetic shower and multiple scattering of low-energy electrons. In water Cerenkov detectors, multi-prong events are not used due to the technical difficulty in separating overlapping rings and the ambiguity in selecting a lepton candidate in a multi-prong event. In tracking calorimeters, multi-prong events are also used based on suitable criteria to select a lepton candidate. One should notice that the analysis of multi-prong

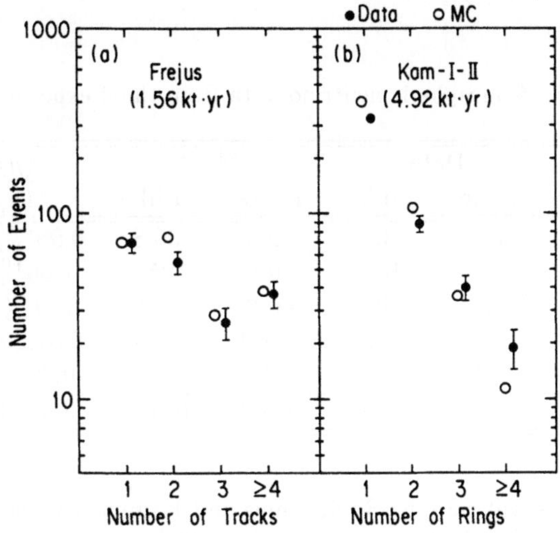

Figure 26: (a); Distribution of the number of charged tracks from the Frejus experiment. (b); Distribution of the number of Cerenkov rings from the Kamiokande experiment.

events is easier in tracking calorimeters than in water Cerenkov detectors. There is another reason to use multi-track events in tracking calorimeters. In tracking calorimeters, the protons recoiled by neutrino interactions are, in some cases, visible; these events are recognized as at least 2 prong events. On the other hand, recoiled protons are invisible in water Cerenkov detectors in most cases.

Table 3 summarizes the results from the four experiments. We focus on the $(\nu_\mu+\bar{\nu}_\mu)/(\nu_e+\bar{\nu}_e)$ ratio and compare the experimental ratio with the MC prediction. The $(\nu_\mu + \bar{\nu}_\mu)/(\nu_e + \bar{\nu}_e)$ ratio has less ambiguities in the prediction, because the ambiguities in the total number of events cancel out. One can see that the $(\mu/e)_{Data}$ from NUSEX and Frejus are in good agreement with the expected values. There is no problem as far as these data are concerned. On the other hand, the $(\mu/e)_{Data}$ from IMB-3 and Kamiokande are smaller than expected.

Fig. 27 shows the $(\mu/e)_{Data}/(\mu/e)_{MC}$ as a function of the momentum. Fig. 28 shows the zenith angle dependence of $(\mu/e)_{Data}/(\mu/e)_{MC}$. In these figures, the error bars are statistical only. One can see that the data from IMB-3 and Kamiokande are in good agreement, and that the small $(\mu/e)_{Data}/(\mu/e)_{MC}$ from IMB-3 and Kamiokande do not show any strong momentum and zenith-angle dependences.

4.4 Details of systematic error on the μ/e ratio

We now look at the systematic errors involved in $(\mu/e)_{Data}/(\mu/e)_{MC}$. Table 4 summarizes the sources of the systematic error on the μ/e ratio from Kamiokande. In the following we describe details of the possible sources of systematic errors using the analysis of Kamiokande [26]. The sources of systematic error in the other experiments are mentioned only briefly.

Table 3: Atmospheric neutrino data from the 4 experiments.

Exp.	Data		νMC[(*)]		$\frac{(\mu/e)_{data}}{(\mu/e)_{MC}}$
	e-like	μ-like	e-like	μ-like	
NUSEX	18	32	20.5	36.8	$0.99^{+0.35}_{-0.25} \pm small$
Frejus[(**)]	57	108	70.6	125.8	$1.06^{+0.19}_{-0.16} \pm 0.15$
(Frejus[***])	56	66	66.8	90.0	$0.87^{+0.18}_{-0.14} \pm 0.12$)
IMB-3	325	182	257.3	268.0	$0.54 \pm 0.05 \pm 0.12$
Kamiokande	159	151	164.9	260.6	$0.60^{+0.07}_{-0.06} \pm 0.05$
Combined	559	473	513.3	691.2	0.63 ± 0.04(stat)

(*) In estimating the event rate, NUSEX, Frejus and Kamiokande used the flux calculation of ref. [5], while IMB-3 used that of ref. [25].
(**) All events.
(***) Fully contained events only.

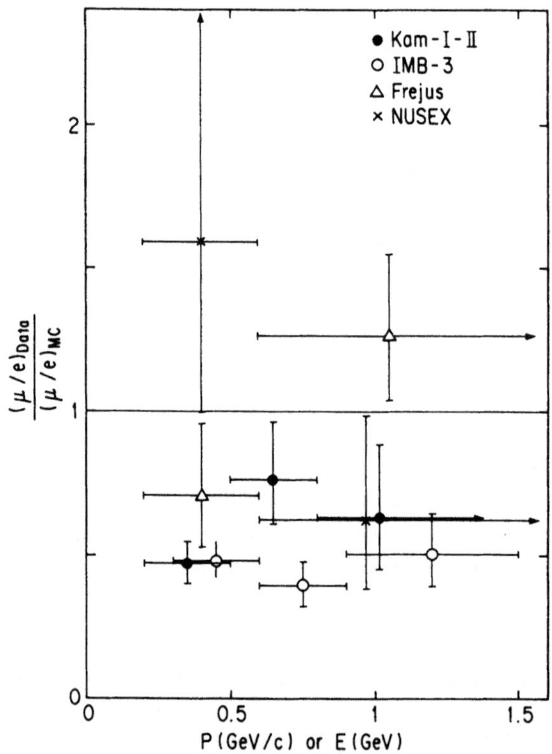

Figure 27: $(\mu/e)_{Data}/(\mu/e)_{MC}$ as a function of lepton momentum (or visible energy of an event) from the 4 experiments discussed in the text.

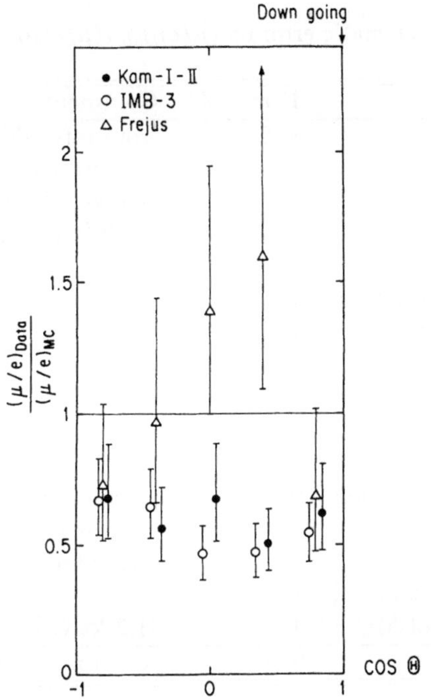

Figure 28: $(\mu/\mathrm{e})_{Data}/(\mu/\mathrm{e})_{MC}$ as a function of zenith angle from Frejus(△), IMB(○) and Kamiokande(•).

The authors in refs. [5] [27] estimate that the uncertainty in the $(\nu_\mu+\bar{\nu}_\mu)/(\nu_e+\bar{\nu}_e)$ ratio is less than 5 %. Kamiokande takes this number as a systematic error of the $(\nu_\mu+\bar{\nu}_\mu)/(\nu_e+\bar{\nu}_e)$ ratio. Indeed, three different flux calculations, (refs. [5] [25] [27]), agree within 3% for the $(\nu_\mu+\bar{\nu}_\mu)/(\nu_e+\bar{\nu}_e)$ ratio.

There is a discrepancy between the calculations of ref. [5] and ref. [27] concerning the $\nu_e/\bar{\nu}_e$ ratio. The discrepancy is as large as 8 % in the $0.3 \sim 1$ GeV energy range. Thus, the Kamiokande collaboration took 8 % for the ambiguity of the $\nu_e/\bar{\nu}_e$ ratio and found that this 8% corresponds to a 2 % error for the μ/e ratio. On the other hand, the $\nu_\mu/\bar{\nu}_\mu$ ratio was calculated to be 1 at low energy of our interest and is stable against changes in the input parameters. Thus, the ambiguity in the $\nu_\mu/\bar{\nu}_\mu$ ratio has only a negligible contribution on the systematic error of the μ/e ratio. Finally, we would mention that the flux calculation in ref. [25] predicts $\nu_e/\bar{\nu}_e = 1$. We do not understand this ratio, and think that authors in ref. [25] might have used some approximation, such as an equal production of π^+ and π^-, in their calculation.

Among the possible systematic errors, the most critical is an error in the μ/e identification, since it affects both the numerator and denominator in the μ/e ratio in opposite direction. Thus, the validity of particle identification was extensively studied in Kamiokande. Electrons (and γ's) produce electromagnetic showers, and low-energy electrons exhibit many multiple scatterings in the detector water. Thus,

Table 4: Source of the systematic error on $(\mu/e)_{Data}/(\mu/e)_{MC}$ from Kamiokande.

Source	Error (%)	Comments
$(\nu_\mu+\bar{\nu}_\mu)/(\nu_e+\bar{\nu}_e)$	<5	from refs. [5] [27]
$\bar{\nu}_e/\nu_e$	2	8 % error in the $\bar{\nu}_e/\nu_e$ ratio changes the μ/e ratio by 2 %.
Particle identification	4	2×(2 % of misidentification)
1-Ring/2-Ring separation	2	
Vertex position determination	1	
Absolute energy calibration	1	
External background	<1	
Ambiguity in the difference of cross section between ν_e and ν_μ	3	
Ambiguity in NC cross section	2	Fraction of NC events in single-ring events is estimated to be 4 %.
γ-ray from exited target nuclei	1	
Statistical fluctuation of MC	4	31.2 kt·yr exposure equivalent
Total	9 %	added in quadrature

the Cerenkov rings produced by electrons (and γ's) are more diffuse than those from muons (and charged pions). Figs. 18(a) and (b) show the Cerenkov pattern of a μ-like and an e-like event, respectively. One can clearly see the difference.

The particle identification program makes use of this difference of the distribution of photoelectrons. (In Kam-II, opening angle of the Cerenkov ring as estimated by the vertex-position reconstruction-program is also used for particle identification. The opening angle should be 42° for electrons (and γ's), but should be smaller than 42° for low-momentum muons and pions, because $\beta < 1$.) The actual discrimination is made based on a maximum-likelihood method. The likelihood (L) is defined as (for full description of this method see ref.[24])

$$L(e \text{ or } \mu) = \prod_{\theta_i < 1.5 \times \theta_{Cerenkov}} Prob_i(e \text{ or } \mu), \tag{2}$$

where i represents a PMT-number and

$$Prob_i(e \text{ or } \mu) = \frac{1}{\sqrt{2\pi}\sigma} \cdot \exp\left[\frac{-(p.e._{\text{obs'd}}(i) - p.e._{\text{exp'd}}(i, e \text{ or } \mu))^2}{2\sigma^2}\right] \tag{3}$$

for $p.e._{\text{exp'd}} > 10$. For $p.e._{\text{exp'd}} < 10$, $Prob_i(e \text{ or } \mu)$ takes into account the Poisson statistics and the pulse-hight resolution of the PMT to single-photoelectron pulses.

Fig. 29 shows the result of the particle identification. One can see that the shapes of the likelihood distributions are similar between the data and the neutrino

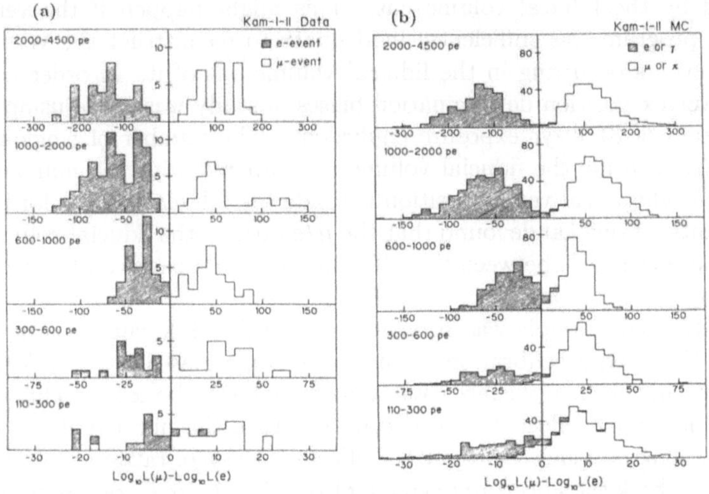

Figure 29: Results from the particle identification for (a) the data and (b) the atmospheric neutrino MC events of Kamiokande. Events indicated by a hatch show electrons or gammas for the MC and events classified as e-events for the data. The event samples are divided into 5 subsamples based on the observed total photoelectron numbers. The horizontal axis shows the difference of e and μ likelihood. Events with $\log L(\mu) - \log L(e) < (>)0$ are $e(\mu)$-like. Some events are classified as $e(\mu)$-like even if $\log L(\mu) - \log L(e) > (<)0$. This is because the final classification to either e- or μ-like is made by this likelihood plus the information of reconstructed opening angle of the Cerenkov ring.

MC events. From these studies, Kamiokande estimates the probability of incorrect particle identification to be 2 %. The quoted systematic error on the μ/e ratio is thus 4 %, since it affects the numerator and the denominator in the μ/e ratio in opposite directions. This 4 % is conservative because the particle identification is applied to both the data and the Monte Carlo events; thus, in principle, most of the systematic biases on $(\mu/e)_{data}/(\mu/e)_{MC}$ coming from particle identification are expected to cancel out.

In the analysis of Kamiokande, events with single Cerenkov rings are used. The selection of single-ring events is made through an interactive graphic display. The separation of single- and multi-ring events are trivial in most of the cases. However, there are some events that are ambiguous in the number of Cerenkov rings. One example of such events is a ν_e CC single pion production with the momentum of the pion near the Cerenkov threshold. In order to estimate the uncertainty resulting from this selection process, the Monte Carlo events of 31 kt·yr exposure equivalent were scanned to select events that are ambiguous in single- and multi-ring separation. This scanning found ± 1.9 % and 0.6 % ambiguity in the number of single-ring electron- and μ-like events, respectively. Adding these numbers in quadrature, Kamiokande obtained a 2 % systematic error in the μ/e ratio coming from the single- and multi-ring separation ambiguity.

There remain some doubts as to whether the μ-like events might be selectively discarded by the fiducial volume cut. This might happen if the vertex reconstruction program was sufficiently inadequate to reconstruct the vertex position of μ-like events occurring in the fiducial volume out of it. In order to check the possible vertex position determination biases, a study was made using the Monte Carlo events of 31 kt·yr exposure equivalent. The number of single-ring events reconstructed inside the fiducial volume is compared with the number of single-ring events whose real vertex position is inside the fiducial volume for both e- and μ-like events. Kamiokande found that the μ/e ratio in the fiducial volume does not change more than 1 % between the MC samples made based on the reconstructed and real vertex positions.

Since the data sample was selected with total visible energy cuts, the uncertainty in the energy calibration might cause some biases in the μ/e ratio. The uncertainty in the energy calibration was estimated to be ± 3 %. This uncertainty has less than a 1 % effect on the μ/e ratio. Also, the uncertainty in the trigger threshold causes a much less than 1 % effect on the μ/e ratio.

We now check the uncertainty coming from the Monte Carlo program. The lepton universality of weak interaction tells us that the cross-section ratio, $\sigma(\mathrm{CC}\,\nu_\mu)/\sigma(\mathrm{CC}\,\nu_e)$, is well understood and is close to 1 at sufficiently high energies, where the effect of the charged lepton mass could be neglected. However, in the energy range of our interest, the masses of leptons must be properly taken into account.

Table 5 summarizes the interaction type of the Monte Carlo fully-contained events. We check the possible changes in the numbers in Table 5 and in the (μ/e) ratio of the Monte Carlo due to uncertainties in the input parameters of the Monte Carlo program. The cross section of CC quasi-elastic interactions can be described using M_A (axial-vector mass), the value of which was determined by accelerator neutrino experiments. The typical magnitude of the measured error and the spread of the central value among the experiments is $5 \sim 10$ %. Thus, Kamiokande took a 10 % error in M_A and found ± 0.8 % error in μ/e from M_A. In a H_2O target, 89 % of the nucleons are bounded within the nucleus. The effect of the ^{16}O nucleus is taken into account by the Fermi-gas model. In this model, the Fermi surface momentum must be introduced by hand based on the experimental data. In order to estimate the uncertainty due to the Fermi surface momentum, Kamiokande changed it by ± 50 MeV/c and obtained a ± 1.0 % change in the μ/e ratio. It is argued [28] that the Fermi-gas model predicts systematically higher lepton momentum in the final state than does the shell model. This effect was checked by simply lowering the charged lepton energy by 20 MeV for both ν_e and ν_μ events. This procedure changed the μ/e ratio by 1.8 %. It can be seen in Table 5 that there are non-negligible contaminations of the CC $n\pi (n \geq 1)$ processes in the single-ring sample. While the estimated error in the μ/e ratio due to the uncertainties in CC quasi-elastic processes is of the order of 1 %, the same error due to the uncertainties in CC $n\pi (n \geq 1)$ processes might be larger. Kamiokande took 5 % for the error of $(\sigma_{\nu_\mu(\bar{\nu}_\mu)}$: CC $n\pi)/(\sigma_{\nu_e(\bar{\nu}_e)}$: CC $n\pi)$, and found that this amount of error translates to a 1.1 % error in the overall μ/e ratio. The comparison between the data and the Monte Carlo simulation in Fig. 3 suggests that the cross-section error in the single-pion channel should be less than 30 %. Kamiokande checked the ambiguity of the μ/e ratio due to the absolute cross-section error of single-pion production channels. They changed the absolute cross section by ±30 % and found that the effect on the μ/e ratio is 1.6 %. By adding all of these errors in quadrature, they obtained a 3 % error on the μ/e ratio due to ambiguities in the CC cross sections.

The contamination of NC events in the single-ring events is estimated to be about 4 %. A 50 % error in some of the cross sections of NC processes changes the μ/e ratio by 2 %. This is the number quoted as a systematic error for NC processes.

Table 5: Summary of the interaction type of the Monte Carlo fully-contained sample of Kamiokande.

Interaction channel(*)	e-like(1-ring) (%)	μ-like(1-ring) (%)	Multi-ring (%)
CC ν_e quasi-elastic	69.9	0.9	3.2
CC ν_e $1\pi^{\pm}$	17.9	0.6	9.8
CC ν_e $1\pi^0$	3.5	0.0	5.0
CC ν_e multi-π	2.2	0.0	6.0
CC ν_μ quasi-elastic	1.4	76.8	1.7
CC ν_μ $1\pi^{\pm}$	0.3	14.8	14.1
CC ν_μ $1\pi^0$	0.4	2.4	8.4
CC ν_μ multi-π	0.2	1.5	14.4
(CC total	95.6	97.0	62.5)
NC quasi-elastic	0.3	0.3	1.0
NC $1\pi^{\pm}$	0.5	1.2	3.1
NC $1\pi^0$	2.5	0.2	18.0
NC multi-π	1.1	1.3	15.4
(NC total	4.4	3.0	37.5)
Total	100	100	100

(*) Pions produced in these interactions might undergo nuclear interactions in the same nucleus. Thus, for example, CC ν_μ 1 π^0 production might have only 1 Cerenkov ring, because the π^0 might be absorbed in the same nucleus.

In neutrino interactions off ^{16}O nuclei, the target nucleus in the final state should be in some excited state. Then, the exited nucleus decay into a ground state emitting about total of 10 MeV gamma rays. (Roughly speaking, nuclei exited above $\sim$ 10 MeV generally emit particles such as protons, neutrons and α's, and decay into a lower energy state. If the energy of the state is less than 10 MeV, gamma-rays are emitted.) Since the extra Cerenkov photons due to these gamma-rays are 'backgrounds' to the Cerenkov photons due to leptons, these gamma rays might have some effect on the particle identification and might change the μ/e ratio. The μ/e ratios of Monte-Carlo-generated events with and without these gamma rays are compared. The ratio changed less than 1 % in these two Monte Carlo samples.

Finally, since the entire Monte Carlo study was carried out based on the sample of 31.2 kt·yr of the detector exposure, there should be a non-negligible error from the statistical fluctuation of the Monte Carlo events, which gives a 4.0 % effect.

Adding all of these errors in quadrature, Kamiokande estimates that the total systematic error on the μ/e ratio is $\pm 9\,\%$.

The Frejus experiment quotes $\pm 14\,\%$ systematic error in the μ/e ratio. They quote a 5 % systematic error in the $(\nu_\mu + \bar{\nu}_\mu)/(\nu_e + \bar{\nu}_e)$ ratio. The Monte Carlo simulation of the neutrino interactions in the detector has an associated systematic error of about 11 %. The largest source of error associated with the Monte Carlo simulation is due to the ambiguity in the trigger threshold. They estimate a 10 % ambiguity in the predicted number of CC ν_e events and a negligible error on the CC ν_μ events. The systematic error due to the event analysis is about 7 %. They estimate the systematic error by performing three independent analyses and measured the spread in the μ/e ratio among the three analyses.

The total systematic error in the μ/e ratio in IMB-3 is estimated to be about $\pm 22\,\%$. The uncertainty in the calculation of the $(\nu_\mu + \bar{\nu}_\mu)/(\nu_e + \bar{\nu}_e)$ ratio is quoted to be less than 5 %. The uncertainties in the Monte Carlo simulation are: 4 % from the 10 % uncertainty of M_A, less than 10 % from the 20 % variation of the Fermi-surface momentum, 7 % from a factor two uncertainty of the cross sections of NC processes. Adding the uncertainties due to the flux, form factor (M_A), Fermi-surface momentum, and neutral-current linearly, they obtained a 20 % overall uncertainty in the expected μ/e ratio. The uncertainty in the measured μ/e ratio due to particle identification was quoted to be 10 %. For other possible sources of systematic error, such as the uncertainties in the detection efficiency through the reduction processes or in the energy calibration, IMB-3 quotes them to be negligible. Adding the 20 % uncertainty in the $(\mu/e)_{MC}$ and 10 % uncertainty in the $(\mu/e)_{Data}$ in quadrature, IMB-3 obtained a 22 % uncertainty in the μ/e ratio measurement. It should be noted that the quoted systematic error in IMB-3 is larger than that of Kamiokande. This could be understood in part because IMB-3 adds some of the sources of the error linearly, while Kamiokande adds all of them in quadrature.

To summarize, the $(\nu_\mu + \bar{\nu}_\mu)/(\nu_e + \bar{\nu}_e)$ ratio is calculated quite accurately, their errors being less than $\pm 5\,\%$. The cross section ratio between ν_μ and ν_e can be accurately estimated from lepton universality of weak interactions. The contribution of neutral-current events is small. All of the other sources of systematic errors are well understood and are small. From these arguments, it is clear that the systematic error could not account for the observed small $(\mu/e)_{Data}/(\mu/e)_{MC}$.

4.5 μ-decay signal

In the case of the imaging water Cerenkov experiments one can check the above results by comparing the fraction of μ-decay-associated events with the MC expectation. Fig. 19 shows (a) a μ-like event and (b) an electron event from the decay of the preceding μ-like event. Since this analysis does not involve the particle identification, it constitutes an independent check. See Table 6 which summarizes the Kamiokande and IMB-3 data. IMB-3 observed for single-ring fully contained events that the fraction of μ-decay associated events, F(Data), was 0.33 ± 0.03 while $F(\text{MC}) = 0.43 \pm 0.01$ was expected from the MC simulation. Kamiokande observed that $F(\text{Data}) = 0.34 \pm 0.03$, while $F(\text{MC}) = 0.46 \pm 0.01$ was expected. These results are consistent with the small μ/e ratio described above; since the fractions (F(Data) and F(MC)) are related to the above $(\mu/e)_{Data}/(\mu/e)_{MC}$ by the approximate relation $(\mu/e)_{Data}/(\mu/e)_{MC} \sim (F(\text{Data})/(1-F(\text{Data}))/(F(\text{MC})/(1-F(\text{MC}))$, the corresponding values of $(\mu/e)_{Data}/(\mu/e)_{MC}$ for the entire momentum range are $0.67^{+0.07}_{-0.06}(stat.)$ for IMB-3 and $0.61 \pm 0.07(stat.)$ for Kamiokande.

It is also important that, for both e- and μ-like events, the expected and observed μ-decay fraction agree well. There is no inconsistency between the particle identification and the μ-decay methods in separating the μ- and e-like events.

Table 6: Summary of the Kamiokande and IMB-3 data on μ-decay.

Experiment and Event sample	Data Total	Data μ-decay	Data fraction of μ-decay	ν MC Total	ν MC μ-decay	ν MC fraction of μ-decay
Kamiokande						
Single ring	329	112	0.34±0.03	452.2	206.8	0.46
μ-like	151	99	0.66±0.05	260.6	190.4	0.73
e-like	178	13	0.07±0.02	191.6	16.4	0.09
($p_e > 100$MeV/c)	(159)			(164.9)		
IMB-3						
Single ring	507	168	0.33±0.03	525.3	223.3	0.43
μ-like	182	117	0.64±0.04	268.0	180.0	0.67
e-like	325	51	0.16±0.02	257.3	43.2	0.17

4.6 Consistency among the experiments

In the last column of Table 3, $(\mu/e)_{Data}/(\mu/e)_{MC}$ from all of the experiments are summarized. There seems to be a large deviation in the central values among the experiments. However, the experimental errors are also large. The lower statistics data of the NUSEX detector, while the data are consistent with no anomaly, does not really exclude the Kamiokande/IMB-3 results. The μ/e ratio calculated using all of the data from Frejus is about 3σ statistically higher than those of Kamiokande and IMB-3. If one uses the fully contained data for Frejus, as in the cases of Kamiokande and IMB-3, the disagreement in the μ/e ratio between (Kamiokande/IMB-3) and Frejus becomes smaller (about a 2σ effect). The significance of the disagreement is reduced if the systematic error is taken into account in this argument. However, a more detailed comparison of the result for the Frejus gas-counter iron-plate detector with the results of the imaging water Cerenkov detectors is subject to the difficulty that the contained event category in the Frejus data includes events of inelastic interactions in addition to the single-prong (largely $\sim 75\,\%$ quasi-elastic) events on which the analyses of the Kamiokande and IMB-3 are based. Furthermore, the energy range covered by the Frejus data is slightly higher than those of the water Cerenkov detectors (see Fig. 25). The trigger threshold of Frejus is about 400 MeV for single-track events at 50 % efficiency [22]. Accordingly, the Frejus data, based on 1.56 kt·yr exposure, do not repudiate conclusively the results from Kamiokande and IMB-3, based on a total exposure of 12.6 kt·yr.

5 Interpretation of the result - neutrino oscillations -

Since it is known that the calculated $(\nu_\mu + \bar{\nu}_\mu)/(\nu_e + \bar{\nu}_e)$ ratio is quite accurate and does not strongly depend on the calculational details, the discrepancy between the calculations and the results from water Cerenkov detectors is very serious, and requires some explanations. One explanation proposed to explain the atmospheric neutrino result was neutrino oscillations [29].*

The probability that a neutrino born as a ν_μ at the source remains a ν_μ at distance L is

$$P(\nu_\mu \to \nu_\mu) = 1 - \sin^2 2\theta \sin^2(\frac{\pi L}{L_v}), \tag{4}$$

where $L_v = 4\pi p_\nu/\Delta m^2$ is called the vacuum oscillation length. θ and Δm^2 represent the mixing angle between two neutrino species and the mass-difference squared $(= |{m_1}^2 - {m_2}^2|)$, respectively. Numerically, this equation is expressed as

$$P(\nu_\mu \to \nu_\mu) = 1 - \sin^2 2\theta \sin^2\left(\frac{1.27\Delta m^2 L}{p_\nu}\right). \tag{5}$$

In the above equation, Δm^2, L and p_ν are in units of $(\mathrm{eV})^2$, km and GeV/c, respectively. The combination of $p_\nu \sim 1\,\mathrm{GeV/c}$ and $L \sim 10 \sim 10^4\,\mathrm{km}$ suggests that the potentially explored range of Δm^2 by atmospheric neutrinos is as low as $10^{-4}\,\mathrm{eV}^2$. This value is more than 2 orders of magnitude smaller than those have been explored by accelerator and reactor experiments. On the other hand, the range of accessible $\sin^2 2\theta$ is limited to be relatively large, due mainly to the low event rate.

*Recently, it was pointed out that the observed small μ/e ratio might be due to signals of proton decay into $e^+\nu\nu$ [31]. In this scenario, the small μ/e ratio is interpreted as being due to an excess of electron-events in the low-momentum ($\lesssim$600 MeV/c) region instead of a deficit of μ-events. If proton decay is the cause of the observed μ/e ratio, then the μ/e ratio above the nucleon decay region must be in agreement with the prediction. See Fig. 27, which shows $(\mu/e)_{Data}/(\mu/e)_{MC}$ as a function of lepton momentum. Kamiokande and IMB-3 observe the small μ/e ratio above the region related to the nucleon decay signal. However, the experimental error in this region is too large to conclude definitely.

If one considers neutrino oscillations between ν_e and ν_x ($x = \mu$ or τ), it is necessary to take into account the effect of the matter in the earth (MSW effect [30]). The MSW effect is due to a phase mismatch in the scattering amplitude between ν_e and ν_x, which is caused by a difference in the coherent forward scattering amplitude between them. It should be noted that the MSW effect never occurs between ν_μ and ν_τ in the earth. As a consequence, oscillation is governed by an effective mixing angle θ_m and an oscillation length L_v, which are written

$$P(\nu_\mu \to \nu_\mu) = 1 - \sin^2 2\theta_m \sin^2(\frac{\pi L}{L_m}), \tag{6}$$

$$\sin^2 2\theta_m = \frac{\sin^2 2\theta}{\sin^2 2\theta + (L_v/L_0 \mp \cos 2\theta)^2}, \tag{7}$$

$$L_m = \frac{L_v}{\sqrt{1 \mp 2(L_v/L_0)\cos 2\theta + (L_v/L_0)^2}} \tag{8}$$

and

$$L_0 = \frac{2\pi}{\sqrt{2}GN} = \frac{1.624 \times 10^7 \text{meters}}{\rho}, \tag{9}$$

where $\rho = N/(6.02 \times 10^{23})$ is the electron density in units of Avogadro's number per cm^3. In the above equations, $\mp$ is for neutrinos $(-)$ and anti-neutrinos $(+)$. Equation 7 shows that the neutrino mixing is enhanced for neutrinos at $L_v/L_0 \simeq \cos 2\theta$, even if the intrinsic mixing angle (vacuum mixing angle), $\sin^2 2\theta$, is small. Numerically, this resonance condition is

$$1.53 \times 10^{-4} \cdot \frac{p(\text{GeV/c}) \cdot \rho(\text{g/cm}^3)}{\Delta m^2(\text{eV}^2)} = \cos 2\theta. \tag{10}$$

For a typical atmospheric neutrino energy of 1 GeV, along with a typical matter density of 10 g/cm^3 in the earth, the resonance condition is satisfied for $\Delta m^2 \sim 1.5 \times 10^{-3}$ eV^2. On the contrary, the MSW effect always suppresses anti-neutrino mixing.

Before looking at the result of a full analysis, the following general arguments could be made:

(1). Because of the large uncertainty in the normalization of the total number of the events, both $\nu_e \leftrightarrow \nu_\mu$ and $\nu_\mu \leftrightarrow \nu_\tau$ oscillations are allowed. (For neutrino oscillations between ν_e and ν_μ, the number of e-like events of the data should be larger than expected and the number of μ-like events of the data should be smaller than expected. Thus, $(\mu/e)_{Data}/(\mu/e)_{MC}$ should be less than 1. Similarly, for neutrino oscillations between ν_μ and ν_τ, the number of e-like events of the data should be equal to the expected number, and the number of the μ-like events of the data

should be smaller than that expected. This is because the threshold energy of charged-current ν_τ interactions is about 3 GeV and most of the ν_τ oscillated from the atmospheric ν_μ do not undergo CC interactions. Thus, $(\mu/e)_{Data}/(\mu/e)_{MC}$ should be less than 1. On the other hand, for neutrino oscillations between ν_e and ν_τ, the number of e-like events of the data should be smaller than that expected, and the μ-like events of the data should be equal to that expected. Thus, $(\mu/e)_{Data}/(\mu/e)_{MC}$ should be larger than 1.)
(2). Because of no significant zenith-angle dependence of $(\mu/e)_{Data}/(\mu/e)_{MC}$, the typical oscillation length of atmospheric neutrinos could be $\lesssim O(100\,\mathrm{km})$ or less. This value of L in Eq. 5 and typical p_ν of $O(1\,\mathrm{GeV/c})$ suggest that

$$\frac{1.27\Delta m^2 L(\lesssim 100\,\mathrm{km})}{p_\nu(\sim 1\,\mathrm{GeV/c})} \gtrsim \frac{\pi}{4}. \tag{11}$$

Therefore,

$$\Delta m^2 \gtrsim O(0.5\times 10^{-2}\,\mathrm{eV}^2). \tag{12}$$

(3). Finally, the large discrepancy between $(\mu/e)_{Data}$ and $(\mu/e)_{MC}$ suggests that the mixing angle between the two neutrino species is large.

Analyses of neutrino oscillations using the Kamiokande result were carried out in [2][26][24]. Fig. 30 shows the allowed regions of the neutrino oscillation parameters from Kamiokande [26]. In their analysis, the momentum and zenith angle dependences of the μ/e ratio are taken into account. It should be noted that the MSW effect is included in obtaining the allowed region of $\nu_\mu \leftrightarrow \nu_e$ oscillations. Also shown in the same figure are the excluded regions from a reactor experiment [32], accelerator experiments [33] [34] [35] and the Frejus experiment [7]. The lack of a clear positive effect in the Frejus result leads to an excluded region. The result from IMB-3 is not shown, because the analysis in terms of neutrino oscillations has not been presented. One can see that large parts of the allowed regions from Kamiokande ($\Delta m^2 > 4\times 10^{-3}$ and $\sin^2 2\theta > 0.3$ for $\nu_\mu \leftrightarrow \nu_e$ oscillations, and $\Delta m^2 > 1\times 10^{-3}$ and $\sin^2 2\theta > 0.45$ for $\nu_\mu \leftrightarrow \nu_\tau$ oscillations) are not seriously excluded by the other experiments. (There are some constraints on the neutrino oscillation parameters from the data on upward-going muons, see sect. 6.)

We now focus attention on the $\nu_\mu \leftrightarrow \nu_\tau$ channel, since the allowed region of the $\nu_\mu \leftrightarrow \nu_e$ channel conflicts with the constraints from the solar-neutrino data if the latter are also to be ascribed to neutrino oscillation. The presently available solar-neutrino data from the ^{37}Cl [36], Kamiokande-II [37], SAGE [38] and Gallex [39] experiments all indicate a smaller solar-neutrino flux than the theoretical prediction. At present, one of the most promising solutions to the solar-neutrino deficit is neutrino oscillations in the solar interior (MSW mechanism [30]). If the MSW mechanism is the solution of the solar-neutrino problem, then $\nu_e \leftrightarrow \nu_\mu$ oscillations are ruled out as a solution of the atmospheric neutrino problem. However, the allowed region for $\nu_\mu \leftrightarrow \nu_\tau$ is not in conflict with any other experimental results.

Neutrino oscillations may also be manifested in the momentum and zenith-angle distributions of the single-ring events, which would reflect a difference in neutrino energy and in path length or of matter traversed. The observed momentum

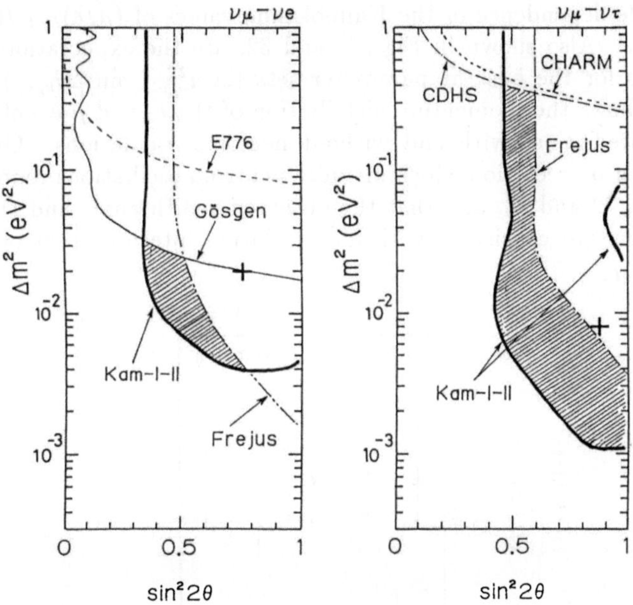

Figure 30: Allowed neutrino oscillation parameters at the 90 % C.L. from the Kamiokande data, for the cases of $\nu_\mu \leftrightarrow \nu_e$ (left), and $\nu_\mu \leftrightarrow \nu_\tau$ (right) [26]. The best fit parameter sets are shown by crosses. All of the other experimental results show excluded regions. The regions marked by a hatch are the regions allowed by the Kamiokande data.

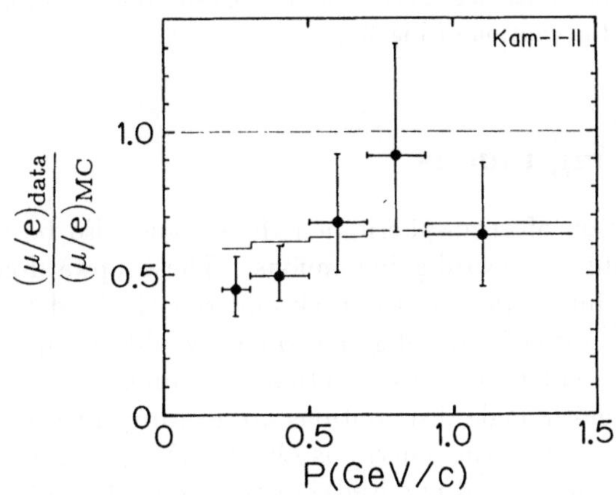

Figure 31: Momentum dependence of $(\mu/e)_{Data}/(\mu/e)_{MC}$ of the Kamiokande data [26]. The circles with $\pm 1\sigma$(stat) vertical bars indicate the values of $(\mu/e)_{Data}/(\mu/e)_{MC}$ in the respective momentum bins. The solid histogram shows the expected $(\mu/e)_{Data}/(\mu/e)_{MC}$ for the neutrino oscillation parameters ($\Delta m^2_{\nu_\mu\nu_\tau}$ in eV^2, $\sin^2 2\theta_{\nu_\mu\nu_\tau}$) = $(0.8 \cdot 10^{-2}, 0.87)$, the best-fit parameter set. The broken line at $(\mu/e)_{Data}/(\mu/e)_{MC} = 1$ shows the case with no neutrino oscillations.

dependence of the Kamiokande values of $(\mu/e)_{Data}/(\mu/e)_{MC}$ is shown in Fig. 31. The zenith-angle dependence of the Kamiokande values of $(\mu/e)_{Data}/(\mu/e)_{MC}$ is shown in Fig. 32. Also shown in Figs. 31 and 32, are the expectations from the MC simulations for the best fit parameter sets $(\Delta m^2_{\nu_\mu\nu_\tau}, \sin^2 2\theta_{\nu_\mu\nu_\tau})$. See also Fig. 33, which shows the momentum distribution of the e- and μ-events, together with the MC predictions with and without neutrino oscillations. One can see that a Monte Carlo prediction which includes neutrino oscillations reproduces the data in Figs. 31, 32 and 33, and that the observed zenith angle and the momentum distributions are consistent with an oscillation interpretation of the small $(\mu/e)_{Data}/(\mu/e)_{MC}$.

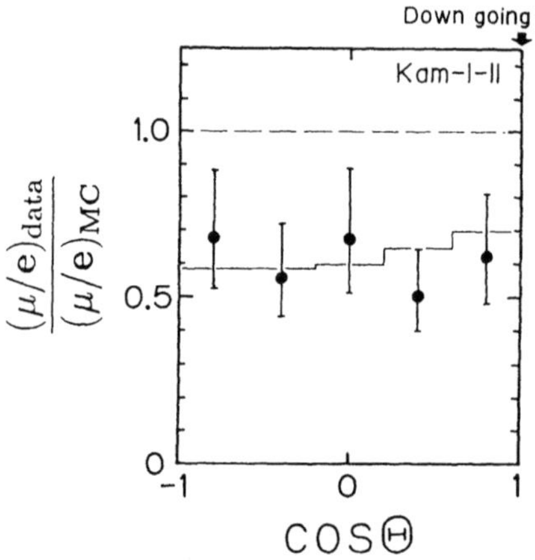

Figure 32: Zenith angle dependence of $(\mu/e)_{Data}/(\mu/e)_{MC}$ of the Kamiokande data [26]. The error bars are statistical only. The histogram and the broken line show the MC predictions, (see the caption of Fig. 31).

6 Upward-going muons

The flavor composition of atmospheric neutrinos could also be studied by measuring the event rate of upward-going muons. These upward-going muons are produced by CC ν_μ interactions in the rock underneath the detector. If ν_μ oscillates to ν_e or ν_τ, the rate of upward-going muons would be suppressed compared with the case of no oscillations. In the actual experiments, the observed number of events, and the zenith-angle distributions are examined with the Monte Carlo simulation. The typical neutrino energy is $O(100\,\mathrm{GeV})$ (see Fig. 34). The typical path lengths vary from (several hundred) km (horizontal) to 13000 km (vertical). This combination suggests the typical accessible Δm^2 to be

$$\frac{1.27\Delta m^2 L(\lesssim 10^4\,\mathrm{km})}{p_\nu(\sim 100\,\mathrm{GeV/c})} \gtrsim \frac{\pi}{4}. \tag{13}$$

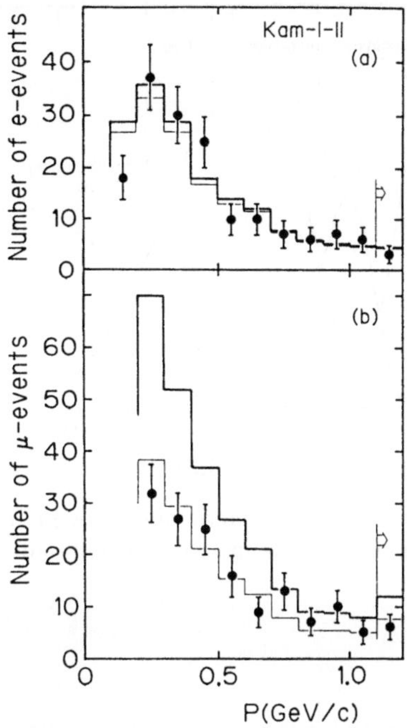

Figure 33: Momentum spectra of (a) e- and (b) μ-events [26]. The histograms show the Monte Carlo predictions without neutrino oscillations (thick) and with neutrino oscillations for $(\Delta m^2_{\nu_\mu\nu_\tau}, \sin^2 2\theta_{\nu_\mu\nu_\tau}) = (0.8 \times 10^{-2}\,\mathrm{eV}^2, 0.87)$ (thin).

Therefore,

$$\Delta m^2 \gtrsim O(0.5 \times 10^{-2}\,\mathrm{eV}^2). \tag{14}$$

This range of Δm^2 is similar to that suggested by the contained event data from Kamiokande (Fig. 30). Thus, the study of upward-going muons could provide an independent check of the fully contained data.

There are at present three experiments which have reported on the flux of upward-going muons. They are the Baksan Underground Scintillator Telescope (Baksan) [40], IMB [41] and Kamiokande [42]. Table 7 summarizes the results from the three experiments. It is clear that the observed fluxes by the three experiments are consistent with the Monte Carlo simulations if one considers the statistical error of the data and the systematic error of the Monte Carlo prediction. The systematic error, which is about 20 %, is dominated by the uncertainty of the neutrino-flux calculation. All three experiments concluded that no evidence for neutrino oscillations was observed in the upward-going muon data.

In order to estimate the flux of upward-going muons, it is necessary to precisely know the flux of atmospheric neutrinos, the neutrino nucleon cross-sections and the dE/dx of muons, all in the range of 1 GeV to 10^5 GeV. Recently, it was argued

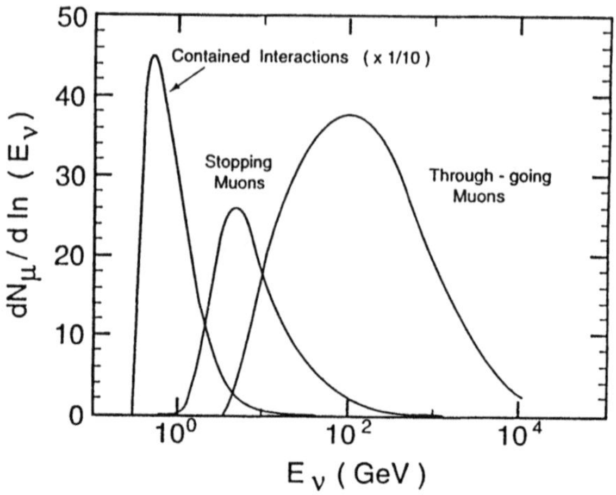

Figure 34: Approximate energies of atmospheric neutrinos giving rise to the fully contained events, as well as the upward stopping and through-going muons in an underground detector [44].

Table 7: Summary of upward-going muons from Baksan,[40] IMB[41] and Kamiokande.[42]

	Zenith angle (deg)	$E_{th}^{(a)}$ (GeV)	Observed number of ev.	Calculated number of ev.
Baksan	180 – 90$^{(c)}$	1.0	421	
IMB	180 – 90	1.8(IMB-1-2)	617(All)	600
		1.0(IMB-3)	532(Exiting)	516
			85(Stopping)	84
Kamiokande	180 – 90	1.7	252	

Observed flux$^{(b)}$ (10^{-13}cm^{-2}sec^{-1}sr^{-1})	Calculated flux$^{(b)}$ (10^{-13}cm^{-2}sec^{-1}sr^{-1})
2.77±0.17±0.09	$2.45^{(d)} \sim 2.75^{(e)}$
2.04±0.13	$1.92^{(d)} \sim 2.20^{(e)}$

(a) Threshold energy for observed muons.
(b) Averaged flux over the zenith-angle range.
(c) The detection efficiency is 50% for zenith angles of 90 – 110 deg.
(d) Volkova's flux [3]
(e) Butkevich's flux [43]

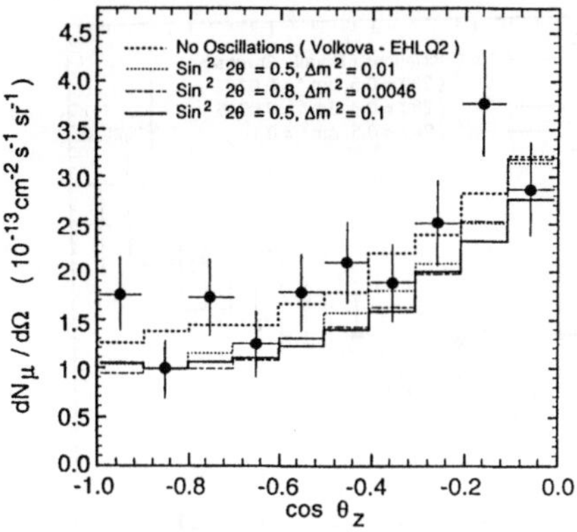

Figure 35: Zenith-angle distribution of the upward-going muons observed by Kamiokande together with the Monte Carlo prediction with the flux of ref. [3] and the neutrino cross sections used by the Kamiokande collaboration. Monte Carlo predictions with neutrino oscillations are also shown. (From [44])

[44] that the present experiments might be underestimating the cross sections of high-energy ($\gtrsim$10 GeV) neutrino interactions. It was also suggested that an independent flux calculation in ref. [45] predicts higher fluxes than that in ref. [3] does. Fig. 35 shows the zenith angle distribution of the Kamiokande data with the flux of ref. [3] and the cross sections used by the Kamiokande collaboration. Monte Carlo predictions with neutrino oscillations are also shown. It is clear that, in this case, Monte Carlo simulations with neutrino oscillations do not reproduce the data well. This conclusion is essentially the same as the conclusion of the three experiments. On the other hand, Fig. 36 shows a similar plot with the flux calculation of ref. [44], as well as the calculated neutrino cross sections based on a recent structure functions of ref. [46]. In this case, a Monte Carlo simulation with neutrino oscillations reproduces the data rather well. This study strongly suggests that we need more studies before making firm conclusions.

In order to avoid the 20 % uncertainty in the absolute flux, IMB compared the rate of upward-going muons which stop in the detector with the rate of those which pass through the detector. In this way the errors from the absolute flux and cross section cancel. Since the medium energy of the stopping muons is calculated to be 6 GeV, as compared to 100 GeV for the through going ones (see Fig. 34), the fraction of stopping muons to through-going ones is a sensitive parameter for neutrino oscillations. Table 7 shows the number of stopping and through-going muons of the data and the Monte Carlo prediction. IMB claimed that the measured data are consistent with the hypothesis of no neutrino oscillations. They obtained an excluded region between 2×10^{-4} and 2×10^{-2} eV^2 for Δm^2 and > 0.5 for $\sin^2 2\theta$, (see Fig. 37).

The data of upward-going muons are potentially very important since they provide independent data on the atmospheric ν_μ/ν_e ratio. However, much more work might be needed to make any definite statement.

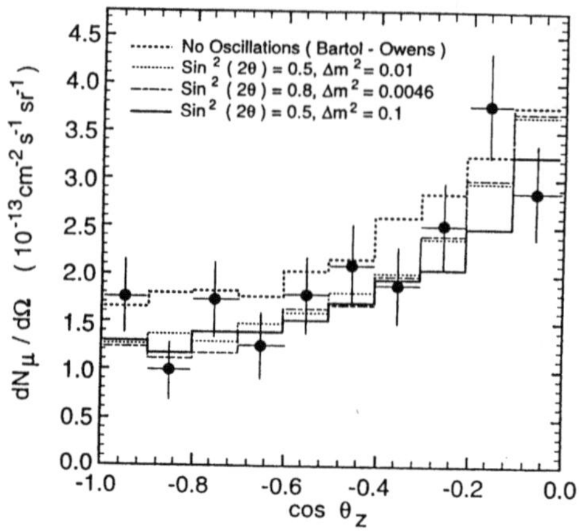

Figure 36: Zenith-angle distribution of the upward-going muons observed by Kamiokande together with the Monte Carlo prediction with the flux of ref. [44] and the calculated neutrino cross sections based on a structure function of ref. [46]. Monte Carlo predictions with neutrino oscillations are also shown. (From [44])

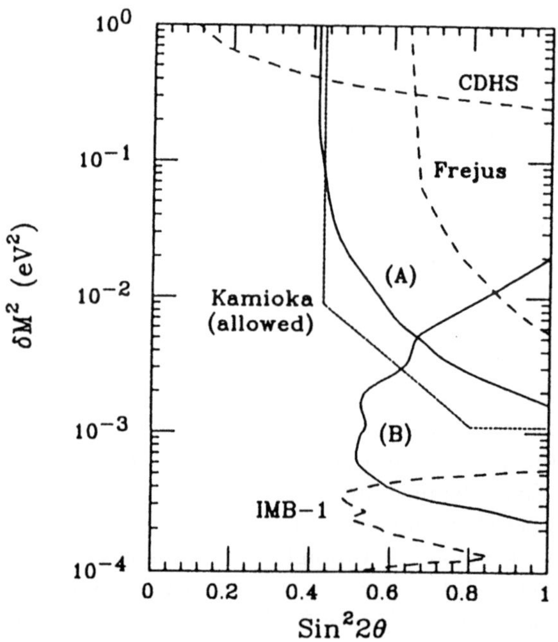

Figure 37: 90 % C.L. limits on $\nu_\mu \rightarrow \nu_\tau$ oscillations from (A) the absolute flux of upward-going muons and (B) the fraction of stopping upward-going muons as obtained by IMB [41]. Excluded regions from Fréjus and CDHS, and the allowed region from Kamiokande are also shown.

7 Future prospect

The results from Kamiokande and IMB-3 on atmospheric neutrinos might be related to as-yet-unaccounted-for physics, such as neutrino oscillations. Further studies are required to obtain a better understanding of what is occurring in atmospheric neutrinos. The following lists the future prospects.

7.1 Other (New) experiments

So far, a small μ/e ratio has been observed only in water Cerenkov detectors. The results from gas-counter iron-plate tracking-calorimeters agree with the prediction. Although the statistics from these experiments are limited compared with that of the water Cerenkov detectors, many physicists worry about this discrepancy. Since both Frejus and NUSEX have already stopped data acquisition, it is necessary to wait for new experiments to start taking data. [47] is a fine-grained tracking calorimeter with the capability of dE/dx measurements. (Frejus and NUSEX did not have the dE/dx information.) The detector starts full operation in the summer of 1993, and is expected to observe ~600 atmospheric neutrino events in 5 years of its operation. With the dE/dx measurement, the detector can distinguish muons from protons, thus, further improving the particle identification efficiency compared with Frejus and NUSEX. This experiment might clarify the possible disagreement between (IMB-3 + Kamiokande) and (Frejus + NUSEX).

Very recently, a preliminary result from Soudan-2 [48] appeared. They observed 26 single-track events in total. Among them, they found 14 e-like and 11 μ-like events, while the expected numbers were 14.5 and 20.8 events, respectively. These numbers can be translated to the μ/e ratio. Then, the Soudan-2 result is

$$\frac{(\mu/e)_{Data}}{(\mu/e)_{MC}} = 0.55^{+0.27}_{-0.19}(\text{stat.}) \qquad (\text{Soudan} - 2,\ \text{Preliminary}). \tag{15}$$

This result is consistent with those from Kamiokande and IMB-3. However, the statistical error is too large to be conclusive. Several more years are probably necessary to obtain a result with sufficiently good statistics from this experiment.

7.2 High-statistics experiments

If the distribution of the μ-like/e-like ratio shows some zenith angle dependence, it could offer additional support for the neutrino-oscillation interpretation of the small μ/e ratio. The systematic uncertainty of the up-down uncertainty of the μ/e ratio should be very small ($\lesssim O(1)\,\%$), since all of the detectors are designed as up-down symmetric. The μ/e ratio observed by the current detectors show no strong zenith-angle dependence, i.e., the small μ/e ratio is independent on the zenith angle within the statistics. Since the only detector which is still operational among the 4 experiments is Kamiokande, and Kamiokande has already been operational for nearly 10 years, one does not expect any great improvements in the statistics with the current detectors. One possible experiment that could improve the statistics

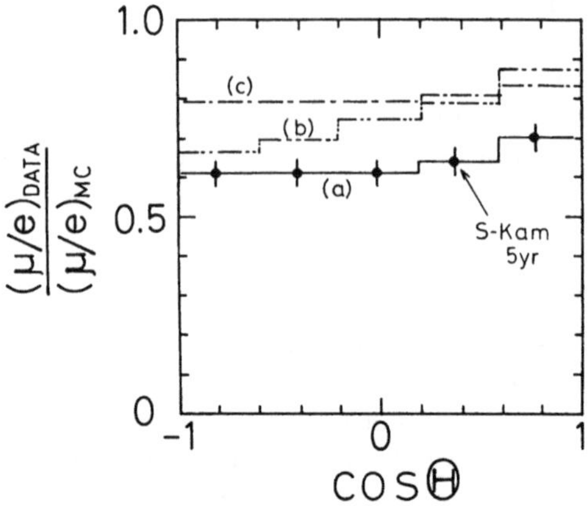

Figure 38: Calculated zenith-angle dependence of $(\mu/e)_{Data}/(\mu/e)_{MC}$ for (a) $(\Delta m^2, \sin^2 2\theta) = (0.8 \cdot 10^{-2}, 0.85)$, (b) $(\Delta m^2, \sin^2 2\theta) = (1.0 \cdot 10^{-3}, 0.85)$, and (c) $(\Delta m^2, \sin^2 2\theta) = (0.8 \cdot 10^{-2}, 0.45)$. The parameters used in (b) and (c) are at the boundary of the allowed region from the Kamiokande result. Also shown are the expected error bars after 5 years of the operation of the Superkamiokande detector.

significantly is Superkamiokande [49]. This is an experiment which has 22 kton of fiducial mass. The detector should observe $\sim$ 1400 single-ring atmospheric neutrino events in one year. Fig. 38 shows the expected error bars on the zenith-angle dependence after 5 years of operation. There might be a chance to observe an up-down asymmetry of $(\mu/e)_{Data}/(\mu/e)_{MC}$.

Finally, a high-statistics experiment with Superkamiokande might be able to distinguish between $\nu_e \leftrightarrow \nu_\mu$ and $\nu_\mu \leftrightarrow \nu_\tau$ oscillations. This might be possible because, in the case of $\nu_e \leftrightarrow \nu_\mu$ oscillations, both μ-like(Data)/μ-like(MC) and e-like(Data)/e-like(MC) should show a characteristic zenith-angle dependence, while in the case of $\nu_\mu \leftrightarrow \nu_\tau$ oscillations only μ-like(Data)/μ-like(MC) should show this type of dependence.

7.3 Better understanding of the systematic error on the μ/e ratio

In Kamiokande, an attempt is being made to reduce the systematic error. In 1990, mirrors were installed around each PMT, thus increasing the light correction by about 25 % with negligible distortion of the event pattern by badly scattered light from the mirror. This improvement should slightly reduce the misidentification probability near to the analysis threshold (205 MeV/c for muons).

An attempt is also being made at Kamiokande and IMB-3 to prove their analysis by a series of beam tests. They constructed a water tank of 11 m in height and 11 m in diameter at KEK in Japan. This tank contains about 1 kton of water and is going to be viewd by $\sim$ 400 PMT's of 20" diameter (Kamiokande). Then after the beam test by the Kamiokande group, the IMB-3 group is going to repeat the same beam test with their 8" PMT's. This test detector is going to be exposed to beams of e, μ, π and possibly ν from the 12 GeV KEK-PS. (A beam test with

Table 8: Summary of the IMB-3 data and the MC predictions for three fluxes (from Table 1 of ref. [8]).

	Data	Lee flux[25]	Gaisser flux[5]	Honda flux[27]
Single ring				
in momentum cuts	236	232.1	302.3	262.0
e-like	139	113.7	150.0	128.7
μ-like	97	118.4	152.2	133.4

neutrinos has not been approved yet.) The exposure is scheduled in 1994. From test experiments, the systematic error of the μ/e ratio resulting from the analysis procedure should become smaller.

7.4 Long-baseline neutrino oscillation experiments

As can be seen from Fig. 30, the allowed Δm^2 of the neutrino oscillations turned out to be $\gtrsim 10^{-3}$ eV^2 with large mixing angles. It might be possible to test the neutrino oscillation hypothesis of the atmospheric neutrino data by employing long-baseline accelerator experiments.

There should be several combinations between the accelerator and the detector. For example, a combination of the Fermilab Main Injector ($\langle E_\nu \rangle \sim 20$ GeV) and the Soudan-2 detector, which is 900 km away from Fermilab, can study Δm^2 in the range of (several $\times 10^{-3}$ eV^2)[50]. Another possibility [51] is a combination of KEK-PS ($\langle E_\nu \rangle \sim 2$ GeV) and the Superkamiokande detector,which is 250 km away from KEK. Although the distance between KEK and Superkamiokande is closer than that between Fermilab and Soudan-2, the range of Δm^2 to be explored is lower, $\Delta m^2 \gtrsim 3 \times 10^{-3}$ eV^2. This small Δm^2 is due to the lower E_ν of the KEK-PS beam (see Eq. 5). It was proposed to construct a new detector for a long-baseline neutrino oscillation experiment using the BNL-AGS. The detectors consist of three massive Cerenkov counters at 1, 3 and 20 km away from AGS. There are other possibilities for long-baseline neutrino oscillation experiments, for example, from CERN-SPS to ICARUS at Gran Sasso Laboratory, from CERN-SPS to Superkamiokande and from Frmilab Main Injector to DUMAND at Hawaii. Because of the importance of the neutrino-mass problem, these possibilities have to be seriously considered. Readers interested in these types of experiments should see, for example, refs. [52], [53].

7.5 Improvements in the flux calculation and its implications

The estimated errors in the calculation of the atmospheric neutrino fluxes are summarized as: (at least) $\pm 20\,\%$ for the absolute normalization, and (less than) $\pm 5\,\%$ for the $(\nu_\mu + \bar{\nu}_\mu)/\ (\nu_e + \bar{\nu}_e)$ ratio, both at $E_\nu \sim 1$ GeV.

So far, all of the arguments have been made based on the μ/e ratio, because of the accuracy in the $(\nu_\mu + \bar{\nu}_\mu)/(\nu_e + \bar{\nu}_e)$ calculation. Both $\nu_e \leftrightarrow \nu_\mu$ and $\nu_\mu \leftrightarrow \nu_\tau$

oscillations could explain the small μ/e ratio. However, if the absolute normalization of the flux calculation is made to an accuracy of better than 10 %, it might be possible to show which of the 2 possibilities is more likely. Table 8, taken from ref. [8], summarizes the situation of the absolute normalization. If one relies on the calculation given in ref. [5], the preferred solution would be the oscillation between ν_μ and ν_τ (e-like(Data) $\simeq$ e-like(MC) and μ-like(Data) $\ll$ μ-like(MC)). On the other hand, if one relies on the calculation given in ref. [25], the preferred solution would be the oscillation between ν_e and ν_μ ((e+μ)-like(Data) $\simeq$ (e+μ)-like(MC)). The flux calculation given in ref. [27] predicts the absolute event rate between the above two calculations. In order to improve the accuracy of the absolute normalization, basic data such as the primary cosmic-ray flux, atmospheric muon flux and proton interactions on various nuclei should also be accurately measured. We hope that the absolute flux can be predicted to an accuracy of better than 10 %. With these improvements, it is expected the current situation concerning the atmospheric neutrino problem can be clarified in the near future.

Finally, we mention the solar-neutrino data. The presently available data from the ^{37}Cl [36], Kamiokande-II [37], SAGE [38] and GALLEX [39] experiments all indicate that the observed solar-neutrino flux is smaller than the theoretical prediction. At present, one of the most promising solutions to the solar-neutrino deficit is neutrino oscillations in the solar interior (MSW mechanism [30]). If the MSW mechanism is the solution of the solar-neutrino problem, the $\nu_e \leftrightarrow \nu_\mu$ oscillations are ruled out as a solution of the atmospheric neutrino problem. $\nu_\mu \leftrightarrow \nu_\tau$ oscillations remain as a possible solution of the atmospheric-neutrino problem.

Thus, for a consistency check of the solar- and atmospheric-neutrino data, it is important to discriminate which of the two possibilities is the real solution to the atmospheric-neutrino problem.

References

[1] K.S.Hirata et al., Phys. Lett. B **205** (1988) 416.

[2] J.G.Learned, S.Pakvasa and T.J.Weiler, Phys. Lett. B **207** (1988) 79; V.Barger and K.Whisnant, Phys. Lett. B **209** (1988) 365; K.Hidaka, M.Honda and S.Midorikawa, Phys. Rev. Lett. **61** (1988) 1537.

[3] L.V.Volkova, Sov. J. Nucl. Phys. **31** (1980) 784; private communication.

[4] T.K.Gaisser, T.Stanev and G.Barr, Phys. Rev. D **38** (1988) 85.

[5] G.Barr, T.K.Gaisser and T.Stanev, Phys. Rev. D **39** (1989) 3532.

[6] M.Aglietta et al., Europhys. Lett. **8** (1989 611.

[7] Ch.Berger et al., Phys. Lett. B **227** (1989) 489; Ch.Berger et al., Phys. Lett. B **245** (1990) 305.

[8] D.Casper et al., Phys. Rev. Lett. **66** (1991) 2561.

[9] M. Honda, in this book.

[10] M.Nakahata et al., J. Phys. Soc. Japan **55** (1986) 3786.

[11] S.J.Barish et al., Phys. Rev. D **19** (1979) 2521.

[12] C.H.Llewellyn Smith, Phys. Rep. **3C** (1972) 261.

[13] S.J.Barish et al., Phys. Rev. D **16** (1977) 3103.

[14] G.L.Fogli and G.Nardulli, Nucl. Phys. **B160** (1979) 116.

[15] D.Rein and L.M.Sehgal, Ann. Phys. **133** (1981) 79.

[16] H.Deden et al., Nucl. Phys. **B85**, (1975) 269.

[17] N.J.Baker et al., Phys. Rev. D **25** (1982) 617.

[18] F.A.Brieva and A.Dellafiore, Nucl. Phys. **A43** (1977) 445.

[19] O.C.Allkofer, K.Carstensen and W.D.Dau, Phys. Lett. **36B** (1971) 425.

[20] G.Battistoni et al., Nucl. Inst. Meth. **A245** (1986) 277.

[21] G.Battistoni et al., Phys. Lett. **133B** (1983) 454.

[22] Ch.Berger et al., Nucl. Inst. Meth. **A262** (1987) 463.

[23] R.Becker-Szendy et al., submitted for publication in Nucl. Inst. Meth.; Boston University Preprint No. BUHEP-92-6.

[24] M.Takita, Doctor Thesis, University of Tokyo (February 1989), ICR-Report-186-89-3.

[25] H.Lee and Y.S.Koh, Nuovo Cimento, **105B** (1990) 883.

[26] K.S.Hirata et al., Phys. Lett. B **280** (1992) 146.

[27] M. Honda et al., Phys. Lett. B**248** (1990) 883.

[28] E.V.Bugaev, G.S.Bisnovatyi-Kogan, M.A.Rudzsky and Z.F.Seidov, Nucl. Phys. **A324** (1979) 350.

[29] Z.Maki et al., Prog. Theor. Phys. **28** (1962) 870; B.Pontecorvo, Zh. Eksp. Theor. Fis., **53** (1967) 1717; V.Gribov and B.Pontecorvo, Phys. Lett. B **28** (1969) 493.

[30] S.P.Mikheyev and A.Yu.Smirnov, Sov. J. Nucl. Phys. **42** (1985) 913; L.Wolfenstein, Phys. Rev. D**17** (1978) 2369, ibid. D**20** (1979) 2634.

[31] W.A.Mann et al., Phys. Lett. B **291** (1992) 200.

[32] K.Gabathuler et al., Phys. Lett. B **138** (1984) 449.

[33] L.Borodovsky et al., Phys. Rev. Lett. **68** (1992) 274.

[34] F.Dydak et al., Phys. Lett. B **134** (1984) 281.

[35] F.Bergsma et al., Phys. Lett. B **142** (1984) 103.

[36] R.Davis, Jr. et al., Phys. Rev. Lett. **20** (1968) 1205; K.Lande, Proc. of the 25th Int'l Conference on High Energy Physics, Singapore, Aug. 1990, p667.

[37] K.S.Hirata et al., Phys. Rev. Lett. **65** (1990) 1297; K.S.Hirata et al., Phys. Rev. Lett. **65** (1990) 1301; K.S.Hirata et al., Phys. Rev. Lett. **66** (1991) 9; K.S.Hirata et al., Phys. Rev. D. **44** (1991) 2241.

[38] A.I.Abazov et al., Phys. Rev. Lett. **67** (1991) 3332.

[39] P.Anselmann et al., Phys. Lett. B **285** (1992) 376; Phys. Lett. B **285** (1992) 390.

[40] M.M.Boliev et al., in Proceedings of the Third International Workshop on Neutrino Telescopes, ed. M.Baldo Ceolin, Venezia, Feb. 26-28, 1991, p235.

[41] R.Becker-Szendy et al., Phys. Rev. Lett. **69** (1992) 1010.

[42] Y.Oyama et al., Phys. Rev. D **39** (1989) 1481; M.Mori, KEK Preprint 92-24, To appear in the Proceedings of the Workshop on High Energy Neutrino Astrophysics, Honolulu, March 23-26, 1992.

[43] A.V.Butkevich, L.G.Dedenko and I.M.Zheleznykh, Sov. J. Nucl. Phys. **50** (1989) 90.

[44] T.K.Gaisser, talk presented at the International Symposium on Neutrino Astrophysics, Takayama, Japan, Oct. 1992, to appear in the Proceedings: W.Frati, T.K.Gaisser, A.K.Mann and Todor Stanev, Preprint BA 92-71, UPR-0218E.

[45] V.Agrawal, T.K.Gaisser, P.Lipari and T.Stanev, draft in preparation for publication.

[46] J.F.Owens, Phys. Lett. B **266** (1991) 126.

[47] D.S.Ayres (Soudan 2 collaboration), Proceedings of the Last Workshop on Grand Unification, North Carolina, USA, April 1989, p28.

[48] P. Litchfield, for The Soudan 2 collaboration, in Proc. of the International Workshop on ν_μ/ν_e Problem in Atmospheric Neutrinos, Laboratori Nazionali del Gran Sasso, Italy, March, 1993, p 114.

[49] M.Koshiba, proceedings of Workshop on Grand Unified Theories and Cosmology, KEK, Japan, December 1983, P24; Y.Totsuka, Proceedings of the Seventh Workshop on Grand Unification / ICOBAN'86, April 1986, Toyama,

Japan, P78; A.Suzuki, Proceedings of the Workshop on Elementary Particle Picture of the Universe, KEK, Japan, February 1987, P136; T.Kajita, Proceedings of the third Workshop on Elementary-Particle Picture of the Universe, Fuji-yoshida, Japan, October 1988, p163, ICR-Report-185-89-2.

[50] Fermilab Letter of Intent P822, "A Long Baseline Neutrino Oscillation Experiment Using Soudan 2", Aug. 1990, proposal submitted March 1991 by the Soudan collaboration.

[51] K.Nishikawa, Preprint INS-Rep.-924, April 1992.

[52] Proceedings of the workshop on Long-Baseline Neutrino Oscillation, Nov. 1991, Fermilab, USA.

[53] S. Parke, talk presented at the Moriond Workshop on Neutrino Physics, Jan. 1993, Villars, Switzerland, Fermilab-Conf-93/056-T (April 1993).

Calculation of Low-Energy Atmospheric Neutrino Fluxes

Morihiro Honda

Institute for Cosmic Ray Research, University of Tokyo

1 Introduction

Calculations of low-energy atmospheric ν–fluxes (100 MeV $\sim$ 3 GeV) have been carried out by many authors. The recent calculations include: Barr et al. (BGS) [1], Bugaev and Naumov (BN) [2], Lee and Koh (LK) [3], and Honda et al. (HKHM) [4]. They agree well in the ratio $(\nu_e + \bar{\nu}_e)/(\nu_\mu + \bar{\nu}_\mu)$ within 5%; A modest agreement of up to 20% is achieved for the $\bar{\nu}_e/\nu_e$ ratio. The calculations, however, do not agree in the absolute flux values by 50%. In this article, the calculation of low-energy atmospheric ν–fluxes and the problem are described based on the work of HKHM.

The atmospheric ν is mainly produced as the decay product of charged π's and K's, which are created by the interaction of cosmic rays in the atmosphere. We therefore need to know the primary flux of cosmic rays, their interactions in the atmosphere, and the decay of mesons in order to calculate the flux of atmospheric ν's. These problem are reviewed in order in the following sections.

2 Primary Flux of Cosmic Rays

Webber and Lezniak have compiled the energy spectrum of primary cosmic rays for hydrogen, helium, and CNO nuclei in a wide energy range of 10 MeV $\sim$ 1000 GeV [5] for three levels of solar activity (Fig. 1). Recently, a similar compilation has been made by Adams et al. [6] for hydrogen and helium nuclei, which agrees well with that by Webber and Lezniak. The uncertainty in the absolute primary fluxes was estimated by Adams et al. as $\sim$ 15% for hydrogen and helium fluxes.

From the compilation of Webber and Lezniak, the chemical composition of cosmic rays is H(proton) $\sim$ 90.6%, He$\sim$ 9.0%, and CNO nuclei $\sim$ 0.4% above $\sim$ 100 MeV/nucleus, and H $\sim$ 95.2%, He$\sim$ 4.5%, and CNO nuclei $\sim$ 0.3% above $\sim$ 2 GeV/nucleus. The portion of other components (Ne, S, Fe, $\cdots$) is so small that they can be neglected in the calculation of the atmospheric ν–fluxes. It is noted that since the atmospheric ν is created through the hadronic interaction of cosmic rays and air nuclei, it therefore depends more on the number of nucleons, rather than the number of nuclei. The contribution of a heavier nucleus cosmic

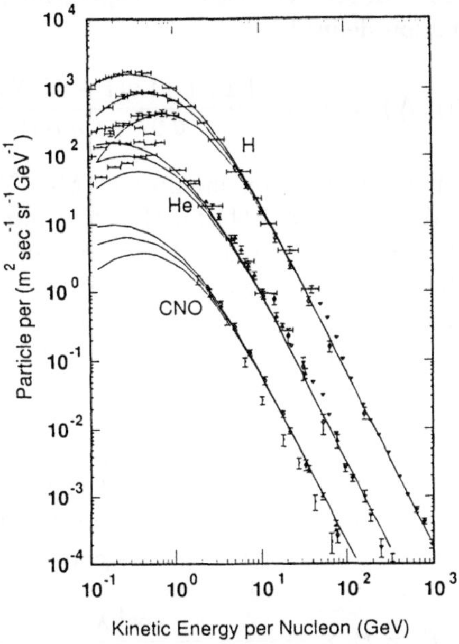

Figure 1: Fluxes of cosmic-ray protons, heliums, and CNOs from a compilation of Webber and Lezniak. The flux of CNOs at low energy ($\lesssim$ 3 GeV) is an extrapolation, while assuming the same dependence on the solar activity.

ray to the atmospheric ν is larger than that of a proton cosmic ray. Heliums carry $\sim$ 15% of the total nucleon in the cosmic rays and the CNO carries $\sim$ 3.6% above $\sim$ 2 GeV/nucleon. This effect is amplified by the effect of the geomagnetic field (cutoff-rigidity), which works depending on the rigidity (= momentum/charge). The rigidity for those nuclei is twice larger than protons when their energy-per-nucleon is the same. The details are explained below.

The flux of low-energy cosmic rays is modulated by solar activity. This is because the solar wind drives back the low-energy cosmic rays which are entering into the solar sphere, and the strength of the solar wind varies with the solar activity. This effect is more evident in the lower energy cosmic rays. The flux at maximum solar activity is two-times smaller than that at minimum solar activity for cosmic-ray protons with $\sim$ 1 GeV (Fig. 1). For $\sim$ 10 GeV cosmic ray protons, however, the flux difference due to solar activity is reduced to $\sim$ 10%.

The primary flux with various solar activities has been parametrized by K. Nagashima et al. [7] as a function of the count of neutron monitor (N) by

$$f(E_k)dE_k = \gamma_i u^{-2.585} M(p,N) \frac{E_k}{u} dE_k, \tag{1}$$

where i stands for the kind of nucleus (=H, He, CNO ..), p for the rigidity in GV ($\equiv$GeV/c/Z), E_k for the kinetic energy per nucleon in GeV, and u for the total energy per nucleon in GeV. γ_i determines the absolute flux value for each component, where, $\gamma_{\rm H} = 10.85 \times 10^3$ m^{2}sec^{-1}sr^{-1}GeV^{-1}, $\gamma_{\rm He} = 5.165 \times 10^2$ m^{2}sec^{-1}sr^{-1}GeV^{-1}, and $\gamma_{CNO} = 3.3 \times 10^{-2}$ m^{2}sec^{-1}sr^{-1}GeV^{-1}, respectively. The function $M(p, N)$ is the modulation function defined by

$$M(p,N) = \exp\left[-\frac{1.15 + 14.9(1 - N/N_{max})^{1.12}}{0.97 + (p/1GV)}\right], \tag{2}$$

and N is the count of the neutron monitor at Mt. Washington with N_{max} = 2465 counts/hour. We take $N = 2445$ for the solar min., $N = 2300$ for the solar mid., and $N = 2115$ for the solar max. The results of parametrization are given in Fig. 1 by the solid lines, which agree well with the data, except for low-energy heliums ($\lesssim$1GeV/nucleon). The offset in low-energy heliums, however, gives only small effects on the calculation of atmospheric ν's due to the small contribution to the total ν yield.

The geomagnetic field determines the minimum energy with which a cosmic ray can arrive at Earth. The analytic study for one propagation of charged particles in a dipole geomagnetic field was originally started by Störmer in his aurora studies, and has been developed further by other authors [8]. The cutoff momentum for protons is approximately given for the dipole geomagnetic field by the Störmer's formula,

$$P_c = P_0 \times \frac{\cos^4 \lambda}{[1 + (1 - \cos^3 \lambda \sin\theta \sin\phi)^{1/2}]^2}, \tag{3}$$

where $P_0 = eRB/c \simeq 59$ (GeV/c), R is the radius of the Earth and B is the strength of the magnetic field at the magnetic equator. λ is the magnetic latitude, θ the zenith angle, and ϕ the azimuth measured clockwise from magnetic north. For the cosmic-ray nucleus, the minimum energy of cosmic rays arriving at Earth is determined by the cutoff-rigidity. We note that the cutoff-rigidity is a function of the entering position on the Earth and the arrival direction (zenith θ and azimuth ϕ). Since the mass/charge ratio of helium or CNO nuclei is twice that of the proton, those nuclei carry lower energy nucleons in the atmosphere than do protons.

The actual geomagnetic field is represented by a multipole expansion by the spherical harmonic function as

$$B_{North} = \frac{1}{r}\frac{\partial V}{\partial \theta}, \quad B_{East} = \frac{-1}{\sin\theta}\frac{\partial V}{\partial \phi}, \quad \text{and} \quad B_{Down} = \frac{\partial V}{\partial r} \tag{4}$$

with the potential function

$$V = R \sum_{n=1}^{n=8} \sum_{m=0}^{m=n} \left(\frac{R}{r}\right)^{n+1} [g_n^m \cos(m\phi) + h_n^m \sin(m\phi)] P_n^m(\cos\theta). \tag{5}$$

Here, R is the radius of the Earth and $P_n^m(x)$ was the associated Legendre function. The expansion coefficient (g_n^m, h_n^m) was compiled and reported by IAGA Division I working Group 1 [9]. Since the geomagnetic field slowly varies with time, the coefficient is reported yearly. The coefficients for 1985 are given in Table 1 up to the 8th order.

The value of the cutoff-rigidity for an actual geomagnetic field can be obtained by a simulation of the orbit of the cosmic ray on a computer. In this simulation, an

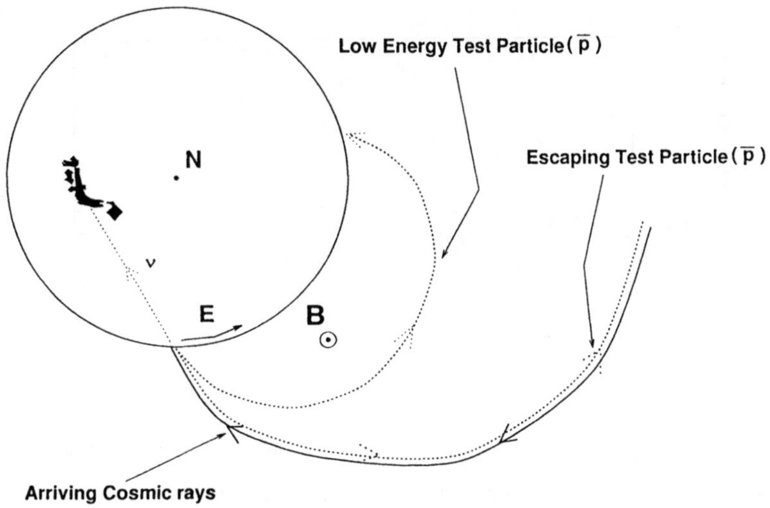

Figure 2: Schematic view of the calculation of the cutoff-rigidity.

antiproton, which has the same mass as protons, but a negatively signed charge, is used as the test particle. We note that the change $e \leftrightarrow -e$ is equivalent to the change of $t \leftrightarrow -t$ in the equation of motion of a charged particle in an electromagnetic field,

$$\frac{\partial \mathbf{p}}{\partial t} = e(\mathbf{E} + \mathbf{v} \times \mathbf{B}), \tag{6}$$

where $\mathbf{p}$ is the momentum, $\mathbf{E}$ the electric field, $\mathbf{v}$ the velocity, and $\mathbf{B}$ the magnetic field. To know the cutoff-rigidity at different positions and for different directions, we launch an antiproton, varying the positions and directions. When the test particle with a given momentum reaches a distance of 10-times the Earth's radius, where the strength of the geomagnetic field decreases to the same level as the interplanetary magnetic field near the Earth ($\sim 3 \times 10^{-8}$ T), it is assumed that the test particle has escaped from the geomagnetic field. Assuming that the momentum distributions of cosmic rays are isotropic in angular space, some cosmic rays which have the same rigidity can arrive at the Earth following the same orbit, but in the inverse direction. The cutoff-rigidity is calculated as the minimum rigidity with which the test particle escapes from the geomagnetic field. A schematic view of the calculation is shown in Fig. 2.

In the calculation of HKHM, they assumed that the ν's keep the direction of the parent cosmic rays (one-dimensional approximation) and calculated the cutoff-rigidity for the arrival direction of the ν's at the ν–detector sites. They found that the magnetic field calculated with up to the 5th order of the expansion gives almost the same result for the cutoff-rigidity as those of higher orders. The cutoff-rigidity at Kamiokande is shown as a contour map in Fig. 3. Integrating over the direction from a ν–detector, we can calculate the rate at which cosmic rays with a rigidity penetrate the geomagnetic shield and create ν's toward the detector (pass–rate). The pass–rate is shown for Kamioka, IMB site, KGF, and Gran Sasso as a function

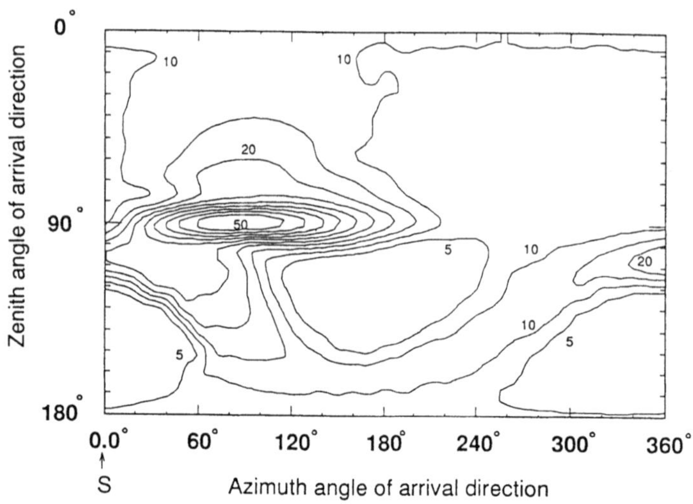

Figure 3: Contour map of the cutoff-rigidity for the ν arrival direction at Kamioka. Azimuth angles of $0°, 90°, 180°$, and $270°$ show the directions of south, east, north, and west, respectively.

of the rigidity in Fig. 4. We note that Stömer's formula is a good approximation for low magnetic latitude, such as at the KGF site. However, the multipole effect becomes important for higher magnetic latitudes, like the IMB site.

3 Interaction of Cosmic Rays

As cosmic rays propagate in the atmosphere, they create π's and K's by interactions with air nuclei. Those mesons create atmospheric ν when they decay, as follows:

$$\begin{aligned} A_{cr} + A_{air} \rightarrow \; & \pi^{\pm}, K^{\pm}, K^{0}, \ldots \\ & \pi^{+} \rightarrow \mu^{+} + \nu_{\mu} \\ & \qquad \mu^{+} \rightarrow e^{+} + \nu_{e} + \bar{\nu}_{\mu} \\ & \pi^{-} \rightarrow \mu^{-} + \bar{\nu}_{\mu} \\ & \qquad \mu^{-} \rightarrow e^{-} + \bar{\nu}_{e} + \nu_{\mu} \\ & \cdot \\ & \cdot \end{aligned} \tag{7}$$

The interactions of cosmic-ray protons and nuclei with air nuclei are hadronic in character. The energy range of the interaction that we need to understand is around $3 \sim 100$ GeV. Although there have been many accelerator proton–nucleus and hellium-nucleus experiments in the 3~30 GeV energy range, not many are applicable to our present purpose. In the calculation of HKHM, they employed the NUCRIN [10][11] Monte Carlo code for hadronic interactions at energies of ≤ 5 GeV, and the LUND code (FRITIOF version 1.6 [12] and JETSET version 6.3 [13]) for energies ≥ 5 GeV. Comparisons of the outputs of those code were

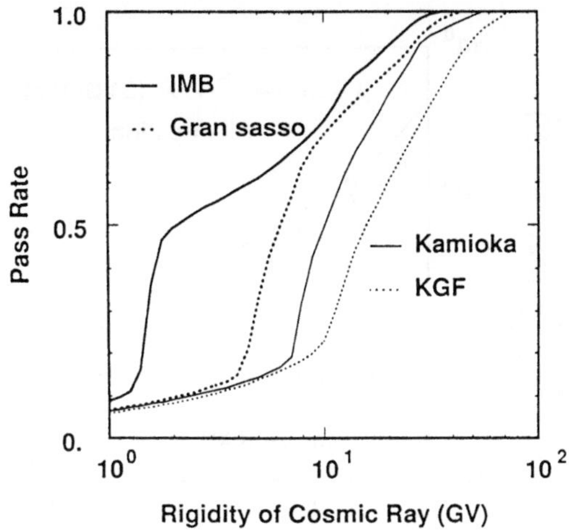

Figure 4: Pass rate for Kamioka, IMB site, Gran Sasso, and KGF.

Table 1: Expansion coefficients g_n^m and h_n^m in equation (5) for a geomagnetic field up to the 8th order for 1985 [9].

n	m	g_n^m	h_n^m	n	m	g_n^m	h_n^m
1	0	-29877	0.0	6	2	50	90
1	1	-1903	5497	6	3	-180	69
2	0	-2073	0	6	4	4	-50
2	1	3045	-2191	6	5	17	-4
2	2	1691	-309	6	6	-102	20
3	0	1300	0	7	0	75	0
3	1	-2208	-312	7	1	-61	-82
3	2	1244	284	7	2	2	-26
3	3	835	-296	7	3	24	-1
4	0	937	0	7	4	-6	23
4	1	780	233	7	5	4	17
4	2	363	-250	7	6	9	-21
4	3	-426	68	7	7	0	-6
4	4	169	-298	8	0	21	0
4	0	-215	0	8	1	6	7
5	1	356	47	8	2	0	-21
5	2	253	148	8	3	-11	5
5	3	-94	-155	8	4	-9	-25
5	4	-161	-75	8	5	2	11
5	5	-48	95	8	6	4	12
6	0	52	0	8	7	4	-16
6	1	65	-16	8	8	-6	-10

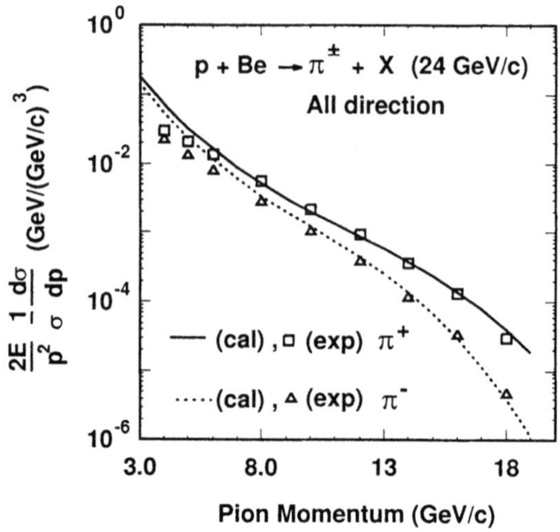

Figure 5: Comparison of the $\pi^{\pm}$-production spectrum in a p-Be collision at $p = 24$ GeV/c between the experiment (Ref. [14]) and the LUND Monte Carlo code. The π's created in the cone of $\leq 7.28°$ are used in the experimental data, and π's in all directions are used in the calculated data.

made with the available experimental data. For the LUND code, a comparison was made of the all–direction π production spectrum in p–Be collisions (24 GeV/c) [14] (Fig. 5); for NUCRIN code, a comparison was made with the π production spectrum at 2.5° in $p+C$ collisions (5 GeV/c) [15] (Fig. 6). The agreement of the LUND code and the experimental data is quite good, except for the very low-momentum region ($\lesssim$5 GeV). Since the energy spectrum of cosmic ray is steep ($\propto E^{-2.5}$), the π-production spectrum in the lower momentum region by nucleons with a fixed energy is not very important. The agreement of the NUCRIN code and the experimental data don't seem to be as good as that of the LUND code. The authors of the NUCRIN code estimated the disagreement in their output and the experimental result as being 10 $\sim$ 20 % [11]. However, they also said that this accuracy is sufficient for the study of hadronic cascade. In the low-energy cosmic-ray interactions, the detailed structure of the π-production spectrum may be smeared due to a flattening of the cosmic-ray spectrum at low energy. We estimate the error of the atmospheric ν calculation resulting from the Monte Carlo code to be less than 10%.

The nature of the atmosphere, such as the chemical compositions and density structure, is important, since it is the target of the hadronic interaction of cosmic rays. The atmospheric composition is given approximately by nitrogen (78%), oxygen (21%), and argon (1%). For the π's and K's propagating in the atmosphere, decay and interaction with the air nuclei are competitive processes. The probabilities of both processes are related to the energy and lifetime of the mesons, as well as to the density of air. When the probabilities for decay and interaction

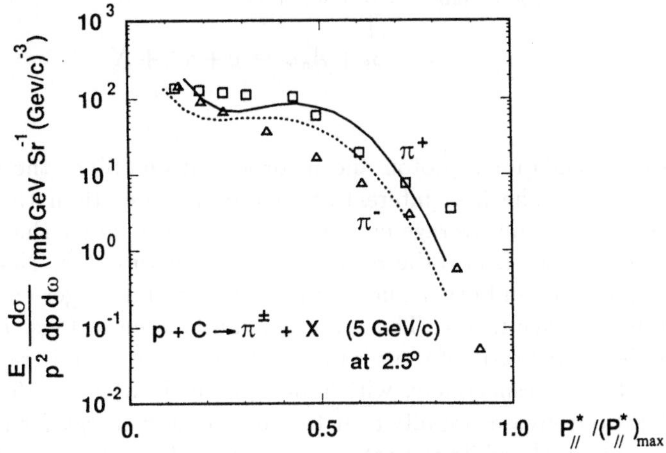

Figure 6: Comparison of the $\pi^{\pm}$-production spectrum in the p-C collision at $p = 5.1$GeV to the direction of 2.5° between the experiment (Ref. [15]) and the NUCRIN Monte Carlo code. P^* denotes the momentum in the total center-of-mass system.

are equal, we have

$$c\tau\frac{E}{mc^2} = \frac{1}{\sigma\rho}, \tag{8}$$

where m is the energy of the meson, ρ the density of air, σ the interaction cross section, and τ the lifetime. Equation (8) is evaluated for the energy with which decay and interaction take place equally as

$$E = \frac{mc^2}{c\tau\sigma\rho} \sim \left\{ \begin{array}{l} 12 \text{ (GeV, for } \pi^{\pm}) \\ 22 \text{ (GeV, for } K_L^0) \\ 90 \text{ (GeV, for } K^{\pm}) \end{array} \right\} \times \frac{\rho_{[sea\ level]}}{\rho}, \tag{9}$$

where $\rho_{[sea\ level]}$ =1.225 kgm^{-3}. For the density structure of the atmosphere, the US standard [16] is commonly used. The interaction mean-free path for cosmic-ray protons is $\sim$ 100 g/cm^2 in column density; the first interaction of vertical cosmic rays takes place at an altitude of 15 $\sim$ 20 km. For cosmic rays entering at an angle the first interaction point is higher than those entering vertically. Since at the altitude where cosmic rays interact with air nuclei, the density of air is $\sim$ 10-times less than that at the sea level, almost all of the secondary mesons created by cosmic rays with an energy of 10 $\sim$ 100 GeV (which are relevant to low-energy atmospheric ν ($\lesssim$ 3 GeV)), decay before the interaction.

A qualitative discussion of the interaction of cosmic rays and air nuclei may be useful. In hadronic interactions, the projectile charge is mainly carried by the leading particle. For proton cosmic rays, the most probable interaction sequence is as follows:

$$\begin{array}{lll} p + A_{air} \rightarrow & n + \pi^+ + X & \\ & \downarrow & \\ & n + A_{air} \rightarrow p + \pi^- + X & \\ & \qquad\qquad \downarrow & \\ & \qquad\qquad\quad \cdots, & \end{array} \qquad (10)$$

In the above p stands for a proton and n for a neutron. Since the energy of the neutrons created by the first interaction in (10) is lower than that of incident protons, the number and average energy of the π^+'s are larger than those of the π^-'s. The excess of π^+ is also the result of charge conservation. Neutrons, which are carried by helium and other nuclei, create π^-'s with a larger energy than the π^+'s due to isospin–symmetry. This situation can be seen in Figs. 7(a) and 7(b), which show the charge ratio of the π's (π^-/π^+) created by cosmic-ray protons and cosmic-ray heliums, respectively, with (energy)/(nucleon) = 5, 10, and 20GeV. The π^-/π^+ ratio decreases rapidly near to the highest π energy for proton cosmic rays, due to excess π^+, while the ratio is near unity for hellium cosmic rays over the entire π energy range. In the case of hellium, the π^+ excess is cancelled by the neutron in the nucleus. Since the proton is dominant in the cosmic-ray composition, there is always an excess of π^+ to π^- by $\sim$ 20%. Consequently, we expect $\sim$ 20% excess of ν_e to $\bar{\nu}_e$.

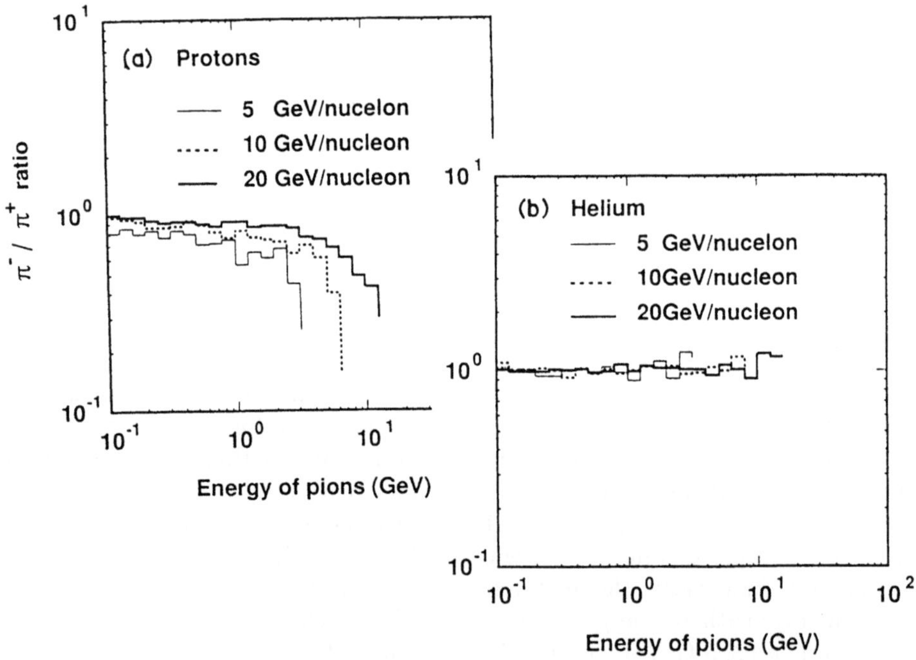

Figure 7: (a) (π^+/π^-) created by vertical cosmic-ray protons with energies of 5, 10, 20 GeV. (b) (π^+/π^-) created by vertical cosmic-ray heliums with energies of 5, 10, 20 GeV/nucleon.

4 Decay of Mesons

Low-energy ν's are produced mainly from the decay of μ's, π's, and K's, which are listed here:

$$
\begin{array}{lllr}
\pi^{\pm} & \rightarrow \mu^{\pm}\nu_{\mu}(\bar{\nu}_{\mu}) & (100\ \%) & \\
\mu^{\pm} & \rightarrow e^{\pm}\nu_{e}(\bar{\nu}_{e})\bar{\nu}_{\mu}(\nu_{\mu}) & (100\ \%) & \\
K^{\pm} & \rightarrow \mu^{\pm}\nu_{\mu}(\bar{\nu}_{\mu}) & (63.5\ \%) & \\
 & \rightarrow \pi^{\pm}\pi^{0} & (21.2\ \%) & \\
 & \rightarrow \pi^{\pm}\pi^{+}\pi^{-} & (5.6\ \%) & \\
 & \rightarrow \pi^{0}\mu^{\pm}\nu_{\mu}(\bar{\nu}_{\mu}) & (3.2\ \%) & (K_{3\mu\nu}) \\
 & \rightarrow \pi^{0}e^{\pm}\nu_{e}(\bar{\nu}_{e}) & (4.8\ \%) & (K_{3e\nu}) \\
 & \rightarrow \pi^{\pm}\pi^{0}\pi^{0} & (1.73\ \%) & \\
 & & & \\
K_{s}^{0} & \rightarrow \pi^{+}\pi^{-} & (68.6\ \%) & \\
 & & & \\
K_{l}^{0} & \rightarrow \pi^{+}\pi^{-}\pi^{0} & (12.37\ \%) & \\
 & \rightarrow \pi^{\pm}\mu^{\mp}\nu_{\mu}(\bar{\nu}_{\mu}) & (27\ \%) & (K_{3\mu\nu}) \\
 & \rightarrow \pi^{\pm}e^{\mp}\nu_{e}(\bar{\nu}_{e}) & (38.6\ \%) & (K_{3e\nu})
\end{array}
\tag{11}
$$

Especially, the decay of charged π's and, consequent, μ decay ($\pi - \mu$ decay):

$$
\begin{array}{ll}
\pi^{\pm} \rightarrow & \mu^{\pm} + \nu_{\mu}(\bar{\nu})_{\mu} \\
 & \downarrow \\
 & \mu^{\pm} \rightarrow e^{\pm} + \nu_{e}(\bar{\nu})_{e} + \bar{\nu}_{\mu}(\nu_{\mu}),
\end{array}
\tag{12}
$$

is dominant among these processes.

When a $\pi^{\pm}$ decays at rest, the energy carried by $\nu_{\mu}(\bar{\nu}_{\mu})$ is $(m_{\pi}^2 - m_{\mu}^2)/2m_{\pi} \sim$ 30 MeV; the $\mu^{\pm}$ carries the rest of the energy. If we ignore the spin of the μ's, each decay particle ($e^{\pm}$, $\nu_e(\bar{\nu}_e)$, and $\bar{\nu}_{\mu}(\nu_{\mu})$) carries 1/3 of the $\mu^{\pm}$'s energy ($\sim$ 37 MeV) on the average in three-body decay. The energy division ratio holds as the average relation when the π's decay in flight, i.e. the π energy is approximately divided into 1/4 to the decay products in $\pi - \mu$ decay on average. We thus expect the flux ratio to be about $(\nu_e + \bar{\nu}_e)/(\nu_{\mu} + \bar{\nu}_{\mu}) = 1/2$ and $\bar{\nu}_{\mu}/\nu_{\mu} = 1$, irrespective of the π spectrum. When the energy of the μ's becomes high ($\gtrsim$5 GeV), however, the μ's tend not to decay in air, but reaching Earth before the decay. In this case, the μ's loose their energy in Earth, and decay after they are almost stopped, or are captured by nuclei in the Earth. These effects reduce the flux of ν_e and $\bar{\nu}_{\mu}$ at $\gtrsim$2 GeV, since there is an excess of π^+ in π production. The ratios $(\bar{\nu}_e + \nu_e)/(\bar{\nu}_{\mu} + \nu_{\mu})$ and $\bar{\nu}_{\mu}/\nu_{\mu}$ decrease at those energies. We note that the μ polarization and μ energy loss in air are important for the precise calculation of atmospheric ν's. Each has $\gtrsim$ 5% effects on the flux and the ratio of atmospheric ν's for $\gtrsim$100 MeV. The energy loss of the μ's in air is taken into account by the Monte Carlo method. The μ polarization is explained as follows.

In the decay process of charged π's, the resulting $\mu^{\pm}$ are fully polarized against (toward) the direction of μ motion in the charged π or in the K rest frame. In sequential μ decay, $\nu_e(\bar{\nu}_e)$ is emitted in the forward direction of the μ's motion, due to the conservation of helicity. Thus, $\nu_e(\bar{\nu}_e)$ resulting from μ decay has a larger energy than does $\bar{\nu}_\mu(\nu_\mu)$ in the π rest frame. Since the π's decay in flight, the polarization is not full. In general, the direction distribution of ν_e is proportional to $(1+\zeta\cos\theta)$, where ζ is the polarization ($\zeta = 1$ is the full polarization) and θ is the angle between directions of ν_e and μ in the CM frame of the μ's.

The μ polarization is calculated in the observer's frame as follows [17]. It is convenient to introduce a 4-vector s, which satisfies $sp = 0, s^2 = -1$, and in the μ rest frame $s = (s^0, \vec{s}) = (0, \vec{\zeta})$, where $\vec{\zeta}$ is the direction vector of the spin. In the observer's frame, s is represented as

$$(s^0, \vec{s}) = (\frac{\vec{\zeta}\cdot\vec{p}}{m_\mu}, \zeta + \frac{\vec{p}(\vec{\zeta}\cdot\vec{p})}{m_\mu(E_\mu + m_\mu)}), \tag{13}$$

where $\vec{p}$ is the μ 3-momentum and E is the μ energy. We denote the quantities in the π rest frame with an asterisk (*), and assume that $\vec{\zeta}^*$ is parallel to p^*. It may be easy to repeat the calculation for the case that $\vec{\zeta}^*$ is antiparallel to $\vec{p}^*$. The expression of s^* is obtained as

$$(s^{*0}, \vec{s}^*) = (\frac{|\vec{p}^*|}{m}, \frac{\vec{\zeta}^* E^*}{m_\mu}). \tag{14}$$

In the frame of the observer, where the π has a velocity of $\vec{u}$, s^* is transformed as

$$(s^0, \vec{s}) = (\gamma[\frac{|\vec{p}|}{m_\mu} + \frac{E^*}{m_\mu}(\vec{u}\cdot\vec{\zeta}^*)], \frac{E^*}{m_\mu}\vec{\zeta}^* + (\gamma - 1)\frac{E^*}{m_\mu u^2}(\vec{\zeta}^*\cdot\vec{u})\vec{u} + \gamma\frac{|\vec{p}|}{m_\mu}\vec{u}), \tag{15}$$

where $\gamma = (1-u^2)^{-1/2}$. The quantity $(\vec{u}\cdot\vec{\zeta}^*)$ can be obtained from the Lorentz transformation of energy:

$$E = \gamma[E^* + |\vec{p}^*|(\vec{u}\cdot\vec{\zeta}^*)]. \tag{16}$$

Equating the expressions for s^0 in equations (15) and (13), and eliminating $(\vec{u}\cdot\zeta^*)$, we obtain the μ polarization along the direction of motion as

$$\vec{\zeta}\cdot\frac{\vec{p}}{|\vec{p}|} = \frac{EE^* - \gamma_\pi m_\mu^2}{\vec{p}\cdot\vec{p}^*} \tag{17}$$

in the observer's frame. Here, E the μ energy, $\vec{p}$ the μ momentum , and γ_π the Lorentz factor of the π in the observer's frame. E^* and $\vec{p}^*$ denote the μ energy and the momentum in the π rest frame, respectively, as before. The above discussion can also be applied to the μ's created by $K^\pm \to \mu^\pm\nu_\mu(\bar{\nu}_\mu)$ decay. For the μ's resulting from $K_{3l\nu}$ decay, the discussion given in Ref. [18] can be applied. Small-angle scattering of the μ's in the atmosphere reduces the μ polarization. This depolarization effect has been evaluated by Hayakawa [17] to be of the order of 21MeV$/vp$, where v and p are the velocity and momentum of the μ's, respectively. Therefore, the depolarization effect may be negligible for μ's which produce ν's having energies $\gtrsim$100 MeV.

If we ignore the effect of μ polarization, the calculated energy of ν_e decreases by $1.5 \sim 2\%$ on the average; therefore, the flux of ν_e is estimated to be systematically smaller by $\sim 5\%$ in the energy range that we are considering

5 The Flux of Atmospheric Neutrino

We can now calculate the flux of atmospheric ν's, combining the study presented in the above sections. Assuming that the ν's keep the direction of the parent cosmic rays (one-dimensional approximation), the flux of the atmospheric ν's integrated over all directions is calculated to be

$$F_\nu(E_\nu) = \sum_i \int \eta^i(E_{cr}, E_\nu)\theta(R^i(E_{cr}) - R_{cut}(\theta, \phi))F^i_{cr}(E_{cr})dE_{cr}d\Omega, \qquad (18)$$

where i stands for the chemical composition of the cosmic ray, $\eta^i(E_{cr}, E_\nu)$ is the ν production spectrum by the i-nucleus cosmic ray with energy E_{cr}, $\theta(x)$ is a function which satisfies $\theta(x) = 1$ for $x > 0$ and $\theta(x) = 0$ for $x < 0$, and $F^i_{cr}(E_{cr})$ is the primary spectrum of the i-nucleus cosmic ray. We note that $\eta^i(E_{cr}, E_\nu)$ is a zenith-angle dependent function, since the density structure along the path of the cosmic ray and its products varies with the zenith angle. As noted before, however, the low energy ν–flux ($E_\nu \lesssim 3$ GeV) has very weak dependence on the density structure of the atmosphere. Practically, we consider $\eta^i(E_{cr}, E_\nu)$ as being a direction-independent quantity.

The calculation method of (18) is different among authors: HKHM employed a Monte Carlo method for all calculations; Bugaev and Naumov (BN) employed an analytic method in calculating $\eta^i(E_{cr}, E_\nu)$ as well as in the integration. The hybrid method, which calculates $\eta^i(E_{cr}, E_\nu)$ by the Monte Carlo method and integrates (18) analytically, is also possible [1]. Since $\eta^i(E_{cr}, E_\nu)$ is a direction-independent quantity, expression (18) could be rewritten with the pass–rate given in Fig. 4, and with the flux of atmospheric ν with constant cutoff-rigidity for all directions as

$$F_\nu(E_\nu) = \int F^*_\nu(R_{cut}, E_\nu)\frac{dP_{pass}}{dR_{cut}}dR_{cut}. \qquad (19)$$

Here, $F^*_\nu(R_{cut}, E_\nu)$ is the ν–flux with a given cutoff-rigidity R_{cut}, and P_{pass} is the pass–rate. This formulation is prefered by the authors who calculated the atmospheric ν by the analytic or hybrid method. $F^*_\nu(R_{cut}, E_\nu)$ for ν_μ is shown in Fig. 8 for several cutoff rigidities (solar mid). The cutoff-rigidity has a very small effect on the ν–flux when it is below ~ 2 GV, while it affects in large at ~ 10 GV. With $R_{cut} = 10$ GV, the atmospheric ν–flux at 1 GeV decreases by 25% from the no-cutoff case; with $R_{cut} = 1.58$ GV, it decrease by 40% from the no-cutoff case.

In this calculation scheme (one-dimensional approximation), the directional dependence of the atmospheric ν can be calculated by this $F^*_\nu(R_{cut}, E_\nu)$ and the cutoff-rigidity for the direction from ν–detector ($R_{cut}(\theta, \phi)$). We have shown the cutoff-rigidity as a function of direction (θ, ϕ) as the contour map for Kamioka (Fig. 3). From the contour map, we find that since there is a large cutoff-rigidity region in the east-horizontal direction, we can expect a low ν–flux intensity from this direction. In the actual case, however, the direction of ν is different from that of the parent cosmic rays, because the mesons are produced in a slightly different direction from that of an incident particle, and again the ν's are created in slightly different directions from those of mesons in their decay. Although we expect that

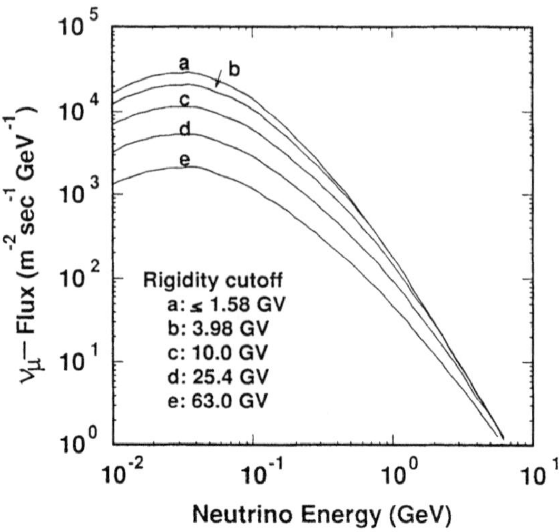

Figure 8: Atmospheric ν_μ–flux calculated with a constant cutoff-rigidity for all directions. The lines below 1.58 GV almost overlap each other.

this effect is rather small, there is another smearing effect of direction in the ν–detector; when a low-energy ν ($\lesssim$3 GeV) creates a lepton by a quasi-elastic process, the lepton has a typical angle of $\sim 30°$ from the ν direction. Therefore, it may be difficult to observe the deficit of the ν–flux from this direction.

Sometimes the observed μ–flux is used to check the calculation of the quantity $F^*_\nu(R_{cut}, E_\nu)$, since the μ–flux is deeply related to the ν–flux in $\pi - \mu$ decay. We note, however, the μ–flux observed at sea level is almost irrelevant to the low-energy cosmic rays which create the ν at low energy ($\lesssim$3 GeV), and an observation at high altitude is needed. Very recently, the μ–flux was measured with good accuracy by the MASS (Matter-Antimatter Superconducting Spectrometer) experiment [19] at several altitudes. Fig. 9 shows a comparison of the observed μ–fluxes and those calculated by HKHM using the same Monte-Carlo code for the ν's. The agreement of experiment and calculation is very good, except for very high altitude ($\lesssim$37 g/cm^2) and the low-momentum region in low-altitude (615 g/cm^2). In the very high altitude, delicate problems, fine structure of atmosphere density, which could be time variant, and low-energy cosmic-ray flux and compositions, are related. Moreover, only a small fraction of atmospheric ν's is created at this altitude. We consider the agreement of the observation and calculation of the high-altitude μ–flux to be satisfactory.

The flux of atmospheric ν's integrated over all directions for Kamioka and the ν_μ–flux for IMB calculated by HKHM are shown in Fig. 10 for solar max., solar mid., and solar min. The peak of the ν spectrum at $\sim$ 30 MeV comes from the ν created by nearly stopping muons. The effect of solar activities can be seen from the figure. It varies with the ν energies; the differences between solar min and solar max are $\sim$ 15% at 100 MeV and $\sim$ 5% at 1 GeV for Kamioka in absolute

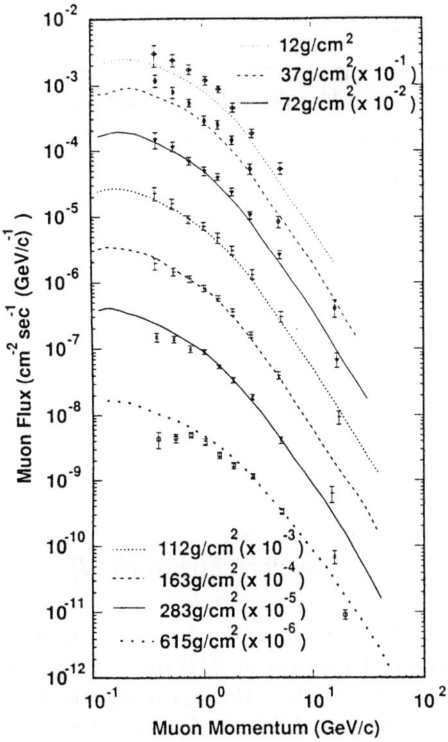

Figure 9: Comparison of the μ–fluxes observered by the MASS (Matter-Antimatter Superconducting Spectrometer) experiment [19] at several altitudes and those calculated by HKHM.

values. This difference is larger for IMB due to the lower cutoff-rigidity ($\sim$ 23% at 100 MeV and $\sim$ 8% at 1 GeV).

The results of other authors, Barr et al.(BGS) [1], Bugaev and Naumov (BN) [2], and Lee and Koh (LK) [3], are compared with HKHM for the $\nu_\mu + \bar{\nu}_\mu$ flux as shown in Fig. 11. Since the spectrum shape of BGS is similar to HKHM, the ratio is nearly constant $\sim$115% for ν energies <1 GeV. Gaisser et al. have shown that this difference between HKHM and BGS could be understand as being a difference in the cutoff-rigidity [20]. BN shows a large deviation from HKHM in ν energies <1 GeV, but this deviation becomes small in the energies $\gtrsim$2 GeV. The flux of LK is 80 $\sim$ 90% of HKHM in the ν energy range 300 MeV $\sim$ 800 MeV, but much smaller in higher and lower energy regions . According to Lee and Koh [3], this difference in the lower energy comes from the different calculation scheme, i.e. a one-dimensional calculation (BGS, HKHM, and BN) compared to a three-dimensional one (LK). However, this does not explain the difference in the higher energy, since the one dimensional approximation becomes good.

In Fig. 12 the ratios $\bar{\nu}_\mu/\nu_\mu$, $\bar{\nu}_e/\nu_e$, and $(\bar{\nu}_e + \nu_e)/(\bar{\nu}_\mu + \nu_\mu)$ expected at Kamioka and IMB are shown for the four calculations. The effect of solar activity on those

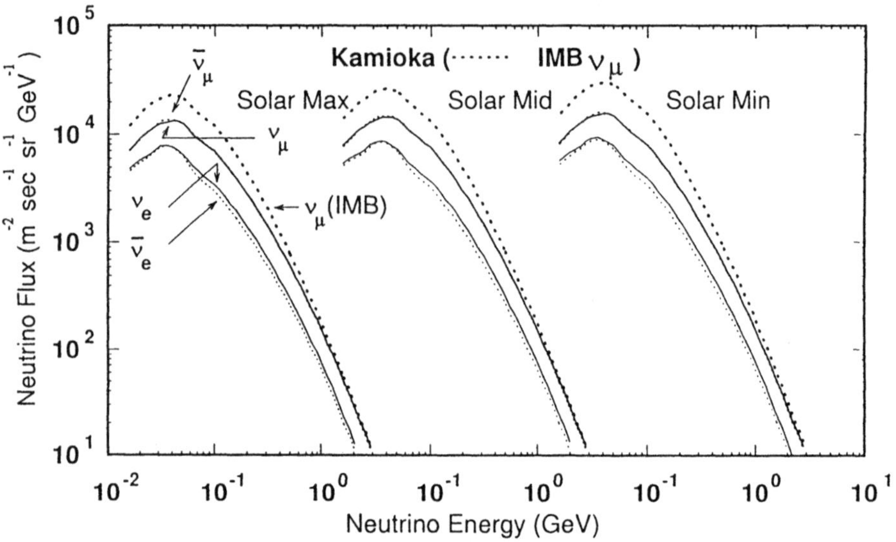

Figure 10: Atmospheric ν–fluxes at Kamioka for solar activity of max, mid, and min by HKHM. For a comparison, the ν_μ–flux at IMB site is shown by a dotted line for each solar activity. The thick solid lines show the flux of ν_μ, the thick dotted lines $\bar{\nu}_\mu$, the thin solid lines ν_e, and the thin dotted lines $\bar{\nu}_e$ for each solar activity and calculation.

ratios is small, $\sim$ 1% for the ratios $\bar{\nu}_\mu/\nu_\mu$ and $(\bar{\nu}_e + \nu_e)/(\bar{\nu}_\mu + \nu_\mu)$, and at most $2 \sim 3$% for $\bar{\nu}_e/\nu_e$ in the energy range 100 MeV $\sim$ 1 GeV. The result for IMB is not shown in LK, since it is identical with that for Kamioka. The $\bar{\nu}_\mu/\nu_\mu$ and $(\bar{\nu}_e + \nu_e)/(\bar{\nu}_\mu + \nu_\mu)$ ratios for $\gtrsim$100 MeV are $\sim$ 1 and $\sim$ 0.5 respectively, as expected from the rough discussion in sect. 4, and are quite similar to each other among different calculations. For $\lesssim$100 MeV, the kinematics of the decay of almost stopped π's and muons is important. In the energy region $\gtrsim$1 GeV, the ratio $(\bar{\nu}_e + \nu_e)/(\bar{\nu}_\mu + \nu_\mu)$ decreases in all calculations. This is because both ν_e and $\bar{\nu}_e$ are predominantly created in muon decay, and muons of $\gtrsim$5 GeV tend not to decay in the air, but to go into the Earth directly. The discussion in sect. 4 assumes that ν_μ, $\bar{\nu}_\mu$, and $\nu_e(\bar{\nu}_e)$ are created in almost the same energy in the π-μ decay. When μ goes into the Earth before its decay, ν's created in the μ decay have lower energies than the ν created in the π decay due to the large energy loss of the μ in the Earth. The discussion in sect. 4 cannot be applied for the ratio of ν species for such a case. Since $\nu_e(\bar{\nu}_e)$'s are created only in the μ decay while ν_μ's are created also in the π decay, the ratio $(\bar{\nu}_e + \nu_e)/(\bar{\nu}_\mu + \nu_\mu)$ becomes lower than 0.5 in the energy $\gtrsim$1 GeV. As protons are dominant in the cosmic ray, π^+ is the main source of atmospheric ν's. Therefore, $\bar{\nu}_\mu/\nu_\mu$ ratio also decreases in higher energies.

It is noted that the difference of $\bar{\nu}_e/\nu_e$ ratio among different calculations is larger than other ratios. Especially that of LK, which is almost equal to 1, is quite

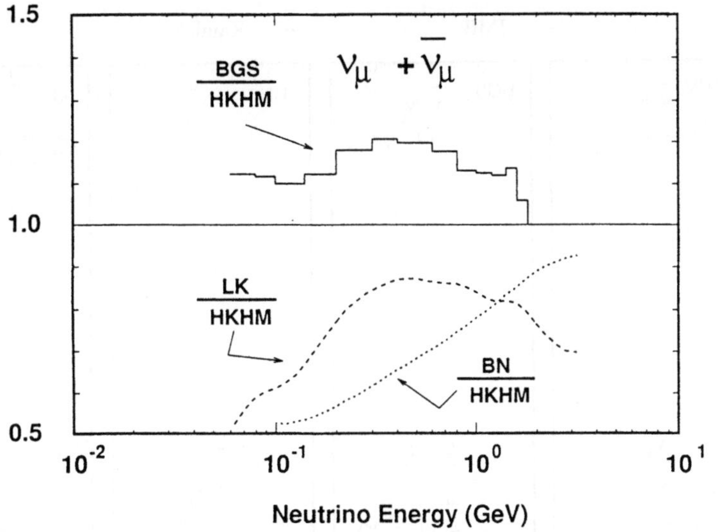

Figure 11: Comparison of $\nu_\mu + \bar{\nu}_\mu$ flux by different calculations. The ratios to HKHM are shown.

different from other calculations ($0.9 \sim 0.7$). The deviation of $\bar{\nu}_e/\nu_e$ from 1 is the result of the proton dominance in the cosmic ray, and the consequent excess of π^+'s. It seems very hard to explain the value of $\bar{\nu}_e/\nu_e \sim 1$ obtained by LK. We note that the ratio $\bar{\nu}_e/\nu_e$ is directly related to the π^-/π^+ ratio in the interaction of the cosmic rays, and therefore to the π production spectrum in the interaction and neutron/proton ratio in low-energy cosmic rays. In the calculations of HKHM, BGS, and BN, the $\bar{\nu}_e/\nu_e$ ratio is smaller for IMB than for Kamioka. This could be understood by the fact that the typical cutoff-rigidity for IMB is $\sim$ 2 GV and $\sim$ 10 GV for Kamioka. With a lower cutoff-rigidity for IMB, the lower energy cosmic rays contribute to the atmospheric ν than at Kamioka. With the higher rigidity cutoff for Kamioka, the portion of the neutron component below 10 GeV/nucleon is larger than that for IMB. All of the differences in the calculation scheme, i.e. the cutoff-rigidity, hadronic interactions, etc, are related to the $\bar{\nu}_e/\nu_e$ ratio.

Table 2: Expected single-ring event number for Kamiokande 6.2 kt yr

	μ-like events	e-like events	e/μ-ratio
BGS	324.0	210.2	0.649
HKHM	292.6	186.9	0.639
BN	199.0	128.3	0.645
LK	249.4	159.1	0.634
Obs.	191	198	1.04 ± 0.10

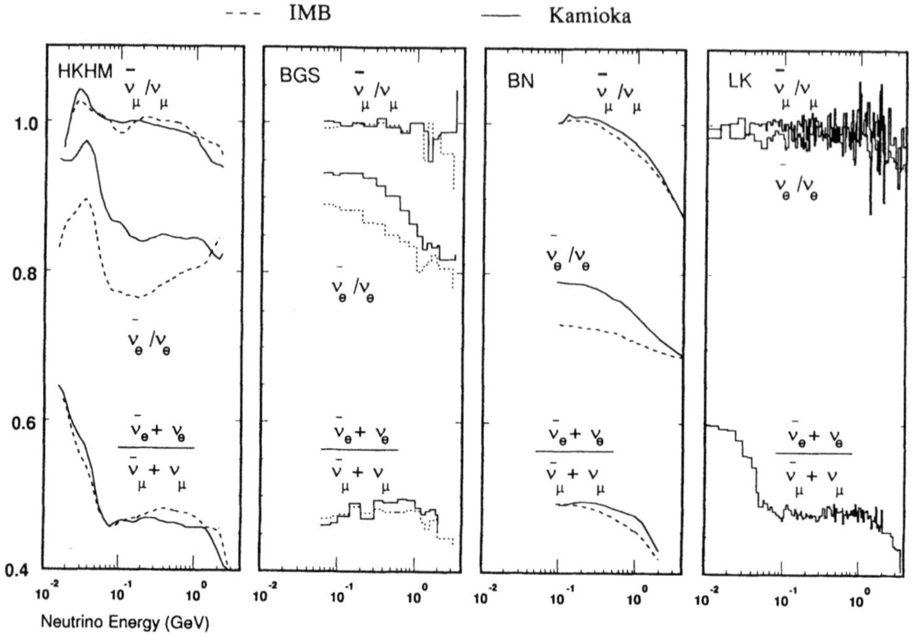

Figure 12: Ratios of $\bar{\nu}_\mu/\nu_\mu$, $\bar{\nu}_e/\nu_e$, and $(\bar{\nu}_e+\nu_e)/(\bar{\nu}_\mu+\nu_\mu)$. The solid lines show the flux ratios for Kamioka and the dashed lines for IMB. For the calculation of Lee and Koh, the result for IMB is not shown, since there is no difference between Kamioka and IMB.

Shown in Table 2 are the expected event numbers calculated for μ-like events and e-like events for Kamiokande with a 6.2 kt yr exposure, as compared with the observed ones [21]. Although the absolute numbers are different among different authors, the ratios of (e-like event)/(μ-like event) lie within a difference of 3%.

6 Conclusion

The knowledge of atmospheric ν's is becoming important, since the flux ratio, ν_μ/ν_e, observed with the water cherenkov type ν-detector shows a significant deviation from the predicted value [22, 23]. Methods and calculations of the flux of atmospheric ν's are presented in this article based on the work of HKHM as well as the results of other authors. The uncertainty in the calculation of atmospheric ν's are mainly originated from the uncertainties of the flux of primary cosmic rays and hadronic interactions. The absolute value of the cosmic ray flux is rather difficult to determine and the uncertainty is estimated $\sim$ 15 %, although the spectrum shape and the relative fraction of chemical components are well determined in $\lesssim$100 GeV. Since available experiments are limited, some properties of hadronic interactions, such as the production spectrum of secondary particles, are not well determined either. Those uncertainties result in the uncertainties in the absolute flux value of atmospheric ν's and in the flux ratio of ν species, especially in $\bar{\nu}_e/\nu_e$ ratio. However, the ratio $(\nu_e + \bar{\nu}_e)/(\nu_\mu + \bar{\nu}_\mu)$ is determined with almost no uncertainties, since the π-μ decay is the main source of ν in this energy region.

The fluxes calculated by different authors are distributed in a wider range beyond the estimated uncertainties. We conclude that the interaction model and/or the flux of primary cosmic rays used by some authors differs more largely than the uncertainties in the interaction model and in the primary cosmic rays including the rigidity cutoff. Although the large differences are seen in the absolute flux value and in some ratios, the agreement of the ratio (e-like event)/(μ-like event) is remarkably good. All calculations agree within 3 % among each other.

Very recently μ-flux was observed in a good accuracy by MASS (Matter Antimatter Superconducting Spectrometer) experiment at high altitude [19]. The μ-flux at high altitude may provide a mean for overall calibration of atmospheric ν's. The agreement of μ-flux observation and calculation of HKHM is very good at the altitude of $\sim$ 100 g/cm^2, where atmospheric ν's are mainly created. This agreement suggests that the flux calculated by HKHM has the accuracy of 10 % in the absolute value.

It is concluded that the anomaly in the ν_e/ν_μ ratio observed by ν-detectors is hard to attribute to the uncertainty of the calculation of atmospheric ν's. Taking the calculation of HKHM, there is a deficit of ν_μ in the atmospheric ν flux. One obvious possibility to explain this deficit is to invoke neutrino oscillations: in this case the determination of the mixing angles of the ν's mass matrix [24] becomes possible by reducing the uncertainties in the calculation of absolute atmospheric ν fluxes.

References

[1] G.Barr, T.K. Gaisser, and T. Stanev (BGS), Phys. Rev. D**39** 3532 (1989).

[2] E.V. Bugaev and V.A. Naumov (BN), Phys. Lett. B**232** 391 (1989).

[3] H. Lee and Y. Koh (LK), Nuovo Cimento B**105** 884 (1990).

[4] M. Honda, K. Kasahara, K. Hidaka, and S. Midorikawa (HKHM), Phys. Lett. B**248** 193 (1990).

[5] W.R. Webber and J.A. Lezniak, Astrophys. Space Sci. 30 361 (1974).

[6] J.H. Adams et al., *in Proceedings of Twenty-Second International Cosmic Ray Conference, Dublin, 1991,* edited by M. Cawley *et al.* (Reprint Ltd, Dublin, Ireland 1991) Vol. 2, p. 631.

[7] K. Nagashima, *et al.*, Nuovo Cimento **12** 173 (1989).

[8] R.A. Alpher, J. Geophys. Res., **55** 437 (1950) and references there in.

[9] D.R. Barraclough et al., (IAGA Division I Working Group 1) J.Geomag. Geoelectr., **37**, 1157 (1985).

[10] K. Hänssget and J. Ranft, J. Comput. Phys. **39** 37 (1986).

[11] K. Hänssget and J. Ranft, Nucl. Sci. Eng., **88** 537 and 551 (1984) J. Comput. Phys. **39** 37 (1986).

[12] B. Nilsson-Almqvist and E. Stenlund, Comput. Phys. Commun. **43** 387 (1987).

[13] T. Sjöstrand and M. Bengtsson, Comput. Phys. Commun. **43** 367 (1987).

[14] T. Eichten et al., Nucl. Phys. B**44** 333 (1972).

[15] J. Papp et al.., Phys. Rev. Lett. **34** 601 and 991 (1975).

[16] S.E. Forsythe, Smithsonian Physical Tables (Smithsonian Institution press, Washington DC, 1969).

[17] S. Hayakawa, Phys. Rev. **108** 1533 (1957).

[18] N. Brene, L. Egrard, and B. Qvist, Nucl. Phys. **22** 553 (1961).

[19] M. Circella et al. *in Proceedings of Twenty-Third International Cosmic Ray Conference, Calgary, 1993,* edited by D.A. Leahy *et al.* (Department of Physcs and Astronomy – The University of Calgary, Canada 1993) Vol. 4, p. 503.

[20] T.K. Gaisser *in Proceedings of Neutrino Astrophysics, Takayama 1993,* edited by Y. Suzuki and K. Nakamura, (Universal Academy Press, to be published), and Private communication.

[21] T. Kajita, *in Proceedings of Neutrino Astrophysics, Takayama 1993,* edited by Y. Suzuki and K. Nakamura, (Universal Academy Press, to be published).

[22] K. S. Hirata *et al.*, Phys. Lett. B**205** 416 (1988).

[23] D. Casper *et al.*, Phys. Rev. Lett. **66** 2561 (1991).

[24] K. Hidaka, M. Honda and S. Midorikawa, Phys. Rev. Lett. **61** 1537 (1988).

Experiments for Solar Neutrinos and Supernova Neutrinos

Yoji Totsuka

Institute for Cosmic Ray Research, University of Tokyo

1 Introduction

Solar energy is believed to be generated by nuclear fusion processes taking place in the core of the Sun. This belief must, of course, be tested experimentally.

The underlying nuclear processes are well known [1]:

[$p-p$ Chain]

$$p+p \rightarrow {}^2\mathrm{H} + e^+ + \nu_e \quad (E_\nu < 0.42\,\mathrm{MeV}) \tag{1.1}$$

or

$$p+e^- + p \rightarrow {}^2\mathrm{H} + \nu_e \quad (E_\nu = 1.44\,\mathrm{MeV}) \tag{1.2}$$

$${}^2\mathrm{H} + p \rightarrow {}^3\mathrm{He} + \gamma$$

$${}^3\mathrm{He} + {}^3\mathrm{He} \rightarrow {}^4\mathrm{He} + 2p$$

or

$${}^3\mathrm{He} + {}^4\mathrm{He} \rightarrow {}^7\mathrm{Be} + \gamma$$

$${}^7\mathrm{Be} + e^- \rightarrow {}^7\mathrm{Li} + \nu_e \quad \left(E_\nu = \begin{matrix} 0.86\,\mathrm{MeV}(90\%) \\ 0.38\,\mathrm{MeV}(10\%) \end{matrix}\right) \tag{1.3}$$

$${}^7\mathrm{Li} + p \rightarrow 2\,{}^4\mathrm{He}$$

or

$${}^7\mathrm{Be} + p \rightarrow {}^8\mathrm{B} + \gamma$$

$${}^8\mathrm{B} \rightarrow {}^8\mathrm{Be}^* + e^+ + \nu_e \quad (E_\nu < 15\,\mathrm{MeV}) \tag{1.4}$$

$${}^8\mathrm{Be}^* \rightarrow 2\,{}^4\mathrm{He}$$

or

$${}^3\mathrm{He} + p \rightarrow {}^4\mathrm{He} + e^+ + \nu_e, \quad (E_\nu < 18.8\,\mathrm{MeV}) \tag{1.5}$$

[CNO Cycle]

$${}^{12}\mathrm{C} + p \rightarrow {}^{13}\mathrm{N} + \gamma$$

$${}^{13}\mathrm{N} \rightarrow {}^{13}\mathrm{C} + e^+ + \nu_e \quad (E_\nu < 1.2\,\mathrm{MeV}) \tag{1.6}$$

$${}^{13}\mathrm{C} + p \rightarrow {}^{14}\mathrm{N} + \gamma$$

$${}^{14}\mathrm{N} + p \rightarrow {}^{15}\mathrm{O} + \gamma$$

$${}^{15}\mathrm{O} \rightarrow {}^{15}\mathrm{N} + e^+ + \nu_e \quad (E_\nu < 1.7\,\mathrm{MeV}) \tag{1.7}$$

$${}^{15}\mathrm{N} + p \rightarrow {}^{12}\mathrm{C} + {}^4\mathrm{He}$$

Processes (1.1) to (1.7) produce electron neutrinos which are called, respectively, *pp*, *pep* and ^{7}Be, ^{8}B, ^{13}N and ^{15}O neutrinos; *pep* and ^{7}Be neutrinos are monoenergetic. Other neutrinos have continuous energy spectra characterized by [2]

$$\frac{dN}{dE_\nu} \propto E_\nu^2 E_e p_e F(-Z, E_e), \tag{1.8}$$

where E_ν is the neutrino energy, E_e and p_e are the (total) energy and momentum of the associated positron, and $F(-Z, E_e)$ is the Fermi function for positron emission with a daughter element of charge Z. The Fermi function is approximately given by [1]

$$F(Z, E_e) = 2(1+\gamma)(2p_e R)^{2\gamma-2}\, e^{\pi\nu} \frac{|\Gamma(\gamma + i\nu)|^2}{|\Gamma(2\gamma+1)|^2}$$

$$\approx \frac{2\pi\nu}{1 - e^{-2\pi\nu}} \quad (\text{for small } Z),$$

where

$$\nu = \frac{\alpha Z E_e}{p_e}, \; \gamma = [1 - (\alpha Z)^2]^{1/2}. \tag{1.9}$$

R is the nuclear radius. Note, however, that the ^{8}B neutrino spectrum is more complicated due to the fact that ^{8}B decays into the broad excited state of ^{8}Be (spin-parity $J^\pi = 2^+$, 2.94 MeV) [2].

The neutrino fluxes for processes (1.1)–(1.7) on Earth have been calculated [3] and are shown in Fig. 1. The solar-neutrino experiment is thus subjected to measuring the absolute flux as well as the energy distribution of solar neutrinos and testing the validity of theoretical calculations.

The detectors that are used for solar-neutrino detection are automatically applicable to other low-energy neutrinos, such as supernova neutrinos, diffuse neutrinos from past supernovae, and possibly neutrinos from terrestrial sources which are thought to be responsible for terrestrial heat.

A massive star with a mass $M \geq 8M_\odot$ ($M_\odot$ denotes the mass of the Sun) evolves to an onion-skin structure with a central iron core of Chandrasekhar mass $5.76\, Y_e^2\, M_\odot$, where Y_e is the number of electrons per nucleon. The iron core radius has been calculated to be about 1,000 km. Once the core mass exceeds the Chandrasekhar mass of about $1.3\, M_\odot$, it becomes unstable and eventually collapses gravitationally. The collapse is halted due to repulsive nuclear forces when the density becomes so high that the nucleons almost 'touch' each other. The core bounces and produces shock waves deep in the core ($\sim 0.7 M_\odot$) which propagate through the dense nuclear matter. This shock, when propagates outward, liberates ν_e, which were produced in neutronization process, $e^- + p \rightarrow \nu_e + n$, in the core and then trapped in the middle of core called neutrino sphere. This process produces the initial ν_e burst of duration $\sim$ 10 ms. The shock wave further heats up the neutrino sphere to a few MeV; thermal neutrinos are then produced and escape from the neutrino sphere. The remaining outer shell of the iron core accretes on this hot protoneutron star and its mass reaches $1.4 - 1.6\, M_\odot$. The heat of the core is quickly carried away by emission of thermal neutrinos dramatically

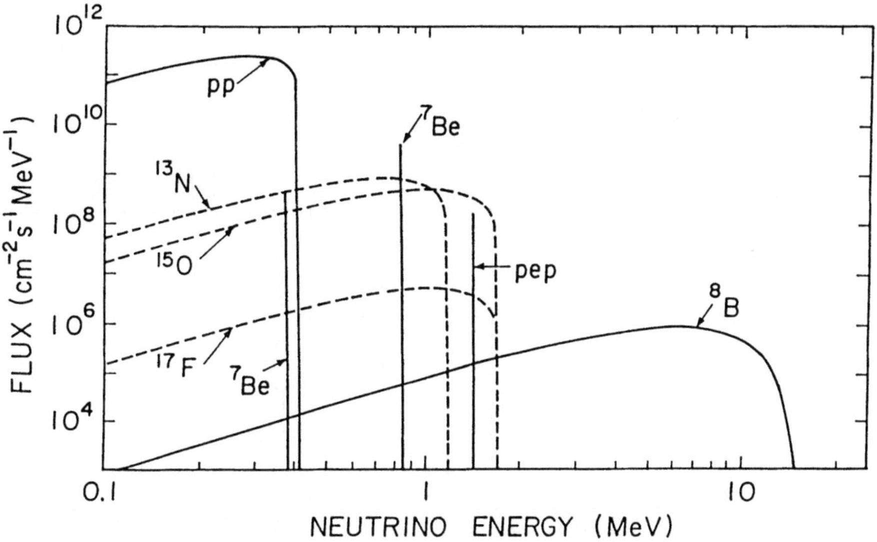

Figure 1: Neutrino fluxes on Earth of various origins as a function of energy. ^{7}Be and *pep* neutrinos are monochromatic and their fluxes are in units of $\mathrm{cm}^{-2}\mathrm{s}^{-1}$.

through the process $e^+ + e^- \to \nu_i + \overline{\nu}_i$ where $i = e$, μ or τ. During the course of cooling, the radius of the protoneutron star (R) shrinks and finally settles down to about 10 km [4][5][6][7][8]. The accretion stops about 0.5 sec after the collapse and the outer shell with $M > 1.7\,M_\odot$ is blown out. Figure 2 shows the resultant neutrino luminosities calculated for initial stellar masses of 12, 25, 50 and 100 $M_\odot$ according to Mayle and Wilson [8]. Only the case of 25 $M_\odot$ shows the luminosities for $\nu_e(L_\nu)$, $\overline{\nu}_e(L_{\overline{\nu}})$ and one of ν_μ, $\overline{\nu}_\mu$, ν_τ and $\overline{\nu}_e(L_\mu)$. The initial sharp peak corresponds to the neutronization burst; remarkably, its luminosity hardly depends on the initial stellar mass. This comes from the fact that the initial collapse occurs at the core mass equal to the Chandrasekhar mass, which is almost independent of the progenitor mass. The total energy carried by the neutronization burst (ν_e) is about 3×10^{51} erg, and that by the thermalization neutrino is 5×10^{52} erg, depending slightly on the progenitor mass [9]. Calculations also show that approximately equal energy is carried off by each species of neutrinos, i.e. $L_{\nu_e} = L_{\overline{\nu}_e} = L_{\nu_\mu} = L_{\overline{\nu}_\mu} = L_{\nu_\tau} = L_{\overline{\nu}_\tau}$. Therefore, the total energy carried by all kinds of neutrinos is about $6\,L_{\overline{\nu}_e} \approx 3 \times 10^{53}$ erg, which, of course, corresponds to the gravitational energy $(3/5)GM^2/R$. The mean energies of the neutrinos have also been calculated: approximately $E_{\nu_e} = 10\,\mathrm{MeV}$, $E_{\overline{\nu}_e} = 15\,\mathrm{MeV}$, and $E_{\nu_\mu} = E_{\overline{\nu}_\mu} = E_{\nu_\tau} = E_{\overline{\nu}_\tau} = 20\,\mathrm{MeV}$ [7][8]. The energy distribution obeys, to a first approximation, the Fermi-Dirac distribution [7][8][10]; namely, the neutrino spectrum on Earth is generally given as

$$\frac{dN_\nu}{dEdt} = 4\pi R^2 \frac{g}{4} \frac{1}{(2\pi)^3} \frac{4\pi E^2}{\exp\left(\frac{1}{T}(E-\mu)\right) + 1} \frac{1}{4\pi D^2}, \tag{1.10}$$

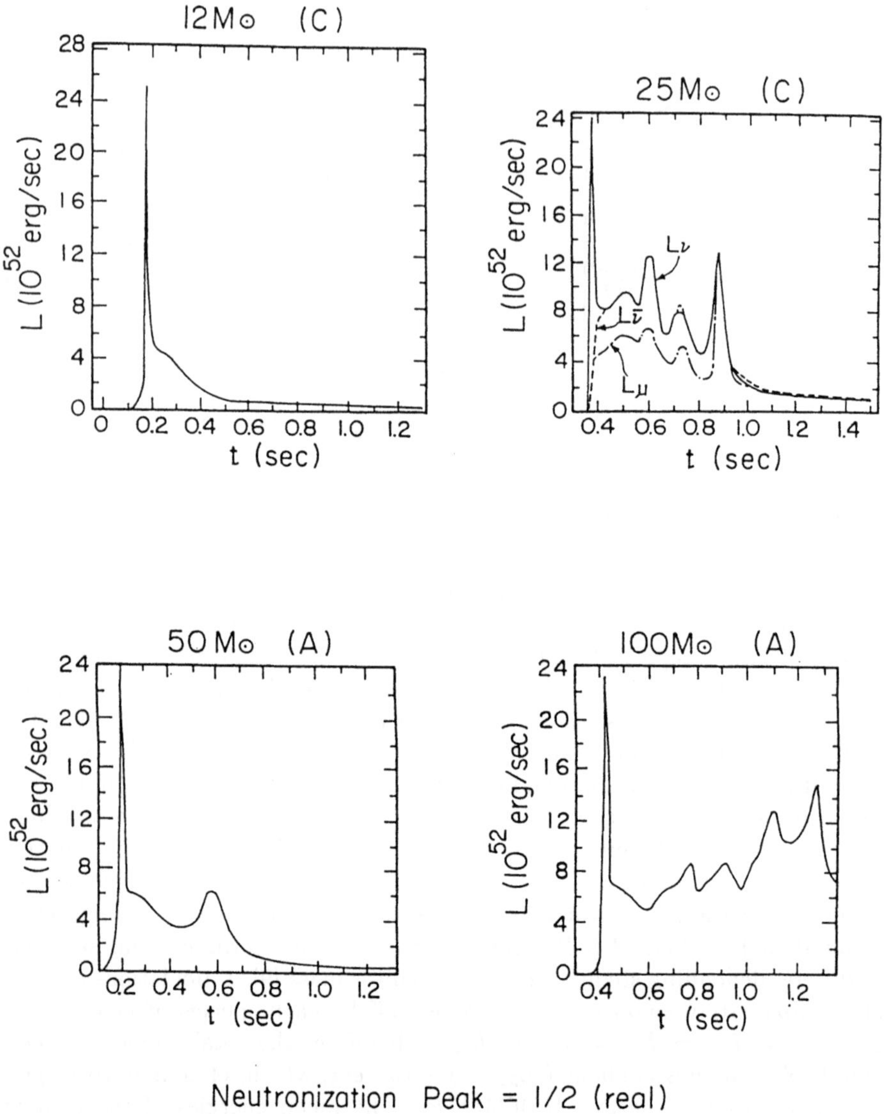

Figure 2: ν_e luminosity (L_ν) for specified progenitor masses. (A) and (C) denote different collapse models. The case for 25 $M_\odot$ shows the $\overline{\nu}_e$ luminosity ($L_{\overline{\nu}}$) and the ν_μ luminosity (L_μ). The neutronization peak is scaled by 1/2. (From [8])

where $g = 1$ is the helicity weight of the neutrino species; E, T, and μ are the neutrino energy and temperature and the chemical potential of the neutrino sphere, respectively; D is the distance from the supernova to Earth; the factor 1/4 comes from the effective solid angle. For $\mu = 0$, the average neutrino energy ($\langle E \rangle$) and the total luminosity per neutrino species (L_{tot}) are given by

$$\langle E \rangle = 3.151\,T \tag{1.11}$$

and

$$L_{tot} = \frac{7}{8}\frac{1}{4}\frac{\pi^2}{30}T^4 \cdot 4\pi R^2 = 1.45 \times 10^{51} \left(\frac{T}{4\,\mathrm{MeV}}\right)^4 \left(\frac{R}{10\,\mathrm{km}}\right)^2 \mathrm{erg/s}. \tag{1.12}$$

The factor of 1/4 appears as before, $(\pi^2/30)T^4$ is the energy density (Stefan-Boltzmann law), and the factor 7/8 is due to the difference between the Fermi-Dirac and Planck distributions.

We now discuss the experiments in the following chapters. In chapter 2 we discuss Real-Time Experiment for Solar Neutrinos. It is subdivided into the Principle of solar neutrino detection(2.1), On-going experiment(2.2), Future solar neutrino experiments(2.3), and Conclusion of this chapter(2.4). Chapter 3 is devoted to Radiochemical Experiment, subdivisions of which are Principle of radiochemical detection of solar neutrinos(3.1), Cross sections(3.2), Expected neutrino capture rates(3.3), Signals from electron-capture decay(3.4), On-going experiments(3.5), and Conclusion of this chapter(3.6). Chapter 4 describes Observation of Supernova Neutrinos, especially neutrinos from SN1987A in the Large Magellanic Clouds. The chapter consists of brief History(4.1), Detector(4.2), Background(4.3), Observation(4.4), Future outlook(4.5) and Conclusion of this chapter(4.6).

We will not discuss the implications based on the observational results concerning solar and supernova neutrinos, but rather concentrate on the basic principles, expected yields and their applications to actual observations.

2 Real-Time Experiment for Solar Neutrinos

2.1 Principle of solar neutrino detection

A real-time experiment is the one which uses conventional counting techniques developed in high-energy physics. It detects electrons with a suitable target material via a charged-current (CC) reaction,

$$\nu_e + A \to e^- + B. \tag{2.1}$$

Since electrons are recorded with high efficiency once the reaction takes place in a detector, the phrase real-times is used. One can measure the energy spectrum of electrons, which is obviously closely related to the neutrino energy spectrum, and possible short-time variation of the neutrino flux, such as day-night effects.

The great difficulty of the real-time experiment lies in how to eliminate the large environmental background caused by natural radioactivities (U, Th, $\cdots$) and from spallation products (β-unstable nuclei such as ^{16}N,^{12}N,^{8}B,^{8}Li,$\cdots$) produced by cosmic-ray muons. The background problem sets the lower limit of the detectable electron energy, which is usually larger than 5 MeV.

The minimum detectable energy of several MeV allows us to detect only ^{8}B neutrinos (see Fig. 1). Therefore, if one is only interested in the origin of solar energy, a radiochemical experiment like ^{71}Ga would be much better than the real-time type, because the process involving ^{8}B (eq.(1.4)) has a branching ratio of a mere 0.02%, i.e. a negligible role compared to the dominant *pp* reaction (eq.(1.1)). However, if the solar neutrino problem, namely the deficit in the ^{8}B neutrino flux, is indeed due to the MSW effect [11], one may expect a sizable distortion in the energy spectrum of the ^{8}B neutrinos. This is exactly where the real-time experiment can do the best. The MSW effect not only distorts the energy spectrum, but also induces a new type of neutrinos, i.e. mu-neutrinos or tau-neutrinos. These neutrinos do not participate in the CC reaction. Therefore, the radiochemical experiment, which detects the daughter element (B), is insensitive to them. On the other hand, the real-time experiment is, by applying various techniques, sensitive to neutral-current (NC) reactions,

$$\nu_i + A \to \nu_i + A' \quad (i \neq e), \tag{2.2}$$

in which one detects product A' in real time. We consider an example later.

It is a challenge to eliminate the environmental background as much as possible and to lower the minimum detectable energy to several 100 keV, which will enable one to observe ^{7}Be neutrinos (monochromatic, 0.861 MeV). An effort is progressing in this direction.

Table 1 lists the real-time experiments currently running or under construction/planning. The target particles are the electron, deuteron or Argon. The relevant reactions corresponding to eq.(2.1) and some relevant comments are described here.

(i) Electron target

The basic process is

$$\nu_i + e \to \nu_i + e, \qquad (i = e, \mu, \tau). \tag{2.3}$$

The cross section is calculable by the well-proven electroweak theory, and is given by

$$\frac{d\sigma}{dy} = \frac{G_F{}^2 m_e E_\nu}{2\pi}\left[A + B(1-y)^2 - Cy\frac{m_e}{E_\nu}\right],$$

where

$$y = \frac{E_\nu - E'_\nu}{E_\nu} = \frac{T_e}{E_\nu}, \quad 0 < y < \frac{2E_\nu}{2E_\nu + m_e}. \tag{2.4}$$

A, B and C are tabulated in Table 2. G_F and m_e are the Fermi coupling constant and electron mass, respectively. E_ν, E'_ν and T_e are the incident and scattered neutrino energies and kinetic energy of the scattered electron, respectively. For $E_\nu \geq 5$ MeV, the total cross section is well approximated as

$$\sigma = \frac{G_F{}^2 m_e E_\nu}{2\pi}\left[A + \frac{B}{3}\right]. \tag{2.5}$$

Table 1: Real-time experiments currently running or under construction/planning.

	T_{min}(MeV)[(a)]	Target	Detection of NC[(b)] via
Kamiokande	6.5	e	$\nu_\mu + e \to \nu_\mu + e$[(c)]
Super-Kamiokande	5.0	e	$\nu_\mu + e \to \nu_\mu + e$[(c)]
SNO	~ 5.0	^{2}H, e	$\nu_\mu + e \to \nu_\mu + e$[(c)] or $\nu_\mu + ^2\mathrm{H} \to \nu_\mu + p + n$, $\mathrm{Cl}(n,\gamma)\mathrm{Cl}$[(d)]
BOREXINO	0.25	e	$\nu_\mu + e \to \nu_\mu + e$
ICARUS	~ 5.0	e, ^{40}Ar	$\nu_\mu + e \to \nu_\mu + e$

(a) minimum detectable kinetic energy of electrons from reaction $\nu_e + A \to e^- + B$.
(b) NC, neutral current.
(c) signal is indistinguishable from charged-current reaction.
(d) γ from (n,γ) reaction is detected to identfy the neutron. No information of ν-directionality or ν-energy is available.

Numerically,

$$\sigma(\nu_e e \to \nu_e e) = 9.5 \times 10^{-44} \left(\frac{E_\nu}{10\,\mathrm{MeV}}\right) \mathrm{cm}^2 \tag{2.6a}$$

and

$$\sigma(\nu_i e \to \nu_i e) = 1.6 \times 10^{-44} \left(\frac{E_\nu}{10\,\mathrm{MeV}}\right) \mathrm{cm}^2, \quad (i \neq e) \tag{2.6b}$$

which are very small. Note that $\sigma(\nu_\mu e) = \sigma(\nu_\tau e) \approx \sigma(\nu_e e)/6$.

The cross section (2.4) has a sharp forward peak. This can be seen after modifying eq.(2.4), as follows:

$$\frac{d\sigma}{d\cos\theta} = 4\frac{m_e}{E_\nu}\frac{(1+\frac{m_e}{E_\nu})^2\cos\theta}{[(1+\frac{m_e}{E_\nu})^2-\cos^2\theta]^2}\frac{d\sigma}{dy}$$

$$\approx 4\frac{m_e}{E_\nu}\frac{\cos\theta}{\sin^4\theta}\frac{d\sigma}{dy} \quad (\text{for } E_\nu \gg m_e),$$

$$y = \frac{2\frac{m_e}{E_\nu}\cos^2\theta}{\left(1+\frac{m_e}{E_\nu}\right)^2 - \cos^2\theta}, \tag{2.7}$$

where θ is the scattering angle of the electron measured from the incident neutrino direction. This feature is very important for improving the signal-to noise ratio over an essentially isotropic background.

The electron kinetic energy (T_e), on the other hand, has an almost flat spectrum ($0 < T_e \leq E_\nu$), as can be seen from eq.(2.4). Therefore, the expected event rate depends sensitively on the minimum detectable kinetic energy (T_{min}). Table 3

Table 2: Parameters in eq.(2.4) for $\nu_e, \overline{\nu}_e, \nu_\mu$ and $\overline{\nu}_\mu$. Those for ν_τ and $\overline{\nu}_\tau$ are the same as that for ν_μ and $\overline{\nu}_\mu$.

ν :	A	B	C
$\nu_e e \rightarrow \nu_e e$	$(g_V + g_A + 2)^2$	$(g_V - g_A)^2$	$(g_V + 1)^2 - (g_A + 1)^2$
$\overline{\nu}_e e \rightarrow \overline{\nu}_e e$	$(g_V - g_A)^2$	$(g_V + g_A + 2)^2$	$(g_V + 1)^2 - (g_A + 1)^2$
$\nu_\mu e \rightarrow \nu_\mu e$	$(g_V + g_A)^2$	$(g_V - g_A)^2$	$g_V^2 - g_A^2$
$\overline{\nu}_\mu e \rightarrow \overline{\nu}_\mu e$	$(g_V - g_A)^2$	$(g_V + g_A)^2$	$g_V^2 - g_A^2$

$g_V = I^3 - 2Q \sin^2\theta_W,\ g_A = I^3.$

I^3 and Q are the third component of weak isospin of the target particle and ite charge, respectively. θ_W is the weak mixing angle ($\sin^2\theta_W = 0.23$).

For electrons,

$I^3 = -1/2,\ \ Q = -1$

$g_V = -0.5 + 2\sin^2\theta_W \approx -0.04,\ \ g_A = -0.5.$

tabulates some numerical values of the integrated cross section over the ^{8}B neutrino spectrum, $\phi_{^8\mathrm{B}}(E_\nu)$ [12],

$$\langle \sigma_e(^8\mathrm{B}, T_{min}) \rangle = \frac{\int_{T_{min}} \phi_{^8\mathrm{B}}(E_\nu)\, dE_\nu \int_{T_{min}}^{E_\nu} \frac{d\sigma}{dT_e}\, dT_e}{\int_0 \phi_{^8\mathrm{B}}(E_\nu) dE_\nu}, \tag{2.8}$$

as a function of T_{min}.

Table 3: Integrated cross sections of $\nu_e e \rightarrow \nu_e e$ averaged over the ^{8}B neutrino spectrum as a function of the minimum detectable energy (T_{min}) of electrons [12].

T_{min} (MeV)	$\langle \sigma_e(^8\mathrm{B}, T_{min}) \rangle (\times 10^{-44} \mathrm{cm}^2)$
0	6.08
4.0	2.48
5.0	1.80
6.0	1.23
7.0	0.790
8.0	0.464
9.0	0.244
10.0	0.110

(ii) Deuteron target

The basic process is the following CC reaction:

$$\nu_e + {}^2\mathrm{H} \rightarrow e^- + p + p. \tag{2.9}$$

The spin-parity and isospin of ^{2}H are $J^\pi = 1^+$ and $T = 1$, respectively. The final pp state is $^1S_0$ and $T = 1$ at low neutrino energy. The cross section is then [13][14][15][16]

$$\frac{d^2\sigma}{dE_e d\Omega} = \frac{1}{\pi} G_F{}^2 \cos^2\theta_c \left(\frac{g_A}{g_V}\right)^2 F(2, E_e)\frac{1}{4\pi}\left(1 - \frac{1}{3}v_e \cos\theta\right)$$
$$\times |I|^2 m_p\, p_p\, E_e\, p_e,$$

where

$$I = \int u_{2p}^*(r) u_d(r) dr. \tag{2.10}$$

Here, θ_c is the Cabibbo angle ($\cos\theta_c = 0.974 \pm 0.001$), $g_A/g_V = 1.25$, F the Fermi function, $v_e = p_e/E_e$, m_p the proton mass, p_p the proton momentum, and $u(r)$ the radial wave function. Note that the CMS motion of the $2p$ system is negligible. $|I|^2$ depends on p_p and can be calculated. The electron kinetic energy is given by

$$T_e = E_\nu - Q - \frac{p_p{}^2}{m_p} \quad (Q = 1.44\,\mathrm{MeV}). \tag{2.11}$$

The last term somewhat smears out the T_e ($FWHM \approx 1.2\,\mathrm{MeV}$ at $T_e = 10.6\,\mathrm{MeV}$) due to a finite spread of the proton momentum (p_p), which is illustrated in Fig. 3 [13].

ν_μ and ν_τ can be detected via the NC reaction,

$$\nu_i + {}^2\mathrm{H} \rightarrow \nu_i + p + n, \quad (i = e, \mu, \tau). \tag{2.12}$$

The neutrons produced in this process are thermalized and captured by Cl,

$$\mathrm{Cl}(n_{thermal}, \gamma)\mathrm{Cl}, \tag{2.13}$$

and emitted γ's are detected. The energies of the γ's consist of several peaks which range from 5 to 8.6 MeV. Their relative intensities are given in Fig. 4; some of them are listed in Table 4 [17].

Since neutrons are thermalized, information concerning E_{ν_i} is completely lost, and one can only measure the total neutrino capture rate for $E_{\nu_i} > 2.22\,\mathrm{MeV}$. The total cross sections of eq.(2.12) have been calculated [15][16]. We list them in Table 5 together with those for the CC reaction, eq.(2.10) [16]. The total NC cross section averaged over the ^{8}B neutrino spectrum $\phi_{^8\mathrm{B}}(E_\nu)$ is [16],

$$\langle\sigma_{NC}(^8\mathrm{B})\rangle = \frac{\int_{2.22\,\mathrm{MeV}} \phi_{^8\mathrm{B}}(E_\nu) dE_\nu \sigma_{NC}(E_\nu)}{\int_0 \phi_{^8\mathrm{B}}(E_\nu) dE_\nu} = 4.72 \times 10^{-43}\mathrm{cm}^2, \tag{2.14a}$$

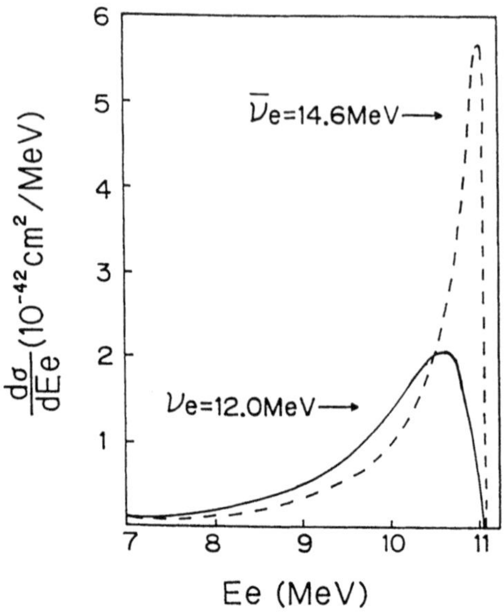

Figure 3: Cross sections of $\nu_e\,{}^2\mathrm{H} \rightarrow e^- pp$ and $\overline{\nu}_e\,{}^2\mathrm{H} \rightarrow e^+ nn$ for incident energies of 12.0 and 14.6 MeV, respectively. The electron energy becomes broad due to two protons in the final state carrying a fraction of the available energy. [13]

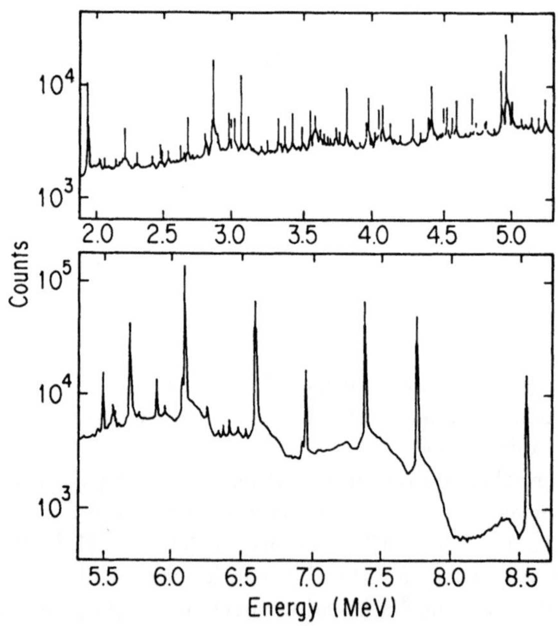

Figure 4: Spectra of gamma rays (sharp peaks in the figure) produced by a capture reaction of the thermal neutrons on Chlorine, Cl(n,γ)Cl [17]. Neutrons are counted by detecting gamma rays from this reaction.

Table 4: Intensity of typical gamma rays from the reaction $^{35}\mathrm{Cl}(n,\gamma)^{36}\mathrm{Cl}$ [17].

E_γ(keV)	Intensity (photons/100 captures)
8578.9	3.39
7790.9	10.22
7414.4	11.58
6978.0	2.74
6627.9	5.89
6620.6	11.20
6111.4	21.66
5714.5	5.89

Table 5: Total cross sections of $\sigma(\nu_e\,^2\mathrm{H} \rightarrow e^- pp) \equiv \sigma(CC)$ and $\sigma(\nu_i\,^2\mathrm{H} \rightarrow \nu_i pn) \equiv \sigma(NC)$ as a function of the neutrino energy (E_ν). ν_i is ν_e, ν_μ or ν_τ [16].

E_ν(MeV)	$\sigma(NC)$ ($\times 10^{-42}\mathrm{cm}^2$)	$\sigma(CC)$ ($\times 10^{-42}\mathrm{cm}^2$)	$\sigma(NC)/\sigma(CC)$
4	0.0307	0.158	0.194
6	0.201	0.624	0.322
8	0.554	1.46	0.379
10	1.10	2.71	0.406
12	1.86	4.37	0.426
14	2.82	6.48	0.435
16	3.99	9.05	0.441
18	5.38	12.1	0.445
20	6.99	15.6	0.448
25	12.0	26.6	0.451
30	18.4	40.9	0.450
40	35.7	80.1	0.446
50	59.1	134	0.441

Table 6: Integrated cross sections for $\nu_e\,^2\mathrm{H} \to e^- pn$ [18] and for $\nu_e\,^{40}\mathrm{Ar} \to e^-\,^{40}\mathrm{K}^*$ [20] averaged over the ^{8}B neutrino spectrum as a function of the minimum detectable energy of electrons (T_{min}).

T_{min} (MeV)	$\langle\sigma_{^2\mathrm{H}}(^8\mathrm{B}, T_{min})\rangle$ ($\times 10^{-42}\mathrm{cm}^2$)	$\langle\sigma_{\mathrm{Ar}}(^8\mathrm{B}, T_{min})\rangle$ ($\times 10^{-42}\mathrm{cm}^2$)
0	1.17	0.785
4.0	1.11	0.460
5.0	1.04	0.292
6.0	0.917	0.142
7.0	0.753	0.0446
8.0	0.560	0.0062
9.0	0.362	—
10.0	0.190	—

We also list in Table 6 the integrated CC cross section averaged over the ^{8}B neutrino spectrum $\phi_{^8\mathrm{B}}(E_\nu)$ [18],

$$\langle\sigma_{^2\mathrm{H}}(^8\mathrm{B}; T_{min})\rangle = \frac{\int_{T_{min}} \phi_{^8\mathrm{B}}(E_\nu)dE_\nu \int_{T_{min}}^{E_\nu} \frac{d\sigma}{dT_e} dT_e}{\int_0 \phi_{^8\mathrm{B}}(E_\nu)dE_\nu}, \tag{2.14b}$$

which should be compared with those given in Table 3. For example, at $T_{min} = 5\,\mathrm{MeV}$, $\langle\sigma_{^2\mathrm{H}}\rangle/\langle\sigma_e\rangle = 58$. Note, however, that this large factor is somewhat compensated by the relative number of target particles when the event rate is calculated, e.g., $N(e^-)/N(^2\mathrm{H}) = 5$ for heavy water. Thus, $Rate(^2\mathrm{H})/Rate(e) = 12$ at $T_{min} = 5\,\mathrm{MeV}$.

(iii) Argon target

The basic process is

$$\nu_e + {}^{40}\mathrm{Ar} \to e^- + {}^{40}\mathrm{K}^*,$$
$$^{40}\mathrm{K}^* \to {}^{40}\mathrm{K} + \gamma. \quad (E_\gamma = 2.09, 1.65\,\mathrm{MeV}, \cdots) \tag{2.15}$$

The Q-value of this reaction is 1.65 MeV (ground state – to – ground state transition). However, since the ground state of ^{40}Ar has $T = 2$ and $J^\pi = 0^+$, the reaction proceeds mostly to the isotopic analog state ($T = 2$, $J^\pi = 0^+$) at 4.38 MeV above the ground state of $^{40}\mathrm{K}(J^\pi = 4^-)$. The angular distribution of the electrons is given by [19]

$$1 + v_e \cos\theta. \tag{2.16}$$

The cross section has been calculated [20] (see section 3.2 for details). Table 6 lists the integrated cross section averaged over the ^{8}B neutrino spectrum as a function of the minimum detectable energy of electrons, T_{min}, $\langle\sigma_{\mathrm{Ar}}(^8\mathrm{B}, T_{min})\rangle$. For example, $\langle\sigma_{\mathrm{Ar}}(^8\mathrm{B}, T_{min} = 5\,\mathrm{MeV})\rangle = 3.79 \times 10^{-43}\mathrm{cm}^2$, i.e. 1/3.6 of $\langle\sigma_{^2\mathrm{H}}(^8\mathrm{B}, T_{min})\rangle$. When

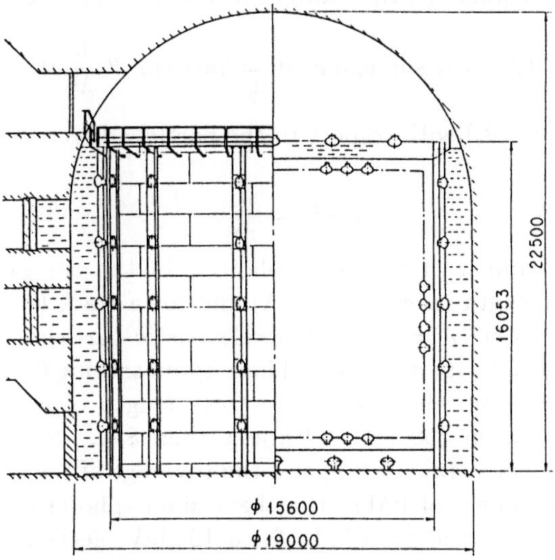

Figure 5: Kamiokande detector. The inner detector has 2,140 t of sensitive volume viewed by 948 PMTs with 20″ diameter. The outer detector contains 1,500 t of water which is used for tagging incoming cosmic-ray muons and for absorbing low-energy gamma rays and neutrons from the surrounding rock.

the event rate is calculated, the number of target particles must be multiplied. For the same weight of liquid Argon and heavy water, $N(\mathrm{Ar})/N(^2\mathrm{H}) = 1/4$. Thus, $Rate(\mathrm{Ar})/Rate(^2\mathrm{H}) = 1/14$ for $T_{min} = 5\,\mathrm{MeV}$ and for the same target weight. Thus, combined with the argument in (ii) above, $Rate(\mathrm{Ar}) = 0.86\,Rate(e)$, which is not very large.

2.2 On-going experiment

There is only one experiment currently running, namely a water-Čerenkov detector called Kamiokande. The target here is the electron, i.e. the basic process is eq.(2.3). It is located 1,000 m underground in the Kamioka zinc mine, which is used to shield the detector against cosmic-ray muons. As shown in Fig. 5, the detector comprises inner and outer parts. The outer part completely covers the inner one, and has about 1,500 t of water. It is equipped with 123 photomultipliers (PMTs) in order to tag still-incoming cosmic-ray muons. Its main role, however, is to absorb gammas and neutrons from the surrounding rock.

The inner part is used for real-time neutrino observations. It is cylindrical and has 2,140 t of pure water (transparent and free from radioactive elements like U, Ra, Rn and Th) viewed by 948 PMTs with 20″ diameter. They cover 20% of the whole inner wall.

A charged particle with velocity v that exceeds the inverse of the refractive index emits Čerenkov photons, the number of which (N_γ), is given by

$$d\,N_\gamma = \alpha\,\sin^2\theta_c\,d\,E_\gamma\,dl = 2\pi\alpha\,\sin^2\theta_c\frac{d\lambda}{\lambda^2}dl. \qquad (2.17)$$

The angle of emission (θ_c) with respect to the particle direction (Čerenkov angle) is given by

$$\cos\theta_c = \frac{1}{nv}, \qquad (2.18)$$

where dN_γ is the number of Cerenkov photons in the energy range ($E_\gamma, E_\gamma + dE_\gamma$) or in the wavelength range (λ, $\lambda + d\lambda$) and for a track length of dl; n is the refractive index of water ($n = 1.34$). For relativistic particles ($v = 1$), $\theta_c = 42°$ and θ_c shrinks as the velocity decreases. Thus, by measuring θ_c, low-energy muons can be distinguished from electrons. The track length of a 10 MeV electron is about 5 cm. N_γ between 350 and 500 nm is then 880. Multiplying it by the PMT's quantum efficiency (about 0.2) and the coverage of the photosensitive area (0.2), the expected number of PMTs that have single-photoelectron signals is 35. Kamiokande actually sees 30 hit PMTs for a 10 MeV electron, which agree well with this estimate.

The number of hit PMTs, (N_{hit}), is found to be a good measure of the electron energies around 10 MeV; $E_e = 30\,N_{hit}$. This $N_{hit} - E_e$ relationship must be calibrated very precisely, because the E_e spectrum is steep at about 10 MeV (see Fig. 7 for the observed energy spectrum). The calibration is carried out with gamma rays produced by Ni(n,γ)Ni, n being thermalized neutrons from ^{252}Cf fissions [21]. The γ energies have several peaks. The dominant ones are 9.00, 7.81 and 6.84 MeV with relative intensities of 0.71:0.16:0.13, respectively. The accuracy in calibration has reached 2% at Kamiokande.

As the minimum detectable energy for electron, (T_{min}), becomes close to or below 10 MeV, the detector suffers from enormous background of various origins. They are: (i) γ's from the rock which were not absorbed in the outer detector. Most of them are eliminated by limiting the fiducial volume to the central 680 t.
(ii) β-particles from ^{16}N,^{12}N,^{6}B,^{8}Li,$\cdots$ produced by break-up of oxygen nuclei by cosmic-ray muons (spallation products). They dominate the event rate in the fiducial volume for $E_e = 8 \sim 17$ MeV. The Kamiokande site is somewhat too shallow from this point of view.
(iii) β and γ from ^{222}Rn and possibly from ^{208}Tl, which dominate at $E_e < 8$ MeV. Since the air is highly contaminated with Rn gas underground in Kamioka ($1,000 \sim 2,000$ Bq/m^3), a strict precaution must be employed to avoid contact of air to the tank water. The present T_{min} of Kamiokande (4.5 Mev trigger and 6.5 MeV analysis threshold) is set by the contamination of ^{222}Rn in the tank water.

The analysis programs consume most of the computer time to eliminate as much background as possible, though elimination is not complete. After background reduction, we take advantage of a forward peak of the $\nu_e - e$ scattering (eq.(2.7)). Figure 6 shows the $\cos\theta_{sun}$ distribution of the events observed during Jan.1987–Apr.1990, in which θ_{sun} is the scattering angle of electrons measured from the Sun. Figure 6a is for $E_e > 7.5$ MeV($T_e > 7.0$ MeV) and Fig. 6b for $E_e > 9.3$ MeV. All of the background in Fig. 6b is due to spallations, while one third of the background

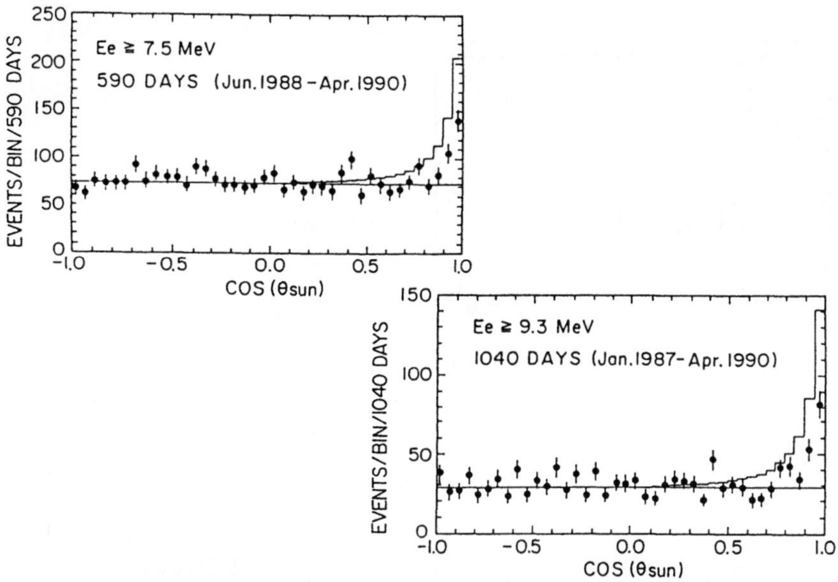

Figure 6: Angular distribution of electrons observed in the Kamiokande detector [21]. θ_{SUN} is defined as angle between electron momentum and a vector pointing from the Sun to the detector. (a) for $E_e \geq 7.5$MeV and (b) for $E_e \geq 9.3$MeV. One sees a flat background plus a forward peak in both figures. The forward peaks are produced by solar neutrinos. Histograms indicate the prediction of Bahcall-Ulrich's SSM.

in Fig. 6a is from ^{222}Rn and the two thirds from spallations. In spite of the large background still remaining, one can clearly see an enhancement in the forward direction, which is, of course, produced by solar neutrinos. The electron energy spectrum is then obtained (Fig. 7). Although the shape is consistent with what one expects from ^{8}B neutrinos, the flux is considerably less than the prediction of the standard solar model recently updated by Bahcall and Pinsonneault (SSM) [3]. For $E_e > 7.5\,$MeV it is found that

$$\frac{Number\ \ of\ \ events\ observed}{Number\ \ of\ \ events\ \ predicted} \equiv \frac{Data}{SSM} = 0.49 \pm 0.04 \pm 0.06, \qquad (2.19)$$

based on 1435 days of observation [22]. Note that the observed event rate is merely one event in 3.5 days. It is obvious that a much larger detector with less background is clearly needed.

A large discrepancy between the theory and observation shown above confirms the solar neutrino problem. Combined with the Homestake experiment [23] and the two Gallium experiments [24][25], the solar neutrino problem could be solved by the MSW effect [11], i.e. the problem in particle physics and not the problem in solar physics.

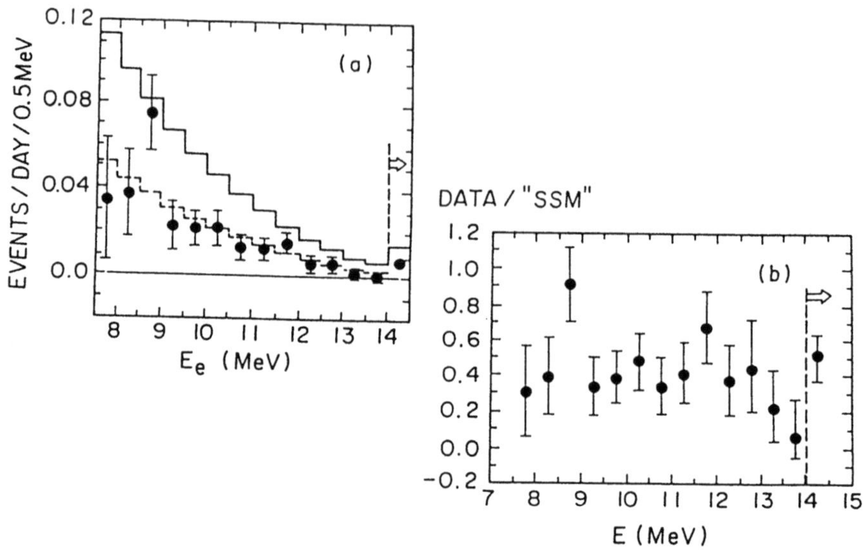

Figure 7: (a) Energy spectrum of electrons scattered by solar neutrinos, observed in the Kamiokande detector [21]. The solid histogram is expected from Bahcall-Ulrich's SSM. The dotted histogram is 0.46× the SSM. (b) Same spectrum normalized by the SSM. One sees a clear depletion of events.

2.3 Future experiments

As mentioned in chapter 1, if the MSW effect is indeed a viable solution for the solar-neutrino problem, the real-time experiment is most suitable to test its validity by accurately measuring any possible distortion of the ^{8}B neutrino spectrum (actually, the electron energy spectrum is measured, *cf.* eq.(2.1)), and also by measuring any possible variation of the ^{8}B neutrino flux (as well as spectrum) between day and night. Moreover, the real-time experiment is able to directly identify $\nu_i(i \neq e)$, the existence of which is clear evidence for the MSW effect (or more precisely, the neutrino oscillations).

Two large underground detectors, SNO and Supekamiokande, are under construction and are expected to be operational in 1995 and 1996, respectively, if everything goes as scheduled. Two other detectors, ICARUS and BOREXINO, are at the stage of prototype tests. The target materials and minimum detectable electron energy (T_{min}) are listed in Table 1. Note that the extremely small T_{min}(0.25 MeV) of BOREXINO. BOREXINO is thus able to measure the flux of ^{7}Be neutrinos, which ought to be almost completely converted to $\nu_i(i \neq e)$ if the MSW effect is a viable solution of the solar-neutrino problem.

Some technical parameters are listed in Table 7. Some comments are in order.

Since Čerenkov light has directionality, Čerenkov detectors are able to point back detected neutrinos to the Sun, especially for an electron target and modestly for an ^{2}H target (distribution, $1 - \frac{1}{3}v_e \cos\theta$) [19]. ICARUS is capable of directly

Table 7: Detector parameters for the five experiments: Kamiokande, Super-Kamiokande, SNO, BOREXINO and ICARUS.

Experiment	Depth (mwe)	Target	Material	Fiducial mass(t)	Detection$^{(a)}$	$\Delta E/E^{(b)}$ (%)	$\Delta\theta^{(b)}$ (deg)
Kamiokande	2,700	e^-	H_2O	680	Č	20	~ 28
Super-Kam	2,700	e^-	H_2O	22,000	Č	14	~ 25
SNO	6,200	$^2H, e^-$	D_2O	1,000	Č	12	~ 25
BOREXINO	4,000	e^-	$B(OCH_3)_3$	100	Scint.	4	–
ICARUS	4,000	$^{40}Ar, e^-$	Ar(*liq*)	4,700	Ioniz. (TPC)$^{(c)}$	5	~ 10

(*a*) Č, Čerenkov detection; Scint., Scintillation detection; Ioniz., Ionization detection (Time projection chamber)
(*b*) for 10 MeV e^- except for ICARUS's $\Delta E/E$ which is for 6 MeV e^-.
(*c*) TPC, Time projection chamber.

Table 8: Purity of target material required for solar-neutrino observation.

	^{238}U	^{232}Th (Bq/kt)$^{(a)}$	^{222}Rn	Remarks
Kamiokande$^{(b)}$	1	1	200	$^{214}Bi \to \beta(3.3MeV)$
Super-Kam	1	1	$1 \sim 10$	$^{214}Bi \to \beta(3.3MeV)$
SNO	0.1	0.01	0.1	$\gamma\,^2H \to pn$ $n\,^2H \to ^3H\gamma(6.3MeV)$
BOREXINO	0.001	0.001	0.001	β, γ from all sources

(*a*) Bq/kt = 1 disintegration/s/1,000 t
(*b*) already achieved.

recording a charged-particle track, and thereby the angular distribution of events ($1 - v_e \cos\theta$ for ^{40}Ar) [19]. On the other hand BOREXINO loses angular information completely due to the isotropic nature of scintillation light. Note that the background can be significantly reduced by the known angular distribution. For example, Kamiokande, with its angular resolution of $\Delta\theta = 28$ deg (for 10 MeV e^-), reduces the isotropic background by a factor of about 20, as shown in Fig. 7. ^2H and ^{40}Ar targets do not exhibit a sharp forward peak, but only modest angular distributions. Therefore, SNO, ICARUS, and especially BOREXINO, must set very strict specifications for radioactive impurities in the target material, which are listed in Table 8. To achieve these specifications is indeed very hard (remember that we are dealing with 100~5,000 t of material). One must first of all invent a measuring device of such low-level radioactivities!

In spite of the seemingly formidable difficulties, these goals will eventually be met, judging from our experience with the Kamiokande detector. SNO, BOREXINO and ICARUS are located sufficiently deep that spallation products are essen-

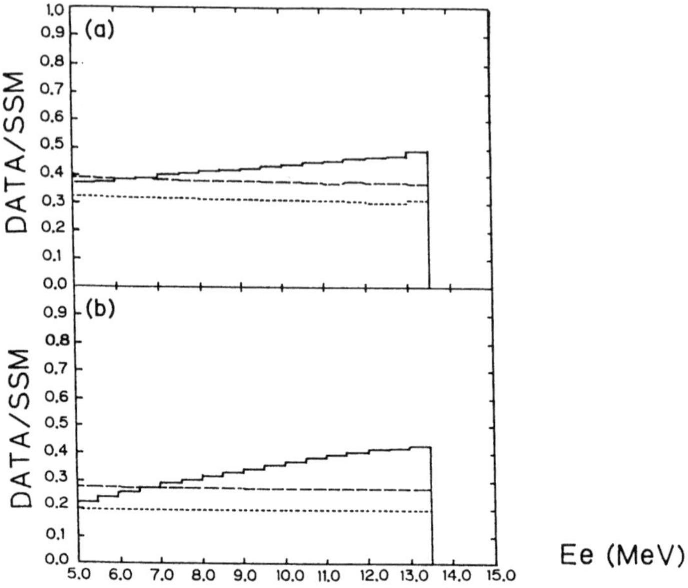

Figure 8: Ratio ($DATA/SSM$) of observed event rate to expected one by Bahcall-Ulrich's SSM as a function of total electron energy (E_e) for (a) Super-Kamio-kande and for (b) SNO. The ratio shown in 13.0-13.5 MeV is actually for $E_e > 13.0$ MeV. Solid, dashed and dotted lines correspond to $(9 \times 10^{-6}\mathrm{eV}^2, 5 \times 10^{-3}), (1 \times 10^{-5}\mathrm{eV}^2, 0.63)$ and $(3 \times 10^{-5}\mathrm{eV}^2, 0.79)$ for the MSW parameters $(\Delta m^2, \sin^2 2\theta)$, respectively. Energy resolutions of both detectors are taken into account.

tially nonexistent, or easily eliminated without any loss of efficiency. Water and scintillator are quite easy to be purified. The remaining technical problems are: (i) to completely isolate the target material from the ambient air, which contains a high level of Radon gas; (ii) to eliminate gamma rays out of PMT glasses, i.e. to use low-radioactive glasses for PMTs; (iii) BOREXINO must use a scintillation technique to receive measurable light signals from electrons of 0.25 MeV. They do not emit Cerenkov lights anyway. The scintillator material considered by the BOREXINO group exhibits a photon yield of $> 5,000$ photons/MeV, which should be compared with about 900 Cerenkov photons/10 MeV in water (see section 2.2).

Now we show the expected event rates for each detector. The relevant formulae and numerical values have already been described in section 2.1. Table 9 lists the expected number of events for various solar neutrinos assuming the standard solar model (SSM) of Bahcall and Ulrich [26]. We expect a great improvement in the statistical accuracy compared to that of Kamiokande.

Figure 8 shows the expected number of events normalized to the SSM prediction (DATA/SSM, see eq.(2.19)) for the three sets of the MSW parameters, which account for the results obtained by Homestake, Kamiokande, SAGE and GALLEX. (a) for Super-Kamiokande and (b) for SNO (CC events only). SNO obviously has a higher sensitivity because T_e represents E_ν rather accurately, i.e. $T_e \approx E_\nu - 1.44$ MeV (see Fig. 3). A few years of operation of both detectors will

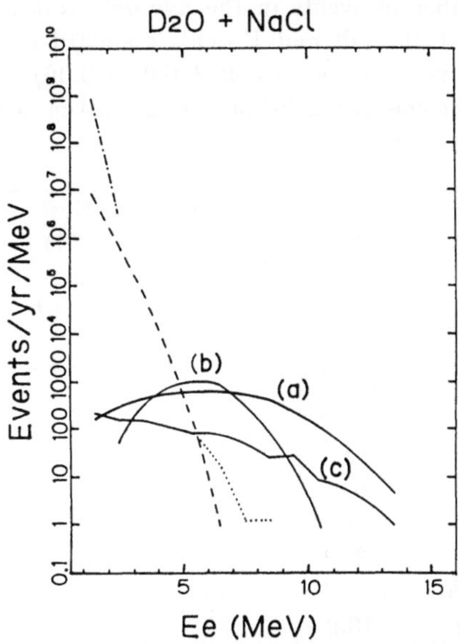

Figure 9: Expected events in the SNO detector as a function of the electron energy. Heavy water is assumed to contain an additional 2.5 t of salt for the reaction Cl(n,γ)Cl to occur with high efficiency [27] (a) signals from $\nu_e\,{}^2\mathrm{H} \to e^- pp$, (b) from $\nu_e\,{}^2\mathrm{H} \to \nu_e pn$, Cl$(n,\gamma)$Cl, and (c) $\nu_e e \to \nu_e e$. Bahcall-Ulrich's SSM is assumed. The dashed curves are the expected backgrounds.

provide statistically a very precise spectrum of ^{8}B neutrinos, which will decisively test the validity of the MSW effect. Figure 9 shows that the absolute values of DATA/SSM significantly differ between Super-Kamiokande and SNO. This is a manifestation of the existence of $\nu_i (i \neq e)$ events in Super-Kamiokande. Thus, a comparison of the electron signals between Super-Kamiokande and SNO will also test the validity of the MSW effect.

SNO has a unique feature to directly observe $\nu_i (i \neq e)$ via the NC reaction with an addition of NaCl in heavy water (eq.(2.12),(2.13)). One expects 31 events/day for SSM (Table 9). Figure 9 shows the expected E_e distribution for SNO with 2.5 t of salt(NaCl) in the heavy water. Some comments are in order:

(i) The visible energy for γ from Cl(n,γ)Cl is less than that for an electron of the same energy.

(ii) The γ signals are isotropic.

(iii) The γ and electron signals coexist and are not separated. One needs alternate observations with/without salt in the D_2O target.

(iv) The γ signals do not represent neutrino energies. Only the number of γ signals is meaningful.

BOREXINO is quite unique to observe ^{7}Be neutrinos which are monochromatic. Since the theoretical uncertainties are expected to be small in the ^{7}Be neutrino flux [26], BOREXINO compares the number of observed events with the

Table 9: Expected number of events in the five detectors for *pep*, ^{7}Be, ^{8}B and hep neutrinos, assumings that Bahcall and Ulrich's standard solar model is correct [26]. Kamiokande's observed event rate was $(0.49 \pm 0.04 \pm 0.06) \times 0.86(ev/d)$ [22]. For the other four experiments the energy resolutions are not taken into account. The third row in SNO represents the NC reaction.

	T_{min} (MeV)	$E_{\nu,min}$$^{(e)}$ (MeV)	*pep* (ev/d)$^{(a)}$	^{7}Be (ev/d)	^{8}B (ev/d)	*hep* (ev/yr)$^{(a)}$
E_ν(MeV)$^{(c)}$			1.44	0.861	< 15	< 18.8
Flux (cm^{-2}s^{-1})$^{(d)}$			$1.4 \cdot 10^{8}$	$4.2 \cdot 10^{9}$	$5.8 \cdot 10^{6}$	$8 \cdot 10^{3}$
Kamiokande	7.0	7.2$^{(b)}$			0.86	
Super-Kam	5.0	5.2$^{(b)}$			67	
	14.5	14.8$^{(b)}$				1.9
SNO	5.0	6.4			33	
	15.0	16.4				3.9
	—	2.22$^{(f)}$			31	
BOREXINO	0.25	0.41$^{(b)}$	3.4	41		
ICARUS	5.0	10.9			11	

(*a*) ev/d = number of observed events/day; ev/yr = number of observed events/year.
(*b*) $T_{min} = E_{\nu\, min} / \left(1 + \frac{m_e}{2E_{\nu\, min}}\right)$, cf. eq.(2.4).
(*c*) Neutrino energy from sources. *pep* and ^{7}Be neutrinos are monochromatic.
(*d*) Neutrino fluxes on Earth.
(*e*) Minimum neutrino energy corresponding to minimum detetable kinetic energy, T_{min}.
(*f*) For the reaction $\nu_i\, {}^{2}\text{H} \rightarrow \nu_i p n$.

theoretical prediction.

2.4 Conclusion of this chapter

It is now possible to observe neutrinos with an energy as low as 5 MeV by applying the real-time technique. Kamiokande is the first experiment to successfully observe ^{8}B solar neutrinos and has shown that the observed flux is smaller than that predicted by Bahcall and Ulrich's standard solar model. However, a single experiment will not solve the solar neutrino problem completely. Fortunately, two much larger detectors are under construction and two more with unique techniques are foreseen in the near future. Armed with these advanced detectors we will eventually be able to solve the long-standing solar-neutrino problem.

Perhaps our efforts will be directed to detecting extremely low-energy electrons with a target mass much larger than that of BOREXINO. We will eventually be able to observe neutrinos of terrestrial origin [28]. This could be called neutrino geophysics and might reveal unexpected surprises, such as the solar-neutrino problem, which we are enjoying so much nowadays.

3 Radiochemical Experiment for Solar Neutrinos

3.1 Principle of low-energy neutrino detection

The second technique applied to detecting solar neutrinos is to take advantage of well-established radiochemistry, though the requirements for the present case are quite stringent.

The basic process used in the radiochemical experiment is the same as that for the real-time one, i.e. a charged-current (CC) reaction,

$$\nu_e + A \rightarrow e^- + B. \tag{3.1}$$

What is detected is now different from real-time observation. The radiochemical experiment tries to pick out produced atoms (B) in eq.(3.1) from a vast amount of target material (A), then to concentrate and purify them, and finally to count their number.

Atom B is radioactive and decays via electron capture, i.e. the inverse reaction of eq.(3.1),

$$B + e^- \rightarrow \nu_e + A. \tag{3.2}$$

The captured electron is previously in the K or L shell of the atom. The resultant vacant state is instantly reoccupied by an electron in the outer shell; this relaxation process emits an Auger electron and/or X-ray, whose characteristic energy corresponds to the K or L energy level. This Auger electron and/or X-ray is detected with a small proportional counter. The proportional counter must be fabricated very carefully in order to avoid contamination due to any background, because the Auger electron and/or X-ray typically has an energy of only $1 \sim 10\,\mathrm{keV}$ and it is not possible to apply a conventional coincidence technique.

The experiment usually exposes the target material for a few weeks, depending on the half-life of atom B. It is thus not possible to observe the possible short-time variation of the solar neutrinos, such as day-night effects.

Also note that radiochemistry is just used to identify the B atoms and to count their number; thus, one obtains no information concerning the neutrino energy spectrum, except for the threshold energy (minimum detectable energy) of neutrinos, ($E_{\nu,min}$), related to the reaction (3.1)

$$E_{\nu,min} = M_B - M_A, \tag{3.3}$$

where $M_B(M_A)$ stands for atomic mass of element $B(A)$.

A crucial point concerning the radiochemical experiment is to select a variety of target elements (A in eq.(3.1)) so that it is possible to observe solar neutrinos from various sources by selecting suitable threshold energies ($E_{\nu,min}$).

^{37}Cl and ^{71}Ga have been successfully used for solar-neutrino observations. R. Davis Jr. first applied the radiochemical technique using ^{37}Cl in 615 t of perchlorethylene C_2Cl_4, and observed the capture reactions due to solar neutrinos which originate mainly from ^{8}B β-decay (eq.(1.4)) [29]. The experiment has been carried out 4,000 mwe underground in the Homestake gold mine. Hereafter, Davis's pioneering work is called the Homestake experiment. It has revealed a deficit in the observed neutrino capture rate after a few years from the initial operation in the mid-1960's. This solar-neutrino problem has since then evolved as one of the central issues of particle physics, i.e. neutrino masses which ought to be related to new physics beyond the standard model.

The radiochemistry technique is unfamiliar to particle physicists. They therefore sometimes are suspicious concerning the results of the Homestake experiment, especially the way of extracting with high efficiency a few atoms out of several hundred tons of liquid.

In 1991-1992 two new radiochemical experiments became operational: they are the long-awaited ^{71}Ga experiments, called SAGE [30] and GALLEX [24].

SAGE is located about 4,700 mwe underground in the Baksan Neutrino Observatory, Russia, and GALLEX is located 4,000 mwe underground in the Gran Sasso Laboratory, Italy. Their technique is essentially the same as that which Davis developed in the Homestake experiment. The important difference from an observational point of view, however, is that their detectors are sensitive to *pp* neutrinos, whose predicted flux has much less uncertainty than that of ^{8}B neutrinos to which the Homestake experiment has the highest snsitivity. Therefore, if they observed a deficit in the neutrino capture rate, it would indicate clear evidence for the particle-physics nature of the solar neutrino problem, namely something peculiar is happening in the neutrino sector. Indeed, both experiments obtained results that showed capture rates lower than the theoretical prediction (but not very significantly lower like the Homestake or Kamiokande results).

The Homestake group is considering a new element, ^{127}I, which could be filled in the existing reaction tank[32]. Its reaction product, ^{127}Xe (see eq.(3.7)), is an inert gas, like ^{37}Ar, the end product from ^{37}Cl. Therefore, the same extraction technique is applicable here.

One could also envisage the possibility of using ^{81}Br [33]. Since neutrino capture converts it into ^{81}Kr (eq.(3.6)), which is again an inert gas, the same technique can be used.

The underlying reactions for ^{37}Cl, ^{71}Ga, ^{81}Br and ^{127}I targets are:

$$\nu_e + {}^{37}\mathrm{Cl} \rightarrow e^- + {}^{37}\mathrm{Ar}, \tag{3.4}$$

$$\nu_e + {}^{71}\mathrm{Ga} \rightarrow e^- + {}^{71}\mathrm{Ge}, \tag{3.5}$$

$$\nu_e + {}^{81}\mathrm{Br} \rightarrow e^- + {}^{81}\mathrm{Kr} \tag{3.6}$$

and

$$\nu_e + {}^{127}\mathrm{I} \rightarrow e^- + {}^{127}\mathrm{Xe}. \tag{3.7}$$

The products ^{37}Ar, ^{71}Ge, ^{81}Kr and ^{127}Xe all decay by electron capture. Their half-lives are 35.0 d, 11.43 d, 2×10^5 yr and 36.4 d, respectively.

We will not discuss ^{81}Br, since an experiment with ^{81}Br will not be realized in the near future. There is a difficulty of counting ^{81}Kr atoms. The long half-life of ^{81}Kr prevents one from adopting a conventional counting technique with a proportional counter, because atoms do not simply decay within a measuring time of about several weeks.

Table 10 lists some characteristics of ^{37}Cl, ^{71}Ga and ^{127}I relevant to solar neutrino detection.

Table 10: Characteristics of the elements ^{37}Cl, ^{71}Ga, ^{81}Br and ^{127}I for the radiochemical observation of solar neutrinos

	$^{37}_{17}$Cl	$^{71}_{31}$Ga	$^{81}_{35}$Br	$^{127}_{53}$I
Average Mass Number	35.45	69.72	79.9	126.90
Abundance Ratio in Number(%)	24.47	39.8	49.3	100
J^π, $T^{(a)}$	$\frac{3}{2}^+, \frac{3}{2}$	$\frac{3}{2}^-$	$\frac{3}{2}^-$	$\frac{5}{2}^+$
ν_e Absorption Threshold Energy (MeV)	0.814	0.233	0.470	0.664
Reaction Product	^{37}Ar	^{71}Ge	^{81}Kr	^{127}Xe
Ground State $J^\pi, T^{(a)}$	$\frac{3}{2}^+, \frac{1}{2}$	$\frac{1}{2}^-$	$\frac{7}{2}^+$	$\frac{1}{2}^+$
Half-life (days)	35.0	11.43	2.1×10^5 yr	36.4
Important Transition Level (J^π, T, Energy (MeV))	$\frac{3}{2}^+, \frac{3}{2}$, $4.993^{(b)}$	ground$^{(c)}$	$\frac{1}{2}^-$ 0.191	$\frac{3}{2}^+$ $0.125^{(d)}$
Auger Electron and X-ray$^{(f)}$ Energy (keV) (=Binding Energy)	2.82 (K) 0.27 (L)	10.4 (K) 1.3 (L) $EC(L/K)$ =0.12	13.4 (K) 1.8 (L) $EC(L/K)$ =0.146	33.2 (K) 5.2 (L)$^{(e)}$ $EC(K/L+M\cdots)$ =7 for 0.203 MeV γ $EC(K/L+M\cdots)$ =4.5 for 0.375 MeV γ
$\log ft$ (s)	5.10	4.37	4.88	not measured for $\frac{3}{2}^+ \rightarrow \frac{5}{2}^+$
Observable neutrino source	^{7}Be, ^{8}B	pp, ^{7}Be, ^{8}B and others	^{7}Be, ^{8}B	^{7}Be, ^{8}B

(a) J^π, spin-parity ; T, isospin.
T is only useful for the Fermi transition.
(b) Isotopic analog state (IAS) of ^{37}Cl.
(c) IAS of ^{71}Ga lies at 8.957 MeV which is above the threshold of particle emission (7.2 MeV).
(d) J^π of other states are not well known.
(e) Electron capture from ^{127}Xe ground state proceeds to excited states ($3/2^+, 1/2^+$) of ^{127}I with the ralative ratio of 54:46.
(f) Binding energy of L is for the L_I shell.

3.2 Cross sections

As described above, the capture rate is the only measurable quantity in the radiochemical experiment. Thus, the cross section of reaction (3.1) must be precisely known. Since complex nuclei are used for the target, any estimate of the cross sections is somewhat uncertain if excited states play a significant role.

We first formulate the cross section for reaction (3.1). The effective Lagrangian density is written as

$$L_{eff} = \frac{G}{\sqrt{2}} \overline{\psi}_B \gamma_\mu (1-\gamma_5) \psi_A \cdot \overline{e} \gamma^\mu (1-\gamma_5) \nu \,, \tag{3.8}$$

where ψ_B, ψ_A, e and ν are fields of nucleus B, A, electron and neutrino, respectively, and $G = G_F \cos\theta_c$, θ_c being the Cabibbo angle ($\cos\theta_c = 0.974 \pm 0.001$). Eq.(3.8) can be approximated at low energies as

$$L_{eff} = \frac{G}{\sqrt{2}} \left[\psi_B^\dagger \psi_A \cdot e^\dagger \nu - \psi_B^\dagger \boldsymbol{\alpha} \psi_A \cdot e^\dagger \boldsymbol{\alpha} \nu + \frac{g_A}{g_V} (\psi_B^\dagger \boldsymbol{\sigma} \psi_A \cdot e^\dagger \boldsymbol{\sigma} \nu - \psi_B^\dagger \gamma_5 \psi_A \cdot e^\dagger \gamma_5 \nu) \right] .$$

But the terms containing $\boldsymbol{\alpha}$ and γ_5 are on the order of $v/c \approx 0.1$, v being the nucleon velocity in a nucleus, and can be neglected compared with the other terms. Then

$$L_{eff} = \frac{G}{\sqrt{2}} \left(\psi_B^\dagger \psi_A e^\dagger \nu + \frac{g_A}{g_V} \psi_B^\dagger \boldsymbol{\sigma} \psi_A \cdot e^\dagger \boldsymbol{\sigma} \nu \right) , \tag{3.9}$$

and the space component of the fields ($\exp(i\boldsymbol{p}\cdot\boldsymbol{r})$) can be set to 1. The total cross section for (3.1) can now be expressed with a standard technique as

$$\sigma = \frac{G^2}{\pi} \left[\langle 1 \rangle^2 + \left(\frac{g_A}{g_V} \right)^2 \langle \boldsymbol{\sigma} \rangle^2 \right] p_e E_e \,, \tag{3.10}$$

where p_e and E_e are the momentum and total energy of the electron. $\langle 1 \rangle^2$ and $\langle \boldsymbol{\sigma} \rangle^2$ are defined as

$$\langle 1 \rangle^2 = \left| \int \psi_B^\dagger \psi_A \right|^2 = \left| \int \psi_B^\dagger \sum_{i=1}^{A} \tau_+^{(i)} \psi_A \, d\boldsymbol{x}_1 \cdots d\boldsymbol{x}_A \right|^2 , \tag{3.11a}$$

where the summation extends over all nucleons (the mass number$=A$),

$$\langle \boldsymbol{\sigma} \rangle^2 = \left| \int \psi_B^\dagger \boldsymbol{\sigma} \psi_A \right|^2 = \sum_{k=1}^{3} \sum_{J_3(B)=1}^{2J_B+1} \left| \int \psi_B^\dagger \sum_{i=1}^{A} \tau_+^{(i)} \sigma_k^{(i)} \psi_A d\boldsymbol{x}_1 \cdots d\boldsymbol{x}_A \right|^2 , \tag{3.11b}$$

In the above the wave functions are normalized such that $|\int \psi^\dagger \psi|^2 = 1$, J_B denotes the spin of the final state B (we will use J_A similarly for the initial state A), and $J_3(B)$ is the third component of $\boldsymbol{J}_B$. $\tau_+^{(i)}$ is the + component ($\tau_+ = \tau_1 + i\tau_2$) of the isospin operator of the i-th nucleon (mass number of the nucleus $= A$), and $\sigma_k^{(i)}$ is the corresponding spin operator. $\langle 1 \rangle^2$ is called the Fermi strength, and can be calculated provided that A and B are in a definite isospin state T (selection rule, $\Delta T = 0$ and $J_A = J_B$). Namely, since

$$T_+ = \sum_{i=1}^{A} \tau_+^{(i)} = T_1 + iT_2 \,,$$

$$\begin{aligned} \langle 1 \rangle^2 &= |\langle B|T_+|A\rangle|^2 = \langle A|T_-T_+|A\rangle \\ &= T(T+1) - T_3(A)^2 - T_3(A) \\ &= T(T+1) - T_3(B)T_3(A). \end{aligned} \quad (3.12)$$

The total Fermi strength summed over all ground and excited states is

$$\sum_{all} \langle 1 \rangle^2 = N_A - Z_A \,, \quad (3.12a)$$

where N_A and Z_A are the number of neutrons and protons in nucleus A, respectively. $\langle \boldsymbol{\sigma} \rangle^2$ is called the Gamow-Teller (GT) strength and cannot be calculated reliably. The selection rule for the GT transition is $J_B - J_A = \pm 1, 0\, (0 \not\to 0)$.

For a transition from ground state to ground state, one can obtain the cross section from the β-decay rate (or electron capture rate) of nucleus B. The decay rate of $B \to A + e^+ + \nu_e$ is calculated in a straightforward way as follows:

$$\Gamma = \frac{G^2}{2\pi^3} \left[\langle 1 \rangle^2 + \left(\frac{g_A}{g_V} \right)^2 \langle \boldsymbol{\sigma} \rangle_A^2 \right] f, \quad (3.13)$$

where

$$f = \int_{m_e}^{E_{max}} F(-Z_A, E_e) p_e E_e (E_{max} - E_e)^2 \, dE_e \,, \quad (3.13a)$$

Z_A being the atomic number of nucleus A. Note that the GT strength in (3.13) is slightly different from (3.11b), such that the sum over $J_3(B)$ is replaced by the sum over $J_3(A)$, and the two strengths are related as

$$(2J_A + 1)\langle \boldsymbol{\sigma} \rangle_A^2 = (2J_B + 1)\langle \boldsymbol{\sigma} \rangle^2. \quad (3.14)$$

With the half-life of B, defined as $t = ln2/\Gamma$, the cross section (σ) is expressed as

$$\sigma = \frac{2J_B + 1}{2J_A + 1} \frac{2\pi^2\, ln2}{(ft)} p_e E_e F(Z_B, E_e). \quad (3.15)$$

The ft value is usually given in units of sec, namely

$$ft = \frac{6163.4 \pm 3.8}{\langle 1 \rangle^2 + (g_A/g_V)^2 \langle \boldsymbol{\sigma} \rangle_A^2} \ (\text{sec}). \quad (3.16)$$

The cross section is now [34]

$$\sigma = \sigma_0 \left(\frac{E_e}{m_e} \right)^2 G(Z_B, E_e) \,, \quad (3.17)$$

where

$$\sigma_0 = \frac{1.206 \times 10^{-42}}{(ft)_{B \to A}} \frac{2J_B + 1}{2J_A + 1} Z_B \ (\text{cm}^2) \quad (3.17a)$$

and

$$G(Z_B, E_e) = \frac{F(Z_B, E_e)}{2\pi\alpha Z_B E_e/p_e} = \frac{F(Z_B, E_e)}{2\pi\nu} \sim \frac{1}{1 - e^{-2\pi\nu}}, \tag{3.17b}$$

Z_B being the atomic number of B. For electron-capture decay, $B + e^- \to A + \nu_e$, the decay rate is calculated in a way similar to eq.(3.10); namely, for K-electron capture,

$$\Gamma_{ec} = \frac{G^2}{\pi}\left[\langle 1\rangle^2 + \left(\frac{g_A}{g_V}\right)^2 \langle \boldsymbol{\sigma}\rangle_A^2\right] E_\nu^2 \, |\psi_K(r=0)|^2 \, \frac{N_K}{2}$$

$$= \frac{G^2}{2\pi^3}\left[\langle 1\rangle^2 + \left(\frac{g_A}{g_V}\right)^2 \langle \boldsymbol{\sigma}\rangle_A^2\right] f_K, \tag{3.18}$$

where ψ_K is the wave function of the K-electron, and N_K is the number of electrons in the K-shell ($N_K = 2$). Since $\psi_K(r)$ is to the first approximation given as

$$\psi_K(r) = \left(\frac{1}{\pi}\right)^{\frac{1}{2}} (\alpha Z_B m_e)^{\frac{3}{2}} \exp(-\alpha Z_B m_e r),$$

$$|\psi_K(0)|^2 = \frac{(\alpha Z_B)^3}{\pi} m_e^3 \tag{3.19}$$

and

$$f_K = 2\pi(\alpha Z_B)^3 E_\nu^2 m_e^3. \tag{3.20}$$

For the capture of an L-electron, ψ_K is replaced by ψ_L. Since $|\psi_L(0)|^2 = |\psi_K(0)|^2/n^3\,(n = 2)$, one expects, naively, that

$$\Gamma_{ec}(L)/\Gamma_{ec}(K) \approx 1/8. \tag{3.21}$$

This relation holds approximately, as can be seen in Table 10. The ft values for ^{37}Ar and ^{71}Ge are given in Table 10. For neutrino capture in ^{127}I one needs the ft value of ^{127}Xe decay $\left(\frac{3}{2}^+ \to \frac{5}{2}^+\right)$, which is not known, because the electron capture proceeds to the excited states of ^{127}I $\left(\frac{1}{2}^+ \text{ and } \frac{3}{2}^+\right)$.

For neutrinos of higher energies, especially for ^{8}B neutrinos, one has to take into account the excited states of the produced nucleus. Their ft values, of course, cannot be obtained with the above method. There are two ways to estimate the Fermi and GT strengths, namely the β decay of the isotopic analog of the relevant nucleus B in (3.1)[34][35] and the forward scattering cross section of the reaction $A(p,n)B$ in the energy range of the proton energy $E_p \approx 100 \sim 200$ MeV [34][36][37]. We discuss the cases for ^{37}Cl, ^{71}Ga and ^{127}I.

(i) ^{37}Cl

^{37}Cl is sensitive to ^{7}Be and ^{8}B neutrinos. Assuming an isospin invariance, the decay $^{37}\text{Ca}\left(J^\pi, T, T_3 = \frac{3}{2}^+, \frac{3}{2}, \frac{3}{2}\right) \to {}^{37}\text{K} + e^+ + \nu_e$ is the mirror reaction to $\nu_e + {}^{37}\text{Cl}\left(\frac{3}{2}^+, \frac{3}{2}, -\frac{3}{2}\right) \to e^- + {}^{37}\text{Ar}$. The excited states of ^{37}K in ^{37}Ca β^+ decay (except for the ground state and $\frac{1}{2}^+$ 1.37 MeV state) decay to ^{36}Ar by proton emission (Fig.

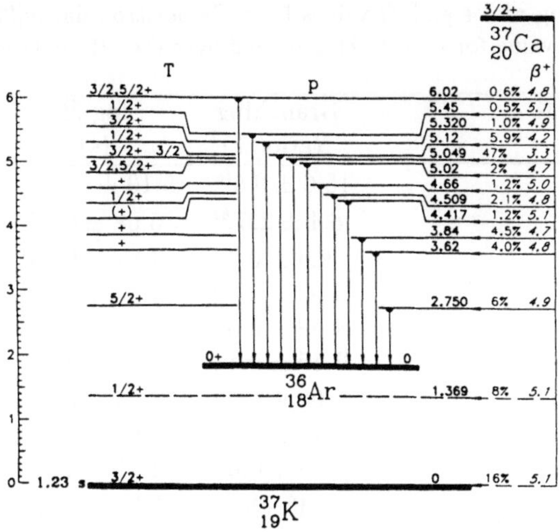

Figure 10: Decay scheme of ^{37}Ca to ^{37}K. Excited states of ^{37}K, except for the ground state and $\frac{1}{2}^+$ 1.369 Mev decay to ^{36}Ar by proton emission. (From [37])

10). The measured branching ratio to each level, together with its excitation energy determined from the measured proton energy, directly provides the relative ft value. Its absolute value is determined assuming $\log ft = 3.30$ for the super-allowed transition ($^{37}\mathrm{Ca}(\beta^+)\,^{37}\mathrm{K}\left(\frac{3}{2}^+, \frac{3}{2}, 5.049\ \mathrm{MeV}\right)$, the transition to the isotopic analog state), which is dominated by the Fermi transition of strength $\langle 1\rangle^2 = 3$ (see eq.(3.1) and (3.16)). A systematic measurement has been carried out by Sextro et al. for the reaction $^{36}\mathrm{Ar}(^3\mathrm{He}, 2n)\,^{37}\mathrm{Ca}$ at an ^{3}He energy of 40 MeV [38]. The result is given in Table 11.

We next consider the forward (p, n) reaction. Goodman *et al.* showed that $d\sigma(0°)/d\Omega$ of the (p, n) reaction for proton energies of $100 \sim 200$ MeV is proportional to the β-decay strength [39],

$$\frac{d\sigma(0°)}{d\Omega} = \left(\frac{E_i E_f}{\pi^2}\right)\frac{k_f}{k_i}\left\{N_\tau^D J_\tau^2\langle 1\rangle^2 + N_{\sigma\tau}^D J_{\sigma\tau}^2\langle\boldsymbol{\sigma}\rangle^2\right\}, \tag{3.22}$$

where E_i, k_i and E_f, k_f are CM energy and momentum of the proton and neutron, respectively, N_τ^D and $N_{\sigma\tau}^D$ are calculable distortion factors in the distorted-wave approximation, and J_τ and $J_{\sigma\tau}$ are proportional constants to be determined experimentally. The results of their measurements on several nuclei are shown in Fig. 11, in which the quantity K is defined as

$$\left[\frac{E_i E_f}{\pi^2}\frac{k_f}{k_i}N_\alpha^D\right]^{-1}, \tag{3.23}$$

N_α^D being N_τ^D or $N_{\tau\sigma}^D$ for a pure Fermi or GT transition. K is now plotted against $\langle 1\rangle^2$ or $\langle\boldsymbol{\sigma}\rangle^2$ deduced from the known β-decay. For a mixed state of Fermi and GT

Table 11: Branching ratios and ft values for ^{37}Ca positron dacay[38]. The last column lists the calculated value for each level averaged over the ^{8}B neutrino spectrum [34].

Energy in ^{37}K (keV)	J^π	Branching Ratio(%)	$\log ft$ (s)	$\langle\sigma_{\rm Cl}(^8{\rm B})\rangle$ (10^{-46}cm^2)
0	$\frac{3}{2}^+$	$15.6 \pm 0.5^{(a)}$	$5.10 \pm 0.01^{(a)}$	596
1,368	$\frac{1}{2}^+$	$8.4 \pm 2.5^{(b)}$	5.08 ± 0.13	418
2,750	$\frac{5}{2}^+$	6.4 ± 0.7	4.87 ± 0.05	429
3,617		4.02 ± 0.39	4.83 ± 0.04	354
3,840		4.48 ± 0.32	4.72 ± 0.03	405
4,417		1.15 ± 0.14	5.12 ± 0.05	126
4,503	$\frac{1}{2}^+$	2.08 ± 0.20	4.84 ± 0.04	231
4,665		1.15 ± 0.46	5.04 ± 0.18	135
5,018	$(\frac{3}{2},\frac{5}{2})^+$	1.9 ± 0.6	4.71 ± 0.14	249
5,047	$\frac{3}{2}^+, T=\frac{3}{2}$	46.74	3.30 (input)	6,375
5,117	$\frac{1}{2}^+$	5.9 ± 0.5	4.17 ± 0.04	799
5,318	$\frac{3}{2}^+$	1.05 ± 0.27	4.85 ± 0.12	153
5,449	$\frac{1}{2}^+$	0.51 ± 0.19	5.11 ± 0.17	78
6,016	$(\frac{3}{2},\frac{5}{2})^+$	0.65 ± 0.09	4.77 ± 0.06	126
			total =	1.05×10^4

(*a*) From mirror decay.
(*b*) Calculated from the sum of all meaured branching ratios and that inferred for the ground-state decay.

transitions, one takes the effective strength

$$\langle eff\rangle^2 = \langle\boldsymbol{\sigma}\rangle^2 + \frac{N_\tau^D}{N_{\sigma\tau}^D}\left(\frac{J_\tau}{J_{\sigma\tau}}\right)^2 \langle 1\rangle^2 , \tag{3.24}$$

using the extracted values of J_τ and $J_{\sigma\tau}$ from the pure states.

What is important in this analysis is that:

(a) There seems to be universal values for J_τ and $J_{\sigma\tau}$ independent of nuclear species, though error bars are large;

(b) Once J_τ and $J_{\sigma\tau}$ are fixed, one is able to estimate the GT strength, assuming that the Fermi strength is given by eq.(3.12).

The measurement for ^{37}Cl$(p,n)^{37}$Ar as carried out by Rapaport et al. [40]. the relative yields of neutrons are shown as a function of the excitation energy in ^{37}Ar in Fig. 12. the incident proton energy was 120 MeV. the resultant ft values are given in Table 12 assuming a value $J_{\sigma\tau}$ = 168 MeV$\cdot fm^3$ and $N_{\sigma\tau}^D = 0.298$, which were derived as shown in Fig. 11 [39]. These values should be compared with those in Table 11, which were deduced from ^{37}Ca β^+ decay. They are in general agreement for energies larger than 5 MeV, though the situation at lower energies are somewhat problematic. Note that the state at 7.65 MeV, which is

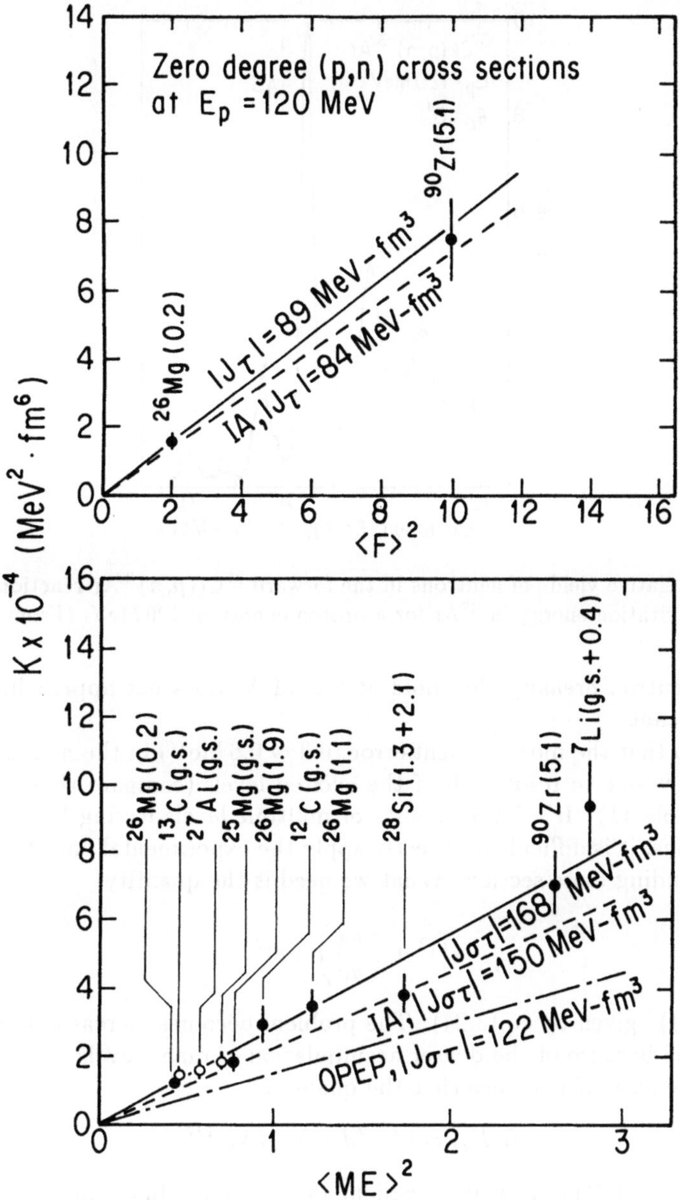

Figure 11: K vs. the strength, $\langle 1 \rangle^2$ or $\langle \sigma \rangle^2$, for a pure Fermi or GT transition. For a mixed state the strength $\langle eff \rangle^2$ is used assuming J_τ and $J_{\sigma\tau}$ shown in the figure. The transitions are labeled by specifying the target nucleus with the excitation energy in the final nucleus given in parentheses. The solid curves represent the fits to ^{26}Mg data. IA and OPEP stand for the impulse approximation and one-pion exchange, respectively. (From [39])

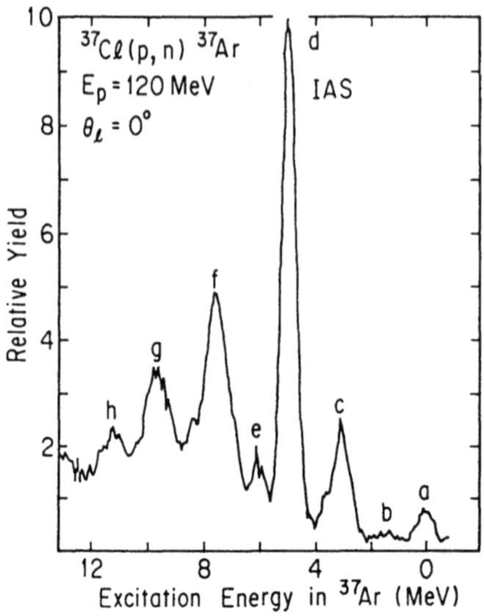

Figure 12: Relative yields of neutrons in the forward $^{37}Cl\,(p,n)\,^{37}Ar$ reaction as a function of the excitation energy in ^{37}Ar for a proton energy of 120 MeV. (From [40])

below the neutron breakup threshold at 8.79 MeV, does not appear in Table 11, but is significant.

Note also that the measurement error ($0.3 \sim 0.6$ MeV) in the neutron energies does not allow one to resolve all of the excited levels (compare these with levels shown in Table 11). If a level consists of multiple levels having both Fermi and GT transitions, it is difficult to directly apply the experimental data to evaluating the corresponding cross section. What we need is the quantity

$$\langle 1 \rangle^2 + \left(\frac{g_A}{g_V}\right)^2 \langle \boldsymbol{\sigma} \rangle^2,$$

and not $\langle eff \rangle^2$ given by eq.(3.24). The problem becomes increasingly serious for heavier nuclei because of the denser accumulation of many levels.

There is empirical evidence that the quantity

$$R(E_p) = |J_{\sigma\tau}/J_\tau|(N_{\sigma\tau}/N_\tau)^{1/2} \qquad (3.25)$$

appearing in eq.(3.24) is mass-independent and linear as a function of the bombarding energy, at least for light elements [41]. E_p is the incident proton Lab-energy. Figure 13 shows the $R(E_p)$ as a function of E_p for the six elements with known GT strengths. Thus, at least for light elements, the contribution from the Fermi transition in eq.(3.24) is about 1/8 of that from the GT transition at $E_p = 150$ MeV.

Figure 14 shows the low-lying levels in ^{37}Ar. Compare them with those in Table 11 [37].

The cross section is now calculated with a formula given by eq.(3.17). For ^8B neutrinos one averages the cross section over the ^8B neutrino spectrum, as was

Table 12: Results of the reaction $^{37}Cl\,(p,n)\,^{37}Ar$ [40]. $T_p = 120\,MeV$, $L = 0$ transitions. Rows 1–8 correspond to peaks a–h in Fig. 12.

Energy in ^{37}Ar (MeV)	$\sigma_{CM}\,(0°)$ (mb/sr)	$\langle\boldsymbol{\sigma}\rangle^2$	log ft
ground state	0.20 ± 0.04	0.034 ± 0.007	5.06 ± 0.09
1.4	≤ 0.05	≤ 0.009	> 5.64
3.1	1.3 ± 0.2	0.23 ± 0.04	4.23 ± 0.08
$4.99^{(a)}$	4.8 ± 0.7	$0.08 \pm 0.06^{(b)}$	3.29 ± 0.02
6.1	0.53 ± 0.1	0.10 ± 0.02	4.60 ± 0.09
7.65	4.7 ± 0.7	0.87 ± 0.13	3.66 ± 0.06
9.65	3.3 ± 0.6	0.64 ± 0.12	3.79 ± 0.08
11.5	2.3 ± 0.4	0.47 ± 0.08	3.92 ± 0.08

(a) Isotopic analog state $\left(\frac{3}{2}^+, \frac{3}{2}\right)$.

(b) $\langle 1 \rangle^2 = 3$ was assumed for this transition.

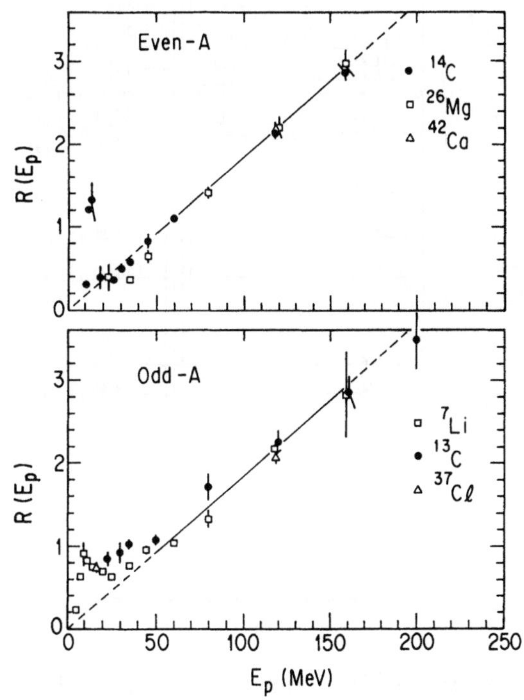

Figure 13: $R(E_p)$ for odd and even A targets. The solid line represents $R(E_p) = (54.9 \pm 0.9\,MeV)^{-1} E_p$, $E_p \geq 50\,MeV$. (From [41])

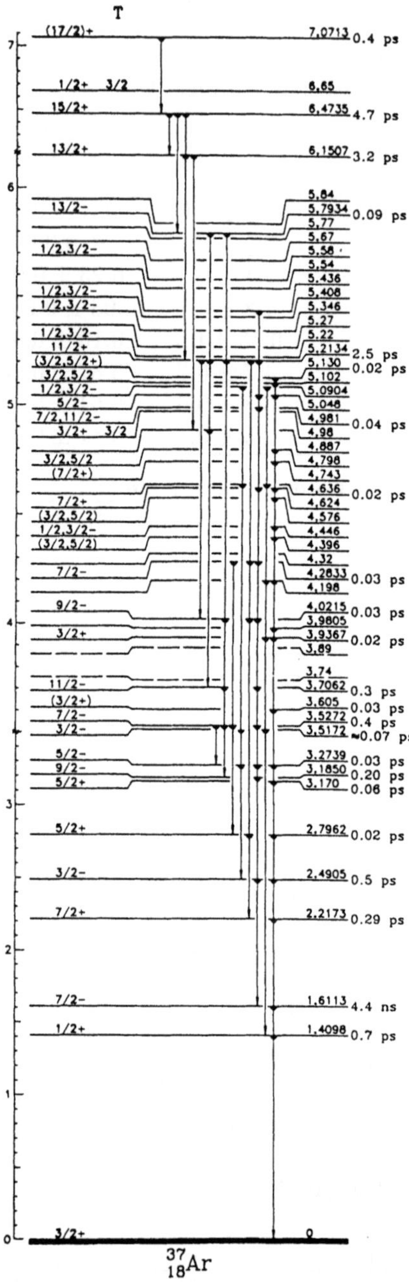

Figure 14: Level scheme of $^{37}_{18}$Ar. (From [37]). Compare this with Table 11.

Table 13: Neutrino capture cross sections averaged over the energy spectra. The units are $10^{-46}\,\mathrm{cm}^2$ [42]. Those in parentheses are the estimated 3σ-uncertainties.

Target	*pp*	*pep*	*hep*	^{7}Be	^{8}B	^{13}N	^{15}O
^{37}Cl	0	16	3.9×10^4	2.4	1.06×10^4	1.7	6.8
		(± 1)	(± 0.4)	(± 0.1)	(± 0.1)	(± 0.1)	(± 0.4)
^{71}Ga	11.8	215	7.3×10^4	73.2	2.43×10^4	61.8	116
	(± 0.8)	(± 47)	$(^{+6.9}_{-3.4})$	(± 6.6)	$(^{+2.14}_{-1.09})$	(± 5.6)	(± 22)
^{127}I $^{(c)}$	0			$\sim 20^{(a)}$	$\sim 2.2 \times 10^{4(b)}$		
				(large)	(large)		

(*a*) Assuming $ft = 10^5$ s for the lowest $\frac{3}{2}^+$ state of ^{127}Xe[43].
(*b*) Calculated in ref. [44].
(*c*) Calculations are not available in vacant positions.

done in chapter 2, namely

$$\langle \sigma_{\mathrm{Cl}}(^8\mathrm{B}) \rangle = \frac{\int_{0.814\,\mathrm{MeV}} \phi_{^8\mathrm{B}}(E_\nu) \sigma_{\mathrm{Cl}}(E_\nu)\, dE_\nu}{\int \phi_{^8\mathrm{B}}(E_\nu)\, dE_\nu} . \tag{3.26}$$

The individual cross section to each level is listed in the last column of Table 10 [30]. Summing all of the contributions, $\langle \sigma_{\mathrm{Cl}}(^8\mathrm{B}) \rangle = 1.05 \times 10^{-42}\,\mathrm{cm}^2$. This does not include the contribution from the excited level of 7.65 MeV found in the (p, n) reaction (Table 12).

Bahcall discussed the uncertainties in the derivation of this cross section, and took as the best estimated value

$$\langle \sigma_{\mathrm{Cl}}(^8\mathrm{B}) \rangle = (1.06 \pm 0.10) \times 10^{-42} \mathrm{cm}^2, \tag{3.27}$$

where the uncertainty is a three-sigma estimate. Note that it does not include a range of violation of the isospin invariance.

The cross sections for various sources of solar neutrinos are listed in Table 13 [42].

(ii) ^{71}Ga

The threshold energy of ^{71}Ga (0.233 MeV) is low enough to be sensitive to the *pp* neutrinos ($E_\nu \leq 0.42$ MeV). The neutrino capture reaction in ^{71}Ga is given by eq.(3.5). The level scheme of the reaction product ^{71}Ge is shown in Fig. 15. The ground-state-to-ground-state transition is possible (GT transition), and is the only transition for *pp* neutrinos. Its ft value is well determined, as is the average cross section for *pp* neutrinos $\langle \sigma_{\mathrm{Ga}}(pp) \rangle$.

For ^{8}B neutrinos, however, the situation is much worse than for ^{37}Cl. There is no precise information concerning the ft values for transitions from ^{71}Ga to excited states of ^{71}Ge. There is one measurement of the ^{71}Ga (p, n) ^{71}Ge reaction

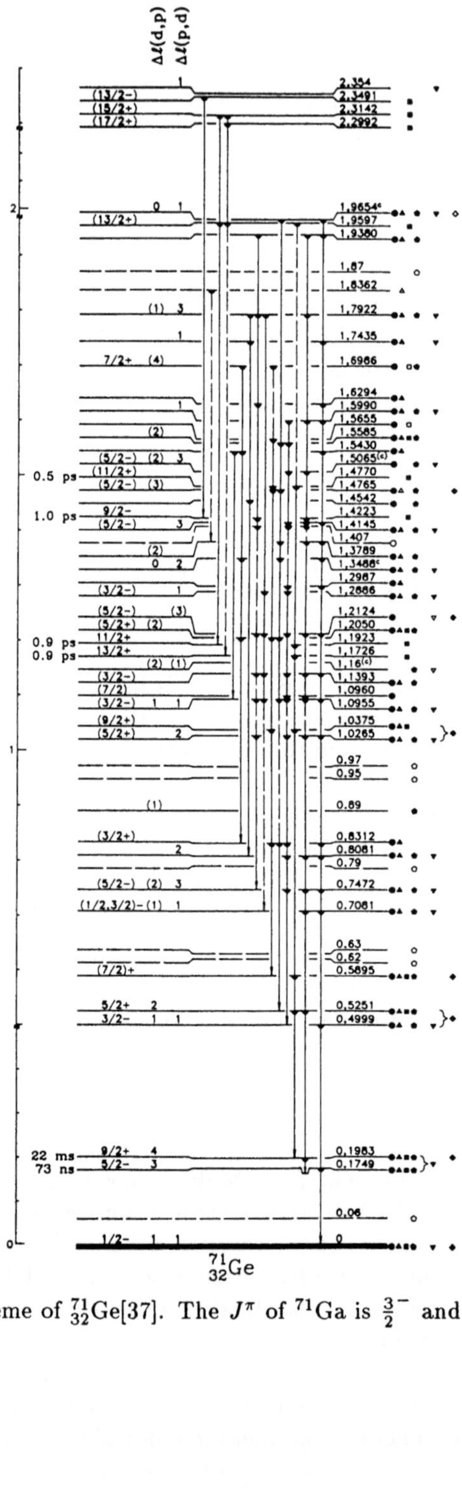

Figure 15: Level scheme of ${}^{71}_{32}$Ge[37]. The J^{π} of ^{71}Ga is $\frac{3}{2}^{-}$ and lies 0.233 MeV below ^{71}Ge.

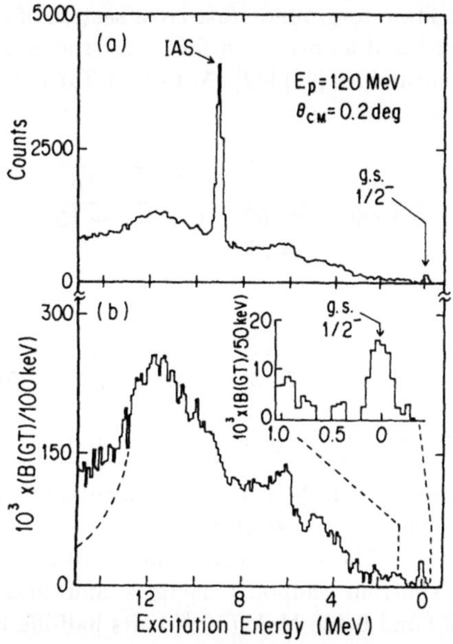

Figure 16: (a) Neutron spectrum in ^{71}Ga$\,(p,n)\,^{71}$Ge $E_p = 120\,$MeV. (b) GT strength in ^{71}Ge as a function of the excitation energy. The isotopic analog state (IAS) at 8.95 MeV is assumed to have the total Fermi strength ($N - Z = 9$, eq.(3.12a)) and has been subtracted. (From [40])

at proton energies of 120 MeV and 200 MeV. Figure 16 shows the GT strength as a function of the excitation energy. There is a strong isotopic analog state (IAS) at 8.95 ± 0.1 MeV, as shown in Fig. 16, which has been subtracted to obtain the GT strength, assuming that IAS has a full Fermi strength ($N - Z = 9$, see eq.(3.12a)). Note that since the particle-emission threshold is 7.4 MeV the GT transition above this energy and the Fermi transition to the IAS are irrelevant. The cross section of the neutrino capture reaction (eq.(3.5)) could be written as

$$\sigma_{\rm Ga} = \sigma_{0,\rm Ga} Q, \tag{3.28}$$

where $\sigma_{0,\rm Ga}$ represents the cross section for the transition to the ground state of ^{71}Ge and Q represents a factor which takes into account the contribution from all of the excited states below the particle-emission threshold (7.4 MeV). For neutrino sources with a continuous spectrum we take the usual average,

$$\langle \sigma_{\rm Ga} \rangle = \frac{\int \phi(E_\nu)\sigma_{\rm Ga}(E_\nu) dE_\nu}{\int \phi(E_\nu) dE_\nu}. \tag{3.29}$$

Table 14 lists $\langle \sigma_{\rm Ga} \rangle$ and Q for various sources of solar neutrinos.

Judging from the factor Q, most uncertainties in $\langle \sigma_{\rm Ga} \rangle$ come from the ^{8}B source. Since the relation between the (p,n) reaction and the corresponding GT strength is

Table 14: Average ground-state-to-ground-state cross section of $\nu_e + ^{71}\mathrm{Ga} \rightarrow e^- + ^{71}\mathrm{Ge}$, ($\langle\sigma_{\mathrm{Ga}}\rangle$) in units of $10^{-46}\mathrm{cm}^2$ and a correction factor for the excited states of ^{71}Ge (Q) for various sources of solar neutrinos [41] [42]. Note that Table 3.4 uses Q different from these.

	pp	*pep*	*hep*	^{7}Be	^{7}Be	^{8}B	^{13}N	^{15}O	^{51}Cr$^{(b)}$
E_ν(MeV)	0.420	1,442	19.80	0.862 89.7%	0.384 10.3%	~ 15	1,199	1,732	0.746 90% 0.426 10%
....	....	....	......	.....	.....	....	....	....	......
$\langle\sigma_{\mathrm{Ga}}\rangle$	11.8	17.0	5,440	74.3	23.2	2,850	58.1	100.1	56 (±6)
Q	1.00	1.27	13.4$^{(a)}$	1.065	1.00	10.9	1.08	1.23	1.06

(a) This was not given in ref. [41], but was calculated as the ratio of the value given in Table 3.4 to that given here.

(b) Calibratoin source. The cross secton has a 10% uncertainty (3σ estimate). The neutrino endpoint energies and branching ratios are 0.746 MeV (90 %) and 0.426 MeV (10 %). Its half-life is 28 days.

purely empirical, one cannot reliably estimate the uncertainties. Looking at Figs. 11 and 13, the experimental errors are at least about 20 ~ 30 %. Bahcall took an estimate of $^{+100}_{-50}$ % (3σ) for the cross sections derived from the (p,n) data [42] .The uncertainty in σ_0 was estimated to be ±6% (3σ), which comes from forbidden corrections [42].

(iii) ^{127}I

The threshold energy of the reaction (eq.(3.7)) is 0.664 MeV. This target is thus sensitive mainly to ^{7}Be and ^{8}B neutrinos (see Table 10). The level scheme is shown in Fig. 17 [43]. An important point is that the transition from ground-state-to-ground-state ($\frac{5}{2}^+ \rightarrow \frac{1}{2}^+$) is first forbidden. ^{7}Be neutrinos produce a transition to the first excited state ($\frac{1}{2}^+$, 0.125 MeV). For ^{8}B neutrinos, higher excited states are expected to be responsible. Unfortunately, there is no information available concerning the ft values for excited states of ^{127}Xe. Haxton[43] and Engel *et al.*[44] estimated the cross sections. Table 13 lists those for ^{7}Be and ^{8}B neutrinos. Obviously, one needs experimental data concerning the transition strengths to the excited states of ^{127}Xe: first of all, $d\sigma/d\Omega\,(0°)$ of $^{127}\mathrm{I}\,(p,n)\,^{127}\mathrm{Xe}$; then, to refine the (p,n) data with artificial neutrinos. Until a reliable estimate of the cross section is achieved, experiments with ^{127}I will not be very quantitative.

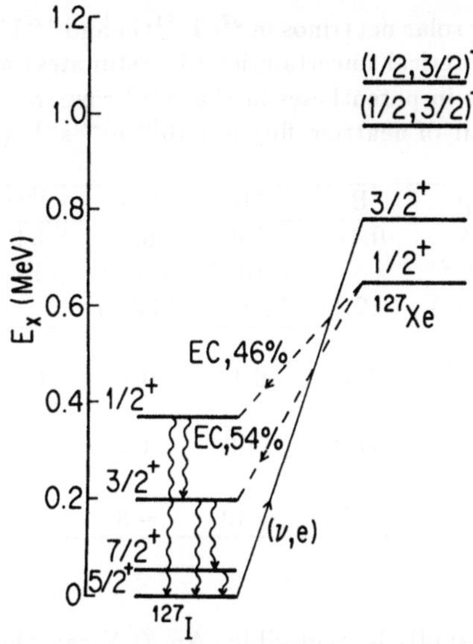

Figure 17: Level scheme of ^{127}I and ^{127}Xe. (From [43])

3.3 Expected neutrino capture rates

Once the average cross section for a capture reaction is known, it is straightforward to calculate the capture rate per target atom,

$$Rate = \sum_i \phi_i \langle \sigma_i \rangle, \tag{3.30}$$

where ϕ_i is the total neutrino flux on Earth from the i-th neutrino source and $\langle \sigma_i \rangle$ is the average cross section discussed in section 3.2. It is conventional to use 10^{-36} capture/s/atom as the unit of the capture rate, which is called SNU (solar neutrino unit). Table 15 presents a summary of the capture rates for ^{37}Cl, ^{71}Ga and ^{127}I for the neutrino fluxes calculated by Bahcall and Ulrich[42]. The underlying model is usually referred to as SSM (Standard solar model). One could also apply the fluxes calculated by different authors [45].

3.4 Signals from electron-capture decay

To actually count the number of produced atoms from a neutrino-capture reaction, one needs to know the properties of the signals from the electron-capture decay of the produced atoms. We have given in Table 10 the binding energies of K- and L_I-shell electrons which are to be liberated as Auger electron and/or X-ray.

K X-ray is emitted from a K-electron capture. Its energy is

$$E_K = E(K) - E(L_i), \tag{3.31}$$

where $E(K)$ is the binding energy of the K-shell electron etc. In this case the K hole is filled by an electron coming from the i-th L shell (i=II and III). The

Table 15: Capture rates for solar neutrinos in ^{37}Cl, ^{71}Ga and ^{127}I in units of SNU (10^{-36} capture/s/atom). The theoretical uncertainties (3σ estimates) are shown in the total capture rates. The numbers in parentheses in the third row are the 3σ uncertainties of the neutrino fluxes. The unit of neutrino flux is $\times 10^{10}$cm^{-2}s^{-1}. (From [42])

	pp	*pep*	^{7}Be	^{8}B	^{13}N	^{15}O	*hep*	Total
Flux	6.0	1.4 $\times 10^{-2}$	0.47	5.8 $\times 10^{-4}$	6.1 $\times 10^{-2}$	5.2 $\times 10^{-2}$	7.6 $\times 10^{-7}$	
	($\pm$0.12)	($\pm$0.07)	($\pm$2.1)	($\pm$3.1)	($\pm$3.0)			
Target								
^{37}Cl	0	0.2	1.1	6.1	0.1	0.3	0.03	7.9 $\pm$2.6
^{71}Ga	70.8	3.0	34.3	14.0	3.8	6.1	0.06	132 $^{+20}_{-17}$
^{127}I			~ 9	~ 13	~ 3			~ 25

contribution from higher shells is negligible. An L X-ray then follows to further fill the vacancy in the L shell.

The other relaxation mechanism is that an Auger electron is dominantly emitted from the L shell, instead of the K X-ray, when a vacancy in the K-shell is filled by an electron from the L shell. The Auger electron is associated with low-energy X-rays in order to fill two vacancies in the L-shell. The energy of the emitted Auger electron is approximately

$$E(KL_iL_j) = E(K) - E(L_i) - E^i(L_j) \qquad (3.32)$$

$$(i = I, II, III),$$

where $E^i(L_j)$ is the binding energy of the L_j shell in an atom with nuclear charge Z ionized in the L_i shell [46].

For example, a $KL_{II}L_{III}$ Auger electron from $^{37}_{18}$Ar electron-capture decay has an energy

$$\begin{aligned} E(KL_{II}L_{III}) &= 2.82 - 0.20 - 0.25 \\ &= 2.37\,(\text{keV}), \end{aligned} \qquad (3.33)$$

where $E^{II}(L_{III})$ is assumed to be $E(L_{III})$ of the $Z = 18$ atom.

Table 16 lists E_K, the binding energies of the K-, L_I- and the L_{II}-shell electrons, the intensity ratios of $K_{\alpha_2}(K - L_{II}$ transition) and $K_{\beta_1'}(K - M_{II,III,IV})$ to $K_{\alpha_1}(K - L_{III})$ and the fraction of K X-ray emission, i.e. $\omega_K = K$ X-ray/(K X-ray + Auger) [47].

Table 17 gives a crude estimate of the intensity ratios of the components in the KL_iL_j Auger electrons, together with the energies of the Auger electrons [47].

For an L-electron capture the Auger-electron emission is more frequent than the L X-ray emission; $1 - \omega_L \sim 0.85$ for $_{53}$I and $1 - \omega_L \geq 0.95$ for $_{17}$Cl and $_{31}$Ga [47].

Table 16: K X-ray energy (E_K), intensity ratios of $K_{\alpha_2}/K_{\alpha_1}$ (K_{α_1}, $K - L_{III}$ transition; K_{α_2}, $K - L_{II}$) and of $K_{\beta_1}'/K_{\alpha_1}$ (K_{β_1}', $K - M_{II,III,IV}$), and the fraction of K X-ray emission in electron-capture decay (ω_K), i.e. ω_K = K X-ray/(K X-ray + Auger electron) [47]. The last three columns list the binding energies of K-, L_I- and L_{II}- shell electrons, respectively (in units of keV) [46].

Element	E_K(keV)	$K_{\alpha_2}/K_{\alpha_1}$	$K_{\beta_1}'/K_{\alpha_1}$	ω_K	$E(K)$	$E(L_I)$	$E(L_{II})$
^{17}Cl	2.6	0.5	0.2	0.06	2.822	0.270	0.202
^{31}Ga	9.2	0.5	0.2	0.45	10.37	1.30	1.14
^{53}I	28.3	0.5	0.3	0.86	33.17	5.19	4.85

Table 17: Approximate intensity ratios of the components in the $K\,L_i\,L_j$ Auger electrons [47] and energies of Auger electrons in units of keV in the parentheses. Note, however, that the Auger electron is always associated with low-energy X-rays.

	^{17}Cl	^{31}Ga	^{53}I
$K\,L_I\,L_I$	1	1	1
	(2.2)	(7.7)	(23)
$K\,L_I\,L_{II}$	3	3	1
	(2.3)	(7.8)	(23)
$K\,L_I\,L_{III}$	0.5	0.5	1
	(2.3)	(7.8)	(23)
$K\,L_{II}\,L_{II}$	0.5	0.5	0.5
	(2.4)	(8.0)	(23)
$K\,L_{II}\,L_{III}$	6	6	3
	(2.4)	(8.0)	(23)
$K\,L_{III}\,L_{III}$	0.5	1	2
	(2.4)	(8.0)	(23)

Experimentally, one expects the following cases in observations:

(a) Both K X-ray and L X-ray are observed. The observed energy (E_{obs}) equals $E(K)$.

(b) The K X-ray escapes and the L X-ray is observed together with subsequent low-energy X-rays. E_{obs} equals $E(L)$.

(c) The KLL Auger electron plus low-energy X-rays are observed. E_{obs} equals $E(K)$.

(d) The LLL Auger electron plus low-energy X-rays are observed. E_{obs} equals $E(L)$.

Thus, the observation exhibits two peaks (called the K and L peaks) in the energy spectrum near to the K and L binding energies.

The relative intensity of the K and L peaks depends on the quantum efficiency of the proportional counter. It thus varies from experiment to experiment. Table 18 lists the mass-absorption coefficients and the mean free paths of X-rays in counter gases of Ar and Xe (at STP), where the X-rays are those produced by deexcitation of Cl, Ga and I after the electron-capture decay of Ar, Ge and Xe, respectively. One can see that the mean-free paths of the K X-rays are too long to give a reasonable quantum efficiency in a small proportional counter ($\sim 1\,\mathrm{cm}^3$). On the other hand, the L X-rays are detectable with high efficiency (possibly except I). Thus, out of the four cases listed above, (b), (c) and (d) are usually realized.

For a Cl experiment, the Auger-electron emission dominates, and the L X-ray energy is below the detection limit. Thus, only case (c) is relevant. One observes an energy deposit of 2.82 keV.

For the Ga experiment, since $\omega_K = 0.45$ and $EC(K)/EC(L) = 8$, one expects 48% for case (c), 39% for case (b) and 13% for case (d). The ratio of the K- and L- peaks are approximately 48:52.

For the I experiment the Auger-electron emission is small (14%). Moreover, since the mean-free path of the L X-ray is 5 mm, the quantum efficiency is about 0.6 for a typical path length of 5 mm in a counter. Thus, one expects that the detection efficiencies for the K- and L- peaks are approximately 14% and 64%, respectively, even if the counter efficiency is 100%.

Table 18: Mass absorption coefficients (μ) and mean-free paths (λ) of $K_{\alpha_1}(K - L_{III}$ transition) and K_{β_1} ($K - M_{III}$ transition) X-rays in counter gases of Ar and Xe (at STP). The X-rays are those produced by deexcitation of Ar, Ge and Xe, respectively.

Element	Energy of X-rays (keV)	in Ar gas $\mu(\mathrm{cm}^2/\mathrm{g})$	in Ar gas λ (cm)	in Xe gas $\mu(\mathrm{cm}^2/\mathrm{g})$	in Xe gas λ (cm)
Cl	$2.6(K_{\alpha_1})$	316	1.8		
	$0.27(K_{\beta_1})$	4.2×10^4	0.013		
Ga	$9.2(K_{\alpha_1})$			200	7.8
	$1.2(K_{\beta_1})$			2.8×10^4	0.0061
I	$28.3(K_{\alpha_1})$			4.9	35
	$4.3(K_{\beta_1})$			340	0.5

3.5 On-going experiments

Three radiochemical experiments are currently operational, as briefly described in section 3.1. They are the Homestake[29], SAGE[30] and GALLEX[31] experiments.

3.5.1 Experiments

The experimental details are summarized in Table 19. However, additional explanations are in order:

(1) The capture rates expected from SSM are $1 \sim 2$ captures/day (9th column).

(2) The yield of the reaction products (N) is given by

$$N = \tau r \left(1 - e^{-\frac{T}{\tau}}\right) , \tag{3.34}$$

here τ, r and T are the decay lifetime of the produced atoms (half-life $t_{1/2} = \tau \ln 2$), the capture rate and the exposure time, respectively. Thus, the exposure time is optimized at $(1 \sim 2)\tau = (1.4 \sim 3)t_{1/2}$ (15th column).

(3) The produced atoms are efficiently extracted by means of the radiochemical technique (12th column). The procedures are as follows:

(i) Homestake [29]

(a) After exposure, the C_2Cl_4 solution is bubbled by 1 tank-volume of He gas ($4 \cdot 10^5\,\ell$).

(b) The gas is passed through a condenser at $-32°C$ and a molecular sieve at room temperature to trap evaporated C_2Cl_4.

(c) The gas is passed through a charcoal trap at liquid-nitrogen temperature to trap argon.

(d) The argon is extracted from a heated charcoal trap and purified; after its volume is measured, it is loaded into a small proportional counter with tritium-free methane.

(ii) SAGE [30][48]

(a) Metallic gallium is stored in $5 \sim 10$ vessels (6 t Ga each) just above the melting temperature (29.8°C). After exposure, an extraction solution (1 kg HCl, 5.2 kg H_2O_2, 68.8 kg H_2O, 70 ℓ in total) is added to each reactor, and then stirred thoroughly.

(b) After a few more processes the extraction solution is siphoned away from each reactor, and produced Ga atoms are concentrated by vacuum distillation.

(c) The argon flow sweeps the Ge as $GeCl_4$ from the concentrated solution into 1 ℓ H_2O.

(d) $GeCl_4$ is further concentrated and converted to GeH_4, which is then led to a proportional counter with additional xenon gas (13th and 14th columns). The entire process is more complicated than that for Homestake, because the product (Ge) is not an inert gas.

(iii) GALLEX [48]

(a) A $GaCl_3$ solution is bubbled by 1,900 m^3 N_2 gas for 20 hours.

(b) $GeCl_3$, which is formed by the carrier Ge or ^{71}Ge produced by the neutrinos, is recovered by scrubbing the gas with 50 ℓ H_2O in three large absorber columns.

(c) The aqueous solution containing $GeCl_4$ is concentrated into 1 ℓ H_2O.

(d) $GeCl_4$ is further concentrated and converted to gas as GeH_4, which is then led to the proportional counter with additional xenon gas (13th and 14th columns). The process after (c) is the same as that for SAGE.

(4) In order to check the extraction efficiency of Ar/Ga, carrier agents are added to the target material before exposure. The amount of carriers is adjusted so as to be compatible with the volume of the proportional counter (10th column). Different

isotopes are alternately used in order to avoid any possible contamination from the previous run (not for SAGE).

Table 19: Parameters of the three radiochemical experiments: Homestake, SAGE and GALLEX.

	1	2	3	4	5	6	7	8
	Underground Depth(mwe)	First Operation (year)	Target					
			Material	Mass(t)	Principal Element	Mass(t)	Target Element	Number of Atoms
Homestake	4,000	1969	C_2Cl_4 (liquid)	615	Cl	526	^{37}Cl	$2.17 \cdot 10^{30}$
SAGE	4,700	1990	Ga (metal)	57[(a)]	Ga	57	^{71}Ga	$1.96 \cdot 10^{29}$
GALLEX	4,000	1991	$GaCl_3$ +HCl +H_2O (liquid)	78.5 3.9 20.5	Ga	30.3	^{71}Ga	$1.04 \cdot 10^{29}$

(a) 30 t for initial runs.

	9	10	11	12	13	14
	Capture[(b)] Rate(day^{-1})	Carrier[(c)]	Amount	Typical[(d)] Extraction Efficiency(%)	Counting gas[(e)]	Proportional Counter Volume(cc)
Homestake	1.48	^{36}Ar,^{38}Ar (alterately)	0.2 cc	> 95%	Ar + CH_4 (93%) (7%)	0.3 – 0.5
SAGE	2.2	natural Ge (Ga-Ge alloy)	160 μg	80 ± 6	GeH_4+Xe (20%) (80%)	0.75
GALLEX	1.2	^{72}Ge,^{74}Ge,^{76}Ge (alternately) (H_2GeO_3 solution)	1 mg	92 ± 2	GeH_4+Xe (30 %) (70 %)	1

(b) Theoretical capture rates given in Table 15.

(c) Isotopes to calibrate the extraction efficiency as well as the counting gas. Different enrichments of isotopes are used to avoid contamination of remaining carriers from previous runs.

(d) Efficiency from the tank to the counter.

(e) The mixing ratio in parentheses is the volume ratio.

	15	16	17	18
	Typical Exposure Time(weeks)	Typical Counting Time(weeks)	Counter Background (day^{-1})	Counter Efficiency for K:L peaks (%)
Homestake	5 ∼ 9	8	0.023	46:0
SAGE	3 ∼ 4	2 ∼ 3	0.10[(f)]	37:34
GALLEX	3	7	0.15[(g)]	35.2:30.6

(f) K region only (^{68}Ge contamination included).

(g) K and L regions (^{68}Ge contamination included).

	19	20	21	22	23
	Check of Hot Chemistry	Extraction efficiency of ionized Atoms (%)	Background		
			Reaction[h]	Cosmic ray[i] μ (SNU)	Evironmental[i] Fast Neutrons (SNU)
Homestake	$C_2Cl_3\,^{36}Cl$, $^{36}Cl^+ \to ^{36}Ar^+$	100 ± 3	$^{37}Cl(p,n)^{37}Ar$	0.4 ± 0.16	small
SAGE	^{71}Ge	99^{+6}_{-8}	$^{71}Ga(p,n)^{71}Ge$	$< 3.3^{(g)}$	small
GALLEX	$^{71}As(EC)^{71}Ge^+$	99.8 ± 3.7	$^{71}Ga(p,n)^{71}Ge$	3.7 ± 1.1	0.15 ± 0.10
	$^{69}Ga(p,n)^{69}Ge^+$	> 99			

(h) Protons are produced by muons, fast neutrons or residnal radioactivities.

(i) Contribution to the Ar or Ge yield from cosmic-ray μ or fast neutrins.

(j) Ref.[50]

3.5.2 Counting

The proportional counters are small ($0.5 \sim 1$ cc) and have extremely low background, because the coincidence technique is not applicable here. Radon gas in the ambient air must be kept from leaking into the counters. The background counting rate is considerably lower for the Homestake counters than for the SAGE and GALLEX counters (17th column).

The counter pulses are subject to a rise-time cut in which only those pulses with a fast rise-time are selected. Figure 18 shows the amplitude of the differentiated pulse (ADP), which is inversely proportional to the rise-time, against the pulse height (energy) for the first 37 days counting of Run 83 of the Homestake experiment. Four events are seen in the region expected from ^{37}Ar [29]. Figure 19 shows the energy spectra of the observed events accumulated in the Homestake experiment. For the first 70-day data one can see a clear peak at 2.8 keV; however, the peak disappears in the second data taken 105-350 days from the end of exposure. Subtracting the second data from the first, one obtains a signal which is perfectly consistent with the Auger electrons from ^{37}Ar electron-capture decay. Figure 20 demonstrates that the observed events indeed decay with the expected half-life of 35 days above the very low background ($\sim$0.023 events/day). Thus, there is no doubt that the Homestake experiment did see ^{37}Ar extracted from the tank.

The SAGE experiment follows what the Homestake experiment developed.

The GALLEX experiment measures the pulse rise-time directly with a transient digitizer. The events observed in the GALLEX counters are displayed in Fig. 21 on the plane of the rise-time vs. energy. The black points are those located in the two regions expected for the K and L peaks. The background is obviously larger in the L-peak region. In any case, the accumulation of events in the allowed regions is clear. Figure 22 shows the energy spectra of the events observed in three

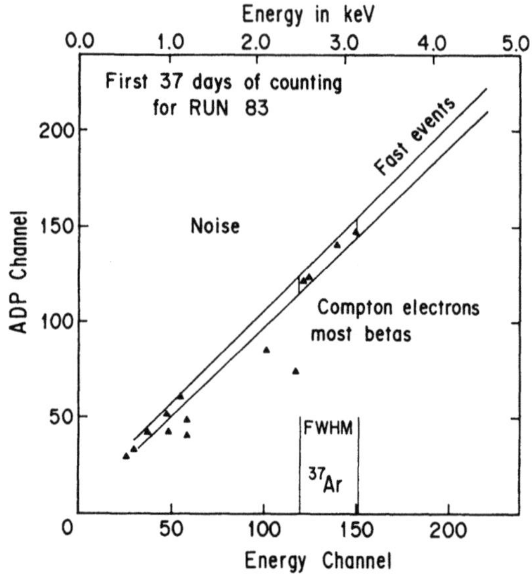

Figure 18: ADP(amplitude of differentiated pulse) versus the energy for events observed in the first 37 days of counting for Run 83 in the Homestake experiment. The pulse rise-time is inversely proportional to ADP. The energy range expected from Auger electrons of ^{37}A electron capture is shown, as well as a selected region of the rise-time. (From [29].

time periods. The first data sample has two clear peaks at the expected K and L peaks. They then decay along with the counting time. The decay feature of the data is demonstrated in Fig. 23. Thus, there is no doubt that GALLEX has observed ^{71}Ge produced in the reaction tank.

3.5.3 Background

The background considered here is ^{37}Ar/^{71}Ge produced by particles other than neutrinos. The only important reaction is ^{37}Cl$(p,n)\,^{37}$Ar or ^{71}Ga$(p,n)\,^{71}$Ge. Protons are produced by cosmic-ray muons still penetrating the rock overburden and by fast neutrons via (n,p) reactions in or near the tank. The background rate must be estimated and subtracted from the observed count rate to obtain the correct capture rate by the solar neutrinos.

We present how the three experiments estimated the background:

(i) Homestake [29]

Three 600 gallon tanks of C_2Cl_4 were exposed at three shallower depths and the ^{37}Ar production rate at each level was measured. Figure 24 shows the results obtained at three depths. The open circle at 4,000 hg/cm^2 (equivalent to mwe for the standard rock) is the estimated production rate scaled according to the muon intensity. An estimate of the production rate was also made by multiplying the measured muon intensity at the experimental site by the calculated ^{37}Ar yield per muon appropriate to the average muon energy. The result is in good agreement

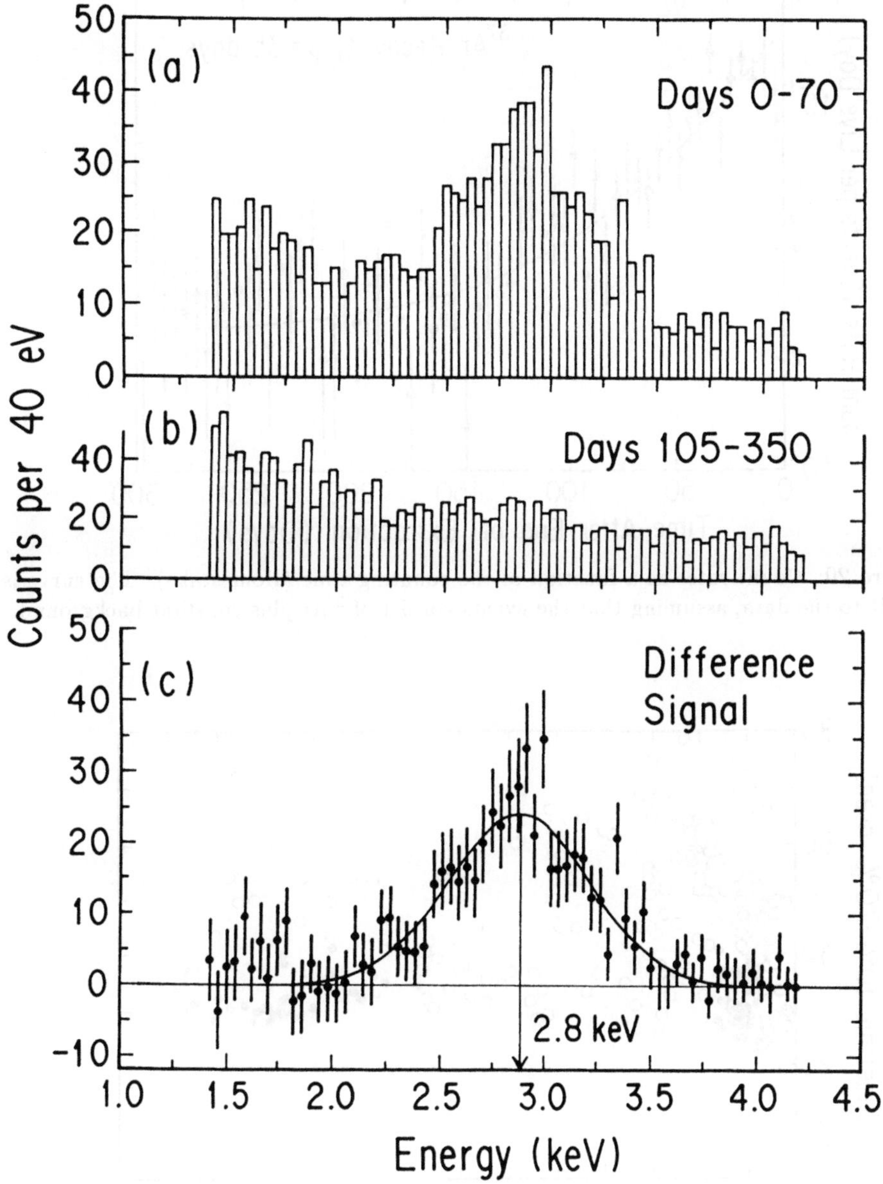

Figure 19: (a) Energy spectrum of all the events observed in the Homestake experiment for the first 70 days of counting. The accumulation of events is seen at around 2.8 keV. (b) Same as (a) for events observed in the 105–350 days of counting. Events near 2.8 keV have disappeared. (c) Energy spectrum of background subtracted events. The signal from Auger electrons of ^{37}Ar electron capture is clearly shown.

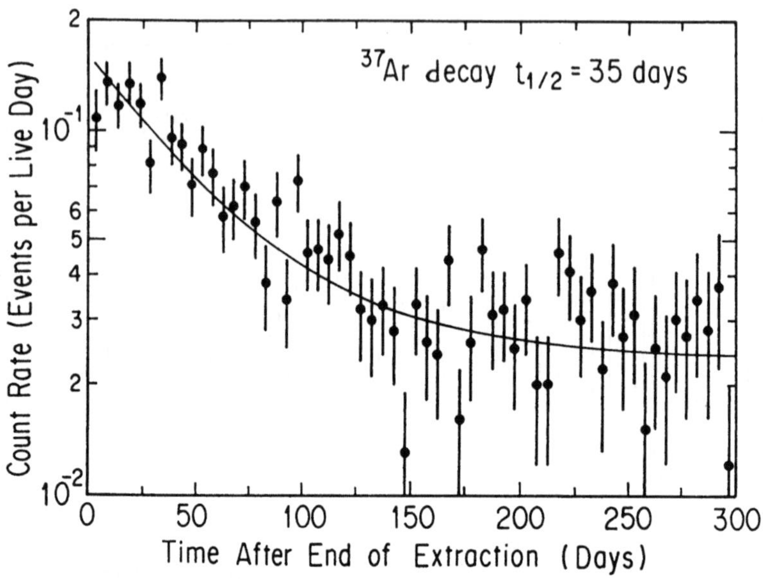

Figure 20: Count rates as a function of the counting time (Homestake). The curve is the fit to the data, assuming that the events consist of ^{37}Ar plus constant background.

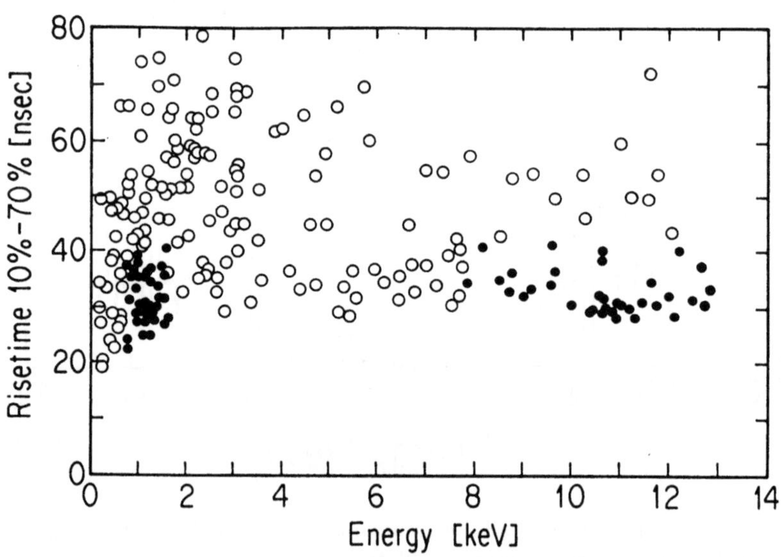

Figure 21: Rise time vs. energy plot of events observed in the GALLEX experiment. The black dots are those that lie in the selected K and L peak regions from ^{71}Ge electron-capture decay. The accumulation of events in these regions is shown.

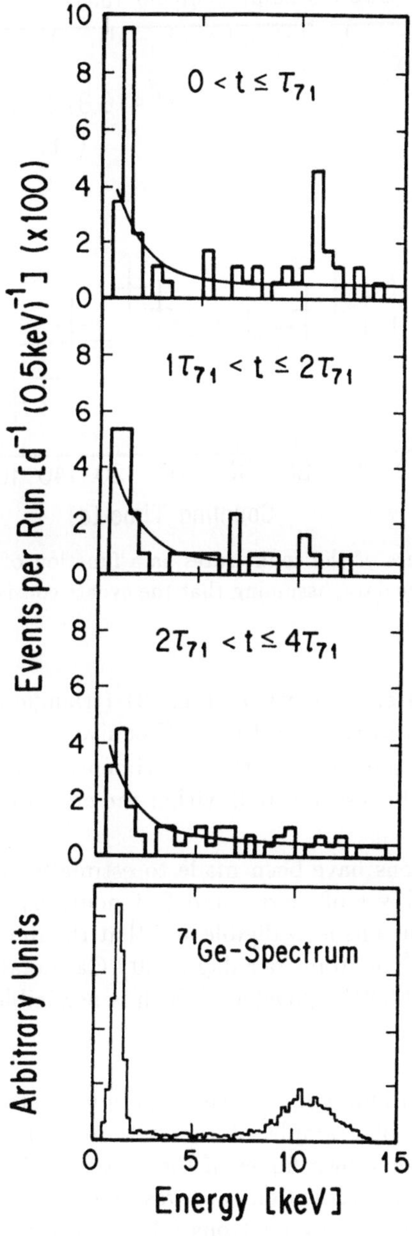

Figure 22: Energy spectra of the events in the rise-time selected region for three different counting periods. τ_{71} is the mean lifetime of ^{71}Ge (16.5 days).

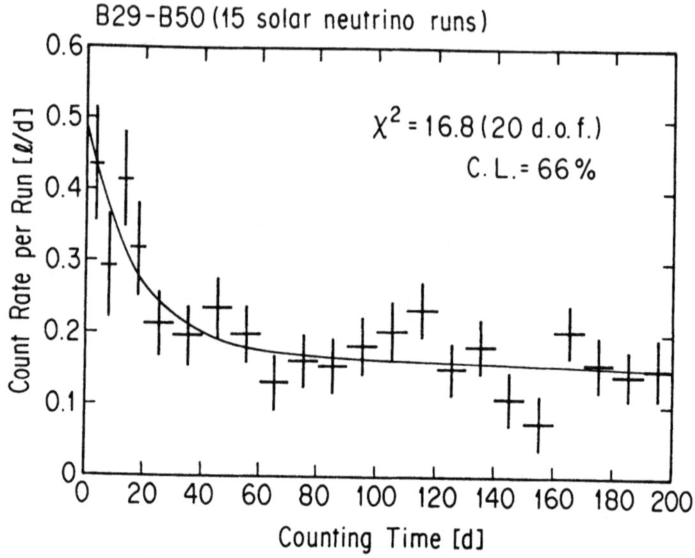

Figure 23: Decay feature of the count rates as a function of the counting time. The curve is the best fit to the data, assuming that the events consist of ^{71}Ge plus a constant background.

with the above estimate, as shown in Fig. 24 (triangle at 4,000 hg/cm^2). The adopted production rate was 0.4 ± 0.16 SNU equivalent.

Fast neutrons were measured at the experimental site and the ^{37}Ar produced by them was found to be less and well within the error of the above estimate.

(ii) SAGE [49]

Extensive calculations have been made to estimate the ^{71}Ge production rate based on the known fluxes of muons and fast neutrons. It was found that the contribution of fast neutrons is negligible and that the muons produce 0.005 ^{71}Ge, 0.013 ^{69}Ge and 0.009 ^{68}Ge atoms per day in 30 t Ga metal. This ^{71}Ge production rate corresponds to 0.6 SNU equivalent, which is negligible compared to the SSN prediction of 132 SNU.

(iii) GALLEX

The integral ^{71}Ge production cross-sections in the detector solution were measured by the irradiation of a sample solution with muon beams and Pu-Be neutron sources. Then, with the known fluxes of the muons and fast neutrons at the experimental site, the ^{71}Ge production rate was estimated to be 3.7 ± 1.1 SNU by muons and 0.15 ± 0.10 SNU by neutrons. They also estimated the background from ^{69}Ge from muons and ^{8}B neutrinos that is falsely attributed to ^{71}Ge to be 1.0 ± 1.0 SNU [24].

The 22nd and 23rd columns of Table 19 summarize the backgrounds, which are a few % of the SSM predictions.

We have not discussed the counter background caused by various sources like

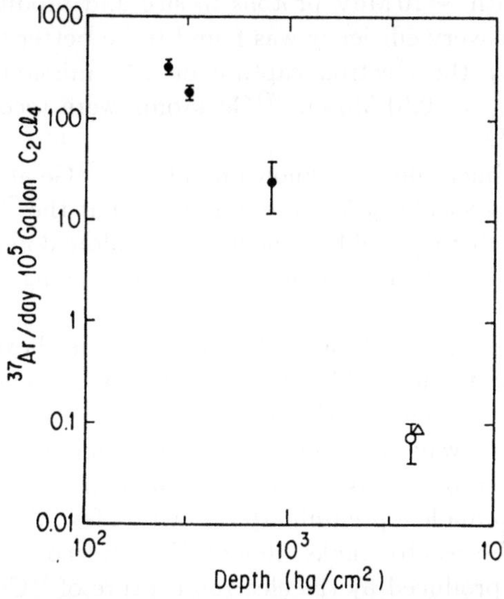

Figure 24: ^{37}Ar production rate by cosmic-ray muons measured at depths of 254, 327 and 819 hg/cm^2 of standard rock. The open circle at 4,000 hg/cm^2 is then estimated according to the depth-intensity relation of the muon intensity. The open triangles are independent estimate based on calculations. (From [29])

electric discharge, material radioactivity and other radioactivity, especially Radon gas in the ambient air. In principle, one can determine this background by counting events long after ^{37}Ar/^{71}Ge atoms have decayed and can subtract its contribution from the data (16th and 17th columns of Table 19). However, one must take any possible precautions to make the background constant in time, since it is difficult to independently estimate it during the measurement. In fact, this causes the largest systematic uncertainty in the SAGE result [30].

3.5.4 Hot chemistry ?

Atoms ^{37}Ar/^{71}Ge produced by solar neutrinos break molecular bonds and are probably ionized to various degrees. The chemical reactions of such hot atoms might be different from normal chemistry. For example, ionized ^{71}Ge atoms might form an alloy Ge-Ga and might not be extracted by the usual procedure. Therefore, the actual hot chemical situation must be simulated and tested.
(i) The Homestake experiment used C_2Cl_4 labeled with ^{36}Cl, which β-decays with a half-life of 3.0×10^5 yr, and $T_{\beta\, max} = 0.71$ MeV, namely $C_2Cl_3{}^{36}Cl \rightarrow C_2Cl_3^+$ (fragments) + $^{36}Ar^+ + e^- + \overline{\nu}_e$, which is nearly equivalent to the reaction $^{37}Cl(\nu, e^-)\,^{37}Ar$. The produced ^{36}Ar is neutralized and can be recovered with the carrier ^{38}Ar if nothing peculiar happens. An experiment was carried out and the ^{36}Ar was, indeed, recovered with $100 \pm 3\%$[29].
(ii) The GALLEX experiment carried out the following tests.

(a) ^{69}Ga$(p,n)^{69}$Ge with $\sim$ 10 MeV protons in situ and produced ^{69}Ge atoms recovered later. The recovery efficiency was found to be better than 99%.
(b) ^{71}Ge produced by the electron capture or β^+ emission from ^{71}As in situ ($t_{1/2}$ = 26.0 hr, $T_{\beta max}$ = 0.81 MeV). ^{71}Ge atoms were recovered with an efficiency of (99.8 ± 3.7)%.
(iii) The SAGE experiment doped a known number of ^{71}Ge atoms in a Ge carrier. The recovery efficiency was 99^{+6}_{-8}%. Note, however, that the ^{71}Ge atoms were not produced in situ; namely they had been neutralized before doping and the test may not have represented the chemical reactions actually taking place in the reactor tanks[30].

The 19th and 20th columns of Table 19 summarize the above discussions. From these results it is almost conceivable that the ^{37}Ar/^{71}Ge atoms produced by solar neutrinos were indeed extracted with high efficiency.

The most convincing way, though subsidiary, to demonstrate the high recovery efficiency of ^{37}Ar/^{71}Ge atoms is to produce them with artificially generated neutrinos. The SAGE group made a preliminary run with a few MCi of ^{51}Cr which were brought into one of the reactor tanks, though they did not obtain a quantitative result. Neutrinos are produced by the electron capture of ^{51}Cr,

$$^{51}\mathrm{Cl} + e^- \rightarrow {}^{51}\mathrm{V} + \nu_e, \tag{3.35}$$

($E_\nu = 0.746\,\mathrm{MeV}, 90.1\,\%;\ 0.426\,\mathrm{MeV}, 9.9\,\%;\ t_{1/2} = 27.7\,\mathrm{d}$).
A similar test is planned by the GALLEX group.

For the Homestake experiment a candidate source could be the electron-capture decay of ^{65}Zn,

$$^{65}\mathrm{Zn} + e^- \rightarrow {}^{65}\mathrm{Cu} + \nu_e, \tag{3.36}$$

($E_\nu = 0.227\,\mathrm{MeV}, 50.75\,\%;\ 1.343\,\mathrm{MeV}, 47.8\,\%;\ t_{1/2} = 224\,\mathrm{d}$).

3.5.5 Results

Figures 25, 26 and 27 show the results from Homestake, SAGE and GALLEX, respectively. The average capture rates are:

Homestake	(1970-1991)	2.28 ± 0.23 SNU,	[29]
SAGE	(1990,1991)	$58^{+17}_{-24} \pm 14$ SNU,	[51]
GALLEX	(1990,1991)	$87 \pm 14 \pm 7$ SNU.	[24][52]

There is an argument that one can take an average of the SAGE and GALLEX results, because their data are consistent with the purely statistical behavior [53]. Then, the average value is

SAGE+GALLEX	(1990, 1991)	$81 \pm 12 \pm 6$ SNU

if the errors are summed in quadrature.

The ratios of the experimental data to the SSM predictions[42], $Data/SSM$, are:

Homestake	(1970-1991)	0.29 ± 0.03
SAGE	(1990,1991)	$0.44^{+0.13}_{-0.18} \pm 0.11$
GALLEX	(1990,1991)	$0.66 \pm 0.11 \pm 0.05,$

or

SAGE+GALLEX	(1990,1991)	$0.61 \pm 0.09 \pm 0.05.$

Recall that Kamiokande gave the result (eq.(2.19)):

Kamiokande	(1987-1992)	$0.49 \pm 0.04 \pm 0.06.$

Therefore, all four experiments have seen fewer neutrino events than that which SSM predicts.

3.6 Conclusion of this chapter

Three radiochemical experiments have been successfully operated. Their threshold energies are 0.814 MeV and 0.233 MeV for ^{37}Cl and ^{71}Ga targets, respectively, which are much smaller than that for Kamiokande (7.0 MeV visible energy). Thus, they can investigate different regions of the solar neutrino spectra. The results are surprising; all of the results gave neutrino capture rates lower than what the standard solar model predicts.

The three experiments have been performed very carefully, and it is convincing that the experimental procedures were thoroughly studied and there was no apparent flaw in them.

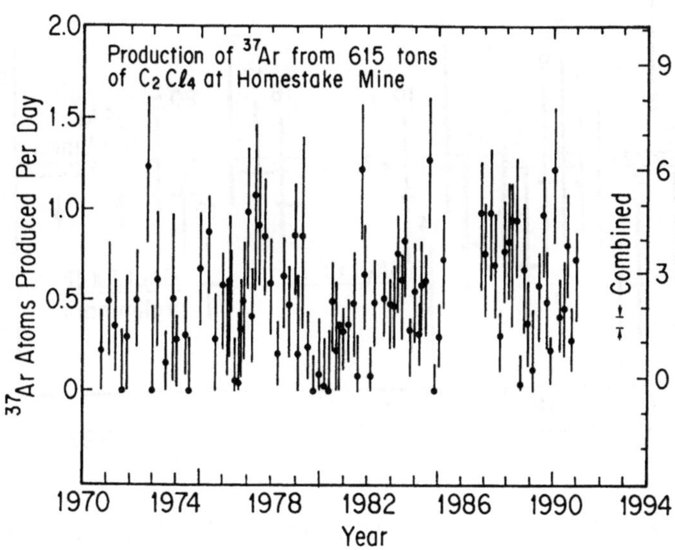

Figure 25: Summary of the Homestake results from 1970 to 1991. The combined ^{37}Ar production rate from all experiments is 0.509 ± 0.031 atoms/day, corresponding to 2.28 ± 0.23 SNU. (From [50])

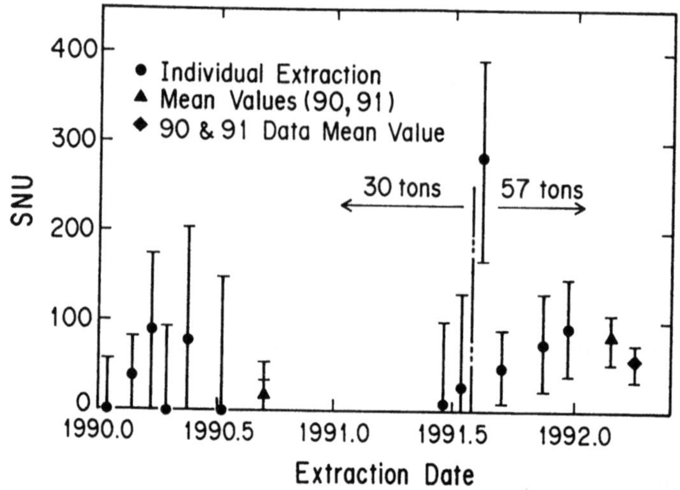

Figure 26: SAGE results from the 1990 and 1991 runs. The first error bar shown for the 1990 mean value is the statistical uncertainty; the second one is the systematic uncertainty assigned to a possible time variation in the background. The target mass has been doubled to 57 tons since the middle of 1991. (From [51])

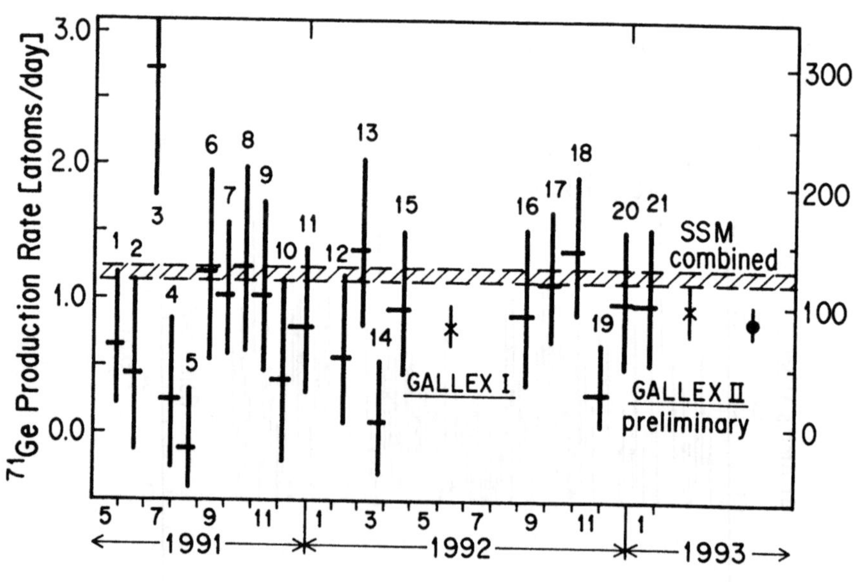

Figure 27: GALLEX results from the 1991, 1992 and Jan,1993, runs. The horizontal dashed lines show the prediction and its uncertainty of the SSM. (From [52])

4 Real-Time Experiment for Supernova Neutrinos

4.1 Brief history

On February 25, 1987, we (the Kamiokande group) received a telefax from S.A. Bludman, saying "Supernova went off in Large Magellanic Clouds. Can you see it? This is what we had been waiting 350 years for!" Within a few hours more information arrived. But it was still too early to definitely identify the supernova as being type I or type II. The type-I supernova is an explosion of a complete star due to uncontrolled nuclear fusion, while a type-II supernova is triggered by the gravitational collapse of the Fe core of a massive star (≥ 8 solar mass). It is this type-II supernova that would leave a neutron star or a black hole after liberation of an enormous amount of energy (3×10^{53} erg) in the form of neutrinos. Therefore, only the type-II supernova is relevant to look for neutrino signals.

It was also frustrating that the time when the stellar collapse actually took place was not definitely determined, because it was believed that the supernova brightened up about a day after the collapse and there was an ambiguity in the time lag of the optical observation. There was a possibility that it had happened well before February 24.

Kamiokande was happily running continuously, though with a 2 minute-interruption every two hours, on February 21-24. We conveyed the supernova news to the shift members in Kamioka on February 25, and asked them to dispatch the magnetic tapes for data up to February 25 to Tokyo as soon as possible (at that time, all of the analyses were being made at the Faculty of Science, University of Tokyo). It never occurred to us that the time of the supernova occurrence would be important. Remember that Kamiokande was searching for proton decay with a sensitivity of 10^{32} years! Who cared about a 1-minute inaccuracy in the clock timing? Even worse, before we realized its importance, our primary power was shut down for a short period on February 26. We lost a chance to recalibrate our computer clock.

At any rate, the tapes arrived on February 27. It was straightforward to analyze the data quickly. The Kamiokande detector had been upgraded to observe solar neutrinos; anticounter layers with a water thickness of ≥ 1.2 m were installed, the water purifier had a new ion-exchange column to remove radioactive elements (^{238}U or ^{226}Ra) in the water, and the detector was made airtight to avoid contamination of ^{222}Rn from the air.

As a result of these efforts, the trigger threshold was lowered to 7.5 MeV (50 percent efficiency). Software work was also in progress. Programs were ready for space reconstruction and the determination of low-energy events. An energy calibration had already been performed with a gamma-ray source from the Ni(n, γ)Ni reaction. We were in quite a good position [54].

Analysis of the data between February 21 and February 24 started from the evening of February 27 and ended the next morning. We simply printed out a two-dimensional plot of time versus energy for each event. It was easy to pick up a cluster of 11 events that had lasted for about 10 seconds from 500 pages of printouts. The events were space-reconstructed and their energies determined. A computer display showed that the events were clear electron signals. A correlation of the cluster with preceding energetic cosmic-ray muons was checked. Muons occasionally produce multiple pions by a violent interaction with oxygen nuclei in water. The pions then break up other ^{16}O and produce a number of radioactive nuclei (^{12}N, ^{12}B, ^{8}B, ^{8}Li and so on) which β-decay in about $10 \sim 20$ seconds, and fake a neutrino burst. It was found that there were no preceding muons that had interacted violently in the water. Therefore, the cluster was most probably produced by an enormous number of neutrinos. The background was completely negligible.

It was quite natural to associate the cluster with supernova SN1987A, because we could not imagine any other physical phenomenon that might emit such an enormous number of neutrinos. This work was completed on February 28. A few days later we learned that the Mt Blanc UNO experiment also observed a cluster of five events within 7 sec, but, strangely enough, the time of occurrence was about 5 hours earlier than Kamioka's observation [55]. We were so confident of our data in that we simply considered the UNO data as being spurious. The UNO people, however, are still confident of their observation. This difference of opinion has not yet been resolved, and will not be resolved until the next supernova in our Galaxy.

It took us another $5 \sim 6$ days to evaluate errors in the energy and angle determinations of the 11 events. The paper was ready on March 6; our result was then delivered to the world.

The IMB experiment, which is another large water Čerenkov detector, searched for a burst of low-energy events after learning of the Kamioka and Mt.Blanc times. They quickly found a cluster of 8 events within 6 sec at the Kamioka time, but none at the Mt Blanc time. Thus, although they confirmed Kamioka's result nicely, the Mt Blanc anomaly remains a mystery.

That the supernova went off in the Large Magellanic Clouds (LMC) has positive as well as negative aspects. It was quite fortunate that the supernova was observed by optical means, which is not always the case in our Galaxy, because the Galactic center is invisible due to thick layers of dust. The data of about 10 events alone would hardly be taken as evidence of the stellar collapse if there were no optical help. The distance to the LMC is known, which enabled us to calculate the neutrino luminosity. The negative aspect is that the LMC is a bit too far (50 kpc), so that the number of burst events was marginal; moreover, we did not observe a definite signal of a short pulse ($\sim 10\,\mathrm{ms}$) expected from neutronization (see chapter 1).

We will not need optical help to observe neutrinos from the next supernova, because we now understand a basic property of the neutrino burst. The next supernova will most probably go off in our Galaxy and will provide much more information concerning stellar collapse and neutrino properties. The supernova rate is believed to be $1/10 \sim 1/30\,\mathrm{yr}$. We can only hope that much better detectors will be ready by that time.

4.2 Detector

Supernova neutrinos have an average energy of 10 $\sim$ 20 MeV. The cross sections are listed for reactions that can be utilized for detecting such low-energy neutrinos:

$$\sigma(\overline{\nu}_e + p \to e^+ + n) = 6.90 \times 10^{-42} \ \ ((E - 1.29)/8.71)^2 \tag{4.1}$$
$$\sigma(\overline{\nu}_e +^2\mathrm{H} \to e^+ + n + n) = 1.23 \times 10^{-42} \ \ ((E - 4.03)/5.97)^2 \tag{4.2}$$
$$\sigma(\overline{\nu}_e + e \to \nu_e + e) = 0.39 \times 10^{-43} \ \ (E/10) \tag{4.3}$$
$$\sigma(\nu_e +^2\mathrm{H} \to e + p + p) = 2.71 \times 10^{-42} \ \ (E - 1.44)/8.56)^{2.25} \tag{4.4}$$
$$\sigma(\nu_e + e \to \nu_e + e) = 0.95 \times 10^{-43} \ \ (E/10) \tag{4.5}$$
$$\sigma(\nu_\mu + e \to \nu_\mu + e) = 0.16 \times 10^{-43} \ \ (E/10) \tag{4.6}$$
$$\sigma(\overline{\nu}_\mu + e \to \overline{\nu}_\mu + e) = 0.14 \times 10^{-43} \ \ (E/10) \tag{4.7}$$
$$\sigma(\nu_e +^{16}\mathrm{O} \to e + {}^{16}\mathrm{F}) = 1.1 \times 10^{-44} \ \ (E - 13)^2 \tag{4.8}$$

Here, the unit is cm^2, and E is the neutrino energy in MeV.

Equations (4.5),(4.6) and (4.7) were derived from Table 2. The cross section for the deuteron target (4.4) has already been discussed in section 2.1(ii). Cross sections (4.2) and (4.4) are apporoximate fits to numerical calcutaions [56][57][15][16]. We tabulate the cross sections for several neutrino energies in Table 20 [16]. Equation(4.1) is a numerical eveluation of the cross section,

$$\sigma(\overline{\nu}_e + p \to e^+ + n) = \frac{G_F^2}{\pi}\left[1 + 3\left(\frac{g_A}{g_V}\right)^2\right]\cos\theta_c\, p_e E_e, \tag{4.9}$$

where $E_e = E - 1.29(\mathrm{MeV})$ for $E \ll m_p$, and θ_c is the Cabibbo angle (see eq.(2.10)). Note $\sigma(\bar{\nu}_e p)$ is larger by two orders of magnitude than $\sigma(\nu_e e)$. Taking into account the relative number of protons and electrons, the ratio of event rates, $N(\bar{\nu}_e p)/N(\nu_e e) = 20 : 1(25 : 1)$, for water (liquid scintillator). Thus, $\bar{\nu}_e$ signals dominate over ν_e signals. However, scattering with electrons has an advantage: a strong directionality of scattered electrons enables us to point back to the position of supernovae. The deuterium target that provides almost free neutrons is interesting: $\sigma(\nu_e\,^2\mathrm{H})$ and $\sigma(\overline{\nu}_e\,^2\mathrm{H})$ are large and especially the large $\sigma(\nu_e\,^2\mathrm{H})$ is good to detect ν_e's from neutronization.

In view of these factors, one may employ either water (hight or heavy) or liquid scintillators as a detector material. Remember that one needs a large amount ($\sim$kt) of material to observe a reasonable number of interactions. Therefore, expensive materials like plastic scintillators are not suited. Table 21 summarizes the materials and their properties. It is interesting that the energy resolution of the water Cerenkov detector (Kamiokande) is as good as that of the liquid scintillation detector (UNO), although the Cerenkov light yield is 100-times less than that of scintillation.

Table 22 shows the four detectors that claimed to detect the supernova neutrinos from SN1987A. Water Čerenkov detectors are generally, much larger than liquid-scintillator detectors, for a good reason. The background rates are, however, are quite different from detector to detector (0.7/day $\sim 1.2 \times 10^{-8}$/ year). Obviously, the Baksan group cannot claim their observation without the help of Kamioka or IMB. The background rate of the Mt Blanc group is rather small

Table 20: Cross sections for neutrino reactions on proton and deuteron targets [16]. The units are $10^{-42}\,\mathrm{cm}^2$.

E(MeV)	$p(\overline{\nu}_e, e^+)n$	$^2\mathrm{H}(\overline{\nu}_e, e^+)nn$	$^2\mathrm{H}(\nu_e, e^-)pp$
4	0.67	—	0.158
6	2.02	0.116	0.624
8	4.10	0.515	1.46
10	6.90	1.23	2.71
....	....	....	....
12	10.4	2.27	4.37
14	14.7	3.63	6.48
16	19.7	5.30	9.05
18	25.4	7.28	12.1
20	31.8	9.55	15.6
....	....	....	....
25	51.1	16.5	26.6
30	75.0	25.0	40.9
35	103	35.1	58.7
40	136	46.5	80.1
45	174	59.2	105
50	216	72.9	134

Table 21: Materials suitable for the detection of supernova neutrinos and their properties.

Material	Detection method	Relative number $p/n/e$	Directionality	Energy Resolution (%)[a]	Angular[a] resolution (deg)
H_2O	Čerenkov	1/0/5	○[b]	22[c]	28[d]
D_2O	Čerenkov	1/1/5	○[b]	22[c]	28[d]
$(CH_2)_n$	Scintillation	1/0/4	×	25[e]	

[a] At 10 MeV.
[b] For electron target.
[c] Parameters of Kamiokande II.
[d] Parameters of Kamiokande II, dominated by multiple Coulomb scattering.
[e] Parameter of UNO.

Table 22: Existing detectors capable of detecting supernova neutrinos as of 1987.

Experiment (technique)	Depth (m.w.e)	Material	Mass (t)	Threshold Energy[a] (MeV)	Observation Time (UT)	Number of Event	Background Rate
Kamioka (Cerenkov)	2,700	H_2O	2,140	7.5	7:35:35	11	1.2 $\times 10^{-8}$/yr
IMB (Cerenkov)	1,570	H_2O	6,800	30	7:35:41	8	5 $\times 10^{-3}$/yr[b]
Mt Blanc (scintillation)	4,000	$(CH_2)_n$	90	5.5	2:52:37	5	0.7/yr
Baksan (scintillation)	850	$(CH_2)_n$	200	10	7:36:12	5	0.7/d

[a] at 50% efficiency
[b] The rate contains edge clipping muons. The non-muon background would be similar to that of Kamioka.

(0.7/year), but as stated before, their observation time is a problem. Two experiments, Kamioka and IMB, are completely free from background, and their observation times are consistent within the quoted errors. Both detectors happen to be water Cerenkov detectors. This is understandable because they are at least 10-times more massive than the scintillation detectors, as shown in Table 22.

Electron/positron energies (E_e) are obtained by counting a number of hit PMT's (N_{hit}) in the Kamiokande/IMB detectors (see sec.2.2). They are calibrated with gamma rays from Ni(n,γ)Ni (~9 MeV) or electrons from $\mu \to e$ decay. Figure 28 shows the correspondence between N_{hit} and E_e for Kamiokande and IMB.

Fig. 29 shows the efficiencies with which low-energy electrons/positrons are detected as a function of E_e. Note that the Kamiokande and Baksan detectors do not have 100% effeciency at higher energies, because some of the events that are produced near to the edge of the volume and directed toward the wall are not accepted.

4.3 Background

Supernova neutrinos come in a burst of about 10-sec duration. In general, a constant background does not pose a serious problem, although it is not negligible. We are interested in the energy range from a few MeV to a few tens of MeV. What kind of constant background exists in this energy range? Figure 30 shows a scatter plot of low-energy events in visible energy versus the angle measured from the direction of the Sun, observed by Kamiokande. Several events are seen in the 20–50 MeV range, which are presumably electrons from decays of very low-energy muons produced by ν_μ (below Čerenkov threshold). However, they are too few to become a serious background for supernova neutrinos. Therefore, the IMB detector is essentially free from background (threshold energy $\approx$ 30 MeV). Below

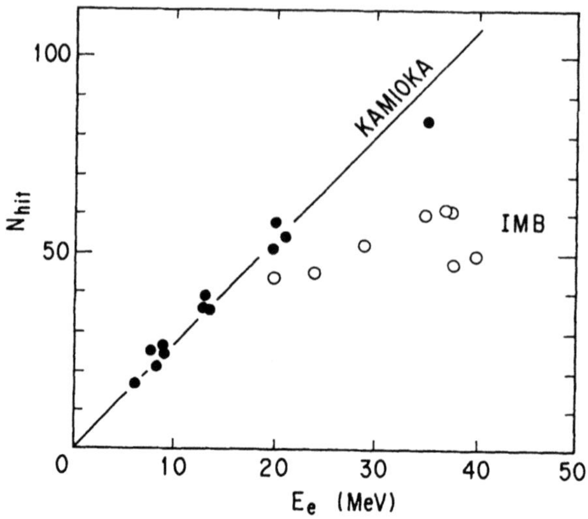

Figure 28: N_{hit} vs. electron energy(E_e) for Kamiokande and IMB. IMB needs much larger corrections to estimate E_e than Kamiokande. The line corresponds to the relation $N_{hit} = 2.6\,E_e$, (E_e in MeV).

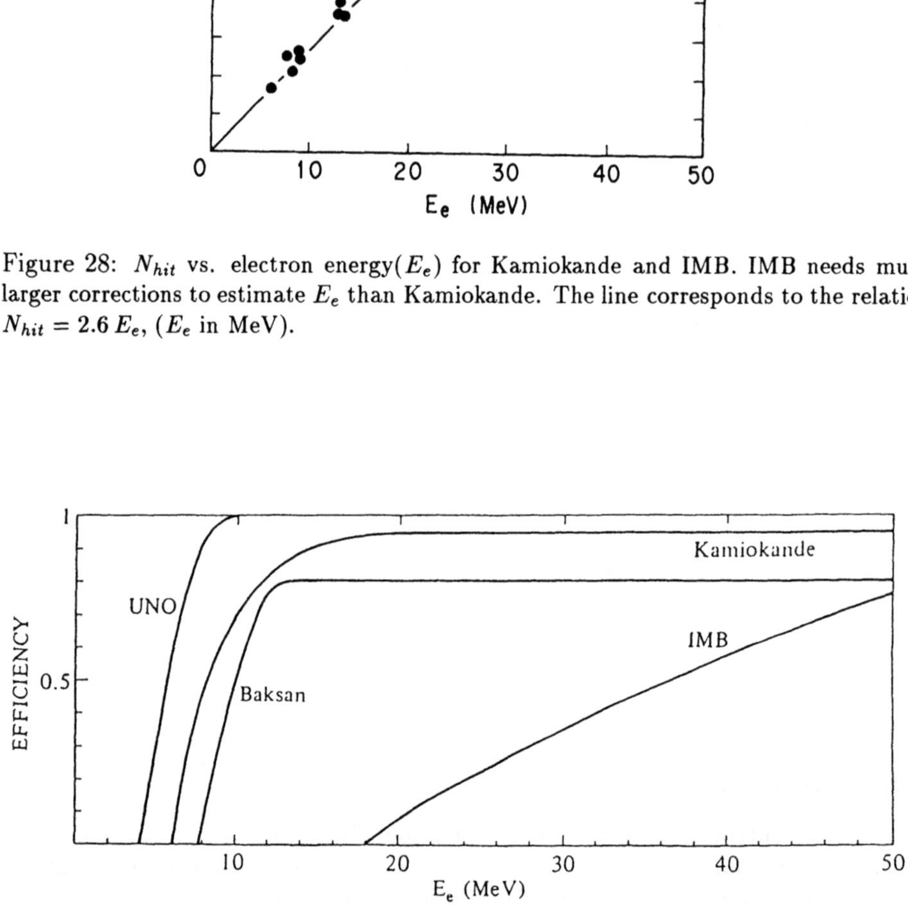

Figure 29: Trigger efficiencies as a function of the electron energy(E_e) for Kamiokande, IMB, UNO and Baksan.

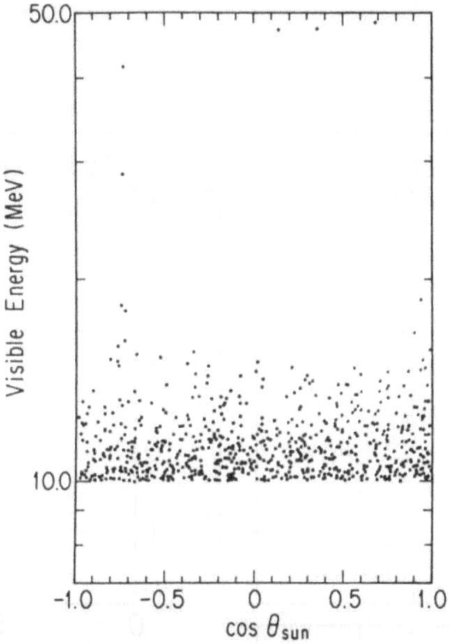

Figure 30: Visible energy and $\cos\theta_{SUN}$ of low-energy events observed by Kamiokande, where θ_{SUN} is the angle measured from the direction from the Sun. They are events still remaining after background refection. The fiducial volume is 680 t in this case. The observations time is 450 days.

20 MeV, however, the background increases quite rapidly. They consist of three types: β-rays of ^{214}Bi in the water, γ-rays unabsorbed in the outer detector from the rock, β-rays of spallation products by cosmic-ray muons like ^{12}N, ^{12}B, ^{8}B and ^{8}Li. In principle, all of them can be eliminated. ^{222}Rn, which are parent elements of ^{214}Bi, can be taken out of the water. A thicker outer detector can absorb external γ-rays. Spallation products are negligible if the detector is located deep underground. However, the existing detectors do not fulfill all of the requirements. A trigger threshold of 7.5 MeV for Kamiokande was, in fact, set due to ^{214}Bi and external γ-rays.

Figures 31(a) and 31(b) show the distributions of the number of events within 10 sec for threshold energies of 7.5 and 11.5 MeV for Kamiokande. Figure 31(c) corresponds to the IMB data[60]. They just follow the Poisson distribution. Note the point corresponding to the neutrino burst. They are far above the expected entries. Therefore, Kamiokande and IMB are free from background for supernova neutrinos, although the trigger thresholds are limited by the background rate.

One must worry about two kinds of burstlike background. One is electric noises coming from electric sparks. It produces a train of pulses and would be a serious problem unless some electronic rejection is made. Kamiokande has a flash ADC (FADC) to directly record the pulse wave forms. Electric noise produces quite distinct wave forms and are easily distinguished from genuine pulses. Moreover, the timing and topological information of PMT signals completely eliminates electric

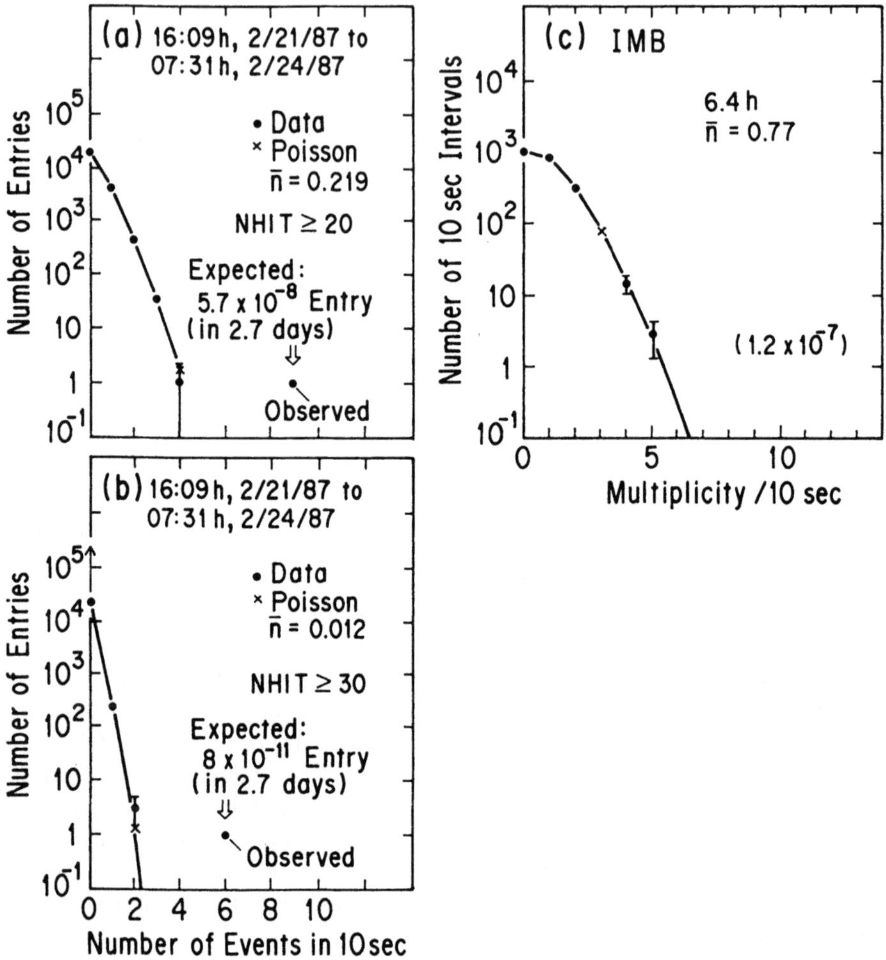

Figure 31: (a) Poission distribution of the number of events with $N_{hit} \geq 20 (E_e \geq 7.5\,\mathrm{MeV})$ per 10 sec in a 2.7 day period. (b) Same for $N_{hit} \geq 30 (E_e \geq 11.5\,\mathrm{MeV})$. (c) Same for the IMB data. The point denoted as observed corresponds to the neutrino burst.

noise. The other source of burstlike events is multiple spallation products produced by a single muon. In fact, Kamiokande observed a burst of 26 low-energy events within 20 sec on January 23, 1987, besides the supernova burst. This event was an obvious spallation event, because a muon with a violent interaction in the detector was found less than 4 msec before the first low-energy event. The location of Kamiokande is sufficiently deep that such events occur very rarely, and it is easy to eliminate them completely. Detectors at much shallower depths will have serious trouble concerning the spallation background due to a much larger muon flux.

4.4 Observation

Once the detector is well understood, observing a neutrino burst is quite an easy task. One first eliminates any electric noise that occurs occasionally, and space-reconstructs each low-energy event. Then, one searches for a cluster of events, say, within 10 sec. When such a cluster is found, a check is made for muons activity preceding the cluster up to 20 sec. One can readily eliminate electrons from multiple $\pi \to \mu \to e$ decays produced by an energetic muon, because the time spread of the cluster is less than about 5 μsec, which is completely different from the supernova time scale. One then estimates the probability that the cluster is of the spallation origin. Some properties of multiple spallation background are as follows: (1) It exhibits an exponential time structure that reflects the known lifetimes of the radioactive fragments from ^{16}O, specifically, an 18±1.2 msec component from ^{12}N and ^{12}B, and also a component with a longer exponential time structure of 1.2±0.5 sec from ^{8}B and ^{8}Li, with relative rates of 2 : 1, respectively; (2) the resultant β-decay electrons with observed energies above 15 MeV occur with less than 4 percent probability; (3) the relative total rate of spallation leading to one or more low-energy electrons is less than 10^{-3} per incident muon; (4) the measured multiplicity distribution of low-energy electron events following an incident muon in time yields a probability of multiplicity ≥ 3 of 3×10^{-3}. For Kamiokande, the rate of muons incident in the detector is 0.3/sec. Therefore, clusters with a multiplicity ≥ 10 occur very rarely ($\leq$ 1/month) and are readily recognized as spallation products. This kind of background is (in any case) negligible for detectors located even deeper underground.

Figure 32 shows the time sequence of all low-energy events (solid lines) and all cosmic-ray muon events (dashed line) within a 45 sec interval centered on 7 : 35 : 35 UT, February 23, 1987, observed by Kamiokande [59]. The event sequence during 0 to 2 sec is shown expanded in the upper-right corner. The left-hand scale (N_{hit}) indecates the event energy ($E_e \approx N_{hit}/2.6$ MeV. See Fig. 28.), while the right-hand scale is the pulse height (in unit of photoelectrons) of the muons. The threshold energy is set to $N_{hit} = 20 (E_e = 7.5\,\text{MeV})$, at which the detection efficiency drops to 50 percent. The burst consists of 12 events. Event number 6 is below the threshold and is consistent with the background. There is a large gap between event number 9 and 10. Events 10, 11 and 12, most probably belong to the burst. The probability that they are background is 4×10^{-5}. Four muons passed the detector within 20 sec prior to the burst; $\mu 1, \mu 2$ and $\mu 3$ are normal through-going muons with

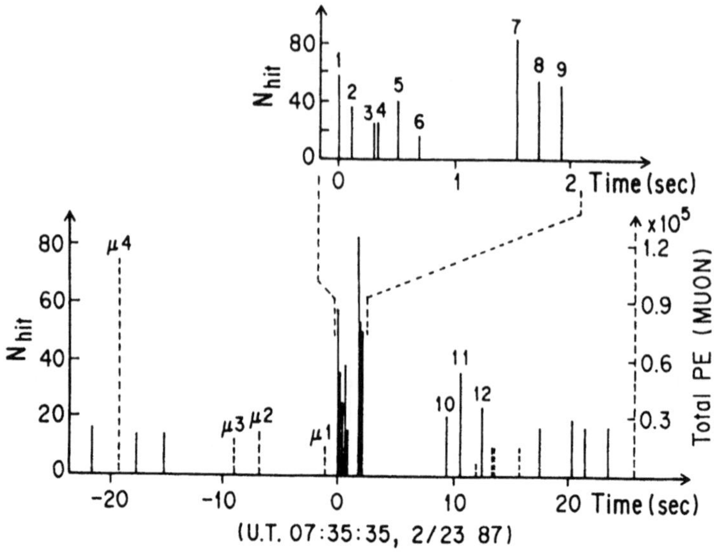

Figure 32: Time suquence of events in a 45 sec intervcal centered on 7 : 35 : 35 UT, Febreary 23, 1987. The solid lines represent low-energy electron events in units of the number of hit PMTs, N_{hit} (left-hand scale). The dashed lines represent muon events in units of the number of photoelectrons (p.e., right-hand scale). The shower energy is approximately given by (p.e.–12,000)/3.6(MeV). The upper-right figure is the 0–2 sec time interval on an expanded scale.

little activity in the detector; $\mu 4$ produced a shower ($E_e \sim 30\,\mathrm{GeV}$), but cannot be a progenitor of the burst; it does not exhibit the features of spallation described above. The burst cannot be due to a statistical fluctuation of constant background, as already shown in Fig. 31. All of the events have a typical Čerenkov pattern produced by electrons. Therefore, we conclude that the burst was produced by neutrinos. It is natural to associate the burst with supernova SN1987A, which became optically visible about 3 hours later.

The IMB group that observed a burst of 8 events within 6 sec on 7 : 35 : 41 UT, February 23, 1987 [60]. Since their threshold energy is 30 MeV (at 50 percent efficiency), spallation is not a problem. The constant background rate in the IMB detector is shown in Fig. 31(c). The burst cannot be a statistical fluctuation of the general background.

The time sequence and energy of the burst events are shown in Fig. 33 for both Kamiokande and IMB. The events observed by IMB, of course, have higher energies due to the higher energy threshold. Note that a 7-sec gap in the Kamiokande data is nicely filled by the two events of IMB. One should not take the gap seriously. The angular distributions for the two data samples are shown in Figs. 34(a) and 34(b) [58][59][60]. The Kamiokande data are consistent with isotropy, while the IMB events are concentrated in the forward direction. This is not due to a possible detector bias[60], but the probability that the distribution is consistent

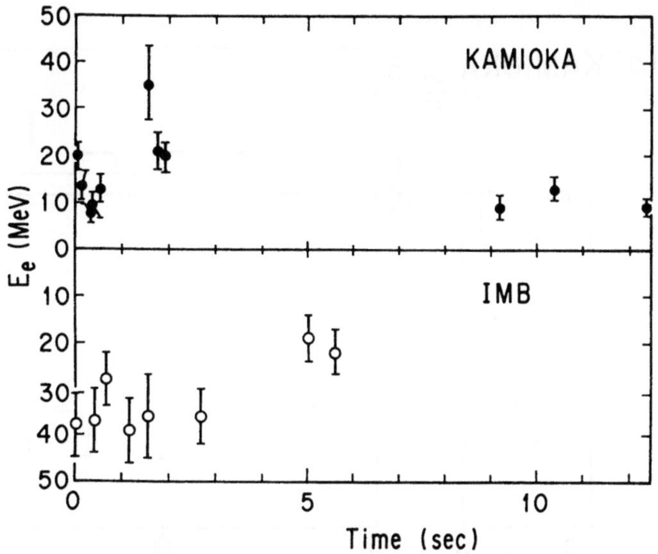

Figure 33: Time equence and energy of the burst evetns observed by Kamiokande and IMB.

with expectation from $\bar{\nu}_e p \to e^+ n$ is 4.5 percent, which is not very small, although not very large [60]. A simple statistical fluctuation is not excluded to explain the IMB angular distribution.

4.5 Future outlook

When will the next supernova occur? This question has been asked many times, especially after SN1987A. The supernova rate has been estimated by several authors, ranging from 1 every 100 years to 1 every 10 years. These days many tend to believe the optimistic rate of about 1 every $10 \sim 30$ years. Thus, one might encounter the next supernova some time around the turn of the century. It will most probably occur near the center of our Galaxy (8.5 kpc from the earth), and will not be seen optically due to thick layers of dust covering the Galactic center. Neutrinos have no problem to punch through and reach Earth. After about a year from the neutrino burst the supernova will become visible by X-rays and gamma rays.

Since optical help is not expected, the direction of the supernova must be obtained by neutrino data alone. This will encourage satellites to search for X-ray or gamma-ray signals afterwards. It is (in principle) possible to determine the supernova direction by neutrinos. One detects elastic scattering $\nu e \to \nu e$. The scattered electron has a strong directionality that points back to the supernova accurately (see sec. 2.1(i)). One expects a scattering rate of about 7/kt for a supernova at 10 kpc away [55]. The angular resolution of a water Čerenkov detector

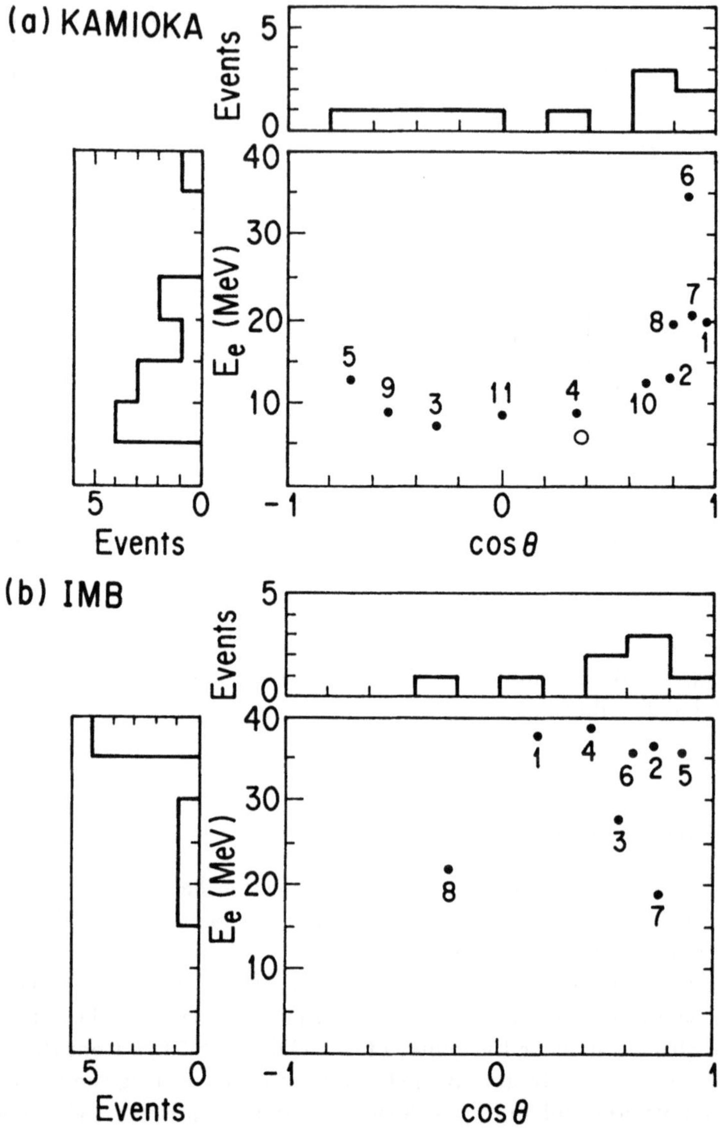

Figure 34: Anglular and energy distributions of the observed electrons. (a)Kamiokande and (b)IMB. The point denoted by an open circle in (a) is below analysis threshold and should not be taken seriously. The number attached to each event is a sequential number in chronological order.

is about $28°/\sqrt{E_e/10\,\mathrm{MeV}}$ (E_e is the electron energy in MeV). Since the average ν_e and $\nu_{\mu,\tau}$ energies are 10 and 20 MeV, respectively, the accuracy of the direction is better than $28°/\sqrt{7W} = 11°/\sqrt{W}$, where W is the detector mass in kt. For example, the Superkamiokande detector[63] will have a total volume of 32 kt, which leads to the directional accuracy of 2°. Another way of determining the direction is to use at least three detectors separated widely and to determine the accurate timing of the neutronization burst. Theory tells us that the pulse width of the neutronization burst is about 10 msec or less. If the separation of the detectors is 10^4 km, then the directional accuracy is about $3 \times 10^5 \times 10^{-2}/10^4 = 0.3\,\mathrm{rad} = 17°$. This is (in fact) the worst case, because each detector would be able to detect more than one event, say N events. Then, the directional acuracy would become $17°/\sqrt{N}$. In this case, one needs large water or scintillation detectors (≥ 10 kt, expected yield ≈ 0.4/kt) or D_2O detectors of kt size.

The observed rate of $\bar{\nu}_e p \rightarrow e^+ n$ is about 130/kt for a supernova 10 kpc away. For example, the Superkamiokande detector will observe 4,000 $\bar{\nu}_e$ events, which are certainly sufficient for a detailed study of stellar collapse, prompt or delayed explosion and so on.

Another important observation in the next supernova will be ν_μ and ν_τ elastic scattering, $\nu_{\mu,\tau} e \rightarrow \nu_{\mu,\tau} e$. The average energy of $\nu_{\mu,\tau}$ is expected to be 20 MeV, which is higher than the average energy of $\nu_e \sim 10$ MeV. Note that $\nu_{\mu,\tau}$ signals are well separated from $\bar{\nu}_e$ by their sharp angular distribution. The only confusion comes from $\nu_e e \rightarrow \nu_e e$ scattering; however, a factor of two difference in the average energies will be able to pick up the $\nu_{\mu,\tau}$ sample, about 2.5 events/kt, from the ν_e background (4.5 events/kt). For example, the Superkamiokande detector will observe 130 ν_e and 80 $\nu_{\mu,\tau}$ events, which are probably large enough to find a two-step energy distribution and to estimate the $\nu_{\mu,\tau}$ flux as well as $\nu_{\mu,\tau}$ energies. Observation of $\nu_{\mu,\tau}$ will have a significant impact on neutrino physics. One could set the $\nu_{\mu,\tau}$ mass limit as low as 50 eV if the $\nu_{\mu,\tau}$ signals are observed in 10 sec. This value is better by a factor of $10^3 - 10^6$ than the present laboratory limits. Should ν_μ or ν_τ turn out to have that high mass, that the neutrino will close the universe would be a likely possibility.

Table 23 shows four detectors being built or planned. All of them are sufficiently large to study a neutrino burst in detail.

4.6 Conclusion of this chapter

The neutrino burst from the supernova SN1987A was clearly observed in coincidence by Kamiokande and IMB. The detection, itself, was easy and straightforward. The total number of observed neutrino events was 11 and 8 for Kamiokande and IMB, respectively. The small data samples are due to the long distance to the LMC (50 kpc) and small size of the detectors ($2 \sim 6$ kt). Nevertheless, they provided important information concerning stellar collapse and neutrino properties.

The supernova watch must continue, but with better detectors. The principal goal of the observation of the next supernova should be to provide a clear answer to the stellar-collapse theory and to set limits on the ν_e and $\nu_{\mu,\tau}$ masses. It is

Table 23: Large underground detectors being built or planned.

	Material	Mass(t)	Thershold Energy(MeV)	Directionality
Super-Kamiokande	H_2O	32,000[(a)]	~ 5	○
SNO	D_2O	1,000	~ 5	○
LVD	$(CH_2)_n$	1,340	~ 5	×
ICARUS	Argon	2,000	~ 5	○

[(a)] A full volume can be utilized for the detection of supernova neutrinos.

essential to have a large detector to observe sufficient numbers of ν_e from the neutronization and $\nu_{\mu,\tau}$. Let us hope that the next supernova will occur in the near future.

References

[1] J.N. Bahcall and R.K. Ulrich, Rev. Mod. Phys. **60**, 297 (1988). See also S. Turck-Chièze, S. Cahen, M. Cassè and C. Doom, Astrophys. J.**335**, 415 (1988).

[2] J.N. Bahcall, Rev. Mod. Phys. **50**, 881 (1978).
^{8}B β-decays into the broad excited state (2^+, 2.94 MeV). Thus eq.(1.2) must further be smeared with the width of this state. See J.N. Bahcall and B.R. Holstein, Phys. Rev. **C33**, 2121 (1986), and E.W. Warburton, Phys. Rev. **C33**, 303 (1986).

[3] J.N. Bahcall and M.H. Pinsonneault, Rev. Mod. Phys. **64**, 885 (1992).

[4] A. Colgate and R.H. White, Astrophys. J. **143**, 626 (1966).

[5] S.E. Woosley, J.R. Wilson and R. Mayle, Astrophys. J. **302**, 19 (1986).

[6] A. Burrows and J.M. Lattimer, Astrophys. J. **307**, 178 (1986).

[7] R. Mayle, Ph.D. Thesis, Univ. of California, Berkeley, 1985; UCRL-53713.

[8] R. Mayle, J.R. Wilson and D.N. Schramm, Astrophys. J. **318**, 288 (1987).

[9] K. Sato and H. Suzuki, Phys. Rev. Lett. **58**, 2722 (1987).

[10] H. Suzuki, Doctor Thesis, Univ. of Tokyo, 1989.

[11] L. Wolfenstein, Phys. Rev. **D17**, 2369 (1978); S.P. Mikheyev and A.Yu. Smirnov, Nuovo Cimento **C9**, 17 (1986).

[12] J.N. Bahcall, Rev. Mod. Phys. **59**, 505 (1987).

[13] F.J. Kelly and H. Überall, Phys. Rev. lett. **16**, 145 (1966).

[14] S.D. Ellis and J.N. Bahcall, Nucl. Phys. **A114**, 636 (1968).

[15] S. Ying, W.C. Haxton and E.M. Henley, Phys. Rev. **D40**, 3211 (1989).

[16] N. Tatara, Y. Kohyama and K. Kubodera, Phys. Rev. **C42**, 1694 (1990).

[17] A.F.M. Ishaq and T.J. Kennett, Canadian J. Phys. **50**, 3090 (1972).

[18] J.N. Bahcall, "Neutrino Astrophysics", Cambridge University Press, 1989.

[19] Note the angular distribution of electrons from reaction $\nu_e {}^{Z-1}A \to e^- {}^{Z}A$ is given by $1 + v_e \, \alpha \cos\theta$, where α is,

$$\alpha = \frac{\langle 1 \rangle^2 - \frac{1}{3}(g_A/g_V)^2 \langle \sigma \rangle^2}{\langle 1 \rangle^2 + (g_A/g_V)^2 \langle \sigma \rangle^2},$$

$g_A/g_V = 1.25$,
$\langle 1 \rangle^2 = |\langle f \parallel \Sigma_i \tau_i{}^+ \parallel i \rangle|^2 = T(T+1) - T_3(f)T_3(i)$, $(spin\ nonflip)$
$\langle \sigma \rangle^2 = |\langle f \parallel \Sigma_i \tau_i{}^+ \sigma_i \parallel i \rangle|^2, (spin\ \ flip)$.
where T is isospin of ${}^{Z-1}A$, and $T_3(i)$ and $T_3(f)$ are the third components of isospin of ${}^{Z-1}A$ and ${}^{Z}A$, respectively. ^{40}Ar has $J^\pi = 0^+$. Thus $\langle \sigma \rangle^2 = 0$ and $\alpha = 1$. For $\nu_e {}^2\mathrm{H} \to e^- pp$, spin flips ($J^\pi; 1^+ \to 0^+$). Thus $\langle 1 \rangle^2 = 0$ and $\alpha = -\frac{1}{3}$.

[20] J.N. Bahcall, M. Baldo-Ceolin, D.B. Cline and C. Rubbia, Phys. Lett. **B178**, 324 (1986).

[21] K.S. Hirata, et al., Phys. Rev. **D44**, 2241 (1991).

[22] Y. Suzuki, Frontiers of Neutrino Astrophysics, ed. Y. Suzuki and K. Nakamura, (Universal Academy Press Inc., 1992), p.61.

[23] R.Davis Jr., Frontiers of Neutrino Astrophysics, ed. Y. Suzuki and K. Nakamura, (Universal Academy Press Inc., 1992), p.47.

[24] P. Anselmann et al., GX27-1993/MPIH-V11-1993/Chem(BNL) C-3991 (1993).

[25] G.T. Zatsepin, Frontiers of Neutrino Astrophysics, ed. Y. Suzuki and K. Nakamura, (Universal Academy Press Inc., 1992), p.71.

[26] J.N. Bahcall and R.K. Ulrich, Rev. Mod. Phys. **60**, 297 (1988). See also S. Turck-Chièze, S. Cahen, M. Cassè and C. Doom, Astrophys. J.**335**, 415 (1988).

[27] "Scientific and Technical Description of the Mark II SNO Detector", SNO-89-15, ed. by E.W. Beier and D. Sinclair.

[28] L.M. Krauss, S.L. Glashow and D.N. Schramm, Nature **310**, 191 (1984).

[29] R. Davis, Phys. Rev. Lett. **12**, 303 (1964); J.K. Rowley, B.T. Cleveland and R.Davis, Jr., AIP Conference Proceedings No.126, p.1 (1985).

[30] A.I. Abazov et al., Phys. Rev. Lett. **67**, 3332 (1991); T.J. Bowles, Invited talk presented at NEUTRINO'92, Granada, 1992.

[31] P. Anselmann et al., Phys. Lett. **B285**, 390 (1992).

[32] K. Lande, Invited talk presented at NEUTRINO'92, Granada, 1992.

[33] G.S. Hurst et al., Phys. Rev. Lett. **53**, 1116 (1984); G.S. Hurst et al., AIP Conference Proceedings No.126, p.152 (1985).

[34] J.N. Bahcall, Rev. Mod. Phys. **50**, 881 (1978).

[35] J.N. Bahcall and C.A. Barnes, Phys. Lett. **12**, 48 (1964).

[36] J. Rapaport et al. Phys. Rev. Lett. **47**, 1518 (1981);

[37] "Table of Isotopes", ed. by C.M. Lederer and V.S. Shirley, John Wiley & Sons, 1978.

[38] R.G. Sextro, R.A. Gough and J. Cerny, Nucl. Phys. **A234**, 130 (1974).

[39] C.D. Goodman et al., Phys. Rev. Lett. **44**, 1755 (1980).

[40] J.Rapaport et al., Phys. Rev. Lett. **47**, 1518 (1981).

[41] T.N. Taddeucci et al., Phys. Rev. **C25**, 1094 (1981).

[42] J.N. Bahcall and R.K. Ulrich, Rev. Mod. Phys. **60**, 297 (1988); J.N. Bahcall, "Neutrino Astrophysics", Cambridge University Press, 1989.

[43] W.C. Haxton, Phys. Rev. Lett. **60**, 768 (1988).

[44] J. Engel, S. Pittel and P. Vogel, Phys. Rev. Lett. **67**, 426 (1991).

[45] S. Turck-Chièze, S. Cahen, M. Cassè and C. Doom, Astrophys. J. **335**, 415 (1988).

[46] J.A. Bearden and A.F. Burr, Rev. Mod. Phys. **39**, 125 (1967).

[47] A.H. Wapstra, G.J. Nijah and R. Van Lieshout, "Nuclear Spectroscopy Tables", North-Holland Publ. 1959.

[48] E. Henrich and K.H. Ebert, Angew. Chem. Int. Ed. Engl.

[49] V.N. Gavrin et al., Proc. of Inside the Sun Conference, Versailles, ed. Berthomieu and Cribier (Kluwer Academic, 1989), p.201. **31**, 1283 (1992).

[50] R. Davis Jr., Frontiers of Neutrino Astrophysics, ed. Y. Suzuki and K. Nakamura, (Universal Academy Press Inc., 1992), p.47.

[51] G.T. Zatsepin, Frontiers of Neutrino Astrophysics, ed. Y. Suzuki and K. Nakamura, (Universal Academy Press Inc., 1992), p.71.

[52] F. von Feilitzsch, Invited talk presented at the International Conference on Neutral Currents Twenty Years Later, Paris, July 1993.

[53] J. Rich and M. Spiro, DAPNIA/SPP 93-01, 1993; Texas/PASCOS 92; Relativistic Astrophysics and Particle Cosmology, ed. C.W. Akerlof and M.A. Srednicki, (Ann. N.Y. Acad. Sci., 1992), p.364.

[54] K.S. Hirata, Doctor thesis, University of Tokyo, 1991; ICRR-Report-239-91-8 (Institute for Cosmic Ray Research, Univ. Tokyo); K.S. Hirata, et al., Ref. [21].

[55] International Astronomical Union Circular (IAU Circular) no.4323 (1987).

[56] J.Weneser, Phys. Rev. **105**, 1335 (1957).

[57] F.J. Kelly and H. Überall, Phys. Rev. Lett. **16**, 145 (1966).

[58] K. Hirata et al., Phys. Rev. Lett. **58**, 1490 (1987).

[59] K.S. Hirata et al., Phys. Rev. **D38**, 48 (1988).

[60] R.M. Bionta et al., Phys. Rev. Lett. **58**, 1494 (1987); C.B. Bratton et al., Phys. Rev. **D37**, 3361 (1988).

[61] M. Aglietta et al., Europhys. Lett. **3**, 1315 (1987).

[62] E.N. Alexeyev et al., Phys. Lett. **B205**, 209 (1988).

[63] Y. Totsuka, Proceedings of International Symposium on Underground Physics Experiments, ed. K. Nakamura, 1990, p.129; Y. Totsuka, ICRR-Report-227-90-20.

Indium Solar Neutrino Experiments

Yoichiro Suzuki and Kunio Inoue

Institute for Cosmic Ray Research, University of Tokyo

1 Introduction

The possibility of detecting low-energy solar neutrinos through inverse β-decay in ^{115}In was first discussed by R. S. Raghavan [1] in 1976. An indium experiment has unique features in that it is possible to measure the energy spectrum of solar neutrinos above 125 keV in real time. Although a variety of ideas concerning indium detectors have been discussed from time to time, and sophisticated ideas have been presented on various occasions, no practical designs for indium detectors have yet been demonstrated. We focus on discussions of problems that stand in the way of realizing a viable experiment in section 3. A survey of past and present development projects of the detector is given in section 4. A practical detector design and detailed background considerations for a dedicated detector for ^{7}Be-pep neutrino detection are given in section 5.

2 Characteristics of a signature and detection principles

Neutrinos are captured by ^{115}In (95.7% natural abundance) through the Gamow-Teller transition, which emits an electron with an energy of $E_{e^-} = E_{\nu_e} - 125$ keV from the $(9/2)^+$ ground state to the $(7/2)^+$ second excited state of ^{115}Sn* with an excitation energy of 614 keV:

$$\nu_e +^{115}\mathrm{In} \rightarrow e^- +^{115}\mathrm{Sn}^*. \tag{1}$$

Subsequently, ^{115}Sn* emits two photons of 116 keV (γ_1) and 497 keV (γ_2) with a half-life of 3.3μsec. Approximately 50% of the γ_1 is internally converted and a 90 keV electron (e_1) is emitted along with the characteristic X-ray. The relevant nuclear energy level diagram is shown in Fig. 1.

The most prominent feature of the indium experiment is that the incident neutrino energy can be directly obtained through the relation $E_\nu = E_e + 125$ keV. This 125 keV minimum-energy threshold enables us to measure the differential energy spectrum of low-energy solar neutrinos mainly from the $p+p \rightarrow d+e^+ + \nu_e$ fusion process. Measurements of monochromatic neutrinos from ^{7}Be$+e^- \rightarrow {}^7$Li$+\nu_e$ and $p+e+p \rightarrow d+\nu_e$ processes are also an exciting possibility. Since the experiment

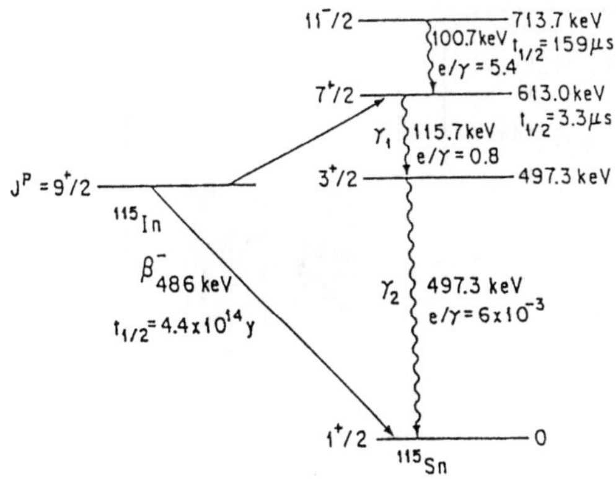

Figure 1: Nuclear level diagram of ^{115}In. The excited state of ^{115}Sn* emits two photons of 116 keV (γ_1) and 497 keV (γ_2) with a half-life of 3.3μsec. The ground state of ^{115}In decays with a life time of 4.4×10^{14}yr to the ground state of ^{115}Sn through β-decay.

can be done in real time, it is very suitable for studying the time variations of solar neutrinos.

The signal can be identified, for example, in the following way (a schematic is shown in Fig. 2): (1) to detect a prompt electron with an energy of $E_{\nu_e} - 125$ keV in a vertex cell; and (2) to detect two decay products of ^{115}Sn* within a sufficiently wide time window to cover $\sim 3.3\mu$sec. The following restrictions may be further applied: the 116 keV or the 90 keV energy deposit for γ_1/e_1 can be found in the vertex cell and a total 614 keV energy deposit can be required in the adjacent cells; The cell size is assumed to be chosen in such a way that the low-energy γ_1 is detected within the vertex cell. (A detailed study concerning the identification algorithm and the detection efficiency in the case for ^{7}Be-neutrino detection are given in section 5.)

The capture rate taken from the calculation of Bahcall and Ulrich [2] is shown in Table 1 with a further discussion of the uncertainties presented in section 3.1. The capture rates of 468 SNU for *pp*-neutrinos and 116 SNU for ^{7}B-neutrinos are large compared to the gallium experiments, in which 71 SNU for *pp*-neutrinos and 34 SNU for ^{7}Be-neutrinos are expected. The expected event rate of the indium experiment for the assumed capture rate in Table 1 implies that for *pp*-neutrino detection; a three-ton detector may yield 0.5 event per day, and therefore 10 tons of indium are required for ^{7}Be-neutrino detection.

3 Problems to be addressed

We now turn to the question of why nobody has succeeded to implement an experiment using indium. There are three major problems: (1) background, (2) detector fabrication, (3) uncertainties in the capture rate. The first and second problems are related to each other, and it is likely that they can be solved simultaneously once we solely aim to measure ^{7}Be-neutrinos, rather than *pp*-neutrinos.

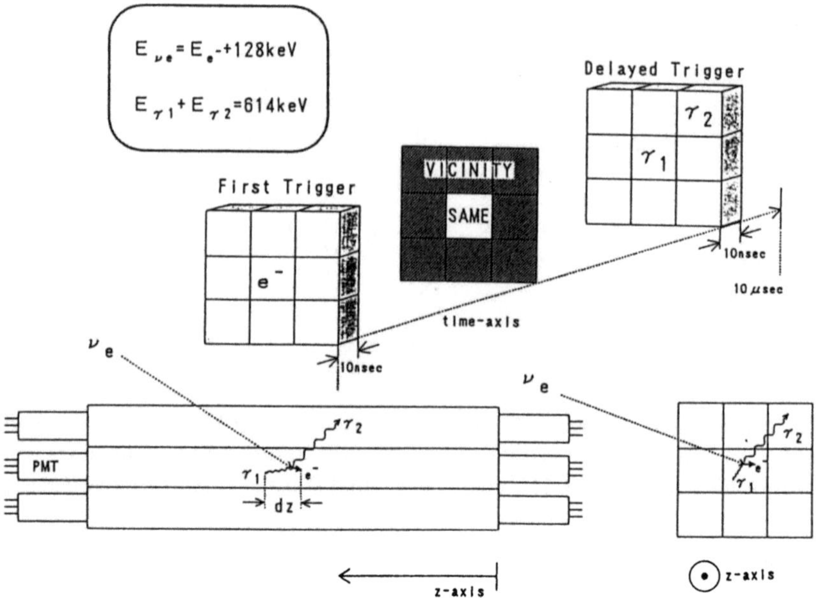

Figure 2: The event identification scheme of the neutrino capture by an indium detector. The real time coincidence of the three signals is the most important characteristics of the detector.

Table 1: Capture rate for ^{115}In from the standard solar model of reference [2]. See text the discussion on errors.

Neutrino Source	Capture Rate (SNU†)
pp	468
pep	8.1
hep	0.05
^{7}Be	116
^{8}B	14.4
^{13}N	13.6
^{15}O	18.5
^{17}F	0.2
Total	639

† SNU (Solar Neutrino Unit): 10^{-36} capture /sec/atom

3.1 Background and detector-fabrication problems

Indium experiments were initially proposed to measure *pp*-neutrinos due to its low threshold energy of 125 keV. There is, however, a large background due to β-decay of radioactive ^{115}In, as shown in Fig. 1. The ground state of ^{115}In undergoes a transition to ^{115}Sn with a half-life of 4.4×10^{14} years. The energy of the electron ($E_{max} = 487$ keV) covers the entire spectrum of what is expected from *pp*-neutrino reaction, of which the end-point energy is 420 keV. The rate of indium β-decay (0.22 decays/g/sec) is 11 orders of magnitudes larger than that of *pp*-neutrinos of 3×10^{-12} captures/g/sec for an assumed capture rate of 640 SNU. The indium β-decay is serious background for a *pp*-neutrino detector. Another dangerous background is stray γ-rays from detector impurities or environmental origins (hereafter we call them environmental γ-rays). Multiple-Compton scattering of those γ-rays mimics the delayed coincidence signature of γ_1 and γ_2 that follow the capture of a solar neutrino.

In order to suppress this type of background, the delayed coincidence signature must be used. Segmentation of the detector is then very effective for minimizing the background; accidental delayed coincidence signals that happened at spatially different points can be rejected as background by segmentation. For an order-of-magnitude estimate of such background, suppose that M(g) is a mass in a unit detector volume in which the signal (both prompt and delayed) can be measured. The rate of the signal is given by $n_s = 3 \times 10^{-12} \times M/\text{sec}$. In the following evaluation we have assumed that the most serious background for the prompt signal is indium β-decay and that for the delayed signal is multiple Compton scattering of stray γ-rays. Although we may further require the energy window for both γ_1 and γ_2 separately, we do not use it in this evaluation and we only use the total-energy deposit of two γ-rays. The rate of the background thus becomes

$$n_{BG} = n_\beta \times n_\gamma \times M \times (\Delta\,\text{keV}) \times 10^{-5}, \tag{2}$$

where $n_\beta = 2.2 \times 10^{-1} \times M$ and the coincidence width of 10μsec and $n_\gamma = 10^{-8}$ /keV/g/sec are used; Δ keV is the energy resolution at around 600 keV. In order to obtain $S/N = 1$, it is required that

$$M \cdot (\Delta\,\text{keV}) \sim 70(\text{keV} \cdot \text{g}). \tag{3}$$

If the detector has a moderate energy resolution, such as $\sim$ 100 keV, $10^6 \sim 10^7$ segmentation is necessary for a 1-ton detector. If we have a superdeeper energy resolution, say 1 keV, then 10^4 segmentation may work. The good energy resolution is certainly effective. Other background for the delayed signal, something like β and γ coincidence or a pile-up of two γ-rays, can be suppressed by an order of 10^8 if a timing resolution of 10nsec is accomplished (which is not difficult to achieve). Those background are therefore not too serious. The only feasible way toward *pp*-neutrino detection is to obtain an extremely good energy resolution for a unit detector size of $\sim$ 100 g in addition to a nano-second timing resolution. Although many attempts have been made, none has been successful.

The ^{7}Be-neutrinos from the $^7\text{Be} + e^- \to ^7\text{Li} + \nu_e$ process have a monochromatic energy of 861 keV (and 383 keV for 10% of the flux), being well above the end-point energy of the indium β-decay. The background due to β-decay can be significantly reduced at this energy and, hopefully, overcome. For a typical energy resolution of 10%, the rate of the background is $\sim 2 \times 10^{-4} \times n_\beta$ (see section 5). Since the solar-neutrino event rate for ^{7}Be is also reduced by a factor of 5, the net effect is $\sim 4 \times 10^{-3}$. Therefore, 10^3 to 10^4 segmentation for a 10-ton detector is needed for the experiment. A large-volume indium detector is needed to compensate for the low capture rate. A reduction of stray γ-rays and a uniform detector quality with reasonably good energy resolution is an essential point for detector development. A detailed study of an experimental configuration and a background evaluation for ^{7}Be-pep detection is described in section 5. If we are aiming at only ^{7}Be-neutrinos, the detector-fabrication problem may also be overcome.

An indium detector is capable of measuring ^{7}Be-neutrinos separately from other solar neutrinos. It is also possible to measure *pep*-neutrinos with a much higher energy of 1440 keV by using the same configuration. The capture rate of these *pep*-neutrinos is roughly 10% of that for ^{7}Be-neutrinos. A simultaneous measurement of both ^{7}Be- and *pep*-neutrinos is very interesting. The ratio of *pep*-neutrinos to *pp*-neutrinos depends on model only weakly [3]. Hence the ration of *pp*-neutrinos to *pep*-neutrinos would give an interesting test for the MSW solutions [4] of the solar-neutrino problem.

3.2 Capture rate and calibration

Another important problem is that the capture rate is uncertain up to 100% due to difficulties in measuring its ft-value directly. Therefore, an indium detector must be calibrated by an empirically reliable method. The most feasible way is to use laboratory neutrino sources, for example strong radio-active sources which emit neutrinos.

The capture cross section to the $(7/2)^+$ state of ^{115}Sn is expressed as

$$\sigma_{capt} = \frac{2\pi^2}{c}\left[\frac{\hbar}{m_0 c}\right]^3 \frac{\ln 2}{(ft)_{inv-\beta}} W_e^2 G(Z, W_e, P_e), \tag{4}$$

where W_e is the total electron energy and $G(Z, W_e, P_e)$ is the modified Fermi function [5]. Both the uncertainty in the ft-value for the transition to the 614 keV state and an additional contribution to the higher excited states (as shown in Fig.1) must be considered. There are no direct measurements of the ft-value between the $9/2^+$ state of ^{115}In and the $7/2^+$ state of ^{115}Sn*, because the γ-transition to its ground state is much faster than the β-decay of ^{115}Sn* that makes a measurement of ft impossible.

There is a systematics that a low-lying 7/2 state appears in odd Sn isotopes. A systematic observation of the allowed Gamow-Teller transitions is made above a mass number of $A = 114$. Only the transition of ^{115}Sn to ^{115}In is not observed. This systematics may give a good basis for predicting the ft-value of the ^{115}In-to-^{115}Sn transition by extrapolating the value from the neighboring nuclei, which is what Raghavan used in his 1976 paper [1]. Here, we briefly recapitulate his estimate. Based on pairing-correlation theory [6], the prediction of the ft-value is $ft = (2I_i + 1)/V_n^2(N)$, where $V_n^2(N)$ is the neutron pair occupation probability.

$V_n^2(N)$ changes by only a few percent for $A > 115$, since $Z = 50$ (Sn) has a proton closed shell. The experimental value are $\log(ft) = 4.4$ for $A = 117, 119, 121$. Therefore, Raghavan predicted that

$$(ft)_{inv-\beta}(^{115}\text{In} \rightarrow ^{115}\text{Sn}) = (ft)_{\beta}(^{117}\text{In} \rightarrow ^{117}\text{Sn}), \tag{5}$$

where $(ft)_{\beta}(^{117}\text{In} \rightarrow {}^{117}\text{Sn}) = 2.5(2) \times 10^4$ sec with $\sim 10\%$ accuracy [1].

The ft-value and its uncertainty can also be estimated with the aid of (p, n) reactions, by which the above prediction can be tested, although this method has its own uncertainty. The result from an experiment carried out by J. Rapaprot et al. [7] is in good agreement with the prediction made by Raghavan [1]. Although the uncertainty of the ft-value assigned by the experiment is of the order of $\sim 20\%$, it has been pointed out [1] that the estimate of the GT matrix element from the (p, n) scattering is uncertain, especially for large mass-number nuclei. The difference between the quantum-mechanical operators for the GT transitions and for (p, n) scattering becomes significant due to the larger contribution from the orbital angular momentum in the large mass-number nuclei than that in the small mass-number nuclei. Although many experiments have been performed in order to examine the reliability of the estimate of GT transition matrix element from the (p, n) scattering data, they have not yielded a conclusive answer. Therefore, Bahcall [1], taking into account the uncertainty of the transition to the excited states above 614 keV, assigned the uncertainty of the cross section as $+100\%$ and -50% for the "3σ" value. A reasonable estimate of the uncertainty of the capture cross section is somewhat between 20 and 30% for 1σ error.

For an indium experiment, it is essential to make an independent calibration using a laboratory neutrino source, or to design an experiment in such a way that it does not depend on the absolute callibration: for example, a measurement of the ratio of the two-line spectra of solar neutrinos.

Conditions required on the calibration sources are different from those for radiochemical experiments. The energy of the prompt electron must be well above the end-point energy of the indium β-decay. The source must have a low γ-ray emission probability, since we must eliminate an additional trigger due to extra γ-rays, since the detector operates in real time. There are three candidates for the laboratory neutrino source [8][9][10]:

(1) $e^- + {}^{65}\text{Zn} \rightarrow {}^{65}\text{Cu} + \nu$ (0.227 MeV 51%)
$e^- + {}^{65}\text{Zn} \rightarrow {}^{65}\text{Cu} + \nu$ (1.343 MeV 48%)
${}^{65}\text{Zn} \rightarrow {}^{65}\text{Cu} + e^+ + \nu$ (0.330 MeV 1%)

(2) $e^- + {}^{51}\text{Cr} \rightarrow {}^{51}\text{V} + \nu$ (0.751 MeV 90%)
$e^- + {}^{51}\text{Cr} \rightarrow {}^{51}\text{V} + \nu$ (0.426 MeV 10%)

(3) $e^- + {}^{37}\text{Ar} \rightarrow {}^{37}\text{Cl} + \nu$ (0.813 MeV 100%).

It is difficult to obtain a strong and handy radioactive source of ^{37}Ar, since Argon is a noble gas and hard to be made in any form of compounds. Although the γ-ray emission from ^{65}Zn and ^{51}Cr is useful to determine the absolute strength of the source, the γ-rays from these sources must be shielded for the detector. Since the high energy γ-rays of 1.1 MeV from ^{65}Zn is difficult to be shielded, ^{51}Cr is the only suitable radioactive neutrino source for the calibration of an indium detector.

The cross section of neutrinos from ^{51}Cr (751 keV) on ^{115}In is estimated to be 2.2×10^{-44} cm^2. Then, ^{51}Cr of 30 MCi gives 100 events/day for a 10-ton indium detector assuming that the detector is placed at 3 m from the source. Considering the production cross section of ^{51}Cr (15.9 barn) from natural Cr by thermal neutrons, we can obtain 33 MCi of ^{51}Cr by irradiating 1.5×10^{14}/cm^2/sec neutrons (at 1 m distance for ~100 MW reactor) on 1 mg of natural Cr. We then require about 900 kg of Cr as a neutrino source. The production of ^{51}Cr neutrino source with the mass of 12.5 kg, which was aimed to use for the calibration of Ga experiment, has already been reported [11]. It was made by using the 35 MW Siloe reactor.

4 Detector development

Historically the development was pursued by Raghavan and his collaborators [12][13]. They made a liquid scintillator from solvent phenethyl alcohol ($C_6H_5CH_2$ CH_2OH) and indium trifluoroacetate ($(CF_3CO_2)_3In$) with 5.1% indium-loading. An energy resolution of $\Delta E \sim 5.16/\sqrt{E}$, where E is in keV, was achieved. However, their liquid scintillator had a serious light-attenuation problem. Although the further development of a similar liquid scintillator along this line was reported [14], the attenuation length was not drastically improved. They obtained 68 cm for 5% indium loading and 28 cm for 10% loading. These earliest attempts are not suitable for either *pp*-neutrino detection (because of its poor energy resolution) nor ^{7}Be neutrino detector (because of its poor light attenuation).

It is recognized that the detector requirement is quite different for *pp*-neutrino detection and for ^{7}Be-neutrino detection. Thus, in the following sections we treat them separately.

4.1 Detector for ^{7}Be-pep neutrino detection

A dedicated indium experiment aiming to detect ^{7}Be-pep neutrinos was first discussed by Drukier and Nest [15] in 1985. Although their idea was to use a coated wire of indium and 15 m^2 of liquid xenon chamber, no experimental test along this line has been performed.

In 1989, researchers from Europe, U.S. and Japan met in Paris to make detailed studies of indium ^{7}Be-neutrino detectors. The technical problems were evaluated; most of them could be generally overcome. Three different types of detectors (scintillation fibers with indium clad, indium plates immersed in a liquid scintillator and indium-loaded liquid scintillators) were studied in detail. Most of them, except for the liquid scintillators, are only in the idea stage, and only a few calculations on their property have been made. The energy resolutions of the various liquid scintillators so far made are given in Table 2. The indium-loaded liquid scintillator brought from a Japanese group [16] provided the best performance. Its energy resolution and characteristics satisfied the necessary requirements to be a

Table 2: A comparison of the indium loaded liquid scintillators dedicated for ^{7}Be-neutrino detection.

Reference	Indium loading (%)	σ/E (%) 734 keV	Size of the detector
L. Pfeiffer et al. [12][13]	5%	8%	10cm cell
N. Booth et al. [14]	5%	18%	1m cell
	10%	30%	1m cell
Y. Suzuki et al. [16][17]	5%	9.5%	1m cell
	7.1%	10.3%	1m cell

practical detector. Since indium-loaded liquid scintillators are the most promising one, we describe in detail their development status.

(i) Indium-loaded liquid scintillators

An indium-loaded liquid scintillator is well suited for ^{7}Be-pep neutrino detectors, since a large apparatus can be made large with a reasonably good energy resolution, and can be constructed relatively cheaply. Based on the argument given in the previous section, the detector unit should have a size of (10 $\sim$ 20 cm) $\times$ (10 $\sim$ 20 cm) $\times$ (100 $\sim$ 200 cm), containing 5 $\sim$10% indium. An attenuation length $\geq$100 cm and a uniform energy resolution better than $\sim 2.7/\sqrt{E}$ (10%@E_ν) over a $\geq$100 cm long cell is crucial for ^{7}Be-detection. A cell of this kind also ensures the containment of 100-keV photons in a vertex cell.

The earliest attempts [12][13][14] are not suitable for ^{7}Be-neutrino detection since they do not have a long light attenuation length. This difficulty results from a well-known problem: how to dissolve polar materials in non-polar liquids. Indium can not be easily dissolved into an organic liquid scintillator.

There are a few possible ways to get around this: (M1) The use of a solvent which has a high solubility for polar compounds with an efficient energy-transfer property. The phenethyl alcohol scintillator is one of such solvents used in this method; (M2) The use of an organic metal of indium and an organic scintillator; (M3) The use of an organic scintillator with a surfactant having hydrophilic and hydrophobic chemical groups. This emulsive scintillator enables one to mix a polar indium compound into a non-polar organic scintillator.

The early developments were based on method (M1). As stated, these liquid scintillators were limited by the light attenuation. Nobody has succeeded in finding suitable combinations of organic indium metals and organic scintillators (M2) to achieve a reasonably good quality. Suzuki et al. [16][17] adopted method (M3), and succeeded to make an indium-loaded liquid scintillator having good qualities.

The first indium-loaded liquid scintillator of Suzuki et al. was made from a commercially available xylene-based emulsive liquid scintillator containing a surfactant (hereafter called EX-H [19]) mixed with $InCl_3 \cdot 4H_2O$ (Indium tri-chloride tetra hydrate) [18]. The EX-H has DPO (2,5-Diphenyloxazole: $C_6H_5 \cdot C_3HNO \cdot C_6H_5$) as a primary flour and bis-MSB (1,4-Bis(2-methylstyryl) benzen: $CH_3C_6H_4CH$:CHC_6H_4 CH: $CHC_6H_4CH_3$) as a secondary flour. A maximum indium loading of 7.5% by weight was obtained.

The indium concentration can be increased by raising the temperature. The density is $0.97 g/cm^3$ and the index of refraction is about 1.5, which is well matched to that of quarts glass.

Inoue et al. [17] extended their work and succeeded to make an indium-loaded liquid scintillator based on 1-2-4 trimethylbenzene (pseudo cumene (PC: $C_6H_3(CH_3)_3$) and toluene. The choice of the solvent is crucial, since xylene has a strong chemical reaction and a low flush point of 27°C. It is therefore not suitable as a practical detector, whereas pseudo cumene is not chemically active and has a flush point of 54°C. The solvability of indium for various combinations of solvents and surfactants is given in Fig. 3. Isopropyl alcohol and commercially available surfactants, (Liponox NC-95 and Nonion-NS210) were used. A maximum indium concentration of 14% was obtained for S500, which is also a commercially available emulsive scintillator. A pseudo-cumene-based scintillator revealed maximum indium concentration of 6%. The indium solvability and quantity used for the scintillator was quite similar for all of the scintillators we have made (except for S500). For a 7.5% indium concentration, 30~40% surfactant was necessary.

The energy response to γ-sources was measured for liquid scintillators with different indium concentrations from 0 to 14%. In Fig. 4 the relative pulse heights of these liquids to EX-H without indium for 477 keV electrons are given. A pseudo-cumene scintillator using isopropyl alcohol as the surfactant has about 1/2 of the light output compared to the other kinds. The better quality was obtained from a combination of pseudo cumene and a commercially available surfactant (NC-95).

The light attenuation length and energy resolutions for a large cell were measured using a cell made of a quartz tube (1 m in length and 5 cm in diameter). Backscattered 477 KeV compton electrons were used for the measurements in which the ^{137}Cs source was moved along the cell. The attenuation length was obtained by fitting the data using a double-expotential formula, PH = $A_1 \exp(-x/\lambda_1) + A_2 \exp(-x/\lambda_2)$. The short component ($\lambda_1$) is typically ~ 20 cm for about 50% of the light ($A_1 \sim 0.5$); the long component (λ_1) is typically ~ 155 cm for about 50% of the light ($A_2 \sim 0.5$) for 7.1% indium loading.

The energy resolution obtained along the 1 m long cell is shown in Fig. 5. They obtained a reasonably uniform energy resolution of 13.1% along the cell for 7.1% loading and 11.7% for 5% loading at 477 keV. If we assume a $1/\sqrt{E}$ dependence of the energy resolution, the resolutions can be expressed as $\sigma/E = 2.8/\sqrt{E}$ and $2.6/\sqrt{E}$ for 7.1 and 5% loading, respectively, where E is in keV. A position resolution of ± 4.5 cm (σ) was obtained from the time difference between the two PMTs attached at each end.

An indium-loaded liquid scintillator made by a new method involving a surfactant has very good qualities: a long light attenuation length, a reasonably good energy resolution of over 1 m cell and the good timing capability. The obtained

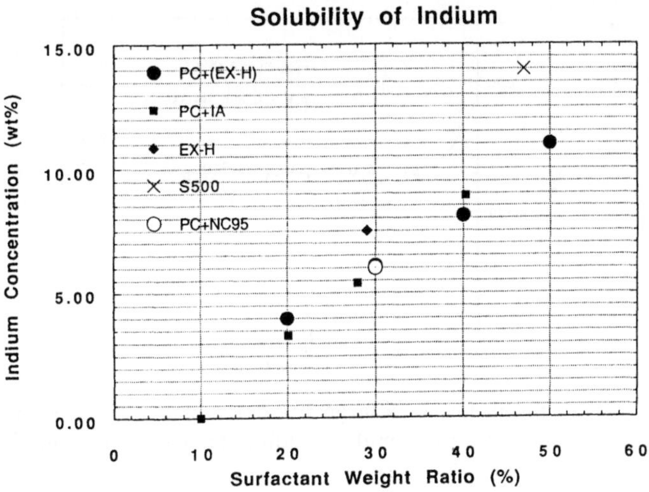

Figure 3: The maximum weight of indium resolved for various combinations of solvents and surfactants in a liquid scintillator while keeping its transparency (the maximum solubility). PC, IA and NC95 stand for 1-2-4-trimethylbenzene (pseudo cumene), isopropyl alcohol and Liponox NC-95 (surfactant), respectively. EX-H and S500 are commercially available emulsive scintillator cocktails. The solubility and the quantity of surfactant are almost proportional and are independent of a kind of combinations of surfactants and scintillators. About 30 to 40% surfactant is necessary to make a 5 to 10% indium loaded liquid scintillator.

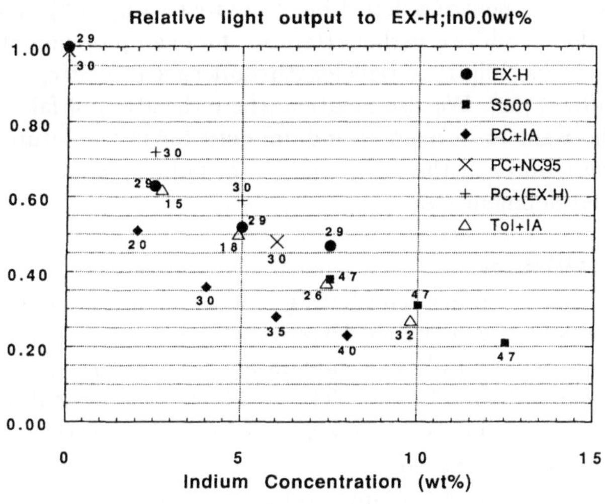

Figure 4: The relative pulse height of various indium-loaded liquid scintillators to those without indium (EX-H and PC+NC95) measured by a backscattered electron from ^{137}Cs. Notations are the same as in Fig. 3. Tol means toluene. The small numbers associated with the symbols show the weight ratio of surfactant in scintillator cocktails.

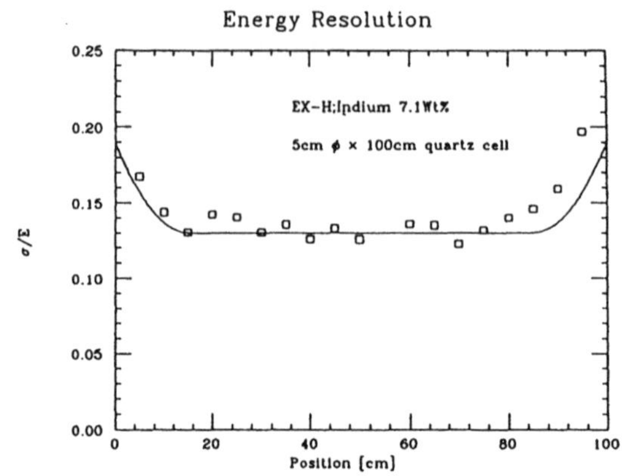

Figure 5: The energy resolution for a large size detector of 100 cm long. The uniform energy resolution is obtained for an emulsive indium-loaded liquid scintillator.

qualities are almost sufficient to satisfy the requirement for the ^{7}B neutrino detector. Liquid scintillators utilizing surfactants have really opened up a possibility to realize a high-quality indium-loaded liquid scintillator suitable for a practical ^{7}Be-solar neutrino detector.

4.2 Detectors developed mainly for *pp*-neutrino detection

So far we have not attempted to review the past and present efforts for an indium detector aimed at *pp*-neutrino detection. Here we pick up subjectively several typical ideas. We do not address earlier attempts, for example, those using indium foils with drift chambers [20][21] or wavelength-shifting scintillating fibers immersed in a liquid scintillator [22]. We also mention where the attempt failed for convenience for people who want to develop an indium detector in the future so that they do not repeat the failures in the past; this list of failures would be a starting point for the next development stage.

(i) Solid scintillators

Solid scintillators (either scintillating crystals or indium-doped solid scintillators) are suitable for *pp*-neutrino detection. They are easily fabricated in small pieces, and are supposed to have a moderately good energy and timing resolution. However, we need millions of pieces. The goal is to find a crystal containing indium and having a large luminescence as well as a low light-absorption coefficient.

(a) scintillating indium crystals

There are a number of indium crystals. Some of them are known to show photo-luminescence. The detailed characteristics, however, are not well known. In general, since they contain indium more than 50% by weight, they are very attractive.

Suzuki et al. [23] have studied the luminescence of InBr, InI and InP at room temperatures and at low temperatures down to 10°K. These crystals were grown by the vertical Bridgeman method to sizes of about 1 cm^3. InBr did not reveal any luminescence, even at 10°K. Though the photo-luminescence was observed for InI at low temperature by laser excitation, it was found that it was the absorption-edge emission. This, as it is, can not be used as a scintillator. InP is very well known as a semi-conductor material. Detector development along this line is discussed later. InP is also known to exhibit photo-luminescence in the infra-red region with various impurity dopings. InP(S) and InP(Zn) were found to have an absorption-edge luminescence at 950 nm, and are thus not suitable for a scintillator. InP(Er), InP(Nd) and pure InP were found to exhibit luminescence at 1.5 μm, and are transparent at this wavelength. A germanium semi-conductor detector was used to detect this luminescence. However, no response was obtained through an irradiation of α particles. A suitable choice of impurity doping may change the situation and may make it useful as a scintillator. Further systematic studies will be necessary to make possible improvement.

L. Gonzalez-Mestres and D. Perret-Gallix [24] pointed out that $InBO_3$, indium orthoborate with trisaled indium ions is one candidate for a scintillating crystal. A crystal with a 65 mm diameter and 2 mm thickness have been made [25], and the powder doped with Tb and Eu has been reported to exhibit cathode-luminescence at 550–600 nm [26]. They have studied single crystals of $InBO_3$ doped with Tb or Ce [24]. They also listed a bunch of candidates of possible scintillating crystals: $InGaO_3$; $NaInSi_2O_6$; $In_2Si_2O_7$; $In_2Ge_2O_7$; In_2O_3. However, none of them, including $InBO_3$, have been studied in detail. Further extensive studies on these materials are needed. They have also proposed the idea of a luminescent bolometer with which the scintillation light and phonons are detected simultaneously at low temperature [27].

The group at ISN Grenoble is also working on indium scintillators [28]. So far, although they found two candidates in which In represents more than 50%, they have had trouble of finding suitable dopants to make these crystals scintillating.

(b) Indium-doped solid scintillators

Extensive studies of indium-doped scintillators: BGO(In), CsI(In) and CsBr(In) have been carried out by Suzuki et al. [23]. BGO and CsI are well known to be good scintillators. Therefore, studies of these crystals are to determine how much indium can be added to them without degrading the scintillator quality. They also tested TlI(In) and TlBr(In). TlBr was found not to have any luminescence. TlI(In) has emission at 700 nm and is transparent at this wavelength up to a 10% indium concentration. However, the amount of light produced was only about 2% of that of NaI(Tl). The eduction of indium was observed in BGO crystal with a 3% indium concentration. Both CsI(In) and CsBr(In) showed good scintillating quality up to a 5% indium concentration. The amount of light for 4% indium concentration is about 30% of that of NaI(Tl); crystals with a 10% concentration were not transparent at all. Both CsI(In) and CsBr(In) with an indium concentration of up to 5% can be promising candidates.

Another well-known solid-state scintillator is plastic. T. Inagaki [29] has developed plastic scintillators with up to a 10% indium concentration with an attenuation length of 29 cm.

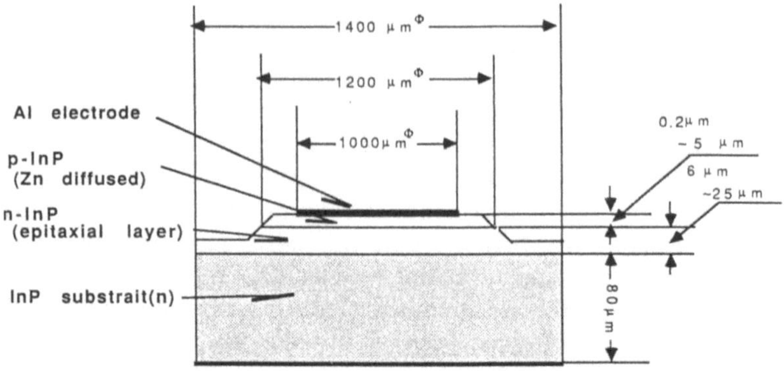

Figure 6: Schematic view of the InP photo diode. The high purity epitaxial layer of ~25 μm is grown above the InP substrate.

(ii) InP as a semi-conductor detector

InP is a high-density (ρ = 4.79 g/cm^3) material with a large atomic number ($\langle A \rangle$ = 80.2); its operability at room temperature is a great asset for it to be a good *pp*-neutrino detector. There are a few previous reports concerning InP photo conductors [30], which, however, give no information on the detection of single particles. InP depletion-layer diodes as a particle detector was first reported by Suzuki et al. [31][32].

Three indium compound semiconductors, namely InP, InAs and InSb, have been well studied, and their characteristics are understood. However, only InP can be operated at the room temperature, because of its large band gap of 1.29 eV. Since the characteristics of InP (mobility, intrinsic resistivity and so on) are just between those of Si and GaAs, InP is expected to reveal a good quality as a solid state detector. Diodes with a low impurity level of 10^{15} cm^{-3} have been fabricated [31][32], but they have a thinner epitaxial layer of 10 μm.

A cross-sectional view of the diode is shown in Fig. 6. High-purity InP layers were grown by liquid-phase epitaxy utilizing a multiple-well horizontal-slider carbon boat on a sulfur-doped $\langle 100 \rangle$ InP wafer (almost etch pit free). The p$^+$-layer of 0.2 ~ 4 μm thickness was obtained by Zn diffusion. The junction depth was controlled by the diffusion temperature and time.

The diameter of the diode was 1.0 mmϕ. The mesa structure was formed by the etchant (1HCl: 1CH_3CHOOH: 1H_2O_2) at 25°C for 3 minutes. Electrodes for the n-InP substrates and the p$^+$-layer are an Au-Ge-Ni alloy metal and an Al metal, respectively.

^{241}Am (60 keV photons) and ^{57}Co (122 keV photons) sources were used to obtain its energy response. The results for ^{241}Am gamma rays are shown in Fig. 7. An escape peak (~ 36 keV) and a 60 keV peak can be seen.

The method of the InP photo conductor has recently been revived by J. C. Lund et al. [33] They used a Fe-doped balk crystal of 2.2 mm thickness. Their energy response is shown in Fig. 8.

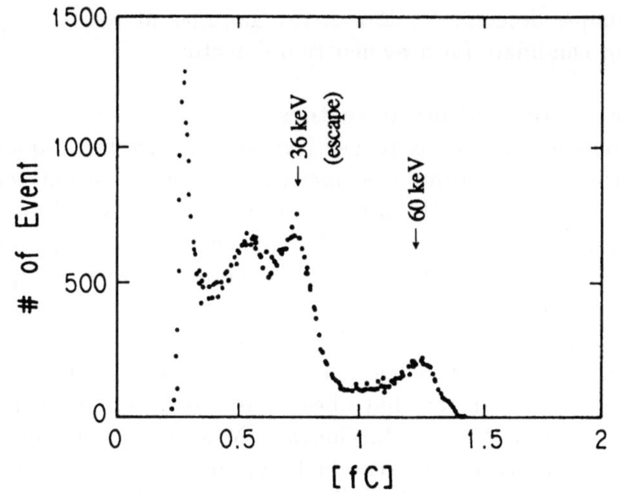

Figure 7: The energy spectrum of InP semiconductor detector for ^{241}Am source. A low energy peak and a 60 keV peak are clearly seen.

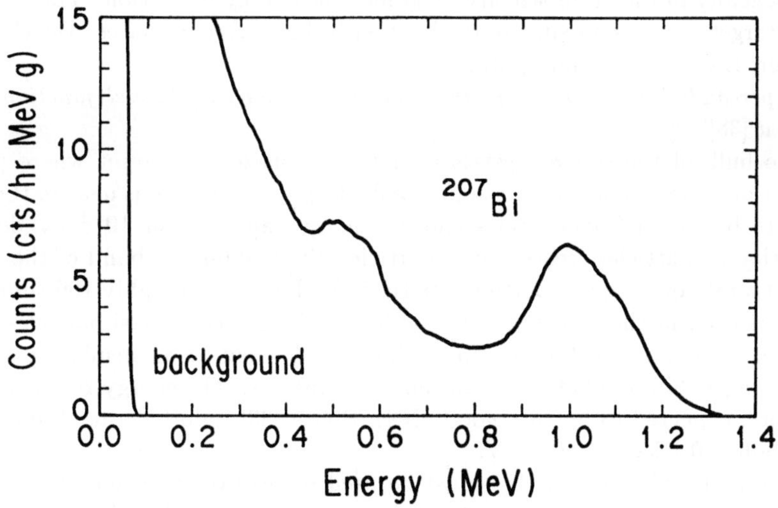

Figure 8: The energy response of the InP photo-conductor developed by Lund et al.[33] irradiated by ^{207}Bi. The 2.2 mm thick detector is operated at a bias of 6 kV.

Indium phosphate solid-state detectors are capable of detecting low-energy photons. If they can be fabricated in larger volume with a slightly better quality, this would be an unique detector for low-energy gammas at room temperature and would be a good candidate for a *pp*-neutrino detector.

(iii) Superconducting indium detectors

Another sophisticated idea is to use low-temperature techniques. Indium is a superconductor and its normal-to-super-conducting transition temperature is 3.408°K. A ^{4}He bath or a ^{3}He-^{4}He dilution refrigerator is needed for such a detector. Two techniques have been extensively pursued: superheated superconducting indium granules and superconducting indium detectors with a superconducting tunnel junction read out.

(a) Superheated superconducting granules (SSG)

This method is explained in detail in other article of this book. Particular applications to indium detectors have been pursued by G. Waysand [34], A. de Bellefon and P. Espigat [35], and L. Gonzalez-Mestres and D. Perret-Gallix [36] [37]. A matrix of superheated superconducting granules is held in a magnetic field. The energy released by particle interaction causes a transition of the granules from the superconducting state to the normal state causing a sufficient change in the magnetic field. A large number of read-out channels are required for this approach. The energy resolution is not yet well known. However, the timing resolution can be expected to be rather good. The major problem associated with SSG development is that very small grains must be made in order to obtain the necessary sensitivity for minimum ionizing particles. However, this results in small electronic signals. Inhomogenity in the grain size also degrades the energy resolution. The SSG is an active target and it is possible to read out with the X-Y coordinates. Development, however, is still in its infancy stage.

(b) Superconducting indium crystal with superconducting tunnel junction (STJ) read out [38][39]

The bulk of the single crystals of indium is used as a target where particle interactions release energy. In a superconducting state, electrons near to the Fermi level are bound in Cooper pairs and an energy gap of order 10^{-3} eV develops. Ionization by particles creates quasi-particles by breaking the bond of the Cooper pairs at a rate of $\sim$1 quasi-particle per 10^{-3} eV. This gives roughly 100-times more electrons than in the case of a semiconductor. The excess quasi-particles can be detected by a superconducting tunnel junction formed by an insulating material comprising a thin film between two super-conductors. The energy resolutions are, therefore, expected to be very good. N.E. Booth [33] reported that the expected resolution is 0.1 keV at 500 keV.

Bulk materials (at least 1g) must be used instead of the usual thin film as a solar neutrino detector. Technical problems were associated with the use of bulk materials. It was difficult to make junctions on the surface of bulk indium. Further, a tunnel junction acts like a semi-permeable membrane. It therefore takes a long time for the signal to pass through, resulting in a small magnitude of the signal current.

A sophisticated idea was proposed by N.E. Booth [40] in order to overcome these problems. If a small piece of another superconductor with a binding energy lower than that of indium is placed between the indium crystal and the junction, the quasi-particles diffuse into it and are trapped. They hit the tunnel junction more frequently and, thus, the signal current increases. This may be called a quasi-particle multiplier. He also elaborated on how to make a tunnel junction on the crystal surface. Actually, he reversed the method; he grows a crystal on the tunnel junction instead of making a tunnel junction on the surface of bulk indium.

However, a serious problem results from the fact that particles (either α-rays or γ-ray) could not produce sufficient quasi-particles, but produce low-energy phonons. This was explained by the fact that for materials which have a low Debye energy, like indium, the quasi-particle recombination rate exceeds the pair-breaking rate [41]. Further studies to overcome these problems are certainly needed.

(iv) Phonon-meadiated detector using InSb [42][43]

The energy deposited by a charged particle goes into the production of phonons in the semiconductor InSb, since the band-gap is only 0.2 eV. Also, the electrons and holes recombine to produce phonons on a time scale of a few μsec, due to its direct band-gap nature. These phonons can be detected by the STJ discussed in a previous section. Goldie et al. [44] fabricated a 2-mm thick crystal of InSb with two STJ's on its surface. Alpha particles were exposed through two small holes on the detector. They obtained a 1 mm spatial resolution through the time difference of the two STJ's. An energy resolution of 1 keV was obtained for 60-keV γ-rays. The InSb seems to be promising, but further study is necessary, especially concerning the best way to resolve two consecutive events in the same crystal. This is crucial for utilizing the three-fold coincidence technique.

4.3 Measurement of the indium β-decay spectrum

Three groups have measured the β-decay spectrum with various qualities, either by liquid scintillators or by InP. The first measurement was performed by L. Pfeiffer et al. [13]. They carried out a sophisticated analysis and obtained the half-life of ^{115}In as 4.41(25) $\times 10^{14}$ yr and the end-point energy of ^{115}In as 482(15) keV. Their results are given in Fig. 9.

The second measurement was carried out by K. Inoue, Y. Suzuki and others [17], who used a 5% indium-loaded liquid scintillator made of pseudo cumene mixed with Liponox-NC 95 in an acrylic container of 10 cm $\times$ 10 cm $\times$ 30 cm containing 150 g of indium. The measured spectrum is shown in Fig. 10 in which the event rate of 36 Hz was observed while 11 Hz was obtained from other detector unit which does not contain indium. The net event rate from the indium β-decay is directly estimated to be 25 Hz, which can be compared with the indium β-decay rate of 28 Hz for 150g of indium.

Indium β-decay was also measured by InP [33]. The authors used a 0.17 cm^2 volume detector accumulated over 117 hours.

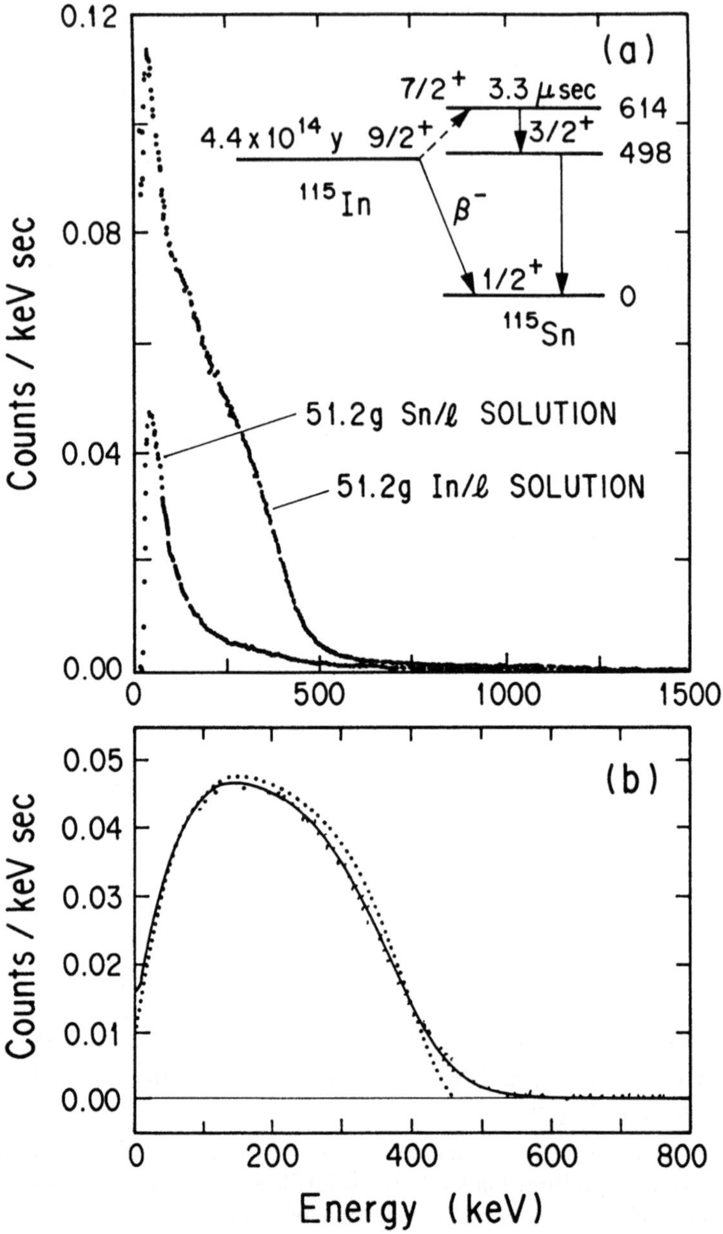

Figure 9: The spectrum of indium β-decay measured by Pfeiffer et al. [13]. (a) The raw β spectrum and background spectrum. (b) The background corrected β spectrum (dots). The solid line is a polynomial fit to the data.

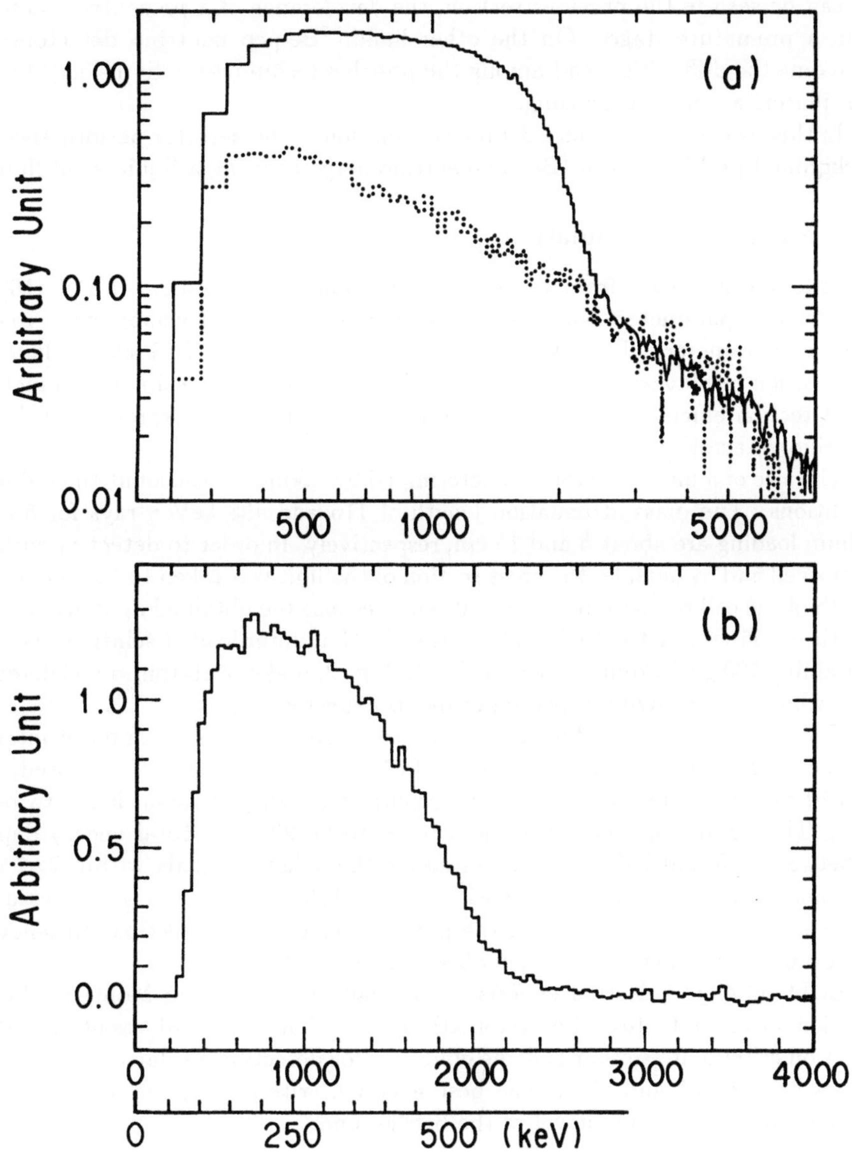

Figure 10: The spectrum of indium β-decay measured by Suzuki et al. [17]. (a) The raw β spectrum and background spectrum. (b) The background subtracted β spectrum. The observed event rate of 25 Hz/150 g agrees well with the predicted indium decay rate of 0.22 decays/g.

5 Design consideration for a ^{7}Be-*pep* neutrino detector

As can be seen in the previous section, the developmen of a *pp*-neutrino detector is in a premature stage. On the other hand, ^{7}Be-pep neutrino detectors may overcome the difficulties, and among the possible techniques indium-loaded liquid scintillators are most promising.

In this section we present a detailed evaluation of the detector performance and background problems for a ^{7}Be-*pep*-neutrino detector using a liquid scintillator.

5.1 Monte-Carlo simulation

We followed the results from a Monte-Carlo simulation carried out in Ref. [16][17]. The detector parameters obtained by the previous detector development were used as input parameters. Particularly, the energy resolution of 10% at 734 keV and the position resolution of ± 5 cm were used. The output of the simulation includes the detection efficiency and the acceptable environmental background level for a neutrino detector.

The size of a unit detector was determined by taking into account the following conditions. The mass attenuation length of 116 and 498 keV γ-rays for 5$\sim$10% indium loading are about 5 and 15 cm, respectively. In order to detect γ_1 within a vertex cell and γ_2 outside, the cross section of the unit was taken to be 10 cm. The length of the cell was assumed to be 100 cm, because the obtained light attenuation length is 155 cm. A total of 14000 cells of 7.5% indium-loaded scintillators, each containing 720 g of indium, assembled as in Fig. 2, makes a 10-ton model detector, with which 0.54 neutrino capture per day is expected.

The 7$\times$7 cells surrounding the vertex cell were taken as a region of interest. For a prompt signal, an energy window of $650\,\text{keV} < E_e < 850\,\text{keV}$ was used. The coincidence-time window between the prompt and delayed signal is set to be 10 μsec. The minimum-energy threshold is set to be 20 keV. Total energy deposit of between 510 and 750 keV is required for the delayed signals within 7$\times$7 cells surrounding the vertex cell. The position of the 116 keV delayed signal is assumed to be less than 15 cm away from the prompt signal. The detection efficiency for this event-selection criteria is then obtained to be 49%.

Eight different selection criteria are actually tested in the Monte-Carlo calculation in order to test the systematic biases. The obtained results for these selection criteria are similar and the efficiency for detecting a solar-neutrino signal ranges between 30 and 67%. The best selection criteria must be determined by considering the rejection power of the background.

5.2 Background for the ^{7}Be-*pep* neutrino detector

The background for the prompt signal (734 keV electron) includes the following:

(P1) Electrons from indium β-decay. While the end-point energy of β-decay is substantially lower than the energy window for prompt signal that concerns us, β-decay still gives a major source of the background due to a tail of the detector's energy resolution. When the β-decay spectrum of $E_{max} = 486\,\text{keV}$

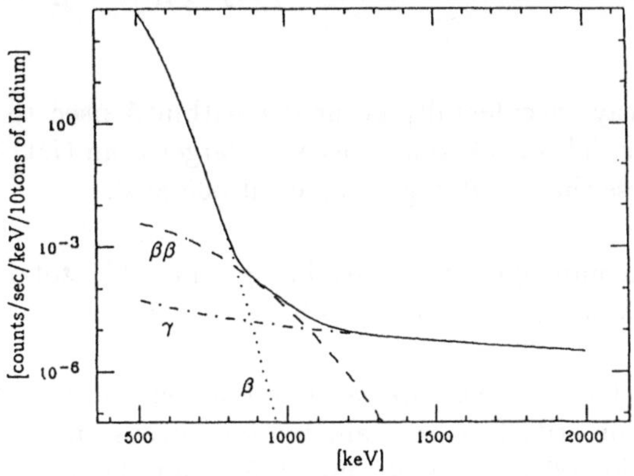

Figure 11: The various backgrounds for the prompt signal of indium ^{7}Be-solar neutrino detector obtained from the detector simulation. The background rate from the smeared tail of indium β-decay is denoted by β. The coincident two β-decays and γ rays from the radio-active impurities (10^{-8} /g/keV/sec at around 500 keV) are denoted by $\beta\beta$ and γ, respectively.

is convoluted with the resolution curve of the detector we expect a rate of 51 Hz in the window between 650 and 850 keV from β-decay for 10 tons of indium (estimated by the Monte-Carlo calculation). This value is compared to neutrino capture of 6.3×10^{-6} Hz. The various background contributions for the prompt signal are shown in Fig. 11.

(P2) Accidental coincidence of the two β-decay which is denoted by $\beta\beta$ in Fig. 11. A coincidence window of 5 ns was assumed. A good timing resolution and fine segmentation can alleviate this background. We estimate that background rate is 0.16 Hz using a Monte Carlo simulation.

(P3) Gamma rays from radioactive impurities in the detector material or environmental origins. See the line indicated by γ in Fig. 11. This background is smaller than 4.4 $\times 10^{-3}$ Hz if the γ-ray rate is less than 10^{-9}/g/keV/sec at around 500 keV. The γ-ray energy was assumed to follow $dn/dE \sim E^{-1.5}$ in the simulation.

(P4) Alpha particles from the impurity of the detector material.

(P5) Spallation products produced by cosmic-ray muons traversing the detector.

Background (P1) is the most serious for a prompt signal. This background is reduced by the additional requirement of the detection of the delayed signals: γ_1 in the vertex cell and γ_2 in the cell surrounding the vertex within 10 μsec. However, there are also many other background sources for the delayed signal.

(D1) Multiple Compton scattering of γ-rays from radioactive impurities in the detector material or of environmental origin. This background invalidates the two-fold coincidence scheme of γ-rays (γ_1 and γ_2) and is therefore most serious.

(D2) Two β-decays accidentally occurring within 5 nsec to mimic the delayed coincidence. This background becomes larger than (D1) when the γ-ray rate becomes less than $\sim 10^{-9}$/g/keV/sec at 500 keV.

(D3) β-decay accompanying Bremsstrahlung γ-ray. The total energy requirement of the delayed signal reduces this background.

Background (D1) and (D2) are taken into account in the Monte-Carlo calculation. The selection criteria were evaluated for vairous environmental γ-ray rates. The event selection effciency is of roughly 50% for the selection criteria we have adopted. The required γ-level must be less than 5×10^{-9}/keV/g/sec in order to make the signal-to-background ratio better than S/N=1. Other selection criteria giving the efficiency between 30 to 60% allow an environmental background level of 10^{-8} to 10^{-10}/keV/g/sec. It is probably safe to say that an indium solar-neutrino experiment can be carried out if the environmental γ-level is reduced to less than a level of 10^{-10}/keV/g/sec.

A similar simulation were carried out by the Munich group and Saclay Group [45]. The Saclay group obtained the allowed background level of $(4 \sim 7) \times 10^{-9}$/keV/g/sec for a 15 cm × 15 cm × 100 cm cell based on the assumption of $\sigma/E = (2.6 \sim 2.8)/\sqrt{E}$ (keV). Their results are in good agreement with each other.

Fig. 12 shows the expected signal obtained by the Monte-Carlo simulations imposing the selection criteria so as to reduce a γ-ray rate of 10^{-10}/g/keV/sec. The solid line indicates the background. In an actual experiment, a background event can be detected by taking an off-timing three-fold coincidence trigger. We are therefore allowed to assume that the background-spectrum can be measured independently and subtracted out. The background-subtracted spectrum is shown in Fig. 12. We can clearly see a peak due to the solar-neutrino signal.

5.3 Acceptable impurity level

A Monte-Carlo simulation was used to estimate the allowed ^{238}U/^{232}Th contamination from the environmental γ-ray background level expressed in units of /keV/g/sec obtained in the previous section. The results obviously depend on the size and configuration of the detector. For the particular configuration used in the previous section, 1ppb of ^{238}U(^{232}Th) contamination in the liquid scintillator was found to be equivalent to 5×10^{-9}/keV/g/sec (2×10^{-9}/keV/g/sec). The calculation was also carried out for the case of ^{40}K contamination; the result was 4×10^{-8}/keV/g/sec/ppb. This means that if we require the background level to be less than 10^{-10}/keV/g/sec, the internal background due to ^{238}U must be reduced down to at least the 0.02 ppb level. This level of the environmental background can be obtained with known technology.

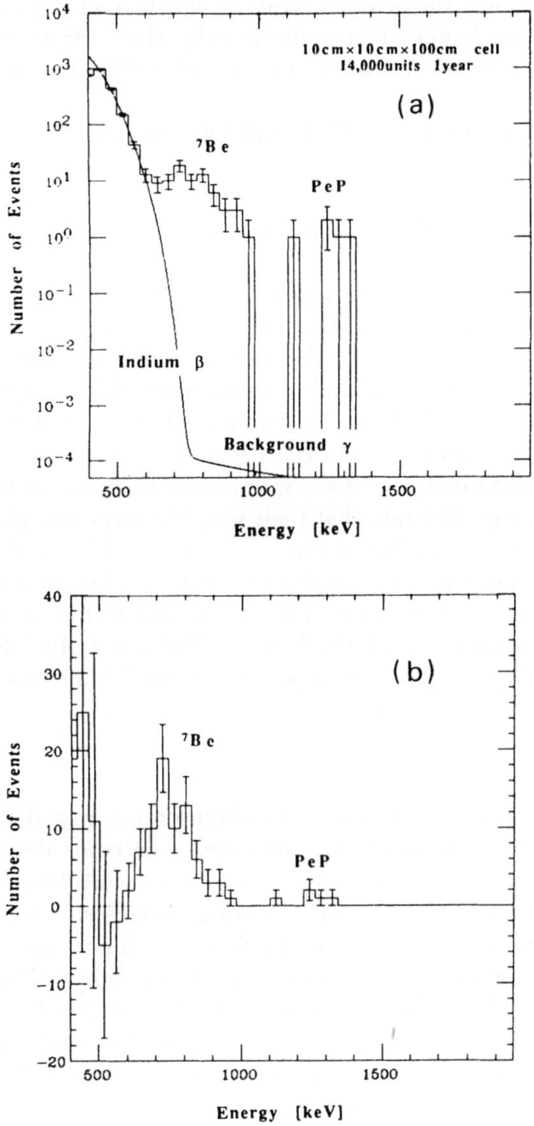

Figure 12: (a) The simulated energy spectrum measured in one year operation of 14000 unit indium detector that contain total of 10 ton of indium. The γ-ray background level of 10^{-10} near 500 keV region is assumed in the simulation. (b) The background subtracted signal for one year operation. The background spectrum can be obtained off-timing coincidence events.

5.4 Time correlated background

Time correlated background should be considered carefully, because the unique three-fold coincidence of the neutrino capture signature of ^{115}In might be mimiced by the series of decays happening simultaneously. Short-life nuclei in decays of the U-Ra and Th series are sources of the time correlated background. For example,

$$^{212}\text{Bi}\ (60.60\ \text{m},\ \alpha(36.0\%)) \rightarrow\ ^{208}\text{Tl}\ (3.053\ \text{m},\ \beta\gamma),$$

and

$$^{218}\text{Po}\ (3.05\ \text{m},\ \alpha) \rightarrow\ ^{214}\text{Pb}\ (26.8\ \text{m},\ \beta\gamma)$$

decays are of most serious in which first α-decay mimics the prompt signal and the subsequent $\beta - \gamma$, or even γ alone, mimics the delayed signals. Fortunately most of those daughter particles further decay within reasonable time which can be used to identify and veto this kind of background. We can estimate the rate of the backgound assuming the U or Th contamination level of 1ppb (although we require less than 0.02 ppb in the previous section). The contamination gives 100 Hz or 500 Hz of above mentioned decays, respectively, in the 10-ton-indium loaded (7%) liquid scintillator. The rate that their second decays take place within 10μsec is 7×10^{-6} Hz. Furthermore this background rate would be reduced to less than 10^{-7} Hz with the aid of a pulse shape discrimination on the prompt α-particles and the restricted energy requirements. The rate of non-time-correlated background from the 1ppb contamination is 4×10^{-6} Hz for ^{238}U and 2×10^{-6} Hz for ^{232}Th. The time correlated background is thus negligible for the ^{7}Be neutrino detection.

6 Summary

It seems to be not possible to make an indium detector to detect *pp* solar neutrinos with the present technology. However, a detector dedicated to observe ^{7}Be-neutrino may be possible. Indium-loaded liquid scintillators, especially those using emulsive scintillators, are most promising. Small-scale test experiments are needed to verify its feasibility and to clarify any technical problems. There still remain a few important problems: (a) capture rate is not well known, calibration by a laboratory neutrino source is necessary. (b) We must reduce any background of internal or external origin. However, the level of the requirement is moderate. (c) A fancy trigger and a read out electronics are necessary. These requirements may increase the cost of the experiments.

References

[1] R. Raghavan, Phys. Rev. Lett. **37**, 259 (1976).

[2] J. N. Bahcall and R. K. Ulrich, Rev. Mod. Phys., **60**, 297 (1988) and references therein.

[3] J. N. Bahcall, "Neutrino Astrophysics", Cambridge University Press (1989).

[4] S. P. Mikheyev and A. Yu. Smirnov, Sov. J. Nucl. Phys. **42**, 913 (1985); L. Wolfenstein, Phys. Rev. **D17**, 2369 (1978); ibid. **D20**, 2634 (1979).

[5] The tables of M. E. Rose, in Beta and Gamma Spectroscopy, edited by K. Siegbahn(North-Holland, Amsterdam, 1955), p875 was used in the calculation made by R. Raghavan's 1976 paper [1].

[6] See references in [1].

[7] J. Rapaport et al., Phys. Rev. Letts. **54**, 2325 (1985).

[8] L. W. Alvares, Physics Notes, Memo No. 767, Lawrence Radiation Lab (1973).

[9] R. S. Raghavan, in Proceedings of Informal Conference on Status and Future of Solar Neutrino Research, edited by G. Friedlander (Brookhaven National Laboratory) Report No. 50879, Vol 2, p1 (1978).

[10] W. C. Haxton, Phys. Rev. **C38**, 2474 (1988).

[11] M. Cribier et al., Nucl. Instr. and Methods, **A265**, 574 (1988).

[12] L. Pfeiffer et al., Phys. Rev. Lett. **41**, 63 (1978).

[13] L. Pfeiffer et al. , Phys. Rev. **C19**, 1035 (1979).

[14] A. G. D. Payne and N. E. Booth, Nucl. Instr. and Meth. **A288**, 632 (1990).

[15] A. K. Drukier and R. Nest, Nucl. Instr. and Methods, **A239**, 605(1985).

[16] Y. Suzuki et al., Nucl. Instr. and Methods, **A293**, 615 (1990).

[17] K. Inoue et al. ICRR-Report-252-91-21, ICRR Univ. of Tokyo, August, 1991.

[18] Dojindo Laboratories.

[19] $InCl_3 \cdot 4H_2O$(Indium tri-chloride tetra hydrate) is also resolved in water [see [16]]. The maximum indium loading of 35% was actually obtained. This indium contained water is very transparent and the density is 1.85g/cm^3. The index of reflection is about 1.7. By adding $CeCl_3$ to the indium loaded water as a wavelength shifter, 360 nm light emission through an 260 nm excitation by a xenon lump was obtained. The indium loaded heavy water can be used as a Cherenkov detector with Ce activator. But this cannot be used as a scintillator since the liquid does not have an energy transfer property.

[20] R. S. Raghavan, in Proc. Conf. on Status and Future of Solar Neutrinos, ed by G. Friedlander(Brookhaven National Laboratory Report BNL-50879, 1979) vol. II, p1.

[21] M. Spiro, in Proc. 4th Moriond Workshop on Massive Neutrinos in Astrophysics and in Particle Physics, ed. by Tran Thanh Van (Editions Frontieres, 1984) p311.

[22] R. S. Raghavan, in Proc. of Neutrino '81, ed. by R. J. Cenece et al. (Univ. of Hawaii, 1981) Vol I, p27.

[23] Y. Suzuki, Report of Grant-in-Aid for Scientific Research, 1988, The Ministry of education, Science and Culture.

[24] L. Gonzalez-Mestres and D. Perret-Gallix, LAPP-EXP-87-03.

[25] K. Oka and H. Unoki, J. of Crystal Growth **64**, 385 (83).

[26] F. J. Avella et al., Soc. Solid State Sci. **114**, 613 (1968).

[27] L. Gonzalez-Mestres, in Low Temperature detectors for Neutrinos and Dark Matter IV, ed. by N. E. Booth and G. L. Salmon, (Editions Frontieres,1992), p471.

[28] M. Avenier et al., Nucl. Phys. **B** (Proc. Suppl.) **28A**, 496, (1992).

[29] T. Inagaki, Proc of the first workshop on solar neutrino detection, KEK Report 86-7, December 1986, ed. by M. Sakuda and Y. Suzuki.

[30] D. R. Kania et al., Appl. Phys. Lett., **44** (1984)105, and references therein.

[31] Y. Suzuki, Proc. of the VIIth Moriond Workshop, Neutrinos and Exotic phenomena in Particle Physics and in Astrophysics, Les Arcs, Savoie, January 23-30, 1988.

[32] Y. Suzuki et al., Nucl. and Instr. and Method, **A275**, 142 (1989).

[33] J. C. Lund et al.,Nucl. Instr. and Meth. **A272**, 885 (1988).

[34] G. Waysand, Proc. of the Moriond workshop on Massive Neutrinos, January 1984, Ed. by J. Tran Thanh Van, p319.

[35] A. de Bellefon and P. Espigat, LPC 86/22.

[36] L. Gonzalez-Mestres and D. Perret-Gallix, LAPP-EXP-84-05-TH-112.

[37] L. Gonzalez-Mestres and D. Perret-Gallix, LAPP-EXP-88-01.

[38] Booth et al.,in Solar Neutrinos and Neutrino Astronomy, ed by M. L. Cherry et al., AIP Conf. Proc. **No.126**, p216 (1985).

[39] N. E. Booth, Sci. Prog. Oxf. **71**, 563 (1987).

[40] N. E. Booth., Appl. Phys. Lett. **50**, 293 (1987).

[41] N. E. Booth, in Proc. Int. Symposium on Neutrino Astrophysics, ed by Y. Suzuki and K. Nakamura, Kamioka/Takayama, October (1992).

[42] C. A. Klein, J. Appl. Phys. **39**, 2029 (1968).

[43] D. J. Goldie, in X-ray Detection by Superconducting Tunnel Junctions, ed. by A. Barone et al., (World Scientific,1991), p.98.

[44] D. J. Goldie et al., in Low Temperature Detectors for Neutrinos and Dark Matter IV, ed. by N.E. booth and G. L. Salmon (Editions Frontieres, 1992), p.245.

[45] A. de Bellefon et al., DPhPE 89-17.

Superheated Superconducting Granules for a Particle Detector

Takeo Ebisu

Department of Physics, Kobe University

Tadashi Watanabe

Department of Information Systems, Tokyo University of Information Sciences

1 Introduction

Recently, cryogenic methods have attracted much attention for the purpose of detecting feeble signals caused by a very small energy transfer induced by solar neutrinos [1][2] or weakly interacting massive particles, such as dark-matter candidates in galactic halos [3]. The principle of detectors is based on the properties of materials expected at low temperatures: 1) Both the specific heat and thermal noise decrease in crystals, 2) superconductivity appears in most metals, and 3) liquid helium shows superfluidity below the lambda point. Since the energy relevant to these properties is of the order of meV, much smaller than that necessary to excite or ionize atoms or molecules, the adoption of only one or two of them could provide novel and promising particle detectors. Feasibility studies have been carried out on the following detection devices: i) bolometers using either ballistic phonons in silicon or rotons in superfluid helium, ii) transition-edge thermometers based on superconductivity, iii) superconducting-tunnel junctions, and iv) superheated superconducting granules (SSGs) [4][5][6].

In this review, an introduction to superheated superconducting granules utilized for a target in a particle detector is given based on the results of the research and development carried out by the present authors. Superconductivity is characterized by perfect conductivity, perfect diamagnetism and the quantization of magnetic flux. During the transition from a state of perfect diamagnetism to that of paramagnetism, the magnetic susceptibility of metals makes a large change from $-1/4\pi$ to a value of the order of 10^{-6}. We can detect this change as an incident particle's signal which is induced in one of the grains embedded in a target. A granule of size less than several 100 μm remains superconductive in an applied

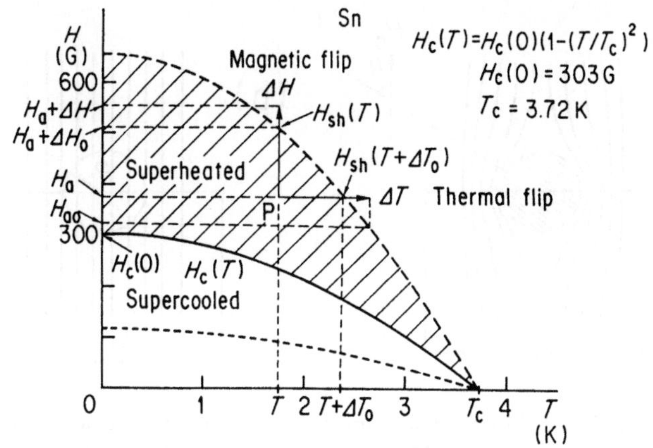

Figure 1: Schematic of the temperature-field ($T - H$) phase diagram for a small tin granule. The state in a hatched region bounded by two curves of $H_{sh}(T)$ and $H_c(T)$ is called a superheated superconducting state. An arrow pointing to the right denotes a thermal flip induced through a rise of temperature, whereas an arrow pointing upward denotes a magnetic flip through an increment of field strength.

field (H_a) which is stronger than the critical field ($H_c(T)$) for a bulky substance, and is weaker than a superheating critical field ($H_{sh}(T)$) at a given temperature (T) (Fig. 1). This is because of the fact that an additional surface energy appears to prevent any macroscopic flux-penetration of the stronger field. A superconducting state in a field beyond the critical field ($H_c(T)$) is called a superheated superconducting state.

As is shown in Fig. 1, two kinds of transitions are possible: a thermal flip induced due to an energy deposit and a magnetic flip due to an increment in the field strength. When an impinging particle deposits a sufficient energy (ΔE) in a granule to heat it by ΔT_0 ($=\Delta E/C$, C:heat capacity), its thermodynamical state steps over the phase boundary at $H_{sh}(T+\Delta T_0)$, smaller than $H_{sh}(T)$ in the temperature-field ($T - H$) diagram. This process is applicable to the observation of a weakly interacting particle losing recoil energy through elastic scattering off a nucleus. Some applications of SSGs have been proposed for the detection of monopoles [7], solar neutrinos [8], dark-matter particles [9] [10], and the double-beta decay [11]. Drukier and Stodolsky [8] originally proposed to measure the flux of solar neutrinos by using the thermal flip of SSGs, which is free from the neutrino-oscillation problem. The recoil energy and the event rate were calculated to be in a range of 1 eV–3 keV and $1 \times 10^{-3} - 5 \times 10^{-2}$ /kg·day, respectively, for neutrinos produced through different nuclear reactions in the Sun. Subsequently, Goodman and Witten [9] and Drukier et al. [10] considered the use of SSGs as detectors for various dark-matter particles in cosmic rays. A recoil energy of 10 eV–100 keV and an event rate of $0.1 - 1 \times 10^4$ /kg·day are expected, depending on their mass and spin.

When a Dirac-type magnetic monopole penetrates a granule, an additional field

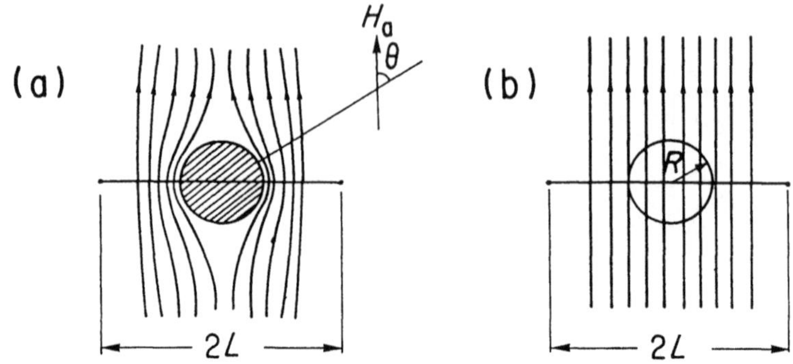

Figure 2: (a) An applied field expelled from inside the spherical granule in a superconducting state, which is known as the Meissner effect. A circular readout coil of diameter 2L is placed coaxially with the applied field (H_a). (b) Collapse of the applied field into the granule upon the transition from a superconducting phase to a normal one. The signal coil encircling the granule detects the difference in the penetrating magnetic flux.

(ΔH) of several 100 gauss is loaded in the granule, due to a magnetic flux of $2\phi_0$ left behind. ϕ_0 denotes the fluxoid quantum. Since the resultant field ($H_a + \Delta H$) exceeds the superheating field ($H_{sh}(T)$), the phase transition occurs through the magnetic cause alone [12].

In both the thermal and magnetic flip, the applied field collapses into a hit granule and the change in magnetic flux can be read out by a surrounding detection coil (Figs. 2a and 2b) connected to charge amplifiers. Since the temperature rise is inversely proportional to the heat capacity, granules of smaller size show a higher sensitivity to an energy deposit. On the other hand, the readout of signals becomes more difficult because the change in the total magnetic flux becomes smaller. Roughly speaking, a large quantity of granules of size 100 μm cooled at several kelvins is available for the detection of monopoles, whereas the size and the temperature should be 1-5 μm and 50-100 mK, respectively, for the detection of solar neutrinos or postulated weakly interacting particles.

In sect. 2, a summary of the thermodynamical and electromagnetical properties of SSG is given. It is shown that the shape and size of a granule have close relations to its magnetic properties, such as metastable superheated superconductivity. The heat capacity and heat conductivity of a granule almost determine its energy sensitivity. Latent heat influences the thermodynamical behavior of a collection of granules.

An SSG-target comprises a collection of small spherical granules which are usually dispersed in some dielectric material as packing material. Materials for SSGs are chosen out of a variety of superconductors, taking into account nuclear spins, critical temperatures, melting points and so on. In sect. 3, the properties of type-I superconductors as well as the properties of dielectric materials are presented and some techniques for producing small spherical granules are introduced.

In sect. 4–7, various features of the magentic flip and the thermal flip of tin granules are described on the bases of the experimental results obtained by the present authors. Some techniques for reading out signals of the phase transition are presented in sect. 4. Discussion is given concerning the transition times and line shapes of output signals which are caused solely by an increment of the applied field. This field increment could be assumed to be due to a pseudo-magnetic monopole.

In sect. 5, the thermal response of tin-SSGs is studied to an energy deposit through the irradiation of strongly interacting α particles. From the results of a measurement is obtained a threshold field-strength which is the smallest field necessary for a granule of a given size to undergo a phase transition with a deposited energy (E) at temperature T. This field is called the phase transition field $H_{pt}(E,T)$. By analyzing H_{pt}, the energy sensitivity ($\partial H_{pt}/\partial E$) of a granule is shown to be inversely proportional to the specific heat, as is expected for a granule size of 150–250 μm and for a temperature range of 2.25–3.00 K. Sample granules of this size turn out to flip upon receiving an energy as small as 1 MeV. Furthermore, the superheated state is found to turn into the normal state with an energy deposit two orders of magnitude less than that required for the entire granule to be uniformly thermalized. This means that the full penetration of an applied field is triggered provided that about 0.1% of the volume of the hit granule is sufficiently heated up. This is a typical case in which the local-heating picture is valid.

A prototype detector is presented in sect. 6. It has a 17 mm-square planar target comprising 190 $\mu m\phi$ tin granules placed at 32 $\times$ 33 matrix points. Four-channel readout coils are attached on each side of the target surface, one in the x-direction and the other in the y-direction. A stability test has been carried out at a temperature of 2.0 K for 650 hours. There were observed signals, each of which is estimated to be induced with an energy deposit of less than 1.75 MeV. Four causes are examined: fluctuations of the applied field and the temperature in the detector; radiation from some radioactive impurity contained in granules; and incident cosmic-ray muons. The flucuations of the field and the temperature as well as the impurity turn out to be too small to induce the observed events. Taking into account both the standard flux of cosmic muons and their ionization-energy loss in the granules, the events can be due to them.

In sect. 7, the thermal response of a tin-SSG of 30 μm in diameter is studied to irradiation of γ-rays with an energy of 60 keV at a temperature of 30 mK in a ^{3}He-^{4}He dilution refrigerator. It is shown that the spectrum of the absorbed energies has two peaks at $\sim$ 30 keV and $\sim$ 55 keV. The former is due to excited electrons from the K shell by the γ-rays, and the latter is due to those from the L shell by either primary γ-rays or a part of secondary X-rays produced by the filling of the K vacancies. The ratio of the counting rates at the two peaks is not very different from the calculated value using the mass-absorption coefficient of the K and L shells at the photon energy. This is a demonstration for obtaining a spectrum of energy deposits by means of SSGs. The superconducting granule of 30 μm in diameter turns out to react to excited electrons with energies less than 10 keV. It is necessary to take into account the specific heats of both the electrons and phonons in the model calculation concerning the absorbed energies, so as to explain the experimental results. This means that a strong coupling between the electrons and phonons still remains at such low temperatures.

In the last section, two plans contemplated by the present authors are outlined: One is an experimental study on the response of tin granules to the irradiation of neutron-rays, which are presumed to be pseudo dark-matter particles. The other is an experiment searching for some dark-matter particle by operating intermittently throughout one year a tin-SSG detector with a 1 gram-weight target .

There have been published several review studies and proceedings of workshops on cryogenic methods for the detection of neutrinos and weakly interacting massive particles [4][5][6][13][14][15][16].

2 Thermodynamics of SSGs

In this section we review the thermodynamics and electromagnetics [17] underlying superheated superconducting granule (SSG) detectors. Each superconducting granule, consisting of the detector target, changes its phase upon receiving an energy deposit of more than a threshold value from a penetrating particle. It is shown how the shape and size of a granule have close relations to its magnetic properties, such as the metastable superheated superconductivity. Both the heat capacity and the heat conductivity of a granule almost determine its energy sensitivity. An argument on latent heat is important to assess the thermodynamical behavior of the collection of granules.

2.1 Critical temperature and critical field

More than 20 elements are known to become superconducting at a temperature below some critical temperature (T_c) in a magnetic field less than the critical value, $H_c(T)$ (see Table 1). The critical field at a given temperature (T) is usually expressed approximately as

$$H_c(T) = H_c(0)[1 - (T/T_c)^2], \tag{1}$$

where $H_c(0)$ is an extrapolated value of $H_c(T)$ to $T = 0$.

The critical temperature is usually defined by the value at which a bulky substance undergoes a transition from a normal phase to a superconducting phase, and vice versa, in no applied field. This temperature is often called the bulk critical-temperature (T_{bc}).

2.2 Shape effect and size effect

The magnetic flux density ($\boldsymbol{B}$) in a material is given as a superposition of the internal magnetic field ($\boldsymbol{H}_{in}$) and the magnetization ($\boldsymbol{M}$),

$$\boldsymbol{B} = \boldsymbol{H}_{in} + 4\pi\boldsymbol{M}. \tag{2}$$

An external field ($\boldsymbol{H}_a$) applied to a material is therein reduced to the field $\boldsymbol{H}_{in}$ by a demagnetizing field ($\boldsymbol{H}_d$),

$$\boldsymbol{H}_{in} = \boldsymbol{H}_a - \boldsymbol{H}_d. \tag{3}$$

The field $\boldsymbol{H}_d$ caused by the magnetization ($\boldsymbol{M}$) is written as

$$\boldsymbol{H}_d = 4\pi N\boldsymbol{M}, \tag{4}$$

where N is the demagnetization coefficient, which depends crucially on the shape of the material used. The values of N for ellipsoidal substances are well-known. It vanishes for a long, thin cylinder placed parallel to the applied field ($\boldsymbol{H}_a$), whereas it is $N = 1/3$ for a sphere.

A small spherical metal has the property of metastable superheated superconductivity, as is described below. We shall make a particle detector of which the target comprises a collection of such spherical granules.

A material in a superconducting state excludes a magnetic flux by virtue of the screening of persistent current circulating on its surface. This is known as the Meissner effect. Since the magnetic flux density (B) vanishes inside the substance in a superconducting phase, the magnetization (M) turns out to be

$$\boldsymbol{M} = -\boldsymbol{H}_a/[4\pi(1-N)], \tag{5}$$

using Eqs. (2), (3) and (4).

For a sphere, Eq. (5) reads

$$\boldsymbol{M} = -(1/4\pi)(3/2)\boldsymbol{H}_a. \tag{6}$$

Thus, the internal field ($\boldsymbol{H}_{in}$) leads to $(3/2)\boldsymbol{H}_a$. From the continuity condition for the tangential components of a magnetic field at an interface, we can easily derive the field strength excerted on the surface of a sphere as

$$B = (3/2)H_a \sin\theta, \tag{7}$$

where θ is the polar angle measured from the direction of the applied field ($\boldsymbol{H}_a$). The field (B) on the surface takes the maximum value of $(3/2)H_a$ at the equator ($\theta = 90°$), and monotonically decreases to take the minimum value of 0 at polar points ($\theta = 0°$, 180°). Note that B exceeds H_a over an equatorial band of angles from $\theta = 42°$ to 138°.

Four typical magnetization curves are depicted for the applied field H_a in Fig. 3. A bulk material takes the maximum magnetization at H_c, which is shown with the broken line (i). On the other hand, the magnetization for a sphere reaches the maximum at $(2/3)H_c$, as depicted by the thin solid curve (ii) in accordance with Eqs. (6) and (7). For fields in the range of $(2/3)H_c < H_a < H_c$, there appears a coexistence of superconducting and normal phases, which is called the intermediate state.

In Fig. 3, the area under the curve represents the energy density ($E(H)$) stored in a material,

$$E(H) = \int M dH, \tag{8}$$

with which the magnetic field is expelled outside. When the material has the shape effect with a nonvanishing demagnetization coefficient, the magnetization takes a

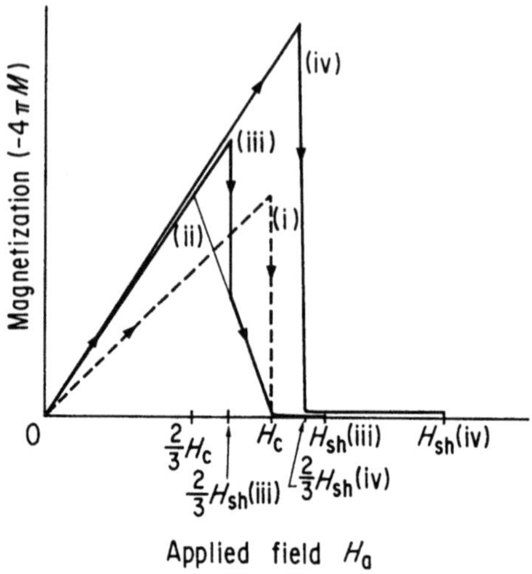

Figure 3: Schematic curves for magnetization. A bulk material takes a maximum magnetization at H_c (i). A bulk sphere takes it at $(2/3)H_c$ (ii). A small-sized sphere takes it at $(2/3)H_{sh}$ ((iii), (iv)).

maximum at a weaker applied field between $(2/3)H_c$ and H_c, though the energy density ($E(H)$) is the same as that for a bulk material.

Since a small-sized substance has a surface energy $E(\mathrm{S})$ in addition to the volume energy $E(H)$ in Eq. (8), it can exclude a stronger applied field due to an increment of stored energies, by taking a superheated superconducting state. The field strength excludable by the superheated state is called a superheating critical field $H_{sh}(T)$, which substitutes for $H_c(T)$ at a given temperatrure (T) and depends not only on the size of a granule, but also on the degree of smoothness of its surface. In Fig. 3, two typical magnetization curves are traced for a small sphere which takes the maximum magnetization at $(2/3)H_{sh}$, instead of $(2/3)H_c$. The bold solid curve (iii) shows the magnetization realized when $H_{sh}(T)$ is less than $(3/2)H_c$, whereas the bold solid line (iv) is for the case that $H_{sh}(T)$ is greater than $(3/2)H_c$. Thus, we should take into account both effects due to the "shape" and "size" of small granules for the study of magnetization.

In a particle detector using SSGs, the change of magnetic flux penetrating a granule is taken as a candidate signal. As can be seen in Fig. 3, a maximal change occurs for a drop of magnetization in curve (iv) from $(-1/4\pi)H_{sh}$ to 0 by neglecting the very small magnetization due to paramagnetism. It is thus desired that the granules used for an SSG-target detector have an $H_{sh}(T)$ greater than $(3/2)H_c(T)$. Otherwise, the penetration of the field is incomplete so that a small fraction of the magnetization remains as is shown in the curve (iii), which leads to the intermediate state.

In the $T-H$ phase diagram (Fig. 4), curves for the superheating (sh) superconducting field $H_{sh}(T)$ and the supercooling (sc) field $H_{sc}(T)$ are shown for 80 and 250 μmϕ tin granules measured in a temperature range from less than 2 K

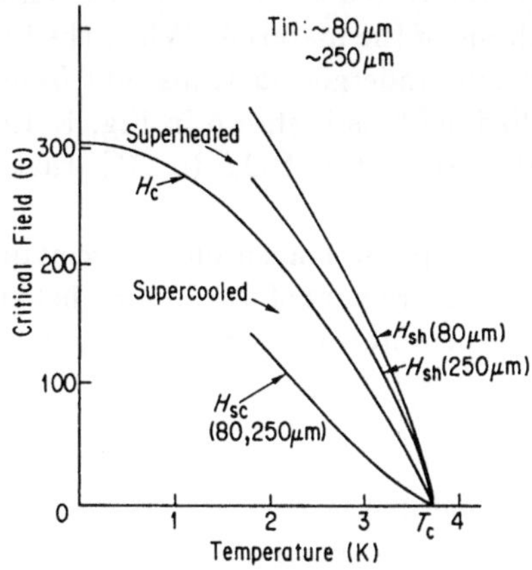

Figure 4: Temperature-field diagram for two tin granules of diameters 80 and 250 μm. The smaller one has a higher superheating superconducting field than the larger one. The supercooling field depends less on the size of the granules.

to its critical temperature ($T_c = 3.7$ K). It is actually seen that the smaller is the size, the higher is $H_{sh}(T)$, because the stored surface energy (E(S)) is the bigger. $H_{sc}(T)$ depends less on the size.

A bulk material takes only two phases, a superconducting phase and a normal conducting phase, whose regions are bounded by the curve $H_c(T)$ in Fig. 4. However, the structure of the phase for a small material is complicated because of a size effect. Its phase shows a hysteresis-like behavior. At a given temperature (T), such a material remains superconducting beyond the curve $H_c(T)$ until the boundary of $H_{sh}(T)$, which is called the superheated superconducting phase. Reversely, a small material remains normal-conducting until the boundary $H_{sc}(T)$ when moved from the outside region of the $H_{sh}(T)$ curve. The phase bounded by curves $H_c(T)$ and $H_{sc}(T)$ is called the supercooled superconducting phase. Both the superheated and supercooled phases are metastable, so that transitions therefrom/thereinto are irreversible.

Ginzburg first theoretically derived that such a superheated superconducting phase can be realized in superconductors of the first kind [18]. In the Ginzburg-Landau (GL) approximation, we have the following relation between H_{sh} and H_c [19]:

$$H_{sh}(T) = 2^{-1/4}\kappa^{-1/2}H_c(T), \tag{9}$$

which is valid near the critical temperature (T_c) and κ is the GL-parameter. For an analysis of experimental results, an empirical formula for the superheating critical field is also useful [20],

$$H_{sh}(T) = H_{sh}(0)[1 - a(T/T_c)^2], \quad 0.5 < a < 1. \tag{10}$$

Receiving an energy deposit (ΔE) from an incident particle, a small granule is heated by ΔT,

$$\Delta T = \Delta Q/CV, \tag{11}$$

where ΔQ is the heat energy converted from ΔE, and C and V are, respectively, the specific heat and the volume of the material. When the temperature rise (ΔT) is sufficiently large, the granule undergoes a transition from the superheated to the normal state in an applied field, as is shown in Fig. 1. Roughly speaking, the energy sensitivity is defined as the ratio of ΔT to ΔQ, that is the inverse of the heat capacity (CV).

The granule can also make a phase transition by receiving an additional field (ΔH), which is either simply an increment of the applied field (H_a) or an increment of the magnetic field left by a penetrating magnetic monopole.

2.3 Specific heats

The specific heats for a material in the normal and superconducting states (C_n and C_s, respectively) both have two components contributed from phonons(p) and electrons(e) as follows:

$$C_n = C_{np} + C_{ne} \tag{12}$$

and

$$C_s = C_{sp} + C_{se}. \tag{13}$$

Since the superconducting state stems from electron pairing and has no direct concern with phonons, no substantial difference appears in the lattice specific heats, (C_{np} and C_{sp}), which are expressed in terms of the Debye temperature T_D as

$$C_{np} = C_{sp} = \alpha(T/T_D)^3, \tag{14}$$

with $\alpha = (12/5)\pi^4 R = 1940$ J/mol·K and $R = 8.3$ J/mol·K.

On the other hand, the electronic specific heats (C_{ne} and C_{se}) have utterly different forms:

$$C_{ne} = \gamma T \tag{15}$$

and

$$C_{se} = \gamma T_c a \exp[-bT_c/T], \tag{16}$$

where γ [mJ/K^{2}mol] takes a value between 0.6 (Ga) and 10 (La, U), and the other two parameters (a and b) are almost independent of the materials, being given as about 9 and 1.5, respectively. The latter one (b) is associated with the energy gap (Δ) generated by the Cooper pair through the relation $bk_B T_c = \Delta$ (= 1 meV), with k_B the Boltzmann constant.

The electronic specific heat (C_{se}) in Eq. (16) decreases more rapidly with decreasing temperature than does the lattice specific heat (C_{sp}) in Eq. (14) in the superconducting state. Provided that almost all of the deposited energies are consumed to increase the electron temperature alone, instead of the phonon temperature, an extraordinarily high energy-sensitivity is anticipated due to Eq. (11).

2.4 Heat conductivity

Heat is conducted with both phonons and conduction electrons with respective conductivities of κ_{sp} and κ_{se} in a superconducting phase,

$$\kappa_{sp} \simeq (T/T_D)^3 \tag{17}$$

and

$$\kappa_{se} \simeq \exp[-bT_c/T], \tag{18}$$

where b is given as 1.45 for tin.

When a sufficient amount of energy is given at a small spot in a superconductor, either the phonons or Cooper pairs are excited. The latter ones become quasiparticles (*i.e.*, conduction electrons) beyond the energy gap (Δ). Phonons and quasiparticles then diffuse and distribute energies inside the material by interacting with each other. However, the process of thermalization is not yet fully understood. When the thermal conduction is large, the deposited energies are effectively conveyed overall to heat the material nearly uniformly before a phase transition takes place. This leads to the picture of "global heating" in which the heat capacity in Eq. (11) turns out to be one for the entire volume of an SSG. On the other hand, for slower thermal diffusion, a small portion of the material has only to be heated for an SSG to change its phase, which leads to the "local heating" model. In this case, the phase transition is triggered by a very small amount of energy with which the entire material could not be sufficiently heated. The heat capacity given in Eq. (11) is taken as that for the partial volume of an SSG. Then, the effectiveness of the deposited energy is amplified. Such amplification is really found in experiments with granules of In and Sn [21], whereas the heating process in Al and Zn is understood in the global-heating picture [22]. The difference between the two pictures becomes obscure for very small materials.

The heat conductivity of the packing material should also be considered, as is briefly referred to in sect. 3, since an actual SSG-target is embedded in a dielectric material.

2.5 Latent heat

Heat is usually absorbed upon transition from a superconducting state to a normal state due to an increase in entropy (an endothermic process). However, an exothermic process becomes possible depending on the difference between the free energies for the two states, as is described below.

From a differential form for the Gibbs free energy (G) at a given temperature (T) and applied field (H_a),

$$dG(T, H_a) = -SdT - MdH_a, \tag{19}$$

where S and M are the entropy and magnetization, respectively, the free energy (G_s) per unit volume of a superconductor is derived as

$$\begin{aligned} G_s(T,H_a) &= G_s(T,0) - \int_0^{H_a} M dH_a \\ &= G_s(T,0) + (1/4\pi)[1/2(1-N)]H_a^2. \end{aligned} \tag{20}$$

For a sphere, Eq. (20) becomes

$$G_s(T,H_a) = G_s(T,0) + (1/8\pi)(3/2)H_a^2. \tag{21}$$

The free energy (G_n) in a normal state, however, is insensitive to the field (H_a), since the magnetization is negligibly small in paramagnetism. It is expressed as

$$G_n(T,H_a) = G_n(T,0). \tag{22}$$

Taking into account that the two states are in a thermal equilibrium at $H_a = H_c$ (the bulk critical field), namely

$$G_s(T,H_c) = G_n(T,H_c), \tag{23}$$

we have the relation

$$G_n(T,0) = G_s(T,0) + (1/8\pi)H_c^2. \tag{24}$$

The heat released during a phase transition from a superheated superconducting state to a normal state under an applied field $H_a > H_c$ is written as (when the heat due to a difference between surface energies of the two states can be neglected)

$$Q = G_s(T,H_a) - G_n(T,H_a) + T(S_s - S_n), \tag{25}$$

where entropy S_s and S_n, respectively, for the superconducting and normal states are given by

$$S_{s,n} = -\left(\frac{\partial G_{s,n}}{\partial T}\right)_{H_a}. \tag{26}$$

Then, for a sphere, Eq. (25) leads to

$$Q = (H_c^2/8\pi)[(3/2)h^2 - 1 - 4t^2/(1-t^2)], \tag{27}$$

where t is a reduced temperature (T/T_c), and h is a reduced applied field (H_a/H_c). The transitions are classified into endothermic and exothermic types for negative and positive Q, respectively.

When the size effect is small, the applied field (H_a) begins penetration at a smaller magnetization and makes an intermediate state (in). For $(2/3)H_c < H_a < H_c$, the free energy (G_{in}) and heat (Q) released are given by, respectively,

$$\begin{aligned} G_{in}(T, H_a) &= G_s(T,0) + \int_0^{\frac{2}{3}H_c} (1/4\pi)(3/2)H_a dH_a \\ &\quad + \int_{\frac{2}{3}H_c}^{H_a} (1/4\pi)3(H_c - H_a)dH_a \\ &= G_s(T,0) + (1/8\pi)H_c^2 - (3/8\pi)(H_c - H_a)^2 \end{aligned} \tag{28}$$

and

$$Q = (H_c^2/8\pi)[2 + (9/2)h^2 - 6h - (12h - 8)t^2/(1 - t^2)]. \tag{29}$$

For both cases above, we have

$$Q > 0\,(\text{exothermic}) \quad \text{for} \quad t < 1/3 \quad \text{at} \quad H_a = H_c. \tag{30}$$

At an ambient temperature satisfying the condition $Q > 0$, there is a possibility that the phase of adjacent granules will turn into the normal one successively, due to heat released by a single granule changing its phase. Such a chain-reaction has been observed in many test-experiments [23]. The authors of ref. [23] called the phenomenon an "avalanche", and expect to utilize it as a way of amplifying originally-small signals.

3 Materials for SSG-targets

An SSG-target comprises a collection of small spherical granules which are usually dispersed in some dielectric material as a packing material. The detection of neutral dark-matter particles with an SSG-detector is based on the phase transition of a granule in which an atomic nucleus of its material receives a recoil energy from an incident particle through elastic scattering [24]. Two cases usually taken for illustration are coherent weak interactions (*e.g.*, Dirac neutrinos or supersymmetric neutrinos) and axial coupling interactions (*e.g.*, supersymmetric neutralinos or Majorana neutrinos). The event rates (R) through the two interactions are given by [24]

$$R(\text{coherent}) \approx 1.2r(N/2)^2 \quad \text{events/kg/day} \tag{31}$$

and

$$R(\text{axial}) \approx 1.2rC\lambda_s^2 J(J+1) \quad \text{events/kg/day}, \tag{32}$$

where $r = 4M_T M_D/(M_T + M_D)^2$, M_T [GeV] is the mass of the target nucleus, M_D [GeV] the mass of an incident particle, N the number of neutrons in the target nucleus, J the total nuclear spin, λ_s^2 the shell-model factor, and C the square hadronic matrix element for photino-nucleon interaction. Axially coupled particles are not detectable with granules, the elements of which have a spinless nucleus.

Materials for SSGs are chosen out of a variety of superconductors, taking into account the nuclear spins, critical temperatures, critical fields ($H_c(0)$) melting points and so on. In this section, the properties of both type-I superconductors and dielectric materials are presented. The techniques for producing small spherical granules are briefly described.

3.1 Properties of type-I superconductors

In Table 1 we list all of the well-known type-I superconducting elements in descending order of the critical temperatures (T_c[K]). Some of their properties are also presented: bulk critical fields ($H_c(0)$[G]) at T=0 K, Debye temperatures (T_D[K]), atomic number (Z), mass number (A), number of neutron (N), the total nuclear spin (J), melting temperatures (T_m [°C]), and natural isotopic abundances (%).

The nuclides with natural abundance of less than 10% are omitted from Table 1. Type-II superconductors (Nb and V), low-T_c superconducting alloys, and high-T_c superconducting oxides are presently left out of consideration for materials of the granule target.

Type-I superconductors have a critical fields $H_c(T)$ expressed by the generic formula $H_c(T) = H_c(0)\ [1 - (T/T_c)^2]$ in Eq. (1). The slope dH_c/dT for a given material decreases with decreasing temperature, as is also seen in the $T - H$ diagram (Fig. 4). The critical values ($H_c(0)$) for 22 elements in Table 1 decrease with decreasing critical temperatures (T_c) with several exceptions. Therefore, at a given temperature (T) the slope (dH_c/dT) is generally larger for materials with lower critical temperatures (T_c). The superheating critical fields ($H_{sh}(T)$) for different materials also show analogous attributes to those of the critical fields ($H_c(T)$) as stated above.

When a granule is placed in an applied field (H_a) which is weaker by ΔH_0 than $H_{sh}(T)$ at temperature T (see Fig. 1), the temperature rise (ΔT) required for a thermal phase transition to occur is estimated to be $\Delta T > \Delta T_0 = \Delta H_0/(dH_{sh}/dT)$. Then, the smaller is the slope (dH_{sh}/dT), the larger is the threshold of a temperature rise (ΔT_0) for a field-deficit (ΔH_0). This means that the energy sensitivity of a granule is low in the region of small dH_{sh}/dT, neglecting the fact that the specific heat is small at low temperatures.

Since materials with a low critical temperature (T_c) have a large slope (dH_{sh}/dT) at a given temperature, granules made from such materials may show a high energy-sensitivity. For this reason, many studies have also been carried out on Cd and Zn-granules.

Table 1: Type-I superconducting elements

	T_c[K]	$H_c(0)$[G]	T_D[K]	Z, A, N	J	T_m [°C]
Tc*	7.92	1410		43, 98, 55		2170
Pb	7.193	803	96.3	82, 208, 126 (52 %)	0	327.5
				82, 206, 124 (24 %)	0	
				82, 207, 125 (22 %)	1/2	
La(b)	6.06	1600	132	57,139, 82	7/2	921
Ta	4.483	830	258	73,181,108	7/2	2996
Hg(a)	4.154	410.9	71.9	80,202,122(30 %)	0	−38.86
				80,200,120(23 %)	0	
				80,199,119(17 %)	1/2	
				80,201,121(13 %)	3/2	
				80,198,118(10 %)	0	
Sn#	3.722	305.50	195	50,120, 70(33 %)	0	231.97
				50,118, 68(24 %)	0	
				50,116, 66(15 %)	0	
In	3.4035	293	109	49,115, 66(96 %)	9/2	156.63
Tl	2.39	171	78.5	81,205,124(70 %)	1/2	302.5
				81,203,122(30 %)	1/2	
Re	1.698	198	415	75,187,112(63 %)	5/2	3180
				75,185,110(37 %)	5/2	
Th	1.368	162	168	90,232,142	0	~1800
Al	1.196	99	420	13, 27, 14	5/2	660.4
Ga	1.091	51	324.7	31, 69, 38(60 %)	3/2	29.78
				31, 71, 40(40 %)	3/2	
Mo	0.92	98	460	42, 98, 56(24 %)	0	2610
				42, 96, 54(17 %)	0	
				42, 95, 53(16 %)	5/2	
				42, 92, 50(15 %)	0	
Zn	0.875	53	309	30, 64, 34(49 %)	0	419.58
				30, 66, 36(28 %)	0	
				30, 68, 38(19 %)	0	
U(a)	0.68	300	206	92,238,146(99 %)	0	1133
Os	0.655	65	500	76,192,116(41 %)	0	2700
				76,190,114(26 %)	0	
				76,189,113(16 %)	3/2	
				76,188,112(13 %)	0	
Cd	0.56	30	209	48,114, 66(29 %)	0	321.1
				48,112, 64(24 %)	0	
				48,111, 63(13 %)	1/2	
				48,110, 62(12 %)	0	
				48,113, 65(12 %)	1/2	
Zr	0.546	47	290	40, 90, 50(51 %)	0	1852
				40, 94, 54(17 %)	0	
				40, 92, 52(17 %)	0	
				40, 91, 51(11 %)	5/2	
Ru	0.49	66	550	44,102, 58(32 %)	0	2250
				44,104, 60(19 %)	0	
				44,101, 57(17 %)	5/2	
				44, 99, 55(13 %)	5/2	
				44,100, 56(13 %)	0	
Ti	0.39	100	425	22, 48, 26(74 %)	0	1675
Ir	0.14	19	420	77,193,116(63 %)	3/2	2457
				77,191,114(37 %)	3/2	
W	0.012	1.07	550	74,184,110(31 %)	0	3387
				74,186,112(29 %)	0	
				74,182,108(26 %)	0	
				74,183,109(14 %)	1/2	

(Tc*: artificial radionuclide, La(b):La with face-centered cubic structure, Hg(a):Hg with rhombohedron structure, Sn#:white tin, U(a): U with hexagonal close-packed structure.)

3.2 Manufacturing spherical granules

Both good sphericity and a defect-free surface are required for small granules to have a sufficiently large $H_{sh}(T)$. The size of grains should be several tens of microns or less to attain an energy sensitivity of the order of 100 eV, for instance. A large quantity of grains weighing from several kilograms to several tons should be provided for a target to gain substantial reaction rates of postulated dark-matter particles. It is further required that every grain embedded in a target be as homogeneous in size as possible to make the same response for the same trigger.

The following methods have been employed so far to make sample granules: (A) Atomizing method: by blowing off melt metal through a nozzle, some kinds of metallic powder are manufactured in large quantity on a commercial basis. The shape of the grains formed by this method is irregular and generally diverse. Almost all of the conditions for SSGs mentioned above are not fulfilled, though it is reported that relatively good sphericity is obtained for some metals, such as tin, copper and bronze.

(B) Ultrasonic method [25]: tiny spherical granules can be formed by spattering colloid metal in an ultrasonic transducer. Their average diameter (ϕ) obtained is thus given semi-empirically as

$$\phi = (\sigma/\rho)^{1/3} f^{-2/3}, \tag{33}$$

where σ is the surface tension of the liquefied metal, ρ the density of the metal, and f the frequency of an ultrasonic wave. Granules of about 30 μm in diameter are expected to be formed with a frequency of 20 kHz according to Eq. (33). For a size of 1 μm, a high frequency of 5 MHz is required. Metallic granules of 20 μm in diameter with rather good surface-quality are produced by this method at some companies in France. The manufacturing rate is said to be 5 kg/hour [26].

(C) Steel's method [27]: this is easily available for metals with lower melting points at laboratories, though not suited for the mass production of grains. The present authors improved and employed this method to obtain sample grains of tin for the measurement of the magnetization and for irradiation experiments. More than 1000 grains for a prototype target are also produced by this method [28].

Oil is heated in a 150 cm-long glass-tube above the melting point of tin, except the bottom part of 10 cm in length (Fig.5). The powder of the metal is thrown into the oil through the mouth of the tube. Then, every grain melts immediately and becomes ball-shaped due to its surface tension during the course of very slow sinking in the oil. Subsequently, it is cooled upon passing through the last 10-cm distance. Lastly, we cleaned the granules in an organic solvent in an ultrasonic-wave washer in order to remove any remnant oil and oxide from their surfaces.

Grains produced through either one of the three ways mentioned above are generally so diverse in size that we must sieve them several times by using a set of wire meshes to collect grains of diameter in the required range.

An actual target comprises a collection of metallic granules which are dispersed in some dielectic material, such as paraffin-wax, Epibond resin, epoxy resin (Emerson and Cummings Stycast 1266), varnish GE7031, Apiezon N grease, Dow-Corning silicon vacuum-grease and so on. The thermal conductivity of these mate-

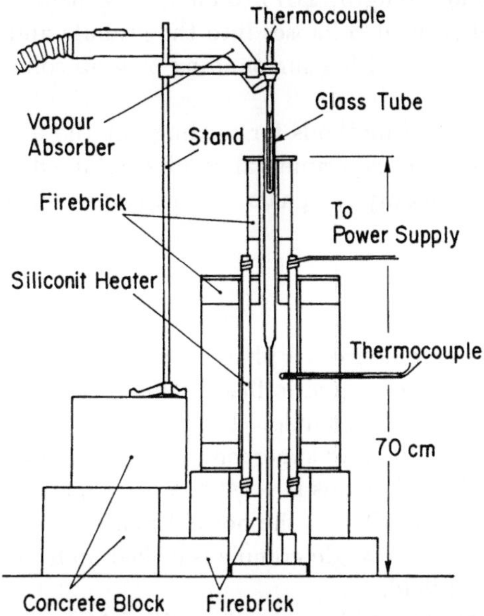

Figure 5: Electric furnace for producing spherical granules after Steel's method.

rials becomes smaller in the order of Epibond 100A, varnish GE7031 and Apiezon N grease at temperatures of around 0.1 K.

Resin is often used as a sort of packing material. After complete solidification, it is impossible to pick up granules again therefrom. On the other hand, varnish GE7031 is resolvable with some solvent, and grease becomes soft again at room temperature.

There have been attempts to arrange granules in a lattice in order to make an analysis of their magnetic interaction easy [29]. One attempt was to make an array of grains on a substrate by virtue of the photolithographic technique from an evaporated superconducting material; a satisfactory result has not yet been obtained. Another method reported [29] is as follows: after composing an array of rectangular pellets of In on a substrate, the pellets were heated above the melting point to make them spherical. The spheres can be made homogeneous in size by weighing the pellets beforehand as rigorously as possible. The superheating fields of granules obtained showed a small deviation from an average value, $\Delta H_{sh}/H_{sh} =$ several %, which is an order of magnitude smaller than that reported so far. This method is remarkable because it can actually arrange equal-sized spheres in a lattice with a definite distance.

4 Responses of tin-SSGs to increment of an applied field

A small spherical granule undergoes a transition irreversibly from a superheated to a normal state, upon receiving either a sufficient energy deposit ΔE (a thermal flip) or an additional field strength (ΔH) to an applied field (a magnetic flip). In both processes the applied field collapses into the granule and any change in the magnetic flux can be read out with a surrounding pick-up coil connected to charge amplifiers.

In this section a couple of methods for reading out the flux changes ($\Delta\phi$) are presented. Subsequently, the experimental results obtained for tin granules are described: the line shapes of output signals and the transition times for increments of the applied field.

4.1 Readout systems

Upon the transition of a granule from a phase of perfect diamagnetism to a phase of Pauli's paramagnetism, the magnetic flux begins to make a macroscopic penetration from the surface of the granule. The interface between the two phases shrinks so as to sweep out the inner superconducting region in a very short time. The magnetic susceptibility of metals is $-1/4\pi$ in the superconducting phase, but takes a typical value of about 10^{-6} in the normal phase. The resultant change in the magnetic flux penetrating a single granule is picked up by a readout coil placed coaxially with the applied field.

The flux change ($\Delta\phi$) induced in a circular coil of $2L$ in diameter is given in a superheating field ($H_{sh}(T)$) as

$$\Delta\phi = \phi_{normal} - \phi_{super} = R^3\pi H_{sh}/L, \tag{34}$$

for a spherical granule of diameter $2R$ (Figs. 2a and 2b). In a rectangular coil having a length sufficiently larger than width (d) it leads to

$$\Delta\phi = 4H_{sh}R^3 d/(d^2 + 4a^2), \tag{35}$$

where a is the distance between the coil plane and the equatorial plane of a flipped granule. The flux change ($\Delta\phi$) increases in proportion to R^3, so that the larger is the size of the granules, the easier it is to read out the signals, though the temperature rise in the whole granule decreases.

Though the change in the susceptibility is large, the resultant flux change is not large. Three methods to detect such a flux change are considered:

(A) Observation of an electromotive force (e.m.f.): The time span (Δt) for the magnetic flux to complete its penetration into a thin cylinder-like material placed parallel to an applied field is calculated to be [30]

$$\Delta t = \sigma\pi R^2, \tag{36}$$

where σ and R are the electric conductivity and the radius in the cross section of the sample rod, respectively. Upon a phase transition, the persistent current circulating around the surface of the rod to prevent any penetration of an applied field turns into a normal conducting current which gradually decays by generating Joule heat. Such a process lasts until the field penetration is completed up to the center of the cylinder. This is the reason why the decay time depends on both σ and the cross section of the rod.

The electromotive force (V[volt]) induced is given on the average as

$$V = d\phi/dt. \tag{37}$$

For the flux change in the circular coil shown in Eq. (34), it reads

$$V = (R/L)(H_{sh}/\sigma). \tag{38}$$

We have alternatives for amplifying the e.m.f. in Eqs. (37) and (38). One is to directly use a charge-sensitive amplifier or a low-noise amplifier. Another is to deal with the voltage induced at a room temperature after receiving it with a transducer [31]. An analysis of the circuit for this amplification is shown in detail in ref. [32]. For lower-noise amplification, there have been studies on FET operated in liquid nitrogen, or on GaAs-FET operated at liquid-helium temperature. Since the transition time of a 10 μm-sized granule is shorter than several hundreds of nanoseconds, the band range of the amplifiers required is wider than the order of 10 MHz.

Method (A) is appropriate for obtaining the time information included in the signal shapes.

(B) Observation of a resonance-frequency change induced in an LC- resonance circuit [33]: Upon a phase transition a resonance frequency of around 100 kHz is obtained in an LC circuit connected to a detection coil encircling the granule. The average of the flux changes induced in a number of granules can be measured using this circuit. Such a measurement is particularly appropriate for studying the collective phase-transitions of small granules, though information concerning individual flips is missed. An "avalanche" of flips can take place on the condition that heat is released during the transition (an exothermic process), as is described in sect. 2.5. Such phenomena are reported to have been observed in a collection of Cd granules by this measuremental method [34].

(C) Observation of very small flux changes with a SQUID: A SQUID-system is needed to observe a very small flux change induced in a granule of diameter less than 10 μm. The impedance of a detection coil should be taken as almost the same as that of the input coil in a SQUID, which is about 2μH in commercially available ones [35].

Each phase-transition can be detected with a SQUID. Though the details of temporal information are generally missed for rapid flux changes, a SQUID-system is sufficiently useful for taking data of rare events. In an experimental study on the magnetic properties of SSG, SQUID-systems are feasible for use in a temperature-cycle, but not in a field-cycle [36].

During the final stage, every signal obtained in (A), (B) and (C) is dealt on personal computers or minicomputers through a GP-IB interface or others [37].

4.2 Line shapes of output signals

In Fig. 6 is presented one of output signals taken by method (A) for the phase transition of a spherical tin granule of 250 μm in diameter at a temperature of 3 K in a field of 92.67 G [38]. The gross features of the line shape are similar

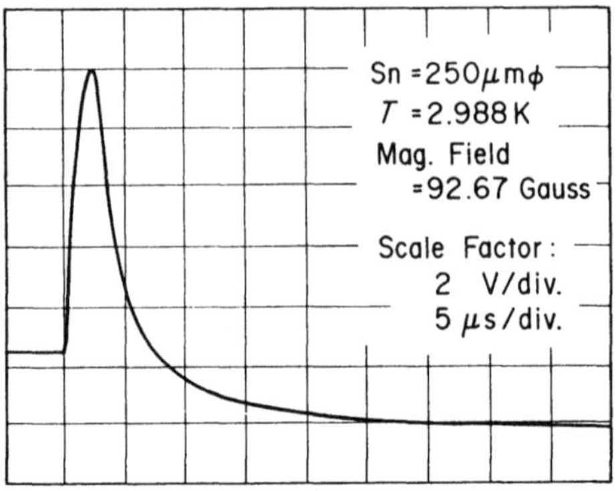

Figure 6: Output for a magentic flip undergone by a granule of 250 μm in diameter at a temperature of 2.988 K in a field of 92.67 G without any irradiation. The x-axis for the time lapse is graduated in 5 μs and the y-axis for the signal amplitude in 2 volts for each division.

to those given in a model calculation by Faber [30]. Though each output signal drops sharply for cylindrical materials, every signal shape obtained for a spherical granule gradually decays with time. Each line shape obtained by de Bellefon also has a long decay tail like that in Fig. 6 [32].

In Fig. 6, the area under the pulse curve, which is given by $\int V dt = \int (d\phi/dt) dt = \phi$, stands for the total flux collapsing into a granule during a phase transition. The collapsing total flux (ϕ) should be given as $\pi R^2 H_{sh}(T)$ for a granule of radius R and a superheating field ($H_{sh}(T)$) at a given temperature. Hence, the shorter is the decay time, the larger is the pulse height. The detection of signals becomes easier for the larger maximum. It is therefore advisable to make the electric conductivity smaller and to make the decay time (Δt) shorter in Eq. (36) by mixing an impurity into the material, unless the superheating field (H_{sh}) becomes so small. It must also be noted that in the case of a narrow pulse-width we need an amplifier which is able to follow an initial rapid rise of such a signal.

High-purity tin (99.9999%) was used in the first stage of the research by the present authors. Lower-purity tin (99.9%, cheaper by one fiftieth) was used afterward. The specific resistance of tin was measured at liquid-helium temperatures by the method of Bean [39]. The results are 7.60×10^{-10} and 5.88×10^{-8} ohm·cm for 99.9999%-purity tin and 99.9%-purity tin, respectively. This difference of two orders of magnitude makes the pick-up of signals much easier.

4.3 Transition times

An experimental study on the transition times from a superheated to a normal state was carried out by using an arrangement of forty tin-granules of 200 μm in diameter (Fig. 7) [40]. Each granule is separated by more than 0.5 mm in a pick-up coil of 1cm in diameter. With this distance, the magnetic interaction among the granules is negligible. The coil was connected to a wide-band preamplifier. A magnetic field is produced with a long superconducting solenoid coil. As the

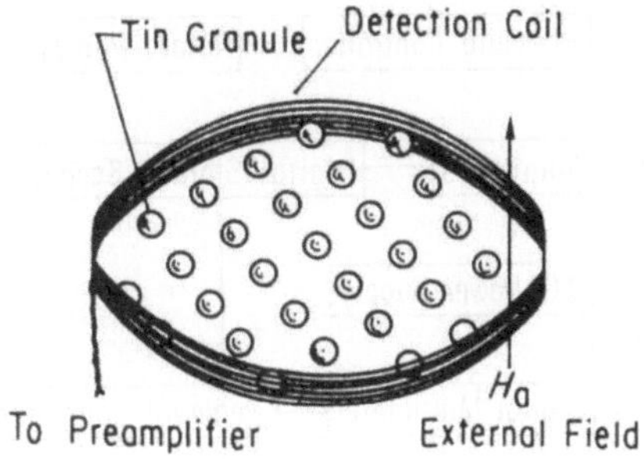

Figure 7: Schematic arrangement of forty tin granules and a pick-up coil.

applied field was gradually increased, magnetic flips were continually observed. The asynchronous appearance of signals is attributed to the difference either in the degree of sphericity, in the size or in the location of sample granules. Patterns of signals were recorded with a transient memoryscope. The transition times are observed to be 5-10 μs, which are consistent with that calculated using a global-heating model of Faber [30]. The risetime of a signal is about 2.5 μs, as can be seen in Fig. 6.

5 Responses of tin-SSGs to alpha-ray irradiations

With a sufficient energy deposit, a superheated superconducting state of a granule changes into a normal state irreversibly through a thermalization process (*i.e.*, a thermal flip). The temperature rise (ΔT) in a material with a heat capacity C and an energy deposit E is estimated to be

$$\Delta T = E/C. \tag{39}$$

The threshold of temperature rise, (ΔT_0), which is the smallest one required for the phase transition to take place, depends on the field strength (H_a) applied to a granule at a given temperature (T). When the field (H_a) is much smaller than the superheating field ($H_{sh}(T)$), a large temperature rise (ΔT) is necessary for the phase transition. Thus, in the experimental study on the thermal flip we have a threshold field strength (H_{a0} in Fig. 1) which is the smallest one for a granule to come out of a superheated state with a given energy deposit (E) at temperature T. This field is called the phase-transition field $H_{pt}(E,T)$. In other words, the difference $\Delta H(= H_{sh}(T) - H_a)$ takes the largest value at $H_a = H_{a0} = H_{pt}(E,T)$ for the thermal flip due to a given energy deposit E.

A change in the magnetic flux at a flip of this kind can also be read out with a surrounding pick-up coil as well as at the magnetic flip described in the preceding section.

This section considers the heat energy deposited in tin-granules of 100-250 μm in diameter through irradiation of α-rays with energies less than 4.7 MeV. Their responses are also considered: the line shapes of the output signals, transition

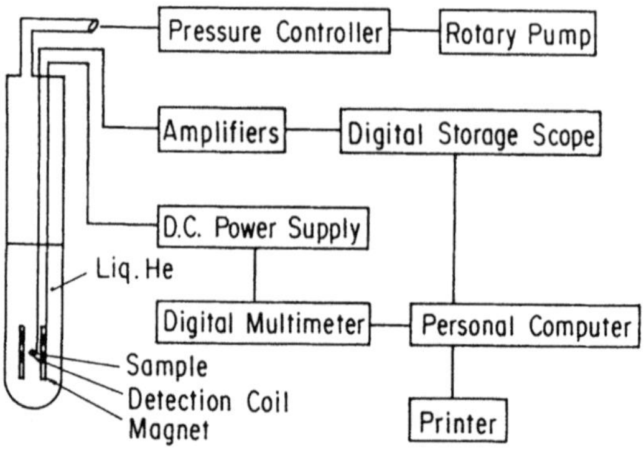

Figure 8: Block diagram for our experiment in which a digital storage scope receives a pulse induced through a phase transition of a granule in applied fields at temperatures lower than the critical point of 3.72 K.

times, and the phase-transition field ($H_{pt}(E,T)$) in the temperature range from 2.25 to 3.00 K [41].

Sample granules of this size turn out to undergo a phase transition while receiving an energy deposit as small as 1 MeV. The energy sensitivity of single tin granules has been ascertained to increase with decreasing temperatures. The superheated states are found to turn normal with an energy deposit two orders of magnitude less than that required for the whole of a granule to be uniformly heated up. This means that the full penetration of an applied field is triggered provided that about 0.1% of the volume of a hit granule is sufficiently heated. This is a typical case in which the local-heating picture is valid.

5.1 Experimental apparatus

Using a similar technique developed by Steel [42], sample granules were produced from tin of 99.9% purity. The sphericity and size of the granules were determined through a microscopic examination and a comparison with a graduated scale, respectively.

The measuring system is schematically shown in Fig. 8. Each signal induced in the detection coil was received in a digital storage scope via wide-band amplifiers and is finally recorded on a personal computer.

The experimental arrangement in the cryostat is depicted in Figs. 9a, 9b and 9c. A sample granule is located in the center of a detection coil comprising a 90 $\mu m\phi$ insulated copper wire wound in 150 turns on a stainless-steel tube of 5 mm in diameter (Fig. 9a). The tube is sheathed with a superconducting magnet, though not drawn, of 100 mm in length and 14 mm in bore-diameter. The magnet produces applied fields of up to 420.0 G with an error less than 0.1 G. The pedestal supporting the material of the measuring device is made of bakelite.

For studying the granule response to radiation a combined device (Fig. 9c) is used, in which another tubular device (Fig. 9b) is inserted with a long brass rod into the detection-coil supporter (a 5 mmϕ stainless-steel tube) in Fig. 9a from

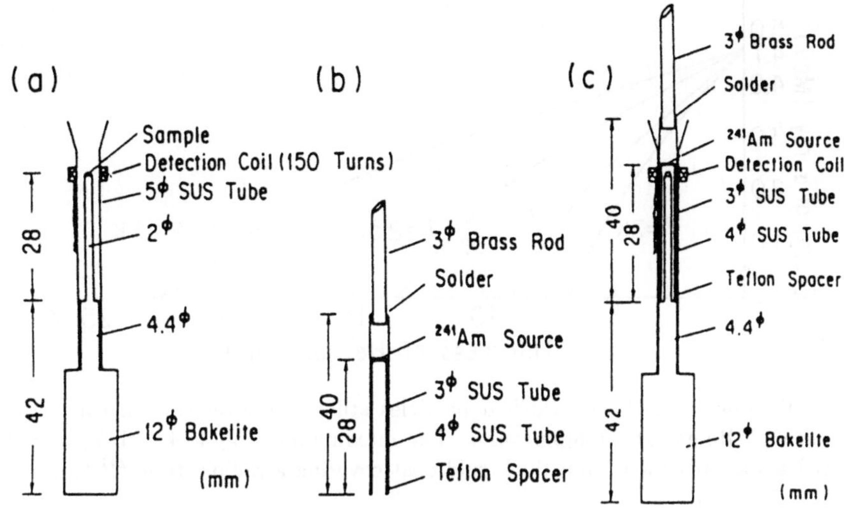

Figure 9: Schematic of the experimental arrangement. (a) A sample granule is placed at the center of a detection coil wound on a 5mm-diam SUS tube. The tube is insetted axially in a superconducting magnet (not depicted), so that signals of the flip can be taken in a changeable field. For measurements in irradiations of α particles a combined device (c) is used, in which a tubular device (b) involving an ^{241}Am source is inserted into the 5mm-diam SUS-tube from the top.

the top. A gold-coated α-ray source of ^{241}Am with activity (0.5 μ Ci $\pm$10%) faces downward on the top of a 3 mmϕ tube. By handling the brass rod from outside the cryostat, the source is placed at the desirable position away from a granule to make definite the path length of α-particles beforehand. Within several minutes after fixing the distance, saturated vapor of helium grows to fill inside both the 3 mmϕ and 4 mmϕ tubes so that the level of liquid helium retreats down to the bottoms of both tubes. Since the thermal conductivity of the vapor is only a factor smaller than that of the liquid, the temperature of the sample granule soon becomes equal to that of liquid helium.

The temperature is controlled at an arbitrary value of between 4.2 and 1.6 K with precision of at worst 1 mK through a vapor-pressure regulator involving a diaphragm device [43]. Alpha particles are emitted from the source cover with an initial energy of 4.7 MeV and hit the granule after losing a definite energy for ionization along the path in helium gas.

5.2 Data-taking during alpha-ray irradiation

In order to obtain the phase-transition field ($H_{pt}(E, T)$), which is the lowest magnetic field necessary for each sample granule to flip in the α-ray irradiation, three quantities should be measured as variables: the temperature (T), the energy (E) of incident α particles, and the applied field (H_a).

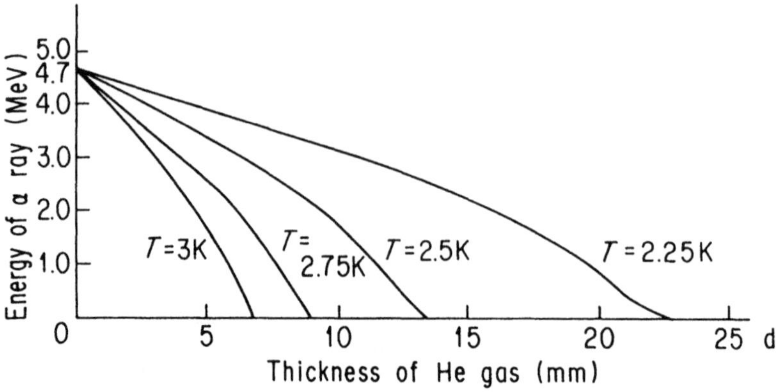

Figure 10: Energy E [MeV] which α particles still carry after traveling a distance d [mm] through He gas at 4 temperatures. The maximum energy is 4.7 MeV, with which α particles start from a 1.5 μm-thick gold coat covering a radioisotope ^{241}Am.

The energy of incident α particles on a granule can be varied by changing the path length in helium gas, though their number also varies thereby, as is discussed in detail in subsection 5.3. The mean range of an α particle having an energy of between 0.1 and 20 MeV in helium gas was calculated by Knoll [44]. By combining his results and the density of saturated helium gas at low temperatures [45], a numerical estimate can be made on the energy (E) which alpha particles still carry after traveling a distance of d in the gas. Four $E - d$ curves are depicted in Fig. 10.

The experimental procedure for obtaining $H_{pt}(E, T)$ is as follows. Firstly, $H_{pt}(E, T)$ is measured by changing the applied field (H_a) stepwise for more than ten values, whereas both the temperature (T) and the mean energy (E) (*i. e.*, the distance between the granule and the α-ray source) are kept fixed. Secondly, $H_{pt}(E, T)$ is determined for another mean energy (E) in the same procedure as mentioned above. Eight energy points between about 1 and 4 MeV are chosen for the measurement. Thirdly, $H_{pt}(E,T)$ is obtained for another temperature. The four temperatures chosen are 3.0, 2.75, 2.5 and 2.25 K.

In the actual measurement, the occurrence of a phase transition was waited for 60 seconds at a given applied field (H_a). This span of time was determined by requiring a sufficiently large number of α particles to bombard the granule, even at the longest distance, in a series of measurements. When no flip occurred during 1 min, the applied field was strengthened a little more. Such a procedure was conducted repeatedly untill the granule flipped.

5.3 Results and discussions

The three sample grains studied were spheres of diameter 145, 165 and 250 μm (A, B and C, respectively).

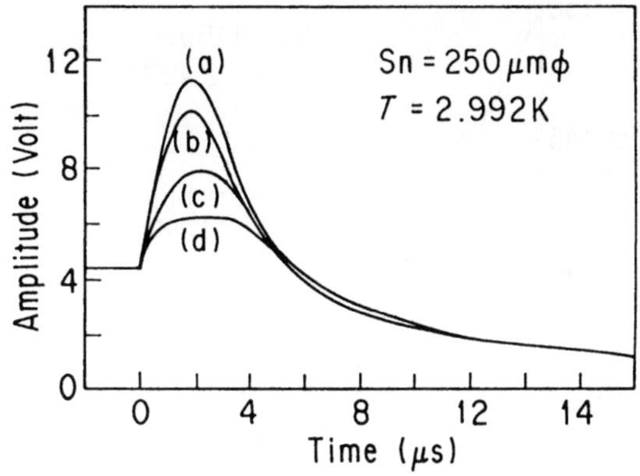

Figure 11: Line shapes of (a), (b), (c) and (d) which were recorded for flips of the 250 μmϕ granule at a temperature of 2.992 K in fields of 92.7, 89.5, 87.8 and 86.0 G, respectively. The signal amplitudes were obtained in volts after amplification. Only the upper signal (a) is induced with the field H_{sh}(= 92.7 G) alone, whereas the other three are induced in the irradiation of the respective fields. The trigger level was set to about 4.4 V.

Signal shapes

Four line shapes of signals are traced in Fig. 11 for a granule of 250 μm in diameter cooled at temperature 3 K. The top pulse (a) is the signal obtained by applying a superheating field (H_{sh}= 92.7G) without any irradiation, whereas (b), (c) and (d) are those induced in the irradiation of α particles with energies of 1–2 MeV at field of 89.5, 87.8 and 86.0 G lower than H_{sh}, respectively. The original energy of 4.7 MeV is reduced to smaller values through the energy loss of ionization. The line shape is not very sensitive to deposited energies, but crucially depends on the strength of the applied field. The pulse heights are observed to diminish with decreasing field. The area under each pulse curve corresponds to the amount of magnetic flux penetrating the granule, but is not always exactly proportional to that because the signal rise is limited, due to the time constant (about 1 μs) of the amplifiers. The tails of the pulses are not much affected by the time constant. The transition time (FWHM) ranges over 4–8 μs. Signal shapes for smaller granules of 145 and 165 μm in diameter are similar to those in Fig. 11, though the transition times are shortened by 1–2 μs.

Every line shape is on the whole similar to the curves given by Faber [46], except for the tail. It drops sharply with time for his cylindrical materials, whereas the signals obtained for the thermal flip of spherical materials gradually approach zero in the same way for the magnetic flip, as is discussed in sect. 4.

Phase-transition fields

As an example, the observational results of the response of a granule of 145 μm in diameter to α-rays at 2.5 K are presented in Fig. 12, in which the circle denotes

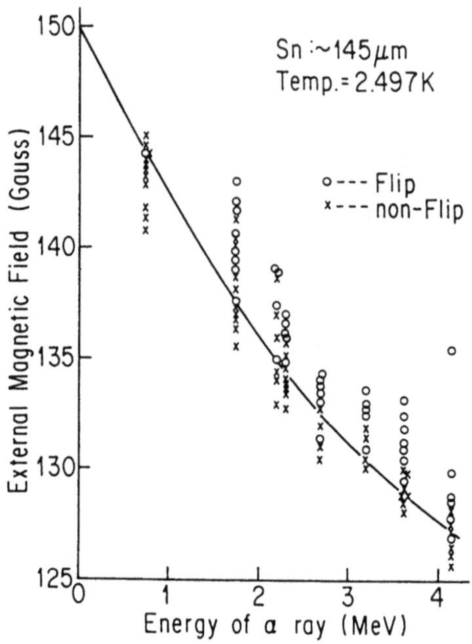

Figure 12: Responses of a 145 $\mu m\phi$ granule to α-ray irradiation at a temperature of 2.497 K. The circle denotes the occurrence of a flip and the cross the non-occurrence during 1 min. The energy of α ray in the abscissa is graduated according to each $E - d$ curve in Fig. 10. The solid line passing through the lowest circles stands for a phase-transition field ($H_{pt}(E,T)$). As the energy approaches zero, H_{pt} should become equal to the superheating field ($H_{sh}(T)$). $H_{sh}(T)$ was measured to be 150 G at 2.5 K for this granule in a separate experiment (Fig.14), so that the $H_{pt}(E,T)$ curve is extrapolated up to a value of 150 G at E=0.

the occurrence of a flip and the cross the non-occurrence during 1 min. Phase transitions were examined in more than ten applied fields for each of eight energy-points of α particles. The field strength on a sphere varies as $(3/2)H_a\sin\theta$, where θ is a polar angle measured from the direction of an applied field (H_a). Hence, the superheated state is more effectively destroyed with an α particle incident near to the equator of a granule than around the poles. It is seen that the circles and the crosses run into each other and that the field strength for the lowest circle is not always larger than that for the top cross at each energy. Such results should be ascribed to the statistical nature as to what point is bombarded during 1 min.

The lowest circles at each energy are taken as events made by an α particle incident near the equator. A solid curve shown in Fig. 12 determines the phase-transition field $H_{pt}(E, T)$ which stands for the lowest field necessary for the transition to occur with an energy deposit of E at a temperature T=2.5 K. The curve is extrapolated up to a point of ($H_a = H_{sh}, E$=0), because $H_{pt}(0,T)$ is identical with the superheating field ($H_{sh}(T)$), as is shown in Fig. 13. The field $H_{sh}(T)$ for this granule was measured for temperatures between 2 K and 3.25 K in a separate experiment. The result is given in Fig. 14; the value of $H_{sh}(T)$ at T = 2.5 K is

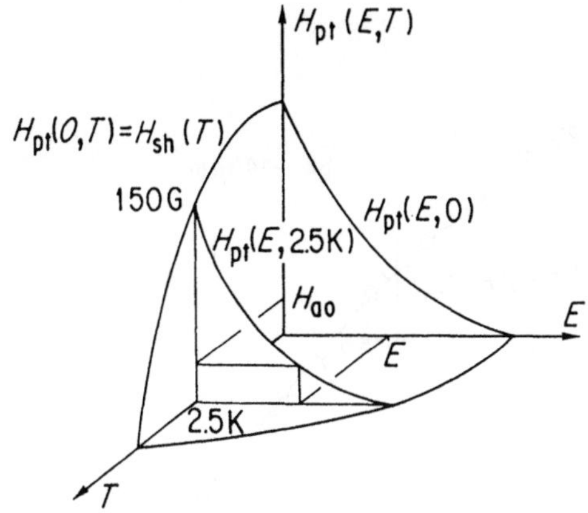

Figure 13: Schematic of the phase-transition field ($H_{pt}(E,T)$) over the $T-E$ plane.

read out to be 150 G.

Flips of the 145 μmϕ granule (A) were measured at incident energies from 4 to 1 MeV or less at temperatures of 2.996, 2.749, 2.497 (shown in Fig. 12) and 2.250 K. The responses of the bigger ones of 165 μm(B) and 250 μm(C) in diameter were studied only at 3.0 K, and the respective phase-transition fields ($H_{pt}(E$,3K)) were obtained.

The smallest response energy was seen to reach 0.75, 0.3 and 0.75 MeV for granule A, B and C, respectively. The measurable smallest energies were determined mainly by the longest distance between a granule and the α-ray source, which was safely taken with the irradiation device shown in Fig. 9c.

Local heating

In Fig. 15, we plot $\Delta H = H_{sh}(T) - H_{pt}(E,T)$ with respect to the energy (E) at four temperatures for a 145 μmϕ granule (A). Every curve is almost linear in the low-energy range. The slopes, $\partial \Delta H/\partial E(= -\partial H_{pt}/\partial E)$ evaluated between 1 and 2 MeV, are given in Table 2. The products of $-\partial H_{pt}/\partial E$ and the specific heat $C(T)$ are also shown there and were found to be insensitive to the temperature. From the conservation of energy we have the following generic relation:

$$C(\partial H_{pt}/\partial E)_T = (\partial H_{pt}/\partial T)_E \cdot A/k\rho v, \tag{40}$$

where A, k, ρ and v are, respectively, the mass number of tin, a conversion factor, the density of the material and the volume which is heated above the threshold temperature (T_0) satisfying the relation $H_{sh}(T_0) = H_a$. In the vicinity of E=0, the derivative in the right-hand side on Eq. (40) can be replaced with $dH_{sh}(T)/dT$. The phase diagram for granule (A) in Fig. 14 shows that the slope (dH_{sh}/dT) takes an almost constant value of 95 G/K in the temperature range over 2.25–3.0 K. Then, the volume (v) is calculated to be 5000 μm^3, which is very small compared with the whole volume (1.6 $\times 10^6 \mu$m^3) of the granule. It is be considered that a thermal nucleus of this volume v is formed for a phase transition.

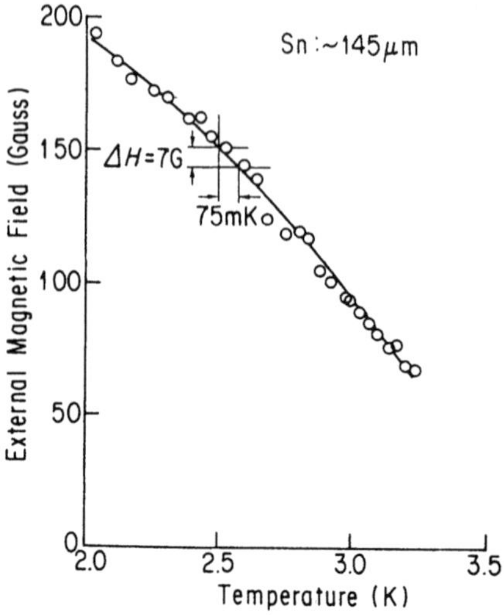

Figure 14: Phase diagram obtained for a tin granule of 145μm in diameter. The curve denotes the superheating critical field ($H_{sh}(T)$).

Table 2: Slopes of the phase-transition fields, $-\partial H_{pt}/\partial E$, and the specific heat ($C(T)$) at four temperatures. Their products are given in the right-hand column.

T [K]	$-\partial H_{pt}/\partial E$ [G/MeV]	C [mJ/mol·deg][a)]	$(-\partial H_{pt}/\partial E) \cdot C$
2.996	3.7	12.9	48
2.749	4.9	10.4	51
2.497	6.3	8.5	54
2.250	6.8	6.9	47

a) interpolated values on the basis of the data given in W.S. Corak, B.B. Goodman, C.B. Satterthwaite and A. Wexler: Phys. Rev. **102** (1956) 656.

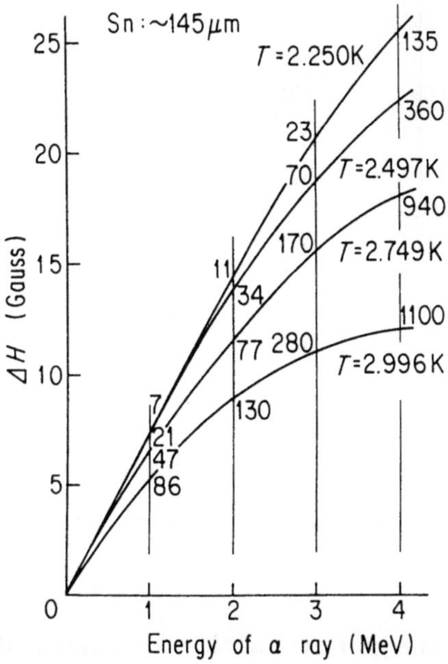

Figure 15: Diagram of $\varDelta H$ defined as $H_{sh}(T) - H_{pt}(E,T)$ to the energy (E) of α particles. The figures given nearby the curves are the calculated numbers of α particles incident on the 145 μmϕ granule during 1 min.

The curves representing the field $H_{pt}(E, 3\mathrm{K})$ for the other granules (B and C) of larger size are analogous to those for granule A at a temperature of 3 K. The volume of the thermal nucleus increases at most by 10% for the largest granule(C) of 250 μm in diameter. Slopes of $-dH_{pt}(E, 3\mathrm{K})/dE$ for granules B and C were evaluated to be 4.5 and 4.0 G/MeV, respectively, between 1 and 2 MeV.

If a deposited energy (E) is entirely used for triggering a phase transition, $\varDelta H$ is expected to be proportional to E, because the slope $(dH_{sh}(T)/dT)$ is almost constant (95 G/K) in the temperature range over 2.25-3.0 K in Fig. 14. The rate of increase in $\varDelta H$ is seen to become smaller in the high-energy range around $E = 4$ MeV than in the vicinity of $E = 0$ in Fig. 15. This means that the incident energy is rather inefficiently used in the high-energy range than in the low-energy one. This is explained qualitatively as follows. The supercurrent flows on a granule at a skin depth of about 0.1 μm, preventing penetration of magnetic flux. Especially in the case of a metastable superheated state, a phase transition of the whole granule is easily induced through a serious disturbance of the flow of a supercurrent. An energy deposit deep inside is less effective. The mean range of α particles is about 3 and 10 μm in tin for their energies of 1 and 4 MeV, respectively. Therefore, a high-energy particle does not work at its face value. For this reason the expected linear increase of $\varDelta H$ is suppressed.

The radioactive source had the shape of a thin disk of 2.4 mm in diameter. The number of α particles bombarding granule (A) per minute was calculated

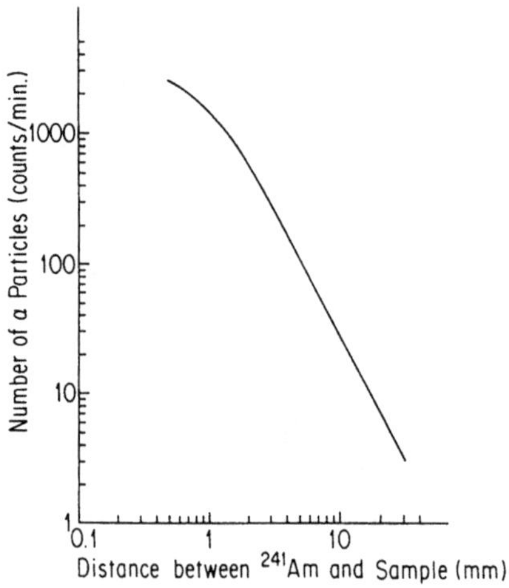

Figure 16: Numbers of α particles incident on a 145 μmϕ granule during 1 min after traveling from the radioisotope.

while taking account of the geometry of the experimental device, and is plotted to the distance between the source and the sample granule in Fig. 16. The figures given nearby the respective curves at each energy in Fig. 15 stand for the number of incident α particles, which is evaluated with the help of the $E - d$ curves in Fig. 10. It increases with energy, since the distance is made smaller. The granule receives only 6–7 α particles for each 10×10 μm^2 area every minute, even for the largest total number of 1100. An energy deposit of 1MeV, for instance, in the small portion is estimated to diffuse all over the granule in the order of 100 ps on the basis of the heat conduction of tin in the normal/superconducting states at temperatures between 2 K and T_c[47]. The difference between the frequency of bombardment (one particle every 10 s) and the diffusion rate (1/100 ps) is so large that the accumulation of deposited energies in such a small portion is impossible. This can be examined more directly in the experimental results. If a small portion is effectively heated through such an energy accumulation due to successive bombardments, $\Delta H(E)$ should be enhanced with energy rather than suppressed. The fact shows the reverse of such an assumption, as is shown in Fig. 15. It is already mentioned that a nucleus for a normal phase is formed in such a small portion. These arguments favor the assumption that the heating is caused by one incident α particle.

Provided that an energy of 1 MeV is distributed uniformly over the entire granule A, it is derived from energetics that the energy can heat the granule only by 0.1 mK at 2.5 K. However, a rise of 75 mK is required to flip the granule at ΔH=7 G, which is derived using the data in Fig. 14. From a difference of more than two orders of magnitude between 0.1 mK and 75 mK, it is concluded that the flip is triggered through a temperature rise of 75 mK in a small portion near to the surface. This is a typical example in which the process of local heating is especially dominant.

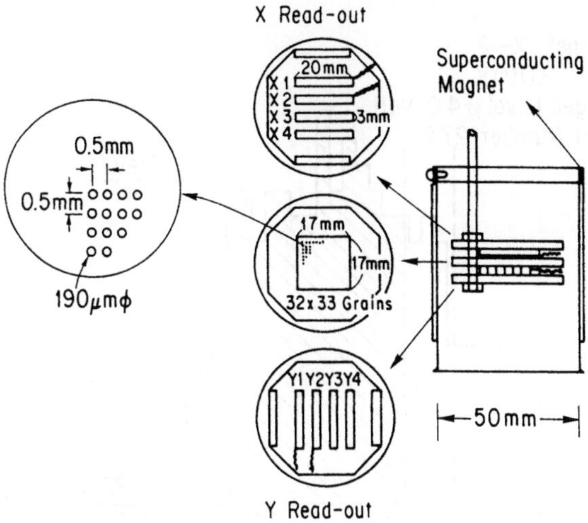

Figure 17: Prototype detector having a planar-type target in 17 mm square, in which 1056 tin grains of 190 μm in diameter are placed at 32 $\times$33 matrix points with 0.5 mm space. Readout coils are made up of 4 channels in each of the x and y-direction sandwiching the target. The coil is a 50 $\mu m\phi$ copper wire wound in 80 turns on an acrylic bar 3 mm wide, 3 mm high and 20 mm long. A plus (minus) signal is induced in a coil when a granule inside (outside) the coil undergoes a phase transition.

Frank et al. [48][49] also studied the response of single grains to gamma- and alpha-ray irradiation. The granules used by them had diameters of 40–60 μm, much smaller than those reported in this section. They reported that the responses of Zn and Al well match a simple picture of global heating, whereas a mixed picture of global and local heating is favorable for Sn.

6 A prototype detector

The structure of a prototype detector and the results of its trial run are described in this section [50]. The detector has a planar-type target consisting of 1056 tin granules of 190 μm in diameter. A stability test was carried out during 650 hours.

The target was 17 mm square and tin grains are placed at 32 $\times$ 33 matrix points with a 0.5 mm space, as is shown in Fig. 17. This space was sufficient to reduce the magnetic interaction between neighboring grains to a negligible amount. Four-channel readout coils in the x-direction were attached on one side of the target surface, and the four y-channel coils on the other side. Each coil had a rectangular shape (3 mm wide and 20 mm long). The flip of a granule induces a positive-sign signal in the coil enclosing this granule, and a small negative-sign signal in the other coils.

For the purpose of illustration, one of the results obtained in the measurement on the x-2 channel with a trigger level of +4.0 volts at temperature 2 K is presented. The distribution of the number of signals exceeding the trigger level is given in

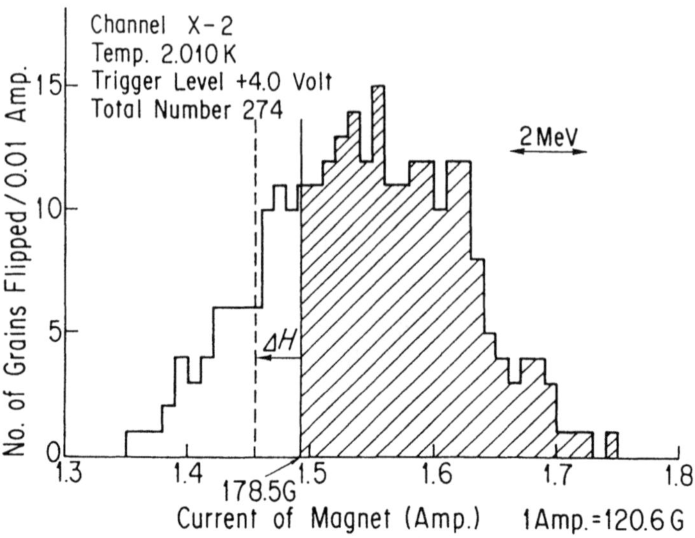

Figure 18: The distribution of the number of flips measured on the x-2 channel coil. When the magnetic field is increased up to 178.5 G shown with a vertical line, one third of the total 274 granules are flipped into the normal states. Then, the external field is lowered by ΔH (a field gap). Stability of the detector was examined by taking 12 field-gaps and by keeping each field constant during 50 hours.

Fig. 18. This was measured by a field-sweep method in which the applied field was increased stepwise with an interval of 1.2 G from 0 to a sufficiently large value. The number of flipped grains in Fig. 18 stands for the average number for every 1.2 G (= 0.01 A) steps obtained through five field-sweeps. The average total number of such signals is 274.

In the stability test, the following procedure was used. At first the field was increased from 0 up to 178.5 G (shown with a vertical line) so that one third of the 274 granules were flipped. That is, every granule included in the white region in Fig. 18 was in the normal-conducting state at that time. Next, the field was lowered by a small value (ΔH) from 178.5 G to the strength shown with a broken line, and was kept at that value for many hours. The field strength is short by more than ΔH for unflipped granules included in the hatched region to be triggered with the magnetic field alone. The field gap (ΔH) can be calculated in terms of the threshold energy (E), which is defined as that required to flip a granule of this size. The relation between ΔH and E is given as E[MeV]=ΔH[G]/3.8 at a temperature of 2.0 K, using the experimental result for sensitivity ($d(\Delta H)/dE$) obtained in the preceding section [51].

The stability of the prototype SSG-detector was examined by keeping the field gap (ΔH) constant at its 12 values for about 50 hours each. The corresponding threshold energies are E=0, 0.25, 0.5, 0.75, 1.0, 1.25, 1.5, 1.75, 2.0, 2.25, 2.5 and 3.0 MeV. The total observation time reached 650 hours. The distribution of obtained

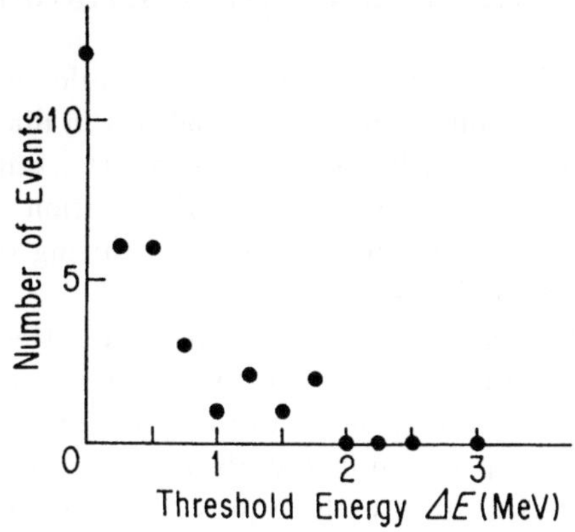

Figure 19: Observed events are plotted with respect to the threshold energy E.

events is given with respect to the threshold energy (E) in Fig. 19. It can be seen that there were events caused by an energy transfer equal to 1.75 MeV or less.

Four causes are considered for the appearance of those events. The fluctuations of either the applied field or the granule temperature are two candidates. From the precision shown by the experimental apparatus used, the former and the latter are estimated to be at most 190 mG and 1 mK, respectively, which correspond to 50 and 10 keV in energy. Therefore, they can only contribute to the phase transitions at the threshold energy E=0. The third candidate for the cause is some radioactive impurity contained in the granules. The radioactivities in the tin used were measured using a low-background gamma-ray detector system in Osaka University [52]. The contamination data of the sample are as follows: U-chain $<$ 3.9 pCi/kg (90% C.L.), Th-chain $<$ 2.1 pCi/kg (90% C.L.), ^{60}Co $<$ 0.42 pCi/kg (90% C.L.), ^{137}Cs= (0.4$\pm$0.3) pCi/kg (1σ) and ^{40}K= (6.2$\pm$2.7) pCi/kg (1σ). Then, the radioactivity due to the U-chain is less than 4.4 nBq and that due to ^{40}K is equal to 7.0 nBq (β ray: 1.3 MeV, γ ray: 1.5 MeV) for a 200 μmϕ granule. It is derived that only 0.1 or 0.2 events are expected at E=0 because of the radioactive contamination for a 50-hour run, though 12 events are recorded. Cosmic-ray muons may cause such flips. An average energy of 300 keV is deposited by a single muon running 0.2 mm in a grain. From the standard flux of secondary cosmic muons, the rate of one event per five hours was calculated in this experiment. Although both the muon flux and the released energy are apparently consistent with the observational results, it should also be noticed that the mechanism of energy loss by muons is different from that by α particles. A muon releases some energy along its lengthy path, and not locally at the surface of a granule as an α particle does, so that the local-heating hypothesis may not be applicable to the case of incident muons.

7 Responses of a tin-SSG to γ-ray irradiation

The response of a superheated superconducting granule to γ-ray irradiation is described in this section. A single tin sphere of 30 μm in diameter was irradiated with 60 keV photons in applied fields of less than the critical field ($H_{sh}(T)$= 325.60 G) at a temperature of 30 mK in a ^{3}He-^{4}He dilution refrigerator [53]. The number of flips of the sample granule was counted during one hour at each ten field-strengths from 315.22 to 324.37 G.

The minimal energy deposit which is necessary for the granule to undergo a phase transition was estimated at each field strength, using both the specific heat in the superconducting state and a working hypothesis of global heating. Then, it is shown that the spectrum of the absorbed energies can be derived by making a differential counting rate with respect to the deposited energies. This is a demonstration for obtaining the spectrum of the energy deposits by means of SSGs. The spectrum has two peaks at ~30 and ~55 keV; the former being due to excited electrons from the K shell by primary γ-rays of 60 keV and the latter due to those from the L shell by either primary γ-rays or a part of secondary X-rays produced by the filling of the K vacancies. The ratio of the counting rates at the two peaks is not very different from the ratio of the mass-absorption coefficient of the K shell to that of the L shell at a photon energy of 60 keV.

Taking account of the original intensity of γ-ray sources, the photon mass attenuation length in intervening absorbers, and the K- and L-edge absorption energies of tin (the granule) and copper (a neighboring detection coil) altogether, this superconducting granule turns out to react to excited electrons with energies of less than 10keV by γ-rays.

In sect. 5, it is shown that superheated superconducting tin granules of diameter 100–200 μm react at temperatures of 2–3 K to an energy deposit down to 1 MeV by α-rays from ^{241}Am [41]. Therefore, the response energy is lowered by two orders of magnitude with a 30 μmϕ granule at 30 mK.

7.1 Experimental apparatus

A 30 μmϕ single tin granule and a 20-turn detection coil of copper wire of 60 μm in diameter were glued together with varnish (GE7031) and completely immersed in liquid helium in a small container made of epoxy resin (Stycast 1266). The container was supplied with liquid helium through a 4 mmϕ SUS tube directly from the ^{3}He-^{4}He mixing chamber (Fig. 20).

The container is wound with a superconducting coil of Nb-Ti-Cu wire of diameter 117 μm in four layers, on which three gamma-ray sources of ^{241}Am(0.5 $\pm$ 0.05 μCi each) are attached at equiangles with each other on a level with the tin granule placed inside. The superconducting magnet was able to produce the field up to 600 G with an error less than 0.1 G. The cooling rate of the refrigerator was about 70 μW at 100 mK, and the temperature in the mixing chamber was monitored with a carbon thermometer inserted in it. A temperature of about 20 mK can be stably maintained in a continuous operation mode, whereas much lower than 15 mK is attainable in a single-operation mode. The structure of the refrigerator is

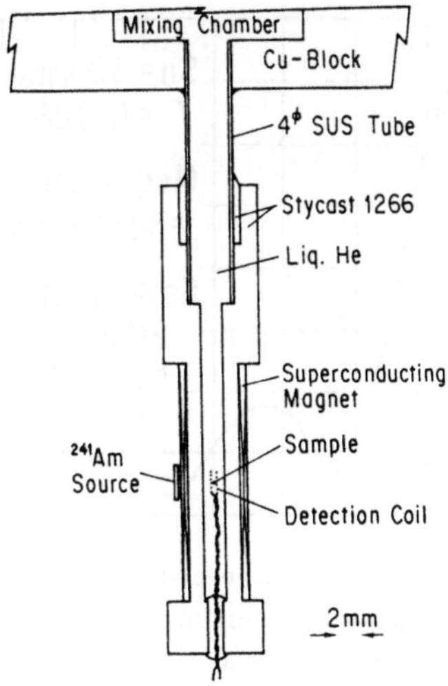

Figure 20: The granule-container made of epoxy resin (Emerson and Cummings Stycast 1266), which is connected through a stainless-steel pipe to the bottom of the mixing chamber. A tin granule and a detection coil glued together with varnish are completely immersed in Liq. He in the container. The coil is connected to amplifiers. The container is wound with a superconducting coil in four layers, on which three γ-ray sources of ^{241}Am are attached (though only one is depicted) at equiangles with each other on a level with the granule placed inside.

outlined in sect. 8.

7.2 Data acquisition and results

Using a sphere of diameter 30 μm singled out from tin powder [54], the superheating superconducting critical field ($H_{sh}(T)$) was measured at temperatures ranging from 30 mK to 3.0 K with no irradiation of γ-rays. Almost all of the data which were obtained several times at each temperature are well described with the conventional formula for the superheating field, $H_{sh}(T) = H_{sho}[1-(T/T_c)^2]$, with $T_c = 3.72$ K. The maximum strength of field $H_{sho} = H_{sh}$ (30 mK) is 327.5 G as a mean value, whereas a field weaker by a 0.6% (about 2.0 G) was obtained during γ-ray irradiation. Therefore, the conventional formula for $H_{sh}(T)$ with H_{sho}= 325.6 G was employed for analyzing the experimental results.

Each signal induced in the detection coil was received in a digital storage scope via wide-band amplifiers. Fig.21 shows one of line shapes for the signals recorded when the applied field was increased beyond $H_{sh}(T)$ at $T = 28$ mK.

Studies on the responses of the sample granule to 60 keV photons were carried out by taking stepwise ten field-strengths smaller than the critical field (H_{sh}) at a temperature of 30 mK. The number of flips during one hour was counted. When no signal was observed during 6 minutes, the field was reset at the same strength and the observation was continued during the next six-minute-period. The applied field was reset for fear that the granule had already flipped for some cause. Therefore,

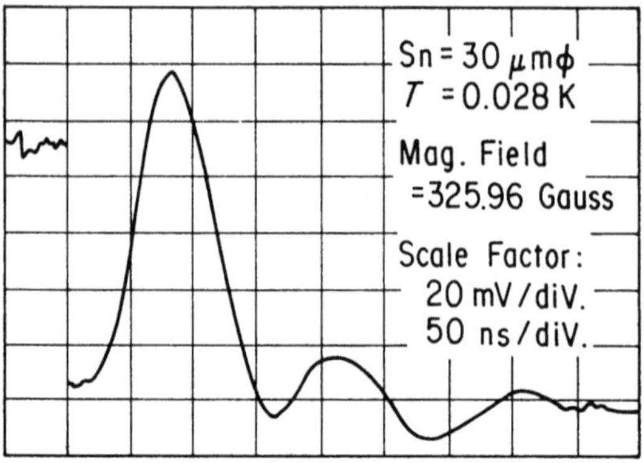

Figure 21: A typical signal shape at a temperature of 28 mK and in a field of 326 G. Its FWHM is 70-80 ns. The amplitude is given in 20 mV/div for the ordinate after amplification and in 50 ns/div for the abscissa.

the field was increased once reaching the critical value 325–326 G to make sure of the granule undergoing a phase transition before the reset. Such observations have been performed at ten field-strengths. The obtained numbers of flips during an hour are given in Table 3 and in Fig. 22.

7.3 Discussions and conclusions

The radioactive isotope ^{241}Am used in this experiment radiates both alpha-rays (5.5 MeV) and gamma-rays (11.9 keV, 0.78%; 13.9 keV, 13.2%; 17.8 keV, 18.1%; 20.8 keV, 4.9%; 26.3 keV, 2.5%; 59.5 keV, 35.3%). By taking into account the original intensity of each source (0.5 μCi) and the radiations absorbed by intervening materials, it turns out that only 59.5 keV γ-rays can reach the granule, as is shown in Table 4. While the initial intensity of 23 photons/min was estimated for the six energy-spectra of γ-rays, only 0.54 photons of 59.5 keV are expected to be absorbed every minute while traversing along the diameter of the granule.

The granule cannot flip because of a deficiency ($\Delta H = H_{sh} - H_a$) in the applied field (H_a). The energy deposit given by an incident γ-ray can compensate for the deficit of field strength less than some critical value (ΔH_0). In an applied field (H_a) beyond the threshold field ($H_{a0} = H_{sh} - \Delta H_0$), in other words, within the value ΔH_0, the counting rate of flips increases with the field strength (H_a) and reaches a maximal, as is expected. The threshold field (H_{a0}) was measured to be around 315 G for 60 keV-photons. From the formula for $H_{sh}(T)$ with H^{*}_{sh0} = 325.6 G, it was derived that a temperature rise of about 640 mK is required for the phase transition to take place in a field of 315 G. Using both the specific heat in the superconducting state and the global-heating picture, a temperature rise of 640 mK is converted into a deposited energy of 65 keV, which coincides roughly with the photon energy.

The maximal counting rate was about 25 flips/hour, whereas the number of absorbed photons during one hour by the granule was evaluated to be 32 or less.

Table 3: Ten applied fields (H_a) and the counting rate N[1/hour] of flips. $H_{sh} - H_a$ stands for the deficit for a granule to flip solely by the applied field. due to the applied field alone. ΔE denotes the necessary energy estimated for the granule to undergo a phase transition due to a temperature rise in each applied field.

	H_a [G]	$H_{sh} - H_a$	N	ΔE [keV]
(0)	325.60 (= H_{sh})			
(1)	324.37	1.23	24	0.9
(2)	323.26	2.34	25	3.2
(3)	322.39	3.21	23	6.2
(4)	321.31	4.29	25	11
(5)	320.39	5.21	26	16
(6)	319.20	6.40	22	24
(7)	318.28	7.32	12	31
(8)	317.24	8.36	7	42
(9)	316.37	9.23	5	51
(10)	315.22	10.38	0	65

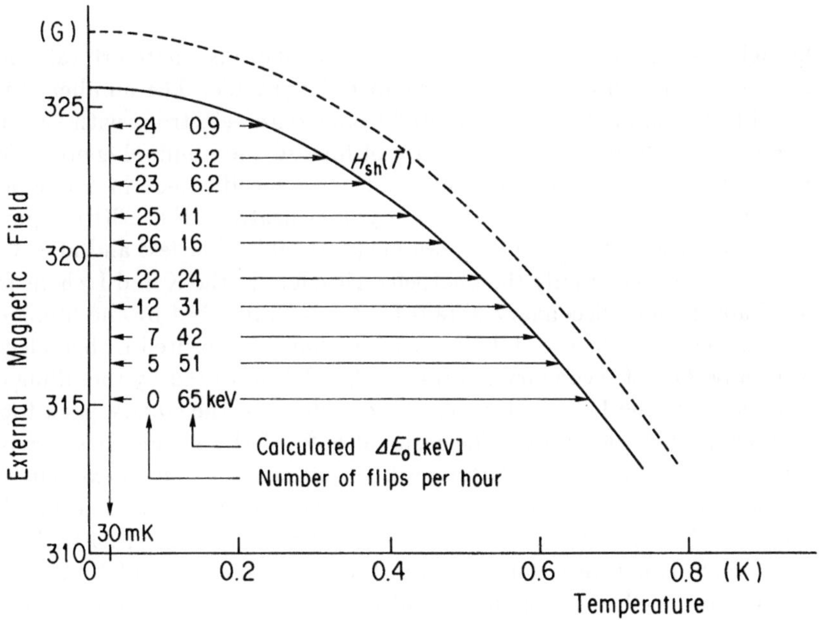

Figure 22: Responses of the 30 μmϕ granule to 60keV photons at a temperature of 30 mK in ten field-strengths. The number of flips counted during an hour are shown. The figures in energy from 0.9 to 65 keV present the smallest value ΔE_0 which is necessary for the granule to flip in each applied field. The smallest energies are estimated by taking into account the specific heat in the superconducting state and a global-heating hypothesis. A curve of broken line stands for the superheating superconducting field $H_{sh}(T)$ in no irradiations.

Table 4: Six energy lines E [keV] of γ-rays from ^{241}Am with relative intensity in % noted. Taking account of both the original intensity 0.5 μCi and the geometrical configuration of the irradiation apparatus, the initial intensity (I_0) of γ-rays was estimated to be 23 photons/min. I_1 [1/min] stands for their remaining intensity after traversing through a solenoid coil (Nb-Ti-Cu) for a superconducting magnet, a cylinder-shaped container (epoxy resin) and a detection coil (Cu) altogether. I_2 [1/min] is the intensity after traversing through a tin granule of 30 μm in diameter. $I = I_1 - I_2$ represents the expectation number of photons absorbed in the granule.

	(1)	(2)	(3)	(4)	(5)	(6)
E [keV]	11.89	13.9	17.8	20.8	26.35	59.54
%	0.78	13.2	18.1	4.87	2.47	35.30
I_1 [1/min]	2.7×10^{-21}	1.0×10^{-10}	2.2×10^{-9}	9.8×10^{-7}	3.0×10^{-4}	4.23
I_2 [1/min]	5.3×10^{-22}	3.8×10^{-11}	1.3×10^{-9}	7.0×10^{-7}	2.5×10^{-4}	3.69
$I = I_1 - I_2$	2.2×10^{-21}	6.2×10^{-11}	0.9×10^{-9}	2.8×10^{-7}	0.5×10^{-4}	0.54

Through the photoelectric effect the absorbed photons excite orbital electrons which subsequently diffuse so as to thermalize the granule. The number of flips is considered to be cumulatively contributed from excited electrons with various energies. Fig. 23(a) shows the counting rate of flips to the minimal energy- deposit calcurated at each field strength, using both the specific heat in the superconducting state and a working hypothesis of global heating. Fig. 23(b) depicts the differential counting rate, which has two peaks at 60 keV or less and at ~30 keV. It is sufficient to consider only the photoelectric effect in the K and L shells for the process of absorption. Secondary X-rays (~ 4 keV) produced by the filling of the L vacancies contribute to the 60 keV peak together with excited L-shell electrons. However, only 20% of secondary X-rays (~ 25 keV) produced by the filling of the K vacancies excite the L-shell electrons to contribute to the 60 keV peak together with excited K-shell electrons. The residual 80% of the secondaries escape out of the granule. Therefore, the latter peak (~ 30 keV) is due to excited K-shell electrons accompanied by such escaping secondary X-rays with an energy of more than 25 keV. The ratio of counting rates at the two peaks reads 1 : 4 with limited statistics, which is not very different from the calculated value of 1 : 2 using the mass-absorption coefficients of the K and the L shells at a photon energy of 60 keV.

It is necessary to take into account the specific heats of both electrons and phonons in the model calculation about the absorbed energies so as to explain the experimental results. This means that a strong coupling between electrons and phonons remains at such low temperatures. Such a calculation is the same as that performed by Freund et al. [55].

The discussion given above remains almost unaltered when a more elaborate

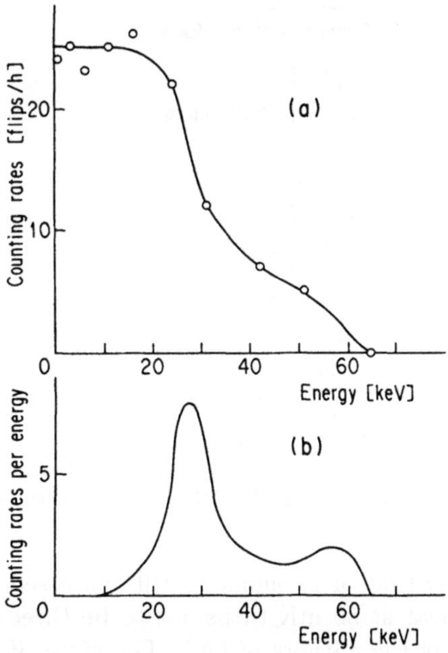

Figure 23: (a) A curve tracing out the counting rates to ten energy-points. (b) A curve representing the differential counting rate with respect to energy deposit. Two peaks should be mainly due to excited electrons from the L and the K shells of tin.

consideration is made on both the other shell's edge-absorption energies and secondary X-rays, not only for tin but also for materials adjacent to the granule. The contribution from kicked electrons through Compton scattering is negligible.

It is concluded that a superconducting granule of diameter 30 μm reacts to electrons with energies of less than 10 keV, which are excited by γ-rays. The energy sensitivity down to 1 MeV was confirmed for superconducting tin granules of diameter 100–200 μm at temperatures 2–3 K by irradiating α-rays in sect. 5 [51]. Therefore, an improvement in the response energy with the tin granule was attained by two orders of magnitude by decreasing the temperature from 2–3 K down to 30 mK and by downsizing the granule by a factor of 3–7.

8 Some plans

The specifc heats of metals become smaller for lower temperatures, so that the sensitivity of granules to deposited energies is generally expected to become higher. Drukier and Stodolsky have pointed out that solar neutrinos with an energy of 420 keV are detectable with tin grains of 2 μm in diameter at a temperature of 50 mK, receiving an average energy of 1 eV [56]. In their calculation the electronic heat capacity (γT) of the normal state is adopted, which likely leads to an overestimate of the heat capacity of granules at such low temperatures. In fact, the specific heat of superconducting electrons diminishes exponentially with decreas-

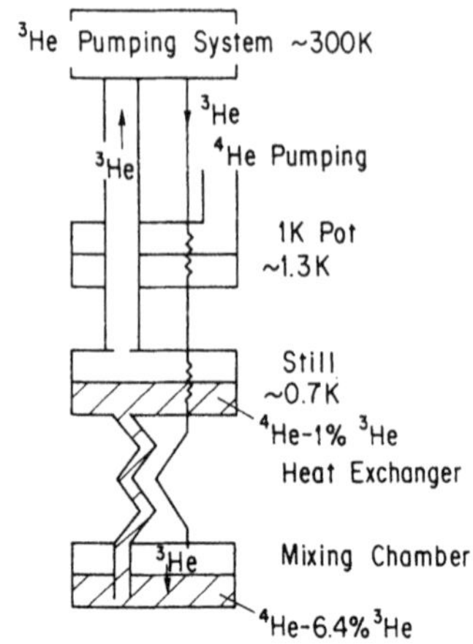

Figure 24: Outline of a ^{3}He-^{4}He dilution refrigerator.

ing temperature, whereas that of phonons is still expressed in the form of αT^3. The electronic specific heat at 50 mK turns out to be three orders of magnitude smaller than that used by the authors of [56]. Therefore, if the heat capacity of superconducting electrons is solely relevant to the phase transition of a metastable superheated state, neutrinos of 420 keV will be detected with tin grains of 30 μm in diameter.

To obtain a sensitivity for an energy deposit of 100 eV, for instance, given by neutrinos or some dark-matter candidates, granules of size 30 μm and an operation temperature of less than 50 mK are required. The present authors are now studying experimentally the response of 30 $\mu m\phi$ tin granules to the irradiation of neutron-rays which can be presumed to mimic pseudo dark-matter particles.

To maintain the temperature of materials below 0.3 K, the ^{3}He–^{4}He dilution refrigerator is practically of use [57]. An outline of it is depicted in Fig. 24. In a stationary operation mode, both the upper half of the mixing chamber and the path to the 1 K pot are filled with almost pure ^{3}He liquid (a concentrated phase), whereas both the lower half of the chamber and the path to the still are filled with almost pure ^{4}He liquid (a dilute phase). Refrigeration is effected in the mixing chamber by ^{3}He atoms moving across the phase boundary from the upper to the lower half. This process is similar to ordinary evaporation in the sense that the upper phase corresponds to liquid and the lower phase to vapor.

The vapor of ^{3}He alone is pumped out from the still, because the vapor pressure of ^{4}He is negligibly small there. Then, ^{3}He atoms ascend from the dilute phase in the mixing chamber, due to the osmotic pressure-gradient, so that the density of ^{3}He is kept constant in the still. Subsequently, ^{3}He atoms dissolve from the concentrated to the dilute phase. The circulation of ^{3}He atoms throughout the system is obtained by means of pumping at room temperature. Incoming gas of ^{3}He is first precooled and liquefied at 1.3 K in a device attached to the ^{4}He-pot (1 K plate). Then, passing by the still at 0.7 K, the liquid enters the mixing chamber through the heat exchanger. It is also required to maintain the still's temperature

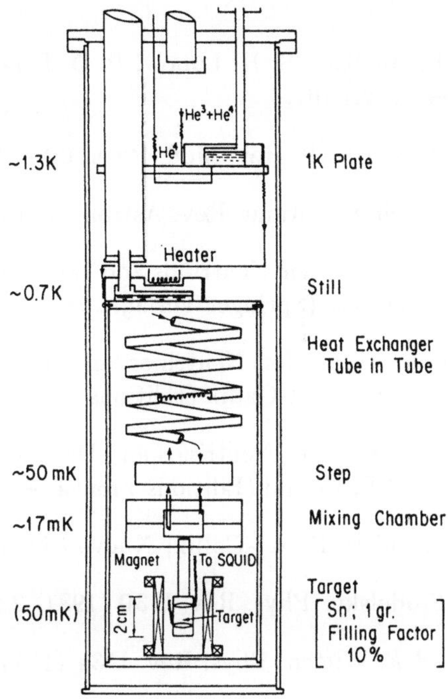

Figure 25: Schematic of the cryogenic-detector system to be constructed by the present authors.

high enough to circulate the gas at a sufficiently high rate.

Fig. 25 shows a schematic of the cryogenic-detector system which is planned to be constructed by the present authors. The circulating rate of ^{3}He atoms in the dilution refrigerator was designed to be around 100 micromoles/sec. The cooling power is 70 microwatts at a temperature of 100 mK; materials of several kilograms will be cooled down to 50 mK. The detector has a cylinder-shaped target of diameter 10 mm and height of 20 mm, which consists of 30 $\mu m\phi$ tin granules embedded in some dielectric material with a filling factor of 10%. The total weight of the granules will be about 1gram. The target is cooled through heat conduction with a metallic rod which connects it to the mixing chamber. Two detection coils, which are each connected to SQUID systems, are wound on the target. After the experimental study on the response of 30 $\mu m\phi$ tin granules to the irradiation of neutron-ray, a stability test of the detector will be made. Next, by running such a detector for two weeks per month throughout one year, the annual variation of signals from some dark-matter particle could be studied.

References

[1] J. N. Bahcall, W. F. Huebner, S. H. Lubow, P. D. Parker and R. K. Ulrich: Rev. Mod. Phys. **54** (1982) 767.

[2] J. N. Bahcall and R. K. Ulrich: Rev. Mod. Phys. **60** (1988) 297.

[3] See, for instance, V. Trimble: Annu. Rev. Astron. Astrophys. **25** (1987) 425.

[4] "Low Temperature Detectors for Neutrinos and Dark Matter" ed. K. Pretzl, N. Schmitz and L. Stodolsky (Springer Verlag, 1987).

[5] "Low Temperature Detectors for Neutrinos and Dark Matter II" ed. L. Gonzalez-Mestres and D. Perret-Gallix (Editions Frontieres, 1988).

[6] "Low Temperature Detectors for Neutrinos and Dark Matter III" ed. L. Brogiato, D. V. Camin and E. Fiorini (Editions Frontieres, 1990).

[7] L. Gonzalez-Mestres and D. Perret-Gallix: Nuovo Cimento **C9** (1986) 573.

[8] A. Drukier and L. Stodolsky: Phys. Rev. **D30** (1984) 2295.

[9] M. W. Goodman and E. Witten: Phys. Rev. **D31** (1985) 3059.

[10] A. K. Drukier, K. Freese and D. N. Spergel: Phys. Rev. **D33** (1986) 3495.

[11] A. F. Pacheco: Mod. Phys. Lett. **A1** (1986) 167.

[12] C. Bernard, A. de Rujula and B. Lautrup: Nucl. Phys. **B242** (1984) 93.

[13] J. R. Primack, D. Seckel and B. Sadoulet: Annu. Rev. Nucl. Part. Sci. **38** (1988) 751-807, "Detection of Cosmic Dark Matter".

[14] "Superconductive Particle Detectors" ed. A. Barone (World Scientific, 1988).

[15] P. F. Smith and J.D. Lewin: Phys. Lett.(Phys. Rep.) **C187#5** (1990) 203–280, "Dark Matter Detection".

[16] "Low Temperature Detectors for Neutrinos and Dark Matter IV" ed. N. Booth and G. Salmon (Editions Frontieres, 1992).

[17] See, for instance, G. K. White: "Experimental Techniques in Low Temperature Physics" 2nd ed. (Oxford Univ. Press, 1968); O. V. Lounasmaa: "Experimental Principles and Methods below 1 K" (Academic Press, 1974); M. Tinkham: "Introduction to Superconductivity" (McGraw-Hill, 1975).

[18] V. L. Ginzburg: J. Exp. Theor. Phys. (USSR) **34** (1958) 113.

[19] "Superconductivity", ed. R. D. Parks (Marcel Dekker, N.Y.,1969).

[20] R. Radebaugh and P. H. Keesom: Phys. Rev. **149** (1966) 209, 217.

[21] N. Perrin and X. C. Zeng: [6] 75; T. Ebisu, N. Nakagawa and T. Watanabe: J. Phys. Soc., Jpn. **60** (1991) 1415.

[22] M. Frank et al.: Phys. Lett. **B230** (1989) 159; Preprint MPI- PAE/Exp.El.225 July 1990.

[23] W. Seidel et al.: Rev. Sci. Instrum. **58** (1987) 1471; L. Gonzales-Mestres and D. Perret-Gallix: [4] 9, [5] 297; F. v. Feilitzsch et al.: [4] 94; P. Freund et al.: [6] 55; M. Chmielowski et al.: [6] 99.

[24] A. Drukier and L. Stodolsky: [8]; M. W. Goodman and E. Witten: [9]; A. K. Drukier, K. Freese and D. N. Spergel: [10]; P. F. Smith and J. D. Lewin: [15] 228.

[25] E. G. Lierke and G. Griehammer: Ultrasonic (1967 Oct.) 246; A. de Bellefon et al.: [14] 241.

[26] L. Gonzalez-Mestres and D. Perret-Gallix: [5] 297.

[27] M. C. Steel: Phys. Rev. **78** (1950) 791.

[28] T. Ebisu and T. Watanabe: [5] 329; T. Ebisu, N. Nakagawa and T. Ebisu: [6] 61.

[29] M. Le Gros et al.: [6] 91.

[30] T. E. Faber: Proc. R. Soc., London **A219** (1953) 75.

[31] See, for instance, Fig. 2 in A. de Bellefon et al.: [4] 59.

[32] A. de Bellefon et al.: [5] 425; Nucl. Instrum. Meth. **A277** (1989) 211.

[33] D. S. McLachlan and J. Feder: Rev. Sci. Instrum. **39** (1968) 1340.

[34] F. v. Feilitzsch et al.: [4] 94; W. Seidel et al.: Rev. Sci. Instrum. **58** (1987) 1471.

[35] See, for instance, O. V. Lounasmaa: "Experimental Principles and Methods Below 1 K", (Academic Press, 1974).

[36] M. Le Gros et al.: [4] 37.

[37] See, for instance, Fig.2 in M. Chmielowski et al.: [6] 99.

[38] T. Ebisu et al.: [6] 61.

[39] C. P. Bean et al.: J. Appl. Phys. **30** (1959) 1976.

[40] T. Ebisu and T. Watanabe: [4] 329.

[41] T. Ebisu et al.:[6] 61; J. Phys. Soc., Jpn. **60** (1991) 1415.

[42] M. C. Steel: Phys. Rev. **78** (1950) 791.

[43] T. Ebisu and T. Watanabe: Rev. Sci. Instrum. **61** (1990) 921.

[44] G.F. Knoll: "Radiation Detection and Measurement" (John Wiley & Sons, Inc., 1979), Chap.2.

[45] G. Klipping: "VDI: Lehrgangshandbuch Kryotechnik" (2nd ed. 1977), Chap.3.

[46] T. E. Faber: Proc. R. Soc., London, **A219** (1953) 75.

[47] K. Mendelssohn: Prog. Low Temp. Phys. **1** (1955) 184.

[48] M. Frank, P. Freund, J. Gebauer, K. Pretzl, A. Singsaas and L. Stodolsky: MPI-PAE/Exp.El. 207 (Aug.1989), to appear in Nucl. Instrum. & Methods.

[49] M. Frank, P. Freund, J. Gebauer, A. Singsaas and L. Stodolsky: Phys. Lett. **B230** (1989) 159.

[50] T. Ebisu et al.: [6] 61.

[51] T. Ebisu et al.: [6] 61; J. Phys. Soc., Jpn. **60** (1991) 1415.

[52] The present authors owe the measurement to Dr. Takayuki Watanabe and Dr. Kenji Matsuoka.

[53] T. Ebisu and T. Watanabe: to appear in Low Temp. Phys. (1993).

[54] The present authors thank Prof. G. Waysand for offering them tin micrograins. A sample granule used in this experiment was selected out of a collection of the grains.

[55] P. Freund, J. Gebauer and L. Stodolsky, [16] 11.

[56] A. K. Drukier and L. Stodolsky: Phys. Rev. **D30** (1984) 2295.

[57] See, for instance, O.V. Lounasmaa: "Experimental Principles and Methods Below 1 K" (Academic Press, 1974) Chap. 3.

Supernova Neutrinos

Hideyuki Suzuki

KEK, National Laboratory for High Energy Physics

1 Introduction

The term 'supernova' was first introduced in 1930's by Baade and Zwicky [1,2] to distinguish the newly recognized extremely luminous nova from the ordinary ones. They are so bright that they are seen even if they occur in other galaxies. In some cases, their maximum absolute magnitude becomes about -19mag, 10^{10} times more luminous than the Sun and 10^4 times than ordinary novae. The time scale of darkening is several months. While about 700 supernovae were observed up to 1990 [3,4], the latest supernovae observed in our Galaxy were the two in the 17th century (Kepler's supernova and the one which left a remnant known as Cassiopeia A) and several other supernovae were recorded before them. Especially the remnant of the supernova in 1054 is now observed as the Crab nebula which contains the Crab pulsar. The nearest supernova observed in this century is SN1987A in the Large Magellanic Cloud.

Today, supernovae are considered as stellar explosions at the death of stars. Moreover, they are also, in some cases, the birth of a compact object such as a neutron star or a black hole. In addition, they drive evolution of the galaxy; the shock wave caused by a supernova explosion propagates through the interstellar space and acts as the trigger to form stars. We owe our lives to supernovae; they scatters away heavy elements synthesized during the stellar life and at the explosion, and these elements constitute heavy elements of the Earth. Cosmic rays are also believed to be accelerated at the supernova explosion.

The plan of this article is as follows. After the classification of supernovae in section 1.1, one of the two types, thermonuclear supernova which is less important as the neutrino source, is briefly described in section 1.2. The rest is devoted to another type, collapse-driven supernovae, and neutrinos from them. Historical overview of the research and the scenario of the collapse-driven supernova explosions are described in section 2.1 and section 2.2 with many figures from numerical studies by various researchers. In section 2.3, we concentrate on neutrinos in and from the supernova core. Section 2 is ended with discussion on the influences of input physics in section 2.4. Section 3 provides the observational results and their implications of the neutrino burst from SN1987A (section 3.1). The search for

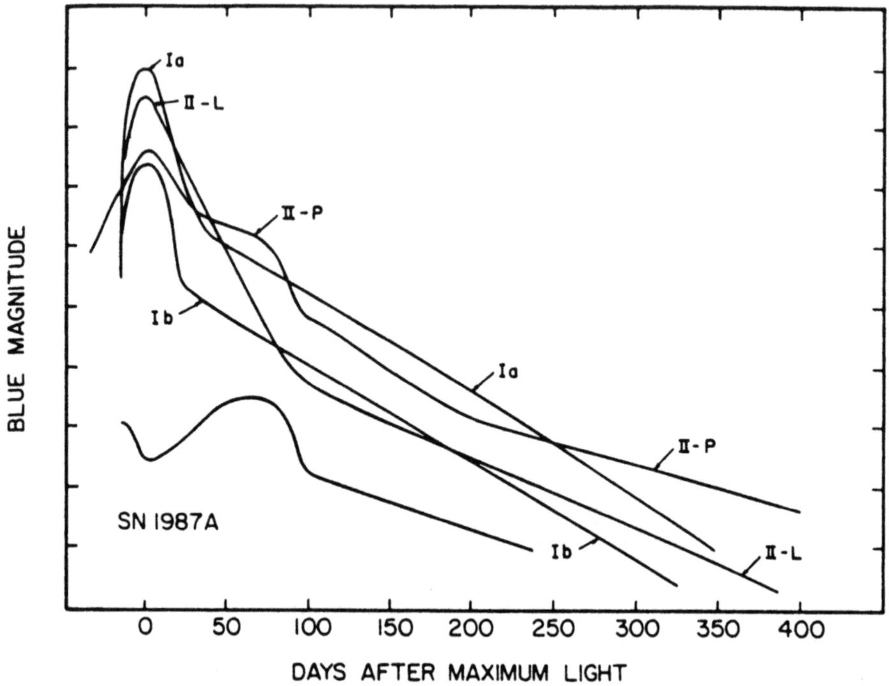

Figure 1: Schematic light curves of various types of supernovae. (taken from Wheeler 1990 [13]) Peculiar behaviour of SN1987A was studied intensively and is now ascribed to the small radius of the progenitor.

the neutrino burst and the relic neutrinos are also reviewed in section 3.2. Basic equations used in numerical simulations of supernova neutrinos are summarised in appendix A.

As further readings, a review on supernova mechanism by Bethe [5], reviews on supernova neutrinos by Burrows [6,7], Mayle [8], Cooperstein [9] and Bahcall [10] and reviews on neutrino astronomy by Koshiba [11] and Totsuka [12], for instance, will be helpful for comprehensive understanding.

1.1 Types of Supernovae

For a long time supernovae have been divided into two classes, Type I and Type II according to the absence or presence of the hydrogen line in their spectra. Recently, however, it is more common to divide each type into subclasses [14,15]. Ordinary Type I (without hydrogen line, with strong silicon line and characterized by homogeneity of the light curve) is now called Type Ia. Type Ia supernovae are observed even in a galaxy consisting only of old stars. On the other hand, Type Ib supernovae, which are characterized by the absence of silicon line and the presence of strong helium line, are observed in the region with young stars as well as Type Ic (without silicon line and with/without weak helium line). Type II supernovae

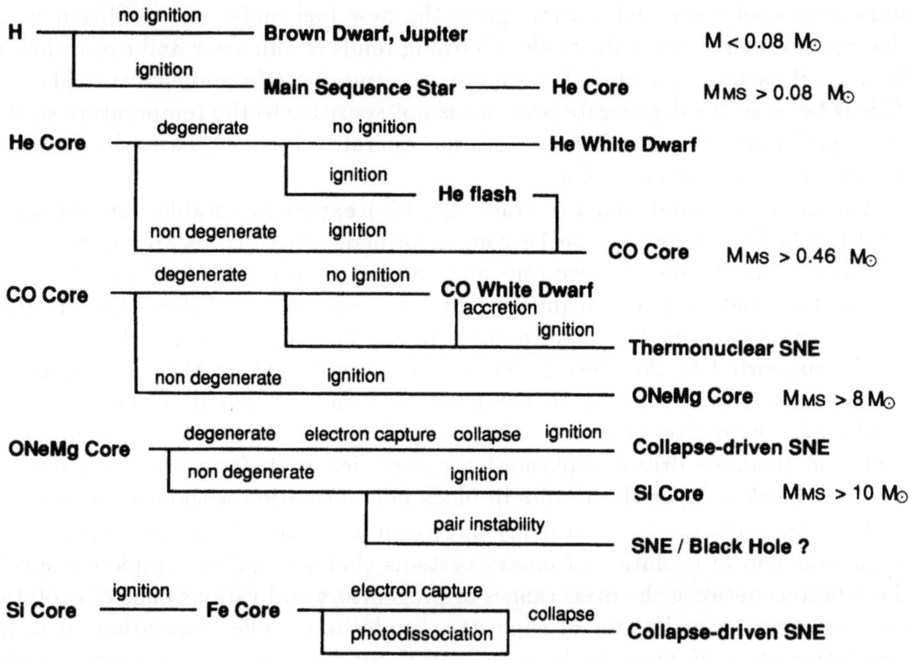

Figure 2: Schematic diagram of the stellar evolution with various main sequence masses ($M_{\rm MS}$).

are subdivided into Type II-P, which has a plateau phase in its light curve, and Type II-L whose light curve declines linearly without a plateau (See Fig. 1). Type II supernovae occur in the region where young stars exist.

In order to understand the above varieties of supernovae, let us recall the theory of stellar evolution. Evolution of stars depends strongly on their masses. The key points are [14,16–18]: 1) Massive stars require large pressure to support their own masses and, hence, their central temperature is higher than that of less massive stars. 2) Since the Coulomb barrier for the nuclear reaction is proportional to Z^2 where Z is proton number of the nucleus, a higher temperature is required for ignition of heavier elements. 3) The stellar core evolves almost independently of the envelope because of the steep density gradient near the core surface. The core mass, however, increases with new ashes from the shell burning unless the envelope is lost by some mass-loss mechanism beforehand. 4) The life of a massive star is much shorter than that of a less massive star because it exhausts the fuel more rapidly to support its large mass. A schematic diagram of stellar evolution and its end is shown in Fig. 2. When the nuclear fuel is consumed at the stellar center, the core of ashes contracts to increase temperature and ashes are ignited as a new fuel. This cycle proceeds from one element to another, from the lightest element H to the most stable element Fe, and a star eventually evolves into the onion-like structure except for cases when the core becomes degenerate. Less massive stars form cool cores and if they ignite the new fuel under the condition where electrons are degenerate, the nuclear burning tends to run away and causes helium flash or, if carbon is ignited, the supernova explosion (thermonuclear explosion). This is because the degenerate pressure is not sensitive to the temperature so that the negative feedback mechanism does not operate which otherwise decreases the reaction rate by core expansion.

On the other hand, massive stars ($\gtrsim 8M_\odot$) experience stable burning stages until the O·Ne·Mg core or the Fe core is formed. These cores are supported by the pressure of degenerate electrons and can be regarded as white dwarfs except for the fact that they are surrounded by large envelopes[1]. When the core mass approaches or exceeds the Chandrasekhar mass, the maximum mass of a star which can be supported by degenerate electron pressure, the O·Ne·Mg core and the Fe core become unstable owing to the pressure deficit caused by electron capture or photodisintegration of nuclei. The core collapses and results in the supernova explosion (collapse-driven explosion) accompanied with formation of a neutron star or a black hole, or else results in black hole formation without explosion.

Unfortunately there are so many uncertainties in the theory of mass loss of a single star and of evolution of binary systems that we cannot completely specify the relations between the mass ranges of progenitors and various ends of evolution such as observational types of supernova explosions. The association to young population stars of Type Ib, Ic and Type II supernovae, however, suggests that their progenitors are short-lived massive stars. It is conceivable that massive stars with the hydrogen envelope explode as Type II supernovae driven by core collapse and that the Type Ib (Ic) supernova also originates from a massive star which has lost the hydrogen envelope (and also helium shell in the case of Type Ic) before explosion [15]. Such a massive progenitor may be a Wolf-Rayet star[2] or a star in a binary system. The difference between II-P and II-L, or, equivalently, the existence and the absence of the plateau phase, may be attributed to the amount of the hydrogen envelope and the explosion energy [20]. On the other hand, Type Ia is considered to be the thermonuclear destruction of a CO white dwarf in a binary system. Of course, there are many other models concerning the linkage of progenitors to the observational types of supernovae; for example, it is possible to regard a single star of the intermediate mass ($6 \sim 8\ M_\odot$) as a progenitor of Type II-L, and to consider it as thermonuclear explosion of a CO core with the hydrogen envelope [15].

In any case, however, it is believed that (not too extremely) massive stars ($8M_\odot \lesssim M \lesssim 30 - 50M_\odot$) explode by means of core collapse and that less massive stars become white dwarfs or explode by means of thermonuclear runaway. The relation between the observational types and the explosion mechanism is complicated with the mass loss and the mass accretion during the stellar life.

---

[1] 'White dwarf' is originally a name of the stars which have high surface temperature (white) and small radius (dwarf). They are supported by the pressure of degenerate electrons and their structures do not depend on the existence of envelopes.

[2] Wolf-Rayet stars are characterized by bright and broad emission lines of helium in their spectra. They are considered as stars which have large mass loss rate and have lost their hydrogen envelopes. [19]

1.2 Neutrinos from Thermonuclear Supernova Explosion

Although both types of supernova explosion accompany the neutrino burst, neutrino burst from the thermonuclear supernova explosion is much weaker than that from the collapse-driven supernova explosion. In the case of thermonuclear supernova explosion [21], the energy source of the explosion is nuclear energy. The total energy released in the thermonuclear supernova explosion can be estimated by assuming that all ^{12}C in a white dwarf burn to ^{56}Fe as

$$E_{\text{thermonuclear}} = \left(m_{\text{C}} - \frac{12}{56} m_{\text{Fe}}\right) c^2 \, \frac{M_{\text{WD}}}{m_{\text{C}}} \sim 3 \cdot 10^{51} \text{erg} \tag{1}$$

where m_{C} and m_{Fe} are the masses of ^{12}C and ^{56}Fe and the mass of the white dwarf, M_{WD}, is set to be the Chandrasekhar mass ($\sim 1.4 M_{\odot}$). The typical energy of emitted neutrinos is $\sim 3T$ (for thermal neutrinos) or $\lesssim 5/6 \cdot \mu_{\text{e}}$ (for ν_{e}'s produced by electron capture[3]), both of which are less than few MeV taking account of the physical condition in the exploding white dwarf (density $\rho_c \lesssim 5 \cdot 10^9 \text{g/cm}^3$, temperature $T \lesssim 1\,\text{MeV}$, electron Fermi energy $\mu_{\text{e}} \lesssim 7\,\text{MeV}$) [22,23]. The neutrino mean free path for scattering off ^{56}Ni is about $2 \cdot 10^8 \text{cm}(\rho_c/5 \cdot 10^9 \text{g/cm}^3)^{-1}(E_\nu/6\text{MeV})^{-2}$ which is longer than the scale of the white dwarf, $5 \cdot 10^7 \text{cm}(M_{\text{WD}}/1.4 M_{\odot})(\rho_c/5 \cdot 10^9 \text{g/cm}^3)^{-1/3}$. This means that the emitted neutrinos escape freely from the white dwarf and they are never thermalized, so that equipartition of thermal energy is not realized. Consequently neutrinos can carry out only a tiny fraction of the released energy (Eq. 1). Nomoto [24] reported that the total energy of the neutrino burst is less than $O(10^{49})$ erg and the time scale of the burst lies in the range from 0.2 sec to 60 sec using his numerical simulations of the thermonuclear supernova explosions [22,25]. In fact the sum of the observed kinetic energy of the ejecta and the radiation energy is comparable to the total thermonuclear energy estimated above. In this way neutrinos do not play an important role in the thermonuclear supernova explosion, so that we will not discuss further it in this article.

2 Collapse-driven Supernova Explosion and Neutrinos

When the core of a massive star ($M_{\text{core}} = 1 \sim 2 M_{\odot}$) collapses to form a neutron star [1], the core density exceeds 10^{14} g/cm^3 and the temperature becomes higher than 10 MeV. Under this condition the typical time scale of weak interaction becomes of the order of $1/\sigma_{\text{weak}} n_{\text{target}} v_{\text{relative}} \sim 8 \cdot 10^{-8}\,\text{sec}\,(T/10\text{MeV})^{-2}$ $(\rho/10^{14} \text{g/cm}^3)^{-1}$ ($\sigma_{\text{weak}} \sim 4 G_F^2 \hbar^2 c^2/\pi \cdot T^2$ [26], $n_{\text{target}} \sim \rho/m_u$ and $v_{\text{relative}} \sim c$ where $G_F \sim 10^{-5} m_{\text{p}}^{-2}$ is the Fermi coupling constant and m_u is the atomic mass).

---

[3]Since the cross section for electron capture on free protons is roughly $\sigma_{\text{e-cap}} \sim \frac{G_F^2 \hbar^2 c^2}{2\pi} E_{\text{e}}^2$ where G_F is the Fermi coupling constant and E_{e} is electron energy, the average energy of emitted ν_{e}'s becomes

$$\frac{\int \sigma_{\text{e-cap}}(E_{\text{e}}) 2 f(E_{\text{e}}) E_\nu n_{\text{p}} \frac{d^3 p_{\text{e}}}{\hbar^3}}{\int \sigma_{\text{e-cap}}(E_{\text{e}}) 2 f(E_{\text{e}}) n_{\text{p}} \frac{d^3 p_{\text{e}}}{\hbar^3}} \sim \frac{\int_0^{\mu_{\text{e}}} E_{\text{e}}^5 dE_{\text{e}}}{\int_0^{\mu_{\text{e}}} E_{\text{e}}^4 dE_{\text{e}}} = \frac{5}{6} \mu_{\text{e}} \quad .$$

Here we use approximations: $E_{\text{e}} \sim E_\nu$, distribution function $f(E_{\text{e}}) = \theta(\mu_{\text{e}} - E_{\text{e}})$ (degenerate electrons), $p_{\text{e}} c \sim E_{\text{e}}$ (relativistic electrons).

Since this time scale is much shorter than the dynamical time scale of the core ($1/\sqrt{G\rho} \sim 0.4\,\mathrm{msec}\ (\rho/10^{14}\mathrm{g/cm}^3)^{-1/2}$)[4], a large number of neutrinos are produced, which are comparable to other constituents such as electrons and photons. Among the constituents of the core, however, neutrinos interact most weakly because they react neither with electromagnetic forces nor with strong forces. This means that it is easiest for neutrinos to move in the core and, hence, it is neutrinos that transport the internal energy of the core and drive the evolution of supernova core. While the gravitational energy released by the core collapse can be estimated roughly as

$$\Delta E_G = -\left(\frac{GM^2_{\mathrm{core}}}{R_{\mathrm{Fe\ core}}} - \frac{GM^2_{\mathrm{core}}}{R_{\mathrm{Neutron\ Star}}}\right) \sim O(10^{53})\ \mathrm{erg},$$

the observed explosion energy (kinetic energy + radiation energy) is only about 10^{51} erg (kinetic $O(10^{51})$ erg, radiation $O(10^{49})$ erg). The gravitational waves may carry out a part of the released energy, but Stark & Piran [27] estimated that this amount is only 10^{51}erg for the case of slowly rotating core. The rest ($O(10^{53})$ erg) is emitted as neutrinos. Thus, in contrast with the case of the thermonuclear supernova explosion, neutrinos play a very important role in the collapse-driven supernova explosion and the total energy emitted as neutrinos is much larger than that in the thermonuclear supernova explosion.

Neutrinos from supernova explosion in the nearby galaxies had been predicted at an observable level and were indeed observed [28, 29]. The first object studied in the (extra-solar-system) neutrino astronomy was the supernova (SN1987A) in the Large Magellanic Cloud. We can '*see*' the supernova core directly by neutrinos instead of the ordinary electromagnetic waves. They bring us information about the dense core which we can never obtain through photons of any frequencies. In addition, neutrino astronomy of the supernova explosion is useful not only for astrophysics but also for nuclear physics and particle physics. That is, the supernova core is a good laboratory for the dense matter and for weakly interacting particles. Since the thermal history of the supernova core is controlled by weakly interacting particles (neutrinos and other exotic particles such as axions if exist) and the core structure is affected much by the equation of state (hereafter EOS) of nuclear matter, observations of the core by neutrinos give us much information about the particle property and the EOS. Moreover, several properties of neutrinos can be also studied by the observation of neutrino burst even if they have little effects on the supernova core. For example, neutrino mass much smaller than core temperature can be studied using the dispersion of the time of flight due to the finite mass [30]. In this way, supernova neutrinos play a unique role in the various fields of natural science.

---

[4]The order of dynamical time scale can be estimated from the equation of motion;

$$m\frac{d^2r}{dt^2} = -G\frac{Mm}{r^2} \quad ,$$

in which one can substitute $M \sim \rho r^3$, and $\frac{d^2r}{dt^2} \sim -\frac{r}{\tau^2_{\mathrm{dyn}}}$, and then obtains $\tau_{\mathrm{dyn}} \sim 1/\sqrt{G\rho}$.

2.1 Historical Overview

At present, collapse-driven supernova explosion is literally considered to be triggered by the core collapse. When the central density exceeds the nuclear density and the EOS stiffens, the collapse is halted and the bounced inner core launches a shock wave into the outer core. Although it is still unsettled whether the neutrino heating of the matter plays a critical role for the explosion, blowing off the envelope by the shock wave is the supernova explosion. At the same time as the shock wave propagates through the outer core and envelope, the bounced inner core settles into a hydrostatic configuration and this protoneutron star cools down to a normal neutron star by emitting neutrinos.

The original idea of this scenario was pointed out by Baade & Zwicky [1] more than 50 years ago. They noticed that a small part of the gravitational energy released at neutron star formation is sufficient to blow off the overlying envelope whose binding energy is very small. Hoyle, Fowler and their collaborators [21, 31, 32] considered thermonuclear burning as the mechanism to convert the released gravitational energy into the explosion energy. Meanwhile, Colgate & Johnson [33] considered the formation of shock wave due to the sudden increase of the pressure. In 1966, Colgate & White [34] performed numerical simulations of the supernova explosion taking into account the energy transfer by neutrinos with very crude method. They pointed out that neutrinos produced in the core can deposit sufficient energy for the envelope to explode. (See also references [35–37].) However, more refined simulations by Arnett [36] and Wilson [38] proved that the neutrino deposition is not sufficient. Aside from the neutrino deposition mechanism, Wilson [38] found purely hydrodynamical explosion by the shock wave in the case of a low mass model ($1.25 M_\odot$).

Situations changed when the neutral current of the weak interaction was discovered. Freedman [39] pointed out that coherent scattering off heavy nuclei is important for neutrino transfer in supernovae. It was expected that the neutral current would enhance the neutrino deposition because the neutrino interaction is now stronger than considered before [39–41]. Sato [42, 43], however, showed that neutrinos are trapped in the core. Mazurek [44, 45] also studied effects of neutrino degeneracy. Refined simulations were performed [46, 47] and it was shown that neutrino deposition is still inefficient to explode the star [47]. On the other hand, the neutrino trapping provides a desirable condition for purely hydrodynamical explosion. Before the establishment of the current picture (neutrino trapping, a cold core, high density EOS in finite temperature and effects of general relativity), the inner core was considered to bounce at lower central density ($\lesssim 10^{14}\mathrm{g/cm^3}$) owing to the thermal pressure of free nucleons [36, 40, 47]. Since the neutrino trapping suppresses the neutronization of nuclei and keeps the entropy low, heavy nuclei do neither drip neutrons nor melt into free nucleons until the density approaches the nuclear density. Collapse to higher density results in a larger released energy and, hence, a larger shock energy, which is desirable for explosion. Brown, Bethe & Baym [48] established the picture that the core bounces and launches a shock wave above the nuclear density. Unfortunately, however, there are some processes to weaken the shock wave such as the endothermic photodisintegration of nuclei and energy loss accompanied with neutrino emission behind the shock front. Various numerical simulations have been performed [49–75] and some of the results show successful prompt explosions, but others show that the shock wave stalls and does not cause explosion. Results depend on adopted EOS, initial models and numerical codes. It is now commonly believed, however, that massive Fe cores, if formed, cannot explode in a purely hydrodynamical way.

In 1984, using numerical simulations with neutrino transfer, Wilson [76] found that the stalled shock wave is revived by neutrino heating and that the core explodes several 100 msec after the bounce. This relatively slow explosion, compared with the prompt one which occurs in the dynamical time scale of about 10 msec, is called delayed explosion. Wilson *et al.* [77] reported that even the massive core of $2.2M_{\odot}$ can explode by this delayed mechanism. However, since no other group has yet reproduced delayed explosion in realistic numerical simulations [64, 66, 67], we cannot regard the delayed mechanism as confirmed.

Aside from the explosion mechanism, the detection of the neutrino burst from supernova explosion and the cooling stage of a protoneutron star is a matter of much interest [78, 79]. In particular, in 1986, Burrows & Lattimer [80] investigated the birth of a neutron star with numerical simulations including simplified neutrino transfer. Experiments which aim at the detection of supernova neutrinos started in 1980's. Two water Čerenkov detectors (KAMIOKANDE-II, IMB) which were originally constructed for the search of nucleon decay observed supernova neutrinos for the first time in 1987 [28, 29]. It was very lucky that SN1987A exploded just after the threshold energy of the KAMIOKANDE-II detector was lowered in order to observe solar neutrinos. The analysis of the observed neutrino burst will be summarized in section 3.1. The most important conclusion is that the theoretical framework of the stellar evolution, the collapse-driven supernova explosion and the neutron star formation is proved to be correct as a whole.

Next let us summarize the physical processes important in each stage of the collapse-driven supernova explosion [14, 48, 81, 82].

2.2 Scenario of the collapse-driven supernova explosion

2.2.1 Pre-collapse stellar model

In Fig. 3, an example of the pre-collapse stellar model is shown. The Fe core[5] is surrounded by shells of lighter elements. This is called onion-skin structure. The size of Fe core is the order of 10^9 cm while the stellar radius is larger than 10^{13} cm, which is comparable to the radius of earth's orbit. The core is supported mainly by the pressure of degenerate electrons and, hence, it can be regarded as a white dwarf. At the boundary between the core and the surrounding shell, the density decreases steeply and, hence, the dynamical time scale of the core is much shorter than that of the envelope. That is, the dynamics of the core is not affected by the envelope. Therefore, we focus on the core hereafter for a while.

[5]Strictly speaking, the "Fe" core consists not only of iron nuclei but also of various nuclei. Isotopes having a neutron excess greater than ^{56}Fe are also included because of the electron capture which is already in progress.

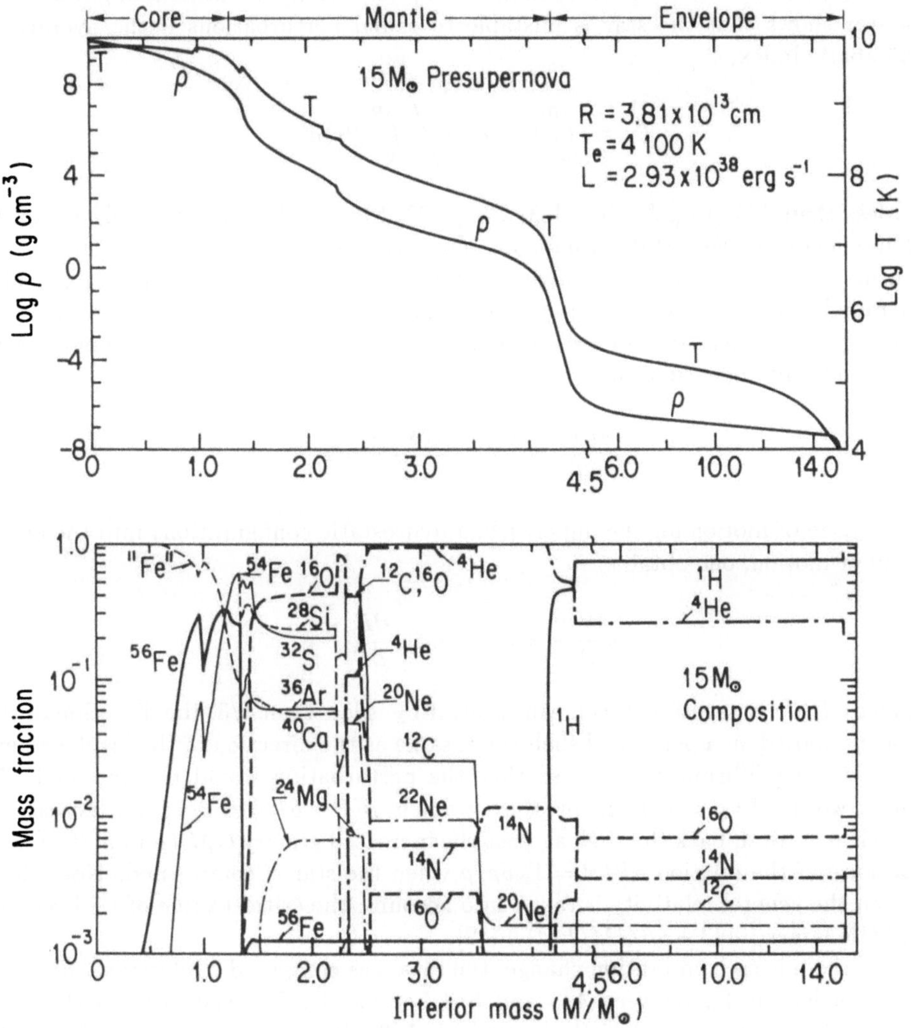

Figure 3: Structure of the pre-collapse star (Woosley & Weaver's $15M_{\odot}$ model taken from [14]). Upper figure: temperature and density profile, lower figure: composition profile. The abscissa is the enclosed mass of each shell. Note that at $4.5M_{\odot}$ the scale of the abscissa is changed.

2.2.2 Onset of collapse

Stability of the core depends on the adiabatic index,

$$\Gamma_S \equiv \left(\frac{\partial \ln P}{\partial \ln \rho}\right)_S, \tag{2}$$

of the matter. Linear perturbation theory for the hydrostatic configuration of a star shows that the star is unstable to radial perturbations if the average of adiabatic index,

$$\overline{\Gamma}_S \equiv \int_0^R \Gamma_S P r^2 dr \Big/ \int_0^R P r^2 dr, \tag{3}$$

is less than 4/3 (Cox & Giuli [83] Chap. 27, Shapiro & Teukolsky [82] Chap. 6, Kippenhahn & Weigert [84] Chap. 38). For intuitive example, consider homologous shrink or expansion of a star with constant adiabatic index, Γ_S, that is, consider a radial perturbation which is proportional to radial position ($r_0 \rightarrow r_0 + \delta r$, $\delta r = \zeta \cdot r$, $\zeta =$ constant). Substituting $\delta P/P = \Gamma_S \delta\rho/\rho$, $\delta\rho/\rho = -3\zeta$ (derived from the continuity equation) and

$$\rho \frac{\partial^2 r_0}{\partial t^2} = \frac{\partial P}{\partial r} - \frac{G M_r \rho}{r_0^2} = 0 \tag{4}$$

(equation of motion for the unperturbed hydrostatic configuration) into the equation of motion, one obtains

$$\rho \frac{\partial^2 \delta r}{\partial t^2} = (3\Gamma_S - 4)\zeta \frac{\partial P}{\partial r} \quad . \tag{5}$$

Since $\partial P/\partial r$ is negative, this means that, if Γ_S is less than 4/3, the direction of the force imposed on a perturbed shell is the same as the direction of the displacement from the equilibrium position so that the perturbation would be amplified. In other words, if $\Gamma_S < 4/3$, the pressure gradient $\partial(P+\delta P)/\partial r$ does not increase so much as to push back the shell against the increased gravity $GM_r(\rho+\delta\rho)/(r+\delta r)^2$ because of the relation $\delta P/P = \Gamma_S \delta\rho/\rho$ when the star is compressed. Note that, when the general relativity is taken into account, the critical value of $\overline{\Gamma}_S$ becomes a little larger ($4/3 + O(GM/Rc^2)$) [82].

If there is no composition change, the ideal gas composed of electrons, photons and nuclei has Γ_S between 4/3 and 5/3 because Γ_S for relativistic particles is 4/3 and Γ_S for non-relativistic particles is 5/3. However, as the core (O·Ne·Mg or Fe) of a massive star grows with ashes of nuclear burning and approaches the Chandrasekhar limit, it becomes unstable owing to two processes; electron capture and nuclear photodissociation. These reactions change the composition of the matter and make Γ_S less than 4/3 in some cases.

Instability caused by electron capture such as

$$^{24}\mathrm{Mg}\ \mathrm{e}^- \quad \rightarrow \quad ^{24}\mathrm{Na}\ \nu_e$$

$$^{56}\mathrm{Fe}\ \mathrm{e}^- \quad \rightarrow \quad ^{56}\mathrm{Mn}\ \nu_e$$

is triggered when the Fermi energy of electrons,

$$\mu_e \sim (3\pi^2 n_e)^{1/3}\hbar c = 11.1\,\mathrm{MeV}(\rho Y_e/10^{10}\mathrm{g/cm^3})^{1/3} \tag{6}$$

where $Y_e \equiv (n_{e^-} - n_{e^+})/n_{\text{baryon}}$ is electron fraction per baryon, exceeds the threshold energy for these reactions (mass difference between the nuclei: $m_{\text{Na}} - m_{\text{Mg}} = 5.5\,\text{MeV}$, $m_{\text{Mn}} - m_{\text{Fe}} = 3.7\,\text{MeV}$). Diminution of electrons causes a deficit in pressure and the core begins to collapse. As collapse proceeds, the consequent increase of μ_e results in further electron capture. If one neglects thermal pressure for simplicity, the pressure comes from relativistic electrons,

$$P_e = \frac{1}{4}(3\pi^2)^{1/3}\hbar c\left(\frac{\rho}{m_{\text{nucleon}}}\right)^{4/3} Y_e^{4/3} \tag{7}$$

and the adiabatic index becomes

$$\Gamma_S = \left.\frac{\partial \ln P}{\partial \ln \rho}\right|_S = \left.\frac{\partial \ln P}{\partial \ln \rho}\right|_{S,Y_e} + \left.\frac{\partial \ln Y_e}{\partial \ln \rho}\right|_S \left.\frac{\partial \ln P}{\partial \ln Y_e}\right|_S = \frac{4}{3}\left(1 + \left.\frac{\partial \ln Y_e}{\partial \ln \rho}\right|_S\right) . \tag{8}$$

Continuation of electron capture implies negative $(\partial \ln Y_e/\partial \ln \rho)_S$ and, hence, Γ_S less than 4/3.

On the other hand, when the temperature exceeds about $5 \cdot 10^9$K, the endothermic photodissociation of nuclei such as

$$^{56}\text{Fe} + \gamma \rightarrow 13\ ^4\text{He} + 4\text{n} - 124.4\,\text{MeV}$$

results in lowering the adiabatic index to be less than 4/3, that is, the increase of the internal energy caused by contraction is exhausted by this endothermic reaction so that the pressure does not increase so much as to suspend the core collapse. For extreme example, consider the case that temperature does not increase at all because of the endothermic reactions even when the core contracts. Since degenerate pressure of relativistic electrons (main component of the total pressure) in finite temperature can be expressed as

$$P_e = K\rho^{4/3}(1 + BT^2\rho^{-2/3}) \tag{9}$$

where

$$K = \frac{(3\pi^2)^{1/3}\hbar c}{4}\left(\frac{Y_e}{m_{\text{nucleon}}}\right)^{4/3} \tag{10}$$

and

$$B = \frac{2\pi^2(3\pi^2)^{-2/3}}{3\hbar^2 c^2}\left(\frac{Y_e}{m_{\text{nucleon}}}\right)^{-2/3}, \tag{11}$$

the adiabatic index becomes

$$\Gamma_S = \frac{4}{3} - \frac{BT^2\rho^{-2/3}}{1 + BT^2\rho^{-2/3}}\left(\frac{2}{3} - 2\left.\frac{\partial \ln T}{\partial \ln \rho}\right|_S\right) . \tag{12}$$

Thus, if $(\partial \ln T/\partial \ln \rho)_S \sim 0$, Γ_S is less than 4/3: the core is unstable.

A relatively less massive core ($M_{\text{main sequence}} \lesssim 12 - 15 M_\odot$ [14, 17, 18]) which is cool and dense becomes unstable by electron capture at first, whereas a hot massive core becomes unstable mainly by photodissociation. When the density and the temperature increase with the core collapse, the two processes cooperate. (As for O·Ne·Mg core, the nuclear burning to iron family nuclei accompanies the collapse and, then, the matter becomes in nuclear statistical equilibrium.) Note that photodissociation does not disintegrate all nuclei into free nucleons in the collapse stage because the temperature of the degenerate core does not become so high. For example, a numerical simulation by Bruenn [62] shows that mass fraction of heavy nuclei is kept to be greater than 0.7 during the collapse stage (See Fig. 4) until they merge to be nuclear matter.

On the other hand, the progress of electron capture during the core collapse depends on the electron capture rate of neutron-rich nuclei. Various treatments of nuclear shell effects in the calculation of the capture rate cause different results [85–87]. The amount of the electron capture before the neutrino trapping determines the trapped lepton fraction which is an important quantity for the success of prompt shock wave propagation as will be discussed in section 2.4.

2.2.3 Neutrino trapping

Neutrino trapping plays very important roles in supernova physics [42–45]. Here we summarize the effect of neutrino trapping. The existence of the neutral current weak interaction implies that the dominant opacity source for ν_e's during the collapse stage is coherent scattering off nuclei, cross section of which is proportional to A^2 [39]. That is, neutrinos with wave length, $\hbar c/E_\nu \sim 20\,\text{fm}\,(E_\nu/10\text{MeV})^{-1}$, longer than the size of nuclei, $1.2A^{1/3}\,\text{fm} \sim 5\,\text{fm}\,(A/56)^{1/3}$, are scattered coherently off A nucleons in the nucleus.

In the collapse stage, only ν_e's are emitted copiously by electron capture. Although ν_e's can escape freely from the core at first, the core becomes opaque for ν_e's as the density exceeds $3 \cdot 10^{10}\text{g/cm}^3$ and the mean free path of ν_e's, $\lambda_\nu \sim 1 \cdot 10^7\text{cm}\,(\rho/3 \cdot 10^{10}\text{g/cm}^3)^{-5/3}\,(A/56)^{-1}\,(Y_e/26/56)^{-2/3}$, becomes shorter than the core radius, $R_{\text{core}} \sim 2.7 \cdot 10^7\text{cm}\,(\rho/3 \cdot 10^{10}\text{g/cm}^3)^{-1/3}\,(Y_e/26/56)^{2/3}$.[6] That is, there appears a 'surface' for neutrinos in the core. Neutrinosphere is usually defined as the surface where the optical depth, $\int_r^\infty \lambda_\nu^{-1} dr$, becomes the order of unity. Roughly speaking, one can consider that the inside of the neutrinosphere is opaque and the outside is transparent for neutrinos. In the opaque core, the diffusion time scale for ν_e's is expressed as $\tau_{\text{diff}} = 3R_{\text{core}}^2/c\lambda_\nu$.[7] When the density

[6] Here we use the following expressions:

$$\lambda_\nu = \frac{1}{\sigma_A n_A}, \quad \sigma_A = \frac{G_F^2 \hbar^2 c^2}{\pi} E_\nu^2 \sin^4\theta_w A^2, \quad E_\nu \sim \frac{5}{6}\mu_e, \quad \mu_e = \left(3\pi^2 \frac{\rho Y_e}{m_u}\right)^{\frac{1}{3}} \hbar c$$

$$n_A = \frac{\rho}{A m_u}, \quad R_{\text{core}} = \left(\frac{3M}{4\pi\rho}\right)^{\frac{1}{3}}, \quad M = M_{\text{Ch}} = 1.256 M_\odot \left(\frac{Y_e}{26/56}\right)^2 .$$

[7] This time scale is derived from the theory of random walk. Since the mean displacement

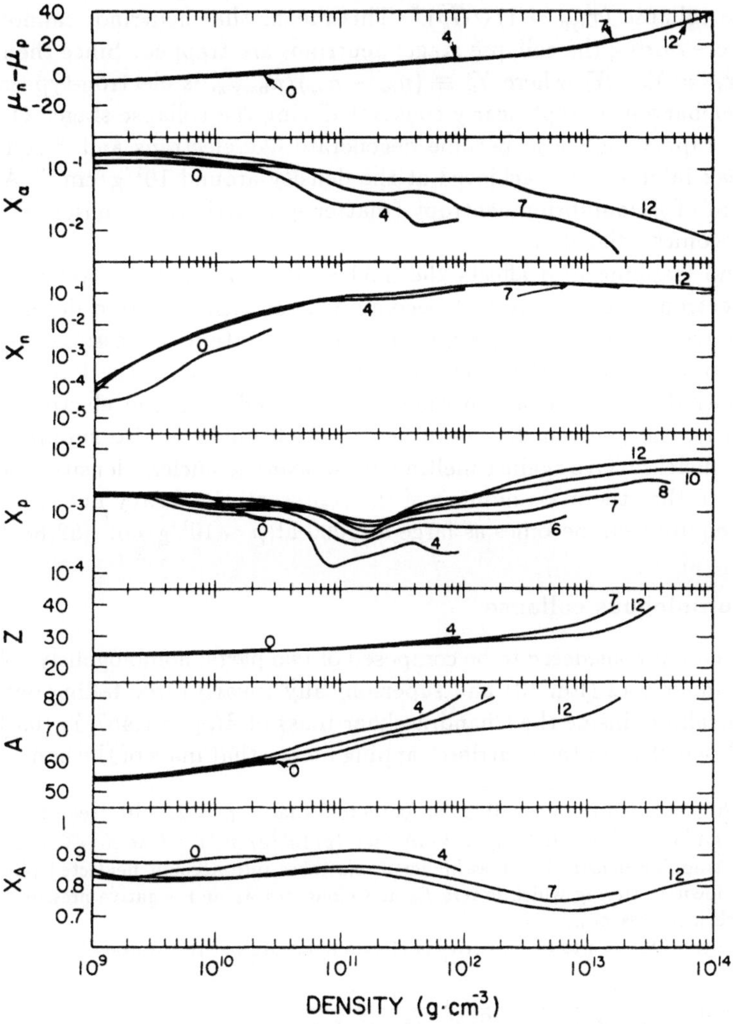

Figure 4: Composition distributions in the collapse stage (taken from Bruenn 1985 [62]). Lines show snapshots at $t = 0\,\text{sec}$ (labelled 0: initial model, $\rho_c = 2.751 \cdot 10^{10}\text{g/cm}^3$), at $t = 39.54\,\text{msec}$ (labelled 4: $\rho_c = 9.022 \cdot 10^{11}\text{g/cm}^3$), at $t = 43.84\,\text{msec}$ (labelled 7: $\rho_c = 1.787 \cdot 10^{13}\text{g/cm}^3$) and at $t = 44.88\,\text{msec}$ (labelled 12: $\rho_c = 3.648 \cdot 10^{14}\text{g/cm}^3$), respectively. $\mu_\text{n} - \mu_\text{p}$ is the difference of chemical potentials of neutrons and protons [in MeV], $X_\alpha, X_\text{n}, X_\text{p}, X_\text{A}$ are mass fraction of α particles, neutrons outside nuclei, protons outside nuclei and heavy nuclei, respectively, Z is charge number of the representative heavy nuclei and A is mass number of the representative heavy nuclei.

---

after N steps of 1-dimensional random walk (1 step $= \lambda_1$) is $\sqrt{N}\lambda_1$, the number of steps required for movement of R is $N_R = (R/\lambda_1)^2$. In 3-dimensional case, we can set $\lambda_1 = \lambda/\sqrt{3}$ and the time scale for neutrinos to diffuse out from the core becomes $N_R\lambda/c = 3R^2/c\lambda$.

exceeds 10^{11}–10^{12}g/cm^3, this diffusion time becomes longer than the time scale of the core collapse ($\tau_{\rm dyn} \sim 1/\sqrt{G\rho}$).[8] This means that neutrinos cannot escape from the core during the collapse stage; neutrinos are trapped. Since then, lepton fraction, $Y_L = Y_e + Y_\nu$ where $Y_\nu \equiv (n_{\nu_e} - n_{\bar{\nu}_e})/n_{\rm baryon}$ is electron-type neutrino fraction per baryon, is kept nearly constant during the collapse stage. Once neutrinos are trapped, ν_e's also become degenerate like electrons and β-equilibrium for the weak interaction is achieved at the density around 10^{12}g/cm^3.[9] After the achievement of β-equilibrium, entropy (matter + neutrinos) is conserved and the collapse becomes adiabatic.

Neutrino trapping also affects the nuclear state as follows. Whereas nuclei capture electrons successively and become neutron-rich, neutrino degeneracy after the neutrino trapping suppresses this neutronization process: conversion of protons into neutrons by the electron capture. It is because the Pauli exclusion principle forbids emission of neutrinos accompanied with the electron capture. Consequently, the neutron drip from neutron-rich nuclei is also suppressed and, hence, the nuclei survive against melting up to near the nuclear density. (See again Fig. 4.) Note that the mass number of the representative heavy nuclei in nuclear statistical equilibrium becomes as large as 1000 at $\rho \sim 10^{14}$g/cm^3 [52,88–90].

2.2.4 Homologous collapse

Collapsing core is considered to be composed of two parts; homologously collapsing (subsonic) inner core (Eq. 13) and supersonically (nearly) free falling outer core. Roughly speaking, inside the Chandrasekhar mass of $M_{\rm Ch} = 1.457(Y_L/0.5)^2 M_\odot$ [10] ($Y_L = 0.30 \sim 0.38$ after the neutrino trapping means that mass of the homologously

---

[8] Although it is found from equations in footnote 6 that $\tau_{\rm diff} = 2.3 \cdot 10^{-1}$sec $(\rho/10^{12}{\rm g/cm^3})$ $(Y_e/26/56)^2$ $(A/56)$ is equal to $\tau_{\rm dyn} = 4 \cdot 10^{-3}$sec $(\rho/10^{12}{\rm g/cm^3})^{-1/2}$ at $\rho \sim 6.5 \cdot 10^{10}$g/cm^3, the realistic trapping density becomes higher if one takes into account neglected physics such as neutrino down scattering which lowers E_ν and elongates λ_ν and negative pressure gradient, $\partial P/\partial r$, which increases $\tau_{\rm dyn}$.

[9] The time scale for ν_e's to be emitted via electron capture on free protons is

$$\tau_{\nu_e\ \rm emission} \sim \frac{1}{\sigma_{\rm e\text{-}cap} n_{\rm p} c} \sim \frac{1}{\frac{4G_F^2\hbar^2c^2}{\pi}\left(\frac{5}{6}\mu_e\right)^2 \frac{\rho}{m_u} X_{\rm p} c}$$

$$\sim 7.4 \cdot 10^{-4}\,{\rm sec} \left(\frac{\rho}{10^{12}{\rm g/cm^3}}\right)^{-\frac{5}{3}} \left(\frac{Y_e}{26/56}\right)^{-\frac{2}{3}} \left(\frac{X_{\rm p}}{10^{-3}}\right)^{-1}$$

where $X_{\rm p}$ is mass fraction of free protons ($\sim 10^{-3}$, see Fig. 4). This time scale becomes shorter than the dynamical time scale at $\rho \gtrsim 2.5 \cdot 10^{11}$g/cm^3. Lower energy neutrinos are equilibrated via $e^-p \leftrightarrow n\nu_e$, $e^-A \leftrightarrow A'\nu_e$ and $\nu_e e^- \leftrightarrow \nu_e e^-$ at higher density. Results of detailed numerical simulations show that β-equilibrium is established at density around 10^{12}g/cm^3 [9,62].

[10] A little strictly speaking, the Chandrasekhar mass including neutrino pressure becomes

$$M_{\rm Ch} = 1.457 M_\odot \left(\frac{Y_L}{0.5}\right)^2 \frac{(Y_e^{4/3} + 2^{1/3} Y_\nu^{4/3})^{3/2}}{Y_L^2},$$

but $(Y_e^{4/3} + 2^{1/3} Y_\nu^{4/3})^{3/2}/Y_L^2$ is in the range between $\sqrt{2/3} \sim 0.816$ and 1 when $\mu_e \geq \mu_{\nu_e} \geq 0$.

collapsing core is $0.5 \sim 0.8M_\odot$.), the infall velocity is proportional to the radius:

$$v(r,t) = a(t) \cdot r \quad , \tag{13}$$

and the velocity of the outer core is of the order of the free fall velocity so that $v \propto 1/\sqrt{r}$. Adiabatic collapse of cores with constant Γ_S was investigated by Goldreich & Weber [91] and Yahil & Lattimer [92, 93] and this double structure is realized in numerical simulations (See Fig. 9.). The essence of the above picture is as follows. Stars with $\Gamma_S < 4/3$ are unstable to radial perturbations and its fastest growing eigen mode is the fundamental mode. Meanwhile the fundamental mode of a star with $\Gamma_S = 4/3$ is such a mode that displacement, δr, from the initial equilibrium position, r_0, has a form $(\xi_0 + \xi_1 t)r_0$ where ξ_0 and ξ_1 are constants. Note that velocity profile of this mode becomes Eq. 13. Since the core which has $\Gamma_S \sim 4/3$ before the onset of collapse becomes unstable ($\Gamma_S \lesssim 4/3$) due to electron capture and/or photodissociation of nuclei and begins to collapse, the collapsing behaviour is similar to the fundamental mode of the star with $\Gamma_S = 4/3$ (homologous collapse). On the other hand, the model of the homologous collapse for the star with $\Gamma_S = 4/3$ shows that the pressure vanishes at the Chandrasekhar mass. That is, it is only the inner core (inside of the Chandrasekhar mass) that collapses homologously. The dynamical time scale for the outer part of the core (outer core/outer mantle) is longer than that of the inner core so that the outer core is left behind the homologous collapse: the outer core is considered to fall freely. Note that the mass of the inner core decreases as Y_L decreases until ν_e's are trapped ($M_{\text{inner core}} \sim M_{\text{Ch}} \propto Y_L^2$). In Figs. 5 and 6 spacial profiles of thermodynamic quantities (density, lepton fraction, electron fraction, entropy per baryon and temperature) at several times in the collapse phase [94] are presented. In these figures, the values of central density are used as time stamps. When the central density becomes 10^{13}g/cm^3, the density of the inner part ($\lesssim 0.4M_\odot$) is greater than $2 \cdot 10^{12}\text{g/cm}^3$. Lepton fraction of this region does not change after that, representing neutrino trapping. On the contrary, electron fraction of the inner region decreases even after that time. This is because the value of Y_e in the β-equilibrium for a given Y_L varies with density (and with temperature).

2.2.5 Core bounce and shock wave formation

When the central density exceeds the nuclear density, the EOS stiffens (Γ_S becomes $5/3 \sim 3$) suddenly because of the hard core component in the nuclear force [48]. The subsonically falling inner core bounces simultaneously (in a time scale for the propagation of sound wave: $\sim$ msec); Since Γ_S is now greater than the critical value ($4/3 + O(GM/Rc^2)$), there is a stable hydrostatic configuration for the inner core and, when the inner core contracts beyond this equilibrium configuration due to the inertia, it bounces back to be the equilibrium configuration like a spring. A pressure wave originated from the center grows to be a shock wave near the boundary between the inner core and the supersonically falling outer core. Unshocked inner core is nearly identical to the homologously collapsing inner core [92] ($M_{\text{unshocked}} = 0.6 \sim 0.9M_\odot$, $R_{\text{unshocked}} \sim$ several $\times$ 10 km) and its

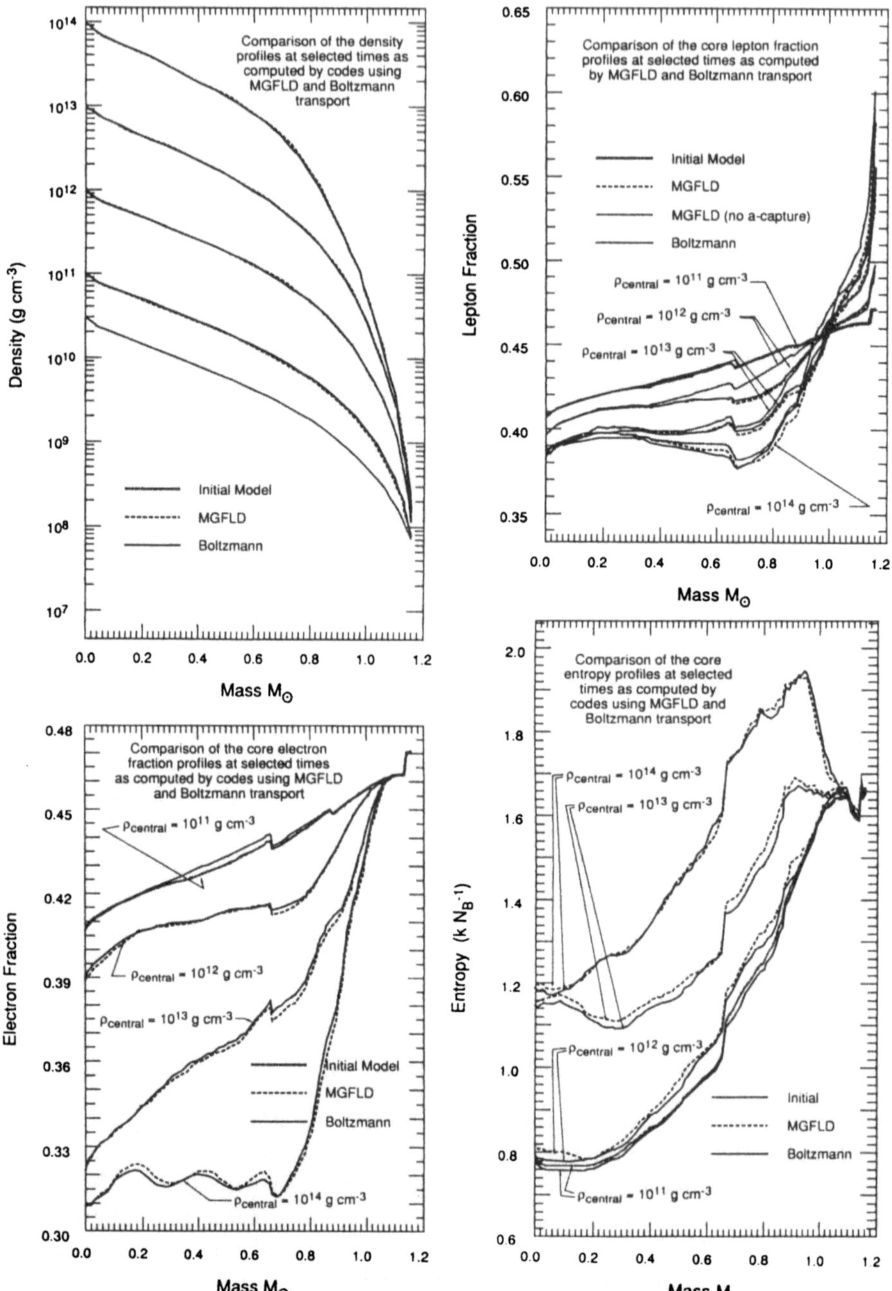

Figure 5: Snapshots of the collapsing core at selected times (taken from Mezzacappa & Bruenn 1993 [94]). Abscissa is enclosed mass. The upper left figure is density profile, the upper right is the lepton fraction, the lower left is the electron fraction, and the lower right is the entropy profile.

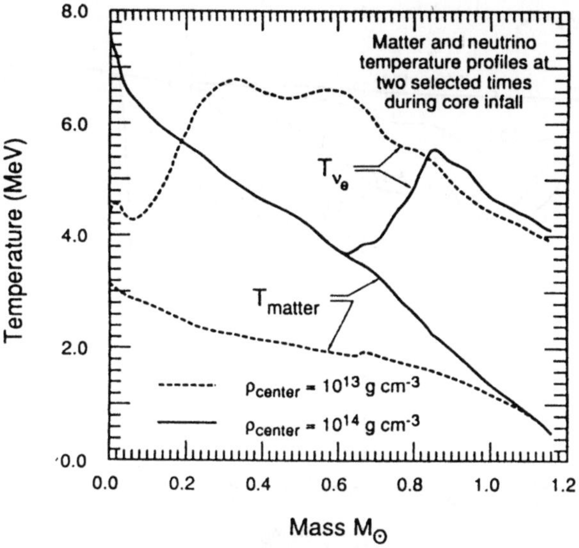

Figure 6: Matter temperature vs. enclosed mass at selected times in the collapse stage (taken from Mezzacappa & Bruenn 1993 [94]).

mass is roughly proportional to the square of the trapped lepton fraction, $Y_{L\ \mathrm{trap}}^2$. Initial energy of the formed shock wave, E_{shock}, is roughly the binding energy of the unshocked core [95];

$$E_{\mathrm{shock}} \sim \frac{GM_{\mathrm{unshocked}}^2}{R_{\mathrm{unshocked}}} \sim \text{several } \times 10^{51}\mathrm{erg} \quad . \tag{14}$$

2.2.6 Core explosion

Since the initial shock energy is greater than the observed explosion energy of $O(10^{51})$ erg, the shock wave can in principle drive out the stellar envelope, resulting in supernova explosion. When the shock wave is propagating through the outer core, however, there are some processes that weaken the outgoing shock. Shock wave must overcome the ram pressure of the supersonically infalling outer core ($P_{\mathrm{ram}} \sim \rho_{\mathrm{infall}}(v_{\mathrm{infall}} - v_{\mathrm{shock}})^2$). Photodissociation of nuclei in the hot shocked region exhausts thermal energy behind the shock front. Simultaneously, since free protons produced by photodissociation of nuclei have a larger electron-capture cross section than nuclei, electron capture in the shocked matter proceeds so much as to diminish the pressure behind the shock front. Moreover, after the shock wave passes through the neutrinosphere, neutrinos produced thermally or by electron capture carry out the internal energy behind the shock front. In Fig. 7, evolution of the lepton fraction profile is shown. One can recognize easily that the lepton fraction near the neutrinosphere drops suddenly (enclosed mass $M = 1 \sim 1.25M_\odot$) when the shock wave passes there and ν_e's produced by the electron capture process become able to escape freely. It results in significant decrease of lepton pressure

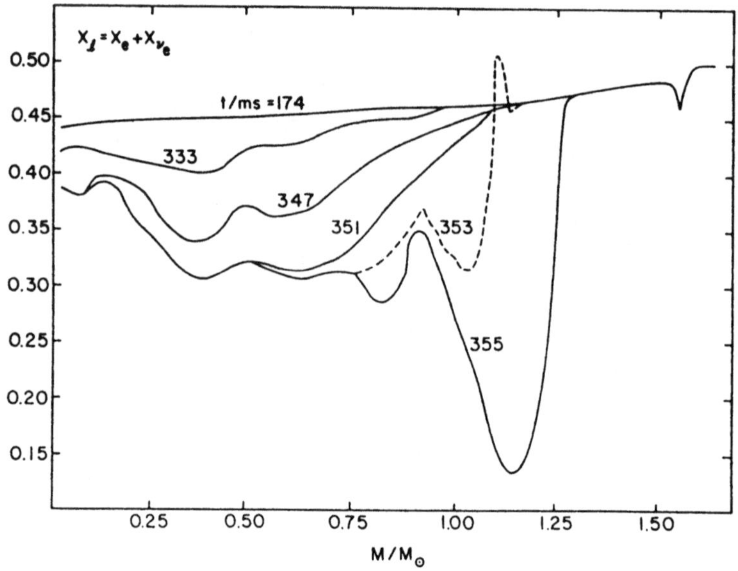

Figure 7: Snapshots of lepton fraction profile at selected times (taken from Mazurek, Cooperstein & Kahana 1980 [50]). Lepton fraction per baryon (Y_L in the text), X_ℓ, is plotted as a function of the enclosed mass. At $t = 351$ msec it is just before the shock formation. Between $t = 353$ msec and $t = 355$ msec, the shock wave passes through the neutrinosphere. This model does not show a successful prompt explosion.

which should push out the shock wave. In this way, these processes all act to weaken the shock wave and the shock wave might stall without driving explosion. Whether the shock wave propagates successfully to the core surface against these obstructions was one of main subjects of current numerical simulations. Fig. 8 shows the evolution of entropy profile. Shock wave is generated around the point $M = 0.75M_\odot$ between $t = 352$ msec and $t = 353$ msec. On shock passing, entropy rises much due to the dissipation.

Prompt Explosion If the shock wave successfully propagates through the outer core without stalling, purely hydrodynamical (core) explosion in the dynamical time scale of $O(10)$ msec is realized [54,56,57,60]. This is called prompt explosion. As an example of the prompt explosion, results of a numerical simulation by Hillebrandt, Nomoto & Wolff [57] are shown in Fig. 9–12. In this model collapse of O·Ne·Mg core ($1.28M_\odot$) initiated by electron capture causes a successful prompt explosion. The corresponding main sequence star has mass of $9M_\odot$ and the simulation was performed for the core of $1.375M_\odot$. According to various numerical simulations [49–75], however, the shock wave can successfully blow off the core surface only when the core is small and cool or the EOS is soft enough. In the case of a small core, the shock wave has only to propagate a small region with dissociating nuclei. Soft EOS leads to sufficient contraction of the inner core to release large gravitational energy. For comparison, Fig. 13 and Fig. 14 show an example

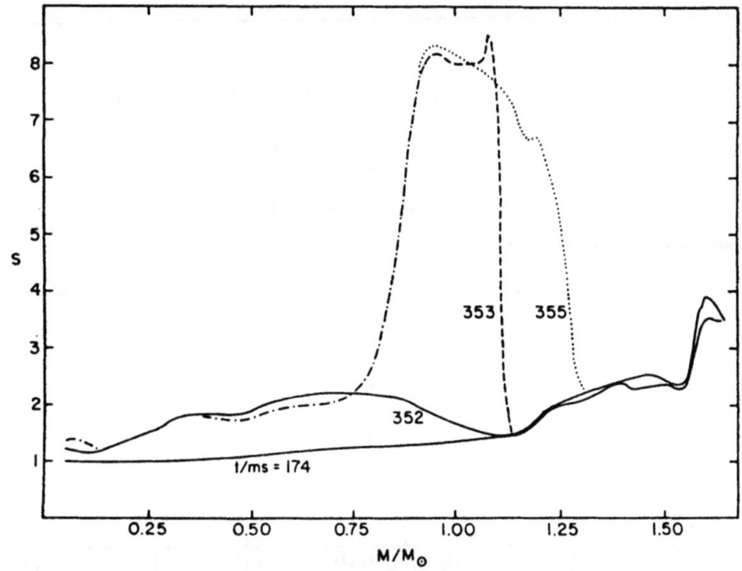

Figure 8: Snapshots of entropy profile at selected times (taken from Mazurek, Cooperstein & Kahana 1980 [50]). Entropy per baryon in unit of Boltzmann constant is plotted as a function of enclosed mass. At $t = 352$ msec it is just before the shock formation.

of unsuccessful prompt explosion (main sequence mass is $20M_\odot$). One can recognize that the outward velocity just behind the shock front decreases and finally becomes zero at about $1.3M_\odot$; the shock wave stalled there. Note that there are some numerical simulations [61, 68, 96] which indicate that the prompt explosion does not occur even for the small core ($1.375M_\odot$ core with $1.28M_\odot$ O·Ne·Mg core) for which Hillebrandt, Nomoto & Wolff reported a successful prompt explosion [57]. Mass range for successful prompt explosion, if exists, is still uncertain. Moreover, it should be noted that recent numerical simulations using current standards for pre-collapse core, for EOS and for the treatment of neutrino down-scattering off electrons seem unable to reproduce prompt explosions.

Delayed Explosion Except for the special cases preferable to the prompt explosion, the shock wave stalls and becomes an accretion shock[11] at the radius of several times 100 km. In that case, supernova explosion does not occur and the star, in turn, collapses to a black hole. Wilson [76], however, discovered from his numerical simulation that the stalled shock wave can be revived by neutrino heating. In Fig. 15, an example of his numerical simulations is shown. Until 0.35 sec after the bounce of inner core, the bounce shock propagates through the outer core and, then, stalls at about 400 km from the center. In the hot region just behind

[11] When the shock front does not move outward in the laboratory frame, it is called accretion shock. Since the supersonically falling matter hits the shock front and, then, merges into the shocked matter behind the shock front, the shock front of an accretion shock moves outward in the fluid frame, of course.

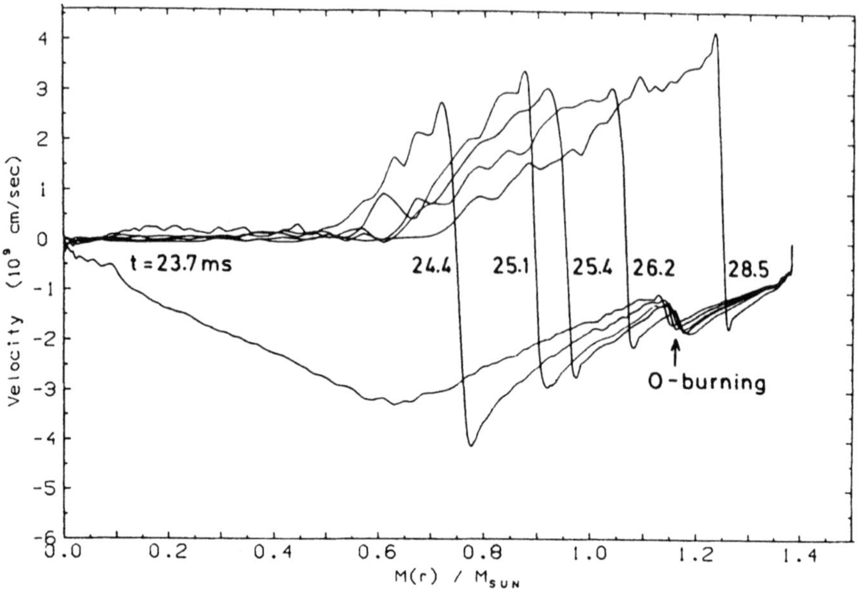

Figure 9: Homologous collapse, shock formation and successful prompt explosion ($M_{\mathrm{main\ sequence}} = 9M_\odot$). Each line is a snapshot of the velocity profile (taken from Hillebrandt, Nomoto & Wolff 1984 [57]). Before the bounce, the inner core ($M(r) < 0.6M_\odot$) collapses homologously ($t = 23.7$ msec). After bounce, the outgoing shock wave is formed and propagates.

the shock front, however, a small part of neutrinos escaping from the inner core is absorbed by free nucleons which are produced by nuclear photodissociation due to the shock passage and hence the matter is heated up (ν_e n $\rightarrow$ p e^-, $\bar{\nu}_e$ p $\rightarrow$ n e^+ around $t \sim 0.5$ sec) [98,99]. In addition, neutrino pair annihilation ($\nu\,\bar{\nu} \rightarrow e^+\,e^-$), especially pair annihilation of high energy ν_μ's and ν_τ's, and electron scattering ($\nu\ e^\pm \rightarrow \nu\ e^\pm$) pour thermal energy into the dilute region ('hot bubble') behind the shock [97,100–102]. Note that neutrino absorption by nucleons cannot contribute to heating of the hot bubble consisting mainly of relativistic particles so much after the shocked matter accreted onto the inner core and the nucleon density became low. Meanwhile the gravitational energy released from the accreted matter is emitted as neutrinos and they also heat up the bubble. Heated bubble expands and pushes out the stalled shock to restart propagation. By this mechanism Wilson and his collaborators found that even a massive core of the progenitor as heavy as $50M_\odot$ can explode. This explosion is named as delayed explosion because the time scale of the explosion (100 msec $\sim$ 1 sec) is much longer than purely hydrodynamical prompt explosion. However, no other group has yet confirmed the validity of delayed explosion owing to the variety of complexities in the supernova core. Compared with the total energy emitted as neutrinos of $O(10^{53})$ erg, the energy required for the explosion is very small, only $O(10^{51})$ erg, and the net energy input in the hot bubble mediated by neutrinos is the difference between the neutrino

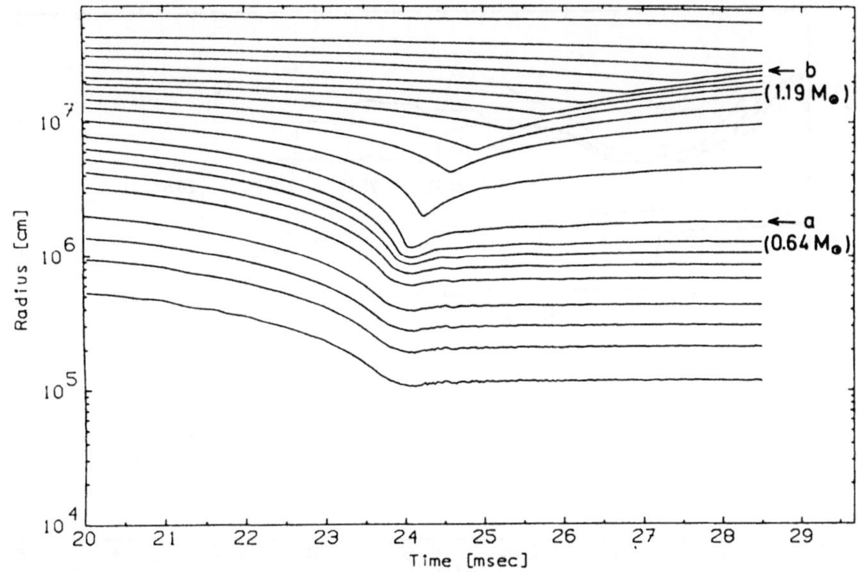

Figure 10: Successful prompt explosion ($M_{\text{main sequence}} = 9M_\odot$). Lines are trajectories of selected mass zones (taken from Hillebrandt, Nomoto & Wolff 1984 [57]). The point 'a' is the boundary of the unshocked inner core. The shock wave successfully propagates to the core surface and the material above the point 'b' will be blown off.

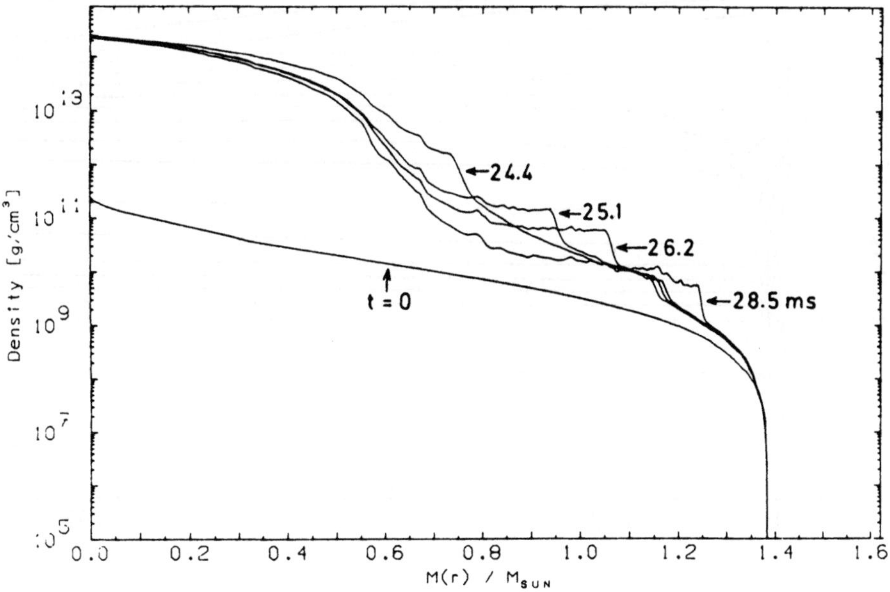

Figure 11: Successful prompt explosion: snapshots of density profile at selected times (initial model and shock propagating phase). $M_{\text{main sequence}} = 9M_\odot$ (taken from Hillebrandt, Nomoto & Wolff 1984 [57])

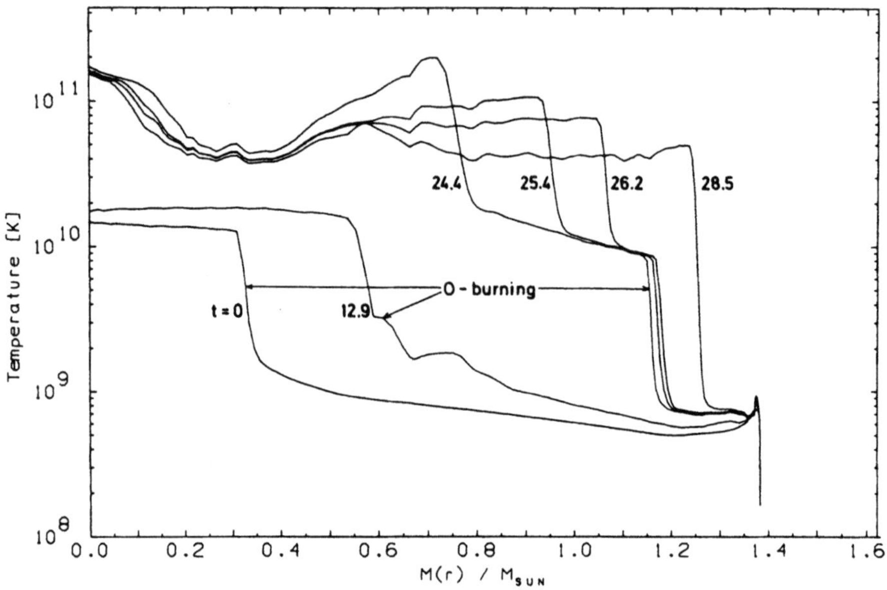

Figure 12: Successful prompt explosion: snapshots of temperature profile at selected times. $M_{\text{main sequence}} = 9M_{\odot}$ (taken from Hillebrandt, Nomoto & Wolff 1984 [57])

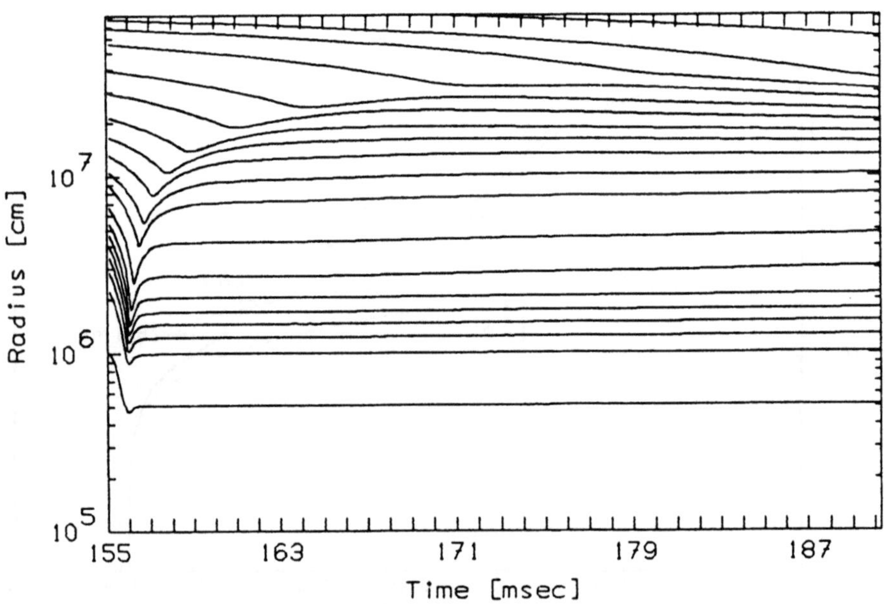

Figure 13: Unsuccessful prompt explosion. ($M_{\text{main sequence}} = 20M_{\odot}$) Lines are trajectories of selected mass zones. The shock wave stalls and cannot expel the envelope (taken from Hillebrandt 1987 [64]).

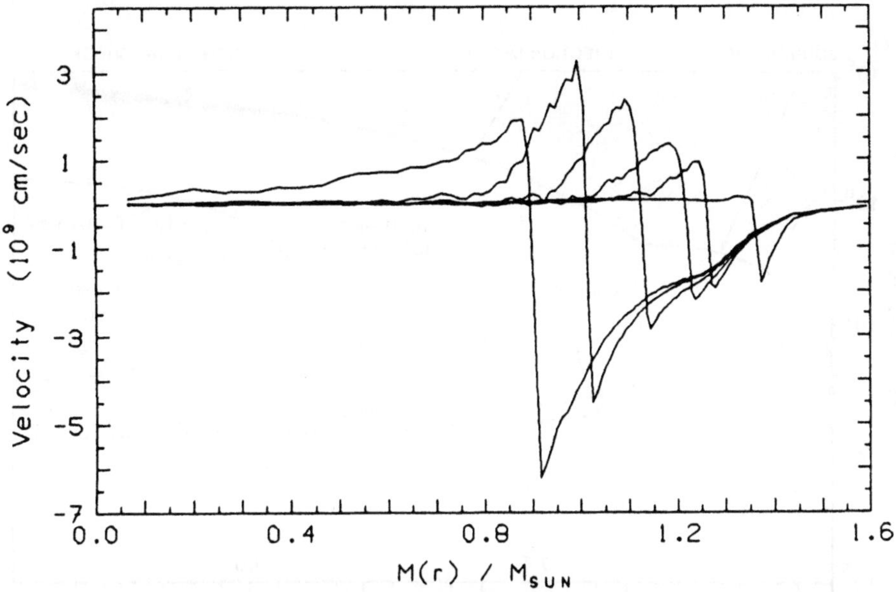

Figure 14: Unsuccessful prompt explosion: snapshots of velocity profile at selected times. ($M_{\text{main sequence}} = 20 M_\odot$ taken from Hillebrandt 1987 [64])

heating and neutrino cooling because the heated bubble emits neutrinos via the reverse processes of heating processes. Heating rate is sensitive to the neutrino luminosity and the shapes of their energy spectra as well as the geometrical condition (intensity of the neutrino flux varies as r^{-2} and the neutrino pair annihilation rate depends on the angle between the two neutrinos of the pair); precise treatment of neutrino transfer is required. Furthermore, the explosion energy obtained in the most of Wilson's numerical simulations is insufficient ($\lesssim 4 \cdot 10^{50}$erg) to explain the observed explosion ($\gtrsim 10^{51}$erg). In these years, convection in the central remnant and/or in the hot bubble draws attention as a candidate for the essence of satisfactory delayed explosion. Many studies including multi-dimensional numerical simulations were/are performed [81, 103–111]. Wilson & Mayle [112] report that convection around the neutrinosphere in the mantle of the core induced by the negative gradient of lepton fraction (see Fig. 7) is indispensable to the successful delayed explosion; it conveys inner hot matter to the neutrinosphere and enhances the neutrino luminosity leading to larger neutrino heating rates behind the shock front. At present, how the core explodes is still an open question.

2.2.7 Protoneutron star cooling

At the same time of the shock propagation, the bounced inner core settles into a hydrostatic configuration in the dynamical time scale of millisecond. The matter of the falling outer core is swept and decelerated by the shock wave, accretes onto the unshocked inner core and forms the shocked outer mantle successively. Thus the protoneutron star which consists of the cool (entropy per nucleon $S \sim O(1)$)

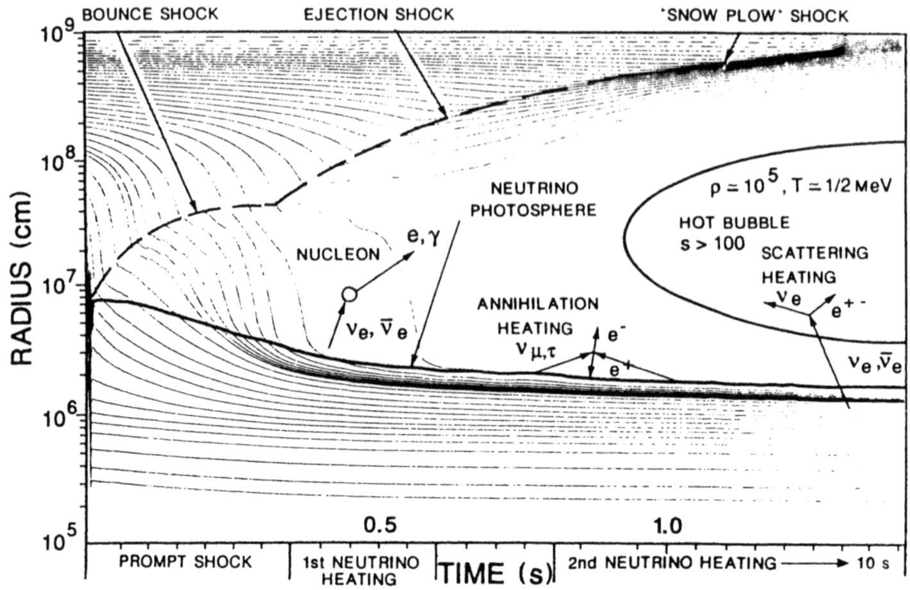

Figure 15: Delayed explosion. Each thin line shows a trajectory of a mass shell in Mayle & Wilson's numerical simulation (taken from Colgate 1989 [97]).

unshocked inner core and the hot ($S \sim O(10)$) shocked mantle grows until the original stellar core explodes and the matter accretion ceases. Since the inner core of protoneutron star contains many leptons and protons ($\sim$ 30%), we cannot call it a *neutron star*. It is neutrinos that carry out trapped lepton number and thermal energy from the protoneutron star in the neutrino diffusion time scale of $O(10)$ sec. Burrows & Lattimer [80, 114] investigated this transition phase from a protoneutron star to a normal neutron star with a simple LTEFLD (local thermodynamic equilibrium flux limited diffusion) neutrino transfer method and we [113, 115] performed more detailed numerical simulations using MGFLD (multigroup flux limited diffusion) scheme for neutrino transfer. Figs. 16–20 show results of our numerical simulation on the quasistatic evolution of a protoneutron star whose initial model was prepared using numerical results (0.4 sec after bounce) of Mayle & Wilson's dynamical simulations [101, 102, 116, 117]. This cooling stage of a protoneutron star can be divided into two stages; the rapid cooling stage of the shocked outer mantle and the succeeding cooling stage of the inner core. At first, the shocked outer mantle cools down and deleptonizes with emitting neutrinos (Figs. 16, 19) because, inside the outer mantle, the density is relatively low and therefore the mean free path of neutrinos is relatively long. The decrease of entropy causes the contraction of the outer mantle with a temperature increase (Fig. 17). Since, initially, there is a region where the temperature gradient is positive, the heat flux flows inwards there and the entropy of the inner core increases. After about one second, the radius of protoneutron star becomes that of the cold neu-

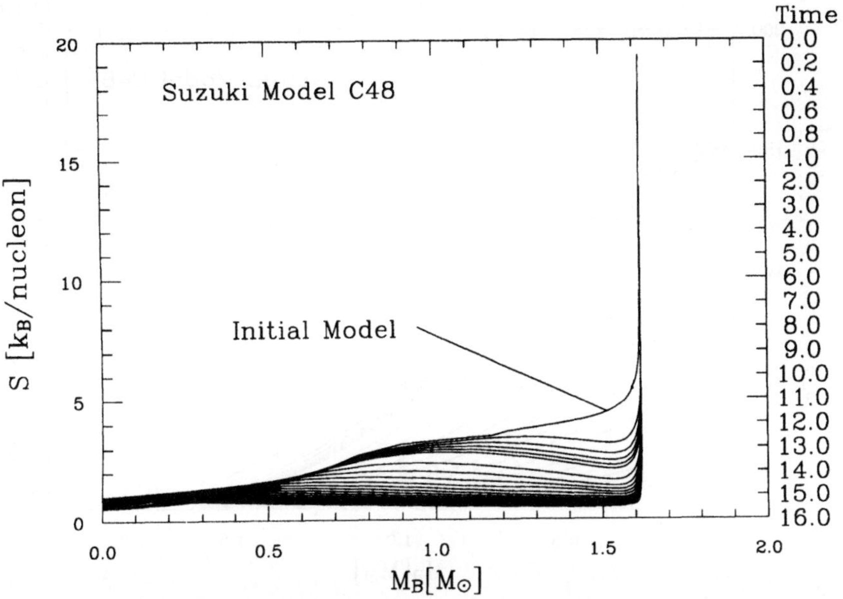

Figure 16: Evolution of the protoneutron star into a 'neutron star'. Snapshots of entropy profile at various times (Suzuki: model C48 [113] $t = 0 \sim 16\,\text{sec}$). Initial profile shows that entropy of shocked outer mantle is much higher than unshocked inner core.

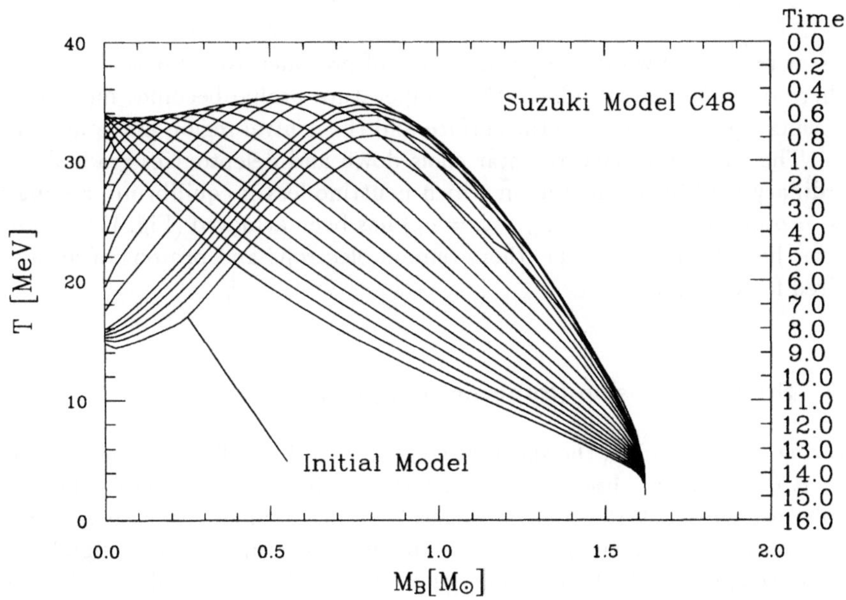

Figure 17: Evolution of the protoneutron star into a 'neutron star'. Snapshots of temperature profile at various times (Suzuki: model C48 [113] $t = 0 \sim 16\,\text{sec}$). Initial peak of temperature is not at the center but in the shocked region.

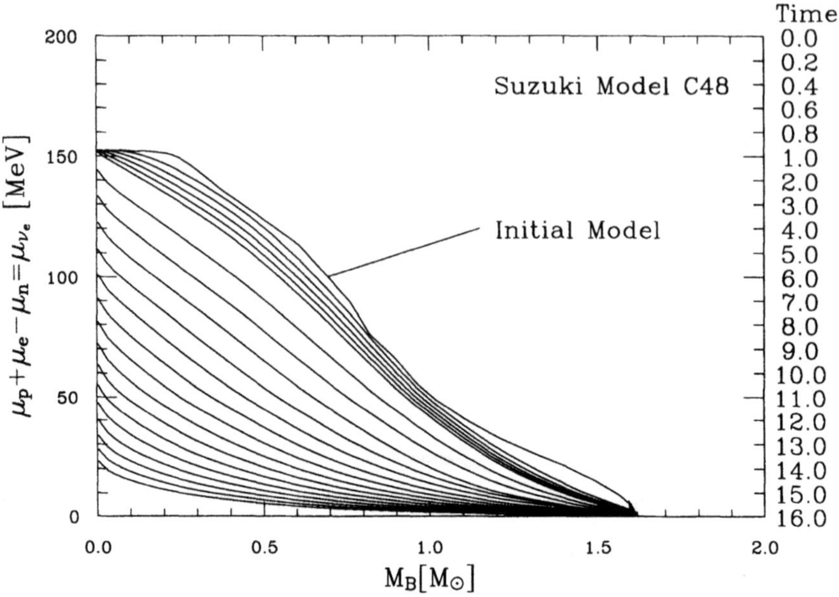

Figure 18: Evolution of the protoneutron star into a 'neutron star'. Snapshots of $\mu_p + \mu_e - \mu_n$ profile at various times (Suzuki: model C48 [113] $t = 0 \sim 16$ sec). In the β-equilibrium state, the chemical potential of ν_e's is equal to $\mu_p + \mu_e - \mu_n$. ν_e's are also degenerate with $\mu_{\nu_e} \sim 150$ MeV $\gg T \sim 20$ MeV in the central region at the early stage.

tron star ($\sim$ 10 km) while the initial radius of protoneutron star is about 100 km. Furthermore, after about 10 sec, the temperature profile becomes monotonically decreasing from the center to the surface and the central region also starts to cool down. The whole protoneutron star cools down and deleptonizes gradually to be an ordinary cold 'neutron star' in which neutrino-less β-equilibrium among neutrons, protons and electrons ($\mu_p + \mu_e = \mu_n$) has been achieved (Fig. 17). Note that even in the cold 'neutron star' there remain electrons and protons of an amount required for β-equilibrium.

2.2.8 Stellar explosion (Supernova explosion)

Following core explosion, the shock wave goes into the stellar envelope and finally blows it off. This is observed as a supernova after the shock wave breaks out the photosphere (not the neutrinosphere). Since, unlike the case in the core, photodissociation and energy loss due to neutrino emission is negligible in the stellar envelope and the binding energy of the envelope is small, the shock wave successfully explodes the star. The propagation time of the shock wave depends on the stellar radius and is in the range from several hours to days. That is, before we recognize the stellar explosion by optical signals, a neutrino burst from the core tells us the occurrence of supernova and the birth of a neutron star.

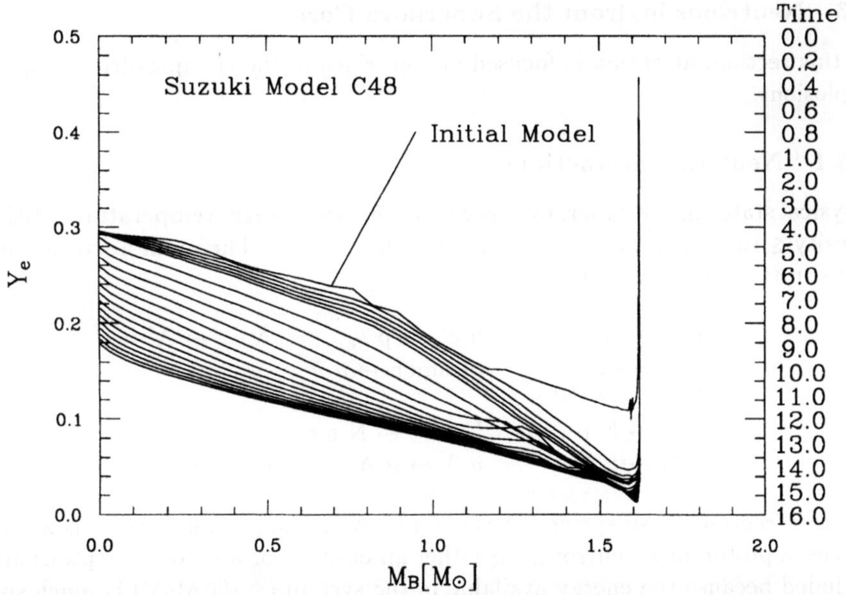

Figure 19: Evolution of the protoneutron star into a 'neutron star'. Snapshots of electron fraction profile at various times (Suzuki: model C48 [113] $t = 0 \sim 16\,\mathrm{sec}$). The trough near the surface was created by the rapid electron capture and ν_e's escape after the shock passage.

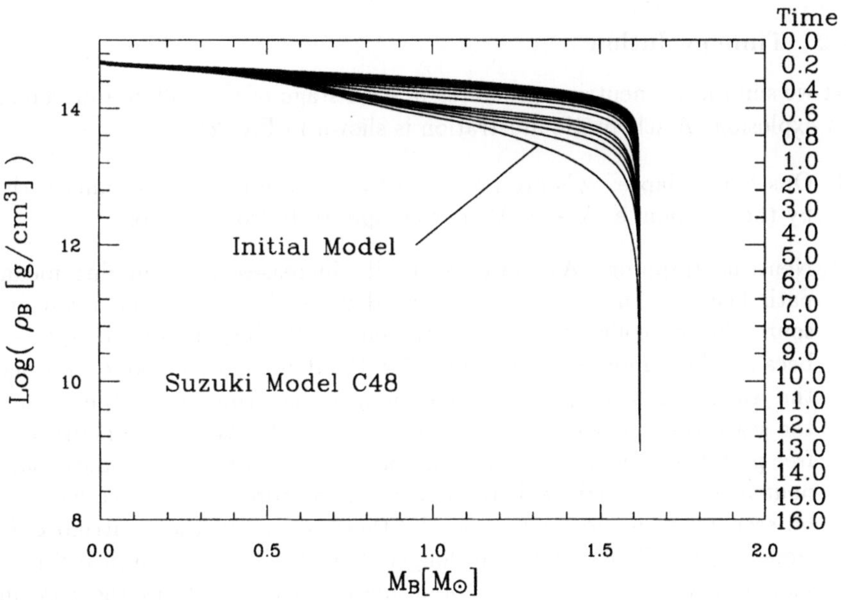

Figure 20: Evolution of the protoneutron star into a 'neutron star'. Snapshots of density profile at various times (Suzuki: model C48 [113] $t = 0 \sim 16\,\mathrm{sec}$).

2.3 Neutrinos in/from the Supernova Core

In this section, attention is focused on neutrinos in the collapse-driven supernova explosions.

2.3.1 Neutrino interactions

Physical states in the supernova core lie in the wide range; temperature $\lesssim$ 100MeV, density $\lesssim 10^{15}$g/cm^3 and electron fraction $0 \lesssim Y_e \lesssim 0.5$. The main neutrino interactions in the supernova core are as follows.

$$
\begin{array}{lll}
\mathrm{p\ e^-} \leftrightarrow \mathrm{n}\ \nu_e, & \mathrm{n\ e^+} \leftrightarrow \mathrm{p}\ \bar{\nu}_e, & \mathrm{A\ e^-} \leftrightarrow \mathrm{A'}\ \nu_e \\
\mathrm{e^- e^+} \leftrightarrow \nu\ \bar{\nu}, & \mathrm{plasmon} \leftrightarrow \nu\ \bar{\nu} & \\
\mathrm{N\ N'} \leftrightarrow \mathrm{N\ N'}\ \nu\ \bar{\nu} & & \\
\mathrm{N\ n} \leftrightarrow \mathrm{N\ p}\ l^-\ \bar{\nu}_l, & \mathrm{N\ p}\ l^- \leftrightarrow \mathrm{N\ n}\ \nu_l & \\
\nu\ \mathrm{N} \leftrightarrow \nu\ \mathrm{N}, & \nu\ \mathrm{A} \leftrightarrow \nu\ \mathrm{A}, & \nu\ l \leftrightarrow \nu\ l
\end{array}
$$

where ν represents all species of neutrinos, A is a heavy nucleus, N is a nucleon (either a proton or a neutron), l is either an electron or a muon. τ leptons are not included because the energy available in the system ($\lesssim$ 100 MeV) is much smaller than the mass of τ leptons. In most of the numerical simulations, muons are also neglected for the same reason and for simplicity. In addition, ν_μ, $\bar{\nu}_\mu$, ν_τ and $\bar{\nu}_\tau$ are often treated as one species which will be denoted as 'ν_μ' in the following. As for the reaction rates of the above interactions, see the references [26, 39, 62, 118–124] for example.

2.3.2 Time evolution

Next we summarize neutrino processes in each stage of the collapse-driven supernova explosion. A schematic illustration is shown in Fig. 21.

1. Onset of collapse: ν_e's are produced by electron capture on nuclei, that is, by the reaction, $\mathrm{e^- A} \to \nu_e \mathrm{A'}$. ν_e's escape freely from the core.

2. Neutrino trapping: As the core density increases, the neutrino mean free path becomes shorter. As mentioned in section 2.2.3, the main opacity source in the collapse stage is coherent scattering off nuclei, $\nu_e \mathrm{A} \to \nu_e \mathrm{A}$ because there remain a lot of nuclei without being dissociated and because the wave length of ν_e's is so much longer than the size of the nuclei that the scattering cross section is proportional to A^2. When the neutrino mean free path becomes shorter than the core size and the core becomes opaque to neutrinos, the neutrinosphere appears in the core as a surface which could be seen by neutrino telescopes. Furthermore when the neutrino diffusion time, $\tau_{\mathrm{diff}} \sim 3R^2/c\lambda_\nu$, becomes longer than the core's dynamical time scale, $\tau_{\mathrm{dyn}} \sim 1/\sqrt{G\rho}$, neutrinos become unable to escape from the core in the collapse stage and can be regarded as trapped. Once ν_e's are trapped, subsequent electron capture and compression build up the degenerate sea of ν_e's. Resultant increase of average energy of ν_e's leads to a larger opacity

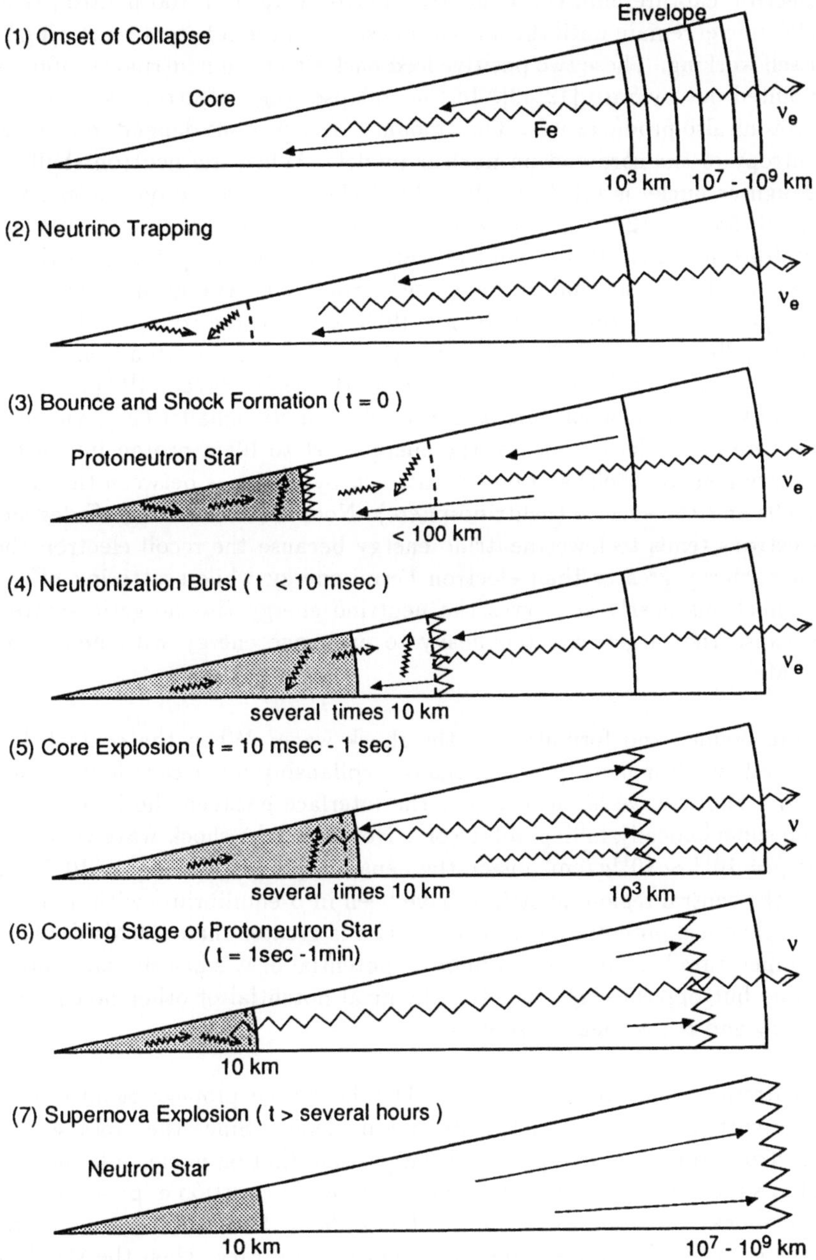

Figure 21: Collapse-driven supernova explosion and neutrinos. Dashed arc represents the neutrinosphere and waved arc represents the shock front. Shaded region is the protoneutron star which has been in hydrostatic configuration. Solid arrow represents the matter motion and waved arrow represents neutrinos.

($\sigma_{\nu \mathrm{A\ scat}} \propto E_\nu^2$). Because of the Pauli principle, degenerate ν_e's prevent the electron capture and, consequently, neutron drip from too neutron-rich nuclei; nuclei remain until the nuclear density and the coherent scattering keeps itself working. These two positive feed back effects on neutrino trapping were pointed out by Sato [42, 43]. In the collapse stage, electron capture on free protons also produces ν_e's. The amount of free protons depends on the initial entropy of the core and on nuclear models. When the neutron shell of the daughter nuclei is filled up, the rate of electron capture on nuclei becomes small [86, 125–127] and the relative importance of electron capture on free protons increases. In Bruenn's numerical simulation [62] this transition happens at density around $2 \cdot 10^{11}\mathrm{g/cm^3}$. Since high energy neutrinos cannot escape from the trapped shell ($\rho \gtrsim 10^{12}\mathrm{g/cm^3}$) in the collapse phase, emergent ν_e flux is composed of low energy ν_e's which are emitted from the shells with $\rho \lesssim 10^{12}\mathrm{g/cm^3}$ ($\langle E_{\nu_e} \rangle \sim 5/6 \cdot \mu_e \sim 30\,\mathrm{MeV}(\rho/10^{12}\mathrm{g/cm^3})^{1/3}\,(Y_e/0.4)^{1/3} \lesssim 30\,\mathrm{MeV}$ via electron capture on free protons or still smaller $\langle E_{\nu_e} \rangle$ via electron capture on nuclei because of the energy cost to lift a proton into a higher energy neutron shell) and down-scattered by electrons between the emitting shells and the surface (neutrinosphere). Note that scattering off degenerate electrons tends to lower neutrino energy because the recoil electron should have energy greater than electron Fermi energy while scattering off nuclei or nucleons nearly conserves the neutrino energy (isoenergetic scattering) because the targets are too heavy to exchange energy with neutrinos ($\sim 10\,\mathrm{MeV}$).

3. Core bounce and formation of the shock wave: When the central density exceeds nuclear density, homologously collapsing inner core halts as a unit and a shock wave is formed near the interface between the inner core and the supersonically falling outer core. Note that the shock wave is generated at $\rho = 10^{12} \sim 10^{14}\mathrm{g/cm^3}$ inside the neutrinosphere ($\rho = 10^{10} \sim 10^{12}\mathrm{g/cm^3}$). At the central region, neutrinos have been in β-equilibrium with matter and the chemical potential of degenerate ν_e's is greater than 100 MeV while the temperature is $O(10)$ MeV. Chemical potential of $\bar{\nu}_e$'s has the same absolute value but opposite sign and the chemical potential of other neutrinos (ν_μ, $\bar{\nu}_\mu$, ν_τ and $\bar{\nu}_\tau$) is (nearly) zero.

4. Neutronization ('break out') burst: The shock wave propagates into the outer core with dissociating nuclei into free nucleons. Since the cross section of electron capture on free protons is larger than that on nuclei, a large amount of ν_e's are produced in the shocked region by the reaction $\mathrm{e^-p} \rightarrow \nu_e\mathrm{n}$. Before the shock wave arrives at the neutrinosphere, however, these ν_e's cannot escape in the dynamical time scale. On the contrary, when the shock wave passes through the neutrinosphere, the emitted ν_e's can escape freely because they are produced outside the neutrinosphere. Moreover, dissociation of the nuclei which have been main obstacles for neutrinos results in considerable decrease of neutrino opacity leading to a recession of the neutrinosphere. The burst of ν_e's thus formed is called the neutronization burst (or 'break

out' burst). The duration of the neutronization burst is the time scale of the shock propagation and, hence, less than 10 msec. While the peak luminosity exceeds 10^{53}erg/sec, the total energy emitted in the neutronization burst is only of the order of 10^{51}erg because of the short duration time. Meanwhile abrupt progress of electron capture around the neutrinosphere at the shock wave passage followed by the neutronization burst of ν_e's forms a deep trough in Y_e/Y_L profile (See Fig. 7.) and weakens the shock wave.

5. Matter accretion onto the protoneutron star, core explosion and the rapid cooling of the outer mantle of the protoneutron star: Since the shocked region is hot ($S \gtrsim O(10)$, $T \gtrsim O(10)$ MeV) and relatively low density, the electron degeneracy is not high and relativistic positrons are also created thermally. Neutrinos other than ν_e's are emitted via reactions $e^+n \to \bar{\nu}_e p$, $e^+e^- \to \nu\bar{\nu}$ and they carry out the thermal energy of shocked matter as well as ν_e's produced via $e^-p \to \nu_e n$ which carry out electron-type lepton number at the same time. A part of shocked outer core accretes onto the hydrostatic protoneutron star and the released gravitational energy is converted into thermal energy and emitted as thermal neutrinos. This accretion phase might last only $O(10)$ msec for the case of prompt explosion while it might last long (during the shock stagnation $\sim O(1)$ sec) for the case of delayed explosion. In addition, the shocked outer mantle of protoneutron star which was formed by the accretion of the shocked matter cools down emitting thermal neutrinos either in parallel with or following the accretion phase. Note that the main opacity sources after nuclear dissociation are scattering off free nucleons ($\nu n \to \nu n$, $\nu p \to \nu p$) or absorption on free nucleons ($\nu_e n \to pe^-$, $\bar{\nu}_e p \to ne^+$). Cooling and neutronization/deleptonization of shocked matter precede those of unshocked inner core. It is interesting that $\bar{\nu}_e$'s and 'ν_μ's carry the thermal energy into the central region at the early stage. The number densities of these neutrinos at the center are less than those in the shocked region due to the ν_e degeneracy or the low temperature at the center and it results in the inward flux of $\bar{\nu}_e$'s and of 'ν_μ's. In the case of delayed explosion, neutrino interactions just behind the stalled shock wave is important for a successful explosion while the emergent neutrino flux is not affected directly by such infrequent interactions above the neutrinosphere. As mentioned in section 2.2.6 the region obtains energy via neutrino absorption on nucleon at first and the generated hot bubble is heated by neutrino pair annihilation and by electron scattering.

6. Cooling stage of the protoneutron star: A protoneutron star cools down and deleptonizes to a normal neutron star emitting thermal neutrinos in the time scale of the neutrino diffusion ($O(10)$ sec). As the temperature lowers, resultant small number of positrons makes the neutrino pair production via e^-e^+ pair annihilation ($e^-e^+ \to \nu\bar{\nu}$) insufficient. Nucleon bremsstrahlung process ($NN' \to NN'\nu\bar{\nu}$), instead, acts as main source of 'ν_μ's during the late stage of protoneutron star cooling. After about 1 minute, the neutron star becomes transparent for neutrinos and the neutrino luminosity drops

suddenly [128].

7. Supernova explosion: After several hours from the core bounce, the shock wave reaches the stellar surface and the star actually explodes. A neutron star has been already formed in the center.

In general, neutrinos diffusing out of the core are once thermalized by the matter even if they are produced by the electron capture process as high energy neutrinos. As the thermalization process, neutrino scattering off electrons plays an important role in addition to the other absorption and emission processes. Though neutrino scattering off nucleons and nuclei (isoenergetic scattering) alone cannot thermalize neutrinos, it acts as a crucial opacity source for neutrinos. Neutrinos are thermalized via absorption/emission and non-isoenergetic scattering processes during many times of isoenergetic scattering. As for the energetics, the total energy carried out by neutrinos is several times 10^{53}erg. Roughly speaking, all species of neutrinos contribute equally. For example, even if one assumes that all protons in the iron core capture electrons and emit ν_e's with mean energy of $\langle E_{\nu_e}\rangle$, the total ν_e's energy amounts only to

$$26\,\frac{M_{\mathrm{Fe\ core}}}{m_{\mathrm{Fe}}}\langle E_{\nu_e}\rangle \sim 1.2\cdot 10^{52}\mathrm{erg}\,\frac{M_{\mathrm{Fe\ core}}}{1.4M_\odot}\,\frac{\langle E_{\nu_e}\rangle}{10\mathrm{MeV}}\quad , \tag{15}$$

which is less than 10% of total neutrino energy [129]. At the collapse stage and in the neutronization burst, the energy carried by ν_e's is only of the order of 10^{51}erg, respectively. The shock-heated matter and the accreted matter emit thermal neutrinos of $O(10^{53})$ erg. Protoneutron star also emits thermal neutrinos of $O(10^{53})$ erg.

An example of the numerical results on the neutrino luminosity curve in a case of successful delayed explosion by Mayle, Wilson & Schramm [129] is shown in Fig. 22. The first peak of the ν_e luminosity is corresponding to the neutronization burst. Without this feature, the behaviour of light curves are similar among the three species of neutrinos. The oscillation of the light curve corresponds to the intermittent mass accretion of the shocked matter during the shock stagnation. This oscillation, if observed, might be an evidence of the delayed explosion. Since matter accretion no longer continues after the core explosion ($t > 0.9$ sec), neutrino luminosity is kept low and monotonous.

In Fig. 23, the evolution of the average energies of the emergent neutrino flux are shown for the same model as Fig. 22. Roughly speaking it is expected that the neutrino mean energies have anticorrelation with the coupling strength between the matter and the neutrino. Since ν_e's interact most strongly with the matter, the neutrinosphere for ν_e's is outermost and coolest, which results in the lowest mean energy. In addition, the rise of the mean energy of $\bar{\nu}_e$ (curves labeled as $\langle\bar{\nu}\rangle_c$ or $\langle\bar{\nu}\rangle_s$) was interpreted by Mayle, Wilson & Schramm as follows. Before the core explosion ($t < 0.9$ sec in this case), the accreting matter containing a lot of protons interacts with $\bar{\nu}_e$'s strongly ($\bar{\nu}_e\mathrm{p} \to \mathrm{ne}^+$ in addition to pair processes and scattering processes) and the mean energy of $\bar{\nu}_e$'s is similar to that of ν_e's which interact with matter via $\nu_e\mathrm{n} \to \mathrm{pe}^-$ instead of $\bar{\nu}_e\mathrm{p} \to \mathrm{ne}^+$. After the core explosion, there is no

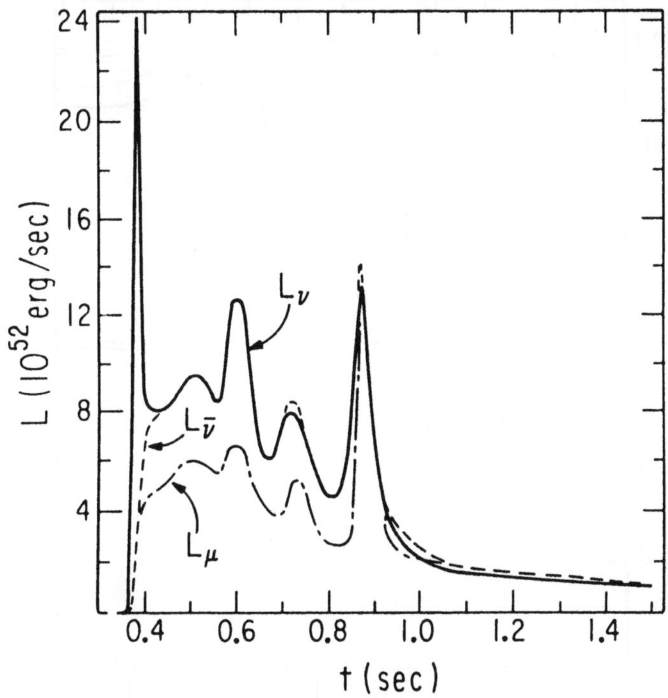

Figure 22: Evolution of the neutrino luminosity in a case of the delayed explosion (Model 25C in Mayle, Wilson & Schramm 1987 [129]). Solid line is ν_e, dashed line is $\bar{\nu}_e$ and dot-dashed line is 'ν_μ'. Note that the neutronization burst of ν_e is drawn one-half its actual height. (The actual peak luminosity is about $5 \cdot 10^{53}$erg/sec.) Gravitational red shift effects are not included in this simulation.

accreting matter and the accreted matter has been neutronized so much that the interaction rates of $\bar{\nu}_e$'s decrease to be similar to those of 'ν_μ's which interact with matter only via pair processes and scattering processes. Hence, the mean energy of $\bar{\nu}_e$'s increases and approaches that of 'ν_μ's.

In Figs. 24–27, the results of other supernova models by Bruenn [66] and by Myra & Burrows [74] are shown. In these simulations, the star could not explode. Before the neutronization (or 'break out') burst of ν_e's, that is, during the collapse phase and at the bounce, there is no significant emission of $\bar{\nu}_e$'s and 'ν_μ's. Furthermore, in contrast with Fig. 23 for the model of successful delayed explosion, the average energies of $\bar{\nu}_e$'s are kept similar to those of ν_e's. This feature is consistent with Mayle, Wilson & Schramm's interpretation mentioned above.

In Figs. 28 and 29, the evolution of the neutrino luminosity and the mean energy during the protoneutron star cooling stage are shown [113]. Since the initial model was constructed referring Mayle & Wilson's results of dynamical simulation at 0.4 sec after bounce [101, 102, 116, 117] and the subsequent matter accretion is assumed not to occur, the origin of time ($t = 0$) should be considered as at the core explosion about 1 sec after the bounce. Fig. 28 shows that the rapid

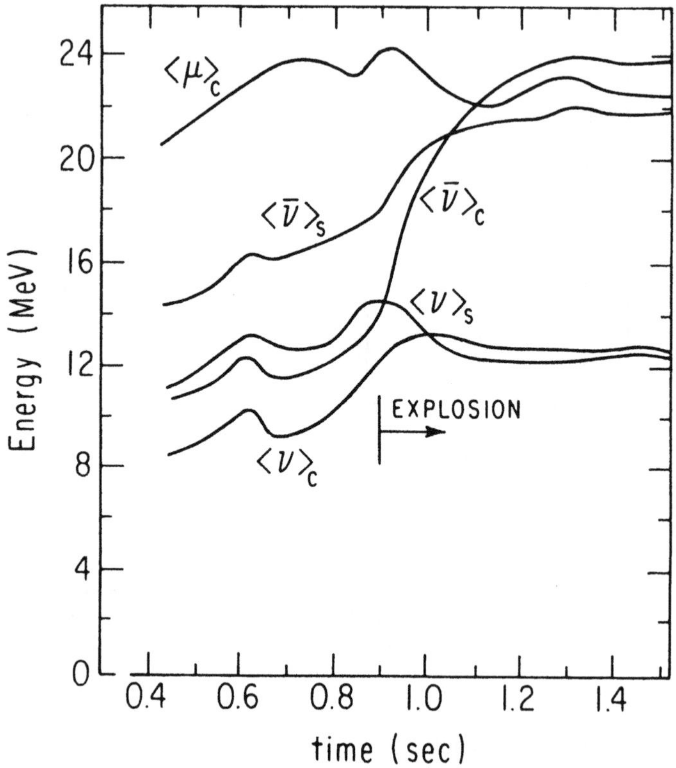

Figure 23: Mean energy of neutrino burst in a case of a successful delayed explosion (Model 25C in Mayle, Wilson & Schramm 1987 [129]). $\langle\nu\rangle$, $\langle\bar{\nu}\rangle$, $\langle\mu\rangle$ are mean energies of ν_e's, $\bar{\nu}_e$'s and 'ν_μ's, respectively. Subscripts 'c' represent the results of evolutionary simulation with multigroup flux limited diffusion scheme. Subscripts 's' represent the results of static analysis using a Boltzmann equation solver. Gravitational red shift effects are not included in this simulation.

cooling stage of the outer mantle ($t \lesssim 1\,\text{sec}$) is followed by the slower cooling stage of the entire protoneutron star. Meanwhile, due to a similar reason to the case of Fig. 23, the average energy of $\bar{\nu}_e$'s approaches to that of 'ν_μ's. That is, the progress of neutronization and cooling leads to a reduction of protons and an increase of neutron degeneracy, resulting in suppression of the reaction $\bar{\nu}_e p \rightarrow ne^+$ which had been distinguishing $\bar{\nu}_e$'s from 'ν_μ's. Note that, Suzuki [115] studied a model (C12) in which the nucleon bremsstrahlung process is ignored but the other input physics is the same as the above. The results showed that insufficient source term for 'ν_μ's results in a weak 'ν_μ' flux; luminosity of 'ν_μ's becomes less than half of ν_e's luminosity or $\bar{\nu}_e$'s luminosity at $t = 10\,\text{sec}$. On the contrary, when we did not ignore the nucleon bremsstrahlung process (model C48), the light curve of 'ν_μ's becomes similar to the light curves of ν_e's and $\bar{\nu}_e$'s (see Fig. 28). Nucleon bremsstrahlung process is very important source for 'ν_μ's and should be included in the study of protoneutron star cooling.

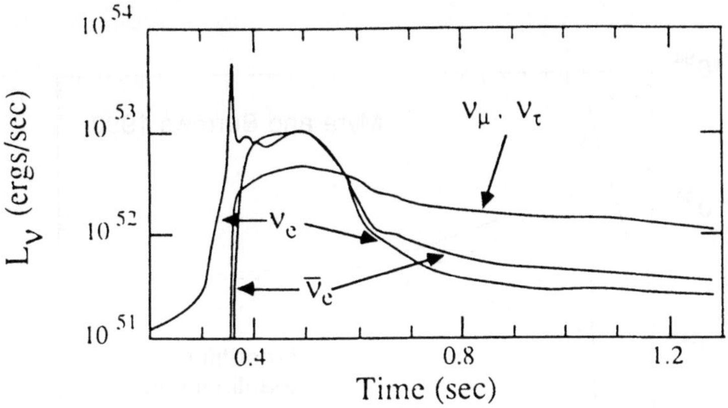

Figure 24: Evolution of the neutrino luminosity at the dynamical stage in the case of $2M_\odot$ Fe core (taken from Bruenn 1987 [66], no prompt explosion). Since a generally covariant formalism is not used in the neutrino transfer, corrections for the red shift and time dilation was made using the red shift parameter at the neutrino emission surface.

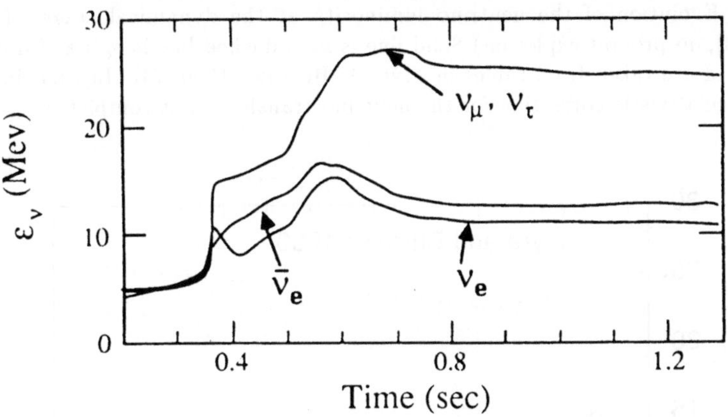

Figure 25: Mean energy of neutrino burst at the dynamical stage. (taken from Bruenn 1987 [66], no prompt explosion). Since a generally covariant formalism is not used in the neutrino transfer, corrections for the red shift and time dilation was made using the red shift parameter at the neutrino emission surface.

In Fig. 30 we compare our results on the average energy of neutrinos with Mayle & Wilson's results [117]. Our simulation showed lower mean energies, especially for 'ν_μ's, than other group's results. Although it might be meaningless to compare the two results in detail because, while Mayle & Wilson studied the dynamical evolution in the time scale of $O(1)$ sec, we studied the quasistatic evolution of the protoneutron star in the time scale of $O(10)$ sec in different manner, one of the origins of the difference is the adopted EOS. We used a stiff EOS provided by Wolff while Mayle & Wilson used a soft EOS (See Fig. 38 for comparison).

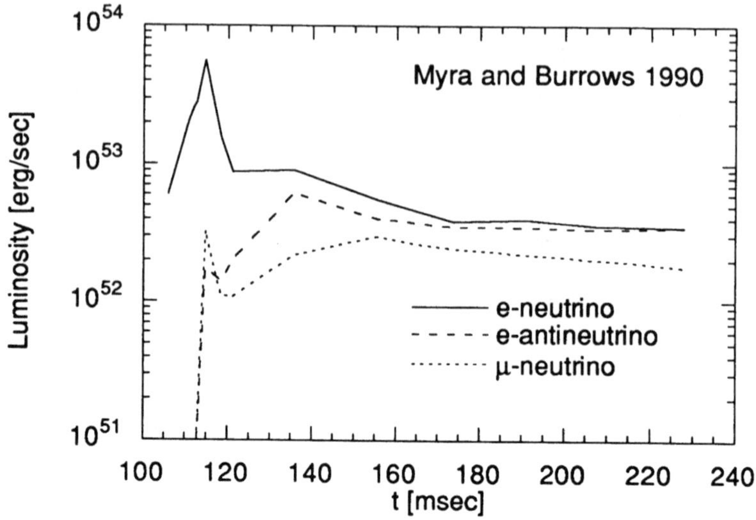

Figure 26: Evolution of the neutrino luminosity at the dynamical stage. (Myra & Burrows [74], no prompt explosion) Solid line is ν_e's, dashed line is $\bar{\nu}_e$'s and dotted line is 'ν_μ's. Data are taken from tables in Myra & Burrows 1990 [74]. In this simulation, the general relativistic correction for the neutrino transfer is not complete.

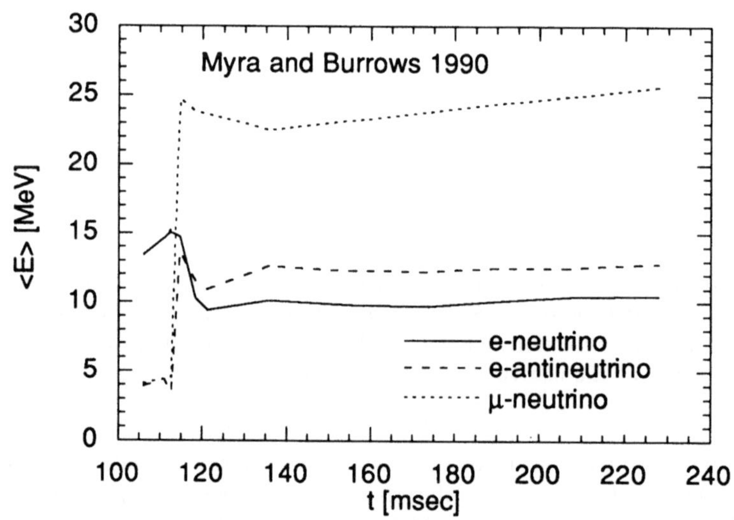

Figure 27: Mean energy of neutrino burst at the dynamical stage. (Myra & Burrows [74], no prompt explosion) Solid line is ν_e's, dashed line is $\bar{\nu}_e$'s and dotted line is 'ν_μ's. Data are taken from tables in Myra & Burrows 1990 [74]. In this simulation, the general relativistic correction for the neutrino transfer is not complete.

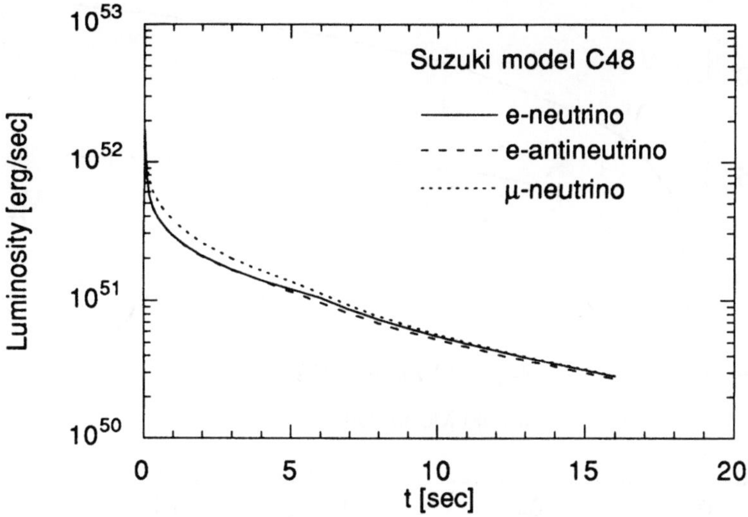

Figure 28: Evolution of the neutrino luminosity at the cooling stage of the protoneutron star (Suzuki: model C48 [113]). Solid line is ν_e's, dashed is $\bar{\nu}_e$'s and dot-dashed is 'ν_μ's, respectively

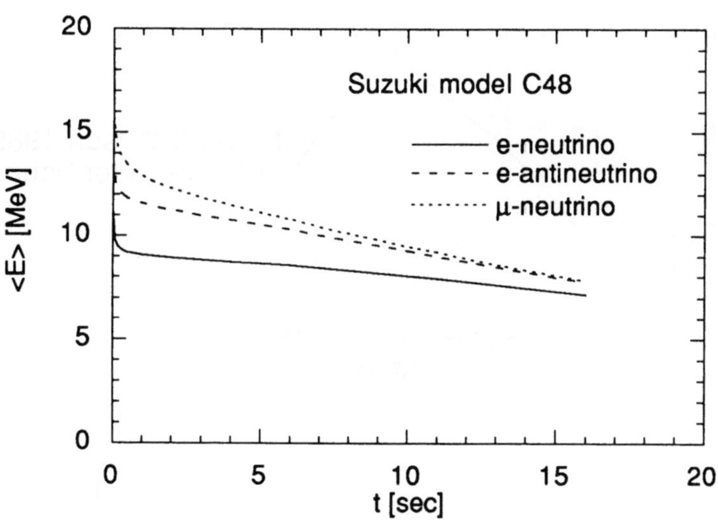

Figure 29: Mean energy of neutrino burst at the cooling stage of the protoneutron star (Suzuki: model C48 [113]). Solid line is ν_e's, dashed is $\bar{\nu}_e$'s and dot-dashed is 'ν_μ's, respectively

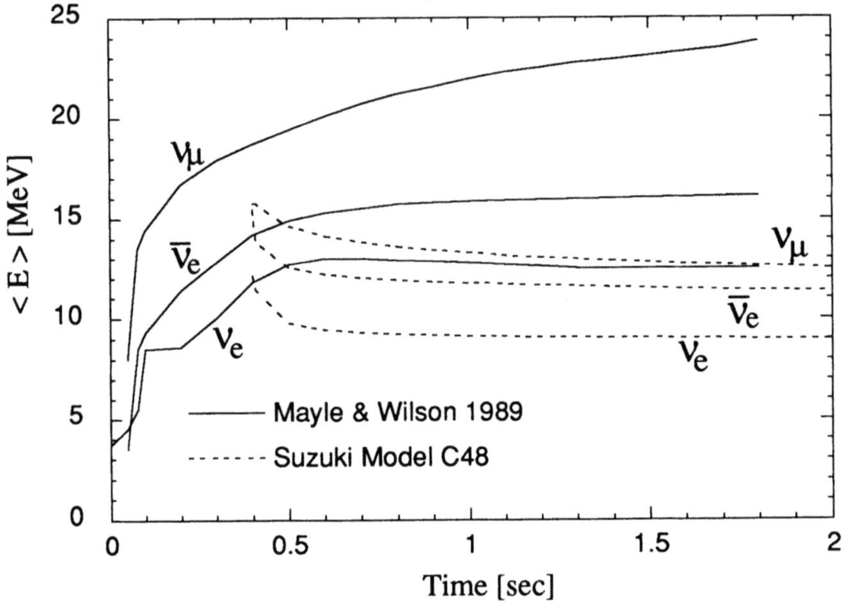

Figure 30: Comparison of neutrino average energies. The hydrostatic initial model for Model C48 (Suzuki) is constructed by referring the numerical results of Mayle & Wilson at 0.4 sec after bounce.

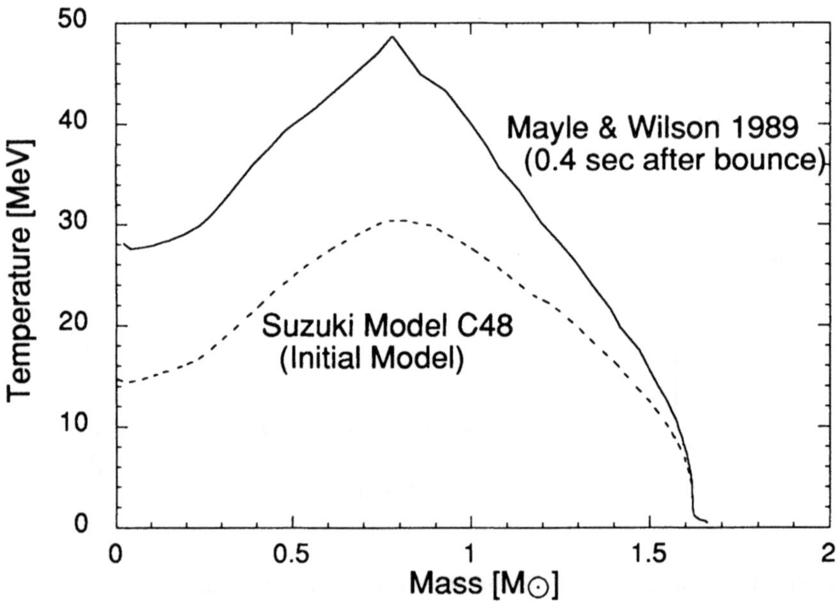

Figure 31: Comparison of temperature profile at 0.4 sec after bounce.

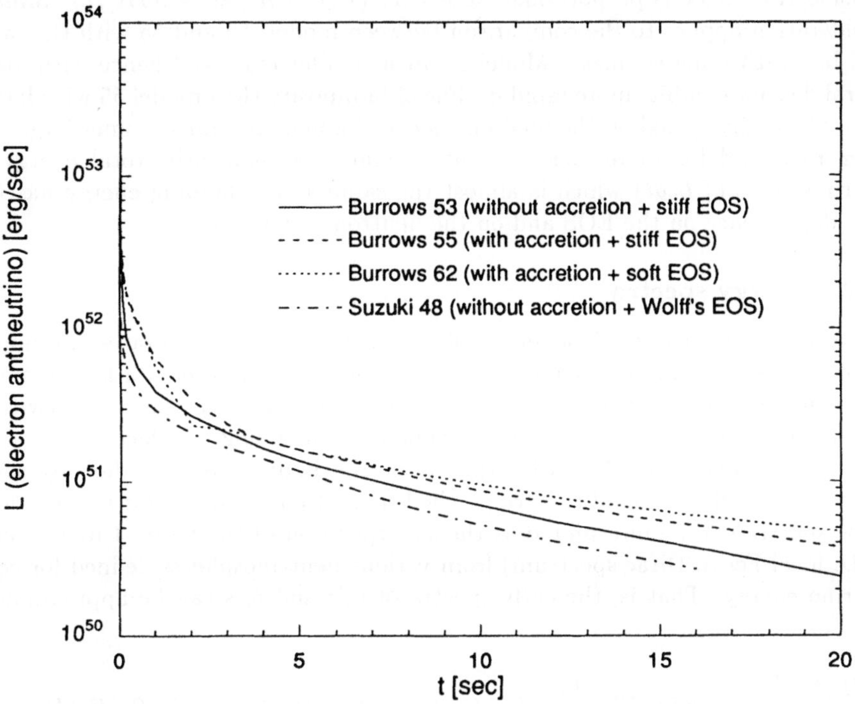

Figure 32: $\bar{\nu}_e$ luminosity vs. time. Data are taken from tables in Burrows 1988 [114]. Model 53: without accretion and with stiff EOS (solid line), Model 55: with accretion and with stiff EOS (dashed line), Model 62: with accretion and with soft EOS (dotted line). As a reference, a result of our numerical simulation (model C48 without accretion, baryon mass = $1.62 M_\odot$) is also plotted by dot-dashed line. Note that our simulation corresponds to the second half of the supernova neutrinos, i.e., about half of total energy has been emitted before our initial model.

Temperature profile is affected much by the adopted EOS as shown in Fig. 31. The low temperature results in the low neutrino mean energy in our simulation.

Meanwhile Burrows [114] investigates the cooling stage of protoneutron star in the case with accretion phase. Fig. 32 shows his results on evolution of $\bar{\nu}_e$ luminosity for models with/without the accretion phase. He assumed the accretion rate $\dot{M}$ is given by $\dot{M}(t) = 0.4 M_\odot/\text{sec} \cdot \exp(-t/0.5\text{sec})$ for models 55 (stiff EOS) and 62 (soft EOS). Initial mass of the protoneutron star was set to $1.3 M_\odot$ for all models. It is clear that, in the accretion phase, neutrino luminosity is kept higher due to the energy supply from accreted matter. Furthermore, in this figure, we can recognize some general dependencies of neutrino burst at the late slow cooling stage on EOS and on neutron star mass (see $t \gtrsim 4\,\text{sec}$ in the figure). Comparing model 55 with 62 having the same mass ($1.5 M_\odot$) but with different EOS, neutrino luminosity decays more rapidly for a model with stiff EOS than soft EOS. This is because of the lower central density in the model with stiff EOS; the neutrino diffusion time scale is proportional to density ($\tau_{\rm diff} \sim R^2/\lambda c \sim n\sigma R^2/c$). Similar consideration applies to the comparison between models 53 and 55 with the same EOS but with different mass. Model 53 with smaller mass and hence with lower central density exhibits more rapid decline of luminosity than model 55 with larger mass. Generally speaking, the binding energy of a neutron star becomes larger for softer EOS and for more massive neutron star. Consequently, total energy of neutrino burst ($\int L_\nu dt$) which is almost the same as the binding energy has the same dependence on the EOS and on the neutron star mass.

2.3.3 Energy spectra

Due to the strong energy dependence of the neutrino interaction rates, an unique neutrinosphere for all energy range cannot be defined. In other words, the emergent neutrino energy spectrum is not so called 'black body' spectrum (with a modification necessary for the case of fermions) from a single surface.

Neutrino spectra in the early stage after the bounce were studied by some groups [130–132] and found to have pinched shape (narrow peak) for $\nu_{\rm e}$'s and $\bar{\nu}_{\rm e}$'s. These spectra can be interpreted as the superposition of the Fermi-Dirac spectra (LFD: local Fermi-Dirac spectrum) from various neutrinospheres defined for each neutrino energy. That is, the early spectra of $\nu_{\rm e}$'s and $\bar{\nu}_{\rm e}$'s can be approximated by

$$\left.\frac{dL_\nu}{dE}\right|_{\rm LFD} = \frac{c}{4}\, 4\pi R_\nu^2(E)\, \frac{4\pi}{(hc)^3}\, f_{\beta\text{-eq}}(\omega(E,R_\nu(E)),R_\nu(E))\, \omega^2(E,R_\nu(E))E \tag{16}$$

where

$$f_{\beta\text{-eq}}(\omega,r) = \frac{1}{\exp[(\omega-\mu_{\beta\text{-eq}}(r))/T(r)]+1} \tag{17}$$

is the neutrino distribution function in β-equilibrium (chemical potential: $\mu_{\beta\text{-eq}}$) with the matter at radius r, $R_\nu(E)$ is the radius of the neutrinosphere for neutrinos with red-shifted energy of E and $\omega(E,r)$ is the corresponding neutrino energy at the sphere. Note that if one substitutes $R_\nu(E) = R$ (constant), $\omega = E$ and exchanges $f_{\beta\text{-eq}}$ with

$$f_\gamma = \frac{2}{\exp(E/T)-1} \tag{18}$$

for photons, dL/dE becomes the normal black body spectrum for photons from a surface with radius R and temperature T. Fig. 33a shows that the early stage spectrum of $\nu_{\rm e}$'s can be interpreted as the LFD spectrum. The main opacity sources for $\nu_{\rm e}$'s and $\bar{\nu}_{\rm e}$'s at this early stage are $\nu_{\rm e}{\rm n} \to {\rm pe}^-$, $\bar{\nu}_{\rm e}{\rm p} \to {\rm ne}^+$ and $\nu{\rm N} \to \nu{\rm N}$ and all of these reactions have cross section proportional to E^2. For low energy neutrinos interacting weakly with the matter, the radius of the neutrinosphere ($R_\nu(E)$) is small and small area of emitting surface ($4\pi R_\nu(E)^2$) results in small flux of low energy neutrinos. For high energy neutrinos, on the other hand, the neutrinosphere resides at outer cool region and, hence, $f_{\beta\text{-eq}}$ for the high energy neutrinos is very small. This is the reason of small flux of high energy neutrinos. Small flux of low energy and high energy neutrinos causes the pinched shape of emergent neutrino spectra.

Meanwhile, the spectrum of 'ν_μ's and the late stage spectrum cannot be fitted by LFD spectrum. For these neutrinos, the main opacity source is isoenergetic scattering off nucleons which cannot thermalize neutrinos even at the neutrinosphere. Fig. 33b shows $\bar{\nu}_e$ spectra at $t = 8\,\text{sec}$ in our numerical simulation. The solid line is our numerical result and the dot-dashed line is LFD spectrum. In order to understand the calculated spectrum, we should take into account the scattering processes more correctly. By referring to the modified black body spectra from the scattering stellar atmosphere [133], we consider the modified Fermi-Dirac spectra (hereafter MFD) as follows:

$$\left.\frac{dL_\nu}{dE}\right|_{\text{MFD}} = \frac{c}{4}\,4\pi R_{\text{th}}^2(E)\,\frac{4\pi}{(hc)^3} \times f_{\text{th}}(\omega(E,R_{\text{th}}(E)),R_{\text{th}}(E))\,\omega^2(E,R_{\text{th}}(E))E\cdot S_{\text{MFD}}(E) \quad , \tag{19}$$

where $R_{\text{th}}(E)$ is the radius of the thermalization-sphere for neutrinos with redshifted energy of E, ω is the corresponding neutrino energy at the sphere, and

$$f_{\text{th}}(\omega,r) = \frac{1}{\exp[(\omega-\mu_{\text{th}}(r))/T(r)]+1} \tag{20}$$

is the thermal neutrino distribution function at that sphere. Note that f_{th} is not necessarily chemically equilibrated one, that is, μ_{th} is the chemical potential of the neutrino which is determined by the temperature $T(r)$ and the neutrino number density but not necessarily equal to the one in the case of β-equilibrium. The thermalization-sphere is defined as the sphere where the effective optical depth ($\tau_{\text{th}}(E,r) = \int_r^\infty \lambda_{\text{eff}}^{-1}(E)dr$) for thermalizing processes (neutrino processes other than isoenergetic scattering)[12] becomes the order of unity. $S_{\text{MFD}}(E)$ represents the effect of scattering above the thermalization-sphere and is defined in terms of the total optical depth, $\tau_{\text{tot}}(E,r) = \int_r^\infty \lambda_\nu^{-1}(E)$, and τ_{th} as

$$S_{\text{MFD}}(E) \equiv \frac{2\tau_{\text{th}}(E,R_{\text{th}})}{\tau_{\text{tot}}(E,R_{\text{th}})+\tau_{\text{th}}(E,R_{\text{th}})} \quad . \tag{21}$$

(See Rybicki & Lightman [133]: Eq. 7.66.) Note that, if the isoenergetic scattering does not dominate ($\lambda_\nu \sim \lambda_{\text{th}}$), the thermalization-sphere is nearly identical to the neutrinosphere ($R_{\text{th}} \sim R_\nu$) and the suppression factor $S_{\text{MFD}}(E)$ is about 1 ($\tau_{\text{th}} \sim \tau_{\text{tot}}$). Furthermore, if non-isoenergetic scattering which can thermalize but cannot

[12]Here λ_{eff} is an effective mean free path for thermalizing processes. When λ_ν is the total mean free path for all processes including isoenergetic scattering and λ_{th} is the mean free path for thermalizing processes, processes other than the isoenergetic scattering, neutrinos interact $N = \lambda_{\text{th}}/\lambda_\nu$ times on the average before one thermalizing interaction. Therefore, the effective mean free path for thermalizing processes becomes $\lambda_{\text{eff}} = \sqrt{N}\lambda_\nu = \sqrt{\lambda_{\text{th}}\lambda_\nu}$.

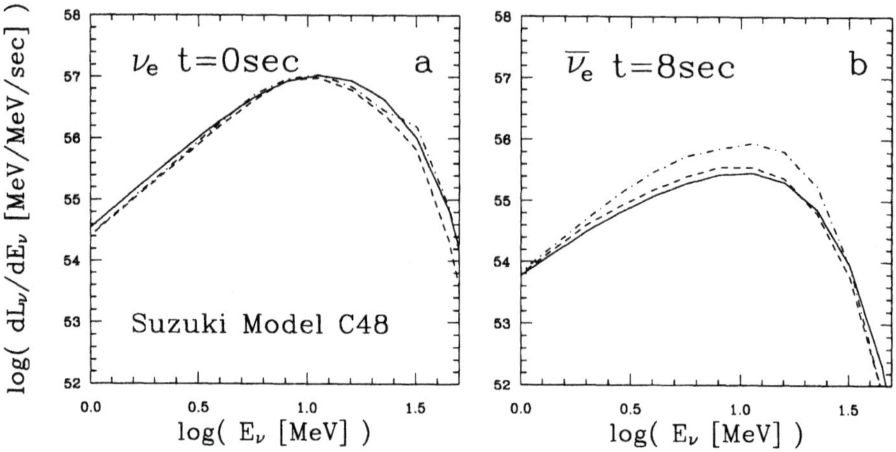

Figure 33: Energy spectra of emergent neutrino fluxes (Suzuki: model C48 [113]). Solid lines are our numerical results using MGFLD, dashed lines are MFD spectra and dot-dashed lines are LFD spectra. a: ν_e at the early stage ($t = 0\,\mathrm{sec}$). b: $\bar{\nu}_e$ at the late stage ($t = 8\,\mathrm{sec}$).

β-equilibrate neutrinos with the matter does also not dominate, thermalization is accompanied with β-equilibration ($\mu_{\rm th}(R_{\rm th}) \sim \mu_{\beta\text{-eq}}(R_\nu)$) and the MFD spectrum approaches the LFD spectrum as the case for ν_e's and $\bar{\nu}_e$'s at the early stage.

In Fig. 33, the MFD spectrum is also shown as the dashed line. It reproduces the numerical result well. Thus, we can understand the spectral shape in terms of the scattered radiation.

In Fig. 34, time evolution of the peak width of the emitted neutrino spectrum during the protoneutron star cooling stage is shown. Using the above discussion on the MFD spectra, we can interpret this evolution in terms of the variation of the relative strength of various channels of neutrino interactions. For example, at the early stage, $S_{\rm MFD}(E)$ for $\bar{\nu}_e$'s does not depend on E so much because the dominant thermalization process is $\bar{\nu}_e p \to n e^+$ while the dominant interaction channels are $\bar{\nu}_e N \to \bar{\nu}_e N$ and $\bar{\nu}_e p \to n e^+$; all of these interactions have cross sections proportional to E^2 so that energy dependence of $S_{\rm MFD}(E) \sim 2\sqrt{\sigma_{\rm tot}\sigma_{\rm th}}/(\sigma_{\rm tot} + \sqrt{\sigma_{\rm tot}\sigma_{\rm th}})$ vanishes. Here we assumed homogeneous models for simplicity. As neutronization and cooling proceed, on the other hand, the thermalization process for low energy $\bar{\nu}_e$'s becomes dominated by $\bar{\nu}_e e^- \to \bar{\nu}_e e^-$ or $NN' \to NN'\nu\bar{\nu}$ while the dominant opacity source is still $\bar{\nu}_e N \to \bar{\nu}_e N$. Then $S_{\rm MFD}(E) \sim 2\sqrt{\sigma_{\rm th}/\sigma_{\rm tot}}$ becomes roughly proportional to $E^{-1/2}$ or $E^{-3/2}$ in the low energy region if one uses naive estimations; $\sigma_{\rm th} \sim \sigma_{\bar{\nu}_e e\ \rm scat} \propto E$ or $\sigma_{\rm th} \sim \sigma_{\rm nucleon\ brems} \propto E^{-1}$. This energy dependence of $S_{\rm MFD}(E)$ (negative power of E) in Eq. 19 causes broad peak of the $\bar{\nu}_e$'s spectrum in the late stage. (See Fig. 34.)

Aside from such analysis of the connection between the neutrino spectra and the physical states of the supernova core as discussed above, parameters which

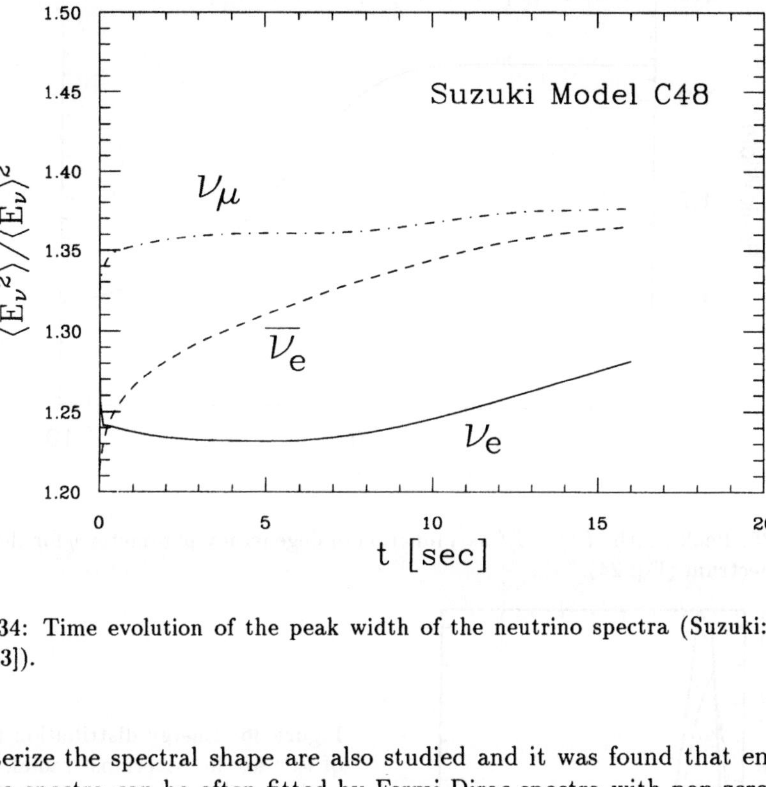

Figure 34: Time evolution of the peak width of the neutrino spectra (Suzuki: model C48 [113]).

characterize the spectral shape are also studied and it was found that emergent neutrino spectra can be often fitted by Fermi-Dirac spectra with non-zero chemical potential [74, 115, 131, 134, 135]. Parameters in a Fermi-Dirac spectrum are temperature T, chemical potential $\mu = \eta T$ where η is degeneracy parameter and radius R; they are related to total luminosity L_ν, average energy $\langle E_\nu\rangle$ and peak width of the spectrum $\langle E_\nu^2\rangle/\langle E_\nu\rangle^2$ as follows:

$$\begin{aligned} L_\nu &= \frac{c}{4}\, 4\pi R^2\, \frac{4\pi}{(hc)^3} \int \frac{E_\nu^3 dE_\nu}{e^{(E_\nu-\mu)/T}+1} \\ &= \frac{c}{4}\, 4\pi R^2\, \frac{4\pi}{(hc)^3}\, T^4 F_3(\eta) \end{aligned} \tag{22}$$

$$\begin{aligned} \langle E_\nu\rangle &= L_\nu \Big/ \frac{c}{4}\, 4\pi R^2\, \frac{4\pi}{(hc)^3} \int \frac{E_\nu^2 dE_\nu}{e^{(E_\nu-\mu)/T}+1} \\ &= T F_3(\eta)/F_2(\eta) \end{aligned} \tag{23}$$

$$\begin{aligned} \langle E_\nu^2\rangle/\langle E_\nu\rangle^2 &= \int \frac{E_\nu^4 dE_\nu}{e^{(E_\nu-\mu)/T}+1} \Big/ \int \frac{E_\nu^2 dE_\nu}{e^{(E_\nu-\mu)/T}+1}\, \langle E_\nu\rangle^2 \\ &= F_4(\eta)F_2(\eta)/F_3(\eta)^2 \end{aligned} \tag{24}$$

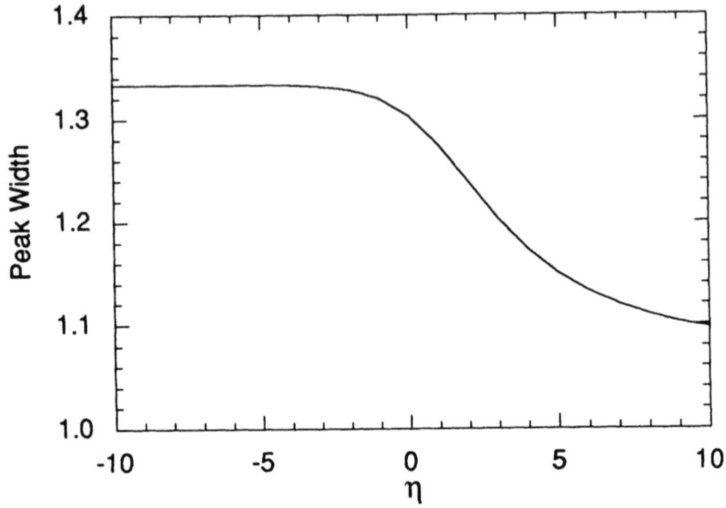

Figure 35: Peak width $\langle E^2\rangle/\langle E\rangle^2$ as a function of degeneracy parameter η for the Fermi-Dirac spectrum (Eq. 24).

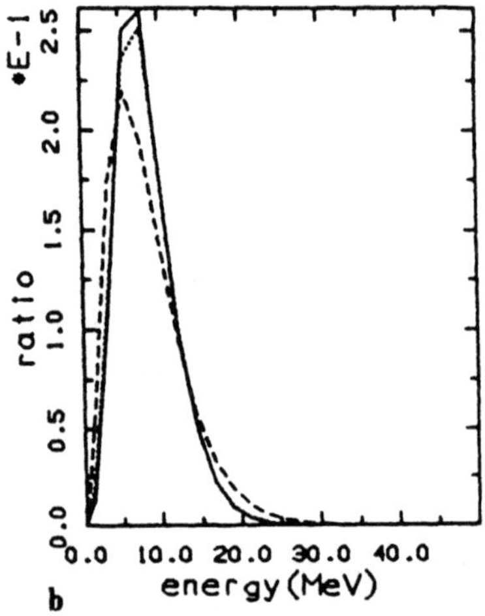

Figure 36: Energy distribution function of ν_e's at the outermost radial position in Janka's Monte Carlo simulation (t = 315 msec after bounce taken from Janka & Hillebrandt 1989 [131]). The solid line is his Monte Carlo result and dotted line is fitting curve by Fermi-Dirac distribution with T = 1.930 MeV and η = 3.504.

where

$$F_n(\eta) \equiv \int_0^\infty \frac{x^n dx}{e^{x-\eta}+1} \quad . \tag{25}$$

In Fig. 35, we plot the peak width as a function of η (Eq. 24). We can calculate η from the peak width if the peak width is less than 4/3, and then T from the average energy and R from the luminosity. Fig. 36 shows the neutrino energy distribution near the surface of the supernova core (solid line) calculated precisely using Monte Carlo simulation by Janka & Hillebrandt [131] with a fitting curve by Fermi-Dirac distribution function (dotted line, T = 1.930 MeV, η = 3.504). Thus neutrino spectra can be parametrized well by η, T and R. One should keep attention,

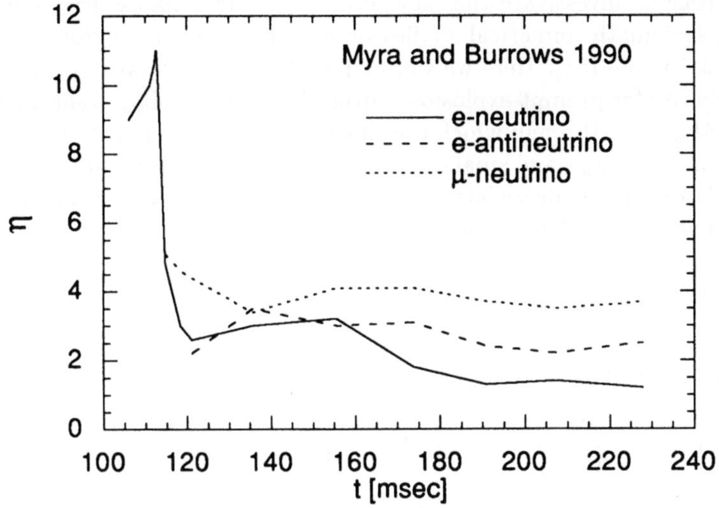

Figure 37: Time evolution of degeneracy parameter in Fermi-Dirac spectral fitting at the dynamical stage of unsuccessful prompt explosion [74]. Solid line is ν_e's, dashed line is $\bar{\nu}_e$'s and dotted line is 'ν_μ's. The core bounces at t = 112.6 msec and the shock breaks out the neutrinosphere at t = 114.8 msec. Data are taken from tables in Myra & Burrows [74].

however, on the point that these parameters have no direct relation to the physical states at a region in the core; they are just fitting parameters. For example, the fitting value of R sometimes exceeds the actual core radius. Fig. 37 shows fitting degeneracy parameters η vs. time during the dynamical stage of an unsuccessful prompt explosion. At the shock breakout (t = 114.8 msec), the spectral shape for ν_e's changes to have moderate degeneracy parameter $\eta = 1 \sim 5$. All species of neutrinos have positive degeneracy parameters corresponding to narrow peaks (pinched shape) (see Fig. 35). Myra, Lattimer & Yahil [134] shows that $\bar{\nu}_e$ spectra both of Bruenn [66] ($2M_\odot$ core model: t = 0.47 sec) and of Mayle, Wilson & Schramm [129] (model 25C t = 0.47 sec) also have narrow peaks corresponding to $\eta = 3.4 \sim 3.6$ but the later spectrum of Bruenn [66] ($t \sim 1$ sec) has broader peak ($\eta \sim 0$). Furthermore Janka [135] calculated neutrino spectra from Wilson's model of a protoneutron star at t = 3.32, 5.77, 7.81 sec using Monte Carlo method and showed that the resultant peak widths correspond to η's ranging from 1.2 to 3.2. Our numerical simulations of evolution of protoneutron star using MGFLD, on the contrary, showed that the peak widths of $\bar{\nu}_e$'s and 'ν_μ's become broader leading to negative η (see Figs. 34 and 35). More studies are required on evolution of spectral shapes.

2.4 Dependence on input physics

There are so many uncertainties in the input physics and in the numerical method that the mechanism of collapse-driven supernova explosion has not yet been clarified. In order to investigate the 'key physics' of the collapse-driven supernova explosions, systematic numerical studies were explored by many groups [51,59,60, 63,67,69–73,75,136–151]. Here we summarise the effects of various input physics qualitatively. As for prompt explosion, initial shock energy $E_{\rm shock}$ and mass of the outer core $M_{\rm outer\ core}$ through which the shock wave must propagate are important quantities. Larger $E_{\rm shock}$ and smaller $M_{\rm outer\ core}$ are preferable to successful prompt explosion. For example, dissociation of all iron nuclei in the outer core on the way of shock propagation exhausts energy [56] of

$$\frac{M_{\rm outer\ core}}{m_{\rm Fe}}\left(30m_{\rm n}+26m_{\rm p}-m_{\rm Fe}\right)c^2 \sim 8\cdot 10^{51}{\rm erg}\left(\frac{M_{\rm outer\ core}}{0.5M_\odot}\right) \quad . \tag{26}$$

On the other hand, roughly speaking, $E_{\rm shock}$ can be expressed as

$$E_{\rm shock} \sim \frac{GM^2_{\rm unshocked\ core}}{R_{\rm unshocked\ core}} \propto M^{\frac{5}{3}}_{\rm unshocked\ core}\rho_c^{\frac{1}{3}} \propto Y_{L\ {\rm trap}}^{\frac{10}{3}}\rho_c^{\frac{1}{3}} \quad , \tag{27}$$

because $M_{\rm unshocked\ core}$ is close to the Chandrasekhar mass which is proportional to $Y^2_{L\ {\rm trap}}$. Note that, since $M_{\rm outer\ core} = M_{\rm Fe\ core} - M_{\rm unshocked\ core}$, large trapped lepton fraction $Y_{L\ {\rm trap}}$ which results in a massive unshocked core (larger $M_{\rm unshocked\ core}$) leads to both larger $E_{\rm shock}$ and smaller $M_{\rm outer\ core}$ [142,144]. At first, we list up the effects of various input physics on the prompt explosion.

- Mass of iron core [56,57,65,71,73,136,144,146,152]
  Small initial iron core mass $M_{\rm Fe\ core}$ leads to small $M_{\rm outer\ core}$. Numerical studies of stellar evolution by Nomoto & Hashimoto [18] and Woosley & Weaver [153] reveal that $M_{\rm core}$ is in the range from $1.20M_\odot$ to $1.95M_\odot$ for stars with $M_{\rm main\ sequence} \lesssim 25M_\odot$. $M_{\rm Fe\ core}$ varies depending upon mass of the main sequence star, reaction rate for $^{12}{\rm C}(\alpha,\gamma)^{16}{\rm O}$, numerical treatment of convection (including overshooting and semiconvection), Coulomb effects in EOS, electron capture rate and so on.

- Softness of EOS for high density matter [53,59–61,71,136–139,148–150]
  If the EOS is soft, the core collapses sufficiently (small $R_{\rm unshocked\ core}$ and high ρ_c in Eq. 27) and releases a large amount of gravitational energy. Although there is no unique parameter representing the "softness" of the EOS, the adiabatic index, $\Gamma \equiv \partial \ln P/\partial \ln \rho$, which corresponds to the slope of $\ln\rho$ – $\ln P$ curve, the nuclear incompressibility at saturation density, $K_0 \equiv 9\partial P/\partial\rho$, which is related to the magnitude of the pressure at the saturation density[13] and the maximum mass of neutron star $M_{\rm NS\ max}$ are often used as such parameters in numerical studies. It is found that the smaller Γ (but greater than 4/3), the smaller K_0 and the smaller $M_{\rm NS\ max}$ (but greater than the maximum mass of observed neutron stars, $1.44M_\odot$) result in the stronger bounce and explosion.

[13] Saturation density is the density at which the free energy of nuclear matter has its minimum for a given proton fraction due to the saturation property of the nuclear matter. For proton fraction ~ 0.5, the saturation density is the so called nuclear density $0.16\,{\rm fm}^{-3} \sim 2.7\cdot 10^{14}{\rm g/cm}^3$.

- Thermal stiffness of the outer core [139, 140, 142]
  Softening of EOS, i.e. pressure decrease due to photodissociation, neutrino loss and so on in the shocked region was modelled by introducing a parameter, thermal stiffness γ_T. It was shown that large γ_T is preferable to the successful shock propagation; that is, photodissociation and neutrino loss are harmful.

- General relativity [52, 53, 60, 138, 147, 149]
  Compared with the model with Newtonian gravity, it was found that general relativity has mainly two effects: smaller homologous core ($\sim$ 10% smaller than Newtonian model) and higher central density. For the model with stiff EOS or with small trapped lepton fraction, small homologous core results in small shock energy and stronger gravity weakens the shock wave; general relativity is harmful in this case. On the contrary, for the model with soft EOS or with large trapped lepton fraction, collapse to higher central density releases much more gravitational energy which overcomes the worse conditions coming from the small homologous core; general relativity is useful in this case.

- Electron capture rate [52, 53, 55, 62, 63, 67, 87, 141, 143]
  Electron capture rate in the collapsing core affects directly the amount of ν_e's which escape from the core before the neutrino trapping, equivalently, the trapped lepton fraction $Y_{L\ \mathrm{trap}}$. Especially, electron capture rate below the trapping density $\sim 10^{12}\mathrm{g/cm}^3$ where neutron-rich nuclei dominate is important. Fuller Fowler & Newman pointed out the importance of nuclear shell effects on electron capture rate. For example electron capture rate of nuclei with protons in $1f_{7/2}$ state drops when neutron number in the nucleus exceeds 40 because the neutron shell $1f_{5/2}$ into which protons in $1f_{7/2}$ state would be converted by the electron capture reaction has been filled up. If one neglects these shell blocking effects, larger electron capture rate results in smaller $Y_{L\ \mathrm{trap}}$ and smaller E_{shock}. Note that average energy of ν_e's emitted by electron capture on nuclei is lower than that by electron capture on free protons because extra energy is necessary for the transition into a nuclear shell with higher energy. Lower ν_e energy also causes larger ν_e loss because of smaller opacity.

- Free proton fraction [70, 144, 154]
  Since electron capture rate on free protons is larger than that on nuclei, the larger the free proton fraction X_p is, the smaller the trapped lepton fraction $Y_{L\ \mathrm{trap}}$ becomes. In particular the abundance of free protons is important where the nuclear shell blocking suppresses the electron capture on nuclei.

- Entropy [59, 71, 144, 145]
  The initial entropy of the iron core depends on the amount of neutrino energy loss during the pre-collapse stellar evolution. The cool core with low entropy contains small amount of free protons and leads to large $Y_{L\ \mathrm{trap}}$.

- Bulk symmetric energy coefficient in the nuclear model [70, 144]
  The bulk symmetric energy coefficient W_{sym} in nuclear models affects the dif-

ference of chemical potentials between neutrons and protons and, hence, the abundance of free protons. Nuclear models with larger W_{sym} have tendency to hold protons in neutron-rich nuclei and, as a result, lead to large $Y_{L\ \mathrm{trap}}$. Bruenn [70] shows that changing W_{sym} from 28 MeV to 40 MeV where standard value is 31.5 MeV causes a drop of about 2 orders of magnitude in X_p for a model of pre-collapse iron core and that $Y_{L\ \mathrm{trap}}$ at the center varies from 0.345 to 0.375.

- Neutrino-electron scattering [62, 67, 69, 70]
  While the main opacity source for ν_e's escaping from the core during the collapse stage is coherent and isoenergetic scattering off nuclei ($\nu_e A \rightarrow \nu_e A$), the non-isoenergetic scattering off electrons ($\nu_e e^- \rightarrow \nu_e e^-$) plays an important role for the determination of $Y_{L\ \mathrm{trap}}$. When ν_e's are scattered by degenerate electrons, recoil electrons should have energy greater than electron Fermi energy; there is energy transfer from ν_e's to electrons. This down-scattering of ν_e's causes the average energy of ν_e's to be lower and the matter entropy to be higher. Both of them leads to smaller $Y_{L\ \mathrm{trap}}$. One should not ignore this energy-non-conservative feature of electron scattering in numerical simulations.

- Nuclear inelastic scattering [75]
  Bruenn & Haxton [75] showed that inelastic scattering off nuclei ($\nu_e A \rightarrow \nu_e A^*$) plays similar roles to electron scattering during the collapse stage because it is also down-scattering.

So far we did not concern the effects of rotation because the ratio of rotational energy to gravitational energy of pulsars at birth, which are considered to be born at the collapse-driven supernova explosions, is estimated to be small (less than $O(10^{-3})$) except for the case of direct formation of millisecond pulsars. On the other hand, rotating core collapse have been also studied [155–162]. It was found that the centrifugal force causes core bounce at lower density leading to smaller amount of gravitational energy release; rotation is harmful for successful prompt explosion [162]. Symbalisty [159] carried out 2-dimensional numerical simulations of core collapse with rotation and magnetic field and he showed only in the case with unrealistically strong magnetic field ($> 10^{15}$ Gauss) and high rotation, there occurs a jet-like explosion of the core. It is also reported that rotating core collapse can cause a large scale circulation flow in the core [158, 159] and that there forms the deformed protoneutron star which is extended in equatorial plane due to the centrifugal force [161]. Janka & Mönchmeyer [163] pointed out that neutrino flux from such a deformed protoneutron star as with a ratio of rotational energy to gravitational energy of about 0.1 varies by a factor of about 2 depending on the direction of observer with respect to the rotational axis.

As for the delayed explosion, intensity of neutrino flux passing through the stalled shock is important. When we approximate the neutrino flux from the inner core as a thermal flux with temperature T_ν and luminosity L_ν, heating rate due to neutrino absorption by nucleons is proportional to neutrino number flux ($\propto T_\nu^3$), absorbed energy ($\propto T_\nu$) and absorption cross section ($\propto T_\nu^2$), that is, proportional to $T_\nu^6 \propto L_\nu T_\nu^2$ [98]. Thus, the efficiency of energy transfer from neutrinos to the matter behind the shock front is sensitive to L_ν and T_ν. Softer EOS makes core collapse further emitting larger amount of binding energy as neutrinos (larger L_ν) and rising core temperature (higher T_ν); soft EOS is preferable to delayed explosion as well as prompt explosion. Hillebrandt [64, 164] performed a numerical simulation of the core collapse for a model of a $20M_\odot$ star which has a similar core to the $15M_\odot$ model of Wilson *et al.* [77]. He attributed the failure of delayed explosion in his simulation to the stiff EOS used and the simplified neutrino transfer scheme (NEFDFLD: Non-equilibrium Fermi-Dirac flux limited diffusion scheme). In Fig. 38, some typical supranuclear EOSs are plotted in the form of density-pressure curve. In the case of EOS for the cold neutron star matter in which all the matter is in the neutrino-less β-equilibrium, one has only to specify the density to determine the other thermodynamic quantities such as pressure. Composition of the matter, that is, proton/electron fraction could be determined by the β-equilibrium condition and temperature could be set to zero. On the contrary, in the case of EOS for the hot matter in the supernova core, one should specify three quantities: temperature T, density ρ_B, and composition Y_e in order to calculate the other thermodynamic quantities. In Fig. 38, we plot pressure as a function of density with given temperature and electron fraction profile which is taken from numerical simulation by Mayle & Wilson 1989 [101, 102, 116, 117]. T profile and Y_e profile used are also depicted in the figure. It is clear that EOS used by Mayle & Wilson (MW89: solid line) is much softer than the other EOSs at the low density. Wolff's EOS which was calculated using Hartree-Fock method [165] is stiff one.

Meanwhile matter accretion onto the protoneutron star during the shock stagnation supplies energy to the neutrino emitting region. Oscillatory enhancement of neutrino luminosity in Fig. 22 is caused by such matter accretion. General relativistic red shift effects for neutrino energy, on the contrary, lower the heating rate [8].

As shown in Figs. 7 and 8 there are regions with negative gradient in Y_L profile or in entropy profile. Such regions might be unstable to convection. Down to the early 1980's, a possibility that lepton-driven convective overturn of the supernova core could cause supernova explosion directly was mainly studied [166–171]. After the appearance of delayed explosion model, convection attracts considerable attention as assistant to the shock revival. Convection underneath the neutrinosphere driven by negative gradient in Y_L profile and/or in entropy profile transports efficiently neutrinos and energy from inner region to the surface; it enhances the neutrino luminosity [77, 81, 96, 103, 106–109, 112, 129, 172–174]. In Wilson & Mayle's numerical simulations, convective enhancement of neutrino luminosity is indispensable to successful delayed explosions. On the other hand, the hot bubble above the neutrinosphere and behind the shock front might be also unstable due to negative entropy gradient [103, 105, 110, 111, 173, 175]. It was found that the convective energy transport itself cannot enhance the shock wave significantly but that induced large scale motion of matter might result in additional neutrino emission due to the released gravitational energy. Multidimensional numerical simulations of such instabilities become current hot topic. Various observations

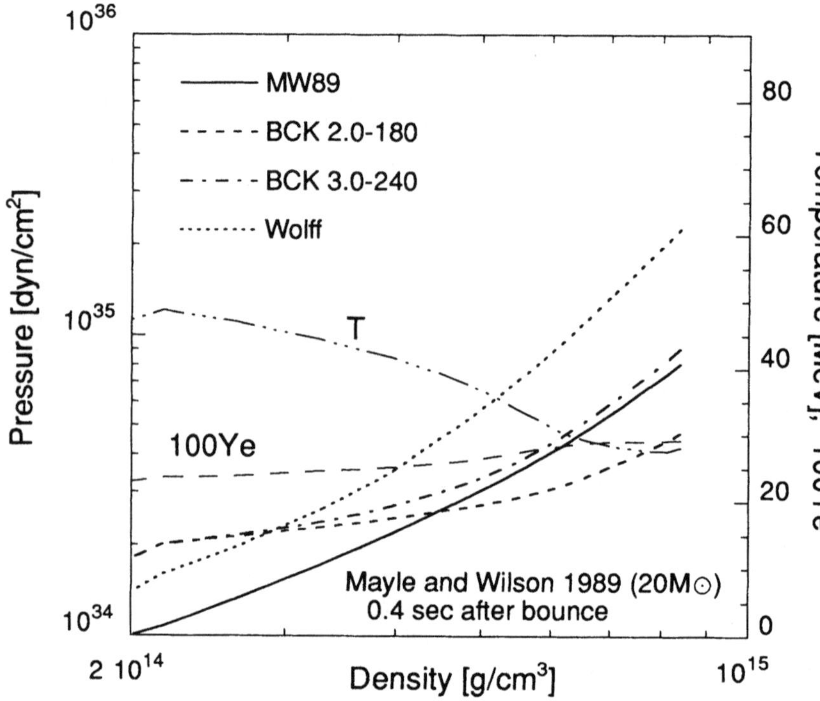

Figure 38: Comparison of several EOSs at supranuclear density. Plotted curves are $P(\rho_B, T(\rho_B), Y_e(\rho_B))$ for 4 EOSs. $T(\rho_B)$ and $Y_e(\rho_B)$ are taken from the numerical result of Mayle & Wilson's dynamical simulation at 0.4 sec after bounce [116]. MW89: Mayle & Wilson [101, 102, 117], Wolff: Hillebrandt & Wolff [165], BCK: parametrized EOS for the cold nuclear matter plus thermal contribution as modelled by Baron, Cooperstein & Kahana [60]. 2.0-180: adiabatic index $\gamma = 2.0$ and incompressibility $K_0^{sym} = 180$ MeV, 3.0-240: $\gamma = 3.0, K_0^{sym} = 240$ MeV. We use the mathematical expression of the BCK EOS for the nuclear pressure in Bethe [5].

of SN1987A in Large Magellanic Cloud using electromagnetic wave (from infrared to gamma) reveal non-spherical features of the explosion. Sato, Shimizu & Yamada [162] consider that convective enhancement of delayed explosion of slowly rotating core could cause the observed non-spherical features.

As one of the neutrino heating processes in the hot bubble, the neutrino pair annihilation process ($\nu\bar{\nu} \rightarrow e^-e^+$) was lighted up by Goodman, Dar & Nussinov [176]. Being much different from other heating processes, however, the efficiency of this process depends strongly on the angular distribution of neutrino momentum; antineutrinos which act as targets for a neutrino have also non-spherical angular distribution outside the neutrinosphere. Since the popular neutrino transfer method, MGFLD, does not take care of the angular distribution, special attention should be paid on the treatment of this process [112]. Janka (using Monte Carlo simulation [177]) and Cooperstein, van den Horn & Baron [178] studied effects of

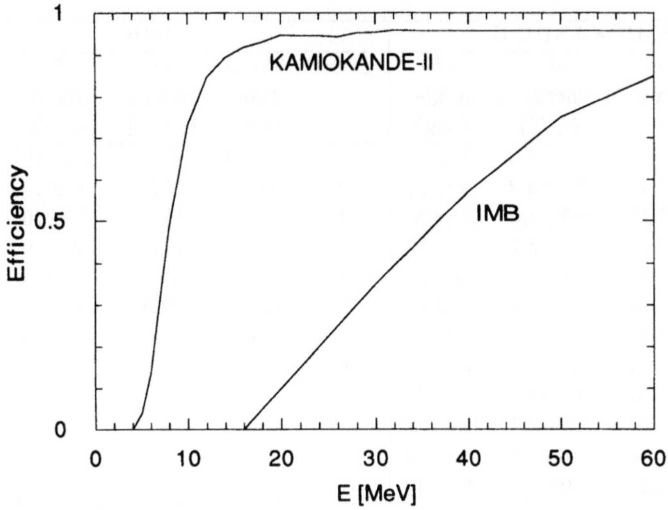

Figure 39: Detection efficiency of the two detectors. The dead time effect is not taken into account in these efficiency curves.

geometry and back scattering, and showed that most of heating occurs near the neutrinosphere and that the heating is less efficient than the estimation of Goodman, Dar & Nussinov.

3 Observation of Supernova Neutrinos

3.1 SN1987A

On 24 February 1987, the supernova 1987A in the Large Magellanic Cloud was discovered by the optical observation [179] as the nearest supernova since the galactic supernova in the 17th century. The observed spectrum shows hydrogen lines [180] so that it was identified as a Type II supernova. Since it was regarded as a collapse-driven supernova explosion and the distance to the supernova was estimated to be only 50 kpc, it was expected that the neutrino burst from SN1987A must have been detected by the underground neutrino detectors before the epoch of optical discovery. As expected, the KAMIOKANDE-II collaboration found that they observed 11 events caused by the neutrino burst from SN1987A on 23 February 1987, 7:35:35 UT(±1min) with their water Čerenkov detector [28]. IMB collaboration also found the detection of the neutrino burst (8 events [29]). Baksan collaboration reported their detection (5 events) [181] much later but their result seems marginal. We present here results of analysis on the data and discuss the validity of current models for the supernova neutrinos. The exciting days around the discovery of neutrino burst from SN1987A are recorded in the review by Koshiba [11].

The detectors of KAMIOKANDE-II and IMB are of the water Čerenkov type,

KAMIOKANDE-II				IMB				
Event	Event time (sec)	$e^{\pm}$ energy (MeV)	$e^{\pm}$ angle (deg)	Event	Event time (sec)	$e^{\pm}$ energy (MeV)	$e^{\pm}$ angle (deg)	$\bar{\nu}_e$ energy (MeV)
1	0.000	20.0±2.9	18±18	1	0.000	38±7	80±10	41±7
2	0.107	13.5±3.2	40±27	2	0.412	37±7	44±15	39±7
3	0.303	7.5±2.0	108±32	3	0.650	28±6	56±20	30±6
4	0.324	9.2±2.7	70±30	4	1.141	39±7	65±20	42±7
5	0.507	12.8±2.9	135±23	5	1.562	36±9	33±15	38±9
*	0.686	6.3±1.7	68±77	6	2.684	36±6	52±10	38±6
6	1.541	35.4±8.0	32±16	7	5.010	19±5	42±20	21±5
7	1.728	21.0±4.2	30±18	8	5.582	22±5	104±20	24±5
8	1.915	19.8±3.2	38±22					
9	9.219	8.6±2.7	122±30					
10	10.433	13.0±2.6	49±26					
11	12.439	8.9±1.9	91±39					

Table 1: Data of the neutrino burst: Event time is measured from 7:35:35±1min UT for KAMIOKANDE-II and 7:35:41.374 UT for IMB. Event angle is with respect to direction away from SN1987A. $\bar{\nu}_e$ energy is estimated assuming that events are due to $\bar{\nu}_e p \rightarrow e^+ n$. Errors of energy do not include the systematic error. Event * in KAMIOKANDE-II data is considered as a background event.

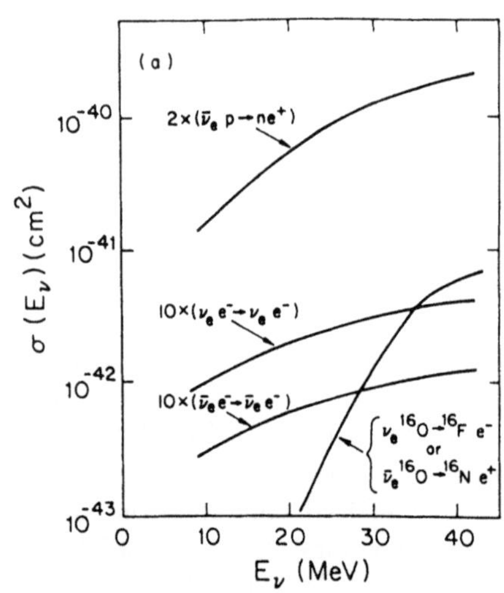

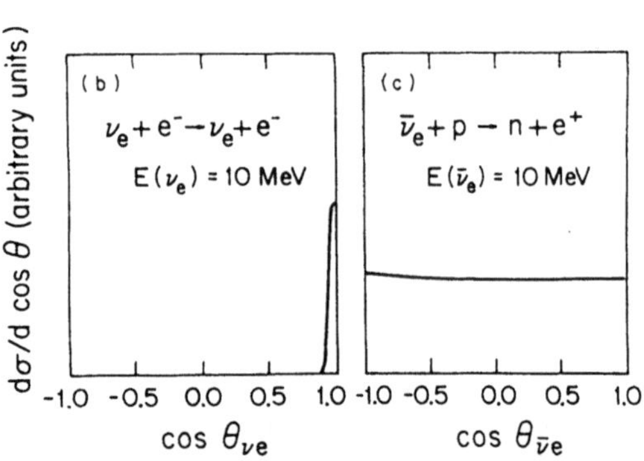

Figure 40: (a) Cross sections for typical neutrino interactions in water Čerenkov detector as functions of incident neutrino energy. (b,c) Angular differential cross section for $\nu_e e^- \rightarrow \nu_e e^-$ (b) and for $\bar{\nu}_e p \rightarrow e^+ n$ (c). (taken from Hirata *et al.* [182])

which were originally constructed for the nucleon decay search while Baksan detector is an underground scintillator telescope aiming at a search for a stellar collapse. Since the Baksan detector has a rather large background rate (0.7/day corresponding to 5 events per 9 seconds), we perform an analysis of the neutrino burst using only the revised data of KAMIOKANDE-II [182] and IMB [183]. Piran & Spergel [184] show that the analysis including Baksan's data does not change the results considerably. The two water Čerenkov detectors can measure energy and direction of electrons or positrons which are produced or scattered by neutrino interactions with water. Note that electrons and positrons are not distinguishable each other. The data of neutrino events observed by the detectors are listed in Table 1. The quoted errors of energy do not include the systematic error which is mainly originated from energy calibration. Since KAMIOKANDE-II detector was under refinement for the observation of solar neutrinos, the threshold energy for detection is significantly lower than IMB's so that lower energy events were observed by KAMIOKANDE-II. The detection efficiency of electrons/positrons $\eta_{\rm detector}$ in each detector is shown in Fig. 39 [183, 185]. We adopt the central values of the reported efficiency of the IMB detector. The detector volumes $V_{\rm detector}$ are 2140m^3 for KAMIOKANDE-II and 6800m^3 for IMB. In addition, since IMB reported the dead time of 13% during the burst, we multiply the detection efficiency by a factor of 0.87. KAMIOKANDE-II detector is almost dead time free. Then, for each neutrino interaction channel, the expected number of events in the detector with electron/positron energy $E_{\rm e}$ and event angle $\theta_{\rm e}$ per unit energy per unit $\cos\theta_{\rm e}$ per unit time, $d^3N_{\rm expect}/dE_{\rm e}d\cos\theta_{\rm e}dt$ can be expressed as

$$\frac{d^3N_{\rm expect}^{{\rm detector}\ (i)}(E_{\rm e},\cos\theta_{\rm e},t)}{dE_{\rm e}d\cos\theta_{\rm e}dt} = n_{\rm target}^{(i)}V_{\rm detector}\eta_{\rm detector}(E_{\rm e})$$

$$\times\ \frac{1}{4\pi d_{SN}^2}\frac{dN_\nu(E_\nu,t)}{dE_\nu}\frac{d\sigma^{(i)}(E_\nu,\cos\theta_{\rm e})}{d\cos\theta_{\rm e}}\Bigg/\frac{dE_{\rm e}^{(i)}(E_\nu,\cos\theta_{\rm e})}{dE_\nu} \qquad (28)$$

where superscript (i) denotes the interaction channel such as $\bar\nu_{\rm e}{\rm p}\rightarrow{\rm ne}^+$ or $\nu{\rm e}^-\rightarrow\nu{\rm e}^-$, $n_{\rm target}^{(i)}$ is the number density of the targets in the water ($n_{\rm p}=2\times\rho_{\rm H_2O}/m_{\rm H_2O}$, $n_{\rm e^-}=10\times\rho_{\rm H_2O}/m_{\rm H_2O}$), d_{SN} is the distance to the supernova ($\sim 50\,{\rm kpc}$), dN_ν/dE_ν is the neutrino number spectrum from the supernova and $d\sigma^{(i)}/d\cos\theta_{\rm e}$ is the differential cross section for the interaction. Note that for neutrinos with energy of $10\sim 30\,{\rm MeV}$ the $\bar\nu_{\rm e}$ absorption by free (not in O nuclei) protons ($\bar\nu_{\rm e}{\rm p}\rightarrow{\rm ne}^+$) has the largest $n_{\rm target}\sigma$ and the second is the interaction $\nu_{\rm e}{\rm e}^-\rightarrow\nu_{\rm e}{\rm e}^-$ (see Fig. 40).

3.1.1 Astrophysical analysis

By analyzing the overall feature of the neutrino events and by comparing it with results of numerical simulations, many studies have been done on validity of the current model for supernova neutrinos, the progenitor, the newly formed neutron star and so on (for example, references [66, 114, 115, 131, 184, 186–198]).

Duration of the burst KAMIOKANDE-II observed events during the time span of 12.4 seconds and IMB 5.6 seconds. This time scale coincides with the neutrino diffusion time scale in the supernova core. This means that neutrinos are trapped once in the core and then diffuse out.

Angular distribution of the events Most of supernova neutrinos are emitted thermally so that all types of neutrinos are emitted equally in the rough. In the water Čerenkov detector, however, $\bar{\nu}_e$ absorption by free protons ($\bar{\nu}_e p \rightarrow e^+ n$) dominates other scattering processes ($\nu e^- \rightarrow \nu e^-$) at neutrino energy around 10 MeV. As for $\bar{\nu}_e$ absorption, since the targets (protons) are considerably heavy for $\bar{\nu}_e$'s with energy of about 10 MeV, the angular distribution of the products (positrons) is nearly isotropic while the scattered electrons in the process $\nu e^- \rightarrow \nu e^-$ have a sharp forward peak (See Fig. 40.). Therefore the angular distribution of the data is expected to be nearly isotropic. For example, expected ratio of the scattering events (forward peak) to the absorption events (isotropic) based on numerical results for the neutrino burst by Mayle (model 15C [81]) is evaluated to be about 0.05 for both the detectors. Actual data (11 + 8 events) include only one event (the first event of KAMIOKANDE-II) which has a large probability to be a scattering event. Thus the angular distribution is consistent with the current model in which all types of neutrinos are emitted thermally. Such a picture that only ν_e's which are inevitably emitted by electron capture processes at the birth of a neutron star are emitted is excluded.

Neutronization burst When the shock wave generated at the core bounce breaks out the neutrinosphere, free protons produced by the dissociation of the nuclei catch electrons and cause a so-called neutronization burst of ν_e's. According to several simulations, the time scale of the neutronization burst is less than 10 msec and the total energy of ν_e's emitted during the neutronization burst is less than several times 10^{51} erg. The expected number of scattering events corresponding to the neutronization burst based on the numerical simulations is less than 0.1 [115, 186]. Although the first event of KAMIOKANDE-II could be regarded as a scattering event due to its direction, we cannot conclude that it is really a scattering event caused by the neutronization burst because of the small expected number of events and the small statistics. Therefore, an assumption that all 19 events are due to $\bar{\nu}_e p \rightarrow e^+ n$ is frequently adopted.

Progenitor mass The progenitor of SN1987A was found to have been a blue supergiant and theoretical studies on the stellar evolution reveal that the progenitor mass is $\sim 20 M_\odot$ [199, 200]. Meanwhile, Wilson's group has done systematic simulations of the supernova explosion using realistic initial models and the multigroup neutrino transfer method. Comparing the expected number of neutrino events based on their simulations with the observed one, we infer that the progenitor mass of SN1987A is less than or equal to $15 M_\odot$. For more massive progenitors than $25 M_\odot$, much more events are expected [115, 186]. This is consistent with the above result obtained from the optical observation. We must, however, mention that this implication on the progenitor mass depends considerably on the models such as pre-supernova stellar evolution and the EOS of the neutron star matter.

Temperature and total energy of $\bar{\nu}_e$ flux Although, strictly speaking, supernova neutrinos do not have black body spectra because of the strong dependence of the interaction rate on neutrino energy, thermally emitted neutrino spectra can be characterized by a neutrino temperature as the zeroth order approximation. Assuming that the time-integrated $\bar{\nu}_e$ spectrum is Fermi-Dirac one with temperature T and zero chemical potential and denoting the time-integrated luminosity (total energy) of $\bar{\nu}_e$ as $E_{\mathrm{tot},\bar{\nu}_e}$, we can express the time-integrated neutrino number spectrum as

$$\int \frac{dN_\nu}{dE_\nu} dt = \frac{E_{\mathrm{tot},\bar{\nu}_e}}{\int_0^\infty \frac{E_\nu^3}{e^{E_\nu/T}+1} dE_\nu} \frac{E_\nu^2}{e^{E_\nu/T}+1} \quad . \tag{29}$$

Substituting Eq. 29 into Eq. 28 we can obtain expected values of various observational quantities for this simplified model. Comparison of expected number and average energy of events with actual data puts constraints on the values of T and $E_{\mathrm{tot},\bar{\nu}_e}$. For example, an application of the maximum likelihood method [201] to the combined data of the two detectors could be as follows. One can calculate expected number and expected average energy of events for each detector as

$$N_{\mathrm{expect}}^{\mathrm{detector}} = \int \frac{d^3 N_{\mathrm{expect}}^{\mathrm{detector}}(E_e, \cos\theta_e, t)}{dE_e d\cos\theta_e dt} dE_e d\cos\theta_e dt \tag{30}$$

$$\langle E_{\mathrm{expect}}^{\mathrm{detector}} \rangle = \int E_e \frac{d^3 N_{\mathrm{expect}}^{\mathrm{detector}}(E_e, \cos\theta_e, t)}{dE_e d\cos\theta_e dt} dE_e d\cos\theta_e dt \Big/ N_{\mathrm{expect}}^{\mathrm{detector}} \quad . \tag{31}$$

Intrinsic fluctuation of the event energy caused by the non-monochromatic flux $\sigma_{\mathrm{expect}}^{\mathrm{detector}}$ becomes

$$(\sigma_{\mathrm{expect}}^{\mathrm{detector}})^2 =$$

$$\int (E_e - \langle E_{\mathrm{expect}}^{\mathrm{detector}} \rangle)^2 \frac{d^3 N_{\mathrm{expect}}^{\mathrm{detector}}(E_e, \cos\theta_e, t)}{dE_e d\cos\theta_e dt} dE_e d\cos\theta_e dt \Big/ N_{\mathrm{expect}}^{\mathrm{detector}} \quad . \tag{32}$$

Since the distribution of the number of events in each detector can be considered as Poisson distribution with mean $N_{\mathrm{expect}}^{\mathrm{detector}}$ and the distribution of the average energy of $N_{\mathrm{obs}}^{\mathrm{detector}}$ events ($N_{\mathrm{obs}}^{\mathrm{KAMIOKANDE-II}} = 11$, $N_{\mathrm{obs}}^{\mathrm{IMB}} = 8$) can be regarded as normal distribution with mean $\langle E_{\mathrm{expect}}^{\mathrm{detector}} \rangle$ and variance

$$\sigma^{\mathrm{detector}} = \sqrt{\frac{(\sigma_{\mathrm{expect}}^{\mathrm{detector}})^2}{N_{\mathrm{obs}}^{\mathrm{detector}}} + (\sigma_{\mathrm{obs}}^{\mathrm{detector}})^2} \tag{33}$$

where $\sigma_{\mathrm{obs}}^{\mathrm{detector}}$ is the error in the average energy of observed events including detector's systematic error ($\sigma_{\mathrm{obs}}^{\mathrm{KAMIOKANDE-II}} = 1.9\,\mathrm{MeV}$, $\sigma_{\mathrm{obs}}^{\mathrm{IMB}} = 5.6\,\mathrm{MeV}$), the joint likelihood function $L(T, E_{\mathrm{tot},\bar{\nu}_e})$ is defined as

$$\begin{aligned} L(T, E_{\mathrm{tot},\bar{\nu}_e}) &= \prod_{i=\mathrm{detector}} \frac{1}{\sqrt{2\pi}\sigma^i} \exp\left[-\frac{1}{2}\left(\frac{\langle E_{\mathrm{obs}}^i \rangle - \langle E_{\mathrm{expect}}^i \rangle}{\sigma^i}\right)^2\right] \\ &\quad \times \frac{(N_{\mathrm{expect}}^i)^{N_{\mathrm{obs}}^i} \exp(-N_{\mathrm{expect}}^i)}{(N_{\mathrm{obs}}^i)!} \quad , \end{aligned} \tag{34}$$

where $\langle E_{\rm obs}^{\rm detector}\rangle$ is the average energy of observed events ($\langle E_{\rm obs}^{\rm KAMIOKANDE-II}\rangle = 15.4\,{\rm MeV}$, $\langle E_{\rm obs}^{\rm IMB}\rangle = 32.8\,{\rm MeV}$). A parameter set which maximises this likelihood function is the best one for reproducing the observed number and average energy of events in the two detectors simultaneously. Our evaluated parameters are $T = 4.0 \pm 0.5\,{\rm MeV}$ and $E_{{\rm tot},\bar{\nu}_e} = 3.4^{+1.4}_{-1.0} \cdot 10^{52}{\rm erg}$ [115]. These values are, of course, in good agreement with the current standard model of supernova neutrinos. Using different likelihood function, Lattimer & Yahil [197] obtained $T = 3.7 \pm 0.4\,{\rm MeV}$, $E_{{\rm tot},\bar{\nu}_e} = (4.8^{+2.1}_{-1.8}) \cdot 10^{52}{\rm erg}$, while Bludman & Schinder [194] gave $T = 3.3 \pm 0.3\,{\rm MeV}$, $E_{{\rm tot},\bar{\nu}_e} = (5.7 \pm 1.7) \cdot 10^{52}{\rm erg}$ and Janka & Hillebrandt [131] gave $T = 3.4^{+1.1}_{-0.8}\,{\rm MeV}$, $E_{{\rm tot},\bar{\nu}_e} = (5.3^{+5.0}_{-2.6}) \cdot 10^{52}{\rm erg}$. A part of the difference between ours and others' comes from the different treatment of low energy threshold of the detectors.

Mass of the neutron star just born When the stellar core collapses to form a neutron star, the binding energy of the neutron star is released as neutrinos, gravitational waves, kinetic energy of the ejecta and radiation(photons). However, as mentioned before, we can consider that the total energy carried out by neutrinos is nearly equal to the binding energy of the neutron star, E_b. We can estimate the former from the observed data of the neutrino burst and, hence, can obtain the mass of the neutron star just born in SN1987A using the relation between neutron star mass and its binding energy [187].

The total energy emitted as neutrinos, $E_{\rm tot}$, can be evaluated roughly as 6 times the total energy emitted as $\bar{\nu}_e$'s, $E_{{\rm tot},\bar{\nu}_e}$. From the previous discussion using the maximum likelihood method, $E_{\rm tot}$ becomes in the range of $1.4 \sim 2.9 \cdot 10^{53}$ erg. When we plot the relation between the mass of the neutron star and its binding energy for various EOSs as in Fig. 41, the region estimated for the case of SN1987A can be superposed on the figure as the horizontal band. This figure implies that if we take the EOS C (Bethe & Johnson I [202]), the gravitational mass of the neutron star is in the range $1.1 \sim 1.6 M_\odot$. This range strongly depends on the adopted EOS. Since the existence of the neutron star with mass of $1.44 M_\odot$ [203] excludes extremely soft EOSs, we consider the EOS F (the maximum mass of the neutron star is $1.46 M_\odot$) [204] as an example of the softest EOS. As for the stiff EOS, we take the EOS M (the maximum mass is $1.96 M_\odot$) [205] as an example of the hardest EOS. Consequently, taking account of the ambiguity of EOSs, we obtain the range of the gravitational mass of the neutron star in SN1987A as $1.0 \lesssim M_G \lesssim 1.7 M_\odot$. Note that the binding energy of the neutron star in SN1987A does not exceed the upper limits for any EOSs, that is, it seems that the black hole was not formed in SN1987A.

Although in the above analysis we take the value of 50 kpc as the distance to the supernova, there are some uncertainties in the estimation of the distance. In order to include the ambiguity concerning the distance, we perform the same analysis for the two limiting values of the distance (46 kpc [208] and 56 kpc [209]). The mass range of the neutron star changes to $0.9 \sim 1.6 M_\odot$ (for 46 kpc) and $1.1 \sim 1.9 M_\odot$ (for 56 kpc), respectively. Even if the distance to the supernova is 56 kpc, the possibility of the black hole formation is small because the binding energy is still less than $3.7 \cdot 10^{53}$erg.

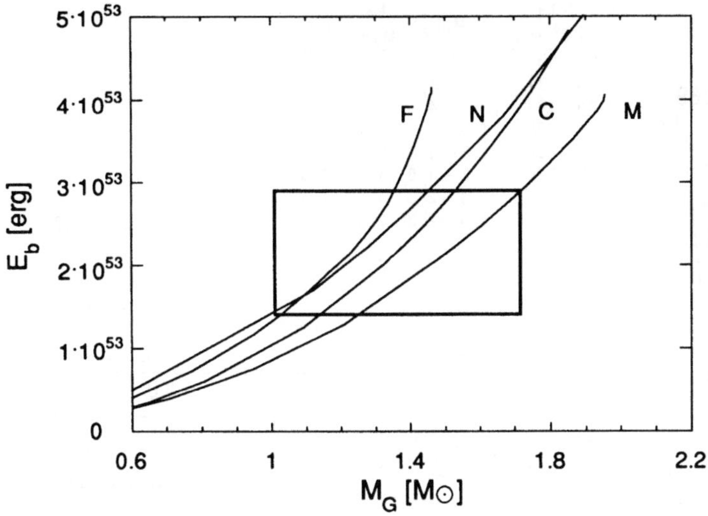

Figure 41: Gravitational mass of neutron star vs. its binding energy for various EOSs: The box represents the estimated region of the binding energy for the case of SN1987A (distance = 50 kpc). The data on the $M_G - E_b$ relations are taken from Arnett & Bowers [206]. The EOS F is Arponen's [204], N: Walecka [207], C: Bethe & Johnson [202] M: Pandharipande & Smith [205].

Time evolution Unfortunately, the small statistics of the observed events prevents us to study the time evolution of the neutrino burst in detail. However, at least it can be seen that high energy events concentrate in the first 3 seconds and that the event frequency becomes low at $t > 3$ sec. It is natural to regard this feature as corresponding to cooling of the protoneutron star. Fig. 42 shows integrated number of $\bar{\nu}_e$ events expected for KAMIOKANDE-II from Burrows' models of protoneutron star cooling [114]. The step like line labelled K II is accumulated histogram of the observed events for KAMIOKANDE-II. It can be seen that the observed time evolution of the neutrino events is well reproduced by appropriate models of standard protoneutron star cooling.

3.1.2 Implications to particle physics

In spite of the limited amount of information obtained from the only 19 events, a large number of implications to particle physics were performed. Since it is beyond the scope of this review to follow all of such implications, we pick up some topics shortly. For more comprehensive understanding, see Bahcall [10], Raffelt [210,211] and references therein, for instance.

What makes neutrinos be important in supernova core is their suitable coupling strength with the matter. While they are produced copiously and hold substantial part of energy in the core, they escape from the core fastest among the constituents. Too weak coupling would prevent their production and too strong coupling would prevent their escape. If there exist other particles having similar coupling strength

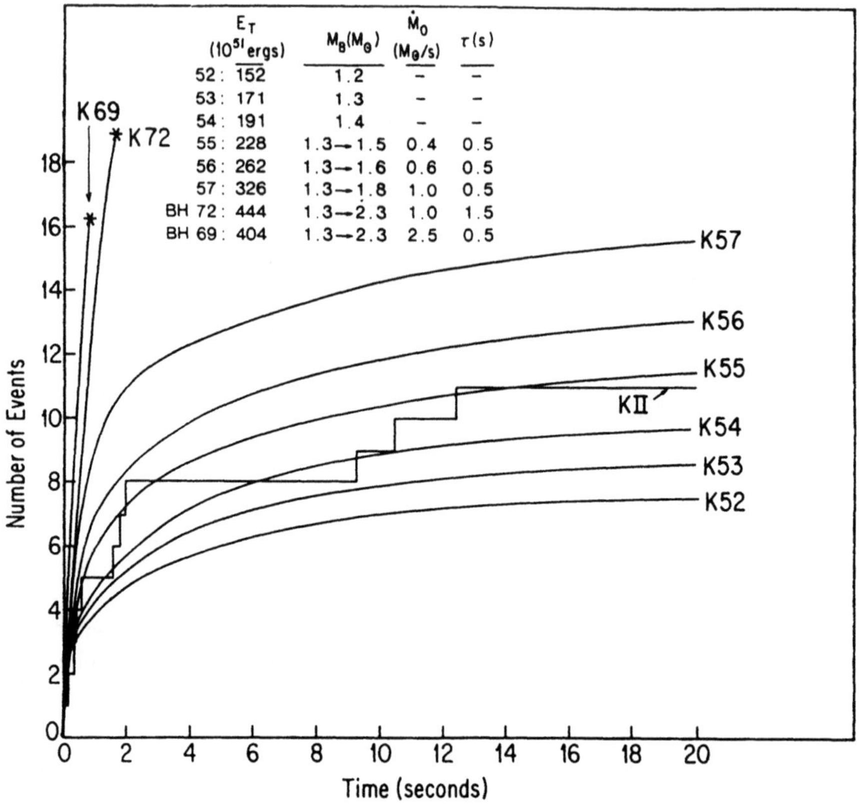

Figure 42: Accumulated number of $\bar{\nu}_e$ events in KAMIOKANDE-II detector for some models of protoneutron star cooling (taken from Burrows [114]). Models 55–72 contain the matter accretion phase and models 69 and 72 result in black hole formation. The distance to the protoneutron star is set to be the distance to the Large Magellanic Cloud, 50 kpc. The actual data for SN1987A is also shown as the line labelled K II.

with the matter, they must have been also emitted from SN1987A as well as neutrinos. Existence of such particles should alter the total energy and the time profile of neutrino burst; for example, leading to smaller number of neutrino events or shorter duration of neutrino burst than the prediction from the standard model of supernova neutrinos. The agreement between the prediction and the observation of neutrinos from SN1987A, conversely, put constraints on existence and property of such particles. For example, one can exclude hadronic axions with mass from $O(10^{-3})$ eV to $O(1)$ eV [212–223]. In addition, if neutrinos are Dirac fermions with finite mass, right-handed neutrinos should exist and they play the same roles as of the above exotic particles. The neutrino observation of SN1987A could exclude mass range from $O(10)$ keV to $O(1)$ MeV [215, 224–231]. On the other hand, irrespective of whether neutrinos are Dirac fermions or Majorana fermions, a finite neutrino mass causes the dispersion in the time of flight from the supernova to the earth; the velocity of higher energy neutrinos is faster than that of lower energy neutrinos. Although it is difficult to reduce the ambiguity coming both from the small number of the observed events and from the variety of supernova models, neutrino events during the 13 seconds indicate that the mass of $\bar{\nu}_e$'s is less than 10–25 eV [186, 189, 190, 198, 232–246], which was comparable to the laboratory limit. A constraint on the neutrino magnetic moment which could also cause the helicity flip is obtained as less than $10^{-12}\mu_B$ [247–250] where μ_B is the Bohr magneton.

3.2 Galactic Supernovae and Neutrino Background originated from Past Supernovae

So far there has been only one supernova (SN1987A) which was observed by neutrinos. In other words, KAMIOKANDE-II [251], IMB [252] and Baksan [253] did not observe neutrino bursts from galactic supernovae during their operation (except SN1987A in the Large Magellanic Cloud, of course) in spite of their sufficient faculties for the successful detection. This negative detection puts a constraint on the galactic (collapse-driven) supernova rate [253] as

$$\text{Supernova rate} < 0.21/\text{year} \quad (90\%\ \text{C.L.}) \tag{35}$$

while the supernova rate [13, 254–256] is estimated to be in the range from 0.01 SN/year to 0.1 SN/year using various astrophysical observations such as stellar statistics [257, 258], statistics of historical supernovae [256], statistics of extra-galactic supernovae [259,260] and pulsar statistics [261]. When the next supernova appears in our galaxy, the next generation of neutrino telescopes such as SUPER-KAMIOKANDE [262,263] and SNO [264] will detect thousands of supernova neutrinos and enable us to investigate details of supernova mechanism, property of high density matter, property of neutrinos and so on. What we can do with the detection of galactic supernova neutrinos was studied by some groups [128,265–269], especially about the possibility for measuring the mass of ν_τ's.

On the other hand, neutrinos emitted from supernovae since the beginning of the universe should form diffuse neutrino background [270–272]. These relic neutrinos might be observed as isotropic background radiation. When one denotes the average emission rate of supernova neutrinos with energy E_ν per unit energy per unit volume at time t as $d^2n_\nu(E_\nu,t)/dE_\nu dt$ (unit: 1/cm^3/MeV/sec), the expected differential number density (unit: 1/cm^3/MeV) of the relic neutrinos from the past supernovae becomes

$$\frac{dn_\nu(E_\nu,t_0)}{dE_\nu} = \int_0^{t_0} \frac{1}{(1+z)^3} \frac{d^2n_\nu(E'_\nu,t)}{dE'_\nu dt} \frac{dE'_\nu}{dE_\nu} dt \tag{36}$$

where $E'_\nu = E_\nu(1+z)$ is the neutrino energy at the time $z(t)$ which has been red-shifted to the present ($t = t_0$) value E_ν and $1/(1+z)^3$ denotes the volume expansion effect. For the simplest example, consider the case of flat matter dominant universe, that is, $dt = -H_0^{-1}(1+z)^{-5/2}dz$ (H_0 is the Hubble constant). Assuming that number density of galaxies $n_{\rm gal}(t) = n_{\rm gal}(t_0)(1+z)^3$, that (collapse-driven) supernova rate $f_{\rm SN}$ has been constant all the time and that the time-integrated neutrino spectrum from a supernova is Fermi-Dirac spectrum with temperature T, zero chemical potential and the total energy $E_{\mathrm{tot},\nu}$, $d^2n_\nu/dE'_\nu dt$ can be written as

$$\begin{aligned}\frac{d^2n_\nu}{dE'_\nu dt} &= n_{\mathrm{gal}}(t)\, f_{\mathrm{SN}}\, \frac{E_{\mathrm{tot},\nu}}{T^4F_3(0)} \frac{E'^2_\nu}{\exp(E'_\nu/T)+1} \\ &= n_{\mathrm{gal}}(t_0)\, f_{\mathrm{SN}}\, \frac{E_{\mathrm{tot},\nu}}{T^4F_3(0)} \frac{E^2_\nu(1+z)^5}{\exp(E_\nu(1+z)/T)+1} \quad . \end{aligned} \tag{37}$$

Then the total number flux of $\bar{\nu}_e$'s at present becomes

$$\begin{aligned} c n_{\bar{\nu}_e} &= c\int \frac{dn_{\bar{\nu}_e}}{dE_\nu} dE_\nu \\ &= \frac{2}{3}\frac{cT^3F_2(0)}{H_0T^4F_3(0)} E_{\mathrm{tot},\bar{\nu}_e}\, n_{\mathrm{gal}}(t_0)\, f_{\mathrm{SN}} \\ &= 44/\mathrm{cm}^2/\mathrm{sec} \left(\frac{H_0}{75\,\mathrm{km/sec/Mpc}}\right)^{-1} \left(\frac{T}{4\,\mathrm{MeV}}\right)^{-1} \left(\frac{E_{\mathrm{tot},\bar{\nu}_e}}{5\cdot 10^{52}\mathrm{erg}}\right) \\ &\times \left(\frac{n_{\mathrm{gal}}(t_0)}{1/50\,\mathrm{Mpc}^3}\right) \left(\frac{f_{\mathrm{SN}}}{1/10\,\mathrm{year}}\right) \end{aligned} \tag{38}$$

and the energy flux becomes

$$\begin{aligned} c u_{\bar{\nu}_e} &= c\int E_\nu \frac{dn_{\bar{\nu}_e}}{dE_\nu} dE_\nu \\ &= \frac{2}{5}\frac{c}{H_0} E_{\mathrm{tot},\bar{\nu}_e}\, n_{\mathrm{gal}}(t_0)\, f_{\mathrm{SN}} \quad . \end{aligned} \tag{39}$$

Note that the average energy of $\bar{\nu}_e$'s at present is

$$\langle E_{\bar{\nu}_e}\rangle = \frac{u_{\bar{\nu}_e}}{n_{\bar{\nu}_e}} = \frac{3}{5} T \frac{F_3(0)}{F_2(0)} \tag{40}$$

where the average energy of $\bar{\nu}_e$'s from the individual supernova is $TF_3(0)/F_2(0)$. The additional factor 3/5 denotes the red shift of neutrino energy due to the cosmic expansion.

KAMIOKANDE-II [251, 273] searched $\bar{\nu}_e$ events in the positron energy range from 19 MeV to 50 MeV in which the background events other than $\bar{\nu}_e$ events due to the relic supernova neutrinos occur very infrequently. $\bar{\nu}_e$ events which exceed the expected background level significantly were not observed during the operation of 1044 days. For the simplified model as mentioned above with $T = 4\,\mathrm{MeV}$, they obtain a flux limit of

$$c n_{\bar{\nu}_e} < 780/\mathrm{cm}^2/\mathrm{sec} \quad (\text{at } 90\%\ \text{C.L.}). \tag{41}$$

This flux limit can be converted to a limit for supernova rate $f_{\rm SN} < 1.8$ supernovae/year/galaxy (at 90% C.L.) for typical values of parameters in Eq. 38. Note that the red shift effects and the high threshold energy for the detection make the high energy tail of neutrino spectrum be important. Since it is considered that the supernova rate relates strongly to the theory of formation and evolution of galaxy, observation of the relic supernova neutrinos is one of the most important targets of new neutrino telescopes such as SUPERKAMIOKANDE under construction.

Acknowledgment

The author would like to thank K. Sato, K. S. Hirata, N. Ishizuka and K. Sumiyoshi for valuable discussions and collaboration. He would also like to express his gratitude to J. Wilson, R. Mayle, R. G. Wolff for kindly providing him with their numerical data and to the editors for useful comments on this article and for encouragement. Numerical Simulations reported were performed on the supercomputer HITAC S820/80 at KEK.

A Basic Equations for Hydrodynamics and Neutrino Transfer

In this appendix, we summarize the basic equations used in numerical simulations of spherical supernova models. General relativistic tranport theory was studied by Lindquist [274] and the expressions of hydrodynamic equations and transfer equations suitable for supernova simulations are presented in Baron *et al.* [275]. (See also references [62, 81, 276–278].) The fundamental equations for hydrodynamics of the matter and neutrino transport include the Einstein equation;

$$R^{\alpha\beta} - \frac{1}{2} g^{\alpha\beta} R = \frac{8\pi G}{c^4} T^{\alpha\beta} \tag{A1}$$

where $R^{\alpha\beta}$ is the Ricci tensor, R is the Ricci scalar, $g^{\alpha\beta}$ is the metric tensor and $T^{\alpha\beta}$ is the energy-momentum tensor of the matter and neutrinos, the equation for energy-momentum conservation;

$$T^{\alpha}{}_{\beta\,;\,\alpha} = 0 \quad , \tag{A2}$$

an equation for baryon number conservation;

$$V_B{}^{\alpha}{}_{;\,\alpha} = 0 \tag{A3}$$

where $V_B{}^{\alpha}$ is the baryon current, equations for lepton number conservation;

$$V_l{}^{\alpha}{}_{;\,\alpha} = 0 \tag{A4}$$

where $V_l{}^{\alpha}$ is the lepton current for each lepton flavour ($l = \mathrm{e}, \mu, \tau$) and an equation for neutrino distribution function;

$$\frac{d\Delta N_\nu}{d\lambda'} = \left(\frac{d\Delta N_\nu}{d\lambda'}\right)_{\rm coll} \tag{A5}$$

where λ' is the affine parameter and ΔN_ν represents the number of neutrinos in a phase volume $\Delta V_{\mathrm{phase}}$. The last equation means that the change of neutrino number along the geodesic is caused only by collision (interaction with the matter and other neutrinos). Neutrino distribution function f_ν is defined as

$$\Delta N_\nu = f_\nu \Delta V_{\mathrm{phase}} \quad . \tag{A6}$$

According to Liouville theorem that the phase volume is kept constant along the geodesic, we get an equation for the evolution of neutrino distribution function;

$$\frac{df_\nu}{d\lambda'} = \left(\frac{df_\nu}{d\lambda'}\right)_{\mathrm{coll}} \quad . \tag{A7}$$

A.1 Equation of motion and Boltzmann equation

Giving a form of the metric and expressions for $T^{\alpha\beta}$, ${V_B}^\alpha$ and ${V_l}^\alpha$, one can write down the above equations in practical forms. For the spherically symmetric case, we consider the comoving frame of the matter (fluid frame) and take the enclosed baryon mass m (atomic mass unit m_u times the number of enclosed baryons) of each shell as a radial coordinate. We use the following metric;

$$\begin{aligned} ds^2 &= g_{\alpha\beta} dx^\alpha dx^\beta \\ &= \mathrm{e}^{2\phi(t,m)} c^2 dt^2 - \mathrm{e}^{2\lambda(t,m)} \left(\frac{G}{c^2} dm\right)^2 - r(t,m)^2 (d\theta^2 + \sin^2\theta d\varphi^2) \quad . \end{aligned} \tag{A8}$$

Baryon current can be written as

$$ {V_B}^\alpha = n_B v^\alpha \tag{A9}$$

where n_B is the baryon number density and v^α is the fluid 4-velocity which is, in the comoving frame, $(\mathrm{e}^{-\phi}/c, 0, 0, 0)$. Eq. A3 leads to

$$\frac{\partial (r^2 \mathrm{e}^\lambda n_B)}{\partial t} = 0 \tag{A10}$$

and equivalently

$$\mathrm{e}^\lambda = \frac{C'(m)}{r^2 n_B} \tag{A11}$$

where $C'(m)$ is a constant independent of t. On the other hand, from Eq. A8, the proper length along the radial direction dr' can be written as

$$dr' = \mathrm{e}^\lambda \frac{G}{c^2} dm \quad . \tag{A12}$$

Since

$$4\pi r^2 dr' m_u n_B = dm \tag{A13}$$

according to the meaning of n_B, $C'(m)$ must be

$$C'(m) = \frac{c^2}{4\pi G m_u} \tag{A14}$$

and hence we obtain the relation

$$e^{\lambda} = \frac{1}{4\pi r^2 \rho_B}\frac{c^2}{G} \tag{A15}$$

where

$$\rho_B \equiv m_u n_B \tag{A16}$$

is the baryon density.

As for the energy-momentum tensor, we assume that it can be divided into that of neutrinos $T_{\nu}{}^{\alpha\beta}$ ($\nu = \nu_e, \bar{\nu}_e, \nu_\mu, \bar{\nu}_\mu, \nu_\tau, \bar{\nu}_\tau$) and that of matter $T_{\rm m}{}^{\alpha\beta}$, that is

$$T^{\alpha\beta} = \sum_{\nu} T_{\nu}{}^{\alpha\beta} + T_{\rm m}{}^{\alpha\beta} \quad . \tag{A17}$$

For neutrinos with distribution function f_ν, the energy-momentum tensor becomes [274]

$$T_{\nu}{}^{\alpha\beta} \equiv \frac{c g_\nu}{h^3}\int f_\nu \, p^\alpha p^\beta \sqrt{-g}\, \frac{dp^m dp^\theta dp^\varphi}{p_t} \quad . \tag{A18}$$

In our case, the freedom of spin $g_\nu = 1$ and

$$\sqrt{-g} = \frac{G}{c} e^{\phi+\lambda} r^2 \sin\theta \quad . \tag{A19}$$

Since it is convenient to express f_ν as a function of (t, m, p, μ) where p is neutrino momentum in the comoving fluid frame and μ is cosine of the angle between the neutrino momentum and the radial direction in the fluid frame, we should express the neutrino 4-momentum in terms of the independent variables (p, μ, φ') in the spherical coordinate which takes the radial direction as its polar axis. To do so, we consider the following tetrad representation;

$$g_{\alpha\beta} = e_\alpha{}^a e_\beta{}^b \eta_{ab} \tag{A20}$$

$$\eta_{ab} = e_a{}^\alpha e_b{}^\beta g_{\alpha\beta} \tag{A21}$$

where $\eta_{ab} = diag(c^2, -1, -1, -1)$. The non-zero components are

$$e_t{}^0 = e^{\phi},\; e_m{}^1 = \tfrac{G}{c^2} e^{\lambda}, \quad e_\theta{}^2 = r,\; e_\varphi{}^3 = r\sin\theta \tag{A22}$$

$$e_0{}^t = e^{-\phi},\; e_1{}^m = \tfrac{c^2}{G} e^{-\lambda}, \quad e_2{}^\theta = \frac{1}{r},\; e_3{}^\varphi = \frac{1}{r\sin\theta} \quad . \tag{A23}$$

With

$$p^0 = p/c, \quad p^1 = p\mu, \quad p^2 = p\sqrt{1-\mu^2}\cos\varphi', \quad p^3 = p\sqrt{1-\mu^2}\sin\varphi', \tag{A24}$$

we obtain

$$p^t = \frac{1}{c}\,\mathrm{e}^{-\phi}p, \tag{A25}$$

$$p^m = \frac{c^2}{G}\,\mathrm{e}^{-\lambda}p\mu, \tag{A26}$$

$$p^\theta = \frac{1}{r}\,p\sqrt{1-\mu^2}\cos\varphi', \tag{A27}$$

$$p^\varphi = \frac{1}{r\sin\theta}\,p\sqrt{1-\mu^2}\sin\varphi'. \tag{A28}$$

Substituting these expressions into Eq. A18, we get

$$T_\nu{}^{tt} = \frac{1}{c^2}\,\mathrm{e}^{-2\phi}u_\nu \tag{A29}$$

$$T_\nu{}^{tm} = \frac{1}{G}\,\mathrm{e}^{-\phi-\lambda}F_\nu \tag{A30}$$

$$T_\nu{}^{mm} = \frac{c^4}{G^2}\,\mathrm{e}^{-2\lambda}p_\nu \tag{A31}$$

$$T_\nu{}^{\theta\theta} = \frac{1}{r^2}\frac{u_\nu - p_\nu}{2} \tag{A32}$$

$$T_\nu{}^{\varphi\varphi} = \frac{1}{r^2\sin^2\theta}\frac{u_\nu - p_\nu}{2} \tag{A33}$$

where the energy density u_ν, energy flux F_ν and pressure p_ν of neutrinos are defined as

$$u_\nu \equiv \frac{1}{h^3}\int f_\nu p c d^3p \tag{A34}$$

$$F_\nu \equiv \frac{c}{h^3}\int f_\nu p c \mu d^3p \tag{A35}$$

$$p_\nu \equiv \frac{1}{h^3}\int f_\nu p c \mu^2 d^3p \ . \tag{A36}$$

The matter other than neutrinos is assumed as perfect fluid and the energy-momentum tensor of the matter is taken as

$$T_{\mathrm{m}}{}^{\alpha\beta} \equiv (\rho_{\mathrm{m}}c^2 + p_{\mathrm{m}})v^\alpha v^\beta - p_{\mathrm{m}}g^{\alpha\beta} \tag{A37}$$

where

$$\rho_{\mathrm{m}}c^2 \equiv \rho_B c^2 + u_{\mathrm{m}}, \tag{A38}$$

and $u_{\rm m}, p_{\rm m}$ are the internal energy density and pressure of the matter, respectively. The non-zero components are

$$T_{\rm m}{}^{tt} = \frac{1}{c^2}\,{\rm e}^{-2\phi}\rho_{\rm m}c^2 \tag{A39}$$

$$T_{\rm m}{}^{mm} = \frac{c^4}{G^2}\,{\rm e}^{-2\lambda}p_{\rm m} \tag{A40}$$

$$T_{\rm m}{}^{\theta\theta} = \frac{1}{r^2}\,p_{\rm m} \tag{A41}$$

$$T_{\rm m}{}^{\varphi\varphi} = \frac{1}{r^2\sin^2\theta}\,p_{\rm m} \quad . \tag{A42}$$

Using the above expressions for the metric tensor and for the energy-momentum tensor, ${}^t{}_m$ component of the Einstein equation (Eq. A1) becomes

$$\frac{2}{c^2 r}\left(\frac{\partial\phi}{\partial m}\frac{\partial r}{\partial t} + \frac{\partial\lambda}{\partial t}\frac{\partial r}{\partial m} - \frac{\partial^2 r}{\partial t\partial m}\right){\rm e}^{-2\phi} = -\frac{8\pi G^2}{c^8}{\rm e}^{\lambda-\phi}\sum_\nu F_\nu \tag{A43}$$

and reduces to

$${\rm e}^{-\phi}\frac{\partial\lambda}{\partial t} = \frac{\partial U}{\partial m}\Big/\frac{\partial r}{\partial m} - \frac{G}{c^4}\frac{4\pi r\sum_\nu F_\nu}{\Gamma} \tag{A44}$$

where

$$U \equiv {\rm e}^{-\phi}\frac{\partial r}{\partial t} \tag{A45}$$

$$\Gamma \equiv \frac{c^2}{G}{\rm e}^{-\lambda}\frac{\partial r}{\partial m} = 4\pi r^2\rho_B\frac{\partial r}{\partial m} \quad . \tag{A46}$$

In the same way, ${}^t{}_t$ component of the Einstein equation becomes

$$\frac{1}{r^2}\left(1+\left(\frac{U}{c}\right)^2 - \Gamma^2\right) + \frac{2}{r}\left(\frac{U}{c^2}{\rm e}^{-\phi}\frac{\partial\lambda}{\partial t} - \frac{c^2}{G}{\rm e}^{-\lambda}\frac{\partial\Gamma}{\partial m}\right) = \frac{8\pi G}{c^4}(\rho_{\rm m}c^2 + \sum_\nu u_\nu) \tag{A47}$$

When we define f_{MG} as

$$f_{MG} \equiv \frac{rc^2}{2G}\left(1+\left(\frac{U}{c}\right)^2 - \Gamma^2\right), \tag{A48}$$

we obtain using Eqs. A43, A47 and A46

$$\frac{\partial f_{MG}}{\partial m} = 4\pi r^2\frac{\partial r}{\partial m}\left(\rho_{\rm m} + \frac{\sum_\nu u_\nu}{c^2} + \frac{U\sum_\nu F_\nu}{c^4\Gamma}\right) \quad . \tag{A49}$$

Here we can define the gravitational mass $m_G(t,m)$ as

$$m_G(t,m) \equiv \int_0^{r(t,m)} 4\pi r^2 \left(\rho_{\rm m} + \frac{\sum_\nu u_\nu}{c^2} + \frac{U\sum_\nu F_\nu}{c^4\Gamma}\right) dr \tag{A50}$$

$$= \frac{rc^2}{2G}\left(1 + \left(\frac{U}{c}\right)^2 - \Gamma^2\right), \tag{A51}$$

and obtain the relations;

$$\frac{\partial m_G}{\partial m} = 4\pi r^2 \frac{\partial r}{\partial m}\left(\rho_{\rm m} + \frac{\sum_\nu u_\nu}{c^2} + \frac{U\sum_\nu F_\nu}{c^4\Gamma}\right) \tag{A52}$$

$$\Gamma = \sqrt{1 + \left(\frac{U}{c}\right)^2 - \frac{2Gm_G}{rc^2}} \ . \tag{A53}$$

The equation of motion can be derived from ${}^m{}_m$ component of the Einstein equation

$$\frac{2}{rc^2}e^{-\phi}\frac{\partial U}{\partial t} - \frac{2c^2}{rG}e^{-\lambda}\frac{\partial \phi}{\partial m}\Gamma + \frac{1}{r^2}\left(1 + \left(\frac{U}{c}\right)^2 - \Gamma^2\right) = -\frac{8\pi G}{c^4}(\sum_\nu p_\nu + p_{\rm m}) \tag{A54}$$

as

$$\frac{\partial U}{\partial t} = e^{\phi}\left[-\frac{Gm_G}{r^2} - \frac{4\pi r G}{c^2}(p_{\rm m} + \sum_\nu p_\nu) + \frac{c^4}{G}\Gamma e^{-\lambda}\frac{\partial \phi}{\partial m}\right] \ . \tag{A55}$$

Note that $\partial\phi/\partial m$ is related to the pressure gradient via the equation

$$T_{\rm m}{}^{\alpha}{}_{m\ ;\ \alpha} + \sum_\nu T_\nu{}^{\alpha}{}_{m\ ;\ \alpha} = 0 \tag{A56}$$

derived from Eq. A2 where

$$T_{\rm m}{}^{\alpha}{}_{m\ ;\ \alpha} = -\frac{\partial\phi}{\partial m}(\rho_{\rm m}c^2 + p_{\rm m}) - \frac{\partial p_{\rm m}}{\partial m} \tag{A57}$$

$$T_\nu{}^{\alpha}{}_{m\ ;\ \alpha} = -\frac{\partial\phi}{\partial m}(u_\nu + p_\nu) - \frac{\partial p_\nu}{\partial m} - \frac{1}{4\pi r^2 \rho_B}$$
$$\times\left[\frac{\Gamma}{r}(3p_\nu - u_\nu) + \frac{1}{c^2}\left\{e^{-\phi}\frac{\partial F_\nu}{\partial t} + 2\rho_B\frac{\partial}{\partial t}\left(\frac{1}{\rho_B}\right)e^{-\phi}F_\nu - 2\frac{U}{r}F_\nu\right\}\right] \tag{A58}$$

Meanwhile, the equation for the energy conservation becomes

$$T_{\rm m}{}^{\alpha}{}_{t\ ;\ \alpha} + \sum_\nu T_\nu{}^{\alpha}{}_{t\ ;\ \alpha} = 0 \tag{A59}$$

where

$$T_{\mathrm{m}}{}^{\alpha}{}_{t\,;\,\alpha} = \rho_B \left[\frac{\partial}{\partial t}\left(\frac{u_{\mathrm{m}}}{\rho_B}\right) + p_{\mathrm{m}} \frac{\partial}{\partial t}\left(\frac{1}{\rho_B}\right) \right] \tag{A60}$$

$$\begin{aligned} T_{\nu}{}^{\alpha}{}_{t\,;\,\alpha} &= \rho_B \left[\frac{\partial}{\partial t}\left(\frac{u_{\nu}}{\rho_B}\right) + p_{\nu} \frac{\partial}{\partial t}\left(\frac{1}{\rho_B}\right) \right. \\ &\quad \left. + \mathrm{e}^{-\phi} \frac{\partial(4\pi r^2 F_{\nu} e^{2\phi})}{\partial m} + \frac{U \mathrm{e}^{\phi}}{r\rho_B}(u_{\nu} - 3p_{\nu}) \right] \quad . \end{aligned} \tag{A61}$$

Next let us derive the general relativistic Boltzmann equation for the neutrino distribution function f_ν from Eq. A7. Regarding f_ν as a function of the coordinate $x^\alpha(\alpha = t, m, \theta.\varphi)$ and the momentum measured in the comoving fluid frame $p^a(a = 1,2,3)$, the left hand side of Eq. A7 becomes

$$\frac{df_\nu}{d\lambda'} = \frac{dx^\alpha}{d\lambda'}\frac{\partial f_\nu}{\partial x^\alpha} + \frac{dp^a}{d\lambda'}\frac{\partial f_\nu}{\partial p^a} \tag{A62}$$

and, from the geodesic equations, we get

$$\frac{dx^\alpha}{d\lambda'} = p^\alpha \tag{A63}$$

$$\frac{dp^a}{d\lambda'} = -\Gamma^a_{bc}\, p^b p^c \quad . \tag{A64}$$

Here Ricci rotation coefficient Γ^a_{bc} is defined as

$$\Gamma^a_{bc} \equiv e_\alpha{}^a e_b{}^\beta e_c{}^\gamma \Gamma^\alpha_{\beta\gamma} + e_\alpha{}^a e_c{}^\beta \frac{\partial e_b{}^\alpha}{\partial x^\beta} \tag{A65}$$

where $\Gamma^\alpha_{\beta\gamma}$ is the Christoffel symbol. Noting that p^0 is not independent of p^1, p^2, p^3 and that f_ν depends only on x^α, p^1, p^2, p^3, substitution of Eqs. A63 and A64 into Eq. A62 and succeeding tedious calculation lead to

$$\begin{aligned} \frac{df_\nu}{d\lambda'} &= p^a e_a{}^\alpha \frac{\partial f_\nu}{\partial x^\alpha} - \Gamma^1_{bc}\, p^b p^c \frac{\partial f_\nu}{\partial p^1} - \Gamma^2_{bc}\, p^b p^c \frac{\partial f_\nu}{\partial p^2} - \Gamma^3_{bc}\, p^b p^c \frac{\partial f_\nu}{\partial p^3} \\ &= \frac{1}{c}\mathrm{e}^{-\phi} p \frac{\partial f_\nu}{\partial t} + \frac{c^2}{G}\mathrm{e}^{-\lambda} p\mu \frac{\partial f_\nu}{\partial m} \\ &\quad - p^2 \left\{ \frac{1}{c}\mathrm{e}^{-\phi}\mu^2 \frac{\partial \lambda}{\partial t} + \frac{c^2}{G}\mathrm{e}^{-\lambda}\mu \frac{\partial \phi}{\partial m} + (1-\mu^2)\frac{U}{cr} \right\} \frac{\partial f_\nu}{\partial p} \\ &\quad + p(1-\mu^2) \left\{ \frac{\Gamma}{r} + \frac{U}{cr}\mu - \frac{1}{c}\mathrm{e}^{-\phi}\frac{\partial \lambda}{\partial t}\mu - \frac{c^2}{G}\mathrm{e}^{-\lambda}\frac{\partial \phi}{\partial m} \right\} \frac{\partial f_\nu}{\partial \mu} \quad . \end{aligned} \tag{A66}$$

On the contrary, the right hand side of Eq. A7, *i.e.* the collision term, can be written in terms of the interaction rate. Denoting the proper time on the comoving fluid frame as t_p, the right hand side becomes

$$\left(\frac{df_\nu}{d\lambda'}\right)_{\rm coll} = \frac{dt}{d\lambda'}\left(\frac{df_\nu}{dt}\right)_{\rm coll}$$
$$= \frac{1}{c}{\rm e}^{-\phi}p\frac{dt_p}{dt}\left(\frac{df_\nu}{dt_p}\right)_{\rm coll}$$
$$= p\frac{1}{c}\left(\frac{df_\nu}{dt_p}\right)_{\rm coll} \quad . \tag{A67}$$

In the comoving fluid frame, the change of f_ν per unit proper time by various weak interactions including absorption, emission and scattering can be expressed using emissivity $j_{\rm emit}$, absorption mean free path $\lambda_{\rm abs}$ and so on as

$$\frac{1}{c}\left(\frac{df_\nu(t,m,p,\mu)}{dt_p}\right)_{\rm coll}$$
$$= j_{\rm emit}(t,m,p,\mu)(1-f_\nu(t,m,p,\mu)) - \frac{1}{\lambda_{\rm abs}(t,m,p,\mu)}f_\nu(t,m,p,\mu)$$
$$+ \int R^{\rm scat}_{\rm in}(t,m,p,\mu,p',\mu')f_\nu(t,m,p',\mu')(1-f_\nu(t,m,p,\mu))dp'd\mu'$$
$$- \int R^{\rm scat}_{\rm out}(t,m,p,\mu,p',\mu')f_\nu(t,m,p,\mu)(1-f_\nu(t,m,p',\mu'))dp'd\mu'$$
$$+ \int R^{\rm pair}_{\rm creation}(t,m,p,\mu,p',\mu')(1-f_\nu(t,m,p,\mu))(1-f_{\bar\nu}(t,m,p',\mu'))dp'd\mu'$$
$$- \int R^{\rm pair}_{\rm annihilation}(t,m,p,\mu,p',\mu')f_\nu(t,m,p,\mu)f_{\bar\nu}(t,m,p',\mu'))dp'd\mu' \tag{A68}$$

where $R^{\rm scat}_{\rm in}(t,m,p,\mu,p',\mu')$ is the scattering rate from the momentum with (p',μ') to (p,μ), $R^{\rm scat}_{\rm out}(t,m,p,\mu,p',\mu')$ is the scattering rate from the momentum with (p,μ) to (p',μ'), $R^{\rm pair}_{\rm creation}(t,m,p,\mu,p',\mu')$ is the pair creation rate for the neutrinos with the momentum (p,μ) and the anti-neutrino with (p',μ'), $R^{\rm pair}_{\rm annihilation}$ (t,m,p,μ,p',μ') is the pair annihilation rate of the neutrinos with the momentum (p,μ) and the anti-neutrino with (p',μ') and $(1-f_\nu)$ denotes the Pauli blocking factor. In the followings, we use a symbolic expression

$$\left(\frac{df_\nu}{d\lambda'}\right)_{\rm coll} = p\left(j_\nu - \frac{1}{\lambda_\nu}f_\nu\right) \tag{A69}$$

instead of Eq. A68 for simplicity.

After all, the general relativistic Boltzmann equation for the spherical model becomes

$$\frac{1}{c}{\rm e}^{-\phi}p\frac{\partial f_\nu}{\partial t} + \frac{c^2}{G}{\rm e}^{-\lambda}p\mu\frac{\partial f_\nu}{\partial m} - p^2\left\{\frac{1}{c}{\rm e}^{-\phi}\mu^2\frac{\partial\lambda}{\partial t} + \frac{c^2}{G}{\rm e}^{-\lambda}\mu\frac{\partial\phi}{\partial m} + (1-\mu^2)\frac{U}{cr}\right\}\frac{\partial f_\nu}{\partial p}$$

$$+p(1-\mu^2)\left\{\frac{\Gamma}{r}+\frac{U}{cr}\mu-\frac{1}{c}e^{-\phi}\frac{\partial\lambda}{\partial t}\mu-\frac{c^2}{G}e^{-\lambda}\frac{\partial\phi}{\partial m}\right\}\frac{\partial f_\nu}{\partial\mu}$$

$$= p\left(j_\nu-\frac{1}{\lambda_\nu}f_\nu\right) \quad . \tag{A70}$$

This equation determines the time evolution, spatial flow, spectral change, and angular distribution of the neutrinos in the supernova core.

Note that integrations of the Boltzmann equation over the momentum space result in

$$\rho_B\left[\frac{\partial}{\partial t}\left(\frac{u_m}{\rho_B}\right)+p_m\frac{\partial}{\partial t}\left(\frac{1}{\rho_B}\right)\right] = -\sum_\nu T_\nu{}^\alpha{}_{t\,;\,\alpha}$$

$$= -e^\phi c^2\sum_\nu\int p\left(j_\nu-\frac{1}{\lambda_\nu}f_\nu\right)\frac{d^3p}{h^3} \tag{A71}$$

$$\frac{\partial\phi}{\partial m}(\rho_m c^2+p_m)+\frac{\partial p_m}{\partial m} = \sum_\nu T_\nu{}^\alpha{}_{m\,;\,\alpha}$$

$$= -e^\lambda c^2\sum_\nu\int p\mu\left(j_\nu-\frac{1}{\lambda_\nu}f_\nu\right)\frac{d^3p}{h^3}\,. \tag{A72}$$

The first equation means that the exchange of energy between the matter and neutrinos via weak interactions cause the change of the entropy of the matter.

As for the lepton number conservation, we define the lepton current for three lepton families ($l = e, \mu, \tau$) as

$$V_l{}^\alpha \equiv \sum_{l'=l^\pm,\nu_l,\bar\nu_l} \ell_{l'}\frac{g_{l'}}{h^3}\int p_{l'}^\alpha f_{l'}\sqrt{-g}\frac{dp_{l'}^m dp_{l'}^\theta dp_{l'}^\varphi}{p_{l't}} \tag{A73}$$

where the spin freedom $g_{l'}$ is 2 for $l^\pm$ and 1 for ν_l and $\bar\nu_l$, lepton number $\ell_{l'}$ is 1 for l^- and ν_l and -1 for l^+ and $\bar\nu_l$. This can be rewritten as

$$V_l{}^t = \frac{1}{c}e^{-\phi}(n_{l^-}-n_{l^+}+n_{\nu_l}-n_{\bar\nu_l}) \tag{A74}$$

$$V_l{}^m = \frac{c}{G}e^{-\lambda}(Fn_{\nu_l}-Fn_{\bar\nu_l}) \tag{A75}$$

where

$$n_{l'} \equiv \frac{g_{l'}}{h^3}\int f_{l'}d^3p \tag{A76}$$

in the particle's number density and

$$Fn_{l'} \equiv c\frac{g_{l'}}{h^3}\int f_{l'}\mu d^3p \tag{A77}$$

is the particle's number flux. The equation for the lepton number conservation Eq. A4 leads to

$$\frac{1}{c}\frac{\partial}{\partial t}\left[r^2 e^{\lambda}(n_{l^-}-n_{l^+})\right]$$
$$=-\frac{1}{c}\frac{\partial}{\partial t}\left[r^2 e^{\lambda}(n_{\nu_l}-n_{\bar{\nu}_l})\right]-\frac{c}{G}\frac{\partial}{\partial m}\left[r^2 e^{\phi}(Fn_{\nu_l}-Fn_{\bar{\nu}_l})\right] \quad . \tag{A78}$$

On the other hand, since integration of the Boltzmann equation over the momentum space after dividing by p yields

$$\int\left(j_\nu-\frac{1}{\lambda_\nu}f_\nu\right)\frac{d^3p}{h^3}=\frac{1}{r^2}e^{-\phi-\lambda}\left(\frac{1}{c}\frac{\partial(r^2e^{\lambda}n_\nu)}{\partial t}+\frac{c}{G}\frac{\partial(r^2e^{\phi}Fn_\nu)}{\partial m}\right), \tag{A79}$$

we obtain

$$e^{-\phi}\frac{1}{c}\frac{\partial Y_l}{\partial t}=-\frac{1}{n_B}\left[\int\left(j_{\nu_l}-\frac{1}{\lambda_{\nu_l}}f_{\nu_l}\right)\frac{d^3p}{h^3}-\int\left(j_{\bar{\nu}_l}-\frac{1}{\lambda_{\bar{\nu}_l}}f_{\bar{\nu}_l}\right)\frac{d^3p}{h^3}\right] \tag{A80}$$

where the lepton fraction per baryon Y_l is defined as

$$Y_l\equiv\frac{n_{l^-}-n_{l^+}}{n_B} \quad . \tag{A81}$$

In addition to the above equations, EOS and weak interaction theory should be used in order to calculate $p_{\rm m}$, j_ν and λ_ν. Then the system of equations is closed. With appropriate boundary conditions, one can calculate the hydrodynamical evolution of the core and the neutrino transfer in the core.

As a summary, variables to be calculated and equations to be used for the calculation are listed in Table 2.

A.2 Multigroup flux limited diffusion scheme

So far we discussed the equations for hydrodynamics and neutrino transfer in the spherically symmetric space-time with minimal assumption. The present facilities for numerical simulations enable us to solve the Boltzmann equation for neutrinos directly with some cost [94, 129]. One should calculate the neutrino distribution function f_ν for various times t, for various position m, for various energy $\omega=pc$, and for various angle μ. Therefore, it is more popular to solve the transport equations in the angle-integrated form, that is, one regards the neutrino distribution as a function of t, m, ω with extra assumptions. Hereafter we derive equations for this multigroup (multi-energy) scheme [47, 62, 67, 115, 275, 279, 280].

For every neutrino energy, let us define the neutrino energy density $u_\nu(\omega)$, energy flux $F_\nu(\omega)$, and the second moment $p_\nu(\omega)$ as

$$u_\nu(\omega)\ \equiv\ \frac{\omega^3}{h^3c^3}\int f_\nu d\mu d\varphi' \tag{A82}$$

Variable	Equation	dependencies
$U(t,m)$	A55	$\phi, r, m_G, p_{\rm m}, p_\nu, \Gamma, \lambda$
$r(t,m)$	A45	ϕ, U
$\lambda(t,m)$	A44	$\phi, U, r, \Gamma, F_\nu$
$\phi(t,m)$	A72	$\rho_{\rm m}, p_{\rm m}, \lambda, j_\nu, \lambda_\nu, f_\nu$
$\Gamma(t,m)$	A46	r, ρ_B
$m_G(t,m)$	A52	$r, \rho_{\rm m}, u_\nu, U, \Gamma, F_\nu$
$\rho_B(t,m)$	A15	r, λ
$n_B(t,m)$	A16	ρ_B
$u_{\rm m}(t,m)$	A71	$\rho_B, p_{\rm m}, \phi, j_\nu, \lambda_\nu, f_\nu$
$\rho_{\rm m}(t,m)$	A38	$\rho_B, u_{\rm m}$
$Y_l(t,m)$	A80	$\phi, n_B, j_\nu, \lambda_\nu, f_\nu$
$p_{\rm m}(t,m)$	EOS	$u_{\rm m}, \rho_B, Y_l$
$T(t,m)$	EOS	$u_{\rm m}, \rho_B, Y_l$
$j_\nu(t,m,p,\mu)$	weak int. theory	T, Y_l, ρ_B, f_ν
$\lambda_\nu(t,m,p,\mu)$	weak int. theory	T, Y_l, ρ_B, f_ν
$f_\nu(t,m,p,\mu)$	A70	$\phi, \lambda, r, U, \Gamma, j_\nu, \lambda_\nu, f_\nu$
$u_\nu(t,m)$	A34	f_ν
$F_\nu(t,m)$	A35	f_ν
$p_\nu(t,m)$	A36	f_ν

Table 2: Summary of transfer scheme using Boltzmann equation.

$$F_\nu(\omega) \equiv c\frac{\omega^3}{h^3c^3}\int f_\nu \mu d\mu d\varphi' \tag{A83}$$

$$p_\nu(\omega) \equiv \frac{\omega^3}{h^3c^3}\int f_\nu \mu^2 d\mu d\varphi' \tag{A84}$$

Integrating the Boltzmann equation Eq. A70 over the angle, we obtain

$$\begin{aligned}
&\frac{1}{c}\mathrm{e}^{-\phi}\frac{\partial u_\nu(\omega)}{\partial t} + \frac{c}{G}\mathrm{e}^{-\lambda}\frac{\partial F_\nu(\omega)}{\partial m} \\
&-\omega\left[\frac{c}{G}\mathrm{e}^{-\lambda}\frac{\partial\phi}{\partial m}\frac{\partial F_\nu(\omega)}{\partial\omega} + \frac{1}{c}\mathrm{e}^{-\phi}\frac{\partial\lambda}{\partial t}\frac{\partial p_\nu(\omega)}{\partial\omega} + \frac{U}{rc}\frac{\partial u_\nu(\omega)}{\partial\omega} - \frac{U}{rc}\frac{\partial p_\nu(\omega)}{\partial\omega}\right] \\
&+\frac{1}{c}F_\nu(\omega)\left(\frac{2\Gamma}{r} + \frac{c^2}{G}\mathrm{e}^{-\lambda}\frac{\partial\phi}{\partial m}\right) + u_\nu(\omega)\left(\frac{2U}{rc} + \frac{1}{c}\mathrm{e}^{-\phi}\frac{\partial\lambda}{\partial t}\right) \\
=\ &\frac{\omega^3}{h^3c^3}\int\left(j_\nu - \frac{1}{\lambda_\nu}f_\nu\right)d\mu d\varphi' \quad .
\end{aligned} \tag{A85}$$

If one integrates the Boltzmann equation after multiplying μ^n, one would obtain an equation including n-th moment, $(n+1)$-th moment, and $(n+2)$-th moment. That is, we need some approximations in order to close the systems of equations. The simplest method is assuming that the neutrino angular distribution in momentum space is nearly isotropic, that is f_ν is assumed to be the form

$$f_\nu(t,m,\omega,\mu) = f_\nu^{(0)}(t,m,\omega) + f_\nu^{(1)}(t,m,\omega)\mu \quad . \tag{A86}$$

This is the case for neutrinos in the inner core where neutrinos are in thermal equilibrium with the matter by virtue of the frequent interactions. Substituting this form into the Boltzmann equation and integrating it over the angle after multiplying μ, we obtain

$$\frac{c^2}{3G}\mathrm{e}^{-\lambda}\left[\frac{\partial u_\nu(\omega)}{\partial m} - \omega\frac{\partial\phi}{\partial m}\frac{\partial u_\nu(\omega)}{\partial\omega} + 3\frac{\partial\phi}{\partial m}u_\nu(\omega)\right] = -\frac{1}{c\lambda_\nu}F_\nu(\omega) \tag{A87}$$

where we consider the static case with dropping the terms containing U and $\partial/\partial t$ and assume that both j_ν and λ_ν are isotropic and we use the relation $u_\nu(\omega) = 3p_\nu(\omega)$ coming from the assumption Eq. A86. In addition, when we introduce a new independent variable

$$\omega_0 \equiv \omega\mathrm{e}^{\phi(t,m)} \tag{A88}$$

which corresponds to the red-shifted energy at infinite distance in place of ω, this result can be written in the form

$$\begin{aligned} F_\nu(\omega) &= -\frac{c\lambda_\nu}{3}\frac{c^2}{G}\mathrm{e}^{-\lambda}\mathrm{e}^{-3\phi}\left(\frac{\partial}{\partial m}u_\nu(\omega)\mathrm{e}^{3\phi}\right)_{t,\omega_0} \\ &= -\frac{c\lambda_\nu}{3}\mathrm{e}^{-3\phi}\Gamma\left(\frac{\partial}{\partial r}u_\nu(\omega)\mathrm{e}^{3\phi}\right)_{t,\omega_0} \quad . \end{aligned} \tag{A89}$$

This is just the Fick's law in the diffusion theory that the flux is proportional to the gradient of density. Although, when the interactions between the matter and neutrinos are so frequent to make the neutrino angular distribution isotropic, this diffusion approximation is valid, it is unphysical to use the above expression of the flux in the limit of free streaming flux. That is, in the case with very long (nearly infinite) mean free path (large λ_ν), the diffusion flux exceeds the causal limit of $cu_\nu(\omega)$. In order to manage this annoying problem, various types of flux limiters are contrived and used [62, 280, 281]. The simplest flux limiter is

$$\Lambda_\nu \equiv \frac{1}{\frac{1}{\lambda_\nu} + \frac{\left|\frac{1}{3}\mathrm{e}^{-3\phi}\Gamma\left(\frac{\partial}{\partial r}u_\nu(\omega)\mathrm{e}^{3\phi}\right)_{t,\omega_0}\right|}{u_\nu(\omega)}} \tag{A90}$$

and the flux is defined using this flux limiter as

$$F_\nu(\omega) = -\frac{c\Lambda_\nu}{3}\mathrm{e}^{-3\phi}\Gamma\left(\frac{\partial}{\partial r}u_\nu(\omega)\mathrm{e}^{3\phi}\right)_{t,\omega_0} \quad . \tag{A91}$$

In the diffusion limit ($\lambda_\nu \to 0$), this flux becomes the diffusion flux Eq. A89 and in the free streaming limit ($\lambda_\nu \to \infty$) this flux becomes $cu_\nu(\omega)$. Multigroup flux limited diffusion (MGFLD) scheme for the neutrino transfer is the transport scheme which solves Eq. A85 with the flux defined as Eq. A91 for various (t, m, ω). Note that MGFLD scheme can calculate neutrino energy spectrum and manage the non-thermal feature of the neutrino energy distribution.

Equations which one must solve in the case of MGFLD scheme are almost the same equations as in Table 2; differences are as follows. In place of an equation for f_ν, one should use equations A85 for $u_\nu(\omega)$, A91 for $F_\nu(\omega)$, some closure relation connecting $p_\nu(\omega)$ to $u_\nu(\omega)$, and definition of the flux limiter such as Eq. A90. u_ν, F_ν, p_ν are now expressed in terms of $u_\nu(\omega)$, $F_\nu(\omega)$, $p_\nu(\omega)$ as

$$u_\nu = \int u_\nu(\omega)d\omega \tag{A92}$$

$$F_\nu = \int F_\nu(\omega)d\omega \tag{A93}$$

$$p_\nu = \int p_\nu(\omega)d\omega \ . \tag{A94}$$

Similarly angular integration of the collision term $(j_\nu - f_\nu/\lambda_\nu)$ in some equations in Table 2 should be rewritten in terms of $u_\nu(\omega)$, $F_\nu(\omega)$, $p_\nu(\omega)$.

Note that we can rewrite Eq. A85 with taking t, m, ω_0 as independent variables instead of t, m, ω. We got

$$\begin{aligned}
&\frac{1}{c}\mathrm{e}^{-\phi}\left(\frac{\partial u_\nu(\omega_0\mathrm{e}^{-\phi})}{\partial t}\right)_{m,\omega_0} + \frac{1}{c}\mathrm{e}^{-\phi}\frac{\partial\phi}{\partial t}\omega_0\frac{\partial u_\nu(\omega_0\mathrm{e}^{-\phi})}{\partial\omega_0} \\
&+\frac{c}{Gr^2}\mathrm{e}^{-\lambda-\phi}\left(\frac{\partial\mathrm{e}^{\phi}r^2F_\nu(\omega_0\mathrm{e}^{-\phi})}{\partial m}\right)_{t,\omega_0} - \frac{1}{c}\mathrm{e}^{-\phi}\frac{1}{\rho_B}\frac{\partial\rho_B}{\partial t}u_\nu(\omega_0\mathrm{e}^{-\phi}) \\
&-\left(r\frac{\partial}{\partial r}\left(\frac{U}{cr}\right) - \frac{4\pi rG\sum_{\nu'}F_{\nu'}(\omega_0\mathrm{e}^{-\phi})}{\Gamma c^5}\right)\omega_0\frac{\partial p_\nu(\omega_0\mathrm{e}^{-\phi})}{\partial\omega_0} - \frac{U}{rc}\omega_0\frac{\partial u_\nu(\omega_0\mathrm{e}^{-\phi})}{\partial\omega_0} \\
=& \ \frac{\omega^3}{h^3c^3}\int\left(j_\nu - \frac{1}{\lambda_\nu}f_\nu\right)d\mu d\varphi' \ .
\end{aligned} \tag{A95}$$

This is the same as Eq. (13) in p.40 of Mayle (1985) [81] except for the term $4\pi rG\sum_{\nu'}F_{\nu'}(\omega_0\mathrm{e}^{-\phi})/\Gamma c^5$.

A.3 More simplified transfer scheme

Furthermore, if one integrates Eq. A85 over the neutrino energy, one obtains the transport equation for energy density of neutrinos. That is, one has only to calculate the energy density as a function of t and m. In this case, however, one should make an additional assumption as for the energy distribution of neutrinos, for example, Fermi-Dirac distribution with chemical potential and the matter temperature (LTE: local thermodynamic equilibrium) or with temperature independent of the matter temperature (NEFD: non-equilibrium Fermi-Dirac). Energy-integrated flux limited diffusion (FLD) scheme which introduces a flux limiter in

Reference	Hydro	ν transfer				EOS
		scheme	ν	GR	NES	$> \rho_{nuc}$
MCK80 [50]	D	LTEFLD	only ν_e			BBP+
HNW84 [57]	GR D	NEFDFLD	only ν_e			HW
Bruenn85 [62]	GR D	MGFLD	ν_e,$\bar{\nu}_e$,'ν_μ'	RS-	direct	FP+
Mayle85 [81]	GR D C	MGFLD	ν_e,$\bar{\nu}_e$,'ν_μ'	RS-	FP	BW+
	GR D	MGFLD	ν_e,$\bar{\nu}_e$,'ν_μ'	GR	FP	BW+
BL86 [80]	GR QS	LTEFLD	ν_e~$\bar{\nu}_e$,'ν_μ'	GR		BCK
MWS87 [129]	GR D	MGFLD	ν_e,$\bar{\nu}_e$,'ν_μ'	RS-	FP	BW+
	S	B	ν_e,$\bar{\nu}_e$,'ν_μ'	no	$R^{\rm scat}_{\rm out}$-	BW+
Bruenn87 [66]	GR D	MGFLD	ν_e,$\bar{\nu}_e$,'ν_μ'	RS-	direct	
MW88 [96]	GR D C	MGFLD	ν_e,$\bar{\nu}_e$,'ν_μ'	RS-	FP	
Burrows88 [114]	GR QS A	LTEFLD	ν_e,$\bar{\nu}_e$,'ν_μ'	GR		LB
Bruenn89 [70,71]	GR D	MGFLD	ν_e,$\bar{\nu}_e$,'ν_μ'	RS-	direct	BCK
WM89 [101,102,117]	GR D	MGFLD	ν_e,$\bar{\nu}_e$,'ν_μ'	GR		MW
JH89 [130]	S	MC	ν_e,$\bar{\nu}_e$,'ν_μ'	no	direct	
Suzuki [113,115]	GR QS	MGFLD	ν_e,$\bar{\nu}_e$,'ν_μ'	GR	direct	HW
MyBu90 [74]	GR D	MGFLD	ν_e,$\bar{\nu}_e$,'ν_μ'	RS-	FP	BCK
Janka [135]	S	MC	ν_e,$\bar{\nu}_e$,'ν_μ'	GR	direct	
MeBr93 [94]	N D	B	only ν_e	no	no	BCK
MeBr93 [284]	N D	B	only ν_e	no	direct	BCK
WM93 [112]	GR D C	MGFLD	ν_e,$\bar{\nu}_e$,'ν_μ'	GR		

Table 3: Summary of computational methods.

the similar way to the MGFLD is also often used in the study of the supernova explosion and neutron star formation [50,57,64,80,106,107,114,282]. Meanwhile a method not using the flux limiter was developed by Cooperstein, van den Horn & Baron [9,178,283]. Their two-fluid method makes a use of shells relative position from the neutrinosphere in the calculation of neutrino flux and pressure from the energy density at the shell. While these energy-integrated method cannot calculate the neutrino spectrum and there are more impreciseness than MGFLD, they can be used for the study of global feature of the neutrino burst.

When one has interest in physics of supernova explosion other than neutrinos, in numerical studies neutrino transfer is sometimes omitted [136–140,142,145–151] or managed by means of very rough methods such as leakage scheme [51–54,59–61,65,68,141,143]. In the leakage scheme, one assumes that neutrinos escape from the shells in some time scales determined by the local condition such as local density. The simplest assumption is that neutrinos escape freely from the shells with density below a given trapping density and that neutrinos are completely trapped in the shells with higher density. Intensive comparisons were made by Bruenn [63] especially between the MGFLD and the leakage scheme.

A.4 Numerical simulations

Here we describe the computational scheme of some numerical simulations referred in this article briefly in tabular format. In the column 'Hydro', GR means general relativistic hydrodynamics, N is Newtonian, D means dynamical simulation, QS is simulation of quasi-static evolution of protoneutron star, S means static analysis, C and A indicate inclusion of convection or matter accretion, respectively. In the column 'ν transfer scheme', B: solving Boltzmann equation directly (neutrino field is a function of (t, m, ω, μ), MGFLD: multigroup flux limited diffusion scheme (neutrino field is a function of (t, m, ω), LTEFLD: local thermodynamic equilibrium flux limited diffusion scheme (neutrino field is a function of (t, m)), NEFDFLD: non-equilibrium Fermi-Dirac flux limited diffusion scheme (neutrino field is a function of (t, m)), MC: Monte Carlo method (neutrino field is a function of (t, m, ω, μ)). In the 'GR' column, RS- means that general relativistic corrections such as red shift effects are incomplete. no: neglecting general relativity. In the 'NES' (neutrino electron scattering) column, FP means adopting Fokker-Planck approximation for the treatment of the reaction ν e $\rightarrow$ ν e, $R^{\rm scat}_{\rm out}$- means that $R^{\rm scat}_{\rm out}$ in Eq. A68 has been averaged over p' and μ'. In the 'EOS' for the supranuclear density column, the basic part of EOS used are FP+: modified Friedman & Pandharipande [285], BBP: Baym, Bethe & Pethick [286] + ideal Fermi gas, BCK: Baron, Cooperstein & Kahana [60], BW+: modified Bowers & Wilson [280], HW: Hillebrandt & Wolff [165], LB: Lattimer & Burrows (1988), MW: Mayle & Wilson [101].

References

[1] Baade, W. and Zwicky, F. 1934, *Phys. Rev.*, **45**, 138.

[2] Baade, W. and Zwicky, F. 1934, *Phys. Rev.*, **46**, 76.

[3] Barbon, R., Cappellaro, E., and Turatto, M. 1989, The Asiago Supernova Catalogue.

[4] IAU circulars, International Astronomical Union.

[5] Bethe, H. A. 1990, *Rev. Mod. Phys.*, **62**, 801.

[6] Burrows, A. 1990, *Ann. Rev. Nucl. Part. Sci.*, **40**, 181.

[7] Burrows, A. S. 1990, in *Supernovae*, edited by Petschek, A. G., (Springer-Verlag, New York · Berlin · Heidelberg · London · Paris · Tokyo · Hong Kong, 1990), p. 143.

[8] Mayle, R. W. 1990, in *Supernovae*, edited by Petschek, A. G., (Springer-Verlag, New York · Berlin · Heidelberg · London · Paris · Tokyo · Hong Kong, 1990), p. 267.

[9] Cooperstein, J. 1988, *Phys. Rep.*, **163**, 95.

[10] Bahcall, J. N. *Neutrino Astrophysics*, chapter 15, p. 421. Cambridge University Press, Cambridge · New York · New Rochelle · Melbourne · Sydney, 1989.

[11] Koshiba, M. 1992, *Phys. Rep.*, **220**, 229.

[12] Totsuka, Y. 1992, *Rep. Prog. Phys.*, **55**, 377.

[13] Wheeler, J. C. 1990, in *supernovae*, edited by Wheeler, J. C., Piran, T., and Weinberg, S., (World Scientific, Singapore · New Jersey · London · Hong

Kong, 1990), volume 6 of *Jerusalem Winter School for Theoretical Physics*, p. 1.

[14] Woosley, S. E. and Weaver, T. A. 1986, *Ann. Rev. Astron. Astrophys.*, **24**, 205.

[15] Wheeler, J. C., Harkness, R. P., and Cappellaro, E. 1987, in *13th Texas Symposium on Relativistic Astrophysics*, edited by Ulmer, M. P., (World Scientific, Singapore, 1987), p. 402.

[16] Clayton, D. D. *Principles of Stellar Evolution and Nucleosynthesis.* The University of Chicago Press, Chicago · London, 1983.

[17] Sugimoto, D. and Nomoto, K. 1980, *Sp. Sci. Rev.*, **25**, 155.

[18] Nomoto, K. and Hashimoto, M. 1988, *Phys. Rep.*, **163**, 13.

[19] Azzopardi, M. A. 1992, in *The Astronomy and Astrophysics Encyclopedia*, edited by Maran, S. P., (Van Nostrand Reinhold, New York, 1992), p. 812.

[20] Shigeyama, T. and Nomoto, K. 1990, *Astrophys. J.*, **360**, 242.

[21] Hoyle, F. and Fowler, W. A. 1960, *Astrophys. J.*, **132**, 565.

[22] Nomoto, K., Thielemann, F.-K., and Yokoi, K. 1984, *Astrophys. J.*, **286**, 644.

[23] Woosley, S. E., Taam, R. E., and Weaver, T. A. 1986, *Astrophys. J.*, **301**, 601.

[24] Nomoto, K. *et al.* 1993, in *Proc. of the International Symposium on Neutrino Astrophysics: Frontiers of Neutrino Astrophysics*, edited by Suzuki, Y. and Nakamura, K., (Universal Academy Press Inc., Tokyo, 1993), number 5 in Frontiers Science Series, p. 235.

[25] Nomoto, K., Sugimoto, D., and Neo, S. 1976, *Astrophys. Sp. Sci.*, **39**, L37.

[26] Tubbs, D. L. and Schramm, D. N. 1975, *Astrophys. J.*, **201**, 467.

[27] Stark, R. F. and Piran, T. 1985, *Phys. Rev. Lett.*, **55**, 891.

[28] Hirata, K. *et al.* 1987, *Phys. Rev. Lett.*, **58**, 1490.

[29] Bionta, R. M. *et al.* 1987, *Phys. Rev. Lett.*, **58**, 1494.

[30] Piran, T. 1981, *Phys. Lett.*, **102B**, 299.

[31] Burbidge, E. M., Burbidge, G. R., Fowler, W. A., and Hoyle, F. 1957, *Rev. Mod. Phys.*, **29**, 547.

[32] Fowler, W. A. and Hoyle, F. 1964, *Astrophys. J. Suppl.*, **9**, 201.

[33] Colgate, S. A. and Johnson, M. H. 1960, *Phys. Rev. Lett.*, **5**, 235.

[34] Colgate, S. A. and White, R. H. 1966, *Astrophys. J.*, **143**, 626.

[35] Arnett, W. D. 1966, *Can. J. Phys.*, **44**, 2553.

[36] Arnett, D. 1967, *Can. J. Phys.*, **45**, 1621.

[37] Schwartz, R. A. 1967, *Ann. Phys.*, **43**, 42.

[38] Wilson, J. R. 1971, *Astrophys. J.*, **163**, 209.

[39] Freedman, D. Z. 1974, *Phys. Rev.*, **D9**, 1389.

[40] Wilson, J. R. 1974, *Phys. Rev. Lett.*, **32**, 849.

[41] Schramm, D. N. and Arnett, W. D. 1975, *Astrophys. J.*, **198**, 629.

[42] Sato, K. 1975, *Prog. Theor. Phys.*, **53**, 595.

[43] Sato, K. 1975, *Prog. Theor. Phys.*, **54**, 1325.

[44] Mazurek, T. J. 1974, *Nature*, **252**, 287.

[45] Mazurek, T. J. 1975, *Astrophys. Sp. Sci.*, **35**, 117.

[46] Bruenn, S. W., Arnett, W. D., and Schramm, D. N. 1977, *Astrophys. J.*, **213**, 213.

[47] Arnett, W. D. 1977, *Astrophys. J.*, **218**, 815.

[48] Brown, G. E., Bethe, H. A., and Baym, G. 1982, *Nucl. Phys.*, **A375**, 481.

[49] Wilson, J. R. 1980, *Ann. N.Y. Acad. Sci.*, **336**, 358.

[50] Mazurek, T. J., Cooperstein, J., and Kahana, S. 1980, in *DUMAND-80: Proceedings of the 1980 INTERNATIONAL DUMAND SYMPOSIUM*, edited by Stenger, V. J., (HAWAII DUMAND CENTER, Honolulu, 1980), volume II, p. 142.

[51] Hillebrandt, W. and Müller, E. 1981, *Astron. Astrophys.*, **103**, 147.

[52] Van Riper, K. A. and Lattimer, J. M. 1981, *Astrophys. J.*, **249**, 270.

[53] Van Riper, K. A. 1982, *Astrophys. J.*, **257**, 793.

[54] Hillebrandt, W. 1982, *Astron. Astrophys.*, **110**, L3.

[55] Bowers, R. and Wilson, J. R. 1982, *Astrophys. J.*, **263**, 366.

[56] Arnett, W. D. 1983, *Astrophys. J.*, **263**, L55.

[57] Hillebrandt, W., Nomoto, K., and Wolff, R. G. 1984, *Astron. Astrophys.*, **133**, 175.

[58] Cooperstein, J., Bethe, H. A., and Brown, G. E. 1984, *Nucl. Phys.*, **A429**, 527.

[59] Baron, E., Cooperstein, J., and Kahana, S. 1985, *Nucl. Phys.*, **A440**, 744.

[60] Baron, E., Cooperstein, J., and Kahana, S. 1985, *Phys. Rev. Lett.*, **55**, 126.

[61] Burrows, A. and Lattimer, J. M. 1985, *Astrophys. J.*, **299**, L19.

[62] Bruenn, S. W. 1985, *Astrophys. J. Suppl.*, **58**, 771.

[63] Bruenn, S. W. 1986, *Astrophys. J. Suppl.*, **62**, 331.

[64] Hillebrandt, W. 1987, in *High Energy Phenomena Around Collapsed Stars*, edited by Pacini, F., (D. Reidel Publishing Company, Dordrecht, 1987), volume C195 of *NATO ASI*, p. 73.

[65] Baron, E., Bethe, H. A., Brown, G. E., Cooperstein, J., and Kahana, S. 1987, *Phys. Rev. Lett.*, **59**, 736.

[66] Bruenn, S. W. 1987, *Phys. Rev. Lett.*, **59**, 938.

[67] Myra, E. S. *et al.* 1987, *Astrophys. J.*, **318**, 744.

[68] Baron, E., Cooperstein, J., and Kahana, S. 1987, *Astrophys. J.*, **320**, 300.

[69] Myra, E. S. and Bludman, S. A. 1989, *Astrophys. J.*, **340**, 384.

[70] Bruenn, S. W. 1989, *Astrophys. J.*, **340**, 955.

[71] Bruenn, S. W. 1989, *Astrophys. J.*, **341**, 385.

[72] Baron, E. and Cooperstein, J. 1991, in *Supernovae: The Tenth Santa Cruz Summer Workshop in Astronomy and Astrophysics*, edited by Woosley, S. E., (Springer-Verlag, New York · Berlin · Heidelberg · London · Paris · Tokyo · Hong Kong · Barcelona, 1991), p. 342.

[73] Baron, E. and Cooperstein, J. 1990, *Astrophys. J.*, **353**, 597.

[74] Myra, E. S. and Burrows, A. 1990, *Astrophys. J.*, **364**, 222.

[75] Bruenn, S. W. and Haxton, W. C. 1991, *Astrophys. J.*, **376**, 678.

[76] Wilson, J. R. 1985, in *Numerical Astrophysics*, edited by Centrella, J. M., LeBlanc, J. M., and Bowers, R. L., (Jones & Bartlett, Boston, 1985), p. 422.

[77] Wilson, J. R. *et al.* 1986, *Ann. N.Y. Acad. Sci.*, **470**, 267.

[78] Burrows, A. and Mazurek, T. L. 1983, *Nature*, **301**, 315.

[79] Burrows, A. 1984, *Astrophys. J.*, **283**, 848.

[80] Burrows, A. and Lattimer, J. M. 1986, *Astrophys. J.*, **307**, 178.

[81] Mayle, R. W. *Physical Processes in Collapse Driven Supernova.* 1985, Ph.D. thesis, Univ. of California, Berkeley. UCRL-53713.

[82] Shapiro, S. L. and Teukolsky, S. A. *Black Holes, White Dwarfs, and Neutron Stars.* John Wiley & Sons, Inc., New York, 1983.

[83] Cox, J. P. and Giuli, R. T. *PRINCIPLES OF STELLAR STRUCTURE.* Gordon and Breach, Science Puplishers Inc., New York, 1968.

[84] Kippenhahn, R. and Weigert, A. *Stellar Structure and Evolution.* Astronomy and Astrophysics Library. Springer-Verlag, Berlin · Heidelberg, 1990.

[85] Bethe, H. A. *et al.* 1979, *Nucl. Phys.*, **A324**, 487.

[86] Fuller, G. M. 1982, *Astrophys. J.*, **252**, 741.

[87] Cooperstein, J. and Wambach, J. 1984, *Nucl. Phys.*, **A420**, 591.

[88] Lamb, D. Q., Lattimer, J. M., Pethick, C. J., and Ravenhall, D. G. 1978, *Phys. Rev. Lett.*, **41**, 1623.

[89] Bonche, P. and Vautherin, D. 1981, *Nucl. Phys.*, **A372**, 496.

[90] Osasawara, R. and Sato, K. 1983, *Prog. Theor. Phys.*, **70**, 1569.

[91] Goldreich, P. and Weber, S. V. 1980, *Astrophys. J.*, **238**, 991.

[92] Yahil, A. and Lattimer, J. M. 1982, in *Supernovae: A Survey of Current Research*, edited by Rees, M. J. and Stoneham, R. J., (D. Reidel Publishing Company, Dordrecht, 1982), p. 53.

[93] Yahil, A. 1983, *Astrophys. J.*, **265**, 1047.

[94] Mezzacappa, A. and Bruenn, S. W. 1993, *Astrophys. J.*, **405**, 637.

[95] Lattimer, J. M., Burrows, A., and Yahil, A. 1985, *Astrophys. J.*, **288**, 644.

[96] Mayle, R. and Wilson, J. R. 1988, *Astrophys. J.*, **334**, 909.

[97] Colgate, S. A. 1989, *Nature*, **341**, 489.

[98] Bethe, H. A. and Wilson, J. R. 1985, *Astrophys. J.*, **295**, 14.

[99] Ray, A. and Kar, K. 1987, *Astrophys. J.*, **319**, 143.

[100] Wilson, J. R. and Mayle, R. W. 1989, in *THE FIFTH MARCEL GROSSMANN MEETING on recent developements in theoretical and experimental general relativity, gravitation and relativistic field theories*, edited by Blair, D. G. and Buckingham, M. J., (World Scientific, Singapore · New Jersey · London · Hong Kong, 1989), p. 217.

[101] Wilson, J. R. and Mayle, R. W. 1989, in *The Nuclear Equation of State, Part A*, (Plenum Press, New York · London, 1989), p. 731.

[102] Mayle, R. W. and Wilson, J. R. 1991, in *Supernovae: The Tenth Santa Cruz Summer Workshop in Astronomy and Astrophysics*, edited by Woosley, S. E., (Springer-Verlag, New York · Berlin · Heidelberg · London · Paris · Tokyo · Hong Kong · Barcelona, 1991), p. 333.

[103] Janka, H.-T. and Müller, E. 1993, in *Proc. of the International Symposium on Neutrino Astrophysics: Frontiers of Neutrino Astrophysics*, edited by Suzuki, Y. and Nakamura, K., (Universal Academy Press Inc., Tokyo, 1993), number 5 in Frontiers Science Series, p. 203.

[104] Colgate, S. A., Herant, M., and Benz, W. 1993, *Phys. Rep.*, **227**, 157.

[105] Herant, M., Benz, W., and Colgate, S. 1992, *Astrophys. J.*, **395**, 642.

[106] Burrows, A. 1987, *Astrophys. J.*, **318**, L57.

[107] Burrows, A. and Lattimer, J. M. 1988, *Phys. Rep.*, **163**, 51.

[108] Wilson, J. R. and Mayle, R. W. 1988, *Phys. Rep.*, **163**, 63.

[109] Burrows, A. and Fryxell, B. A. 1992, *Science*, **258**, 430.

[110] Shimizu, T., Yamada, S., and Sato, K. 1993, *Publ. Astron. Soc. Japan*, **45**, L53.

[111] Yamada, S., Shimizu, T., and Sato, K. 1993, *Prog. Theor. Phys.*, **89**, 1175.

[112] Wilson, J. R. and Mayle, R. W. 1993, *Phys. Rep.*, **227**, 97.

[113] Suzuki, H. 1993, in *Proc. of the International Symposium on Neutrino Astrophysics: Frontiers of Neutrino Astrophysics*, edited by Suzuki, Y. and Nakamura, K., (Universal Academy Press Inc., Tokyo, 1993), number 5 in Frontiers Science Series, p. 219.

[114] Burrows, A. 1988, *Astrophys. J.*, **334**, 891.

[115] Suzuki, H. *Neutrino Burst from Supernova Explosion and Proto Neutron Star Cooling.* 1990, Ph.D. thesis, Univ. of Tokyo.

[116] Wilson, J. 1989, private communication.

[117] Wilson, J. 1990, in *WEAK INTERACTIONS AND NEUTRINOS - 1989*, edited by Singer, P. and Eilam, G., (North-Holland, Amsterdam, 1990), volume 13 of *Nucl. Phys. B (Proc. Suppl.)*, p. 380.

[118] Dicus, D. A. 1972, *Phys. Rev.*, **D6**, 941.

[119] Yueh, W. R. and Buchler, J. R. 1976, *Astrophys. Sp. Sci.*, **39**, 429.

[120] Yueh, W. R. and Buchler, J. R. 1976, *Astrophys. Sp. Sci.*, **41**, 221.

[121] Schinder, P. J. *et al.* 1987, *Astrophys. J.*, **313**, 531.

[122] Maxwell, O. V. 1987, *Astrophys. J.*, **316**, 691.

[123] Itoh, N. *et al.* 1989, *Astrophys. J.*, **339**, 354. Errutum in Astrophys. J. 360 (1990) 741.

[124] Braaten, E. and Segel, D. 1993, *Phys. Rev.*, **D48**, 1478.

[125] Fuller, G. M., Fowler, W. A., and Newman, M. J. 1980, *Astrophys. J. Suppl.*, **42**, 447.

[126] Fuller, G. M., Fowler, W. A., and Newman, M. J. 1982, *Astrophys. J.*, **252**, 715.

[127] Fuller, G. M., Fowler, W. A., and Newman, M. J. 1982, *Astrophys. J. Suppl.*, **48**, 279.

[128] Burrows, A., Klein, D., and Gandhi, R. 1992, *Phys. Rev.*, **D45**, 3361.

[129] Mayle, R., Wilson, J. R., and Schramm, D. N. 1987, *Astrophys. J.*, **318**, 288.

[130] Janka, H. T. and Hillebrandt, W. 1989, *Astron. Astrophys. Suppl.*, **78**, 375.

[131] Janka, H. T. and Hillebrandt, W. 1989, *Astron. Astrophys.*, **224**, 49.

[132] Giovanoni, P. M., Ellison, D. C., and Bruenn, S. W. 1989, *Astrophys. J.*, **342**, 416.

[133] Rybicki, G. B. and Lightman, A. P. *RADIATIVE PROCESSES IN ASTROPHYSICS.* John Wiley & Sons, New York · Chichester · Brisbane · Tronto · Singapore, 1979.

[134] Myra, E. S., Lattimer, J. M., and Yahil, A. 1988, in *Proc. of Supernova 1987A in the Large Magellanic Cloud*, edited by Kafatos, M. and Michalitsianos, A., (Cambridge University Press, Cambridge, 1988), p. 213.

[135] Janka, H.-T. *Neutrino Transport in Type-II Supernovae and Protoneutron Stars by Monte Carlo Methods.* 1991, Ph.D. thesis, Technische Universität München.

[136] Van Riper, K. A. 1978, *Astrophys. J.*, **221**, 304.

[137] Van Riper, K. A. and Arnett, W. D. 1978, *Astrophys. J.*, **225**, L129.

[138] Van Riper, K. A. 1979, *Astrophys. J.*, **232**, 558.

[139] Lichtenstadt, I., Sack, N., and Bludman, S. A. 1980, *Astrophys. J.*, **237**, 903.

[140] Van Riper, K. A. 1980, *Astrophys. J.*, **240**, 658.

[141] Epstein, R. L. and Pethick, C. J. 1981, *Astrophys. J.*, **243**, 1003.

[142] Takahara, M. and Sato, K. 1982, *Prog. Theor. Phys.*, **68**, 795.

[143] Bludman, S. A., Lichtenstadt, I., and Hayden, G. 1982, *Astrophys. J.*, **261**, 661.

[144] Burrows, A. and Lattimer, J. M. 1983, *Astrophys. J.*, **270**, 735.

[145] Lichtenstadt, I. and Bludman, A. 1984, *Astrophys. J.*, **276**, 746.

[146] Takahara, M. and Sato, K. 1984, *Prog. Theor. Phys.*, **71**, 524.

[147] Takahara, M. and Sato, K. 1984, *Prog. Theor. Phys.*, **72**, 978.

[148] Takahara, M. and Sato, K. 1985, *Phys. Lett.*, **156B**, 17.

[149] Van Riper, K. A. 1988, *Astrophys. J.*, **326**, 235.

[150] Takahara, M. and Sato, K. 1988, *Astrophys. J.*, **335**, 301.

[151] Janka, H.-T., Zwerger, T., and Mönchmeyer, R. 1993, *Astron. Astrophys.*, **268**, 360.

[152] Mazurek, T. J. 1982, *Astrophys. J.*, **258**, L13.

[153] Woosley, S. E. and Weaver, T. A. 1992, in *The Structure and Evolution of Neutron Stars*, edited by Pines, D., Tamagaki, R., and Tsuruta, S., (Addison-Wesley Publishing Company, Redwood City, 1992), p. 235.

[154] Arnett, W. D. 1980, *Ann. N.Y. Acad. Sci.*, **336**, 366.

[155] LeBlanc, J. M. and Wilson, J. R. 1970, *Astrophys. J.*, **161**, 541.

[156] Müller, E., Hillebrandt, W., and Rozyczka, M. 1980, *Astron. Astrophys.*, **81**, 288.

[157] Tohline, J. E., Schombert, J. M., and Boss, A. P. 1980, *Sp. Sci. Rev.*, **27**, 555.

[158] Müller, E. and Hillebrandt, W. 1981, *Astron. Astrophys.*, **103**, 358.

[159] Symbalisty, E. M. D. 1984, *Astrophys. J.*, **285**, 729.

[160] Mönchmeyer, R. and Müller, E. 1989, *Astron. Astrophys.*, **217**, 351.

[161] Janka, H.-T. and Mönchmeyer, R. 1989, *Astron. Astrophys.*, **226**, 69.

[162] Sato, K., Shimizu, T., and Yamada, S. 1993, in *Proc. of the International Symposium on Neutrino Astrophysics: Frontiers of Neutrino Astrophysics*, edited by Suzuki, Y. and Nakamura, K., (Universal Academy Press Inc., Tokyo, 1993), number 5 in Frontiers Science Series, p. 191.

[163] Janka, H.-T. and Mönchmeyer, R. 1989, *Astron. Astrophys.*, **209**, L5.

[164] Hillebrandt, W. 1985, in *Proc. 35th Workshop on Nuclear Astrophysics*, edited by Hillebrandt, W., Müller, E., and Thielemann, F. K., (MPA, Garching, 1985), 199, p. 59.

[165] Hillebrandt, W. and Wolff, R. G. 1985, in *Nucleosynthesis: Challenges and New Developments*, edited by Arnett, W. D. and Truran, J. W., (The Univ. of Chicago Press, 1985), chapter 10, p. 131.

[166] Epstein, R. I. 1979, *Mon. Not. R. Astron. Soc.*, **188**, 305.

[167] Colgate, S. A. and Petschek, A. G. 1980, *Astrophys. J.*, **236**, L115.

[168] Bruenn, S. W., Buchler, J. R., and Livio, M. 1979, *Astrophys. J.*, **234**, L183.

[169] Livio, M., Buchler, J. R., and Colgate, S. A. 1980, *Astrophys. J.*, **238**, L139.

[170] Smarr, L. *et al.* 1981, *Astrophys. J.*, **246**, 515.

[171] Lattimer, J. M. and Mazurek, T. J. 1981, *Astrophys. J.*, **246**, 955.

[172] Arnett, W. D. 1987, *Astrophys. J.*, **319**, 136.

[173] Bethe, H. A., Brown, G. E., and Cooperstein, J. 1987, *Astrophys. J.*, **322**, 201.

[174] Burrows, A. and Fryxell, B. 1993, in *Proceedings of the International Symposium on ORIGIN AND EVOLUTION OF THE ELEMENTS*, edited by Kubono, S. and Kajino, T., (World Scientific, Singapore · New Jersey · London · Hong Kong, 1993), p. 165.

[175] Miller, D. S., Wilson, J. R., and Mayle, R. W. 1993, *Astrophys. J.*, **415**, 278.

[176] Goodman, J., Dar, A., and Nussinov, S. 1987, *Astrophys. J.*, **314**, L7.

[177] Janka, H.-T. 1991, *Astron. Astrophys.*, **244**, 378.

[178] Cooperstein, J., van den Horn, L. J., and Baron, E. 1987, *Astrophys. J.*, **321**, L129.

[179] IAU circular No. 4316, International Astronomical Union, 1987.

[180] IAU circular No. 4317, International Astronomical Union, 1987.

[181] Alexeyev, E. N. *et al.* 1988, *Phys. Lett. B*, **205**, 209.

[182] Hirata, K. S. *et al.* 1988, *Phys. Rev.*, **D38**, 448.

[183] Bratton, C. B. *et al.* 1988, *Phys. Rev.*, **D37**, 3361.

[184] Piran, T. and Spergel, D. 1988, in *DARK MATTER: Proc. of the XXIIIrd Rencontre de Moriond*, edited by Audouze, J. and Tran Thanh Van, J., (Editions Frontières, Cedex, 1988), p. 453.

[185] Hirata, K. S. Observation of supernova neutrinos at KAMIOKANDE. Master's thesis, Univ. of Tokyo, 1988.

[186] Sato, K. and Suzuki, H. 1987, *Phys. Rev. Lett.*, **58**, 2722.

[187] Sato, K. and Suzuki, H. 1987, *Phys. Lett. B*, **196**, 267.

[188] Mayle, R. W. and Wilson, J. R. Calculations of stellar collapse, neutrinos, and SN1987a. preprint, 1987.

[189] Arafune, J. and Fukugita, M. 1987, *Phys. Rev. Lett.*, **59**, 367.

[190] Burrows, A. and Lattimer, J. M. 1987, *Astrophys. J.*, **318**, L63.

[191] Kahana, S. H., Cooperstein, J., and Baron, E. 1987, *Phys. Lett.*, **B196**, 259.

[192] Krauss, L. M. 1987, *Nature*, **329**, 689.

[193] Bahcall, J. N. *et al.* 1987, *Nature*, **327**, 682.

[194] Bludman, S. A. and Schinder, P. J. 1988, *Astrophys. J.*, **326**, 265.

[195] Blinnikov, S. I., Imshennik, V. S., and Nadyozhin, D. K. 1988, *Astrophys. Sp. Sci.*, **150**, 273.

[196] Hillebrandt, W. and Höflich, P. 1989, *Rep. Prog. Phys.*, **52**, 1421.

[197] Lattimer, J. M. and Yahil, A. 1989, *Astrophys. J.*, **340**, 426.

[198] Loredo, T. J. and Lamb, D. Q. 1989, *Ann. N.Y. Acad. Sci.*, **571**, 601.

[199] Nomoto, K., Shigeyama, T., and Hashimoto, M. 1987, in *ESO Workshop on the SN1987A*, edited by Danziger, I. J., (ESO, Garching, 1987), number 26 in ESO Conference and Workshop Proceedings, p. 325.

[200] Woosley, S. E., Pinto, P. A., and Ensman, L. 1988, *Astrophys. J.*, **324**, 466.

[201] Frodesen, A. G., Skjeggestad, O., and Tøfte, H. *Probability and Statistics in Particle Physics.* Universitetsforlaget, Bergen-Oslo-Tromsø, 1979.

[202] Bethe, H. A. and Johnson, M. B. 1974, *Nucl. Phys.*, **A230**, 1.

[203] Taylor, J. H. and Weisberg, J. M. 1989, *Astrophys. J.*, **345**, 434.

[204] Arponen, J. 1972, *Nucl. Phys.*, **A191**, 257.

[205] Pandharipande, V. R. and Smith, R. A. 1975, *Nucl. Phys.*, **A237**, 507.

[206] Arnett, W. D. and Bowers, R. L. 1977, *Astrophys. J. Suppl.*, **33**, 415.

[207] Walecka, J. D. 1974, *Ann. Phys.*, **83**, 491.

[208] Tammann, G. A., Sandage, A., and Yahil, A. 1980, in *Physical Cosmology*, edited by Balian, R., Audouze, J., and Schramm, D. N., (North-Holland, Amsterdam · New York · Oxford, 1980), p. 53.

[209] De Vaucouleurs, G. 1979, *Astrophys. J.*, **227**, 380.

[210] Raffelt, G. G. 1990, *Mod. Phys. Lett. A*, **5**, 2581.

[211] Raffelt, G. G. 1990, *Phys. Rep.*, **198**, 1.

[212] Ellis, J. and Olive, K. A. 1987, *Phys. Lett.*, **B193**, 525.

[213] Mayle, R. *et al.* 1988, *Phys. Lett.*, **B203**, 188.

[214] Mayle, R. *et al.* 1989, *Phys. Lett.*, **B219**, 515.

[215] Raffelt, G. and Seckel, D. 1988, *Phys. Rev. Lett.*, **60**, 1793.

[216] Engel, J., Seckel, D., and Hayes, A. C. 1990, *Phys. Rev. Lett.*, **65**, 960.

[217] Turner, M. S., Kang, H.-S., and Steigman, G. 1989, *Phys. Rev.*, **D40**, 299.

[218] Brinkmann, R. P. and Turner, M. S. 1988, *Phys. Rev.*, **D38**, 2338.

[219] Ishizuka, N. and Yoshimura, M. 1990, *Prog. Theor. Phys.*, **84**, 233.

[220] Burrows, A., Ressell, M. T., and Turner, M. S. 1990, *Phys. Rev.*, **D42**, 3297.

[221] Burrows, A., Turner, M. S., and Brinkmann, R. P. 1989, *Phys. Rev.*, **D39**, 1020.

[222] Turner, M. S. 1988, *Phys. Rev. Lett.*, **60**, 1797.

[223] Mayle, R. *et al.* 1993, *Phys. Lett.*, **B317**, 119.

[224] Gaemers, K. J. F., Gandhi, R., and Lattimer, J. M. 1989, *Phys. Rev.*, **D40**, 309.

[225] Grifols, J. A. and Massó, E. 1990, *Phys. Lett.*, **B242**, 77.

[226] Pérez, A. and Gandhi, R. 1990, *Phys. Rev.*, **D41**, 2374.

[227] Turner, M. S. 1992, *Phys. Rev.*, **D45**, 1066.

[228] Gandhi, R. and Burrows, A. 1990, *Phys. Lett.*, **B246**, 149. Erratum in Phys. Lett. B261 (1991) 519.

[229] Burrows, A., Gandhi, R., and Turner, M. S. 1992, *Phys. Rev. Lett.*, **68**, 3834.

[230] Pantaleone, J. 1991, *Phys. Lett.*, **B268**, 227.

[231] Pantaleone, J. 1992, *Phys. Rev.*, **D46**, 510.

[232] Bahcall, J. N. and Glashow, S. L. 1987, *Nature*, **326**, 476.

[233] Arnett, W. D. and Rosner, J. L. 1987, *Phys. Rev. Lett.*, **58**, 1906.

[234] Hillebrandt, W. *et al.* 1987, *Astron. Astrophys.*, **177**, L41.

[235] Midorikawa, S., Terazawa, H., and Akama, K. 1987, *Mod. Phys. Lett.*, **A2**, 561.

[236] Huzita, H. 1987, *Mod. Phys. Lett.*, **A2**, 905.

[237] Roos, M. 1987, *Europhys. Lett.*, **4**, 953.

[238] Kolb, E. W., Stebins, A. J., and Turner, M. S. 1987, *Phys. Rev.*, **D35**, 3598. Errata and Addendum in Phys. Rev. D36 (1987) 3820.

[239] Spergel, D. N. and Bahcall, J. N. 1988, *Phys. Lett.*, **B200**, 366.

[240] Burrows, A. 1988, *Astrophys. J.*, **328**, L51.

[241] Chiu, H.-Y., Chan, K. L., and Kondo, Y. 1988, *Astrophys. J.*, **329**, 326.

[242] Abbott, L. F., De Rújula, A., and Walker, T. P. 1988, *Nucl. Phys.*, **B299**, 734.

[243] Cowsik, R. 1988, *Phys. Rev.*, **D37**, 1685.

[244] Adams, E. N. 1988, *Phys. Rev.*, **D37**, 2047.

[245] Evans, D., Fong, R., and Collins, P. D. B. 1988, *Astron. Astrophys.*, **189**, L10.

[246] Avignone III, F. T. and Collar, J. I. 1990, *Phys. Rev.*, **D41**, 682.

[247] Nötzold, D. 1988, *Phys. Rev.*, **D38**, 1658.

[248] Goldman, I. *et al.* 1988, *Phys. Rev. Lett.*, **60**, 1789.

[249] Lattimer, J. M. and Cooperstein, J. 1988, *Phys. Rev. Lett.*, **61**, 23. Errata in Phys. Rev. Lett. 61 (1988) 2633.

[250] Barbieri, R. and Mohapatra, R. 1988, *Phys. Rev. Lett.*, **61**, 27.

[251] Hirata, K. S. *Search for Supernova Neutrinos at KAMIOKANDE-II.* 1991, Ph.D. thesis, Univ. of Tokyo.

[252] Dye, S. T. *et al.* 1989, *Phys. Rev. Lett.*, **62**, 2069.

[253] Alekseev, E. N. *et al.* 1993, *JETP*, **77**, 339.

[254] van den Bergh, S. 1991, *Phys. Rep.*, **204**, 385.

[255] van den Bergh, S. and Tammann, G. A. 1991, *Ann. Rev. Astron. Astrophys.*, **29**, 363.

[256] Tammann, G. A. 1993, in *Proc. of the International Symposium on Neutrino Astrophysics: Frontiers of Neutrino Astrophysics*, edited by Suzuki, Y. and Nakamura, K., (Universal Academy Press Inc., Tokyo, 1993), number 5 in Frontiers Science Series, p. 255.

[257] Bahcall, J. N. and Piran, T. 1983, *Astrophys. J.*, **267**, L77.

[258] Ratnatunga, K. U. and van den Bergh, S. 1989, *Astrophys. J.*, **343**, 713.

[259] Muller, R. A. *et al.* 1992, *Astrophys. J.*, **384**, L9.

[260] Cappellaro, E. *et al.* 1993, *Astron. Astrophys.*, **273**, 383.

[261] Lyne, A. G., Manchester, R. N., and Taylor, J. H. 1985, *Mon. Not. R. Astron. Soc.*, **213**, 613.

[262] Totsuka, Y. 1990, in *Proc. of International Symposium on UNDERGROUND PHYSICS EXPERIMENTS*, edited by Nakamura, K., ICRR, Univ. of Tokyo, (ICRR, Univ. of Tokyo, Tanashi, Tokyo, 1990), p. 129.

[263] Takita, M. 1993, in *Proc. of the International Symposium on Neutrino Astrophysics: Frontiers of Neutrino Astrophysics*, edited by Suzuki, Y. and Nakamura, K., (Universal Academy Press Inc., Tokyo, 1993), number 5 in Frontiers Science Series, p. 135.

[264] Ewan, G. T. *et al.* 1993, in *Proc. of the International Symposium on Neutrino Astrophysics: Frontiers of Neutrino Astrophysics*, edited by Suzuki, Y. and Nakamura, K., (Universal Academy Press Inc., Tokyo, 1993), number 5 in Frontiers Science Series, p. 147.

[265] Minakata, H. and Nunokawa, H. 1990, *Phys. Rev.*, **D41**, 2976.
[266] Acker, A., Pakvasa, S., and Raghavan, R. S. 1990, *Phys. Lett.*, **B238**, 117.
[267] Seckel, D., Steigman, G., and Walker, T. P. 1991, *Nucl. Phys.*, **B366**, 233.
[268] Krauss, L. M. *et al.* 1992, *Nucl. Phys.*, **B380**, 507.
[269] Balantekin, A. B. and Loreti, F. 1992, *Phys. Rev.*, **D45**, 1059.
[270] Bisnovatyi-Kogan, G. S. and Seidov, Z. F. 1984, *Ann. N.Y. Acad. Sci.*, **422**, 319.
[271] Krauss, L. M., Glashow, S. L., and Schramm, D. N. 1984, *Nature*, **310**, 191.
[272] Woosley, S. E., Wilson, J. R., and Mayle, R. 1986, *Astrophys. J.*, **302**, 19.
[273] Zhang, W. *et al.* 1988, *Phys. Rev. Lett.*, **61**, 385.
[274] Lindquist, R. W. 1966, *Ann. Phys.*, **37**, 487.
[275] Baron, E. *et al.* 1989, *Astrophys. J.*, **339**, 978.
[276] May, M. M. and White, R. H. 1967, *Meth. Comp. Phys.*, **7**, 219.
[277] Mezzacappa, A. and Matzner, R. A. 1989, *Astrophys. J.*, **343**, 853.
[278] Harleston, H. and Vishniac, E. T. 1992, *Phys. Rev.*, **D45**, 4458.
[279] Bruenn, S. W., Buchler, J. R., and Yueh, W. R. 1978, *Astrophys. Sp. Sci.*, **59**, 261.
[280] Bowers, R. L. and Wilson, J. R. 1982, *Astrophys. J. Suppl.*, **50**, 115.
[281] Levermore, C. D. and Pomraning, G. C. 1981, *Astrophys. J.*, **248**, 321.
[282] Cooperstein, J. and Baron, E. A. 1990, in *Supernovae*, edited by Petschek, A. G., (Springer-Verlag, New York · Berlin · Heidelberg · London · Paris · Tokyo · Hong Kong, 1990), p. 213.
[283] Cooperstein, J., van den Horn, L. J., and Baron, E. A. 1986, *Astrophys. J.*, **309**, 653.
[284] Mezzacappa, A. and Bruenn, S. W. 1993, *Astrophys. J.*, **410**, 740.
[285] Friedman, B. and Pandharipande, V. R. 1981, *Nucl. Phys.*, **A361**, 502.
[286] Baym, G., Bethe, H. A., and Pethick, C. J. 1971, *Nucl. Phys.*, **A175**, 225.

Detection of VHE γ-Rays and Neutrinos

Tadashi Kifune

ICRR, University of Tokyo

Masaki Mori

Department of Physics, Miyagi University of Education

1 Introduction

Observation of Galactic γ-rays was first proposed by Hayakawa and Morrison at the end of 1950's. Cosmic rays collide with interstellar matter producing γ-rays and, thereby, the observation can give an estimate of the amount of interstellar matter together with the existence of high energy protons. Unfortunately, the expected γ-ray flux is small when estimated from the known cosmic ray intensity. However, a more intense and more detectable flux of γ-rays might be expected if a considerable amount of antiprotons exist in the Galaxy as suggested by Morrison. At that time, radio wave detection had just started to uncover high energy phenomena in the universe. One mechanism of radio wave emission is synchrotron radiation, and in this case it was concluded that high energy electrons as well as magnetic fields exist in various objects at far distances from us. In addition to the high energy processes due to electrons, observation of γ-rays reveals high energy phenomena including high energy protons at far places in the universe.

Gamma-rays, photons with energies higher than ~ 1 MeV, interact with high energy protons and other elementary particles. γ-rays are produced as a result of interactions of energetic elementary particles, such as from a decay of the π^0 meson. Fundamental mechanisms for producing γ-rays are given in Section 2.

The purpose of this article is to describe high energy γ-ray astronomy, the last frontier of electromagnetic radiation, and the possibility of detecting high energy neutrinos. The observational results are, generally speaking, more controversial in the higher energy region. However, remarkable progress is being made at GeV energies, also with evidence of γ-ray detection at $\sim$ TeV ($= 10^{12}$ eV) energy on a sound basis of statistical significance. An observation window has been now opened up to as high as the VHE (Very High Energy ; $\geq 10^{11}$ eV) region. The recent status in this VHE region is discussed as well as γ-ray astronomy at GeV energies which is indispensable for understanding the VHE region. The energy region of UHE (Ultra High Energy ; $\geq 10^{13}$ eV) is also discussed. The detection method for high energy γ-rays above 100 MeV is explained in Section 3.

A number of unexpected/important results are being reported from observations with the Compton Gamma-Ray Observatory (GRO) Satellite, launched in April 1991 and in operation while this paper is prepared. Some of its results on γ-ray pulsars and on extra-galactic objects are mentioned in Sections 4 and 5. By using the improved data of high energy γ-rays, a better understanding is obtained on the sites where high energy elementary particles are generated and their energetic interactions take place, for example, at pulsars, active galaxies and other locations, which will be described in Section 5.

Below $\sim$ 100 MeV γ-ray interactions are somewhat different from those at higher energies; γ-rays from π^0 decay are not the main production mechanism and nuclear transitions cause line γ-rays around 1 – 10 MeV. The detection method is also different because the dominant interaction of γ-rays is the Compton effect and pair creation, respectively below and above $\sim$ 100 MeV. The results at a few MeV energy are not discussed in this article, such as the unexpected results on gamma ray bursts [1].

The last section (6) is devoted to high-energy neutrino astronomy. When γ-rays are produced from neutral pions, neutrinos are also produced from charged pions, generated in the same hadronic collisions. Thus high-energy γ-rays and neutrinos are closely related in their astrophysical presence. However, the detection methods are completely different; in order to filter the overwhelming flux of cosmic rays, neutrino detectors must be placed deep underground/underwater. Though we have no indication of the existence of extraterrestrial high-energy neutrinos up to now, there is a good hope that next-generation experiments now underway will see positive signals.

2 Production mechanism of high energy γ-rays

If γ-rays of GeV energies are of thermal origin, the temperature must be higher than 10^{13} K $\sim$ 1 GeV, the same order as the nucleon mass. A thermal equilibrium at such high temperatures results in copious production of antinucleons and, inversely, annihilation of nucleon and antinucleon pairs into γ-rays. However, no evidence has been found, at least inside our Galaxy, to suggest the abundant antiprotons expected from such a process. A state of thermal equilibrium is known to exist in compact objects, X-ray stars and possibly γ-ray bursters, with a high temperature of keV to a few MeV. The thermalized energy is given to electrons when protons accrete onto the surface of a neutron star releasing gravitational potential energy. The available kinetic energy is calculated to be $\leq$ 10 MeV per particle. The thermal γ-rays with energies higher than hundreds of MeV seem unlikely to be produced at such compact objects.

The VHE and UHE γ-rays are thus considered to be from nonthermal radiation by energetic particles. Such progenitor particles are known as cosmic rays. Cosmic ray nucleons can produce high energy γ-rays in collisions with other protons in interstellar space or at the place where the cosmic rays are created,

$$p + p \rightarrow p(n) + p(n) + \pi^+ + \pi^- + \pi^0 + ... \tag{1}$$

Neutral pions produced in the reaction decay into two γ-rays. The energy given to the secondary particles and γ-rays is typically about one tenth of the energy of the progenitor cosmic ray. In the energy region much higher than 1 GeV, the energy spectrum of the γ-rays from pion decays tends to have the same spectral shape as that of the progenitor protons, $I_p(E_p), \propto E_p^{-\alpha}$.

A high number density of soft photons is expected around active objects such as pulsars where particle acceleration may take place, and high energy protons may coexist with a high density of soft photons. In this circumstance, pions can be created through proton-photon collisions,

$$p + photon \rightarrow p + \pi^0, n + \pi^+, \tag{2}$$

if the proton energy E is high enough that the total energy in the center of mass system $W = \sqrt{4\varepsilon E + M^2c^4}$ is larger than $\sqrt{M + \mu} \cdot c^2$, where ε, M and μ are the photon energy, proton mass and pion mass.

The number of such target photons can be enhanced by energetic electrons. Electrons radiate photons by synchrotron radiation in a magnetic field, and also scatter soft photons into higher energies through the inverse Compton effect. If the electron energy is very high and/or if the magnetic field is strong enough, the energy of radiated photons can be as high as γ-rays. The two processes have a similar relationship between the energy of emitted photons and the reaction rate to the electron energy. The energy k (k_{sync} and k_{ic} for the synchrotron and inverse Compton process, respectively) of the emitted photon, in units of TeV, is approximately given as

$$\begin{aligned} k_{sync} &= a \cdot BE^2 = 6 \cdot 10^{-9} \cdot BE^2, \\ k_{ic} &= b \cdot E^2 = 2 \cdot 10^{-3} \cdot \frac{\varepsilon}{10^{-4}\,\mathrm{eV}} E^2, \end{aligned} \tag{3}$$

where B is the magnetic field in gauss, E the energy of the progenitor electron in TeV, and ε the energy of ambient photons in eV.

In general, as is seen from the above expressions, inverse Compton effect is more efficient in producing higher energy photons. TeV electrons can create GeV γ-rays through the inverse Compton effect in collisions with microwaves, while synchrotron radiation needs magnetic fields of $\sim 10^8$ gauss. High energy γ-rays, in turn, can supply more electrons through pair creation of electrons and positrons. In a system of energetic particles in conjunction with either high photon number density or dense matter, the particles produce secondary pions, γ-rays and higher numbers of electrons and soft photons. The secondaries then cause other reactions. The whole process cosists of complicated successive interactions. Such a cascade of processes is believed to occur in a variety of systems, such as pulsars, X-ray stars and active galactic nuclei. The luminosity and the energy spectrum of the high energy γ-rays depends on the detailed structure of the cascade process.

The energy spectrum of particles non-thermally accelerated, cosmic rays, is characterized by a power law $\sim E^{-\gamma}$. It is known that radiated photons from electrons with such a spectrum have also a power spectrum. The energy loss per electron or the emission power increases in proportion to $10^{-3} \cdot B^2 E^2$ for synchrotron radiation, and the energy transferred to photons is given by the product of the number of electrons and the emission power. Thus, the energy spectrum of photons $I(k)$ is

$$\begin{aligned} I(k) &= 10^{-3} \cdot B^2 E^2 \cdot E^{-\gamma} dE \\ &= 10^{-3} \cdot a^{\frac{\gamma-3}{2}} \cdot B^{\frac{\gamma+1}{2}} \cdot k^{-\frac{\gamma-1}{2}} dk. \end{aligned} \tag{4}$$

The radiated photons have a power spectrum $\propto k^{-\alpha} dk$ when the number of progenitor electrons is $\propto E^{-\gamma} dE$. In the inverse Compton case, the emission power is independent of B but depends on the number density of ambient photons, ρ. By putting the power equal to $P_0 E^2 \approx \rho c \sigma_T \cdot k$, and $k = b \cdot E^2$, where σ_T is the Thomson cross section, we obtain the relation,

$$\begin{aligned} I(k) &= P_0 \cdot E^2 \cdot E^{-\gamma} dE \\ &= P_0 \cdot b^{\frac{\gamma-3}{2}} \cdot k^{-\frac{\gamma-1}{2}} dk. \end{aligned} \tag{5}$$

Thus, we obtain $\alpha = \frac{\gamma-1}{2}$ for both processes.

3 Detection methods of high energy γ-rays

Gamma-rays do not penetrate the atmosphere and celestial γ-rays must be observed outside the atmosphere. We need a larger detection area for higher energy γ-rays since the flux falls off rapidly with energy that limits the highest energy γ-rays to be detected with a detector. Observations with satellite-borne detectors, SAS 2 (launched in 1972) and COS B (in 1975 – 1982), have been done in the energy range below 10 GeV = 10^{10} eV, and have detected γ-ray emission along the Galactic disc as well as several γ-ray stars.

In order to detect higher energy γ-rays, we use the atmosphere as a radiator material which converts γ-rays into electron-positron pairs to increase the detection area. The pair can develop, if the γ-ray energy is sufficiently high, into an extensive air shower which can be observed with a ground-based telescope.

The VHE γ-rays are observed with optical telescopes (Čerenkov telescopes) by detecting the Čerenkov light emitted by the shower particles in the upper atmosphere. The Čerenkov light is emitted within $\sim 1°$, which results in a lateral spread of ~ 100 m on the ground. The Čerenkov telescope enjoys a large effective detection area by using a small area detector, due to this lateral spread of Čerenkov light. In the UHE region, the shower particles survive to reach the ground level, and an array of particle detectors (air shower array) is used for the detection of the shower particles. The array has the advantage that it can be operated day and night throughout the year, whereas the Čerenkov telescope is useful only on moonless clear nights.

The first attempt with a ground-based telescope was made for the Crab Nebula at ~ 1 TeV $= 10^{12}$ eV by the Russian group in Crimea. It was stimulated by the discovery of radio emission from the nebula. The radio emission is considered to be synchrotron radiation from energetic electrons, and it was attempted to find evidence for energetic protons in the nebula which are likely to emit high energy γ-rays.

About two decades ago, evidence for γ-ray emissions was found from an X-ray star, Cyg X-3, at ~ 1 TeV by the Crimean Group. Early in the 1980's, a γ-ray signal from Cyg X-3 was reported in the higher energy region of 1 PeV $= 10^{15}$ eV. With "detections" of γ-rays in TeV energies from several other stars, these results encouraged many groups to begin hunting for energetic γ-rays, as well as theoretical work to study the mechanism for γ-ray emission beyond 10^{12} eV.

Ground-based detectors are heavily contaminated with cosmic ray background. The usual way of distinguishing γ-ray events is to use the ballistic nature of γ-ray propagation. A signal should appear as a peak of the counting rate in the direction of the source or as a time varying signal over a uniform and steady but overwhelming cosmic ray background. A correlation between the data acquisition times and the pulsar period is looked for, if the assumed source is a pulsar.

In the detection of γ-rays, even at intermediate energies of MeV to GeV, thick material is necessary to absorb γ-rays of high penetrating power, and it is not easy to achieve angular resolution much better than 1°. The improvement of this resolution is of great importance in γ-ray astronomy.

Recently, a new type of Čerenkov telescope (the imaging Čerenkov telescope) has been developed (see for an example [2]). This telescope has an angular resolution as good as or better than 0.1° and also the capability of distinguishing cosmic ray background. Since extensive air showers (EAS) have an extended structure, emitted Čerenkov light is also extended and has a finite image size of $\sim 1°$ when observed by a telescope, as is explained in Fig. 1. In addition, the image is located at a direction shifted from the true direction of the γ-rays (indicated as "a" and "b" and chosen at the center of the field of view of the telescope, "O", in the figure), since shower electrons are at a finite height in the atmosphere. However, the image of Čerenkov light from γ-ray showers is elongated with an elliptical shape, and the true direction of γ-rays lies on the line extended from the major axis of the ellipse. The imaging telescope measures such characteristics of Čerenkov image, and the true direction is determined within the accuracy of a few arc minutes. In the case of showers initiated by cosmic ray protons, the image has, due to the hadronic interaction, a wider and more irregular shape compared with that of γ-ray showers, enabling us to reject typical cosmic ray shower events. The sensitivity is thus considerably improved at TeV energies, such as those at which γ-rays from the Crab pulsar/nebula were detected.

Satellite-borne detectors

The satellite-borne detectors for γ-rays below 10 GeV are a combined appa-

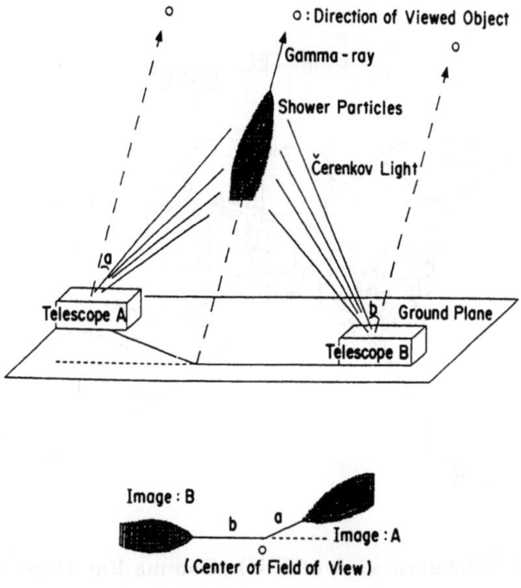

Figure 1: Čerenkov light image viewed with imaging Čerenkov telescope.

ratus of calorimeter and track detector, the former measuring the photon energy and the latter determining the arrival direction of γ-rays. An incident γ-ray is converted into an electron-positron pair at the top of the detection system. A typical detection area is one thousand cm^2. A total weight sufficient to absorb the electron-photon shower cascaded from the pair electrons limits the size of the detector on board the space vehicle. The opening angle of the pair of electrons and the spatial resolution of the tracking detectors limit the angular resolution to not much better than 1°. Fig. 2 shows a picture of GRO launched in 1991, and Fig. 3 is a schematic view of the EGRET detector of GRO.

The BATSE (Burst and Transient Source Experiment) detector is an all sky monitor of γ-rays of 20 keV to 10 MeV. It consists of eight crystal scintillation detectors, and the detection area is 2000 cm^2. The COMPTEL (Compton Telescope) detects 1 to 30 MeV γ-rays with two layers of scintillation detectors. A γ-ray Compton-scattered by one of the seven upper liquid scintillation detectors is then detected by one of the fourteen lower NaI(Tl) crystal scintillation detectors. Low background operation is obtained by using anticoincidence, time of flight and pulse shape techniques. The OSSE (Oriented Scintillation Spectrometer Experiment) detector consists of four large collimated scintillation detectors with the effective area of $\sim 700 \times 4$ cm^2 for detecting 0.1 to 10 MeV γ-rays. The typical energy resolution is 8% and 3.2% FWHM at 622 keV and 6.1 MeV, respectively. The EGRET (Energetic Gamma-Ray Experiment Telescope) detector has an effective area of ~ 1600 cm^2, about ten times the area of the earlier experiments, SAS 2 and COS B, and detects 20 MeV – 30 GeV photons. Incoming γ-rays are

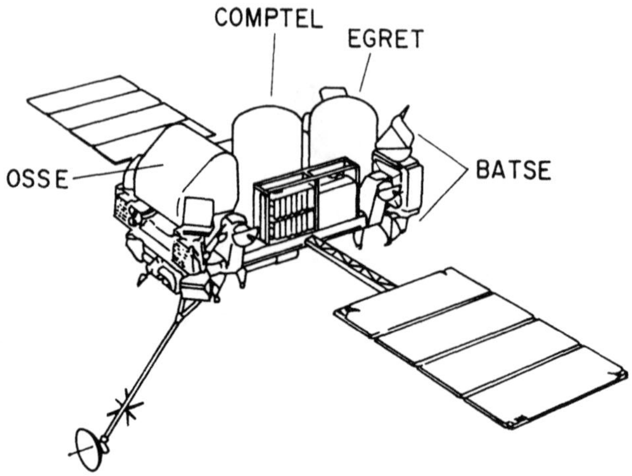

Figure 2: Schematic view of GRO (Gamma Ray Observatory) [3].

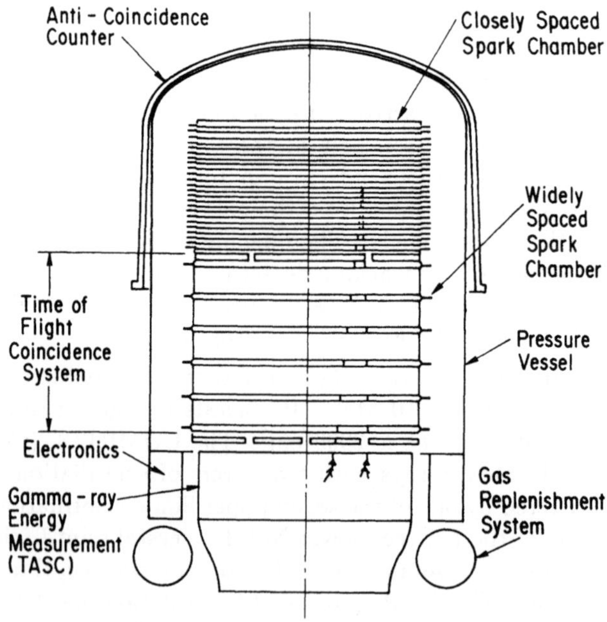

Figure 3: Schematic arrangement of EGRET apparatus of GRO [3].

converted to an electron-positron pair in the upper part of the spark chamber module. The tracks of pair electrons are determined by the lower spark chambers, and the energy is measured by the scintillation crystal calorimeter at the end of the instrument.

Čerenkov telescope

The Čerenkov telescope is useful for detecting γ-rays at energies $\sim 10^{12}$ eV. It is an apparatus composed of mirror for collecting the light and a camera of fast photomultiplier tubes set at the focal point of the mirror. A merit of the Čerenkov telescope is in its large detection area with a one-mirror telescope, utilizing the fact that the lateral distribution of Čerenkov light is approximately flat extending to about 150 m allowing, as a result, $\pi \times (150\text{ m})^2$ of detection area to be covered by one telescope. Nevertheless, the telescope must be of a large size, compared with a telescope for optical astronomy, in order to collect faint Čerenkov light of ~ 50 photons per m^2 from a 1 TeV γ-ray. However, the surface quality of the mirror need not be as good as required in optical astronomy. The telescope usually consists of a number of small mirrors sacrificing surface quality to achieve a total mirror area typically of the order of 10 m^2.

The telescope of the Whipple Observatory in the United States (shown in Fig. 4) has a spherical mirror of diameter 10 m. The surface quality of the mirror is of an accuracy of $0.2° \sim 0.3°$. It was the first imaging Čerenkov telescope, and in the focal plane of the telescope is set a camera of 37 photomultiplier tubes. Each photomultiplier tube has a pixel size of 0.4° separated from the neighboring ones by 0.5° (now the number of tubes has increased to 109 with a pixel size of 0.2° by the Project called GRANITE), aiming at a circular region of the sky of 0.25° in diameter (pixel size). The threshold energy of detectable γ-ray is about 300 GeV. Another imaging telescope has just started operation in 1992 in a collaboration of Japan and Australian groups (CANGAROO) aiming at the objects in the southern hemisphere [4]. A photograph of the telescope brought from Japan and installed in Australia is shown in Fig. 5, which is cooperated with another telescope of Adelaide University shown in Fig. 6. The telescope from Japan has a pixel size of $0.12° \times 0.12°$, to take advantage of the good surface quality of the mirror with a blur better than one minute of arc.

A simultaneous observation with multiple imaging telescopes has just been started by the Whipple group as well as by CANGAROO. The true direction can be inferred event by event as the intersection of the two major axises fitted to the elliptical images of the two telescopes in simultaneous operation, as explained in Fig. 7. A better angular accuracy and background rejection factor is expected with the "stereoscopic observation" of two telescopes than those obtained from a single telescope.

EAS arrays

When the γ-ray energy is higher than about $10^{13} \sim 10^{14}$ eV, the EAS reaches ground level, and the measurement is made by an array of particle detectors. The EAS arrays have been used for cosmic ray observations, and the current largest array dedicated to UHE γ-ray detection is the CASA Array (shown in Fig. 8) in the United States. The distance between two adjacent scintillation counters,

Figure 4: A photograph of the Imaging Čerenkov telescope of the Whipple Observatory.

Figure 5: Japanese telescope (diameter of 3.8 m) of the CANGAROO project operated in Australia.

Figure 6: BIGRAT Telescope of Adelaide University.

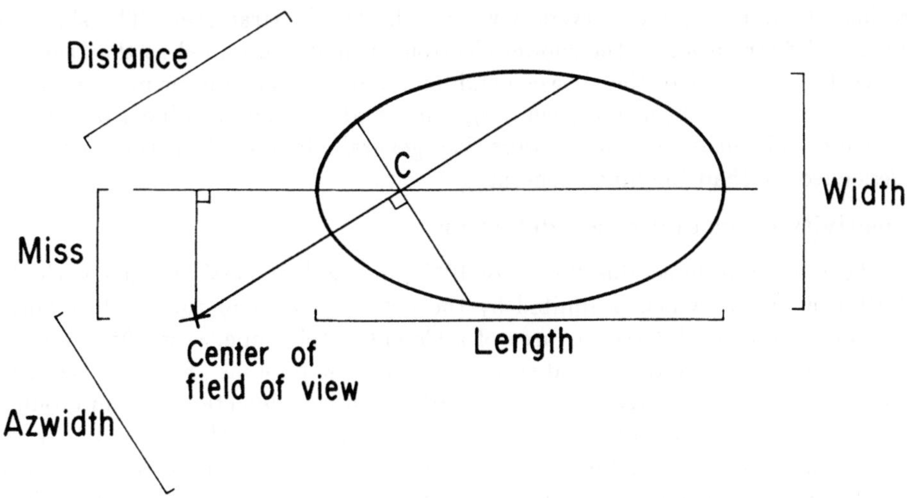

Figure 7: Simplified sketch of the Čerenkov image and definition of imaging parameters. (From Weekes et al. [24])

appearing as a number of dots in the figure, is 15 m, and the total number is 1024. The shower particles come down with the light velocity c bunched in a thin pancake-shaped disk perpendicular to the arrival direction. The arrival direction of each EAS is, therefore, determined from the arrival time at each detector. In Fig. 9, the pulse height (a) and arrival time (b) at all the detectors is shown by the height of the vertical bars. The arrival direction of the measured event is perpendicular to the "slope" connecting the tops of the bars in Fig. 9(b). The accuracy of angular determination is $0.5° \sim 1°$.

The detection sensitivity is usually limited by the statistics of background events of cosmic rays. An accumulation of data for improving the statistics has a limitation determined by experimental systematic errors. The rate of cosmic rays is, for example, sensitive to the pressure and temperature of the atmosphere to an extent as large as $\sim 10^{-2}$ per mb or degree. Thus, it is necessary to separate out the background events to get better sensitivity with the EAS array. The selection of γ-ray signals has been attempted by measuring the number of muons, a decay product of pions or kaons, because γ-ray showers contain fewer muons among the shower particles than proton showers which include hadronic interactions. The method, however, in spite of its long history yet remains to be established. Muon detection is usually done by using a thick absorber material which only muons can penetrate. Many EAS arrays lack or are poorly supplied with muon detectors. The CASA Array has the largest area with 2500 m^2 of muon detector, and is in the closest position to proving the method. It is sometimes argued that events are biased towards γ-ray showers if we use the "age" parameter. The shape of the lateral distribution of the shower electrons, usually described as the "age", is related to the height of the shower maximum and, thus, to the primary particle. However, there has been no general agreement about the usefulness of the age parameter in identifying γ-ray showers. It is generally believed that this parameter is less effective than the muon content.

Sensitivity of ground-based detection

The minimum detectable flux is $\sim 10^{-8}$ cm^{-2} s^{-1} at GeV energies with the EGRET detector, which is limited by the detection area of the satellite-borne detector. The typical detection area of a Čerenkov telescope, $\sim \pi 150^2$ $m^2 \approx 10^9$ cm^2, is determined by the lateral spread of Čerenkov photons. Recent large EAS arrays have also an effective area $\approx 10^9$ cm^2. Even this detection area in ground-based observation is not sufficient to detect the small flux of VHE and UHE γ-rays. Most of the current upper limits or the observed fluxes so far reported have been near the minimum detectable flux as limited by cosmic ray background. The observation time on one object is typically about 10^5 s (several nights) and $3 \cdot 10^6$ s (during 1 year operation), respectively for Čerenkov telescopes and EAS arrays. A typical field view of 1° half-angle corresponds to a solid angle $\sim 10^{-3}$ sr. The cosmic ray flux is about 2×10^{-5} and 2×10^{-10} cm^{-2} s^{-1} sr^{-1} at 10^{12} eV (1 TeV) and 10^{15} eV (1 PeV), respectively. The sensitivity is estimated by equating the γ-ray flux to the statistical fluctuation, $\sim \sqrt{N}$, of cosmic ray counts N. Thus, we obtain the minimum detectable flux to be $10^{-11} \sim 10^{-10}$ $(10^{-14} \sim 10^{-13})$ cm^{-2} s^{-1} for 1 TeV (PeV) γ-rays.

(a)

(b)

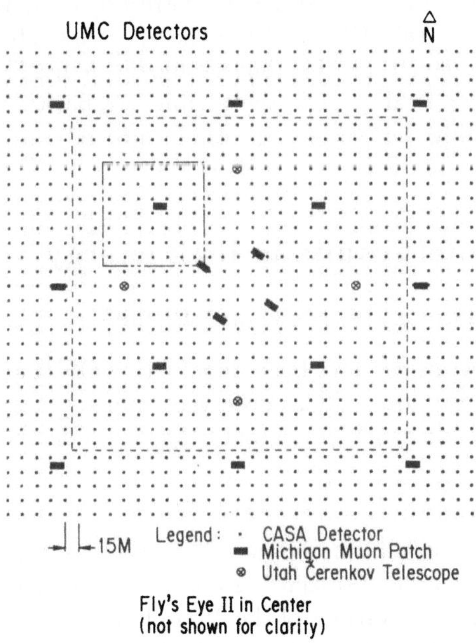

Figure 8: CASA array of the University of Chicago. (a) Photograph and (b) Schematic drawing to describe the layout of array elements of the scintillation detectors. (Courtesy of J. W. Cronin, University of Chicago)

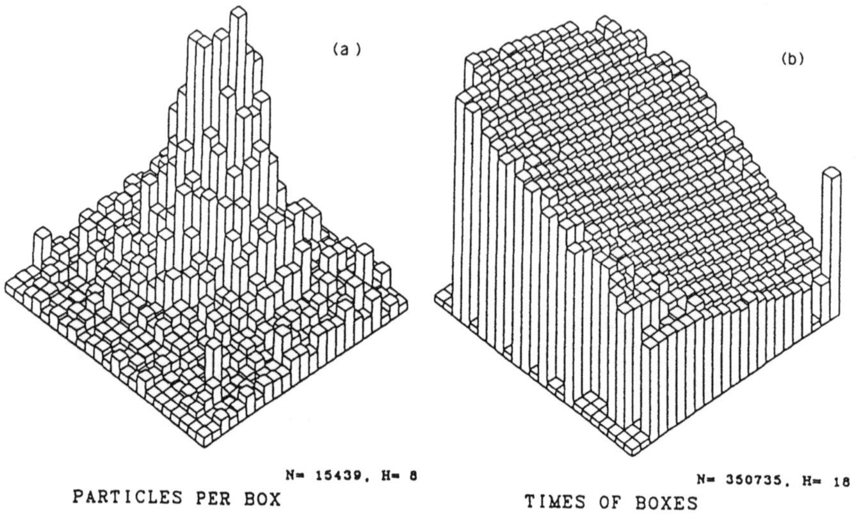

Figure 9: Pulse height and arrival time distribution in each detector of the CASA array. (Courtesy of J. W. Cronin)

Assuming that the integral energy spectrum of γ-rays has a power index of 1, a flux of $10^{-8} \sim 10^{-7}$ cm^{-2} s^{-1} at 1 GeV correspond tos $10^{-11} \sim 10^{-10}(10^{-14} \sim 10^{-13})$ cm^{-2} s^{-1} at 1 TeV (PeV). Since the spectrum steepens with increasing energy, sensitivity better than the above values is generally required. The imaging Čerenkov telescope can reduce the cosmic ray background by the combined effects of distinguishing the image shape and improved angular resolution, and the minimum detectable flux is improved to as low as $\sim 10^{-12}$ cm^{-2} s^{-1} at $\sim$ 1 TeV.

4 Observational results in high energy γ-rays

4.1 Energies between 100 MeV and 10 GeV

Satellite-borne observations have been done with COS B and GRO, following the OSO-3 and SAS 2 satellites. The observed intensity of γ-rays is plotted as a function of Galactic longitude (integrated over latitudes between $-5°$ and $5°$) in Fig. 10 [5].

Diffuse component

The general structure of the intensity variation is explained by the uniform cosmic rays in the Galactic disc and the distribution of interstellar matter, which is inferred from radio observations of the 21 cm line of the H I atom and millimeter line of the CO molecule to trace column density of H I and molecular hydrogen H_2.

The total luminosity of the Galactic disc above 100 MeV is $(1-2) \times 10^{39}$ erg s^{-1} [5]. The observed differential energy spectrum of the diffuse γ-rays at $\geq$ 100 MeV shown in Fig. 11 is well explained by γ-rays from the decay of π^0 mesons produced by cosmic ray protons in collisions with interstellar matter. (At lower energies,

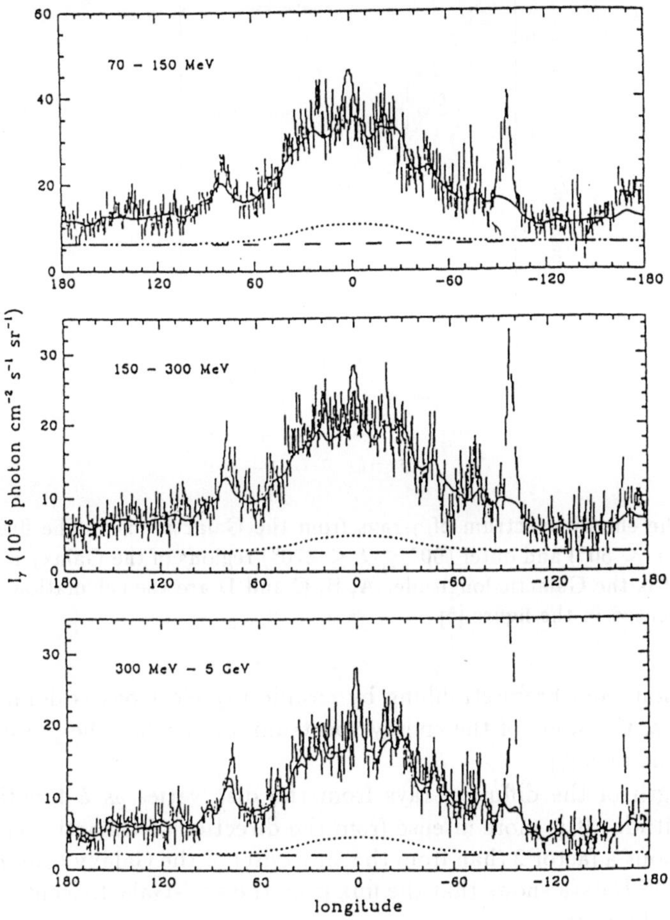

Figure 10: Diffuse γ-ray distribution from the Galactic disc. The intensity is plotted as a function of galactic longitude (integrated over latitudes between $-5°$ and $5°$) for three energy bins indicated in the figure [5].

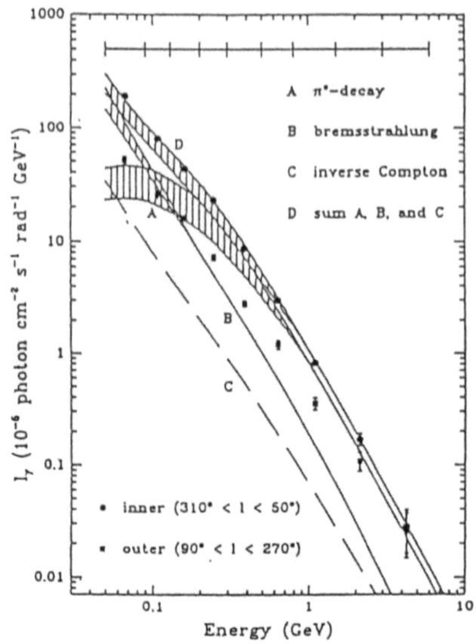

Figure 11: The energy spectrum of γ-rays from the Galactic disc. The flux from the inner ($310° \leq \ell \leq 50°$) and outer ($90° \leq \ell \leq 270°$) regions of the Galaxy is separately shown, where ℓ is the Galactic longitude. A, B, C and D are the calculations from each process as indicated in the figure [5].

the contribution from bremsstrahlung by cosmic ray electrons is dominant). At higher energies, the slope of the energy spectrum approaches the power index of -2.75 of cosmic rays [6].

The strength of the diffuse γ-rays from the disc varies as a function of the Galactic longitude. It is more intense from the direction of Galactic center (inner Galaxy) by about a factor 2 than from the anticenter of the Galaxy (outer Galaxy), though the COS B data shows that the flux from the anti-Galactic center decreases less steeply with energy.

Galactic point-like sources

In addition to the diffuse γ-rays due to cosmic ray interactions with interstellar matter, several point sources have been found with the COS B satellite [7] (Table 1). Identified with known sources which are observed in other wavelengths are the Crab and the Vela pulsars. Recently, X-ray observations with the ROSAT satellite detected a periodic signal of 0.237 sec from the X-ray star 1E0630+178 [8], which has been suspected to be the counter part of the γ-ray source Geminga (2CG195+04). It was discovered that the COS B and GRO data are pulsed with the same period of 0.273 sec [9], and the long standing "mystery" of identifying Geminga has been solved. The EGRET detector also observed a periodic signal of 0.102 sec from 2CG342−02, which was then identified with a radio pulsar PSR1706−44. Thus, the number of γ-ray pulsars now amounts to four. The other point sources, possibly γ-ray pulsars, are not yet identified with any known objects.

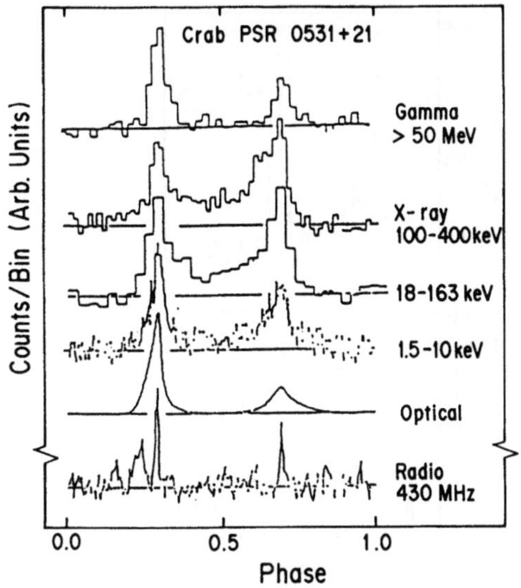

Figure 12: γ-ray signals from the Crab as a function of the phase of the pulsar spin period, together with X-ray, optical and radio signals [11].

The intensity variation of the Vela and the Crab with the period of 33 msec and 89 msec, respectively, observed with the COS B satellite, is shown in Fig. 12 and 13. Both pulsars as well as the Geminga have double peaks, whereas PSR1706−44 shows a single broad peak in γ-rays. In the case of the Crab pulsar, the γ-ray pulse is seen at the same phase as the radio, optical and X-ray peaks. Interpulse emission can be seen between the two peaks in the case of the Vela, amounting to about 50 % of the total, while the interpulse intensity is less than 15 % in the case of the Crab. Unpulsed emission from the Crab is about 45 % around 400 MeV and has a steeper spectrum than the pulsed one [10]. It is not clarified yet whether this has originated from the pulsar or the nebula.

Geminga, first detected by SAS 2 and confirmed by COS B, is the second most intense point source next to the Vela pulsar. The radiation in different wavelengths has a small flux or just an upper limit in radio, optical light and X-ray emission from the object. The source has its largest luminosity in γ-ray.

One of the COS B sources, 2CG353+16, is from the direction of a molecular cloud ρ Oph, suggesting that the excess of γ-rays are produced by cosmic rays colliding with the dense matter in the cloud. This mechanism works more effectively if intense cosmic rays exist near the molecular cloud. An association of supernova remnant (SNR) with a molecular cloud is a candidate for such a type of γ-ray source. A SNR has been found within the error circle of 2CG006−00, 2CG065+00 and 2CG078+01 (of course, also in 2CG135+01 and 2CG263−02).

The Durham group studied several COS B sources, and found that out of 9 unidentified sources, 6 can be explained as due to cosmic ray interactions [12].

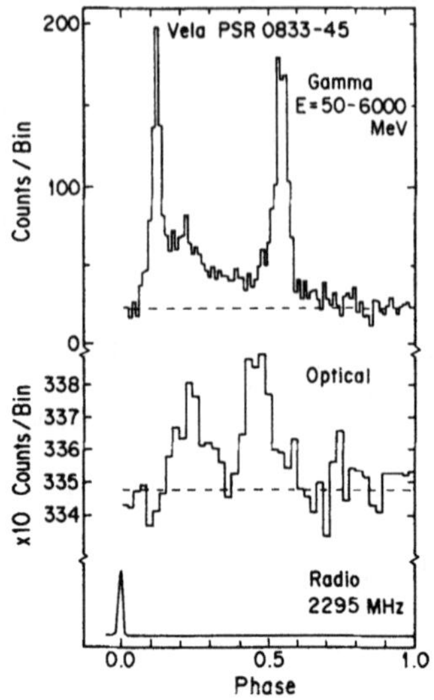

Figure 13: Gamma-ray signals from the Vela pulsar are plotted as a function of the phase of the pulsar spin period and compared with the optical and radio signals [11].

The required enhancement factor of cosmic ray intensity was estimated to be ~10, ~4 and ~1.5 for 2CG135+01, 2CG121+04, and 2CG353+16, respectively. These γ-ray sources are expected to be spatially extended. By analyzing the COS B data carefully for the point-like nature of the sources, Mayer-Hasselwander and Simpson [13] reported that 8 sources out of 20 in the 2CG catalogue remain point-like and new 9 sources are newly added. A different analysis by Grenier, Hermsen and Pollock [14] lists 12 sources as the revised candidates for point-like sources.

The limiting flux of the COS B search is $\sim 1 \times 10^{-6}$ cm^{-2} s^{-1} for $E > 100$ MeV, and the luminosity of $\sim 10^{35}$ erg s^{-1}, assuming the distance is typically ~ 5 kpc. From these values, the total number of γ-ray sources in the galaxy is estimated to be ~ 160 [15].

Extra-Galactic sources

One source, 2CG289+64, in the COS B catalogue appears at high Galactic latitude and is likely to be identified with QSO 3C273. In addition to 3C273, GRO reported that a total of fourteen Active Galactic Nuclei (in June 1992 and more being found) emit γ-rays higher than hundreds of MeV (circulated as GRO news).

Very and ultra high energy observations

VHE and UHE γ-rays have been searched for from various types of objects such as radio and X-ray pulsars and supernova remnants. The number of point sources reported from VHE γ-ray observations exceeds 10 and there are a few from UHE. However, the situation is still controversial, and it is hard to conclude how many out of those sources are established as emitters of high energy γ-rays. The

Table 1: The second COS B catalogue

name	ℓ	b	error radius	flux$^{(a)}$	spectrum$^{(b)}$	note	
2CG 006−00	6.7	−0.5	1.0°	2.4	0.39 ± 0.08	W28$^{(c)}$	
2CG 010−31	10.5	−31.5	1.5°	1.2			
2CG 013+00	13.7	+0.6	1.0°	1.0	0.68 ± 0.14		
2CG 036+01	36.5	+1.5	1.0°	1.9	0.27 ± 0.07		
2CG 054+01	54.2	+1.7	1.0°	1.3	0.20 ± 0.09		
2CG 065+00	65.7	0.0	0.8°	1.2	0.24 ± 0.09	G65.1+0.6$^{(d)}$	
2CG 075+00	75.0	0.0	1.0°	1.3			
2CG 078+01	78.0	+1.5	1.0°	2.5		G78.2+2.1$^{(c)}$	
2CG 095+04	95.5	+4.2	1.5°	1.1			
2CG 121+04	121.0	+4.0	1.0°	1.0	0.43 ± 0.12		
2CG 135+01	135.0	+1.5	1.0°	1.0	0.31 ± 0.10		
2CG 184−05	184.5	−5.8	0.4°	3.7	0.18 ± 0.04	PSR0531+21	(a)
2CG 195+04	195.1	+4.5	0.4°	4.8	0.33 ± 0.04	Geminga	
2CG 218−00	218.5	−0.5	1.3°	1.0	0.20 ± 0.08		
2CG 235−01	235.5	−1.0	1.5°	1.0			
2CG 263−02	263.6	−2.5	0.3°	13.2	0.36 ± 0.02	PSR0833−45	
2CG 284−00	284.3	−0.5	1.0°	2.7			
2CG 288−00	288.3	−0.7	1.3°	1.6			
2CG 289+64	289.3	+64.6	0.8°	0.6	0.15 ± 0.07	3C273	
2CG 311−01	311.5	−1.3	1.0°	2.1			
2CG 333+01	333.5	+1.0	1.0°	3.8			
2CG 342−02	342.9	−2.5	1.0°	2.0	0.36 ± 0.09		
2CG 353+16	353.3	+16.0	1.5°	1.1	0.24 ± 0.09	ρ Oph	
2CG 356+00	356.5	+0.3	1.0°	2.6	0.46 ± 0.12		
2CG 359−00	359.5	−0.7	1.0°	1.8			

Flux for $E \geq 100$ MeV in 10^{-6} photons cm^{-2} s^{-1}, assuming E^{-2} spectra.
(b) Intensity ($E \geq 300$ MeV) / Intensity ($E \geq 100$ MeV), assuming E^{-2} spectra for both intensities.
(c) Supernova remnants may be associated with this source [16].
(d) Supernova remnant G65.1+0.6 was recently found within the error circle [17].

Crab nebula is so far the only object commonly accepted as firmly established. Bearing this warning in mind, a list of the sources is presented in Table 2. Some details about the sources such as Crab nebula/pulsar, Cyg X-3, Her X-1 and some others will be discussed in the following. The flux and luminosity given in the table is the averaged one over the period if the detected signal is periodic. The group who made the detection is also indicated in the column for convenience, but the list is not complete (see Weekes [18] for the review with a more complete list of references). An extragalactic object, Centaurus A, was reported to be a VHE γ-ray source [19] in the mid-1970's. A BL Lac object, Markarian 421, one of the active galaxies detected with EGRET detector, was discovered to be TeV γ-ray source [20].

The objects can be divided into two groups: periodic and non-periodic sources. The evidence of detection may be influenced by the use of the technique of period analysis applied to periodic sources [21]. The imaging telescope has been found useful in detecting non-periodic sources [22]. "Careful application" of period analysis to the data of the imaging Čerenkov telescope is necessary to put observed evidence on a more sound basis.

Crab nebula and pulsar

A large number of observations on this unique source have been done in various ways to try to find the pulsed signal with the 33 msec period, the unpulsed excess of events and transient bursts.

Among them, the detection of a persistent unpulsed flux by the Whipple group has a high significance of 20σ and was confirmed by repeated observations. The stable nature of the unpulsed signal is shown in the Table 3. Thus, the result is considered to be the first clear detection of VHE γ-rays. The flux is $dN(E)/dE = 2.5 \times 10^{-10}(E/0.4\,\mathrm{TeV})^{-2.4\pm0.5}$ photons $\mathrm{cm}^{-2}\ \mathrm{s}^{-1}\ \mathrm{TeV}^{-1}$ between 0.4 and 4 TeV. The integrated flux above 400 GeV is $(7.0 \pm 0.4) \times 10^{-11}\ \mathrm{cm}^{-2}\ \mathrm{s}^{-1}$ [23].

The observation was done with an imaging Čerenkov telescope. The observed image of the Čerenkov light is characterized by several parameters. The image from the γ-rays is elongated towards the true direction of the source. Experimental errors, however, affect the orientation of the elongation axis and the observed angular distance from the axis to the source direction is defined as "miss" (Fig. 7). The "miss" indicates how close the arrival direction of each event is to the assumed source. The "width", the angular width of the image perpendicular to the elongation axis, has smaller values for γ-rays than for protons. A parameter, "azwidth", a combination of "width" and "miss", is defined as the width perpendicular to the line connecting the true direction and the center of the image, and was used in analyzing the Whipple data of the Crab. The distribution in Fig. 14 shows simulated γ-rays (bold line) and protons (dashed line). Simulation indicates that keeping smaller values of "azwidth" can reject 98% of protons while retaining more than 50% of γ-ray events. Two independent runs of observation (on- and off-source runs) were made, the former having the target object, the Crab, at the center of field of view and the latter pointing at a region of the sky away from the object of interest. A comparison of the observed events was done as a function of "azwidth" between on- and off-source runs. A small excess of on-source events

Table 2: Source List for VHE γ-Rays

source	D (kpc)	period	flux (cm^{-2} s^{-1})	threshold	luminosity (erg s^{-1}) (detection group)
Crab pulsar	2	33 ms	8×10^{-12}	~ 1 TeV	~ 10^{34} ?
Crab nebula			1×10^{-11}	~ 1 TeV	2×10^{34}
Vela pulsar	0.4	89 ms	7×10^{-12}	~ 1 TeV	5×10^{32}(Pachmari)
Vela nebula			8×10^{-11}	~ 200 GeV	5×10^{32} ? (Grindlay et al.)
PSR1937+21	5	1.6 ms	2×10^{-11}	~ 1 TeV	2×10^{35}
PSR1509-58	4.2	150 ms	4×10^{-11}	1.5 TeV	5×10^{34}
			2×10^{-11}	1.5 TeV	(Potchefstroom)
			$< 7 \times 10^{-11}$	0.3 TeV	(Durham)
PSR0355+54		156 ms	8.6×10^{-12}	1.3 TeV	1.3×10^{34}
PSR1953+29	3	6.1 ms	1×10^{-12}	~ 1 TeV	3×10^{35}
PSR1957+20	0.9	1.6 ms	$< 1 \times 10^{-11}$	4.3 TeV	$< 3 \times 10^{34}$ (Pachmari)
				~ 1 TeV	$< 3 \times 10^{34}$ (Durham)
			(3.5 ± 0.8) $\times 10^{-14}$	0.5 PeV	(8.3 ± 1.9) $\times 10^{33}$ (KGF)
Cyg X-3	11	12.6 ms ?	5×10^{-11}	~ 1 TeV	3×10^{36}
		4.8 hrs	2×10^{-14}	~ 1 PeV	1×10^{36}
Her X-1	5	1.24 s	3×10^{-11}	~ 1 TeV	3×10^{35}
			3×10^{-12}	~ 1 PeV	2×10^{37}
4U0115+63	3.5	3.6 s	7×10^{-11}	1 TeV	6×10^{35}
			4×10^{-10}	400 GeV	1.7×10^{36} (Durham)
Vela X-1	1.8	283 s	2×10^{-11}	1 TeV	3×10^{34}
				300 GeV	3.6×10^{34} (Durham)
		9.0 day	9×10^{-15}	3 PeV	2×10^{34}
Cen X-3	8		(1.3 ± 0.9) $\times 10^{-10}$	2 TeV	1.4×10^{35}
1E2259+586	4	6.979 s	2×10^{-11}	0.4 TeV	1.4×10^{33} (Durham)
			$< 2 \times 10^{-10}$	0.6 TeV	$< 2 \times 10^{33}$ (Whipple)
			$< 3 \times 10^{-11}$	0.6 TeV	$< 3 \times 10^{32}$ (Whipple, imaging)
LMC X-4	50	13.5 s	2×10^{-10}	~ 1 TeV	~ 10^{37} (Durham)
			$< 1 \times 10^{-10}$	~ 1 TeV	$< 5 \times 10^{36}$ (Durham)
		1.4 day	3×10^{-15}	10 PeV	1×10^{38}
SMC X-1	60		2 % of cosmic ray	~ 1 TeV	~ 3×10^{37} (Durham)
4U1626-67		7.68 s	8×10^{-11}	300 GeV	(Durham)
Sco X-1			2×10^{-11}	300 GeV	(Potschefstroom)
			6×10^{-10}	1000 GeV	(Durham)
Cen A			1×10^{-11}	~ 300 GeV	3×10^{40} (Grindlay et al.)
					$< 2.5 \times 10^{41}$ (Durham)
Mkn 421	~ 10^5		1.2×10^{-11}	> 500 GeV	(Whipple)

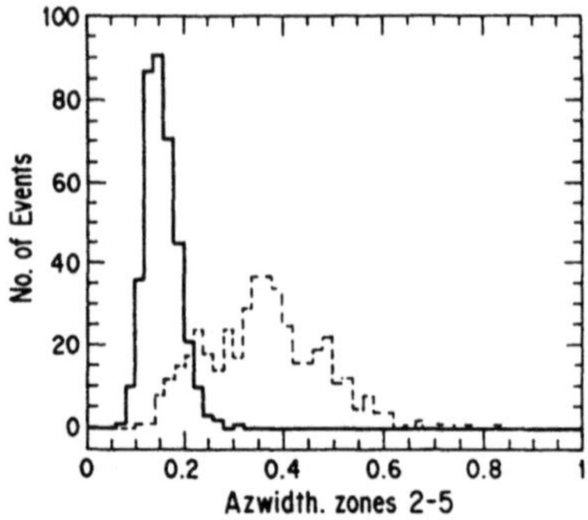

Figure 14: Simulated distribution of γ-rays (bold line) and proton events (dashed line) as a function of azwidth [24].

Table 3: Summary of Crab nebula detection by the Whipple Group

date	no image cut	azwidth cut	
	on / off ratio	on / off ratio	flux (10^{-11} cm^{-2} s^{-1})
Nov 1988	1.017 ± 0.004	1.27 ± 0.02	7.7 ± 0.7
Dec 1988	1.002 ± 0.004	1.31 ± 0.03	6.1 ± 0.5
Jan 1989	1.013 ± 0.005	1.28 ± 0.03	7.3 ± 0.8
Feb 1989	1.019 ± 0.008	1.31 ± 0.05	8.1 ± 1.1
Mar 1989	1.019 ± 0.008	1.26 ± 0.05	6.0 ± 1.2

relative to the off-direction is regarded as being from the Crab nebula. When the data is plotted as a function of "miss", the excess appears as an increase of events below 0.2° (Fig. 15), indicating that the excess is due to the Crab.

The Durham group [25] reported a pulsed signal with a 33 msec period from data of 156,342 events during 103 hr from 1982 September to 1983 November. There are 6587 events in the phase bin of the main pulse of the Crab. The expected number is 6224, based on the 74,684 events between phases 0.48 – 0.96 for which the COS B data suggest no γ-ray emission. The excess is 0.223% $\pm$ 0.054% of the background level of cosmic ray protons. This corresponds to the time averaged pulsed flux of $(7.9 \pm 1.8) \times 10^{-12}$ cm^{-2} s^{-1} above 1000 GeV.

The Whipple data, however, appears uniform when it is plotted as a function of the phase. The upper limit is 7×10^{-12} cm^{-2} s^{-1} at 99 % confidence level above a 400 GeV γ-ray energy. The γ-ray flux from the Crab pulsar/nebula is summarized in Fig. 16 together with the COS B result.

Cyg X-3

At the time of the radio outburst in 1973 of this source, a sporadic increase of counting rate was detected with the Čerenkov telescope of the Crimean Observa-

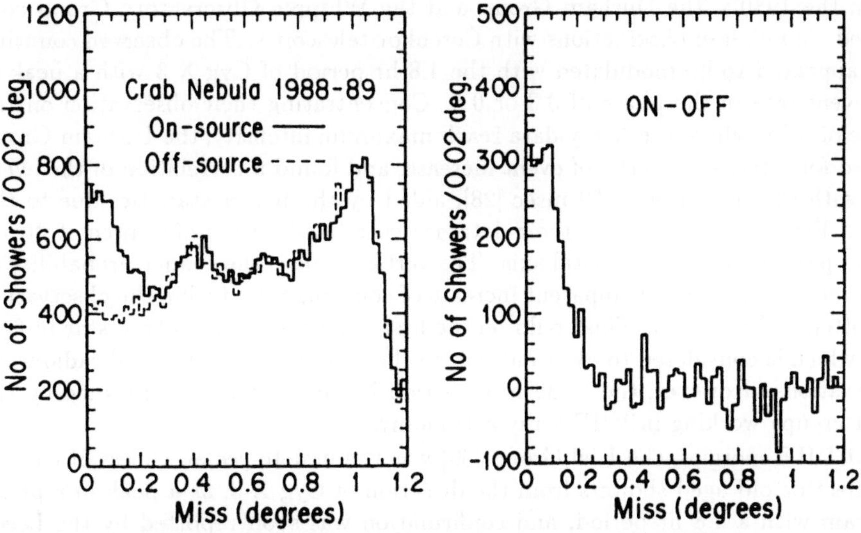

Figure 15: The data from the Crab (on-source) and an off-source direction observed by the Whipple Group as a function of the "miss" parameter [23].

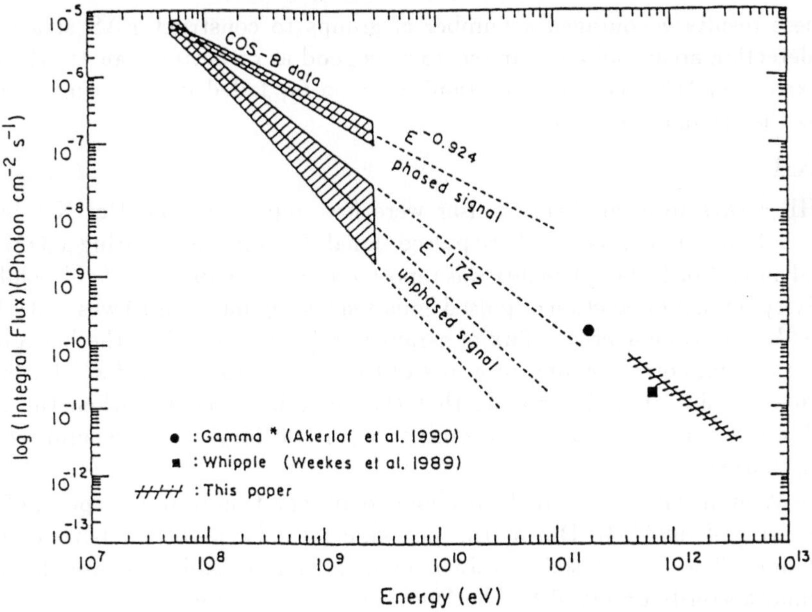

Figure 16: The γ-ray intensity from the Crab. (From Vacanti et al. [23])

tory, the first report of VHE γ-rays from a point source [26, 27].

In the 1980's, the Durham Group and the Whipple Observatory Group conducted a number of observations with Čerenkov telescopes. The observed counting rate appeared to be modulated with the 4.8 hr period of Cyg X-3 with a peak of the event rate in the phase of 0.3 or 0.6. Concentrating their observation on the phase of 0.65, where the X-ray data reach maximum intensity, the Durham Group waited for a transient burst of event increase, and found a correlation of the event rate with the period of 12.59 msec [28], aided by the better statistics due to the burst. Fig. 17(a) shows how their data are correlated with 12.59 msec of "test" period plotted on the horizontal axis. The vertical axis is the chance probability of observed correlation. An apparent increase of counting rate during the observation is plotted in Fig. 17(b). This could be the first evidence for a neutron star in Cyg X-3, which is considered to exist but has not been seen with X-ray and radio wave observations. However, the 12.59 msec period has not yet been confirmed by the other groups working in VHE γ-ray astronomy.

At UHE energies, the Kiel Group [30] was the first to see an excess of counts, by selecting old aged showers from the direction of Cyg X-3, as a peak in a phase diagram with a 4.8 hr period, and confirmation was soon reported by the Leeds Group [31]. The Kiel result also suggested that the showers from Cyg X-3 contained as many muons as normal cosmic ray proton showers, in contrast to the muon-poor nature expected from the standard model of γ-ray interactions. A possible detection of a weaker flux was reported by selecting muon-poor showers [32], which is in accordance with normal interaction of γ-rays.

These results encouraged a number of groups to construct EAS arrays with large detection areas and angular accuracy as good as or better than 1°. However, these arrays of "the second generation" have so far failed in detecting a flux as high as the original Kiel result.

Her X-1

VHE γ-rays from an X-ray pulsar were first reported from Her X-1 by the Durham Group, detecting a 1.24 sec pulsed signal that increased during a 3 minutes interval on 17 April 1982. The flux was reported to be 1.2×10^{-9} cm^{-2} s^{-1}, while the sensitivity (for a 3σ detection of pulsed emission lasting for 10 min) was $\sim 3\times10^{-10}$ cm^{-2} s^{-1}. The time-averaged flux of γ-rays was (0.84 ± 0.4)% of the background cosmic ray flux, corresponding to a flux of $(3\pm1.5)\times10^{-11}$ cm^{-2} s^{-1} for energy $\geq$ 1000 GeV [33]. It is interesting that the detection occurred when the X-ray period of the pulsar changed to becoming longer contrary to the general trend of getting shorter.

The Whipple Group has made 275 hours of observation from the spring of 1984 to the spring of 1987 [34]. During this period, eight episodes of burst-like emission were found. The first episode on June 11 in 1984 is notable because this burst was simultaneously observed by the Durham Group in Dugway. The DC excess was 165 ± 88, and the probability of random noise creating the periodic feature as observed was calculated to be 1.5×10^{-6}. This episode was also interesting because the Rayleigh power (see section 2.4) was peaked at the frequency shifted from the X-ray pulsar period of 1.238 sec by (0.16 ± 0.02)%. The same amount of

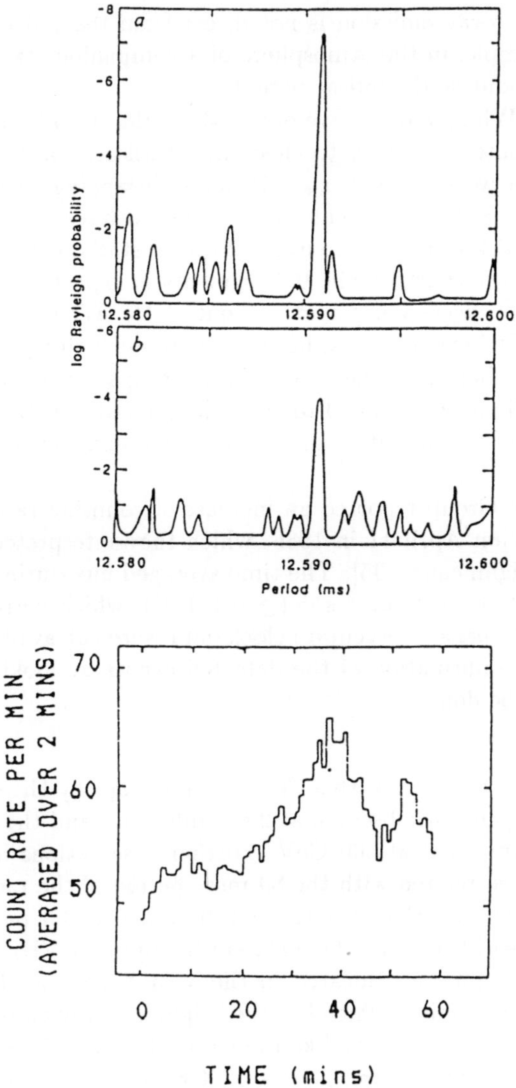

Figure 17: (a) The chance probability of the discovery of the 12.59 msec pulsar in Cyg X-3 by the Durham Group is plotted as a function of test period. The top and the bottom figures are from 1983 September 12 and October 2 observations, respectively. (b) The increase of event rate on 1983 September 12. (From Chadwick et al. [28, 29]).

shift was also observed by the Los Alamos Group at PeV energy, and also by the Haleakala Group at TeV energy in May and July of the same year, 1986. The shift, if true, suggests that γ-ray emission is not direct from the pulsar but might be reproccessed, for example, in the atmosphere of a companion star, whose motion and geometry can modulate the pulsar period.

According to the Whipple data, five out of the eight episodes have a duration time of about 30 minutes. In order to check the statistical basis of these occurrences, the total data were divided into 578 non-overlapping segments of about 30 minutes duration. Fifteen segments were found to have Rayleigh power, P, larger than 6.0 when 6.4 segments are expected from statistical fluctuations, and 3 segments had P larger than 8 while 0.9 segments is expected. The number of segments in which DC excess was found is 7 out of 15. When all these episodes are summed, a weak DC excess exists, however, without high significance.

It should be noted that, after the success of imaging analysis in the case of the Crab, the same technique was applied to the Whipple data on Her X-1, resulting in the disappearance of the periodic signal rather than enhancing the significance of positive evidence.

The Tata Institute Group reported an increase of counting rate from Her X-1 at 2147UT – 2201UT on April 11 in 1986, which they interpreted as a burst of γ-ray events, of 42σ significance [35]. The time averaged flux during the burst-like event increase was 1.8×10^{-8} cm^{-2} s^{-1} (≥ 0.4 TeV), which corresponds to the luminosity of 1.8×10^{37} erg s^{-1}. Accurate clock data were not available during the observation, and the confirmation of the detected events by looking for 1.24 sec periodicity could not be done.

Vela pulsar

This source was first observed with a Čerenkov telescope by Grindlay et al. [36]. An excess of counting rate was observed at 2σ significance, and the 3σ upper-limit flux was 8.5×10^{-11} cm^{-2} s^{-1} at 300 GeV. Further observations have been done by looking for signals correlated with the 89 msec period of the radio and optical pulsar. Bhat et al. [37] reported a persistent emission at 0.3 periods from the optical peak. It is also likely that the Vela emits unpulsed VHE γ-rays similar to the Crab. The Vela pulsar is located in the Vela X nebula. It is of current importance to know if the object which has the brightest γ-ray emission in the 100 MeV – 1 GeV also emits VHE γ-rays like the Crab. The Vela X extends to $\sim 1°$, however with a compact synchrotron nebula of X-rays. A search for an unpulsed signal is necessary with the improved current technique.

Other X-ray and radio pulsars

There is evidence of VHE γ-ray emission from X-ray pulsars, such as Vela X-1, Cen X-3, SMC X-1, 1E2259+63, and from some radio msec pulsars, as listed in Table 2. Almost all the results are based on a weak correlation of the observed data with the known periodicity established by radio or X-ray observations.

Supernova 1987A

This supernova that exploded recently in 1987, SN1987A in the Large Magellanic Cloud (LMC), may have resulted in a birth of a pulsar, which could be very active with a short spinning period. The possibility of this newly born pulsar emitting intense VHE and UHE γ-rays was indicated by several authors, and a number of observations to prove this prediction have been attempted with no positive answers, except a probable detection of a burst of TeV γ-rays about 300 days after the supernova explosion [38]. It is interesting to note that the time of the occurrence coincides with the time when the X-ray satellites, Ginga and Kvant, detected a burst of X-rays. The suspected burst of TeV γ-rays suggests that particle acceleration at least up to 10 TeV took place in the shock created by the supernova ejecta.

The long-term upper limit, $\leq 2 \times 10^{37}$ erg s^{-1}, of γ-ray luminosity gives the upper-bound on cosmic rays from the supernova as 10^{39} erg s^{-1}, which is still much larger than the total luminosity of the Galactic cosmic rays, $10^{37} \sim 10^{38}$ erg s^{-1} above 10^{15} eV. Thus, we conclude that a supernova just born is not an extraordinarily strong emitter which can supply a considerable fraction of UHE cosmic rays in the short time of several years after the presumed birth of a pulsar.

Diffuse γ-rays from the Galactic disc

Energetic γ-rays are produced by cosmic ray interactions with interstellar protons. The column density of the interstellar matter along the line of sight in the Galaxy is roughly 1 proton cm$^{-3} \times 3 \cdot 10^{22}$ cm $= 3 \cdot 10^{22}$ protons cm^{-2}. Since the total cross section of proton - proton reactions is $\sim 10^{-26}$ cm^2, the relative flux of energetic γ-rays, estimated from the product of the column density and the cross section, is no higher than 10^{-4}.

An estimation can be also done empirically from the γ-ray data in the lower energy region. The energy spectrum of diffuse γ-rays is considered to have the same power index as cosmic ray, i.e., $\propto E^{-1.7}$. The γ-ray flux detected with the COS B satellite is (7 or 2) $\times 10^{-4}$ cm^{-2} s^{-1} sr^{-1} from the general direction of the Galactic center (GC) or the anticenter direction. The spectrum, when extrapolated with the same index as the cosmic ray spectrum, gives 5×10^{-10} cm^{-2} s^{-1} sr^{-1} at 1 TeV and 3×10^{-15} cm^{-2} s^{-1} sr^{-1} at 1 PeV from GC. This intensity is $10^{-5} \sim 10^{-6}$ of the cosmic ray flux.

The estimated flux is a factor of $10^2 \sim 10^4$ smaller than the detection sensitivity at 1 TeV $\sim$ 1 PeV, pointing out again the importance of distinguishing γ-rays from the cosmic ray background. Though a current imaging Čerenkov telescope can improve the sensitivity by a factor of 10, its minimum detectable flux is still about one order of magnitude larger than the expected flux. Either the operation of multiple imaging telescopes or a telescope with further improvements is necessary to detect the diffuse γ-rays from the Galactic disc.

4.2 Comments on periodic signals

Most of the evidence for VHE and UHE γ-ray signals are from the time correlation between pulsar period and observed event rates. Consider a set of data, $(t_1, t_2, .., t_i, .., t_N)$, to examine if t_i appears to correlate with a period p, where t_i is the time when the i-th event occurred. A random distribution of t_i can produce a spurious periodicity, and the standard method of evaluating this probability is the Rayleigh test.

Let us define the phase of each event by $a_i = 2\pi\nu t_i$, a product of the acquisition time, t_i, and $2\pi\nu$, where ν is $1/p$. The function $P = |R|^2/N$ is called Rayleigh power, where N is the number of events and R is $\sum(\cos a_i + i \sin a_i)$ summed over all the observed events. If the events are distributed at random over the phase of the assumed period, R grows as $\propto \sqrt{N}$ with increasing event number, and the probability of R being larger than a certain value R_0 is given by e^{-R_0}. The correlation is likely to be true when the Rayleigh power is found to be large, because its occurrence probability is too small to be a chance coincidence of random background events. The largest value of the Rayleigh power is thus searched for by varying the assumed period ("test period") near the predetermined or the assumed spin period of the neutron star. An example of such a search is shown in Fig. 17(a).

Any data set has a finite duration time and this causes an effect on the analysis of periodic correlation. When the duration time of observation is T, the fundamental frequency of the Fourier transform is given by T^{-1}, which implies that the neighbouring test periods located within $\Delta p = \nu^{-2}\Delta\nu = p^2 T^{-1}$ are not independent from each other. Thus, "the independent number of trials" in the Rayleigh test is given by $\sim \Delta f \cdot T$, where Δf is the frequency range in which the search is made. This effect of "over-sampling" is usually corrected for by multiplying the probability calculated from the Rayleigh power by this number of trials.

It is obvious that the assumed period must be accurately known compared with Δp, but this is not always the case. It is difficult to deduce a period from the high energy γ-ray data itself, so the period from radio or X-ray observations is usually assumed. Since the pulsar period varies with time, it is necessary to use the most recent and accurate period, if available. In the case of binary systems, rotational period of the pulsar is Doppler shifted due to orbital motion. This effect must be corrected for accurately before any periodicity is looked for. Failure to use rigorous orbital correction procedures may result in a failure to obtain a periodic signal [21].

For each celestial object, observation is usually done near its culmination in a ground-based experiment detecting VHE and UHE γ-rays, causing every day a gap of no observation. This can produce a systematic effect of artificial periodicity when observations span many days. For an instance, the Cyg X-3 period of 4.8 hrs variation is very close to one fifth of 24 hrs. The beat frequency between one sidereal day and this period is one cycle per several months.

The effect of the observation gap is not simple and obvious; let us illustrate another trap following the work by Lewis, Lamb and Biller [39]. The considered case is a data set consisting of two segments with a gap of length T. Both segments have the number of counts N and duration time of $T_s \ll T$. It is assumed that we have γ-ray events with periodicity coherent between the two segments. The Rayleigh power, R, is calculated as $\frac{1}{2N}\left(|R_1 + e^{2\pi i \Delta\nu T} R_2|^2\right)$, where R_1 and R_2 correspond to the amplitudes of the first and second segments, respectively. By taking the searched range of frequency as having a width of $\frac{1}{T}$, the independent number of trials is made to be one, so that the extra factor with phase $2\pi i \Delta\nu T$ multiplying R_2 scans the full range of phase from 0 to 2π. Then the exact alignment of phase between R_1 and R_2 can occur within the searched frequency range, i.e., $|R_1 + R_2| = |R_1| + |R_2|$, to maximize the Rayleigh power. This mechanism would increase the Rayleigh power, R, larger than the "true one". The true chance probability is thus larger than the calculated value of e^{-R}. Detailed calculations and a Monte Carlo simulation tell us that the underestimation of chance probability is by more than a factor of 10 and depends on various experimental conditions such as the number of segments.

Based upon the arguments introduced above, Lewis, Lamb and Biller [39] concluded that there had been no evidence of coherent periodic emission persistently lasting for more than a day in the data for 4U0115+63, 1E2259+586, and Her X-1.

Most of the evidence had been from periodic correlations in the data when a transient burst, a sudden increase of event rate, was detected. We have to observe a greater number of bursts if we want to confirm the preceding results. The "signal" to noise ratio is typically 10% for a burst and, when the source is quiescent, it is lower than 1%, so that significant correlation is hard to observe. If we accumulate data during a quiescent time for many days, we have to consider the effect described above as well as other systematic errors. The analysis of time varying features is a useful method to apply to the data with a low signal to noise ratio. However, in order to avoid confusion, it is desirable that we improve the detection method, reduce cosmic ray background and obtain a better signal to noise ratio.

5 Implications and prospects of VHE gamma-ray astronomy

5.1 γ-ray pulsars

The two pulsars of the Crab and Vela are found as γ-ray pulsars in the COS B data. The luminosity from the Crab and Vela is plotted as a function of energy band in Fig. 18(a) and (b), respectively. The luminosity is at its maximum in the γ-ray region in the case of the Vela pulsar. In order to explain these features, Ruderman et al. developed a pulsar outergap model [40, 41, 42].

According to them, a mechanism for forming a gap of electric potential is active in the outer region of the pulsar magnetosphere, where electron-positron pairs are created. The energetic electrons and positrons undergo at and near the gap a cascade of processes such as synchrotron radiation and the inverse Compton effect. The resultant radiation can be adjusted to be consistent with the observed emission from the Crab and Vela pulsars. Geometrically, the outergap is formed so as to yield two peaks of γ-rays in the phase diagram of the pulsar period as are observed for the Crab and the Vela. The radius to the light cylinder, at which place the rotating velocity of the pulsar magnetosphere reaches light velocity, increases for a longer period of neutron star rotation, and the outergap will be located at weaker magnetic field. This effect tends to shift the peak of emission toward higher energies, and a larger fraction of the spin-down energy of the pulsar is used for maintaining the gap and for radiating γ-rays. The discovery of a pulsed signal in Geminga [9], which has its most of energy output in γ-rays above 100 MeV, is in general agreement with this view. However, it had been predicted by Ruderman and Cheng [42] that the mechanism for maintaining the outer gap ceases at around

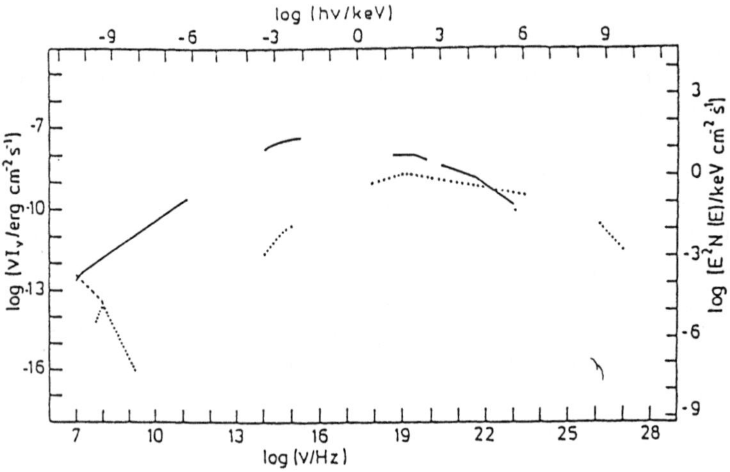

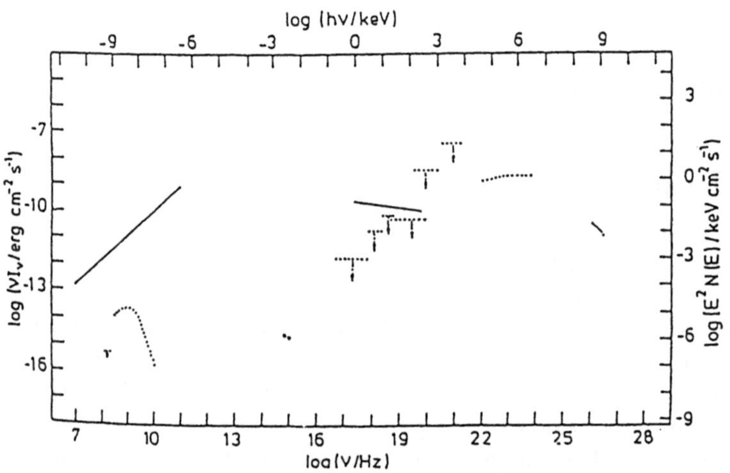

Figure 18: The luminosity per energy is plotted as a function of frequency for (a) the Vela and (b) Crab Pulsar. The frequency is converted into units of keV and shown in the upper abscissa. Solid and dotted lines indicate the emission from nebula and pulsar, respectively. Recent Whipple result is not included [11].

Table 4: γ-ray pulsars

Pulsar	0531+21	0833-45	1706−44	1E0630+178
SNR/nebula	Crab	Vela X		
COS B	2CG184−05	2CG263−02	2CG342−02	2CG195+04 (Geminga)
p (sec)	0.033	0.089	0.102	0.237
$\dot{p}$ (10^{-15} s/s)	423	125	93	11.4
distance (kpc)	2	0.4	1.5	0.038 – 0.38
E (erg/s)	5×10^{38}	7×10^{36}	3.4×10^{36}	3.5×10^{34}
dynamic age (years)	1230	11200	17300	3.2×10^{5}

a 0.12 sec period, which is apparently in contradiction with the discovery of the 0.237 sec period of Geminga.

In the outer gap model, energetic electrons up to 10^{13} eV are considered to cause $\sim 10^{9}$ eV γ-rays. It is not certain if these electrons radiate 10^{12} eV unpulsed γ-rays in the pulsar magnetosphere, or it may be the case that the electrons are transported into the nebula region, possibly reaccelerated by a shock mechanism and then radiate 10^{12} eV γ-rays. Whatever the true mechanism is, it is likely that the TeV γ-ray emission is closely related to the pulsar structure near the light cylinder [43].

The characteristics of four γ-ray pulsars are listed in Table 4. The PSR 1706−44 shows a single broad γ-ray pulse in the phase diagram, while the other three pulsars have double peaks.

5.2 Young supernova remnants and pulsars

A pulsar with a shorter period stores a larger amount of rotational energy and emits more luminous radiation per unit of time. A larger voltage for particle acceleration is induced from the faster rotation of the pulsar magnetic field. Thus, young pulsars, which have short periods, are probably the sites where particles are accelerated to high energies and γ-rays are produced. The observed γ-ray pulsars are consistent with this general view. In Table 5 are listed some of the COS B sources associated with SNR and pulsars. The young pulsars are likely to be accompanied by SNR which are still bright enough to be detected, so searching for pulsars in SNR would be useful for identifying γ-ray sources. Or unknown VHE γ-ray sources may be in the SNR which contains a pulsar, since unpulsed VHE γ-rays from the Crab may suggest that energetic electrons in the nebula of the SNR are the γ-ray source.

No optical or radio nebula of a SNR is found around the γ-ray pulsars Geminga. However, if electrons as energetic as $10^{13} \sim 10^{14}$ eV are generated at the pulsar causing high energy γ-ray radiation, a halo of synchrotron X-rays may accompany these γ-ray sources. The high energy electrons would emit synchrotron radiation in the X-ray region in a magnetic field at least as high as $\sim 1\mu$ gauss of the interstellar value. Such an example was observed at Cyg X-3 in the data of the X-ray satellite EXOSAT [45]. Young pulsars like the Crab are likely to have intense

Table 5: Associations of SNR and γ-ray source

γ-ray source	SNR	pulsar
2CG184−05	Crab	PSR0531+21
2CG263−02	Vela	PSR0833−45
2CG006+00	W28	PSR1758−23
2CG078+01	γ-cigny	
2CG311−01	G311.5−0.3	
321.0−1.2	G320.4−1.2	PSR1509−58
327.5−0.5	KES27	
2CG065+0.6	G65.1+0.6	PSR1953+29 ?

References are D'Amico [44] and Landecker, Clutton-Brock and Purton [17] for 2CG065+00.

radio and X-ray radiation. The number of SNR, radio and X-ray pulsars in the Galaxy so far detected is about $\geq$ 150, 300 and 30, respectively. The Crab had been for a long time the only source which had a SNR, radio and X-ray pulsar, but recently some more such objects were discovered. These Crab-like pulsars are likely to be γ-ray sources. Two examples are PSR1509−58 (with a period of 150 ms and named MSH 15−52 as a SNR) and PSR0540−69 (in the LMC, 50.3 ms period with $\dot{p} \sim 9.5 \times 10^{-12}\ \mathrm{s\,s^{-1}}$). A search in SNR and COS B sources with radio waves found PSR1338−62 (associated with SNR G308.7+0.0) and PSR1758−23 (associated with W28) to be also radio pulsars. The latter pulsar PSR1758−23, which can be associated with a SNR, is within the error circle of COS B source 2CG006−00.

The cataloged number of SNR and pulsars in the Galaxy must be smaller than the true number. The total number of SNR in the Galaxy is, for example, estimated by Helfand et al. [46] to be $\sim$ 360 to $\sim$ 590 compared with the cataloged number of 155. There are a number of supernova remnants in which X-rays are emitted via synchrotron radiation presumably from very high energy electrons, similar to the Crab. Recent searches for X-ray pulsars in these nebula by the ROSAT and Ginga satellites found evidence of a pulsed signal in the Vela nebula which has been known to be a SNR and radio pulsar but not an X-ray pulsar. The SNR nebula RCW 103 was also found to contain a pulsar of 69 ms period.

The characteristics of the Crab and the Vela pulsar, simultaneously SNR and γ-ray sources, are listed in Table 6. Some potential γ-ray sources which have Crab-like features are also listed in Table 7.

Radio pulsars of short periods, PSR1055−52 (197 ms, $d = 920$ pc), PSR1620−26 (11 ms, X-ray binary), PSR1821−24 (3 ms, isolated), PSR1855+09 (5 ms) and PSR0021−72A,B (4.5 ms and 6.1 ms), have been studied by the Durham Group with no positive results.

Among the known $\sim$ 30 X-ray pulsars, there are a few single pulsars, such as the Crab, PSR1509−58 (MSH 15−52) and PSR0540−69 in LMC. All of the others are close binary systems. Several X-ray stars from a close binary system have been

Table 6: SNR, pulsar, COS B and γ-ray source

Pulsar SNR/nebula COS B	0531+21 Crab 2CG184−05	0833−45 Vela X 2CG263−02
p (sec)	0.033	0.089
$\dot{p}$ (10^{-15} s/s)	423	125
distance (kpc)	2	0.4
E (erg/s)	5×10^{38}	7×10^{36}
dynamic age (years)*1	1230	11200
$L_{optical}$ (erg/s)	4×10^{33} (pulse)	1×10^{29}
L_r (erg/s) *2	2×10^{30}	1×10^{29}
L_X (erg/s) *3 nebula	2.3×10^{37}	4×10^{32}
point	1×10^{36}	~ 1% of nebula
L_γ (erg/s) *4	2×10^{35}	4×10^{34}
$L_{VHE\gamma}$ (erg/s)		
pulsed	$\sim 10^{34}$?	5×10^{32} ?
unpulsed	2×10^{34}	$\sim 10^{32}$?

(*1) $p/(2 \cdot \dot{p})$
(*2) radio in 10^7 to 10^8 Hz.
(*3) X-ray in 0.2 to 4 keV. X-ray luminosity is from [47].
(*4) γ-ray in 0.05 to 10 GeV.

Table 7: Crab-like source

Pulsar SNR	1509−58 MSH 15−52	0540−69 in LMC	1951+32 CTB 80	1E2259+586 CTB109 (SNR 109.1−1.0)
p (sec)	0.150	0.050	0.0395	6.979
$\dot{p}$ (10^{-15} s/s)	1490	~ 9500	6	620
distance (kpc)	4.2	50	2.5	3.6 − 5.2
E (erg/s)	2×10^{37}	1.5×10^{38}	4×10^{36}	$\sim 10^{32}$
age (years)	1550	1660	10^5 ? 582?(*5)	$(1.2 - 1.7) \times 10^4$
$L_{optical}$ (erg/s)	$< 1 \times 10^{32}$ (pulse)		$< 1 \times 10^{33}$	$< 4 \times 10^{32}$
L_r (erg/s) *1	4×10^{28} (pulse)		$< 2 \times 10^{30}$	$< 1 \times 10^{28}$
L_X (erg/s)*2 nebula	1.7×10^{35}	8×10^{36}	2.2×10^{34}	1.2×10^{35}
point	1.9×10^{34}	2.4×10^{36}	1.0×10^{34}	1.3×10^{34}
L_γ (erg/s) *3	$< 1.8 \times 10^{35}$		$< 9 \times 10^{35}$	$< 1.8 \times 10^{35}$
$L_{VHE\gamma}$ (erg/s) *4	5×10^{34} ?	$< 2 \times 10^{37}$	—	3×10^{35}? 1.4×10^{33} ?

(*1) radio in 10^7 to 10^8 Hz.
(*2) In (1.2 − 20) keV X-ray for CTB109 and in 0.2 to 4 keV for the other sources.
(*3) γ-ray in 0.05 to 10 GeV.
(*4) VHE γ-ray.
(*5) Historical supernova SN1408 (1404)?
X-ray data are from Trussoni et al. [48] (MSH15−52), Ogelman and Buccheri [49] (CTB 80) and Koyama et al. [50] (CTB 109). The VHE γ-ray data is from Nel et al. [51] (MSH15−52) and from the Durham group (PSR0540 and CTB109).

Table 8: Active galactic nuclei which emit high energy gamma-rays

Name	flux ($E > 100$ MeV) (10^{-6} cm^{-2} s^{-1})	Δ_1 error (arcmin)	Δ_2 shift (arcmin)	red shift z	Luminosity (0.1 – 5 GeV) 10^{48} erg s^{-1}
QSO 3C279	0.6 – 4.9	5	5	0.528	0.3 - 2
QSO 3C273	0.24 ± 0.06	30	30	0.158	0.008
QSO 0208−512	1.0 ± 0.1	9	4	1.03	2
QSO 1633+382 (4C38.41)	0.8 ± 0.1	~15	4	1.81	6
QSO 0528+134	0.8 ± 0.1	9	4		
QSO 0836+710 (4C71.07)	0.15 ± 0.04	30	35	2.17	1.1
QSO 2251+15 (3C454.3)	1.3 ± 0.2			0.859	0.8
QSO 2230+11	0.4 ± 0.1	24	23	1.037	1.0
BL Lac 1101+384 (MKN421)	0.11 ± 0.03	13	5	0.031	0.0001
BL Lac 0537−441	0.3 ± 0.1	36	26	0.894	0.2
BL Lac 0716+714	0.20 ± 0.06	24	28		
QSO 0202+149				0.852	
BL Lac 0235+164				0.524	
QSO 0420−014				0.633	

reported to be VHE and UHE γ-ray emitters. However, it is obvious that further careful experimental studies remain to be done before we draw a conclusion.

5.3 γ-Rays from active galactic nuclei

Energetic γ-rays (> several hundred MeV) from a number of active galactic nuclei (AGN), star-like objects in the center of active galaxies such as quasars, have been detected by the Compton GRO satellite. They have energy spectra of a power law, and are seemingly of non-thermal origin. The γ-ray data provide us with more direct clues than softer photons for understanding energetic non-thermal processes which take place in and near AGN. The fourteen AGN are listed in Table 8.

The Δ_1 and Δ_2 in the table are the uncertainty of the γ-ray source position and the difference between the known position of the galaxy and the estimated location of the γ-ray source. The given luminosity assumes uniform emission.

The contents of the table was from the presentation by R. C. Lamb (in the Workshop "Towards a Major Atmospheric Čerenkov Telescope" held at Ecole Polytechnique, Palaiseau, Paris in June 1992), which added three new sources.

A very energetic process of particle acceleration is expected in AGN, for which the energy source is considered to be a massive black hole of $\sim 10^8 M_\odot$. The high energy particles which may produce γ-rays are presumably accelerated by energy from the accretion flow onto the black hole. However, high energy γ-rays, if produced, cannot easily escape the central core of the AGN engine. This is because there is in the central region an intense radiation field of soft photons, which can interact with γ-rays to create electron and positron pairs and thus absorb energetic γ-rays. The typical spectrum of radiation from AGN is such that it has a bump in ultra violet (UV) wavelengths, presumably due to the thermal emission from the accretion disc. High energy γ-rays with energy k can create electron-positron pairs when colliding with UV photons if $k \cdot 10$ eV $> (m_e c^2)^2$, so the γ-rays of hundred GeV could hardly escape out of the AGN core. Similarly, X-ray photons can cause absorption of even GeV γ-rays. High energy nucleons also suffer from energy attenuation by producing pions in their collisions with photon fields if their energy E is larger than $M\,\mu\, c^4\,/\,2\varepsilon$, which is $10^{13} \sim 10^{14}$ eV for $\varepsilon \sim 10^3$ eV, where M, μ and ε are nucleon and pion mass and the energy of ambient photons, respectively. Thus, it is usually considered that high energy γ-rays are not from AGN.

The objects in Table 8 are blazers, OVV (Optical Violently Variable) galaxies or BL Lac type galaxies, which are considered to have jets toward us. This suggests that γ-rays are emitted from jets from a region far out from the core of the AGN. Dermer et al. [52] proposed an efficient and economical mechanism in which γ-ray emission collimated within a cone along the jet direction. According to their model, inverse Compton scattering of accretion disk photons by relativistic electrons in the jet has such a kinematical relation that the γ-ray emission is focussed into the directions in which the jet is observed in the radio measurement as superluminal. The BL Lac object, Mkn 421, from which TeV γ-rays were recently detected, is suspected to be accompanied by components that exhibit apparent superluminal movement of $v = 1.6 \sim 21.9$ [53].

In the case that accelerated protons are the energy source of the powerful radiation of AGN, neutrons, which must also exist there, may, before decay, escape out to 1 – 100 pc from the central engine. A considerable portion of the total energy can be thus transferred into the peripheral region and high energy γ-rays from there would be less absorbed and be able to escape out to intergalactic space. VHE γ-rays may be thus emitted from AGN (for examples, Sikora, Begelman and Rudak [54], Kirk and Mastichiadis [55], Mastichiadis and Protheroe [56] and Atoyan 1992 [57]). The emission of high energy γ-rays and their allowed maximum energy is determined as a result of complicated cascade processes around the central engine. The conditions which allow high energy γ-ray emission seem to depend on the size of the central core and on the intensity of ambient photons or the total luminosity of AGN. The high energy γ-rays, if emitted from AGN core, would be useful for inferring these parameters as well as the mechanism of particle acceleration in the AGN.

The observed spectrum shape would depend on the distance to the source and the intensity of intergalactic infrared and microwave photons, because γ-rays with energies of 0.1 – 1000 TeV suffer from considerable absorption in the collision with these background photons. Fig. 19 shows the absorption calculated by Stecker, Jager and Salamon [58]. The dotted line is the extrapolation of the flux from 3C279 detected with the Compton GRO, and curves below it correspond to various assumed intensities of the radiation of intergalactic space. It is seen that absorption

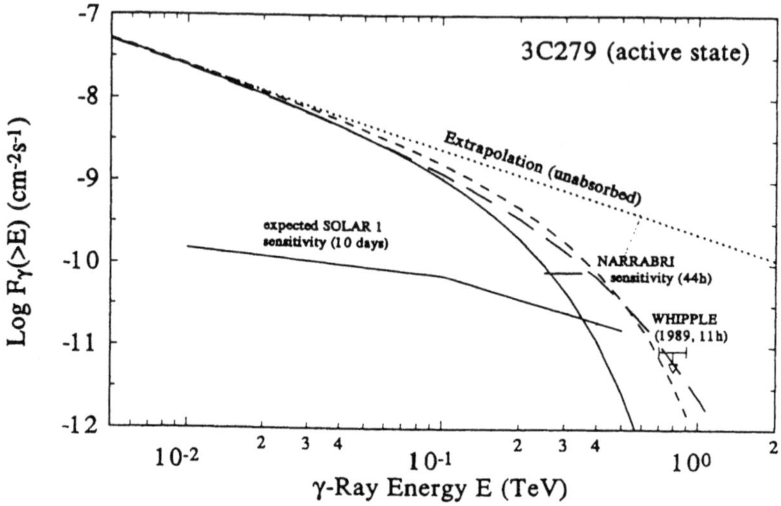

Figure 19: Spectrum of very high energy γ-rays from 3C279 plotted against γ-ray energy taking into account the absorption due to inter-galactic infrared photons. The dotted line is the extrapolation of the flux in the active state detected with the Compton GRO, and the two dashed curves are the attenuated cases with two different estimates of infrared intensity in inter-galactic space. The upper limit by the Whipple group and the sensitivities of some other detectors are also shown [58].

happens somewhere between 0.1 to 1 TeV of γ-ray energy for the distance of 3C279 ~ 1.6 Gpc (red shift ~ 0.54). The TeV γ-rays were detected by the Whipple Group from Mkn 421 [20].

5.4 Acceleration site of cosmic rays

A notable feature of cosmic rays is in their energy, extending to 10^{20} eV with a non-thermal spectrum, making it the most energetic radiation observed in the present universe. However, these charged particles propagate in the Galaxy by diffusion, and it is hard to infer the sources of cosmic rays directly from the observed arrival direction, unlike the case of neutral radiation such as radio waves, optical light and X-rays. The general characteristics, such as the total production rate of cosmic rays, are consistent with a supernova origin. As the supernova ejecta expands, a shock is created in interstellar space. It is considered that the first order Fermi acceleration takes place in the turbulent region, which explains the observed cosmic ray power spectrum with the index of about -2.7.

However, this acceleration near SNR has a limitation on its maximum accelerated energy of around $10^{14\sim15}$ eV, where the cosmic ray spectrum is known to bend (usually called "knee") and to become steeper at higher energies. The origin of cosmic rays at the energies above the knee has for a long time been debated as well as the causes of the knee. One view point, since the discovery of radio pulsars, is that pulsars accelerate the cosmic rays of energies around the knee. The point sources of VHE and UHE γ-rays are likely a direct indication of such acceleration sites of cosmic rays. An alternative view is that the cosmic rays at the higher energies are due to a shock in a very large extended region like the termination shock of the Galactic wind. The detection of γ-ray point sources can be a clue for solving this long-standing problem.

The map of the γ-ray emission in the Galaxy is useful for inferring the general distribution of cosmic rays in the Galaxy. The distribution of cosmic rays is likely to follow the distribution of SNR as the origin of cosmic rays as well as determined by their propagation in the Galaxy.

If a supernova, as an origin of cosmic ray particles, is embedded in an interstellar dense cloud, the interaction of cosmic rays and the dense matter in the cloud will cause an enhanced emission of γ-rays. Applying this mechanism to the COS B data, the production rate of γ-ray energy from such a source was estimated by Morfill and Tenerio-Tagle [59] to be $\dot{E}_\gamma = 10^{34} \sim 10^{35}$ erg s^{-1}. The γ-ray luminosity at TeV energy can be estimated to be $10^{32} \sim 10^{33}$ erg s^{-1}, since the power exponent -2.7 of cosmic rays implies that the energy dependence of the energy production rate of γ-rays is $\sim E^{-0.7}$.

The total number of SNR in the Galaxy can be estimated from the explosion rate of SN which is assumed to be one per 50 yrs. The enhanced intensity, f, relative to the ambient cosmic rays is related to the age and the size of the SNR, and, given the volume of the Galactic disc, the total volume fraction with the enhancement, f, is estimated to be $\sim 0.2/f$. The number of clouds in the Galaxy is assumed to be about 4000. The SN and the clouds are homogeneously distributed in the Galaxy, so the number of γ-ray sources is therefore $N_\gamma \sim 700/f \sim 7 - 70$, for $f \sim 10$ to 100 depending on the distance from the center of the nebula. The candidates for this case would be W28, W44, IC443, Cyg loop and ρ Oph [59].

The recent detection of VHE γ-rays from Mkn 421 may imply that this AGN has also a high activity of high energy cosmic rays. The existence of a number of such objects can explain a considerable portion of the cosmic ray flux near the knee [60]. It is interesting to note that such a model of an extra-galactic origin of cosmic rays was also suggested by the COS B result of the map of the γ-ray emission in the Galaxy.

5.5 Diffuse γ-rays from galaxies

Diffuse VHE γ-rays from the Galactic disc are yet below detection level. However, the luminosity from the whole disk above 100 GeV is as large as 10^{37} erg s^{-1}. This estimation is based on the extrapolation of the luminosity 10^{39} erg s^{-1} of the COS B result to higher energies with a power law of $E^{-0.75}$. The luminosity within a field of view of 10^{-2} sr (half angle of 3°) is 10^{35} erg s^{-1}, comparable or even larger than the value observed from the Crab nebula, though it is not easy to identify because of the uniform incidence of γ-rays in the field of view.

The above total luminosity from a galaxy at the distance of 100 kpc yields a flux of $\sim 10^{-11}$ cm^{-2}s^{-1}, which is near the current sensitivity of an imaging Čerenkov telescope . If the distance is farther, the emission appears like a point source. The upper limit on the luminosity from the nearest galaxy, LMC (at a distance of 50 kpc), is $\sim 10^{37}$ erg s^{-1}, as large as the limit for SN1987A. This indicates that the cosmic ray activity in LMC is no stronger than in our Galaxy. However, the LMC is of a smaller size than our Galaxy, and the cosmic ray flux from the LMC is in fact expected to be much smaller than this limit. Observations in other wavelengths show, for example, that the total diffuse X-rays from the Galactic disc is $\sim 10^{38}$ erg s^{-1} for a 2–11 keV region and $\sim 2 \times 10^{35}$ erg s^{-1} from the LMC. Optical and X-ray luminosity varies from one galaxy to another, and higher activities than our Galaxy appear in some galaxies. There might be some galaxies which contain more intense cosmic rays so that the VHE γ-ray emission would be above the detection level.

6 High-energy neutrino astronomy

Since photons cannot pass through a shield of a few hundreds of grams of matter, only neutrinos carry information of active objects in the Universe behind those shields. In addition, if we discover a high-energy neutrino source, it will be firm proof that the cosmic accelerator works with hadronic processes, not well-known electromagnetic ones. It can be a clue for solving the long-standing problem: acceleration of cosmic rays. Thus the observation of cosmic high-energy neutrinos has a possibility of opening a entirely new window of astronomy.

6.1 Production of high-energy neutrinos

High-energy neutrinos are produced in decays of pions (or kaons) generated in interactions of accelerated particles with matter, i.e., *pp*-collisions (eq.(1)) or $p\gamma$-collisions (eq.(2)), and high-energy γ-rays are the decay products of neutral pions generated in the same reactions:

$$\begin{array}{cccccccc} p + p \,\mathrm{or}\, \gamma \rightarrow & \pi^0 & + & \pi^+ & + & \pi^- & + \cdots . \\ & \downarrow & & \downarrow & & \downarrow & \\ & 2\gamma & & \mu^+ \nu_\mu & & \mu^- \bar{\nu}_\mu & \\ & & & \downarrow & & \downarrow & \\ & & & e^+ \nu_e \bar{\nu}_\mu & & e^- \bar{\nu}_e \nu_\mu & \end{array} \tag{6}$$

In contrast to *pp*-neutrinos, the production of $p\gamma$-neutrinos has a threshold effect; a proton must have energy greater than $\sim 100(E_\gamma/3\mathrm{keV})$ TeV. Thus $p\gamma$-neutrinos are important only in the ultra-high-energy region.

Let us consider the condition under which *pp*-collisions produce neutrinos efficiently [61]. The density of the target matter (assumed to be hydrogen: n_H) should be low enough so that the pions will decay before interacting within the target:

$$\gamma_\pi \tau_\pi \ll 1/n_H \sigma_{\pi N} c \quad \mathrm{or} \quad n_H \gg 10^{18} (E_p/\mathrm{TeV})^{-1}\,(\mathrm{cm}^{-3}), \tag{7}$$

while the column density (z) should be large enough so that the accelerated protons interact in the target:

$$N_A z \gg 1/\sigma_{NN} \quad \mathrm{or} \quad z \gg 40\,\mathrm{g\,cm}^{-2}, \tag{8}$$

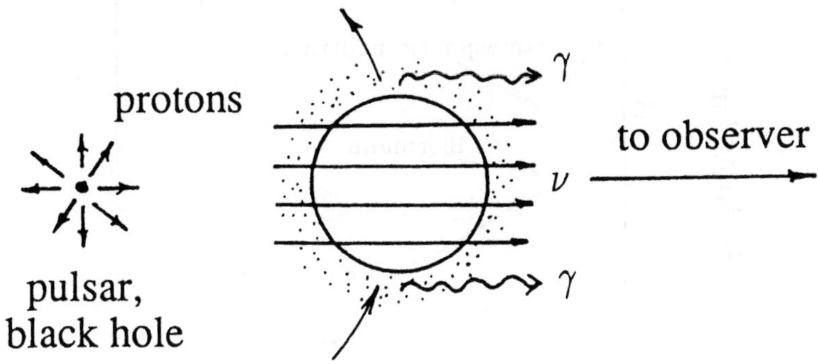

Figure 20: A binary system as a source of high-energy neutrinos and γ-rays [61].

where E_p is the energy of an accelerated proton, $\gamma_\pi = E_\pi/m_\pi \sim 10^4 E_p/\mathrm{TeV}$ is the Lorentz factor of a pion, $\tau_\pi = 2.6 \times 10^{-8}\,\mathrm{s}$ is the lifetime of a charged pion, and $\sigma_{\pi N}(\sigma_{NN}) \approx 2 \times 10^{-26}(4 \times 10^{-26})\,\mathrm{cm}^2$ is the cross section of a pion (nucleon) with a nucleon and N_A (cm) is Avogadro's number.

Here we briefly summarize the relation between the source luminosity and the neutrino flux observed at the Earth. The neutrino spectrum dN_ν/dE at a distance d from the source is related to the accelerator luminosity $\mathcal{L}_p$ as

$$\int E \frac{dN_\nu}{dE} dE = \frac{1}{4\pi d^2} \epsilon_\nu \mathcal{L}_p, \tag{9}$$

where ϵ_ν accounts for the fact that only a finite fraction of the proton energy goes into neutrino production. When the proton spectrum takes a power-law form of index -2 ($\propto E^{-2}$), the neutrino spectrum is given by

$$\frac{dN_\nu}{dE} = \frac{1}{4\pi d^2} \frac{\epsilon_\nu \mathcal{L}_p}{\ln \gamma_{max}} E^{-2} \tag{10}$$

$$\approx 10^{-10} \left(\frac{d}{10\mathrm{kpc}}\right)^{-2} \left(\frac{\mathcal{L}_p}{10^{38}\mathrm{erg\,s^{-1}}}\right) \left(\frac{E}{\mathrm{TeV}}\right)^{-2} \mathrm{cm^{-2}s^{-1}TeV^{-1}}, \tag{11}$$

where $\gamma_{max} = E_{p,max}/E_{p,min}$ is the maximum Lorentz factor attained in the accelerator and, under optimal conditions, $\epsilon_\nu = 0.11(0.094, 0.041, 0.029)$ for $\nu_\mu(\bar{\nu}_\mu, \nu_e, \bar{\nu}_e)$ (for a E^{-2} spectrum [62]).

6.2 Sources of high-energy neutrinos

Naturally the high-energy γ-ray sources discussed in the previous section can also be high-energy neutrino sources [61]. Thus we do not discuss them again, but here we point out the differences in the production mechanism for the case of a binary system, for example.

Fig. 20 shows a binary system comprised of a fast active pulsar and a companion star. The companion star acts as a target of the particles accelerated by the pulsar. γ-rays are produced in collisions with the atmosphere and neutrinos are produced

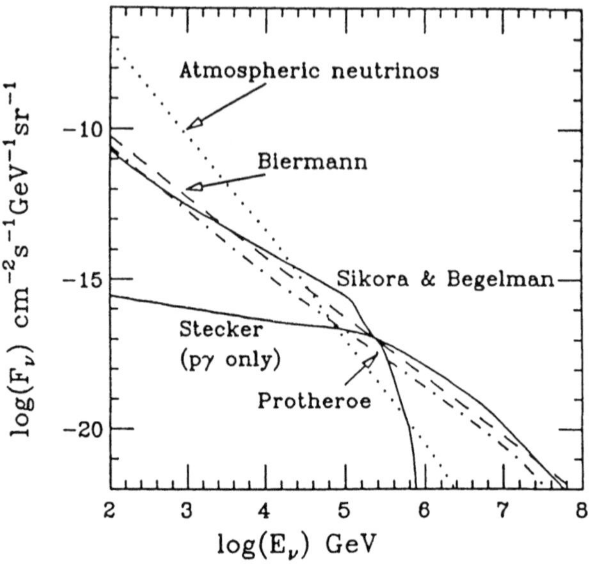

Figure 21: Predicted flux of neutrinos from Active Galactic Nuclei, according to various models, compared with the background of atmospheric neutrinos [65].

from particles bombarding the star. Then the neutrino flux is about δ_ν/δ_γ times the γ-ray flux, where δ_ν and δ_γ are the durations (or phase widths) of neutrino and γ-ray pulses. From the observed phase width $\delta_\gamma \sim 0.1$, we obtain $\delta_\nu/\delta_\gamma < 5$ (see Fig. 20). Considering the estimate given in the previous subsection, a high-energy γ-ray source observed with a flux of 10^{-11} cm^{-2} s^{-1} above 1 TeV can produce less than 0.1 upward-going muon events per year above 1 TeV in a 1000 m^2 detector.

In addition to γ-ray sources, there could be "hidden sources" of intense high-energy neutrinos in which the accompanied γ-rays are absorbed by the surrounding matter. However, because some kind of electromagnetic radiation must inevitably accompany the neutrino flux, the neutrino flux from that kind of source is limited by the observation of X-rays and other frequencies [61].

Recently a promising source of ultra-high-energy neutrinos has been discussed intensively: active galactic nuclei (AGN). A scenario written by Biermann et al. [63] assumes a massive black hole as an accelerator at the center of the AGN which is the source of the jets. These jets terminate in "hot spots" in the radio robes which is a dense cloud of photons. Termination shocks of the jet are supposed to accelerate protons and they bombard the photon cloud; thus $p\gamma$-neutrinos are produced, but the cloud is optically thick and γ-rays are unlikely to escape the source. Stecker et al. [64] pointed out that the radio-quiet AGN's could also be a powerful emitter of neutrinos. Suppose a massive black hole surrounded by an accretion disk in a very dense cloud of photons. High-energy protons accelerated in the shocks in the accretion disk produce pions from collisions with photons in the cloud and neutrinos are emitted via decays of these pions. A bump of 10 eV ultraviolet photons is reported in the NGC4151 spectrum, showing the presence of a cloud of dense photons. All of the high-energy photons generated in the beam dump cascade down to X-ray energy and should account for the observed X-ray spectrum. Even if the neutrino fluxes from individual AGN's are small, the combined flux from hundreds of radio-quiet sources can be larger than the atmospheric neutrino background at energies greater than $\sim$ 100 TeV. Fig. 21 shows the predicted flux of neutrinos from AGN's based on various models [65].

Other type of sources are also proposed: collisions of ultra-high-energy ($\sim 5 \times 10^{19}$ eV) cosmic-rays and the 2.7 K background radiation [66], weakly interacting massive particles constituting the dark matter of the Universe [67], and so on.

6.3 Detection of high-energy neutrinos

Upward-going muons

The most efficient way to detect high-energy neutrinos is to use the Earth as a muon-neutrino target; they are observed as upward-going muons in an underground detector [68]. The number of muons expected in a detector of a unit area is

$$N_\mu = \int_{E_{th}}^{\infty} \frac{dN_\nu}{dE} P_\nu^{obs}(E, E_{th}) dE, \tag{12}$$

where $P_\nu^{obs}(E, E_{th})$ is the probability to observe a neutrino of energy E as a muon of energy greater than E_{th} through an interaction with the Earth and is given by

$$P_\nu^{obs}(E, E_{th}) = N_A \int_{E_{th}}^{E} \frac{d\sigma_{\nu N}(E, E_\mu)}{dE_\mu} R_\mu(E_\mu, E_{th}) dE_\mu, \tag{13}$$

where $d\sigma_{\nu N}/dE_\mu$ is the charged-current cross section of a neutrino with a nucleon producing the muon of energy E_μ and R_μ is the range of the muon. In Fig. 22 is shown the total charged-current cross section of ν_μ and $\bar{\nu}_\mu$ with a nucleon and their observation probability in rock. R_μ for rock can be approximated by

$$R(E_\mu, E_{th}) = 2.5 \times 10^5\,\mathrm{g\,cm^{-2}} \cdot \ln \frac{E_\mu + 510\,\mathrm{GeV}}{E_{th} + 510\,\mathrm{GeV}}, \tag{14}$$

in the GeV–TeV range. (For details, see ref.[69].) If the neutrino energy is much higher than E_{th}, which is 1 $\sim$ 100 GeV depending on the thickness of the detector, P_ν^{obs} does not depend on E_{th}.

For the flux given in eq.(11), we can estimate the number of muons as 0.15 $(d/10\mathrm{kpc})^{-2}(\mathcal{L}_p/10^{38}\mathrm{erg\,s^{-1}})$ events per year above 1 TeV in a 1000 m^2 detector.

The ultra-high-energy neutrino flux is attenuated by interactions inside the Earth [70]. Fig. 23 shows the shadowing factor

$$\frac{dS}{d\Omega} = \exp[-N_A \sigma_{\nu N}(E_\nu) L], \tag{15}$$

integrated over azimuth, for several values of E_ν (L is the distance traveled through the Earth). However, this feature can be used to estimate the energy spectrum of neutrinos from a point source only when such a source is found.

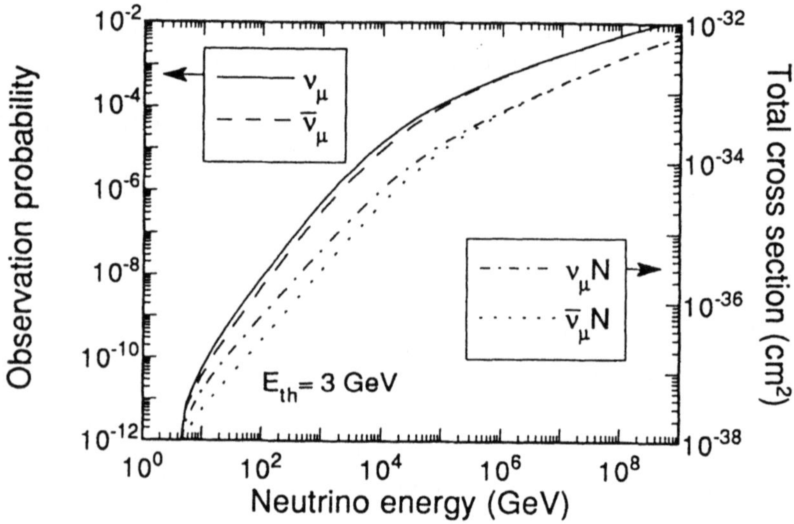

Figure 22: Total charged-current cross section of ν_μ and $\bar{\nu}_\mu$ with a nucleon and their observation probability in rock.

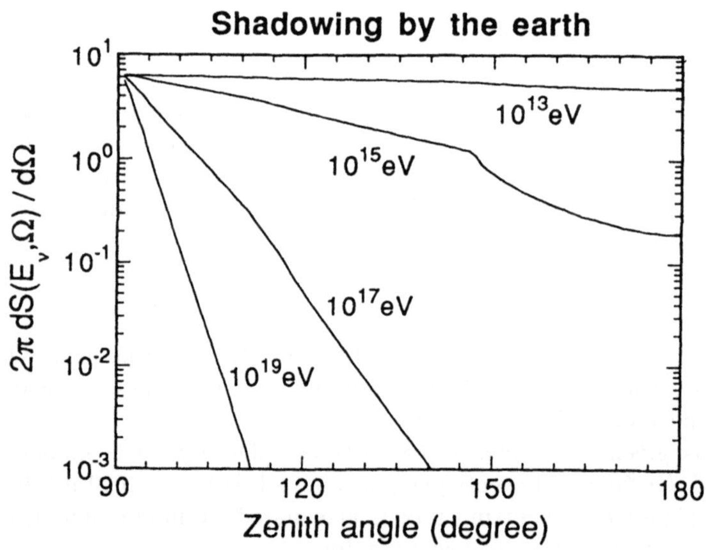

Figure 23: Shadowing factor (eq.(15)) for a detector on the Earth's surface, integrated over azimuth, as a function of the zenith angle for neutrinos of 10^{13}, 10^{15}, 10^{17} and 10^{19} eV incident on the Earth.

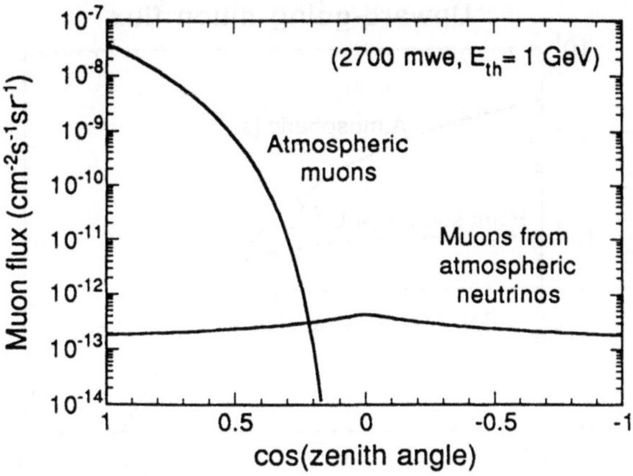

Figure 24: Anglular distribution of cosmic-ray muons and muons from atmospheric neutrinos for a 2700 mwe (meter water equivalent) deep experiment.

Other methods of detection

If a neutrino interaction happens inside the detector volume, it can be easily identified by its cascade shower. In general this probability is small; however, there is a resonant reaction (often called "Glashow resonance") $\bar{\nu}_e + e^- \rightarrow W^-$ which has a peak cross section of $\sim 10^{-31}$ cm^2 at 6.3 PeV [71]. This type of event can be detected in a large volume detector [72].

Since extensive air showers generated by cosmic-ray nuclei cannot reach surface detectors at large zenith angles, *horizontal air showers*, if observed, can be a signature of high-energy neutrinos [73]. The diffuse flux from AGN's is already limited marginally by the existing data, but the use of larger surface detectors are encouraged [74].

Radio and acoustic detection of high-energy neutrinos via neutrino-induced cascades in deep ice/water was suggested many years ago [75]. The possible large detection area is the advantage; however, the high detection threshold (>several PeV) made the idea unattractive. Recent predictions of a higher energy neutrino flux from AGN's may revive these methods [72].

Backgrounds

The major background for the search of high-energy neutrino sources comes from the atmospheric neutrinos produced by the interactions of cosmic ray nuclei; it amounts to $\sim 2 \times 10^{-13}$ cm^{-2} s^{-1} sr^{-1} for muons above several GeV [69]. The angular distributions of cosmic-ray muons and muons from atmospheric neutrinos is shown in Fig. 24. Therefore, in order to obtain meaningful results above this background, a detector of area larger than several hundreds of square meters is required.

For a discovery of a point source, the atmospheric neutrino background, which distributes almost uniformly in the celestial sphere, can be reduced by imposing

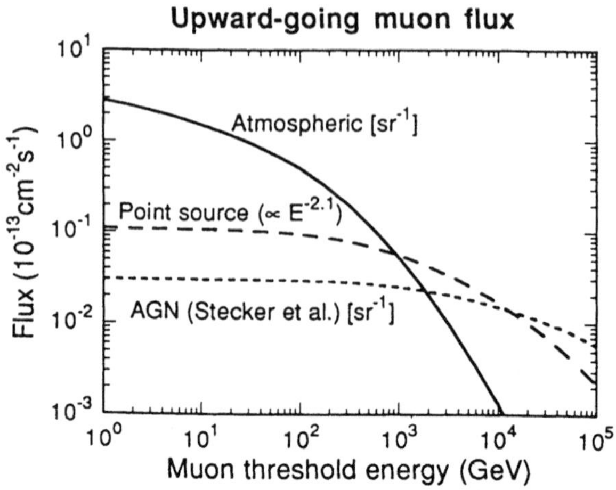

Figure 25: Energy threshold dependence of the muon flux from neutrinos from the atmosphere, a point source with the spectrum of $4 \times 10^{-6}(E_\nu/\mathrm{GeV})^{-2.1}\ \mathrm{cm}^{-2}\,\mathrm{s}^{-1}$ ($E_\nu \leq 10^6\,\mathrm{GeV}$) and AGN's [64].

an angular cut. Since the mean scattering angle of a muon from the direction of the parent neutrino is $\sim 2.6(E_\nu/100\,\mathrm{GeV})^{1/2}$ degrees [61], an angular resolution of one degree or better is required. However, for diffuse sources, such as the ensemble of distant AGN's, we cannot dismiss the atmospheric neutrino background.

Fig. 25 illustrates the energy threshold dependence of the muon flux, showing the advantage of having a higher threshold. As the signal-to-noise ratio increases with energy, the diffuse flux from AGN's may appear above the atmospheric ones.

6.4 Present status and prospects for experimental searches

The first experiments to search for muons induced by extraterrestrial neutrinos were carried out at the Kolar Gold Field in India [76] and the East Rand Proprietary Mine in South Africa [77] in the 1960's. The former experiment used neon flash tubes and plastic scintillators, and the latter used flash tubes and liquid scintillators. The detectors were arranged deep underground, 7,500 mwe (meter water equivalent) and 9,000 mwe, respectively, to observe neutrinos from nearly horizontal directions in both cases. These two experiments observed neutrino-induced muons consistent with the expectations from the atmospheric neutrinos and gave upper limits on the muon flux from point sources on the order of $\sim 10^{-13}\ \mathrm{cm}^{-2}\ \mathrm{s}^{-1}$.

A three-layer liquid scintillation counter array has been built at the Baksan Neutrino Observatory in Russia [78]. Upward-going muons are differentiated from down-going ones by the time-of-flight technique. Recently, larger underground detectors, IMB [79] and Kamiokande [80], originally built to search for proton decay, were used to search for high-energy neutrino sources. These two are based on the water Cherenkov technique and the direction of a particle crossing the

Table 9: Large underground/undersea experiments

Name	Location	Depth (mwe)	Area (m^2)	Data taking	Ref.
Baksan	Baksan, Russia	850	300	1978~	[78]
IMB	Cleveland, U.S.A.	1,570	300	1982~1991	[79]
Kamiokande	Kamioka, Japan	2,700	150	1983~	[80]
MACRO	Gran Sasso, Italy	3,400	1,300	1989~	[81]
LVD	Gran Sasso, Italy	3,600	480	1991~	[82]
Super-Kamiokande	Kamioka, Japan	2,700	600	1996?	[83]
NT-200	Lake Baikal, Russia	1,000	2,000	1993/1994?	[84]
DUMAND-II	Hawaii, U.S.A.	4,760	20,000	1993?	[85]
NESTOR	Pylos, Greek	3,500	20,000	1994?	[86]
AMANDA-I	South Pole	1,000	20,000	1994?	[87]

detector is unambiguous – an advantage for identifying upward-going events.

All experiments carried out have reported no positive results up to now. However, the total number of upward-going muon events accumulated so far is only ~ 2000 events, too small to draw a strong negative result; the given upper limits are far above the predicted flux. In Table 9 are listed major underground/undersea experiments currently taking data or in preparation. (See ref.[72] for a more extensive list.) The next generation of experiments aims to have sensitivities more than an order-of-magnitude higher than present ones. All the new projects in Table 9 are based on the water Cherenkov technique mainly for economical reasons.

The DUMAND project in Hawaii has been active since 1975, defining the problems, doing the background studies needed and creating the technology necessary for working in the deep ocean. Now the construction is underway for nine tethered strings, each having 24 optical modules (Fig. 26). The first 3 strings will be laid from the 4760 m deep site to the shore station 30 km distant in 1993 [85].

The NESTOR project has carried out demonstration tests in the Mediterranean sea off the southwest coast of Greece, and the first "tower", a 12-story hexagonal floor having 7 omni-directional optical modules, is under construction (Fig. 27). They hope to have seven towers with an effective area of 10^5 m^2 in about 1996 [86].

The AMANDA project aims to put optical detectors in the clear deep ice at the South Pole. Below several hundred meters in depth the ice is bubble-free due to pressure. The test at 800 m depth in 1991–1992 showed that the attenuation length of light is consistent with measurements in the laboratory (> 18 m). The advantages are that ice will form a mechanical support and electronics can be set near the detector. They plan to construct a nine string array in arround 1994 (Fig. 28) [87].

Even if these projects are realized, however, the recent estimates of the high-energy neutrino signal requires even larger area, a detector of order 10^5 m^2 [88]. Planned detectors are designed to be easily expanded, but new detection techniques

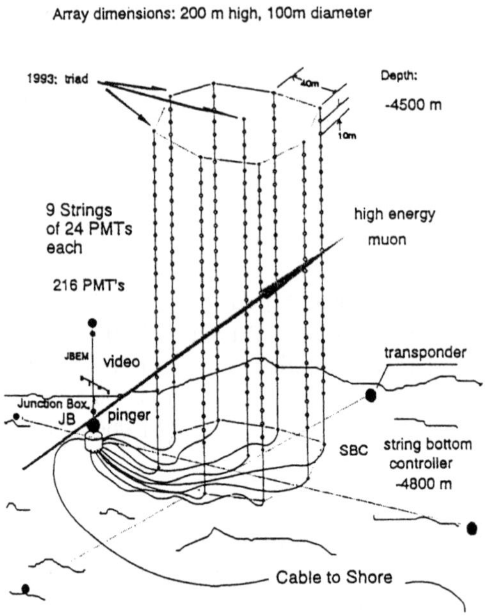

Figure 26: Overall view of the DUMAND-II detector [85].

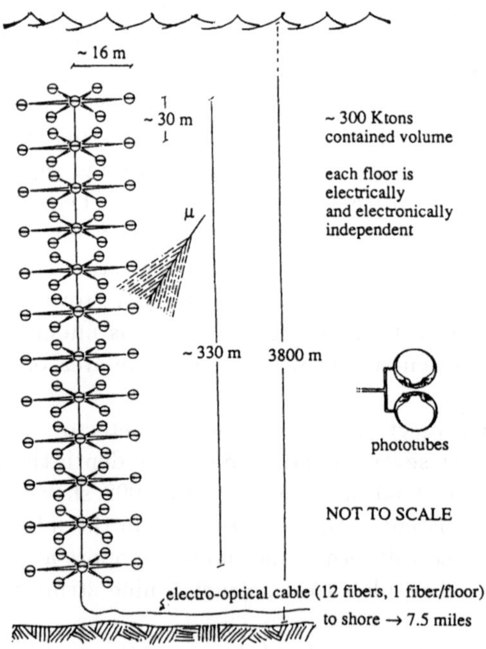

Figure 27: Schematic view of the NESTOR detector [86].

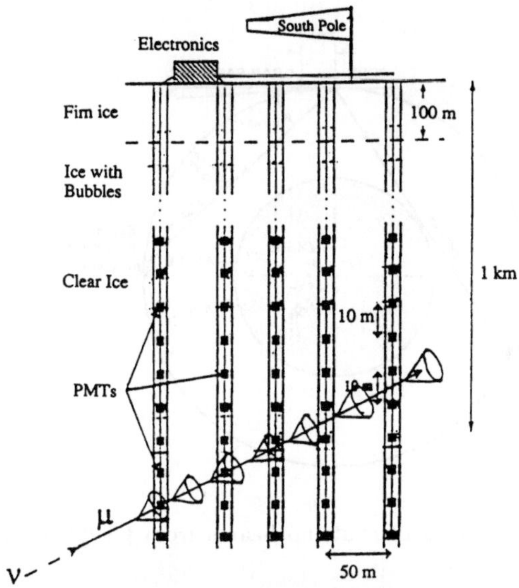

Figure 28: Schematic view of the AMANDA detector [87].

to cover a larger area with reasonable cost are highly desired.

6.5 Comments on long-baseline experiments

Recent indications of a deficit in the ν_μ flux of atmospheric neutrinos have motivated new searches for neutrino oscillations with a small neutrino mass-squared difference ($\Delta m^2 < 1$ eV^2). Since the oscillation probability is given by

$$P = \sin^2 2\theta \sin^2(1.27 \Delta m^2 L/E), \tag{16}$$

where Δm^2 is in eV^2, E is the neutrino energy in GeV and L is the distance from the production point to the detector in km, a large L/E ratio is a key factor for such experiments.

There are several proposals to use large underground/undersea detectors for long-baseline neutrino oscillation experiments performed with neutrino beams from accelerators [89]. Since the energy spectrum of accelerator neutrinos can be known better than atmospheric neutrinos and can be controlled, one can measure oscillation parameters with high precision. For example, DUMAND is located 6,482 km away from Fermilab and Gran Sasso, NESTOR and SuperKamiokande are 732, 1676 and 8752 km away from CERN, respectively, and so L will be very long compared with previous oscillation experiments (Fig. 29). In addition, if the oscillation is mainly between ν_μ and ν_e and L is larger than ~ 1000 km, one might see matter effects from the Earth.

Usually accelerator beams are designed with long decay tunnels so that the neutrinos result from pion and kaon decay and are mostly ν_μ. The mean energy

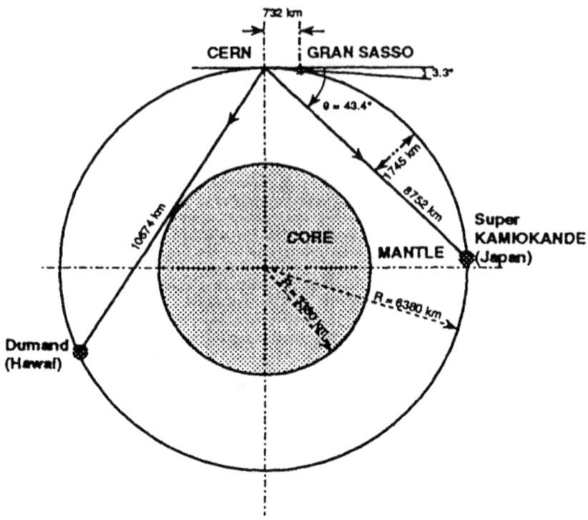

Figure 29: Various possibilities of neutrino beams from CERN towards existing large neutrino detectors [90].

of the neutrinos are several tens of GeV, if the proton accelerators in Fermilab or CERN are used. Neutrino oscillations are searched for via the disappearance of ν_μ and the appearance of other flavors. Thus, in principle, these experiments will be sensitive down to $\Delta m^2 \sim 10^{-3}$ eV^2 and will cover the range indicated by the atmospheric ν_μ deficit.

The main difficulty of these experiments comes from the high cost of aiming the beam down to point toward the distant detectors. Also, the event rate at large distances becomes a problem. Another difficulty is the identification of neutrino flavors; large detectors are not designed to be fine-grain, as a tradeoff made to achieve a great size. The best one can do may be to distinguish "muon-full" events and "muonless" events; their ratio is a measure of charged-current to the neutral-current interaction ratio [91].

Anyway, next-generation deep underground/underwater experiments will search for neutrino oscillations with atmospheric neutrinos as an inherent background; if evidence of the ν_μ deficit becomes stronger, using neutrino beams from accelerators will be a good option for these experiments.

References

[1] Meegan, C. A., *et al.*, 1992, *Nature*, **355**, 143.

[2] Fegan, D., 1993, *Towards a Major Čerenkov Detector*, eds. Fleury, P. and Vacanti, G. (Editions Frontieres), p.1.

[3] Kanbach, G. *et al.*, 1988, *Space Sci. Rev.*, **49**, 69.

[4] Patterson, J. R. and Kifune, T., 1992, *Australian and New Zealand Physicist*, **29**, 58.

[5] Bloemen, H., 1989, *Ann. Rev. Astron. Astrophys.*, **27**, 469.

[6] Dermer, C. D., 1986, *Astron. Astrophys.*, **157**, 223.

[7] Swanenburg, B. N., *et al.*, 1981, *Astrophys. Journ. Lett.*, **243**, L69.

[8] Halpern, J. P. and Holt, S. S., 1992, *Nature*, **357**, 222.

[9] Bertsch, D. L., *et al.*, 1992, *Nature*, **357**, 306.

[10] Clear, J., *et al.*, 1987, *Astron. Astrophys.*, **174**, 85.

[11] Bignami, G. F., and Hermsen, W., 1983, *Ann. Rev. Astron. Astrophys.*, **21**, 67.

[12] Houston, B., Riley, P. A., and Wolfendale, A. W., 1983, *Space Sci. Rev.*, **155**.

[13] Mayer-Hasselwander, H. A., and Simpson, G., 1990, *Proceednings of 21st International Cosmic Ray Conference*, Adelaide, **1**, 261.

[14] Grenier, I. A., Hermsen, W., and Pollock, A. M. T., 1990, *AIP Conference Proceedings 220, High Energy γ-ray Astronomy*, ed. Matthews, J. (University of Michigan, Ann Arbor MI).

[15] Bignami, G. F., Caraveo, P., and Maraschi, L., 1978, *Astron. Astrophys.*, **67**, 149.

[16] Plock, A. M. T., 1985, *Astron. Astrophys.*, **150**, 339.

[17] Landecker, T. L., Clutton-Brock, M., and Purton, C. R., 1990, *Astron. Astrophys.*, **232**, 207.

[18] Weekes, T. C., 1988, *Phys. Rep.*, **160**, 1.

[19] Grindlay, J. E. *et al.*, 1975, *Astrophys. Journ. Lett.*, **197**, L9.

[20] Punch, M. *et al.*, 1992, *Nature*, **358**, 477.

[21] Chadwick, P. M., *et al.*, 1993, *Towards a Major Čerenkov Detector*, ed. Fleury, P. and Vacanti, G. (Editions Frontieres) p.53.

[22] Lamb, R. C., 1993, *Towards a Major Čerenkov Detector*, ed. Fleury, P. and Vacanti, G. (Editions Frontieres) p.43.

[23] Vacanti, G. *et al.*, 1991, *Astrophys. Journ.*, **377**, 467.

[24] Weekes, T. C. *et al.*, 1989, *Astrophys. Journ.*, **342**, 379.

[25] Dowthwaite, J. C., *et al.*, 1984, *Astrophys. Journ. Lett.*, **286**, L35.

[26] Vladimirski, B. M, Stepanian, A. A. and Fomin, V. B., 1973, *Proceedings of 13th International Cosmic Ray Conference*, Denver, **1**, 456.

[27] Stepanian, A. A. *et al.*, 1977, *Proceedings of 15th International Cosmic Ray Conference*, Plovdiv, **1**, 135.

[28] Chadwick, P. M., *et al.*, 1985, *Nature*, **318**, 642.

[29] Chadwick, P. M., *et al.*, 1987, *Very High Energy γ-ray Astronomy*, ed. Turver, K. E. (D. Reidel Publishing Company), p.115.

[30] Samorskii, M. and Stamm, W., 1983, *Astrophys. Journ. Lett.*, **268**, L17.

[31] Lloyd-Evans, J. *et al.*, 1983, *Nature*, **305**, 284.

[32] Kifune, T. *et al.*, 1986, *Astrophys. Journ.*, **301**, 230.

[33] Dowthwaite, J. C., *et al.*, 1984, *Nature*, **309**, 691.

[34] Lamb, R. C., *et al.*, 1988, *Astrophys. Journ. Lett.*, **328**, L13.

[35] Vishwanath, P. R., *et al.*, 1989, *Astrophys. Journ.* , **342**, 489.

[36] Grindlay, J. E. *et al.*, 1975, *Astrophys. Journ.*, **201**, 82.

[37] Bhat, P. N, *et al.*, 1987, *Astron. Astrophys.*, **178**, 242.

[38] Bond, I. A., *et al.*, 1988, *Phys. Rev. Lett.*, **61**, 2292.

[39] Lewis, D. A., Lamb, R. C., and Biller, S. D., 1991, to appear in *Astrophys. Journ.*.

[40] Cheng, K. S, Ho, C. and Ruderman, M., 1986, *Astrophys. Journ.*, **300**, 500.

[41] Cheng, K. S, Ho, C. and Ruderman, M., 1986, *Astrophys. Journ.*, **300**, 522.

[42] Ruderman, M. and Cheng, K. S., 1988, *Astrophys. Journ.*, **335**, 306.

[43] Kwok, P. W., Cheng, K. S, and Lau, M. M., 1991, *Astrophys. Journ.*, **379**, 653.

[44] D'Amico, N., 1983, *Space Science Rev.*, **36**, 195.

[45] Kifune, T. *et al.*, 1987, *Mon. Note R. Astron. Soc.*, **228**, 243.

[46] Helfand, D. J., *et al.*, 1989, *Astrophys. Journ.*, **341**, 151.

[47] Seward, E. D., and Harnden Jr., E. R., 1984, *X-ray Astronomy*, Bologna, p.297.

[48] Trussoni, E. *et al.*, *Astron. Astrophys.*, **234**, 403.

[49] Ogelman, H., and Buccheri, R., 1987, *Astron. Astrophys.*, **186**, L17.

[50] Koyama, K. *et al.* 1989, *Publ. Astron. Soc. Japan*, **41**, 461.

[51] Nel, H. I. *et al.*, 1990, *Astrophys. Journ.*, **368** 585.

[52] Dermer, C. D., Schlickeiser, E., and Mastichiadis, A., 1992, *Astron. Astrophys.*, **256**, L27.

[53] Zhang, F. J., and Bååth, L. B., 1990, *Astron. Astrophys.*, **236**, 47.

[54] Sikora, M., Begelman, M. C., and Rudak, B., 1989, *Astrophys. Journ. Lett.*, **341**, L33.

[55] Kirk, J. G., and Mastichiadis, A., 1989, *Astron. Astrophys.*, **213**, 73.

[56] Mastichiadis, A., and Protheroe, R. J., 1991, *Mon. Not. R. Astron. Soc.*, **246**, 279.

[57] Atoyan, A. M., 1992, *Astron. Astrophys.*, **257**, 476.

[58] Stecker, F. W., De Jager, O. C., and Salamon, M. H., 1992, *Astrophys. Journ. Lett.*, **390**, L49.

[59] Morfill, G. E., and Tenorio-Tagle, G., 1983, *Space Sci. Rev.*, **36**, 93.

[60] Halzen, F., 1992, presentation at the Workshop *"Towards a Major Atmospheric Čerenkov Telescope"*, Ecole Polytechnique, Palaiseau, Paris.

[61] Berezinsky, V. S., Castagnoli, C. and Galeotti, P., 1985, *Nuovo Cim.*, **8C**, 185; Berezinsky, V. S., 1991, *Nucl. Phys. (Proc. Suppl.)*, **19**, 375.

[62] Lee, H. and Bludman, S. A., 1985, *Astrophys. Journ.*, **290**, 28. See also Stenger, V. J., 1984, *Astrophys. Journ.*, **284**, 810.

[63] Mannheim, K., Stanev, T. and Biermann, P. L., 1992, *Astron. Astrophys.* **260** L1; Biermann, P.L., 1992, ref.[65], p.86.

[64] Stecker, F. W. *et al.*, 1991, *Phys. Rev. Lett.*, **66**, 2697; *ibid.*, 1992, **69**, 2738 (Erratum).

[65] *Proc. of the Workshop on High Energy Neutrino Astronomy*, Hawaii, U.S.A., eds. Stenger, V. J., Learned, J. G., Pakvasa, S. and Tata, X., (World Sicentific Singapore, 1992).

[66] Hill, C. T. and Schramm, D. N., 1983, *Phys. Lett.*, **131B**, 247 and references therein.

[67] Berezinsky, V. S., 1993, *Nucl. Phys. (Proc. Suppl.)* **31**, 413 and references therein.

[68] Markov, M. A., 1960, *Proc. 1960 International Conf. on High energy Physics*, Rochester, U.S.A., p.578.

[69] T. K. Gaisser, 1990, *Cosmic Rays and Particle Physics* (Cambridge University Press), Chapter 8.

[70] Honda, M. and Mori, M., 1987, *Prog. Theor. Phys.*, **78**, 963; Reno, R. H. and Quigg, C., 1988, *Phys. Rev.*, **D37**, 657.

[71] Glashow, S. L., 1960, *Phys. Rev.*, **118**, 316; F. Wilczek, 1985, *Phys. Rev. Lett.*, **55**, 1252.

[72] Learned, J. G., 1993, *Nucl. Phys. (Proc. Suppl.)*, **33A,B**, 77.

[73] Baltruisaitis, R. M. *et al.*, 1985, *Phys. Rev.*, **D31**, 2192.

[74] Halzen, F. and Zas, E., 1992, *Phys. Lett.*, **B289**, 184.

[75] Learned, J. G., 1979, *Phys. Rev.*, **D19**, 3293 and references to earlier work by Askaryan and Bowen therein; Halzen, F., Zas, E. and Stanev, T., 1992, *Phys. Rev.*, **D45**, 362.

[76] Achar, C. *et al.*, 1965, *Phys. Lett.*, **18**, 196; Krishnaswamy, M. R. *et al.*, 1971, *Proc. Roy. Soc. Lond.*, **A323** 489.

[77] Reines, F. *et al.*, 1965, *Phys. Rev. Lett.*, **15**, 429; Crouch, M. F. *et al.*, 1978, *Phys. Rev.*, **D18**, 2239.

[78] Alexeyev, E. N. *et al.*, 1979, *Proc. 16th Int. Cosmic Ray Conf.*, Kyoto, vol.10, p.276; Boliev, M. M. *et al.*, 1981, *Proc. 17th Int. Cosmic Ray Conf.*, Paris (Centre d'Etudes Nucleaires, Saclay), vol.7, p.106; Boliev, M. M. *et al.*, 1991, *Proc. Third Int. Workshop on Neutrino Telescopes*, Venice, ed. M. Baldo Ceolin, p.235.

[79] Svoboda, R. *et al.*, 1987, *Astrophys. Journ.*, **315**, 420.

[80] Oyama, Y. *et al.*, 1989, *Phys. Rev.*, **D39**, 1481; Mori, M., 1992, ref.[65], p.221.

[81] Ahlen, S. *et al.*, 1993, *Nucl. Inst. Meth. in Phys. Res.*, **A324**, 337.

[82] Bari, G. *et al.*, 1989, *Nucl. Inst. Meth. in Phys. Res.*, **A277**, 11.

[83] Totsuka, Y., 1990, Preprint ICRR-Report-227-90-20 (presentation at the Int. Symp. on Underground Physics Experiments, Tokyo, April 1990).

[84] Belolaptikov, I. A. *et al.*, 1991, *Nucl. Phys. (Proc. Suppl.)*, **19**, 388.

[85] *DUMAND II Proposal, Preprint HDC-2-88* (Hawaii DUMAND Center, University of Hawaii, 1988); A. Okada, 1992, ref.[65], p.277.

[86] Resvanis, L. K., 1992, ref.[65], p.325.

[87] Barwick, S. *et al.*, 1992, ref.[65], p.291.

[88] Barwick, S. *et al.*, 1991, *Proc. Third Int. Workshop on Neutrino Telescopes*, Venice, ed. Baldo Ceolin, M., p.437.

[89] Bernstein, R. H. and Parke, S. J., 1991, *Phys. Rev.*, **D44**, 2069.

[90] Rubbia, C., 1993, Preprint CERN-PPE/93-08 (presentation at the Conversaciones de Madrid, Madrid, 5 September 1992).

[91] Stenger, V. J., 1992, to appear in *Proc. of the 2nd NESTOR International Workshop*, Pylos, Greece, 19–22 October 1992, ed. Resvanis, L. K.; Resvanis, L. K., 1993, to appear in *Proc. Fifth Int. Workshop on Neutrino Telescopes*, Venice, 2–4 March 1993, ed. Baldo Ceolin, M.

Particle Acceleration Mechanisms in Astrophysics

Fumio Takahara

Department of Physics, Tokyo Metropolitan University

1 Introduction

The origin and acceleration mechanisms of cosmic rays have been one of the most important problems in astrophysics. Recently, they have also received much attention in connection with the advent of observations of high-energy γ-rays and neutrinos from astronomical objects. During the past few years two distinct views have been extensively discussed. One is the direct acceleration mechanisms due to an electric field around rotating neutron stars; the other comprises statistical acceleration mechanisms, called Fermi acceleration, which work in a turbulent interstellar medium and astrophysical shocks.

As for the former, two basic mechanisms were proposed by Ostriker and Gunn [1] as well as by Goldreich and Julian [2] after the discovery of pulsars in 1967. Pulsar models have been constructed in terms of rotating neutron stars with a strong dipole magnetic field. Ostriker and Gunn studied the case in which the axis of the dipole magnetic field is inclined from the rotation axis. Pulsars emit magnetic dipole radiation, and particle acceleration occurs in the strong electromagnetic field of the waves. Alternatively, Goldreich and Julian examined the case in which the axis of the dipole magnetic field is aligned with the rotation axis. Although the resulting fields are stationary, an electrostatic potential is generated on the surface of the star due to unipolar induction. This potential produces an electric field which accelerates charged particles. Since then, much effort has been devoted to studies of these mechanisms in connection with pulsar models. Moreover, interest in these models has occasionally been revived when new observations appeared, such as high-energy γ-rays from Cyg X-3 and Her X-1 [3] and the appearance of SN1987A [4], although observational evidence for very high-energy γ-rays from point sources has recently become less convincing [5]. The application of a unipolar induction mechanism to accretion disks has also been made in connection with jet production in active galactic nuclei [6] and high-energy γ-ray sources [7]. However, direct acceleration mechanisms have not been accepted with wide consensus, since several problems have been noticed concerning them. For example, there is no agreement on what is the correct pulsar energy-loss mechanism. A currently popular model is that pulsars lose their rotation energy mostly through relativistic winds, which involve a large number of particles [8]. In this case, the energy of individual particles would be significantly reduced. It should also be noted that direct acceleration mechanisms have difficulty in producing the characteristic power-law spectrum of cosmic rays. We review the basic concept and problems of direct acceleration mechanisms in sections 2, 3 and 4.

The Fermi acceleration mechanism has become much more promising, since it was recognized that shock waves can provide an efficient place for this mechanism. Several authors took this approach in the late 1970s [9][10][11][12]. The most important characteristic of shock acceleration is that it can produce a universal power-law spectrum irrespective of the shock speed; the power-law index is determined by the compression ratio. It is also stressed that fairly high efficiency is achieved for the conversion of shock energy into accelerated particles. Moreover, evidence for shock acceleration has been found from in-situ spacecraft observations for shocks in the solar system, such as the Earth's bow shock, which is generated by a collision of the solar wind with the Earth's magnetic field (see reviews e.g., [13][14][15]). These observations greatly enhance the belief that shock acceleration also occurs for astrophysical shocks.

The observed cosmic-ray spectrum is shown in Fig. 1. We consider three energy regions: below $\sim 10^{16}$ eV, between $\sim 10^{16}$ and $\sim 10^{19}$ eV, and above $\sim 10^{19}$ eV. When we apply shock acceleration to the origin of cosmic-rays, we must first identify their acceleration sites for each energy region. The view is now widely accepted that supernova remnants are responsible for cosmic rays of energies less than the knee region, which lies at about 10^{16} eV. However, for energies higher than 10^{14} eV, since the acceleration time scale of protons becomes long compared with the lifetime of the shock, Fermi acceleration in supernova remnants becomes less efficient [16]. This is one of the quantitative problems that must be solved. In astrophysics, shocks are seen rather ubiquitously, and there are several candidates for the origin of high-energy cosmic rays. For example, if our Galaxy blows a galactic wind, it may generate a shock by colliding with intergalactic gas [17]. Observations of extragalactic radio sources often reveal structures which appear to be shocks. Among others, it is widely believed that jets emanating from active galactic nuclei collide with intergalactic gas, and thus generate shocks which can be recognized as bright hot spots in the radio lobes [18][19][20]. Shocks resulting from accretion onto neutron stars have also been examined [21][22][23].

The simplest version of shock acceleration theory adopts a test-particle approximation, i.e., it assumes that the shock front is an a-priori given discontinuity of the background fluid flow, which convects the scattering centers of cosmic rays, and that cosmic rays move among these scattering centers without influencing the shock structure. It adopts the diffusion approximation for particle transport. Further, it assumes a steady, non-relativistic and plane-parallel shock with the magnetic field direction parallel to the shock normal. Various modifications have been extensively studied during this decade (for reviews see [24][14][15]). In this paper we review the basic mechanism of shock acceleration for the most simplified case in section 5. In section 6 we discuss various modifications to the simplest version, such as oblique shocks, relativistic shocks, and nonlinear modifications of the shock structure due to accelerated particles. In section 7, applications to various astrophysical sites are examined in order to identify the origin of cosmic rays.

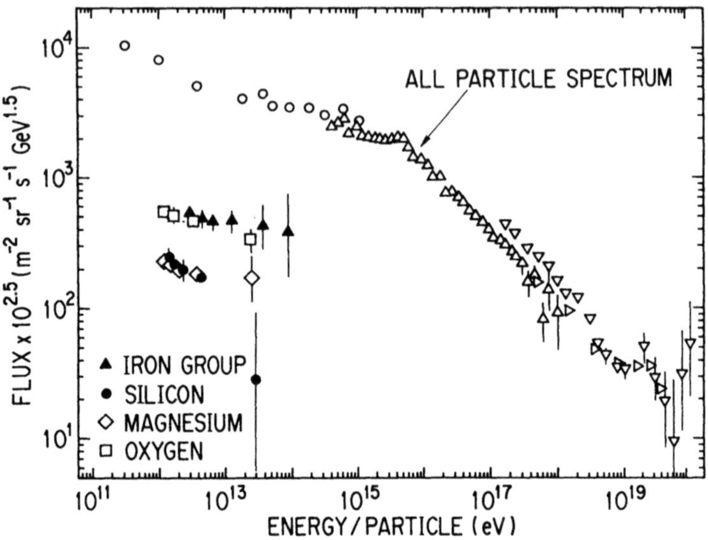

Figure 1: Observed spectrum of cosmic rays (from [25]).

2 Acceleration by Strong Waves

Radio pulsars are believed to be rotating neutron stars with a strong dipole magnetic field. If they are placed in a vacuum, and if the dipole axis is inclined from the rotation axis, they emit electromagnetic waves at a frequency which coincides with that of the rotation through magnetic dipole radiation. The emission mechanism is essentially that of a rotating magnet. The electromagnetic fields are so strong that charged particles gain very high energy [1]. The fields can be calculated in terms of classical electromagnetism. We define the radius of the light cylinder as $r_{\rm lc} = c/\Omega$, where c is the speed of light and Ω is the angular frequency of rotation of the neutron star. The radius of the light cylinder represents the distance within which the magnetosphere can corotate with the star. By adopting cylindrical coordinates with radial coordinate r, for $r < r_{\rm lc}$ the field is dominated by the stellar dipole field, while for $r > r_{\rm lc}$ it is dominated by that of the magnetic dipole radiation.

If the dipole axis is perpendicular to the rotation axis, the energy loss rate by the magnetic dipole radiation $L_{\rm md}$ is given by

$$L_{\rm md} = \frac{2\mu^2\Omega^4}{3c^3} = 9.6 \times 10^{38} B_*^2 R_*^6 P_*^{-4} \text{ erg s}^{-1}, \tag{1}$$

where $\mu \equiv \frac{1}{2}BR^3$ is the magnetic dipole moment. Typical values of the surface magnetic field at the magnetic pole $B_* = B/10^{12}$G the radius $R_* = R/10^6$cm and the rotation period $P_* = P/10$ msec are inserted for a neutron star. The rotation period and angular frequency are related by $P = 2\pi/\Omega$. The wavelength of the emitted waves is given by $\lambda = c/\Omega$, which is equal to the radius of the light cylinder $r_{\rm lc}$. The waves are linearly polarized in the equitorial plane and circularly polarized in the polar directions.

Since the magnetic field decreases as r^{-3} for $r < r_{\rm lc}$ and as r^{-1} for $r > r_{\rm lc}$, the typical field strength for $r > r_{\rm lc}$ becomes $B(R/r_{\rm lc})^3(r_{\rm lc}/r)$. To be specific, we consider the wave fields in the equatorial plane, which can be written as

$$E_\theta = B_\varphi = 2.2 \times 10^8 B_* R_*^2 P_*^{-2} \frac{\sin(\Omega(t - r/c))}{r/R} \ \mathrm{erg}^{1/2}\ \mathrm{cm}^{-3/2}. \quad (2)$$

We define the strongness of the waves in terms of a strongness parameter (the ratio of the cyclotron frequency (ω_c) in the wave field to the wave frequency Ω) as $\nu \equiv \omega_c/\Omega$, which results in

$$\nu = 3.4 \times 10^9 \frac{Z}{A} \frac{B_* R_*^2 P_*^{-1}}{r/R}, \quad (3)$$

where Z and A are the charge and mass number of the particle. The physical meaning of ν can be understood as follows. Consider a charged particle of a charge e at rest in its initial state. If $\nu \ll 1$, the momentum that a particle attains due to the electric field of the wave is on the order of eB/Ω. Thus, for $\nu \ll 1$ it is less than mc, and the particle remains nonrelativistic. On the contrary, if $\nu \gg 1$ the particle becomes relativistic before the wave field changes its sign and it experiences many Larmor rotations in one wave period through the action of the Lorentz force. This means that the particle motion is not a mere linear response to the changing wave field, but becomes highly nonlinear while attaining a relativistic energy.

The equation of motion of charged particles is given by

$$\frac{d\vec{p}}{dt} = Ze(\vec{E} + \frac{1}{c}\vec{v} \times \vec{B}). \quad (4)$$

As an illustration, we first consider the motion in the plane waves which propagate in the z-direction and are linearly polarized in the x-direction [26]. We take $E_x = B_y = B_0 \sin\chi$, where $\chi \equiv \Omega(t - z/c)$ is the phase of the wave. The equations of motion are written as follows:

$$\frac{dp_x}{dt} = Ze(1 - \frac{v_z}{c})B_0 \sin\chi, \quad (5)$$

$$\frac{dp_y}{dt} = 0 \quad (6)$$

and

$$\frac{dp_z}{dt} = Ze\frac{v_x}{c}B_0 \sin\chi. \quad (7)$$

Noting that

$$\frac{d\chi}{dt} = \Omega(1 - \frac{v_z}{c}), \tag{8}$$

we obtain from Eq. (5)

$$p_x = -\nu mc(\cos\chi - \cos\chi_0), \tag{9}$$

where we impose the condition that the particle is at rest at $\chi = \chi_0$.

It is convenient to use the following supplementary equation for the Lorentz factor γ:

$$\frac{d\gamma}{dt} = \frac{Ze}{mc^2} v_x B_0 \sin\chi. \tag{10}$$

From Eqs. (7) and (10) we obtain $\gamma = p_z/mc + 1$, where the integration constant is determined by the condition that the particle is at rest at $\chi = \chi_0$. Inserting the result for p_x into the equation for p_z, we obtain

$$p_z = \frac{\nu^2}{2} mc(\cos\chi - \cos\chi_0)^2. \tag{11}$$

While the particle energy oscillates with the wave phase, during most of the period the particle can be accelerated to a highly relativistic energy of $\gamma \sim \nu^2$. The essential features of the particle motion are the same for elliptically polarized waves as those for linearly polarized waves. The maximum Lorentz factor of particles can become as high as 10^{16} for plane waves with $\nu = 10^8$, a typical value attained at the light cylinder for typical parameters of $B_* = P_* = R_* = 1$. This corresponds to 10^{25} eV for protons and is five orders of magnitudes larger than the maximum energy of observed cosmic rays.

However, a particle cannot be accelerated to this energy, since the phase does not actually change very much. Since the wave phase that the particle feels proceeds as

$$\frac{d\chi}{dt} = \frac{\Omega}{\gamma}, \tag{12}$$

the period of the particle motion becomes

$$T = \frac{1}{\Omega}\int_0^{2\pi} \gamma d\chi = P\{1 + \frac{1}{4}\nu^2(1 + 2\cos^2\chi_0)\}, \tag{13}$$

which is much longer than the rotation period of the neutron star; in many cases it is larger than, or comparable to the age of the pulsar itself. This argument implies that the phase the particle feels takes a nearly constant value as soon as it becomes relativistic, which is called phase-locking. Moreover, in realistic situations, the assumption of infinite trains of plain waves is not satisfied; a correct treatment of spherical waves in which the field strength decreases with distance from the pulsar is thus needed.

To determine the effect of spherical waves, we first consider a perturbation analysis of plane waves [27]. Since acceleration occurs while the phase change is small as discussed above, we can expand the phase variable as $\chi = \chi_0 + \chi_1$. We then obtain from Eq. (11)

$$\gamma = \frac{\nu^2}{2}\chi_1^2 \sin^2 \chi_0. \tag{14}$$

Combining this with Eq. (12), the distance that the particle moves to the propagation direction is approximated as

$$z \approx \int c dt = r_{\rm lc}\frac{\nu^2 \chi_1^3}{6} \sin^2 \chi_0. \tag{15}$$

By eliminating χ_1 from these two equations we obtain

$$\gamma = (\frac{3z\nu \sin \chi_0}{\sqrt{2} r_{\rm lc}})^{2/3}. \tag{16}$$

Because ν behaves as r^{-1} for spherical waves, a perturbation analysis based on the constant ν may be, at best, correct for z with an order of the radius of particle injection r_0, which is taken here as $r_{\rm lc}$. Thus, for spherical waves the maximum Lorentz factor becomes on the order of $\nu_0^{2/3}$, where ν_0 is the strongness parameter at the light cylinder. An analytical study for the case of spherical waves under some reasonable approximations leads to a result that is very similar to the above [1] [26], such as

$$\gamma = \{\frac{3}{\sqrt{2}}\nu_0(1 - r_0/r) \sin^2 \chi_0\}^{2/3}. \tag{17}$$

Thus, for $r \gg r_0 = r_{\rm lc}$ and $\chi_0 = \pi/2$ we obtain

$$\gamma_{\rm max} = 2.8 \times 10^5 (\frac{Z}{A})^{2/3} B_*^{2/3} R_*^2 P_*^{-4/3}. \tag{18}$$

A numerical study of the particle motion essentially confirms this magnitude for the maximum attained energy [28].

Although this mechanism predicts a fairly large value of $\gamma_{\rm max}$, there seems to be some problems. If we try to explain ultra-high-energy γ-rays from Cyg X-3, this model fails. From Eqs. (1) and (18) we obtain $L_{\rm md} = 4.3 \times 10^{22} \gamma_{\rm max}^3$ erg s^{-1} for $A = Z = 1$. If we take $\gamma_{\rm max} = 10^7$, in order to produce γ-rays of 10^{16} eV, we obtain $L_{\rm md} = 4.3 \times 10^{43}$ erg s^{-1}. If we tentatively take a rotation period of 10 msec and the rotation energy of the neutron star as 2×10^{50} erg, the rapid rotation would decay on a time scale of less than a year, which contradicts the reported observations [3].

A more serious problem of strong wave acceleration is that strong waves are unlikely to be propagated into the neutron-star environment. If the neutron star is surrounded by a certain amount of plasma instead of a vacuum, the propagation of such low-frequency waves is prevented. The plasma frequency is given by $5.6 \times 10^4 n_e^{1/2}$ s^{-1} with an electron number density of n_e. Although the density is hard to estimate, the model discussed in the next section predicts that n_e is much larger than 1 cm^{-3}. Even if the wave propagation becomes possible, by evacuating the surrounding plasma by waves themselves, the waves will most likely suffer from various instabilities [29]. Although the detailed processes are still an open problem, one prevailing view is that the instabilities lead to a conversion of the wave energy into the production of a relativistic magnetohydrodynamical wind comprising electron-positron pairs and magnetic fields[30][8]. We thus conclude that the particle acceleration due to strong waves seems to be less probable.

3 Unipolar Induction

Most pulsar models utilize a strong electric field induced by unipolar induction in a rotating magnet to explain the acceleration of particles which are responsible for pulsed emission over a wide range of frequencies. Unipolar induction is a familiar notion in classical electromagnetism, in which an axially symmetric rotating magnet generates an electrostatic potential on its surface. Since it is generated by the aligned component of the magnetic field with the rotation axis, almost all papers have concentrated on stationary aligned rotators, following a paper by Goldreich and Julian [2].

We assume that a point dipole is located at the center of the neutron star, and that the magnetic dipole axis aligns with the rotation axis. Figure 2 illustrates the considered situation. Since neutron stars are a good electric conductor, the electric field is given by $\vec{E} = -\vec{v} \times \vec{B}/c$, where $\vec{v} = \vec{\Omega} \times \vec{r}$ inside the star. Assuming that the star has no net charge and that it is surrounded by vacuum, we can calculate the electrostatic potential in the spherical coordinates as

$$\Phi = \frac{\mu\Omega}{c}(\frac{\sin^2\theta}{r} - \frac{2}{3R}) \tag{19}$$

inside the star, and

$$\Phi = -\frac{2\mu\Omega R^2}{3cr^3}\frac{3\cos^2\theta - 1}{2} \tag{20}$$

outside the star. Thus, a quadrupole electric field is generated in the surrounding vacuum region. The corresponding surface-charge density is given by

$$\sigma = -\frac{\mu\Omega\cos^2\theta}{2\pi c R^2}. \tag{21}$$

The large potential difference given above can be used for particle acceleration if the induced electric field has a component along the magnetic field lines, as is the case for the potential given above, since particles move along a magnetic field line due to a strong synchrotron loss of perpendicular momentum. The electric field perpendicular to the magnetic field leads to a drift motion, and is not used for acceleration. It is also to be noted that if the magnetic field lines are closed, they eventually become an equipotential line through a redistribution of the electric charge. They are not then used for particle acceleration, either. Actually, the field lines which emerge near to the equator are closed, thus making a corotating magnetosphere. Thus, strong particle acceleration occurs if the magnetic field line

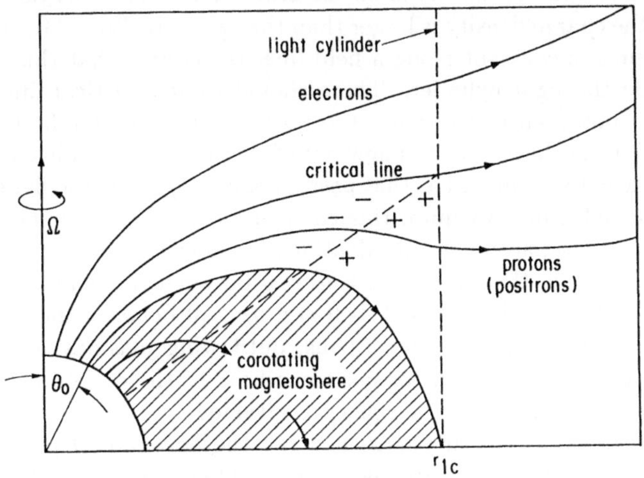

Figure 2: A schematic view of rotating magnetic neutron stars based on the model of ref. [2].

extends to infinity from the surface. Such a case is realized for field lines which emerge near to the poles, and become open as they approach infinity.

Thus, not all of the potential difference between the surface of a star and infinity is available for particle acceleration. If we assume that the dipole shape of the magnetic field remains a good assumption inside the light cylinder, we can find the location of the field line which separates the open and closed field lines as follows. Denoting the polar angle of this boundary field line at the surface of the neutron star by θ_0, we obtain $\sin^2\theta_0 = R\Omega/c$ by assuming that the field line which passes $\theta = \theta_0$ at the surface goes through $\theta = \pi/2$ at the light cylinder. The neutron star has a corotating magnetosphere comprising field lines with $\theta > \theta_0$ at the stellar surface. The open field lines are possible only for $0 \leq \theta \leq \theta_0$ at the surface of the star, along which particle acceleration and a flow of charged particles occurs.

Though a deformation of the magnetic field lines surely occurs due to the generated electric current, we assume for simplicity that the magnetic field lines are not greatly changed from that of the original dipole field. If a sufficient number of charged particles exist, the local electric field vanishes in the plasma rest frame which is corotating with the neutron star, and the electric field in the inertial frame is given by

$$\vec{E} = -(\vec{\Omega} \times \vec{r}) \times \vec{B}/c, \tag{22}$$

instead of that given by the gradient of Eq. (20). The electric field is sustained by the charge density,

$$\rho_E = \frac{1}{4\pi}\mathbf{div}\vec{E} = -\frac{\vec{\Omega}\cdot\vec{B}}{2\pi c} = -\frac{\mu\Omega}{2\pi c r^3}(3\cos^2\theta - 1). \tag{23}$$

The typical number density is on the order of 10^{14} cm^{-3} at the stellar surface. As long as the charge density is larger than that given by Eq. (23), the electrostatic potential remains constant along a field line; we assume that this is realized in a region inside the light cylinder. There should exist a critical line on which the electrostatic potential is zero since the potential is zero at infinity and particles with both signs of charges should flow out of the star to maintain charge neutrality. This line is usually identified as that passes $\cos\theta = 1/\sqrt{3}$ at the light cylinder from Eq. (20) for fields in a vacuum; this condition coincides with the condition that $\rho_E = 0$ in the above, although this identification does not seem to be very accurate. If $\vec{\Omega}$ is parallel to $\vec{\mu}$, then $\Phi < 0$ and $\rho_E < 0$ between the poles and these critical lines at the light cylinder, while $\Phi > 0$ and $\rho_E > 0$ between the critical lines and the equator at the light cylinder. Thus, outflowing plasmas comprise electrons near to the poles, and protons or positrons near to the equator. Alternatively, if $\vec{\Omega}$ is anti-parallel to $\vec{\mu}$, all of the signs are reversed.

To keep the neutron star charge neutral, the flow rate of particles with both signs should be equal. The available potential difference is given by half the difference between the pole and the boundary line which divides open and closed lines on the stellar surface; or

$$\Delta\Phi = \frac{1}{2}\frac{\mu\Omega}{cR}\sin^2\theta_0 = 3.3\times 10^{16} B_* R_*^3 P_*^{-2}\ \text{V}. \tag{24}$$

Here, the numerical value is given in units of Volt. If plasmas are completely charge separated, each particle is accelerated to an energy of $e\Delta\Phi$. The energy loss rate due to this mechanism is simply estimated by assuming that the charge density is given by Eq. (23), and that plasmas flow at the velocity of light. We obtain the energy-loss rate due to a unipolar induction L_{UI} as

$$L_{\rm UI} = \Delta\Phi\rho_E\pi(R\theta_0)^2 c = 7.2\times 10^{38} B_*^2 R_*^6 P_*^{-4}\ \text{erg s}^{-1}, \tag{25}$$

which turns out to be 3/4-times that of strong waves.

While the basic idea of this mechanism is believed to play some role in radio pulsars, there seem to be some problems which have not yet been resolved. Although the maximum energy given by Eq. (24) is sufficiently large to explain the claimed detection of ultra-high-energy γ-rays from compact sources, such as Cyg X-3 and Her X-1, it is difficult to explain those from several slow rotators, such as Vela X-1, which has a rotation period of 283 s. The maximum energy turns out to be on the order of GeV in this case.

Apart from the issue of high-energy gamma-ray emission, we note here some problems with this model in relation to pulsars. One of them is the source of accelerated particles. For parallel rotators there are electrons over the entire surface of the neutron star, and there seems to be no source of protons, as can be seen in Eq. (21). On the other hand, there are no electron sources for anti-parallel rotators. To maintain a steady state, no net charge should flow out of the central star; this means that both positive and negative particles should flow out from the neutron star, and that the electric current must somehow be closed. Although there has been much discussion concerning this current closure problem, no consensus has yet been reached on this very well-formulated problem of classical electromagnetism [31]. A widely discussed possibility is that electron-positron pairs are self-generated by cascades in the strong electric fields, and that only those produced particles are accelerated [32]. In this model, however, there is no room for proton acceleration, which is an indispensable ingredient of models for ultra-high-energy γ-ray production. Another problem is whether plasmas are completely charge separated or are quasi-neutral. For the latter case, the attained energy of individual particles is significantly reduced. As a result, the energy corresponding to the given potential difference is shared by many particles to produce a relativistic wind as discussed in the previous section [30][8].

We close this section by estimating the effect of the radiation loss which is sometimes important. If accelerated particles move along curved magnetic field lines, they suffer from a curvature radiation loss. The typical frequency and emission rate of curvature radiation have been estimated to be

$$\nu_c \approx \frac{c}{2\pi R_c}\gamma^3 \text{ Hz} \tag{26}$$

and

$$mc^2\frac{d\gamma}{dt} \approx -\frac{ce^2}{R_c^2}\gamma^4 \text{ erg s}^{-1}, \tag{27}$$

respectively, where R_c is the radius of curvature. These formula are an analog to those of synchrotron radiation, and were obtained by replacing the Larmor radius by the radius of curvature. Equating the energy loss rate and the acceleration rate, we obtain

$$\gamma_{\max} \approx (\frac{R_c^2 R\Omega B}{ec})^{1/4} \approx 8 \times 10^7 R_*^{3/4} P_*^{-1/4} B_*^{1/4} \tag{28}$$

for $R_c \sim R$. The typical energy of curvature radiation becomes

$$h\nu \approx 10^{13} R_*^{5/4} P_*^{-3/4} B_*^{3/4} \text{eV}. \tag{29}$$

If $\gamma_{\max}mc^2$ is smaller than $e\Delta\Phi$, the maximum attainable energy is limited by $\gamma_{\max}mc^2$, which is the case for electrons. For protons, both energies have comparable magnitudes.

4 Accretion Disks

Some other models for particle acceleration in compact objects have also been proposed. Instead of using the rotation energy of neutron stars, they involve gravitational energy release during the course of accretion onto compact objects, such as neutron stars and black holes. While X-ray stars and active galactic nuclei are now believed to be powered by accretion, their basic models have largely relied on thermal processes. Generally, the models involving direct acceleration processes discussed in this section are not well incorporated in the standard models of accretion-powered objects.

To devise a model of very high-energy gamma rays from galactic X-ray sources, Chanmugan and Brecher [7] proposed a model involving unipolar induction in accretion disks. This mechanism was originally proposed by Lovelace [6] regarding the production of jets in active galactic nuclei (AGNs). If an accretion disk is threaded by a large-scale poloidal magnetic field, it produces a radial component of an electric field in the disk. For simplicity, we assume that magnetic field has only a z-component which behaves as r^{-m} in cylindrical coordinates, and that rotation velocity is Keplerian. The electric field has a radial component of $E_r = -\Omega(r)rB_z/c$, and the charge density is

$$\rho_E = -\frac{1}{4\pi r}\frac{d}{dr}(r\Omega(r)B_z(r)). \tag{30}$$

If the field lines are open and reach to infinity, particle acceleration will occur. This is the case in the model of [6], in which the available potential difference is obtained by integrating the electric field given above with $m = 1/2$,

$$\Delta\Phi \approx 6.3\times10^{19}\frac{M}{10^8 M_\odot}\frac{B_I}{10^4 G}\sqrt{\frac{r_I}{r_g}}\ln\left(\frac{r_{II}}{r_I}\right)\ \mathrm{V}, \tag{31}$$

where $r_g \equiv 2GM/c^2$ is the Schwarzschild radius of the central gravitating object with mass M; B_I is the magnetic field at $r = r_I$, and r_I and r_{II} are the inner and outer radii of the accretion disk, which can act as a unipolar inductor. It is conventional to take $r_I = 3r_g$, thus identifying it with the radius of the innermost stable circular orbit around the Schwarzschild black hole. Since this potential difference amounts to the energy of the highest-energy cosmic rays, several people have proposed that nearby active galactic nuclei are the sources of such particles.

However, this model suffers from several serious problems. First, even if particle acceleration might occur, the energy loss through interactions with ambient photons in AGNs prevents the acceleration of protons to such a high energy, as was discussed by Sorrell [33]. This point will be described in section 7, where shock acceleration in AGNs is discussed. The second problem concerns a more intrinsic point of the model. The assumption that all of the magnetic field lines are open seems to be rather unlikely. The charge density given by Eq. (30) has the same sign for $m > -1/2$ as long as the magnetic field keeps the same orientation, which is quite conceivable if a dominant poloidal field exists at all. If a flow with a single charge occurs, an electric current does not close, and an electric field will be generated to prevent such a flow. We should then consider an alternative situation in which all of the field lines are closed. If the surfaces at the different radii of the disk are connected by a magnetic field line with a potential drop somewhere along the line, the potential difference might be used for particle acceleration. However, the most likely outcome is magnetohydrodymanical processes which lead to the formation and deformation of magnetic loops due to the differential rotation in the disk, while maintaining a vanishing electric field in the plasma frame. The energy stored in the magnetic field will eventually be released through magnetic activity similar to that on the solar surface. The final point is that it is very difficult for the accretion disk to possess such large-scale poloidal magnetic fields at all. If there exists such a field, the accretion itself is prevented, since the energy density of the magnetic fields dominates over that of the Keplerian motion at the outer radii. The energy density of the magnetic field behaves as r^{-1}, while that of the rotation energy goes as r^{-3}, even for constant density. Frozen-in poloidal fields can be convected from the surrounding, only when the magnitude is far short of the above estimate. Dynamo theory predicts that a self-generated magnetic field is dominantly toroidal, and is of no use in the above mechanism.

In the model of Chanmugan and Brecher [7], by applying the same mechanism to galactic X-ray sources the resultant energy of particles turns out to be relatively low. According to conventional accretion-disk theory, the magnetic pressure, can be at most comparable to the fluid pressure, and B_I is roughly proportional to the inverse square root of M. As a result, the potential difference goes as the square root of M, and significantly decreases for a stellar-mass object. Moreover, the inner radius of the disk lies far from the stellar surface due to the effect of a stellar magnetic field. For a neutron star of $1M_{\odot}$ with a surface magnetic field of 10^{12} G, the inner radius $r_{\rm I}$ is located at an Alfven radius of $\sim 10^8$ cm and $B_{\rm I} \sim 10^6$ G. Thus, the maximum energy becomes on the order of 10^{15} eV.

We now discuss other models concerning the particle acceleration around neutron stars. Many of the alleged very high-energy γ-ray sources belong to binary X-ray sources which are powered by gravitational energy release by accretion onto neutron stars. It is very probable that accreting plasmas drastically change the magnetospheric structure of neutron stars. According to the standard model of such sources (e.g.[34]), an accretion flow once stops at the Alfven radius where the magnetic pressure due to the stellar field equals the incoming momentum flux of accreting matter. Plasmas accrete onto the stellar surface at the magnetic poles along the field lines. In this picture, the electrostatic potential produced by unipolar induction is almost completely shielded by accreting plasmas. The number density of accreting plasmas is much higher than that of a completely charge-separated plasma expected in the model of [2]. Thus, there seems to be little room for particle acceleration in this picture.

Nevertheless, several authors have proposed models for particle acceleration by compromising the model of isolated pulsars and that of accreting X-ray sources. Katz and Smith [35] considered the possibility of particle acceleration by plasma turbulence excited in the magnetosphere. Cheng and Ruderman [36] proposed a model in which some of the open field lines penetrate into accretion disks. If the accretion disk rotates faster than the neutron star, a vacuum gap appears between the stellar magnetosphere and accretion disk, where a large potential drop exists. If $\vec{\mu}$ is parallel to $\vec{\Omega}$, protons are accelerated towards the disk and electrons towards the polar-cap region. If $\vec{\mu}$ is anti-parallel to $\vec{\Omega}$, the flow direction is reversed. They also suggested that the small period differences between X-rays and high-energy γ-rays reported for Her X-1 are due to a difference in the rotation period of the central neutron star and the outer edge of the accretion disk. Alternatively, Kazanas and Ellison [21][37], Kiraly and Meszaros [38], and Gaisser et al. [23] proposed models in which shock acceleration is involved; we briefly discuss these models in section 7. A detailed discussion of various models for particle acceleration in Cyg X-3 is given in [39].

5 Fermi Acceleration by Shock Waves

5.1 Fermi acceleration

In contrast to direct acceleration mechanisms, statistical acceleration mechanisms have proved to explain the salient features of the observed properties of cosmic rays. Statistical acceleration is often referred to as Fermi acceleration, named after the first proposer of the statistical mechanism [40][41]. Fermi's original proposal utilized the collisions of particles with randomly moving magnetized interstellar clouds; head-on collisions lead to an increase in the particle energy, while rear-end collisions decrease it; a small difference of their collision probabilities results in a net acceleration of particles. For simplicity, we consider only relativistic particles. Denoting the velocity of a cloud by V, and the particle momentum by p, a head-on collision leads to $\Delta p = 2(V/c)p$, while a rear-end collision gives $\Delta p = -2(V/c)p$. The former collision rate is $(1+V/c)/t_c$, while the latter rate is $(1-V/c)/t_c$, where the mean collision time is denoted by t_c. We thus obtain

$$\frac{dp}{dt} = 4(\frac{V}{c})^2\frac{p}{t_c}. \tag{32}$$

Thus, the momentum increases with time as

$$p = p_0 \exp\frac{t}{t_a}, \tag{33}$$

where $t_a = t_c(c/2V)^2$ gives the acceleration time. By incorporating the escape probability of particles with the escape time from the galactic disk t_e, the age distribution of particles $n(t)$ follows the equation

$$\frac{dn(t)}{dt} = -\frac{n(t)}{t_e}, \tag{34}$$

where t means the age from the injection at the momentum p_0. A solution is found to be

$$n(t) = n_0 \exp(-\frac{t}{t_e}). \tag{35}$$

Combining Eqs. (33) and (35), we obtain

$$\frac{dn}{dp} \propto p^{-(1+\frac{t_a}{t_e})}. \tag{36}$$

Thus, the number spectrum turns out to have a power-law shape. This mechanism is sometimes called second-order Fermi acceleration, since dp/dt is proportional to the square of the velocity of the clouds. A more formal formulation, which uses the Fokker-Planck equation, has also been established, in which a dispersion of the momentum change around the mean momentum change is incorporated (e.g., [42]). The numerical factor in the acceleration time is determined more accurately.

A critical problem with this mechanism has long been recognized. If we adopt $V = 10$ km/s and $t_c = 10^9$ s for the mean separation between clouds of 10 pc, based upon observations, the acceleration time is about 10^{10} yrs. This is much longer than the escape time of cosmic rays 10^7 yrs deduced from the observed isotope ratios of cosmic rays. Thus, the spectrum becomes very steep, and effectively no acceleration would occur. This problem might be remedied by utilizing very active locations, such as supernova remnants, as being the acceleration regions, since the cloud velocities there may be much larger than those of interstellar clouds. In this case, however, the spectral index would differ among individual cosmic-ray sources reflecting the difference in the cloud velocities; the superposed spectrum would not have a power-law behavior.

Fermi acceleration was resurrected after people realized that shock waves can also accelerate charged particles statistically with quite a high efficiency, and that the resultant power law index is universal irrespective of the velocity of shocks [9][10][11][12]. Although this acceleration mechanism is often called first-order Fermi acceleration by shock waves, the acceleration rate is actually proportional to the square of the shock velocity, as is shown later. Shock waves are seen in many places, from supernova remnants to radio galaxies, and there now seems to be a wide consensus that cosmic rays of energies less than about 10^{16} eV are accelerated by shock waves in supernova remnants [43]. One important fact is that this mechanism has been confirmed by in-situ observations of the solar system (e.g. [13][14]). For example, Earth's bow shock is generated by the collision of the solar wind with the Earth's magnetosphere; there exist shocks called "corotating interacting regions" which are generated by the collisions of fast and slow streams in the solar wind. In the vicinities of these shocks high-energy particles are observed by spacecraft, though these particles are predominantly still non-relativistic. Motivated by these observations, several mechanisms for particle acceleration by shock waves have been proposed. Apart from Fermi acceleration, it is also recognized that a single encounter of a particle with a shock front also leads to particle acceleration if the magnetic field is quasi-perpendicular to the shock normal, which is called the drift mechanism. Mechanisms which take account of the internal structure of collisionless shock waves have also been proposed. As a dissipation mechanism of the formation of collisionless shock waves, particle acceleration has been recognized as being an essential ingredient of the physics of collisionless shocks [15].

In this paper, however, we concentrate on the Fermi mechanism, since it is difficult to accelerate particles to relativistic energies and to produce a power-law spectrum by a single-encounter acceleration. For brevity, we often refer to Fermi acceleration by shock waves simply as "shock acceleration". During the last decade, many authors have studied shock acceleration from various point of views, and numerous papers have been published. We refer to the excellent reviews by Drury [24], Blandford and Eichler [14], and Jones and Ellison [15]. We now describe the basic features of shock acceleration in the simplest version. In the next section, we treat some of the current issues, which go beyond the simplest case. The topics include relativistic shocks, oblique shocks and nonlinear treatments. In section 7 we estimate the maximum energy of accelerated particles in various astrophysical sites, such as supernova remnants, radio galaxies and accretion shocks near to compact objects.

5.2 Basic mechanism: individual particle approach

The basic mechanism of shock acceleration can be understood in its simplest version by using an individual-particle approach or another approach using a diffusion-convection equation. Both the former and the latter, which give the same results, are described in this and the next subsection, respectively. The simplest version adopts the test-particle approximation, i.e., it assumes that the shock front is an a-priori given discontinuity of the background fluid flow, which convects the scattering centers of cosmic rays, and that cosmic rays move among these scattering centers without influencing the shock structure. It also considers a plane-parallel and non-relativistic shock moving with a velocity of V_1, and assumes that the magnetic fields play no role, except that their fluctuations efficiently scatter particles giving rise to a nearly isotropic distribution. This situation may be realized if the direction of the magnetic field is parallel to the shock normal, which is referred to as parallel shock. The individual-particle approach adopted by Bell [11] is useful for intuitively understanding the basic mechanism.

A schematic view is depicted in Fig. 3. We work in the shock rest frame, where the fluid velocity upstream and downstream is given by V_1 and V_2, respectively. It is assumed that since there exist some particles with momentum much greater than that of the thermal particles, they can cross the shock front back and forth, and that their angular distribution is nearly isotropic in each rest frame of the background fluid. Scattering is assumed to be elastic and that no energy change occurs in the fluid frames. Here, we assume for simplicity that accelerated particles have already attained relativistic momenta.

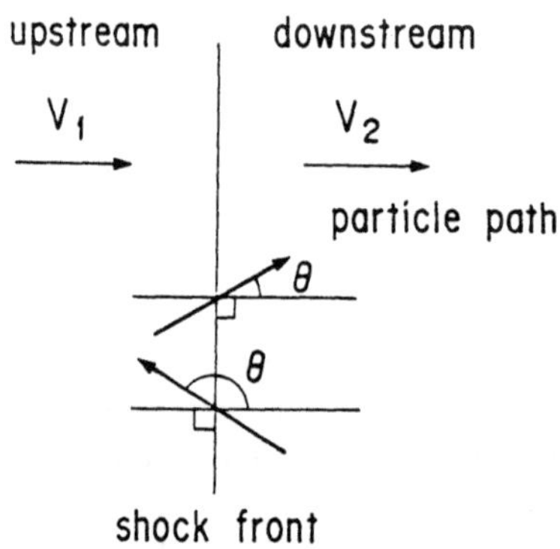

Figure 3: A schematic view of shock acceleration.

Consider a particle which crosses the shock front from upstream to downstream and then crosses the front in the reverse direction after it has been scattered by collisions in the downstream. A cycle is completed when it recrosses the front after suffering scattering upstream. The particle energy in the shock frame E, and those in the upstream and downstream fluid frames E^u and E^d, are related according to

$$E = E^u(1 + V_1 \cos\theta/c) = E^d(1 + V_2 \cos\theta/c), \tag{37}$$

for a non-relativistic shock velocity, where we neglect terms with higher orders of V/c. The angle θ is measured from the shock normal, which is directed from upstream to downstream. We denote the energy of the particle after completing k cycles by E_k. Noting that E^u and E^d do not change for elastic scattering and that E does not change when traversing the shock front, the ratio of E_{k+1} to E_k is given by

$$\frac{E_{k+1}}{E_k} = \frac{1 + (V_1 \cos\theta_{k+1}^A - V_2 \cos\theta_k^A)/c}{1 + (V_1 - V_2)\cos\theta_k^B/c}. \tag{38}$$

Here, superscripts A and B denote the forward and backward crossings, respectively, and subscript k corresponds to the cycle number. The mean values of $\cos\theta^A$ and $\cos\theta^B$ are 2/3 and −2/3, respectively (note that the flux is proportional to $\cos\theta$ and that the average should be taken with a weight of $\cos\theta$). Thus, we obtain the average ratio as

$$\frac{E_{k+1}}{E_k} = 1 + \frac{4}{3}(V_1 - V_2)/c. \tag{39}$$

Physically, this increase is ascribed to the fact that the shock wave represents a convergent flow. The upstream and downstream fluids approach each other; in the upstream rest frame, the downstream fluid and particles are observed to approach; at the same time, in the downstream rest frame, the upstream fluid and particles are observed to approach. Thus, at first glance the acceleration becomes first order in shock velocity.

After a particle crosses the shock front back and forth n times, the energy E_n becomes

$$E_n = E_0\{1 + \frac{4(V_1 - V_2)}{3c}\}^n \approx E_0 \exp\{\frac{4n(V_1 - V_2)}{3c}\}. \tag{40}$$

Particles escape only through convection with the downstream fluid; the escape probability can be estimated as follows. Denoting the number density of accelerated particles by n, the incident flux of particles from the upstream to the front is $cn/4$, and the average flow rate is nV_2 downstream. Then, the escape probability is given by $4V_2/c$. The probability of crossing the front n times is given by

$$A_n = \frac{4V_2}{c}(1 - \frac{4V_2}{c})^n \approx \frac{4V_2}{c}(\frac{E_n}{E_0})^{-\frac{3V_2}{V_1 - V_2}}. \tag{41}$$

By combining these two equations we obtain the energy spectrum as

$$N(E) \propto A_n \frac{dn}{dE_n} \propto E^{-\mu}, \tag{42}$$

where

$$\mu = \frac{3V_2}{V_1 - V_2} + 1 = \frac{r+2}{r-1}, \tag{43}$$

and $r \equiv V_1/V_2$ is the compression ratio. Thus, the resultant spectrum depends only on r; this spectrum becomes universal in the sense that it is determined only by the compression ratio. In the strong shock limit (for infinite Mach number) the compression ratio is $(\gamma+1)/(\gamma-1)$, where γ is the adiabatic index. For a gas with $\gamma = 5/3$, the compression ratio becomes 4, which is usually realized in astrophysical situations. In this case the resultant power-law index becomes 2, which is near to the observed value 2.7 for galactic cosmic rays with energy in the range of 10^9 eV to 10^{16} eV. The difference between 2 and 2.7 can be explained in terms of propagation effects. While the source spectrum is universal with an index of 2, the escape time from the galactic disk decreases with energy, which results in a spectral steepening [14]. This is most simply explained as follows. The simplest form of the continuity equation for cosmic rays in the Galaxy is written as

$$\frac{dn}{dt} = q - \frac{n}{t_e}, \tag{44}$$

where q denotes the injection term, which here describes the production of cosmic rays at the sources. If q is proportional to $E^{-2.0}$, and if t_e is proportional to $E^{-0.7}$, we then obtain the steady state spectrum of cosmic rays in the Galaxy, which turns out to be proportional to $E^{-2.7}$.

5.3 Basic mechanism: diffusion-convection equation approach

Another method to derive the above results is based on the diffusion-convection equation, which is written as

$$\frac{\partial f}{\partial t} + V\frac{\partial f}{\partial x} = \frac{\partial}{\partial x}\kappa\frac{\partial f}{\partial x} + \frac{1}{3}\frac{\partial V}{\partial x}p\frac{\partial f}{\partial p}. \tag{45}$$

Here, f denotes the isotropic part of the distribution function, V the velocity of the background fluid, and κ the diffusion coefficient. The positive direction of x is taken as that of the fluid velocity. We skip a derivation of this equation on microscopic bases, which is fairly complicated. We refer interested readers to [14] and [44]. The physical meaning of each term is rather clear; the second term on the left-hand side, the first and second terms on the right-hand side represent the convection with a fluid, diffusion and acceleration, respectively. It should be noted that momentum is measured in the fluid frame, so that the particle flux at the given momentum in the shock frame is given by

$$F(x,p) = -\kappa\frac{\partial f}{\partial x} - \frac{1}{3}pV\frac{\partial f}{\partial p}, \tag{46}$$

where the second term represents the Compton-Getting effect; the flux is not simply fV, since the x-component of the momentum as measured in the shock frame, is different from that in the fluid frame by VE/c^2.

The diffusion-convection equation has been widely used for studies of the acceleration and propagation of cosmic rays, and is sufficiently accurate if the particle distribution function is nearly isotropic and if the fluid motion is non-relativistic [14]. We take the coordinate as $x = 0$ at the shock front, $x < 0$ for the upstream, and $x > 0$ for the downstream, respectively. The steady state solution of this equation is easily obtained as follows; both upstream and downstream, since the fluid velocity V is constant, we integrate it with respect to x once, thus obtaining

$$Vf - \kappa(x,p)\frac{\partial f}{\partial x} = Vg(p), \tag{47}$$

where $g(p)$ is an arbitrary function of p. The solution of this equation is given by

$$f(x,p) = g_1(p)\exp\left(\int_0^x \frac{V}{\kappa(x,p)}dx\right) + g(p), \tag{48}$$

where g_1 is another arbitrary function of p. Assuming that no particles are convected from far upstream, we take the boundary condition $f = 0$ at $x = -\infty$, and obtain for upstream ($x < 0$)

$$f_1(x,p) = g_1(p)\exp\left(\int_0^x \frac{V_1}{\kappa(x,p)}dx\right) = g_1(p)\exp\left(\frac{xV_1}{\kappa_1}\right), \tag{49}$$

where the second equality holds if $\kappa = \kappa_1 = \text{const.}$ On the other hand, for downstream $x > 0$, f should remain finite; we thus obtain

$$f_2(x,p) = g_2(p), \tag{50}$$

where $g_2(p)$ is also an arbitrary function of p. The matching conditions at the shock front between f_1 and f_2 are given by the continuity of the number density, and the particle flux at all momenta given by Eq. (46). This leads to

$$g_1(p) = g_2(p) \tag{51}$$

and

$$\kappa\frac{\partial f_1}{\partial x} + V_1 p\frac{1}{3}\frac{\partial f_1}{\partial p} = V_2 p\frac{1}{3}\frac{\partial f_2}{\partial p}. \tag{52}$$

The second equation is also obtained by integrating the diffusion-convection equation over a small spatial width including the front.

From these conditions, we obtain

$$(r-1)p\frac{dg}{dp} = -3rg, \tag{53}$$

where $r = V_1/V_2$ and $g = g_1 = g_2$. Integration of this equation leads to

$$g(p) \propto p^{-3r/(r-1)}. \tag{54}$$

Thus, the number spectrum becomes a power law with an index of $-3r/(r-1)+2 = -(r+2)/(r-1)$, which is the same as that obtained by the individual-particle approach. Conservation of the total number of particles is ensured by including the injection term neglected in Eq. (45). It is implicitly assumed that injection occurs only at the shock front $x = 0$, monoenergetically with $p = p_0$.

5.4 Acceleration time scale

While the steady state solution obtained above continues up to infinite momentum, the upper cutoff should exist in realistic situations; it should be determined by energy losses and a finite time scale available for acceleration. Although time dependent solutions can be obtained in principle by Laplace transformation, it is somewhat complicated and then we have recourse to an argument utilizing the comparison between the various time scales. The mean residence time in the upstream per cycle Δt_1 is obtained by dividing the total number of particles in the upstream by the number flux crossing the front; the former is obtained by an integration of Eq. (49) and turns out to be $\kappa_1 g/V_1$, and the latter is $gc/4$. We therefore obtain $\Delta t_1 = 4\kappa_1 c/V_1$. The mean residence time in the downstream Δt_2 should be estimated only for particles which return to the front; after some manipulation, we obtain $\Delta t_2 = 4\kappa_2 c/V_2$ [24]. Thus, the average time for a particle to complete a cycle by crossing the shock front back and forth is given by

$$\Delta t = \frac{4}{c}\left(\frac{\kappa_1}{V_1} + \frac{\kappa_2}{V_2}\right), \tag{55}$$

and κ_i denotes the diffusion coefficient; the subscripts 1 and 2 refer to the upstream and downstream, respectively.

The acceleration time scale is estimated by

$$t_{\rm acc} = \frac{p}{\Delta p}\Delta t = \frac{3}{V_1 - V_2}\left(\frac{\kappa_1}{V_1} + \frac{\kappa_2}{V_2}\right). \tag{56}$$

Thus, the acceleration time scale depends on the numerical values of the diffusion coefficient and the shock velocity. An estimate of the diffusion coefficient for ions is not so straightforward. Usually, we assume that magnetic field fluctuations with a wavelength coincident with the Larmor radius of the ion resonate with ion motion and change the direction of the ion by a small amount; this is referred to as pitch angle scattering. The resultant diffusion coefficient is written as

$$\kappa = \frac{c r_{\rm L}}{3}\left(\frac{B}{\delta B}\right)^2 \equiv \frac{c r_{\rm L}}{3}\xi, \tag{57}$$

where $r_{\rm L} \equiv mc^2\gamma/eB$ is the Larmor radius of a particle, and δB denotes the amplitude of the magnetic field fluctuations (e.g., [14]). The maximum energy attained by a specific shock wave is estimated by equating $t_{\rm acc}$ to the life time of the shock, the time scales of energy losses and particle escape, which will be discussed in section 7.

6 Current Problems of Shock Acceleration

We have shown in the previous section that Fermi acceleration by shock waves can produce the universal power law spectrum. We will show in the next section that the acceleration time is reasonably short for various candidate acceleration sites. The analysis in section 5, however, is made under a number of simplifying assumptions such as plane-parallel geometry, infinite boundaries and non-relativistic speed of the shock front as well as the test particle approximation and diffusion approximation. As for the effect of the magnetic field, we implicitly assumed that field direction is parallel to the shock normal. During this decade much work has been done on extensions of the shock acceleration theory to relax these assumptions and to apply them to realistic astrophysical situations. Since it is beyond the scope of this article to review all aspects of these developments, here we treat some selected topics related to the origin of cosmic rays.

6.1 Oblique shocks

In this subsection, we discuss the case where the direction of the magnetic field is not parallel to the shock normal. Since the average magnetic field is expected to cross the shock front obliquely in most situations, it is very important to investigate the consequences of the obliqueness. If the momentum and energy fluxes of the fields are neglected, compared with those of the fluid in determining the jump conditions of the shock front, it is straightforward to obtain the magnetic field configuration. There are several frames in which particle motion is described. A schematic view is shown in Fig. 4. In the inertial frame of the upstream fluid, the shock front propagates to the minus x direction with velocity V_s. We denote the magnetic field strength and the angle between the field and the shock normal in the upstream by $B_{\rm up}$ and $\Phi_{\rm up}$, respectively. The magnetic field is taken to be in the $x-y$ plane. In this frame, the physical quantities are time dependent as the shock front moves. By making a Lorentz transformation to the minus x direction with the velocity V_s, we move to the shock frame where the velocity field of the fluid is stationary. Lorentz transformation to the shock frame gives

$$B^{\rm s}_{1x} = B_{\rm up}\cos\Phi_{\rm up}, \tag{58}$$

$$B^{\rm s}_{1y} = \Gamma_{\rm s} B_{\rm up}\sin\Phi_{\rm up} \tag{59}$$

and

$$E^{\rm s}_{1z} = -\Gamma_{\rm s}\frac{V_{\rm s}}{c} B_{\rm up}\sin\Phi_{\rm up}, \tag{60}$$

where $\Gamma_{\rm s} = 1/\sqrt{1 - V_{\rm s}^2/c^2}$. Here and hereafter, superscript 's' means quantities in the shock frame and subscript 1 means those in the upstream. It is to be noted that the angle between the direction of the field and the shock normal in the upstream becomes

$$\tan\Phi^{\rm s}_1 = \Gamma_{\rm s}\tan\Phi_{\rm up}. \tag{61}$$

The jump conditions for fields at the shock front are the continuity of the normal component of the magnetic field and the perpendicular component of the electric field. We have $B^{\rm s}_{2x} = B^{\rm s}_{1x}$ and $E^{\rm s}_{2z} = E^{\rm s}_{1z}$, where subscript 2 refers to quantities in the downstream. Imposing the condition that the electric field should disappear in the downstream fluid frame gives $B^s_{2y} = rB^s_{1y}$, where r is the compression ratio. The particle motion in the shock frame is still complicated since the energy of a particle is not conserved due to the existence of the electric field.

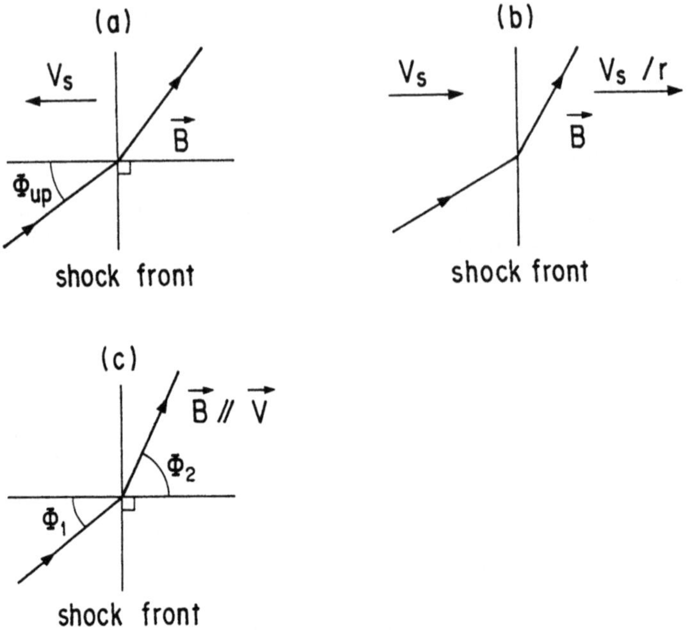

Figure 4: A schematic view of an oblique shock. Figures (a), (b) and (c) correspond to inertial frame of the upstream fluid, shock frame and de Hoffman-Teller frame, respectively.

We can work in more convenient frames by making another Lorentz transformation to the y direction. Note that, in the shock frame, the intersection of an individual magnetic field line with the shock front moves along the shock front with a phase velocity of $V_t = -V_s\Gamma_s \tan\Phi_{up}$. Oblique shocks are divided into two cases depending on whether $|V_t| > c$ or not. Shocks with $|V_t| > c$, or equivalently $V_s/c > \cos\Phi_{up}$, are called superluminal and it is convenient to make a Lorentz transformation with a velocity of c^2/V_t. In this frame, the magnetic field becomes perpendicular to the shock normal and the electric field makes particles drift to the direction of the shock normal. If the cross field diffusion is unimportant, and then the guiding center of particles is constrained to move on a fixed magnetic field line, particles experience an adiabatic compression and drift acceleration at the shock front only once and then are convected away with the fluid. Thus, no statistical acceleration will occur for superluminal shocks.

In the opposite case for $|V_t| < c$, or equivalently $V_s/c < \cos\Phi_{up}$, it is possible to move to the frame in which the magnetic field is parallel to the fluid velocity both in the upstream, and in the downstream, by making a Lorentz transformation with a velocity of V_t. This case is called subluminal. The results are

$$V_{1x} = rV_{2x} = \frac{V_s}{\Gamma_t}, \tag{62}$$

$$V_{1y} = V_{2y} = -V_t, \tag{63}$$

$$B_{1x} = B_{2x} = \frac{B_{up}\cos\Phi_{up}}{\Gamma_t}, \tag{64}$$

and

$$B_{1y} = \frac{B_{2y}}{r} = \Gamma_s B_{up}\sin\Phi_{up}, \tag{65}$$

where $\Gamma_t = 1/\sqrt{1 - V_t^2/c^2}$. It can been seen that the magnetic field line is parallel to the stream line. The angle between a magnetic line of force and the shock normal in this frame is given by

$$\tan\Phi_1 = \frac{V_{1y}}{V_{1x}} = \frac{B_{1y}}{B_{1x}} = \Gamma_t\Gamma_s\tan\Phi_{up} = \frac{\tan\Phi_{up}}{\sqrt{1 - \frac{V_s^2}{c^2\cos^2\Phi_{up}}}} \tag{66}$$

and

$$\tan\Phi_2 = \frac{V_{2y}}{V_{2x}} = \frac{B_{2y}}{B_{2x}} = r\tan\Phi_1, \tag{67}$$

for upstream and downstream, respectively. This frame is referred to as the de Hoffman-Teller frame and no electric field appears, thus the energy of a particle is conserved.

For subluminal shocks, statistical acceleration occurs if magnetic field fluctuations exist. It may seem to be paradoxical that particle acceleration occurs in spite of the vanishing electric field. Particle acceleration occurs because field fluctuations carry an electric field which eventually leads to a change of the energy of particles, although no electric field appears in the background fluid. While an extension of the treatment in parallel shocks to that in oblique shocks seems to be rather simple at first glance, there remain many problems to be clarified. Analysis based on the individual particle approach [45], and that on the diffusion-convection equation [24], show that the same power law index is obtained as that for the parallel shock if an isotropic distribution of particles is realized. As is discussed later, the most important consequence of oblique shocks is that the acceleration time can be significantly shortened compared with parallel shocks [45][46].

For oblique shocks, we should take account of the interaction of particles with the magnetic discontinuity at the shock front. The ratio of the magnetic field strength in the downstream to that in the upstream turns out to be

$$b = \sqrt{1 + (r^2 - 1)\Gamma_s^2\sin^2\Phi_{up}} \tag{68}$$

We assume first an adiabatic interaction, that is, assume that the magnetic moment of the particle is conserved. Denoting the pitch angle cosine by μ, the magnetic moment is given by $p^2(1-\mu^2)/B$. Using the conservation of energy and magnetic moment of the particle in the de Hoffman-Teller frame, it is easily shown that in the shock interaction, upstream particles with $0<\mu<\mu_0$ are reflected and those with $\mu_0<\mu$ are transmitted to the downstream, where $\mu_0=\sqrt{1-1/b}$. Downstream particles with $\mu<0$ are transmitted to the upstream. The Liouville theorem assures that the phase space density is conserved and that the number density turns out to be the same at the downstream and upstream, if the particle distribution is isotropic in the fluid frames.

When we work with the diffusion-convection equation, the boundary condition relating the particle flux at a given momentum at the front is needed, and it is also the same as that for the parallel shock. Thus the only modification to the parallel shock is the replacement of κ by $\kappa_{\|}\cos^2\Phi+\kappa_{\perp}\sin^2\Phi$ in Eq. (45), where $\kappa_{\|}$ and $\kappa_{\perp}$ are the diffusion coefficient along the magnetic field and the cross field diffusion coefficient, respectively. Usually a form of $\kappa_{\perp}=\kappa_{\|}/\xi^2$ is adopted. Since the number spectrum given by Eq. (54) is independent of the diffusion constant, the spectral index is the same as that for the parallel shock. The acceleration time scale turns out to be

$$t_{\mathrm{acc}}=\frac{3r\kappa_0}{(r-1)V_s^2}\cos^2\Phi_1\{\xi+\frac{\tan^2\Phi_1}{\xi}+\frac{r}{\cos^3\Phi_1}(\frac{1}{1+r^2\tan^2\Phi_1})^{3/2}(\xi+\frac{r^2\tan^2\Phi_1}{\xi})\},\tag{69}$$

where $\kappa_0=cr_{\mathrm{L}}/3$. Thus, the ratio of the acceleration time for oblique shocks to that for the parallel shock is given by

$$\frac{\cos^2\Phi_1}{1+r}\{1+\frac{\tan^2\Phi_1}{\xi^2}+\frac{r}{\cos^3\Phi_1}(\frac{1}{1+r^2\tan^2\Phi_1})^{3/2}(1+\frac{r^2\tan^2\Phi_1}{\xi^2})\}.\tag{70}$$

This factor is significant even for a mild obliqueness; for example, for $r=4$, and $\xi\gg1$, a reduction factor of 18 and 470 is realized for for $\cos\Phi_1=0.5$ and 0.1, respectively. However, this reduction in the acceleration time is not large if the magnetic field is turbulent enough to give $\xi\approx1$; for $\xi=1$, the reduction factor is only 2.3 and 2.5 for $\cos\Phi_1=0.5$ and 0.1, respectively.

For given r and Φ_1, the shortest acceleration time is realized at around

$$\xi_{\min}=\tan\Phi_1\{\frac{r^3+\cos^3\Phi_1(1+r^2\tan^2\Phi_1)^{3/2}}{r+\cos^3\Phi_1(1+r^2\tan^2\Phi_1)^{3/2}}\}^{1/2}\tag{71}$$

and it turns out to be

$$\begin{aligned}t_{\mathrm{acc,min}} &= \frac{6r\kappa_0}{(r-1)V_s^2}\frac{\tan\Phi_1}{\cos\Phi_1(1+r^2\tan^2\Phi_1)^{3/2}}\\ &\quad\times\{r+\cos^3\Phi_1(1+r^2\tan^2\Phi_1)^{3/2}\}^{1/2}\\ &\quad\times\{r^3+\cos^3\Phi_1(1+r^2\tan^2\Phi_1)^{3/2}\}^{1/2}.\end{aligned}\tag{72}$$

It is noted that since we should take $\xi\geq1$ to have a physically meaningful diffusion coefficient, these two equations are not applicable if $\xi_{\min}<1$. As a numerical example, for $r = 4$ and $\cos\Phi_1 = 0.5$, $t_{\rm acc,min}$ turns out to be $5.7\kappa_0/V_s^2$ at $\xi_{\rm min} = 2.6$, which should be compared to the acceleration time for the parallel shock $20\xi\kappa_0/V_s^2$ given by Eqs. (56) and (57). For $\cos\Phi_1 = 0.1$, $t_{\rm acc,min}$ is $1.2\kappa_0/V_s^2$ at $\xi_{\rm min} = 13.7$. As $\cos\Phi_1$ approaches 0, the limiting case of subluminal shocks ($\cos\Phi_{\rm up} = V_s/c$), $t_{\rm acc,min}$ approaches 0 in proportion to $\cos\Phi_1$; this is due to a vanishing diffusion coefficient in this limit, and is certainly unphysical. If we require the extent of particles in the upstream (κ/V_s) be at least an order of the Larmor radius, ξ should be smaller than c/V_s. In this limit, the acceleration time becomes an order of r_L/V_s, which is by a factor of c/V_s shorter than the conventional estimate. Anyhow, an order of magnitude reduction in the acceleration time is quite possible for oblique shocks. Another merit is that this is realized for a reasonable degree of the amplitude of the magnetic field fluctuations. This has an important consequence on the estimation of the maximum energies attained in specific objects as will be discussed in the next section.

The individual particle approach made by [45] reached the same conclusion as above. He incorporated the effects of reflection and transmission of particles at the shock front with the adiabatic approximation. It may seem to be surprising at first sight that the power law index remains to be the same in spite of an enhanced acceleration rate due to the particle-shock interaction. The reason for this is that as the speed of the fluid along the magnetic field is increased by the obliqueness, the escape probability is also increased to just compensate for the above effect. An unresolved issue is how good are the assumptions of adiabaticity and isotropy of the particle distribution. The first assumption is not justified since the the thickness of the shock front is small compared to the gyro radius of particles. To incorporate non-adiabaticity one should trace every particle orbit, which is practically impossible for studying statistical acceleration. Calculations of the single interaction have been performed in several works and large deviations from adiabaticity were found [47]. But the often stated point is that the adiabatic treatment gives a good approximation in the statistical sense. This is demonstrated by the Monte Carlo approach by Takahara and Terasawa [48], though further study is needed. As for the isotropy of particle distribution, the nature of the scattering may play an important role. If a large angle isotropic scattering is adopted [48], particles tend to be easily isotropized. Alternatively, if a pitch angle scattering is adopted [49], higher degrees of anisotropy and density jumps at the shock front are found. A deviation from the isotropy leads to a change of a resultant energy spectrum, and [49] found much flatter spectra than the canonical one. Finally we note that for oblique shocks, relativistic effects appear at relatively low shock velocity since the velocity along the magnetic field is magnified [50].

6.2 Relativistic shocks

There are several places where relativistic shocks are likely to exist. Radio observations, using very long baseline interferometers of the central cores of radio loud quasars and radio galaxies have revealed that the emission structure changes with time so that peaks of radio emission move with an apparent velocity larger than light. These phenomena are called superluminal motions and are most convincingly interpreted in terms of relativistic beams emanated from the central engine [51]. It is also expected that fast pulsars lose their rotation energy by relativistic winds as was stated in sections 2 and 3. The composition of relativistic beams and winds is uncertain; it may be either a usual proton-electron plasma or electron-positron pair plasma, although for pulsar winds the latter composition is more likely. While the essential mechanism of particle acceleration by relativistic shocks is quite similar to that for non-relativistic shocks, it is more difficult to treat theoretically. First, internal energy of the background fluid itself becomes relativistic in the downstream and it is uncertain what kind of thermal state the fluid takes there. It is unlikely for the fluid to attain a complete thermal equilibrium and so the ratio of fluid velocity in the upstream to that of the downstream depends on the manner of how the downstream plasma is relaxed. For example, in the limit that both upstream and downstream fluids are relativistic, the velocity ratio becomes 3 in the strong shock limit from the Rankine-Hugoniot relation. Unless we can solve the shock structure self-consistently, we must somehow assume a value of the velocity ratio. Second, for relativistic shocks, the particle distribution becomes intrinsically anisotropic since both background fluid and accelerated particles move near the speed of light. Then, we cannot make use of the diffusion approximation and more sophisticated approaches are required.

For parallel shocks, in which the direction of the magnetic field is parallel to the shock normal, an analytic approach incorporating anisotropic distribution was developed by Kirk and Schneider [52][22]. Alternatively, Monte Carlo simulations have been done by several authors [53][54][55]. A review of these results is given by [55]. The particle spectrum takes a power law form and becomes flatter as the shock velocity and the velocity ratio increase. The spectral index depends on the nature of the scattering. Large angle scattering leads to a flatter spectrum than pitch angle scattering because the majority of upstream particles cross the front with a large angle to the shock normal in the case of the pitch angle scattering. As for the acceleration time, t_{acc} becomes short by a factor of a few compared to that expected from Eq. (56). A much larger reduction of t_{acc} claimed by [54] was not confirmed by [55] and is thus in question. For oblique shocks, where the direction of magnetic field is not parallel to the shock normal, most relativistic shocks become superluminal. Cross field diffusion is necessary to realize Fermi acceleration for oblique relativistic shocks.

6.3 Geometrical effects

All results presented above are obtained only for rather idealized cases. We have assumed steady shocks and infinite and plane-parallel geometries. In realistic situations, however, relevant regions are neither plane-parallel nor infinite and shocks have a finite lifetime. Then, the mean free path of accelerated particles should also affect the resultant spectral shape. As an example of this, we discuss the case of spherical shock waves. For expanding spherical shock waves, analytical and numerical studies have shown that the resultant power law index is the same as that for plane-parallel shocks only when the mean free path of the accelerated particles is much shorter than the shock radius. The spectrum becomes steeper as the mean free path becomes longer, because the average velocity of particles moving away from the shock front in the downstream increases owing to the diffusion caused by non-uniform particle distribution, thus the escape probability increases. Numerically the index becomes larger by about $60\kappa/(RV_s)$ [56][57], which means that the mean free path should be shorter than $0.05RV_s/c$ in order to restrain the difference from the universal power law spectrum below 0.1. These results are confirmed by a Monte Carlo simulation [58].

The above argument can also be applied to plane-parallel shocks with finite boundaries. Unless the mean free path is much shorter than the size of the region, a spectral steepening should occur. This gives a stronger constraint on the maximum energy of protons accelerated by shocks than for the infinite boundary. One interesting case is a hypothetical accretion shock which occurs near neutron stars or black holes. In this case the downstream is the region between the shock front and the central star. Particles are removed from the downstream region only when they hit the stellar surface or the horizon of the black hole. Thus particles can easily cross the shock front back to the upstream even when the mean free path of accelerated particles is fairly long. A preliminary numerical simulation by Takahara shows that a flatter spectrum than the canonical one becomes possible.

6.4 Nonlinear modifications of shocks

Several important problems remain to be discussed. These include particle injection and the conversion efficiency of shock energy into particle acceleration, the generation of magnetic fluctuations which are responsible for the scattering of particles, and modifications of the shock structure due to the pressure of accelerated particles. The latter two problems have been extensively studied over the last decade. Nonlinear treatment is indispensible in clarifying the injection mechanism, the conversion efficiency, and excitation and damping of magnetic fluctuations which determine the mean free path of particles as well as the modification in the shock structure. First, the diffusion coefficient, which is governed by magnetic field fluctuations, should be self-consistently calculated while it is a priori given in the test-particle approximation. Second, accelerated particles can diffuse to the upstream from the downstream, which would weaken the shock strength by increasing the upstream pressure. On the other hand, relativistic particles may be regarded as a gas with the specific heat ratio of 4/3, which would result in the increase of the shock compression ratio to 7 and a significant flattening ($\mu = 1.5$) of the resultant spectrum if the cosmic ray pressure dominates that of the background gas.

The present status of these issues is excellently summarized in two review papers [14][15]. Since a detailed examination is beyond the scope of this paper, we here just give a very brief summary. Most studies have adopted the two-fluid approach consisting of a background fluid and a cosmic ray fluid. They utilize the equations of continuity of mass, momentum and energy of the background fluid as well as the cosmic ray hydrodynamic equation. The shock structure becomes complicated by including the gas subshock and a change of the compression ratio. For a Mach number greater than a few, solutions seem to be unphysical in the sense that they give rise to an extremely high efficiency of acceleration. Jones and Ellison discussed that this is an intrinsic weakness in the two fluid approach [15]. The essential point is that the two-fluid approach does not incorporate the conservation of cosmic ray particle number and lacks spectral information. The alternative approach adopted by Eichler [59][14] incorporates the escape of the highest energy particles, which may be one of the useful methods to treat the problem. A Monte Carlo approach by Ellison et al. includes the injection of seed particles for acceleration from a thermal pool, and solves the shock structure and particle distribution iteratively, though the mean free path of particles is a priori given [60]. The standpoint of Eichler and Ellison is that the particle acceleration is not a byproduct of shocks but an essential part of the formation of collisionless shocks. Much remains to be studied in the nonlinear treatment of shock acceleration.

7 Astrophysical Sites of Shock Acceleration

In this section, we review possible sites of the acceleration of cosmic rays based on Fermi acceleration by shock waves. Observationally, the cosmic ray spectrum is divided into three energy regions as is seen in Fig. 1. From $\sim 10^9$ eV to $\sim 10^{16}$ eV, the spectrum is described by a single power law with an index of 2.7. At about 10^{16} eV the spectrum steepens to another power law spectrum with an index of 3.1 up to $\sim 10^{19}$ eV. Above $\sim 10^{19}$ eV the spectrum again flattens and continues to above 10^{20} eV. The first component is generally considered to be produced by supernova remnants. As for the second one, a thorough understanding of their acceleration sites is still forthcoming, although most people regard that they are of galactic origin. Finally, the highest energy component is most probably of extragalactic origin. Although the composition of cosmic rays at energies higher than about 10^{13} eV is not yet well determined, we assume that protons are the main component while the fraction of heavier particles may increase with energy. Before discussing the individual acceleration sites, we give a general formula used to estimate the maximum energy attained in the shock acceleration.

7.1 Maximum energy attained by shock acceleration

The maximum energy attained by shock acceleration in an astrophysical environment is determined by equating the acceleration time given by Eq. (56) with the shortest one among the lifetime of the shock, escape time of the particles and energy loss time. The lifetime of the shock may be estimated as the time scale in which the shock traverses the region considered. We denote this time scale by $t_{ev} = R/V_s$, where R is the size of the acceleration region. Equating t_{ev} and t_{acc}, and adopting the compression ratio of 4, we obtain the maximum energy as

$$E_{max} = Am_pc^2\gamma_{max} = \frac{3ZeBR}{20\xi}\frac{V_s}{c}. \tag{73}$$

If the shock is stationary, as is the case for a hypothetical galactic wind termination shock discussed in subsection (7.3), then at first sight t_{ev} may be replaced by the real life time of the shock. Actually, however, t_{ev} should be replaced by the escape time of particles, which results in nearly the same maximum energy as above [61]. By equating t_{acc} with the escape time R^2/κ, we obtain the maximum energy only $\sqrt{20}$ times larger than the above. Actually, the restriction due to particle escape might be much more severe when the spectral steepening by geometrical effects is taken into account, as discussed in the previous section.

The value of ξ depends on the strength of magnetic field fluctuations, and the level of magnetic field fluctuations is still an open problem. Observations of shock acceleration in the solar system suggest that ξ takes a value of order 10. On the other hand, oblique shocks significantly reduces the acceleration time scale as was discussed in the section 6. For a reasonable guess of an order of 10 for ξ, t_{acc} can be reduced by a few orders of magnitude, and then the maximum energy can be so much larger than that given by Eq. (73). Keeping all these uncertainties in mind, we here choose the estimate given by Eq. (73) with $\xi = 1$ as a canonical estimate. As is shown below, however, in many cases the required maximum energy is one or two orders of magnitude larger than the canonical estimate, which may suggest that acceleration by oblique shocks plays an important role. Energy losses become important in restricting the maximum energy of protons through synchrotron radiation in strong magnetic fields and pion and electron-positron pair production by photo reactions in the existence of copious photons. These are discussed later for relevant cases.

7.2 Supernova remnants

Supernova remnants have been accepted as a prime source of cosmic rays for a number of reasons including energetics and chemical composition. Eq. (73) gives the maximum energy of about 10^{14} eV for $B = 10\mu$G, $R = 3$ pc and $V_s = c/30$. This agrees with the estimate given by [16] which has been frequently referred to. Supernova remnants evolve through many stages. In the initial stage, where the mass of ejecta dominates over that of the entrained mass of interstellar matter, the shock is propagated with a constant velocity. In the stage where the latter dominates over the former, the shock is decelerated with time as $t^{-3/5}$ while the entrained mass increases. This is called the Sedov phase. Cosmic rays can be accelerated only in the constant velocity phase, since in the Sedov phase the shock velocity begins to decrease and the acceleration time scale increases with the time from the explosion. So there remains a problem that a theoretical estimate is short of the observed break energy by one or two orders of magnitudes. For parallel shocks, much more intense magnetic fields or a more tenuous environment is needed to realize a reasonable maximum energy.

To overcome this difficulty, some possibilities have been proposed. The composition difference is an obvious possibility; at energy around 10^{15-16} eV, heavy ions may dominate over protons since irons can be accelerated up to an energy 26 times higher than protons. Although there is no evidence for a change in the composition up to about $\sim 10^{13}$ eV, it remains an open problem as to whether a change occurs at higher energies. An alternative possibility is that oblique shocks can accelerate more efficiently as was discussed above.

7.3 Multiple shocks versus galactic wind termination shock

The origin of cosmic rays in the energy region between 10^{16} eV and 10^{19} eV has remained fairly uncertain. The Larmor radius of protons with an energy 10^{18} eV, and $B = 3\ \mu$G, is 300 pc and comparable to the thickness of the galactic disk; since the spectrum breaks downward at about 10^{16} eV from the index of 2.7 to 3.1, they are most likely of galactic origin. Two possibilities have been suggested; one is the first order Fermi acceleration by multiple shocks and the other is the acceleration by the galactic wind termination shock.

The first one proposed by Axford utilizes the fact that the interstellar space is filled with many shocks generated by supernova remnants [43]. Although cosmic rays at this energy region are not efficiently confined and accelerated in a single supernova remnant, they can interact with many shocks and can be accelerated in principle. Although at first glance this appears to be the second order Fermi acceleration, actually it is first order, because all shocks are expanding and all the collisions are head-on as far as the Larmor radius is shorter than the size of the supernova remnants. The resulting spectral index depends on the average shock speed, the mean distance between neighbouring shocks and the cosmic ray residence time within the galaxy and is not unique. However, Axford argued that a reasonable choice of these quantities can reproduce the observed index.

An alternative explanation is the acceleration by the galactic wind termination shock advocated by Jokipii and Morfill [17]. Although the very existence of the galactic wind is a speculative one, some of the interstellar gas may be heated to blow out to the intergalactic space and the collision of the wind with the intergalactic matter may generate the termination shock. The velocity is an order of 300 km/s which is determined by the depth of the gravitational potential, the magnetic field is an order of the galactic field, and the position of the shock is expected to be an order of mean galactic separation. The maximum energy of protons attained by the galactic wind termination shock is estimated by Eq. (73) to be 10^{17} eV for $V_s = 300$ km/s, $B = 3\ \mu$G and $R = 100$ kpc. The maximum energy is higher by a factor of 5 because the termination shock is stationary and particle escape limits the maximum energy as discussed in subsection (7.1). Assuming that the shock is highly oblique, [17] argued that irons can be accelerated up to 10^{19-20} eV. Both models have not been examined in detail and do not seem to be fully convincing at present.

7.4 Active galactic nuclei and radio galaxies

Charged particles with an energy above 10^{19} eV cannot be confined within the galactic disk and their origin is supposed to be extragalactic. Hillas [61] discussed previously various possibilities of the acceleration of highest energy cosmic rays and concluded that the shock acceleration is rather inefficient. However, some authors have recently argued that Fermi acceleration by relativistic or mildly relativistic shocks in active galaxies is the most promising candidate for the origin of highest energy cosmic rays [20][54][18][19]. In order to obtain $E_{\max} = 10^{20}$ eV, eBR should be larger than 2×10^{21}eV for $\xi \approx 1$ and $V_s/c \approx 1/3$. In practical units, BR should be larger than about 2G · pc. This condition is more stringent than the condition that Larmor radius is shorter than the size of the acceleration region. Again we note that for oblique shocks this condition may be relaxed by a few orders of magnitudes. In radio galaxies or radio loud quasars, there seem to be several places where the above conditions are satisfied ranging from the very compact cores where $B = 10^3$ G for a size of 10^{-3} pc, to the extended lobes of 100 kpc size with $B = 10^{-5}$ G [51]. In extended lobes there exist structures called hot spots where radio emission is enhanced. These hot spots are considered to be an interaction region of a beam with the intergalactic medium and contain mildly relativistic shocks. These hot spots have a characteristic size of a few kpc and a characteristic magnetic field of a few times 10^{-4} G.

One of the problems with active galaxies being the origin of the highest energy cosmic rays is that they suffer from energy losses due to pion production, and electron-positron pair production through collisions with ambient photons. For protons of an energy 10^{20} eV, pion production and electron-positron pair production occur mainly by collisions with photons of energy around $\epsilon = 10^{-2}$ eV and 10^{-5} eV, respectively. The average cross section of pion production is an order of 10^{-28} cm^2. While the cross section of electron-positron pair production is an order of 10^{-26} cm^2, the energy loss per collision is much lower. We can roughly estimate the time scale of energy loss as $t_{\rm loss} = 1/n_\gamma \sigma c$, where the effective cross section σ is taken to be 3×10^{-28} cm^2. The number density of target photons is estimated by

$$n_\gamma = L_\epsilon / 4\pi c R^2 \epsilon, \tag{74}$$

where L_ϵ is the luminosity in a logarithmic interval around an energy of ϵ. A typical energy spectrum of radio galaxies, or radio quasars, in the relevant energy region is nearly flat so that L_ϵ/ϵ is a constant in a rough approximation. For a typical luminosity of $L_\epsilon = 10^{43}$ erg s^{-1} at $\epsilon = 10^{-3}$ eV, the condition of $t_{\rm loss} > t_{\rm ev}$ leads to the constraint that R should be larger than 10 pc for $V_s = c/3$. This eliminates the compact cores of strong extragalactic radio quasars which have more than $L_\epsilon > 10^{44}$ erg s^{-1} as a candidate of the origin of 10^{20} eV protons. Other places such as jets, extended lobes and hot spots are not excluded by this restriction. Among others, hot spots are the most promising because they in fact contain mildly relativistic shocks. This argument in turn implies that compact cores are powerful emitters of very high energy γ-rays and neutrinos [62][63]. A substantial fraction of energy injected for particle acceleration will be converted to this high energy emission.

Next, we examine the energetics of the highest energy cosmic rays. The energy density of cosmic rays above 10^{19} eV in the Galaxy is about 10^{-20} erg cm^{-3}, which is 8 orders of magnitude smaller than that in GeV range. If these very high energy cosmic rays are produced in the Galaxy, the required luminosity is 10^{36} erg s^{-1} for a volume of 10^{67} cm^3 and a confinement time of 3×10^3 yrs. On the contrary, if these cosmic rays are accelerated in radio galaxies and uniformly distributed in the intergalactic space, the required cosmic ray luminosity is dependent on the space density, and lifetime of the radio galaxies, and the way of propagation of the cosmic rays. Because of the energy loss due to collisions with microwave photons, protons of 10^{20} eV can traverse only about 100 Mpc, and radio galaxies within this distance contribute to the observed flux. In other words, the life time of cosmic ray protons of 10^{20} eV is about 3×10^8 yrs. The space density of radio galaxies depends very much on radio luminosity and radio frequency. A crude estimate is that there are 10^{-5} and 10^{-7} radio galaxies per 1 Mpc^3 at the radio luminosity of 10^{40} and 10^{42} erg s^{-1} at GHz range, respectively. When we adopt the former density, the required cosmic ray luminosity is 10^{41} erg s^{-1} for each radio galaxy, even if the mean life of cosmic rays is taken to be 10^{10} yrs. Thus the required cosmic ray luminosity should be an order of magnitude larger than the observed total radio luminosity. A more optimistic estimate by [18] is mainly due to an overestimate of the space density of radio galaxies. However, since our Galaxy lies in the local supercluster, an enhancement of the number density of radio galaxies is also expected, which will somewhat alleviate this requirement.

The strength of the intergalactic magnetic field is not known. If the intergalactic magnetic field is an order of 10^{-9} G, the Larmor radius becomes as large as 100 Mpc for 10^{20} eV protons and they propagate straight from nearby sources. Thus we expect an anisotropy due to nearby radio galaxies, such as Cen A and Virgo A. The distance to Cen A is about 5 Mpc and the expected flux is 3×10^{-11} erg s^{-1} cm^{-2} for the cosmic ray luminosity of 10^{41} erg s^{-1}. This should be compared to the average flux of cosmic rays above 10^{19} eV of 3×10^{-11} erg s^{-1} cm^{-2} sr^{-1}. Thus we expect strong anisotropies in the direction of individual radio galaxies. Although we do not have enough data at 10^{20} eV, the observed anisotropy is not apparently strong in the energy region of 10^{19} eV; this may be due to scattering off magnetic fields in the intergalactic space and the galactic halo. For 10^{19} eV protons, the Larmor radius becomes 10 kpc for the halo field of 1 μG and 10 Mpc for the intergalactic field of 10^{-9} G, respectively. Thus the arrival direction may be broadened for the majority of the cosmic rays above 10^{19} eV. It is particularly important to determine the arrival direction of the highest energy cosmic rays in order to test the hypothesis of radio galaxy origins or other models.

7.5 Compact sources

Finally we briefly discuss shock acceleration near compact objects. For neutron stars with the surface magnetic field of $B = 10^{12}$ G, Eq. (73) shows that the maximum energy amounts to more than 10^{19}eV, if we assume that an accretion shock is formed at the stellar surface with the free fall velocity of $\sqrt{2GM/R}$. If the shock is formed away from the stellar surface, the maximum energy decreases with $R_s^{-5/2}$. However, since the synchrotron cooling time

$$t_{\mathrm{syn}} = \frac{8\pi m_p^2 c^3}{B^2 \sigma_T} (\frac{m_p}{m_e})^2 \frac{1}{E} \tag{75}$$

is much shorter than the acceleration time

$$t_{\mathrm{acc}} = \frac{20cE\xi}{3eBV_s^2}, \tag{76}$$

the maximum energy is limited by

$$E_{\max} \approx 10^{14}(\xi B_*)^{-1/2}\frac{V_s}{c}\mathrm{eV}. \tag{77}$$

Several models based on accretion shocks have been proposed to explain very high-energy γ-rays from Cyg X-3 and similar sources. Since those γ-rays from magnetic poles are opaque to electron-positron pair production against strong magnetic fields and X-ray photons, most models incorporate particle acceleration in other regions. Kazanas and Ellison proposed spherical accretion shock onto non-magnetic compact objects [21] [37]. Magnetic fields are assumed to be in rough equipartition with the kinetic energy density of accreting plasmas. The maximum attainable energy of protons is estimated by equating $t_{\rm acc}$ with synchrotron loss time or advection time scale of an order of R/V_s. It turns out to be an order of 10^{15} eV for stellar mass objects and may be up to 10^{20} eV for AGNs if shocks are formed near the Schwarzschild radius. For the latter case, however, pion and electron-positron production through collisions with soft photons prevents particles from attaining such a high energy as was discussed in the previous subsection.

Alternatively, Kiraly and Meszaros [38] proposed a model for shock acceleration in magnetized neutron stars. In their model, corotating beams are produced along open field lines in the magnetic poles and a shock is generated in the plasma around the Alfven radius through the interaction of the corotating beams with the accreting matter. The maximum energy of protons is also an order of 10^{15} eV. Another model proposed by [23] studied particle acceleration in shocks by pulsar winds. For this kind of shocks, arguments given in the previous subsections can apply, and a large value of BRV_s is needed to accelerate protons to high energies.

Although shock models involving accreting plasmas around compact stars have some attractive features, details remain to be investigated [39]. Those include the question of shock formation during accretion onto black holes, geometrical effects, effects of finite mean free path and effects of inelastic processes in high energy density environment. Geometrical effects have been discussed in subsection 6.3, and it was suggested that a flatter spectrum may be realized even for a large mean free path. Inelastic processes are important for emission of high-energy γ-rays and neutrinos as well as neutron production.

8 Summary

In this paper we reviewed various mechanisms for the acceleration of high-energy cosmic rays. The main conclusions are summarized as follows:

1) It should be stressed that direct acceleration mechanisms around compact objects do not explain the power law spectrum which is the most important feature of cosmic rays.
2) Acceleration by strong waves is unlikely to occur. In the current view, the propagation of waves is prevented and the wave energy is converted to relativistic MHD winds consisting of electron-positron pairs and magnetic fields.

3) Acceleration by unipolar induction around neutron stars may occur and plays a certain role in pulsars, though the most likely outcome is also an MHD wind. It is still an open question as to whether the acceleration of protons occurs in binary systems such as Cyg X-3. It is also to be noted that it cannot explain very high-energy γ-rays from slow rotators such as Vela X-1.
4) Unipolar induction in accretion disks seems to be unlikely since large scale strong poloidal magnetic fields in accretion disks are difficult to be realized.
5) Shock acceleration has been most extensively studied and widely accepted as the most plausible acceleration mechanism of galactic cosmic rays. The most important feature is that it can reproduce the universal power law spectrum under appropriate approximations.
6) Current topics on basic ingredients of shock acceleration theory are briefly reviewed including oblique shocks, relativistic shocks and nonlinear modifications. These are important to estimate the maximum attainable energy of protons and the efficiency of conversion of shock energy into relativistic particles. **7)** Supernova remnants are responsible for cosmic rays below 10^{16} eV, though a simple estimate of the maximum energy gives one or two order of magnitudes lower energy. This discrepancy may be solved by the shorter acceleration time for oblique shocks.
8) Astrophysical origin of cosmic rays between 10^{16} and 10^{19} eV remains to be open, though it is probably of galactic. The origin of the highest energy cosmic rays above 10^{19} eV is likely to be extragalactic. A possible acceleration site is mildly relativistic shocks in radio galaxies, especially in hot spots in radio lobes.

References

[1] J.P. Ostriker, J.E.Gunn: Astrophys.J **157**, 1395 (1969).

[2] P. Goldreich, W.H. Julian: Astrophys.J **157**, 869 (1969).

[3] D. Eichler, W.T. Vestrand: Nature **307**, 613 (1984).

[4] T. Nakamura, Y. Yamada, H. Sato: Prog.Theor.Phys. **87**, 1065 (1987).

[5] J. Matthews: In Proc. Internat. Symp. "Astrophysical Aspects of the Most Energetic Cosmic Rays, Kofu (1991, World Scientific) p.162.

[6] R.V.E. Lovelace: Nature **262**, 649 (1976).

[7] G. Chanmugan, K. Brecher: Nature **313**, 767 (1985).

[8] C.F. Kennel, F.V. Coroniti: Astrophys.J. **283**, 694 (1984).

[9] G.F. Krymskii: Sov.Phys.-Dokl. **23**, 327 (1977).

[10] W.I. Axford, E. Leer, G. Skadron: In Proc. 15th Internat. Cosmic Ray Conf., Plovdiv, **11**, 132 (1977).

[11] A.R. Bell: Mon.Not.R.Astron.Soc. **182**, 147 (1978).

[12] R.D. Blandford, J.P. Ostriker: Astrophys.J.Lett. **221**, L29 (1978).

[13] T. Terasawa, M. Sholer: Science **244**, 1050 (1989).

[14] R. Blandford, D. Eichler: Phys.Rep. **154**, 1 (1987).

[15] F.C. Jones, D.C. Ellison: Spa.Sci.Rev. **58**, 259 (1991).

[16] P.O. Lagage, C.J. Cesarski: Astron.Astrophys. **125**, 249 (1983).

[17] J.R. Jokipii, G.E. Morfill: Astrophys.J. **312**, 170 (1987).

[18] F. Takahara: Prog.Theor.Phys. **83**, 1071L (1990).

[19] W.H. Ip, W.I. Axford: In Proc. Internat. Symp. "Astrophysical Aspects of the Most Energetic Cosmic Rays", Kofu (1991, World Scientific) p.273.

[20] P.L. Biermann, P.A. Strittmatter: Astrophys.J. **322**, 643 (1987).

[21] D. Kazanas, D.C. Ellison: Nature **319**, 380 (1986).

[22] J.G. Kirk, P. Schneider: Astron.Astrophys. **201**, 177 (1988).

[23] T.K. Gaisser, A.K. Harding, T. Stanev: Astrophys.J. **345**, 423 (1989).

[24] L.O'C. Drury: Rep.Prog.Phys **46**, 973 (1983).

[25] T. Hara et al.: Proc. 17th Internat. Cosmic Ray Conf., Bangalore, **9**, 198 (1981).

[26] J.E. Gunn, J.P. Ostriker: Astrophys.J. **165**, 523 (1971).

[27] K.O. Thielheim: Fund.Cos.Phys. **13**, 357 (1989).

[28] H. Laue, K.O. Thielheim: Astrophys.J.Suppl. **61**, 465 (1986).

[29] C. Max: Phys.Fluids **16**, 1277 (1973).

[30] M.J. Rees, J.E. Gunn: Mon.Not.R.Astron.Soc. **167**, 1 (1974).

[31] F.C. Michel: Rev.Mod.Phys. **54**, 1 (1982).

[32] M.A. Ruderman and P.G. Sutherland: Astrophys.J. **196**, 51 (1975).

[33] W.H. Sorrell: Astrophys.J. **323**, 647 (1987).

[34] P. Ghosh, F.K. Lamb: Astrophys.J. **234**, 296 (1979).

[35] J.I. Katz, I.A. Smith: Astrophys.J. **326**, 733 (1988).

[36] K.S. Cheng, M. Ruderman: Astrophys.J.Lett. **337**, L77 (1989).

[37] D. Kazanas, D.C. Ellison: Astrophys.J. **304**, 178 (1986).

[38] P. Kiraly, P. Meszaros: Astrophys.J **333**, 719 (1988).

[39] A. Mitra: Astrophys.J **370**, 345 (1991).

[40] E. Fermi: Phys.Rev. **75**, 1169 (1949).

[41] E. Fermi: Astrophys.J. **119**, 1 (1954).

[42] D.B. Melrose: Plasma Astrophysics Vol.2 (1980, Gordon and Breach).

[43] W.I. Axford: In Proc. Internat. Symp. "Astrophysical Aspects of the Most Energetic Cosmic Rays", Kofu (1991, World Scientific) p.406.

[44] F.C. Jones: Astrophys.J. **361**, 162 (1990).

[45] M. Ostrowski: Mon.Not.R.Astron.Soc. **233**, 257 (1988).

[46] J.R. Jokipii: Astrophys.J. **313**, 842 (1987).

[47] R.B. Decker: Spa.Sci.Rev. **48**, 195 (1988).

[48] F. Takahara, T. Terasawa: In Proc. Internat. Symp. "Astrophysical Aspects of the Most Energetic Cosmic Rays", Kofu (1991, World Scientific) p.291.

[49] M. Ostrowski: Mon.Not.R.Astron.Soc. **249**, 551 (1991).

[50] J.G. Kirk, A.F. Heavens: Mon.Not.R.Astron.Soc. **239**, 995 (1989).

[51] M.C. Begelman, R.D. Blandford and M.J. Rees: Rev.Mod.Phys. **56**, 255 (1984).

[52] J.G. Kirk, P. Schneider: Astrophys.J. **315**, 425 (1987).

[53] J.G. Kirk, P. Schneider: Astrophys.J. **322**, 256 (1987).

[54] J.J. Quenby, R. Lieu: Nature **342**, 654 (1989).

[55] D.C. Ellison, F.C. Jones, S.P. Reynolds: Astrophys.J. **360**, 702 (1990).

[56] G.F. Krymskii, S.I. Petkhov: Sov.Astron.Lett. **6**, 124 (1980).

[57] V.L. Prishchep, V.S. Ptuskin: Sov.Astron. **25**, 446 (1981).

[58] F. Takahara: In Proc. "Joint Varenna-Abastumani Internat.School and Workshop on Plasma Astrophysics", Sukhumi 1986, ESA SP-251, p.253.

[59] D. Eichler: Astrophys.J. **229**, 419 (1979).

[60] D.C. Ellison: In Proc. Internat. Symp. "Astrophysical Aspects of the Most Energetic Cosmic Rays", Kofu (1991, World Scientific) p.281.

[61] A.M. Hillas: Ann.Rev.Astron.Astrophys. **22**, 425 (1984).

[62] K. Mannheim, P.L. Biermann: Astron.Astrophys. **221**, 211 (1989).

[63] F.W. Stecker et al.: Phys.Rev.Lett. **66**, 2697 (1991).

MIX
Papier aus verantwortungsvollen Quellen
Paper from responsible sources
FSC® C105338

If you have any concerns about our products,
you can contact us on
ProductSafety@springernature.com

In case Publisher is established outside the EU,
the EU authorized representative is:
Springer Nature Customer Service Center GmbH
Europaplatz 3, 69115 Heidelberg, Germany

Printed by Libri Plureos GmbH
in Hamburg, Germany